FIELD ROBOTICS

FIELD ROBOTICS

Proceedings of the 14th International Conference on Climbing and Walking Robots and the Support Technologies for Mobile Machines

University Pierre et Marie Curie (UPMC), Paris, France, 6 – 8 September 2011

Editors

Philippe Bidaud
Universite Pierre et Marie Curie, France

Mohammad O. Tokhi
University of Sheffield, UK

Christophe Grand
Universite Pierre et Marie Curie, France

Gurvinder S. Virk
University of Gävle, Sweden

World Scientific

NEW JERSEY • LONDON • SINGAPORE • BEIJING • SHANGHAI • HONG KONG • TAIPEI • CHENNAI

Published by

World Scientific Publishing Co. Pte. Ltd.
5 Toh Tuck Link, Singapore 596224
USA office: 27 Warren Street, Suite 401-402, Hackensack, NJ 07601
UK office: 57 Shelton Street, Covent Garden, London WC2H 9HE

British Library Cataloguing-in-Publication Data
A catalogue record for this book is available from the British Library.

The front cover image has been provided by: Sami Ylonen, Helsinki University of Technology, Helsinki, Finland.

FIELD ROBOTICS
Proceedings of the 14th International Conference on Climbing and Walking Robots and the Support Technologies for Mobile Machines

ISBN-13 978-981-4374-27-9
ISBN-10 981-4374-27-X

Printed in Singapore by B & Jo Enterprise Pte Ltd

PREFACE

Robotics has been an exciting field in engineering and natural sciences for many decades; it has held considerable fascination for researchers and scholars and many important contributions have been made by industrial robots in various manufacturing tasks such as assembly, welding, painting, and material handling. In recent times, we have been witnessing the emergence of new service robots which are intended to perform a variety of tasks in new environments such as search and rescue, surveillance, exploration, security missions and surgical operations as well as provide assistance to various users.

Field robotics aims to bring technologies that allow autonomous systems to assist and/or replace humans performing tasks that are difficult, repetitive, unpleasant, or take place in hazardous environments. These robotic systems will bring sociological and economic benefits through imporved human safety, increased equipment utilisation, reduced maintenance costs and increased production.

CLAWAR 2011 is the fourteenth in a series of international conferences on Climbing and Walking Robots and the Support Technologies for Mobile Machines. The aim of the conference is to provide an open forum where researchers, scientists, engineers and practitioners from throughout the world can come together to present and discuss the latest achievements, future challenges and exciting applications for mobile service machines in general, and climbing and walking robots in particular. The proceedings of CLAWAR 2011 include state of the art research and development findings presented during the CLAWAR 2011 conference in 119 technical presentations by authors from 28 countries covering the five continents.

The editors would like to thank members of the International Program Committee, International Advisory Committee and National Organising Committee for their efforts in reviewing the submitted papers, and the authors in addressing the comments and suggestions of the reviewers in their final submissions. It is intended that the CLAWAR 2011 proceedings will be a valuable source of reference for research and development in mobile robotics.

P. Bidaud, M. O. Tokhi, C. Grand, G. S. Virk

CONFERENCE ORGANISERS

University Pierre et Marie Curie
Paris, France

CLAWAR Association

A non-profit making membership based professional organization serving the robotics community
www.clawar.org

CONFERENCE COMMITTEES

General Chairs and Programme Chairs

of Fourteenth International Conference on Climbing and Walking Robots and the Support Technologies for Mobile Machines

P. Bidaud (General Chair) – University Pierre et Marie Curie, France
M. O. Tokhi (Program Chair) – University of Sheffield, UK
C. Grand (Program Vice-Chair) – University of Tsukuba, Japan
G. S. Virk (IAC Chair) – University of Gävle, Sweden

International Advisory Committee (IAC)

of Fourteenth International Conference on Climbing and Walking Robots and the Support Technologies for Mobile Machines

H. Levent Akin – Turkey
Manuel Armada – Spain
Yvan Baudoin – Belgium
Karsten Berns – Germany
Bryan Bridge – UK
Anibal T de Almeida – Portugal
H. Fujimoto – Japan
Krzysztof Kozlowski – Poland
Giovanni Muscato – Italy
Lakmal Seneviratne – UK
Ming Xie – Singapore

International Programme Committee (IPC)

of Fourteenth International Conference on Climbing and Walking Robots and the Support Technologies for Mobile Machines

National Organising Committee

of Fourteenth International Conference on Climbing and Walking Robots and the Support Technologies for Mobile Machines

F. B. Amar
C. Chevallereau
S. Doncieux
S. Lacroix
B. Gas
A. Godin
T. Hamel
J. Morillon
M. Mouaddib
J. B. Mouret
F. B. Ouezdou
V. Pasqui
F. Plumet
B. Thuilot
S. Zeghloul

CONTENTS

Preface v
Conference organisers vi
Conference committees vii
Table of contents xi

Section–1: Assistive Robots

Human interface of a wearable robot for walking on a step 3
T. Kagawa, Y. Oda and Y. Uno

Interactive person following for social robots: hybrid reasoning based on fuzzy and multiple- objectives decision making approaches 11
C. Granata, P. Bidaud and A. Buendia

Development of a passive-type robotic wheelchair 27
D. Chugo, T. Higuchi, Y. Sakaida, S. Yokota and K. Takase

Design of an efficient back-drivable semi-active above knee prosthesis 35
M. Awad, K. S. Tee, A. Dehghani, D. Moser and S. Zahedi

Section–2: Autonomous Robots

Myon: Concepts and design of a modular humanoid robot which can be reassembled during runtime 45
M. Hild, T. Siedel, C. Benckendorff, M. Kubisch and C. Thiele

Design and real-time PID control approach of a two-wheeled vehicle with an extended rod 53
A. C. Rosendo, S. A. Agouri, M. O. Tokhi and K. M. Goher

Distribution of reaction forces during movement walking robots 62
I. Ion, I. Simionescu, A. Curaj, V. Luige, A. Vasile

Online discovery of locomotion modes for wheel-legged hybrid robots: A transferability-based approach 70
S. Koos and J.-B. Mouret

A dynamic based path tracking controller for a fast rover with independent steering and drive 78
M. Krid and F. B. Amar

Navigation with obstacle avoidance of an autonomous sailboat 86
C. Petres, M. A. Romero-Ramirez, F. Plumet

Extrinsic RGB-D camera calibration for legged robots 94
M. A. Hoepflinger, C. D. Remy, M. Hutter and R. Siegwart

Sailing robot performance: Maximum speed tracking vs energy efficiency 102
Y. Briere

A wind-independent control strategy for autonomous sailboats based on Voronoi diagram 110
K. Xiao, J. Sliwka and L. Jaulin

Section–3: Biologically Inspired Systems and Solutions

Proposed omnidirectional locomotion strategy for traveling-wave-type mobile robot (TORoIII) 127
T. Ogawa, M. Yamamoto and T. Nakamura

Energy based gait control of an under actuated robotic fish 135
D. T. Roper, S. Sharma, R. Sutton and P. Culverhouse

Balance recovery of a humanoid robot using cognitive sensorimotor loops (CSLs) 142
M. Kubisch, C. Benckendorff and M. Hild

Controlling human-like locomotion of a biped by a biologically motivated approach 149
J. Zhao, T. Luksch and K. Berns

Modelling of flexible manipulator systems using bacterial foraging algorithms 157
H. Supriyono, M. O. Tokhi and B. A. Md Zain

Development of an omnidirectional wall-climbing robot by using traveling wave propagation 166
T. Nakamura and T. Hayakawa

Study of ant locomotion in surface transitions for climbing robot design 174
A. Lozano, G. Peters, D. Liu and K. Waldron

Biological-inspired locomotion with the multi-segmented inspection robot KAIRO-II ... 182
L. Pfotzer, S. Bohn, C. Birkenhofer and R. Dillmann

Object localisation using active whiskers on underwater autonomous walking robots ... 190
T. Rooney, M. J. Pearson, J. Welsby, I. Horsfield, R. Sewell, S. Dogramadzi

Sensory coupled action switching modules (scasm) for modeling and synthesis of biologically inspired coordination ... 196
B. Rutter, B. Taylor, R. Quinn, W. Lewinger, J. Bender, R. Ritzmann and M. Blumel

Section–4: Co-operative Robot Systems

Machine learning for heterogeneous multi-robots systems in logistic application frame ... 207
T. Wang, D. M. Ramik, C. Sabourin and K. Madani

Over the hill and far away: Aerial/ground cooperation for long range navigation ... 215
R. Boumghar and S. Lacroix

Leader/neighbour trajectory following by mobile robots ... 223
T. K. Patel, T. P. Sattar and H. L. Rodriguez

Section–5: Education, Intelligence and Learning for CLAWAR

A student-built ball-throwing robotic companion for hands-on robotics education ... 233
J. Oberländer, T. Harbaum, G. Kurz, N. Ahmed, T. Kos-Grabar, A. Hermann, A. Rönuau and R. Dillmann

A real-time robot vision approach combining visual saliency and unsupervised learning ... 241
D. M. Ramík, C. Sabourin and K. Madani

Generation and testing of gait patterns for walking machines using multi-objective optimization and learning automata ... 249
J. L. D. Santos and C. L. N. Júnior

Learning fast walking patterns for a NAO robot using Qwalking ... 257
J. Ma and S. A. Cameron

Section–6: Flexible Manoeuvring Systems

Fuzzy control scheme for two overhead cranes with three degrees of freedom 267
R. Algazaly and M. O. Tokhi

Measurement of flexed posture for driving flexible mono-tread mobile track using flexible displacement sensor 275
T. Haji, S. Myoraku, T. Kinugasa, T. Akagi, K. Yoshida, H. Amano, M. Iribe, K. Tokuda, R. Hayashi and K. Osuka

Nonlinear two DoF dynamic modelling of a TRMS using an adaptive network based inference system 283
A. Rahideh, M. H. Shaheed and S. A. Niknam

Hybrid fuzzy control of a flexible manoeuvring system 292
M. Omar, F. Aldebrez and M. O. Tokhi

Section–7: Guidance and Navigation

Avoiding tracking error with estimation techniques in visual servoing 303
C.-A. Díaz-Hernández, J.-L. Muñoz-Lozano and J. López-Coronado

Robust planar position control of an autonomous airship 311
S. Samaali, L. Beji and A. Abichou

On-field test of a low cost, high precision augmented global navigation satellite system for agriculture applications 321
F. Bonaccorso, L. Cantelli, D. Longo, C. D. Melita and G. Muscato

Mobile robot environment map building using a RGB-D sensor 329
E. Georgiou and J. Dai

Section–8: HMI, Tele-presence and Tele-operated Robots

Human-Robot: From interaction to relationship 339
M. Grandgeorge and D. Duhaut

A software architecture for efficient teleoperation of a semi-autonomous walking robot 347
P. Skrzypczyński

On-line incremental 3d human body reconstruction for HMI or AR applications 355
L. Almeida, F. Vasconcelos, J. P. Barreto, P. Menezes and J. Dias

Remote control system for rescue robot FMT: Removal algorithm of obstacles in the images ... 363
K. Tokuda, T. Hirayama, T. Kinugasa, T. Haji and H. Amano

Modular disposal teleoperated robot ... 371
H. L. Rodriguez and T. Sattar

Section–9: Innovative Design of CLAWAR

Wall climbing robots using electro-adhesion technology ... 381
G. Griffiths and M. Studley

Compact climbing robot rolling on flexible magnetic rollers, for generator inspection with the rotor still installed ... 388
W. Fischer, G. Caprari, R. Siegwart and R. Moser

PARASWIFT – Hybrid climbing and base-jumping robot for entertainment ... 397
L. Geissmann, M. Denuder, D. Keusch, L. Pfirter, D. Röthlisberger, M. Ritter, P. Thoma, R. Siegwart, W. Fischer, G. Caprari, J. Weber and P. Beardsley

CY-MAG^{3DE}: Magnetic climbing inspection robot ... 407
F. Rochat, P. Schoeneich, O. Marti, H. Bleuler and F. Mondada

Climbing robot with thermal glue ... 415
F. Rochat, V. Hirshmann, T. Barras, H. Bleuler and F. Mondada

TREMO: An inspection climbing inchworm base on magnetic switchable device ... 421
F. Rochat, R. Beira, H. Bleuler and F. Mondada

Design of a novel climbing robot — CliffRider: The abseiling face inspection robot ... 429
J. Cappelli and A. H. Göktogan

Hardware independent platform for modular robotics based on protothreads and Contiki loader ... 437
M. Bachratý, P. Hubinský and A. Vitko

Static analysis of a novel biped walking and climbing robot ... 445
G. Figliolini and P. Rea

Section–10: Innovative Sensing and Actuation

High-accuracy measurement system for secure steering of large vehicles ... 455
H. Montes, C. Salinas, J. Sarria and M. Armada

The use of vision systems for building terrain maps for mobile robots 463
J. Kowal, B. Karwat and A. Sioma

Hybrid fuel cell power for long duration robot missions in field environments 471
J. Thangavelautham, D. Gallardo, D. Strawser and S. Dubowsky

Self-configuring localization algorithms in wireless sensor networks: Problems and issues in mobile robot applications 479
M. M. Sanaatiyan and H. Hassanpour

Section–11: Locomotion

A spare method for obstacle sensing in legged locomotion 491
Q. Bombled and O. Verlinden

Modular-joint design for efficient walking and modeling of a bipedal robot TPINOKIO 499
T.-C. Wee, A. Astolfi and M. Xie

High compliant series elastic actuation for the robotic leg ScarlETH 507
M. Hutter, C. D. Remy, M. A. Hoepflinger and R. Siegwart

Steering control for a 3D biped robot considering ground contact and stability 515
T. Wang and C. Chevallereau

Passive dynamic quadruped walker with flat feet and ankle springs 523
T. Kinugasa, H. Komatsubara, K. Yoshida, S. Fujimoto, M. Iribe and K. Osuka

Gait generation for the humanoid NAO based on human walking parameterization 531
J.-C. P. Lamarche, O. Bruneau and J.-G. Fontaine

Compliant joints increase the energy efficiency of bipedal robot 539
A. Haq, Y. Aoustin and C. Chevallereau

On-board energy consumption estimation for a six-legged walking robot 547
A. Roennau, F. Sutter, T. Kerscher and R. Dillmann

Kinematic and dynamic analysis of a hexapod pronking gait robot and control actions 555
S. Soyguder and H. Alli

Adaptive walking gait for locomotion on terrain with non-uniform slope 563
C. Neves, R. Ventura and J. Sequeira

Energy control of periodical oscillations of biped robot in the frontal plane ... 571
A. Telesh, F. Palis, A. Rudskyy, A. Melnykov, T. Dynnik, M. Konyev and U. Schmucker

Limit cycles walking of biped robot based on total energy control and virtual constraints ... 579
A. Telesh, F. Palis, A. Rudskyy, A. Melnykov, T. Dynnik, M. Konyev and U. Schmucker

Prototype passive walker development which has capabilities to measure the state parameters of its own ... 587
M. Iribe, H. Furukawa, Y. Sugimoto, K. Osuka and T. Kinugasa

Modeling of quadrupedal running gaits ... 595
K. J. Waldron

Analyzing energy-efficient kinematic configuration in hexapod robots ... 605
D. Sanz-Merodio, E. Garcia and P. G. de-Santos

Bio-inspired locomotion controller design for legged robots ... 615
R. Zaier

A distributed hybrid controller for walking over uneven terrain ... 623
L. Palmer III and M. Palankar

On the step handling of virtual slope walking ... 631
M. Zhao

Dynamic obstacle-crossing of a wheeled robot with double-wishbone suspension ... 642
J. C. Fauroux and B. C. Bouzgarrou

Development of a personal mobility vehicle for rough terrain ... 650
S. Nakajima

Design, simulation and experimental tests of a hybrid rover for overpassing obstacles ... 658
G. Castelli and E. Ottaviano

Sensor-based control of a skid-steering mobile robot combined with feet for moving on all terrain ... 666
T. Okada, S. Tezuka and Y. Sasaki

Obstacle crossing with the wheel-legged robot HyLoS based on a robustness criterion on the frictional contacts ... 676
P. Jarrault, C. Grand and P. Bidaud

The simplest parkour model: Experimental validation and stability analysis ... 684
A. W. Long , R. D. Gregg, and K. M. Lynch

Section–12: Medical and Biomedical Robots

Robotic assisted intramedullary nailing of femoral shafts ... 695
E. Cerny, K. Bouazza-Marouf and D. Kerr

Force characterisation of one axis for an adhesive foot in a climbing robot for minimally invasive abdominal surgery ... 705
A. M. Lopez, M. Khazravi, R. Richardson, A. Dehghani, R. Roshan, T. Liskiewicz, A. Morina, A. Neville and D. G. Jayne

Phase-dependent knee model identification in gait motion using an active knee rehabilitation orthotic device ... 715
L. Saint-Bauzel, O. Unluhisarcikli and C. Mavroidis

Section–13: Modelling and Simulation of CLAWAR

Modelling and analysis of an auto-reconfiguring modular micro mobile robot ... 725
J. Hodgins and D. Zhang

Modelling of two-wheeled robotic wheelchair with moving payload ... 731
A. M. Almeshal, M. O. Tokhi and K. Goher

Dynamic stability improvement of walking robots ... 742
L. Vladareanu, I. Ion, A. Curaj and S. A. Dumitru

Compliance for a cross four-bar knee joint ... 750
A. Hamon and Y. Aoustin

Walking direction control of a six-legged robot by optimal servo system ... 758
H. Uchida and T. Kameyama

Virtual modelling, prototyping and simulation of transfemoral active prosthetic limb ... 766
Z. W. Lui, M. Awad, A. A. Dehghani and N. Messenger

Mobile vibrating robot for movement on metal surface ... 774
S. F. Jatsun, A. A. Cherepanov, S. B. Rublev and A. S. Yatsun

Quadruped robot optimization using a genetic algorithm ... 782
M. F. Silva

Stability augmentation of slip-like legged locomotion exploiting hip actuation 790
S. Pouya, R. Moeckel, F. Peuker, A. Seyfarth and A. J. Ijspeert

Section–14: Perception and Sensor fusion

Sensory system evaluation methodology for automatic fruits harvesting 801
C. Salinas, R. Fernández, H. Montes and M. Armada

A distance-dependent people detection system for a wheeled mobile domestic robot 809
X. Wang and X. Clady

Obtaining geometry of stairs using monocular vision system 817
K. Walas

Laser scan matching for self-localization of a walking robot in man-made environments 825
P. Skrzypczyński

Dynamic calibration of a gyroscope using a compound pendulum 833
K. S. Tee, M. Awad, A. Dehghani, D. Moser and S. Zahedi

A probabilistic model of equilibrium perception 839
S. N'Guyen, J. Droulez and A. Berthoz

An autonomous system for ground-based cloud tracking using the mean-shift algorithm 847
M. D. O'Toole, S. A. Wormald, D. Kerr, J. M. Coupland and A. P. Sandford

Obstacle detection using integration of omni-directional camera and inertial sensor 855
Y. Guo, S.-H. Ieng, M. Romero, R. Benosman and B. Gas

Section–15: Planning and Control

Behavior-based adhesion control system for safe adherence of wall-climbing robots 865
D. Schmidt and K. Berns

Stochastic controllers for robust underwater mobile manipulation 873
F. Bonsignorio

Planning of motion trajectories for the control of torque-unit manipulator 881
K. Yoshida, Y. Katayama, T. Kinugasa, K. Fukuda and K. Osuka

A goal driven perspective to generate complex humanoid motion synthesis 889
J. Salini, V. Padois, A. Ibanez, P. Bidaud and A. Buendia

Section–16: Service Robots

A leg design for a biped humanoid service robot 901
T. Li and M. Ceccarelli

TRALOC – Teleoperated snake-robot with tracks on each segment, for searching victims in collapsed buildings 909
T. Bleiker, J. Darbre, S. Droll, F. Heer, D. Hunziker, T. Winandy, W. Fischer and R. Siegwart

Pangolin: A variable geometry tracked vehicle with independent track control 917
M. Yusof and T. Dodd

Design of mobile base for a self-reconfigurable intelligent swarm fixture system 925
L. Xiong, M. Zoppi, R. Molfino and L. M. D. L. Girard

Modularity and parallel kinematics: An original design methodology applied to high precision 933
M. Richard and R. Clavel

CLAWAR systems as enablers of intelligent buildings operation, construction and design: A holistic view 941
F. Bonsignorio

TUBULO II: Magnetic climbing inchworm for inspection of boiler tubes 949
F. Rochat, P. Schoeneich, O. T.-D. Nguyen, H. Bleuler and F. Mondada

Safety for emerging service robots 956
G. S. Virk and S. Moon

Author index 965

SECTION–1

ASSISTIVE ROBOTS

HUMAN INTERFACE OF A WEARABLE ROBOT FOR WALKING ON A STEP *

TAKAHIRO KAGAWA, YUKI ODA AND YOJI UNO

Department of Mechanical Science and Engineering,Nagoya University, Furo-cho, Chikusa-ku, Nagoya, Japan

The goal of our study is to develop a human interface of a wearable robot for walking on a step. Our system consists of (1) sensing of a step from a movement of a walker, (2) detecting a foot placement state related to a edge of a step and (3) generating gait patterns of climbing and stepping down. The gait pattern is determined on the basis of the estimated step height. We examined the feasibility of the proposed sensor and control systems for two kinds of steps with different height.

1. Introduction

Paraplegics with spinal cord injuries cannot keep standing and walking, and they use a wheelchair for mobility. Life in a wheelchair often results in secondary complications. Several researches have attempted to develop gait reconstruction systems for paraplegics. Feed-forward gait controls with functional electrical stimulation [1] or a wearable robot [2] make paraplegics walk. However, the adjustability of gait pattern according to the user's intention is required to walk under various environments. A human interface of the system is important to control movements of the paralyzed joints based on the user's intension [3]. In previous studies of wearable robotics, EMG signals are used for human interface to detect the user's intension [4]. However, voluntary EMG signals are not detected from leg muscles of paraplegics.

We have addressed the development of sensor and control systems for Wearable Power-Assist Locomotor (WPAL) shown in Figure 1 which assists leg movements of paraplegics. The stride length in walking is controlled from the distance of a preceding movement of a walker, which is measured by angle-acceleration sensors [5]. A swing movement is triggered on the basis of the stability of zero moment point (ZMP) which is evaluated by ground reaction force [6]. It is more desirable to walk on various environments such as a step, stairs and rugged terrain. As a first approach to improve the locomotor functions, we developed a human interface and control system for walking on a step. We

---

* This study was supported by Grants-in-Aid for Scientific Research (B) (No. 21300092) and for Young Scientists (B) (No. 20760165). This study was also supported by Electro-Mechanic Technology Advancing Foundation.

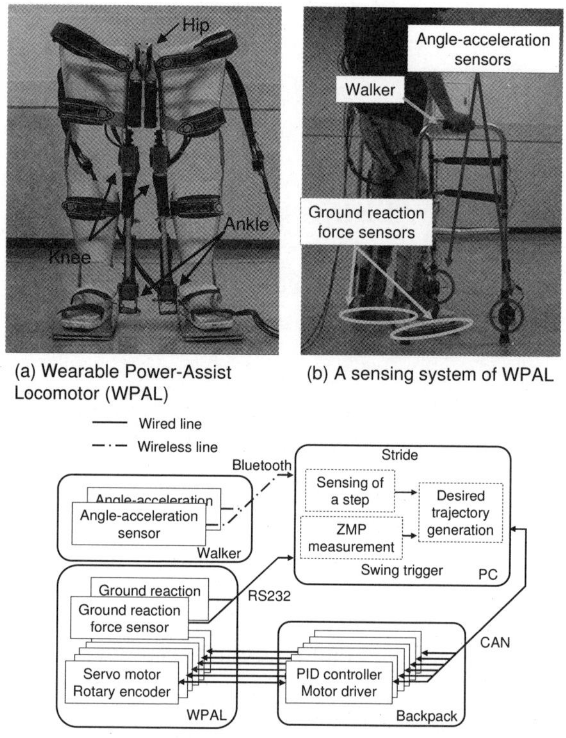

(a) Wearable Power-Assist Locomotor (WPAL)

(b) A sensing system of WPAL

(c) A diagram of the sensing and control system

Figure 1 Sensing and control system of wearable robot WPAL (Wearable Power-Assist Locomotor). (a) shows WPAL which has six actuators on hip, knee, and ankle joints. (b) shows walking with WPAL. The user uses a walker and the movement is measured by angle-acceleration sensors. Ground reaction force is measured by sensors attached to the sole of WPAL. (c) shows a signal processing diagram of the sensor and control system.

addressed three problems to assist walking on a step; (1) sensing of a step from a movement of a walker, (2) avoiding backward balance losses while climbing and stepping down and (3) generating gait patterns of climbing and stepping down.

2. Sensing and control systems for walking on a step

2.1. *Wearable robot and human interface*

Figure 1 shows a wearable robot WPAL which has 6 actuators assisting extension and flexion movements of hip and knee joints and plantarflexion and dorsiflexion movements of ankle joints. The ranges of joint motion are ± 30 deg (hip extension and flexion), 0–120 deg (knee flexion), and ± 30 deg (ankle dorsiflexion and plantarflexion). WPAL can support walking, sitting-to-standing

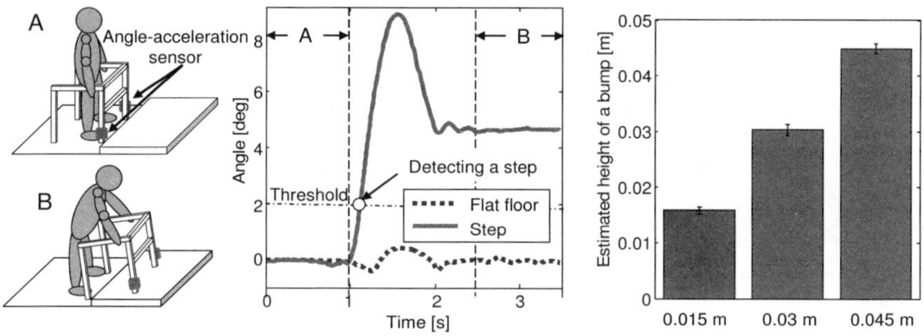

(a) Detection of a step and estimation of the step height based on the angle of the walker

(b) Estimated height of steps

Figure 2 Detection of a step and estimation of the height using angle-acceleration sensors attached to the walker. (a) shows the angle profiles of walker while walking on flat floor and on a step. (b) shows the estimation results of the height of steps.

and standing-to-sitting movements. Figure 1 (b) shows the walking movement with WPAL assisting leg movements. To maintain the upright posture, arm support with a walker is required. Voluntary movements of a user are measured by angle-acceleration sensors and ground reaction force sensors. The angle-acceleration sensors are attached to the front frame of the walker. The ground reaction force sensors are attached to the foot soles of WPAL [6].

Figure 1(c) shows a signal processing diagram of the sensing and control system. A main controller (PC) determines the desired trajectories of the actuators, and motor drivers control rotational angle of the actuators using PID feedback control. The angle and acceleration of the walker movement and the force sensor signals are collected in the PC. The 3-axis ground reaction force, center of pressure (COP) and ZMP are calculated from the force sensor signals. On the basis of sensor signals, desired trajectories of the actuators are determined. A leg swing movement is triggered according to ZMP [6]. Our prototype system is not stand alone and WPAL is wired for PC and external power supply. The PC was used to collect the sensing and control data while walking on a step. We did not use batteries for the power supply to avoid a fall due to running out of power. In the following sections, the methods of the sensing of a step and the gait pattern generation are described.

2.2. *Sensing of a step*

A user moves a walker before climbing or stepping down. Detection of the step and estimation of the step height are performed on the basis of the angle of the walker. Figure 2 (a) shows the angular profiles of the walker related to a horizontal axis when the user moves the walker before climbing a step (a solid line) and walking on a level ground (a dashed line). Existence of a step is detected according to the angle threshold. The height h of the step is estimated by the angle of the walker on the step. According to the step height, it is

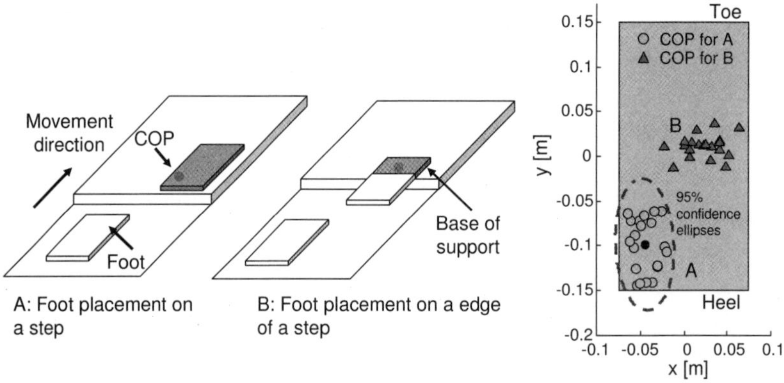

Figure 3 Foot placement in landing of a first step while climbing on a step.

distinguished whether the intended movement is climbing ($h > 0$) or stepping down ($h < 0$). Figure 2(b) shows estimated height by measured angle of the walker for three conditions of the height (0.015, 0.03, 0.045 m). The estimation errors and SDs were within 1 mm.

A foot placement of the first step during climbing is important because a backward falling would be induced according to the relationship between foot and edge of the step (A and B in Figure 3). The position of COP must locate within the area of the plantar surface contacting the top of the step. Therefore, foot placement related to the edge can be detected by COP measured by the ground reaction force sensor. The right part of Figure 3 shows the distribution of the COP positions in landing trials of the first step. The circles and triangles indicate COP positions of landing on the top and the edge of the step, respectively. When the whole foot was on the step (A), the COP was located posterior in the sole. When the foot was on the edge (B), the COP was located more anterior. The difference of the distributions of the COPs at landing was evaluated using 95% confidence ellipse of A shown by the dashed line. The whole foot landing is judged when the COP at landing is located in the ellipse.

2.3. *Gait pattern generation for climbing and stepping down*

Figure 4 (a) shows the definition of joint angles of the robot. θ_1 and θ_3 denote angles of knee joints, and θ_2 and θ_4 denote angles of ankle joints. θ_5 is hip joint angle between left and right thigh links. For the initial and terminal states, joint angles are 0 deg (a standing posture) as shown in Figure 4 (b). While climbing, the robot raises the right foot on the top of the step. Then, the robot raises the left foot on the step. The sequence of stepping down from the step is as well as that of climbing consisting of the two-step procedures.

In a gait patterns generation, trajectories of toe position and foot angle of the swing leg related to a horizontal axis are determined to avoid a collision between the swing foot and the step. These trajectories are calculated on the

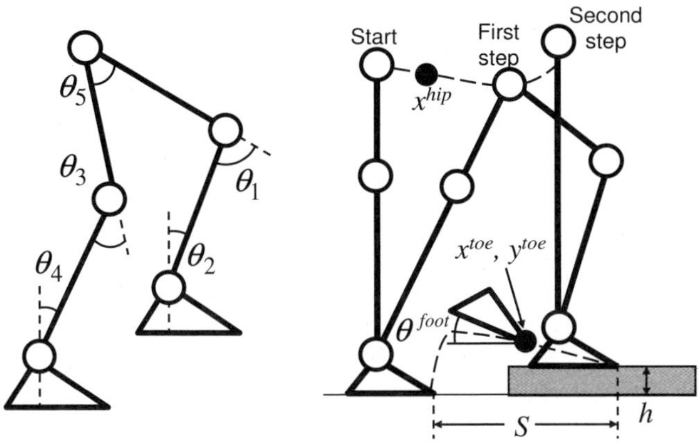

(a) Joint angle

(b) Postures and trajectories while stepping on a bump

Figure 4 Definition of joint angles and variables for trajectory generation. (a) shows the joint angles of the robot. (b) shows the variables to calculate desired trajectories of joint.

basis of minimum jerk (change of acceleration) trajectory with via-point constraints [7] which are specified by the estimated height of the step. In addition, the stance knee joint angle is kept at maximum extension position (0 deg). Once the horizontal hip joint position is determined, the joint angle can be calculated using inverse kinematics equations with the toe position, foot angle and knee joint angle.

The trajectory of hip position is determined so that the sum total of the squared joint angular jerk is minimized. The cost function is given by

$$J = \sum_{i=1}^{5} \sum_{j=0}^{N-1} \left\{ \frac{d^3}{dt^3} \theta_i(t_j) \right\}^2, \tag{1}$$

where N denotes the number of sampling. Avoiding rapid acceleration or deceleration of joint movement, the power of actuators can be decreased and balance losses of the user can be avoided. In addition, sufficient horizontal velocity of hip at toe-off in the second step would be required to avoid backward balance loss because the hip position is posterior to the leading foot and the gravity decelerates the progressional movement. Assuming that the dynamics of the body center of mass movement can be simplified as a linear inverted pendulum model and that the horizontal hip position is equal to the horizontal center of mass position, the horizontal velocity of the hip must satisfy the following condition to avoid backward balance loss [8].

$$\dot{x}^{hip}(t^*) > -\omega \, x^{hip}(t^*), \tag{2}$$

where t^* denotes the time at toe-off, and ω indicates natural frequency of the inverted pendulum. Its value was given by $\sqrt{g/l}$ = 3.5 (1/s). To reduce the optimization variables, the hip trajectory is represented by a M-polynomial function $x^{hip}(t) = a_0 + a_1 t + \ldots + a_M t^M$. Then the boundary conditions for the start point ($t = t_0$) and end point ($t = t_{N-1}$) are given by

$$x^{hip}(t_0) = x_0^{hip} \quad \dot{x}^{hip}(t_0) = \ddot{x}^{hip}(t_0) = 0, \tag{3}$$

$$x^{hip}(t_{N-1}) = x_f^{hip} \quad \dot{x}^{hip}(t_{N-1}) = \ddot{x}^{hip}(t_{N-1}) = 0, \tag{4}$$

From equation (3), we obtain $a_0 = x_0^{hip}$, $a_1 = a_2 = 0$. From equation (4), a_3, a_4, a_5 are represented by the coefficients over the 6th order terms as follows

$$a_3 = -\frac{1}{2}\sum_{k=6}^{M}(k-4)(k-5)t_f^{k-3}a_k + \frac{10(x_f^{hip} - x_0^{hip})}{t_f^3}, \tag{5}$$

$$a_4 = \sum_{k=6}^{M}(k-3)(k-5)t_f^{k-4}a_k - \frac{15(x_f^{hip} - x_0^{hip})}{t_f^4}, \tag{6}$$

$$a_5 = -\frac{1}{2}\sum_{k=6}^{M}(k-3)(k-5)t_f^{k-5}a_k + \frac{6(x_f^{hip} - x_0^{hip})}{t_f^5}. \tag{7}$$

The coefficients over the 6th order terms of the polynomial function were determined so that the sum total of the squared joint angular jerk was minimized subject to the condition at toe-off (equation 2). Since the coefficients of over the 9th order terms were almost 0 in preliminary calculations, the 8th order polynomial function was used to determine the hip trajectory. Due to the nonlinearity of the cost function, iterative procedure was required to calculate the optimal coefficients corresponding to the step height h. For the online generation for specified value of h, the relationship between the values of the coefficients and the height h were approximated by linear regressions. We confirmed that the optimal coefficients were constant or monotonically increase or decrease in relation to the height, and that the trajectories with approximated coefficients were little different from those with the optimal coefficients.

Upper and middle figures in Figure 5 show gait patterns for climbing and stepping down with $h = 0.06$ m, respectively. As shown in (a) and (d), the joint angle profiles are smooth. The stick pictures of the first step, shown in (b) and (e), and the second step, shown in (c) and (f), show that the collision between the swing foot and the step does not appear. The lower figures in Figure 5 show horizontal velocity and position of hip during the second step. The solid and dashed lines indicate the trajectories with and without the condition to avoid backward balance loss, respectively. The trajectory without the condition shows that horizontal hip velocity is almost zero and much smaller than right-side value of equation (2). On the other hand, the trajectory with the condition shows that hip position moves forward and the velocity is sufficiently large at toe-off.

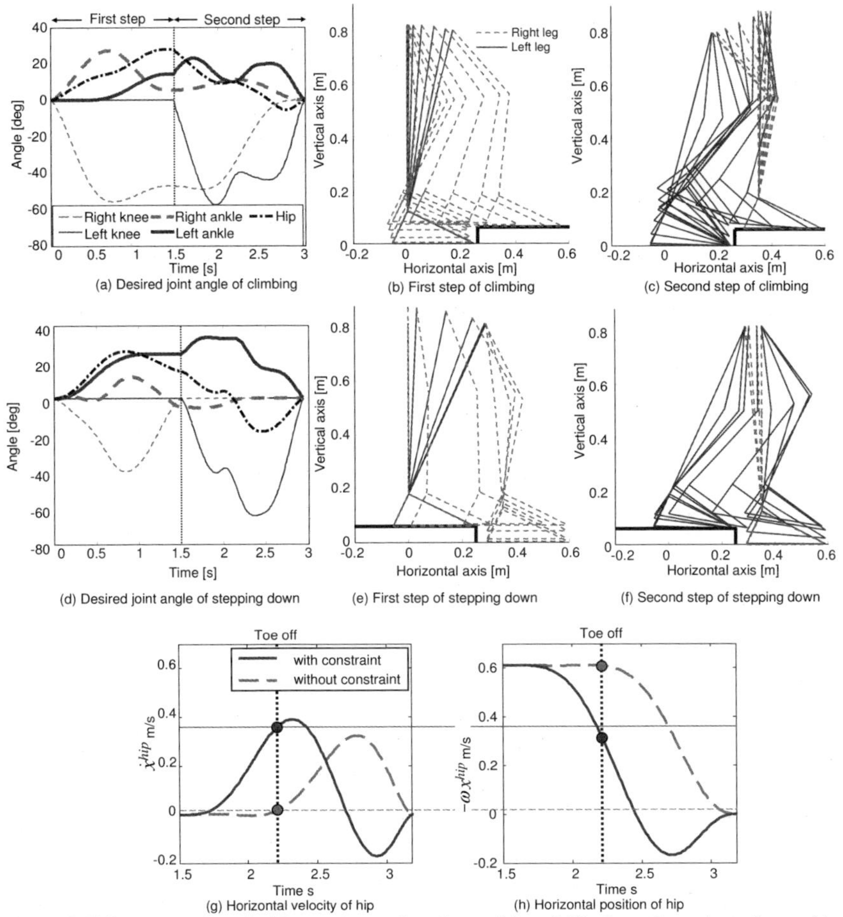

Figure 5 Gait patterns for climbing and stepping down. (g) and (d) show the trajectories on hip position. The solid and dashed lines show the profiles of the trajectories with and without a constraint to avoid backward falling, respectively.

3. Experiment

We performed the experiments of walking on steps to evaluate the feasibility of our system. To examine the walking on steps with different height, a user walked on each of steps with the height of 0.03 m and 0.06 m. Figure 6 (a) and (b) show the pictures of walking on a step with the height of 0.06 m. Our system enabled to assist the leg movement while climbing and stepping down for the steps. Figure 6 (c) shows COP trajectories during the second steps. The COPs at landing located within the 95% confidence ellipse and they moved forward during the second step, indicating that the backward balance losses were avoided during climbing steps with our system.

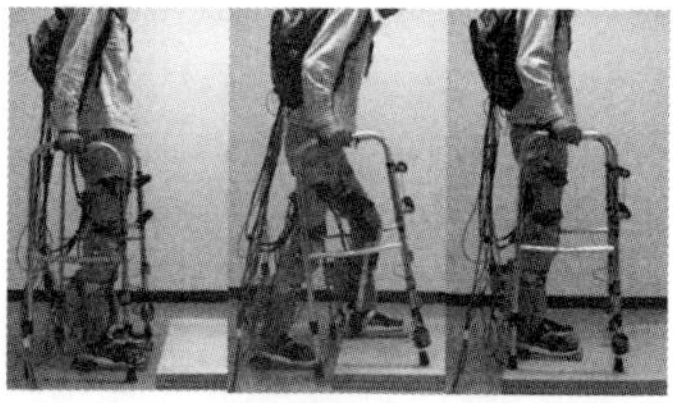

(a) Climbing a step

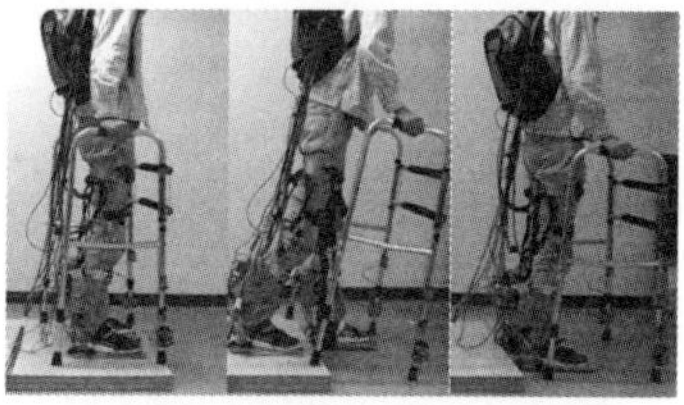

(b) Stepping down from a step

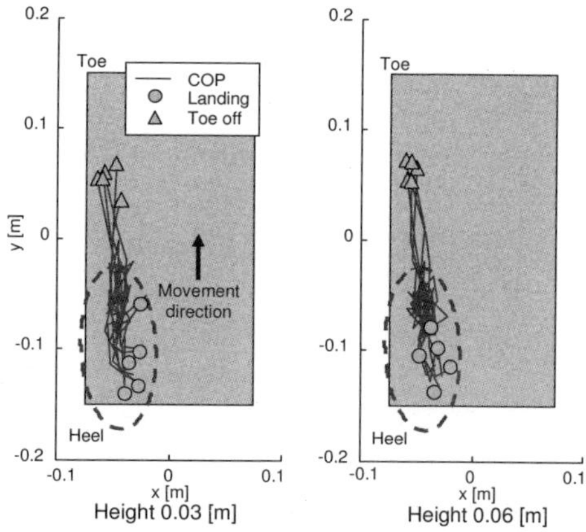

(c) COP trajectories in the right sole during the second step to climb a step

Figure 6 Experimental results of walking on a step with WPAL. (a) and (b) show the pictures while climbing and stepping down for a step with 0.06 m height, respectively. (c) shows the COP trajectories while climbing a step.

4. Conclusion

We proposed a sensing and control system for walking on a step with a wearable robot. We demonstrated the feasibility of our system to assist climbing and stepping down for steps using the wearable robot.

The maximum height of a step for assistance with WPAL was 9 cm. This limitation was due to the range of motion in the hip actuators and can be overcome by improvement of the drive mechanism of hip actuators. For practical implementation, a main controller should be embedded to a microprocessor, and a battery system with fail safe would be required. In addition, the effectiveness of our system for paraplegics should be examined.

References

1. D. Popovic, et. al, *Medical Engineering & Physics*, **25**, 67 (2003).
2. T. Kagawa and Y. Uno, *Proc. IEEE International Symposium on Robot and Human Interactive Communication*, 633 (2009).
3. P. Veltink, et. al, *Archives of Physiology and Biochemistry*, **109**, 1 (2001).
4. H. Kawamoto and Y. Sankai, *Advanced Robotics*, **19-7,** 717, (2005).
5. T. Kagawa and Y. Uno, *Proc. IEEE/RSJ International Conference on Intelligent Robots and Systems*, 4067 (2009).
6 H. Kitamura et. al, *Emerging Trends in Mobile Robotics*, 1047 (2010).
7. T. Flash and N. Hogan, *Journal of Neuroscinece*, **5-7**, 1688 (1985).
8. T. Kagawa and Y. Uno, *Human Movement Science*, **29-6**, 964 (2010).

INTERACTIVE PERSON FOLLOWING FOR SOCIAL ROBOTS: HYBRID REASONING BASED ON FUZZY AND MULTIPLE-OBJECTIVES DECISION MAKING APPROACHES

C. GRANATA* and P. BIDAUD

Institut des Systemes Intelligent et de Robotique
Universite Pierre et Marie Curie(CNRS - FRE 2507)
4, place Jussieu 75252 Paris cedex 05, France
E-mail: consuelo.granata@robosoft.fr
E-mail: philippe.bidaud@upmc.fr
**Robosoft (Bidart, France)*

A. BUENDIA

SpirOps Headquarters
8, passage de la bonne graine 75011 Paris,France
E-mail: axel.buendia@spirops.com

In this study we propose an innovative decision making engine for interactive person following for a mobile social robot. The developed engine combines 1. A Strategy controller that selects the strategy to apply 2. A Multiple-Objective system that provides to the robot the way to perform this strategy. We show the performances of this system by two sets of experiments (in simulation environment): the first one is based only on the observed data; in the second one we predict the state of the user with a Kalman filter and the state of the robot with a predictive model. We demonstrate that the proposed reasoning system is able to regulate the robot behavior in complex and critical situations. The suggested predictors improve the system performances considerably.

Keywords: Interactive person following; Fuzzy logic; Multiple-Objective Decision Making; SpirOps AI.

1. Introduction

Personal robotic systems can be used in our households, offices, hospitals, shopping malls, and others in-door cluttered and human-populated environments. One of their main functions is to be placed near the person with whom they must stay connected, this taking into account the movements

of the person, the environment in which it operates and the tasks that the robot is supposed to achieve for any individual. The synthesis of interactive behavior and especially following people in a robust way is one of the central issues for these systems. Person following behavior synthesis in an indoor environment is particularly challenging. Solving this problem in full autonomy requires first that the robot can perceive its fixed and moving environment, localize, recognize and take into account both the context and the person activity. In addition, it may react to possible interactions with the user (inputs from different interfaces: audio, tactile, video, etc). It must also consider the state of the person, predict its movements and anticipate its path if necessary. This interactive following process may well not simply be based on the control of the distance[1] and/or of the relative speed[2] with respect to the person. It must exploit appropriated reasoning mechanisms. Gockley et al[3] . present an interactive approach of person following discussing two methods, direction-following and path-following, and they propose as ideal behavior a hybrid approach, with the robot automatically selecting which method to use. A similar approach is described by Yuan et al.[4] : they propose a combination of direction-following, path-following and parallel-following behaviors depending on the current state of the environment and on the relative position of the followed subject. Kruijff et al[5] . discuss various interaction strategies observed in independent Wizard-of-Oz studies. One of their conclusions is that in person following it is very important to take into account the human behavior.
In this paper we propose a decision making engine combining two modules: 1. A Strategy controller that selects a particular strategy to apply 2. A Multiple-Objective system that provides to the robot the way to perform this strategy. This reasoning mechanisms allows the robot to achieve the assigned task even in case of noisy and uncertain perception of the user and of the environment. The presented system is conceived to be implemented on the social mobile robot Kompaï (Figure 1).

2. Methodology

2.1. *The Fuzzy Inference System*

Deterministic approaches for decision making systems are suitable when the available information are precise. The use of precise mathematical models in a situation which is imprecise by its nature, will hardly yield an acceptable result. The fuzzy approach is most suitable when the knowledge of the environment and of the context is imprecise and the solution requires approximate reasoning. For this reason we adopt a decision making en-

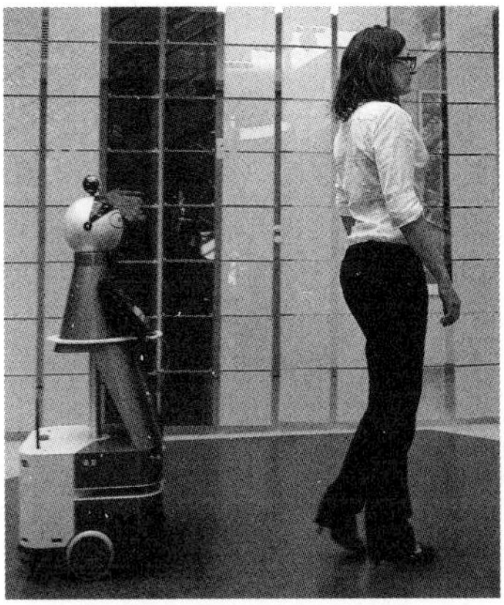

Fig. 1. Kompaï, the mobile robot for which the decision making engine presented in this paper is conceived.

gine that combine a powerful Strategy controller and a Multiple-Objective optimization system in order to regulate the mobile robot behavior when the robot actions cannot be planned in a deterministic manner because of the involved dynamics and uncertainties.

2.2. *Strategy controller*

In order to build an interactive person following behavior for a mobile robot, we need to define different strategies that the robot has to select dynamically according to the available information about the user and the environment. In this paper we will restrict to 3 strategies: GoToUser, GlobalSearch and LocalSearch. The decision of the strategy to apply is made through fuzzy rules (an example is shown in Figure 2). These rules assign a degree of interest to each strategy. The degree of interest is computed as the product of different factors (the user detection confidence, the time elapsed from the last visit of a place, the distance to the position to reach etc...) and it can take any value between 0 and 1. In the developed framework, the computation of the interests is left to the SpirOps AI decision making toolbox which is described in Section 2.5.

2.2.1. *GoToUser*

This strategy is activated when the decision making engine determines that the robot can approach to the user. This decision is taken when the robot has sufficiently high confidence on the user detection. As shown in the left image on Figure 2, the user detection confidence is calculated on the basis of the number of detections (n) during the last 10 laser scans.

2.2.2. *GlobalSearch*

When the robot has to start an interaction with the user and it doesn't know the current user position, it must search for him in the environment. In the known robot workspace (the robot integrates a SLAM algorithm [a] which allows it to quickly establish the mapping of the environment) we have selected some points where the robot can go to search for the user. The criteria used to choose the next point among the predetermined points is based on the time information of the last visit: the chosen point is the place which has been visited less recently. This behavior is activated if the robot didn't detect the user for too long time (see GlobalSearch rule in Figure 2 where It is the time elapsed from the last successful detection).

2.2.3. *LocalSearch*

This strategy is triggered when the robot is going close to the user and it loses him (see LocalSearch rule in Figure 2). It consists of a local search around the last detected position of the user. For this function we have developed two different strategies: the first one consists of searching the user by moving to the known points (the same points used in the GlobalSearch strategy) selected on the basis of the proximity to the last detected position of the user; the second one is based on estimating the user position using a Kalman filter.

2.3. *Multiple-Objective Decision Making*

The purpose of the Strategy controller is to select the action to execute, i.e. what the robot has to do in order to stay connected to the user. But the decision about how execute the selected strategy is a different problem that we treat by a Multiple-Objective Decision Making (MODM) approach. In other words we use MODM approach to regulate the local behavior of the robot while it is executing a selected strategy. MODM methods deal with the process of making decisions in the presence of multiple objectives

---

[a]The SLAM software used is based on the Karto®robotic navigation software under license from SRI International®.[6]

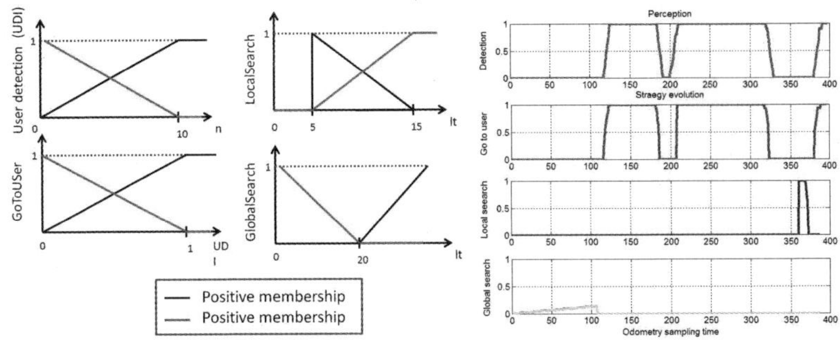

Fig. 2. Left: rules used to associate the user detection to the strategy to apply. Right: evolution of interests associated to each strategy with respect to the user detection (right figure).

usually conflicting. In MODM, the alternatives are not predetermined but instead a set of Objective Functions is optimized subject to a set of constraints. The most satisfactory and efficient solution is sought; the solution is highly dependent on the preferences of the decision-maker and must be a compromise. We have developed a control system based on MODM formalism, in which we consider the coordination of multiple behaviors as an optimization problem with a number of relevant Objective Functions.

2.4. *Objectives*

We define Objectives as basic robot navigation behaviors. In our system we have 3 Objectives (Turn, GoForward and PathFollow) that are blended to form a more complex behaviors. The Objectives are competitive and often conflicting, i.e. the improvement in one of them can be entail a deterioration in another one. Each Objective is characterized by a parameter named Attribute and by an Objective Function. An Attribute is a measurable quantity whose value represents the degree of achievement of the respective Objective. The Objective Function evaluates the desirability of the associated Objective.

2.4.1. *Turn*

It corresponds to the robot capability to turn on spot. The aim of this behavior is to minimize the angular error (the angular position of the point to reach in the robot frame). This angular error θ_r^d is the Attribute of the Turn Objective. The Turn Objective Function can be formulated as follows:

$$O_{Turn}(V, \omega) = (0, \frac{\omega}{||\omega_{MAX}||^2} \theta_r^d)$$

Under constraints on the angular speed of the robot.

$$\omega \leq \omega_{MAX}$$

$$\omega^c - \frac{d\omega}{dT} \leq ||\omega|| \leq \omega^c + \frac{d\omega}{dT}$$

Where ω^c is the current angular speed of the robot.

2.4.2. *GoForward*

It corresponds to the robot capability to go forward with a certain linear speed. The aim of this basic behavior is to minimize the quadratic error between the current robot position (P^c) and the desired position (P^d):

$$\epsilon_d = ||P^c - P^d||^2$$

The Attribute of the GoForward Objective is this quadratic error ϵ_d. The GoForward Objective Function can be formulated as follows:

$$O_{GoForward}(V, \omega) = (\frac{V}{V_{MAX}} \epsilon_d, 0)$$

The constraint to satisfy are

$$V \leq V_{MAX}$$

$$V^c - \frac{dV}{dT} \leq ||V|| \leq V^c + \frac{dV}{dT}$$

Where V^c is the current robot linear speed.

2.4.3. *PathFollow*

It is an Objective that allows the robot to follow a path computed by the PathPlanning algorithm. The PathPlanning algorithm is a function provided by the module of SLAM implemented on our platform. The algorithm used for the path following is the control law proposed by Morin and Samson in.[7] The Attribute associated to this Objective is the angular error θ_e (the angle characterizing the orientation of the robot chassis with respect to the tangent of the path in the projection of the robot control point). In order to minimize this error, the PathFollow function provides the vector of desired robot speeds (V,ω), respecting the non-holonomic constraints of the robot. The Objective Function can be formalized as follows:

$$O_{PathFollow}(V, \omega) = (V^C, -V^C * (\frac{tan(\theta_e)}{l} + \frac{k*d}{cos(\theta_e)}))$$

Where V^C is a constant linear speed defined *a priori* for the PathFollower, k is the Morin-Samson gain and d is the distance from the robot to the path to follow.

2.4.4. *Combination of Multiple-Objectives*

We propose a weighting method to combine the Multiple-Objectives in order to satisfy simultaneously all Objectives as well as possible. We associate to each Objective Function a weight w. The weights are computed as Parameters of the selected Strategy by the SpirOps AI toolbox taking into account the position of the point to reach with respect to the robot and the presence of obstacles close to robot (Figure 3). The obstacles are detected by the laser scan embedded on the robot. This method is based on scalar vector optimization and can be formalized as follows:

$$\hat{x} = argmin_{x \in X}(\sum_{i=1}^{n} w_i O_i(x)) = w^T O(x)$$

where w_i are normalized weights and $\hat{x}$ is a nondominated solution if it is a unique solution or if $w_i > 0$ for all i (see[8]).

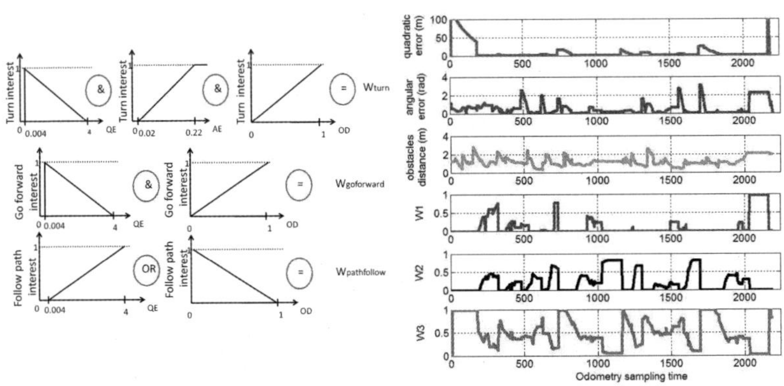

Fig. 3. Left: rules used to regulate the weights for Multiple-Objectives Decision Making (QE= quadratic error, AE= angular error, OD= obstacles distance)). Right: evolution of weights in function of the the position of the point to reach and of the environment.

2.5. *Spirops AI*

SpirOps AI tools[9] provide an open environment for relatively simple incremental design of autonomous decision processes. A graphical editor is used to edit independent needs or goals called Drives. This decision process pulls data from the host application as lists of typed objects described by attributes, including the entity that uses the decision process itself. The decision-making is executed by a simple call. It outputs a report of all the Desires it has selected to be acted upon. A Drive is an action producer. Each Drive represents an independent module responsible for a reflex, a need or a goal. It produces Desires as possible intentions of action. Each Desire includes a target, an action, and the different parameters used to produce a complex action. A Drive is a decision making unit. It has to compute its adequacy with respect to the pursued goal and its opportunity in the current context, which leads to the computation of the degree of interest. The interest can be seen as a fuzzy value, and the Drive as an extended 0+ fuzzy rule.[10] For each sensed object, the Drive will compute the degree of interest and the parameters for each action it produces.

3. Robot test bed and simulation environment

Our approach has been studied to be integrated with the Kompaï robot, a differential non-holonomic mobile robot which is equipped with all necessary sensors for autonomous navigation: laser sensor for localization and obstacle detection, US and IR sensors for obstacle detection and 2 cameras. The low-level control is managed by the embedded Microcontroller (Emtrion SH7780 SBC). For high-level control and user interfaces, the Kompaï uses a Tablet PC running Windows Vista. The implemented software architecture has been tested in simulation. In order to have a realistic robot behavior we use a dynamic simulation environment (Ageia PhysX[11]). The simulated robot kinematics model corresponds to the real platform. The laser sensor is simulated considering a realistic noise of the real sensor. The software architecture is programmed in MRDS (Microsoft Robotics Dev Studio) platform in *CSharp* language. In order to take into account the detection problems (very common in unstructured, cluttered and dynamic environments as homes, offices, hospitals etc) we have implemented a detector based on the simulated laser sensor installed on the robot. As we do not dispose of a simulated human entity, we use a square box to simulate the human and we have implemented a simple segment detector in order to simulate the user detection and localization.

4. Experiments I

4.1. *Exiting through the door*

The first critical case analyzed is exiting through the door. The user leaves a room and he can turn on the right or on the left. The walls presence involves the loss of detection by the user. The strategy applied here is to send the robot to the predefined point closest to the last user detected position. The performance of this strategy depends from the choice of the points where the robot can go to search the user. If the point closest to the last detected position is on the opposite direction with respect to the user walking, the robot cannot find locally the user and it has to start a global search. An example is showed in Figure 4 where the yellow point is the start user position, the orange point is the end user position and the robot is represented in different colors according to the selected strategy (violet when it goes close to user, blue when it makes a local search, green when it makes a global search). In this example the robot loses the user, it chooses the wrong direction and it has to explore the environment with a global search. The robot spends about 12 s of local search and one minute of global search to find the user again.

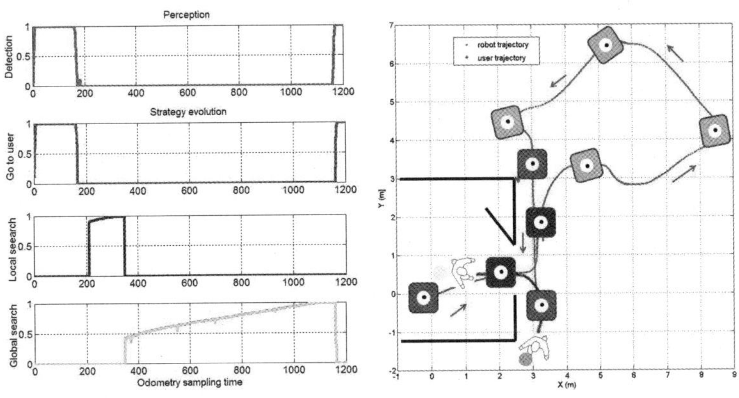

Fig. 4. Left: the strategies selected by the decision-making engine. Right: the trajectories of the user and of the robot.

4.2. *Turning around the table*

The second critical case analyzed is walking around the table. An example is showed in Figure 5 where we use the same colors notation explained in

Section 4.1. The robot loses the user localization because the user is hidden by the table. It applies first the local strategy (for 13 s) that is not fruitful, then the global strategy that allows it to find the user again after about 30 s.

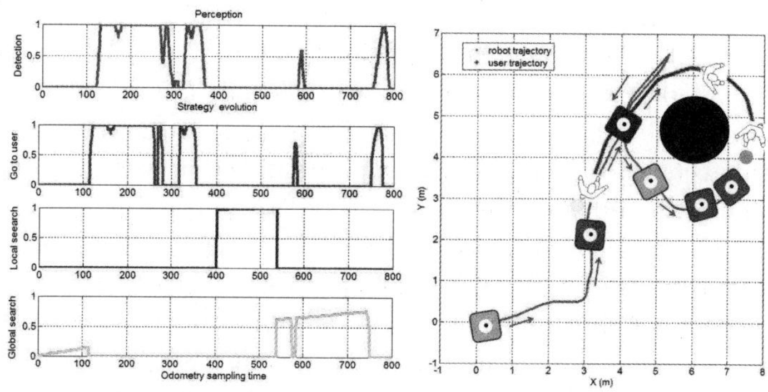

Fig. 5. Left: the strategies selected by the decision-making engine. Right: the trajectories of the user and of the robot.

5. Experiments II

In order to ameliorate the robot behavior, we have introduced in our system two prediction steps: a Kalman filter to predict the trajectory of the user and a predictive model to predict the trajectory of the robot. We demonstrate the improvement of the system analyzing: the simple case of the user walking straight and the same critic cases as the experiments described in Section 4.

5.1. *Predictive models*

5.1.1. *Prediction of user position*

In order to predict the user state we use a Kalman filter that is an optimal estimator of the state for discrete time linear systems in stochastic context. Assuming that the human moves with a constant velocity in a certain time interval and that the noise of our system is a Gaussian noise we can describe the human motion in polar coordinates by the following discrete state variable equations.

$$x_{k+1|k} = F_k x_{k|k} + w_k$$
$$y_k = C_k x_k + v_k$$

$$x_k = \begin{bmatrix} \rho_k \\ \theta_k \\ \dot{\rho_k} \\ \dot{\theta_k} \end{bmatrix}; K_k = \begin{bmatrix} 1 & 0 & \Delta t_k & 0 \\ 0 & 1 & 0 & \Delta t_k \\ 0 & 0 & 1 & 0 \\ 0 & 0 & 0 & 1 \end{bmatrix}; C_k = \begin{bmatrix} 1 & 0 & 0 & 0 \\ 0 & 1 & 0 & 0 \end{bmatrix} \tag{1}$$

where ρ_k and θ_k are the polar coordinates at the sampling interval k; $\dot{\rho_k}$ and $\dot{\theta_k}$ are the velocities of ρ_k and θ_k at the sampling time k; F_k is the state transition matrix; C_k is the observation matrix and w_k and v_k are respectively the system noise due to mis-modeling and the observation noise due to the sensors errors.

The use of a human state predictor allows to fix in a lot of cases the problems of leak of detection. We use the Kalman filter in the local search strategy in order to minimize the time spent by the robot to search the user. The new local search strategy consists on send the robot in the point where the user is estimated be after 5 s from the last detection. The filter is able to compute this point taking in to account the user trajectory and speed observed before the lost of detection.

5.1.2. *Prediction of robot position*

The robot Kompaï used in our application is a differential non-holonomic mobile robot.

We make the following hypothesis:
1. The robot considered is made up of a rigid body with non deforming wheels;
2. The vehicle moves without slipping on a plane.
So, according to the motion principle of a rigid body, the kinematics model of the robot can be described by the following system:

$$\begin{bmatrix} \dot{x} \\ \dot{y} \\ \dot{\theta} \end{bmatrix} = \begin{bmatrix} cos(\theta) & 0 \\ sin(\theta) & 0 \\ 0 & 1 \end{bmatrix} \begin{bmatrix} V \\ \omega \end{bmatrix}$$

Considering: a discrete-time representation of this system, a sampling period T, a sampling instant k and applying the Euler's approximation, we obtain the following discrete-time model for the robot motion:

$$\begin{cases} x(k+1) = x(k) + v(k)cos\theta(k)T \\ y(k+1) = y(k) + v(k)sin\theta(k)T \\ \theta(k+1) = \theta(k) + w(k)T \end{cases}$$

We use the robot predictive model and the output of the Kalman filter described in Section 5.1.1 to define a new Objective: GoForwardFast. The aim of this basic behavior is to minimize the error ϵ_p between the predicted position of the user (at time k+T) and the predicted position of the robot (at time k+T).

$$\epsilon_p = ||P_u(k+T) - P_r(k+T)||$$

Where k is the current sampling instant, T is the sampling period, $P_u(k+T)$ and $P_r(k+T)$ are respectively the predicted position of the user and the predicted position of the robot at next sampling time. The Attribute of the GoForwardFast Objective is this error ϵ_p. The GoForwardFast Objective Function is formulated as follows:

$$O_{GoForwardFast}(V, \omega) = (V\epsilon_p, 0)$$

The constraint to satisfy are the same as for the GoForward Objective.

The weight associated to this Objective is computed combining two rules: the first one based on the user detection confidence, the second one based on the error ϵ_p (Figure 6).

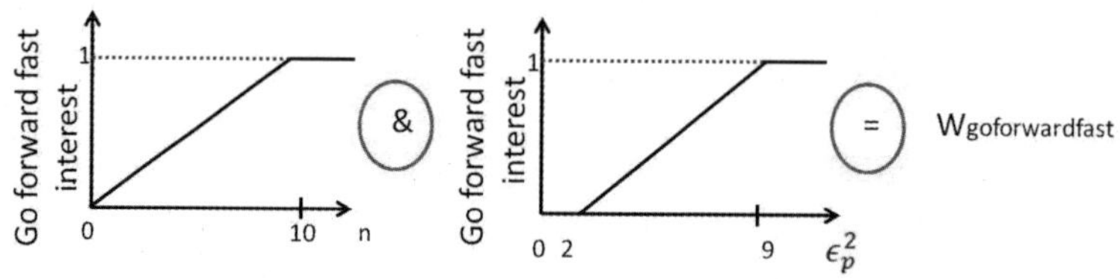

Fig. 6. Rules used to compute W4 (= Wgoforwardfast). n is the number of user detection in the last 10 laser scan, and ϵ_p^2 is the squared Attribute of GoForwardFast Objective.

In order to demonstrate the positive effect of the introduction of GoForwardFast Objective, we have reproduced for the both systems (with and without GoForwardFast Objective) the following experiment: the user walks straight with a speed of 0.7 m/s and he remains all the time in the field of view of the robot (he is never hidden). As showed in Figure 7, when the quadratic distance between the predicted user position and the predicted robot position exceeds 2m, the linear speed remains constant in the case of $w4 = 0$ and increases otherwise. The speed increment allows the robot to stay connected with the user. In fact in the case of $w4 \neq 0$ the quadratic distance between the predictions is $\leq$ 5m, in the case of $w4 = 0$ this distance increases to more than 10m.

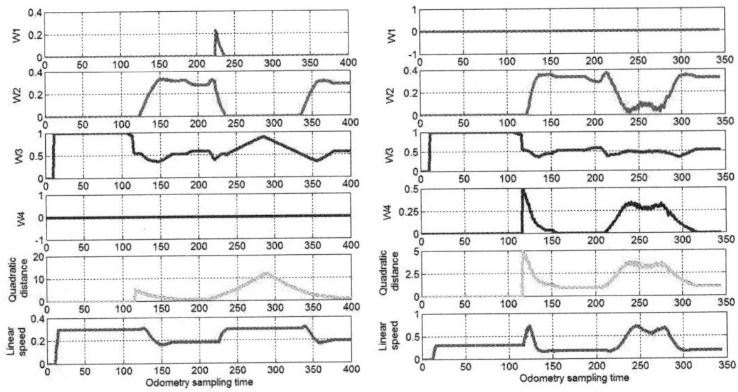

Fig. 7. The computed weights and the linear speed of the robot following an user that moves straight (left considering $w4 = 0$; right considering $w4 \neq 0$)

5.2. *Exiting through the door*

The robot looses the user detection when the user leaves the room, but by predicting his position it can take the right direction where to search locally for the user. As showed in Figure 9 (where we use the colors notation explained in Section 4.1), introducing the prediction step, the robot found again the user after 8 s of local search and without the need to do a global search (as it had to do in the same situation no filtered). It means that even if it looses sight of the person, it is able to estimate its position.

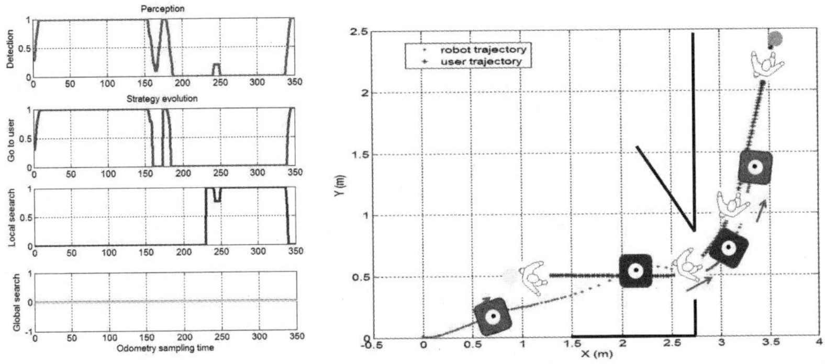

Fig. 8. Left: the strategies selected by the decision-making engine. Right: the trajectories of the user and of the robot.

5.3. *Turning around the table*

As descried in Section 4.2, the user is hidden by the table and he turns around. Introducing the prediction step we significantly improve the robot performances: in the example showed in Figure 9 (colors notation described in Section 4.1), in about 6 s of local search the user is detected again.

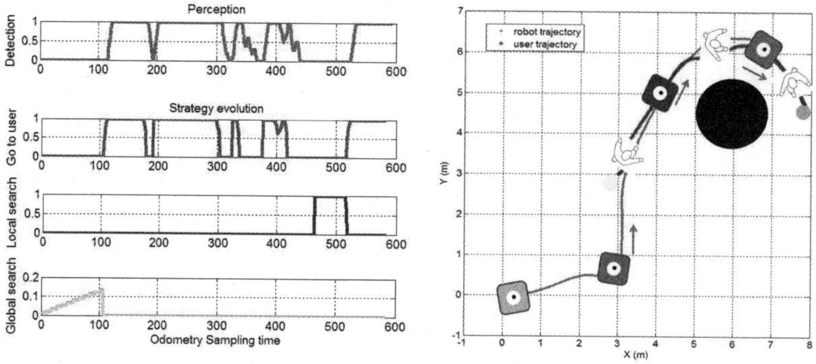

Fig. 9. In the left side, the strategies selected by the decision-making engine. In right side, the trajectories of the user and of the robot.

5.4. *Anticipation*

The predictive filter introduction increases the local search strategy performances, but has also an anticipating role in the robot trajectory. In fact, when the robot estimates a certain position of the user, it computes autonomously the trajectory to follow and it can decide to take a shortcut when it is more convenient. In Figure 10 (colors notation described in Section 4.1), an example of shortcut is showed. The robot detects that the human is passing around an obstacle and it decides to reach him directly after the obstacle.

6. Conclusion and perspectives

In this study we have proposed a decision making engine for interactive person following by a social mobile robot. This engine combines a Strategy controller that selects the strategy to apply by fuzzy rules and a Multiple-Objective system that provides to the robot the way to perform this strategy. We proposed a set of Objectives combined by fuzzy rules in order to generate a smooth robot movement taking into account the geometric shape of the environment, the constraints of the robot and the state of the user.

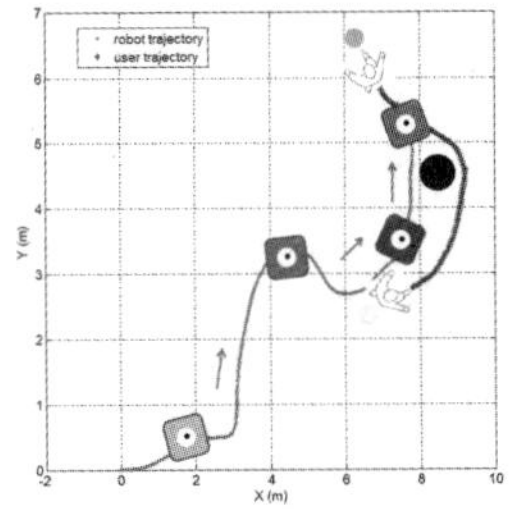

Fig. 10. The robot takes a shortcut to reach the user.

We proposed also a set of Strategies activated by our decision making engine in order to manage the robot movement around the followed person and to choose the action to execute when the person is not detected. The use of fuzzy rules allows to have acceptable behaviors compensating the uncertain and noisy perception. Finally we have introduced a prediction step to improve the system performances. We demonstrate in simulated application, that the robot is able to reasoning and to decide the task to achieve in order to maintain the person following also in critic difficult cases. For future work we intend to implement our decision making engine in the real robot with a people detector based on laser and camera data[12] . We will also sophisticate our system using various inputs issue from the robot multimodal interface (voice recognition, touch screen and camera) in order to enrich the robot behaviors.

7. Acknowledgements

This work was supported by the Domeo[13] project AAL 2008-2008-1-159.

References

1. R. C. Luo, Y. J. Chen, C. T. Liao and A. C. Tsai, Mobile robot based human detection and tracking using range and intensity data fusion. Advanced Robotics and Its Social Impacts. ARSO 2007. IEEE Workshop on, (2007).
2. J. Satake and J. Miura, Robust stereo-based person detection and tracking for a person following robot. In Proceedings of the IEEE ICRA 2009 workshop on people detection and tracking, (2009).
3. R. Gockley, J. Forlizzi and R. Simmons. Natural person-following behavior for social robots. In Proceedings of Human-Robot Interaction, (March 2007).
4. F. Yuan, M. Hanheide and G. Sagerer. Spatial context-aware person-following for a domestic robot. In: International Workshop on Cognition for Technical Systems, Munich, Germany (2008).

5. G.-J. M. Kruijff, H. Zender, P. Jensfelt and H. I. Christensen. Situated Dialogue and Spatial Organization: What, Where... and Why?, International Journal of Advanced Robotic Systems, vol.4, number 2 (2007).
6. http://www.kartorobotics.com/
7. Morin P. and Samson C. Motion control of wheeled mobile robots. Handbook of Robotics, Springer Verlag, (2008).
8. V. Chankong. Multiobjective Decision Making: Theory and Methodology, vol 8, North-Holland, (1983).
9. Buendia A. Spirops AI, a behavior production tool. [Online]. available: http://www.spirops.com/SpirOpsAI.php.
10. Magnus P.D. Forallx an introduction to formal logic. University at Albany, State University of New York, (2005).
11. Craighead J., Murphy R., Burke J. and Goldiez B. A survey of commercial and open source unmanned vehicle simulators. In Proceedings of ICRA 2007 : IEEE/Int. Conf. on Robotics and Automation, pp. 852-857, Roma,Italy, (April 2007).
12. X. Wang, X. Clady, C. Granata. A Human Detection System for Proxemics Interaction. ACM/IEEE International Conference on Human-Robot Interaction (Late Breaking Results). Lausanne, Switzerland, (2011).
13. http://www.aal-domeo.org/

DEVELOPMENT OF A PASSIVE-TYPE ROBOTIC WHEELCHAIR

D. CHUGO* and T. HIGUCHI

School of Science and Technology, Kwansei Gakuin University,
Sanda, Hyogo 6691337, Japan
**E-mail: chugo@kwansei.ac.jp*

Y. SAKAIDA

RIKEN-TRI Collaboration Center for Human-Interactive Robot Research,
Nagoya, Aichi 4630003, Japan

S. YOKOTA

Faculty of Science and Engineering, Setsunan University,
Neyagawa, Osaka 5728508, Japan

K. TAKASE

Graduate School of Information and System, The University of Electro-Communications,
Chofu, Tokyo 1828585, Japan

We propose a novel intelligent wheelchair based on the passive robotics. Our proposed assistive wheelchair consists of a frame, casters, wheels and servo brakes. Controlling a torque of the servo brake, our wheelchair changes its dynamics and realizes the reference tracks. Our system requires no actuators, and its mechanism is simple and low cost. There is no risk by malfunction of servomotors and patients can use it intuitively because they use our wheelchair passively with their own intentional force. Our key ideas are two topics. One is the development of the passive-type assistive wheelchair for practical use. In previous researches, assistive wheelchairs use actuators and their systems are complexity. Without actuators, our system realizes the simple mechanism which is easy to implement for a general standard wheelchair. The other key topic is proposing a human adaptive motion control algorithm, which changes the dynamics of our wheelchair according to reference tracks and input driving force by its user. Using proposed scheme, the user can drive our wheelchair with natural feeling. We test our proposed assistance system by the experiments with our prototype and verify its effectiveness.

Keywords: Assistive Wheelchair; Passive Robotics; Servo Brake.

1. Introduction

Wheelchairs are widely used by many people who are mobility impaired in daily activities. However, many accidents often occur to wheelchair users and its risk is serious. More than 80[%] of wheelchair accidents are caused by environmental problems.[1] Especially, the inclination of sidewalk has high risk for wheelchair user. In Japan, it is permitted to incline the sidewalk up to 5[deg] (Fig.1).[2] This inclination leads the wheelchair user to run out from the sideway to the roadway and it causes the accident between the wheelchair and the car. Therefore, the driving assistance system for wheelchair is important.

(a) Inclination of the sidewalk

(b) The example of typical sidewalk

Fig. 1. Risks of the sidewalk

In previous works, a lot of assistive technologies for wheelchairs are developed. In general, many handicapped people traditionally use power wheelchairs[3] and previous researchers have tried to realizing assistance functions by adding wheels with actuators and controlling them based on the robot technology such as motion control technology,[4] sensing technology and computational intelligence.[5,6] These intelligent wheelchairs provide many functions, such as suitable motion, obstacle avoidance and navigation; thus, they provide a maneuverable system. However, many wheelchair users have the upper body strength and dexterity to operate a manual wheelchair. For these wheelchair users, its cost is too expensive and is not acceptable.

On the other hand, a manual wheelchair without servomotors, which consists of a frame, wheels, casters and hand brakes, is commercially available and widely used. Its mechanism is simple and low cost. There is no risk by malfunction of servomotors and patients can use it intuitively because they use these wheelchairs passively with their own intentional force. Of course, these wheelchairs cannot assist to drive dynamically as powered wheelchairs. Therefore, there is no well-adapted wheelchair assistance system for users who are healthy and have the upper body strength.

In this paper, we propose a passive-type robotic wheelchair with servo brakes. We use the concept of passive robotics.[7] This system moves passively based on external force without actuators. The system uses servo brakes, which can change brake torque, and realizes the desired motion changing the physical parameters using servo brakes according to the applied force and reference tracks. The passive robotics system requires no actuator and its mechanism is simple, therefore, the system will be low cost. This characteristic is suitable for wheelchair assistance system for users who have the upper body strength. In the research area of assistive robotics, the passive robotics concept has been used for the walker and its performance is useful.[8,9] Thus, we adapt this concept for the wheelchair assistance system.

2. System Configuration

2.1. *Assistance Mechanism*

Fig.2 shows our prototype. Our proposed wheelchair utilizes a powder brake, which is one of a servo brake. Comparing the walker, the wheelchair requires larger brake torque and we choose the powder brake as Fig.2(c) (ZKG-YN50, Mitsubishi Electric Corp.), which can generate enough brake torque for stopping 4[km/h] moving wheelchair with a 100[kgf] user by 1[sec]. Our system has compatibility with a general wheelchair, which fulfills Japan Industrial Standard (JIS T9201). This means the user can built our system into their wheelchair without special construction. Our wheelchair has a battery on its body and can work more than 12 hours with one charge. Fig.1(b) shows our prototype. Our prototype can lift up the patient of 180[cm] height and 150[kg] weight maximum, and it can assist him during walking using actuated wheels.

Fig.3 shows its controller. Our wheelchair has two powder brakes with the tension controller on each wheel. Our system can measure the rotational velocity using encoders on each wheel.

2.2. *Kinematics*

Our wheelchair can move based on the intentional force applied by its user because our system has no actuator. Thus, to control its motion, the following condition has to be satisfied.[10]

$$\begin{cases} f_h + f_b \geq 0 & (f_h \geq 0) \\ f_h + f_b < 0 & (f_h < 0) \end{cases} \tag{1}$$

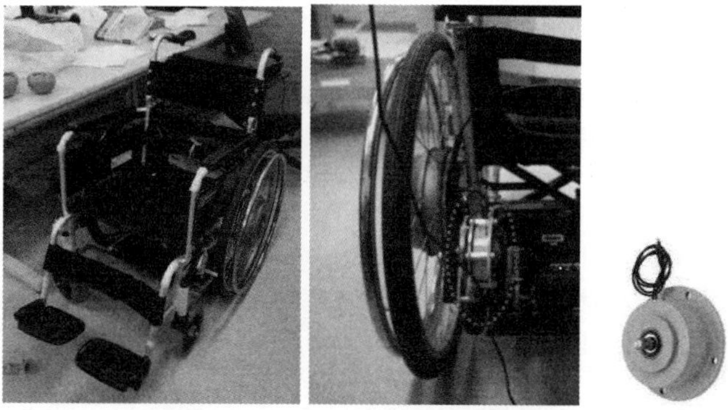

(a) Overview (b) Installed brake (c) Powder brake

Fig. 2. Our Prototype

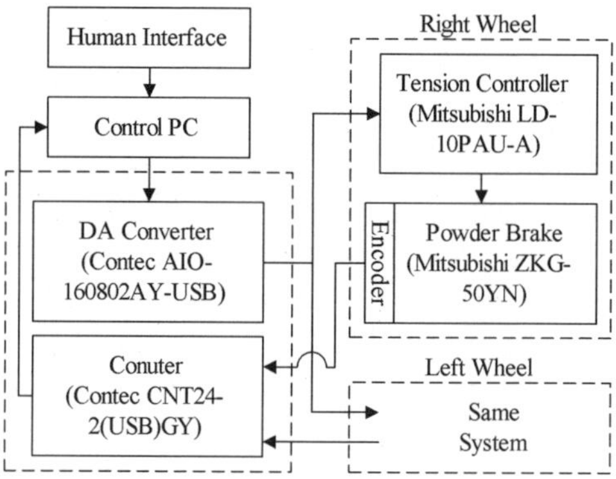

Fig. 3. Overview of our controller

where f_h is the applied force by its user and f_b is brake force generated by the servo brakes.

The kinematics of our wheelchair system is shown in Fig.4. We assume the center of mass m of the wheelchair is in the middle between the wheels. The dynamics of our wheelchair based on the force and moment $\tau_h = [f_h, n_h]^T$ applied by its user and the force and moment $\tau_b = [f_b, n_b]^T$ generated by servo brakes is expressed as follows:

$$\mathbf{M}\ddot{\mathbf{q}} + \mathbf{D}\dot{\mathbf{q}} = \mathbf{E_h}\tau_\mathbf{h} - \mathbf{E_b}\tau_\mathbf{b} \tag{2}$$

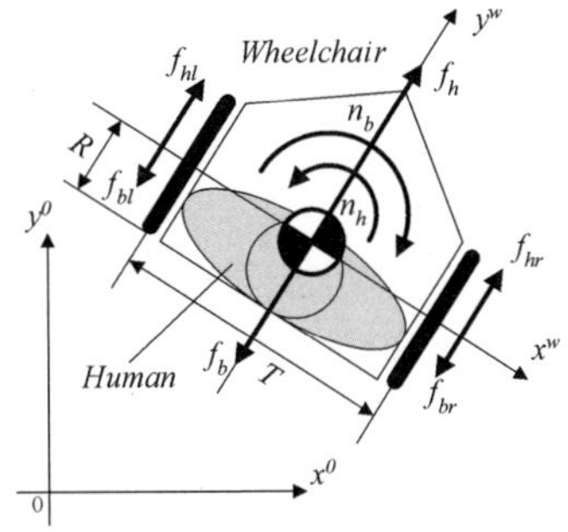

Fig. 4. Kinematical model of our wheelchair

$$\mathbf{M}=\begin{bmatrix} m & 0 & 0 \\ 0 & J & 0 \\ 0 & 0 & m \end{bmatrix},\quad \mathbf{D}=\begin{bmatrix} D & 0 & 0 \\ 0 & D_\theta & 0 \\ 0 & 0 & D \end{bmatrix},\quad \mathbf{q}=\begin{bmatrix} x^0 \\ 0 \\ y^0 \end{bmatrix},\quad \mathbf{E_h},\mathbf{E_b}=\begin{bmatrix} \cos\theta & 0 \\ 0 & 1 \\ \sin\theta & 0 \end{bmatrix}$$

where D is the damping coefficient, D_θ is the damping coefficient around the center of mass, J is the inertial moment. $\dot{q}$ and $\ddot{q}$ are the velocity and acceleration of the wheelchair.

Using (2), we can control the wheelchair motion by the brake torque. The brake torque of each wheel $t_b = [t_r, t_l]^T$ is shown in (3).

$$\mathbf{t_b}=\begin{bmatrix} t_r \\ t_l \end{bmatrix}=\begin{bmatrix} Rf_r \\ Rf_l \end{bmatrix}=\begin{bmatrix} R/2 & R/T \\ R/2 & -R/T \end{bmatrix}\begin{bmatrix} f_b \\ n_b \end{bmatrix} \tag{3}$$

2.3. *Control Algorithm*

For realizing the suitable controller for our wheelchair, we utilize the control algorithm developed by Hirata.[9,10] The dynamics equation of the wheelchair is defined by (2) and we consider a following its apparent dynamics as (4),

$$\mathbf{M}_d\ddot{\mathbf{q}}+\mathbf{D}_d\dot{\mathbf{q}}=\mathbf{E_h}\tau_\mathbf{h} \tag{4}$$

$$\mathbf{M}_d=\begin{bmatrix} m_d & 0 & 0 \\ 0 & J_d & 0 \\ 0 & 0 & m_d \end{bmatrix},\quad \mathbf{D}_d=\begin{bmatrix} D_d & 0 & 0 \\ 0 & D_{\theta d} & 0 \\ 0 & 0 & D_d \end{bmatrix}$$

where $\mathbf{M}_d$ is the apparent inertia matrix of the wheelchair and $\mathbf{D}_d$ is the apparent damping matrix. The wheelchair has a non-holonomic constraint expressed as (5).

$$\dot{x}^0 \sin\theta - \dot{y}^0 \cos\theta = 0 \tag{5}$$

By (2) and (4), we can derive the brake force and moment $\tau_b = [f_b, n_b]^T$ using $\dot{q}$ and $\ddot{q}$ as (6).[11]

$$\begin{aligned}\tau_b &= \begin{bmatrix} f_b \\ n_b \end{bmatrix} \\ &= \begin{bmatrix} \frac{m_d - m}{\cos\theta}\ddot{x}^0 + \frac{(m_d - m)\dot{\theta}\tan\theta + D_d - D}{\cos\theta}\dot{x}^0 \\ -(J_d - J)\ddot{\theta} - (D_{\theta d} - D_\theta)\dot{\theta} \end{bmatrix}\end{aligned} \tag{6}$$

When we derive the brake force and moment $\tau_b = [f_b, n_b]^T$ from (6) and specify the brake torques of the powder brake of the wheelchair in (3), the wheelchair can move as if it has the apparent dynamics expressed by $\mathbf{M}_d$ and $\mathbf{D}_d$.

Our developed wheelchair moves with low speed and acceleration, and the influences of acceleration will be minor. Therefore, in practical use, we assume $\mathbf{M}_d$ is equal to $\mathbf{M}$ and modify (6) as (7). In this paper, we use (7) for our controller.

$$\tau_b = \begin{bmatrix} f_b \\ n_b \end{bmatrix} = \begin{bmatrix} \frac{D_d - D}{\cos\theta}\dot{x}^0 \\ -(D_{\theta d} - D_\theta)\dot{\theta} \end{bmatrix} \tag{7}$$

3. Experiments

We test our prototype in the experiment. In this experiment, the user goes straight with our wheelchair on the test load with inclination as Fig.5. In this condition, the wheelchair moves to the direction shown by the dashed arrow and it is difficult to go straight for wheelchair user.

In this experiment, we utilize a standard PID controller to our system, and for simplify the problem, we set $D = D_\theta = J = 0$.

As the result, the user can drive straight with our system as Fig.6. On the other hand, without our assistance system, it is difficult to go straight by the inclination as Fig.7. Fig.8 shows the velocity of each wheel during the experiment. In Fig.8(b), the velocities are almost same and this means that the wheelchair goes straight comparing with Fig.8(a) which shows the case without proposed assistance.

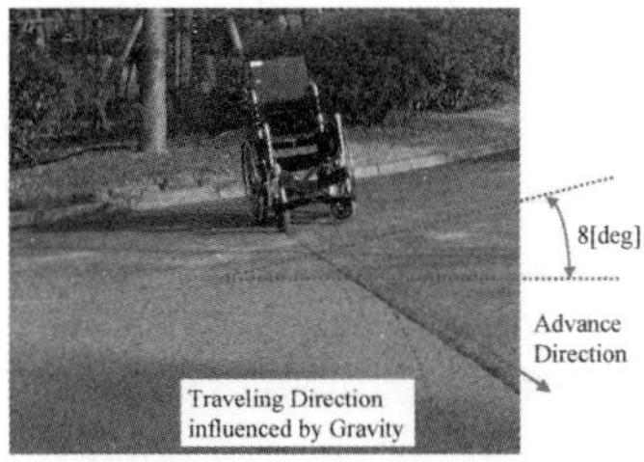

Fig. 5. Experimental condition

Fig. 6. Test run with our assistance system

Fig. 7. Test run without our assistance system.

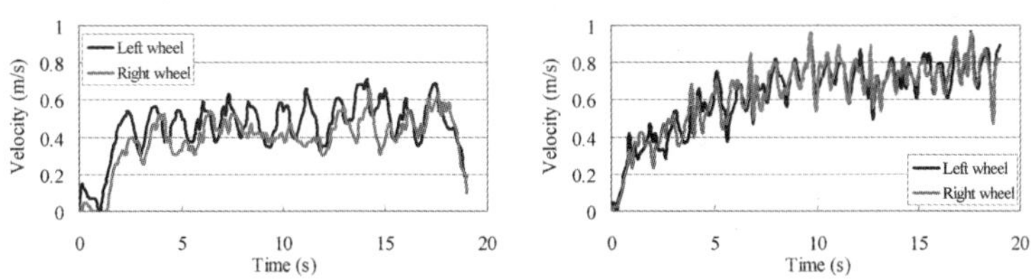

(a) Without our assistance system (b) With our assistance system

Fig. 8. Risks of the sidewalk

4. Conclusion

In this paper, we develop a novel assistive wheelchair for healthy users who have enough upper body strength. For realizing low cost and easy to use practically, we propose the passive-type wheelchair with servo brakes. Controlling a torque of the servo brake, our wheelchair changes its dynamics and realizes the reference tracks. Utilizing the human adaptive motion control algorithm, the subjects can move with our wheelchair even in non-flat ground.

In our future work, we will design the assistance algorithm based on the moving will of the user and implement it in our prototype.

Acknowledgments

This work was supported in part by Kawanishi Memorial ShinMaywa Education Foundation.

References

1. National consumer affairs center of japan, considering the safety of self-propelled manual wheelchairs *NCAC News* **14**2002.
2. A sidewalk guideline for smooth mobility, 2nd edition, japan institute of construction engineering (2008), ISBN: 4802893922.
3. Jwii, yamaha motor co., ltd. (2011), `http://www.yamaha-motor.jp/wheelchair/unit/jw2/`.
4. D. P. Miller and M. G. Slack, *Autonomous Robots* **2**, 77 (1995).
5. N. I. Katevas, N. M. Sgouros, S. G. Tzafestas, G. Papakonstantinou, P. Beattie, J. M. Bishop, P. Tsanakas and D. Koutsouris, *IEEE Robotics and Automation Magazine* **14**, 60 (1997).
6. Y. Murakami, Y. Kuno, N. Shimada and Y. Shirai, Collision avoidance by observing pedestrian's faces for intelligent wheelchairs, in *Proc. IEEE/RSJ Int. Conf. on Intelligent Robots and Systems (IROS2001)*, (Hawaii, US, 2001).
7. A. Goswani, M. A. Peshkin and J. Colgate, Passive robotics: An exploration of mechanical computation, in *Proc. IEEE Int. Conf. on Robotics and Automation (ICRA'90)*, (Ohio, US, 1990).
8. A. J. Rentschler, R. A. Cooper, B. Blaschm and M. L. Boninger, *J. of Rehabilitation Research and Development* **40**, 423 (2003).
9. Y. Hirata, A. Hara and K. Kosuge, *IEEE Trans. on Robotics* **23**, 981 (2007).
10. Y. Hirata, A. Muraki and K. Kosuge, Motion control of intelligent passive-type walker for fall-prevention function based on estimation of user state, in *Proc. IEEE Int. Conf. on Robotics and Automation (ICRA2006)*, (Florida, US, 2006).
11. P. Prautsch, T. Mita and T. Iwasaki, *Trans. of IEEJ* **120D**, 372 (2000).

DESIGN OF AN EFFICIENT BACK-DRIVABLE SEMI-ACTIVE ABOVE KNEE PROSTHESIS

MOHAMMED AWAD[*+], KIAN SEK TEE[*], ABBAS DEHGHANI[*]

[*]*Institute of Engineering Systems and Design, School of Mechanical Engineering, University of Leeds, Leeds, LS2 9JT, United Kingdom.*

[+]*Mechanical Engineering, Faculty of Engineering, Ain Shams University, Cairo, Egypt.*

DAVID MOSER, SAEED ZAHEDI

Chas A Blatchford & Sons Ltd, Lister Road, Basingstoke, Hants, RG22 4AH, United Kingdom.

This paper presents the design and development of an electrical above knee prosthesis, which works as a passive knee prosthesis in part of gait cycle phases and as an active knee prosthesis during other portions. During the passive mode, the system works as a non-holonomic system, and the dynamic coupling between the thigh segment and the knee prosthesis is used to control the prosthesis. Therefore, this knee prosthesis is designed to be back-drivable in passive mode. In order to present a proper design for the knee prosthesis, the mechanism synthesis and analysis for the proposed back-drivable semi-active knee prosthesis are covered in this paper.

1. Introduction

In the past, walkers, crutches, open socket peg legs, wheelchairs were used to help amputees to walk; but with limitations. Therefore, there was a need to develop a more efficient prosthesis to restore the lost locomotive functions for the amputees. Hence, the research in prostheses field has started in MIT since 70's of the last century [1-4]. Furthermore, commercial adaptive passive prostheses were developed; such as Otto Bock C-Leg, Blatchford smart IP+[5], and Ossur's Rheo Knee. These passive prostheses interact with the amputee's healthy segments according to the passive dynamic walking concept which was introduced by McGeer [6]. This concept provides an excellent technique to generate a natural gait for a biped by using inertia between segments. Therefore, an amputee's hip is considered the main engine source to control the inertia interaction between thigh and prosthesis knee as shown in figure 1. This technique is called voluntary control for the prostheses. However, adding more energy at hip to control produces uncomfortable walking experiences for the amputee due to extra pressure between stump-socket interface. Moreover, there are some periods during level ground walking, ramp and stair climbing which require a generated energy by knee. Therefore, there is a need for an actuator during certain gait periods.

This research aims to design and develop a back-drivable semi-active above knee prosthesis for more efficient gait. Whilst other researchers have developed fully actuated prostheses by using pneumatic actuator [7] or by using DC electrical motors [8, 9], these prostheses do not get the benefit of using the dynamic coupling during some periods to reduce the consumed energy and the actuator size. In addition to that, Ossur developed a commercial power knee which is full actuated. And whilst there is another knee prosthesis [10] that could perform as a semi-active device by using a hydraulic actuator, the effect of the dynamic coupling between the amputee's thigh and the knee prosthesis has not been studied in it. Moreover, the proposed design focuses on using a DC electric motor in which has less noise and higher efficiency in comparison to hydraulic system.

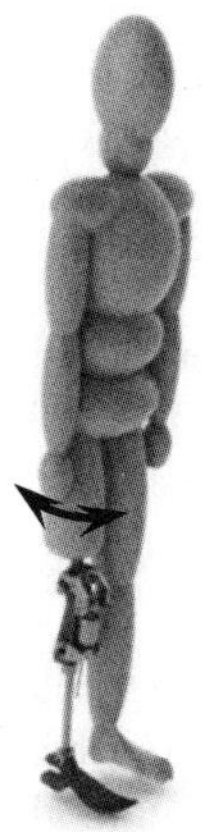

Figure 1: Above knee amputee movement mechanism

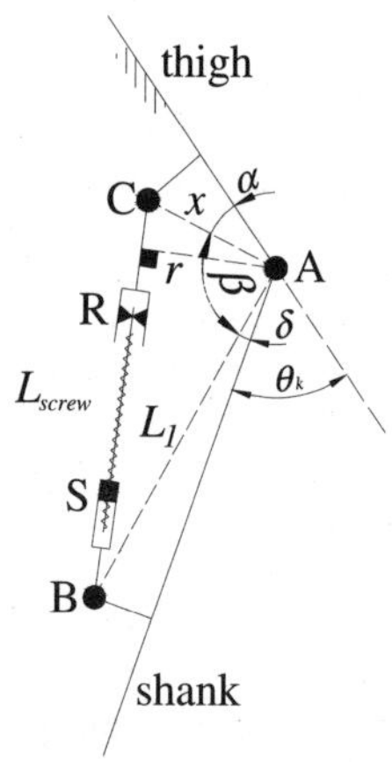

Figure 2: Schematic diagram for closed loop kinematic chain

2. Mechanism Design Synthesis and Analysis

2.1. *Mechanism Selection*

To determine which mechanism is suitable for developing the back-drivable prosthesis, open and closed kinematics chain were studied. The open kinematic chain has the ability to deliver constant transmission ratio with respect to the knee angle. This configuration could be implemented by connecting a Dc motor with a geared harmonic to drive the knee prosthesis axis directly. However, the harmonic drive has rather high back-driving resistance. On the other hand, closed kinematic chain developed with low friction and high lead angle ball screw could back-drive easily. Therefore, closed kinematic chain was chosen.

2.2. *Mechanism Analysis*

The first step in accomplishing an efficient design is to find all possible assembly modes and the most significant parameters which affect on the mechanism performance. The assembly modes were investigated by finding the closed form equation (1) of the mechanism shown in figure 2. Hence, $\partial r/\partial\beta = 0$ was calculated in order to find β^* which define the angle $\measuredangle CAB$ at maximum torque arm r_{max}. Therefore, there were four answers for β^* as shown in equation (2) which shows that the mechanism could be assembled in four different configurations as shown in figure 3. Furthermore, the knee angle at maximum transmission ratio for the mechanism is calculated according to equation (3). In order to find which configuration has better transmission ratio, the torque arm r was calculated to these configurations as shown in figure 4. It is clear from figure 4 that assembly modes 2 and 4 have better torque arm variation which allows better transmission ratio. Furthermore, the effect of the β^* parameter on the mechanism performance was studied in figure 5 which shows that the smaller β^* the better transmission can be achieved. However, if $\beta^* < \theta_k^*$, the singularity would happen at small knee angles. Therefore, β^* range should be within $\theta_k^* < \beta^* < 90^o$, and should be chosen smaller as possible.

$$r = \frac{xL_1}{\sqrt{L_1^2+x^2-2xL_1 cos\beta}} \sin\beta \tag{1}$$

$$\beta^* = \left\{ \begin{array}{l|l} +\tan^{-1}\dfrac{\sqrt{L_1^2 - x^2}}{x} & if\ L > x\ mechanism\ configuration\ (a) \\ -\tan^{-1}\dfrac{\sqrt{L_1^2 - x^2}}{x} & if\ L > x\ mechanism\ configuration\ (b) \\ +\tan^{-1}\dfrac{\sqrt{x^2 - L_1^2}}{L_1} & if\ x > L\ mechanism\ configuration\ (c) \\ -\tan^{-1}\dfrac{\sqrt{x^2 - L_1^2}}{L_1} & if\ x > L\ mechanism\ configuration\ (d) \end{array} \right\} \tag{2}$$

$$\theta_k^* = \pi - [(\alpha \pm \delta) + \beta^*] \tag{3}$$

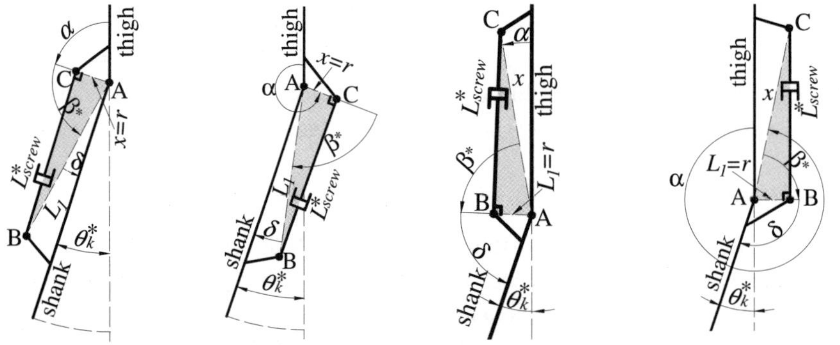

(a) assembly mode 1 (b) assembly mode 2 (c) assembly mode 3 (d) assembly mode 4

Figure 3: Four possible assembly modes for the knee prosthesis.

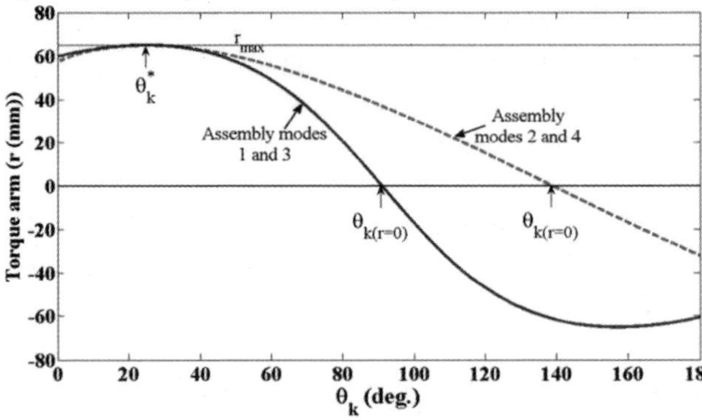

Figure 4: The torque arm r variation according to the knee angle for the four possible modes.

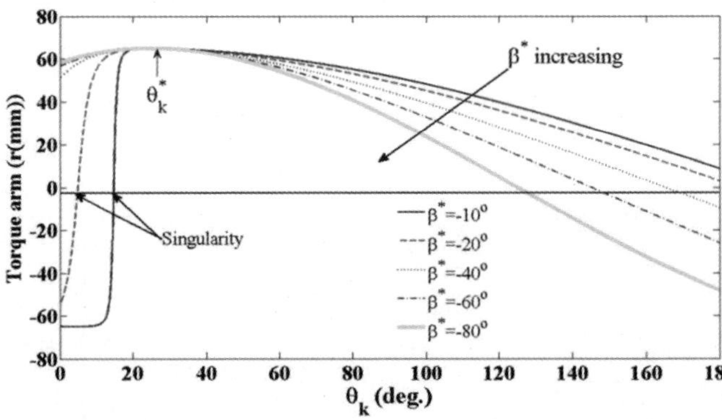

Figure 5: The torque arm r variation according to the knee angle at different mechanism dimensions.

3. Dynamic Coupling Model

As explained before, this research focuses in designing a back-drivable prosthesis applying the dynamic coupling effect between the thigh and the prosthesis to help amputee to walk and to reduce the consumed power from the actuator. Therefore, a simple dynamic model for a simplified prosthesis as shown in figure 6 was deduced. This mode was simplified to 4 degree of freedom system in sagittal plane. The hip joint could move in X_0Y_0 plane by x_h and y_h respectively, and the thigh and prosthesis shank could rotate by θ_h and θ_k respectively. The shank and foot are considered as one link in this model for simplification. Furthermore, the effect of all other active segments in human body such as trunk and healthy leg are seen in hip kinematic data x_h and y_h. Therefore, the hip and knee torques were derived by using Lagrangian formulation as shown in equations (4) and (5) respectively.

On the other hand, if both hip and knee are fully controlled and actuated, the system behaves like holonomic system. However, it could be used the dynamic coupling effect between the hip and the knee to control the knee at some extend during walking. The dynamic coupling effect for the knee is clear in equation (5) in M_{41}, M_{42}, M_{43}, M_{44}, and V_{cen43}. Therefore, the hip kinematics data could be used to control the knee angle acceleration and speed in case of passive mode or non-holonmonic system as shown in equation (6). This technique is used to control the passive prosthesis during swing phase, and it is called voluntary control which done by amputee to control the prosthesis. Therefore, a gait analysis kinematics data for healthy hip subject was used to solve equation (6) under no ground reaction forces, and figure 7 shows the knee angle performance under the hip movement without the need for actuator at knee in comparison to knee angle of a healthy subject.

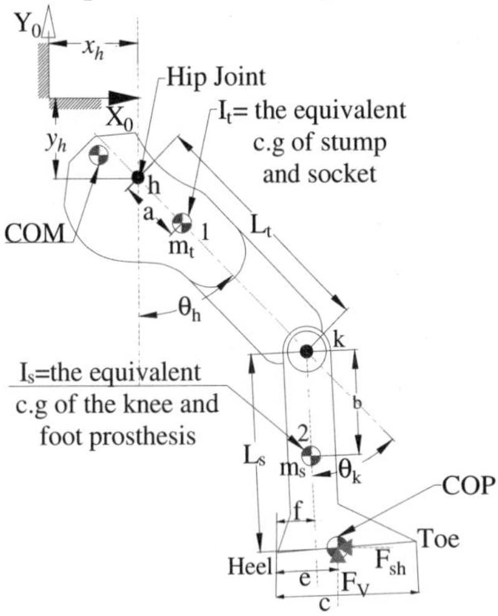

Figure 6: Simplified diagram for the dynamic model of knee prosthesis.

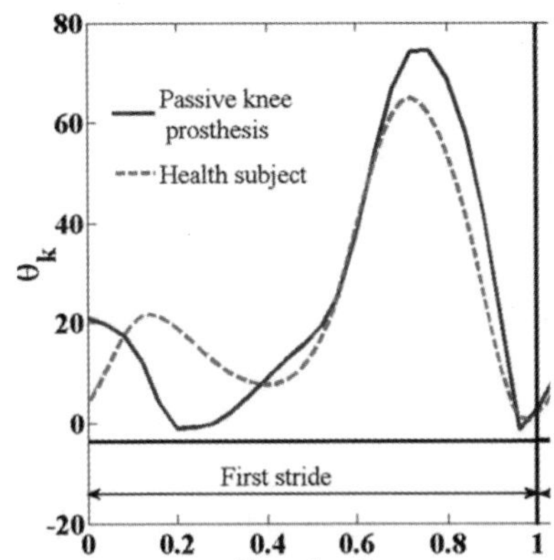

Figure 7: The performance of passive knee under dynamic coupling effect.

$$\tau_h = M_{31}\ddot{x}_h + M_{32}\ddot{y}_h + M_{33}\ddot{\theta}_h + M_{34}\ddot{\theta}_k + V_{cen34}\dot{\theta}_k^2 + V_{cor3}\dot{\theta}_h\dot{\theta}_k + G_3 + \frac{\partial x_{cop}}{\partial \theta_h}F_v + \frac{\partial y_{cop}}{\partial \theta_h}F_{sh} \quad (4)$$

$$\tau_k = M_{41}\ddot{x}_h + M_{42}\ddot{y}_h + M_{43}\ddot{\theta}_h + M_{44}\ddot{\theta}_k + V_{cen43}\dot{\theta}_h^2 + G_3 + \frac{\partial x_{cop}}{\partial \theta_k}F_v + \frac{\partial y_{cop}}{\partial \theta_k}F_{sh} \quad (5)$$

$$\ddot{\theta}_k = -M_{44}^{-1}[M_{41}\ddot{x}_h + M_{42}\ddot{y}_h + M_{43}\ddot{\theta}_h + V_{cen43}\dot{\theta}_h^2 + G_3 + \frac{\partial x_{cop}}{\partial \theta_k}F_v + \frac{\partial y_{cop}}{\partial \theta_k}F_{sh}] \quad (6)$$

4. A Back-Drivable Semi-Active Knee Prosthesis Mechanical Design

4.1. *Design Considerations and Specifications*

The design has been carried out according to the considerations from sections 2 and 3. Those considerations could be summarized in the following points:

1. $\theta_k^* = 25^o$ to cover level ground walking at high speed.
2. $\beta^* = 66^o$ because it is limited to the mechanism dimensions.
3. The prosthesis weight affects on both equations (4) and (6). Therefore, the heavy prosthesis requires more hip torque generated by the amputee, but lighter prosthesis reduces the effect of the dynamic coupling in non- holonomic mode. Therefore, there is a need to find the suitable prosthesis weight to perform natural walking.
4. The back-driving effect which achieved by using a ball screw with high lead helix angle.

4.2. *Mechanical Design*

The prosthesis was designed by using of a ball screw which has 95.5% efficiency in forward direction and 95.2% efficiency in back-driving direction. The prosthesis was driven by a 42V DC motor in active mode which is connected to the ball screw through a timing belt. Figure 8a shows the CAD model for the knee prosthesis. The prosthesis was designed to cover the working range for most human tasks. The extreme positions during fully knee extension flexion are shown in figures 8a and 8b when the knee angle is zero and 104°.

On the other hand, the stress analysis was tested by developing a finite element model where a real human ground reaction loads and knee angle trajectory for level ground walking are imported to the model. This model is divided into frames and the motion kinematics and ground loads are set at each individual frame. Hence, the motion analysis at each joint and link were done to import those motion loads and boundary conditions to the model. Figure 8c shows the maximum von Mises stress which happens at frame number eight

(a) (b) (c)

Figure 8: CAD model and FEA of the back-drivable semi-active knee.

5. Conclusions

An investigation into the design of an efficient back-drivable semi-active prosthetic system has been carried out. Investigations have accordingly provided the key design parameters of the mechanism which affect the prosthesis performance. Furthermore, an analytical work was carried out on the dynamic coupling effect between the hip and the knee. Among these it was noticed that the prosthesis weight has an important effect on the passive dynamic walking performance.

Acknowledgments

The corresponding author (M. Awad) would like to acknowledge the Egyptian Government and Egyptian Cultural Bureau in the UK for their financial support.

Appendix

1. The parameters of dynamic equations (4), (5) and (6) are the following:

$M_{31} = m_t a \cos\theta_h + m_s[L_t \cos\theta_h + \; b\cos(\theta_h + \theta_k)]$

$M_{32} = m_t a \sin\theta_h + m_s[L_t \sin\theta_h + \; b\sin(\theta_h + \theta_k)]$

$M_{33} = I_t + m_t a^2 + m_s[L_t^2 + \; b^2 + 2L_t b\cos\theta_k] + I_s$

$M_{34} = M_{43} = I_s + m_s[b^2 + L_t b\cos\theta_k]; M_{41} = m_s b\cos(\theta_h + \theta_k)$

$M_{42} = m_s b\sin(\theta_h + \theta_k)\,; M_{44} = I_s + m_s b^2$

$V_{cen34} = -m_s b L_t \sin\theta_k\,;\; V_{cen43} = \; m_s b L_t \sin\theta_k\,;\; V_{cor3} = -2m_s b L_t \sin\theta_k$

$$G_3 = m_t g a \sin\theta_h + m_s g[L_t \sin\theta_h + b\sin(\theta_h + \theta_k)]$$
$$G_4 = m_s g b \sin(\theta_h + \theta_k)$$
$$\frac{\partial x_{cop}}{\partial \theta_h} = -(e-f)\sin(\theta_h+\theta_k) + L_t\,cos\theta_h + L_s cos(\theta_h+\theta_k)$$
$$\frac{\partial x_{cop}}{\partial \theta_k} = -(e-f)\sin(\theta_h+\theta_k) + L_s cos(\theta_h+\theta_k)$$
$$\frac{\partial y_{cop}}{\partial \theta_h} = (e-f)\cos(\theta_h+\theta_k) + L_t\,sin\theta_h + L_s sin(\theta_h+\theta_k)$$
$$\frac{\partial y_{cop}}{\partial \theta_k} = (e-f)\cos(\theta_h+\theta_k) + L_s sin(\theta_h+\theta_k)$$

References

1. W.C. Flowers, "A Man-Interactive Simulator System for Above-Knee Prosthetics Studies", Department of Mechanical Engineering, PhD Thesis, MIT, 1972.
2. W.C. Flowers, "Use of an Amputee-computer Interactive Facility in Above-Knee Prosthesis Research", Proceedings of the 1974 Annual Conference -Volume 1, Association for Computing Machinery (ACM), New York, 1974.
3. D. R. Lampe, "Design of A Magnetic Particle Brake Above-Knee Prosthesis Simulator System", Department of Mechanical Engineering, PhD Thesis, MIT, 1976.
4. B. Hunter, "Design of self-contained, active, regenerative computer controlled above-knee prosthesis", Department of Mechanical Engineering, PhD Thesis, MIT, 1981.
5. S. Zahedi, A. Sykes, S. Lang and I. Cullington, "Adaptive prosthesis- a new concept in prosthetic knee control", Roboica, Cambridge University press, vol. 23, pp.337-344, 2005
6. T. McGeer, "Passive walking with knees", IEEE International Conference on Robotics and Automation, Cincinnati, OH, USA, May 13-18, 1990.
7. F. C. Sup and M. Goldfarb, "Design a Pneumatically Actuated Transfemoral Prosthesis", Proceedings of IMECE2006, ASME International Mechanical Engineering Congress and Exposition, Nov. 5-10, 2006, Chicago, Illinois.
8. K. Fite, J. Mitchell, F. Sup, and M. Goldfarb, "Design and Controlled of an Electrically Powered Knee Prosthesis", IEEE International Conference on Rehabilitation Robotics, Noordwijk, The Netherlands, June 12-15, 2007.
9. Bedard et al., "Actuated Leg Prosthesis for Above-Knee Amputees", United States Patent, US 7,314,490 B2, Jan. 1, 2008.
10. B. Lambrecht and H. Kazerooni, "Design of a Semi-Active Knee Prosthesis", IEEE International Conference on Robotics and Automation, Kobe, Japan, May 12-17, 2009.

SECTION–2

AUTONOMOUS ROBOTS

MYON: CONCEPTS AND DESIGN OF A MODULAR HUMANOID ROBOT WHICH CAN BE REASSEMBLED DURING RUNTIME

Manfred Hild, Torsten Siedel, Christian Benckendorff, Matthias Kubisch, and Christian Thiele

Neurorobotics Research Laboratory
Humboldt-Universität zu Berlin
Unter den Linden 6
10099 Berlin, Germany
{hild|siedel|benckend|kubisch|thiele}@informatik.hu-berlin.de

We describe the design concepts of the modular humanoid robot MYON, which can be disassembled and reassembled during runtime. The body parts are fully autonomous in a threefold sense: they all possess their own energy supply, processing power, and a neural network topology which allows for stand-alone operation of single limbs. The robot has especially been designed for robustness and easy maintenance. It exhibits a combination of an endoskeleton with an exoskeleton, the latter of which can manually be detached without the need for technical equipment. One of the essential parts is a novel flange which firmly connects the body parts mechanically, whilst at the same time relaying the power supply lines and sensorimotor signals. We also address the details of the antagonistic and compliant actuation system which not only protects the gears against high impact forces but also enables biologically inspired joint control.

Keywords: Humanoid robot; Modular architecture; Power autonomy; Antagonistic actuation.

1. Introduction and Design Concepts

Humanoid robots are highly complex systems and as such prone to damages and malfunctioning. This is especially true if not only an upper torso with head and arms is used, but the full body, e.g. to test and analyze sensorimotor loops for walking motions and stable standing. There are several approaches to remedy the situation, depending on the experimental settings that are planned to be addressed with the humanoid platform. If full autonomy is not needed, then energy supply and processing power can be placed outside the robot and the bare skeleton can be optimized for maximal mechanical robustness. Also, the type of the actuators plays a crucial

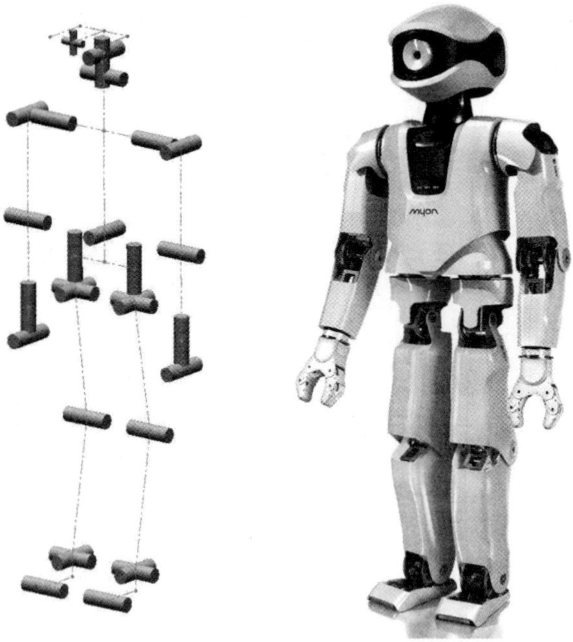

Figure 1. The humanoid robot MYON. Left: Schematic diagram of the joints' positions and orientations. Right: Image of the functional robot including the exoskeleton shells. The exoskeleton is a structural element which prevents the inner endoskeleton from external torsional loads. Furthermore, the electronic components are protected from mechanical impacts and the user from crushing injuries.

role. Pneumatic-driven humanoid robots can withstand a drop from more than one meter height onto the ground without any problem, although it has to be noted, that pneumatic actuators have much longer response times than electric motors.[1] Hence, if the robot needs to be mobile within the experimental setting then electric actuators and on-board batteries have to be used. The robot MYON, which is shown in Figure 1, has been designed for experiments on artificial language evolution[2] and therefore needs to be able to autonomously wander around, recognize and manipulate different objects, and communicate with other robots. Moreover, the robot is also to be used as a research platform for biologically inspired behavior control, using tight sensorimotor loops and antagonistic joint actuation. As recently stated by Migliore et al.,[3] the vast majority of walking robots still loses energy by ignoring the potential benefit of using passive elastic components at their joints. On MYON, elastic components could be incorporated in a very compact way along with the antagonistic actuation system. This not only opens up the research field of energy-efficient walking behaviors, but in the first instance protects the gears against high external impact forces.

In the following, we outline the overall system architecture and focus on the robot's modularity and the antagonistic actuation system. We additionally address the system's power autonomy, the compliance of the actuation system, and the flange design which allows the robot's body parts to be detached and re-attached during runtime.

2. Overall Properties

The robot has especially been designed for robustness and easy maintenance. It exhibits a combination of an endoskeleton with an exoskeleton, the latter of which can manually be detached without the need for technical equipment. All in all, the robot is 1.25 m tall and weighs 15 kg, including the shells (see Table 1). It consists of six body parts (head, torso, arms, and legs) which are fully autonomous in terms of energy supply, processing power, and neural network topology. This way, single body parts can

Table 1. Overview of the robot's mass, degrees of freedom (DOFs), and number of actuators. Except for the eye, a single type of actuator is used for all joints. Joints which need a large amount of torque, e.g. the knee, are driven by several actuators in parallel.

Module	Mass (kg)	Joint	DOFs (number)	Actuators (number)	Actuators (type)
Head	1.4	Eye	4	4	Micro servo
		Neck	3	3	RX-28
Arm (2x)	1.1	Shoulder	1	1	RX-28
		Elbow	1	1	RX-28
		Wrist	1	1	RX-28
Gripper (2x)	0.2	Fingers	1	1	RX-28
Torso	2.5	Shoulder (2x)	1	2	RX-28
		Waist	1	1	RX-28
		Leg (2x)	1	1	RX-28
Leg (2x)	3.0	Hip	2	5	RX-28
		Knee	1	3	RX-28
		Ankle	2	5	RX-28
		Toe	1	–	passive
Shells (total)	2.5		–	–	
Total	15.0		32	48	

be operated stand-alone using the distributed local processing nodes and special neuro-modules[4] as control structures. One of the essential parts is a novel flange which firmly connects the body parts mechanically, whilst at the same time relaying the power supply lines and sensorimotor signals.

3. System Architecture

The inner system architecture is shown in Figure 2. Each body part exhibits an *Energy Module* built of four Lithium-ion polymer batteries (3.55 Ah at 14.8 V, nominal) and an analog circuit which guarantees lossless energy balancing between all body parts. They are connected using a multi-core bus, the so-called *Extended Spinal Cord*, which transfers energy, sensorimotor data at a rate of 4.5 MBaud, and a control signal which is used to switch the robot on and off. Processing is done by 25 distributed processing nodes. They are called *AccelBoard3D*, since they also possess a 3-axis acceleration sensor, despite the Cortex-M3 RISC processor running at 72 MHz. Actuators and additional sensors are connected to each *AccelBoard3D*. The robot MYON does not exhibit a central processor, but the head features an *Field Programmable Gate Array* (FPGA) for image processing (see *BrainModule* in Figure 2).

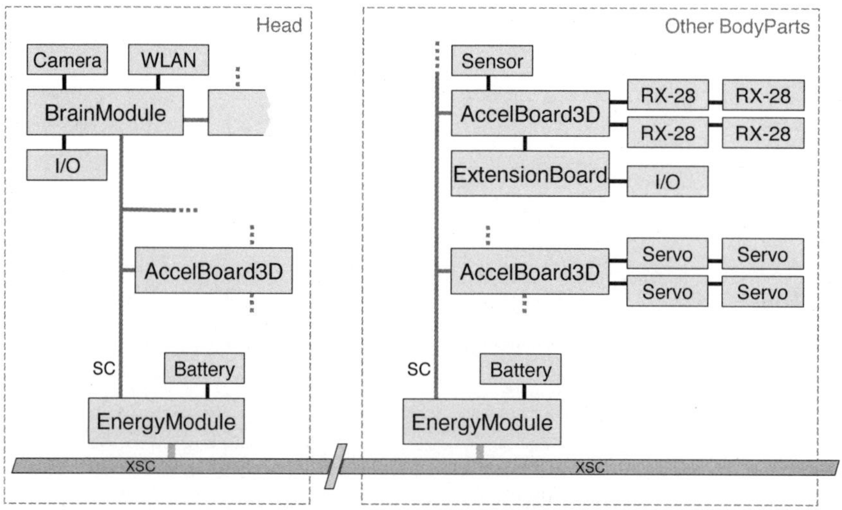

Figure 2. System architecture of the robot MYON. Components within each body part are connected via the so-called *Spinal Cord* (SC), whereas the body parts are connected by the *Extended Spinal Cord* (XSC) which includes several additional lines for energy transfer as well as a control line for quick shutdown of the robot.

4. Runtime Modularity

The body parts of the robot MYON can be detached and reattached during runtime, since every body part is fully autonomous, having its own power supply, computational power, as well as a special firmware which allows for hot plugging and respects the topological structure of the underlying recurrent neural control network. Figure 3 shows the robot's torso and the battery pack which can be inserted and removed easily. The flange design can be seen in Figure 4 in more detail. It consists of two interlocking rings which guarantee a mechanically firm connection. This is very important, e.g. for the legs, where large forces have to be transmitted between the limb and the torso. The electrical connection is placed in the middle of the flange. It relays the *Extended Spinal Cord* (XSC) as shown in Figure 2.

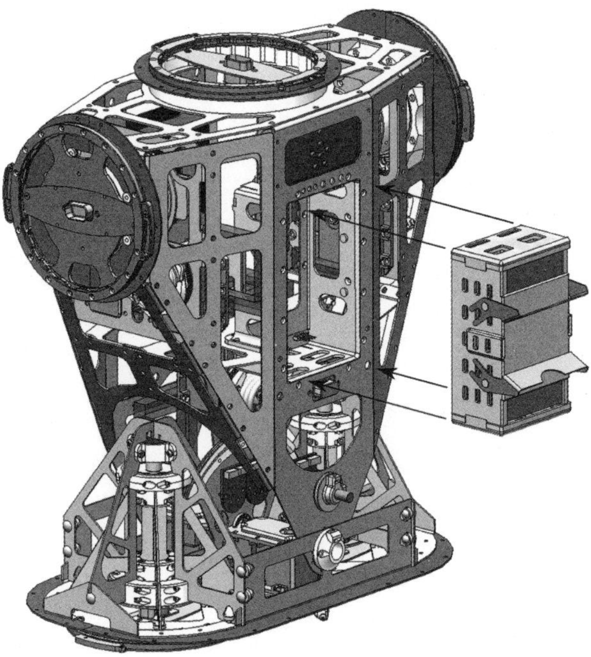

Figure 3. Construction of the robot's torso, exhibiting the flanges for the other body parts. The torso is made of aluminum and therefore only weighs 2.5 kg including actuators, electronics and batteries. The design of the battery pack allows for easy insertion and removal using a single hand.

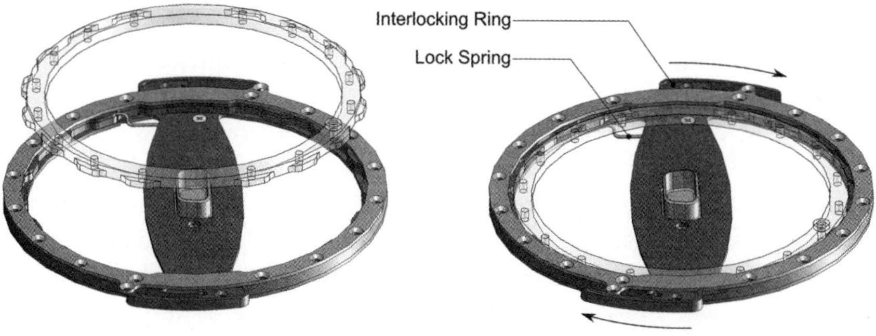

Figure 4. Detailed view of the novel flange which firmly connects the robot's body parts mechanically, whilst at the same time relaying the power supply lines and sensorimotor signals. The flange can be controlled using a single hand and allows for detachment and reattachment of body parts during runtime.

5. Power Autonomy

The power autonomy of the robot's body parts was a crucial design goal, since it enables stand-alone experiments with single limbs. An example of such an experiment is the stand up motion of a single leg, as presented by Hild and Kubisch.[5] Also the balance recovery reported by Kubisch et al.[6]

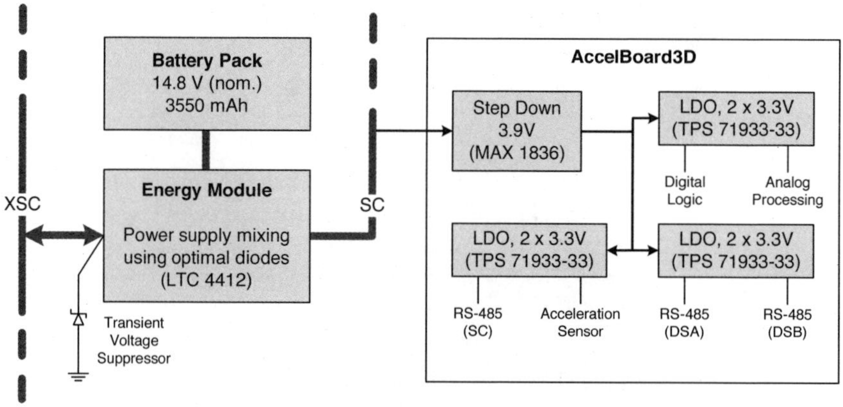

Figure 5. Power distribution within the robot. The *Extended Spinal Cord* (XSC) contains power lines from all body parts, whereas the *Spinal Cord* (SC) locally provides power within each body part (also see Figure 2). The SC power is predominantly recruited from the local battery pack, but other body parts can provide additional power if needed, e.g. during a knee bend. In order to save energy, and also to achieve maximum robustness against system failures, each *AccelBoard3D* processing node uses a step down converter (3.9 V) and six short-circuit-proof, low-drop-out voltage regulators (3.3 V each) for the different functional parts of the circuit, as indicated.

uses only part of the robot, namely the torso with two legs. As the details of the power distribution reveal (see Figure 5), each body part can be driven by a mix of local power from the body part's own battery pack and the power from all other body parts which may more distant to the energy-consuming actuators, but which also may be not as exhausted as the local supply. Depending on the type of full body motion, a single battery pack is able to power the whole robot for twenty minutes or even longer. Surely, power can also be supplied via unconnected flanges (if not all body parts are used) or via an extra connector below the torso's battery pocket (see Figure 3).

6. Antagonistic Actuation

The robot exhibits 32 degrees of freedom which are driven by 48 actuators (see Table 1), i.e., several actuators (all of the type Robotis RX-28) drive the same joint, allowing for biologically inspired, antagonistic control strategies. Each actuator is equipped with a torsion spring, as shown in Figure 6. This spring provides a series elasticity which balances forces between actuators that drive the same joint. The spring stiffness is chosen so that high impact forces are damped enough not to break the actuator's gear in. Figure 7 shows a close-up of the ankle pitch joint, which is driven by four actuators.

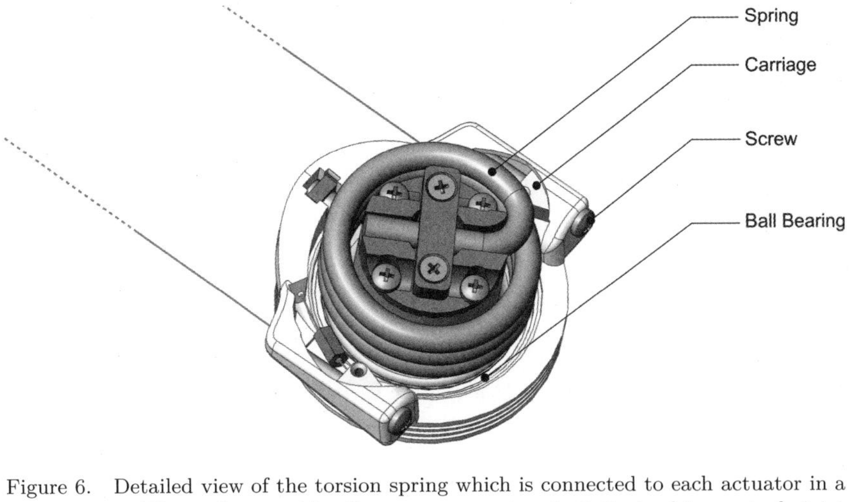

Figure 6. Detailed view of the torsion spring which is connected to each actuator in a compact way. The spring provides the necessary series elasticity to drive a single joint with multiple actuators in an antagonistic way. Forces are transmitted via wire ropes (also see Figure 7) which are clamped by a screw. Additionally, the spring also protects the actuator's gear from unforeseen high external impact forces.

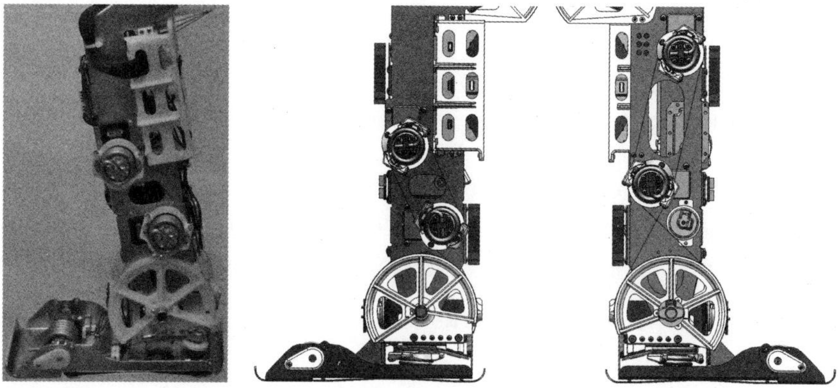

Figure 7. Detail of the ankle joint. Left: Photograph of the foot and leg without exoskeleton shells. Middle and Right: Schematic side views showing the four actuators (small circles) which drive the ankle pitch joint (big circle) via wire ropes.

7. Summary

We have described the modular humanoid robot MYON, which can be disassembled and reassembled during runtime, since all body parts are fully autonomous. The robot is robust and easy to maintain due to its exoskeleton and special flanges. In addition, MYON exhibits an antagonistic actuation system which enables researchers to study biologically inspired control strategies.

Bibliography

1. K. Hosoda, Y. Sakaguchi, H. Takayama and T. Takuma, Pneumatic-driven jumping robot with anthropomorphic muscular skeleton structure, in *Autonomous Robots*, 2010.
2. L. Steels and M. Spranger, How experience of the body shapes language about space, in *Proc. of the 21st Internat. Joint Conf. on Artifical Intelligence (IJCAI'09)*, 2009.
3. S. A. Migliore, L. H. Ting and S. P. DeWeerth, Passive joint stiffness in the hip and knee increases the energy efficiency of leg swinging, in *Autonomous Robots*, 2010.
4. M. Hild, Neurodynamische Module zur Bewegungsteuerung autonomer mobiler Roboter, in *PhD thesis, Humboldt-Universität zu Berlin*, 2007.
5. M. Hild and M. Kubisch, Self-Exploration of Autonomous Robots Using Attractor-Based Behavior Control and ABC-Learning, in *Proc. of the 11th Scandinavian Conf. on Artificial Intelligence*, (Trondheim, Norway, 2011).
6. M. Kubisch, C. Benckendorff and M. Hild, Balance Recovery of a Humanoid Robot Using Cognitive Sensorimotor Loops (CSLs), in *Proc. of the 14th Internat. Conf. on Climbing and Walking Robots*, (Paris, France, 2011).

DESIGN AND REAL-TIME PID CONTROL APPROACH OF A TWO-WHEELED VEHICLE WITH AN EXTENDED ROD

A C Corredor Rosendo[1], S A Agouri[2], M O Tokhi[2] and K M Goher[3]
[1]Departamento Electrónica y Circuitos, Universidad Simon Bolivar, Venezuela
[2]Department of Automatic Control and Systems Engineering,
The University of Sheffield, United Kingdom
[3]Department of Mechanical and Industrial Engineering
Sultan Qaboos University, Oman, E-mail: kgoher@squ.edu.om

This paper presents the design and control of two wheeled inverted pendulum robot. The aim of the design is to provide a prototype for research purposes. The design of the robot incorporates the use of different components to read the tilt angle and apply the proposed control algorithm. A simple PID controller was used to control the robot posture. The control algorithm provided successfully balances the robot in the upright position with minimal vibrations around the set point.

1. Introduction

The research on balancing two-wheeled robots has gained momentum over the last decade in a number of robotics laboratories around the world. This is due to the inherent unstable dynamics of such systems. The control quality of such robots is characterized by the ability to balance on its two wheels and provide spin on the spot. This additional manoeuvrability allows easy navigation on various terrains, turn at sharp corners and traverse small steps or curbs. These capabilities have the potential to solve a number of challenges in industry and society. For example, a motorized wheelchair utilizing this technology would give the operator greater manoeuvrability and thus access to places most able-bodied people take for granted. Small carts built utilizing this technology allow humans to travel short distances in a small area or factories as opposed to using cars or buggies which are more polluting. The rapid increase of the aged population in countries like Japan has prompted researchers to develop robotic wheelchairs to assist an elderly to move around, Takahashi et al. (2000). Two-wheeled machines have different applications due to their advantages which arise from their special design. For example, a two wheeled vehicle may be safer for the occupants while simultaneously being more agile to navigate narrow city streets. Furthermore, the reduced volume and lower mass of this configuration would increase fuel efficiency and overall functionality.

Choi *et al.* (2008) developed a wheeled inverted pendulum and investigated the interaction between human and the robot through applying a pulling or pushing force. They developed a control algorithm able to move and stabilize the robot in coordination with the external force of a human. Imamura *et al.* (2008) discussed the configuration of a baggage transportation system by an inverted pendulum robot and realized a navigation system. Jeong and Takahashi (2007) discussed the mobile control and the standing and sitting motions of an inverted pendulum type assistant robot aiming at the coexistence of safety and work capability. Jingtao *et al* (2007) discussed the mechanical design and dynamic modelling of a differential drive two-wheeled inverted pendulum mobile robot, which has the advantage of high manoeuvrability on slopes and narrow spaces and is able to carry human beings as well as other goods. Nawawi *et al.* (2006 and 2007) designed and developed a two wheeled inverted mobile robot for use as flexible platform, comprising an embedded unstable linear plant intended for research and teaching purposes. Through this work, issues such as selection of actuators and sensors, signal processing units, modelling and control scheme have been addressed and discussed. Zheng *et al* (2007) discussed the mechanical design and control system of a miniature autonomous surveillance robot ("BMS-1") for indoor reconnaissance tasks. The design of two-wheeled robotics machines with an extended rod has first addressed in the work by Goher and Tokhi (2009 and 2010), they have presented novel design of such machine with an additional degree of freedom by incorporating a linear actuator to the intermediate body.

The aim of this work is to design a suitable control strategy in order to test the developed prototype. This work focuses on testing the machine with mainly two types of sensors; an accelerometer and a gyroscope, in addition to complementary filters (high-pass filter and low-pass filter) to obtain a precise tilt angle from the upright balanced position of the two-wheeled platform.

2. Development of the Real Prototype

This section will cover the two main parts of the two-wheeled platform, the hardware and the software, and how they are interfaced with each other to show a working robot structure.

The work introduces a two wheeled vehicle with an additional feature and a linear translational degree of freedom. The design has two main wheels driving the vehicle which allow it to manoeuvre around. Other two auxiliary wheels are attached to the front and rear of the robot the robot safety to prevent it from falling off. The other main part of the vehicle is the intermediate body (IB) of the vehicle which is composed of two co-axial parts connected by a linear actuator with a payload attached to the end of the upper part. The linear actuator allows the payload to move up and down along the vehicle intermediate body (IB), and this adds an additional translational degree of freedom. The robot thus is four degrees of freedom system. A real test rig, shown in Figure 1 has been designed and developed by (Goher *et al*) at the Automatic Control and Systems Engineering department, University of Sheffield, based on the developed design.

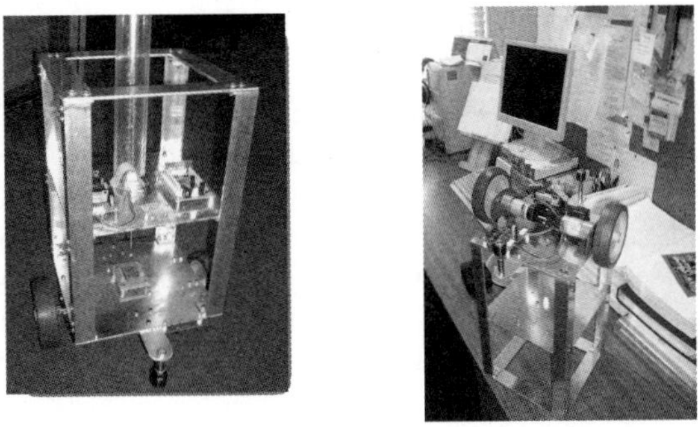

Fig. 1 Real test rig, ACSE department, University of Sheffield

A. Mechanical components and parts

This comprises an Aluminum structure with four right-handed angles in each side. Those four angles are used to support the base plate attached to the two main deriving motors, the intermediate plate carrying the linear actuator and the upper plate. The base plate of the test rig is supported by means of two auxiliary rollers attached at both sides of the base plates. In the real development of the test rig, the rollers are used to the replace the two auxiliary wheels; depending on wheels will add constraints to the entire system in manoeuvring and rotating specifically in rough terrains.

B. *Hardware components and accessories*

a. Two DC permanent magnet motors are used drive the vehicle. Each motor is provided with a built-in shaft encoder with 360 counts/shaft revolution. In order to provide the attached payload with the required sliding motion, a screw thread driver is used. The linear actuator used is able to hold a weight of more than 1.3 kg while travelling through a maximum stroke of 48mm.

b. Atmel TM ATMEGA32 Micro-Controller: ATMEGA32 is a high performance, low power AVR® 8-bit micro-controller manufactured by ATMELTM Corporation. The micro-controller has a CPU speed of 16 MHz and 32 Kbytes self-programmable flash memory. It has eight 10-bit Analogue to Digital Converter (ADC) channels. Moreover it has four Pulse Width Modulation (PWM) channels, two 8-bit timers/counters and one 16-bit timer/counter.

c. ANALOG 3-Axis Accelerometer ADLX335: The ADLX335 is a small, low power, complete 3-axis accelerometer system with signal conditioned voltage outputs. It has 300 mV/g sensitivity for each axis;

x-axis, y-axis and z-axis. The output signals are analogue voltages that are proportional to acceleration. It can measure the static acceleration of gravity in tilt-sensing applications as well as dynamic acceleration from motion, shock and vibration. The sensor is a poly-silicon surface-micro-machined structure built on top of a silicon wafer.

d. Analogue Digital 75% Gyroscope: MLX90609-N2: The MLX90609-N2 Angular Rate Sensor is a full gyroscopic system with high resolution and dynamic range. It shows two output signals; analogue and digital (SPI) outputs which are proportional to the angular rate perpendicular to the device surface. This gyroscope typical has 26.67 mV/°/s sensitivity. The working principle is a z-axis rate-sensing device (yaw-rate sensing), positive output voltage for clockwise rotation around the axis normal to the device top, and negative for counterclockwise rotation.

e. Motor Driver Board MD25-H Bridge Motor Driver: A 12v 2.8A Dual H-Bridge motor controller is utilized. This motor driver is designed to work with the EMG30 gear motors; the MD25 will drive two motors. It can be controlled by serial or the I2C Interface. There are two modes of operation; direct individual control of the motors or the ability to send a speed and a turn commands. The MD25 is a robust I2C or serial dual motor drive designed for use with EMG30 motors.

C. *Software packages*

a. WIN AVRTM: WIN AVRTM is a package of open source software development tools that enables the programmer to write in C, C++ or assembly codes for the ATMELTM AVR series micro-controllers. WIN AVRTM has a GNU C compiler (GCC) and debugging tools.

b. PONYPROG: This program is a serial device programmer application. PonyProg is an open source program that can be used to download the compiled codes (hex-files) to the micro-controller via the RS232 interface. It can be used in several types of micro-controller boards; programming of ATMELTM AVR micro-controllers.

3. Control Strategy

A simple PID controller was selected to be implemented on this device. The PID controller code is summarized in three main parts; a proportional term, a derivative term and an integral term. Error measurement is based on the angle referring to the vertical balance position of the robot using an accelerometer and a gyroscope. For the integral term, it is important to create an anti-windup function so this term will not exceed a maximum value causing an unstable behaviour. It is important to explain the sensors fusion; this is based on a simple Kalman filter function. The accelerometer is accurate for long time scales because the variations in the data are not drastic in a long period of time.

However, the gyroscope which is accurate for short time scales because of the drift in time. In this algorithm, the tilt angle will be given by both sensors. As the main problem of the gyroscope is the drift in time, it could be solved by using the output data from the accelerometer. In summary; the accelerometer returns the correct angle in the long term. However, it is adversely affected by short term conditions.

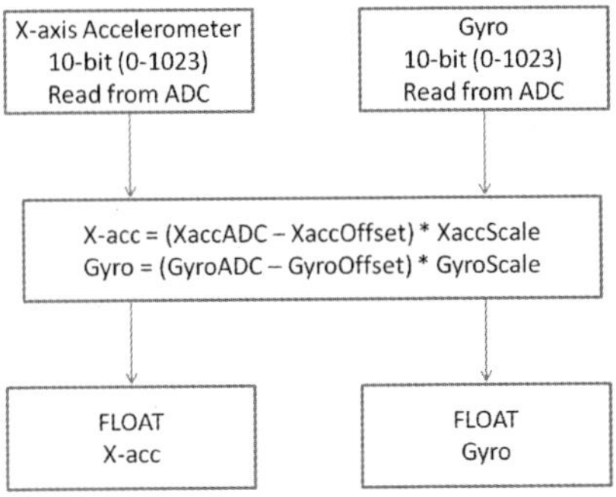

Fig. 2 Diagram of the algorithm of accelerometer and gyroscope

A simple solution is implementing a filter for the measurements of both sensors, shown in Figure 3. A low-pass filter is used for the accelerometer output and a high-pass filter is used for the gyroscope output. Combining these two filtered signals results in generating a reasonable angle measurement. The constants are the complimentary filter coefficients that are chosen, the small value from the filter coefficients means that only long-terms changes to the acceleration will affect the estimated value of the angle.

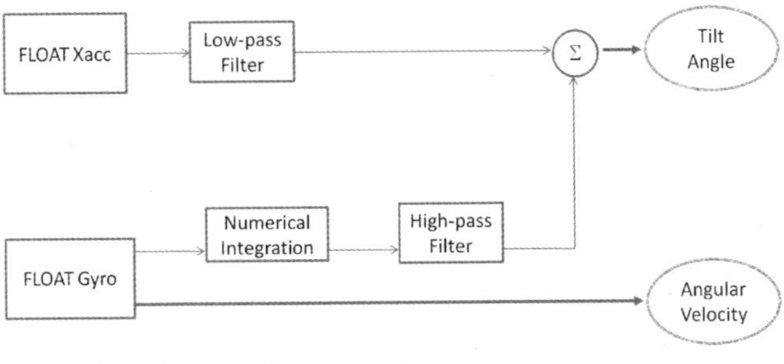

Fig. 3 Digram of filter for accelerometer and gyroscope

The equation for the filter is:

$$Angle = \left(Coef_{1\,filter}\right)\times\left(Angle + Measure_{gyroscope}\times dt\right)+\left(Coef_{2\,filter}\right)\times X_{acc} \qquad (1)$$

$$\tau = \frac{adt}{1-a} \qquad (2)$$

Where:
τ = constant
dt = Sample period
a = Filter coeffecient

Then, the tilt angle of the intermediate body can be calculated using the following equation:

$$y = a \times (y) + (1 - a) \times (x) \quad (3)$$

Where:
y = Tilt angle
a = Filter coeffecient
x = Acceleration measured by the accelerometer

The filter coefficient could be known once the time constant and simple frequencies were chosen.

In this case, to design the filter, the method of trial and error was chosen in order to check the sensors while the sampling period was used to obtain the filter coefficient. As the gyroscope has a bias (drift) of 5°/sec, a constant less than 2.5 seconds is needed to ensure that there is no deviation greater than few degrees in either direction in which the platform can move. If the time constant is greater than 2.5 seconds, only pass through the filter accelerations related to the gravity of the platform. However, for this particular time, the gyroscope measurements provide important errors as many deviations occurred during that time. Moreover, with a small time constant, noise in the horizontal acceleration passes through the filter which will cause errors in the tilt angle.

The following set of parameters has been used: $\tau = 0.45\ s$, $dt = 0.01 s$, $a = 0.97$

The τ defines the limit at which the gyroscope and the accelerometer will work properly. If time interval is less than half a second (0.45s), the gyroscope integration becomes superior and the noise from the horizontal accelerations is filtered. For a longer period of time; more than half a second, the average measure of the accelerometer will have a greater weight from the gyroscope measurements which at this point may have the drift over time.

This method was used to solve the problems presented by the two sensors; the noise in the accelerations from the accelerometer, horizontal accelerations due to gravity effect and the deviation in time (drift) due to the gyroscope. Furthermore, the tilt angle will be quickly calculated and with no delays, which improves stability of the platform. Finally, this method doesn't use excessive memory of the micro-controller.

4. Simulation Results

a. Using accelerometer only

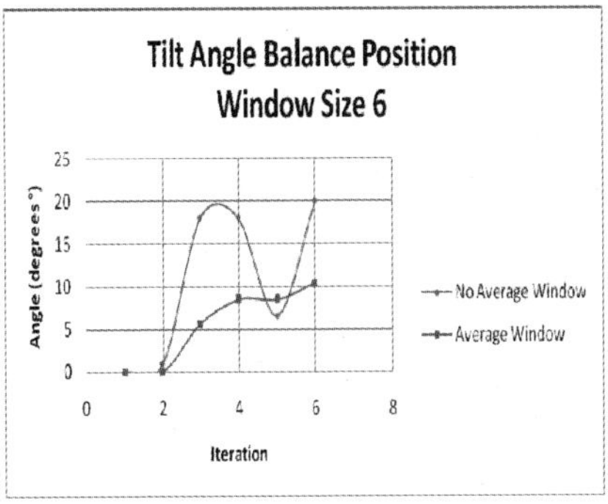

Fig. 4 Tilt angle, window size 6

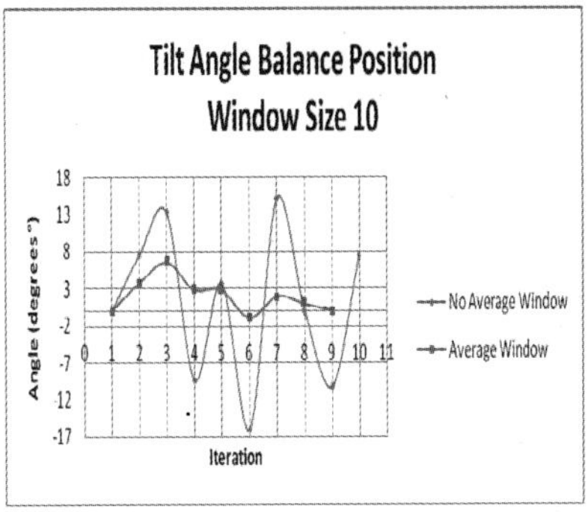

Fig. 5 Tilt angle, window size 10

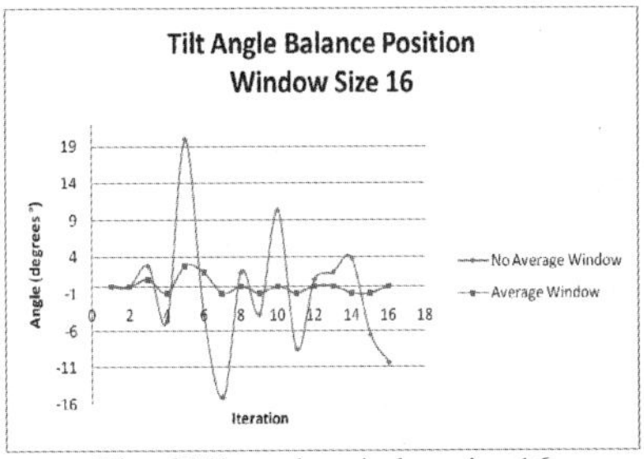

Fig. 6 Tilt angle, window size 16

From the previous graphs, shown in Figures 4, 5 and 6, it can be concluded that the greater the size of the window the smoother platform behaviour without sudden changes to its acceleration. This results in smoother results for the tilt angle which leads to greater consistency in the values of the speed of the wheels and in the angular position of the intermediate body. When the sample frequency is changing to a higher value it will indicate a faster entry into the interrupt routine. Also, it is important to note that even though the changes are smooth with large values of window size, it produces a lag because this method spends time moving and obtaining the average of the values from the accelerometer. This could cause a lag in the feedback controller that can be solved by changing the sampling frequency.

b. Using gyroscope only

It's important to notice that in cases where the gyroscope has been used, different positions of the platform can't be used to obtain results, only the balancing position. This is because when the robot is stationary, the gyroscope will show a constant value because the angular velocity is not changing. Due to this phenomenon, the shown graph in Figure 7 is only for the tilt angle when the robot is trying to balance. As shown in this Figure, it can be concluded that the robot is close to achieve the balance position. On the other hand, after a few measurements, the angle starts increasing without control, leading to losing the balance condition after a period of time.

a. Using accelerometer and gyroscope

This graph shows the angle of the platform when it is trying to achieve a balance position. As explained before, because the gyroscope is one of the sensors used, only the balancing position can be analyzed. An acceptable behaviour of the two-wheeled platform has been obtained by using both sensors. The angles of the robot are limited within a range of ±5°, which is considered an acceptable deviation from the target upright balance position. After analyzing the results shown in Figures (4) – (8); using accelerometer only, gyroscope only and using both accelerometer and gyroscope sensors, it can be concluded that using both sensors has resulted in improving the results in terms of limiting the tilt angle to a small value. Furthermore, the drift problem has disappeared due to using the accelerometer for calculating the inclination angle of the platform.

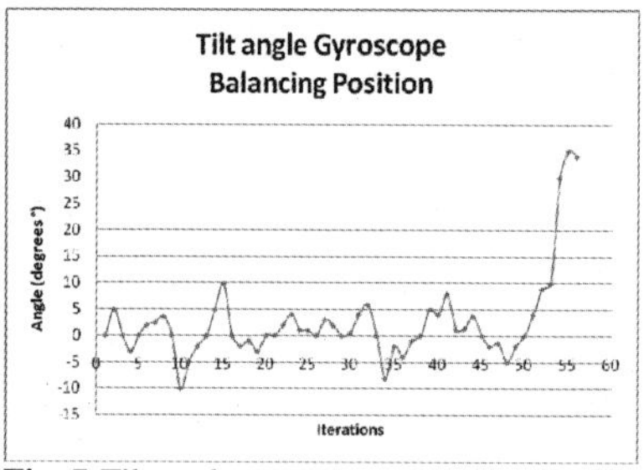

Fig. 7 Tilt angle measurement using gyroscope only

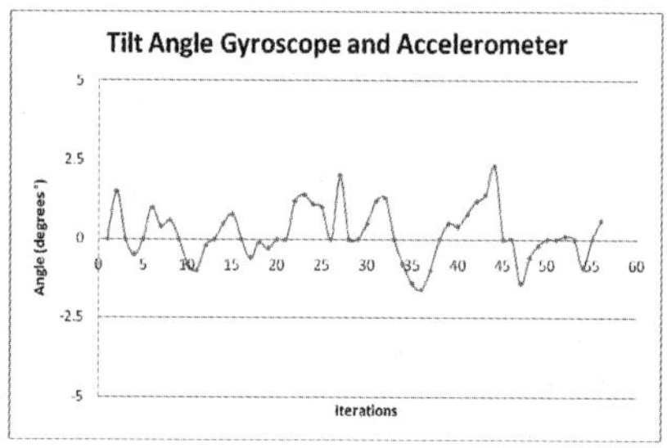

Fig. 8 Tilt angle measurement using gyroscope and accelerometer

5. Conclusions

A two wheeled vehicle with an additional degree of freedom along it intermediate body has been presented. A PID based control scheme control strategy has been developed and implemented on the system in order to simultaneously achieve balance condition with a limited overall system movement in the horizontal plane. Two sensing devices have been used to test the applicability of the developed control algorithm; accelerometer and gyroscope. The experimental results presented in this paper represented a first step in testing the machine for balancing in upright position.

References

1. Choi D and Jun-Ho Oh, "Human-friendly Motion Control of a Wheeled Inverted Pendulum by Reduced-order Disturbance Observer", 2008 IEEE International Conference on Robotics and Automation, Pasadena, CA, USA, May 19-23, 2008.

2. Goher K. M. and Tokhi M. O., Development of a Novel Design of Two-Wheeled Vehicle with an Extended Rod, Journal of Selected Areas in Robotics and Control (JSRC), December Edition, 2010, 1(2), 6 – 16.
3. Goher K M and Tokhi M.O., A Two-Wheeled Vehicle with an Extended Rod, Mechanical Design Clues and Dynamic Modelling, in the Proceedings of the 8th IEEE International Conference on Cybernetic Intelligent Systems, Birmingham, UK, September 9-10., (2009).
4. Imamura R, Takei T, Yuta S, "Sensor Drift Compensation and Control of a Wheeled Inverted Pendulum Mobile Robot", 978-1-4244-1703-2/08/$25.00 ©2008 IEEE.
5. Jeong S H and Takahashi T, "Wheeled Inverted Pendulum Type Assistant Robot: Inverted Mobile, Standing, and Sitting Motions" Proceedings of the 2007 IEEE/RSJ International Conference on Intelligent Robots and Systems, San Diego, CA, USA, Oct 29 - Nov 2, 2007.
6. Jingtao Li et al, "Mechanical Design and Dynamic Modelling of a Two-Wheeled Inverted Pendulum Mobile Robot", Proceedings of the IEEE International Conference on Automation and Logistics, August 18 - 21, 2007, Jinan, China.
7. Nawawi S. W., Ahmad M. N. and Osman J. H. S., "Development of a Two-Wheeled Inverted Pendulum Mobile Robot", the 5th Student Conference on Research and Development - SCOReD 2007, 11-12 December 2007, Malaysia.
8. Nawawi S.W, Ahmad M.N, Osman J.H.S, "Control of Two-wheels Inverted Pendulum Mobile Robot Using Full Order Sliding Mode Control", Proceedings of International Conference on Man-Machine Systems 2006, September 15-16 2006, Langkawi, Malaysia.
9. Zheng C et al, "Mechanical Design and Control System of a Miniature Autonomous Surveillance Robot", Proceedings of the 2007 IEEE International Conference on Mechatronics and Automation, August 5-8, 2007, Harbin, China.
10. Tsai C. C., Lin S.C. and Luo W.L. "Adaptive Steering of a Self-balancing Two-wheeled Transporter", Proceedings of 2006 CACS Automatic Control Conference, St. John's University,Tamsui, Taiwan, Nov. 10-11, 2006.

DISTRIBUTION OF REACTION FORCES DURING MOVEMENT WALKING ROBOTS

ION ION[1], ION SIMIONESCU[2], ADRIAN CURAJ[3], VLADAREANU LUIGE[4] AURELIAN VASILE[3]

[1]Manufacturing Technology Department [2]Mechanism and Robot Theory Department, [3]Automatic Control and System Engineering Department, POLITEHNICA University of Bucharest,[4] Institute of Solid Mechanics, Romanian Academy

Romania E-mail: ioni51@ yahoo.com

An important feature of the walking robots is that the ground's configuration never influences or affects their movement too much. Such a feature makes this locomotion type turn into an attractive solution for many applications that require movement on an unplanned land, namely having an uneven configuration. As uneven land covers an endless variety it is difficult to match all the types and cases of walking on such grounds. The movement of the walking robots can be divided in two modes:- under condition of the static stability; - under condition of the dynamic stability. The loss of static or quasi-static stability may be produced in two cases, namely:- the vertical projection of the gravity center is out of support polygon;- one or more tangential components of reaction forces from support points are greater than the friction forces The movement of the walking robot under dynamical stability conditions may be realized when the forward average speed is greater than a certain limit. In the work are analyzed the possibilities of determination of forces and the limit conditions for the stable displacement of the novel modular walking robots MERO 2.

1. Introduction

Distribution of reaction forces from the supports of the legs is one of the important problems that must be resolved to organize movements on land legs walking robots with relief complicated. Friction cones of the supports are circular and can be oriented arbitrarily and the points of support - whether they are more than three - are not covered by a plan.

When the number of support points is greater than three, the problem of determining the distribution of forces is statically indeterminate. To calculate these forces is necessary to know the layout and terrain feature and the positions, dimensions and materials of components of the robot (kinematic and organological size, modules of elasticity).

Choosing an element of this set is based on the travel restrictions imposed on the system. In some cases, the set of admissible solutions can be null if data movement is impossible to achieve, or may be formed from a single element and then there is only a possibility of realization of movements, without being able to take into account additional restrictions.

For a given configuration of the system displacement, the forces of reaction of the supports are determined unequivocally. The leadership of walking robots with many legs, an optimal distribution of reaction forces of the supports can be taken into account in determining the stepping strategy.

Moreover, this strategy can be set so that the forces or torque engine are minimal power consumption is minimized so. These criteria for optimizing the movement of walking robots can be taken individually or several at once.

2. Distribution of forces in the mechanisms of the robot walking displacement

The walking robots moving on a terrain, strewn with many obstacles, which can be convex or concave, there is a danger that the position of these robots are not stable.

One of conditions imposed on the motion of walking machines is the stability. The movement of the legged robots can be divided in two modes:

- under condition of the static stability;
- under condition of the dynamical stability.

The main difference between robots which walked under the static stability and under the dynamical stability conditions originates from the fact that during statically walking, the vertical projection of the gravity center of the robot must lies into the supporting polygon, where as during the dynamical walking, this condition can be not satisfied.

The problem of quasi-statical stability analysis in condition of arbitrary step when the accelerations of points of component elements are much smaller than gravity acceleration is identical with the problem of stability analysis when the robot does not walk. The inertial forces are neglected and the walking can be controlled in a kinematics way.

The investigation of statically stability is based on the notion namely *hardening configuration*. Hardening configuration is a term used to indicate the rigidly structure of the robot, obtained through the shutting off the driving motors.

The position of the walking robot is stable if the hardening configuration is in posture of stable equilibrium under the action of gravity forces.

The hardening configuration is statically stable if this accomplished the following conditions:

The vertical projection of the gravity center of the robot in its entirety (platforms, connecting elements and leg chains, control system and driving, pregnancy, etc.) must be inside the *supporting polygon*. The supporting polygon is the minimum convex area which is obtained by connecting all support points. A body at rest in a gravitational field, subject to ideal connections, has the differential of the gravity center elevation equal with zero.

Vertical projection of the care center of gravity G of the robot is inside the support polygon if all the distances d_i, measured from this point to a P_iP_{i+1}, sides are positive, assuming that the supports on the perimeter of the polygon are numbered clockwise (Fig.1).

Magnitudes of these distances are calculated with the following:

$$d_i = \frac{(\xi_G - \xi P^i)(\eta P^{i+1} - \eta P^i) - (\eta_G - \eta P^i)(\xi P^{i+1} - \xi P^i)}{\sqrt{(\xi P^{i+1} - \xi P^i)^2 + (\eta P^{i+1} - \eta P^i)^2}}$$

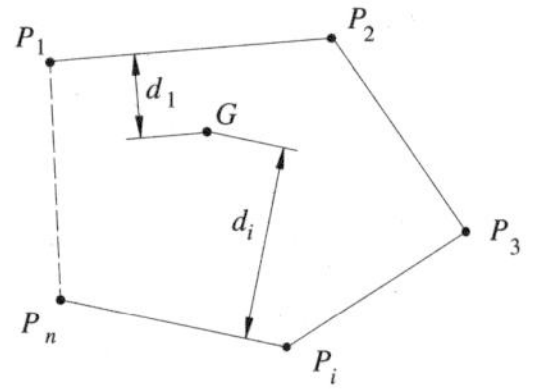

Figure1 Polygon support of a walking robots

For establishing the stable positions of a walking robot it is necessary to determine the forces distribution in the shifting mechanisms.

Determination of the real forces distribution in the shifting mechanisms of a walking locomotion system which moves in rugged land at low speed is necessary for the analysis of stability. The position of a walking system depends on the following factors:

- the configuration of walking mechanisms;
- the masses of component elements and their position of gravity centers;
- the values of friction coefficients between terrain and feet;
- the stiffness of terrain;
- the shape of terrain surface.

The active surface of the foot is relatively small and it is considered that the reaction force is applied in the gravity center of this surface. The reaction force represents the resultant of the elementary forces, uniformly distributed on the foot sole surface. The gravity center of foot active surface is called *theoretical contact point.*

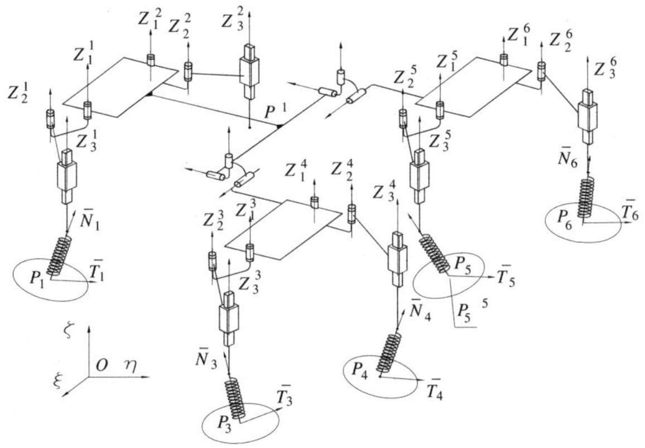

Figure 2 The reaction forces at points of contact leg with the ground surface - model

- normal component $\overline{N}$, perpendicular on the surface of terrain in the theoretical contact point;

- tangential component $\overline{T}$,the magnitude of $\overline{T}$ vector cannot be greater than the product of the magnitude of the normal component N by the frictional coefficient μ between foot sole and terrain. If this magnitude is greater than the friction force, then the foot slips along the support surface down to the stable position, where the magnitudes of this component decrease under the above-mentioned limit.

Therefore, the problem of determining the stable position of a walking robot upon some terrain has not a unique solution. For every foot is available a field which covers all contact points in which the condition $\mu T \leq N$ is true. The equal sign corresponds to the field's boundary.

To determine the stable position of the walking robot which leans upon n legs, on some shape terrain, it is necessary to solve a nonlinear system, which is formed by:

- Equations that express the feet contact the ground surface:

$$X_{0Pi} = \xi(u_i, v_i);\ Y_{0Pi} = \eta(u_i, v_i);\ Z_{0Pi} = \zeta(u_i, v_i),\ i = \overline{1,\ n}, \tag{1}$$

- the transformation matrix equation

$$\left\| \begin{matrix} 1 \\ X_{0Pi} \\ Y_{0Pi} \\ Z_{0Pi} \end{matrix} \right\| = \mathbf{A}\mathbf{A}_0\mathbf{A}_1^i\mathbf{A}_2^i\mathbf{A}_3^i \left\| \begin{matrix} 1 \\ X_{4Pi} \\ Y_{4Pi} \\ Z_{4Pi} \end{matrix} \right\|,$$

where:

A is the transformation matrix of coordinate of a point from the system $O_0X_0Y_0Z_0$ of the robot platform to the system $O_1\xi\eta\zeta$;
$\mathbf{A}_i$ is the Hartenberg – Denavit transformation matrix of coordinates of a point from the system $O_{i+1}X_{i+1}Y_{i+1}Z_{i+1}$ of the element $\boldsymbol{i}$ to the system $O_iX_iY_iZ_i$ of the element (***i*-1**) [3],

- The balance equations

$$\sum_{i=1}^{n} \overline{R_i} + \overline{F} = 0;\ \sum_{i=1}^{n} \overline{M}_{(R_i)} + \overline{M} = 0, \tag{2}$$

which expressed the equilibrium of the forces and moments system which acted on the elements of walking robot.

The system is compatible for $n = 3$ support points.

If the number of feet which are simultaneous in the support phase is bigger than three, the system is undetermined static and is necessary to take into consideration the deformations of the mechanical structure of the walking robot and terrain.
In case of a quadrupedal walking robot, the hardening configuration is a six fold hyperstatical structure. To determinate the force distribution, one must use a specific method for indeterminate static systems.

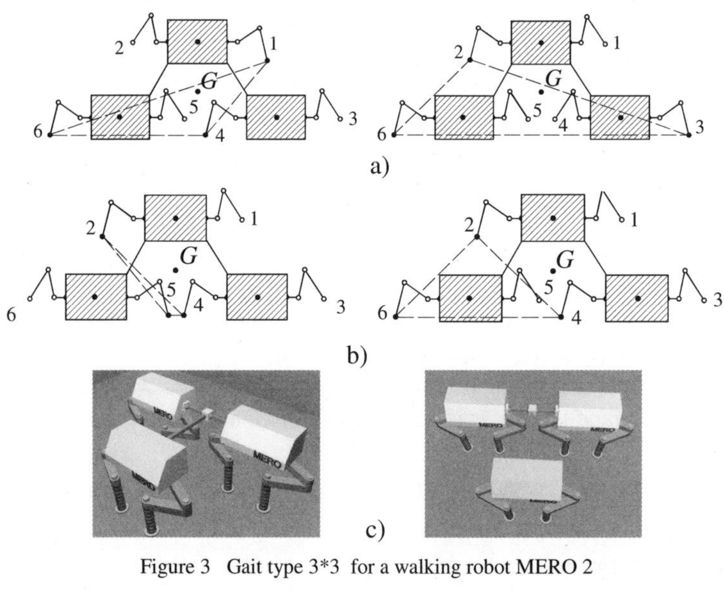

Figure 3 Gait type 3*3 for a walking robot MERO 2

Motion control walking robots using moving blindly walking may split into two complementary processes, namely: the control and active control outfit. The control outfit is used to control the inclination to maintain the body in a desired direction so that the vertical projection of center of gravity of the robot to find support within the polygon, to achieve static stability conditions. Active control is provided via contact sensors and transducers.

A special problem arises during the robot movement (fig.3) is that of preserving stability when one or more legs is in transfer phase. Geometry support polygon can change and the risk that the center of gravity to move outside it. Resolving this situation is based on how your machine is configured and running strategy chosen.

2.1. *Stepping over isolated obstacle while preserving static stability*

The presence of the obstacles on the support surface may lead to changing the direction, the height of the body's movement and its orientation in space as well as the appropriate reorganization of the sequel of watching the regime of the legs' movement.

This paragraph will follow, in compliance with [3], [2], [4], the case when to prevent the obstacles, it is enough only to change the sequel of following and maybe the height of the body's center of gravity (mass point) preserving the same the other movement features of the robot.

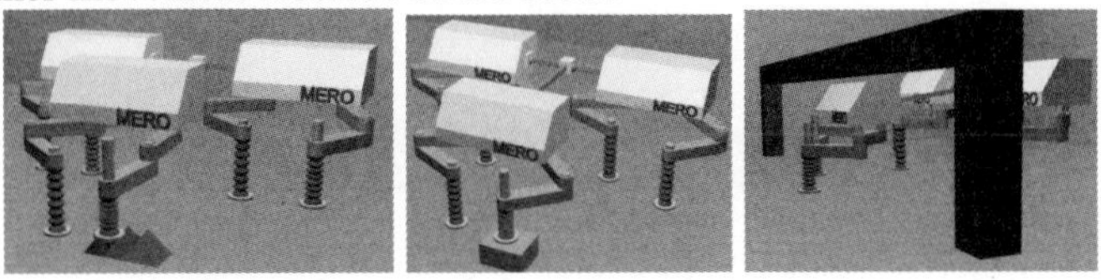

Figure 4 Stepping over isolated obstacles
for a walking robot MERO 2 (Computer graphics simulation)

Like before we will assume that the legs' suspension points are symmetrically placed to the vertical plane including the body's center and the speed parallel to it.

We will name the obstacle 'isolated' if we can include it in the field between two vertical planes, erect on the body's movement direction so that the follow paths beyond this field lack the points forbidden to advance.

On the support area the above mentioned planes separate the area called the *obstacle's zone* where it is forbidden to place any follow points. Its ends are erect on the follow paths.

The zone's width depends on the ratio of the obstacle's dimensions and the geometrical parameters characteristic to the robot as well as its control system.

It occurs the problem of establishing the sequence of follow, which enables the robot to step over the area without disturbing its static stability.

Conclusion/ Application. For version 2 MERO modular walking robot consists of three modules below shows the procedure for determining the reaction forces. (Fig.5)

Feet are modeled by points in the centers of gravity of their surfaces. The land is shaped by moving robot plane passing through the supports. Stability modular walking robots, bipedal three modules, is discussed during the journey, when it focuses only on three legs.

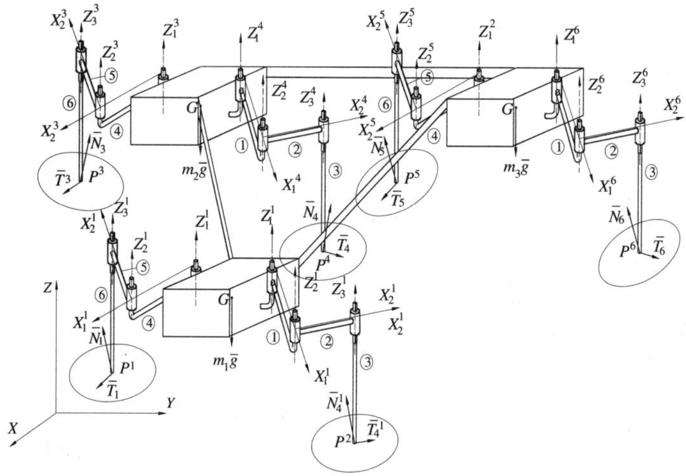

Figure 5 The Hartenberg – Denavit coordinate systems and the reaction forces at points of contact with the ground leg

Equations which is defined planes land, and passing through points P1, P4 and P6 (Fig. 5) are:

$$l_1(x - x_{P1}) + m_1(y - y_{P1}) + n_1(z - z_{P1}) = 0;$$
$$l_4(x - x_{P4}) + m_4(y - y_{P4}) + n_4(z - z_{P4}) = 0;$$
$$l_6(x - x_{P6}) + m_6(y - y_{P6}) + n_6(z - z_{P6}) = 0, \tag{3}$$

where: l_i, m_i şi n_i, $i = 1, 4, 6$, are coefficients of the normals to planes executives support. On the structure of the robot act reinforced a system of forces, ie forces and weight of the load components. Torsorul this system is the point of reduction A of components

$$\overline{F} = F_x\overline{i} + F_x\overline{j} + F_x\overline{k} \text{ şi } \overline{M} = M_x\overline{i} + M_x\overline{j} + M_x\overline{k}, \tag{4}$$

where: F_x, F_y, F_z, M_x, M_y, M_z projections on the axes are fixed system components (the resulting net force and moment)

The reaction forces of the supports P^1, P^4 şi P^6 are: $\overline{R_i} = \overline{N_i} + \overline{T_i}$, $i = 1, 4, 6$,

Where, the normal components $\overline{N_i}$ of support planes are perpendicular and tangential components (frictional forces) $\overline{T_i}$ are contained in planes of support The equations of equilibrium of forces and moments are:

$$F_x + N_1 l_1 + T_1 u_1 + N_4 l_4 + T_4 u_4 + N_6 l_6 + T_6 u_6 = 0;$$
$$F_y + N_1 m_1 + T_1 v_1 + N_4 m_4 + T_4 v_4 + N_6 m_6 + T_6 v_6 = 0; \quad (5)$$
$$F_x + N_1 n_1 + T_1 w_1 + N_4 n_4 + T_4 w_4 + N_6 n_6 + T_6 w_6 = 0;$$

$$\begin{Vmatrix} 0 & -z_A & y_A \\ z_A & 0 & -x_A \\ -y_A & x_A & 0 \end{Vmatrix} \begin{Vmatrix} F_{1x} \\ F_{1y} \\ F_{1z} \end{Vmatrix} + \begin{Vmatrix} M_{1x} \\ M_{1y} \\ M_{1z} \end{Vmatrix} + \mathbf{M}_1\mathbf{R}_1 + \mathbf{M}_4\mathbf{R}_4 + \mathbf{M}_6\mathbf{R}_6 = 0, \quad (6)$$

where: u_i, v_i, w_i directors of the straight lines are the coefficients of the support components of the tangential reaction forces of the supports,

$$\mathbf{M}_i = \begin{Vmatrix} 0 & -z_{Pi} & y_{Pi} \\ z_{Pi} & 0 & -x_{Pi} \\ -y_{Pi} & x_{Pi} & 0 \end{Vmatrix} \quad \mathbf{R}_i = \begin{Vmatrix} N_{ix} + T_{ix} \\ N_{iy} + T_{iy} \\ N_{iz} + T_{iz} \end{Vmatrix}, \; i = 1, 4, 6.$$

These lines are contained in the underlying support planes, so

$$l_i u_i + m_i v_i + n_i w_i = 0, \quad (7)$$

$$u_i^2 + v_i^2 + w_i^2 = 1, \; i = 1, 4, 6. \quad (8)$$

System of equations (5), (6), (7) and (8) resolve uncertainties in relation to N_1, N_4, N_6, T_{1x}, T_{1y}, T_{4z}, T_{4x}, T_{4y}, T_{4z}, T_{6x}, T_{6y}, T_{6z}.

If one or more of the conditions $\sqrt{T_{ix}^2 + T_{iy}^2 + T_{iz}^2} < \mu N_i$, $i = 1, 4, 6$,

is not satisfied at some point, the robot moves by sliding to a position where all inequalities above are verified (μ is the coefficient of friction between feet and ground).

References

1 Denavit J., Hartenberg R.S., *A Kinematic Notation for Lower Pair Mechanisms Based on Matrices*, Journal of Applied Mechanics, Tr. ASME, 1955, Vol. 77

2 Ion I., Vladareanu L., Simionescu I. and Vasile A., *"The Gait Analysis for Modular Walking Robot MERO Walks on the Slope"*, ISI Proceedings of the 9th WSEAS International Conference on Automation & Information: Theory and Advanced Technology (ICAI '08), Published by WSEAS Press pag. 222-229, ISBN: 978-960-6766-77-0, (2008

3 Simionescu I., Ion I. and Ciupitu L. *Mechanism of Industrial Robots* Published by AGIR Press Bucharest -In Romanian. (Tome1, 2008 and Tome 2 ,2010)

4 S.M. Song, and J.K. Waldron, *An Analytical Approach for Gait Study and its Application on Wave Gait,* International Journal of Robotics Research, Vol. 6, No. 2, 1987, pp. 60-71

ONLINE DISCOVERY OF LOCOMOTION MODES FOR WHEEL-LEGGED HYBRID ROBOTS: A TRANSFERABILITY-BASED APPROACH

S. KOOS* and J.-B. MOURET

ISIR, Université Pierre et Marie Curie-Paris 6, CNRS UMR 7222.
Paris, 75005, France
**E-mail: koos@isir.upmc.fr*
http://people.isir.upmc.fr/koos/

Wheel-legged hybrid robots promise to combine the efficiency of wheeled robots with the versatility of legged robots: they are able to roll on simple terrains, to dynamically adapt their posture and even to walk on uneven grounds. Although different locomotion modes of such robots have been studied, a pivotal question remains: how to automatically adapt the locomotion mode when the environment changes? We here propose that the robot autonomously discovers its locomotion mode using optimization-based learning. To that aim, we introduce a new algorithm that relies on a forward model and a stochastic multi-objective optimization. Three objectives are optimized: (1) the average displacement speed, (2) the expended energy and (3) the transferability score, which reflects how well the behavior of the robot is in agreement with the predictions of the forward model. This transferability function is approximated by conducting 20 experiments of one second on the real robot during the optimization. In the three investigated situations (flat ground, grass-like terrain, tunnel-like environment), our method found efficient controllers for forward locomotion in 1 to 2 minutes: the robot used its wheels on the flat ground, it walked on the grass-like terrain and moved with a lowered body in the tunnel-like environment.

Keywords: Hybrid robot; Learning; Multi-objective Evolutionary Algorithm; Transferability

1. Introduction

Wheel-legged hybrid robots aim at combining the efficiency of wheeled robots with the versatility of legged robots:[1–3] by adding wheels at the end of legs, they can act like wheeled robots on simple terrains and adapt their posture to the shape of an uneven ground; they can also stop their wheels and be equivalent to a classic legged robot. Several papers deal with the control of such wheel-legged robots[1–3] but most of them describe single controllers that can take advantage of the legs to adapt the posture of the robot.[2,3] A few papers investigate different locomotion modes, such as rolling with passive wheels and walking.[1] Nevertheless, none of them tackles one of the most important questions: *how should the robot select its locomotion mode?* And, since there are an infinity of possible situations and an infinity of hybrid locomotion modes, *how to discover the best controller in an unforeseen situation?* The present paper introduces a new algorithm to answer these two questions. In the typical scenario (see Fig. 1), the robot first moves on a flat terrain, in which using the wheels is *a priori* efficient; the robot then encounters a tall grass field in which its wheels are not working anymore: it has to find a new locomotion mode; finally, it leaves the grass field and enters a tunnel with a low ceiling that should never be hit: a new adaptation is required. This scenario illustrates three situations, but the goal of the present paper is to introduce a general adaptation algorithm that could be used in any situation and for any robot.

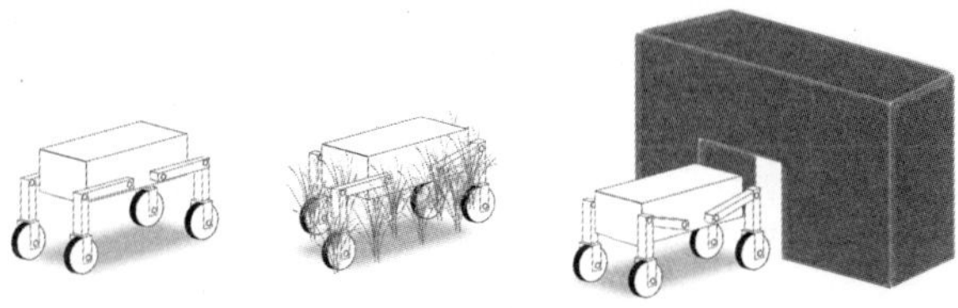

Fig. 1. Typical scenario made of three situations : (left) uniform flat ground, (center) grass-like terrain, (right) tunnel-like terrain.

2. Proposed Approach

2.1. *Overview*

Online adaptation of a mobile robot to its environment can be split in three different phases, as pictured on Fig. 2, left: (1) the robot detects that the environment has changed or that its current behavior is not efficient anymore; (2) the robot enters an adaptation phase, during which new efficient locomotion modes are looked for; (3) an efficient behavior is next selected and exploited, while no new situation is encountered. We here focus on the adaptation phase (phase 2).

To discover locomotion modes, the robot can *learn* a new strategy by itself. Such a situation is a typical use case of reinforcement learning[4] or optimization-based learning,[?] but these algorithms require a large number of trials on the real robot, making the learning phase long[a] and potentially dangerous. The Estimation-Exploration Algorithm[6] proposes a faster alternative which relies on an automatically learned, internal dynamic model of the robot: using it, a star-shaped quadruped robot only needed to perform 15 simple actions to learn a new walking behavior after the loss of a leg. However, each time a disagreement is detected between the forward model and the reality, the EEA requires to learn a new forward model from scratch; we think that this complex step is inefficient if the disagreement stems from a change of the environment (most parts of the forward model should still be reliable) and not from a change in the morphology of the robot. Additionally, actions performed by the robot are not goal-directed: the robot can spend a long time to model a part of its morphology which is useless for its goal.

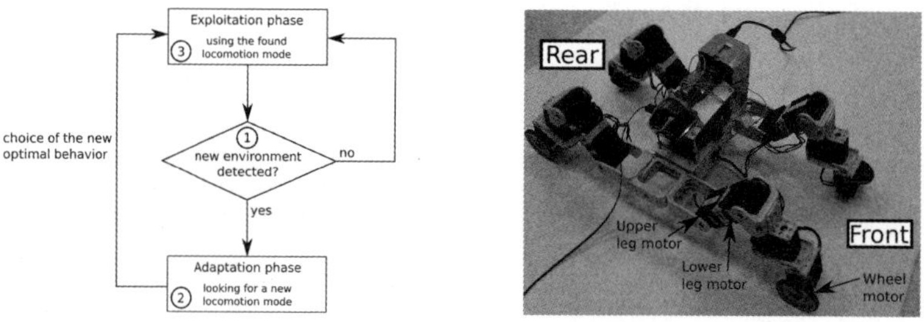

Fig. 2. Left, the three different phases of our adaptation algorithm. Right, the 12-DOF quadrupedal wheel-legged robot used for the experiment.

To overcome these limitations, we propose that the adaptation phase relies on a similar dynamic forward model, except that it will be provided by the robot's designer. In our approach, the learning algorithm does not modify this model but it discovers what potentially interesting behaviors are not properly working in reality in order to avoid them. We take inspiration from the transferability approach,[7] by looking for good compromises between performance objectives and the transferability measure, an approximate function that reflects how well the reality matches the prediction of the forward model for a given controller. This last function is learned by transfering a few well-chosen controllers on the real robot during the optimization and then by comparing the behaviors in simulation and in reality.

[a]about 3 hours to learn a quadruped locomotion pattern for the Aibo robot[4] and about 20 minutes for a snake-like robot.[?]

2.2. *Algorithm*

The several locomotion modes brought into play by hybrid wheel-legged robots can be interpreted as different compromises between performance and expended energy. For instance, walking behaviors are more versatile than rolling behaviors, but leads to higher energy consumption. We consequently formulate the learning problem as a multi-objective optimization process with four objectives to be optimized:

(1) average displacement speed of the robot;
(2) expended energy to perform the behavior[b];
(3) approximated transferability measure $\hat{T}$;
(4) behavioral diversity objective.

This last objective allows to maintain behavioral diversity among the population, which efficiently enhances exploration of the controller state space.[?,?] To quantify the diversity of a controller from the already transferred ones, we define a behavioral diversity value as follows. Let $\mathfrak{C}_T$ be the set of the already transferred controllers and b_{dist} a behavioral distance defined in simulation, the behavioral diversity value $diversity(c)$ for a given controller c is:

$$diversity(c) = \min_{c_i \in \mathfrak{C}_T} b_{dist}(c,\ c_i)$$

The behavioral distance function b_{dist} is based on a set of behavioral features defined by the user and that describe the behaviors of controllers in simulation. The distance between two controllers is then computed as the Euclidean distance between their vectors of behavioral features.

The relation that links the behavioral features in simulation and the exact transferability measure is interpolated Inverse Distance Weighting (IDW) by interpolating. An outline of the learning algorithm is pictured on Fig. 3:

A. For each individual x of the population, the four objective values are computed in simulation (average speed, expended energy, approximated transferability and behavioral diversity).
B. The controllers are optimized via a multi-objective evolutionary algorithm (MOEA) during N generations.

[b]Walking uses more energy than rolling

C. Every N generations, the controller of the current population with the highest behavioral diversity is transferred onto the real robot:
   C1. the behavior of the robot is logged;
   C2. the corresponding exact transferability value is computed based on the simulated and real behaviors;
   C3. the approximated transferability measure is updated with the new data by IDW interpolation; back to step A.

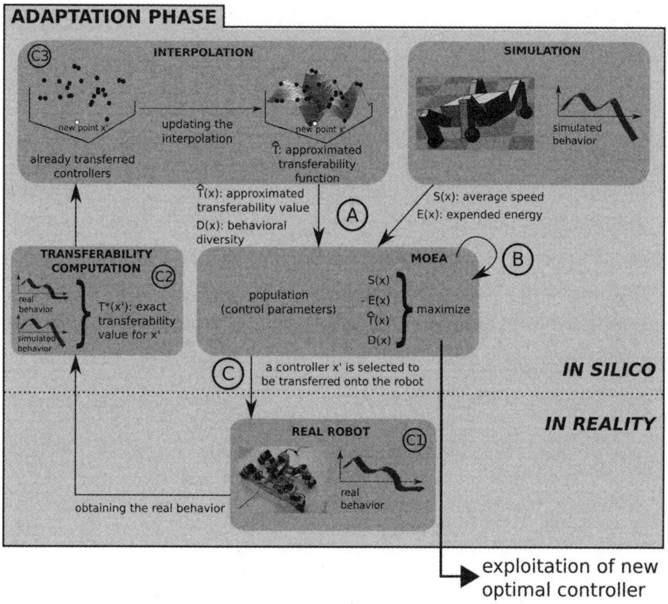

Fig. 3. Outline of the adaptation phase. The four objectives are computed for the controllers (step A), which are optimized during N generations with a multi-objective evolutionary algorithm (MOEA, step B). The controller with the highest behavioral diversity is then transferred onto the real robot and the generated data allows to update the approximated transferability function (steps C1 to C3).

3. Application

We tested this new approach on a 12-DOF quadrupedal wheel-legged hybrid robot (Fig. 2, right) inspired by the Hylos robot.[2] Each leg of the robot includes 4 Dynamixel AX-12+ Robot Actuators from Bioloid Kit. In the experiment, the two upper motors and the wheel motor of each leg are controlled. The control signals of all the motors depend on three optimized real parameters $(p_1, p_2, p_3) \in [-1, 1]^3$: p_1 is linked to the posture of the robot, p_2 to the amplitude of legs' movements and p_3 to the speed of wheels.The behavioral features b^i are directly derived from these parameters and their values are indicated for each case in the following equations. Concerning the leg motors, the desired angular position α^d of a motor i at time t is obtained by:

$$\alpha^d(i,t) = \frac{5\pi}{48} \cdot p_1 - \begin{cases} \frac{5\pi}{24} \cdot p_2 \cdot sin(2\pi t - \phi(i)) & \text{, if } p_2 > 0 \quad (b^1 = p_1,\ b^2 = p_2) \\ 0 & \text{, otherwise} \quad (b^1 = p_1,\ b^2 = 0) \end{cases}$$

The phase angle $\phi(i)$ is 0 for the upper leg motors of each leg and $\pi/2$ for the lower leg motors of each leg (for orientation, see Fig. 2, left). Both motors of one leg consequently have the same control signal with different phases. New angle positions are sent every 0.05 seconds.

The four wheel motors are controlled with the same speed value v defined as follows depending on the p_3 parameter (v_{max} = 6 rad/s):

$$v = \begin{cases} p_3 \cdot v_{max} & \text{, if } p_3 > 0 \quad (b^3 = p_3) \\ 0 & \text{, otherwise} \quad (b^3 = 0) \end{cases}$$

The average speed and the expended energy are computed in a simulation model based on Open Dynamics Engine (ODE)[c]. Whatever the real environment looks like, it always simulates the displacement of the robot on a flat ground during one second. Energy is crudely approximated by summing the angular movement of each degree of freedom. The displacement speed of the robot in reality is measured during one second with a motion tracking system but further work will rely on on-board visual odometry.[10] The exact transferability measure T^* is the opposite of the absolute variation between the average speeds measured with the forward model and on the robot.

The four objectives are simultaneously optimized with NSGA-II, a state-of-the art multi-objective evolutionary algorithm[11d]. This stochastic algorithm finds an approximate set of all Pareto-optimal trade-offs. The size of population is set to 40 and the optimization process stops after 200 generations. The approximate transferability measure is updated every ten generations (N = 10), by transferring the controller of the current population with the highest behavioral diversity, i.e. we perform 20 experiments

---

[c] `http://www.ode.org`

[d] This work has been implemented within the Sferes$_{v2}$ framework.[?] The source code is available at: `http://www.isir.fr/evorob_db`

by run[e]. At the end of the optimization, we extract from the final non-dominated set the solutions whose transferability values are higher than -1 cm/s. We then select the controller, which minimizes the distance to an ideal controller with an average speed of 18 cm/s and an energy value of 25.[7] This ideal controller corresponds to the most efficient rolling behavior on flat ground.

4. Results

We investigated a scenario made of three situations to which the robot has to adapt: (1) a flat ground without any constraint, (2) a grass-like environment in which wheels are blocked and, (3) a tunnel-like environment in which robot's movements are blocked if $p_1 < 0.7$. To obtain statistical results, we performed 5 experiments in each situation. Results (figure 4) show that: (1) the robot autonomously chose to use its wheels when it was put on a flat ground, (2) it learned to walk when its wheels were unavailable and, (3) it lowered its body when it encountered the tunnel. Each learning phase required 1 to 2 minutes with a recent multi-core computer[f] (including the tests on the real robot). Table 1 reports quantitative results: the rolling behavior found on flat ground cannot be used on grass (average speed is null) and is often useless in the tunnel (average speed is low: many controllers did not work), but it is the most energy-efficient controller; the walking mode requires more energy but it appears more versatile; the controllers optimized for the tunnel use slightly more energy than those optimized for flat ground only because the robot has to lower its body. *These measures show that adapting the locomotion mode was always useful and often mandatory for the robot to move forward.*

Table 1. Average speed for each controller and in each situation (5 trials), standard deviations are in brackets; energy is in arbitrary unit.

	flat ground (cm/s)	grass (cm/s)	tunnel (cm/s)	energy (E)
learned on flat ground	17.1 (0.3)	0.0 (0)	8.6 (9.9)	-26 (3)
learned in grass	13.1 (1.9)	13.1 (1.9)	8.3 (7.6)	-33 (1)
learned in tunnel	15.9 (1.7)	0.0 (0)	15.9 (1.7)	-27 (2)

[e]At generation 0, a random controller is transferred.
[f]Intel Xeon E5520 dual quad-core at 2.3 GHz

Fig. 4. Typical selected behaviors: (left) flat ground: rolling behavior; (center) grass-like: walking behavior; (right) tunnel: rolling behavior with a lowered body.

References

1. G. Endo and S. Hirose, Study on roller-walker (multi-mode steering control and self-contained locomotion), in Proc. of *IEEE-ICRA'00*, 2000.
2. C. Grand, F. Benamar, F. Plumet and P. Bidaud, *The International Journal of Robotics Research* **23** (2004).
3. P. Jarrault, C. Grand and P. Bidaud, Large Obstacle Clearance Using Kinematic Reconfigurability for a Rover with an Active Suspension, in *Proc. of CLAWAR'10*, 2010.
4. N. Kohl and P. Stone, Policy gradient reinforcement learning for fast quadrupedal locomotion, in *Proc. of IEEE-ICRA'2004*, 2004.
5. A. Sproewitz, R. Moeckel, J. Maye and A. J. Ijspeert, *The International Journal of Robotics Research* **27**, 423 (2008).
6. J. Bongard, V. Zykov and H. Lipson, *Science* **314**, 1118 (2006).
7. S. Koos, J.-B. Mouret and S. Doncieux, Crossing the Reality Gap in Evolutionary Robotics by Promoting Transferable Controllers, in *Proc. of GECCO'2010*, 2010.
8. S. Doncieux and J.-B. Mouret, Behavioral diversity measures for evolutionary robotics, in *Proc. of IEEE-CEC'10*, 2010.
9. J.-B. Mouret and S. Doncieux, *Evolutionary Computation* (2011, to appear).
10. D. Nistér, O. Naroditsky and J. Bergen, *Journal of Field Robotics* **23**, 3 (2006).
11. K. Deb, S. Agrawal, A. Pratab and T. Meyarivan, A Fast Elitist Non-Dominated Sorting Genetic Algorithm for Multi-Objective Optimization: NSGA-II, in *Proc. of PPSN VI*, 2000.
12. J.-B. Mouret and S. Doncieux, Sferes v2: Evolvin' in the multicore world, in *Proc. of IEEE-CEC'10*, 2010.

A DYNAMIC BASED PATH TRACKING CONTROLLER FOR A FAST ROVER WITH INDEPENDENT STEERING AND DRIVE

M. KRID* and F. BEN AMAR**

ISIR - Institut des Systèmes Intelligents et Robotiques,
Universit Pierre et Marie Curie Paris 6, CNRS UMR 7222
PARIS, 75252 cedex 05, France
**E-mail : krid@isir.upmc.fr*
***E-mail : amar@isir.upmc.fr*

The paper presents a path tracking controller for a fast rover which has independent front and rear steering and four independent driven wheels. The paper develop first the dynamic model of the vehicle moving on a horizontal plane. Next, the model is linearized by assuming that the side slip angles remain small and a linear lateral slippage model introducing the lateral contact stiffness is used. The kinematics of the vehicle is also linearized by assuming that the rover remains close a reference path and with a small relative heading angle. The linearized model permits the design of a Linear Quadratic Regulator LQR which is tested by using Matlab-Simulink/Adams co-simulation. Simulation results, obtained with numerous reference paths, show a good control accuracy and appears to be robust with respect to environment and robot state changes.

Keywords: Mobile robot, path tracking, wheel slippage, LQR controller.

1. Introduction

Outdoor mobile robots has to perform operations more and more far and more and more quickly. Therefore we are interested in the design and control of fast rovers which are able to move autonomously in natural environment at high velocity (up to 10m/s) following a reference path referred by RTK-GPS points. Generally, vehicle dynamics is neglected when designing path tracking controller[1],[2] however purely kinematic based approaches are clearly incongruous for solving the problem of motion control of fast rovers.

Many researches focused on path following of autonomous vehicle in presence of sliding have been developed[1].[3] A kinematic model was used in[4] where sliding effects are introduced as additive unknown parameters. The sliding angle was estimated on-line. Robust adaptive controller was designed in[5] and a sliding mode control is defined in.[6] High-gain controller was designed in[7] to solve the trajectory tracking problem. A fuzzy controller was developed in[8] where several levels of competences and behaviors are implemented. These methods, as well as developed in[2],[4] utilize wheel velocities and/or steering angles as control inputs, and a kinematic model of the vehicle for designing a controller which stabilizes some tracking errors. Since the vehicle dynamics is neglected, these approaches reduce hugely the control performance (accuracy, stability, robustness...).

The paper develops a dynamic model which takes into consideration the lateral slippage of wheels, the wheel torques distribution and independent front and rear steering. This paper is organized as follow. Section II starts with a description of the vehicle dynamic model with 3-DOF and using a linear slippage model. Then, the vehicle kinematic model is presented, introducing the relative tracking errors with respect to a reference path. Finally, we combine the two kinematic and dynamic models, thus providing a global linear state model. This model is used for designing a feedback controller based on quadratic optimization. This is developed in section III. The simulation results are given in the last section IV, including the performance of the proposed tracking controller.

2. Modeling

This section describes the vehicle dynamics model which will be used for control design. We consider the double tracks model (handling model) illustrated in Figure (1). The vehicle has 3 DOF : longitudinal, lateral and yaw motions[9] of the chassis. The dynamic model is given by the following equation of motion.

$$\begin{aligned}
\dot{V}_x M &= M V_\psi V_y + (F_{x1} + F_{x2}) \cos\beta_f + (F_{x3} + F_{x4}) \cos\beta_r \\
&- (F_{y1} + F_{y2}) \sin\beta_f - (F_{y3} + F_{y4}) \sin\beta_r && (1) \\
\dot{V}_y M &= -M V_\psi V_x + (F_{x1} + F_{x2}) \sin\beta_f + (F_{x3} + F_{x4}) \sin\beta_r \\
&+ (F_{y1} + F_{y2}) \cos\beta_f + (F_{y3} + F_{y4}) \cos\beta_r && (2) \\
\ddot{\psi} I_z &= a\left((F_{x1} + F_{x2}) \sin\beta_f + (F_{y1} + F_{y2}) \cos\beta_f\right) && (3) \\
&+ d\left((F_{x2} - F_{x1}) \cos\beta_f + (F_{y1} - F_{y2}) \sin\beta_f\right) \\
&- b\left((F_{x3} + F_{x4}) \sin\beta_r + (F_{y3} + F_{y4}) \cos\beta_r\right) \\
&+ d\left((F_{x4} - F_{x3}) \cos\beta_r + (F_{y3} - F_{y4}) \sin\beta_r\right)
\end{aligned}$$

We denote by F_x and F_y the longitudinal and lateral tire forces, V_x and

V_y are the longitudinal and lateral velocities of the chassis center of mass, $\dot{\psi}$ is the yaw rate, I_z is the yaw-inertia moment, M is the car mass, d is the half-width of the wheel-base, a and b are the distances from the center of gravity to the front and rear axles, β_f and β_r are front and rear wheel steering angles.

2.1. *Lateral force model*

Except for the gravity and aerodynamique forces, the main external forces applied on the vehicle are produced by the tires. The lateral forces acting on the front and rear tire are assumed to be a linear function of the side-slip angle,[10] i.e.

$$F_{y_{(f,r)}} = C_{(f,r)}\alpha_{(f,r)} \tag{4}$$

where C_f and C_r are, respectively, the tire cornering stiffness of the front and rear tires.

The side-slip angle $\alpha_{f,r}$ is defined as the angle between the wheel velocity and the longitudinal axis of the wheel itself. Assuming that side-slip angles are quite small (less than 10° in practice), we can use a linear approximation, that is :

$$\alpha_f = \frac{V_y + a\dot{\psi}}{V_x} - \beta_f, \alpha_r = \frac{V_y - b\dot{\psi}}{V_x} - \beta_r \tag{5}$$

2.2. *Robot dynamics model*

We focus in this paper on the lateral dynamic and path tracking. Henceforth, the longitudinal velocity V_x will be considered constant ($\dot{V}_x = 0$). Combining the double tracks model (2) and the linear tire model (4) and (5), we can write the expression for the lateral acceleration :

$$\begin{aligned} M\dot{V}_y = {} & 2\frac{C_f + C_r}{V_x}V_y + \left(2\frac{aC_f - bC_r}{V_x} - MV_x\right)\dot{\psi} \\ & + (2F_{xf} - 2C_f)\,\beta_f + (2F_{xr} - 2C_r)\,\beta_r \end{aligned} \tag{6}$$

Angular acceleration about the yaw axis can be written by combining (3), (4) and (5)

$$\begin{aligned} I_z\ddot{\psi} = {} & 2\frac{aC_f - bC_r}{V_x}V_y + \left(2\frac{a^2C_f - b^2C_r}{V_x}\right)\dot{\psi} \\ & + a\,(2F_{xf} - 2C_f)\,\beta_f - b\,(2F_{xr} - 2C_r)\,\beta_r + d\,(\Delta F_{xf} + \Delta F_{xr}) \end{aligned} \tag{7}$$

where ΔF_{xf} and ΔF_{xr} are the longitudinal force differences between the right and left wheels, respectively, on the front and rear axle.

Let's consider ΔF_{xf} and ΔF_{xr} as new control inputs. This leads to the the following state system :

$$\begin{pmatrix} \dot{V}_y \\ \ddot{\psi} \end{pmatrix} = [a] \begin{pmatrix} V_y \\ \dot{\psi} \end{pmatrix} + [b] \begin{pmatrix} \beta_f \\ \beta_r \\ \Delta F_{xf} \\ \Delta F_{xr} \end{pmatrix} \tag{8}$$

where $[a]$ is 2×2 matrix and $[b]$ is 2×4 matrix.

2.3. *Robot-path kinematic model*

Figure 2 illustrates the path tracking problem considered in the current study. ψ and ψ_{ref} are the robot heading and the heading of the local tangent of the prescribed path with respect to a fixed arbitrary reference direction. The rate of change of ψ_{ref} with respect to the curvilinear distance measured along the path defines the path curvature k. e_ψ represents the heading error of the robot. e_n is the lateral deviation of the robot center of mass with respect to the prescribed path.

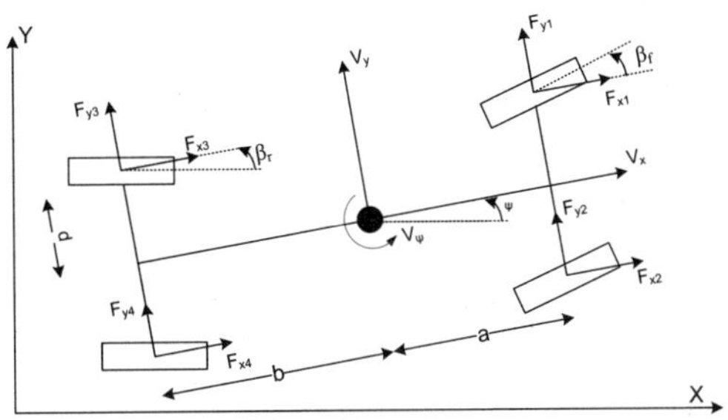

FIGURE 1. Double tracks model.

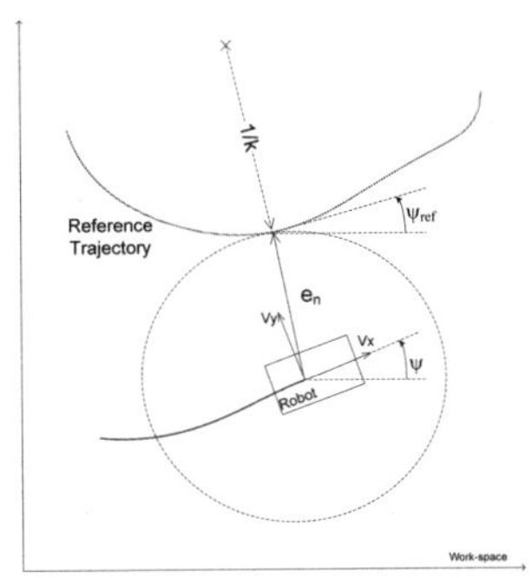

FIGURE 2. The reference and actual path.

Sign conventions for both e_ψ and e_n are denoted in Figure (2). With reference to the same figure and by assuming that the angular deviation is small, the following kinematical relationships are derived :

$$\begin{aligned} \dot{e}_n &= V_x \sin e_\psi + V_y \cos e_\psi \\ &= V_x e_\psi + V_y \end{aligned} \tag{9}$$

$$\begin{aligned}\dot{e}_\psi &= \dot{\psi} - \dot{\psi}_{ref} \\ &= \dot{\psi} - \frac{V_x \cos e_\psi - V_y \sin e_\psi}{1/k + e_n}\end{aligned} \tag{10}$$

In case of straight line tracking ($k = 0$), this last equation is not singular and is equivalent to $\dot{e}_\psi = \dot{\psi}$, since $\dot{\psi}_{ref} = 0$ for a straight line path. Assuming that $e_n << 1/k$, equation (10) can be linearized as follow :

$$\dot{e}_\psi = \dot{\psi} - V_x k + V_x k^2 e_n \tag{11}$$

Assuming a constant curvature, the derivative with respect to time of this latter equation gives

$$\ddot{e}_\psi = \ddot{\psi} + V_x^2 k^2 e_\psi + V_x k^2 V_y \tag{12}$$

3. Design control

In this section, we define a path tracking control of the robot using linear state feedback. Combing the kinematic model of the path-tracking errors (9) and (12) with the dynamic model of the vehicle (8) results in a new state-space model :

$$\dot{X} = AX + BU \tag{13}$$

where $X = \begin{bmatrix} V_y \ \dot{\psi} \ e_\psi \ \dot{e}_\psi \ e_n \end{bmatrix}^T$ is the state vector and $U = \begin{bmatrix} \beta_f \ \beta_r \ \Delta F_{xf} \ \Delta F_{xr} \end{bmatrix}^T$ is the inputs vector, A is 5×5 matrix and B is 5×4 matrix.

In this study, the main objective of the control design is to improve the steering behavior of the robot in the presence of lateral and yaw disturbances. Therefore, the controller should minimizes the side-slip angle β of the robot center mass through minimizing lateral velocity V_y and its yaw rate (thereby decreasing the probability of lateral skidding). To achieve good tracking performance, the cost function must include a term that penalizes both the heading angle deviation e_ψ and the lateral offset e_n,

$$J = \lim_{t \to \infty} \frac{1}{t} \int_0^t \left(X^T Q X + U^T R U \right) dt \tag{14}$$

The second term on the right-hand penalizes the weighted-magnitude of the control inputs vector. This will prevent the use of too high control commands and the resultant control actions will be smooth. Q and R are the nominal state-weighting and the input-weighting matrices.

4. Simulation results and discussion

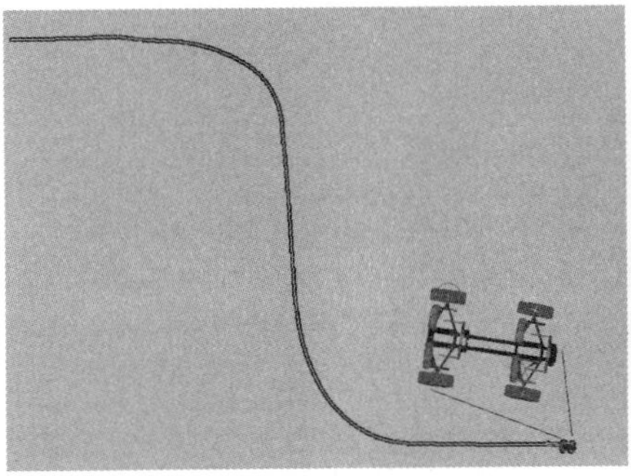

FIGURE 3. simulation views.

FIGURE 4. ADAMS views.

In this section, the LQR controller presented in the previous section has been tested using co-simulation between ADAMS for the mechanical model and Matlab for the control one. The ADAMS model consider a mechanical model with 16 DOF, including 4 independent double wishbone suspensions. The classical Z and O paths are used for testing the controller, they are plotted respectively on (Fig.5) and (Fig.9). In the simulation, the LQR controller gains are calculated on-line, as they depend on the longitudinal velocity V_x and the reference path curvature k. The path curvature was filtered to remove the abrupt variations. During the first 3 seconds of simulation, the system is open-loop until reaching the reference longitudinal velocity. For the first path, we give the control inputs for the front and rear steering angles. The time evolution of lateral deviation and orientation errors are shown on figure (7) and figure (8). For the second path, we give on figure (10) the differential forces ΔF_{xf} ΔF_{xr} for the front and the rear axles.

Results show that the proposed LQR control based on the dynamics model is able to track the reference path with quite good accuracy and quite small and smooth inputs. The new inputs controllers, ΔF_{xf} and ΔF_{xr}, seems to increase the rank of the controllability matrix what permits to reduce the steering angles.

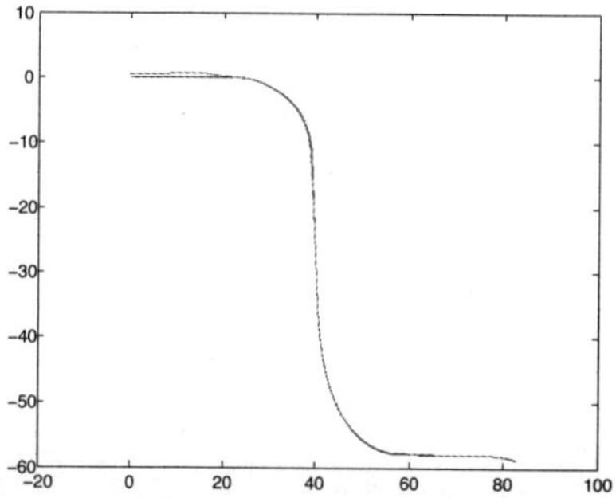

FIGURE 5. The reference (blue) and actual (red) paths for the 'Z' path.

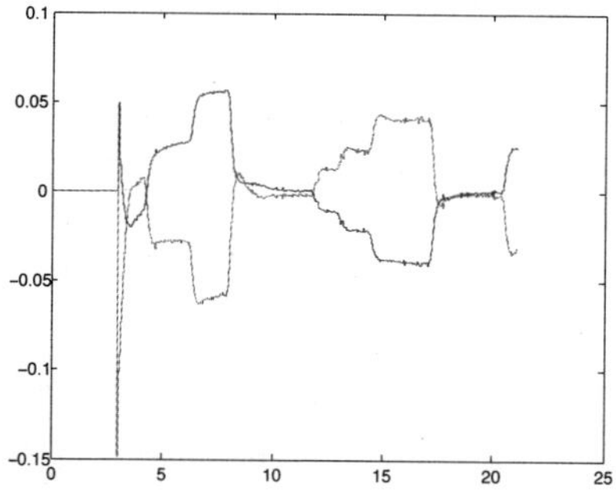

FIGURE 6. Steering angles : front (red) and rear (blue) for the 'Z' path [rad].

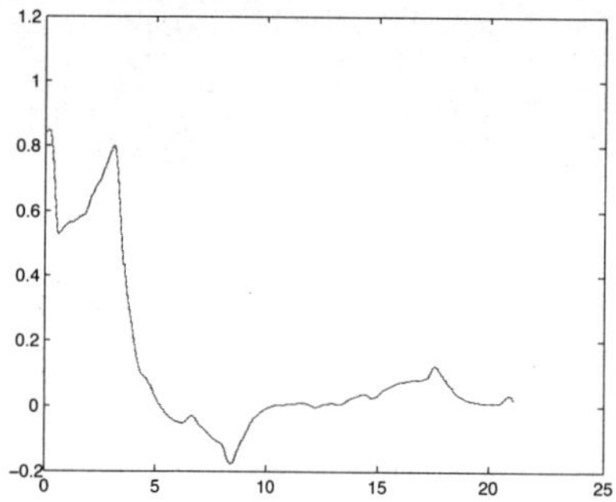

FIGURE 7. Lateral deviation for the 'Z' path [m].

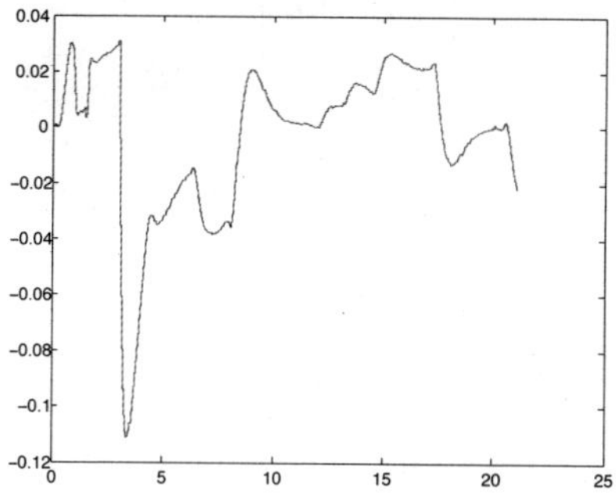

FIGURE 8. Yaw deviation for the 'Z' path [rad].

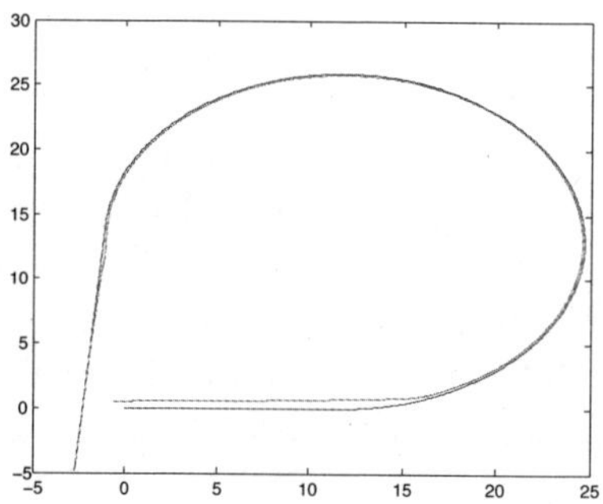

FIGURE 9. The reference (blue) and actual (red) paths for the 'O' path.

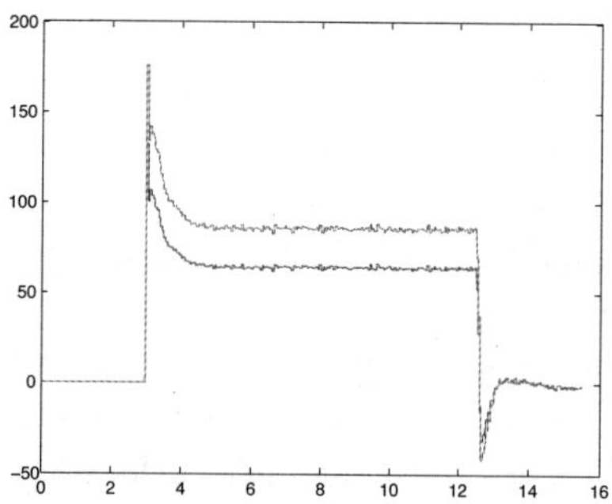

FIGURE 10. The differential forces on front (red) and rear (blue) axle for 'O' path [N].

References

1. H. Fang, R. Lenain, B. Thuilot and P. Martinet, Trajectory tracking control of farm vehicles in presence of sliding, in *Intelligent Robots and Systems, 2005. (IROS 2005). 2005 IEEE/RSJ International Conference on*, aug. 2005.
2. S. L. Tan and J. Gu, Investigation of trajectory tracking control algorithms for autonomous mobile platforms : theory and simulation, in *Mechatronics and Automation, 2005 IEEE International Conference*, july-1 aug. 2005.
3. N. Bouton, R. Lenain, B. Thuilot and P. Martinet, Backstepping observer dedicated to tire cornering stiffness estimation : application to an all terrain vehicle and a farm tractor, in *Intelligent Robots and Systems, 2007. IROS 2007. IEEE/RSJ International Conference on*, 29 2007-nov. 2 2007.
4. C. Cariou, R. Lenain, B. Thuilot and P. Martinet, Adaptive control of four-wheel-steering off-road mobile robots : Application to path tracking and heading control in presence of sliding, in *Intelligent Robots and Systems, 2008. IROS 2008. IEEE/RSJ International Conference on*, sept. 2008.
5. H. Fang, R. Lenain, B. Thuilot and P. Martinet, Robust adaptive control of automatic guidance of farm vehicles in the presence of sliding, in *Robotics and Automation, 2005. ICRA 2005. Proceedings of the 2005 IEEE International Conference on*, april 2005.
6. J.-M. Yang and J.-H. Kim, *Robotics and Automation, IEEE Transactions on* **15**, 578 (jun 1999).
7. M. L. Corradini, T. Leo and G. Orlando, *International Journal of Robotic Systems* **19**, 177(April 2002).
8. S. Yang, H. Li, M.-H. Meng and P. Liu, *Fuzzy Systems, IEEE Transactions on* **12**, 436 (aug. 2004).
9. D.-C. Liaw, W.-C. Chung, C.-C. Song and C.-S. Hsieh, A feedback linearization design for the control of vehicle's lateral dynamics, in *SICE-ICASE, 2006. International Joint Conference*, oct. 2006.
10. S. Jung and T. Hsia, Explicit lateral force control of an autonomous mobile robot with slip, in *Intelligent Robots and Systems, 2005. (IROS 2005). 2005 IEEE/RSJ International Conference on*, aug. 2005.

NAVIGATION WITH OBSTACLE AVOIDANCE OF AN AUTONOMOUS SAILBOAT

Clément Pêtrès, Miguel-Angel Romero-Ramirez, Frédéric Plumet

UPMC Univ Paris 06, ISIR (Institut des Systèmes Intelligents et de Robotique)
CNRS - UMR 7222, 4 place Jussieu, 75005 Paris, France
E-mail: author@isir.upmc.fr

This paper describes a reactive path planning and sail trimming method based on potential fields for autonomous sailboat. The originality of our method is to build a local potential around the boat location to take upwind and downwind constraints into account. Simulations, using a boat specific dynamic simulator, show that the proposed method can successfully drive an autonomous sailboat toward a predefined way-point under time varying wind conditions while avoiding obstacles. This method is computationally efficient and is suitable for on-board real-time implementation.

Keywords: path planning, obstacle avoidance, potential field, autonomous sailboat

1. Introduction

Environmental monitoring, global warming, meteorological and ecological studies in particular need better characterization of ocean processes and consequently requires a broad spectrum of observational tools to understand the complex dynamic coupling between the Earth's oceans and the atmosphere. Thus, considerable progress have been observed in the development and use of autonomous marine platform such as drifters, gliders, autonomous underwater vehicles (AUV), and autonomous surface vehicles (ASV). Each of these platforms provides various capabilities for payload, endurance, range, communication, mobility, and autonomy. Sailboat robots could be an efficient solution for long term missions and semi-persistent presence in the oceans since they rely on renewable energy like solar and wind energy.[1,2]

In the recent years, numerous methods have been proposed for the path planning of autonomous sailboats : based on logic rules,[3,4] state machine,[5–7] neural control[8] or ray-tracing techniques.[9] Potential fields methods, due to Khatib[10] and Krogh,[11] have been successfully employed for robot control

and motion planning of mobile robots (see for example[12–15] among many others) but never for sailboat robots.
The originality of our method is to adapt this popular potential field approach to reactive path planning of a sailboat by building a so-called local potential around the boat location to take upwind and downwind sailing constraints into account. An hysteresis potential relative to the cost of tackling and jibing is also proposed, which makes our method easy to tune according to the vehicle specifications.

2. Sailboat model and performance curves

The sailboat thrust force heavily rely on the wind (speed and direction) and sail angle as well as on the underlying complex dynamic properties of the ship. The components of the hydrodynamic resistance of the hull have been experimentally identified using the towing tank ($150m \times 5m \times 3m$) of *Ecole Centrale de Nantes*. The hull, without keel and rudder, was mounted on a six DoF force-torque sensor with non restricted heave and sway (Fig. 1) and the resulting forces and torques have been recorded under various traction conditions, with and without heel and leeway for a set of ship velocities.[16] Then, the parameters of a specific computational fluid dynamics (CFD)

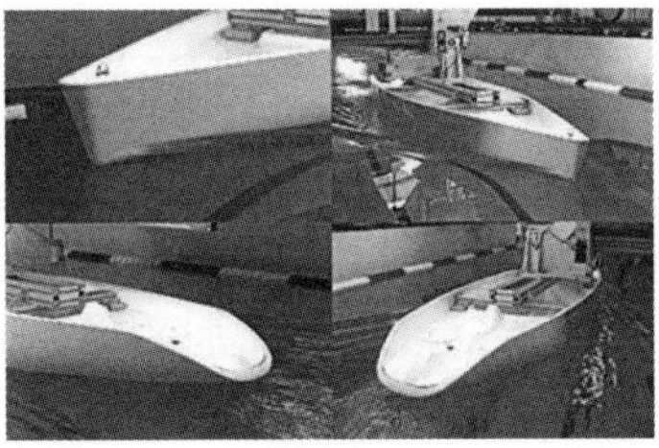

Fig. 1. Hull attached to a 6 components force-torque sensor in the towing tank of Ecole Centrale de Nantes

software have been tuned to match the experimental data recorded in the towing tank and simulations have been re-run for a wide and high sampling range of ship velocities, heel and leeways angles.

Finally, a numerical simulator have been derived using the classical, solid mechanism Newton-Euler's laws with the external aerodynamic force acting on the sail and hydrodynamic force acting on the hull, keel and rudder coming from tabulated values coming from the above mentioned CFD simulations results.

This numerical model is used to compute the speed performance curves of the sailboat (see left side of Fig. 2). These curves correspond to the steady-state maximum velocity of the vehicle for a given wind speed (10 knots, coming from east), for each headings φ relative to the wind direction and for a set of different boom angles. The speed profile of the vehicle, which corresponds to the maximun velocity of the sailboat over all the possible boom angles for each heading, is depicted by the convex hull of all the polar curves. These curves exhibit two "prohibited" sectors: an *upwind no-go zone* for $-45° \leq \varphi \leq 45°$ and a *downwind no-go zone* for $150° \leq \varphi \leq 210°$. A high potential will be affected to these upwind and downwind sailing regions in the path planner presented in the next section. This simulator is also used

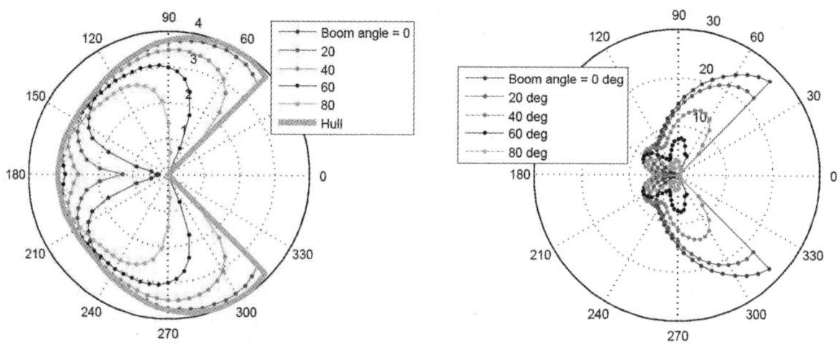

Fig. 2. Speed and heel polar diagram of our sailboat (wind speed: 10kt)

to derive the heel polar diagram of the sailboat (see right side of Fig. 2). These curves correspond to the heel of the boat for the same conditions (wind speed and direction, heading and set of different boom angles). The heel tends to increase as we go upwind and reaches a local minimum around 120 degrees. This well-known characteristics of sailboats will be exploited in the path planning algorithm to favor trajectories with low heel.

This numerical simulator is also used to simulate the behaviour of our prototype under different environmental conditions and tests the reliability of the path planner (see section 4).

3. Path planning

Our potential field approach is decomposed in two parts, a global potential field and a local potential field. The global potential field is relative to the goal point and to the obstacles. The local potential field is relative to the wind direction.

Global potential

In order to attract the vehicle towards the goal point P_{goal} a linear potential P_g is built for every point P of the map as follows:

$$P_g = G_g.dist(P, P_{goal})$$

where G_g is the desired attractive gradient and $dist(P, P_{goal})$ is the Euclidean distance between P and P_{goal}.

A repulsive potential P_o relative to the obstacles located in P_{obst} is built for every point P of the map as follows:

$$P_o = \frac{k}{dist(P, P_{obst})} \quad \text{where } k \text{ is a tunable scalar}$$

Local potential

The local potential is built in the neighborhood of the boat location. It is rebuilt after each displacement of the boat according to the wind direction and the latest tack.

In order to take the upwind no-go zone into account a so-called upwind potential P_{up} is built dynamically for each point P_w inside a window (size W_{size}) centered on the boat location P. The magnitude of P_{up} is a function of the angle ϕ between the direction of $\overrightarrow{PP_w}$ and the wind direction TWA. It is linear inside the upwind no-go zone sector defined by the angle ϕ_{up} (see Fig. 3) and null outside:

$$\begin{cases} P_{up} = G_{up}.dist(P_w, P) & \text{if } 0 < |\phi| < \phi_{up} \\ P_{up} = 0 & \text{elsewhere} \end{cases}$$

where G_{up} is the desired gradient inside the upwind no-go zone sector.

In order to take the downwind no-go zone into account a so-called downwind potential P_{down} is built similarly to P_{up}:

$$\begin{cases} P_{down} = G_{down}.dist(P_w, P) & \text{if } 0 < |\phi - \pi| < \phi_{down} \\ P_{down} = 0 & \text{elsewhere} \end{cases}$$

where G_{down} is the desired gradient inside the downwind no-go zone sector.

In order to take the cost of gybing and tacking into account a so-called hysteresis potential P_h is built dynamically for each point P_w inside the window centered on the boat location P. The magnitude of P_h is a function of the angle ϕ and the latest tack. For instance, if the boat navigates with the wind coming from its left side, P_h will be defined at the right side of the wind direction to prevent the boat to tack or gybe too frequently (see

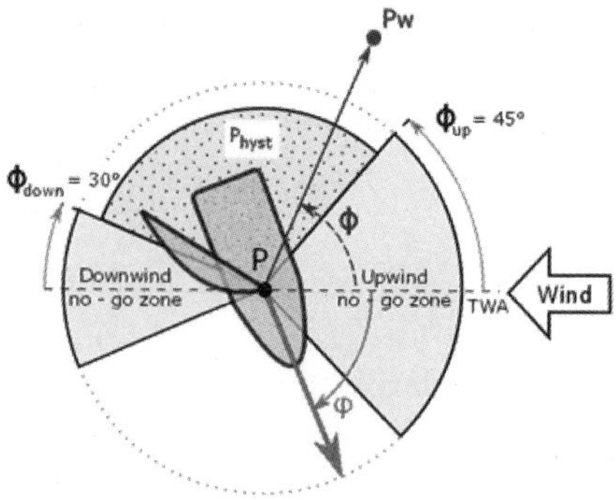

Fig. 3. Scheme of the upwind and downwind no-go zones. The dotted region corresponds to the region where $P_h \neq 0$

Fig. 3). In this example, P_h is defined as follows:

$$\begin{cases} P_h = G_h.dist(P_w, P) \text{ if } \phi_{up} < \phi < \pi - \phi_{down} \\ P_h = 0 \qquad\qquad\qquad\quad \text{elsewhere} \end{cases}$$

where G_h is the desired gradient of P_h. This way, only one parameter has to be set up to fit the cost of gybing and tacking.

In order to take the speed and the heel characteristics of the boat a weighted speed ws is built as follows:

$$ws(v, h) = v.\left(\frac{h_{lim} - min(h, h_{lim})}{h_{lim}}\right)^{\frac{1}{2}}$$

where v is the boat velocity, h_{lim} is an user defined maximum heel value. The weighted speed polar diagram of our sailboat is given in Fig. 4. The

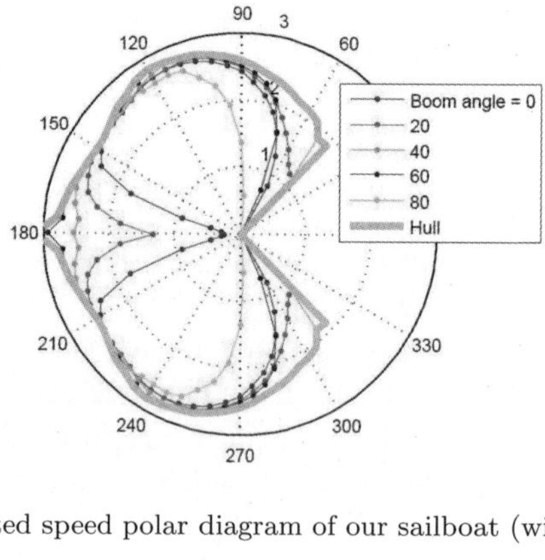

Fig. 4. Weighted speed polar diagram of our sailboat (wind speed: 10kt)

potential P_{ws} associated to the weighted speed is defined outside the no-go zones as follows:

$$P_{ws} = G_{up}.\frac{ws_{max} - ws}{ws_{max}}$$

where ws_{max} is the maximum weighted speed over all the headings (the maximum of the hull).

Gradient descent
Once the local potential has been built and added to the global potential, a gradient descent is carried out on the overall potential $P_t = P_g + P_o + P_{up} + P_{down} + P_h + P_{ws}$ to find the new heading for the boat. The gradient descent algorithm is stopped when the boat arrives in the vicinity of the goal point. Precisely, the stop criteria is defined as follows: $dist(P, P_{goal}) < \epsilon$, with $\epsilon = W_{size}$.

4. Results

For the sake of clarity, the environment in the following simulations is assumed to be static (obstacles are fixed) and completely known (obstacles are mapped without uncertainty). Besides, the wind is assumed to be constant in direction as well as in velocity. However, since our local potential is incrementally built as a function of instantaneous wind conditions and visible obstacles around the boat location, this reactive path planning approach holds for dynamic environments and can be implemented without any change on real embedded systems.

Upwind navigation example
Fig. 5 depicts a tricky manoeuver for a sailboat: sailing directly upwind while avoiding obstacles. In this simulation, default parameters for the potential have been set up as follows: $G_g = 3$, $k = 100$, $G_{up} = 10$, $G_{down} = 5$, $G_h = 2$, $\phi_{up} = 45°$, $\phi_{down} = 30°$, $h_{lim} = 25°$. As we can see, the proposed path planner provides coherent trajectories in terms of heading constraints for a sailboat. Trajectories follow admissible headings regarding the no-go zones and regarding the weighted speed profile of the vehicle. The number of gybes and tackles may be tuned by the hysteresis potential to suit the real cost of the manoeuvres. Besides, attractiveness of the goal point and repulsiveness of obstacles are easily tunable in this potential field framework.

Numerical path planning validation
In this section, realism of the solutions given by the path planing algorithm is improved by using the dynamic model of the sailboat described

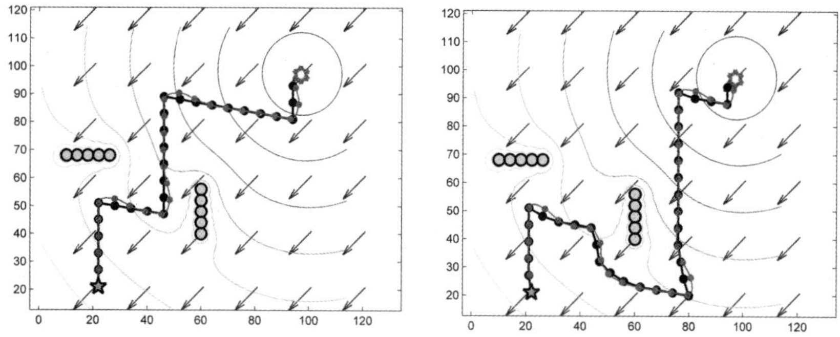

Fig. 5. Left: a trajectory computed with $P_{ws} = 0$ with the wind coming from North-East. Right: a trajectory in the same conditions computed with $P_{ws} \neq 0$

in section 2. In the simulations so far the vehicle dynamics has been neglected. Resulting paths extracted from the start to the goal point may not be physically admissible for a given sailboat. Introducing the sailboat dynamics would greatly improve the reliability of the overall path planning method. At each step of the path planning algorithm, the numerical simulator is run and the resulting position and heading of the sailboat will be used as the current state for the next step.

The low level control loop for the boom angle and for the rudder angle is a simple PID controller that makes the heading converging to the heading reference given by the high level path planner.

A comparison between a trajectory computed using only the high level path planner and a trajectory resulting from the coupling of the path planner with the numerical simulator is given in Fig. 5. While easily done when neglecting the vehicle dynamics, this manoeuver is far from immediate in practice. The inertia of the boat may impact such a turn in the trajectory, as depicted by the red curves.

5. Conclusion and future works

In this work, a reactive path planning method for the navigation of an autonomous sailboat has been presented. This path planning algorithm, based on the potential field approach, reacts to changing wind conditions as well as detected obstacles in real-time by recalculating the heading periodically. This method have been successfully tested and validated in simulation under various wind conditions, using a computational efficient numerical simulator specifically tuned for the boat that will be used in the next future.

References

1. N. Cruz and J. Alves, Autonomous sailboats: An emerging technology for ocean sampling and surveillance, in *MTS/IEEE OCEANS 2008*, 2008.
2. P. Rynne and K. von Ellenrieder, A wind and solar-powered autonomous surface vehicle for sea surface measurements, in *OCEANS 2008*, 2008.
3. M. Neal, A hardware proof of concept of a sailing robot for ocean observation, *IEEE Journal of Oceanic Engineering* **31** (2006).
4. R. Stelzer and T. Proll, Autonomous sailboat navigation for short course racing, *Robotics and Autonomous Systems* **56**, 604 (2008).
5. Y. Briere, F. L. Cardoso Ribeiro and M. Vieira Rosa, Design methodologies for the control of an unmanned sailing robot, in *8th Conf. on Manoeuvring and Control of Marine Craft*, 2009.
6. G. Elkaim and C. Boyce, Energy scavenging and aerodynamic performance of a rigid wing propulsion system for an autonomous surface vessel, in *ION Global Navigation Satellite Systems Conference*, 2009.
7. H. Erckens, G.-A. Busser, C. Pradalier and R. Siegwart, Avalon: Navigation strategy and trajectory following controller for an autonomous sailing vessel, *IEEE Robotics Automation Magazine* **17**, 45 (2010).
8. C. Sauze and M. Neal, A neuro-endocrine inspired approach to long term energy autonomy in sailing robots, in *Proc. TAROS 2010 : Towards Autonomous Robotic Systems*, 2010.
9. R. Stelzer, K. Jafarmadar, H. Hassler and R. Charwot, A reactive approach to obstacle avoidance in autonomous sailing, in *3rd Int. Robotic Sailing Conference*, 2010.
10. O. Khatib, Real-time obstacle avoidance for manipulators and mobile robots, *International Journal of Robotics Research* **5**, 90 (1986).
11. B. Krogh and C. Thorpe, Integrated path planning and dynamic steering control for autonomous vehicles, in *IEEE Int. Conf. on Robotics and Automation*, 1986.
12. J. Barraquand, B. Langlois and J.-C. Latombe, Numerical potential field techniques for robot path planning, *IEEE Transactions on Systems, Man and Cybernetics* **22**, 224 (1992).
13. H. Haddad, M. Khatib, S. Lacroix and R. Chatila, Reactive navigation in outdoor environments using potential fields, in *IEEE Int. Conf. on Robotics and Automation*, 1998.
14. S. Ge and Y. Cui, Dynamic motion planning for mobile robots using potential field method, *Autonomous Robots* **13**, 207 (2002).
15. S. Shimoda, Y. Kuroda and K. Iagnemma, Potential field navigation of high speed unmanned ground vehicles on uneven terrain, in *IEEE Int. Conf. on Robotics and Automation*, 2005.
16. E. Jacquin, Y. Roux, P.-E. Guillerm and B. Alessandrini, Toward numerical vpp with the full coupling of hydrodynamic and aerodynamic solvers for acc yacht, in *17 Chesapeake Sailing Yatch Symposium*, 2005.

EXTRINSIC RGB-D CAMERA CALIBRATION FOR LEGGED ROBOTS

MARK A. HOEPFLINGER, C. DAVID REMY, MARCO HUTTER, ROLAND SIEGWART*

Autonomous Systems Lab, Institute of Robotics and Intelligent Systems, Swiss Federal Institute of Technology (ETHZ), Zurich, Switzerland, markho@ethz.ch

This paper describes a method to identify the extrinsic parameters of RGB-D cameras mounted on legged robots. Since the calculation of the parameters is based on the detection of the robots feet in the camera images, no special calibration objects are required. Therefore the method is simple to use and can even be applied automatically. Experiments demonstrate the precision and the robustness of the method.

1. Introduction

The main advantages of legged robots is their ability to move in very rough and unstructured terrain where wheeled systems usually would fail. While simple execution of stable walking patterns can already cope with substantial variation in the ground level, careful foothold selection becomes indispensable when it comes to large obstacles or gaps in the ground. Successful step planning in such an environment requires accurate knowledge of the robot position as well as an appropriate map of the environment. Within the DARPA Little Dog Challenge, several groups have impressively pushed the state of the art in traversing highly unstructured terrain [1,2,3]. However, they could rely on accurate maps of the environment that were known before execution as well as on real time state observation by external sensors – information that is not available in most real application.

Along the lines of the little dog challenge, we developed ALoF [4], an improved quadruped robot that has enough payload capability to carry on board sensors for localization and mapping. The appropriate usage of such sensors (for example cameras, laser scanners or inertial measurement units) requires the identification of its mounting position and orientation. This identification is known as extrinsic camera calibration problem. Common methods for extrinsic calibration of RGB

* This research was supported by the Swiss National Science Foundation through the National Centre of Competence in Research Robotics.(NCCR).

cameras are often based on the solution of a point correspondence problem, where the correspondence between points with known 3D coordinates and their projection in the image is found. Even if specially designed calibration objects simplify the usage of such methods [5], they are usually still very impractical and involve manual interaction. Other methods, that could in principle be applied to extrinsically calibrate the camera, are based on the detection of geometrical patterns, like lines, circles or ellipses [6, 7], but require multiple views of the pattern. Further methods use motion analysis for the calibration (such as the method developed by F. Pomerleau et al, based on [8]) and are difficult to be applied on legged robots.
Therefore this paper describes a method for the extrinsic calibration of RGBD cameras mounted on legged robots. Because the positions of the front robot feet with known 3D coordinates and their corresponding position in the RBG image are used to identify the mounting pose, no additional calibration objects or sensors and no manual interaction are required to identify the 6 degrees of freedom of the camera.

2. System Description

The RGB-D camera (Microsoft Kinect for Xbox 360) is mounted on ALoF, a four legged robot with a total weight of about 15 Kg and a length of half a meter [4]. Each of the legs has three degrees of freedom. The mechanical design allows hip abduction/adduction, hip flexion/extension as well as knee flexion/extension with a large range of motion (Table 1). This enables the execution of special maneuvers like crawling, recovering from tipping over, but also setting the robot in the calibration configuration (Fig 1) without manual interaction. In this configuration, the front legs are stretched, so that the spherical feet, made of blue rubber material, are visible by the RGB-D camera.
The calibration method has been implemented as a ROS [9] node and applies functions of the OpenCV [10] and ROS PCL libraries.

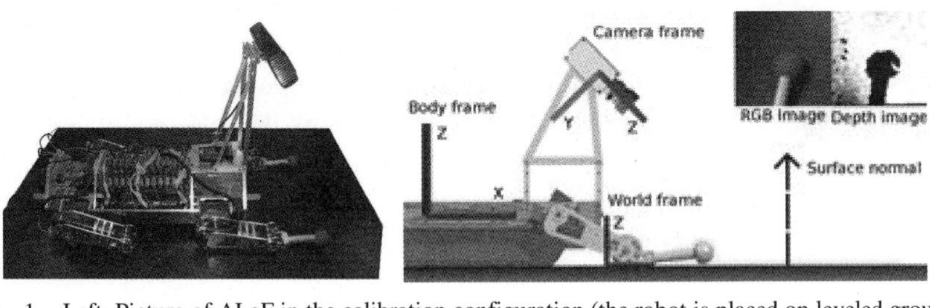

Fig. 1: Left: Picture of ALoF in the calibration configuration (the robot is placed on leveled ground with the front feet in the range of the RGB-D camera).
Right: Introduction of the coordinate frames used for the calibration.

Parameter	Value	Parameter	Value
Overall Length	557 mm	Total Mass	15 Kg
Typical Height	400 mm	Main Body Mass	10 Kg
Overall Width	386 mm	Range of motion for hip flexion/extension	±180 deg
Length of Thigh Segment	150 mm	Range of motion for knee flexion/extension	+90/-160 deg
Length of Shank Segment	150 mm	Range of motion for hip abduction/adduction	+45/-90 deg
Foot diameter	40 mm		

Table 1: Hardware characteristics of the walking robot ALoF

3. Method

The approach presented in this paper allows to identify the extrinsic camera parameters without manual interaction or the need of any special calibration object. Instead, the robot feet are used as reference point with known 3D position. Together with an estimation of the surface normal of the leveled ground in front of the robot, all the extrinsic camera parameters can be determined.
It is assumed, that the intrinsic camera calibration is done in prior and that the RBG and the depth image are aligned correctly. The current implementation of the method is also based on assumptions about the actual robot (such as the shape and color of the robot feet), but the principle can be applied whenever parts of the robot at known positions can be tracked precisely in an image provided by the RGB camera.
The calibration process can be separated in two parts. In the first part, the camera position and orientation with respect to the world frame is identified. This allows to estimate the camera height above ground and its roll and pitch angle. In a second step, the camera pose in the robot body frame is computed.

The pitch and the roll angle of the camera could have been measured by the three axis accelerometer (Kionix KXSD9) that is incorporated in the Kinect. But, on one hand, the exact pose of the accelerometer inside the Kinect device was unknown and on the other hand, not all RGB-D sensors incorporate additional accelerometers. Therefore, our calibration method does not rely on accelerometer measurements. Instead, estimates of the surface normal of the leveled ground in front of the robot are used to identify the camera roll and pitch angles as well as the height over leveled ground.

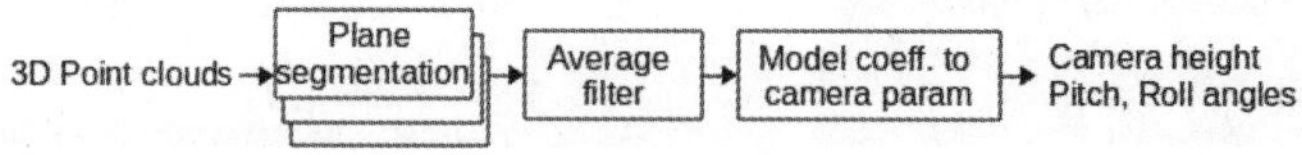

Fig. 2: Diagram illustrating the steps to compute the camera height above ground and its roll and pitch angle

The estimation of the surface normal is done by matching planes (RANSAC plane matching algorithm [9]) to a set of point clouds acquired from the depth sensor. The computed coefficients of the surface normal are then averaged and applied to calculate the camera pose in the world frame.

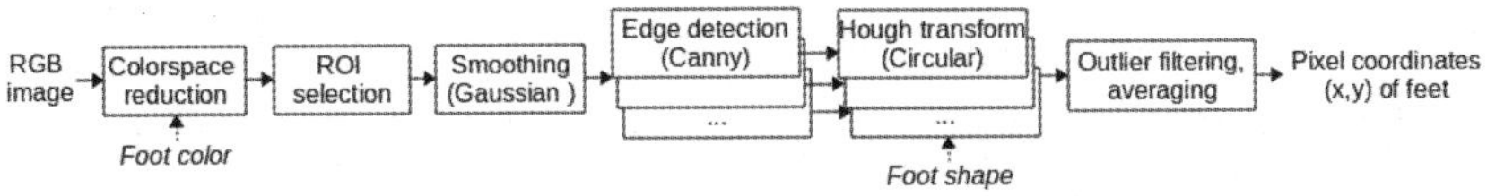

Fig. 3: Diagram illustrating the steps to compute the pixel coordinates of the feet in the RGB image

The assessment of the displacement within the robot body frame is based on the search of the point correspondence of the known robot foot position and its 2D projection in the image. Because of the limited accuracy, precision and eventually sparse depth data at the foot location, precise foot tracking was not possible using only the depth data. Therefore the tracking has been performed on RGB images.

First, under the assumption of having blue feet, the color-space of the RGB images has been reduced. Then the region of interest (ROI) has been set to the lower half of the image where the feet are visible in the calibration configuration of the robot. To lower noise effects, the image has been smoothed with a 5x5 pixel wide Gaussian kernel. Finally, canny edge detection with a set of different thresholds was performed and a circular hough transform was applied to detect the spherical feet in the image. The set of detected circles possibly contained false positives, which, in a next step, are removed by a statistical outlier detection filter. Finally, from an average of the inlier, the estimated foot position in pixel coordinates has been computed (Fig. 3).

Knowing the foot geometry, it would have been possible to compute the remaining camera parameters without using depth image data. In this approach the 2D points from the RGB image are projected to the depth image. Because the depth information in the range of the feet was sparse (Fig. 5, left), the computed 2D position could not directly be mapped to the point cloud data. Instead, linearly distributed sample points through the center of the detected feet in the image have been projected to the point cloud (Fig. 5, right). Using a linear regression (ordinary least squares) on the co-linear sample points, the 3D foot position could be calculated.

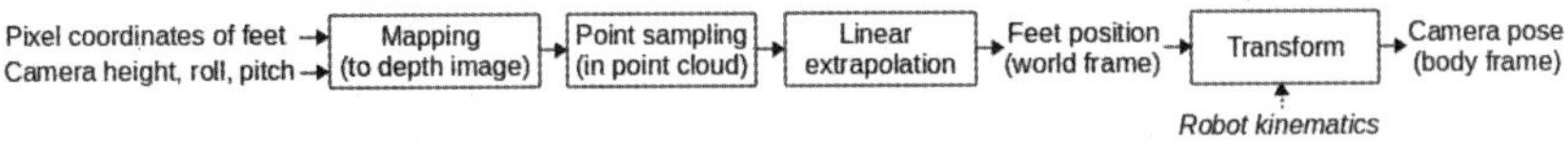

Fig. 4: 3: Diagram illustrating the steps to compute the camera pose in the robot body frame

The foot position served to compute the yaw angle of the camera in the robot body frame and finally, considering the known kinematics of the robot to identify the position of the body frame with respect to the camera frame (Fig. 5).

4. Experiments

To evaluate the performance of the camera calibration, experiments have been performed on three different artificial terrain types. The robot has been placed on a homogeneous white and black surface and on a floor with a textured wooden surface (Fig. 5, left side). The white surface was selected to test the calibration under conditions where the reflectivity as well as the contrast between the robot feet and the surroundings compared to the other surfaces is large. For each type of surface, the calibration routine has been executed 100 times.
Due to the missing specification of the geometry of the Kinect sensor, the position of the camera within the housing could only be estimated. Therefore, the camera position has been measured with a measuring tape (1 mm resolution), which resulted in a ground truth with low accuracy. The pitch and the roll angle of the camera have been quantified by using the Kinect accelerometers. Again, missing specifications about the mounting position/orientation and the application of a sensor that is not very accurate resulted in a imprecise reference.

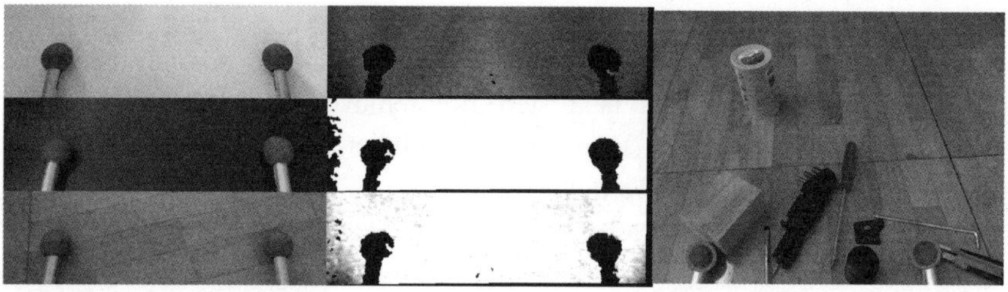

Fig. 5: Left side: Sample images of the different surfaces aquired from the RGB camera.
Middle images: The corresponding depth images (the color has been normalized for visualization).
Right side: Sample image of the foot detection in a disturbed scene.

5. Results:

The following box plots (Fig. 6 and Fig. 7) illustrate the lower/upper quartile (first and third quartile) and the median of the samples. The whiskers of the plot point out the samples that are still within +/- 1.5 times the interquartile range (difference of the third and the first quartile). Samples outside this range are marked as outliers.

Fig. 6 shows plots of the identified camera parameters, that are based on the calculation of the surface normal of the terrain. The computed camera height above ground varies most for the calibrations performed on the white surface (about +/- 1.25 mm). The most precise estimations have been achieved on the black surface (50% of the samples are within about +/- 0.12 mm around the median). Even when considering the outliers, the samples vary by less than approx. +/- 0.75 mm from the median value.

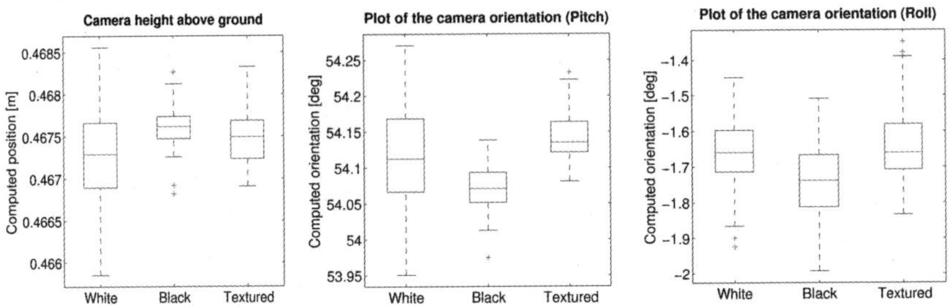

Fig. 6: Box diagram of the identified camera height above ground as well as pitch and roll angle.

The plot in the middle of Fig. 6 presents the computed pitch angle. Again, the estimations based on measurements on the white surface are the most imprecise. Still, 50% of the samples are within +/- approx. 0.05 degrees around the median. While the values measured on the black surface are distributed more symmetrically, the precision is comparable to the one of the measurements performed on the textured surface. The precision of the roll angle measurements is nearly the same for the different surfaces (50% of the samples are within about +/- 0.06 deg around the median, the maximum variation is approx. 0.5 deg).

The median value of the samples differs for the measurements on the discriminative surfaces. The maximal difference is less than 0.1 degrees for the computed angles and 0.5 mm for the height.

The identification of the other extrinsic parameters is based on estimations of the foot position within the RGB image. Therefore it is expected that the optical properties of the surface have an even stronger influence on the quality of the estimation.

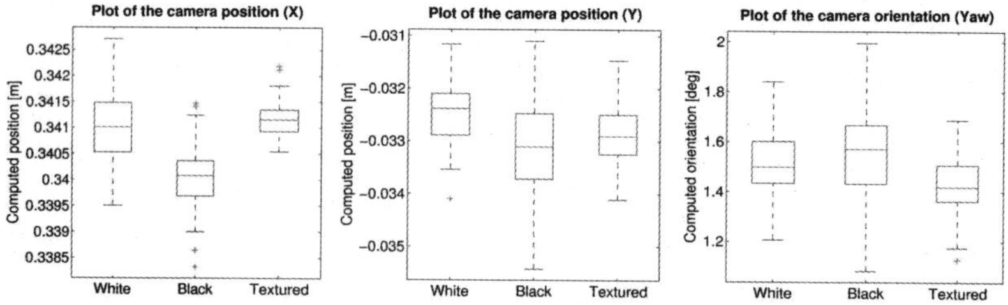

Fig. 7: Box diagram of the identified camera position (X/Y) and yaw angle (robot body frame).

The evaluation of the computed position is visible in Fig. 7, in the left and middle plot. The precision of the measurements performed on the textured surface was the highest. For the X coordinate of the position, the standard deviation is approximately 0.3 mm. Measurements recorded on the black surface where the most imprecise: The biggest variation of all 100 measurements was about 3 mm (and up to 4.3 mm for the estimation of the y component). This also resulted in imprecise yaw estimations (Maximal difference of all samples including outliers: 0.91 degrees for the black surface, 0.56 degrees for the textured surface, and 0.63 degrees for white surface). Still, the standard deviation of the measurements taken on the white or the textured surface is less than 0.13 degrees.

The accuracy of the results seems to be in a reasonable range (Table 2), but due to imprecise and missing reference values, no quantitative comparison was possible.

It can be assumed that the precision of the calibration method could even be improved by the means of applying constant illumination conditions, adding calibration markers to the feet, incorporating additional sensors or simply by processing more RGB-D images per calibration step.

	Mean			**Standard deviation**			**"Ground truth"**
	White	Black	Textured	White	Black	Textured	
X [mm]	341.1	340.1	341.2	0.657	0.575	0.342	340.0
Y [mm]	-32.5	-33.1	-32.8	0.552	0.942	0.587	-30.0
Z [mm]	*402.3*	*402.6*	*402.5*	*0.572*	*0.247*	*0.318*	*400.0*
Roll [deg]	*-1.66*	*-1.74*	*-1.63*	*0.096*	*0.107*	*0.109*	*-1.30*
Pitch [deg]	*54.11*	*54.07*	*54.14*	*0.071*	*0.029*	*0.032*	*54.60*
Yaw [deg]	1.51	1.56	1.42	0.113	0.168	0.126	n.a.

Table 2: Mean and standard deviation of the identified camera parameters, and the 'ground truth'..

6. Conclusion

The paper presented a method to calibrate the extrinsic camera parameters of a RGB-D camera mounted on a legged robot. The calibration is performed while the robot is placed on leveled ground and requires no additional equipment.

Experiments on three different ground surfaces indicate the precision of the method. Standard deviations of around 0.5 mm for the camera position estimations and of about 0.1 degrees for the camera angles indicate a good precision for calibrations performed on the employed white and textured surface.

References

1. J. R. Rebula, P. D. Neuhaus, B. V. Bonnlander, M. J. Johnson, and J. E. Pratt, "A controller for the littledog quadruped walking on rough terrain," in Proceedings of the IEEE International Conference on Robotics and Automation, 2007
2. J. Buchli, M. Kalakrishnan, M. Mistry, P. Pastor, and S. Schaal, "Compliant quadruped locomotion over rough terrain," in IEEE International Conference on Intelligent Robots and Systems, 2009
3. J. Zico Kolter, Mike P. Rodgers, and Andrew Y. Ng, "A control architecture for quadruped locomotion over rough terrain", in IEEE International Conference on Robotics and Automation, 2008, pp. 811-818
4. C. D. Remy, O. Baur, M. Latta, A. Lauber, M. Hutter, M. Hoepflinger, C. Pradalier, R. Siegwart, "Walking and crawling with ALoF – A robot for autonomous locomotion on four legs", in 12th International Conference on Climbing and Walking Robots , 2010
5. J. A. Hesch, A. I. Mourikis, and S. I. Roumeliotis, "Mirror-Based Extrinsic Camera Calibration",
6. X. Meng and Z. Hu, "A New Easy Camera Calibration Technique Based on Circular Points", Pattern Recognition, Vol. 36, pp. 1155-1164, 2003
7. C. Yang, F. Sun and Z. Hu: "Planar Conic Based Camera Calibration", ICPR, 2000
8. F. Pomerleau, S. Magnenat, F. Colas, M. Liu and R. Siegwart, "Tracking a Depth Camera: Parameter Exploration for Fast ICP", submitted to International Conference on Robotics and Automation, 2011
9. http://www.ros.org
10. http://sourceforge.net/projects/opencvlibrary/

SAILING ROBOT PERFORMANCE: MAXIMUM SPEED TRACKING VS ENERGY EFFICIENCY

YVES BRIERE

University of Toulouse / ISAE, 10 Avenue Edouard Belin
31055 Toulouse, France

Automated control of Autonomous Sailing Robots (ASR) is a recent challenge. In this paper we first describe the prototype which has been designed at ISAE to the purpose of crossing the Atlantic Ocean autonomously. The performance of an ASR is closely dependent to the actual sailing conditions: wind, current, waves, etc. We present a new control solution which is aimed to constantly adapt to the best expected performance. Simulation and experimental results are given.

1. Introduction

In the field of automated oceanographic measurement there is an increasing need of large scale and long term data gathering. Among the variety of measurement systems (drifting buoys, moored buoys, ships, etc.) there is a growing attention to a new family of autonomous device: Autonomous Sailing Robot [1,2]. An ASR, as considered in this document, is supposed to sail autonomously during weeks or months in open seas, facing rough seas and strong or opposite winds. Man-machine interaction is consequently limited only to high level trajectory or mission planning. Every short term or middle term decision must be taken autonomously by the robot. Autonomous control of Sailing Robot is a nice challenge well adapted to fuzzy control [3, 4, 5, 6], neural networks for identification and control of rudder [7] and sail [8], and machine learning strategies [9,10]. Although open sea can be reasonably considered to be a very well structured environment (the ocean is flat and there are virtually no obstacles) there is a need to address high level tasks as routing and obstacle avoidance [11, 12, 13].

In the next chapter of this article we give a description of our prototype. In chapter 3 we address the intermediate problem of performance optimization: the adaptive sail control strategy is presented.

2. The Iboat II prototype

2.1. *The boat*

A prototype has been developed at ISAE [2, 5, 8]. Iboat II is the current version and is supposed to demonstrate the feasibility of long term navigation in open sea. The final version (Iboat III) will be launch for a transatlantic journey.

Figure 1 : Photography of the first prototype sailing.

The prototype is made of fiber glass and carbon. The resulting structure is consequently very light. Sail surface has been chosen to be very small: 0.8 m^2 to be compared to the 4m^2 expected for this kind of boat. Performance won't be optimal for small and medium wind but it will be much more robust in case of strong wind. The boat being very light it "lies down" when wind is too strong.

The sail rotates thanks to a single actuator (see Figure 1). Two rudders are actuated by another single actuator. There is no other actuator (like a propeller).

2.2. *Systems and energy*

Stepper motors with integrated control are used. They are controlled via a simple CAN protocol (see below).

Sensors are very like the ones used in a real boat. The main sensor is a compass that gives boat's heading (H): the angle between the longitudinal axis of the boat and the magnetic North. This sensor gives also the rate of turn and roll angle.

A wind sensor is placed on top of the mast (mast orientation is corrected). It gives both speed and heading of the wind (WH and WS). The wind measured is the one experienced by the moving boat: real wind speed and direction is affected by how fast the boat is moving. The last sensor is a conventional GPS giving position and speed over the ground.

Sensors and actuators are all connected to a microcontroller via a common CAN bus. This bus is very safe and compatible with the NMEA2000 marine protocol allowing us to include commercial marine sensors. The Microcontroller is an Infineon C167 with 2Mb RAM and 2Mb Flash and a 25MHz.clock.

An HF radio modem (868 MHz 10km range) is used for ground/sea medium range communication (less than 5km). This transmission system is very well fitted for development and research. For long range communication an SBD Iridium protocol is used.

Energy is stored in a Lithium Polymer 720Wh battery. A total of 0.66m2 of Photovoltaic Panels is mounted on the roof allowing for 80W of peak power and an average of 13Wh/h. The average power consumption of our prototype is 7.68W, close to the theoretical average power of 13W delivered by the solar panels. In case of low sunlight period the battery can last for approximately 4 days. Reducing average energy is a major issue for Iboat III:

- Compass and GPS will be replaced by PCB mounted components and the average energy divided by a factor of ten.
- The very robust ultrasonic wind sensor will be used only as a backup and replaced for normal operation by a conventional wind vane.
- The microcontroller power can be easily reduced by a factor of two.

The expected average power after optimization will be approximately 1.5W without actuators. It is of major importance to reduce the use of actuator in order to save energy.

2.3. *Control*

Design, test and validation of control algorithm are conducted on a Matlab/Simulink based simulation model [5, 8]. C code is then automatically generated and compiled for the Infineon processor. This “software in the loop” approach results in a very stable and reliable code.

3. Sail control Strategy

3.1. *Sail Control Principle and performance criteria*

As said before it is of major importance to reduce the use of actuator in order to save energy. We address in this paper the problem of sail actuator only. It is clear that an accurate tuning of the sail will give better performance than an approximate tuning. The point is to find the tradeoff between performance and energy saving.

In sailing boats the complex task of sails tuning is always done by the crew members in a very complex multi criteria optimization process: he reads speed sensors, he feels the boat's movements, he watches carefully the shape of the sail, etc.... In our robot this tasks has to be performed by the computer: the sail is constantly tuned at the optimal angle to maximize speed.

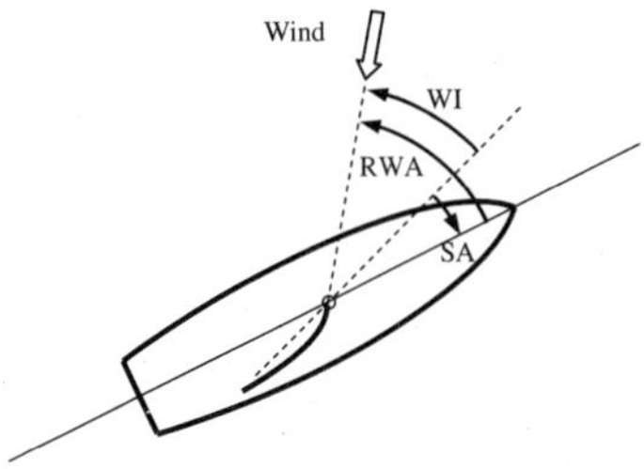

Figure 2 : SA: Sail Angle relative to the boat's longitudinal axis; RWA: Wind Angle Relative to the boat's longitudinal axis; WI: Wind Incidence, angle between sail and wind directions.

The main concern is to find, for each Relative Wind Angle (RWA), which is the optimal wind incidence (WI) to produce maximum forward velocity (Figure 2). In our robot we will simply maintain the Sail Angle near the supposed optimal angle with a threshold in order to minimize actuation time as explained in Figure 3.

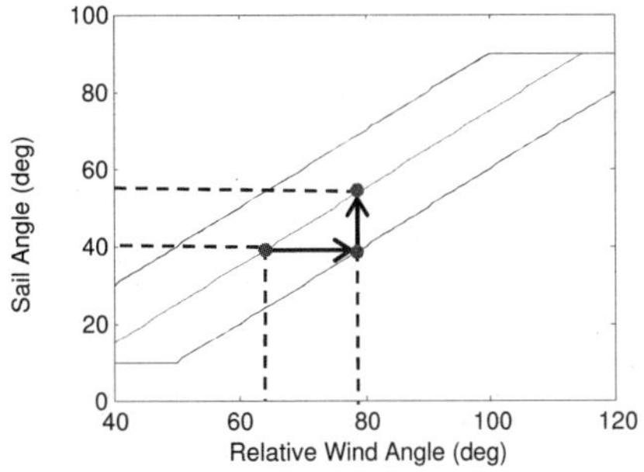

Figure 3: Initial Sail Angle is maintained at 40°. When the Wind Angle is greater than 78° the corresponding Wind Incidence is greater than the minimum of 10°. The Sail Angle is consequently set at the optimal value of 54° corresponding to the optimal Wind Incidence of 25°. The Sail Angle is unchanged while the Wind Incidence remains into the acceptable range of [10° 40°].

From earlier experimental tests it has been found that a Sail Angle less than 10° produced very small forward velocity and Sail Angle bigger than 40° caused an excessive lateral drift. Several tests were conducted in order to explore the Wind Incidence continuum between 10° and 40° (Figure 4). The final goal is, for each Relative Wind Angle value, to find the acceptable Sail Angle range [SA_{min} SA_{max}].

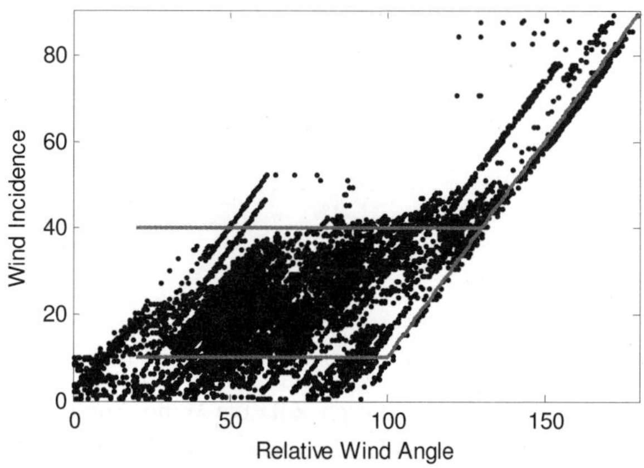

Figure 4: data gathering with Iboat. WI was constrained in the [10° 40°] interval. Each dot represents a measure. Sampling rate: 0.2s; Experiment total duration: 3hrs. The trajectory was pseudo-random and measured wind was between 5kts and 20kts.

3.2. *Recursive Performance Analysis*

Figure XX shows an example of boat speed (S) according to the Wind Incidence (WI) for the particular condition (RWA=70°,WS=13kts). The red line is the best second order fitting according to least square criteria. More generally and for each possible pair (RWA,WS) a second order polynom is recursively obtained.

The (RWA,WS) space is divided in a discrete set of subspace by steps . RWA varies between 50° and 120° by steps of 10 degrees and WS varies between 5kts and 30kts by steps of 5kts. There is consequently 35 second order (a,b,c) polynoms:

$$S = a_{RWA,WS}WI^2 + b_{RWA,WS}WI + c_{RWA,WS}$$

For a new measurement it is first decided in which (RWA, WS) cell it belongs. The Recursive least Square algorithm runs along with the estimation of cross correlation Q^{-1}. The full algorithm is as follows:

$$p = [a_{RWA,WS} \quad b_{RWA,WS} \quad c_{RWA,WS}]^T$$
$$X = [WI^2 \quad WI \quad 1]^T$$
$$p = p + \frac{Q_{RWA,WS}}{(1 + p^T . Q_{RWA,WS} . p)} . (S - p^T . X)$$

Those polynoms are used to predict, for each possible (RWA, WS) cell the optimal Wind Incidence tuning and the corresponding maximum velocity as shown in Figure 5. The polynom is also used to predict which interval of Wind Incidence guarantee a performance near the optimal.

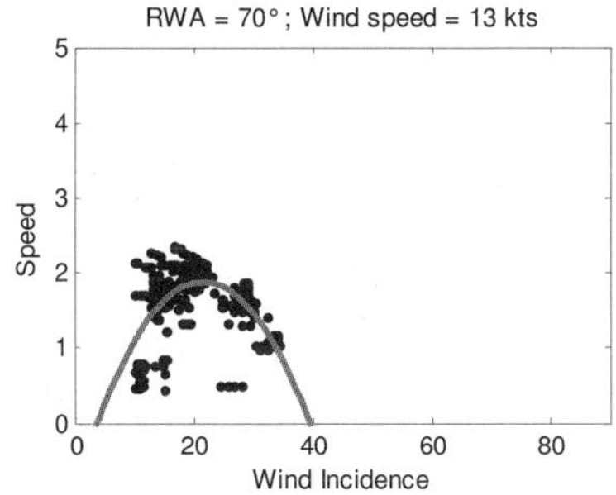

Figure 5: Measured boat speed respectively to the wind incidence for Relative Wind Angles of 70° and wind speed of 13kts (dots). Best second order fitting (red line).

3.3. *Adaptive Sail Control Strategy*

The initial minimum and maximum Wind Incidence were set constant at 10° and 40°. From several hours of field tests and data processing the new Wind Incidence / Relative Wind Angle was obtained. Figure 6-1, 6-2 shows typical recursive analysis result (here for the wind force of 13kts). After linear interpolation the Sail Angle (minimal, optimal and maximal) according to the Relative Wind Angle is as on Figure 7.

The adaptive tuning of the minimum and maximum incidence for each Relative Wind Angle leads to a narrower interval for small RWA and larger for large RWA. The overall performance of the sailing boat is increased by approximately 15% while the sail actuator duty cycle remains close to 5%. In other words the performance gain is sensible for almost the same power.

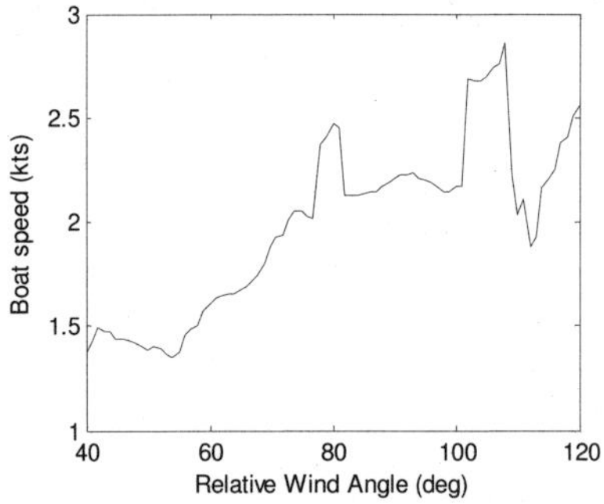

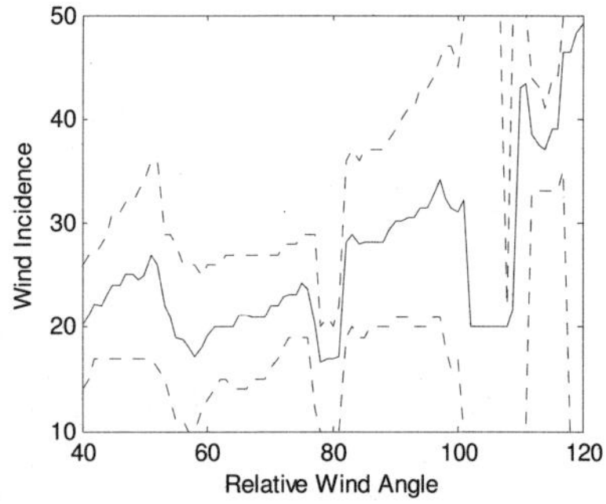

Figure 6-1: expected optimal speed relatively to RWA for a wind speed of 13kts. Figure 6-2: Wind Incidence tuning relatively to RWA for a wind speed of 13kts (cont.: optimal, dashed: optimal minus 10%).

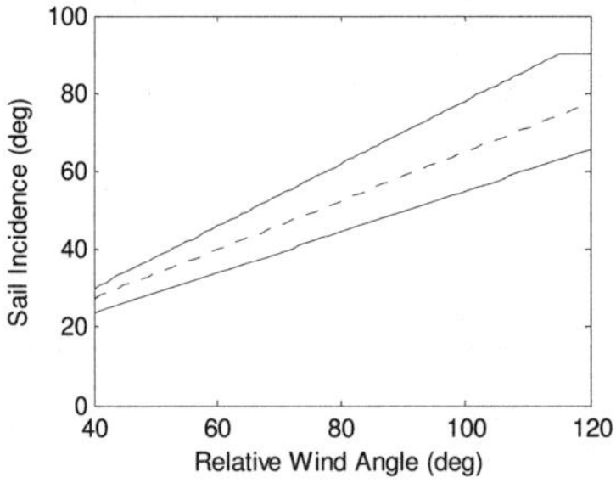

Figure 7: Minimal, optimal and maximal sail angle dependence on the Relative Wind Angle.

4. Conclusion

The presented algorithm has been successfully tested online in simulation and offline with real test datalogs. It was clearly demonstrated that starting from a suboptimal sail parameterization and recursively learning from experience the algorithm converges to an optimal set of parameters. In current works we address the low level rudder control strategy.

The challenge to autonomously cross the Atlantic Ocean using only the wind as source of propulsion is still open [1, 2] and will be the ultimate validation of the autonomous sailing robot concept.

5. References

[1] M. Neal, A hardware Proof of Concept of a Sailing Robot for Ocean Observation, IEEE Transaction on Oceanic Engineering, 2006

[2] Y. Briere, F. Bastianelli, M. Gagneul, Challenge Microtransat, CETSIS05, Nancy France oct. 25-27 2005

[3] J. Abril, J. Salom, O. Salom, Fuzzy Control of a Sailboat, International Journal of Approximate Reasoning 1997, 16:359-375, 1995 Elsevier

[4] E.C. Yeh, J.C. Bin, Fuzzy Control for Self Steering of a Sailboat, Singapore International Conference on Intelligent Control and Instrumentation, 1992, Proc. 17-21 Feb 1992, vol2 p1339-1344

[5] Briere, Yves and Cardoso Ribeiro, Flavio Luis and Vieira Rosa, M.A (2009) Design methodologies for the control of an unmanned sailing robot. In: 8th Conference on Manoeuvring and Control of Marine Craft (MCMC'2009), 16-18 Sept 2009, Guarujá, Brazil.

[6] R. Stelzer, T. Pröll,R. I. John, Fuzzy Logic Control System for Autonomous Sailboats, in 2007 IEEE International Conference on Fuzzy Systems, London, UK

[7] A. Tiano, A. Zirili, C. Yang, C. Xiao, A Neural Autopilot for Sailing Yachts, 9th Mediterranean Conference on Control and Automaiton, June 27-29 2001, Croatia

[8] Briere, Y. (2007) IBOAT: an unmanned sailing robot for the MICROTRANSAT challenge and ocean monitoring. In: TAROS 2007 : Towards Autonomous Robotic Systems, 3-5 Sep 2007, University of Wales, Aberystwyth, UK.

[9] M.L. van Aartrijk, J. Samoocha, Learning to Sail in 'Proceedings of European Symposium on Intelligent Technologies, Hybrid Systems and their implementation on Smart Adaptive Systems (EUNITE 2003), Verlag Mainz, Aachen, 2003

[10] P. J. Sterne, Reinforcement sailing, Master of Science Artificial Intelligence School of Informatics University of Edinburgh 2004

[11] Sauze, C. and Neal, M. (2010), "A Raycast Approach to Collision Avoidance in Sailing Robots", In proceedings of the 3rd International Robotic Sailing Conference, Kingston, Ontario, Canada, June 2010, Pages 26-33.

[12] Stelzer, R. ; Jafarmadar, K. ; Hassler, H. and Charwot, R. (2010), "A Reactive Approach to Collision Avoidance in Autonomous Sailing", In proceedings of the 3rd International Robotic Sailing Conference, Kingston, Ontario, Canada, June 2010, Pages 34-40.

[13] Guo,Y. and Romero Ramirez,M.A. and Ieng,S.H. and Plumet,F. and Benosman,R. and Gas,B. (2011). Reactive Path Planning for Autonomous Sailboat using an Omni-Directional Camera for Obstacle Detection. IEEE International Conference on Mechatronics (ICM 2011-13-15 April, 2011, Istanbul). Istanbul,Turkey

A WIND-INDEPENDENT CONTROL STRATEGY FOR AUTONOMOUS SAILBOATS BASED ON VORONOI DIAGRAM

K. XIAO

University of Minnesota,
Minneapolis, USA
E-mail: kaixiao1988@gmail.com
www.umn.edu

J. SLIWKA* and L. JAULIN

ENSTA-Bretagne, OSM, 2 rue Francois Verny,
Brest, 29200, FRANCE
**E-mail: jan.sliwka@ensta-bretagne.fr*

This paper proposes a new control strategy based on Voronoi diagram for autonomous sailboats. This strategy applies the method of Voronoi diagram to decompose the space and uses a lookup table to make the boat sail as planned. The contribution of this paper is that for the first time it proposes an approach to control autonomous sailboats without the data from wind sensors (with data from GPS and compass only). Since the sensor for detecting wind direction can break down at any time, this new strategy, independent of wind direction, is much more reliable. A simulation is proposed to illustrate the principle and the effectiveness of the approach. Two real experiments on buggy and sailboat are also performed for validation.

Keywords: robotics, autonomous sailboat, control, wind-independent, Voronoi diagram.

1. Introduction

This paper puts up a new control strategy that is independent of wind sensor data for autonomous sailboats. It is motivated mainly by the *micro-transat challenge,* which focuses on the development of small autonomous sailboat robots that could go cross the Atlantic ocean. The challenge is quite meaningful since the unmanned sailboats powered by wind have a great potential application in future oceanographic observations and long term offshore operations (see,[1,2] and[3]).

Due to the harsh environment of the ocean, all components of such robots should be robust with respect to all situations (heavy weather, waves, salt water, low level of energy, long trip, ...). Sensors on autonomous sailboats could be divided into two types: One type is *reliable sensors,* which could survive to all situations. Such sensors include the GPS (providing the position and speed of the boat with a good accuracy), the compass (measuring the north direction with a rather good accuracy in ocean), the gyrometers (giving the rotational speed of the boat) and accelerometers (measuring the roll and pitch of the robot). All these sensors are low energy consumers, can be enclosed inside a waterproof tank and can survive for years. The other type is *unreliable sensors,* which have a high probability to break down in case of heavy weather. Anemometer (device that is used for measuring wind speed), weather vane (which returns the direction of the wind) and dynamometer (which measures the forces on the sail or the rudder) are considered as unreliable. They are directly in contact with aggressive natural elements (wind, waves, salt) and can fail at any time. For simplicity, anemometers and weather vane are called wind sensors in the following passage.

Traditionally, to control the boat, it is necessary to get data from unreliable sensors in order to know where the wind comes from and how strong it is (see e.g.,[4],[56]). However, as said above, sensors for measuring the speed and direction of the wind are vulnerable and could not sustain for long periods ([7]). Therefore, we intend to take a new control strategy that could operate without the knowledge of wind direction and speed.

In this paper, we come up with a new and simple control strategy that is wind-independent based on the Voronoi diagram. We have noticed that for every sailboat there exists a certain no-go-angle θ_{no} on the polar speed diagram ([8]). It means that no matter where the wind comes from, if we provide the sailboat with m ($m \geq 2$) possible directions whose angular separations are greater than the no-go-angle, the sailboat could definitely follow one of them. Our new strategy is based on a lookup table which defines m ($m \geq 2$) possible directions for each zone of the sea. We have introduced the method of Voronoi diagram to implement the decomposition of the sea. As a result, we don't need any data about the wind direction and speed to control our sailboat.

Section 2 provides the basics of polar speed diagram, Voronoi diagram and presents the sailboat to be considered. Section 3 explains how our control strategy works. Section 4 provides a simulation of the strategy. Two real tests on buggy and sailboat are presented on Section 5 and Section 6 concludes the paper.

Fig. 1. sailboat with wind sensor

2. Preliminaries

2.1. *Polar Speed Diagram*

As we know, if the boat is pointed too close to the wind (unless they are backed), the sails will be luffing ("flapping") in the breeze and cannot generate any effective power, only making noise like a flag. Different boats, of course, have different performance characteristics. These characteristics depend on such variables as hull design, keel design, rigging, sails, etc.

The polar diagram of a sailboat is the set S of all pairs (θ,v) that can be reached by the boat, in a cruising mode. The area of direction that cannot be reached is called a no-go-zone. The size of the no-go zone (no-go-angle) will differ based on the performance characteristics of the particular sailboat. For example, racing sailboats can usually sail much closer to the wind (i.e., fewer degrees off the wind direction) than cruising yachts. This is known as "pointing higher."

However, for a given sailboat, the no-go-angle is certain and could be retrieved on its polar diagram. It means that the boat could definitely follow one of m ($m \geq 2$) predefined directions that have an included angle greater than its no-go-angle, whatever the wind direction is.

2.2. *Voronoi Diagram*

A Voronoi diagram is a special kind of decomposition of a metric space determined by distances to a specified discrete set of objects in the space, e.g., by a discrete set of points. In the simplest case, we are given a set of points S in the plane, each point corresponding to one Voronoi site. Each point p has a Voronoi cell, $V(p)$ consisting of all points closer to p than to any other point. The segments of the Voronoi diagram are all the points in the plane that are equidistant to the two nearest points.

In general, the set of all points closer to a point p of S than to any other point of S is the interior of a (in some cases unbounded) convex polytope, called the Dirichlet domain or Voronoi cell for p. The set of such polytopes tessellates the whole space, and is the Voronoi tessellation corresponding to the set S. If the dimension of the space is only 2, then it is easy to draw pictures of Voronoi tessellations, and in that case they are sometimes called Voronoi diagrams.

Given a set of $S \triangleq \{p_i \in \mathbb{R}^n\}$, the Voronoi cell associated with point p_i is the set

$$V(p_i) \triangleq \{x \mid d(x, p_i) \leqslant d(x, p_j), \forall i \neq j\}$$

and the Voronoi diagram of the set S is given by the union of all of the Voronoi cells.

Another way to define $V(p_i)$ is in terms of the intersection of halfplanes. Given two sites p_i and p_j, the set of points that are strictly closer to p_i than to p_j is just the open halfplane whose bounding line is the perpendicular bisector between p_i and p_j. Denote this halfplane $h(p_i, p_j)$. It is easy to see that a point q lies in $V(p_i)$ if and only if q lies within the intersection of $h(p_i, p_j)$ for all $j \neq i$. In other words,

$$V(p_i) \triangleq \cap h(p_i, p_j), (\forall j \neq i).$$

Since the intersection of halfplanes is a (possibly unbounded) convex polygon, it is easy to see that $V(p_i)$ is a (possibly unbounded) convex polygon.

There are a number of algorithms for computing Voronoi diagrams. Of course, there is a simple $O(n^2 \log n)$ time algorithm, which operates by computing $V(p_i)$ by intersecting all the bisector halfplanes $h(p_i, p_j), (\forall j \neq i)$. However, there are much more efficient ways like the sweepline algorithm developed by Steven Fortune[9], which has $O(n \log n)$ worst-case running time.

Voronoi diagram has already been applied on Robotics. For example, it was employed on a path planning strategy for mobile robots on a two-dimensional map, especially for avoidance of obstacles en route (see e.g.[10]).

2.3. *Sailboat*

In order to simulate our control strategy, we need a sailboat dynamics model to implement it. We have introduced here the following state equations based on those in.[11] According to our measurement and test, they are a good approximation of the dynamics of the sailboat represented on Figure 2

$$\begin{cases} \dot{x} &= v\cos\theta + p_1 a\cos\psi \\ \dot{y} &= v\sin\theta + p_1 a\sin\psi \\ \dot{\theta} &= \omega \\ \dot{v} &= \frac{f_s\sin\delta_s - f_r\sin u_1 - p_2 v}{p_9} \\ \dot{\omega} &= \frac{f_s(p_6 - p_7\cos\delta_s) - p_8 f_r\cos u_1 - p_3\omega + p_{11}\,\mathrm{rand}(t)}{p_{10}} \\ f_s &= p_4 a\sin(\theta - \psi + \delta_s) \\ f_r &= p_5 v\sin u_1 \\ \gamma &= \cos(\theta - \psi) + \cos(u_2) \\ \delta_s &= \begin{cases} \pi - \theta + \psi & \text{if } \gamma \leq 0 \\ sign(\sin(\theta - \psi)).u_2 & \text{otherwise.} \end{cases} \end{cases}$$

The state vector $x = (x, y, \theta, v, \omega)^{\mathrm{T}}$ is composed with

- the coordinates x, y of the inertial center G of the boat
- the orientation θ of the boat
- the tangential speed v of G
- the angular velocity ω of the boat around G

The internal variables are

- the thrust force f_s of wind on the sail
- the force f_r of water on the rudder
- the maximum possible angle of the sail δ_s at the current length of mainsheet (γ is to judge whether the sail is tight or not)

The parameters p_i, a, ψ are assumed to be known exactly. They are

- the drift coefficient p_1
- the tangential friction coefficient p_2 between the boat and water
- the angular friction coefficient p_3 between the boat and water
- the sail lift coefficient p_4 and the rudder lift coefficient p_5
- p_6, p_7, p_8 could easily be seen from Figure 2
- the mass of the boat p_9 and the angular inertia of the boat p_{10}
- the speed a and direction ψ of the wind

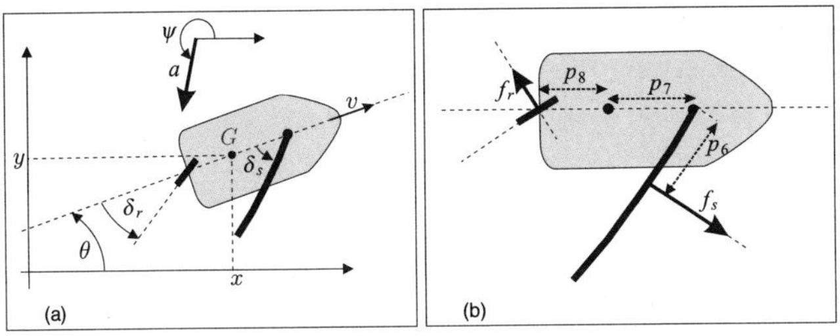

Fig. 2. Sailboat to be observed

The sailboat has two inputs: $u_1 = \delta_r$ is the angle between the rudder and the sailboat and u_2 is the current length of the mainsheet which limits δ_s. This model is similar to that described in,[11] except that here, *(i)* the control is not any more the sail angle, but the length of the mainsheet, which is more realistic, *(ii)* we added a random input on $\dot{\omega}$ that represents the random wave effect on the direction of the boat.

It should be noted that these state equations are introduced only for simulation and presentation purpose. They are not assumed as excellent models, just serving as an example. Any other dynamic models for sailboat could also be applied. Moreover, we simulate the behavior of the boat using Euler approximation.

3. Control Strategy

3.1. *Algorithm*

Our control algorithm is divided into two parts, one offline and the other online. The offline part is performed by the PC and needs human intervention while the online part is done by the microcontroller embedded in the sailboat.

3.1.1. *The offline part*

The offline part is mainly a planning step and needs human help. It takes time to draw the Voronoi diagram with our current technique and thus it is not suitable to be performed on the microcontroller. The PC could do the job and gives out the computed Voronoi diagram to the embedded microcontroller. The offline part is performed at the beginning of one certain voyage.

step1: According to start and end points and obstacles en route, decompose the space by giving reference points that define the Voronoi diagram.

step2: Compute the Voronoi diagram.

step3: Assign each Voronoi site with m ($m \geq 2$) directions.

step4: Generate the C code for the microcontroller.

Remark 3.1. The angular separations between the m directions in one Voronoi site must be greater than the no-go-angle of the boat in order that the boat could follow at least one of them.

Remark 3.2. The effectiveness and efficiency of the strategy depends largely on the setting of reference points and directions. It needs to be done by an experienced sailor.

Remark 3.3. All the directions should ultimately lead to the target in some way.

Remark 3.4. To avoid obstacles some reference points should appear before them and directions associated with these sites should point outward from the obstacles.

3.1.2. *The online part*

The online part of the algorithm is the action step and performed by the embedded microcontroller. It takes the Voronoi diagram given by the PC and controls the action of the sailboat in real-time. The steps are as follows:

step1: Check which Voronoi site the sailboat is in.

step2: Choose the first direction assigned to the current site.

step3: Tune the sail so that the boat gets maximum speed on the target direction and control the rudder so that the boat goes on the target direction.

step4: If the boat fails to follow the current given direction, choose the next one of the current site.

step5: Go to *step3.*

step6: If it fails again, go to *step2.*

Remark 3.5. To determine whether the sailboat follows the desired direction, there could be many criteria using the sensor data about speed, direction or position of the boat. After comparison with other means, we chose the most well-proven way combining the data from GPS sensor and

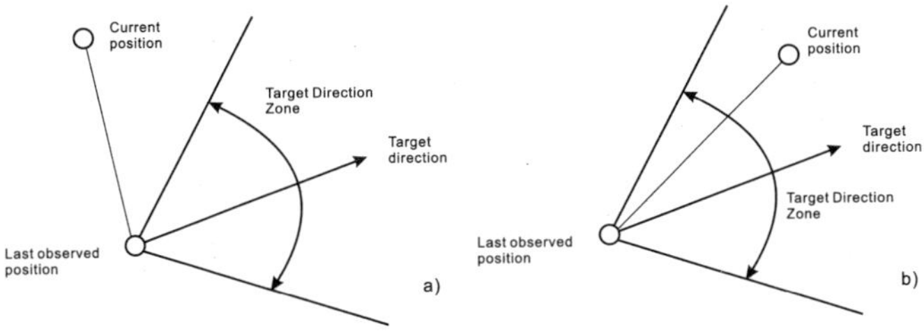

Fig. 3. a) the boat doesn't follow desired direction b) the boat follows target direction

compass which are quite accurate in our scale of problem. The criteria is that if the actual direction of the boat during our observation time (the current position with respect to the last one observed) is within a certain zone around the target direction, then we assume the boat is successful in following it (shown on Figure 3).

3.2. *Tuning the Sail*

Because the wind direction is totally unknown, we cannot rely on any existing knowledge to control the angle of sail. As a result, we need to tune the sail to find the best sail angle so that the boat could fully utilize the power of the wind and follow the given direction. In our current strategy, we have two tuning functions. The first is to find the global optimal angle through a traversal technique, namely it tries every possible angle of sail and finds the best one. This function will be called when the given direction is changed, which means a totally different wind direction with respect to the boat will appear and therefore the best sail angle would be quite different from the last one. The second is to change the sail angle a little once at a time to try to find the local optimal angle. This function helps maintain the stability of the boat while searching for the best sail angle and avoids the high cost of time the first one needs.

3.3. *Controlling the Boat Direction*

In our control strategy, it is quite important for the boat to follow the given direction. For that purpose, a simple but efficient method for controlling the rudder should be applied, since the sailboat's direction is controlled mainly by the rudder. For most of the sailboats, the rudder is efficient enough to make the boat follow the target direction as long as it doesn't belong to the no-go-zone of polar diagram. In our current strategy, the controller

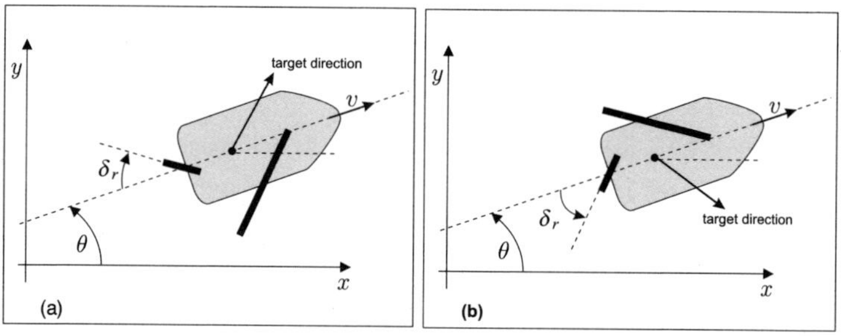

Fig. 4. rudder control strategy

for the rudder is one kind of proportional (proportional to the difference between desired direction and current boat direction). Figure 4 shows how the controller works with different target directions.

4. Simulation

Applying the control strategy and the sailboat model on SCILAB, we conducted a simulation of the new control strategy. We chose the Brest Gulf as our simulation environment because we are familiar with its conditions. Again, it only serves to present our strategy and any other environment could be chosen for the test. We imported the boundary of Brest Gulf into the simulation program (the black lines represent the shoreline). According to our destination and desired path, we decomposed the map into 6 Voronoi sites (the red lines represent the boundary of Voronoi zones) and set the corresponding reference points (the red circles in the figure) along the desired path. To simplify the simulation without losing generality, we assigned each site with only two given directions (the two arrows associated with each reference point, with the darker one representing the first direction and the lighter one representing the second direction) that point to the destination in some way and make the boat avoid obstacles. Since the no-go-angle for most sailboats (including ours[12]) is less than 60 degrees, the included angle between the two directions of each Voronoi site should be greater than 60 degrees. The boat is represented by the black polygon. For simplicity, we used the straightforward algorithm previously described in Section 2.2:

$$V(p_i) \triangleq \cap h(p_i, p_j), (\forall j \neq i)$$

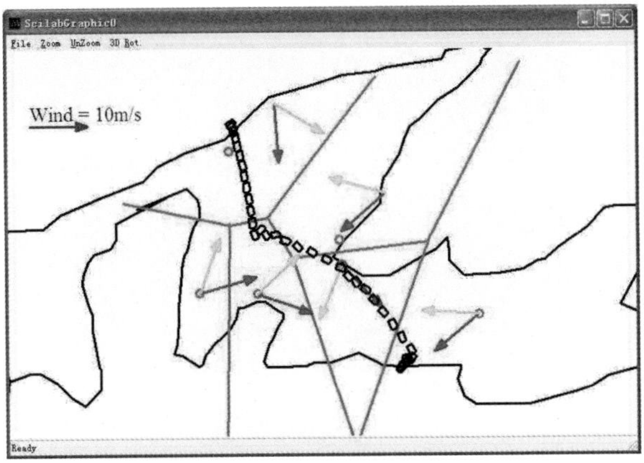

Fig. 5. simulation result

to compute the Voronoi diagram which costs $O(n^2 \log n)$ time. Due to its slowness, we computed the diagram once and stored it in the memory.

According to the real sailboat built by us[12], the parameter vector has been chosen as

$$\mathbf{p} = (0.1, 100, 500, 1500, 70, 1.1, 1.4, 2, 1000, 2000)^{\mathrm{T}}.$$

The result is satisfactory: the boat goes from the starting point to the target without hitting obstacles in any given wind direction. One trajectory of the boat is shown on Figure 5.

The SCILAB code of the simulation as well as movies illustrating our simulation can be downloaded at

`http://www.ensieta.fr/jaulin/kai.html`

5. Testcase

5.1. *With Buggy*

In order to easily implement and see the result of our new control strategy, we tested first on a general ground robotic platform called buggy. There are some small differences between controlling the buggy and the sailboat since the buggy could follow any direction without difficulty. So we ignore the step3 and step5 of the online part that tune the sail. Instead, we just give the buggy a fixed speed. The steering servo which controls the direction of the buggy could be controlled as the rudder of the sailboat. And if we need to see what would happen when it fails to follow one direction, we

Fig. 6. buggy

could just move it in the wrong direction by hand. We chose a football field for test so that the GPS data would be quite accurate. At first, we intended the buggy to go around the field following some kind of spiral trajectory that directs the buggy to the center of the field. So we chose 4 reference points and decomposed the field into 4 sites. The sites and directions belonging to each site could be seen on the left half of Figure 7. The actual trajectory of the buggy is shown on the right half of the figure. The result was quite satisfactory, with the buggy following given direction quite well and it followed a spiral trajectory on the field as intended.

5.2. *With Sailboat*

After satisfactory results on buggy, we did a real test with our sailboat on Lake Tycolo near Brest, France. Further information about our sailboat could be seen in.[12] The only difference is that we unmounted the wind sensor (see Figure 8). Lake Tycolo is a small lake with good eyesight so that the behavior of the sailboat could be better watched during the test. For our test we intended the boat to travel from the bank of lake to the middle of it without hitting any boundary of the lake. According to our strategy, we decomposed the lake into 4 Voronoi sites. Again for simplicity, we assigned each site with only two directions whose angular seperation is greater than no-go-angle (60 degrees for our testcase). Then, we implemented the online part of the control strategy described above on a PIC microcontroller embedded in our sailboat.

The result of the test is quite satisfactory. The boat could tune its sail so that it could find the best sail angle for the moment, although it does not have any information about the wind. The angle the boat finds by itself

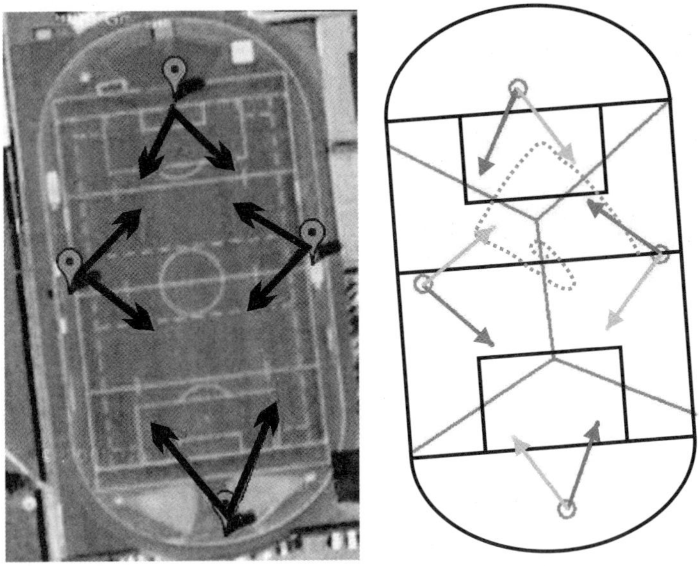

Fig. 7. test with the buggy

Fig. 8. sailboat without wind sensor

matches very well with human expectations. Moreover, the boat's trajectory is shown on Figure 9, which is almost as intended. The SCILAB code for

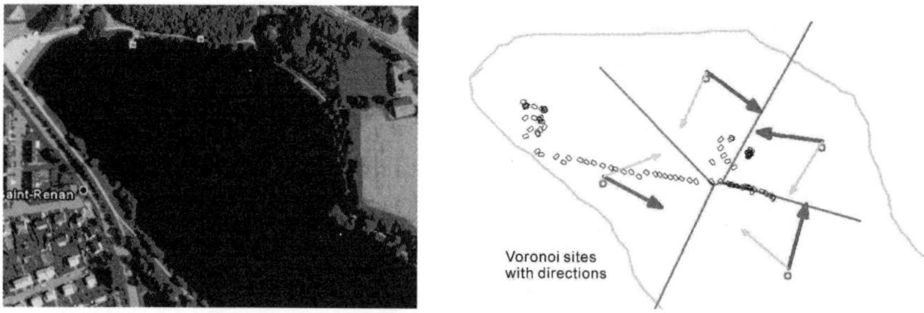

Fig. 9. test with the sailboat robot

the online part, the C code for the offline part of the strategy as well as the video showing our experiment can be downloaded at

`http://www.ensta-bretagne.fr/jaulin/kai.html`

6. Conclusion

In this paper, a new control strategy for autonomous sailboats based on Voronoi diagram has been presented with several advantages.

- The strategy's biggest advantage is that it is *wind-independent*, thus *much more reliable.* As shown above, since the boat could definitely follow one of the two given directions that ultimately point to the target, the boat will undoubtedly get to the target no matter where the wind comes from. Thus, the strategy would be much more reliable than most of others that depend highly on the data from wind sensors which are not so reliable especially in harsh environment. And due to the simplicity of the strategy, the new strategy would have little chance to fail.
- The strategy is *simple and easy to implement.* Due to the Voronoi diagram and the look-up table method, the strategy is quite simple and can be applied easily without putting a high demand on the hardware. Moreover, thanks to the simplicity of the strategy, the program costs few time to execute. In our real test for the sailboat, the computation time for one loop is at the millisecond level. With some advanced table look-up method and algorithms for generating Voronoi diagram, it could be even faster.

Although the strategy has been illustrated only on sailboats, it could have many other areas of application where some unreliable data are necessary for a reliable control. Some potential application areas include space exploration robots, autonomous submarines, etc.

Some future work could help improve the strategy:

- Although the strategy works in our test, its effectiveness and efficiency depend largely on the quality of Voronoi diagram and the two given directions assigned to each Voronoi site. As a result, we currently need an experienced sailor to construct the Voronoi diagram and assign the two given directions. This would greatly limit its application. Future work could focus on finding some techniques to generate the Voronoi diagrams automatically. Many methods such as fuzzy logic, expert system and evolutionary algorithms could be turned for and there are a lot of interesting topics to discover.
- Another important improvement for our strategy is on the computation of Voronoi diagram. To compute the Voronoi diagram takes time. Our current solution is to compute it once and for all in the initialization stage of the program and store it in memory. But it has brought about another problem: in this way the Voronoi diagram could not be updated in real-time. Although the problem does not matter in most cases, it would limit the application of the strategy in some areas. The problem could be largely mitigated by applying some advanced algorithms for computing Voronoi diagram. Now we used a simple, reliable but time-consuming algorithm. There exist a great number of more efficient algorithms for computing Voronoi diagram and Delaunay triangulation. Applying them in our strategy could largely reduce the initialization time for the whole system and could even incorporate the computation in every loop so that the Voronoi diagram could be updated in real time.

References

1. P. F. Rynne and K. D. von Ellenrieder, *Marine Technology Society* **43**, 21 (2009).
2. N. A. Cruz and J. C. Alves, *Journal of the Oesterreichische Gesellshaft fuer Artificial Intelligence (Austrian Society for Artificial Inteligence)* **27**, 25 (2008).
3. Y. Briere, Iboat: An autonomous robot for long-term offshore operation, in *The 14th IEEE Mediterranean Electrotechnical Conference*, 2008.
4. C. Sauze and M. Neal, An autonomous sailing robot for ocean observation, in *in proceedings of TAROS 2006*,).

5. R. Stelzer and T. Proll, *Robotics and Autonomous Systems* **56**, 604 (2008).
6. M. Neal, *IEEE Journal of Oceanic Engineering* **31**, 462 (2006).
7. P. Herrero, L. Jaulin, J. Vehi and M. A. Sainz, *International Journal of Control Automation and Systems (to appear)* (2009).
8. J. Abril, J. Salmon and O. Calvo, *International Journal of Approximate Reasoning* **16** (1997).
9. S. J. Fortune, *Algorithmica* **2**, 153 (1987).
10. S. Garrido, L. Moreno, M. Abderrahim and F. Martin, Path planning for mobile robot navigation using voronoi diagram and fast marching, in *IEEE/RSJ International Conference on Intelligent Robots and Systems*, (Beijing, China, 2006).
11. L. Jaulin, Modlisation et commande d'un bateau 'a voile, in *CIFA2004 (Confrence Internationale Francophone d'Automatique), In CDROM*, (Douz (Tunisie), 2004).
12. J. Sliwka, P. Reilhac, R. Leloup, P. Crepier, H. D. Malet, P. Sittaramane, F. L. Bars, K. Roncin, B. Aizier and L. Jaulin, Autonomous robotic boat of ensieta, in *2nd International Robotic Sailing Conference*, (Matosinhos, Portugal, 2009).

SECTION–3

BIOLOGICALLY INSPIRED SYSTEMS AND SOLUTIONS

PROPOSED OMNIDIRECTIONAL LOCOMOTION STRATEGY FOR TRAVELING-WAVE-TYPE MOBILE ROBOT (TOROIII)

TERUYOSHI OGAWA, MASAYUKI YAMAMOTO, AND TARO NAKAMURA

Faculty of Science and Engineering, Department of Precision Mechanics
Chuo University
1-13-27 Kasuga, Bunkyo-ku, Tokyo 112-8551, Japan

An omnidirectional movement mechanism that can move a robot in a narrow complicated passage is needed. However, existing mechanisms cannot achieve stable operations. We noted that a snail uses traveling waves and can achieve a stable operation because of a large landing area. Therefore, we developed a traveling-wave-type mobile robot (TORoIII) using a snail's locomotion mechanism. However, the directions of the robot were restricted by the number of units, i.e., the directions corresponded to the number of units. In this study, we propose a new perfectly omnidirectional locomotion strategy for TORoIII. In addition, we propose odometry based on kinematics to estimate locomotion track. We experimentally confirmed utility of these proposed methods.

1. Introduction

Many robots that work in narrow indoor spaces, such as a factory or a hospital, have recently been developed. They require an omnidirectional movement mechanism that does not require a large space to turn. Conventionally, an omniwheel is used to achieve the omnidirectional movement [1]. However, an omniwheel cannot achieve sufficient friction force with the ground by using mechanism that slides in the axial direction. On the other hand, a snail's locomotion mechanism uses traveling waves [2]-[5]. A snail can move forward by propagating traveling waves from its tail to head. An advantage of this movement mechanism is that it can achieve stable operation because of a large landing area. For this reason, we previously developed traveling-wave-type omnidirectional mobile robot (TORoIII) [6] [7], which uses the locomotion mechanism of a snail. However, the robot could move in only as many directions as it had units. In this study, we propose a new perfectly omnidirectional locomotion strategy for TORoIII. In addition, we propose odometry based on kinematics to estimate locomotion track. We also confirmed utility of these proposed methods by experiments.

2. The locomotion mechanism of the snail

The common belief is that the snail moves by a traveling wave locomotion mechanism. This produces a traveling wave in the gastropod's foot that propagates in the direction of locomotion [3]. The state of the traveling wave seen from a section of the gastropod's foot is shown in Fig. 1, and the state of the traveling wave when moving forward is shown in Fig. 2. The snail has two kinds of muscles: the anterior oblique and the posterior oblique muscles. These produce a traveling wave that propagates from tail to head. When the traveling wave occurs, the part that is not forming the traveling wave in the gastropod's foot remains in contact with the ground. The ground generates a frictional force with the grounded foot of the gastropod.

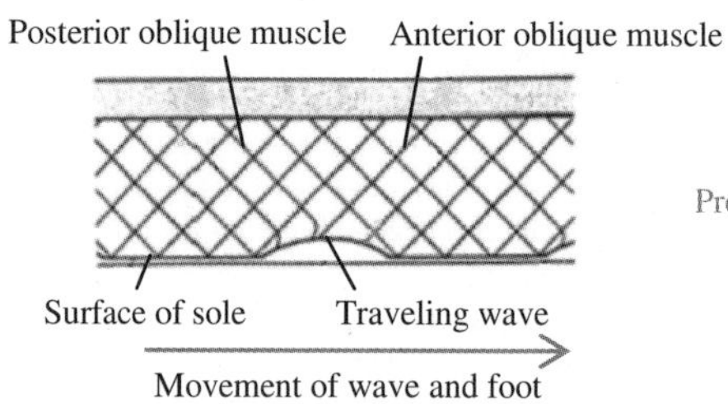

Fig. 1 Cross section of a gastropod's foot [4]

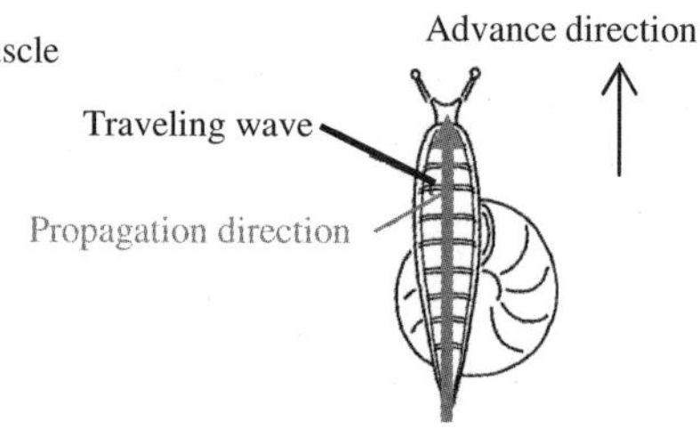

Fig. 2 Propagation of traveling wave [5]

3. Traveling-wave-type omnidirectional mobile robot (TORoIII)

3.1. *Overview*

TORoIII is shown in Fig. 3, and its specifications are shown in Table 1. As shown in Fig. 3, the structure of the robot consists of eight units that are connected by a gear-linkage mechanism.

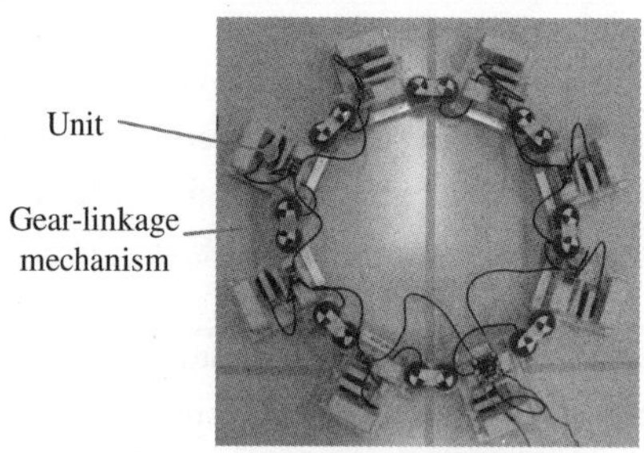

Fig. 3 TORoIII

Table 1 Specifications of TORoIII

Maximum diameter [mm]	755
Weight [kg]	3.6
Number of unit	8

3.2. *Mechanism of the unit*

The design of the units is shown in Fig. 4, and its specifications are shown in Table 2. The unit has two servo motors, denoted as servomotor I and servomotor II. Servomotor I can control the contraction and extension of the unit by a crank mechanism. In addition, a friction plate covered with a rubber sheet moves up and down with the contraction and extension of the unit, respectively. During extension, the friction plate contacts the ground and the unit is fixed to the ground; during contraction, the friction plate leaves the ground and the unit can be easily moved. Servomotor II can control the angle between the units. Units are connected by gear-linkage mechanism. The unit can rotate around point of meshing of gears by rotating the active linkage with servomotor II.

3.3. *System configuration of the robot*

Table 3 shows the system configuration of the robot. The system controls the robot with PC using serial communication. The motor can be controlled by sending the data of the angle by serial communication. We can also measure the angle of the motor because potentiometer is equipped with the motor.

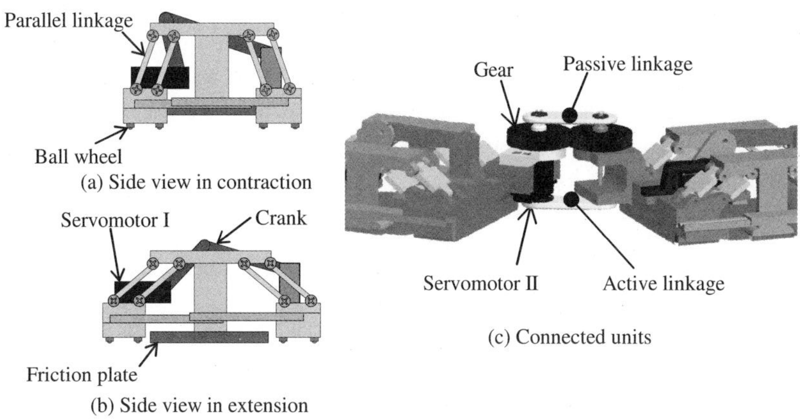

(a) Side view in contraction

(b) Side view in extension

(c) Connected units

Fig. 4 Mechanism of the unit

Table 2 Specifications of the unit

Lenghth [mm]	maximum	200
	minimum	160
Width[mm]	144	
Wight [g]	450	

Table 3 System configuration of the robot

PC	OS	Windows XP SP3
	Software	Visual Studio 2010
	Program language	C++
	Sampling time [ms]	20
Serial communication method		TTL
Actuator		PRS-DE07MS (Pirkus, Inc.) x16

4. Modeling the shape of the robot

To develop omnidirectional locomotion strategy and odometry, we modeled the shape of the robot, as shown in Fig. 5. The robot can be modeled in an octagonal shape consisting of contraction links and joints. The shape of the robot can be expressed by solving the coordinates of the joints by kinematics, as Eq. (1).

$$\mathbf{P}_i = \begin{bmatrix} x_i \\ y_i \end{bmatrix} = \begin{bmatrix} x_{i-1} + L_{i-1}\cos(\theta_{def} + \sum_{j=0}^{i-1}\theta_j) \\ y_{i-1} + L_{i-1}\sin(\theta_{def} + \sum_{j=0}^{i-1}\theta_j) \end{bmatrix} \quad (1)$$

$$\left(\begin{array}{l} \mathbf{P}_0 = \begin{bmatrix} x_0 \\ y_0 \end{bmatrix}, \mathbf{P}_1 = \begin{bmatrix} x_1 \\ y_1 \end{bmatrix} \\ \theta_{def} = \arctan 2\{(y_1 - y_0), (x_1 - x_0)\} \\ i = 2,3,\cdots, N-1 \end{array} \right)$$

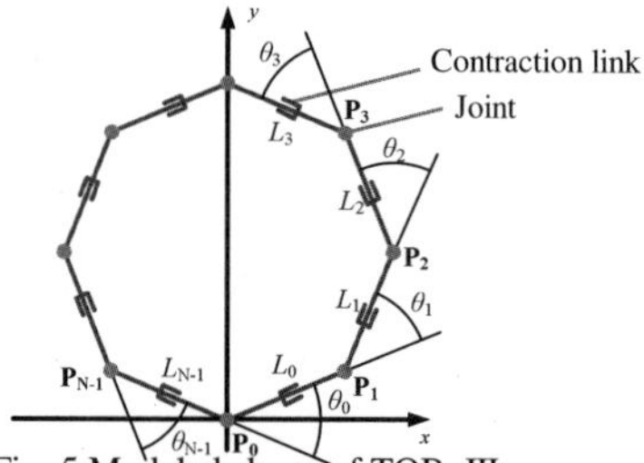

Fig. 5 Modeled shape of TORoIII

Here, θ is the angle between the units, L is the length of each unit, $\mathbf{P}(x,y)$ is coordinate of a joint, and N is the number of units in the robot, as shown in Fig. 4.

5. Omnidirectional locomotion strategy

An overview of the omnidirectional locomotion strategy for the robot is shown in Fig. 6. First, we define a wave propagation line with a propagation velocity *Vw*, propagation angle *Dw* and wavelength *Lw*. Then, we propagate the wave propagation line. If the wave propagation line and the joint do not overlap, the joint remains in its default position as shown in Fig. 6(a). At this point, the current coordinates of joints $\mathbf{P_o}$ are as expressed as Eq. (2).

$$\mathbf{P}_{\mathbf{o}i} = \mathbf{P}_{\mathbf{d}i} \quad (2)$$

Here, $\mathbf{P_d}$ denotes the coordinates of the joints when the robot is a regular octagon and all the units are in extension.

If the wave propagation line and joint overlap, then the joint moves in the direction of the wave propagation line as shown in Fig. 6(b). At this point, the current coordinates of the joints are expressed as Eq. (3).

$$\mathbf{P}_{\mathbf{o}i} = \mathbf{P}_{\mathbf{d}i} + \mathbf{M}_i \quad (3)$$

$$\mathbf{M}_i = \begin{bmatrix} Vw\cos(Dw)\times(t - t_{oi}) \\ Vw\sin(Dw)\times(t - t_{oi}) \end{bmatrix}$$

Here, t_o is the point at which the wave propagation line and joint overlap and t is the current time.

The joint stops after moving by one wavelength. The coordinates of the joints after moving are expressed as Eq. (4).

$$\mathbf{P}_{\mathbf{o}i} = \mathbf{P}_i + \begin{bmatrix} Lw\cos(Dw) \\ Lw\sin(Dw) \end{bmatrix} \tag{4}$$

When all the joints complete their movement, TORoIII comes to a stop. In addition, locomotion speed V is expressed as Eq. (5). Here, L_{toro} is the maximum diameter of the model.

$$V = Lw/\{(L_{toro} + Lw)/Vw\} \tag{5}$$

However, one exception in the omnidirectional locomotion strategy of TORoIII is that when the wave propagation line is parallel to the unit, the unit moves in a grounded state. Therefore, the joint cannot move correctly. To avoid this grounded state, the joint moves along a sine trajectory, as shown in Fig. 6(e). The coordinates of the moving joint are expressed as Eq. (6). Here, B is a constant.

$$\mathbf{P}_{\mathbf{o}i} = \mathbf{P}_i + \mathbf{M}_i \pm \mathbf{M}_{\mathbf{o}i} \tag{6}$$

$$\mathbf{M}_{\mathbf{o}i} = \begin{bmatrix} A\sin\left\{(\dfrac{\pi Vw}{Lw})(t - to_i)\right\} \times \cos(Dw + \dfrac{\pi}{2}) \\ A\sin\left\{(\dfrac{\pi Vw}{Lw})(t - to_i)\right\} \times \sin(Dw + \dfrac{\pi}{2}) \end{bmatrix}$$

$$A = B\sin\left\{\frac{N}{2}\mathrm{mod}(Dw, \frac{2\pi}{N})\right\}$$

By changing the direction of the wave propagation, TORoIII can move in an arbitrary directions. By repeating this motion, TORoIII can follow an arbitrary path.

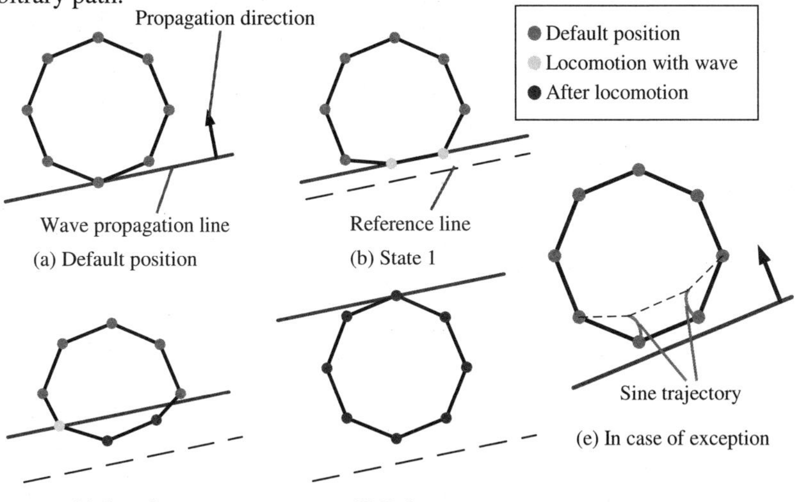

Fig. 6 Omnidirectional locomotion strategy

6. Odometry based on kinematics

Odometry is performed using the shape of the robot as modeled in Section 4. Initially, when a unit of the robot is in extension, the unit achieves a large frictional force with the ground because the friction plate, covered with a rubber sheet, is in contact with the ground. Therefore, we assume that a unit in extension is fixed to the ground and that the joint in the model is fixed to a world coordinate system when two units adjacent to the joint are in extension. Odometry is performed by solving kinematics from this fixed joint in real time. However, because more than one fixed joint exist, the position and the attitude of the robot is decided by solving kinematics from all fixed joints and averaging them, as Eq. (7). Here, *n* is the number of the fixed joints.

$$\mathbf{P}_{\mathbf{od}i} = \begin{bmatrix} x_{\mathrm{od}\,i}(t) \\ y_{\mathrm{od}\,i}(t) \end{bmatrix} = \begin{bmatrix} \frac{1}{n}\sum_{j=1}^{n} x_{ij}(t) \\ \frac{1}{n}\sum_{j=1}^{n} y_{ij}(t) \end{bmatrix} \tag{7}$$

By this method, we can predict the locomotion track of the robot.

7. Locomotion experiments

7.1. *Experimental environment*

We conducted experiments to evaluate the omnidirectional locomotion strategy and odometry. Fig. 7 shows the experimental environment. All the experiments were conducted on a floor where the friction coefficient with the robot was 0.85. The locomotion tracks were analysed using movie analysis software (MOVIAS Pro: Nac Image Technology Inc.). As shown in Fig. 6, two gauge marks were placed on the robot and the midpoint was used for the evaluation of the locomotion track.

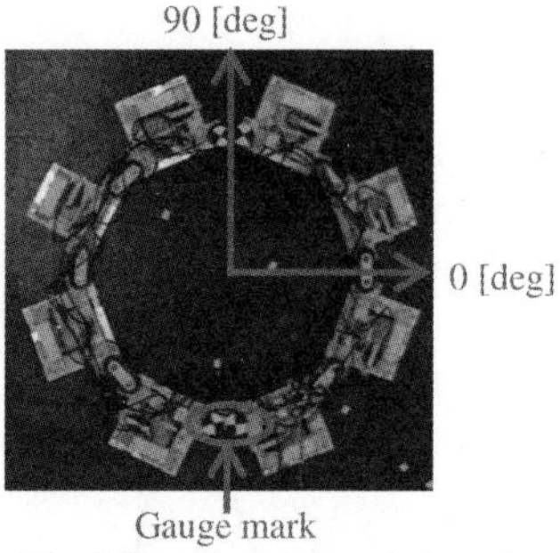

Fig. 7 Experimental environment

7.2. *Experiment to verify locomotion direction*

We experimentally verified whether the robot can move in any direction using the omnidirectional locomotion strategy. In this experiment, we changed the propagation angle of the wave propagation line from 0° to 45° in 5° increments. We set the wavelength of the wave propagation line to 40 mm and propagated the wave propagation line five times at each propagation angle. The results are shown in Fig. 8.

This figure shows that TORoIII can move along a desired track in a range of locomotion directions from 0° to 45°. Furthermore, because TORoIII is a regular octagon, it is symmetric in other directions. Therefore, we confirmed that TORoIII can move omnidirectionally using our proposed strategy.

7.3. *Experiment to verify locomotion velocity*

We experimentally verified whether the locomotion velocity can be controlled by the omnidirectional locomotion strategy. In this experiment, we changed the propagation velocity of the wave propagation line from 20 mm/s to 300 mm/s and set the wavelength of the wave propagation line to 40 mm. The results are shown in Fig. 9.

From this figure, we can see that the locomotion velocity can be controlled. However, as the propagation velocity increases, the amount of deviation from the desired value increases. We think that this deviation is observed because the motor cannot follow the desired angle as this strategy does not consider dynamics.

7.4. *Long-distance locomotion experiment*

We experimentally verified whether the robot can follow an arbitrary path using the omnidirectional locomotion strategy and whether the locomotion track can be predicted by odometry. The results are shown in Fig. 10.

The figure shows that the robot can move along the desired path. Furthermore, the deviation between the desired path and the experimental track can be predicted by odometry. However, some deviation between the final goal of the experimental track and odometry is observed. We believe that this is a result of slippage which odometry does not consider because it is assumed that the extending unit is fixed to the ground.

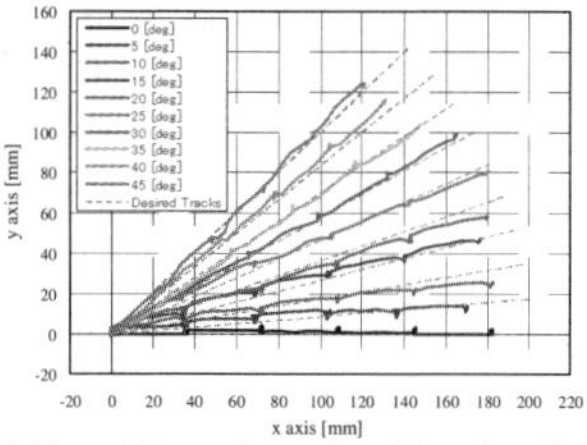

Fig. 8 Experimental result of locomotion direction

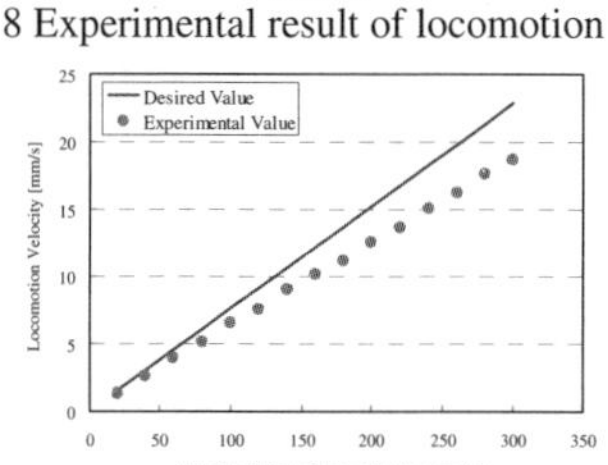

Fig. 9 Relationship between propagation velocity and locomotion velocity

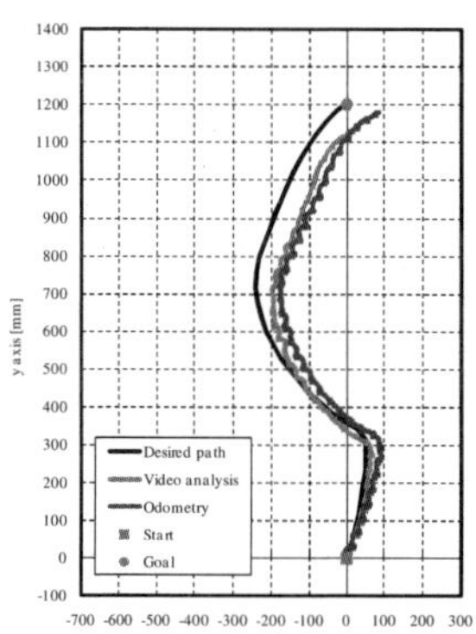

Fig. 10 Result of long-distance experiment

8. Conclusion

In this study, we proposed a novel, perfectly omnidirectional locomotion strategy for TORoIII and odometry based on kinematics to predict the locomotion track of a robot. The conclusions are as follows:

1) Locomotion velocity and locomotion direction of the robot can be controlled by the omnidirectional locomotion strategy.
2) As the locomotion distance increases, the locomotion track of the robot deviates from the desired path. However, this deviation can be estimated by odometry.

In the future, we will propose a path-tracking method using the odometry.

Reference

[1] J. Urbano, K. Terashima, T. Miyoshi and H. Kitagawa, "Velocity Control of an Omni-directional Wheelchair Considerring Use's Comfort by Suppresing Vibration", IEEE/RSJ 2005 International Conference on Intelligent Robots and Systems, pp. 3169-3174, 2005

[2] Brian Chan, N.J.Balmforth, and A.E.Hosoi, "Building a better snail: Lubrication and adhesive locomotion", Physics of Fluids, pp. 113101:1-10, 2005

[3] Ryousen Fujiwara, Hirohisa Morikawa, and Shunishi Kobayashi, "The Mechanism of Pedal Locomotion of Gastropod", The Japan Society of Mechanical Engineers, JAPAN, Vol. 67, No. 658, pp. 1934-1940, 2001.

[4] H.D.Jones, "Circulatory pressures in Helix pomatia", L. Comp. Biochem. Physiol., 39A, 289, 1971

[5] Ryousen Fujiwara, Hirohisa Morikawa, Yuya Hukuya, Hiroshi Sakai and Shunichi Kobayashi, "Pedal-Like Locomotion Mechanism Modeled on Pedal Crawling of Terrestrial Gastropod", The Japan Society of Mechanical Engineers, JAPAN, Vol. 70, No. 695, pp215-221, 2004

[6] Taro Nakamura and Kuniaki Sato, "Locomotion Strategy for an Omni-directional Mobile Robot Using Traveling Waves Propagation" IEEE International Conference on Robotics and Automation, pp. 3769-3774, 2010

[7] Taro Nakamura and Kuniaki Sato, "Development of an omni-directional mobile robot using traveling waves based on snail locomotion", Journal of Industrial Robot, Vol. 35, No. 3, pp. 206-210, 2008

ENERGY BASED GAIT CONTROL OF AN UNDER-ACTUATED ROBOTIC FISH

D. T. ROPER*, S. SHARMA, R. SUTTON

School of Marine Science and Engineering, University of Plymouth
**E-mail: daniel.roper@plymouth.ac.uk*

P. CULVERHOUSE

School of Computing and Mathematics, University of Plymouth

The recent explosion of interest in sub sea research and industries has led to a demand for efficient versatile submarine robots to perform interaction and survey tasks. Whilst most of these robots employ conventional propeller or thruster based locomotion, there has been a growing interest in the development of biologically inspired robotic swimmers (or robotic fish). In a hope to gain some of the manoeuvrability and efficiency advantages characteristic of their biological counterparts.

Early robotic fish achieved their gait, by directly controlling the relative angel between each vertebrae, using multiple active actuators.

Following observations that many biological locomotory gaits, utilize spring dynamics to create efficient oscillatory motion with minimal active input, there has been a recent trend towards the development of under-actuated robotics utilizing dynamics to achieve harmonic locomotory gaits, examples can be found in McGeer's passive dynamic walkers, or MIT's compliant swimming devices.

For a harmonic dynamic system, the path is dependent on the total energy in the system. By controlling the total energy in the system the gait can be controlled by varying a single degree of freedom.

This paper aims to explore the potential for energy based control to produce an effective propulsive gait.

Keywords: Under-Actuated, Robotic Fish, Gait Control, Energy Based Control

1. Introduction

In recent years there has been a growing interest in the development of biomimetic robotic swimmers or *robotic fish*.[1] Typically constructed as a planar multi body chain, robotic fish emulate the moving body wave employed by their biological counterparts to generate locomotive thrust. Like terrestrial biomimetic locomotors, there has been a recent trend towards under actuated design of robotic fish such as the three link robotic dolphin outlined in,[2] where only one of the joints was actively controlled. Or the compliant swimmers outline in,[3] again with only one active actuator. These swimmers take advantage of an affect demonstrated in,[4] whereby harmonic stimulation of a passive body can create an effective forward swimming gait.

Unlike with terrestrial biomimetic locomotion where the reaction forces of the solid ground can easily be calculated, with aquatic locomotion the reaction forces created by the fluid are difficult to calculate, and subject to highly chaotic dynamics.

2. Energy Based Control

Unlike a classical geometric control approach to gait generation, which would seek to directly control the geometry of the robot, *Energy based control*, (EBC) models the system as a harmonic oscillator, then simply seeks to control the total amount of energy in the system. For simple examples of EBC the reader is directed to.[5,6] EBC is particularly attractive for gait generation, as it is likely to result in near harmonic motion which is ideal for most locomotive patterns. Further more the direct control of energy is likely to result in a good energy efficiency, findings presented in[7] demonstrated that a simple harmonic model of a human walking gait could predict step frequencies for individuals under a variety of load conditions.

3. Introduction to Model

In this paper a three link free floating planer multi body chain with sprung joints will represent a robotic fish, as shown in Fig: 1. This model is inkeeping with observations made in[8] that fish use springs in parallel with active muscle movement to produce efficient harmonic swimming gaits. The angle of each body is given relative to the global frame, giving the control space state vector as,

$$\mathbf{q} = \begin{pmatrix} \theta_1 \\ \theta_2 \\ \theta_3 \end{pmatrix}$$

By taking moments from the centre of mass of each body, it is possible to find the x and y co-ordinates of the centre of mass of each body relative to the centre of mass of the whole system,

$$\mathbf{x} = \mathbf{M}\cos(\mathbf{q}),\ \mathbf{y} = \mathbf{M}\sin(\mathbf{q}) \tag{1}$$

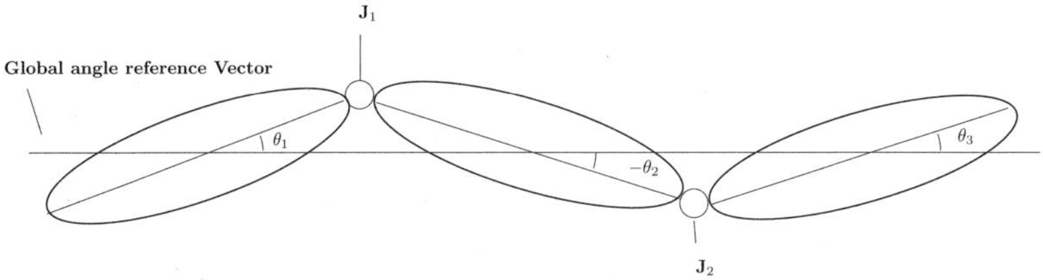

Fig. 1. Simple three link flexible multi body structure, each body posses a θ angle relative to global reference vector, the three bodies are joined by two sprung joint $\mathbf{J}_1$ and $\mathbf{J}_2$

where $\mathbf{M}$ is the matrix of moments of the system at $\mathbf{q} = \mathbf{0}$. Therefore using Newton's second law the reaction forces in the the joints can be calculated as.

$$\mathbf{Rf} = \begin{pmatrix} m_1 & 0 & 0 \\ m_1 & m_2 & 0 \end{pmatrix} \left[\mathbf{M} \frac{d^2}{dt^2} \left(\cos(\mathbf{q}) \ \sin(\mathbf{q}) \right) \right. \tag{2}$$

$$\left. - \left(\frac{\mathbf{F}_x}{\mathbf{m}} + A_x \ \frac{\mathbf{F}_y}{\mathbf{m}} + A_y \right) \right]$$

Where $\mathbf{m}$ is the mass vector, $\mathbf{F}_x$ and $\mathbf{F}_y$ are the x and y components of the external force vector, A_x and A_y are the x and y components of the global acceleration vector. The second differential can be evaluated as,

$$\frac{d^2}{dt^2} \begin{pmatrix} \cos(\mathbf{q}) \\ \sin(\mathbf{q}) \end{pmatrix}^T = \begin{pmatrix} -\ddot{\mathbf{q}} \sin(\mathbf{q}) - \dot{\mathbf{q}} \cos(\mathbf{q}) \\ \ddot{\mathbf{q}} \cos(\mathbf{q}) - \dot{\mathbf{q}} \sin(\mathbf{q}) \end{pmatrix}^T \tag{3}$$

The moments on each body can then be evaluated as,

$$\boldsymbol{\tau}_{total} = \mathbf{B}\tau_{in} + \mathbf{T} \left(-\sin(\mathbf{q}) \ \cos(\mathbf{q}) \right) \mathbf{Rf} \tag{4}$$

$$-\mathbf{Kq} + \boldsymbol{\tau}_{ext}$$

where $\mathbf{B}$ is the input vector which for a single actuator design will be $(1, -1, 0)^T$,$\mathbf{K}$ is the system spring matrix, $\boldsymbol{\tau}_{ext}$ is the vector of external torques and $\mathbf{T}$ is the transform matrix given as,

$$\mathbf{T} = \begin{pmatrix} l_{1r} & 0 \\ l_{2f} & l_{2r} \\ 0 & l_{3f} \end{pmatrix}$$

Where the l values represent the distance from the joints to the relative centers of mass. The system spring matrix $\mathbf{K}$ will be given as,

$$\mathbf{K} = \begin{pmatrix} k_1 + \epsilon & -k_1 & 0 \\ -k_1 & k_1 + k_2 & -k_2 \\ 0 & -k_2 & k_1 \end{pmatrix}$$

where k_1 and k_2 are the spring constants of the first and second joint respectively, ϵ is a numerically insignificant value to ensure non singularity. By substituting $\boldsymbol{\tau}_{total} = \mathbf{I}\ddot{\mathbf{q}}$ it is possible to derive an equation for the system in the form,

$$\mathbf{H}(\mathbf{q})\ddot{\mathbf{q}} = \mathbf{C}(\mathbf{q})\dot{\mathbf{q}} - \mathbf{K}\mathbf{q} + \mathbf{D}(\mathbf{q}, \boldsymbol{\tau}_{ext}, \mathbf{F}) + \mathbf{B}\tau \tag{5}$$

Where $\mathbf{H}$ is equal to $\mathbf{I}$ minus the coefficient of the $\ddot{\mathbf{q}}$ component of the reaction force induced moment, and $\mathbf{C}$ is the coefficient of the $\dot{\mathbf{q}}$ component. $\mathbf{D}$ is the total disturbance torques, including the disturbance forces contribution to the reaction forces and $\mathbf{B}$ is the input matrix $(1 \ -1 \ 0)^T$.

The System can be approximated as the state space system,

$$\begin{pmatrix} \dot{\mathbf{q}} \\ \ddot{\mathbf{q}} \end{pmatrix} = \begin{pmatrix} \mathbf{0} & \boldsymbol{I}_3 \\ -\mathbf{H}^{-1}\mathbf{K} & \mathbf{H}^{-1}\mathbf{C} \end{pmatrix} \begin{pmatrix} \mathbf{q} \\ \dot{\mathbf{q}} \end{pmatrix} + \begin{pmatrix} \mathbf{0} \\ \mathbf{H}^{-1}\mathbf{D} \end{pmatrix} + \begin{pmatrix} \mathbf{0} \\ \mathbf{H}^{-1}\mathbf{B} \end{pmatrix} \tau \tag{6}$$

Where $\boldsymbol{I}_3$ is the three by three identity matrix and $\mathbf{0}$ is the matrix of zeros. Discretization of equation (6) in the time domain gives,

$$\mathbf{Q}_{t+\Delta t} = \mathbf{e}^{\mathbf{A}\Delta t}\mathbf{Q} + \mathbf{A}^{-1}(\mathbf{e}^{\mathbf{A}\Delta t} - \boldsymbol{I})(\mathbf{B'}\tau + \mathbf{D'}) \tag{7}$$

Where $\mathbf{Q}$ is the modified state variable, $\mathbf{B'} = \mathbf{H}^{-1}\mathbf{B}$ and $\mathbf{D'} = \mathbf{H}^{-1}\mathbf{D}$, and $\mathbf{A}$ is the compositional matrix from equation (6).

4. Energy Control of the System

If the system is planar and orthogonal to the action of gravity, then the total energy can be is calculable as the sum of the rotational and linear kinetic energy and the energy stored in the springs. given by;

$$E = \frac{1}{2}\left(\dot{\mathbf{q}}^T\left(\mathbf{I} + \mathbf{M}^T\mathbf{m}\mathbf{M}\right)\dot{\mathbf{q}} + \mathbf{q}^T\mathbf{K}\mathbf{q}\right) \tag{8}$$

Which can be expressed as;

$$E = \frac{1}{2}\begin{pmatrix} \mathbf{q} \\ \dot{\mathbf{q}} \end{pmatrix}^T \begin{pmatrix} \mathbf{K} & \mathbf{0} \\ \mathbf{0} & \mathbf{I} + \mathbf{M}^T\mathbf{m}\mathbf{M} \end{pmatrix} \begin{pmatrix} \mathbf{q} \\ \dot{\mathbf{q}} \end{pmatrix} \tag{9}$$

Where $\mathbf{G}$ is the compositional matrix from equation (9).

Ignoring the added mass of the surrounding fluid and isolating the system from all external forces and losses. Every energy level will represent a distinct harmonic phase cycle. By appropriate selection of spring constants these phase cycles can be shaped into useful forward swimming gaits. The amplitude of which will be determined by the total amount of energy in the system.

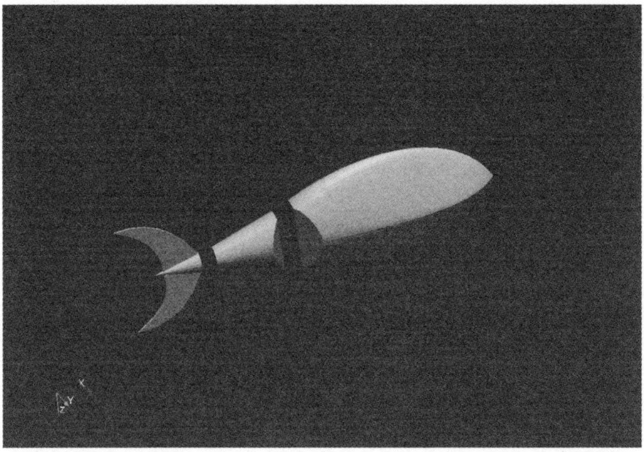

Fig. 2. Simulation geometry, three body chain one meter in length

Ignoring all disturbances from the fluid and combining equations (7) and (9)

$$E_{t+\Delta t} = (\mathbf{e}^{\mathbf{A}\Delta t}\mathbf{Q})^T\mathbf{G}\mathbf{e}^{\mathbf{A}\Delta t}\mathbf{Q} + (\mathbf{e}^{\mathbf{A}\Delta t}\mathbf{Q})^T\mathbf{G}\mathbf{A}^{-1}(\mathbf{e}^{\mathbf{A}\Delta t} - \boldsymbol{I})\mathbf{B}'\tau \qquad (10)$$

$$+(\mathbf{A}^{-1}(\mathbf{e}^{\mathbf{A}\Delta t}-\boldsymbol{I})\mathbf{B}'\tau)^T\mathbf{G}\mathbf{e}^{\mathbf{A}\Delta t}\mathbf{Q}+(\mathbf{A}^{-1}(\mathbf{e}^{\mathbf{A}\Delta t}-\boldsymbol{I})\mathbf{B}')^T\mathbf{G}\mathbf{A}^{-1}(\mathbf{e}^{\mathbf{A}\Delta t}-\boldsymbol{I})\mathbf{B}'\tau^2$$

Since the system has no internal losses, then the natural response should result in no net change in energy, therefore,

$$E_{target} = E_t + \mathbf{e}^{\mathbf{A}\Delta t}\mathbf{Q}^T\mathbf{G}\mathbf{A}^{-1}(\mathbf{e}^{\mathbf{A}\Delta t} - \boldsymbol{I})\mathbf{B}'\tau \qquad (11)$$

$$+(\mathbf{A}^{-1}(\mathbf{e}^{\mathbf{A}\Delta t}-\boldsymbol{I})\mathbf{B}'\tau)^T\mathbf{G}\mathbf{e}^{\mathbf{A}\Delta t}\mathbf{Q}+(\mathbf{A}^{-1}(\mathbf{e}^{\mathbf{A}\Delta t}-\boldsymbol{I})\mathbf{B}')^T\mathbf{G}\mathbf{A}^{-1}(\mathbf{e}^{\mathbf{A}\Delta t}-\boldsymbol{I})\mathbf{B}'\tau^2$$

Forms a quadratic in τ the solutions of which represents the torque required in the absence of fluid interaction to achieve the desired total energy by the next time iteration in each direction. By selecting the solution of lest magnitude the system will move towards the closest harmonic phase cycle at the desired energy level. A proportional energy error controller is achieved.

5. Methods

The system was simulated in OPENFOAM® version 1.7.0 available from OpenCFD® using the pimpleDyMFoam solver, an additional multi-body chain solver was written to solve the body motion.

An approximately tuna shaped geometry as in Figure (2), was used for the simulation. Total length of the model was one meter, The flow was solved in a domain six meter by four meters by four meters. In a fluid with kinematic viscosity $10^{-6}m^2s^{-1}$ and density 10^3Kgm^{-3}.

A dynamic time step was used with a limitation on the maximum Courant number being 0.5. The mesh was produced using snappyHexMesh, with six layers of refinement on an original mesh size of components sized 0.2m.

6. Results

Preliminary results have shown that the employment of this control strategy can produce an effective forward swimming gait although to date the velocities achieved have been limited. Figure (3) shows that the tail is achieving horizontal movement with a positive angle of attack throughout the majority of the tail beat cycle.

At the time of writing this the best results to date have resulted in a net forward velocity of approximately 2.6% of body-length per second. By comparing the energy expenditure to overcome the rigid body drag at this velocity, a propulsive energy efficiency of $\approx 36\%$ was calculated. This compares very well against low speed propulsion efficiencies put forward in[10] where low speed propulsion efficiencies for a robotic fish of $3-5\%$ were put reported.

Evaluating the Strouhal number of the resultant swimming gait gave $St \approx 0.66$ which is significantly higher than the range for efficient swimming given as $St = 0.25 - 0.35$.[9]

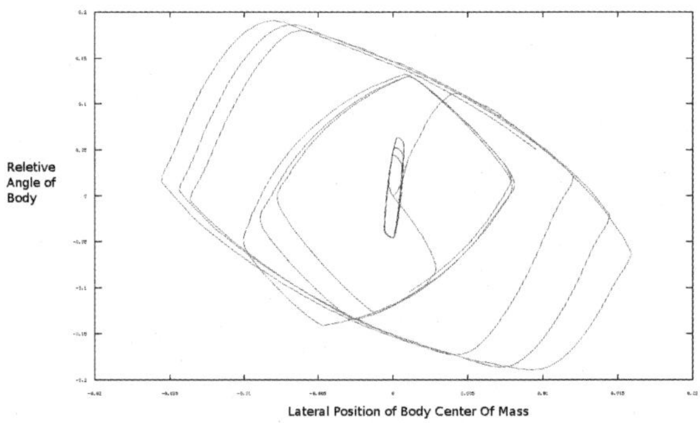

Fig. 3. Phase plot of forward swimming gait, horizontal displacement along the x axis against relative rotation on the y, for the three bodies, blue = head, green = midsection, red = tail

7. Conclusions

Although to data the swimming gait achieved have not been within the optimal optimal range of Stroahal number, and hence resulted in a relatively low propulsion efficiency, the results put forward in this paper do indicate that a harmonic energy based gait production tactic can produce an effective forward swimming motion.

It is hoped that further optimization will result in more efficient locomotion at more practical velocities.

Further improvements may also be possible through alterations to the geometry. Addition of a dorsal fin should result in additional rotational dampening on the forward section of the body reducing energy losses through sway.

References

1. D. Roper, S. Sharma, R. Sutton, and P. Culverhouse, "A review of developments towards biologically inspired propulsion systems for autonomous underwater vehicles," *Proc. IMechE Part M: J. Engineering for the Maritime Environment*, vol. 225, 2011, pp. 77-96.
2. M. Nakashima, Y. Takashi, and K. Ono, *Bio-mechanisms of Swimming and Flying*. Springer, 2004, ch. Three-Dimensional Maneuverability of the Dolphin Robot, pp. 79–92.
3. P. V. y Alvardo, "Design of biomimetic compliant devices for locomotion in liquid environments," Ph.D. dissertation, Massachusetts Institute of Technology, USA, January 2007.
4. D. N. Beal, "Propulsion through wake synchronization using a flapping foil," Ph.D. dissertation, Massachusetts Institute of Technology, USA, June 2003.
5. M. Iwashiro, K. Furuta, and K. J. Aström, "Energy based control of pendulum," in *Proceedings of IEEE international Conference onControl Aplications*, Dearborn, MI, September 1996, pp. 715–720.
6. Y. Dong, Z. Wang, Z. Feng, D. Wang, and H. Fang, "Energy-based control for a class of under-actuated mechanical systems," in *Image and Signal Processing, 2008. CISP '08. Congress on.* Sanya, China: IEEE, 2008, pp. 139–143.
7. K. G. Holt, J. Hamill, and R. O. Andres, "The force-driven harmonic oscillator as a model for human locomotion," *Human Movement Science*, vol. 9, pp. 55–68, 1990.
8. D. A. Pabst, "Springs in swimming animals," *American Zoologist*, vol. 36, pp. 723–735, 1996.
9. M. S. Triantafyllou and G. S. Triantafyllou, "An Effcient Swimming Machine,"Scientific American, vol. 272, pp. 64–71, 1995.
10. C. M. Watts "A Comparison Study of Biologically Inspires Propulsion Systems for an Autonomous Underwater Vehicle," Ph.D. dissertation, University of Glasgow, UK, September 2009.

BALANCE RECOVERY OF A HUMANOID ROBOT USING COGNITIVE SENSORIMOTOR LOOPS (CSLs)

Matthias Kubisch, Christian Benckendorff, and Manfred Hild

Neurorobotics Research Laboratory, Humboldt-Universität zu Berlin, Germany
{kubisch|benckend|hild}@informatik.hu-berlin.de

Recovering balance from unknown disturbances can be considered a complex sensorimotor ability of humanoid robots. We present so-called Cognitive Sensorimotor Loops (CSLs), discuss their properties and show how they can be used for motion generation and balance recovery on robots. Their behavioral abilities will be demonstrated on a single robot leg controlled by CSLs and we will show that a complex stand-up motion can emerge from the interplay of independent joint controllers. Furthermore, we explain how CSLs can be used to help a robot to adapt to changing slopes and recover balance after disturbance.

Keywords: Balance Recovery, Sensorimotor Control, Humanoid Robots

1. Introduction

Balance control is a crucial skill for natural and artificial systems alike. So, in order to accomplish balancing tasks, modern robots possess rich sensory feedback, but their performance still does not reach the robustness of biological individuals. Although the fusion of different sensor modalities may eventually lead to an optimal solution, we try to approach the problem from a different, more minimalist perspective.

In the paper at hand we will extend the notion of Cognitive Sensorimotor Loops (CSLs)[1] and use them for motion creation and balance recovery on the modular humanoid robot *Myon*.[2] For this, the structure is as follows: On the basis of the CSL described in Section 2, we will show how a single robot leg can perform a complex stand-up motion with the use of multiple independent joint controllers (Section 3). Here, we also demonstrate that the CSLs enable a humanoid robot to stabilize its upright standing and recover balance after disturbances, e. g. pushes or changing slopes. In the final section, we give a brief overview on future research.

2. Cognitive Sensorimotor Loops (CSLs)

Connecting sensory inputs to motor outputs, which in turn cause perceptible changes in sensory data, is usually termed a closed sensorimotor loop. The basis for our experiments on robot motion is the simple but yet effective structure depicted in Figure 1 (left). It uses the joint angle $\varphi(t)$ as input, and outputs the actuator's driving voltage $u(t)$. The discrete time update rule is given by $u(t) = -g_i\varphi(t) + \left[g_i\varphi(t) + g_f u(t)\right] z^{-1}$ with $u(0) = 0$, where z^{-1} is the unit delay operator. The left half, comprising the input paths $-g_i$, g_i and the unit delay, forms a differentiator with negative sign. Hence, there is no need for absolute sensory values: it is sufficient to provide a velocity signal and remove the lower input pathway. The right half can either function as leaky integrator, ideal integrator, or integrator with additional feedback. Joint angle $\varphi(t)$ and motor input $u(t)$ are supposed to have the same sense of rotation, therefore, if $u(t) > 0$ then the motor accelerates to get $\dot{\varphi}(t) > 0$.

Depending on the parameters g_i and g_f, a CSL can operate in different modes which equip a robot with distinct behaviors like contraction and release. We will refer to them as *behavioral modes* (see Table 1 for an overview).

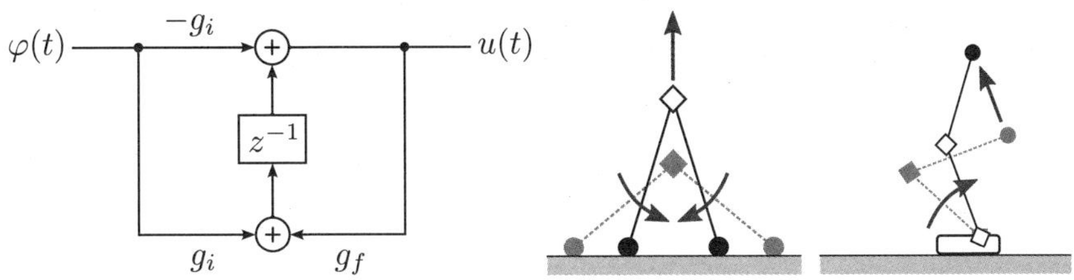

Figure 1. Left: Structure of the CSL. Right: CSL operating in contraction mode for different morphologies which are subjected to gravity.

For example, by setting g_i positive and the feedback to $0 \leq g_f < 1$ the CSL will operate in *release mode*. The actuator's driving voltage is proportional to the negative angular velocity which will slow down the motion, acting as additional fluid friction. On the other hand, if we turn the sign of g_i the control loop will support the current motion. Thereby, the angular velocity is positively fed back.[3]

If we choose to set $g_i > 0$ and $g_f = 1$ the sensorimotor loop will try to hold the current position. Beginning with a steady state, we can think of $u(t)$ as the angular displacement. The CSL henceforth operates like a position controller with the set point put to the last known angular position at rest.

In the following, the behavioral mode of major interest is *contraction mode.* It works against all external forces, e. g. pushes or gravity. For this, the velocity is negatively fed back and further amplified due to the feedback parameter g_f larger than one. See the CSL's behavior for two different morphologies subjected to gravitational forces in Figure 1 (right). A particular property of contraction mode appears when the physical system passes an unstable fixed point, e. g. one can think of an inverted pendulum controlled by CSL which is approaching the upright position. In this case, the velocity and therefore the CSL's output decrease until zero-crossing and finally change the sign. This makes the joint motor turn its rotational direction and eventually stabilize this unstable fixed point.

In contraction mode, the value of $u(t)$ can be used to implicitly detect what is happening to the physical system. When $u(t)$ is almost zero, the physical system is on the transient into a steady state, i. e., an unstable fixed point. Otherwise, if it diverges and tends to grow out of bounds (here: $[+1, -1]$), the physical system is situated in a stall situation while maximally driving its motors. In the same way, the CSL will experience attached loads and can be used to distinguish *up* and *down* while it is actively opposing gravity. That is why we call this sensorimotor loop *cognitive.*

Table 1. Overview of possible behavioral modes. Each parameter configuration exhibits a distinct behavior, see Figure 2 below for illustration.

Release	Hold Current Position	Contraction	Support Motion
$g_i > 0$ $0 \leq g_f < 1$	$g_i > 0$ $g_f = 1$	$g_i > 0$ $g_f > 1$	$g_i < 0$ $g_f = 0$

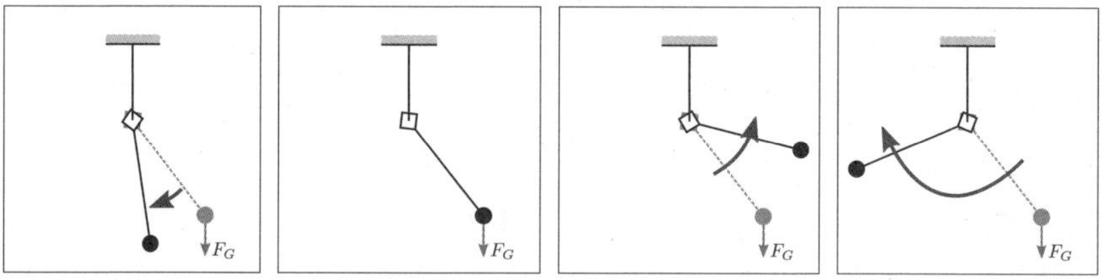

Figure 2. Different behavioral modes of the CSL form nature-like motions; From left to right: release, hold current position, contract, or support an externally induced motion.

The motion of a physical system, e. g. a robot leg, is usually bounded by joint limits. Desirably, these limits should be estimated during self-exploration[1,4–6] and learned for future use. But for now, a simple way to cope with joint limits is to manually identify the maximal angular positions and set the CSL into release mode when a joint limit is reached. Otherwise, the sensorimotor loop in contraction mode will *sense* the mechanical barrier as the direction of maximal resistance and work against it.

3. Single Leg Stand-up and Balance Recovery

In what follows, we concentrate on the control of a single leg of the modular humanoid robot Myon with the proposed Cognitive Sensorimotor Loops. Due to the robot's distributed processing nodes and its onboard power supply, a single leg already constitutes a fully autonomous robot. The programming code is deployed onto the robot. Also, there is no further connection to external computers. For the experimental run described below, we only use a fixed set of parameters for the CSLs. When the hip, knee, and ankle joints are each controlled by local CSLs in contraction mode, then a complex behavioral sequence emerges, as outlined in Figures 3 and 4.

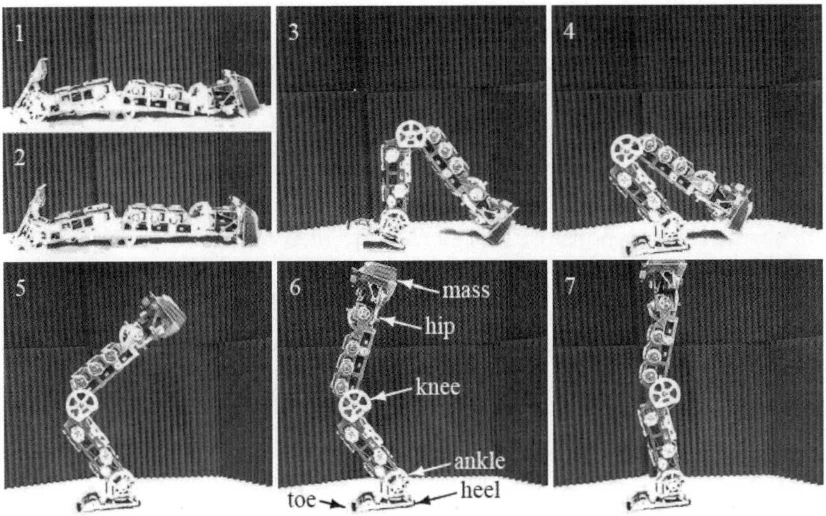

Figure 3. A single leg of the modular humanoid robot Myon standing up using the proposed CSLs. Each of the three joints is locally controlled by a contracting CSL whilst no direct communication takes place between the controllers. Complex behavior emerges from the interplay via momentum.

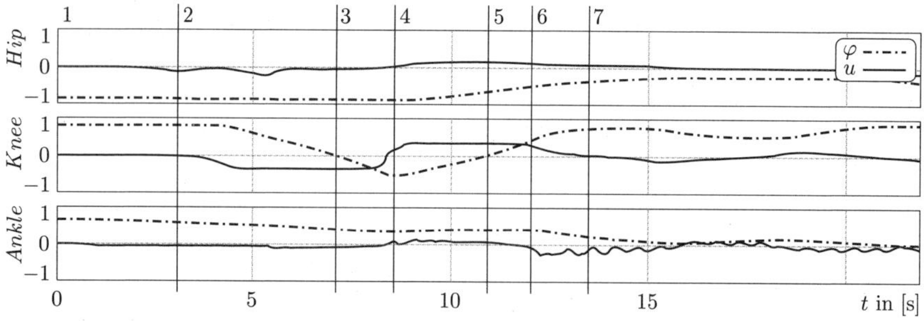

Figure 4. Joint angles of Myon's leg and the control outputs in the stand-up motion. The numbered vertical lines correspond to the snapshots of Figure 3. An experimental run takes about 20 seconds. In the end (7) the leg is never at rest but performs a slight swinging motion in upright position.

First, only ankle, knee, and the mass touch the ground, so the hip joint experiences load and starts to contract (1–2). This lifts the knee from the ground, so the knee joint also starts to contract (3). At some point the leg tilts over onto toe and heel, so that the mass is in the air (4). This makes the knee joint turn its rotational direction, it now works against gravity (5). The other joints are well compensating the forces, so the leg ends up in a fully upright position (6–7) and stays there balancing even if it is pushed or the ground is tilted. In summary, it can be stated that a leg controlled by CSLs is able to stand up and balance without any changes in control, e. g. without switching of target values or adjustment of parameters.

To demonstrate the CSLs' ability to balance more complicated robot morphologies, we did tests with another configuration of Myon as shown in Figure 5 (right). Here, two CSLs are used for the left and right hip roll joint, which are responsible for lateral movements of the legs. When used in contraction mode, the CSLs will create a motion of the legs similar to the one depicted in the middle panel of Figure 1. This enables the robot to successfully recover balance after moderate disturbances of different kinds and also stabilizes its upright posture even when the ground is tilted.

When using contraction mode only, the hip actuators of the two legs will work against each other all the time. This behavior is unintended, once the robot is in a fully upright position and already well balanced. To avoid useless heating-up of the motors we propose crossed inhibitory connections between the two controllers. This is done by modulating the feedback g_f, effectively changing one CSL smoothly from contraction to release mode when both operate in opposing directions. For this, the output $u(t)$ is been split into positive and negative parts denoting $u_+(t) = \max(0, u(t))$ and

$u_-(t) = \min(0, u(t))$, which are connected to $i_\pm$ of the other CSL, cf. Figure 5 (left). These inputs inhibit joint contractions either in positive or negative sense of rotation.

Beyond that, it has come to light that for this body configuration even a single driven hip roll joint is sufficient for balancing an upright posture while withstanding gravity and slight disturbances.

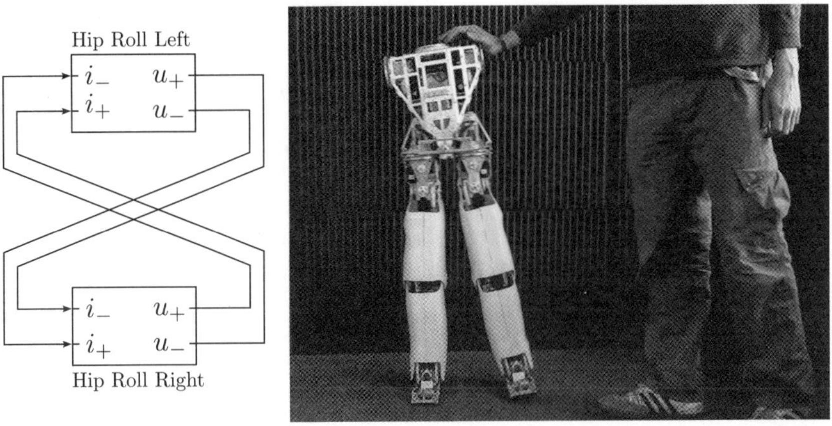

Figure 5. Left: Schematic view of crossed inhibitory connections between two CSLs controlling left and right hip motors of a humanoid robot. Right: Lower body part of Myon during balancing motion. Head and Arms are not attached. Contracting CSLs on the hip joints make the body withstand pushes or adapt its upright posture to tilted grounds.

4. Summary and Outlook

We have explained the functional principle of so-called Cognitive Sensorimotor Loops (CSLs) and have shown their benefits for the creation of stand-up motions for humanoid robots. We demonstrated how complex multi-joint motions emerge from the interplay of local joint controllers using a strictly reduced set of sensory values, namely only position and velocity information of each joint and no additional communication between the controllers.

We also approached the problem of balance recovery in a model-free manner and will take the next step to stabilize movements and disturbances within the sagittal plane making use of the arms. Further work will focus on the generation of model-free walking patterns using reactive limit cycle walking methods[7] incorporating the proposed CSLs.

Acknowledgments

This work has been partly supported by the European research project ALEAR (FP7, ICT-214856). The authors send special thanks to Mario Weidner for his contributions and suggestions regarding the stand-up of the single Myon leg.

Bibliography

1. M. Hild and M. Kubisch, Self-Exploration of Autonomous Robots Using Attractor-Based Behavior Control and ABC-Learning, in *Proceedings of the 11th Scandinavian Conference on Artificial Intelligence*, (Trondheim, Norway, 2011).
2. M. Hild, T. Siedel, C. Benckendorff, M. Kubisch and C. Thiele, Myon: Concepts and Design of a Modular Humanoid Robot Which Can Be Reassembled During Runtime, in *Proceedings of the 14th International Conference on Climbing and Walking Robots*, (Paris, France, 2011).
3. A. Schneider, Local Positive Velocity Feedback for the Movement Control of Elastic Joints in Closed Kinematic Chains: A Modelling and Simulation Study of a 2DoF Arm and a 3DoF Insect Leg (2006), Dissertation.
4. M. Hild, M. Kubisch and S. Höfer, Using Quadric-Representing Neurons (QRENs) for Real-Time Learning of an Implicit Body Model, in *Proceedings of Robotica*, (Lisbon, Portugal, 2011).
5. M. Kubisch, M. Hild and S. Höfer, Proposal of an Intrinsically Motivated System for Exploration of Sensorimotor State Spaces, in *Proceedings of the 10th International Conference on Epigenetic Robotics*, (Örenäs Slott, Sweden, 2010).
6. M. Toussaint, A Sensorimotor Map: Modulating Lateral Interactions for Anticipation and Planning, *Neural Computation* **18** (2006).
7. J. H. Solomon, M. Wisse and M. J. Hartmann, Fully Interconnected, Linear Control for Limit Cycle Walking, *Adaptive Behavior* **18** (2010).

CONTROLLING HUMAN-LIKE LOCOMOTION OF A BIPED BY A BIOLOGICALLY MOTIVATED APPROACH

J. ZHAO*, T. LUKSCH and K. BERNS

Robotics Research Lab, University of Kaiserslautern,
Kaiserslautern, 67655, Germany
**E-mail: zhao@cs.uni-kl.de*
http://rrlab.cs.uni-kl.de

Controlling biped robots in a human-like way is still in the air in robotics. Due to the existing of shortcomings of technical control methods, researchers began to try to find help from biomechanics, neuroscience, human movement analysis and other biological fields. This paper presents a biologically inspired method of controlling the locomotion of biped robots based on the neurological and biomechanical research of human walking. The control concept looks inside of the human motion control to generate an efficient, robust control approach for bipedal dynamic walking. The framework of the control approach consists of a hierarchical architecture of control network, motor patterns and reflexes that works locally and distributed to exploit the inherent dynamics of of the system. In this paper, the approach is validated on a simulated anthropomorphic biped with 21 DoFs. Several locomotion like standing, walking initiation and cyclic walking have been deeply studied. This paper will study the biomechanical aspect and motion analysis of termination of gait and its application in a biped robot. The corresponding reflexes and motor patterns will be also introduced in this paper.

Keywords: Biped Locomotion; Motor Patterns; Reflexes; Behavior-Based Control; Gait Termination.

1. Introduction

In last past decades, there are two different approaches in controlling the biped robot. The first one is called technical control approach. Relying on the collaboration of mechanical engineering, physical and mathematical modeling, classical control theories, it is widely applied in the control of robots. Though the considerable achievements over last decades, the main drawbacks and the restrictions are still clearly existing.[1]

On the other hand, the control of human motions has been evolved and optimized over millions of years in sense of energy optimization and adaptation on different terrains. From the observation of human motions, one can find that there are features existing inherently in human dynamics. Thus the application of biological concepts of human walking expands the solution on controlling biped robots.[2–5]

2. Control Architecture

Hierarchical Control Layout

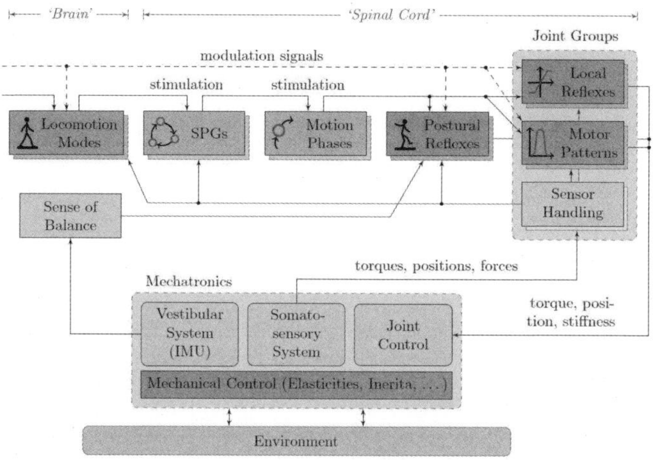

Fig. 1: Hierarchical organization of control units and the control systems, the system mechatronics and the environment[1]

Figure 1 shows the hierarchical categorization of control units based on the control levels. The locomotion modes in highest level are responsible for switching modes among different walking behaviors. *Spinal pattern generators* (SPGs) are generating the patterns for varying muscle activities. Motion phases are defined as the controller of stimulating or inhibiting lower control units. Postural reflexes regulate the posture of each joint in the current motion phase or current task. The relative lower control units, e.g. motor patterns, are to generate the motion commands for muscle groups in a way of feed-forward only directly influencing the local adjacent joints. For local and postural reflexes, the feedback control is thus implemented: local reflexes only influence locally adjacent joints based on data of neighboring sensors and introduce a tight sensor-actor coupling.

Behavior-Based Control Architecture

For controlling above mentioned units, behavior-based architectures have been established as a preferred control approach for robotic systems acting in situations and environments without any modeling effort. The behavior-based control architecture **iB2C** (**i**ntegrated **B**ehavior-**B**ased **C**ontrol) is well suited for implementing the aspired hierarchical layout, command fusion, or stimulation and inhibition mechanisms. Further details on **iB2C** and the implementation of the control system introduced above can be found in.[6]

3. Walking Scenarios

The locomotion of walking, as the standing, initiation of walking, walking rhythmic have been studied. Nevertheless, the research about termination of human gait has received little attention. The termination of rhythmic walking should be finished with accurate control to minimize the possibility of a weakly controlled stop and probability of falling forward or backward.

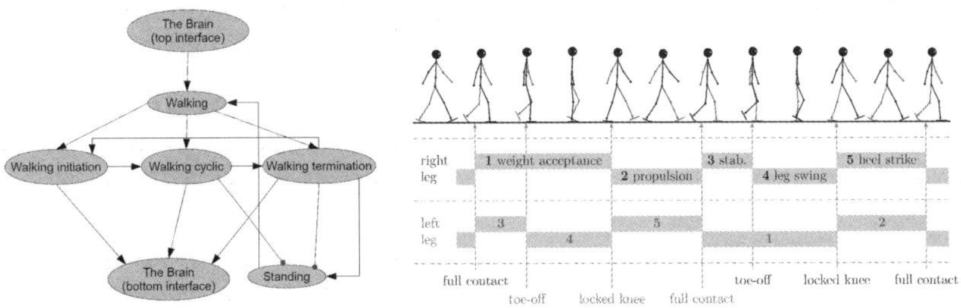

(a) State machine in the brain group

(b) Walking phases for both legs

Fig. 2: Walking modes and phases

Based on the approach mentioned previously, the commands of top level e.g. the brain should be firstly studied. Figure 2(a) shows an overview of flowchart of locomotion activities on the top level. The experiment starts from a stable standing position which contains reflexes to maintain the stability and the optimization of standing. After triggering the walking command, the walking initiation will be firstly enabled. Once the behavior of initiation is satisfied, the cyclic walking then becomes active and inhibits the standing mode at the same time (inhibiting connection with the dotted red end). The termination of gait needs to be enabled if the stop command is given. Thus another criterion should be considered consequently. That is, if the command can be directly and immediately performed by trunks and legs depends on the current walking phase of both legs. The details will be discussed later in this section. After the termination accomplished, it will automatically enable the standing mode and the loop can therefore lively perform.

As shown in Fig. 2(b), the cyclic walking can be categorized into five different phases for each leg. Termination of human gait is thus defined as the transient period starting from repetitive gait to a full stop. In the case of symmetrical feet position in termination, the gait is going to be stopped in 2 steps after transmission of the stop command if it is synchronized with the activity of heel contact (in Fig. 2(b)). Otherwise, it will produce the delay of termination of the gait. Stopping the gait before the heel contact may arise inconvenient or extra movement which is called a poorly controlled stop. The total course of leg behaviors is illustrated in the Fig. 3(a). The gait termination starts from the heel contact of either foot, and then the contralateral leg involves in the activity of swinging. After the foot of swing leg touches the ground, the last step is thus stimulated. At the last step, the swing leg has to be propelled with adequate power to find the proper position to place the foot to fit for the symmetrical or quasi-symmetrical feet placement for next activity like restarting the walk.

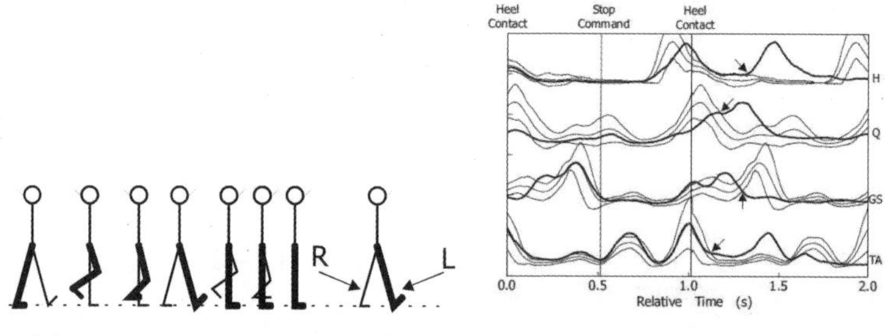

(a) Termination process of both legs (b) EMG experiments in gait termination

Fig. 3: Phases and EMG signals in the gait termination[7]

In the Fig. 3(b), electromyography experiment measures the activities of following muscles: tibialis anterior (TA), gastrocnemius and soleus (GS), quadriceps (Q) and hamstrings (HS). The biomechanical findings in[7] show that, the walking termination process consists of reduction of the toe-off for propelling the leg during the stance phase of the step before last step and the increase of the braking force at the last step of the terminal foot placement. The increasing activity in hamstrings to function as hip extensor to prevent the trunk from an over-swinging and falling forward.

4. Reflex and Motor Control for Gait Termination

As mentioned above, the gait termination is controlled by hierarchical units on the spinal cord and muscle level. The spinal cord coordinates with the control units while the reflex and motor controllers are generating the control commands for distributed joint controllers. Thus the following reflexes and motor patterns working during last step of gait termination can be observed.

- **The Leg Propel Motor Pattern** is intended to provide additional energy so the swing leg can be moved forward to synchronize the trunk motion. The motor pattern *Leg Propel* creates a plantarflexion of the ankle which is corresponding to the muscular activity of the gastrocnemius and soleus before toe-off. Compared to the motor pattern *Leg Propel* in cyclic walking, the output magnitude and duration of this activity are in a relative smaller scale. This is due to the decrease of forward velocity at the last step which needs less energy input than that in the course of cyclic walking.
- **The Lock Ankle Pitch Angle Reflex and Motor Pattern** are activated when the ankle pitch angle of stance leg is approaching the neutral angle of ankle in standing phase. Due to the body weight resting on the foot after touch-down, the leg leans to rotate about the heel contact point. Normally the tibialis anterior muscle functions to prevent the toes from too much dorsiflexion. This activity would generate resisted torques and enough stiffness to prevent the shank of stance leg from overswinging. It works also for swing leg after its touch-down event, for keeping the leg staying at its neutral posture.
- **The Hold Foot Posture Reflex** tends to control the pitch angle of ankle of the swing leg during the last step of gait termination. Since without controlling the posture of the pitch angle of ankle, the foot is inclined to be of plantarflexion. This phenomenon would induce a very uncomfortable landing of the foot which is prone to lead to a failure of gait termination. Therefore, after triggering the toe-off event during the swing of the leg at the last step, it is necessary to ensure the plantar surface to be almost parallel to

the walking surface. During this process, the resisted torque will be generated to prevent the plantarflexion of the ankle. Additionally, after the foot being of parallel to the walking surface, the stiffness of it should also be triggered.

- **The Lock Knee Reflex** is stimulated as the knee angle of stance leg accesses to the neutral angle of knee in the standing phase. Due to the inertia of the trunk and thigh, they intend to rotate about the knee joint if the ankle has been locked stiff enough. For keeping the knee to stay at the neutral angle, additional torque will be generated.
- **The Lock Hip Pitch Angle** is intended to activate when the trunk is accessing to the neutral pitch angle of hip in standing phase. It produces the torques to keep the trunk stopping rotating about the hip join in pitch axis.
- **The Lock Arm Swing Reflex** controls the swing of both arms. When the swing leg has touched with ground, the reflex should be stimulated immediately. The function of arm swing during cyclic walking is planned to stabilize the upper trunk to compensate possible trunk rotation disturbing the walking. After the happening of touch-down event, it is no need to stimulate the arm to swing any more. However, due to the existing of arm inertia, it is yet necessary to produce a small torque to compensate the rotating of the arm about the should joint and then lock it at the neutral position.
- **The Active Knee Swing Motor Pattern** provides the last propulsive energy for the swing leg during the swing phase. Since the stride distance is much shorter than normal stride in cyclic walking, the leg can be driven forward due to its inertia without energy input anymore. The knee swing will drive the foot to achieve at the standing position in a short time.
- **The Touch-down Reflex** of swing leg activates when the ground contact of the swing leg has reached at the threshold. It is intended to be the signal to switch the termination state for swing leg. If the foot of swing leg has touched down with more part than threshold, the swing leg will enter into next termination state. Then, as mentioned above, *lock ankle pitch angle reflex* and *hip pitch angle reflex*, *lock knee angle reflex* and *lock arm swing reflex* will start working on prevent the upper trunk from overswinging. The swing leg of last step during termination begins to stiffen its knee joint and ankle joint to be able to restrain the overswinging and consequent oscillations.

5. Experiments and Results

The experiments are performed on the framework of MCA2[a]. The corresponding physics engine is Newton Game Dynamics[b]. The experiments are mainly focused on the vertical force measured on the plantar surface of each foot during the cyclic walking and gait termination. From the existing test scenarios, we will set the speed of cyclic walking at 0.8 m/s. Firstly, we will measure the vertical forces on the foot before (e.g. foot full contact) and during *leg propel* of both cyclic walking and gait termination. The simulated biped is tested on above mentioned environment for five times at the given speed. Figure 4(a) and 4(b) show the average value of vertical forces on both feet of different experiments. Figure 4(a) illustrates the vertical forces imposed on the foot at the cyclic walking. The first peak indicates the ready process to get enough propulsion for the swing leg, while the next peak denotes the last big propulsion of leg propel before toe-off. Figure 4(b) has the same course as the Fig. 4(a). However it has comparative shorter time and average value. Besides, the duration of leg propel has been also decreased. Those explain that at the last step of gait termination, the propulsive force generated by leg propel is reduced compared to that of cyclic walking. The video of gait termination can be found on http://agrosy.informatik.uni-kl.de/en/robot-gallery/biped/.

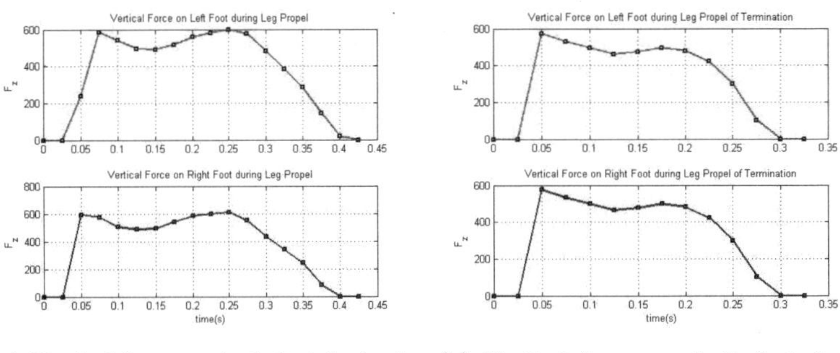

(a) Vertical forces on both feet during leg propel at cyclic walking

(b) Vertical forces on both feet during leg propel in gait termination

Fig. 4: Vertical forces on both feet at cyclic walking and gait termination

[a]http://rrlib.cs.uni-kl.de/

[b]http://newtondynamics.com/forum/newton.php

6. Conclusion and Outlook

In this paper, a biologically motivated control concept for biped locomotion has been addressed. The concept of a hierarchical control architecture, reflexes and motor patterns based on the findings of biomechanical and neuroscientific research, has been transferred into the design of a biped robot. Most control units are from biomechanical and muscles analysis of certain locomotion, but can also be derived by using mathematical optimization.[8] The complete network of walking behaviors, including standing, walking initiation, cyclic walking and walking termination has been generated.

Future work will emphasize on the theoretical study of stability analysis. Mathematically simplified and approximated model, combined with the behavior-based control, which improves the robustness and skill of walking of the biped in different environment, will also be established. Several higher walking behaviors, such as curve walking, stepping on and off stairs, will be the further walking scenarios for the biped to test the adaptation of the control methods.

References

1. T. Luksch, Human-like control of dynamically walking bipedal robots, PhD thesis, University of KaiserslauternOctober 2009.
2. S. Collins and A. Ruina, A bipedal walking robot with efficient and human-like gait, in *Proceedings of the 2005 IEEE International Conference on Robotics and Automation (ICRA)*, (Barcelona, Spain, 2005).
3. J. Pratt, Exploiting inherent robustness and natural dynamics in the control of bipedal walking robots., in *PhD thesis*, (MIT Press, Cambridge, 2000).
4. B. Vanderborght, B. Verrelst, R. V. Ham, M. V. Damme, P. Beyl and D. Lefeber, Torque and compliance control of the pneumatic artificial muscles in the biped "lucy", in *Proceedings of the 2006 IEEE International Conference on Robotics and Automation (ICRA)*, (Orland, FL, USA, 2006).
5. S. Miyakoshi, G. Taga, Y. Kuniyoshi and A. Nagakubo, Three dimensional bipedal stepping motion using neural oscillators, in *Proceedings of the International Conference on Intelligent Robots and Systems (IROS98)*, 1998.
6. M. Proetzsch, T. Luksch and K. Berns, The behaviour-based control architecture iB2C for complex robotic systems, in *Proceedings of the 30th Annual German Conference on Artificial Intelligence (KI)*, (Osnabrück, Germany, 2007).
7. P. Vanitchatchavan, Termination of human gait, in *Systems, Man and Cybernetics, 2009. SMC 2009. IEEE International Conference on*, October 2009.
8. T. Luksch, K. Berns, K. Mombaur and G. Schultz, Using optimization techniques for the design and control of fast bipeds, in *Proceedings of the 10th International Conference on Climbing and Walking Robots (CLAWAR)*, (Singapore, 2007).

MODELLING OF FLEXIBLE MANIPULATOR SYSTEMS USING BACTERIAL FORAGING ALGORITHMS

H. SUPRIYONO, M.O. TOKHI and B.A. Md ZAIN
Department of Automatic Control and Systems Engineering, The University of Sheffield, UK
Email: herusupriyono@gmail.com

This paper presents current work on application of bacterial foraging algorithms for modelling of a single-link flexible manipulator. Bacterial foraging algorithms (BFAs) with adaptable chemotactic step size are used to develop a single-input single-output model characterising a flexible manipulator from torque input to hub-angle response. A laboratory scale single-link flexible manipulator rig is used for development and validation of the models. The performances of the adaptive BFAs are compared to that of standard BFA. The comparison is made based on convergence to optimum value, the optimum value achieved, and time-domain responses of the developed models.

1. Introduction

Compared to rigid manipulators, the use of flexible manipulators have several advantages such as lower energy consumption, smaller actuator required, safer operation due to reduced inertia, compliant structure, possible elimination of gearing, less bulky design, and low mounting strength and rigidity requirements [1,2]. These advantages allow flexible manipulators to be used in various applications such as for reducing the space-launch cost in space exploration [3], handling hazardous waste material in hazardous plants where access to underground storage is limited [4], etc.

Several methods have been proposed for modelling of flexible manipulator systems such as by solving the partial differential equation (PDE) representation, and by using numerical analysis based on finite difference (FD) and finite element (FE) methods to solve the PDE [1]. Flexible manipulator systems may also be modelled using system identification techniques. For this approach, a model of the system is derived based on input-output data pairs collected at input-output terminals of the system. As a consequence, the accuracy of the model highly depends on accuracy of the collected data [5]. With system identification techniques, firstly, qualitative operation which defines the structure of the model such as type and order of the differential/difference equation that relates the input to the output has to be performed. This step is referred to as characterization. Then, numerical values of the structural parameters, which minimize the residual/error between actual system and the model, are determined. This is referred to as identification.

A number of estimation methods have been used by researchers such as least mean squares (LMS) [6] and recursive least squares (RLS) [6-8]. However, the risk of getting trapped at local minima is their main disadvantage [9,10]. To overcome this, intelligent optimisation approaches may be used in the identification step. The use of BFA [11] for modelling a single-link flexible manipulator has not been reported yet. This paper presents approaches for modelling of a single-link flexible manipulator using improved BFAs proposed in [12] using real-world data collected from an experimental rig.

2. The modelling process

2.1. *Flexible manipulator*

The flexible manipulator rig considered in this work is a laboratory-scale facility depicted in Figure 1. A shaft encoder with a resolution of 2048 pulses/revolution, manufactured by Heidenhain is used for measuring the hub-angle of the manipulator. A precision interface circuit PCL 818G is used to interface the flexible manipulator system with a computer [1,13]. The physical parameters of the flexible arm used in this work are as follows: length l = 960 mm, width w = 19.008 mm, thickness h = 3.2004 mm, mass density per unit volume ρ = 2710 $K_g m^{-3}$, the second moment of inertia I = 5.1924×10^{-11} m^4, the Young modulus E = 71×10^9 Nm^{-2}, moment of inertia I_b = 0.04862 $K_g m^2$, and I_h = 5.86×10^{-4} $K_g m^2$. In this work, the impact of gravity is neglected.

Figure 1. The laboratory-scale single-link flexible manipulator rig.

2.2. *Preliminary experimentation*

A random torque signal of $\pm$ 0.3 Nm maximum amplitude was applied as input to excite the flexible manipulator system. The experimental data thus obtained is depicted in Figure 2. From the total 6400 pairs of input/output data, the first 3400 data pairs are used for modelling and the remaining 3000 data pairs are used for validating the model.

2.3. *Model structure formulation*

Various model structures have been reported in the modelling of flexible manipulator systems, i.e. autoregressive moving average with exogenous input (ARMAX) and nonlinear ARMAX (NARMAX) [6] and autoregressive moving average (ARMA) [7, 14]. Considering the simplicity, performance and computation costs, here the ARMA structure is chosen to model the flexible manipulator system. The general ARMA model structure by neglecting the noise term can be formulated as [5]:

$$\hat{y}(k) = -\sum_{i=1}^{N} a_i \times y(k-i) + \sum_{j=0}^{M} b_j \times u(k-j) \tag{1}$$

where $a_i, (i = 1,2,3, \dots, N)$, $b_j, (j = 1,2,3, \dots, M)$, are the output and input coefficients, $N \geq M$ represents the order of the model, y is measured output, and u is input. Thus, the modelling process using ARMA model comprises determining the numerical values of coefficients a_i and b_j which result in minimum error between measured and estimated outputs, y and $\hat{y}$.

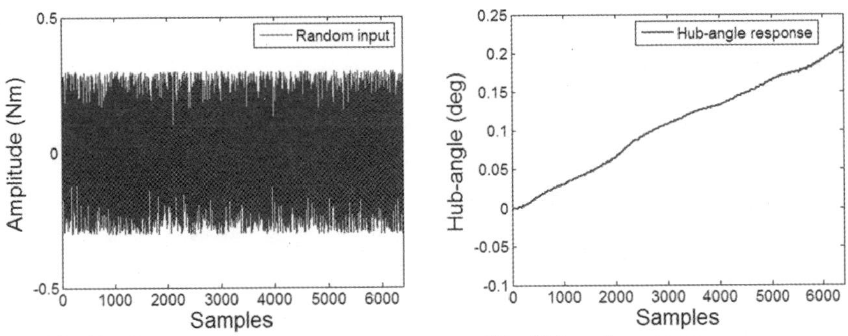

(a) Random torque input (b) Hub-angle response

Figure 2. Experimental input/output data of the flexible manipulator.

2.4. *ABFA-based modelling*

In the original BFA (Passino, 2002), here referred as standard BFA (SBFA), the mechanism of foraging can be subdivided into four steps, namely chemotaxis, swarming, reproduction, and elimination and dispersal. Chemotaxis is such a foraging behaviour that bacteria try to climb up the nutrient concentration optimally by implementing a type of biased random walk with constant value.

Making chemotactic step size adaptable is one way to improve BFA. The chemotactic step size can be made adaptive, i.e. the value of chemotactic step size changed to follow certain conditions. A step size adaptation technique makes the chemotactic step size dependent on the nutrient value; if the nutrient value is high then the step size is big and if the nutrient value is low then the step size is small. Realization of such a mechanism tends to an adaptive BFA (ABFA) that will be faster in convergence and will also be able to reach the global optimum. Three adaptation mechanisms for chemotactic step size based linear, quadratic and exponential function of cost function have been proposed and validated using benchmark functions [12]. In this work, the coefficient values of ARMA model, a_i and b_j, are determined using the ABFAs proposed in [12]. The model prediction error is used to set up the cost function to be optimised by ABFA. In this manner, the mean squared error (MSE) is used as the cost function J;

$$J = \frac{1}{n}\sum_{k=1}^{n}\left(e(k)\right)^2 \tag{2}$$

where n represents the number of data points, and $e(k)$ is the model prediction error;

$$e(k) = y(k) - \hat{y}(k) \tag{3}$$

$y(k)$ is the actual output and $\hat{y}(k)$ is the predicted output. A block diagram of the optimisation mechanism is depicted in Figure 3.

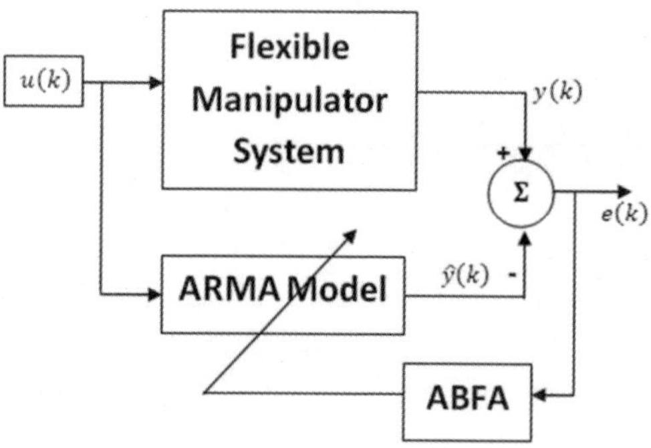

Figure 3. ABFA-based ARMA modelling of the flexible manipulator modelling.

2.5. *Model Validation*

Following the modelling phase, the validation phase is carried out. Firstly the resulted model must be stable. Secondly, the model shall predict the system response to given excitation. To validate the predictive capability of the model the experimental data is split into two sets: estimation set (modelling set) and test set (prediction set). Normally, the data set is divided into two halves. Finally, if the developed model is adequate, the errors (residuals) should contain no information about the past residuals or dynamics of the system. It can be verified by showing that the residuals are uncorrelated with all linear and nonlinear combinations of past inputs and outputs. These types of tests are very complex in nature and are difficult to realise in practice. However, the correlation tests described as below are used [15].

$$\begin{aligned} \phi_{\varepsilon\varepsilon}(\tau) &= E[\varepsilon(t-\tau)\varepsilon(t)] = \delta(\tau) \\ \phi_{u\varepsilon}(\tau) &= E[u(t-\tau)\varepsilon(t)] = 0, \ \ \forall\tau \end{aligned} \tag{4}$$

The correlations will never be exactly zero for all lags and the 95% confidence bands are used to indicate if the estimated correlations are significant or not.

3. Results and discussion

For hub-angle modelling, a 6th order ($N = 6, M = 5$) ARMA model was selected after investigations with various model orders for best results. The first 3400 data points were used in the modelling phase. The convergence plots in Figure 4(a) shows that ABFAs were able to converge faster to the optimum value than the SBFA. The optimum cost function values achieved were 14.323×10^{-7}, 3.5950×10^{-7}, 4.1589×10^{-7} and 1.1807×10^{-7} with SBFA, LABFA, QABFA and EABFA respectively. Thus, it can be noted that the ABFAs outperformed SBFA in the ARMA-based hub-angle modelling process. The time-domain hub-angle responses depicted in Figure 4(b) show that the responses of all models resulted by BFAs were able to mimic the actual system output.

The hub-angle models were validated using the remaining 3000 pairs of input – output data. The cost function values for hub-angle validating process achieved were 4.2637×10^{-5}, 1.2273×10^{-5}, 1.1292×10^{-5} and 1.1836×10^{-5} for SBFA-based model, LABFA-based model, QABFA-based model and EABFA-based model respectively. These cost function values suggest that ABFAs-based models achieved better validating results than SBFA-based model. The time-domain hub-angle responses for the validating phase depicted in Figure 4(c) show that all the models were able to predict the measured output well. It was noted from pole-zero analysis and correlation tests that all models were stable; all poles of the transfer functions were inside the unit circle with one or more zeros outside the unit circle indicating non-minimum phase behaviour. Correlation tests depicted in Figure 5 – Figure 8 show that all the functions were within the 95 % confidence interval. The discrete transfer functions constructed from the optimal parameters resulted by the algorithms with sample time of 0.001 seconds are given as:

$$H(z)_{SBFA} = \frac{-0.0028z^5+0.0000135z^4-0.003z^3-0.0017z^2+0.0046z+0.0011}{z^6-0.2281z^5-0.2285z^4-0.0739z^3-0.1651z^2-0.1654z-0.1347} \tag{5}$$

$$H(z)_{LABFA} = \frac{-0.000137z^5+0.0028z^4+0.000237z^3+0.000152z^2+0.000595z+0.000924}{z^6-0.1322z^5-0.2964z^4-0.0265z^3-0.3079z^2-0.1400z-0.0965} \tag{6}$$

$$H(z)_{QABFA} = \frac{0.0000124z^5-0.0020z^4+0.0015z^3+0.000364z^2-0.0002098z-0.0015}{z^6-0.1810z^5-0.2307z^4-0.2065z^3-0.1336z^2-0.1569z-0.0901} \tag{7}$$

$$H(z)_{EABFA} = \frac{-0.000316z^5+0.000553z^4-0.000202z^3+0.000342z^2+0.000529z+0.000357}{z^6-0.1843z^5-0.1628z^4-0.1137z^3-0.362z^2-0.1444z-0.0326} \tag{8}$$

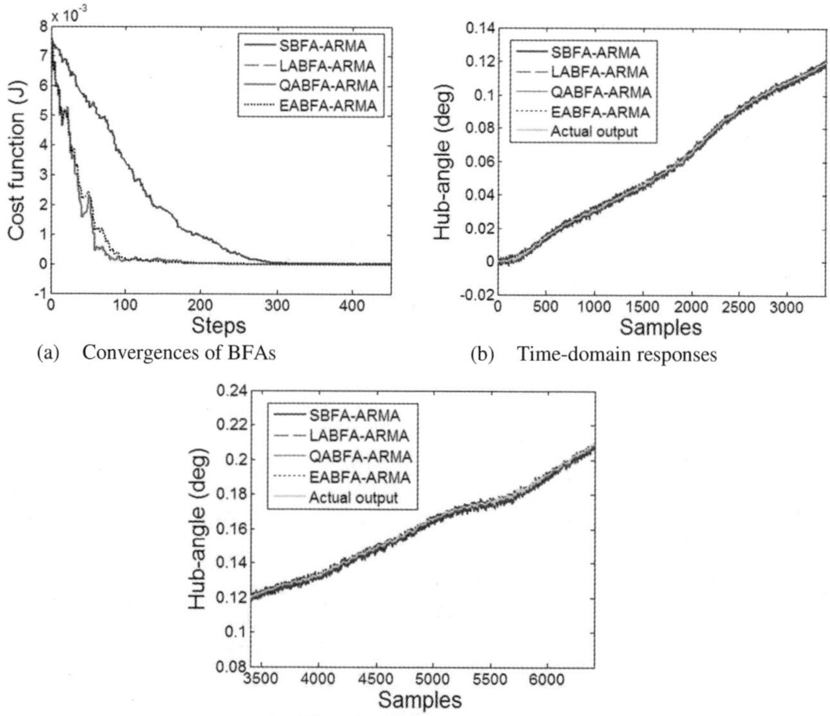

(a) Convergences of BFAs

(b) Time-domain responses

(c) Time domain responses in validation

Figure 4. Simulation results in modelling and validation phase.

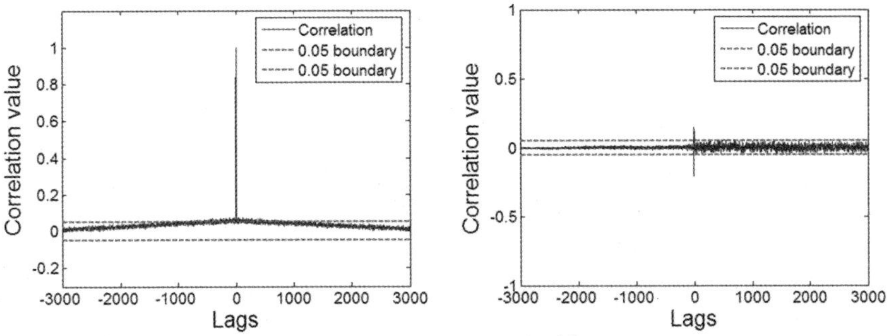

(a) $\phi_{\varepsilon\varepsilon}(\tau)$ (b) $\phi_{u\varepsilon}(\tau)$

Figure 5. Correlation tests for SBFA-ARMA

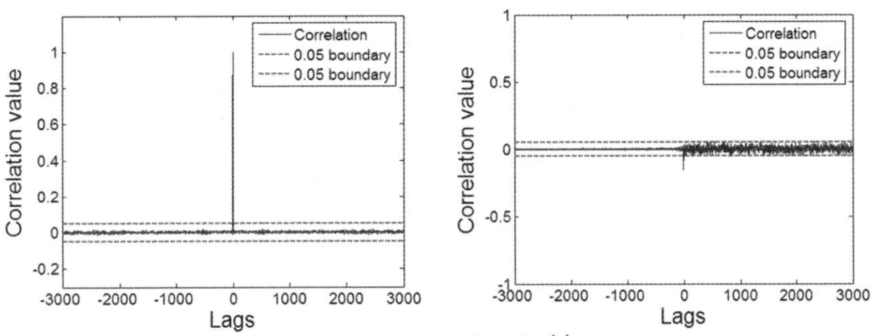

(a) $\phi_{\varepsilon\varepsilon}(\tau)$ (b) $\phi_{u\varepsilon}(\tau)$

Figure 6. Correlation tests LABFA – ARMA

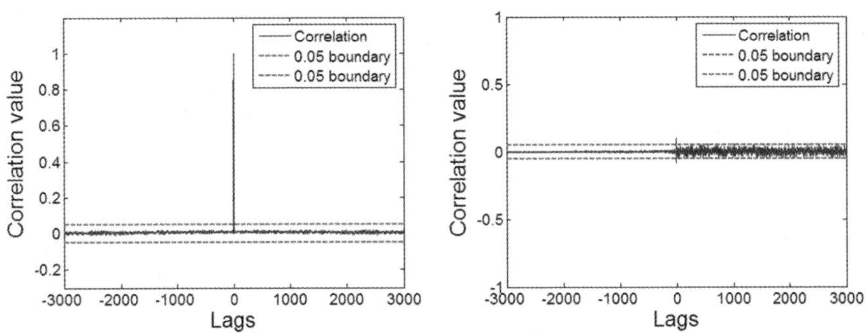

(a) $\phi_{\varepsilon\varepsilon}(\tau)$ (b) $\phi_{u\varepsilon}(\tau)$

Figure 7. Correlation tests for QABFA – ARMA

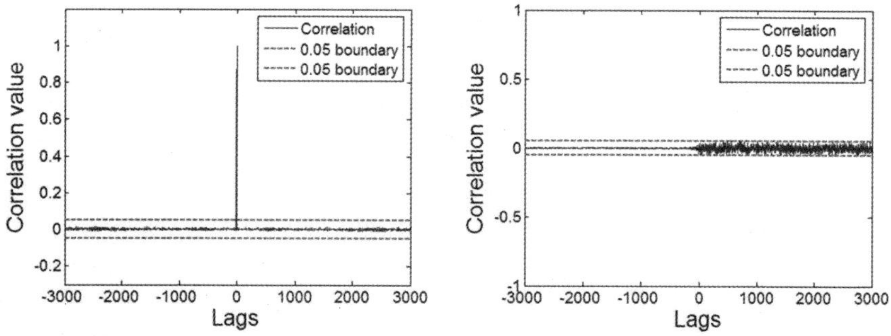

(a) $\phi_{\varepsilon\varepsilon}(\tau)$ (b) $\phi_{u\varepsilon}(\tau)$

Figure 8. Correlation tests for EABFA – ARMA

4. Conclusion

In this work, BFAs with adaptable chemotactic step size have been adopted for modelling a single-link flexible manipulator system. Input-output data pairs have been collected from an experimental single-link flexible manipulator rig and used in developing linear models of the system from input torque to hub-angle. The performances of ABFAs have been assessed in comparison to SBFA. It has been demonstrated that the ABFAs converge faster and achieve better optimum value, and thus better model than SBFA. These modelling approaches can be used in other areas where input-output data pairs are available. ABFAs are general optimisation techniques and can also be used in other areas including control of multi-joint flexible manipulators. Future work, will consider combination of BFAs and other intelligent methods such as NN and FL for modelling of flexible manipulator systems.

Acknowledgement

Heru Supriyono acknowledges the financial support of National Education Department of Republic of Indonesia, and Muhammadiyah University of Surakarta (UMS), Indonesia.

References

1. A. K. M. Azad, *PhD thesis, Department of Automatic Control and Systems Engineering, The University of Sheffield, UK.* (1994).
2. W. J. Book and M. Majette, *Tran. ASME J. Dyn. Sys., Meas. and Cont.* **105**, 245 (1983).
3. M. Yamano, A. Konno and M. Uchiyama, *Proc. IEEE Int. Conf. Rob. Aut.* **482** (2000).

4. M. Jamshidi, T. M. Akbarzadeh and N. Vadie, *Tech. Rep. WERC 1207, New Mexico Waste-Management Education and Research Consortium, USA*, (1998).
5. L. Ljung, *System identification: theory for the user, 2nd edition*, Englewood Cliffs, NJ: Prentice hall, (1999).
6. M. H. Shaheed and M. O. Tokhi, *Robotica.* **20**, 93 (2002).
7. B. A. Md Zain, M. O. Tokhi and S. Md Salleh, *Proc. 2009 Third UKSim Eur. Symp. Comp. Mod. and Sim.* **373** (2009).
8. H. Poerwanto, *PhD thesis, Department of Automatic Control and Systems Engineering, The University of Sheffield, UK,* (1998).
9. C. Goodwin, and K. Sin, *Adaptive filtering prediction control*, Prentice-Hall Inc. Englewood Cliffs, New Jersey, USA, (1984).
10. S. Haykin, *Adaptive filter theory*, Prentice-Hall Inc., Englewood Cliffs, New Jersey, USA, (1991).
11. K. M. Passino, *IEEE Cont. Sys. Mag*. June, **52** (2002).
12. H. Supriyono and M. O. Tokhi, *Proc. Sec. Int. Conf. Comp. Intel., Com. Sys. Net. 2010 (CICSyN2010).* **72** (2010).
13. M. O. Tokhi and A. K. M. Azad, *Robotica.* **15**, 283 (1997).
14. M. S. Alam and M. O. Tokhi, *J. Low Freq. Noise, Vib. and Act. Con.* **26**, 57 (2007).
15. S. A. Billings and W. S. F. Voon, *Int. J. Con.* **44,** 235 (1986).

DEVELOPMENT OF AN OMNIDIRECTIONAL WALL-CLIMBING ROBOT BY USING TRAVELING WAVE PROPAGATION

TARO NAKAMURA and TORU HAYAKAWA

Faculty of Science and Engineering, Department of Precision Mechanics
Chuo University 1-13-27 Kasuga, Bunkyou-ku, Tokyo 112-8551, Japan

A wall-climbing robot can access high places and approach disaster sites that a human cannot. In this study, an effective omnidirectional wall-climbing robot is developed using the locomotion mechanism of snails. A snail moves by generating travelling waves with its muscle elasticity and propagating the wave in the direction of movement. Since this locomotion mechanism is in contact with a large ground area, it is extremely stable. This is advantageous for climbing unstable walls. The proposed robot uses a locomotion mechanism that consists of several slider cranks for locomotion and a centrifugal fan to adhere to the wall by suction. The robot can be moved from the floor to a wall to the ceiling using these two locomotion mechanisms. In addition, the robot moves omnidirectionally by rotating the locomotion mechanism. We confirmed the robot's abilities on various walls. The experiment results showed good performance.

1. Introduction

Recently, robots that can replace humans in many activities, contributing to safety, reducing costs, and increasing efficiency have been developed. For example, a wall-climbing robot has the ability to access high places and approach disaster sites that a human cannot. Climbing unstable walls requires high stability. Since a snail can crawl on non-horizontal planes, we adopted its locomotion mechanism for the robot developed in this study.

The proposed robot uses a locomotion mechanism that consisted of several slider cranks for locomotion [1], and a centrifugal fan to adhere to the wall by suction [2-3]. The robot can be moved from the floor to a wall to the ceiling using these two locomotion mechanisms. In addition, the robot moved omnidirectionally by rotating the locomotion mechanism.

2. Locomotion mechanism of a snail

A snail moves by propagating travelling waves. By extending and contracting its muscles, a small part of the snail's body leaves the ground surface and sends the aperture between the snail's body and the surface forward like a wave, thereby advancing the snail. Figure 1 shows how a snail propagates the wave.

The advantage of this locomotion is that it is extremely stable. There are three reasons for this stability. First, because the contact area is large, the load is dispersed. Second, because the area in transition is small, the change in the load of the contact area is also small. Third, the aperture created during movement is small; thus, there is little possibility of slipping off a surface[4-5].

The locomotion of a snail is advantageous for surfaces like walls. Therefore, we adopted travelling wave propagation as the developed wall-climbing robot's locomotion mechanism.

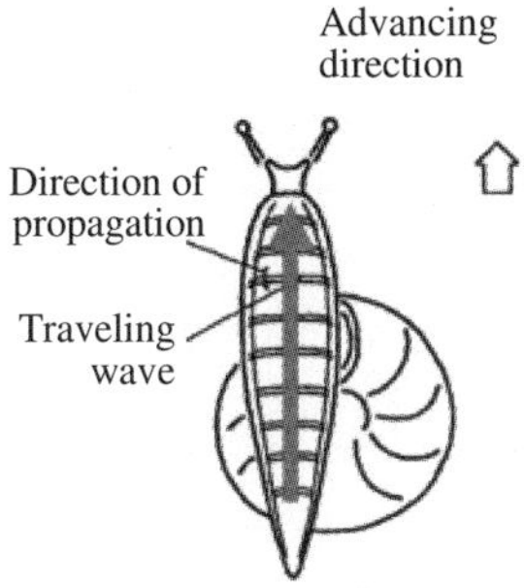

Figure 1. Snail locomotion mechanism. [1]

3. Design of the wall-climbing robot

Figure 2 shows an overview of the wall-climbing robot. Then, Table I shows specifications of this robot. The robot is composed of two identical locomotion units connected by an active joint part using motors. Since this robot has two units, it can move from the floor to a wall to the ceiling. Each locomotion un it consists of two parts: one part contains several slider cranks for locomotion and the other contains a fan for suction.

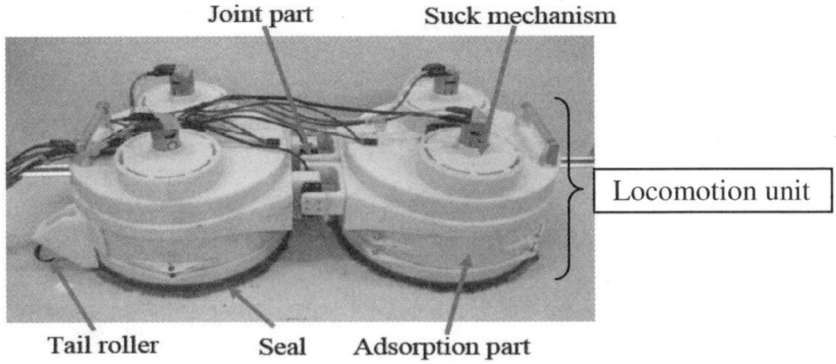

Figure 2. Proposed Omnidirectional wall-climbing robot.

Table 1 Specifications of the robot

Width [mm]	245
Depth [mm]	530
Height [mm]	190
Weight [kg]	4.6

3.1. *Locomotion mechanism unit using several slider cranks*

When the travelling wave is created, tracks of the point on the surface that is draw the elliptical orbit. We achieved an elliptical orbit by using a slider–crank chain. The slider–crank chain transforms the rotary motion of a crank into an elliptical locomotion at the head of the arm, as shown in Fig. 3. Different phases are then added to the motion along the elliptical orbit, and the wave is propagated by a combination of the differences in phases. Figure 4 shows the wave propagates by the coordinated action of slider cranks [1]. Power is transmitted to the adjoining slider cranks via gears. The locomotion mechanism unit is composed of seven worm drive gears that are driven by a single gear motor, as shown in Fig. 5.

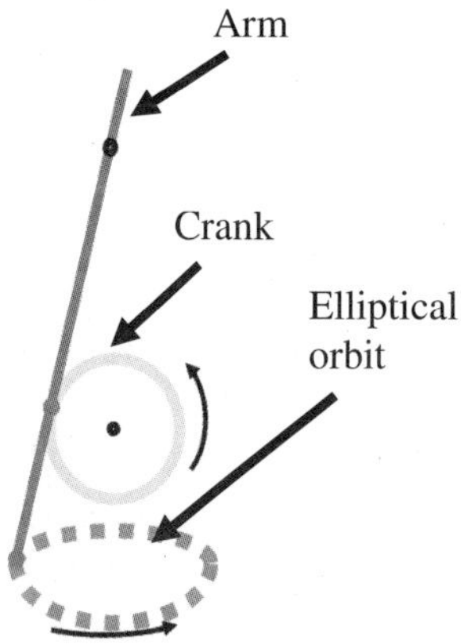

Figure 3. Motion of the slider

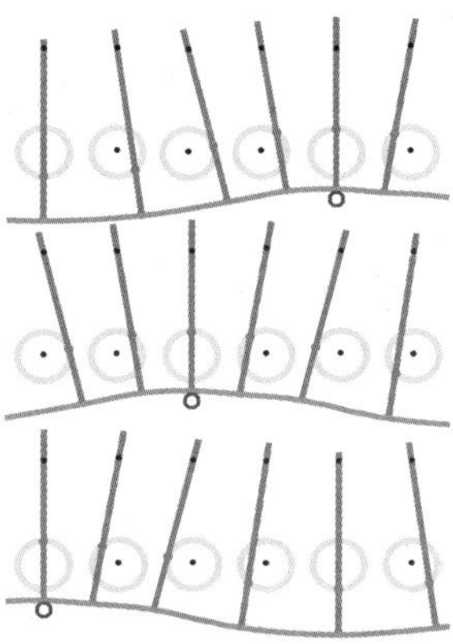

Figure 4. Slider cranks for the locomotion mechanism.

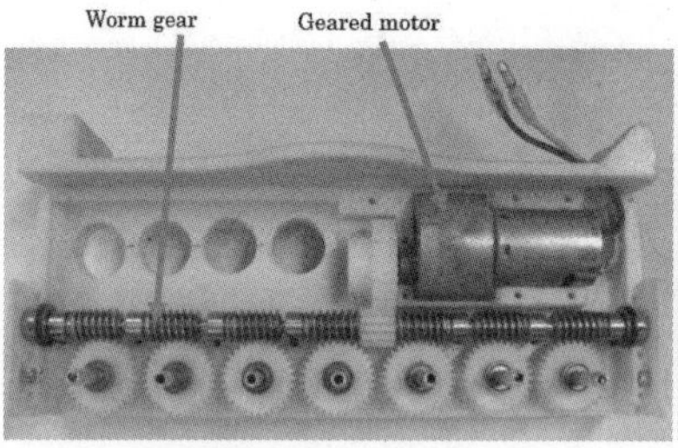

(a) Inner structure

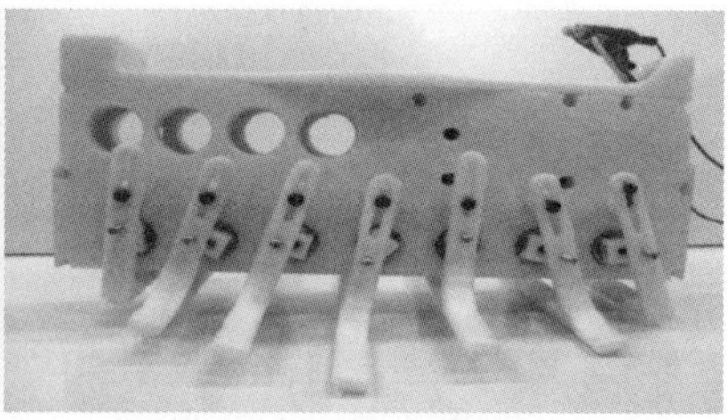

(b) Multi slider clanks

Figure 5. Locomotion mechanism.

3.2. *Suction mechanism using a centrifugal fan*

Figure 6 shows schematic view of the adsorption system. Suction achieved using a centrifugal fan was adopted as an effective mechanism for holding the robot to a wall [2-3]. The fan discharges internal air and generates a pressure differential, which allows the robot adhere to a wall. Thus, by discharging an amount of air greater than that flowing in, the adhering condition is maintainable, even if some openings appear. This is advantageous in an environment where openings occur, such as on coarse or grooved surfaces.

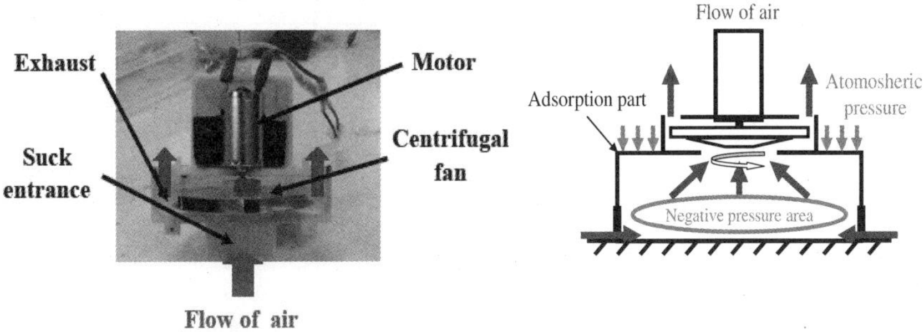

(a) Cross section of suck mechanism (b) A structure of suck mechanism

Figure 6 Schematic view of the adsorption system

3.3. *Omnidirectional locomotion mechanism*

The robot's omnidirectional locomotion mechanism is shown in Fig. 7. As noted in the figure, the locomotion mechanism in the adsorption part can be rotated by servomotors.

Figure 8(a) shows the omnidirectional locomotion method of the robot. The robot can be moved in the desired direction by rotating both locomotion mechanisms mentioned in 3.1. Further, Figure 8(b) shows the turning locomotion method of the robot. The robot can turn right and left by symmetrically arranging the rotation angle of the unit.

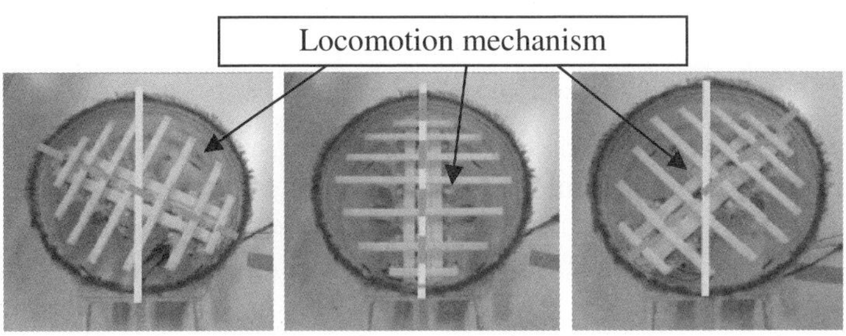

(a) Reverse rotation (b) Default position (c) Normal rotation

Figure 7. Omnidirectional locomotion mechanism.

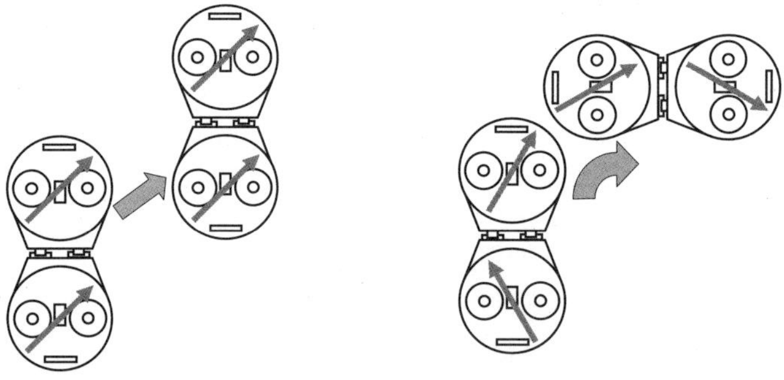

(a) Omnidirectional motion (b) Turning motion

Figure 8. Omnidirectional locomotion method

3.4. *Lifting motion for moving from the floor to a wall to the ceiling*

The robot's lifting motion for moving from the floor to walls to the ceiling is shown in Fig. 9. Since the motor is attached to lift arm, it can lift the other locomotion unit. The lifted unit adsorbs other walls and ceilings. Further, The other unit supports the unit lifted by adsorbing the floor.

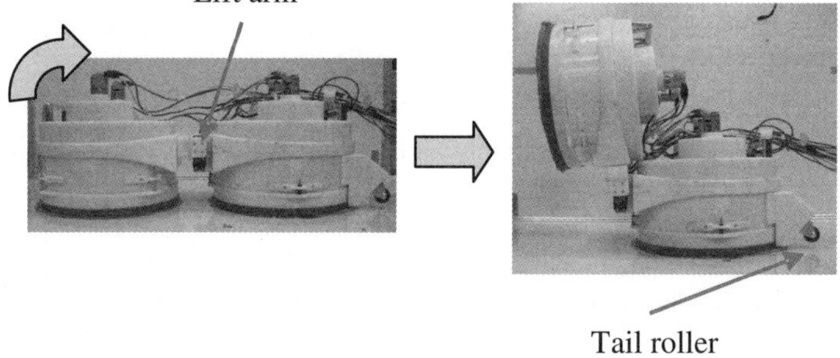

Figure 9 Overview of lifting motion

4. Experimental Results and discussion

The developed wall-climbing robot was tested and the climbing velocity was timed. Figure 10 shows the robot climbing on various wall surfaces. The robot successfully climbed on (1) a smooth surface, (2) a glass window, (3) 100 × 100 mm smooth square tiles (4) and 100 × 100 mm square tiles with grooves 3 mm wide and 2 mm deep. Figure 11 shows the climbing velocities of the robot on various wall surfaces. From these figures, we can see that the robot climbs faster on rough tiles, because its friction coefficient is greater than that of other walls.

Figure 12 shows omnidirectional locomotion trajectory of the robot on the horizontal plane. We confirmed that the robot can move in the desired direction, as shown in Fig.8(a). Figure 13 shows turning locomotion trajectory of the

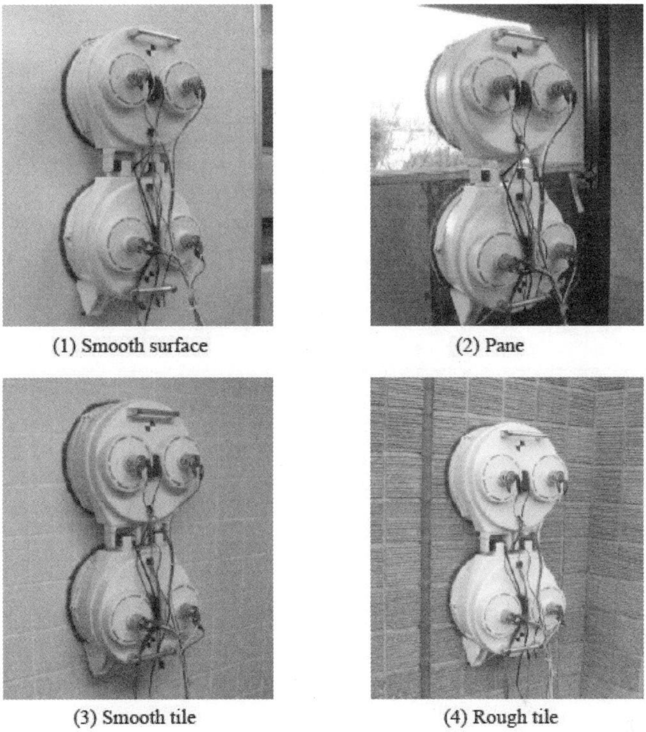

Figure10. Exhibit of robot climbing various wall surfaces.

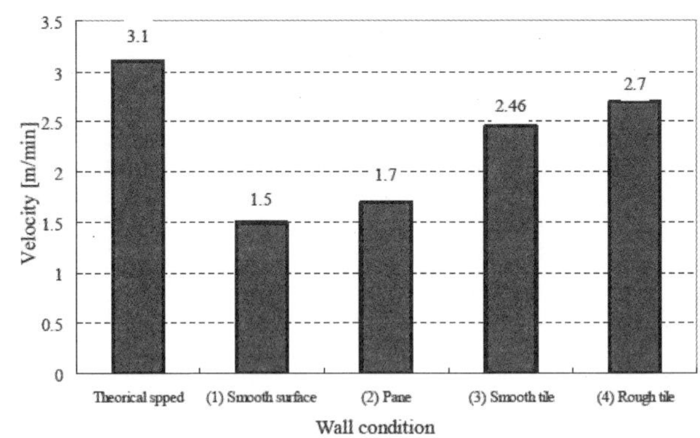

Figure 11. Experimental climbing velocities of the robot on various wall surfaces.

robot on the horizontal plane. From this figure, we can see that the robot can move on the desirable circle track precisely. In addition, we examined how the robot moved from the floor to the wall, and Fig. 14 demonstrates the transfer between these planes.

5. Conclusions

We proposed and developed the omnidirectional wall climbing robot which can move from the floor to walls to the ceiling. The experimental results showed good performance. In the future, we are going to apply to move on the ceiling. And the battery and the controller will be installed in the robot.

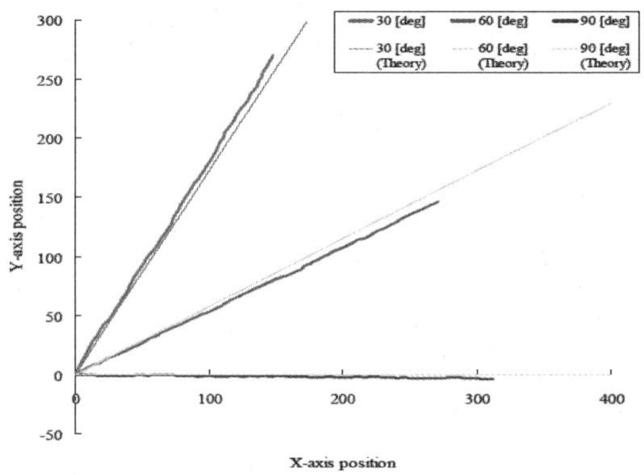

Figure 12. Tracking of robot locomotion on the horizontal plane.

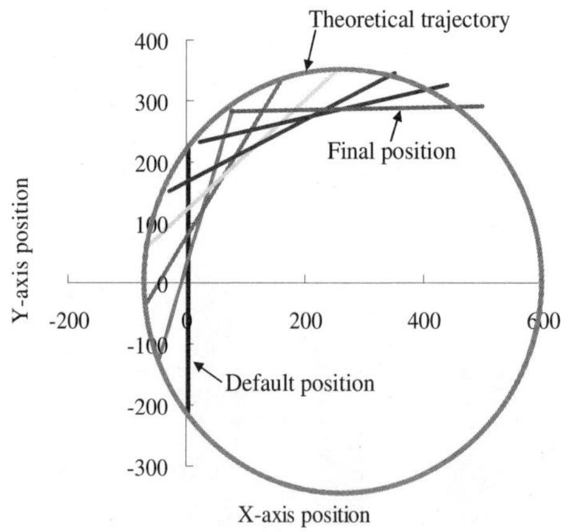

Figure13 Trajectory of turning motion in horizontal plane

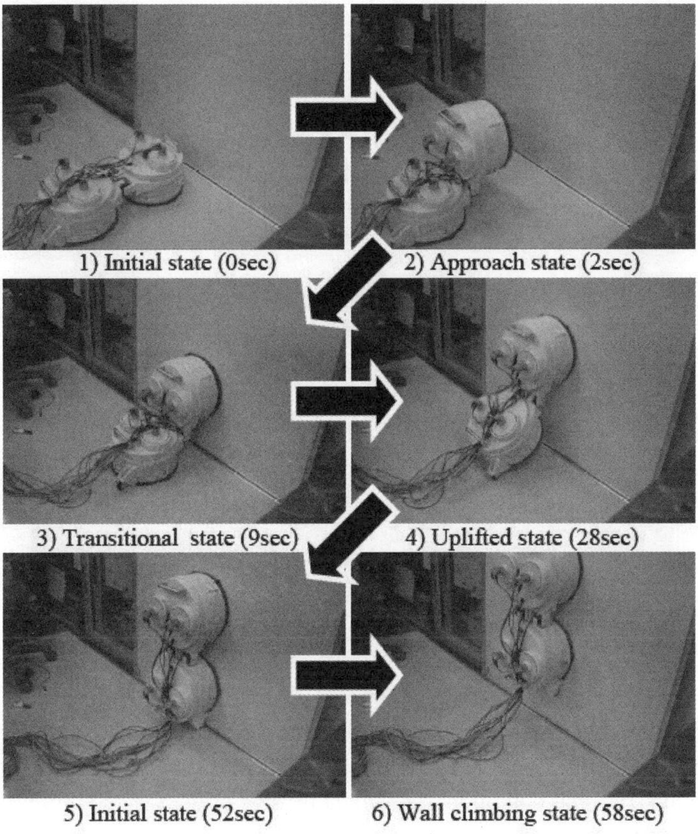

Figure 14. Demonstration of robot movement from floor to wall.

References

(1) T. Hayakawa and T. Nakamura, Development of a wave propagation type wall climbing robot using slider cranks, Proceedings of 12th International Conference on Climbing and Walking Robots and the Support Technologies for Mobile Machines, pp.225-232, 2009

(2) Daijun Xu et.al, Suction Ability Analyses of a Novel Wall Climbing Robot,2006 *Proceedings of IEEE International Conference on Robotics and Biomimetics*, pp.1506-1511, 2006

(3) Jizhong Xiao and Ali Sadegh,Modular Wall Climbing Robots with Transition Capability, 2005 *Proceedings of IEEE International Conference on Robotics and Biomimetics*, pp.246-250, 2005.

(4) Taro Nakamura and Kuniaki Satoh, "Development of an Omni-Directional Mobile Robot Using Travelling Waves Based on Snail Locomotion", *International Journal of Industrial Robot* Vol.35 No.3 , pp.206-210, (2008)

(5) Taro Nakamura and Kuniaki Satoh, "Locomotion Strategies for an Omni-Directional Mobile Robot Using Traveling Waves Propagation", *Proceedings of* 2010 *IEEE International Conference on Robotics and Automaton*, pp3769-3774., 2010.

STUDY OF ANT LOCOMOTION IN SURFACE TRANSITIONS FOR CLIMBING ROBOT DESIGN

AUSTIN LOZANO, GREGORY PETERS, DIKAI LIU & KENNETH WALDRON

Centre for Autonomous Systems (CAS), University of Technology Sydney (UTS), 15 Broadway, Ultimo, NSW 2007, Australia

Many climbing robots designed based on simple representations of the arthropod body have proven to be able to move on relatively flat surfaces and simple environments. However these robots are not able to navigate in complex environments such as steel bridges. This paper presents experimental studies on ant locomotion, particularly focusing on ant leg gait and posture when traversing a variety of complex surfaces, with the aim of providing climbing robot designers a better understanding of biological ant locomotion through a complex terrain.

1. Introduction

Ever increasing standards in occupational health and safety have resulted in preventing human workers from working in hazardous and dangerous environments. One such job is the vital periodical inspection for paint degradation, rust and condition assessment of steel bridges. Such bridges feature complex geometry, confined spaces, difficult access coupled with precarious heights and the constant danger of moving vehicles.

Recently a UTS research group developed a proof-of-concept prototype legged climbing robot, with the intention of performing steel bridge inspection exercises [7]. This robot demonstrated feasibility, however identified further research, particularly in the areas of surface transition technique. The success of steel bridge inspection by robots is largely determined by the robot's ability to safely and effectively navigate the complex structure which consists of girders, beams and trusses. There is currently no such robot capable of reliable navigation in these complex environments. For designers of legged climbing robots it is of critical importance to understand how the robot will move within the complex environment.

This preliminary study aims to provide a better understanding of how a biological system, such as arthropods, works transferring among different surfaces In this research, the Australian Meat Ant (*Iridomyrmex Purpureus*) is chosen for study of locomotion and body posture in transitions among various surfaces. The ant was selected for analysis as they are known for capability and consistent maneuverability throughout entire complex structures. Previous studies have focused on the ability of arthropods to transition over rough ground terrain [3][4][8] and large gaps [1]. Many control strategies for such terrain have been explored, of which some suggest avoiding deliberately planned reference trajectories [4]. Surface transition remains a substantial research challenge.

A series of surface transitions are studied with the analysis of the leg coordination and body posture. This information may be used in climbing robot design and development, preluding a future study of the underlying control techniques.

1.1. *Experimental Setup*

The ants were contained within a clear Perspex enclosure that was 100mm in length, 20mm high and 20mm deep. This small enclosure provided a confined space to allow capture of walking gaits, surface transitions and climbing behavior. The ants were filmed using a high speed camera (150fps – 1000fps) fitted with a macro lens with limited depth of field. The enclosure was illuminated with high intensity lighting. The rear wall of the enclosure was fitted with a 1mm square grid pattern for measurements. Small neodymium magnets were positioned in various forms to generate a range of obstacles.

2. Surface Transition

In this experimental study of ant locomotion in various surface transitions, 72 transition sequences through 21 transitions were captured. These are presented through several key frames, where the moving legs are colored[a]. It is found that normal and slow movements by ants utilise individual leg sequences. Ants also perform the tetrapod gait, which has previously been proven to be undesirable in robotic applications in respect to stability, speed and power consumption [2][6]. At high speeds, ants adopt the tripod gait.

2.1. *Transition between horizontal and vertical surfaces*

Observations of ants transitioning from a horizontal to a vertical surface demonstrate that ants assume a rearing stance. The front legs and middle legs are placed onto the vertical surface whilst the rear legs compress to prepare to push

[a] Leg abbreviations and colours for leg sequence figures: L1 [aqua blue], left front leg; L2 [cerulean blue], left middle leg; L3 [dark blue], left rear leg; R1 [yellow], right front leg; R2 [peach], right middle leg; and R3 [red], right rear leg.

the body upwards. This sequence is shown in Figure 1 through the 11 frames (a-k) detailing step by step, the ant's method in scaling the wall from a horizontal rest position [5].

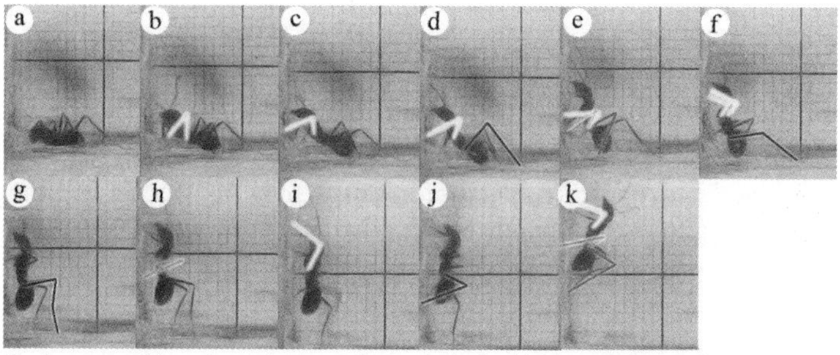

Figure 1 Surface transition from horizontal to vertical: (a) Ant prepares the transition; (b) L1 prepares; (c) L1 reaches vertical surface; (d) R1 & L3 move simultaneously. R1 attaches to vertical wall (e) R2, L2, R3 move as one. Middle pair attaches to vertical surface & R3 compresses for rearing; (f) Front pair reach up surface & L3 starts compressing; (g) L3 moves close to vertical surface while ant pulls body up wall using front legs; (h) L2 moves further up vertical surface; (i) Ant reaches further up by moving L1 while pushing with R3; (j) Ant places L3 onto vertical surface while pushing with R3; (k) Ant brings R3 onto vertical surface. L2 & R1 move simultaneously up.

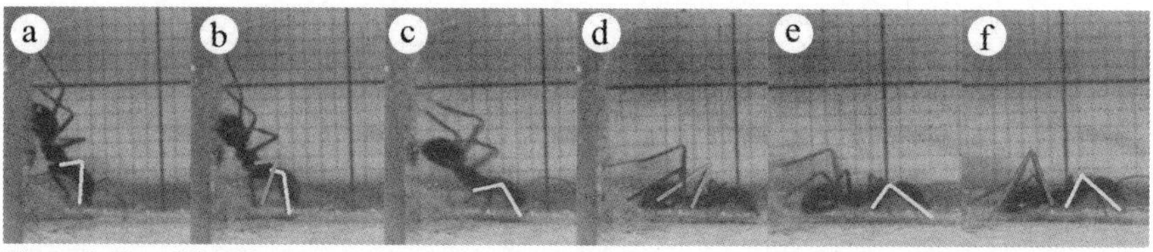

Figure 2. Transition from vertical to horizontal surfaces: (a) L1 leg is placed on horizontal; (b) L2 is placed on horizontal, then L1 is placed over it; (c) R1 & R2 are placed onto horizontal surface. R2 is placed over R1 leg; (d) L2 & L3 legs are placed onto horizontal surface; (e) R3 leg pushes off wall and R1 is placed further forward; (f) R3 on horizontal and L1 is placed forward

Surface transition from a vertical plane back to a horizontal plane is also shown in Figure 2.It is noted that the ant requires a significant number more steps to navigate up the vertical surface than to navigate down to the horizontal surface. In Fig. 1 the ant is required to position its centre of gravity towards the vertical surface before it can continue to climb up whereas in climbing down, the centre of gravity is easily controlled.

2.2. *Vertical Walking*

Ant climbing vertically upwards (Figure 3) and downwards (Figure 4) are studied. There is a significant number more steps involved in the ant walking up the wall as opposed to walking down it. This again is due to the nature of controlling the ant's centre of gravity. Therefore more time and steps are required for ensuring the ants body is in the correct position to progress up the wall. Recordings and analysis of ants walking inverted along the ceiling of the enclosure show that the locomotion gait sequences employed are indistinguishable from vertical climbing. It is recommended that the similar control methods for walking up the wall are employed when walking upside down.

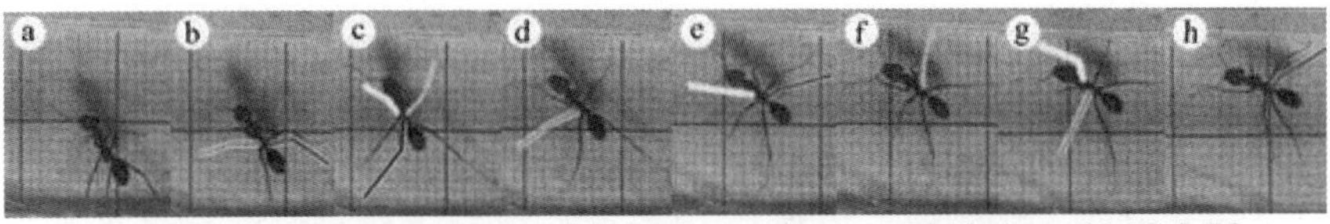

Figure 3. Ant locomotion sequence up a wall: (a) Ant resting on vertical plane; (b) L2 & R3 move forward; (c) L1, L3 and R2 perform tripod gait; (d) L2 moves forward; (e) L1 & R3 move body forward; (f) R2 leg is used; (g) L2 & R1 used; (h) R3 brings ant to start position.

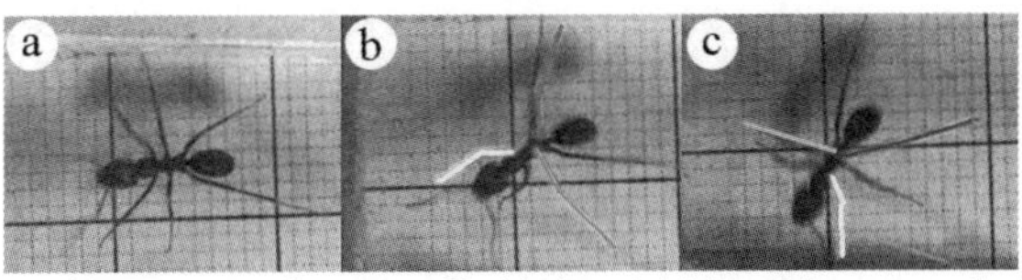

Figure 4.Ant locomotion sequence down a wall: (a) Ants initial position on vertical surface; (b) Ant uses standard tripod gait for movement; (c) Second part of the tripod gait executed.

2.3. *Transition between inverted and vertical surfaces*

Planar transition from a vertical position to an inverted position (Figure 5) and the reverse transition from an inverted to a vertical plane (Figure 6) are recorded. These two sequences involve the similar number of steps to complete. The major difference between the two is that when the ant moves to the ceiling position, at first one leg is positioned onto the inverted surface, followed by pairs of legs and finally the last leg is used. The technique is reversed when climbing down from the ceiling onto the wall. That is firstly the front leg pair is positioned onto the wall, followed by individual placement of the middle legs, and finally the rear leg pair.

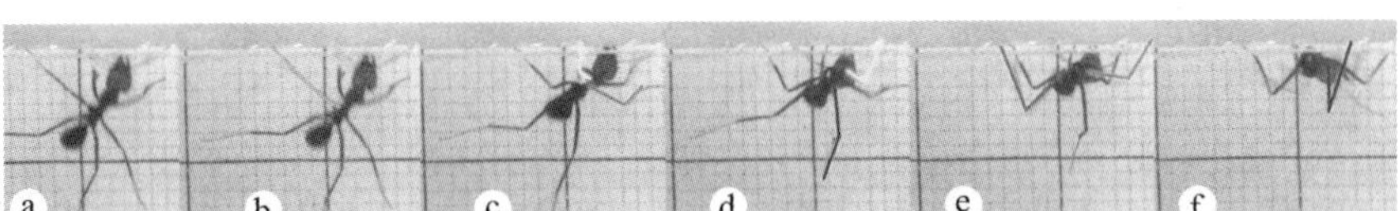

Figure 5.Surface transition from vertical to inverted: (a) Initial position before transition; (b) R2 placed on ceiling surface; (c) R1 is next on the ceiling & L2 placed further up vertical surface; (d) L1 is next placed on the ceiling, L3 moves up wall; (e) L2 & R3 is next placed onto the ceiling; (f) Lastly L3 is placed on the ceiling.

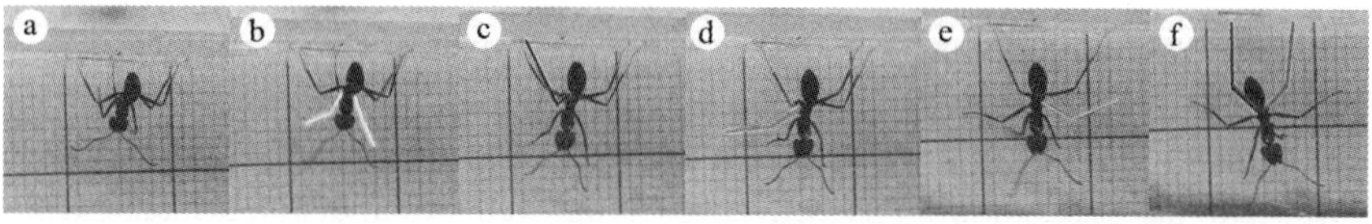

Figure 6.Surface transition from inverted to vertical: (a) Ant begins to rear for transition; (b) Firstly front legs are placed onto vertical surface; (c) R3 leg placed closer to vertical surface; (d) R2 leg place onto vertical surface; (e) L2 leg next placed onto the vertical surface; (f) R3 & L3 legs finally placed onto vertical surface.

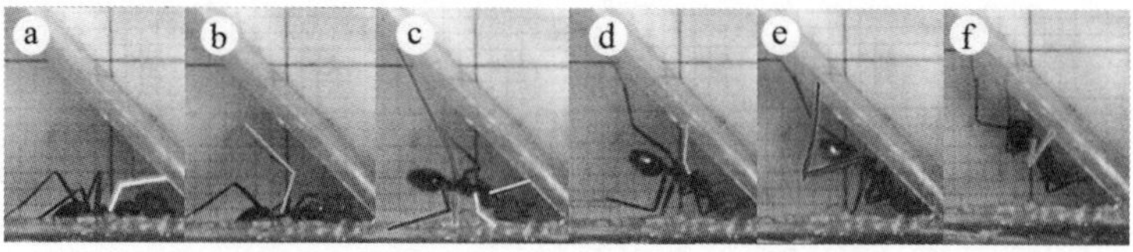

Figure 7. Transition from horizontal to an angled plane: (a) R1 is placed onto angled plane; (b) R2 leg is placed onto a side wall; (c) R3 onto angled plane as L3, L2 L1 pushes body up and R1 pulls body onto angled plane; (d) R2 is placed onto angled plane; (e) L3 is placed onto the angled plane with together L1 and R3; (f) Lastly L2 leg is placed onto the angled surface.

2.4. *Angled Transition*

Transitioning from horizontal to a 45° plane was analyzed (Figure 7). It is observed that the ants in this experiment always utilised the side walls during transition, with the ant never executing direct planar transition from horizontal to the angled plane. The possible reason could be that the ant body finds it easier to adjust its centre of gravity through twisting its body around compared to direct planar transition. Direct planar transition would require the ant's rear legs to compress while the front and middle legs actively position the body. Climbing robots facing this style of surface transition will find this approach difficult.

3. Climbing Barriers

A range of obstacles / barriers are built in this experiment. These barriers simulate common geometry on steel bridge structures, including rivet heads, webs, gusset plates and stringers. The ants were observed to easily pass over the barriers without any problems. A small single obstacle (1.8mm in height) showed no noticeable difference to walking on normal horizontal and flat plane. A wall with the height of 3.6mm was also evaluated as shown in Figure 8. In both the cases, the ant body was pitched when approaching the barrier. The main difference is that the ant slows down as it approaches the higher obstacle.

Figure 8. Locomotion over a 3.6mm obstacle: (a) Ant L1 to top surface of bump; (b) R1 leg placed on bump; (c) L3 forward to prepare pushing body over bump; (d) R2 leg on bump while rear legs push; (e) L3 placed before bump and L1 placed over bump; (f) L3 and L1 to push forward; (g) R3 to bump and R1 forward; (h) L2 to push body over bump and L1 forward; (i) L2 placed over L1 to provide stability as L3 pushes body forward; (j) L1 reaches more forward with L3 pushing; (k) R3 to horizontal surface; (l) R2 placed forward then R1 placed forward and finally L3 makes it onto horizontal surface.

Figure 9 shows the experiment with ants climbing an edge. Steps a-d position the ant to a suitable position before the ant begins 'rolling' its body over the top of the wall (frame 3), Frames f-j show the ant making adjustments in its legs to reposition the body (frame k). Frames l-o show the ant in motion down the remaining wall section. This transition was achieved through leg placement and body stance.

Finally, there may be cases where climbing robots are required to transition a gap in an environment, here simulated by two large walls (Figure 10). The ants which were studied tended to walk down the flats instead of trying to cross it directly, or would rather use the side walls to get to the other side of the gap. Similar to the angled transition, the reason may be that positioning the ant body is difficult for a direct gap transition, especially when the gap is larger than the reach of its front legs. This may prove why the ant chooses to utilise the side wall or walk directly down and up the flats.

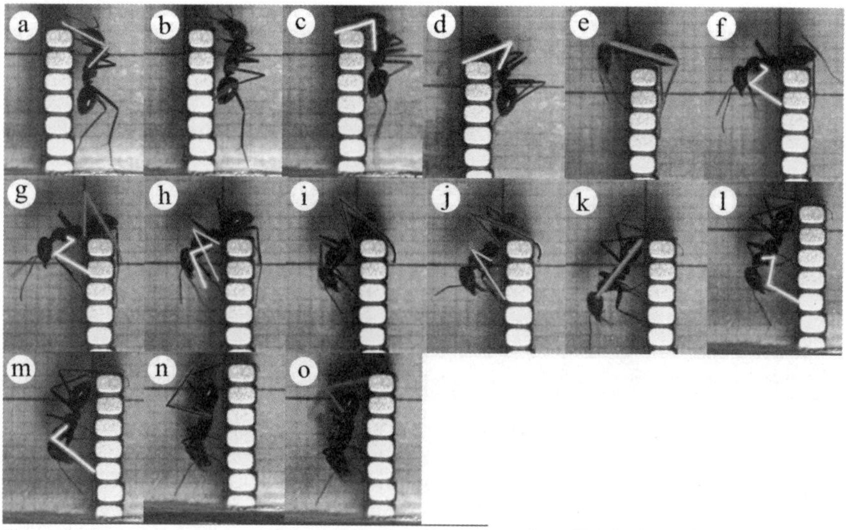

Figure 9. Locomotion over an edge: (1) R2 placed onto top edge; (2) L3 placed further forward on the vertical surface; (3) R1 placed over the edge; (4) L1 placed over the edge; (5) Rear legs push whole body over the edge. The body pivots over the edge; (6) L1 leg placed down the edge; (7) R1 placed onto the opposite edge while R3 leg placed on top edge as support; (8) L1 and R2 legs are placed further forward; (9) L3 placed onto edge for further support; (10) R2 leg place further forward as body pivots over edge; (11) Body of ant continues to pivot over the edge; (12) L1 placed further forward; (13) R1 placed further forward; (14) L3 leg detaches from the edge support and placed onto the opposite side; (15) R3 leg detaches from edge and placed onto surface

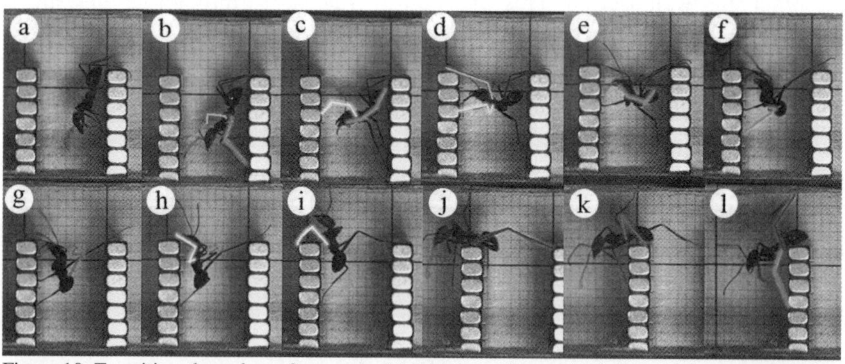

Figure 10. Transition through an 8mm gap: (a) Initial position; (b) L2 forward for support. R2 to side wall; (c) Body flexes while R1 to opposite side plane; (d) L1 to opposite side while R2 to top edge of opposite side; (e) Ant body flexes to provide better access to leg positions; (f) L2 to opposite surface; (g) R3 to side wall; (h) L1 to top edge of opposite side; (i) R1 over edge of opposite side; (j) L3 pushes off wall to help body to other side of the gap; (k) R3 to top edge of wall; (l) L3 placed on side wall for extra support while L2 placed further down opposite side.

4. DISCUSSION & FUTURE WORK

This paper has provided preliminary observation of locomotion techniques used by ants in various surface transitions within complex environments. Successful surface transition is directly linked to the ant's body posture. Control of the ant bodies three segments, i.e. head, thorax and abdomen, provides the key to effective locomotion as they prepare the body form prior to transition.

Analysis of an ant's centre of gravity during surface transitions will help understanding of effective ant locomotion in a complex environment, which is one of the future research works. Analysis on fall recovery techniques used by ants is another further work.

How the observed techniques used by ants in surface transition can be used in design of climbing robots is challenging because of limitations in actuation units (e.g. high weight torque ratio of motors), sensing and control.

5. REFERENCES

1. Cruse, H. Blaesing, B. 'Stick insect locomotion in a complex environment: climbing over large gaps', Journal of Experimental Biology, vol. 207, pp. 1273-1286, 2004
2. Federle, W.E., 'Locomotion and adhesion: dynamic control of adhesive surface contact in ants', *Arthropod Structure & Development*, vol. 33, no. 1, pp. 67-75, 2003
3. Holk, C. 'The control of the body position in the stick insect (Carausius morosus), when walking over uneven surfaces', Journal of Biological cybernetics, vol. 24, pp. 25-33, 1976
4. Koditschek, D. Full, R. Buehler, M. 'Mechanical aspects of legged locomotion control' Journal of Arthropod Structure & Development, vol. 33, pp. 251-272, 2004
5. Lozano, A., Peters, G. and Liu, D. 'Analysis of an Arthropodal System for Design of a Climbing Robot', accepted by the 28th International Symposium on Automation and Robotics in Construction, June 29- July 2, 2011, Seoul Korea.(To be presented)
6. P. Gonzalez de Santos, M. A. Jimenez and M. A. Armada, 'Dynamic Effects in Statically Stable Walking Machines', Journal of Intelligent and Robotic Systems, vol. 23, no. 1, pp. 71-85, 1998
7. Peters, G., Pagano, D., Liu, D,. Waldron, K, 'A prototype climbing robot for inspection of complex ferrous structures', 13th International Conference on Climbing and Walking Robots and the Support Technologies for Mobile Machines CLAWAR 2010
8. Zollikofer, C. 'Stepping Petterns in Ants – Influence of Body Morphology', Journal of Experimental Biology, vol. 192, pp. 107-118, 1994

BIOLOGICALLY-INSPIRED LOCOMOTION WITH THE MULTI-SEGMENTED INSPECTION ROBOT KAIRO-II

L. Pfotzer*, S. Bohn, C. Birkenhofer, R. Dillmann

Forschungszentrum Informatik an der Universität Karlsruhe (FZI)
Haid-und-Neu-Str. 10-14, 76131 Karlsruhe, Germany
**E-mail: pfotzer@fzi.de*
www.fzi.de

This paper describes how to adapt the locomotion of a caterpillar and an inchworm to a wheel-driven multi-segmented inspection robot. The idea is to use alternative biologically-inspired locomotion methods to traverse unstructured or soft grounds, where wheel-driven robots would normally abandon. We have evaluated the implemented locomotion patterns in simulation and outdoors on a grass area and compared the effectivity with the traditional wheel-driven movement.

Keywords: Inspection Robot, Multi-Segmented Robot, Biologically-Inspired Locomotion, Caterpillar Locomotion, Inchworm Locomotion

1. Introduction

Transferring biological locomotion principles to technical robot systems is a very common approach. It is promising to adapt the locomotion of crawling insects because evolution has optimized their movement for thousands of years. There are already a few attempts using the locomotion of caterpillars or inchworms to move biologically-inspired robots.

Chen et al.[1] have defined a multi-segmented robot model consisting of two different types of modules for moving inchworm-like robots. In this approach a finite-state machine is used to control the contraction or extension of every module. Another approach[2] shows the development of an inchworm-like robot prototype for search and rescue applications. Their robot is biologically inspired, but due to technical restrictions they have implemented a more simple kinematic with fewer degrees of freedom (DOF).

In this work we will transfer biologically-inspired locomotion methods to the wheel-driven robot KAIRO-II. Originally, KAIRO-II was developed for the inspection of sewers, and thus it is specialized to move through channel-like environments. When asked to perform common inspection and maintenance tasks in unstructured environments, wheel-driven robots reach their limits. For example, in sandy environments the wheels sink into the ground and slip. Through additional locomotion patterns we want to increase KAIRO-II's flexibility and overcome this deficit of wheel-driven robots.

2. The Multi-Segmented Inspection Robot KAIRO-II

Fig. 1. The multi-segmented inspection robot KAIRO-II.

KAIRO-II is a multi-segmented sensor carrier, consisting of drive segments with wheels, which are connected through joint segments (see Fig. 1). Each drive segment has two DOF and each joint segment has three DOF. All joint angles are measured with optical encoders, giving both absolute and incremental data. By measuring the currents of the wheel motors, it is possible to detect ground contact.[3]

The hierarchical control software is implemented in the robotic framework MCA2 (Modular Controller Architecture)[4] and includes base control, motion execution and maneuver control. The complete base control for regulating motors and reading sensors is implemented by a UCoM[5] (Universal Controller Module) in every segment. Above the base control the motion execution handles sensor data and calculates the joint angles of the robot kinematics. At the top-level the maneuver control plans the motion of the robot using a virtual rail. There is an initial set of elementary maneuvers defined to achieve basic movements for inspection. Every maneuver is implemented as a collection of states inside a finite state machine.[3]

3. Biologically-Inspired Locomotion with KAIRO-II

In order to realise biologically-inspired locomotion we added the maneuvers *Omega* and *Caterpillar* to the adaptive control of KAIRO-II. The implementation details for both locomotion methods are described in the following paragraphs.

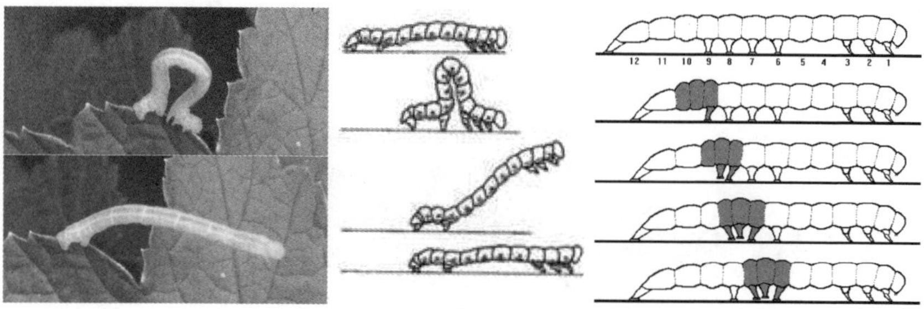

Fig. 2. The left figure shows the *Omega*(Ω) motion of a geometer moth inchworm.[6] The schematic in the middle shows an *Omega* motion sequence and the schematic on the right shows the linear contraction and extension of a *Caterpillar*.

3.1. *Omega*(Ω) *Locomotion*

The first motion is very common to inchworms and is described as *Omega*(Ω) locomotion.[6] Figure 2 (left) shows a geometer moth inchworm moving forward by contracting the whole body to an Ω-shaped vertical loop and then stretching themselves to a rectilinear position. The schematic in the middle of Fig. 2 shows the *Omega* motion sequence consisting of two general steps: Raising and lowering. To transfer this motion to KAIRO-II we implemented these two steps as states of the *Omega* maneuver. At the raising state KAIRO-II move up the middle segment while pinning the robot's head at its current position. The maximum raising height depends on the total number of segments. To ensure static stability the first and the last segment have to stay on ground. After the middle segment has reached the maximum height the state machine switches to the lowering state. At this state the tail of the robot is pinned at its current position and the middle segment is moved down, so that KAIRO-II returns to the rectilinear starting configuration. To realize a continuous locomotion, we have to alternate the raising and lowering states.

3.2. *Caterpillar Locomotion*

The second locomotion method is adapted from a caterpillar, which moves by linear contraction and extension of small groups of body segments. To achieve a forward locomotion, the contraction starts at the tail and moves like a wave towards the head (see Fig. 2 right). With KAIRO-II it is not possible to shrink or extend segments linearly. Therefore we use the resulting contraction of raising one joint segment to approximate a nearly similar motion. We added the four states *Start*, *Raise*, *Lower* and *End* to the state machine of the *Caterpillar* maneuver (see Fig. 3).

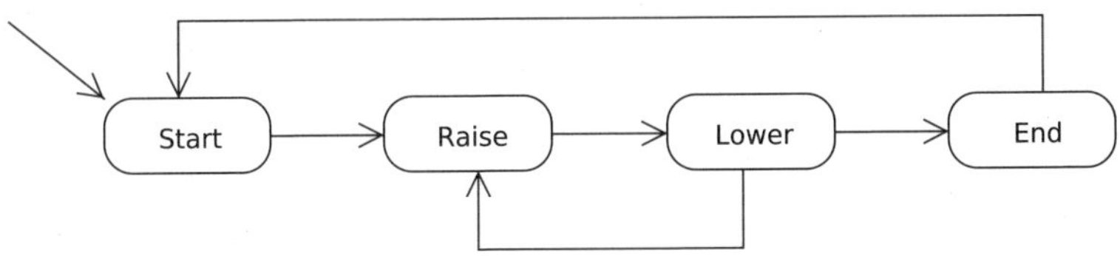

Fig. 3. State machine of the *Caterpillar* maneuver.

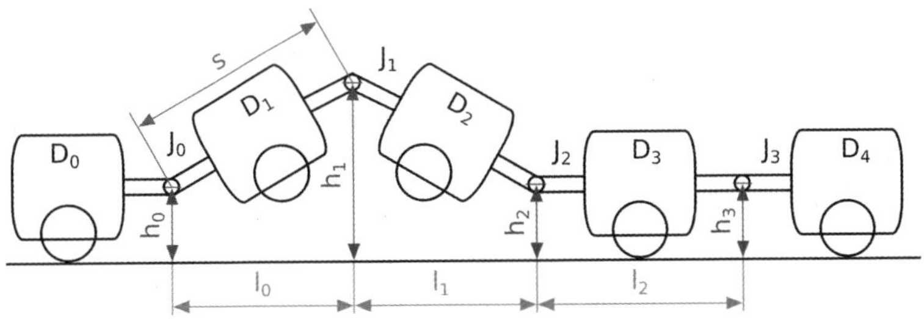

Fig. 4. Starting of the *Raise* state of the *Caterpillar* maneuver with KAIRO-II.

The locomotion starts with lifting up the first drive segment to shrink the overall robot length. We propagate this contraction in a wave-like motion through all segments to move the robot without using its wheels. To realise such a wave-like motion while keeping the robot length constant, we calculate a vertical balancing motion for raising every segment. Figure 4 illustrates this situation: The joint segment J_1 has reached the upper position at height h_1 and we want to raise the joint segment J_2. We can gradually shift the peak from J_1 to J_2 while keeping the length $L = l_0 + l_1 + l_2$ constant by simultaneously changing the heights h_1 and h_2. The motion is calculated as follows:

In the first step we calculated the length L between J_0 and J_3 with Eq. 1. Because the joints J_0 and J_3 should stay at their lower position, we fixed h_0 and h_3 to zero. Moreover, we decided that it is save to move the joints up to a maximum height of 150 mm and fixed $h_1 = 150\,\text{mm}$ and $h_2 = 0\,\text{mm}$ as starting condition. So we can calculate the length L which should stay constant.

$$L = \sqrt{s^2 - (h_0 - h_1)^2} + \sqrt{s^2 - (h_1 - h_2)^2} + \sqrt{s^2 - (h_2 - h_3)^2} \quad (1)$$

In the second step we calculated the vertical balancing height h_1 in respect to the height h_2 with Eq. 1 as well. To achieve this, we fixed h_0 and h_3 to zero again and used the length L calculated above. Then, we resolved the Eq. 1 to h_1. The resulting equation for h_1 is very complex, but looks similar to a parable. Our implementation approximates the function with a polynomial of the degrees three for a faster calculation (see Eq. 2).

$$h_1 = -0.00000794h_2^3 - 0.00181056h_2^2 + 0.48173159h_2 + 150.0173769 \quad (2)$$

It is now possible to calculate the balancing height h_1 while moving upwards the joint segment J_2. If the joint segment J_2 reaches its upper position the state machine will change to the *Lower* state and move down the joint segment J_1. Therefore we used Eq. 2 to calculate the balancing height h_2 in respect to h_1. If the wave reaches the last joint segment, this segment could be lowered without considering any other joints. After finishing the *End* state the whole *Caterpillar* motion cycle starts again.

4. Evaluation and Results

To evaluate the newly added *Omega* and *Caterpillar* motion, we tested them both in simulation and with KAIRO-II. In addition to measuring the driven distances at our laboratory, we evaluated the locomotion methods at outdoor tests.

4.1. *Simulation of a Nine-Segmented Robot*

In our first experiment we simulated the *Omega* and the *Caterpillar* motion. We decided to use a nine-segmented robot for simulation in contrast to the five-segmented KAIRO-II to show the scalability of our approach. By simulation we could illustrate that our added motions, based on the existing kinematics and virtual rail, enable the robot to move either like an inchworm or a caterpillar (see Fig. 5). Especially the *Caterpillar* locomotion looks very natural and smooth because of the implemented balancing motion for neighboured joint segments.

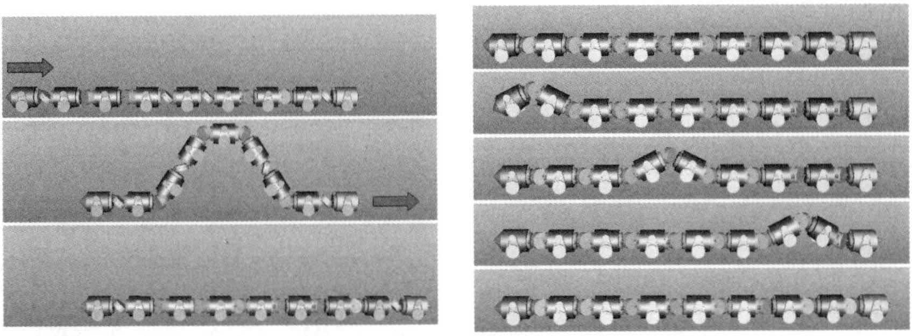

Fig. 5. Simulation of the *Omega* motion (left) and the *Caterpillar* motion (right) using a nine-segmented robot.

4.2. *Evaluation of the Omega(Ω) Motion with KAIRO-II*

Evaluating the *Omega* maneuver, we calculated the theoretical maximum moving distance *mmd* (see Eq. 3) and compared it to the measured values of experiments with KAIRO-II (see Fig. 6).

$$mmd(n) = 2\left\lfloor\frac{n-3}{2}\right\rfloor s - 2\sqrt{\left(\left\lfloor\frac{n-3}{2}\right\rfloor s\right)^2 - h_{\max}} \tag{3}$$

The *mmd* function depends on the number of drive segments n the robot consists of whereby s correspond to the distance between two joint segments and h_{max} is the maximum raising height of the middle drive segment. With a five-segment configuration, we can achieve a theoretical maximal moving distance of 280.5 mm per cycle, where the segment length s is fixed by construction to 330 mm and the maximum raising height we set to $h_{max} =$ 270 mm. We moved the robot five steps and measured the covered distances shown in Table 1. The average value of 281 mm is nearly similar to the calculated maximal moving distance. Overall KAIRO-II moved 1405 mm with five steps.

Table 1. Measured distances KAIRO-II moved using *Omega*(Ω) motion.

cycle	1	2	3	4	5	ø	calculated
distance [mm]	275	275	285	290	280	281.0	280.5

4.3. *Evaluation of the Caterpillar Motion with KAIRO-II*

In our third experiment we evaluated the *Caterpillar* motion with KAIRO-II (see Fig. 6). We measured the moved distance after every cy-

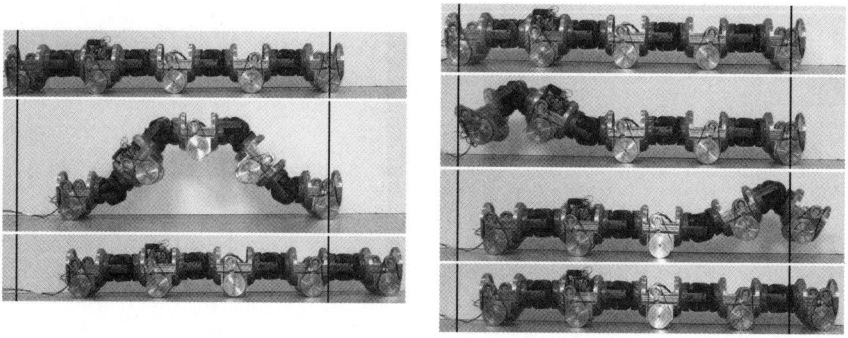

Fig. 6. The *Omega* motion (left) and the *Caterpillar* motion (right) with the five-segmented robot KAIRO-II.

cle (see Table 2) and compared the average of 67.1 mm with the theoretical maximum moving distance of 72.1 mm. This could be calculated by Eq. 3, if we set $n = 5$ and $h_{max} = 150$ mm. The difference between the theoretical and observed distances is due to the lower number of segments. With the *Caterpillar* motion, there are three segments which have no ground contact for a short time. This results in tipping back of KAIRO-II with its five-segment configuration because only two segments have ground contact. This problem will not occur if the robot has six or more segments. Overall KAIRO-II moved 480 mm with seven steps.

Table 2. Measured distances KAIRO-II moved using *Caterpillar* motion.

cycle	1	2	3	4	5	6	7	ø	calculated
distance [mm]	70	65	65	70	75	65	60	67.1	72.1

4.4. *Outdoor Evaluation of Both Locomotion Methods*

Moreover, we took KAIRO-II to an outdoor grass area and evaluated the newly implemented locomotion methods in this unstructured environment (see Fig. 7). First we found out that it is not possible to move wheel-driven through the ground. Moreover, KAIRO-II moved with the *Omega* motion about 10 m and 3.5 m with the *Caterpillar* motion. Although we could not observe any problems by moving through the grass area with the newly locomotion methods. Currently, tests on sandy and more wavy grounds are in progress. So far, our results indicate that it is feasible to use the *Omega* and *Caterpillar* maneuvers as alternative locomotion methods, also in areas where driving with wheels is not possible with KAIRO-II.

Fig. 7. Outdoor test with KAIRO-II in an unstructured environment.

5. Conclusion and Prospect

In this work we have presented an approach to transfer biologically-inspired locomotion methods to KAIRO-II. To achieve this, we extended the adaptive control by two maneuvers and implemented the necessary states. We illustrated how to implement a vertical balance motion to compensate the lack of contracting segments linearly. KAIRO-II now has the possibility to move in environments such as sandy grounds or grass areas where wheel-driven inspection robots usually have difficulties. Through the adaptive implementation of the control software it is feasible to switch between any existing maneuver on the fly, including the new *Omega* and *Caterpillar* locomotion methods. Our approach can handle an arbitrary number of segments.

Speeding up the locomotion of large chain-type robots by running multiple waves through the robot will be investigated in future work, especially for the *Caterpillar* motion.

References

1. I.-M. Chen, S. H. Yeo and Y. Gao, Locomotive gait generation for inchworm-like robots using finite state approach, in *Robotica*, 2001.
2. E. A. Avila, A. M. Melendez and M. R. Falfan, An inchworm-like robot prototype for robust exploration, in *Electronics, Robotics and Automotive Mechanics Conference*, CERMA, 2006.
3. C. Birkenhofer, Adaptive Steuerung eines mehrsegmentigen Inspektionsroboters, PhD thesis, FZI Karlsruhe, 2010.
4. K. Uhl and M. Ziegenmeyer, MCA2 – An Extensible Modular Framework for Robot Control Applications, in *Proceedings of CLAWAR2007, 10th International Conference on Climbing and Walking Robots*, 2007.
5. K. Regenstein, Modulare, verteilte Hardware-Software-Architektur für humanoide Roboter, PhD thesis, Karlsruher Institut für Technologie. 2010.
6. J.-S. Koh and K.-J. Cho, Omegabot : Biomimetic inchworm robot using SMA coil actuator and smart composite microstructures (SCM), in *IEEE International Conference on Robotics and Biomimetics*, ROBIO, 2009.

OBJECT LOCALISATION USING ACTIVE WHISKERS ON UNDERWATER AUTONOMOUS WALKING ROBOTS

T. Rooney*, M.J.Pearson, J. Welsby, I. Horsfield, R. Sewell, S. Dogramadzi

Bristol Robotics Laboratory, University of the West of England,
Bristol, BS161QD, UK
**E-mail: thomas.rooney@live.uwe.ac.uk*
www.brl.ac.uk

A low-cost, biologically inspired underwater walking robot (see Fig. 1) has been designed and built to covertly explore the seabed and to determine properties of submerged objects in obscure and inaccessible underwater locations. This paper focuses on a preliminary evaluation of an artificial active whisker to instrument this platform. Results demonstrate that both range and bearing to objects contacted by the whisker can be determined using simple data driven heuristics.

Keywords: Biomimetic; whisker; underwater robots; legged locomotion

1. Introduction

Adopting legged locomotion for traversing the seabed has a number of operational advantages; firstly, the platform can maintain its position without expending energy[1] ; secondly, the typically unstructured terrain of the sea bed can be scaled efficiently; and thirdly, movement generates a low acoustic signature which, for applications such as mine clearance or littoral warfare, would be beneficial. One of the problems to overcome using legged marine locomotion is how to sense the environment around the platform. The proximity to the sea bed inhibits the effectiveness of acoustic (sonar) based sensing[2] whilst the motion of the legs will disturb the sediment thus impairing the performance of any machine vision based solution.

Taking inspiration from nature we find that many marine mammals use facial whiskers to explore the seabed and catch prey using their sense of touch.[3] Similarly, many other marine animals, such as lobsters and shrimps, also employ touch to explore their environment using active (moving) antennae. One of the principal differences between mammalian whiskers and the antennae of invertebrates is that the sensory afferents, or mechanore-

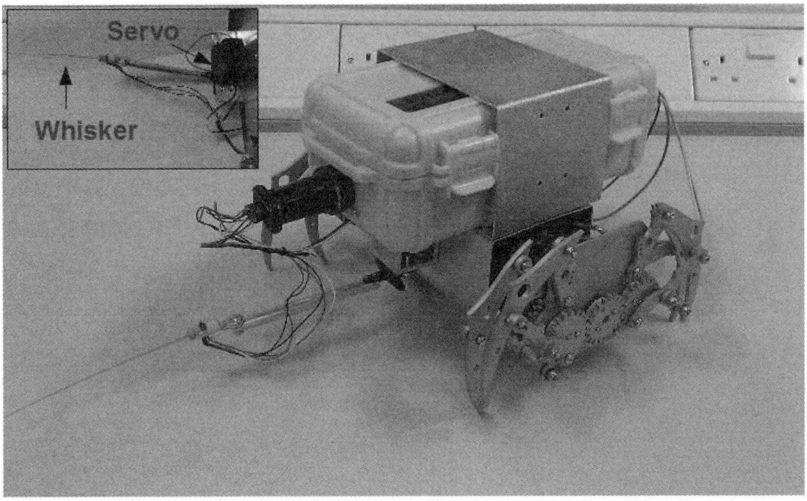

Fig. 1. photo of machalobster showing detail of whisker attachment

ceptors, of the invertebrate antennae are distribution along the length of the appendage. In mammalian facial whiskers the mechanoreceptors are all located at the base with the whisker shaft itself composed of, effectively, dead hair cells.[4] From an engineering perspective developing a touch sensor based on the facial whiskers of marine mammals would, therefore, be a less expensive and potentially more robust solution than a fully instrumented active antennae.[5] However, the derived sensory information will be less rich and may require more processing to determine the location and other properties (such as surface form or texture) of any contacted objects.

Here we report on the preliminary evaluation of a bio-inspired active whisker based sensor for determining the bearing and range to objects encountered by a legged underwater robot. The design of the robotic platform, hereafter referred to as MechaLobster (shown in Fig. 1), is described in the methods section along with the whisker sensor and the experimental setup used to evaluate its performance in both air and water. The results focus on investigating data driven approaches for determining heuristics for bearing and range estimation following whisker contacts when operating in both media.

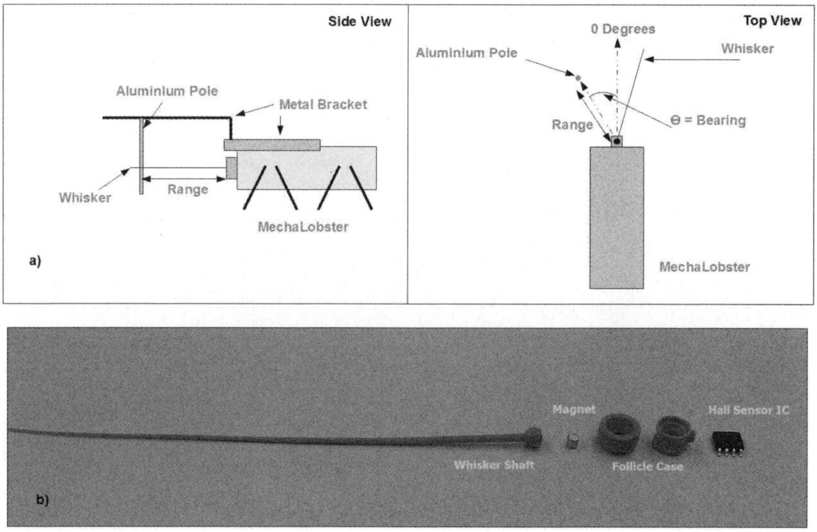

Fig. 2. **a)**schematic of experimental setup and **b)** exploded view of whisker sensor

2. Method

Mechalobster measures 170x550x260mm and is composed of; a watertight, negatively buoyant controller and battery housing (OtterBox); two modified MFA drill motors (1:148 gearbox) driving the Klann linkage[6] inspired leg mechanism; and a waterproof Traxxas 2065 servo to actuate a flexible, 150mm long whisker instrumented with a Hall effect based sensor. The whisker is sensitive to deflections of the shaft in 2 axes; however, here we constrain our evaluation to deflections measured only around the dorsal-ventral axis of the robot, i.e., in the same plane as the actuated degree of whisker motion. The deflections of the shaft are converted into a proportional voltage by the Hall Sensor at the base and are sampled at 250Hz using the 10-bit ADC module of a dsPIC30F4011 micro-controller located in the watertight controller housing. This sensory information, along with the current angle of the whisker drive servo, is packaged and transmitted via an ER400TRS radio module and logged using a remote PC. This 433Mhz radio signal can reasonably penetrate up to half a metre of water, which allows real-time sensory monitoring when the robot is located at the bottom of a laboratory based testing tank.

The experimental setup for generating the preliminary data set was as follows: (Shown in Fig.2a) A 400mm long metal bracket that can be rotated around the dorsal-ventral axis of the robot was attached to the top of MechaLobster directly above the centre of rotation of the whisker drive servo. An aluminium pole was then clamped to this bracket at various points along its length such that the whisker will make contact with the pole at specific distances along its own length and at a particular bearing with respect to the centre of servo rotation. Sinusoidal drive patterns of various frequencies and magnitudes (referred to as whisking) were then applied to the whisker servo such that the whisker would contact and deflect against the aluminium pole which, in turn, was set at various ranges and bearings. The first data set was gathered with the platform out of the water before being repeated with the platform immersed in water.

3. Results

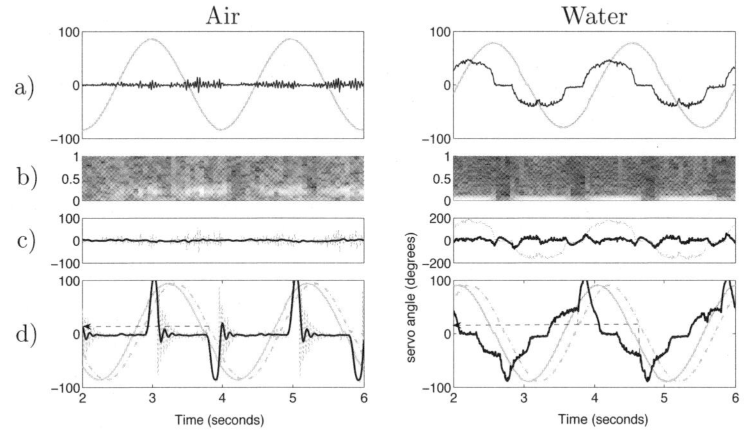

Fig. 3. Determining bearing to contact from active whisker data collected in air and water, **a)** Time series of 'free-whisking'; Grey traces are the demanded servo drive signals and black are the scaled and overlaid whisker deflection signals. **b)** Spectrogram of free-whisking data; Frequency range normalised to half sample rate (i.e., 125Hz). **c)** Black traces are low-pass (air) and high-pass (water) filtered versions of the original grey data. **(d)** Contacting a pole located at 15 degrees bearing; dashed grey traces are lag compensated whisker angles, dashed arrows indicate inferred bearing to contact.

The self-motion induced whisker signals in air were dominated by the resonant frequency of the whisker (25Hz, see Fig.3b). In water, the dominant time series feature is the derivative of the driven whisker motion, i.e.,

0.5 Hz. Whisking in air can be adequately cleaned using a simple low-pass filter (8th order Butterworth, see Fig. 3c), however, an equally simple high-pass filter applied to water based whisking was found to be inappropriate (see Fig. 3c). In previous works[7] a bio-inspired de-correlation adaptive filter model of the cerebellum was used to compensate for similar 're-afferent' noise and would be a good candidate for this application. For subsequent work reported here, the air whisking data was low-pass filtered and the water based signals remained unfiltered. A visual inspection of the data set whilst whisking into a pole aligned 15 degrees from the centre line of the robot was made to determine the phase lag introduced by the filter and the response of the servo drive (see Fig. 3d). In air, this was found to be 36 degrees whilst in water a 60 degree lead was found to be appropriate for reliably determining bearing to contact (as indicated by the blue arrows). In future work the actual whisker angle will be measured using a shaft encoder to derive more accurate bearing estimates. To determine range to

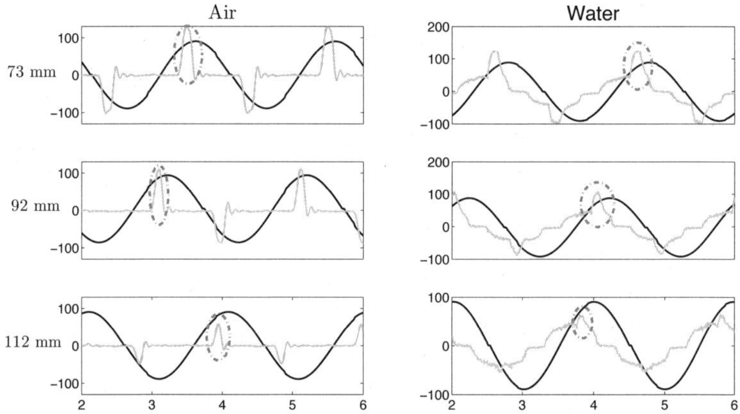

Fig. 4. Determining range to contact from active whisker data collected in air and water into a pole set at 3 different radial distances from the base of the whisker. Axes same as figure 3, dashed ellipses indicate contact incidences

contact, a data set was gathered whisking in air and water at 0.5Hz into a vertical pole, set at a fixed bearing, at three different radial distances from the base of the whisker (73mm, 92mm and 112mm, see Fig. 4). A simple feature that can be used to determine range to contact has been shown as the peak magnitude of the whisker deflection during the whisk.[8] This heuristic is born out in both air and water whisking by a visual inspection of the data set in Fig. 4.

4. Discussion and future work

These encouraging results support further investigation of active whiskers for underwater sensing. It is now known that information for contact localisation is contained in the data but that more experiments are needed. The next phase of work will be to gather a larger control set with which to train a feature based template classifier (based on this preliminary set) to autonomously extract both range and bearing information. This will then be used to orient and direct the movement of the legged robot as it explores to build a tactile map of the seabed. The control of the whisker motion is also an area of particular interest as the classification of more detailed surface features, such as texture[9] or orientation, will require fine motor control to constrain the sensory range.[10]

References

1. G. Packard, R. Stokey, R. Christenson, F. Jaffre, M. Purcell and R. Littlefield, Hull inspection and confined area search capabilities of remus autonomous underwater vehicle, in *OCEANS 2010*, Sept 2010.
2. C. F and J. T. Cobb, Sonar processing for short range, very-high resolution autonomous underwater vehicle sensors, in *OCEANS 2005*, 2005.
3. Z. C, *Science* **293**, 29 (2001).
4. S. Ebara, K. Kumamoto, T. Matsuura, J. E. Mazurkiewicz and F. L. Rice, *J Comp Neurol* **449**, 103 (2002).
5. M. Pearson, I. Gilhespy, C. Melhuish, B. Mitchinson, M. Nibouche, A. Pipe and T. Prescott, *International Journal of Advanced Robotic Systems* **2**, 335 (2005).
6. J. Klann, Klann linkage (2010), `http://mechanicalspider.com/`.
7. S. R. Anderson, M. J. Pearson, A. Pipe, T. Prescott, P. Dean and J. Porrill, *Robotics, IEEE Transactions on* **26**, 1065 (2010).
8. M. Evans, C. Fox, M. Pearson, N. Lepora and T. Prescott, Whisker-object contact speed affects radial distance estimation, in *IEEE International conference on Robotics and Biomimetics(ROBIO)*, (IEEE, 2010).
9. N. Lepora, M. Pearson, B. Mitchinson, M. Evans, C. Fox, A. Pipe, K. Gurney and T. Prescott, Bayes novelty detection for a moving, whiskered robot, in *IEEE International conference on Robotics and Biomimetics(ROBIO)*, (IEEE, 2010).
10. M. Pearson, B. Mitchinson, J. Welsby, T. Pipe and T. Prescott, Scratchbot: Active tactile sensing in a whiskered mobile robot, in *From Animals to Animats 11*, (Springer, 2010).

SENSORY COUPLED ACTION SWITCHING MODULES (SCASM) FOR MODELING AND SYNTHESIS OF BIOLOGICALLY INSPIRED COORDINATION*

BRANDON L. RUTTER, BRIAN K. TAYLOR AND ROGER D. QUINN

Biologically Inspired Robotics Laboratory, Mechanical and Aerospace Engineering Dept., Case Western Reserve University, 10900 Euclid Avenue Cleveland, OH 44106, USA

WILLIAM A. LEWINGER

Institute for Perception, Action and Behaviour, University of Edinburgh, EH8 9AB, Edinburgh, UK

JOHN A. BENDER AND ROY E. RITZMANN

Department of Biology, Case Western Reserve University, 10900 Euclid Avenue Cleveland, OH 44106, USA

MARCUS BLÜMEL

Department of Animal Physiology, Institute for Zoology, University of Cologne, 5023 Cologne, Germany

Biological inspiration has long been pursued as a key to controlling legged robots. However, concise descriptions of system function and architecture and the use of biologically-based elements in control design remains difficult. Recent discoveries in insect locomotion systems based on the study of the neural mechanisms of stick insect leg coordination have made it possible to coordinate joints in robotic legs based on known neural pathways. Our work with these models has led to Sensory Coupled Action Switching Modules (SCASM); a control concept that can describe systems with adaptable, sensory-driven coordination. We use SCASM as a framework to describe and develop neurally-based control for robotic models of insect legs.

1. Introduction

Insects have served as inspiration for the design of legged machines ranging from the highly abstracted Whegs™[1], through less abstract but still highly simplified systems like RHex[2] and MechaRoach[3,4], to more flexible and complex

* This work is supported by the U.S. AFOSR under grant FA9550-07-1-0149.

systems, including Robot II[5], the TUM walking machine[6], the Tarry series[7], the Lauron series[8], and BILL-Ant-p[9].

As engineers wish to know more detail about animal neuromechanics, modeling has become more important as a tool to aid biologists. Many robotic models of legged animals have been behaviorally based. However, the behavioral approach has drawbacks with respect to biology[10] – examining a black box does not generally help to understand the behavior generating mechanisms, and understanding these biological mechanisms can unlock the design and control principles that engineers seek.

Recent developments in our understanding of insect walking systems, encapsulated in the study of the neural mechanisms of stick insect leg coordination[11], have enabled the construction of robotic model legs based on known neuromechanical properties of biological systems. Systems here and in[12] were constructed mainly for modeling insect step generation, particularly to address questions about the role of higher-level influences on transitions between widely varying behaviors in local leg control systems.

The implementation of controllers capable of generating insect stepping behavior in the manner described by[11], as in[12,13], led to the development of Sensory Coupled Action Switching Modules (SCASM) as a control concept. Here, we present a more general description of SCASM along with diagramming tools to describe, understand and design systems with this concept.

The basic function, neuromechanical nature, and resilience of SCASM-controlled steppers have been demonstrated in[12-14]. SCASM provides a set of system representations that can be used to aid sensory-driven controller design for particular cyclic behaviors in a multi-degree-of-freedom (DOF) system. We present the original basis of SCASM in biology, then develop the concept more generally and present tools for its representation. The associated design process is outlined for a hypothetical minimal stepper.

2. The Basis in Biology

The model controller in[11] was the basis for developing a leg motion control system consisting of independent joint controllers. Each joint controller is a bi-stable pattern generator that switches between joint flexion and extension. To make this decision, each controller has access to specific sensor data, but no direct information about the state of the other controllers.

If the joint pattern generators are implemented as finite state machines as in[11], each pattern generator's behavior can be expressed as a set of Boolean decision rules. In the case of the stick insect the appropriate rules for each joint can be derived almost completely from the literature; only threshold values have to be adjusted for a given leg geometry. The stick insect model controller described in[11] has been applied to several robotic legs in[12-15].

3. The SCASM System Concept

A SCASM control system consists of a set of action switching modules (ASM's), each of which has some direct influence on a subset of the entire system to be controlled. In the stick insect leg in Figure 1, the blue module only controls the action of the femur-tibia (FTi) joint. Each ASM has an associated set of actions. An action describes some effect on the dynamics of the system subset associated with its module. In this example, the FTi joint has 1 DOF and 2 actions: joint flexion (FLX) and extension (EXT). The module is then a control unit that switches between these actions, making it an ASM. The dynamics of an action consist of neural/control behavior, plant mechanics and environmental interaction. The action output is the resulting torque and/or movement at that joint.

The switching between actions at a module is influenced by sensory signals, which can be affected by the actions at other modules. These influences couple the modules via sensory feedback, making the system a set of Sensory Coupled Action Switching Modules (i.e., SCASM). The system's behavior depends on the strength and sense of the inter-module connections, the control dynamics of each action, the system's physical dynamics, and the environment.

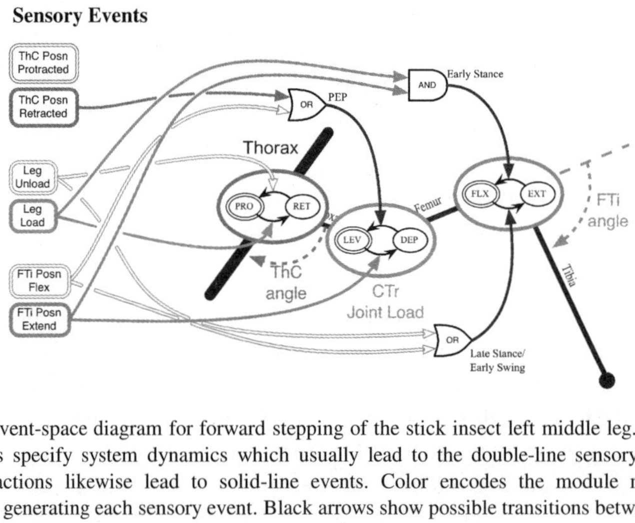

Figure 1. Event-space diagram for forward stepping of the stick insect left middle leg. The double-line actions specify system dynamics which usually lead to the double-line sensory events, and solid-line actions likewise lead to solid-line events. Color encodes the module most directly involved in generating each sensory event. Black arrows show possible transitions between actions.

Influences on action switching for the stick insect leg are represented as sensory events on the left in Figure 1. As an example, "FTi joint is extended", would encourage CTr depression and, in combination with leg load, FTi flexion.

4. Using the SCASM System Concept to Construct a Control Method

In this section we present steps that may be used to specify sensory influences that, when combined with appropriate action dynamics, will generate a desired behavior for a SCASM system. With proper sensory influences, reasonable target behavior can often be achieved with simple action dynamics[12-14].

For this exercise we use a minimal leg-like mechanism capable of producing forward or reverse motion using 2 DOF: protraction & retraction at a "Thorax-Coxa" (ThC) joint for forward/back motion; levation & depression at a "Coxa-Trochanter" (CTr) joint for raising/lowering the foot (Figure 2). We have used this process for 3-DOF robot legs[12-15] and it can be used for higher dimensions; we present the minimal case for brevity and conceptual clarity.

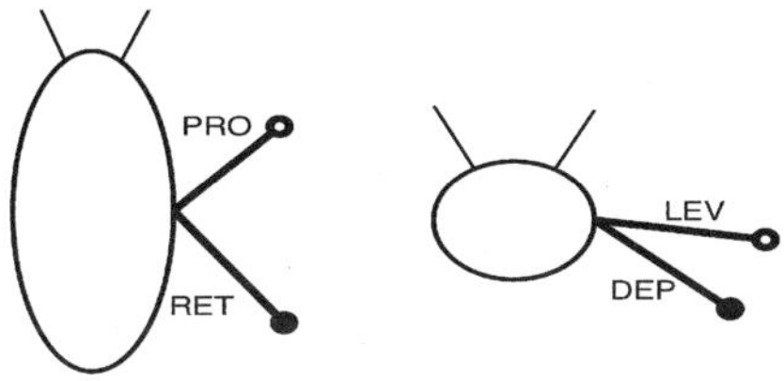

Figure 2. The minimal stepper; top view on the left showing protraction-retraction. Rear view on the right showing levation-depression. The open-circle foot shows the protracted-levated leg at the end of swing (for forward stepping); closed-circle shows the retracted-depressed leg at the end of stance.

Synthesis of a SCASM controller can be attained through the following steps:

Structural decomposition rule: *Choose modules that represent the physical system.* A SCASM controller takes account of physical plant mechanics implicitly, largely through the choice of modules, actions and sensory events. In simple systems such as this, start with one module per DOF. Here, we choose ThC for forward/backward leg motion, and CTr for upward/downward motion.

Start with minimal set of system actions. For each module, choose as few simple actions as possible for the associated system subset. It is probably a good idea to concentrate on "few" to begin with; 2 is a good goal for a 1-DOF subsystem. In this example it is natural to choose Protraction/Retraction as the actions for the ThC module, and Levation/Depression for CTr (Figure 4 right).

Start with simple sensory events that sense the limits of these actions; i.e. sensory events that these actions should eventually produce. Some potential events for our 2 example actions are shown in Figure 3. The minimal set of sensory events should consist of things expected to happen in every cycle. Better performance often results from choosing limit events *related to environmental interaction*, which provides a degree of automatic sensory-driven adaptation of system behavior to environmental variation. With this in mind, we choose "ground contact" for depression. Obstacle collision is not expected in every step, so for protraction the recommended choice is "Protracted Angle". For the other actions in these modules we start with the opposing sensory events, "no ground contact" for levation and "retracted angle" for retraction.

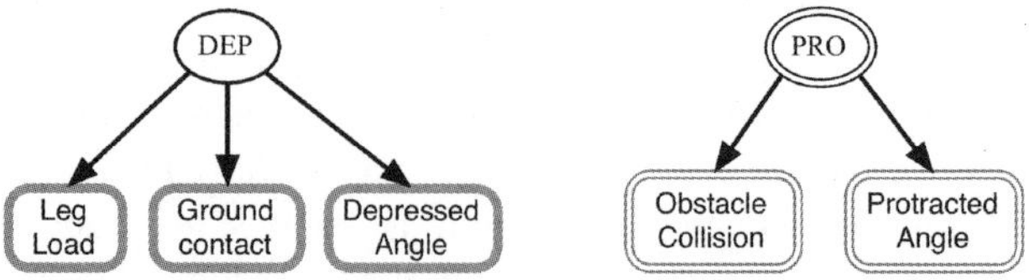

Figure 3. Sensory events that could result from depression at CTr joint or protraction at ThC joint.

In the controller implementation associated with this design step, one must choose real sensory conditions (usually thresholds operating on raw or processed sensory data) that represent these sensory events.

Construct global state diagram(s) for the system: If there are too many states for this to be easy (more than 3), make a sub-global state diagram using the three modules most likely to be important. It may also be useful to choose a decomposition based on one module that is highly involved with the global state description. **Describe and label these global states**, or at least the global states that are in the target behavior (Figure 4 left).

Choose a desired progression of global states for a behavior. This is done for forward stepping in the bold red arrows of Figure 4. To a degree, the use of transient global states (e.g., pre-stance and pre-swing) may provide both a more sensory-driven behavior and a more robust system overall. Choosing only "swing" and "stance" both represents something like central coupling between the modules, and ignores the complexity associated with these transitions, which could hinder the system's ability to cope with e.g., changes in substrate height.

Construct the event-space diagram for the behavior. Start in one of the desired global states; preferably one that takes a large percentage of overall cycle time. Identify the limit sensory events this state leads to. Choose events (ideally at least one from a module that does *not* change actions in the next global state) to trigger the appropriate action switching to reach the next global state. Iterate until finished, as shown in Figure 5 and described below.

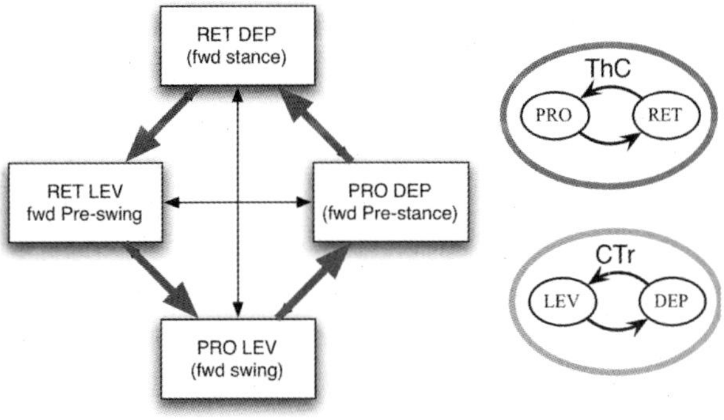

Figure 4. Global state diagram for minimal forward stepping; bold arrows show nominal state progression.

Start in stance. Next state changes from DEP to LEV at CTr, choose the ThC signal being generated in stance (RET-> ThC Retracted) and have it promote CTr Lev. (line 1).
Now in pre-swing. Next state changes RET to PRO at ThC. Choose CTr signal being generated in Pre-swing (LEV → No Ground Contact) and have it promote ThC PRO (line 2)
Now in swing. Next state changes LEV to DEP at CTr, choose ThC signal being generated in swing (PRO-> ThC Protracted) and have it promote CTr DEP (line 3)
Now in pre-stance. Next state changes PRO to RET at ThC, choose CTr signal being generated in pre-stance (DEP -> Ground Contact) and have it promote ThC RET (line 4), returning us to stance and completing the cycle.

Note that this design process provides a high-level set of control rules for a particular behavior; a more general representation of the sensory flow in a system can be derived from these as desired. Once a set of modules and input events has been chosen, potential system behaviors are generated by changing the strength and sense of the connections between them.

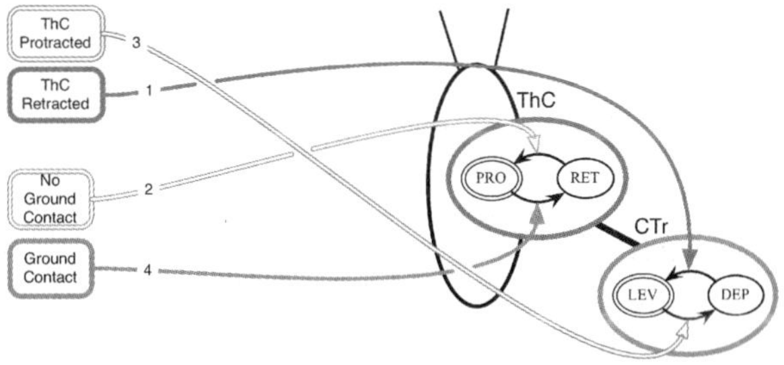

Figure 5. Event-space diagram for forward stepping in the minimal SCASM stepper. The switching event lines are labeled according to the construction steps in the text.

5. Conclusions

The control organization that is at least partially responsible for the generation of stepping in stick insect legs, as described by[11] suggested a control concept dubbed Sensory Coupled Action Switching Modules (SCASM). Tools have been developed to facilitate the use of this concept in the description and synthesis of neural control. The informational content and use of event-space diagrams and global state diagrams has been presented and demonstrated. Additionally, the steps necessary to construct a SCASM controller for a system and a particular behavior have been identified and presented though the example of the minimal SCASM stepper.

A controller developed using the SCASM concept can be implemented as an augmented finite state machine if switching between actions is discrete, with each action considered a state. The transition function is implemented by the process(es) which detects switching events, with the possible addition of a switching timer internal to the module. In animal nervous systems action switching is not discrete. This feature can be included in a SCASM controller.

SCASM is especially useful when developing a model of a legged animal's control system with incomplete data; it can suggest hypothetical pathways that biologists can later research. But, the SCASM concept/representation need not be limited to the control of legged systems; it can be used to describe any system with sub-elements that perform a set of actions, in which the sensory coupling between these elements can override existing central coupling. SCASM is a potential tool to describe or design any multi-dimensional controller that goes through a coordinated set of states in concert with the environment.

Acknowledgments

We thank Nicole Doorly and Luther Palmer for additional help in the clarification of the event-space and global state diagrams.

References

1. T. J. Allen, R. D. Quinn, R. J. Bachmann, and R. E. Ritzmann, "Abstracted Biological Principles Applied with Reduced Actuation Improve Mobility of Legged Vehicles," in *IEEE/RSJ international conference on intelligent robots and systems (IROS 2003)*, Las Vegas, NV, 2003, pp. 1370.
2. R. Altendorfer, N. Moore, H. Komsuolu, M. Buehler, H. B. Brown, D. McMordie, U. Saranli, R. Full, and D. E. Koditschek, *Autonomous Robots,* 11, Nov 2001.

3. M. Boggess, R. Schroer, R. Quinn, and R. Ritzmann, "Mechanized Cockroach Footpaths Enable Cockroach-like Mobility," in *International conference on robotics and automation; 2004 IEEE*, New Orleans, La., 2004, pp. 2871.
4. T. E. Wei, R. D. Quinn, and R. E. Ritzmann, "A CLAWAR That Benefits From Abstracted Cockroach Locomotion Principles," in *Climbing and walking robots Conference*, Madrid, Spain, 2004, pp. 849.
5. K. S. Espenschied, R. D. Quinn, R. D. Beer, and H. J. Chiel, *Robotics and autonomous systems,* 18, 1996.
6. F. Pfeiffer, H. J. Weidemann, and J. Eltze, "The TUM Walking Machine. - In: Intelligent Automation and Soft Computing," in *Trends in Research, Development and Applications.* vol. 2: TSI Press, 1994, pp. 167.
7. A. Buschmann, "Home of Tarry I & II: design of the walking machine Tarry II," 2000.
8. B. Gassmann, K.-U. Scholl, and K. Berns, "Behavior control of LAURON III for walking in unstructured terrain," in *Intl. Conference on Climbing and Walking Robots (CLAWAR '01)*, Karlsruhe, Germany, 2001, pp. 651.
9. W. A. Lewinger, M. S. Branicky, and R. D. Quinn, "Insect-inspired, actively compliant robotic hexapod," in *International Conference on Climbing and Walking Robots (CLAWAR)*, London, U.K., 2005.
10. B. Webb, *Behavioral and Brain Sciences,* 24, 2001.
11. Ö. Ekeberg, M. Blümel, and A. Büschges, *Arthropod structure & development,* 33, 2004.
12. B. L. Rutter, W. A. Lewinger, M. Blümel, A. Büschges, and R. D. Quinn, "Simple Muscle Models Regularize Motion in a Robotic Leg with Neurally-Based Step Generation," in *International Conference on Robotics and Automation (ICRA '07)*, Rome, 2007.
13. W. A. Lewinger, B. L. Rutter, M. Blümel, A. Büschges, and R. D. Quinn, "Sensory Coupled Action Switching Modules (SCASM) generate robust, adaptive stepping in legged robots," *CLAWAR 2006: 9th International Conference on Climbing and Walking Robots* Brussels, Belgium, 2006.
14. W. A. Lewinger and R. D. Quinn, "A Hexapod Walks Over Irregular Terrain Using a Controller Adapted from an Insect's Nervous System," *International Conference on Intelligent Robots and Systems (IROS)* Taipei, Taiwan, 2010.
15. B. L. Rutter, W. A. Lewinger, B. K. Taylor, M. Wilson, M. Blümel, Ö. Ekeberg, A. Büschges, R. E. Ritzmann, and R. D. Quinn, *2006 Neuroscience Meeting Planner,* 2006.

SECTION–4
CO-OPERATIVE ROBOT SYSTEMS

MACHINE LEARNING FOR HETEROGENEOUS MULTI-ROBOTS SYSTEMS IN LOGISTIC APPLICATION FRAME

T. Wang, D.M. Ramik, C. Sabourin and K. Madani

Signals, Images, and Intelligent Systems Laboratory (LISSI/EA 3956), Paris Est University, Senart Institute of Technology,
Avenue Pierre Point, 77127 Lieusaint, France
E-mail: wangting0310@yahoo.com.cn, dominik.ramik@u-pec.fr, sabourin@u-pec.fr, madani@u-pec.fr
www.u-pec.fr

In this paper, we propose machine learning for heterogeneous multi-robots systems in the logistic application frame. The team's robot is composed by a humanoid robot and three wheeled robots. The goal of the humanoid robot is to lead the wheeled robots according to environment (e.g. obstacle, narrow path, etc..). The aim of the wheeled robots, which move in rigid formation, is to act together in order to carry a cumbersome load. Machine learning allows to give automatically path planning.

Keywords: Machine Learning, Multi-Robot Systems; Robot Formation, Logistic Application.

1. Introduction

In the future, numerous robots (eg. wheeled robots, legged robots, humanoid robots, network sensors, etc..) will be integrated in human environment. These robots will be used mainly to provide physical service to humans in autonomous manner. These services will be the result of the work of one robot, or the result of the cooperation between several robots allowing to carry out complex tasks. Effectively, many applications, like warehouse management, industrial assembling, military applications, daily tasks, can benefit from multi-robot systems[1].[2] In short, we think that the self-organization between heterogeneous robot will play a key role in the further applications, as well in home as in industrial domains. But it must be pointed out that because of news advanced technologies, namely in wireless communications, the "intelligence" will be distributed.

For industrial applications, the logistic domain in warehouse environment can benefit of the coordination of several mobile robots. For example, multi-robots systems are used by KIVA systems,[3] where a lot of small robots are used to transport some objects. This approach seems very interesting but there are some limitations in it. All of robots have individual behaviors and all of robots are controlled by a supervisor. In this context, this work focuses on the design of new approach based on multi-robot systems and more especially the collaboration between heterogeneous robots for logistic tasks. The proposed infrastructure is based on a network robots which includes one humanoid robot, wheeled robots, cameras, and remote computer. All devices (robots, cameras, computers, etc..) can communicate between them by using wireless network. The goal of the humanoid robot is to lead the wheeled robots according to the environment (e.g. obstacle, narrow path, etc..), and wheeled robots are used to carry the load. The camera allows to give complementary information about the environment.

The next section presents the problem formulation. Section 3 describes only a short overview of the control strategy. More precise details about the control strategy, namely about the ANFIS controller and the path planning, may be found in several other papers (see for example[456]). Section 4 presents the simulation results and the experimental validation. The last section contains conclusion and potential future extensions of our work.

2. Problem Formulation

The Fig.1 shows an example of the virtual environment used during simulation. In this virtual environment, robots team is composed of three KheperaIII [a], a humanoid robot Nao [b], a camera and a virtual robot supervisor. The environment (unknown environment for the robots) is decomposed into 21 rooms (see Fig. 1) where each room is cut into 100 states, and each state is a square with side of 30 centimeters. The total area of the environment is equal to $6 \times 12m^2$.

The three wheeled robots move in rigid formation (eg a line, column, etc..). The aim of these three robots is to act together in order to carry a cumbersome load. The humanoid robot perceives environment and it leads the wheeled robots. It sends to them high level control as "go forwards", "turn left" and so on. When the humanoid see one or more obstacles, it needs more information and it asks help to supervisor. The supervisor ob-

[a]http://www.k-team.com/
[b]http://www.aldebaran-robotics.com/

Fig. 1. Virtual environment of the multi-robot system.

tains from a camera, located at the top of environment, a picture of this environment. In fact, this camera may be considered like a virtual robot which is located above the worlds with the height $5m$. The field of view of the camera is $3 \times 3m^2$. The camera can be moved by a supervisor from 1 to 7 along with y-axis and from 1 to 3 along with x-axis among the points of the intersection of two dotted lines for exploiting the whole environment step by step (see Fig.1 right). By using image processing and learning process, the supervisor compute a path planning for the rigid formation. It sends to the humanoid robot a set of actions in order to control the three wheeled robots.

3. Control strategy overview

Based on the previous description, the main problem is how to compute the path planning (to find action like "forwards", "left", and so on) in order to move the formation from an initial point to a goal point in a constraint environment (obstacle avoidance, narrow path, etc..). Designing an on-line approach for a real-time application, we have based our concept on two levels. The first one allows us to control a rigid formation of robots by using high level control (eg "forwards", "left", etc..), and the second is based on a learning process including the on-line image processing. It must be noticed that we consider that all wheeled robots move in synchronous manner in order to keep a rigid formation.

3.1. *Control formation*

The state of the formation is given by the position of the reference robots (e.g. the robot number 1) and the orientation of the formation. We define 8 possible orientations (see Fig. 2.a) dependent directly of the angle θ: $\theta = 0°$

for the formation 1, $\theta = 45°$ for the formation 2, $\theta = 90°$ for the formation 3, and so on. For each orientation, we define height possible actions (from 1 to 8) (see Fig. 2b). These actions correspond to the directions allowing to move the formation of robots. Two other actions allow to do a rotation of the formation (Action 9 and 10 are the clockwise and anti clockwise rotation action respectively).

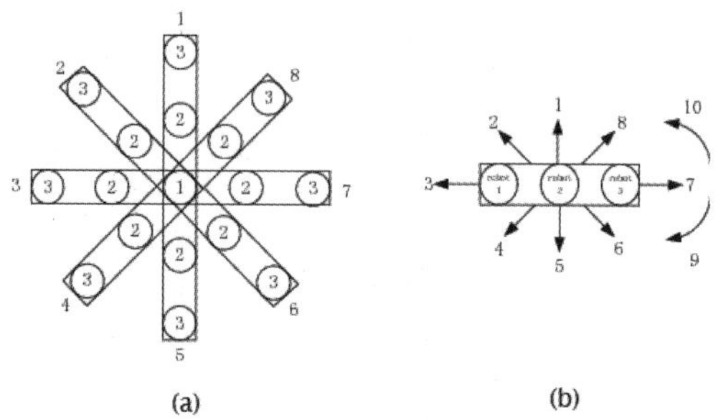

Fig. 2. Schematic description of the 8 formations and 10 actions.

3.2. *Path planning*

In order to get automatically numerical information allowing to represent environment, we have developed an approach based on image processing. This procedure uses the following process:

- Taking the photo of the environment,
- Convert the RGB image into a binary image,
- The last step allows to describe environment by a binary matrix.

After the image processing, it is possible to get the equivalent quantified environment in which 1 represents obstacles and 0 is free path. Based on this quantification of the real environment, we try to find the best way in order to move robots from an initial point to a goal point. To solve this problem, one solution consists to use a reinforcement learning (e.g. Q-learning). The goal of the reinforcement learning algorithm is to find the action which maximizes a reinforcement signal where the reinforcement signal provides an indication of the interest of last chosen actions. Consequently, when the supervisor have found a solution, it is possible to lead the wheeled robots in the constraint environment.

4. Results

This section firstly presents results about simulation and secondly experimental validation. Simulations have been performed by using software Webots and controllers have been designed with the software Matlab. For the experimental validation, we have use real robots KhperaIII and the humanoid Nao, a wireless camera and a computer for the remote control. The camera is put on the 2 m height.

4.1. *Simulation*

In this simulation, we present results when the robots have to move from the room (2,2) to room (1,7) (see Fig 1) by using the following path: (2,2) $\rightarrow (2,5) \rightarrow (3,5) \rightarrow (2,6) \rightarrow (1,6) \rightarrow (1,7)$.

When there is no obstacle, all robots go forwards like in the rooms (2,2) or (2,5). But when there are obstacles, the path planning process describes in section 3.2 is used. And thanks to the proposed approach, the robot team can move automatically in an unknown complex environment. To illustrate this part, we show the obtained results when the robot have to cross the room (2,6). The Fig. 3 shows the image processing described in section 3.2. After the learning process, the result allowing to control the formation is $A = \{3, 3, 9, 2, 2, 2, 2\}$ where each number corresponds to action (see Fig. 2). The Fig. 4 shows a snapshot when the robot team cross this room. It must be pointed out that the wheeled robot keep a rigid formation.

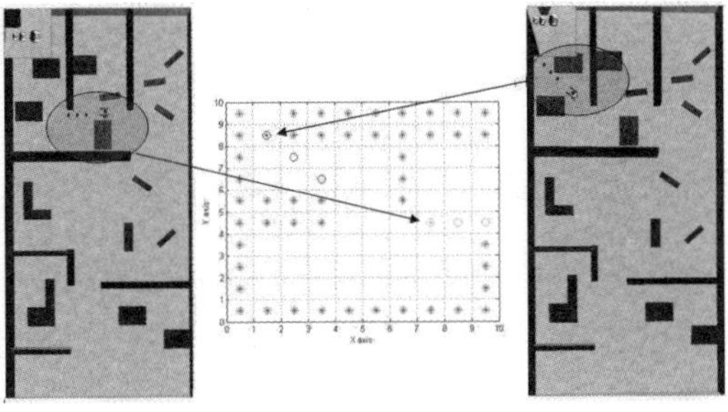

Fig. 3. Illustration of the path planning task carried out by the top camera and remote computer: start point (left), goal point (right).

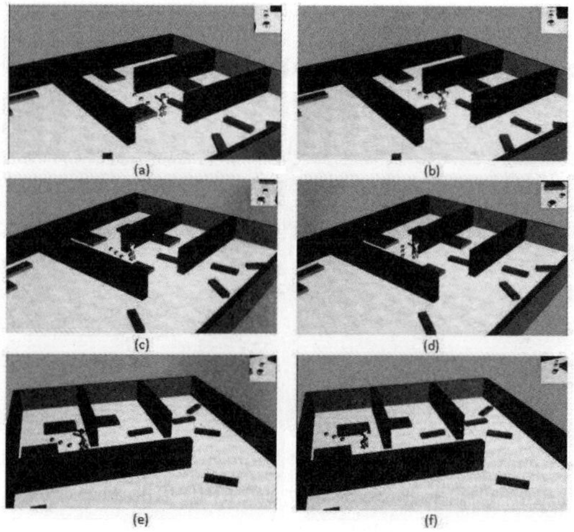

Fig. 4. Snapshot of simulations results.

4.2. *Experimental validation*

Now, we present the results of two preliminary experiments in the real environment to prove the feasibility of our strategy.

The first experiment consists to an on-line control of the formation. In this example, the humanoid robot perceives environment and it sends to wheeled robot high level control as "go forwards", "turn left" and so on. The Fig. 5 shows a snapshot of this experimental validation. According to the position of the wheeled robots, the humanoid robot Nao stops the formation, sends the order "clockwise" in order the formation crosses the narrow path by keeping the rigid formation. The goal is, not only to show that the humanoid Nao plays the role of a supervisor through a remote control, but to prove that this robot is able to perceive environment and to take adequate decision to guide wheeled robot (more information about strategy used to perceive environment can be found in the following reference[6]).

The Fig. 6 describes the second experiment which consists firstly to compute the path planning and secondly to control the rigid formation of the robot. During this experimentation, we use two KhperaIII, a camera and a computer for the remote control of the both camera and robots. The camera put on the 2 m height. The real environment is made up of a 1.5 m wide and 2.1 m long rectangular floor and some paper play the role of obstacles. The environment is separated into 35 states and each state is a 30 cm side square. Image processing is execute in C , path planning and the robots control is executed in Matlab. The connection between robots and computer is WIFI, and robots can received the real-time command transmission from computer. The results of experiment have showed in Fig. 6. The initial position is in Fig.6(a). Robots begin to turn left and to go to the entrance of obstacles (see Fig. 6(b)). Then robots go forwards for three states. This process is showed in 6(c), and Fig.6(d). Robots turn right in Fig.6(e), and go forwards in the last step. The final position has showed in Fig.6(f).

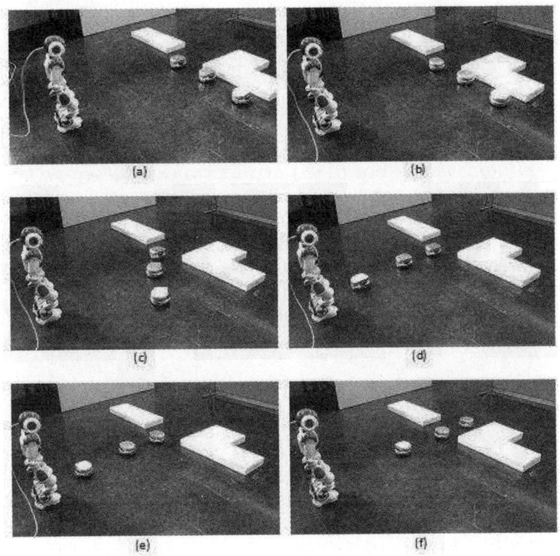

Fig. 5. First experiment: the robot Nao control wheeled robot according to environment.

5. Conclusion

In this paper, we have presented the design of new approach based on multi-robot systems about the collaboration between heterogeneous robots. This infrastructure was based on a network robots which includes one humanoid robot, wheeled robots, cameras, and remote computer. The proposed approach, which was based on machine learning for heterogeneous multi-robots systems, has allowed us to solve complex task for logistic applications. Furthermore, the preliminary results obtained during experimentation prove the feasibility of the presented strategy for real applications. Future works will focus on the improvement of the proposed method,

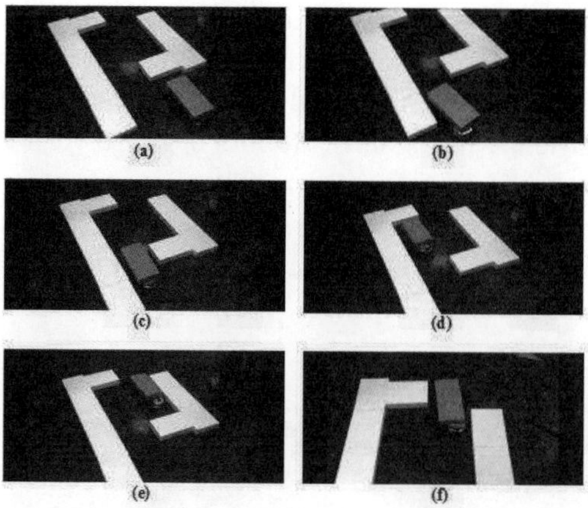

Fig. 6. Second experiment: Snapshot of experimental results where the two robots move in formation by transporting a load.

namely on more complex formation and dynamic environment. We will investigate realistic problems in the domain of industrial application. And for the long term, we will study more general approach for the self-organization between heterogeneous robot as well in home as in industrial domains.

References

1. Cao, Y. U., Fukunaga, A. S., and Kahng, A. (1997). Cooperative mobile robotics: Antecedents and directions. Autonomous Robots, 4:727
2. Parker, L. E. (2008). Multiple mobile robot systems. In Siciliano, B. and Khatib, O., editors, Springer Handbook of Robotics, pages 921941. Springer Berlin Heidelberg.
3. P. R. Wurman, R. D'Andrea, M. Mountz, *Coordinating Hundreds of Cooperative, Autonomous Vehicles in Warehouses*, (AI Magazine, Vol 29, pp:9-20, 2008).
4. Ting Wang, Fabien Gautero, Christophe Sabourin and Kurosh Madani A Neural Fuzzy Inference based adaptive controller using learning process for nonholonomic robots. The international Work-Conference on Artificial Neural Networks (IWANN), 2011.
5. Ting Wang, Christophe Sabourin, Kurosh Madani. Strategy Based on Machine Learning for the Control of a Rigid Formation in a Multi-Robots Frame. 8th International Conference on Informatics in Control Automation and Robotics (ICINCO), 2011. 4 pages.
6. Dominik M. Ramk, Christophe Sabourin, Kurosh Madani, A Cognitive Approach for Robots Vision Using Unsupervised Learning and Visual Saliency. (IWANN), 2011. Accepted Publication.

OVER THE HILL AND FAR AWAY: AERIAL/GROUND COOPERATION FOR LONG RANGE NAVIGATION

R. Boumghar[1,2,3] and S. Lacroix[1,2]

1: CNRS; LAAS; 7 avenue du colonel Roche, F-31077 Toulouse, France
2: Université de Toulouse ; UPS, INSA, INP, ISAE; LAAS; F-31077 Toulouse, France
3: Magellium Toulouse,24 rue Hermès - BP12113, 31521 Ramonville Saint-Agne, France
E-mail: firstname.name@laas.fr

This paper presents a strategy for rover navigation in initially unknown or poorly known environments. The strategy consists in determining the areas in which information is relevant to gather for the rover to reach the goal. The approach relies on a probabilistic reasoning on the currently available information on the environment, and on the models of the vehicle perception and motion abilities. The interest of perceiving a given area is assessed by analysing the way navigation costs are propagated during the search of the optimal path to reach the goal, taking into account the vehicle perception and motion models. Results illustrate the proposed strategy, and show its pertinence when compared to more classical navigation strategies. The extension towards multiple-robot coordination is straightforward, and is exemplified in the case where an aerial robot assists the rover by acting as a scout, aiming at optimizing the overall mission time.

Keywords: Autonomous navigation; navigation strategies; Air/ground robot cooperation

1. Problem statement

The problem of path finding for a ground rover (UGV) in an unknown or partially known environment is a key one in robotics. Its difficulty comes from the fact that the required information to define optimal paths, namely *terrain traversability*, is known with a varying degree of certainty. The efficacy of a given path not only depends on the information held at the decision time, but also on the information on the environment it will allow to acquire: shortest paths may be less efficient than longer ones that allow gathering more reliable information. The path finding problem to reach a given goal can thus be turned into "finding the areas where information is required to reach the goal".

Several studies[1,2] focus on the maximization of the information gain. Gancet[3] proposed an approach for perception and path planning in which confidence and utility of perception are depicted in a single robot navigation situation. Similarly, utility is defined in the paper of Nabbe[2] to extract points of interest by introducing "hallucinated maps". Some contributions exploit the available information in the context of multiple robot missions. Burgard[4] proposes an algorithm for multi-robot exploration using target points selected on frontiers between known and unknown areas. Marjovi[5] guides the exploration with goal points placed on frontiers between sensed and unknown environment and selecting them by choosing the lowest cost provided by the A* algorithm called for each point considered on the frontiers. The utility is defined by a sum of distance of other robots to a frontier point and is used to rule the target points distribution.

We consider the problem of long range navigation in poorly known environments. Our approach combines the information gain and utility maximization principles: it assesses the areas that are relevant to perceive *in order to reach the goal.* The approach is instantiated in a cooperation scheme, in which the UGV is assisted by an UAV (unmanned aerial vehicle) that senses the environment ahead the rover. The navigation problem then turns out to concurrently find sub-goals to reach for the UGV and relevant areas to explore for the UAV.

2. Environment Model

Our approach relies on an information-based decision process, that exploits the available information on the environment and a model of the robot perception abilities. The environment model is a traversability uncertainty grid dedicated to a UGV that is endowed with various motions modes (namely fast navigation on flat terrains, slower navigation on rough terrains). Each pixel of the grid encodes the probabilities $P(C_i)$ that a given navigation mode can be applied, plus a probability that it is an untraversable obstacle (cf. Fig. 1): to each pixel is associated a probability distribution[a] over the N_c terrain states C_i (in our experiments, $N_c = 3$, and the considered classes are flat, rough and obstacle).

A navigation graph is defined from the terrain model, after a quad-tree decomposition to reduce the number of nodes. Quasi-uniform regions are extracted using a criteria based on the standard deviation $\sigma_{P(C_i)}$ of the partial probabilities of the region pixels. For this purpose, the condition

[a]The way such a terrain model is built is out of the scope of the paper.

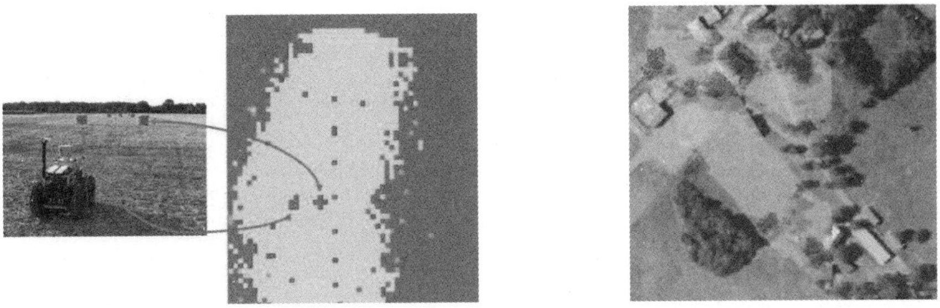

Fig. 1. Left: traversability map built on the basis of range data by a UGV. The hay balls correspond to the clusters of red pixels obstacles, single isolated pixels correspond to $1cm$ thin poles not visible in the picture. Right: traversability map extracted from a monocular sequence of a UAV[6], superimposed on an orthoimage of the terrain.

$|P(C_i) - \mu_{P(C_i)}| < N * \sigma_{P(C_i)}$ is used, where $\mu_{P(C_i)}$ is the mean of the partial probabilities $P(C_i)$ of a region, and N defines the homogeneity of the region (Fig. 2).

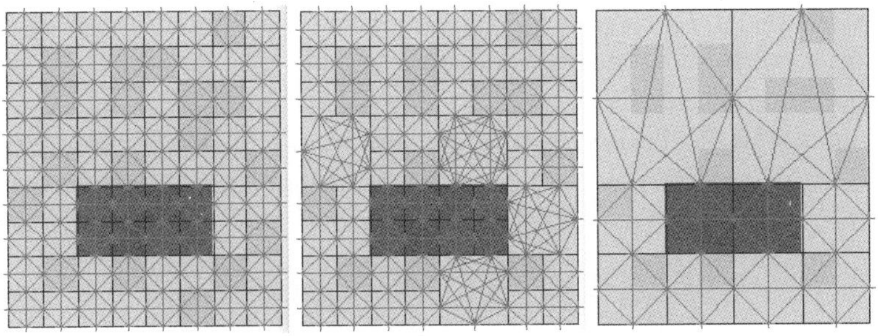

Fig. 2. Three navigation graphs defined on the same environment model. Left: graph defined without any quad-tree decomposition; center: quad-tree decomposition using a strict homogeneity criterion ($N = 0$); right: quad-tree decomposition with $N = 1.5$. The number of nodes are respectively 161, 142 and 27.

3. Assessing relevant areas to perceive

The A* algorithm is used on the navigation graph to find optimal paths for the UGV, but more importantly, we exploit the costs propagated during the path search to assess alternative paths. This information is coupled with the current traversability information and a model of the mapping processes to define the areas that are relevant to perceive to reach the navigation goal.

Navigation Cost A navigation cost is associated to the traverse of each region. The intuitive cost expectation $F_c = \sum_i^{N_c} c_i P(C_i)$, where c_i are the costs associated to each terrain class C_i, can not be used straightforwardly because of the obstacle class – to which an infinite cost should be associated.

We define the traverse cost of a region as follows:

$$F_c = \Big(w_0 + \sum_{i=1}^{N_c-1} \big(w_i(1 - \sum_{j=1}^{i} P(C_j))\big)\Big) * \Big(1 + log\big((1 - \Big(\frac{P(C_{N_c})}{P^{\lim}(C_{N_c})}\Big)^2)^4\big)^2\Big) \quad (1)$$

in which the terrain classes are ordered according to their increasing nominal costs (C_{N_c} is the obstacle terrain class), $P^{\lim}(C_{N_c})$ is the probability over which the terrain is considered as untraversable, and the weights w_i are defined on the basis of the terrain nominal costs c_i so that $c_i = \sum_{j=0}^{i-1} w_j$. In Eq. (1), the first member is akin to a cost expectation, whereas the second member assesses the risk of traversing a region.

Navigation Success Potential The minimum path depends on the current available information on the terrain, the current robot position and the goal position. In some cases such as illustrated by Fig. 3, a slight difference on these information (here the start position) implies large differences in the minimal path solution: in this case, gathering information at the right of the left side of the obstacle would tell the robot which way to go.

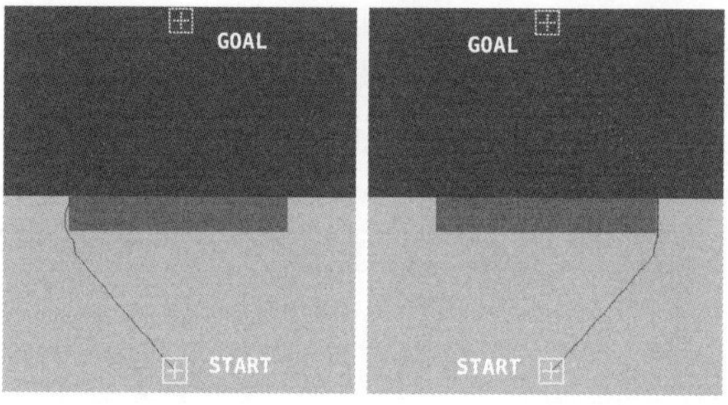

Fig. 3. Two slightly different start positions lead to two very different optimal solutions. Here only the lower part of the environment is well known and a wall in the middle forces the robot to choose a solution on the left or on the right. These two alternatives drive the robot towards two different areas of the frontier between known and unknown terrain.

Contributions on terrain exploration[4] introduce frontier points between well known and unknown areas. In order to assess which areas on these information frontiers are relevant for the navigation task, one need to determine if they correspond to useful path alternatives.

After an **A*** minimal path search, a cost is defined for every visited and open nodes of the graph. The set of evaluated nodes $\mathcal{G}_i = \{\mathcal{N}_i\}$ define an area bordered by a set of open nodes $\mathcal{G}_o = \{\mathcal{N}_j^o\}$ where the search stopped before reaching the goal. We introduce the navigation success potential (NSP) for all the nodes belonging to these sets to evaluate how promising are path alternatives. The NSP, noted $\mathcal{P}$, is calculated from the open nodes on every node of $\mathcal{G}_i$ by accounting how many other paths back-point through it. Each count for an alternative is weighted by the inverse of its estimated cost at the end of the search $F_c(\mathcal{N}_j^o)$ with the heuristic h of the open node it leads to:

$$\forall \mathcal{N}_i \in \mathcal{G}_i, \;\; \mathcal{P}_i = \sum_{j=\{\mathcal{G}_o^i\}} \frac{1}{F_c(\mathcal{N}_j^o) + h(\mathcal{N}_j^o)} \tag{2}$$

Figure 4 illustrates the NSP on the examples of Fig. 3, by displaying the value of the NSP on each node i and their respective incident edge. The highest value of the NSP shown as the lightest point in these figures is at the start point where the maximum number of alternatives are possible. As the path search evolves the NSP develops according to the number of alternatives offered by each node, the navigation cost to the open nodes it leads to and how close the alternatives are to the goal.

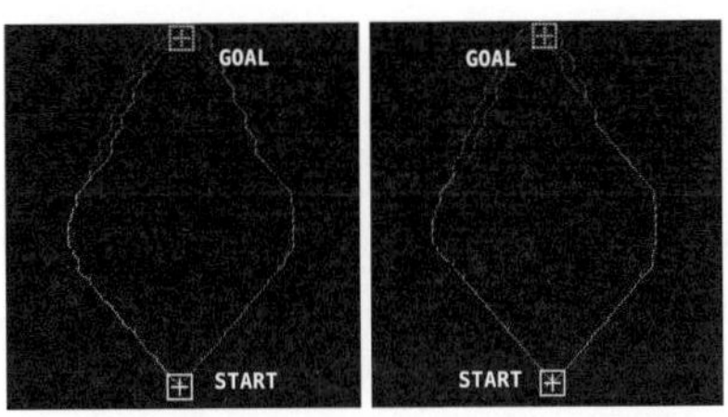

Fig. 4. Navigation Success Potential on the examples of Fig. 3. The NSP highlights the second alternative.

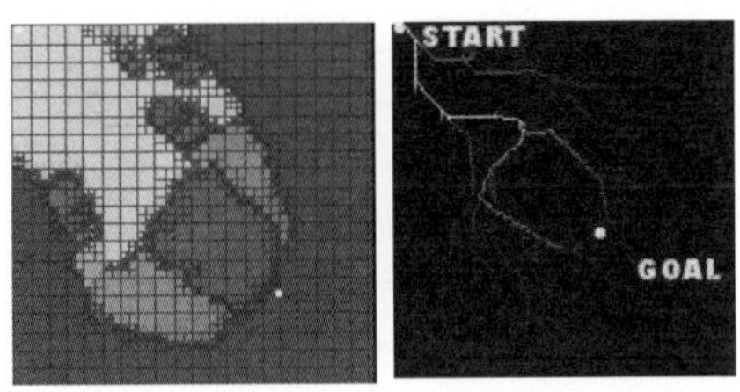

Fig. 5. NSP evaluated after an A* path search which exhibits the relevant alternatives around an obstacle (mapped in the left figure).

4. Cooperative Navigation Strategy

In an aerial/ground cooperative navigation mission where the UGV has to reach a given goal, the NSP by itself is not sufficient to decide where the

UAV must gather more information for the navigation task. It is combined with the predicted information gain that can be assessed with a model of the UAV mapping ability:

$$M_s = \begin{pmatrix} P(C_1|C_1) & \dots & P(C_1|C_{N_c}) \\ \dots & P(C_i|C_i) & \dots \\ P(C_{N_c}|C_1) & \dots & P(C_{N_c}|C_{N_c}) \end{pmatrix} \quad (3)$$

This model allows to predict the evolution of the Shannon entropy H for each perceived area. The utility to perceive areas with non-zero NSP is then expressed as:

$$\mathcal{U}_i = \mathcal{P}_i(H^- - H^+) \quad (4)$$

Where H^- and H^+ resp. denote the current and predicted entropy of the area. Figure 6 shows a step of the map discovery during a navigation mission with its corresponding NSP and navigation utility.

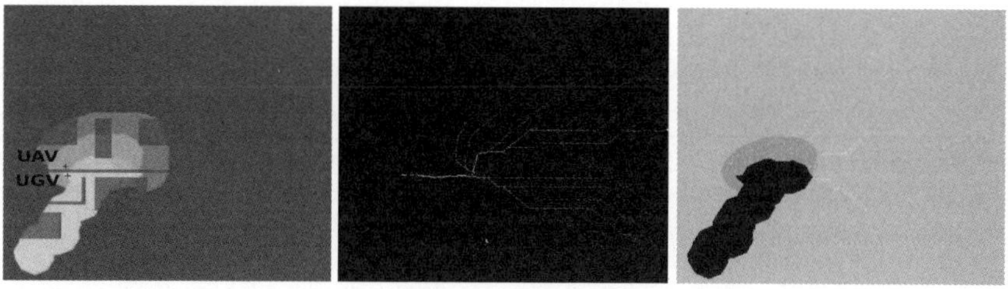

Fig. 6. A cooperative navigation step: traversability map, NSP, perception utility.

Once areas relevant to perceive are determined, the UAV is directed to perceive them on the basis of an optimisation of a utility / cost ratio, while satisfying communication constraints to ensure that it can transfer the information it acquires back to the UGV. This comes to extend the viewing horizon as in Ref. 2, but using an approach that explicitly reasons on the available information on the environment and the robot perception abilities to define the areas of interest.

5. Results

Extensive simulation results have been made to assess the efficiency of the developed strategy and to compare it to more classical strategies. The environment perception is simulated according to the sensor model (cf. Eq.

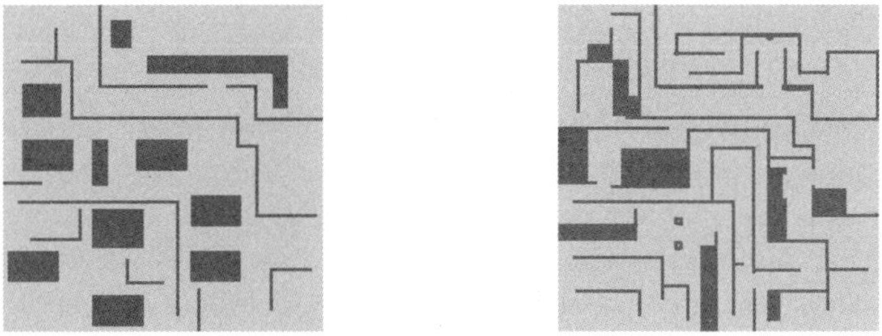

Fig. 7. The two maps used in simulation. Left: an environment with easy detours, right: an environment with many dead-ends.

(3)). The UGV is endowed with a sensor giving good perception within a 10m circular footprint and is not able to see behind obstacles, while the UAV is endowed with a larger (30m) bird-view sensor able to see all around the overflown obstacles, but with a lower quality perception.

Table 1. Comparison of Final Track Length averaged over 100 runs of different navigation strategies on the two maps of Fig. 7.

	Mono-UGV Simple Planner	UGV/UAV Overhead	UGV/UAV Scout	UGV/UAV Cooperative Approach
Final Track Length (m)				
Detour map	27930.22	12933.24	11906.09	**6178.56**
Dead-ends map	53182.93	18576.60	14293.10	**9371.88**
Explored Zone Percentage (%)				
Detour map	14.94	56.20	40.70	**65.48**
Dead-ends map	27.42	56.61	47.97	**79.53**

Table 1 shows the simulation results on two different maps shown in Fig. 7, comparing four different navigation strategies:

- Mono-UGV: The UGV alone with an A* planner.
- UGV/UAV Overhead: the UAV supports the UGV by providing overhead perception.
- UGV/UAV Scouting: the UAV supports the UGV by scouting 50 meters ahead of the current optimal path.
- UGV/UAV Cooperative Approach: the UAV uses the utility $\mathcal{U}$ to plan observation paths every time it meets the UGV to share in-

formation. Both robots start at the same position and gives each other rendez-vous points to exchange data.

The proposed approach yields paths 48% shorter than the UGV/UAV Scout strategy, and is 77% more efficient than the mono-robot strategy. Furthermore the explored zone is in general 60% wider, this results from the optimization given to the UAV paths. Our approach is the closest to the 2500m track generated by an A* planner on the totally well known dead end map. These results show that the main benefits of using NSP coupled with the information gain are mission time reduction and energy savings.

6. Conclusion

Our cooperation strategy exhibits clear benefits: the navigation success potential and information gain prediction help optimizing the UGV path, by exploring the environment where relevant information to reach the goal can be obtained. This approach to navigation somehow unifies land exploration and path finding to clear a short path to the goal.

More simulations are on their way to qualify the strategy on several maps with different obstacle densities and different kind of difficulties. The algorithms have also been integrated on actual robot architectures using the Morse simulation environment[7], and large scale experiments are currently under way.

References

1. S. J. Moorehead, R. Simmons and W. L. Whittaker, Autonomous exploration using multiple sources of information, in *Robotics and Automation, 2001. Proceedings 2001 ICRA. IEEE International Conference on*, 2001.
2. B. Nabbe and M. Hebert, *The International Journal of Robotics Research* **26** (2007).
3. J. Gancet and S. Lacroix, Pg2p: A perception-guided path planning approach for long range autonomous navigation in unknown natural environments, in *IEEE/RSJ International Conference on Intelligent Robots and Systems, Las Vegas (USA)*, Oct. 2003.
4. W. Burgard, D. Fox, M. Moors, R. Simmons and S. Thrun, *ICRA* (2000).
5. A. Marjovi, J. Nunes, L. Marques and A. de Almeida, Multi-robot exploration and fire searching, in *IEEE/RSJ International Conference on Intelligent Robots and Systems*, Oct. 2009.
6. S. Bosch, S. Lacroix and F. Caballero, Autonomous detection of safe landing areas for an UAV from monocular images, in *IEEE/RSJ International Conference on Intelligent Robots and Systems, Beijing (China)*, Oct. 2006.
7. G. Echeverria, N. Lassabe, A. Degroote and S. Lemaignan, Modular open robots simulation engine: Morse, in *ICRA, Shangai (China)*, May 2011.

LEADER/NEIGHBOUR TRAJECTORY FOLLOWING BY MOBILE ROBOTS

TEJAS K PATEL AND TARIQ P SATTAR

Centre for Automated and Robotic NDT, London South Bank University,
Department of Engineering and Design, 103 Borough Road, London, SE1 0AA,
patelt2@lsbu.ac.uk

HERNANDO LEON RODRIGUEZ

Industrial Engineering Department, Faculty of Engineering, Nueva Granada Military University, Bogotá - Colombia, hernando.leon@unimilitar.edu.co

A control scheme and a sensor system are developed to enable a group of robots to work collaboratively. The control scheme maintains a desired configuration of a group of robots while they follow the trajectory of the leader robot. One robot is a pre-defined leader and other robots in the team are followers which follow the trajectory of the leader while maintaining a desired separation distance from neighbouring robots. Simulation results are presented to show the performance of a novel control scheme. The scheme requires measurement of the distance between each robot and a pre-designated neighbour which it will follow. It also requires measurement of the angle between the orientations of the two robots. A real-world test rig is described which uses the Nintendo Wii motion-plus remote controllers to obtain the necessary sensor data. The Nintendo Wii motion-plus controller provides very cheap but sophisticated electronics (a 3-axis accelerometer, a 3-axis gyroscope, an Infra-Red camera and Bluetooth transmitter/receiver) that is used as a sensor system on each robot to enable sensing of the motion trajectory of the leader robot or its nearest neighbouring robot.

1. Introduction

Collaborative robots have been studied in recent years. The use of multiple robots has many uses including surveillance or for search and rescue during natural disasters where the use of simple cheap robots which can be deployed in large numbers would assist in the rescue of human beings by being able to search in unsafe places that are more difficult than using existing methods.

A leader-follower flocking system is reported in (1), the authors investigate a system where a few robots are leader robots "who have global knowledge (a desired trajectory), while majority of the members do not have global knowledge". The follower robots determine the flocking algorithm based on estimation of the position of the flocking centre. They also look at the possibility of followers being able to move away from the flocking group and lead to a group splitting.

We are primarily interested in using multiple robots in the task of conducting Non-Destructive Testing (NDT) where the use of robots working as a team would speed up the inspection of large areas. For example the NDT of large steel plates used in the construction of very large infrastructure is a time consuming task. A large team of robots searching for defects with a predefined configuration would speed up the inspection of these plates.

This paper presents work being done to develop a Distributed Robots System (DRS), a group of collaborating mobile robots where individual robots (follower robots) in the group respond to rotations and movements by a pre-defined leader robot (2; 3). Each robot in the group maintains a given fixed distance relative to its neighbouring robot while following the leader's trajectory. This distance can be adjusted depending on the size of defects to be found by the robots i.e. to detect large defects the robots can be further apart while for small defects they are closer to each other.

The follower robots in the group use wireless sensors to collate the information necessary about the rotations and motion of the robot designated as its nearest neighbour (which will depend on the configuration which the robots are required to maintain and thus could be in front of the robot or to its side, etc.). Each robot in the group follows the designated nearest neighbour. Thus a chain of robots is constructed leading ultimately to the leader robot.

The leader robot can be controlled remotely by an operator or be set up to move autonomously, the follower robot measures the angle of its neighbouring robot and distance between itself and its neighbouring robot to compute the trajectory of the robot, which it then tries to match. This paper is divided into the following sections. Review of the related work is presented in section 2, followed by the configurations we have tested in section 3, a overview of our simulation model is given in section 4 followed by simulation results in section 5 and details on how the test rig is being developed are given in section 6 followed by the conclusion in section 7.

2. Related Work

In recent years collaborative robots have become an increasingly active research area due to the advantages of using multiple robots to tackle complex tasks with simple robots or to speed up the performance of tasks. An algorithm has been proposed in (4) to improve the efficiency of a leader-follower flocking robot system. The aim is to move the follower robots into formation as fast and efficiently as possible given an input from the leader robot. The robots are then required to follow the motion of the leader robot while maintaining the formation. The follower robots are claimed to get into formation in 1.5 seconds and begin following the leader robot thereafter. Our control scheme has similar goals, however we do not require the follower robots to find and get into formation with the leader robot autonomously as we put the robots into formation prior to the movement by the leader robot.

Another leader-follower system (5) presents a strategy aiming to remove the risk of a follower robot losing the leader robot by using inter-robot communication using ad-hoc WLAN communication. Simulation results show a group of robots moving past some obstacles during which one following robot loses the group (using traditional methods). With the use of the inter-robot communication method, the follower robot re-locates the group and continues following the leader robots motion.

With traditional methods of control, the distance between the robot which loses the group and the leader robot increases constantly during the simulation while the distances between the leader and other robots in the group remain under control. The use of ad-hoc inter-robot communication maintains the distances between the leader robot and all other robots in the group.

Another method for co-operative robotics (6) uses language and talking to each other with experiments in which real and simulated robots play language games to ground concepts for effective communication about their world with the use of lexicons. The authors get the robots to play several different games such as "where we are", "go to", "how far", "what direction", "where is there" and "how long". In these games the robots need to associate words with locations (lexicons) in order to build a map of their environment and complete their task.

3. Group Configurations

In this work we have performed simulations of four robot configurations, see Figure 1Figure 4. The configurations are a Horizontal Chain of robots, Vertical Chain of Robots, Cross Configuration and an Array Configuration. The number of robots in each configuration has been varied with the array configuration being simulated for up to 25 robots in the group.

Figure 1: Horizontal Chain of Robots

Figure 2: Vertical Chain of Robots

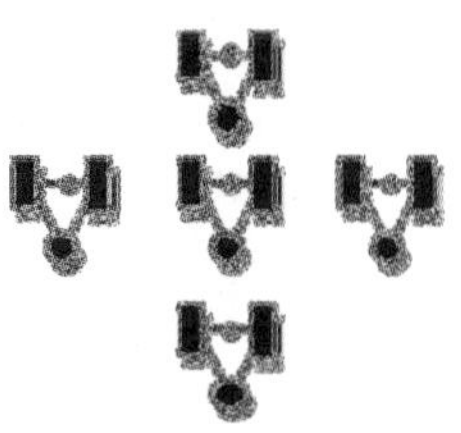

Figure 3: Cross Configuration

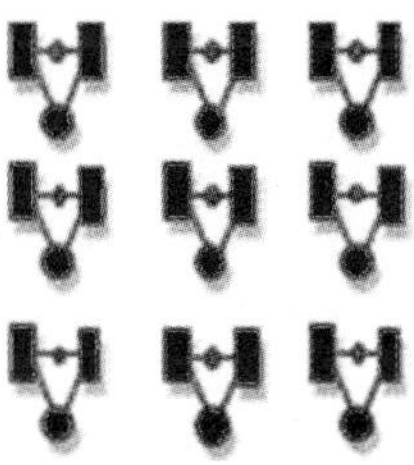

Figure 4: Array Configuration

4. Simulation Model

Kinematics previously presented in (2) has been used for the control scheme. Each robot has two actively driven wheels actuated by two independently controlled motors and one passive wheel. A sensor system measures the distance *d* between a robot and a designated nearest neighbour and also the angle between the orientations of the two robots. These signals are used as feedback by the digital control scheme to maintain a desired distance between the robots and a zero angle between the two orientations i.e. maintains the same heading.

We are using a multi loop multi input and single output *(ML-MISO)* PID controller reported in (2; 3) to follow the motion trajectory of the leader robot. The ML-MISO controller consists of inner and outer loops. The inner loops correct for errors between the motion of the leader robot and the follower robot *(tracking error).* The outer loop corrects for differences in individual robot motion dynamics to provide for greater accuracy in following the leader robot.

The ML-MISO PID controller can be fine tuned to ensure it gives the best computed value of control *'R'* to be sent to the following robot, enabling it to calculate the amount it needs to move and rotate to match the movement of the lead robot. The simulation results in section 5. show the effect of the outer control loop and how it ensures that the following robots maintain the desired distance. Each robot has dynamic functions for the rotation and motion respectively, the rotation dynamic block employs a feedback loop to ensure the robot has rotated to the desired angle. The output from this block is sent to the robot trajectory block which along with the data from the motion length *'r'* and previous position of the robot computes the new position of the robot at each sampling time.

The new position of the robot is then sent to the controller at each sampling time which along with the position of robot 2 and the desired distance calculates the value of 'r' that the robot 1 has moved by. The second robot uses the rotational information from the rotational dynamic block of robot 1 to rotate by the same angle and along with the information of the length 'r' from the controller the second robot is able to accurately follow changes in angle and position by the first robot.

5. Simulation Results

Simulation results are presented in figures Figure 5 - 16. They show the accurate performance of the controller with different robot configurations (array configuration, vertical chain and a horizontal chain), with different numbers of robots in a given configuration, desired trajectories and desired distance between the robots.

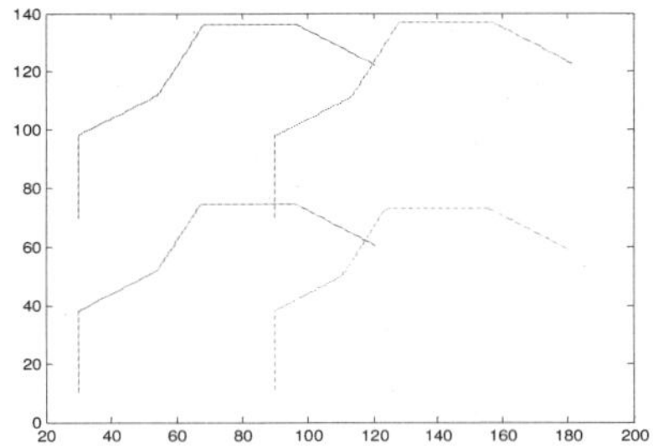

Figure 5: Graph of the X-Y position coordinates of four robot array configuration with distance of 20cm between robots

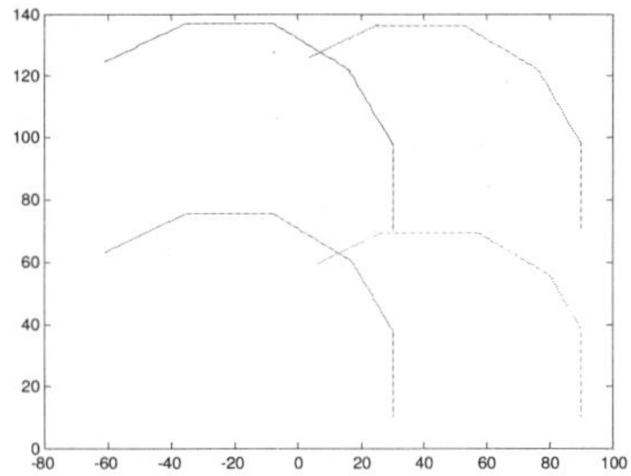

Figure 6: Graph of the X-Y position coordinates four robot array configuration with distance of 20cm and a different trajectory

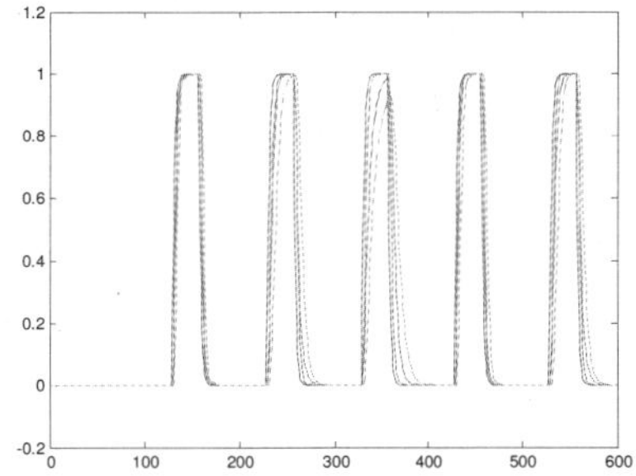

Figure 7: Motion pulse for the simulation in figure 5 with the following robots tracking the motion of the leader robot

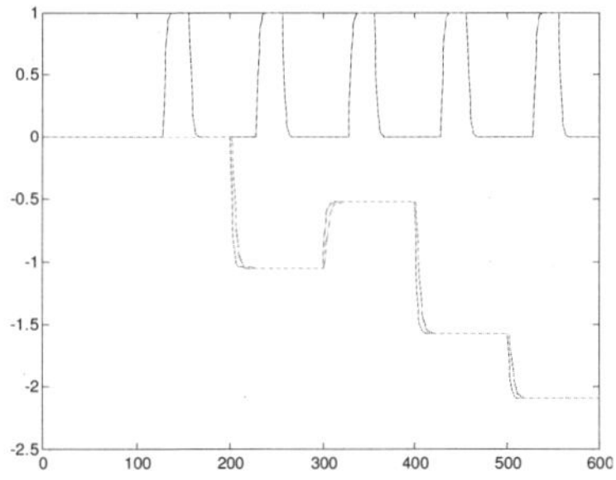

Figure 8: Angular motion graph for the simulation in figure 5 showing the following robots tracking the rotation of the leader robot

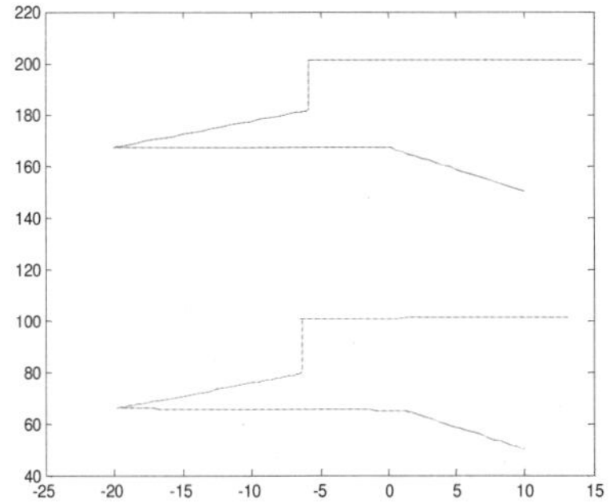

Figure 9: Graph of the X-Y position coordinates two robot vertical chain with the distance set to 100cm

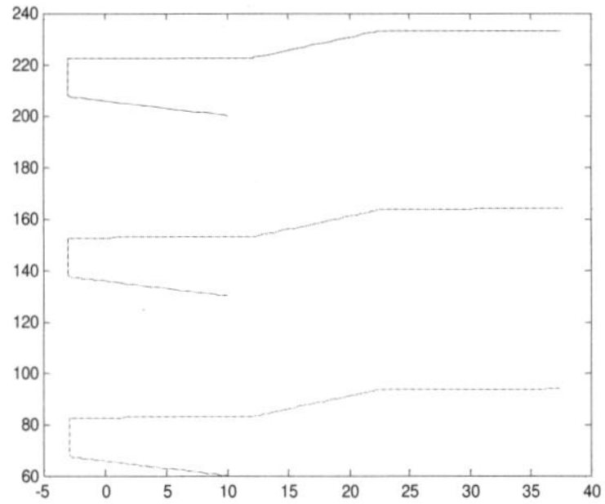

Figure 10: Graph of the X-Y position coordinates three robot vertical chain with the distance set to 70cm

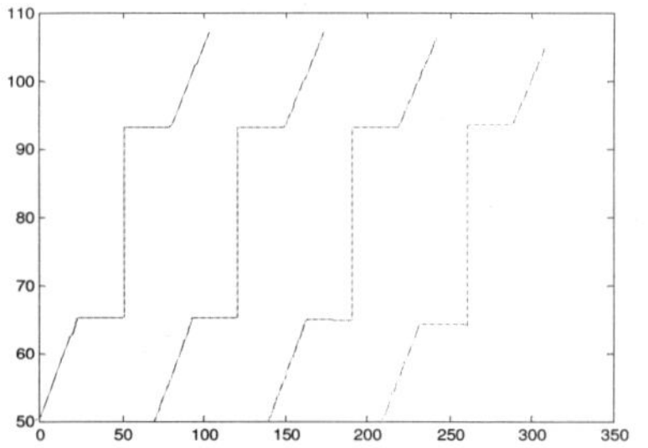

Figure 11: Graph of the X-Y position coordinates four Robot Horizontal Chain

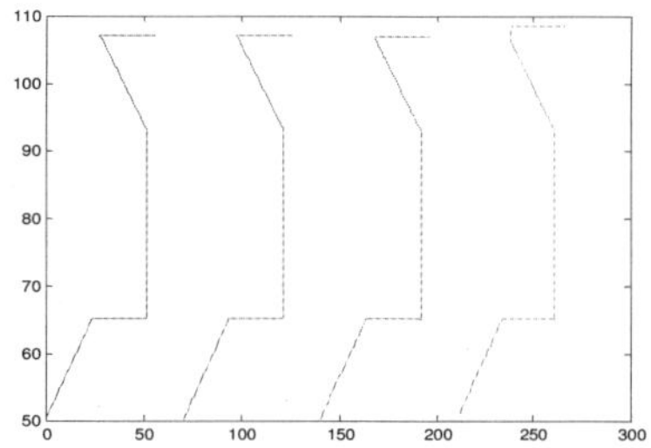

Figure 12: Graph of the X-Y position coordinates four Robot Horizontal Chain

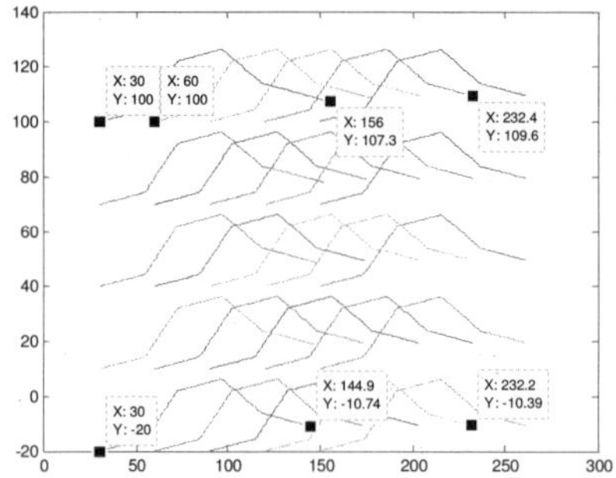

Figure 13: Graph of the X-Y position coordinates motion trajectories of a 5 by 5 array configuration distributed robots

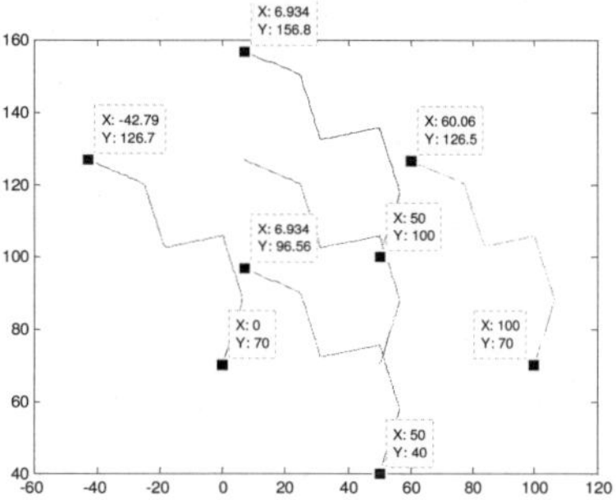

Figure 14: Graph of the X-Y position coordinates motion trajectories of a cross configuration simulation

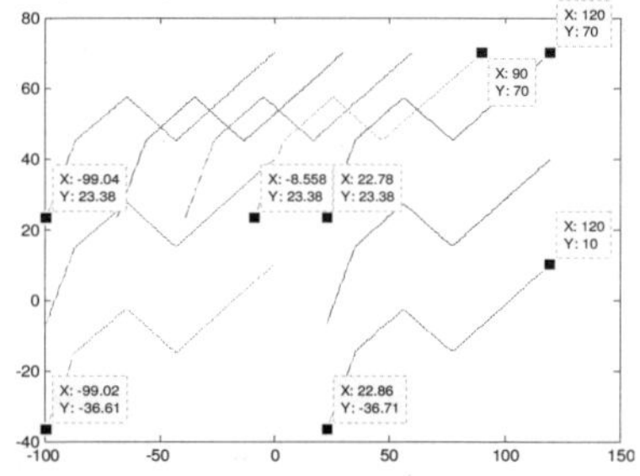

Figure 15: Graph of the X-Y position coordinates motion trajectories for N-Configuration simulation

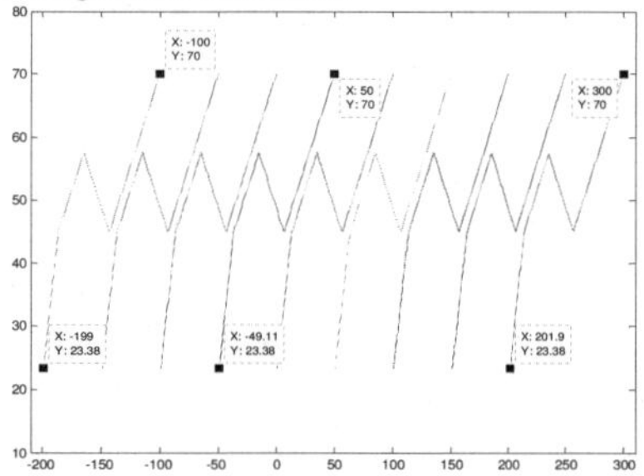

Figure 16: Graph of the X-Y position coordinates motion trajectories for horizontal N-configuration simulation

We conducted simulations for array configurations of various sizes such as 2 by 2, 3 by 4 up to 5 by 5 for which we have presented the motion trajectory, Figure 13. Although we have not simulated array sizes greater than 5 by 5, we are confident that the system will be able to maintain the overall configuration during the motion. The desired distance between the robots was set at 30cm and from the graph in Figure 13 it can be seen that the system has largely maintained the distance while maintaining the overall configuration and having all the robots match the motion trajectory of the lead robot.

Figure 14 shows the motion trajectories for a five robot cross configuration distributed robots system, leader robot is shown with the blue trajectory. The distance for this simulation has also been set at 30cm for the vertical connection and 50cm for the horizontal connection, the graph shows that again the configuration has been maintained by the system throughout the motion. The number of robots for this configuration can be increased in the four directions or a multiple cross configuration can be created by joining two cross configurations together side by side.

Figure 15 and Figure 16 show the motion trajectories for an N-configuration. In this configuration there are a central group of robots positioned in a horizontal chain with robots positioned vertically at both ends of the chain. Dynamic differences in robot dynamics and inner control loop positioning errors were incorporated into the system but the DRS managed to keep the motions of all the robots largely error free with all the robots tracking the leader robot accurately. It can be said from the results that the DRS operates correctly at any heading angle and maintains the desired distance between the robots which can be set to any value required. The number of robots in the system does not affect system operation. The errors emanating from the dynamic differences between the robots have been minimised by our control architecture resulting in a near zero error tracking of changes by other robots in the DRS.

6. Test Rig

A Test rig for the system is currently being developed, which involves using wireless sensors mounted on Amigobot's. Amigobot's have been developed by Adept Mobile Robots Inc. The robots are equipped with sonar sensors, gyroscopes and wi-fi communication. They have the ability to navigate autonomously in any environment using maps of the environment. We have also acquired a Pioneer 3-DX robot which has all the equipment of the amigobot with the addition of an on-board computer and a stereoscopic camera also developed by Adept Mobile Robots (7).

Our system is being tested to verify the results obtained in simulations. Tests are being done using different configurations, distance between robots, the number of robots, *(constrained to 5 for the test rig)*, and different motion trajectories. We are using a Nintendo Wii motion plus controller with the Wii sensor bar, the combination of the two will provide the angles and distances required for the control of our system. The Wii controller uses a high speed infrared vision camera to simultaneously track up to five infrared LED's, has 6 degrees-of-freedom using a 3-axis accelerometer, a 3-axis gyroscope, and Bluetooth technology to communicate with the Wii. It uses the standard Bluetooth HID protocol and therefore appears as a standard input to any Bluetooth host (8).

The Wii sensor system consists of two infra-red LED clusters mounted on the leader robot with a receiver mounted on the follower; this measure's the distance between the robots and with the aid of the accelerometer provides the angle of the leader robot. The information from the Wii controller is transmitted to a remote computer via Bluetooth communication. This data is managed to compute the necessary motion trajectory by the following robots to track the leader robot. Instructions will then be transmitted to the following robot to autonomously follow the leader robot.

7. Conclusions

Simulations of the Distributed Robots System (DRS) show a high degree of accuracy which would enable the system to be used for Non-Destructive Testing (NDT) where the speed and accuracy matters. As the system uses multiple robots the time required to carry out of the NDT would be reduced, also if required multiple methods could be employed like ultrasound, visual inspection or eddy currents. Our simulations have been carried out using up to 25 robots; however our ML-MISO PID control with inner-outer loop control system has maintained the level of accuracy achieved while using fewer robots in the group. We are confident that we can operate the system with more than 25 robots in the group, with the accuracy still being maintained.

Future work will attempt to validate the simulation results with the test rig based on a group of five Amigo robots and the Wii gyroscope/infrared vision system.

8. References

1. *Leader-Follower Flocking: Algorithms and Experiments.* **Gu, Dongbing and Wang, Zongyao.** 5, s.l. : IEEE Transactions, 2009, Vol. 17. 1063-6356.

2. *Distributed Mobile Robots.* **Sattar, Tariq, Patel, Tejas and Rodriguez, Hernando Leon.** Istanbul : World Scientific, 2009. Clawar. Vol. 2009.

3. *Distributed Robot Systems.* **Patel, Tejas K and Sattar, Tariq P.** Nagoya : World Scientific, 2010.

4. *Fast and Efficient Formation Flocking for a Group of Automous Mobile Robots.* **Xiong, Naixue, et al.** 3, s.l. : IEEE, 2008, Vol. 08. 978-1-4244-1694-3/08.

5. *Multi-robot Formation Control Using Leader-follower for MANET.* **Zhang, Yi, et al.** Guilin, China : IEEE, 2009. 978-1-4244-4775-6/09.

6. *Robots, Communication, and Language: An Overview of the Lingodroid Project.* **Schulz, Ruth, et al.**

7. Research Robots. *Adept Mobile Robots.* [Online] 2011. http://www.mobilerobots.com/ResearchRobots/ResearchRobots.aspx.

8. **WiiBrew.** Wiimote. *WiiBrew.* [Online] 3 March 2011. [Cited: 2011 March 10.] http://www.wiibrew.org/wiki/Wiimote.

SECTION–5

EDUCATION, INTELLIGENCE AND LEARNING FOR CLAWAR

A STUDENT-BUILT BALL-THROWING ROBOTIC COMPANION FOR HANDS-ON ROBOTICS EDUCATION

J. Oberländer*, T. Harbaum, G. Kurz, N. Ahmed,
T. Kos-Grabar, A. Hermann, A. Rönnau, R. Dillmann

FZI Forschungszentrum Informatik, 76131 Karlsruhe, Germany
**E-mail: oberlaender@fzi.de www.fzi.de/ids*

We describe the Cataway (Catapult Segway), a Segway RMP-based mobile robotic platform which can autonomously throw a ball at a given target, intended to be used as a companion for a human beach ball player. The platform's throwing mechanism and software architecture were designed, implemented and tested as part of a students' lab course in practical mobile robotics. The concept allows students to gain practical experience with an actual mobile robot through a motivating scenario and a gentle learning curve. We present results of a simple, yet effective machine learning approach which allows the robot to learn to hit its target from any position.

Keywords: machine learning; education; entertainment robots

1. Introduction

The platform was developed in the course of a students' mobile robotics lab course held at the FZI Research Center for Information Technology, an institution that is strongly tied to the Karlsruhe Institute of Technology (KIT). The aim of the course was to enhance the mobile platform using simple means to be able to visually identify a target surface, throw a ball at it and learn how to throw to get the best results. The platform represents a prototypical "robotic beachball playmate", to be enhanced and improved in consecutive lab courses. By reducing the throwing task to learning an approximate linear relationship between the target size in the camera image, the throwing angle, and the resulting vertical distance from the target center, we obtain a simple, easy-to-implement learning scheme. In this paper, which is the result of joint work with the students who participated in the lab course, we show how we achieve good performance hitting the target center using this approach.

2. Related Work

In 1987 Aboaf *et al.* described a task-level approach to robot learning for the application of throwing a ball onto a rectangular surface.[1] The authors demonstrated that learning at the task level does not require a perfect model of the robot's behavior at lower levels in order to achieve good results. The approach learns how to throw the ball from a fixed position, by adjusting the release velocity and locally estimating the parameters of the underlying model using feedback from previous throws. The approach is similar to our approach, but uses a much more complex model and a fixed distance to the target. Also, the authors rely on a human to manually determine the position where the ball hit the target.

In 2010, the Italian Institute of Technology presented an iCub[2] humanoid robot capable of learning the skill of archery.[3] Having been programmed to hold a bow and release the arrow, a regression-based learning algorithm was used to learn the optimal relative positions of the robot's hands in order to hit the target. The algorithm uses camera feedback and considers only the N best hits and the relative hand positions that initiated them. The relative hand positions are then modified under the assumption that they can be linearly approximated within a small neighborhood of the given observations. Experimental results showed that the algorithm needed ten throws to converge on the target center, but the learned result is only valid for a given target distance. In our work we also use a linear interpolation scheme to minimize the vertical error. However, we will show that our linear model can provide a reasonably good global approximation, enabling us to throw from any given distance, rather than a fixed position.

In 2010 researchers presented a stationary robot designed to throw a ball into a basket which moves randomly in the robot's field of view.[4] A stereo camera system is used to locate the target in 3D space. The algorithm estimates the optimal release speed of the end effector and its angle towards the target. The release speed is related to the target distance by a cubic polynomial whose coefficients are determined offline using a least squares method. This method is successful at a distance of 3 m about 99 % of the time. While this approach is the only one in which the relative position between the robot and the target is variable, it does not perform any online learning, and does not receive feedback on the throwing results.

3. Hardware/Software

The robot's hardware is based on a Segway RMP-200, a self-balancing two-wheeled robotic platform. It is equipped with a PC from which we control the platform's behavior. A self-constructed catapult is mounted on top of the platform. The catapult can be automatically triggered by using a motor which is connected to the PC by a UCoM (Universal Controller Module). In order to perceive the environment the robot is equipped with a Firefly MV USB FMVU-03MTC camera.

Fig. 1. Used robot platform and developed catapult

The software is based upon the Modular Controller Architecture MCA2,[5] a robot programming framework which was developed at FZI. MCA2 structures the program into individual modules which can be grouped hierarchically. Each module can have a number of controller and sensor inputs and outputs. Different modules can be connected by creating edges from one module's output to another module's input. From the provided code base, students can quickly add their modules to control the Segway, the camera and the catapult in a straightforward manner.

4. Implementation

As the central idea of this work is to try and obtain a good vertical placement by simple, easy-to-implement means, we make a number of simplifying assumptions. First, we assume that the target will be located in the same position throughout all trials and is detected in the camera image relatively easily. The robot, however, may be placed at any arbitrary spot within reach of the throwing mechanism. Since the height at which we hit the target depends on the distance to the target and the Segway's pitch angle at the time the catapult is released, the task is simplified to finding the right pitch angle. The Segway platform can then be controlled using a fixed movement sequence which ensures that a wide range of pitch angles is covered in a repeatable manner (see Fig. 2). There are, then, two ways of trying to throw the ball at the right time: The simplest way is to release the catapult at a precise moment in time after the start of the movement sequence. The alternative way is to monitor the Segway's pitch angle directly and trigger the catapult release when the desired pitch angle is reached.

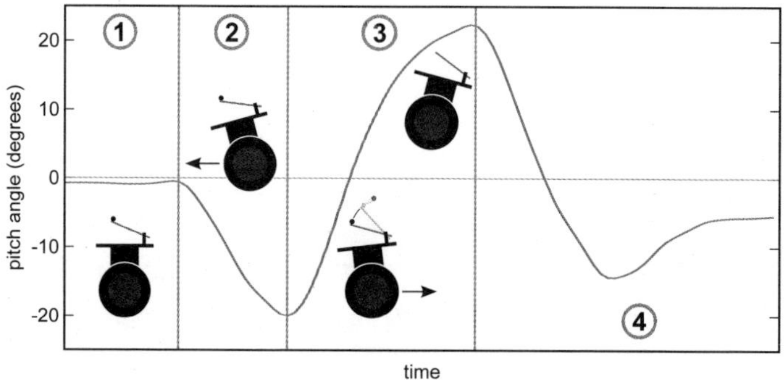

Fig. 2. Catapult mechanism and movement sequence of the Segway. From the starting position (1), the Segway first accelerates away from the target (2) and then towards it (3), before coming to a rest again (4). The catapult is triggered in phase 3, either by picking a fixed launch time, or by directly observing the Segway's pitch sensor and waiting for a given threshold. The underlying curve shows the approximate pitch angle over the course of the sequence.

The vision component operates as follows. The source image, as obtained from the camera, is binarized by an HSV filter to detect the black target area. If necessary, morphological operations can be used to remove misclassified pixels. Then the largest connected component is determined and assumed to be the target. Its center point and the four corners are extracted. The ball is detected by removing everything except the target and considering the connected components of the inverted image. If there are two components, the smaller one is assumed to be the ball (this simple approach means that a ball of any color can be used as long as it is different from the target area). The segmentation result is shown in Fig. 3.

The ball position is subsequently transformed by a perspective transform mapping the target area to the square $(-1,-1)$ – $(1,1)$. A perfectly centered hit would thus end up at position $(0,0)$, and the throwing behavior can be optimized easily by minimizing the absolute offsets. While the horizontal error is minimized by rotating the Segway, the vertical error is minimized by adjusting the moment at which to release the catapult. The horizontal and vertical errors can therefore be optimized independently. In

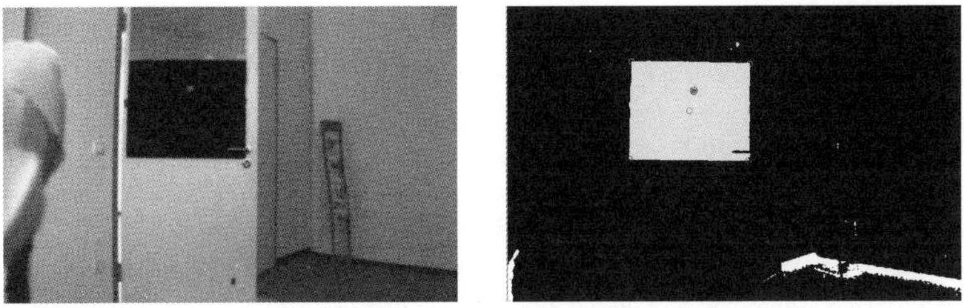

Fig. 3. Original image and segmented image. The target corners and centroid are marked as well as the ball, which in this example was thrown a little too high.

this paper we only concern ourselves with the vertical error.

Initial experiments indicated that the catapult hardware only reached a fairly low level of precision (a mean vertical error of about 3.2 cm, the 90th percentile being at 6.3 cm; cf. Fig. 4). This, along with the given setup (target location, distance, and ball velocity), encouraged us to reduce the task to learning a simple linear relationship between the target size in the camera image, the catapult release pitch, and the resulting vertical position. While this is not physically correct, the results presented below show it to be a reasonable assumption for our task.

This assumption means that the ideal throwing behavior can be learned with a simple Reinforcement Learning scheme, without requiring any complex knowledge of the underlying model. We simply collect a number of samples $(s_i, t_i, v_i)_{i=1}^n$, where s_i represents the height of the target area in the camera image, t_i represents the chosen release point (either a fixed time value or a desired pitch), and v_i represents the resulting vertical ball position on the target. We can then perform linear regression to fit a plane with

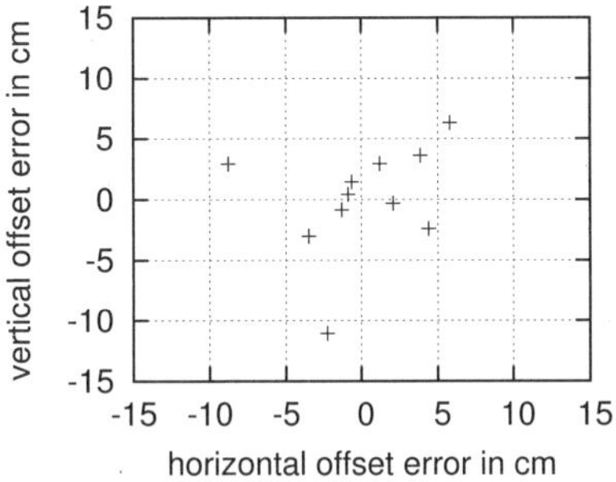

Fig. 4. Offset error observed while the Segway was stationary at a fixed position approximately 1 m from the target. The graph shows the imprecision of the catapult mechanism.

minimum squared error in v_i by solving the overdetermined linear equation

$$\mathbf{X}b = y \quad \text{with} \quad \mathbf{X} = (1\ s_i\ t_i)_{i=1}^n \text{ and } y = (v_i)_{i=1}^n \ , \tag{1}$$

obtaining the solution

$$\hat{b} = (\mathbf{X}^T\mathbf{X})^{-1}\mathbf{X}^T y \ . \tag{2}$$

The line where the resulting plane $v = b_1 + s \cdot b_2 + t \cdot b_3$ intersects $v = 0$ allows us to determine a linear function relating the target size to the optimal timing. To make the plane fit robust against outliers, a RANSAC scheme is used for the plane fit.

In order to see whether a time-triggered release of the catapult could provide enough accuracy, we analyzed the repeatability of our movement sequence. The results, depicted in Fig. 5, show the Segway's behavior over 33 trials. The plots show that the pitch angle at a fixed point in time may vary by up to 10 degrees. As this would result in extremely weak performance, we chose instead to trigger the catapult release upon reaching a specified pitch angle. The Reinforcement Learning scheme then learns the optimal pitch angle form a series of trials using the plane fitting approach described above.

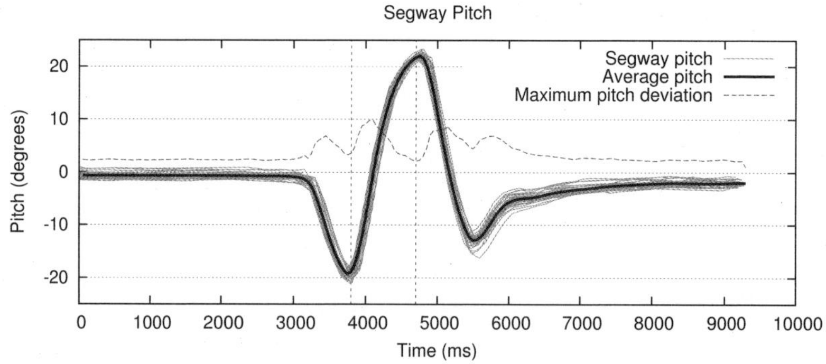

Fig. 5. Pitch analysis of 33 trials of repeating the same movement sequence on the Segway. The part of the sequence where the ball is thrown is between approx. 3800 ms and 4700 ms (indicated by the vertical dashed lines).

A plane fit requires only three trials, but due to hardware and timing imprecisions, giving noisy results and potential outliers, we perform six initial trials with randomly selected pitch angles. Further trials use the pitch angle calculated from the learned data, but the additional observations are added to the set of observations to improve the robustness of the plane

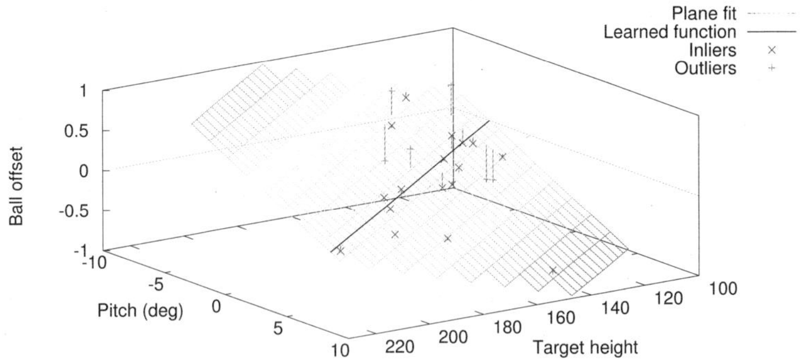

Fig. 6. RANSAC fit of a plane over 24 observed trials, with six outliers. The plane approximates a linear relationship between the height of the target in the camera image, a chosen pitch at which to throw the ball, and the resulting vertical offset of the ball on the target. The line with zero offset represents the optimal pitch angle for a given target height.

fit. The learning scheme can be made adaptable, e. g. to adjust to catapult wear and tear, by only keeping a fixed number of the most recent trials in memory. Fig. 6 illustrates the process of determining the desired pitch angle from the observed data.

With the given learning scheme, one can simply conduct a number of trial throws for arbitrarily chosen target distances, and the robot will learn the desired function automatically.

5. Results and Conclusion

The final experiment spanned 24 observed trials (and four more in the early exploration phase where the robot missed the target altogether). Once three observations were available, the Reinforcement Learning scheme using the plane fitting approach described above could propose a pitch angle for each of the following trials. In trials 4–6, the proposed angle was further perturbed by Gaussian noise in order to better explore the parameter space and have a better basis for the plane fit. Fig. 7 shows the evolution of the absolute error in vertical offset over the course of the trial period. The regression line illustrates how the error shrinks; after the seventh trial, the error is never more than about 16 cm, still dominated by noise and outliers. For the last seven trials the results look quite positive, with a mean error of 3.25 cm, a median error of 2.81 cm, and a maximum error of 9.05 cm.

We have presented an approach that allows a mobile robot with a catapult mounted on top to learn how to hit a target from different positions.

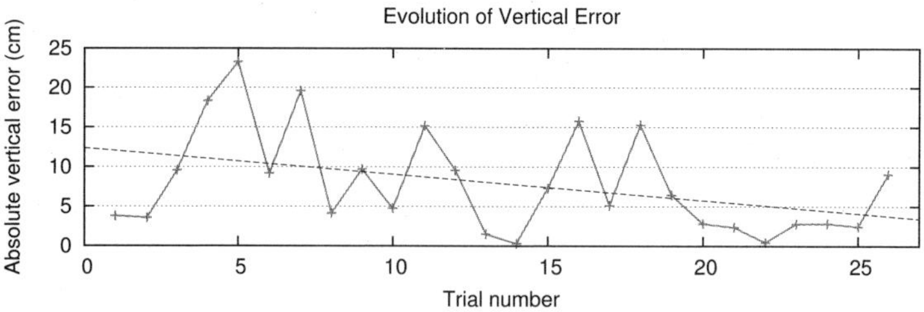

Fig. 7. Evolution of the vertical offset error over the course of 24 trials, using the Reinforcement Learning scheme based on a simple RANSAC plane fit. The first six trials are part of the exploration phase, where the chosen pitch angles are randomized.

A vision algorithm and a machine learning scheme based on Reinforcement Learning were implemented which give encouraging results even with a very simple underlying model: it simply postulates a global linear relationship between the target distance, the pitch angle, and the resulting position on the target surface. Considering the obstacles resulting from mechanical imprecision and inaccurate timing on a non-realtime system, a high precision could be achieved.

Further work might include improvements to the mechanical components of the system. So far only the vertical accuracy is considered and automatically optimizing the horizontal accuracy could be added to the system. Hitting a moving target could provide additional challenges in the future: implementing the visual detection of actual beach ball rackets can bring us closer to being able to play ball with a human.

References

1. E. W. Aboaf, C. G. Atkeson and D. J. Reinkensmeyer, *Task-level robot learning: Ball throwing*, Tech. Rep. 1006, Massachusetts Institute of Technology (1987).
2. G. Sandini, G. Metta and D. Vernon, The iCub cognitive humanoid robot: An open-system research platform for enactive cognition, in *50 Years of AI*, ed. M. L. et al. (Springer-Verlag, 2007).
3. P. Kormushev, S. Calinon, R. Saegusa and G. Metta, Learning the skill of archery by a humanoid robot iCub, in *10th IEEE-RAS International Conference on Humanoid Robots (Humanoids 2010)*, 12 2010.
4. J.-S. Hu, M.-C. Chien, Y.-J. Chang, S.-H. Su and C.-Y. Kai, A ball-throwing robot with visual feedback, in *Intelligent Robots and Systems (IROS), 2010 IEEE/RSJ International Conference on*, 10 2010.
5. K. Uhl and M. Ziegenmeyer, MCA2 – an extensible modular framework for robot control applications, in *Proceedings of CLAWAR2007, 10th International Conference on Climbing and Walking Robots*, 7 2007.

A REAL-TIME ROBOT VISION APPROACH COMBINING VISUAL SALIENCY AND UNSUPERVISED LEARNING

DOMINIK MAXIMILIÁN RAMÍK
CHRISTOPHE SABOURIN
KUROSH MADANI

Signals, Images, and Intelligent Systems Laboratory (LISSI / EA 3956), Paris Est University, Senart Institute of Technology, Avenue Pierre Point, 77127 Lieusaint, France

Facing the complexity of the real world, no robot can rely solely on pre-programmed knowledge and human-prepared data. If we are to create a truly autonomous robot, learning on-the-fly in the real environment is a must. The aim of this paper is to present a novel hybrid approach to unsupervised real-time learning of objects in context of mobile robotics. We develop on techniques inspired by human visual system and by research on how human infants learn. We combine our contribution on salient object detection with state-of-the-art object recognition algorithms in order to acquire both fast learning and recognition capabilities for a humanoid robot. To test it, verification on the MSRA Salient Object Database benchmark is carried out as well as several experiments with learning generic objects in a real office environment using humanoid robot Nao.

1. Introduction

In close future, humans will share their environment with autonomous mobile robots (wheeled, legged or humanoid ones). These robots will be used to provide physical services to humans. In this context, the human-robot interactions will require of robots the same abilities as those of humans, mainly in perception and recognition tasks. In this paper, we focus on a fundamental skill of humanoid robots: the learning process based visual perception. Any fully autonomous robot needs perception to navigate in space and recognize objects and environment in which it evolves. However, most contemporary approaches to robot navigation, mapping, exploration and similar problems (see [1] or [2] for an instance) depend on human-labeled data when it comes to object recognition. However, given the complexity of the real world, no intelligent system can rely solely on pre-programmed knowledge and human-prepared data. What is crucial is the ability to self-organize and to learn on-the-fly in real environment (e.g. human-made environment for service robots) and in real time. In this paper, we present an unsupervised approach enabling a mobile robot, on one hand, to identify salient parts of the environment in the camera video stream, and on the other hand, to learn objects from the perceived, unlabeled (i.e. raw) images. The proposed approach allows also for direct recognition of previously learned objects. To demonstrate capabilities of our approach, we apply it to a humanoid robot Nao, thus enabling the robot to learn without human intervention new objects encountered while moving in a real indoor environment.

The next paragraph will explain foundations of the presented system, followed by detailing on our approach to visual saliency calculation and salient object extraction in paragraph three. Paragraph four gives a description of learning methods we propose to learn previously extracted objects. The fifth paragraph presents results that we have achieved and the last paragraph contains discussion and potential future extensions of our work.

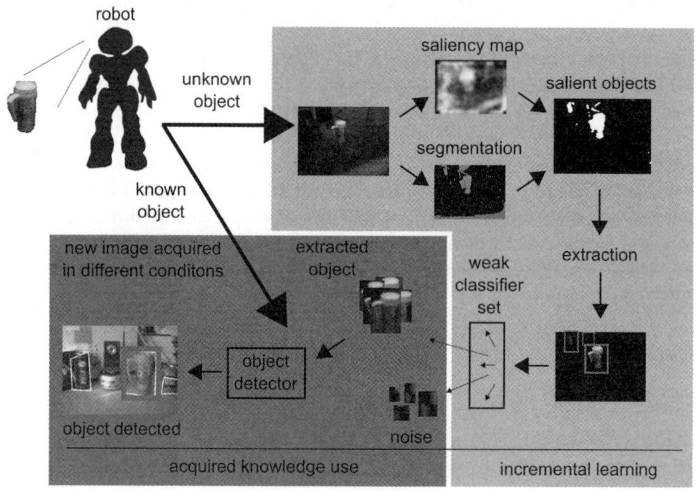

Figure 1. An overview of the entire proposed system. An unknown object is incrementally learned and the acquired knowledge is provided to the robot for further object detection.

2. Overview of the Proposed System

An overview of our system is depicted on Fig. 1. At first, robot takes images while encountering previously unseen objects. Saliency map of each image is calculated and salient objects are extracted. Same objects found on successive images are grouped over the whole sequence of images. Only objects with sufficiently high frequency of occurrence in the sequence are learned by the object detector. Having the object detector learned, known objects can be detected directly from the input image when seen again, without recurrent saliency calculation.

The presented approach extracts objects of interest from raw image data by means of visual saliency. This is a biologically inspired technique proved to play an important role in human learning process [3], especially in distinguishing important information from the irrelevant one. Building on existing work on the field of visual saliency (notably [4], [5] and [6]), we propose a novel fast salient object detection algorithm. It makes use of photometric invariants, which allows increasing the robustness in comparison to the traditional visual saliency. It serves to gather sample images of objects for learning the robot. To enable the system to recognize previously seen objects in real-time, we use a fast object recognition method, SURF [7].

For experiments in real environment, we used humanoid robot Aldebaran Nao to learn presented objects. Images were taken with robots monocular color CMOS camera and processed by a remote PC communicating with the robot via network connection. This heterogeneous structure allows on one hand use of necessary computing resources dedicated for image processing and learning. On the other hand it eventually allows for simple accessing of learned knowledge by other robots connected to the same network to promote robot-robot collaboration and shared cognition.

3. Visual Saliency Driven Object Extraction

We propose a novel visual saliency detector composed of two independent parts, which can be computed in parallel. The first part captures saliency in terms of hybrid distribution of colors (i.e. a global saliency characteristic). The second part calculates local characteristics using a center-surround operation. Their resulting saliency maps are eventually merged together using a translation function, resulting in the final saliency map.

The global saliency map M is computed following Eq. (1) as a hybrid combination of intensity saliency map M_L (Eq. (2)) and chromatic saliency map $M_{\phi\theta}$ (Eq. (3)). C stands for color saturation. As a color space, we use a spherical interpretation of RGB, which allows us to work with photometric invariants (see [8] for more details).

$$M(x, y) = \frac{1}{1+e^{-C}} M_{\phi\theta}(x, y) + \left(1 - \frac{1}{1+e^{-C}}\right) M_L(x, y) \quad (1)$$

$$M_L(x,y)=\left\|I_{\mu L}-I_L(x,y)\right\| \tag{2}$$

$$M_{\phi\theta}(x,y)=\sqrt{(I_{\mu\phi}-I_{\phi}(x,y))^2+(I_{\mu\theta}-I_{\theta}(x,y))^2} \tag{3}$$

The saliency map M is combined with result of center-surround feature detector D [5], resulting in the final saliency map M_{final} (Eq. (4)), which is used to extract salient objects themselves from the image.

In order to obtain its coherent parts we apply a fast segmentation algorithm [9] on the original image. This segmentation algorithm uses equally a spherical interpretation of RGB color space and is particularly robust to difficult real-world illumination conditions such as directed light, shadows and reflections. Subsequently we extract only those segments that are salient enough. A more detailed description can be found in [10]. On a common PC our algorithm needs in average 100ms per 320x240px image. In the MSRA Salient Object Database* benchmark [5], we have achieved results close to those of authors of [5], but with the advantage of real-time processing and native capacity to detect multiple salient objects on image at once and not only the most salient one.

$$M_{final}(x,y)=\begin{cases} D(x,y);\ if\ (M(x,y)<D(x,y) \\ M(x,y)*D(x,y);\ otherwise \end{cases} \tag{4}$$

4. Learning and Recognition

Both processes of learning and recognition are carried out on-line. The following algorithm describes our learning workflow. In the first time, the algorithm classifies each found fragment, and in a second time, the learning process is updated (on-line learning).

```
acquire image
extract fragments by salient object detector
for each fragment F
    if(F is classified into one group)
         populate the group by F
    if(F is classified into multiple groups)
          populate by F the closest group by Euclidian distance of
     features
if(F is not classified to any group)
create a new group and place F inside
```

* http://research.microsoft.com/en-us/um/people/jiansun/SalientObject/salient_object.htm

```
select the most populated group G
    use fragments from G as learning samples for object detection
algorithm
```

For grouping of extracted salient objects, we employ a combination of four weak classifiers, which classify each object by means of its features such as area, aspect ratio, chromaticity distribution and texture uniformity. More details on the online classification process are described in [10]. Groups obtained by this process contain at last images of known objects, which can serve as a learning sample database for a detection algorithm. For future detection of the classified objects, any suitable real time recognition algorithm can be used. For demonstration, we have used the SURF detector [7], as it provides fast detection and is robust to partial occlusions and perspective deformations. In case of this particular detector, we use fragments acquired as matching templates. To preserve the real-time operation of detection even in presence of high numbers

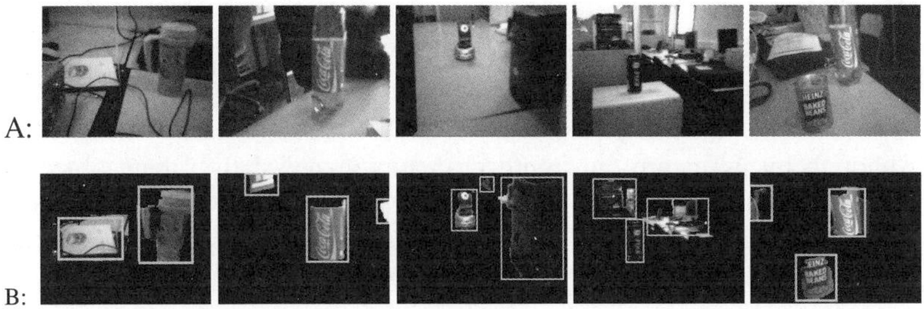

Figure 2. Sample images from training sequences. Fragments containing salient objects detected by our algorithm are marked by rectangles. A: raw robot camera images. B: salient objects found and segmented.

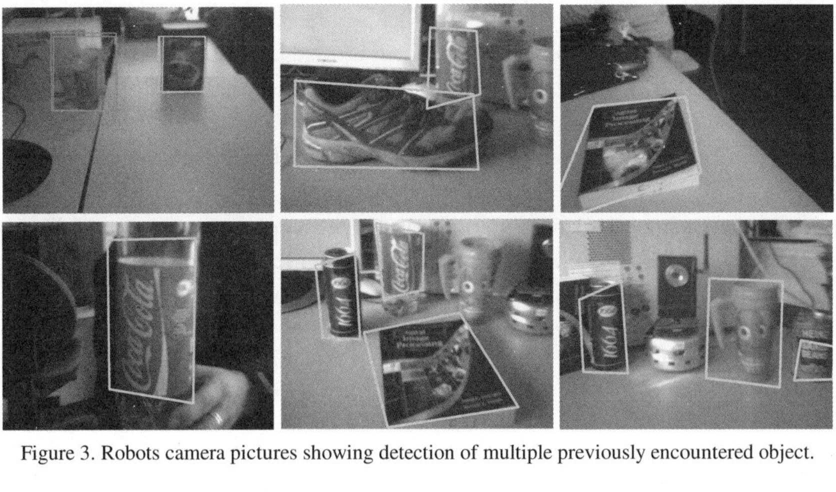

Figure 3. Robots camera pictures showing detection of multiple previously encountered object.

of templates, we pre-extract key points from each template in advance. In detection stage, we match first in several parallel threads templates with the greatest number of key-points (i.e. containing more visual information) and stop this process when another image from camera arrives. This gives us opportunity to test up to few tens template matches per frame. Further important speed-up can be achieved using parallel computation power of CUDA-like architectures on modern GPUs.

5. Results

For experimental verification, we have collected ten common objects with different surface properties to be explicitly learned and presented them in a wide variety of light conditions, poses, distances and visual contexts (Fig. 2). The number of images acquired for each object varied between 100 and 600 for learning image sequences and between 50 and 300 for testing sequences, with multiple objects occurring on the same scene. Note that the high number of learning images was taken primarily in order to test our saliency detection and fragment grouping algorithm. The learning process itself required generally significantly less samples acquired to perform sufficiently well.

Although by its nature our approach learns any visually salient object in robot's environment, for the sake of clarity the following numbers represent only the rates achieved for ten particular "target" objects we have chosen for the robot to learn.

During learning, objects were correctly extracted from 82% of the images acquired by the robot. The subsequent grouping of fragments has achieved on the same dataset success rate of 96%, i.e. only 4% of image fragments (usually bearing close visual resemblance to other similar objects on the scene) were placed into a wrong group, which contained a fragments from a different object. Using the SURF detector, learned objects were then successfully recognized, when encountered again, in 89.3% of cases. Several objects were exposed in robots field of view, presented in visual context different to the learning time; see Fig. 3.

To demonstrate real-time abilities of our system, we have successfully run experiments, where the Nao robot was required to learn a presented object and then to find it in its environment, to track and to follow it. Images from a video acquired during those experiments are shown on Fig. 4.

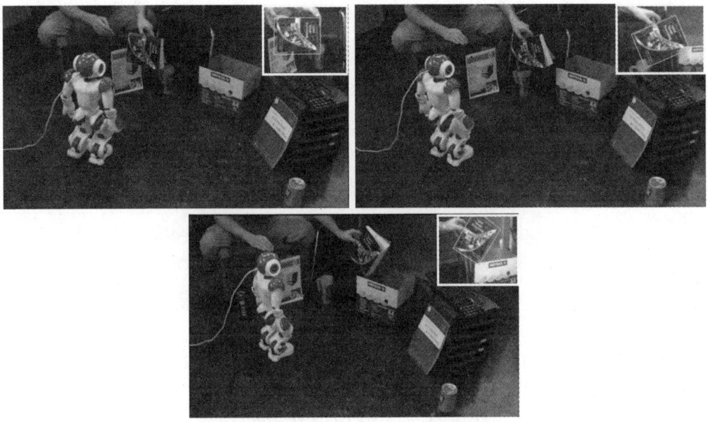

Figure 4. Images from tracking a previously learned moving object. Robot camera picture is shown in upper right corner of each image.

6. Conclusion and Future Work

In this paper, we present a novel approach to unsupervised robot learning based on visual perception. The technique allows real-time learning of objects from unlabeled images and their further detection. We demonstrate its benefits in mobile robotic applications, enabling a robot to learn and to detect objects that surround it without human supervision. To achieve this, we suggest a novel algorithm for visually salient object detection. This algorithm is robust to difficult illumination conditions and due to its low complexity it can be run in real-time on contemporary machines. Our approach is verified on Nao humanoid robot in a real office environment with encouraging results, allowing the robot to learn, search for and track objects in a completely autonomous manner.

In the future, other object detection algorithms, like receptive field co-occurrence histograms [11], could be adopted and objects with different surface properties could be learned by algorithms that best suit their nature in a "mixture of experts" manner. As to our visual saliency detector, the center-surround feature detector could be supplied by or replaced by an interesting approach of spectral residua detection published in [12]. An open question is, whether our technique, instead of learning individual objects, could be adopted for use in context of robot navigation for place learning and recognition.

Our approach allows learning to recognize new objects, but it provides no semantic level clues about the object. We are greatly interested in whether this learning system could be extended in semantic way by an unsupervised interaction with existing information sources such as the Internet to anchor the newly learned object automatically into an existing ontology to finally allow the robot to understand the nature of the particular object and be able to reason about it.

References

1. S. Vasudevan, S. Gachter, V. Nguyen, and R. Siegwart, "Cognitive maps for mobile robots-an object based approach," in Robotics and Autonomous Systems. Amsterdam: North-Holland Publishing Co., vol. 55, pp. 359-371, (2007)
2. O. M. Mozos, P. Nejsfelt, H. Zender, G. M. Kruijff, and W. Burgard, "From Labels to Semantics: An Integrated System for ConceptualSpatial Representations of Indoor Environments for Mobile Robots," in Proceedings of the ICRA-07 Workshop on Semantic Information in Robotics (SIR). Rome, Italy, pp. 33-40, (2007)
3. Brand, R. J., Baldwin, D. A., Ashburn, L. A.: Evidence for `motionese': modifications in mothers' infant-directed action. In : Developmental Science, vol. 5, pp.72-83 (2002)
4. R. Achanta, S. Hemami, F. Estrada, and S. Susstrunk, "Frequency-tuned Salient Region Detection," CVPR (2009)
5. T. Liu, N. Zheng, X. Tang, and H. Shum, "Learning to detect a salient object," IEEE CVPR. Minnesota (2007)
6. Jeremy M. Wolfe, Todd S. Horowitz, "What attributes guide the deployment of visual attention and how do they do it", Nature Reviews Neuroscience 5, 495-501 (2004)
7. H. Bay, T. Tuytelaars, and L. Van Gool, "SURF: Speeded up robust features", European Conference on Computer Vision, 404-417, (2006)
8. R. Moreno, M. Graña, and E. Zulueta, "RGB colour gradient following colour constancy preservation," Electronics Letters, vol. 46(13), 908-910 , (2010)
9. R. Moreno, M. Graña, D. M. Ramík and K. Madani, "Image Segmentation on the Spherical Coordinate Representation of the RGB Color Space", submitted to IET-IPR, 2011
10. D. M. Ramík, Ch. Sabourin, K. Madani, "A Cognitive Approach for Robots' Vision Using Unsupervised Learning and Visual Saliency", IWANN, (2011) accepted paper
11. S. Ekvall and D. Kragic, "Receptive field cooccurrence histograms for object detection," IROS, pp. 84-89, (2005)
12. X. Hou and L. Zhang, "Saliency Detection: A Spectral Residual Approach," in *IEEE CVPR 2007*, pp. 1-8, (2007)

GENERATION AND TESTING OF GAIT PATTERNS FOR WALKING MACHINES USING MULTI-OBJECTIVE OPTIMIZATION AND LEARNING AUTOMATA

Jeeves Lopes dos Santos and Cairo Lúcio Nascimento Júnior

Laboratory of Intelligent Machines - LMI, Division of Electronics Engineering, Aeronautical Institute of Technology - ITA, São José dos Campos, São Paulo, Brazil
E-mail: jeeves@ita.br, cairo@ita.br

This article presents a gait synthesis methodology for a three-legged robot that considers multi-objetives, where the designer must assign a degree of relevance of each performance measurement available (velocity, smoothness of the locomotion, maximum actuator torque and energy consumption) in the gait characteristics. To this end, the position in time of each leg actuator is described by a periodic function that is found by using a reinforcement learning technique called Learning Automata. During the training, the SimMechanics Toolbox of MATLAB/Simulink is used to simulate the robot built using the Bioloid Comprehensive Kit, an educational robot kit manufactured by Robotis. After the training fase, the solution is applied to the real robot and the response is then evaluated and compared with the simulated robot response. In the case studies, two gait synthesis were performed with different desired characteristics. It is shown that proposed solution generates a quite satisfactory gait for the real robot in both cases.

Keywords: Mobile Robotics, Walking Machines, Legged Robots, Reinforcement Learning, Learning Automata.

1. INTRODUCTION

All research about walking machines is direct or indirect related with the leg coordination problem. The machine legs have to be synchronized such that its body moves in the desired direction with the desired gait characteristics. This means that all leg actuators have to be coordinated, which is a difficult problem considering the robot dynamics and the number of functions involved.

This research presents a methodology to generate the gait pattern of a walking machine using multi-objetive optimization and a reinforcement learning technique known as Learning Automata. To this end, a simulated environment using MATLAB R2009b/Simulink/SimMechanics Toolbox (MathWorks, http://www.mathworks.com) is used during the training and the result is tested in a real robot built with the educational robotics kit BIOLOID Comprehensive Kit.

This paper is organized as follows: section 2 presents the robot physical composition, section 3 mathematically describes the problem discussed in this article, section 4 shows how to use the Learning Automata technique to solve the previously discussed problem. In section 5 a three-legged robot is used to illustrate the proposed solution. Finally, concluding remarks and some possibilities for future work are given in section 6.

2. THE ROBOT MORPHOLOGY

The robot presented in this paper (Fig. 1) has three legs (1, 2 and 3) with three joints each (A, B and C). Each joint has one angular degree of freedom which is driven by a small servomotor located in the joint. The angular velocity and angular position of each joint are controlled by the local controller of the servomotor that receives a reference signal sent from the master controller via a wired serial communication network. Therefore, in this article, solving the joint coordination problem corresponds to finding a reference signal for each joint in each robot leg.

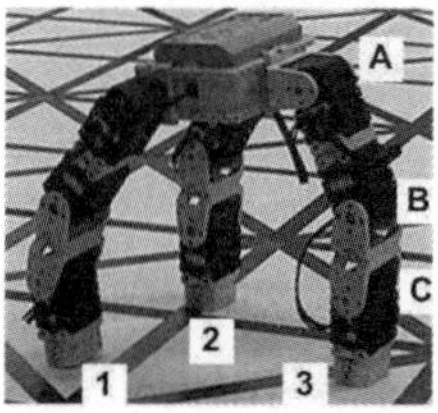

Fig. 1. Real tripedal robot with three actuators per leg (A, B and C).

With: 1) 1.05 Kg, 2) 11 cm of length (x axis), 3) 22.6 cm of height (y axis) and 4) 19 cm of width (z axis), the real robot was built using the BIOLOID Comprehensive Kit manufactured by ROBOTIS. This kit contains a set of components that can be arranged in various ways enabling the assembly of robots with legs, claws and/or wheels.

3. PROBLEM FORMULATION

To coordinate the walking machines actuators, a reference signal is used by the local controller in each actuator of all legs. The signal of actuator a of leg p is characterized as a periodic function in time $f_p^a(t)$ which is defined by a period T and a set of NE points linearly interpolated between them.

In order to simplify the problem, the robot symmetry can be used to minimize the number of variables that need to be adjusted as used in Santos et. al.,2010.[1] Therefore the legs that are symmetric in relation of the robot's Center of Gravity (CG) may be grouped. Thus, legs which belong the same group can share the same functions $f_p^a(t)$, but with a certain lag $\phi(p)$.

By analyzing the symmetries of the tripedal robot used, one can see that there are two groups ($Ng = 2$) of legs. Using the identification presented on Fig. 1, the first group contains the legs 1 and 2 and the second one has leg 3. Thus, it is necessary to define: 1) three periodic functions for the three actuators per group ($NA = 3$), 2) the time period T and 3) the lag between the two legs on the first group ($N_\phi = 1$). Then, considering functions described by four points ($NE = 4$), it is necessary to define 26 variables ($Nv = Ng\ NE\ Na + N_\phi + 1$).

So, the propose of the methodology used in this research is find a value of the Nv variables which describe all periodic functions $f_p^a(t)$ for the robot actuators in order to provide a gait with some desired characteristics.

To this end, a degree of relevance must be assigned (through the vector $F = [F_V\ F_I\ F_\tau\ F_{Ec}]$) to each of the four performance measurements obtained by the simulated robot:

(1) Velocity: The F_V value quantifies the relevance of the maximization of the average linear velocity ($\overline{V}$) of robot's CG in the positive x axis;

(2) Smoothness of the robot locomotion: The F_I value quantifies the relevance of minimizing the variations of linear and angular velocities observed on the robot's CG.

The rates of linear velocity variations (I_V) and angular velocity variations (I_W) are defined for measure these variations. I_V and I_W are obtained by Eq. 1 and Eq. 2, where: 1) $V(i,j)$ and $W(i,j)$ correspond to CG linear and angular speeds on axis x, y or z ($i = 1$, 2 or 3 respectively) at time of sampling j ($j = 1$ to N); 2) $\overline{V(i)}$ and $\overline{W(i)}$ is the average of $V(i, 1 : N)$ and $W(i, 1 : N)$, respectively.

$$I_V = \sqrt{\frac{\sum_{i=1}^{3}(\sum_{j=1}^{N}(V(i,j) - \overline{V(i)})^2)}{N}} \tag{1}$$

$$I_W = \sqrt{\frac{\sum_{i=1}^{3}(\sum_{j=1}^{N}(W(i,j) - \overline{W(i)})^2)}{N}} \tag{2}$$

(3) Maximum torque required: the F_τ value is the relevance for the minimization of the maximum instantaneous torque applied by the actuators (τ_{max});

(4) Energy consumption: the relevance of the minimization of energy consumption (measured by the sum of rotational kinetic energy observed in all actuators - Ec) is assigned through F_{Ec}.

4. METHODOLOGY

To find each variable value, this research uses a technique of reinforcement learning[2] called Learning Automata (LA).[3] The LA uses stochastic automata to describe the knowledge associated with the system and the experience to improve this knowledge.

To describe the knowledge, the designer needs to specify a finite number of possibilites for each variable. This variable is associated with an automaton which is described by a vector whose elements store the odds of success when the possible values associated with them are choused to be the variable value.

Thus, each variable is represented by one LA as described above which is called as Finite Action-set Learning Automata (FALA). The set of FALA, as used in this research, is known as game of FALA and provide a parallel learning that makes the training process faster.[4]

To improve the knowledge, in other words, to adjust the probabilities in the game of FALA, the algorithm 4.1 are used.

In order to aid the search for a solution that takes into account the degree of relevante assigned by the designer for the multi-objetives available, the reinforce signal R has some other features besides such signal encourage the desired performance:

(1) R favors an evolution of performance achieved, in others words, the value of R is positive if the observed performance in the current iteration exceeds a certain average obtained in a historic of previous iterations;

(2) R stimulates only the "good" results, this means that there will be a adjust the odds of the selected variables when the quality of the response obtained is considered positive. In other words, the algorithm uses a linear reward-inaction scheme (L_{R-I});[5]

Algorithm 4.1 Algorithm used to adjust the odds in the game of FALA.

repeat
    Select a solution to be tested by using the probability associated with the possible values. As an example, if a specific value has a probability of success of 25%, there are 25% of chance to select this value;
    Apply the solution in a simulated robot;
    Quantify the quality of the response obtained through the performance measurements and its degrees of relevance assigned by the designer;
    Adjust the probability of success assigning a reforce signal R;
    Check the knowledge convergence;
until Convergence is true

More details about the adjustment of probabilities estimation of success can be seen on appendix of this paper.

5. CASE STUDY

To evaluate the coordination methodology here presented, two case studies are showed with the tripod described above where different gait characteristics are desired by the designer. In both cases: 1) the periodic function for each actuator is described by four points ($NE = 4$), 2) each point is selected from 20 possibilities and 3) the time period T is selected from 20 possibilities linearly spaced between 0.76 s and 1.00 s.

The first case used a $F = [1\ 0\ 1\ 1]$ while the second one used a $F = [1\ 2\ 1\ 1]$. After the learning process convergence, which took place with 11458 and 20290 iterations respectively, the methodology defined the reference functions shown in Fig. 2 and no lag between the two legs on the first group in both cases. As result, the obtained periodic functions have the performance presented in Table 1 on the simulated robot[a]

Comparing the results, when the smoothness of the locomotion is considered, there was a significative decrease of $\overline{V_x}$ to allow a gait with minor variations in the linear and angular velocities in the robot's CG. Furthermore, a smoothness locomotion has a tendency to require less power from the actuators. This fact explain why there were a decrease of τ_{max} and Ec.

[a]The gait obtained for the simulated and real robot are shown in the videos available at:
F_1 - simulated: ftp://labattmot.ele.ita.br/ele/jeeves/videos/CLAWAR2011_3ps1.wmv
F_1 - real: ftp://labattmot.ele.ita.br/ele/jeeves/videos/CLAWAR2011_3pr1.wmv
F_2 - simulated: ftp://labattmot.ele.ita.br/ele/jeeves/videos/CLAWAR2011_3ps2.wmv
F_2 - real: ftp://labattmot.ele.ita.br/ele/jeeves/videos/CLAWAR2011_3pr2.wmv

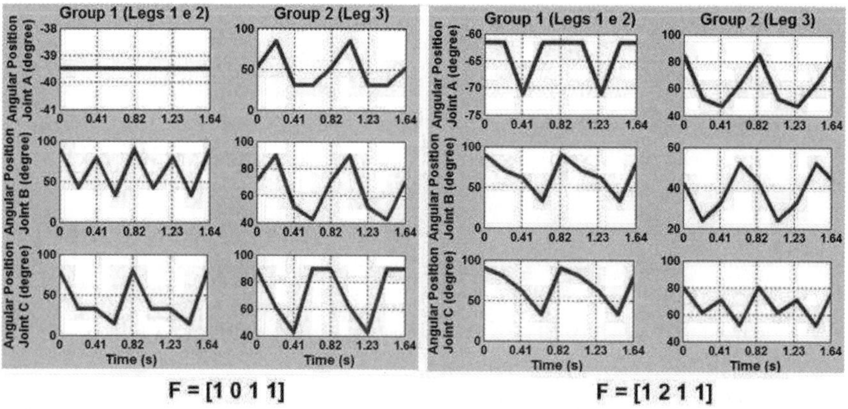

Fig. 2. Periodic functions obtained by the learning process for the joints A, B and C for the two groups of legs.

Table 1. Performance measurement of the solution obtained by the learning process for the simulated tripedal robot.

F	T	$\overline{V_x}$	I_V	I_W	τ_{max}	Ec
[1 0 1 1]	0.82	$28.21\frac{cm}{s}$	$0.40\frac{rad}{s}$	$1.99\frac{m}{s}$	$52.07Nm$	$1.81J$
[1 2 1 1]	0.84	$11.62\frac{cm}{s}$	$0.22\frac{rad}{s}$	$0.93\frac{m}{s}$	$17.14Nm$	$1.55J$

When the periodic functions in Figure 2 was used by the real robot[a]. (Fig. 3), the walking speed was 18.59 cm/s and 7.11 cm/s which is about 66% and 61% of the velocities observed during the simulation with $F = [1\ 0\ 1\ 1]$ and $F = [1\ 2\ 1\ 1]$, respectively. These differences of performance were caused by imperfections in the models used in the simulation, especially in the iteration between the robot and the environment such as the frictional forces.

Since the real robot has not been equipped with the proper sensors, the maximum torque generated by each servomotor, the energy consumed by the leg actuators and the variation rates for the linear and angular CG's velocities can not be measured.

6. CONCLUSIONS AND FURTHER WORK

This paper presented a solution to the problem of gait synthesis for a three-legged robot with three joints in each leg for multi-objetives assigned by the designer. Although only a single case (tripedal robot) is presented, this methodology has been successfully applied to generate gaits for the follow-

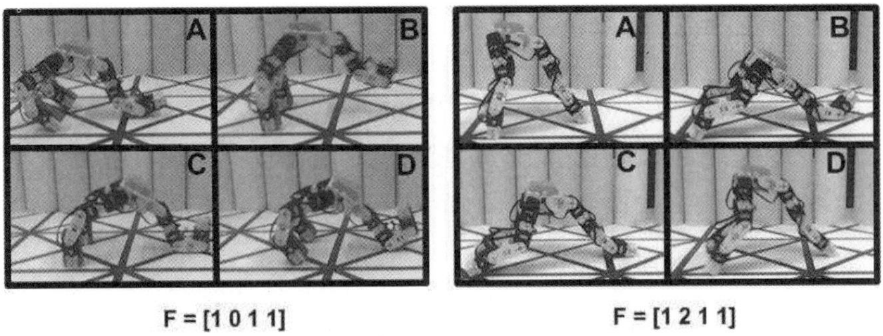

Fig. 3. Behavior observed in the real Tripedal robot with the periodic functions showed in Fig. 2.

ing cases:: 1) a quadrupedal robot, 2) a quadrupedal robot with unactuated wheels in its feet, and 3) a bipedal robot. The results obtained in these case studies show that the proposed methodology achieves the desired results.

Some possible directions for further work are:

(1) improve the simulation models used to describe the friction between the robot feet and the surface where the robot moves around;
(2) put sensors on the real robot;
(3) continue learning using the real robot after obtaining the solution using the simulation environment;

Appendix A.

The adjustment of probability estimation of success associated with the selected solution is accomplished by assigning a reforce signal R whose value is given by Eq. A.1:

$$R = \begin{cases} R_G \text{ if } R_c \geq R_G \\ R_c \text{ if } R_G > R_c > 0 \\ 0 \text{ if } R_c \leq 0 \text{ or if robot falls.} \end{cases} \tag{A.1}$$

The R_c value is obtained by (A.2), where R_P is the lower limit determined by designer and f_R represents the reinforcement function.

$$R_c = f_R\left(\frac{R_G - R_P}{2}\right) + \frac{R_G + R_P}{2} \tag{A.2}$$

The f_R function is calculated from (A.3), where:

(1) $\overline{I_W^{Nc}}$, $\overline{I_V^{Nc}}$, $\overline{\tau_{max}^{Nc}}$ and $\overline{Ec^{Nc}}$ correspond to I_W, I_V, τ_{max} and Ec averages of the last Nc iterations, with Nc determined by designer;
(2) $\overline{V_+^{Nc}}$ corresponds to the average velocity in the last Nc iterations that the average velocity of the robot were positive;
(3) C_{fr} is calculated by Eq. A.4. This value makes f_R tends to 0 when the values of the vector J tend to the average of the last Nc iterations.

$$f_R = \frac{1 + F_V \frac{\overline{V_x}}{\overline{V_+^{Nc}}}}{1 + F_I \left(\frac{I_V}{\overline{I_V^{Nc}}} + \frac{I_W}{\overline{I_W^{Nc}}} \right) + F_\tau \frac{\tau_{max}}{\overline{\tau_{max}^{Nc}}} + F_{Ec} \frac{Ec}{\overline{Ec^{Nc}}}} - C_{fr} \tag{A.3}$$

$$C_{fr} = \frac{1 + F_V}{1 + 2F_I + F_\tau + F_{Ec}} \tag{A.4}$$

To adjust the probabilities associated with each parameter selected Eq. (A.5) is used, where: 1) P_{k-1} corresponds to each previous probability estimation of success of the estimates associated with the selected variables and 2) P_K is the probability estimation of success after the adjustment.

$$P_K = P_{K-1} \left(1 + \frac{R}{100} \right) \tag{A.5}$$

After adjusting the probabilities of the parameters used to determine the reference functions, the probabilities in each column are normalized such that their sum is 1.

In this article, the coefficients used in the test are: 1) $Nc = 20$, 2) $R_P = -20\%$ and 3) $R_G = 20\%$.

References

1. J. Santos, C. N. Jr. and L. F. W. Barbosa, Desenvolvimento de um sistema de aprendizado para o controle do caminhar de um robô utilizando aprendizado por reforço, in *CBA 2010*, sep 2010.
2. R. S. Sutton and A. G. Barto, *Reinforcement Learning: An Introduction* (The MIT Press, March 1998).
3. K. S. Narendra and M. A. L. Thathachar, *Systems, Man and Cybernetics, IEEE Transactions on* **SMC-4**, 323 (july 1974).
4. M. A. L. Thathachar and P. S. Sastry, *Networks of Learning Automata: Techniques for Online Stochastic Optimization* (Springer-Verlag New York, Inc., Secaucus, NJ, USA, 2003).
5. K. S. Narendra and M. A. L. Thathachar, *Learning automata: an introduction* (Prentice-Hall, Inc., Upper Saddle River, NJ, USA, 1989).

LEARNING FAST WALKING PATTERNS FOR A NAO ROBOT USING QWALKING

J. MA[1] and S. A. CAMERON[2]

Oxford University Department of Computer Science,
Wolfson Building, Parks Road, Oxford, OX1 3QD, UK
http://www.cs.ox.ac.uk
E-mail: jie.ma@cs.ox.ac.uk[1], *stephen.cameron@cs.ox.ac.uk*[2]

QWalking is a model-free biped locomotion method that uses Q-Learning to find optimal walking patterns for a humanoid robot. Previously tested in a simulated environment, here we investigate the implementation of QWalking on a real NAO robot to yield fast locomotion patterns. We compare related work with respect to locomotion on the NAO robot and discuss the software infrastructure in order to use QWalking with it. Experimental results show that it develops fast walking patterns that are competitive to the current fastest approach published for this robot, with the advantage of not requiring precise dynamic models or extensive state sensing.

Keywords: QWalking, Biped Locomotion, NAO Robot, Walking, Reinforcement Learning, Q-Learning

1. Introduction

Biped locomotion has been important in the robotics domain for the last few decades. We proposed a model-free biped locomotion approach called QWalking and it has been successfully implemented and evaluated within a RoboCup 3D simulation[1,2]. Here we explore the implementation of QWalking on a real NAO robot, and the experimental results of learning and executing fast locomotion patterns using QWalking.

The NAO robot (*Fig.* 1) is a 22 dof humanoid device that has been widely used in the RoboCup Standard Platform League (SPL) since 2008, and several papers related to the NAO have recently been published[3–5]. Here we review related locomotion methods that have been implemented and evaluated on NAO robots. The official Aldebaran omnidirectional walk provided in the Aldebaran ALMotion module is based on walking pattern generation using the Inverted Pendulum (IP) method[6]. The implementation details of this particular locomotion method have just been published by Gouaillier et al. at the end of 2010 in[7]. In this implementation, a foot planner defines the 2-dimensional positions of feet, according to which an IP method can calculate the positions of the Centre of Mass (CoM). With two sets of positions of the foot and the CoM, inverse kinematics can then be used to generate the actual commands for the actuators. The feedback from the sensors were also used to make the walking skill more stable. The reported walking speed of their method is 10 cms^{-1}.

Kulk and Welsh investigated how adding stiffness can further improve the walking speed and energy-efficiency over that of the unidirectional Aldebaran walk in[8]. According to their experiments, the speed of their low-stiffness walk method is about 14 cms^{-1}, which is about 60% faster than that of the official unidirectional Aldebaran walk at 8.7 cms^{-1}.

Fig. 1. The photo shows the NAO robot used in this paper as the experimental platform for evaluating the QWalking algorithm in the real world.

The team rUNSWift (Tay et al.) adopted the Bézier curve approach to control the length and the height of the gaits using a Centre of Mass (CoM) trajectory[9]. In their original approach, the walking speed was only 3 cms^{-1}, but after using the feedback data from the Inertial Measurement Units (IMUs) and Force Sensitive Resistors (FSRs) they finally increased the walking speed to about 16 cms^{-1}. The success of implementing this walking algorithm in the ALMotion module finally led them to win world 2nd place in RoboCup SPL 2010 in Singapore.

At the end of 2010, Graf and Röfer from the world champion team B-Human published their 3D-LIPM locomotion approach in[10]. Their walking algorithm was also based on the Inverted Pendulum method[6], but by using the dynamic phase duration and sensor feedback they successfully improved the speed and the stability of the original method. The maximum walking speed of their 3D-LIPM method is 28 cms^{-1}, which is reported as the fastest omnidirectional walk implemented on the NAO robot so far.

Under a ZMP-based method, Strom et al. reported the walking speed of their algorithm was 10.5 cms^{-1} in[11].

2. QWalking Implementation

QWalking is a model-free biped locomotion method that uses Q-Learning to find optimal walking patterns while avoiding falling down, where *optimal* could mean fastest[1], or the most energy-efficient[2]. In this paper fast walking patterns are investigated in particular, in which a robot is assumed to have limited knowledge of its dynamics model.

As currently QWalking takes about 20 hours to converge using a simulated robot operating at faster than real speed, it is not practical to do the learning directly on a real NAO robot. Instead, a commercial simulator called Webots[12] was used in this paper to learn walking patterns on the simulated NAO robot, and we then investigated the result of executing the refined walking patterns on the real robot. The manufacturer Aldebaran also officially support Webots, which minimises the gap between the Webots simulation and the real robot.

Q-Learning is a standard learning mechanism in which a table of *Q-values* are manipulated until they converge. Q-values are indexed according to the current system state, s, and subsequent action a, as $Q_t(a, s)$ for the appropriate Q-value at time t. The idea is that the Q-values should converge so that the best action to take in any given state can be estimated according to a given reward function. Thus three core elements of implementing QWalking on a real robot are discussed here, namely reward methods, state determination, and action execution.

Reward methods. The rewards used for QWalking are defined according to the following Q-value update equations. Either the robot falls down when executing the walking pattern a:

$$Q_{t+1}(a, s) = (1 - \alpha_t) Q_t(a, s) + \alpha_t \left[\rho R_t^- + \gamma \max_{a'} Q_t(a', s')\right] \quad (1)$$

or alternatively if the walking pattern a is successful:

$$Q_{t+1}(a, s) = (1 - \alpha_t) Q_t(a, s) + \alpha_t \left[(1 - \rho) R_t^+(s) + \gamma \max_{a'} Q_t(a', s')\right] \quad (2)$$

As we are interested in fast locomotion patterns, the average speed $\bar{v}$ of an action is used for the positive rewards $R_t^+(s)$, and the corresponding negative reward for failure is $R_t^- = -|\bar{v}_{\max}|$ where $|\bar{v}_{\max}|$ is the maximum robot speed. The other parameters are described in §3.

Space Determination. As in[1,2] we use a simplified space model to very roughly indicate the ground projection of the centre of the head (GPCoH) as the robot is assumed to have only a rough knowledge where the GPCoH is relative to its feet; thus we have 3 states corresponding to the GPCoH being in front of/above/behind the feet. In practice we can use a built-in Force Sensitive Resistor to guess the rough position of the GPCoH, but a more direct method is to use the ALMotion module in the NAO software framework (see *Fig.*2) to get the position data.

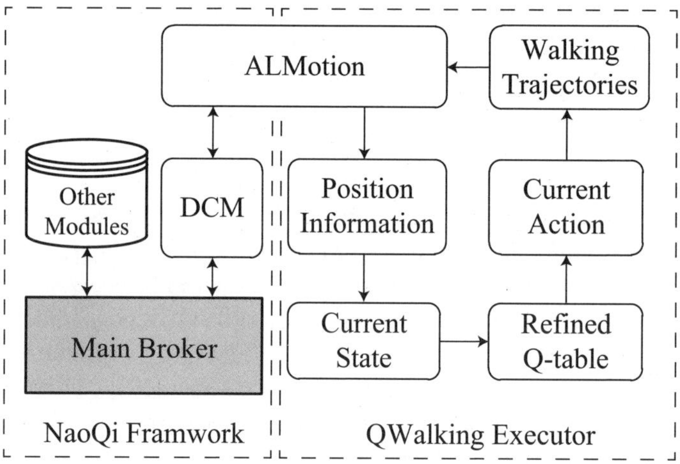

Fig. 2. The QWalking decision architecture used, where the QWalking executor and the related NaoQi framework are shown. The QWalking executor contains the procedure to get the current action based on the refined Q-table that is learned in the simulated Webots environment. Each action contains two trajectories, one for the swinging foot and one for the hip. ALMotion transforms the position trajectories into the eventual control sequence using IK and sends it to the DCM interface.

Action Execution. In QWalking, an action is a 4-dimensional tuple $\langle T_c, D_s, y_{sd}, y_{ed} \rangle$ of parameters that generates the trajectories of the swinging foot and the hip using cubic spline interpolation, where T_c and D_s are the gait period and the gait length and y_{sd} and y_{ed} are two length parameters for the swinging leg. The key to executing an action in QWalking is the mapping from the position trajectories to the corresponding angles of a set of joints (servos). In order to solve this problem Inverse Kinematics was used in QWalking (and in all the other walking methods reviewed here).

QWalking was implemented in Python on the NAO as it is a cross-platform programming language, and it can be easily used both in the Webots and on the real robot without re-compiling. This ensures that the actions, to the largest extent, are executed on both platforms in the same way. The QWalking decision architecture on a real NAO robot is given in *Fig*.2.

3. Experimental Results

In this section we investigate how QWalking was used in a real NAO robot to find fast walking patterns.

Learning QWalking in Webots. We assume both for the Webots-based simulator and in the real world that the ground is even and torque limits are implemented; thus the values of the main parameter range $T_c \in [0.15s, 0.30s]$ was increased from that used with the RoboCup 3D simulation[1,2] and we kept the other ranges of the 3 main action parameters the same, namely: $D_s \in [0.02m, 0.20m]$, $y_{sd} \in [0, 0.5D_s]$ and $y_{ed} \in [0, 0.5D_s]$. For these intervals, the action space was discretised with linear increments of 0.1. Since $y_{ed} > y_{sd}$[13] the number of the combinations of these are therefore ${}^6C_2 = 15$. Therefore we have $16 \times 19 \times 15 = 4560$ actions for each decision.

The same learning rate configuration that was used for the 3D simulation was employed: a floating Q-learning learning rate α_t, which varies from 0.8 to 0.2 during the learning process, and the same discount rate $\gamma = 0.5$. The average speed $\bar{v}$ of an action was used as the positive rewards $R_t^+(s)$ and the negative reward was changed to $-|v_{\max}| = -0.3\text{ms}^{-1}$ corresponding to the maximum published speed that a real NAO robot can achieve. A small punishment rate $\rho = 0.1$ was used directly according to the findings in[1].

In Webots, the experiment lasted for about 21 hours with 300 episodes on a standard PC, and each episode consisted of 200 trials, with the progression shown in *Fig*.3 showing that convergence was achieved after about 100 episodes. The converged average walking speed $\bar{v}$ is above 22 cms^{-1}. Compared with the learning results shown in[1], the walking speed that is obtained in the Webots simulation was much smaller, and it is mainly due to the more realistic physical models used in Webots.

Executing QWalking on a Real Robot. The real NAO robot executes the gait using the QWalking decision architecture shown in *Fig*.2, using the action with the maximum Q-value according to the current state, which

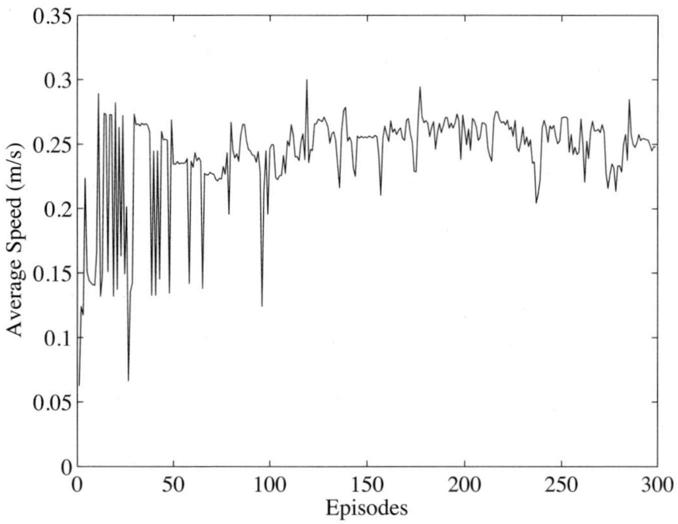

Fig. 3. The progression of the QWalking learning process in the Webots simulation.

the QWalking executor determines according to the position information obtained from the ALMotion module in the NaoQi system. ALMotion is then used to transform the position trajectories required into the eventual control sequence using IK and to send it to the DCM interface. A real NAO robot executing the first QWalking gait is illustrated in *Fig.*4.

In our experiment, a real robot successfully yielded walking using the refined Q-table learned within Webots, with the walking speed as high as 22cms^{-1}. This speed was compared with a number of recent published methods, and QWalking is competitive to the current fastest approach published — 3D-LIPM — and faster than all the other methods reviewed here (see *Fig.*5).

4. Conclusions

QWalking is a novel biped locomotion method that does not requires precise knowledge of dynamic models. Here we investigated the implementation of QWalking on a real NAO robot, and in §3 we showed that we could generate a walking speed as high as 22cms^{-1}. Moreover, QWalking does not require dynamic models to develop the trajectories so it can deal with wear and imprecise physical models, while all the other methods including 3D-LIPM method in *Fig.*5 need the precise position of the Centre of Mass (CoM)— this feature alone differentiates QWalking from other locomotion methods. From here we intend to includes the ability to change direction

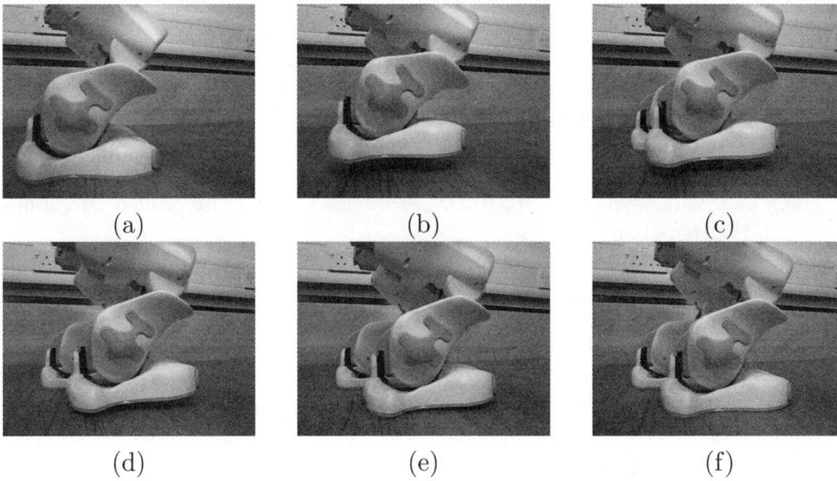

Fig. 4. A real NAO robot executing a QWalking gait is illustrated. The time between two frames is about 1/20 seconds. The right foot of NAO is the swinging foot, and the left foot supports the body.

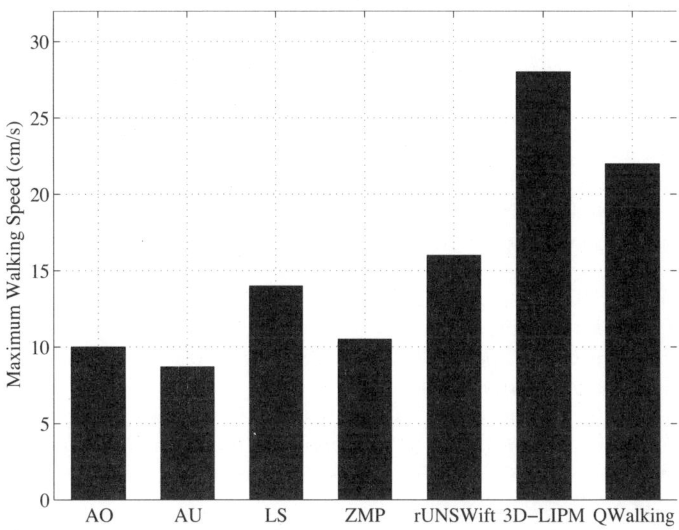

Fig. 5. This bar chart compared QWalking with other recent published locomotion approaches. The walking speed that QWalking yielded on a real NAO robot is competitive to the current fastest approach, 3D-LIPM, and it is faster than the other methods reviewed in this paper.

whilst walking, and to further refine the learn walking patterns on the real robot, both for initial motions and in order to deal with changes over time.

References

1. J. Ma and S. Cameron, Learning to walk: A model-free biped locomotion approach, in *Towards Autonomous Robotic Systems (TAROS)*, (Londonderry, United Kingdom, 2009).
2. J. Ma and S. Cameron, Learning robust and energy-efficient biped walking patterns using QWalking, in *12th International Conference on Climbing and Walking Robots and the Support Technologies for Mobile Machines (CLAWAR)*, (Istanbul, Turkey, 2009).
3. M. Clerx, Using a hierarchy of reactive behaviors to simplify NAO control, bachelor thesis, Universiteit Maastricht 2009.
4. E. Hashemi and M. G. Jadidi, Dynamic modeling and control study of the NAO biped robot with improved trajectory planning, in *International Conference on Advanced Computational Engineering and Experimenting*, 2010.
5. Çetin Mericli and M. Veloso, Biped walk learning on NAO through playback and real-time corrective demonstration, in *9th Int. Conf. on Autonomous Agents and Multiagent Systems (AAMAS 2010)*, 2010.
6. S. Kajita, F. Kanehiro, K. Kaneko, K. Yokoi and H. Hirukawa, The 3D linear inverted pendulum mode: a simple modeling for a biped walking pattern generation, in *Proc. IEEE/RSJ International Conference on Intelligent Robots and Systems*, 29 Oct.–3 Nov. 2001.
7. D. Gouaillier, C. Collette and C. Kilner, Omni-directional closed-loop walk for NAO, in *IEEE-RAS International Conference on Humanoid Robots*, (Nashville, TN, USA, 2010).
8. J. Kulk and J. Welsh, A low power walk for the NAO robot, in *2008 Australasian Conference on Robotics & Automation*, eds. J. Kim and R. Mahony (Canberra, Australia, 2008).
9. A. Tay, Walking NAO omnidirectional bipedal locomotion, PhD thesis, School of Computer Science and Engineering, The University of New South Wales 2009.
10. C. Graf and T. Röfer, A closed-loop 3D-LIPM gait for the RoboCup standard platform league humanoid, in *4th Workshop on Humanoid Soccer Robots, A workshop of the 2009 IEEE-RAS Intl. Conf. On Humanoid Robots (Humanoids 2009)*, (Paris, France, 2010).
11. J. Strom, G. Slavov and E. Chown, Omnidirectional walking using ZMP and preview control for the NAO humanoid robot, in *RoboCup 2009: Robot Soccer World Cup XIII*, eds. J. Baltes, M. Lagoudakis, T. Naruse and S. Shiry (Springer, 2010) pp. 378–389.
12. O. Michel, Webots: Symbiosis between virtual and real mobile robots, in *Virtual Worlds*, ed. J.-C. Heudin, Lecture Notes in Computer Science, Vol. 1434 (Springer Berlin / Heidelberg, 1998).
13. Q. Huang, J. Yang, Z. Yu, W. Xu, J. Li and K. Li, Measurement of human walking and generation of humanoid walking pattern, in *IEEE International Conference on Robotics and Biomimetics (ROBIO)*, Dec. 2007.

SECTION–6

FLEXIBLE MANOEUVRING SYSTEMS

FUZZY CONTROL SCHEME FOR TWO OVERHEAD CRANES WITH THREE DEGREES OF FREEDOM

R. ALGAZALY AND M. O. TOKHI

Department of Automatic Control and Systems Engineering, University of Sheffield, UK

This paper presents development of fuzzy logic control scheme for positioning tow overhead cranes acting in the same work space and minimizing the sway of the payloads when the system is in motion. Proportional-Derivative (PD)-type fuzzy logic controllers are developed to move the payload to the target location to achieve accurate positioning and minimum sway. The system is realized within the Visual Nastran (vn4d) software environment, and integrated with Matlab-Simulink for analysis and control purposes. Simulation results are then presented in time domain to assess the effectiveness of the control approach. It is demonstrated that accurate positioning and significant reduction in the system sway is achieved.

1. INTRODUCTION

In a large number of applications, only one overhead crane acts in the work space. However, in some applications the use of one crane is inadequate if the task is to manipulate and place more than one payload at the same time. In such cases a multiple set of cranes will be required, and therefore it will be necessary to add a third degree of freedom to the dynamics of the cranes. Although there are common features in the control approaches for multi-cranes and single crane systems, the control of cooperative cranes is more difficult to achieve than that of a single crane. The added difficulty of a multi-crane system is that each crane may become a moving obstacle in the path of another crane. The control challenge in such a system is to avoid collision and accurately position the payloads with minimum vibration. Researchers have developed many control methods to address the problem of positioning and load swing of overhead cranes, some used open-loop control such as input shaping control and optimal control and others used closed-loop control such as state feedback, nonlinear control, fuzzy control and adaptive control. In [1] a Delayed Feedback Signal (DFS) and proportional-derivative (PD)-type fuzzy logic controller was used to suppress the sway angle of the rope for a two-dimensional overhead gantry crane system. The dynamic model of the system was derived using the Euler-Lagrange formulation. Performances of both controllers were examined in terms of sway suppression, disturbances cancellation, time response specifications and input force. The design of a fuzzy controller for a practical overhead crane to ensure precise positioning and zero sway angles was considered in [4], and that the energy in the input and states was minimized. The fuzzy system was chosen to represent the dynamics of the overhead crane, and a guaranteed cost fuzzy control design was proposed for the fuzzy descriptor system. [3] Used the inertia theorem and presented a method that accelerated transportation and minimized the payload swing of the overhead crane. The method did not need a complex dynamic model for an overhead crane system, but only used trolley position and swing angle data to design the proposed fuzzy projection controller. The feasibility and effectiveness of the proposed scheme were also compared to the conventional PD and nonlinear coupling methods. [5] Presented a new fuzzy controller for anti-swing and position control for an overhead crane based on Single Input Rule Modules (SIRMs). The trolley position and velocity, the rope swing angle and angular velocity were selected as the inputs, and the trolley acceleration as the output. Each input was given with a SIRM and a dynamic importance degree. The control system proved to be asymptotically stable to the destination. The controller was robust with different rope lengths and had generalization ability for different initial positions. Considerable amount of research has been conducted in field of collision avoidance. [2] Presented a fuzzy controller with two rule bases to minimize the sway of the payload attached to an overhead crane. One rule base was for positioning control of the payload and the other is for reducing the sway angle; the simulation results showed that the controller performed well on giving both accurate positioning and suppressing the sway of the payload. In [6] a scheme for motion control and collision avoidance of a vehicle was presented. The method is based on designing a desired trajectory according to the distance between the vehicle and the fixed or moving obstacle. They used a sliding mode controller which controls motion of the vehicle to track the desired trajectory. The approach was appropriate for real time implementation due to less expensive computations. Their Simulation results show the effectiveness of the scheme in avoiding the collision.

2. SYSTEM SETUP

A system comprising two overhead cranes working together in a 10m by 10m workspace is developed. The system is realized within the Vn4D software environment and integrated with Matlab-Simulink for analysis and control purposes. The task of each crane is to move and position its specific payload to a specific target location. The payloads are aluminum rods with 2.3 cm in diameter, and 70 cm in length. Each overhead crane consists of a hook hanging on a pendulum-like lift line. The hook is lowered and raised in vertical (z) direction. Both the rail and the trolley are capable of motion in the y direction. The trolley can move along the rail in the x direction. In this way the hook and the payload attached to it can move in three dimensions. Fig. 1 shows a schematic diagram of the crane model. The upper crane (Crane-2) is modified by allowing the two rails (Rail 2-1, Rail 2-2) to move vertically up by approximately 1 meter above their original position. This will permit the cranes to swap positions after picking up their payloads. The swap process will be necessary if the payloads are desired to be placed differently from their initial arrangement prior to pickup. As such, the control protocol will need to be modified so that collision-free motion is achieved.

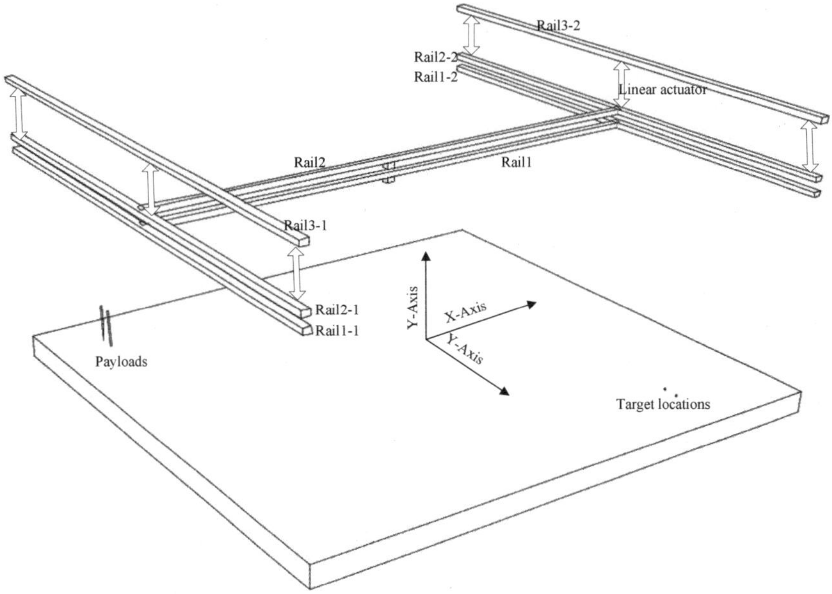

Figure 1. Schematic diagram of the overhead crane with 3DOF.

There are six actuators to connect the Rails 2-1 and 2-2 with the new fixed Rails 3-1 and 3-2. The distance between the fixed and movable rails is set to 0.9 m. The task of the linear actuators is to move the rails 2-1 and 2-2 upward. This will allow the whole crane to move vertically up by approximately 1 meter above its original position. The operating requirements of the system are as follows:

- Each crane moves from its resting position to the position of a specific pair of rods.
- The crane then extends its hook down in order to attach itself to a pair of rods.
- Each rod is hoisted onto the train.
- If Crane-1's target location is further away than Crane-2's, then a swap process is required to avoid collision between Crane-1 and the payload of Crane-2.
- The swap process is achieved by moving Crane-2 upward for approximately one meter, and subsequently letting Crane-1 move first.
- The crane is to move to its target at the desired speed, with minimum sway while avoiding any collisions with the other crane.
- The crane is to place down the pair of rods at the target position with minimum vibration.

3. THE CONTROL DESIGN SCHEME

Four independent fuzzy-PD controllers are used to control each crane; two for positioning control (one for x direction and one for y direction) and two for suppressing the sway (x and y directions). The anti-sway controller is to reduce the payload sway. The positioning controller aims to position the crane at the exact position of the payload during the grabbing phase, and at the destination position during the payload dropping phase. Further controller is used for maintaining a safe distance between the two cranes. These controllers are set to be active only if this separation is less than the safe distance. The controller's scaling factors are tuned using genetic algorithm (GA). Each crane has two inputs: driving force in the x direction (Fx) and driving force in the y direction (Fy). There are nine main outputs. These are the actual positions of the cranes in the x and y directions, the sway angles of the payloads in the x and y directions and the distance between the cranes. These outputs are used for the simulation and system analysis purposes. Fig. 2 shows the Matlab-Simulink structure of the system.

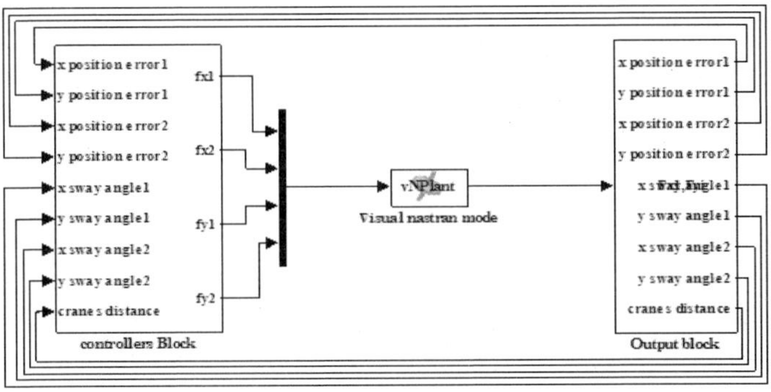

Figure 2. Simulink implementation of the two overhead cranes

4. IMPLIMETATION AND RESULTS

Figure 3 shows the results of the positioning control for Crane-1, where the payload was initially positioned at (3.75 m,4 m) on the coordinate plane. The controller calculates the distance between the trolley and the payload during the first phase, and then between the trolley and the target location after picking up the payload. This was used as an error which the control process minimizes. Thus when the error becomes zero, the trolley would have reached the required position.

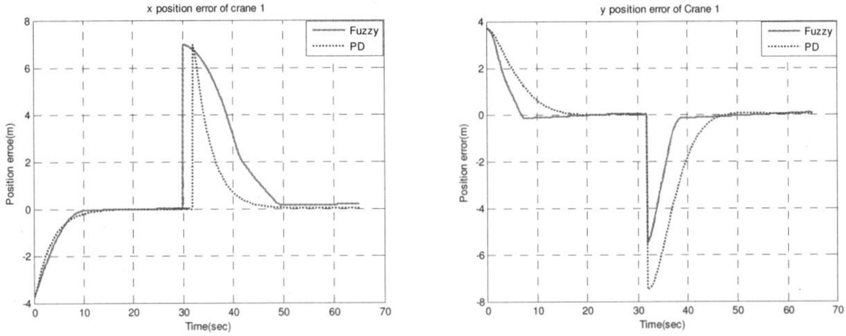

Figure 3 Position error of Crane-1

As Crane-1 is required to pass Crane-2 and its payload, this means that the y position error for Crane-1 will be greater than that with using the PD controllers. However, the x position error remains the same. Figure 4 shows the positioning control results for Crane-2, which indicates that the control system achieved an accurate positioning of the payload within reasonable time. As can be seen from Figure 4, there was a slight delay (2 sec) before the crane began its journey to the target location. This is due to the swap process and also to allow Crane-1 to move first.

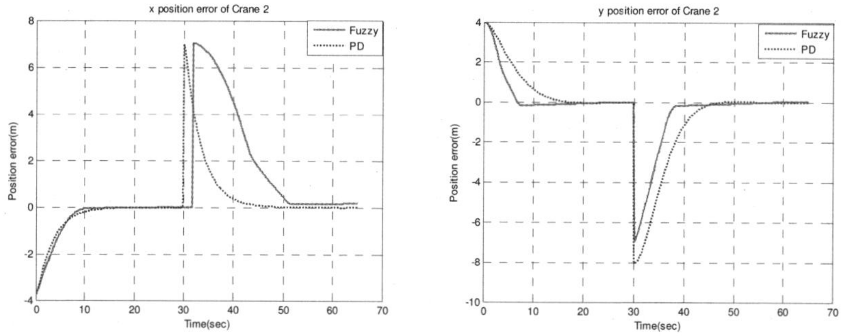

Figure 4 Position error of Crane-2

Figure 5 illustrates the sway angle of the hook attached to Crane-1 in the x and y directions. The maximum angle in the x direction was less than one degree when the crane first moved from its rest position to where the payload was located. It should be noted that the sway angle prior to the payload pickup was damped to zero within 10 seconds. After picking up the payload, the crane reached the target location within 30 seconds. It should be noted that the maximum sway angle was approximately 2 degrees and this was reduced to zero within 30 seconds. The sway angle in the y direction, as noted, was acceptable. The angle was around zero degrees during the first phase, and after picking up the payload and moving to the target location, the maximum sway angle in the y direction was approximately 1 degree. This was suppressed in 30 seconds to around zero.

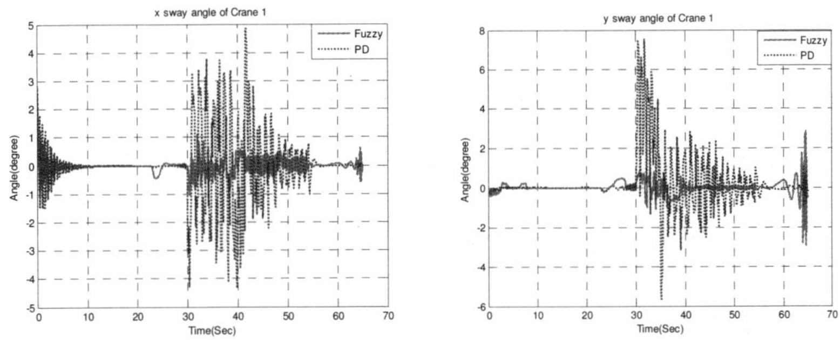

Figure 5 Sway angle of Crane-1

Similarly, Figure 6 shows the sway angle results for Crane-2. it also shows that the sway in both the x direction and y direction were significantly less with the fuzzy controllers than that achieved with the PD controllers – this evidence of the superior damping capabilities that the fuzzy controllers has over the PD type controllers. In Figure 6, it can be seen that the sway angle at approximately 65secs is greater with the fuzzy controller than with the PD controller. This is due to the scaling factors tuning.

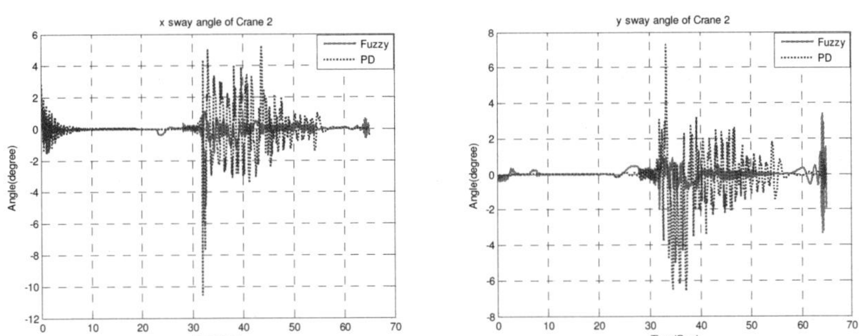

Figure 6 Sway angle of Crane-2

5. CONCLUSION

A system of two overhead cranes acting in the same workspace has been designed using the Vn4D software environment and integrated with Matlab-Simulink for analysis and control purposes. Four independent PD-Fuzzy controllers have been designed for positioning each crane and controlling the sway angle of the payloads. The simulation results demonstrate an excellent performance of the system with accurate positioning, minimum sway angle and reduced vibration of payloads, and a collision-free motion. Furthermore, the control protocol allows the payloads to be placed in any desired target arrangement.

References

1. M. A. Ahmad, Active sway suppression techniques of a gantry crane system. *European Journal of Sciences Research, **27**(3),* 322-333 (2009)
2. Mahfouf, M., Kee, C. H. , Abbod, M. F., and Linkend, D. A., Fuzzy logic Based Anti-sway control design for overhead cranes, neural computing and applications, 9 , 2000, pp. 38-43
3. C. Y. Chang & K. H. Chiang, Fuzzy projection control law and its application to the overhead crane. *Mechatronics, **18**(10),* 607-615 (2008)
4. Y. J. Chen, W. J. Wang, & C. L. Chang, Guaranteed cost control for an overhead crane with practical constraints: fuzzy descriptor system approach. *Engineering Application of Artificial Intelligence, **22**(4-5),* 639-645 (2009)
5. Yi, J., Yubazaki, N., & K. Hiora, Anti-swing and positioning control of overhead traveling crane. *Information Sciences*, ***155**(1-2), 19-42* (2003).

Durali, M., Javid, G.A., Kasaiezadeh, A. (2006), Collision avoidance maneuver for an autonomous vehicle, 9th IEEE International Workshop on Advanced Motion Control, pp. 249-254

MEASUREMENT OF FLEXED POSTURE FOR DRIVING FLEXIBLE MONO-TREAD MOBILE TRACK USING FLEXIBLE DISPLACEMENT SENSOR

Takafumi Haji[1], Shu Myoraku[1], Tetsuya Kinugasa[1], Tetsuya Akagi[1], Koji Yoshida[1], Hisanori Amano[2], Masatsugu Iribe[3], Kenichi Tokuda[4], Ryota Hayashi[5] and Koichi Osuka[6]

[1] *Okayama Univ. of Science, 1-1, Ridai-cho, Okayama, 700-0005, JAPAN*
[2] *National Research Institute of Fire and Disaster,* [3] *Osaka Elector-Communication Univ.,* [4] *Wakayama Univ.,* [5] *Kagoshima Univ.,* [6] *Osaka Univ.*

We have proposed Flexible Mono-tread mobile Track (FMT) as a mobile mechanism for hazardous environment such as disaster area. FMT has only one track which can flex in three dimension. Generally speaking, one has to teleoperate robots under invisible condition. Flexed posture of FMT decides turning radius and direction, and then, it is very important to know its posture, accordingly. FMT has a vertebral structure composed of vertebrae as rigid body and intervertebral disks made by flexible devices such as rubber cylinder and spring. Since the intervertebral disks flex in three dimension, it is not easy for traditional sensors such as potentiometers, rotary encoders and range finders to measure its deformation. The purpose of the paper, therefore, is to measure flexed posture of FMT while driving by using a new flexible displacement sensor.

Keywords: Measurement, flexible mono-tread mobile track, flexible displacement sensor

1. Introduction

It is expected that robotics is applied to some mission on rough terrain such as rescue activity and planetary explorer, etc. Tracked vehicles are effective against such environment because contact pressure can be distributed wider. In order to improve mobility, some novel mechanism such as serpentine type tracked vehicle connected several small tracks serially have been proposed. We have also proposed Flexible Mono-tread mobile Track (FMT). FMT has only one track which can flex in three dimension. Its fundamental mobility have been evaluated using a prototype, "WORMY".[1] The vehicle should be teleoperated to avoid risk of hazardous environment. Flexed posture of FMT decides turning radius and direction, and thereby it is very important to know its posture. FMT has a vertebral structure composed of vertebrae as rigid body and intervertebral disks made by flexible devices. Since the intervertebral disks flex in three dimension, it is not easy for traditional sensors such as potentiometers, rotary encoders and range finders to measure its deformation.

The purpose of the research, therefore, is to measure flexed posture of FMT using a novel flexible displacement sensor developed by Akagi et al.[7] We have already proved that FDS has been able to measure the flexed posture of FMT,[8] however the measurement has been achieved only for flexion, that is, FMT cannot drive in the measurement. In the paper, we realize flexed posture measurement and visualization for FMT while driving.

2. Flexible Mono-Tread Mobile Track (FMT)

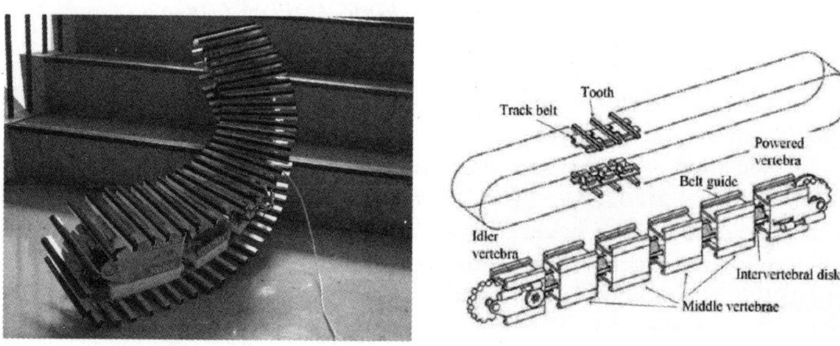

Fig. 1. Prototype of FMT called "WORMY"

Figure 1 shows the prototype of FMT, "WORMY". WORMY has vertebral structure composed of six rigid boxes called vertebrae and five flexible devices such as rubber cylinder and spring called inter-vertebral disks. The body is flexed by two pairs of 2.5 [mm] stainless wire pulled by DC motor connected by warm gear system. Each joint has stoppers which limit the maximum flexion angle. The maximum flexion angle of each joint is 14 [deg] for retro-flexion and 18 [deg] for lateral flexion, and thereby, total flexed angle of WORMY is 70 [deg] for retro-flexion and 90 [deg] for lateral flexion. The retro-flexion means that FMT flexes about pitch axis and the lateral flexion about yaw axis.

2.1. *Flexible Displacement Sensor*

Some methods have been proposed to measure the flexible structure.[2–6] In addition, Akagi et al. have proposed novel flexible displacement sensors using mechanism potentiometer.[7] We introduce the flexible potentiometer so as to get over the problem. Figure 2 shows a photograph and a cutaway drawing of the flexible displacement sensor. The sensor consists of two fixed electrodes (that is an anode and a cathode), a slide electrode (a device located at centre of Fig.2) and a nylon string coated with carbon (a black

Fig. 2. Flexible displacement sensor (FDS)

line in Fig.2, we call it "NSCC"). It works as a flexible potentiometer by sliding the slide electrode along NSCC while holding the electrical contact.

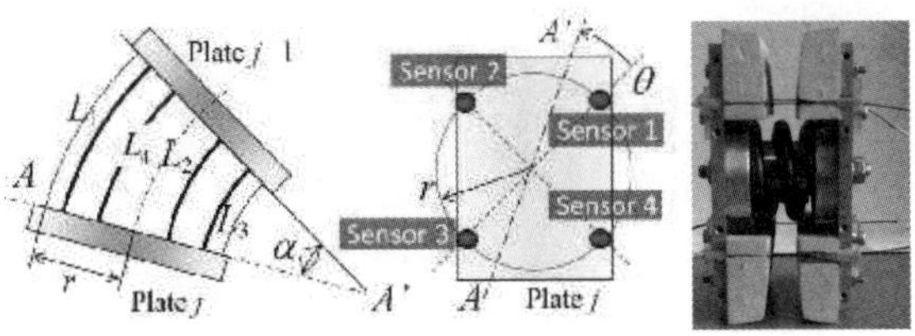

Fig. 3. Model and implementation of FDS

Flexed posture of one joint is measured by four FDSs arranged in parallel at the position of 90 [deg] on a joint of FMT, as is shown in Fig.3.

We assume that flexed shape of NSCC is a segment of a perfect circle. Let j be number of joint. We define flexed angle α_j as the maximum angle between the plates enclosing a intervertebral disk. Then, there exist the lines in the plates which form the angle α_j as shown in Fig.3 (left). The plane which includes the lines have a line segment as the intersection with the plate j; the line segment is indicated A-A' in Fig.3. "Direction angle of flexion" θ_j is defined by the angle on the plate from a base line to the line segment A-A'. The base line is defined as the line which includes the two points on which Sensor 1 and 3 are attached (see Fig.3 (right)). From the geometrical relationship, we can derive Eq.s 1, 2 and 3 for direction angle of flexion θ_j and flexed angle α_j.

$$\theta_j = \begin{cases} 0, & \Delta L_{1j} = \Delta L_{3j} \\ \theta_{0j}, & \Delta L_{2j} \leq \Delta L_{4j} \\ \theta_{0j} + \pi, & \Delta L_{1j} > \Delta L_{3j} \& \Delta L_{2j} > \Delta L_{4j} \\ \theta_{0j} - \pi, & \Delta L_{1j} < \Delta L_{3j} \& \Delta L_{2j} > \Delta L_{4j} \end{cases} \tag{1}$$

$$\theta_{0j} = \tan^{-1} \frac{\Delta L_{4j} - \Delta L_{2j}}{\Delta L_{1j} - \Delta L_{3j}}, \ \Delta L_{1j} \neq \Delta L_{3j} \tag{2}$$

$$\alpha_j = \frac{\sqrt{(\Delta L_{4j} - \Delta L_{2j})^2 + (\Delta L_{3j} - \Delta L_{1j})^2}}{2r_j} \tag{3}$$

Counter clockwise direction is defined as positive for the both angle. i indicates number of sensor. In the figures and equations, L_{ij} indicates each length of four FDSs, ΔL_{ij} each displacement and r_j the distance between each FDS and the centre of sensor axis. r_j equals to 22.5 mm in experiment.

Fig. 4. Measurement for true value

Accuracy of static measurement have been evaluated. We made FMT flex until the limitation angle in the direction of left, right(lateral flexion) and upward(retro-flexion), measured its joint angles of flexion α_j, and compared with true values. The true value was obtained from photograph for each posture as shown in Fig.4. Figure 5 shows the measured angle (green line), the true value (blue line) and error as difference (red line) versus joint number. The left figure shows the results of the left (dash line) and right (solid line) flexion, and its average error. The right figure shows the result of the upward retroflexion. The error was staying in the area ± 1.5 [deg].

2.2. *GUI*

In order to operate conveniently, it is necessary to construct a graphical user interface (GUI). GUI is realized as a three dimensional model imitating WORMY using OpenGL. The model is composed of five rigid boxes

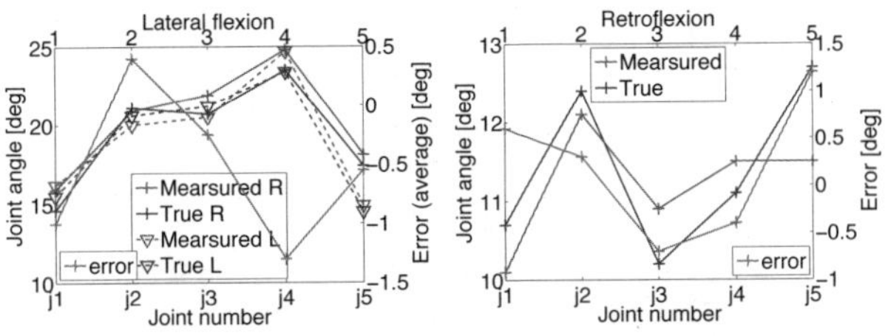

Fig. 5. Accuracy of measurement for flexed angle α_j

connected by ball joints (Fig.6, left). The sensor signal is detected by TPIP2 board (SANRITZ Automation co. Ltd.) and sent through WiFi. After the raw data were transformed to the flexed angle and the direction angle in TPIP2, they are given to OpenGL. In the space of OpenGL, position and attitude vectors of each boxes, $\boldsymbol{p}_j$ and $\boldsymbol{\varphi}_j$ are given by the flexion angle α_j and direction angle θ_j as following equations:

$$\boldsymbol{p}_j = (\sin\alpha_j \cos\theta_j,\ \sin\alpha_j \sin\theta_j,\ \sin\theta_j \sin\alpha_j), \tag{4}$$

$$\boldsymbol{\varphi}_j = (\sin\theta_j,\ \cos\theta_j,\ 0), \tag{5}$$

where j is joint number corresponding to that of WORMY.

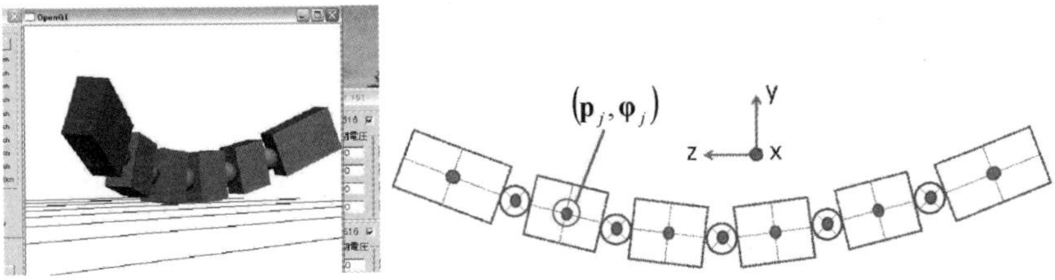

Fig. 6. 3D model of WORMY using OpenGL and its coordinate

3. Measurement of Flexed Posture

It has been difficult to detect flexed posture while driving for our previous result,[8] since signal lines could not wired compactly and fixed electrodes were not reliable. The wiring and contacts are improved, as is shown in Fig.7.

Fig. 7. Improvement of FDS: the previous version (above) and the latest (below)

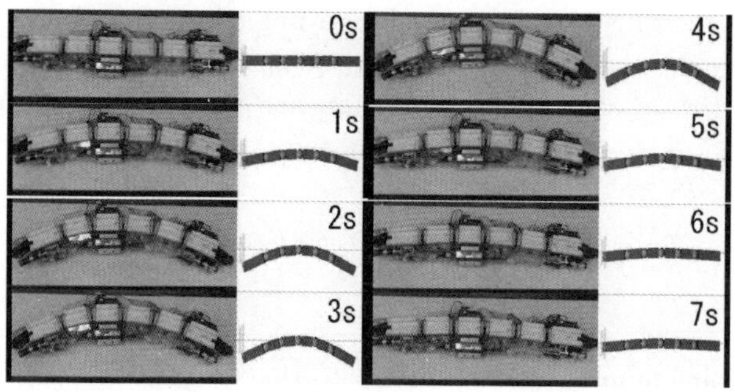

Fig. 8. Measurement without track belt: sequential photograph of WORMY and screen shots of GUI

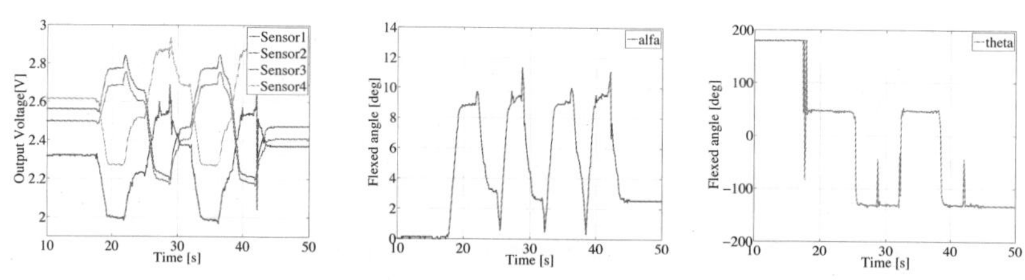

Fig. 9. Voltage of four FDSs at the 3rd joint of WORMY without the track belt, flexion angle α_3 and direction angle θ_3 from the left

3.1. *Measurement without Track Belt*

Firstly, flexed posture of FMT removing its track belt was measured. Figure 8 shows sequential photographs of FMT (odd columns) and screen shots of GUI (even columns) at the same time in case that WORMY flexed to the right. The right end of WORMY in each photograph and each screen shot is defined as the front, and thereby, the right direction of lateral flexion is downward direction of the figure. The GUI model expresses WORMY appropriately, and thus the implemented sensors functioned well. Figure 9 shows voltage of four FDSs at the 3rd joint of WORMY, flexion angle α_3 and direction angle θ_3 from the left. The latter two are calculated by the former data. Since the direction angle θ in case of the right flexion is defined as the origin, θ_3 is chattering between 0 [deg] and 360 [deg]. Hence we can also confirm that FDS functioned well.

3.2. *Measurement while driving*

Finally, we measured the flexed posture of WORMY while driving. Sequential photographs and screen shots of the test are appeared in Fig.10. We can find that the flexed posture of GUI are similar to that of WORMY. Figure 11 shows voltage of four FDSs at the third joint of WORMY, flexion angle α_3 and direction angle θ_3 from the left. WORMY was driving without visible vibration, nevertheless we can see that all joints of GUI model were vibrating while driving. From the left figure of Fig.11, the signal of all FDSs is including vibrational component. Since this vibration was not appearing in the previous result, the vibration is caused only by driving. We are predicting that the vibration comes from vibration of NSCC of which fixed electrode is free end. The vibration is especially find at the second

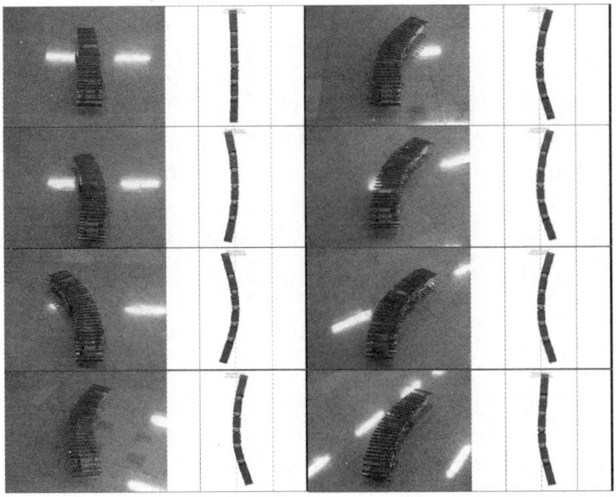

Fig. 10. Sequential photographs of WORMY while driving and screen shots of GUI

and fourth joints. These joints were flexing more widely and rapidly, and then the phenomenon was more apparent than the others. However the

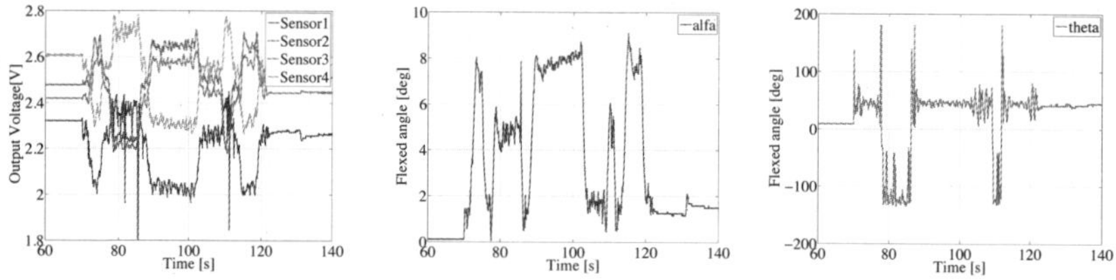

Fig. 11. Voltage of four FDSs at the 3rd joint of WORMY while driving, flexion angle α_3 and direction angle θ_3 from the left

GUI model could project the flexed posture of WORMY with no difficulty of operation even though it was including some vibration. And then, the measurement system using FDS for vertebral structure of FMT is effective.

4. Conclusion

We have realized a measurement system for flexed posture of FMT using a novel flexible displacement sensor. The sensors were implemented in five joints of our prototype and the flexed posture while driving could be detected. Moreover, the measured posture was visualized by OpenGL in three dimensional space. The result validates the effectiveness of the flexible displacement sensor. Future works are (autonomous) control of FMT, etc.

Acknowledgment

This work was supported in part by the Ministry of Education, Culture, Sports, Science and Technology of Japan through a Financial Assistance Program of the Social Cooperation Study and Chugoku Industrial Innovation Centre.

References

1. T. Kinugasa, et al. Proc. of IROS,1642(2008).
2. H. Okubo, et al., Trans. of JSME(C), 59(565), 2676(1993).
3. D. B. Camarillo, et al., Proc. of IEEE ICRA, 2940(2008)
4. K. Kure, et al., Proc.of IEEE IROS, 1012(2006).
5. Gentile et al., United States Patent, 5086785(1992).
6. S. Morimoto and T. Yamashita, Biomechanism 12, 223(1994).
7. I. Hamamoto, et al., Proc. of SICE2006, 1943(2006)
8. T.Kinugasa, et al., Proc. of IEEE SSRR (2010)

NONLINEAR TWO DoF DYNAMIC MODELLING OF A TRMS USING AN ADAPTIVE NETWORK BASED INFERENCE SYSTEM (ANFIS)

A RAHIDEH, M H SHAHEED and S A NIKNAM
School of Engineering and Materials Science, Queen Mary University of London London E1 4NS, UK

In this paper dynamic modeling approach for a highly nonlinear aerodynamic test rig, a twin rotor multi-input-multi-output system (TRMS) using a neuro-fuzzy approach is presented in terms of 2 DoF motion of the system. The modelling process is realized in an adaptive-network-based fuzzy inference system (ANFIS) framework to utilize the learning and adaptation capability of neural networks and human like expert knowledge of fuzzy logic. However, the TRMS is a 2 DoF system, a 2 DoF model is therefore necessary from control perspective as well as to truly characterize its dynamics. As such a 2 DoF model of the system is further developed. The developed models optimally represent the behaviour of the TRMS in 2 DoF domains as validated by three convincing tests. It is noted that the approach presented here has the potential to be used to model similar nonlinear and complex air vehicles.

1. Introduction

Control approaches for aerospace systems are normally developed on the basis of linear models or linearized version of nonlinear models restrained to a safe but specific operating region. However, over the years there has been a great demand for air vehicles such as UAVs and helicopters to be used in many complex applications with the requirement to follow unprecedented course of action within a very limited space. These include applications in war zones, search and rescue operations in complex terrain, and unmanned surveillance in hostile environment. As a result, in developing new design and control paradigms for air vehicles, emphatic consideration is given to aircrafts to be agile (i.e., high angle of attack), low observable (stealth), multi-functional etc. All these imply that linear models within a specific operating boundary cannot adequately describe aircraft dynamics and very reliable nonlinear models need to be adopted (1, 2). In addition, to address real-time control issues, it is necessary to develop such nonlinear models and test them using data extracted from real systems taking all the DoFs into account. The use of artificial intelligence like neural networks (NNs) and Fuzzy Logic (FL) look promising in this respect.

The application of NNs in tuning the fuzzy membership functions results in a neuro-fuzzy approach, a suitable candidate for dynamic modelling and control of nonlinear systems. In this way, by combining the advantages of NNs such as learning capabilities, optimization abilities and connectionist structure with the expert knowledge based humanlike if-then rules of FL their individualistic disadvantages could be addressed. Thus, the design of effective nonlinear models for a complex system like the TRMS using neuro-fuzzy approaches becomes feasible. In this perspective, the main goal of this study has been to use the advantages of the ANFIS proposed by Jang [3] in 2 DoF dynamic modelling of the TRMS. ANFIS can serve as a basis for constructing a set of fuzzy if-then rules with appropriate membership functions to generate the predetermined input-output pairs [3]. The objective of ANFIS is to realise the process of fuzzy reasoning using the structure of an NN and to optimally find the parameters of FL through the optimization nature of NNs. ANFIS can automatically identify the optimum fuzzy rules and tune the membership functions through learning capabilities of NNs.

The increased use of ANFIS in modelling nonlinear dynamic systems in various applications has been motivated by the requirements of industrial automation in recent years [3-7]. 1 DoF modelling of a TRMS using ANFIS has also been reported in literature [8-9]. Toha et al. [8] have modelled the 1 DoF vertical motion of a TRMS using ANFIS and Omar et al. [9] have assumed decoupled vertical and horizontal channels and have presented two separate ANFIS 1 DoF models using the same structures but different membership functions. In this study both 1 DoF and 2 DoF dynamic modelling of the system are investigated. It is noted that the TRMS is a 2 DoF system with significant cross-coupling and therefore, to address effective control of a nonlinear system like the TRMS, 2 DoF modelling is not only advantageous but essential. The paper is organized as follows: In Section 2, the TRMS is briefly introduced. A basic theoretical overview of the main components of ANFIS is presented in Section 3. Experimentation including the choice of the excitation signal, data extraction and pre-processing is described in Section 4. The procedure of the modelling and the obtained results are presented and discussed in Section 5 and 6 respectively. Finally, the paper is concluded in Section 7.

2. The System

The aerodynamic test-rig, the TRMS, considered in this investigation is shown in Figure 1. It is an experimental test-bed representing complex air vehicles with high nonlinearity and cross-coupling between axes. Its characteristics are, to certain degree, similar to that of a helicopter and of a UAV. From control perspective, it poses formidable challenges [10-11]. The system is developed on purpose by the Feedback Instruments Ltd [12] for modelling and control experiments. The main functional unit of the TRMS is a beam pivoted on its base. The beam can rotate freely in both horizontal and vertical planes. Its propellers (the main and tail) are driven by two permanent magnet DC motors. The system is designed to represent its 1 DoF or 2 DoF experiments and to provide a high-order, nonlinear, MIMO system with significant cross coupling.

3. Neuro-fuzzy modelling framework

ANFIS architectures and learning algorithms for the Sugeno fuzzy model have been thoroughly described by Jang [3]. Figure 2 illustrates a two-input, two-rule ANFIS with a single output where A_i 's ($i = 1,2$) are the membership functions (e.g. triangular, trapezoidal or Gaussian) of the first input, B_j 's ($j = 1,2$) are the membership functions of the second input, each circle node labelled Π is any *T-norm* operator (e.g. multiplication), w_k 's are the outputs of the second layer and return the *T*-norm result of the inputs, $\overline{w}_k = w_k / \sum_k w_k$ are the outputs of the third layer obtained from the circle node labelled N, f_k 's ($k = 1,2$) are linear functions of the inputs ($f_k = p_k x + q_k y + r_k$) which represent the 'then' part of the if-then rules where p_k, q_k and r_k are constant coefficients to be determined, and finally $f = \sum_k \overline{w}_k f_k$ is the network output. Alternatively, it is possible to combine the third and forth layers to form a network with four layers.

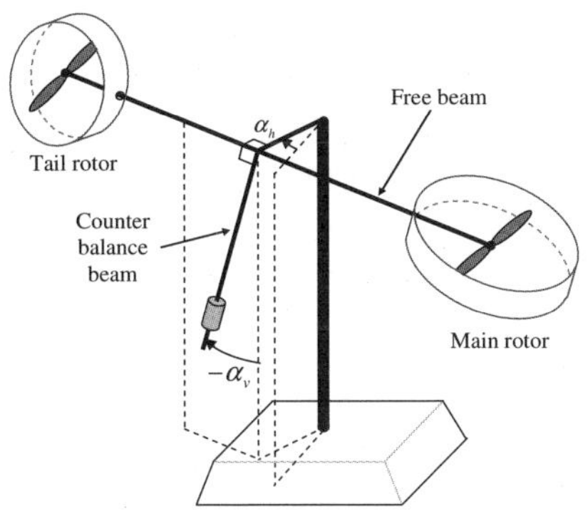

Figure 1. Illustration of the TRMS.

Having defined the inputs and outputs of the system to be modelled and extracted the required input-output data, the type and number of membership functions for each input need to be determined. The objective is to find the optimal values for the linear parameters of the ANFIS (p_k 's, q_k 's and r_k 's) and the nonlinear parameters of the membership functions so that the error between the real and model output is minimized.

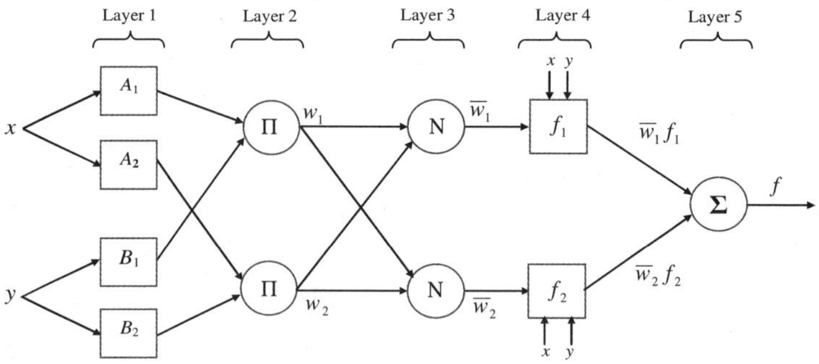

Figure 2. ANFIS architecture for the two-input two-rule Sugeno fuzzy model [3].

4. Experimentation

The experimentation was carried out in a suitable laboratory environment, as a very small blow of wind can move the body of the TRMS, especially in the horizontal plane, influencing the mined data. It is intuitively assumed that the body resonance modes of the TRMS lies in a low frequency range of 0-1 Hz, while the main rotor dynamics are at significantly higher frequencies [1]. As such the input excitation signals were designed with frequency ranging from 0.01 Hz to 1 Hz covering amplitude levels between –2.5 and 2.5 V that are the minimum and maximum applicable voltage to the system. In this study, the system was excited with various input signals with different shapes (e.g. sine and square waveforms), different possible amplitudes and different necessary frequencies. According to the dynamics of the system the input signals with higher frequencies cause no considerable movement in the TRMS beam due to the fact that the dominant time constant of the system (5s in the case of main rotor) is significantly higher than the period of the higher frequencies. Collected data were then pre-processed to address high-frequency disturbances above the frequency of interests, occasional bursts, outliers, missing data, trends etc. In the case of TRMS, some repeated parts from the extracted data have been removed as they will only increase the training burden without any benefit. Some data have been removed since during the data collection the beam hit the restrictors and bounced back (restrictors are used to retain the beam angles in a specific range). The data with some abnormalities such as sudden spikes have been recorded again.

5. Modelling of the system and model validation

A nonlinear autoregressive with external input model structure (NARX) model structure was selected for the system. In the ANFIS-based NARX model structure the inputs to the model are the past (and current) control inputs of the real system, $[u(t-d), u(t-d-1), \ldots, u(t-d-m)]$ and also the past observed outputs, $[y(t-1), y(t-2), \ldots, y(t-n)]$, where d is some multiple of the sampling period, and m and n are the input and output lag spaces. The model of the TRMS in terms of its 2 DoF dynamics as shown in Figure 6 has 6 inputs and two outputs as shown in Fig. 3 where $u_v(k)$ and $u_h(k)$ are the voltage of the main and tail rotors respectively; $\alpha_v(k)$ and $\alpha_h(k)$ are the pitch and yaw angles of the beam. The inputs are the main motor voltage at time k and $(k-1)$, the tail motor voltage at time k and $(k-1)$ and the pitch and yaw angles at time $(k-1)$. The outputs are the estimated pitch and yaw angles at the current time. Note that in this study two separate models with 6 inputs and 2 outputs are used in the case of 2 DoF modelling.

It is necessary for the model to be in a good agreement with the observed data and it can represent the 'true system' adequately. The extracted model is validated using overlapping of the actual and model responses both in time and frequency domains. However, for further validation, a more convincing correlation tests are also carried out.

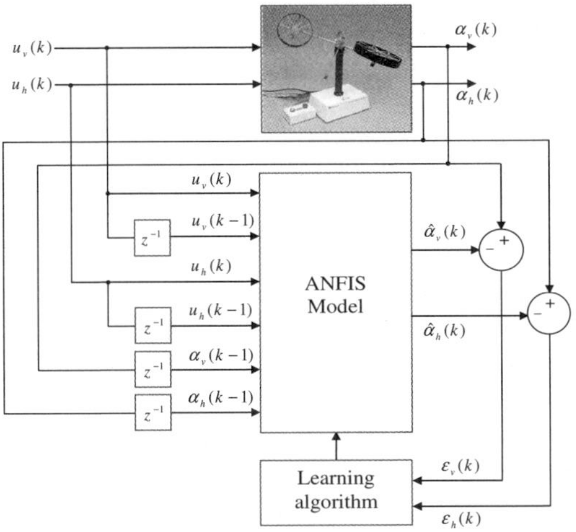

Figure 3. 2 DoF ANFIS based model with 6 inputs and 2 outputs.

If the model of a system is adequate then the residuals or prediction errors $\varepsilon(t)$ should be unpredictable from (uncorrelated with) all linear and nonlinear combinations of past inputs and outputs. This can be tested by means of the following correlation functions [13]:

$$\begin{aligned}
&\phi_{\varepsilon\varepsilon}(\tau) = E[\varepsilon(t-\tau)\varepsilon(t)] = \delta(\tau) && \\
&\phi_{u\varepsilon}(\tau) = E[u(t-\tau)\varepsilon(t)] = 0 && \forall\,\tau \\
&\phi_{u^2\varepsilon}(\tau) = E[(u^2(t-\tau)-\overline{u}^2(t))\varepsilon(t)] = 0 && \forall\,\tau \\
&\phi_{u^2\varepsilon^2}(\tau) = E[(u^2(t-\tau)-\overline{u}^2(t))\varepsilon^2(t)] = 0 && \forall\,\tau \\
&\phi_{\varepsilon(\varepsilon u)}(\tau) = E[\varepsilon(t)\varepsilon(t-1-\tau)u(t-1-\tau)] = 0 && \tau \geq 0
\end{aligned} \tag{1}$$

Where, $\phi_{\varepsilon\varepsilon}(\tau)$ represents auto-correlation of prediction errors and $\phi_{u\varepsilon}(\tau)$ indicates the cross-correlation function between the input and prediction errors and $\delta(\tau)$ is an impulse function.

The first two linear correlation tests in equation (1) alone are not sufficient to validate nonlinear models. Thus, higher-order correlation tests are included in this study. In general, if the correlation functions in equation (1) are within the 95% confidence intervals, $\pm 1.96/\sqrt{N}$, where, N is the total number of data points, the model is regarded as satisfactory.

6. Results and Discussion

In order to model the TRMS in terms of its 2 DoF dynamics using the ANFIS technique, two separate models each with six inputs and one output have been considered to cover the nonlinearities of the system. Figs. 4 and 5 show a part of training results for the pitch and yaw angles of the model respectively. Figs. 6 and 7 illustrate the test results. Due to the space limitation only the first set of the test results are presented. The number of membership functions in this case is set to be 2 for each input. Therefore the total number of membership functions is 12. Similarly, the Gaussian combination function is used for all membership functions. Table 1 lists the characteristics of the 2 DoF ANFIS based model. The RMS errors of the training and test results are listed in Table 2. Their PSD tests are depicted in Figs.8 and 9. Fig. 10 shows the correlation validation test results for the pitch angle channel. Similar results have been obtained in the case of yaw angle channel.

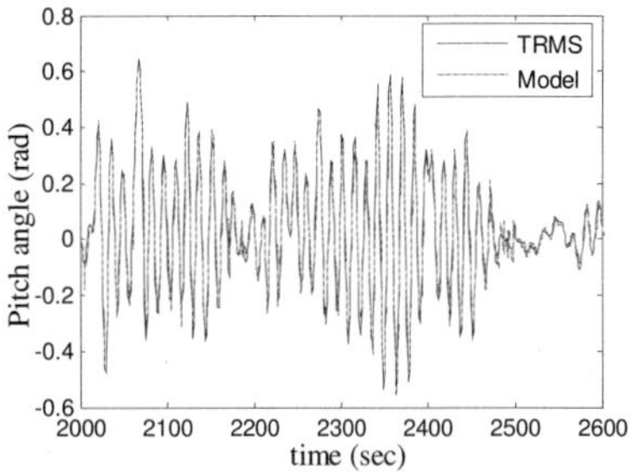

Figure 4. Training result of the pitch angle

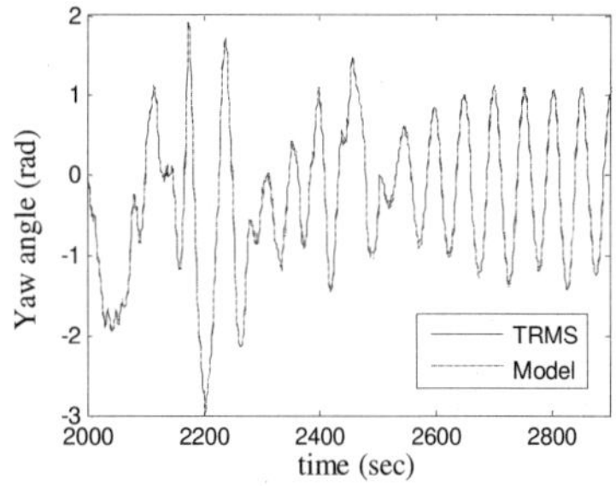

Figure 5. Training results of the yaw angle

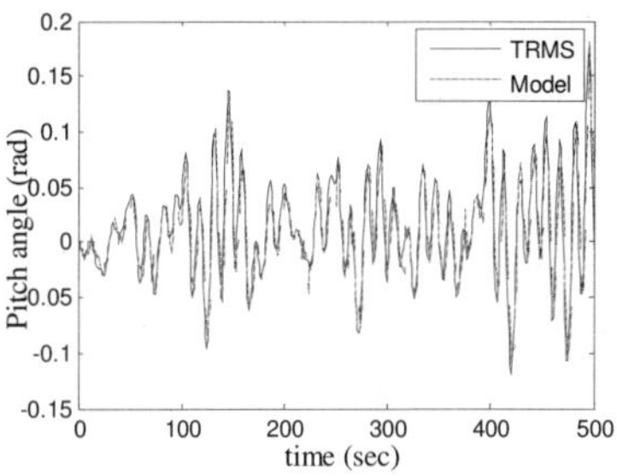

Figure 6. Test results of the pitch angle

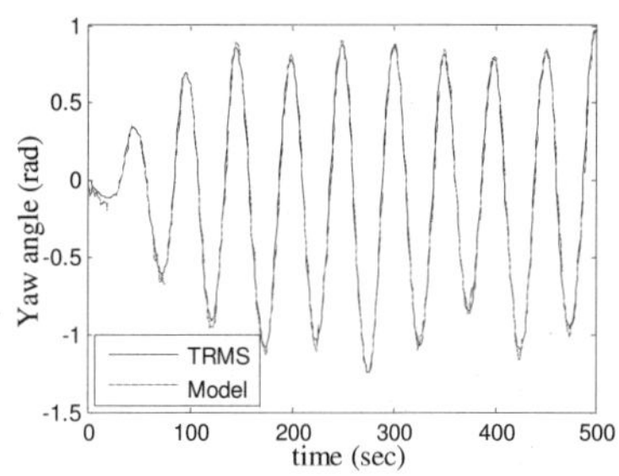

Figure7. Test results of the yaw angle

Table 1. RMS errors of training and tests of the 2 DoF model

RMS error	RMS error of test	
	Set1	0.000825
0.000626	Set2	0.001068
	Set3	0.000756

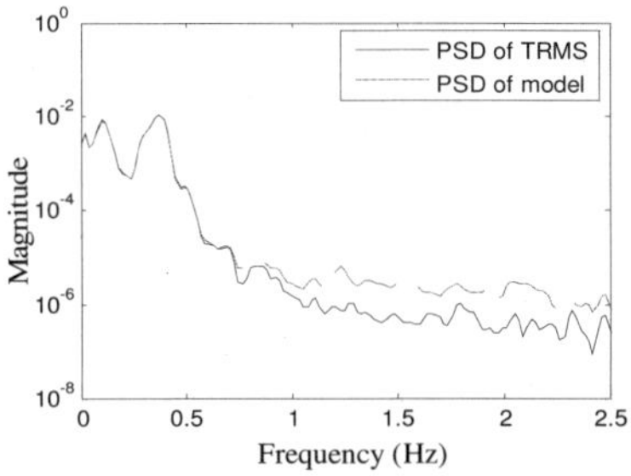

Figure 8. PSD of the pitch angle.

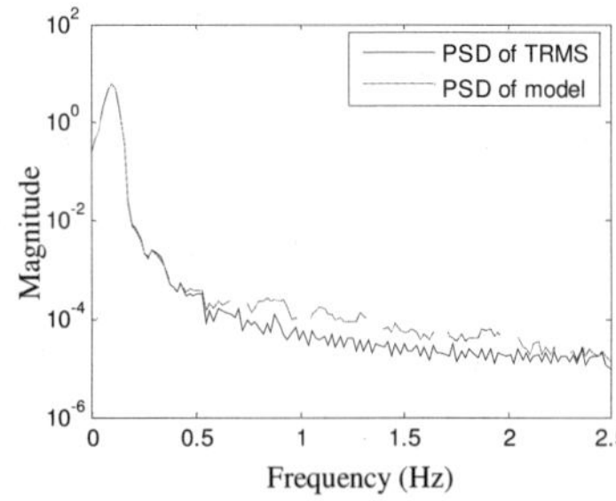

Figure 9. PSD of the yaw angle.

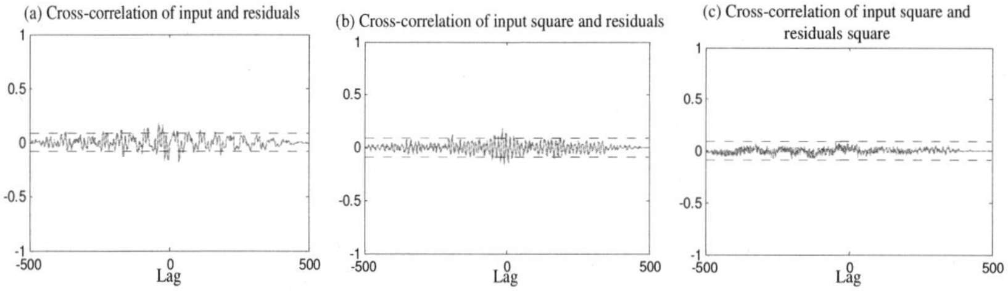

Figure 10. (a) Cross correlation of input and residuals, (b) Cross correlation of input square and residuals, (c) Cross correlation of input square and residuals square.

6. Conclusion

In this paper dynamic modelling approach for a highly nonlinear aerodynamic test rig, a twin rotor multi-input-multi-output system (TRMS) using a neuro-fuzzy approach is presented in terms of 2 DoF motion of the system. The modelling process is realized in an adaptive-network-based fuzzy inference system (ANFIS) framework to utilize the learning and adaptation capability of neural networks and human like expert knowledge of fuzzy logic. However, the TRMS is a 2 DoF system, a 2 DoF model is therefore necessary from control perspective as well as to truly characterize its dynamics. As such a 2 DoF model of the system is further developed. The developed models optimally represent the behaviour of the TRMS in 2 DoF domains as validated by three convincing tests. It is noted that the approach presented here has the potential to be used to model similar nonlinear and complex air vehicles.

References

1. S. M., Ahmad., M. H. Shaheed, A. J. Chipperfield and M. O.Tokhi, Non-linear modelling of a 1 DOF TRMS using Radial Basis Function Networks. *Proceedings of the IMechE, Part-G: Journal of Aerospace Engineering*. **216**, 197-208 (2002).
2. M. H. Shaheed, Feedforward Neural Network based Nonlinear Dynamic Modelling of a TRMS using RPROP Algorithm, *Aircraft Engineering and Aerospace Technology*. **77 (1)**, 13-22(2005).
3. J. S. R., Jang, ANFIS: Adaptive–Network–Based Fuzzy Inference System. *IEEE Transactions on Systems, Man and Cybernetics*. **23(3)**, 665 – 685(1993).
4. M Hayati, A. Rezaei, M. Seifi, and A. Naderi, Modelling and Simulation of Combinational CMOS logic circuits using ANFIS, Microelectronics Journal, **41**, 381-387 (2010)

5. M. Buragohain and C. Mahanta, A Novel Approach for ANFIS modeling based on full factorial design, Applied Soft Computing, **8**, 609-625 (2007).
6. A. Mellit and S. A. Kalogirou, ANFIS-based modeling for photovoltaic power supply system, A case study, Renewable Energy, 36, 250-258 (2011).
7. Z. X. Hou and H. Q. Li, Nonlinear system identification based on adaptive neural fuzzy inference system. *Conference on Communications, Circuits and Systems*. China, Vol. **1-4**, pp. 2067-2069(2006).
8. S.F. Toha, M. O. Tokhi and Z., Hussain, ANFIS Modelling of a Twin Rotor System, UKACC'08, Manchester, UK, September, 2-4, (2008).
9. M. Omar, M. A. Zaidan, and M. O., Tokhi, Modelling of Flexible Manoeuvring System Using ANFIS Techniques. *12th International Conference on Computer Modelling and Simulation. UKSim*, Cambridge, England, pp. 33-38 (2010).
10. A. Rahideh and M. H. Shaheed, Mathematical Dynamic Modelling of a Twin Rotor Multiple Input-Multiple Output System. *Proc. IMechE, Part I: Journal of. Systems and Control Engineering*, **221**, pp. 89-101(2007).
11. A. Rahideh, M. H. Shaheed and H. J. C.Huijberts, Dynamic Modelling of a TRMS using Analytical and Empirical Approaches. *Control Engineering Practice (Elsevier Journal)*. **16**, 241-259(2008).
12. Feedback Instruments Ltd, *Twin Rotor MIMO System* 33-220 user manual (1998).
13. S. A. Billings and W.S.F. Voon, Correlation based validity tests for nonlinear models. *Int. Journal of Control*, **44 (1)**, 235-244(1986).

HYBRID FUZZY CONTROL OF A FLEXIBLE MANOEUVRING SYSTEM

M. OMAR, F. M. ALDEBREZ AND M. O. TOKHI

Department of Automatic Control and Systems Engineering, University of Sheffield, UK

This paper addresses the input tracking control problem of a highly nonlinear flexible system namely a twin rotor multi-input multi-output (MIMO) system (TRMS) in the vertical plane motion. The TRMS can be considered as an aerodynamic experimental model, representing the control challenges of complex air vehicles. A hybrid control scheme comprising a fuzzy PD-like control and a conventional integrator is developed by adding the outputs of both control actions. To satisfy the tracking characteristics of the manoeuvring system, an efficient combination of the control actions is developed using a smooth nonlinear activation function which continuously regulates contribution of the integral term in the controller output. The effect of the activation function greatly reduces response overshoot when the system is exposed to abrupt changes during its manoeuvre. The control system performance is tested in real-time situation using the TRMS. The obtained results prove the efficiency of the proposed scheme in terms of time domain performance specifications of the system.

Keywords: Activation function, flexible systems, hybrid fuzzy control, twin rotor system.

1. Introduction

Advances in control theory has provided a wide variety of artificial intelligence design techniques in addition to the more traditional proportional, integral, derivative (PID) approaches which have been applied to control of flexible manoeuvring systems. Robust optimal control of flexible structures with active feedback techniques requires good models of the base structure, and knowledge of uncertainties of these models. Such information may not be easy to acquire for certain systems. Moreover, the flexible structure in theory is characterised as a distributed-parameter system with infinite number of modes, which makes it difficult to control, while controller design for flexible spacecraft is in general based on an approximated finite-dimensional model by truncating the infinite number of modes to a finite number through neglecting the higher frequency modes. Fuzzy systems can approximate any continuous function on a compact set to any accuracy. It has been known that fuzzy-logic-based modelling and control could serve as a powerful methodology for dealing with imprecision and nonlinearity efficiently. On the other hand, classical control laws are designed for linear systems and they provide a preferable cost over benefit ratio. However, the presence of nonlinear effects limits their performances. Therefore, hybridization of fuzzy logic techniques and classical PID control schemes will exploit the beneficial features of both categories.

The twin rotor multi-input multi-output (MIMO) system (TRMS) is a laboratory scale set-up designed for control experiments by Feedback Instruments Ltd. [1]. It is reported in the literature that various hybrid control structures have been proposed for several applications [2 – 12]. In some applications, PID control and fuzzy logic control are combined by a switch [8 – 11]. In other applications such as [12] a fuzzy switching method between fuzzy control and conventional PID control is used to achieve smooth control during switching. However, this implies aggregation of a second rule base in addition to the main fuzzy controller rule base. Both approaches are associated with significant drawbacks that can be referred to either the switching operation which degrades smoothness of the system response or the aggregation of a second rule base which implies large computational time and higher memory requirement.

This paper investigates the design of a novel hybrid control scheme that comprises a fuzzy PD-like control and a conventional integrator in which a smooth nonlinear activation function is used for on-line regulation of contribution of the integral term.

2. Experimental Setup

The TRMS is a laboratory scale set-up designed for control experiments by Feedback Instruments Ltd. [4]. It is an aero-dynamical system similar to a helicopter as shown in Figure 1. It consists of a beam pivoted on its base in such a way that it can rotate freely in both its horizontal and vertical planes producing two rotating movements around yaw and pitch axes, respectively.

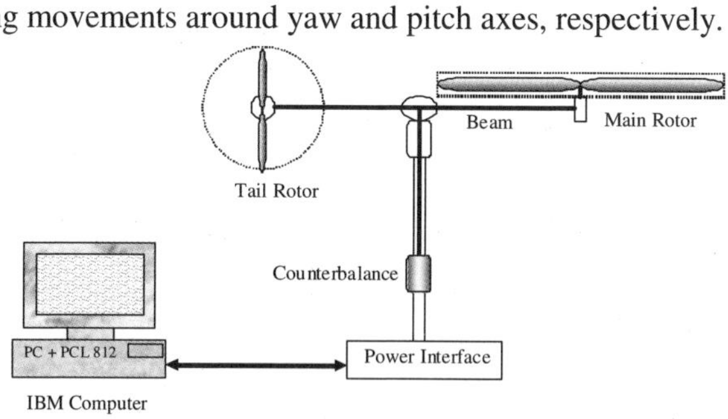

Figure 1. Schematic diagram of the TRMS

At both ends of the beam, pivoting on its base, there are two propellers driven by DC-motors. The articulated joint allows the beam to rotate in such a way that its ends move on spherical surfaces. There is a counter-weight fixed to the beam to determine a stable equilibrium position. The system is balanced in such a way, that when the motors are switched off the main rotor end of the beam is lowered down due to its weight.

In a typical helicopter, the aerodynamic force is controlled by changing the angle of attack of the blades. However, in the TRMS the pitch angle of the blades is fixed and the aerodynamic force is controlled by varying the speed of the motors. Therefore, the control inputs are the supply voltages of the DC-motors. The system is interfaced with a personal computer through a data acquisition board, PCI-1711.

3. Hybrid Fuzzy Control Scheme

Several approaches have been employed for controlling flexible systems. The current study focuses on the design and implementation of hybrid control schemes comprising fuzzy PD-like control and a conventional integrator for set-point tracking of the vertical motion of the TRMS. The proposed control scheme is shown in Figure 2.

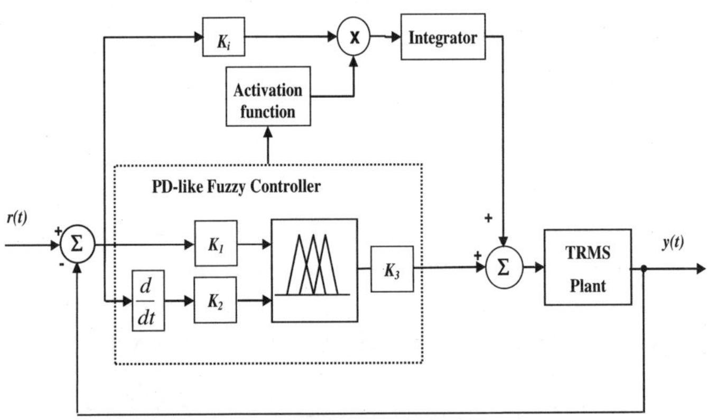

Figure 2. Hybrid fuzzy PD-like control scheme

The sigmoid function is a bounded function with limits at $-\infty$ and $+\infty$ of 0 and 1 respectively. The sigmoid function can be expressed as:

$$f(x) = \frac{1}{1 + e^{-a(x-c)}} \tag{1}$$

where
a determines the slope of the curve.
c determines the point of deflection.

The value of the gain a determines the slope of the function at the origin, and can change the functional behaviour from that of a slowly rising transition as $a \to 0$ to a unit step as $a \to \infty$. The effect of the term c is to shift the response curve in the control plane without changing the shape of the curve. This bias term can therefore be used to provide scaling of the sigmoid function. According to the behaviour of the sigmoid function, control state variables that are small in magnitude will have medium to large gain applied, thereby providing enhanced sensitivity to fine-scale control adjustments. In the large control state regime, the function suppresses large changes in control output, thereby stabilizing the control response. The sigmoid control function therefore provides a control stage that provides enhanced control sensitivity to control regions that are inherently controllable, while suppressing control actions that tend to diverge. The adjustable parameter a controls the slope of the sigmoid, which translates into adjustment of the sigmoid control function rise and decay characteristics.

4. Implementation and Results

In this investigation a hybrid control scheme comprising fuzzy PD-like control and a classical integrator is developed and implemented in real-time. Single-input single-output control is considered for vertical motion of the TRMS. The implementation is realized with Matlab. The combination of fuzzy controller and classical integrator is performed by using an activation function. In this scheme the contribution of the classical control part is regulated by the fuzzy controller through the activation function. It is well known that the integral term introduces oscillations and overshoots during the transient state which makes the settling time rather long. Therefore, the developed activation function is used to suppress integral action during the transient state and retrieve it back during the steady-state in such a way that to influence the system response during abrupt manoeuvres. This approach is achieved by connecting the output of the fuzzy PD-like controller to the input of the activation function as shown in Fig. 1. The output of the activation function is then multiplied by the integral gain Ki so that the integral gain will be tuned online and the integral term is continuously regulated by the fuzzy PD-like term through a sigmoid activation function. For the fuzzy control, appropriate control rules between the two inputs (control error and change of error) and one output of the fuzzy inference system are defined based on the rule-base arrays shown in Table 1.

Table 1: Rule-base array for vertical motion

e / Δe	**NB**	**NS**	**Z**	**PS**	**PB**
NB	NS	NB	NS	NB	NB
NS	Z	Z	Z	Z	NB
Z	NS	NS	Z	PS	PB
PS	PM	NS	PS	Z	PB
PB	PM	PB	PB	PB	PB

The two fuzzy input sets (tracking error and its derivative) of the PD-like fuzzy controller are passed on to a Gaussian type membership function to generate input fuzzy variables which are then used to evaluate the fuzzy rules and generate output fuzzy variables. Output fuzzy variables are then defuzzified to generate output for the rotor. The inputs and output universes of discourse of the PD-like fuzzy controller are normalized within the range [–1, 1]. The input gains K_1 and K_2 were used to map the actual inputs (tracking error and its derivative) of the fuzzy system to the normalized universe of discourse. Similarly, K_3 is the output gain that scales the output of the controller. Centre of gravity (COG) defuzzification is used in the design of the fuzzy control systems throughout this investigation.

The proposed control scheme was implemented on the actual system and its performance was compared with the developed controller without the activation function. It can be seen in Figs. 3 and 4 that by introducing the developed activation function, overshoot and settling time were greatly reduced. It can be clearly noted that the overshoot was reduced from 31% to 12.87%. Also the settling time was reduced from 7.5 sec to 3.5 sec. Fig. 5 demonstrates the effect of the activation function on the integral coefficient where it can be clearly seen that the activation function reduced the integral coefficient by more than 50% at rising and falling edges of the square-wave input and maintained the original value of the coefficient otherwise. Consequently, the output of the integral term was also suppressed at the abrupt changes of the reference input as shown in Fig. 6. As a result of the integral action suppression during abrupt transients, the output of the hybrid controller, which comprised fuzzy PD-like and integral term, was also reduced as shown in Fig. 7 and the unwanted excessive output was avoided resulting in better time domain specification of the system response.

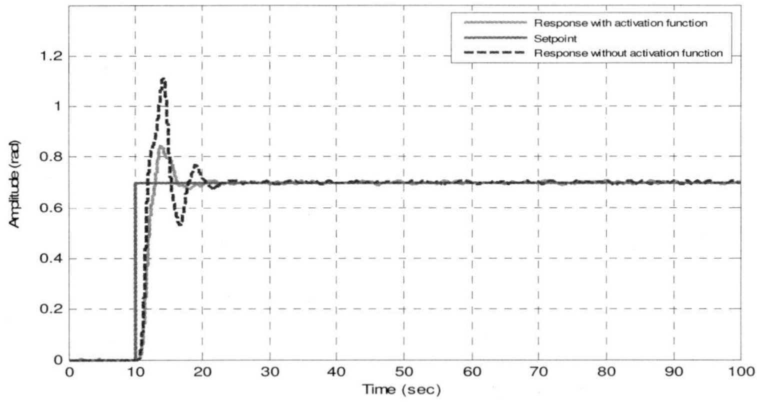

Figure 3. Closed-loop response to step input.

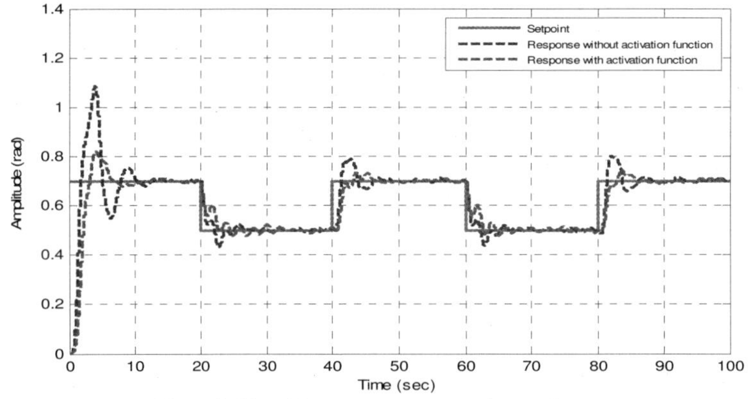

Figure 4. Closed-loop response to square wave input.

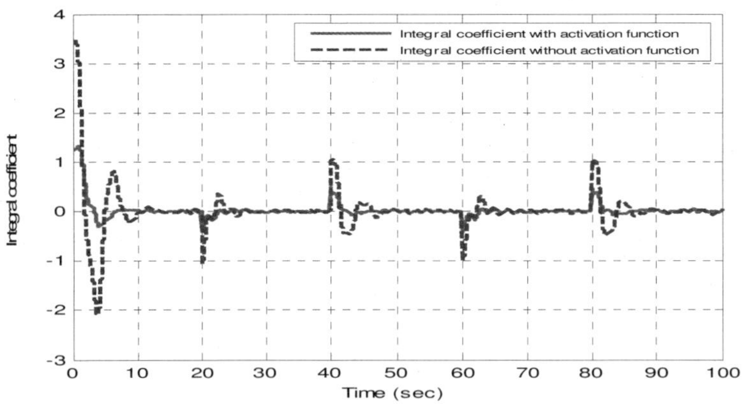

Figure 5. Effect of the activation function on the integral coefficient.

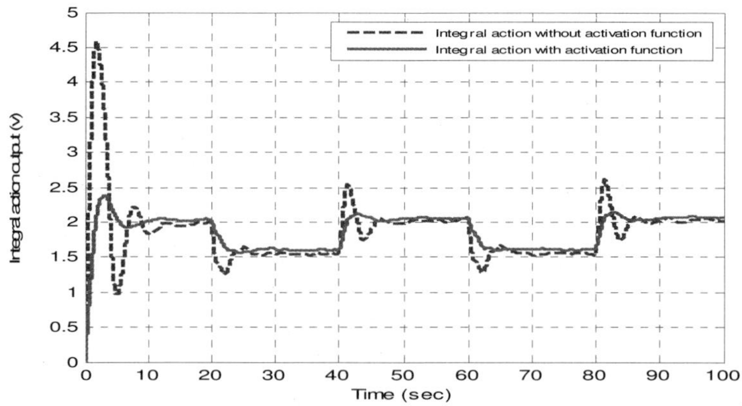

Figure 6. Effect of the activation function on the integral action.

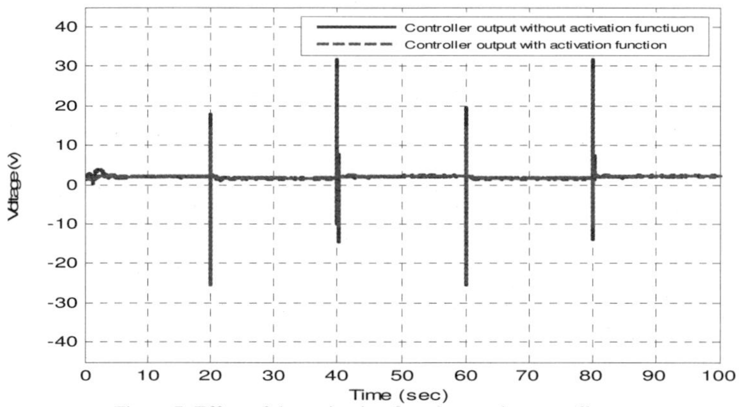

Figure 7. Effect of the activation function on the controller output.

5. Conclusion

In this investigation, a hybrid fuzzy control scheme has been developed and implemented for set-point tracking control of the TRMS in vertical plane motion. The control scheme comprised a fuzzy PD-like controller and a conventional integrator. A nonlinear sigmoidal activation function has been used to regulate the contribution of the conventional integrator into the controller output. The performance of the developed control strategy including activation function has been assessed in comparison to the control strategy which does not include activation function. The obtained results revealed that the developed control scheme can greatly improve the system response in terms of time domain specification.

References

1. Feedback Instruments Ltd., Twin Rotor MIMO System Manual 33-007-0, Sussex, UK (1996).
2. D. P. Kwok, P. Tam, C. K. Li, P. Wang, "Linguistic PID controllers," Proceedings of the 11th World Congress, Tallin, Estonia, 7, pp. 192-197, 1990.
3. T. Brehm, K. S. Rattan, "Hybrid fuzzy logic PID controller," Proceeding of the IEEE National Aerospace and Electronics Conference, Vol:2, pp. 807-813, 1993.
4. W. Li, "Design of hybrid fuzzy logic proportional plus conventional integral-derivative controller," IEEE Trans. Fuzzy Systems 6(4), pp. 449-463, 1998.
5. W. Li, X. G. Chang, F. M. Wahl, S. K. Tso, "Hybrid fuzzy P+ID control of manipulators under uncertainty," Mechatronics (9), pp. 301-315, 1999.
6. L. Xiaoyin, Q. Belmin, "Fuzzy-PID controller," IEE TENCON'93, Beijing, pp. 296-299, 1993.
7. L. Reznik, O. Ghanayem, A. Bourmistrov, "PID plus fuzzy controller structure as a design base for industrial applications," Engineering Applications of Artificial Intelligence, 13 , pp. 419-430, 2000.
8. R. Ketata, , D. D. Geest, A. Titli, "Fuzzy controller: design, evaluation, parallel and hierarchical combination with a PID controller," Fuzzy Sets and Systems (71), pp. 113-129, 1995.
9. N. Matsunaga, S. Kawaji, "Fuzzy hybrid control for DC servomotor," Trans. Inst. Electrical Eng. Japan D (111), pp. 195-200, 1991.
10. A. Otsubo, K. Hayashi, S. Murakami, M. Maeda, "Fuzzy hybrid control using simplified indirect inference method," Fuzzy Sets and Systems (99), pp. 265-272, 1998.
11. M. Parnichkun, C. Ngaecharoenkul, "Kinematics control of pneumatic systems by hybrid fuzzy PID," Mechatronics (11), pp. 1001-1023, 2001.
12. M. J. Er, Y. L. Sun, "Hybrid fuzzy proportional integral plus conventional derivative control of linear and nonlinear systems," IEEE Trans. On Industrial Electronics 48(6), pp. 1109-1117, 2001.

SECTION–7

GUIDANCE AND NAVIGATION

AVOIDING TRACKING ERROR WITH ESTIMATION TECHNIQUES IN VISUAL SERVOING

C. A. DÍAZ-HERNÁNDEZ*, J. L. MUÑOZ-LOZANO and J. LÓPEZ-CORONADO.

Department of System Engineering and Automatic, Polytechnical University of Cartagena,
Cartagena,30202, Spain
*E-mail: *ca.al.di.he@gmail.com, joselu.mlozano@upct.es, jl.coronado@upct.es*

This paper focuses the visual servoing of a mobile robot in dynamic environments. We assume a target with maneuvering capabilities, and thus it can be hidden from the camera by the obstacles in the scene. These two problems must be taken into account in the control law to ensure correct servoing. The control law must consider the target movement to reduce tracking error as small as possible. Moreover, the control law should consider visual loss management (reconstruction of the visual signal in case of occultation), and collision avoidance, estimating the obstacle motion. We present a strategy to avoid the tracking error due to the movement of the target itself.

Keywords: Visual servoing, estimation, mobile robot

1. Introduction

This paper is concerned with the visual servoing of non-holonomic mobile robot in dynamic environments, eventually, with the presence of moving obstacles. Visual servoing is a technique which uses feedback information extracted from a vision sensor to control the motion of a robot.

Under the main problem of controlling a mobile robot in dynamic environments, a problem to solve is the command calculation taking into account an estimate of the target movement with the aim of reducing the tracking error.

In this article, we focus on the problem of taking into account the target movement in the control law. We consider the tracking of a mobile target by a mobile robot. In the literature there are several solutions to estimate target motion: in Ref. 1 a visual controller is presented, that is based on a classical method in automatic control to cancel tracking errors consists of compensating the target motion through an integral term in the control law. This scheme allows the controller to cancel the tracking errors only

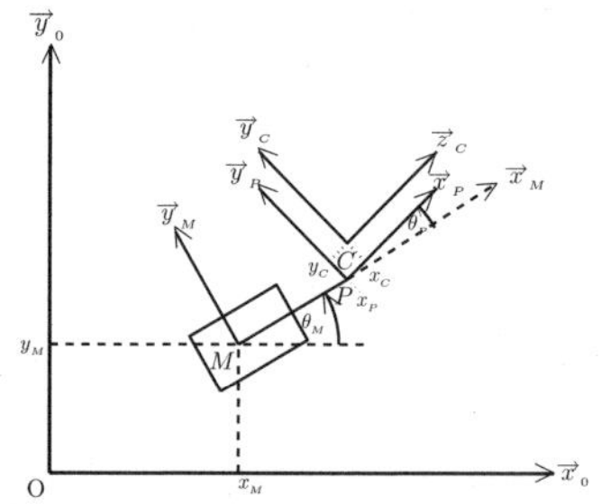

Fig. 1. The mobile robot with pan-platform.

if the target has a constant velocity. Other approach, as in Ref. 2 that is based on feed forward control, estimates directly the velocity through the image measurements and the camera velocities.

This article is structured as follows. In Sec. 2 we present the model of the serving robot. In Sec. 3 we propose a first approach to control law that takes into account the target movement. Validation of the precedent control law is discussed in Sec. 4. Finally, in Sec. 5 we present the conclusions.

2. Robot Model

We consider a robotic platform consist of a non-holonomic robot equipped with a camera on a pan-platform (see Fig. 1). The configuration of the mobile base is described by a vector $\mathbf{Q} = [x_{\mathrm{M}}, y_{\mathrm{M}}, \theta_{\mathrm{M}}]^{\mathbf{T}}$, where the pair $(x_{\mathrm{M}}, y_{\mathrm{M}})$ denotes the coordinates of the robot reference point $\mathbf{M}$ with respect to the scene frame $\mathbf{R}_{\mathrm{O}}(\mathbf{O}, \vec{x}_{\mathrm{O}}, \vec{y}_{\mathrm{O}}, \vec{z}_{\mathrm{O}})$ and the angle θ_{M} represents the orientation of the mobile base to the axis $(\mathbf{O}, \vec{x}_{\mathrm{O}})$. $\mathbf{R}_{\mathrm{M}}(\mathbf{M}, \vec{x}_{\mathrm{M}}, \vec{y}_{\mathrm{M}}, \vec{z}_{\mathrm{M}})$ denotes the frame linked to the mobile base. $\mathbf{P}$ denotes the center of rotation of the pan-platform and x_{P} is the abscissa of $\mathbf{P}$ in $\mathbf{R}_{\mathrm{M}}$. $\mathbf{R}_{\mathrm{P}}(\mathbf{P}, \vec{x}_{\mathrm{P}}, \vec{y}_{\mathrm{P}}, \vec{z}_{\mathrm{P}})$ and $\mathbf{R}_{\mathrm{C}}(\mathbf{C}, \vec{x}_{\mathrm{C}}, \vec{y}_{\mathrm{C}}, \vec{z}_{\mathrm{C}})$ represent respectively the frames attached to the pan-platform and the camera. θ_{P} defines the orientation of $\mathbf{R}_{\mathrm{P}}$ respect to $\mathbf{R}_{\mathrm{M}}$. The camera's optical center, denoted $\mathbf{C}$, has the coordinates $[x_{\mathrm{C}}, y_{\mathrm{C}}, z_{\mathrm{C}}]^{\mathbf{T}}$ in $\mathbf{R}_{\mathrm{P}}$. Vector $\vec{z}_{\mathrm{C}}$ matches up with the optical axis of the camera.

Defining the vector $\mathbf{q} = [l_{\mathrm{M}}, \theta_{\mathrm{M}}, \theta_{\mathrm{P}}]^{\mathbf{T}}$, where l_{M} denotes the curvilinear abscissa of $\mathbf{M}$, the velocity screw of the camera with respect to $\mathbf{R}_{\mathrm{O}}$ is expressed in $\mathbf{R}_{\mathrm{C}}$ as:

$$\mathbf{T}_{\mathrm{C}} = \mathbf{J}_{\mathrm{q}}\, \dot{\mathbf{q}}, \tag{1}$$

where

$$\mathbf{J}_{q} = \begin{pmatrix} 0 & 0 & 0 \\ -\sin\theta_{\mathbf{P}} & x_{\mathbf{C}} + x_{\mathbf{P}}\cos\theta_{\mathbf{P}} & x_{\mathbf{C}} \\ \cos\theta_{\mathbf{P}} & -y_{\mathbf{C}} + x_{\mathbf{P}}\sin\theta_{\mathbf{P}} & -y_{\mathbf{C}} \\ 0 & -1 & -1 \\ 0 & 0 & 0 \\ 0 & 0 & 0 \end{pmatrix},$$

and

$$\dot{\mathbf{q}} = [\mathbf{v}_{\mathbf{M}}, \omega_{\mathbf{M}}, \omega_{\mathbf{P}}]^{\mathbf{T}}.$$

The null rows of the Jacobian matrix $\mathbf{J}_{q}$ define the restricted movements of the camera. Thus, the velocity screw is reduced, by retaining only the controllable degrees of freedom of the camera, as expressed in Eq. (2).

$$\mathbf{T}_{\text{red}} = \mathbf{J}_{\text{red}}\,\dot{\mathbf{q}}. \tag{2}$$

where

$$\mathbf{J}_{\text{red}} = \begin{pmatrix} -\sin\theta_{\mathbf{P}} & x_{\mathbf{C}} + x_{\mathbf{P}}\cos\theta_{\mathbf{P}} & x_{\mathbf{C}} \\ \cos\theta_{\mathbf{P}} & -y_{\mathbf{C}} + x_{\mathbf{P}}\sin\theta_{\mathbf{P}} & -y_{\mathbf{C}} \\ 0 & -1 & -1 \end{pmatrix}.$$

In the pin-hole camera model, scene points are projected onto the image plane through a perspective projection. Thus, if $\underline{x}_{g} = [x_{g}, y_{g}, z_{g}]^{\mathbf{T}}$ represents the coordinates of a point g of the scene expressed in $\mathbf{R}_{\mathbf{C}}$, then, its projection G in the image plane in metric coordinates is $\underline{X}_{G} = [X_{G}, Y_{G}, Z_{G} \equiv \mathrm{f}]^{\mathbf{T}}$ computed by:

$$\underline{X}_{G} = \frac{\mathrm{f}}{z_{g}}\,\underline{x}_{G}, \tag{3}$$

where f represents the focal length of the camera.

Let $\mathbf{s}$ the vector associated with the observed visual pattern. The speed of the visual features, $\dot{\mathbf{s}}$, in the image is related to $\mathbf{T}_{\mathbf{C}}$ by:

$$\dot{\mathbf{s}} = \mathbf{L}_{s}\,\mathbf{T}_{\mathbf{C}}, \tag{4}$$

where $\mathbf{L}_{s}$ denotes the interaction matrix which is defined by Samson and Espiau.[3] This matrix depends on the nature of the visual characteristics considered (points, lines, etc.) and depth of the target from the camera.

In the case of a point g in $\mathbf{R}_{\mathbf{C}}$, Chaumette[4] defines $\mathbf{L}_{s}$ as:

$$\mathbf{L}_{s} = \begin{bmatrix} \frac{-1}{z_{g}} & 0 & \frac{X_{G}}{z_{g}} & X_{G}Y_{G} & -(1 + X_{G}^{2}) & Y_{G} \\ 0 & \frac{-1}{z_{g}} & \frac{Y_{G}}{z_{g}} & 1 + Y_{G}^{2} & -X_{G}Y_{G} & -X_{G} \end{bmatrix}. \tag{5}$$

In order to match the dimensions of $\mathbf{L}_{\mathbf{s}}$ and $\mathbf{T}_{\mathrm{red}}$, we remove the columns of $\mathbf{L}_{\mathbf{s}}$ corresponding to the not allowed movement of the camera. Thus, the reduced interaction matrix is written as:

$$\mathbf{L}_{\mathrm{red}} = \begin{bmatrix} 0 & \frac{X_G}{z_g} & X_G Y_G \\ \frac{-1}{z_g} & \frac{Y_G}{z_g} & 1 + Y_G^2 \end{bmatrix}. \tag{6}$$

In our case, the target is modeled by the four-corners of a square, thus $\mathbf{s}$ is defined by a vector containing the coordinates (X_i, Y_i) of these points, they are indexed as $i = 1, \ldots, 4$. The final interaction matrix for the chosen pattern is:

$$\mathbf{L} = \left[\mathbf{L}_{\mathrm{red},1}^{\mathbf{T}}, \mathbf{L}_{\mathrm{red},2}^{\mathbf{T}}, \mathbf{L}_{\mathrm{red},3}^{\mathbf{T}}, \mathbf{L}_{\mathrm{red},4}^{\mathbf{T}}\right]^{\mathbf{T}}. \tag{7}$$

3. Control Law

The vision task can be expressed as the regulation to zero of the task function:[5]

$$\mathbf{e}(\mathbf{q}, t) = \mathbf{C}\,(\mathbf{s}(\mathbf{q}, t) - \mathbf{s}^*), \tag{8}$$

where $\mathbf{e}$ has the same dimension as $\mathbf{q}$, and $\mathbf{s}^*$ denotes the reference value for $\mathbf{s}$; $\mathbf{C}$ is the combination matrix and allows to enclose more information to compute the command than the number of camera's degrees of freedom. A practical solution is to define $\mathbf{C} = \widehat{\mathbf{L}}_{\mathbf{s}}^{+}$, where $\widehat{\mathbf{L}}_{\mathbf{s}}^{+}$ is the left pseudo-inverse of an estimated of $\mathbf{L}_{\mathbf{s}}$.

We want that $\mathbf{e}$ decrease exponentially; therefore, in close loop, it must to check the law:

$$\dot{\mathbf{e}} = -\lambda\,\mathbf{e}, \tag{9}$$

where $\lambda > 0$. From Eq. (8), the derivative of $\mathbf{e}$ is expressed as:

$$\dot{\mathbf{e}} = \frac{\partial \mathbf{e}}{\partial \mathbf{q}}\,\dot{\mathbf{q}} + \frac{\partial \mathbf{e}}{\partial t} = \mathbf{C}\,\mathbf{L}_{\mathbf{s}}\,\mathbf{J}_{\mathrm{red}}\,\dot{\mathbf{q}} + \frac{\partial \mathbf{e}}{\partial t}, \tag{10}$$

where $\frac{\partial \mathbf{e}}{\partial t}$ represents the variation of $\mathbf{e}$ due to target movement.

Taking into account the Eqs. (9) and (10), Chaumette[4] deduces the next control law:

$$\dot{\mathbf{q}} = \mathbf{J}_{\mathrm{red}}^{-1}\left(-\lambda\,\mathbf{e} - \widehat{\frac{\partial \mathbf{e}}{\partial t}}\right). \tag{11}$$

Chaumette[4] estimates $\frac{\partial \mathbf{e}}{\partial t}$ by the recurrence:

$$\left(\widehat{\frac{\partial \mathbf{e}}{\partial t}}\right)_{k+1} = \left(\widehat{\frac{\partial \mathbf{e}}{\partial t}}\right)_{k} + \mu\,\mathbf{e}_k, \tag{12}$$

where $\left(\widehat{\frac{\partial \mathbf{e}}{\partial t}}\right)_0 = 0$ and μ is a predefined gain. This estimator, type integrator, is based solely on successive measurement values of $\mathbf{e}$, which is also based on visual characteristics.

Bensalah[2] has proposed a measure of the apparent movement of the target in the image based on the difference between total variation of $\mathbf{e}$ and the change of $\mathbf{e}$ due to camera motion.

$$\left(\widehat{\frac{\partial \mathbf{e}}{\partial t}}\right) = \hat{\dot{\mathbf{e}}} - \left(\widehat{\frac{\partial \mathbf{e}}{\partial \mathbf{q}}}\right) \mathbf{T}_{\mathrm{C}}. \tag{13}$$

In the discrete domain (period Δt), it is expressed as:

$$\left(\widehat{\frac{\partial \mathbf{e}}{\partial t}}\right)_k = \frac{\mathbf{e}_k - \mathbf{e}_{k|k-1}}{\Delta t} - \left(\widehat{\frac{\partial \mathbf{e}}{\partial \mathbf{q}}}\right) (\mathbf{T}_{\mathrm{C}})_{k-1}, \tag{14}$$

where $\mathbf{e}_{k|k-1}$ is the predicted value for $\mathbf{e}_k$ computed from $\mathbf{e}_{k-1}$ and the applied velocity screw at $k-1$ instant. Equation (14) is the measure equation for the evolution models of $\frac{\partial \mathbf{e}}{\partial \mathbf{q}}$, which are the basis for a Kalman filter.

4. Results

The command law has been evaluated for tracking tasks, in which a mobile robot must tracks a mobile square pattern. The coordinates of the corners of the pattern have been used as visual information. The target robot with the visual pattern follows a straight line path with a constant velocity.

$\mathbf{s}^*$ is set to $(-0.1, 0.1, -0.1, -0.1, 0.1, -0.1, 0.1, 0.1)^{\mathbf{T}}$. This pattern will be observed by the camera only when 1) the axis $(\mathbf{C}, \vec{z}_{\mathrm{C}})$ be normal and centered to visual pattern, and 2) the camera's optical center is located to reference distance $d^* = 2m$ from the pattern plane.

The visual servoing has been evaluated, taking into account that $\widehat{\frac{\partial \mathbf{e}}{\partial t}} = 0$ and estimating the value of $\widehat{\frac{\partial \mathbf{e}}{\partial t}}$ through the Eqs. (12) and (14).

Figures 2, 4, and 6, show the temporal evolution of $\mathbf{s}$. In all these graphs, we make comments on the left superior point in the image pattern. $\mathbf{s}^*$ is drawn as *reference point. Starting point* and *ending point* correspond to the observed points at initial and final instant of simulation, respectively. The line, connecting the starting and ending points, represents the temporal evolution of the point. We encircle the ending point and reference point to highlight the error.

Figures 3, 5, and 7 display the error $\epsilon = \mathbf{s} - \mathbf{s}^*$ for $\underline{X}_1$. The vertical and horizontal error components are plot in dash and solid lines, respectively. A single point has been represented, because the absolute value of the vertical and horizontal components of each point has the same behavior and it make easier the visualization.

4.1. *Visual Tracking without Target Movement Estimation*

The control law has been evaluated taking into account that the target movement is not meaningful. Under this configuration, the task is not correctly realized, because not all the error components converge to zero, specifically those components related with the horizontal error observed in the image plane, which converge in magnitude to a same value. It was observed that the magnitude of this error is directly related with the lineal velocity of the target.

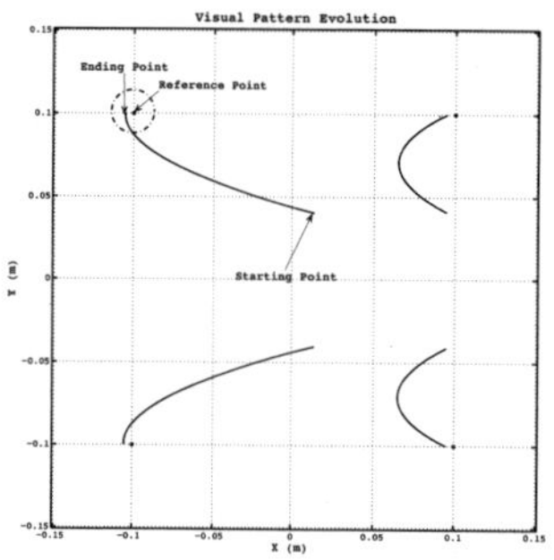

Fig. 2. Evolution of $\mathbf{s}$ considering $\frac{\partial \mathbf{e}}{\partial t} = 0$. Note the error observed between the *reference point* and the *ending point* is encircled, the error is mainly on the horizontal components of $\mathbf{s}$.

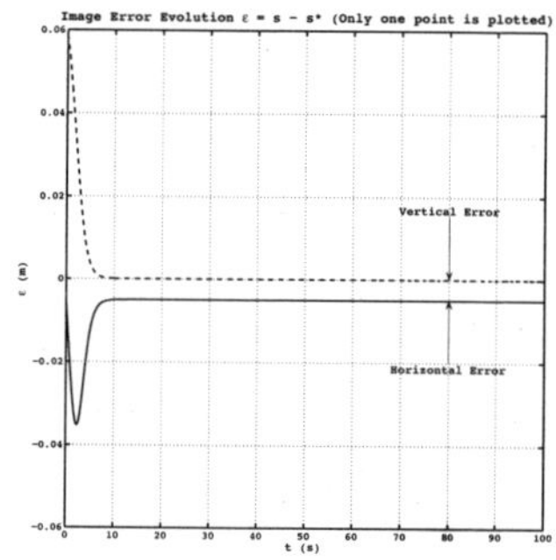

Fig. 3. Temporal evolution of error ϵ considering $\frac{\partial \mathbf{e}}{\partial t} = 0$. Note, at steady state the vertical component of error are null, however, its horizontal component converges to a constant value and the magnitude is proportional to target velocity in $\mathbf{R}_{\mathrm{O}}$.

As it was mentioned previously, Fig. 2 shows that the error is presented only in the abscissas of the image coordinates; correspondingly, Fig. 3 shows that the vertical component of the error tends to zero, and it implies that the desired distance d^* has been reaching, however the error in the horizontal component of ϵ indicates that the line that passes through the point $\mathbf{C}$ and the center of the square pattern does not coincide with the axis $(\mathbf{C}, \vec{z}_{\mathrm{C}})$. The error of the vertical components tends to zero, however, the magnitude of the horizontal error tends to a constant value, it indicates that the image plane is parallel to the pattern plane.

4.2. *Visual Tracking with Target Movement Estimation*

The control law expressed by Eq. (11) has been evaluated using estimators for the target velocity proposed by Chaumette and Bensalah, expressed by Eqs. (12) and (14) respectively. In both cases the error observed in the image, ϵ, tends to zero and, thus, the task function expressed by Eq. (8) is correctly executed. This result can be observed in Figs. 4–5 and Figs. 6–7.

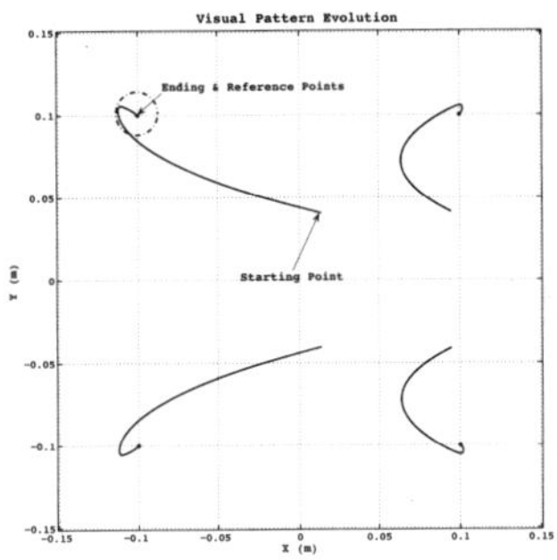

Fig. 4. Evolution of $\mathbf{s}$ using the Chaumette estimator. Note, differently to Fig. 2, the *ending point* coincides with the *reference point.*

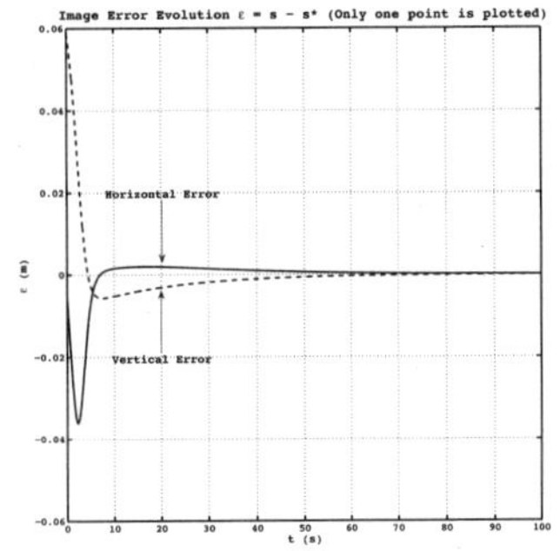

Fig. 5. Temporal evolution of ϵ using the Chaumette estimator. Note, differently to Fig. 3, also the horizontal component of error converges to zero, although we observe oscillations on both error componentes.

Although in both cases the task function is correctly executed, the dynamic obtained is different; the Bensalah estimator allows the error, ϵ, tends to the final value faster than Chaumette estimator. Figures 5 and 7 show the dynamic of the error in both cases. The effect of $\widehat{\frac{\partial \mathbf{e}}{\partial t}}$ calculated with the Chaumette estimator arise that $\mathbf{s}$ exceeds $\mathbf{s}^*$; in Fig. 4 this effect is shown by the curve generated by $\mathbf{s}$ in the proximity of $\mathbf{s}^*$ and in Fig. 5 when ϵ overcomes the t axis in the graph.

The fast convergence of Bensalah estimator is mainly due to it takes into account only information from the instant t and $t-1$, avoiding the effect of the error observed at the beginning of the task, which is generally larger in magnitude.

5. Conclusion

In this paper we have presented the controller for a robotic system, composed by a non-holonomic mobile base and a camera mounted on a pan-platform, from inverse kinematics. The controller exploits the properties of visual servoing method. The robotic system allows the driver to perform tasks of visual tracking of a moving target by a non-holonomic robot in dynamic environments.

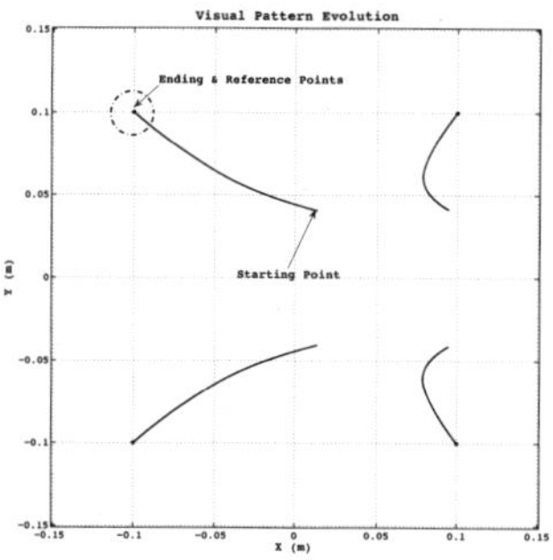

Fig. 6. Evolution of $\mathbf{s}$ using the Bensalah estimator. Note, similar to Fig. 4, the *ending point* coincides with the *reference point*, but, in this case, $\mathbf{s}$ do not exceeds $\mathbf{s}^*$.

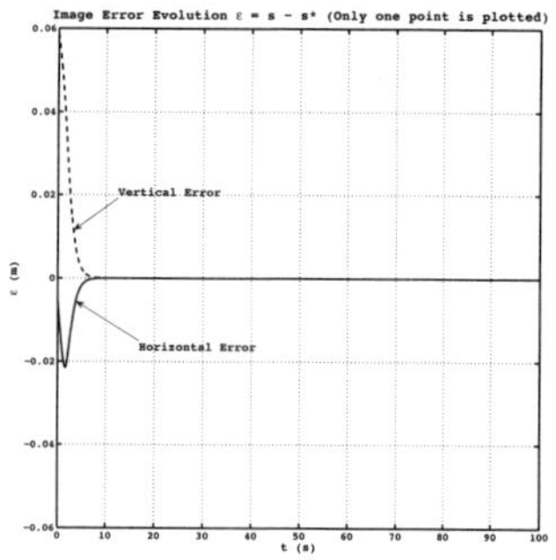

Fig. 7. Temporal evolution of ϵ using the Bensalah estimator. Note, differently to Fig. 5, the error in both components converge to zero, however, in this case its convergence is more rapid than that of Chaumette's estimator.

In order to increase the accuracy of the tracking system, in this article, we have focused on the compensation of the command using an estimate of the movement of the target itself. The application of this compensator allows an efficient execution of the robotic tasks; the results show that the estimator proposed by Bensalah allows fast convergence and an accurate estimate of the target velocity and therefore the convergence of the task function. The simulated results show the feasibility, effectiveness and proper functioning of the controller. In the future, we will extend this work to make the target occultation management and obstacle avoidance method.

Acknowledgments

This work has been supported by Spanish Administration under grants **TEC2006-11730-C03-03** and **TEC2009-10639-C04-03**.

References

1. Chaumette F. and Santos A., Tracking a moving object by visual servoing, in *Proc. of 12th World Congress IFAC*, July 1993.
2. Bensalah F., Estimation du mouvement par vision active, PhD thesis, Université de Rennes 1, (France, 1996).
3. Espiau, B., Chaumette, F. and Rives, P., *IEEE Transactions on Robotics and Automation* **8**, 313 (1992).
4. F. Chaumette and S. Hutchinson, *Handbook of Robotics: Visual servoing and visual tracking* (Springer, 2008), ch. 24, pp. 563–583.
5. Samson, C., Espiau, B. and Le Borgne, M., *Robot Control: The Task Function Approach* (Oxford University Press, 1991).

ROBUST PLANAR POSITION CONTROL OF AN AUTONOMOUS AIRSHIP

S. SAMAALI and L. BEJI[†] *

[†] *Laboratory IBISC EA 4526, University of Evry,*
40 rue du Pelvoux, 91020 Evry, France,
E-mail: beji@iup.univ-evry.fr
www.ibisc.univ-evry.fr

A. ABICHOU

LIM Laboratory, Polytechnic School of Tunisia,
La Marsa, 2070, Tunisia
E-mail: Azgal.Abichou@ept.rnu.tn;sarrasam@yahoo.fr

The nonlinear planar dynamic model of an underactuated autonomous airship and its position/orientation control are addressed, for moderate-amplitude sinusoidal perturbation in three directions. The dynamic model, even if it is reduced into a plan, it presents some strong coupling terms due to the non coincidence of buoyancy and gravity centers and the presence of added masses while two control acting on the airship model. A nonlinear control law is proposed that guaranteed the convergence of the airship in the neighborhood of a desired point in presence of perturbations. Numerical results are presented to show the effectiveness of the proposed nonlinear control law.

Keywords: Position control; Sinusoidal disturbances; Backstepping methodology; Lyapunov theorem.

1. Introduction

Many researches focus on studying the control of underactuated autonomous vehicles (terrestrial, aerial, naval) with no disturbances such as.[2–5,9,10] One cites the control algorithms for the airship disturbance-free dynamic model, such as in[10] and.[9] The nature of the control varies from system to another, according to its dynamic. The position and the attitude control of underactuated autonomous airship with disturbances haven't been considered yet. Compared to unmanned surface vessels[8] and the references

*Corresponding author

therein, the autonomous airship dynamic model, even if it is reduced into a plan, it presents some strong coupling terms. This is due to the non coincidence of buoyancy and gravity centers, and the presence of added masses. From view point the tracking/stabization objectives, the airship dynamic model can't take the output feedback form nor a strict feedback form. Hence, the stabilization analysis becomes a challenge problem. In absence of disturbances, the airship is small-time locally controllable at any equilibrium point. Therefore, the airship can be asymptotically stabilized at any attitude and orientation by only three controls. In[10] a time varying control was proposed to stabilize asymptotically the airship in the neighborhood of a desired point. They proved that the airship fails the Brockett's necessary condition. Consequently it can't be asymptotically stabilized by a smooth time-invariant feedback. So, they succeed to construct a time-varying control that guarantees its asymptotic stabilization. In[8] the problem of stabilization of underactuated surface vessel with various disturbances such as waves and wind is treated. They have shown that a smooth time varying feedback law stabilizes the position of the vessel in neighborhood of zero and its validity in presence and absence of disturbances. In[7] the authors dealt with the spin-axis stabilization control for underactuated rigid spacecraft under sinusoidal disturbances. They studied the spin-axis stabilization and constructed the controllers for both axisymmetric spacecraft and asymmetric spacecraft separately by combining the Lyapunov direct method with the LaSalle's theorem. In this paper, we overcome the difficulties of underactuation through an adequate change of variables, proposed to position and velocity states.

The paper will be structured as follows: In the first section, we detail the dynamic model of the underactuated autonomous airship in the horizontal plan in presence of the exosystem. Section 3 will present our control law according to the objective ensuring the convergence of the airship in neighborhood of a desired point. In order to prove the performance of the proposed controller, section 4 deals with numerical simulations. Some suggestions will conclude the paper.

2. The airship Kinodynamics

The nonlinear kinodynamic model that describes the motion of the airship in the horizontal plan P_{xy}, which be decomposed of two parts: kinematics and dynamics. The kinematic part allows us to determine the relation between the local speeds referred to the local frame, and the global speeds with respect to the earth frame. Let $\xi(x, y, \psi)$ denotes the position/orientation vector of the airship in the earth fixed frame. With respect to a local frame, attached to the airship center of mass, the variables u, v and r are the velocities in surge, sway and yaw respectively (see figure 1). The dynamic

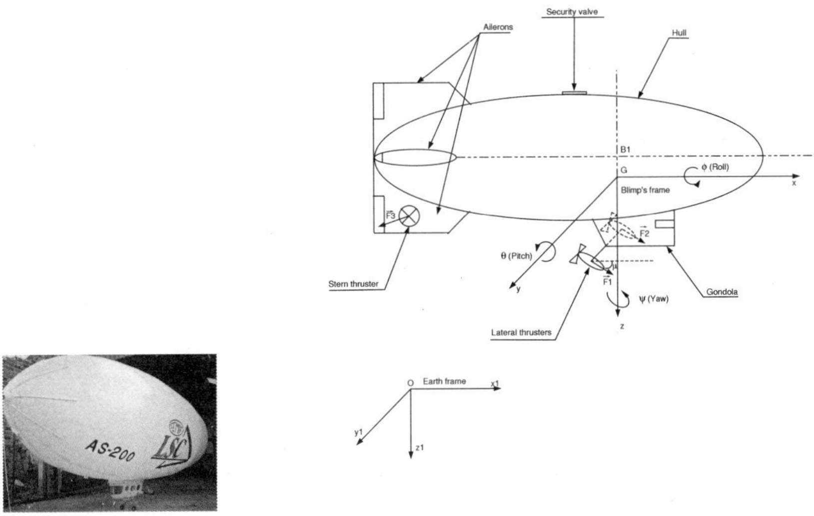

Fig. 1: The airship parametrization

part makes a synthesis of the exterior forces acting on the airship including drag forces and sinusoidal disturbances, acting in the three directions. Hence, combining kinematic and dynamic parts, the system of equations is as follows:

$$\begin{aligned}
\dot{x} &= c_\psi u - s_\psi v; \quad \dot{y} = s_\psi u + c_\psi v; \quad \dot{\psi} = r \\
\dot{u} &= \frac{1}{m_x}[x_u u + m_y v r + \tau_1] + \delta_1 \\
\dot{v} &= \frac{1}{m_y}[y_v v - m_x u r + \tau_2] + \delta_2 \\
\dot{r} &= \frac{1}{J_z}[(m_x - m_y) v u + N_r r - G o_{3x} \tau_2] + \delta_3
\end{aligned} \tag{1}$$

where $\delta_1 = \frac{d_1}{m_x}$, $\delta_2 = \frac{d_2}{m_y}$ and $\delta_3 = \frac{d_3}{J_z}$ show the three dimension disturbances, assumed to be sinusoidal with completely known and constant phase, amplitude, and frequency, and satisfy: $\ddot{d}_1 + \omega_1 d_1 = 0$, $\ddot{d}_2 + \omega_2 d_2 = 0$ and $\ddot{d}_3 + \omega_3 d_3 = 0$. The other parameters such as m_x, m_y and J_z justify the airship inertia and added mass effects. The x_u, y_v and N_r are time invariant aerodynamic damping parameters. The two control inputs are

$\tau_1 = |F_1 + F_2| cos\mu$ and $\tau_2 = |F_3|$. According to the considered plan motion $\mu = 0$, and there is no reverse motion. Consequently, the two acting forces are assumed to be equal in amplitudes. So, any airship yaw-motion is asserted only by τ_2 acting on the tail. Accordingly, we will consider the problem of controlling the position and the elimination of the effects of the sinusoidal external power.

3. Exponential Asymptotic Stabilization

After the adequate coordinate transformation on the system's states, the section is divided into two subsections; in the first one, we will write the error dynamic model that present the difference between the airship model and the reference model, and in the second subsection we will use the backstepping method[1] that provides a nonlinear control law that realizes the control objective. Before proceeding to the control design steps and in order to make our study system in a diagonal form, we propose the following coordinate transformation: $x_1 = x + \varepsilon c_\psi$, $y_1 = y + \varepsilon s_\psi$ and $v_1 = v + \varepsilon r$. where $\varepsilon = \frac{J_z}{Go_{3x} m_y}$. Under these transformations, system (2) takes the following form:

$$\begin{aligned} \dot{x}_1 &= u c_\psi - v_1 s_\psi; \quad \dot{y}_1 = u s_\psi + v_1 c_\psi; \quad \dot{\psi} = r \\ \dot{u} &= \tau_u \\ \dot{v}_1 &= \alpha v_1 + \beta r + \gamma u r + \lambda v_1 u + \delta_2 + \varepsilon \delta_3 \\ \dot{r} &= \tau_r \end{aligned} \tag{2}$$

where τ_u and τ_r being considered as new controls to be designed later,

$$\begin{aligned} \tau_u &= \frac{1}{m_x}[x_u u + m_y v r + \tau_1] + \delta_1 \\ \tau_r &= \frac{1}{J_z}[(m_x - m_y) v u + N_r r - Go_{3x} \tau_2] + \delta_3 \end{aligned}$$

and the new system's parameters $\alpha = \frac{y_v}{m_y}$, $\beta = \varepsilon(-\frac{y_v}{m_y} + \frac{N_r}{J_z})$, $\gamma = -\frac{m_x}{m_y} - \frac{\varepsilon^2(m_x - m_y)}{J_z}$ and $\lambda = \frac{\varepsilon(m_x - m_y)}{J_z}$. The δ-perturbations appear as a combination in v_1 dynamics.

3.1. *System of errors*

Our objective is to determine the inputs τ_u and τ_r which allow the following system of errors to converge asymptotically in the neighborhood of 0_{IR^6}, in order to enable the airship to reach the desired position $(x_{1r}, y_{1r}, \psi_r, 0, 0, 0)$.

The corresponding systems of errors can be easily written as,

$$\begin{aligned} \dot{X}_e &= RU; \quad \dot{\psi}_e = r \\ \dot{u} &= \tau_u \\ \dot{v}_1 &= \alpha v_1 + \beta r + \gamma ur + \lambda v_1 u + \delta_2 + \varepsilon\delta_3 \\ \dot{r} &= \tau_r \end{aligned} \tag{3}$$

where the following notations are considered.

$$R = \begin{pmatrix} c_\psi & -s_\psi \\ s_\psi & c_\psi \end{pmatrix}; \; X_e = \begin{pmatrix} x_{1e} = x_1 - x_{1r} \\ y_{1e} = y_1 - y_{1r} \end{pmatrix}; \; U = \begin{pmatrix} u \\ v_1 \end{pmatrix} \tag{4}$$

U is the linear velocity vector, X_e is the position error vector, with x_{1r} and y_{1r} are desired virtual positions and we will introduce v_{1r} is a desired virtual sway velocity.

3.2. *Control law design*

In this section we seek to design a position control law for the system of errors (3). For the sake of clarity, we split our design conception into four steps.

step 1. In order to stabilize the vector position X_e, we assume u and v_1 as virtual controls. We start by defining a Lyapunov function candidate as

$$V_1 = \frac{1}{2} X_e^T X_e \tag{5}$$

and computing its time derivative (T the transpose), we obtain:

$$\dot{V}_1 = X_e^T RU \tag{6}$$

We suggest as desired expressions for the virtual controls $[\alpha_u, \alpha_v]^T$: $(\alpha_u, \alpha_v)^T = -R^T K X_e$, with K is a positive definite matrix. Substituting these virtual controls into (6) yields after development,

$$\dot{V}_1 = -X_e^T K X_e \tag{7}$$

step 2. Since the components of the vector U are not true controls, we need to introduce new error variables z_u and z_v defined as: $z_u = u - \alpha_u$, $z_v = v_1 - \alpha_v$. Then, the derivative of the Lyapunov function is rewritten as:

$$\dot{V}_1 = X_e^T(-KX_e + RZ_u) \tag{8}$$

where $Z_u = (z_u, z_v)^T$. In the following we would derive z_u to zero, so we added a term to V_1 such that,

$$V_2 = V_1 + \frac{1}{2} z_u^2 \tag{9}$$

The time derivative of V_2 can be expressed as

$$\dot{V}_2 = -X_e^T K X_e + z_v(y_{1e}c_\psi - x_{1e}s_\psi) + z_u(\tau_u + ku + x_{1e}c_\psi + y_{1e}s_\psi) \tag{10}$$

Setting

$$\tau_u = -c_1 z_u - c_2 z_u^3 - ku - x_{1e}c_\psi - y_{1e}s_\psi + f_u, \tag{11}$$

where f_u is a design variable for subsequent use, the time derivative of V_2 becomes,

$$\dot{V}_2 = -X_e^T K X_e + z_v(y_{1e}c_\psi - x_{1e}s_\psi) - c_1 z_u^2 - c_2 z_u^4 + z_u f_u \tag{12}$$

where c_1 and c_2 are positive constants. So far, the controlled subsystem of error dynamics equations has been transformed to,

$$\begin{aligned} \dot{X}_e &= R\alpha(X_e) + RZ_u \\ \dot{z}_u &= -c_1 z_u - c_2 z_u^3 - x_{1e}c_\psi - y_{1e}s_\psi + f_u \end{aligned} \tag{13}$$

step 3. Before starting, we perform some manipulations on the sway error dynamics equation. Then $\dot{z}_v$ can be expanded as

$$\dot{z}_v = (\beta + \gamma u)r + \lambda v_1 z_u + \chi_{v1} \tag{14}$$

where

$$\chi_{v1} = \lambda v_1 \alpha_u + (k + \alpha)v_1 + \delta_2 + \varepsilon\delta_3 \tag{15}$$

In order to stabilize z_v and ψ_e, we consider the velocity in yaw as an auxiliary control, assuming $(\beta+\gamma u) \neq 0$ which is a natural requirement, because u converges to zero and $\beta \neq 0$. We note α_r as a virtual control of r. Choosing the following Lyapunov function

$$V_3 = V_2 + \frac{1}{2}(z_v^2 + \psi_e^2) \tag{16}$$

and substituting the equation (14) into the time derivative of V_3, we obtain,

$$\begin{aligned} \dot{V}_3 = &-X_e^T K X_e - c_1 z_u^2 - c_2 z_u^4 + z_u f_u + \lambda v_1 z_v z_u + ((\beta + \gamma u)z_v + \psi_e)\alpha_r \\ &+ z_v(y_{1e}c_\psi - x_{1e}s_\psi + \chi_{v1}) \end{aligned}$$

setting

$$\alpha_r = -c_3((\beta + \gamma u)z_v + \psi_e) \tag{17}$$

with $c_3 > 0$ then we have,

$$\begin{aligned} \dot{V}_3 = &-X_e^T K X_e - c_1 z_u^2 - c_2 z_u^4 + z_u f_u + \lambda v_1 z_v z_u - c_3((\beta + \gamma u)z_v + \psi_e)^2 \\ &+ z_v(y_{1e}c_\psi - x_{1e}s_\psi + \chi_{v1}) \end{aligned}$$

step 4. Since the variable r is not the true control, we introduce a new error variable $z_r = r - \alpha_r$. So we will concept the command τ_r to stabilize the following subsystem

$$\begin{aligned} \dot{z}_v &= (\beta + \gamma u)(z_r + \alpha_r) + \lambda v_1 z_u + \chi_{v1} \\ \dot{\psi}_e &= z_r + \alpha_r \\ \dot{z}_r &= \tau_r - \dot{\alpha}_r \end{aligned} \tag{18}$$

So, the study dynamic equations are transformed as:

$$\begin{aligned} \dot{X}_e &= -KX_e + RZ_u \\ \dot{z}_u &= -c_1 z_u - c_2 z_u^3 - x_{1e} c_\psi - y_{1e} s_\psi + f_u \\ \dot{z}_v &= (\beta + \gamma u) z_r - c_3(\beta + \gamma u)^2 z_v - c_3(\beta + \gamma u)^2 \psi_e + \lambda v_1 z_u + \chi_{v1} \\ \dot{\psi}_e &= z_r - c_3 \psi_e - c_3(\beta + \gamma u) z_v \\ \dot{z}_r &= \tau_r - \dot{\alpha}_r \end{aligned} \tag{19}$$

we choose

$$V_4 = V_3 + \frac{1}{2} z_r^2 \tag{20}$$

and taking the time derivative, we obtain

$$\begin{aligned} \dot{V}_4 = &-X_e^T K X_e - c_1 z_u^2 - c_2 z_u^4 + z_u(f_u + \lambda v_1 z_v) - c_3((\beta + \gamma u) z_v + \psi_e)^2 \\ &+ z_v(y_{1e} c_\psi - x_{1e} s_\psi + \chi_{v_1}) + z_r[\tau_r + (\beta + \gamma u) z_v + \psi_e + c_3((\beta + \delta u)^2 + 1) r \\ &+ c_3(\beta + \gamma u)(\chi_{v_1} + \lambda v_1 z_u)] \end{aligned}$$

Taking

$$\begin{aligned} \tau_r = &-c_4 z_r - c_5 z_r^3 - (\beta + \gamma u) z_v - \psi_e - c_3((\beta + \gamma u)^2 + 1) r \\ &- c_3(\beta + \gamma u)(\chi_{v_1} + \lambda v_1 z_u) \end{aligned}$$

with c_4 and c_5 are strictly positive constants. Consequently $\dot{V}_4$ has the new following form:

$$\begin{aligned} \dot{V}_4 = &-X_e^T K X_e - c_1 z_u^2 - c_2 z_u^4 + z_u(f_u + \lambda v_1 z_v) - c_3((\beta + \gamma u) z_v + \psi_e)^2 \\ &+ z_v(y_{1e} c_\psi - x_{1e} s_\psi + \chi_{v_1}) - c_4 z_r^2 - c_5 z_r^4 \end{aligned}$$

choosing $f_u = -\lambda v_1 z_v - c_6 z_u z_v^2$, where $c_6 > 0$ is a constant. Under this consideration the derivative of V_4 becomes,

$$\begin{aligned} \dot{V}_4 = &-X_e^T K X_e - c_1 z_u^2 - c_2 z_u^4 - c_4 z_r^2 - c_5 z_r^4 - c_6 z_u^2 z_v^2 - c_3 \psi_e^2 - c_3(\beta + \gamma u)^2 z_v^2 \\ &- 2c_3(\beta + \gamma u)\psi_e z_v + z_v(y_{1e} c_\psi - x_{1e} s_\psi) + z_v \chi_{v_1} \end{aligned}$$

In the above expression we remark that the last three terms have uncertain signs. So we will examine as being all positive. For the analysis we will use the inequality of Young, with the quantities ϵ_i, $i = 1....6$ are positive constants.

$$z_v(y_{1e}c_\psi - x_{1e}s_\psi) \leq \frac{1}{4\epsilon_1}|y_{1e}|^2 + \epsilon_1|z_v|^2 + \frac{1}{4\epsilon_1}|x_{1e}|^2 + \epsilon_1|z_v|^2 \tag{21}$$

$$-2c_3\psi_e z_v(\beta+\gamma u) \leq \frac{1}{\epsilon_2}\gamma^2|u|^2|\psi_e|^2 + \epsilon_2 c_3^2|z_v|^2 + \frac{1}{\epsilon_2}\beta^2|\psi_e|^2 + \epsilon_2 c_3^2|z_v|^2 \tag{22}$$

Now, we will expand the expression of $z_v\chi_{v_1}$, we obtain the following expression,

$$z_v\chi_{v_1} = \lambda z_v v_1 \alpha_u + (k+\alpha)z_v v_1 + z_v(\delta_2 + \varepsilon\delta_3) \tag{23}$$

Then, we will bound the three terms of (23) one by one as follows. After replacing α_u by its expression we have,

$$\lambda z_v v_1 \alpha_u \leq \frac{1}{4\epsilon_3}k^2|v_1|^2|y_{1e}|^2 + \epsilon_3\lambda^2|z_v|^2 + \frac{1}{4\epsilon_3}k^2|v_1|^2|x_{1e}|^2 + \epsilon_3\lambda^2|z_v|^2 \tag{24}$$

$$(k+\alpha)z_v v_1 \leq \frac{1}{4\epsilon_4}(k+\alpha)^2|v_1|^2 + \epsilon_4|z_v|^2 \tag{25}$$

$$z_v(\delta_2 + \varepsilon\delta_3) \leq \frac{1}{4\epsilon_5}|\delta_2 + \varepsilon\delta_3|^2 + \epsilon_5|z_v|^2 \tag{26}$$

To proceed further, we consider the maximum values for the time varying quantities, that will be noticed as v_{1max} and u_{max} which verify: $|u|^2 \leq u_{max}^2$, $|v_1|^2 \leq v_{1max}^2$. Taking into account the results from (21) to (26), the time derivative of V_4 in (21), will be increased by,

$$\dot{V}_4 \leq -\varpi_1 x_e^2 - \varpi_1 y_e^2 - c_1 z_u^2 - c_2 z_u^4 - \varpi_3 z_v^2 - \varpi_2 \psi_e^2 - c_4 z_r^2 - c_5 z_r^4 - [c_6 z_u^2 - \zeta] z_v^2 + \mu \tag{27}$$

where the terms: $\varpi_1 = k - \frac{1}{4\epsilon_1} - \frac{1}{4\epsilon_3}k^2 v_{1max}^2$, $\varpi_3 = c_3(\beta + \gamma u_{max})^2$, $\varpi_2 = c_3 - \frac{1}{\epsilon_2}(\beta^2 + \gamma^2 u_{max}^2)$, $\mu = \frac{1}{4\epsilon_4}(\alpha + k)^2 v_{1max}^2 + \frac{1}{4\epsilon_5}(\delta_2 + \varepsilon\delta_3)^2$ and $\zeta = 2\varepsilon_1 + 2c_3^2\varepsilon_2 + 2\lambda^2\varepsilon_3 + \varepsilon_4 + \varepsilon_5$ are all positive.

Ramark. *The coefficient $c_6 z_u^2 - \zeta$ must be positive. Consequently $|z_u|$ must be above $\sqrt{\frac{\zeta}{c_6}}$, which holds for large c_6 and small ζ.*

Consequently, the final study system is defined by:

$$\begin{aligned}
\dot{X}_e &= -KX_e + RZ_u \\
\dot{z}_u &= -c_1 z_u - c_2 z_u^3 - x_{1e}c_\psi - y_{1e}s_\psi + f_u \\
\dot{z}_v &= (\beta+\gamma u)z_r - c_3(\beta+\gamma u)^2 z_v - c_3(\beta+\gamma u)^2\psi_e + \lambda v_1 z_u + \chi_{v_1} \\
\dot{\psi}_e &= z_r - c_3\psi_e - c_3(\beta+\gamma u)z_v \\
\dot{z}_r &= -c_4 z_r - c_5 z_r^3 - (\beta+\gamma u)z_v - \psi_e
\end{aligned} \tag{28}$$

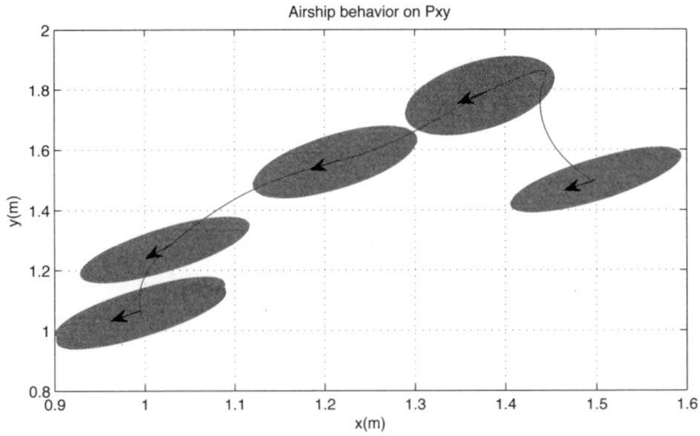

Fig. 2: Stabilization response into the navigation plan (x, y) including disturbance rejections.

So, from the inequality (41), we can deduce that:

$$\dot{V}_4 \leq -\varpi_1 x_e^2 - \varpi_1 y_e^2 - c_1 z_u^2 - \varpi_3 z_v^2 - \varpi_2 \psi_e^2 - c_4 z_r^2 + \mu \leq 2\nu V_4 + \mu \tag{29}$$

where $\nu = min\{\varpi_1, c_1, \varpi_2, \varpi_3, c_4\}$.
By using the comparison lemma, we can yield

$$V_4(t) \leq V_4(0)e^{-2\nu t} + \frac{\mu}{2\nu}, \qquad t \geq 0 \tag{30}$$

4. Simulations

In this section, we present numerical simulations showing the efficiency of the construct command for the stabilization of the underactuated model in presence of sinusoidal disturbances. Before starting, we will present the system parameter values(IS units) used for simulations: $m = 9.0718$, $m_x = 10.2$, $m_y = 16.32$, $J_z = 27.63$, $X_u = -10$, $Y_v = -10$, $N_r = -10$, $GO_{3x} = 3$. We will consider the sinusoidal disturbances torques defined as follows:$d_1(t) = 0.1sin(0.7t)$, $d_2(t) = 0.1sin(0.8t)$ and $d_3(t) = 0.1sin(0.9t)$. The simulations results were obtained with gains chosen as: $k = 4, c_1 = 10, c_2 = 5, c_3 = 6, c_4 = 6, c_5 = 3$ and $c_6 = 1$. The initial velocities, position and orientation are defined as: $x(0) = 1.5$, $y(0) = 1.5$, $\psi(0) = 0$, $u(0) = 0$, $v(0) = 0$ and $r(0) = 0$. Also, we choose as desired attitude and velocities: $x_d = 1$, $y_d = 1$, $\psi_d = 0$, $u_d = 0$, $v_d = 0$ and $r_d = 0$. Figure 2 shows that the control performance is quite well in presence of sinusoidal disturbances. It is clear that the positions x and y converge exponentially to the desired point. A converging results are obtained to ψ-motions. Further, the velocities converge to zero which confirm the stabilizing nature of the (τ_u, τ_r) controllers.

5. Conclusions

Throughout the paper, a depth theoretical analysis for the position/orientation control problem of an underactuated airship into the P_{xy} navigation plan was presented. The model integrates perturbations caused by wind like sinusoidal behavior. The stabilizing two control inputs are determined explicitely and led to an exponential convergence, obtained step by step using the backstepping approach. The simulation results confirm the robustness of the proposed control scheme.

References

1. H.K.Khallil, *Nonlinear Systems*, second ed. Prentice-Hall, Upper Saddle River, 1996.
2. Morin, P.,and Samson, C.(1995) Time-Varying Exponential Stabilization of the Attitude of a Rigid Spacecraft with Two Controls. In Porc. 34th IEEE Conf. on Decision and Control, pages 3988-3993,New Orleans, LA, Dec.
3. Pettersen, K.Y. and Egeland, O. (1999) Time-varying Exponential Stabilization of the Position and Attitude of an Underactuated Autonomous Underwater Vehicle, IEEE Trans. on Automatic Control, Vol.44, No.1.
4. Pettersen, K.Y. and Egeland, O. "Exponential Stabilization of an underactuated surface vessel" Proc of the 35th IEEE Conf on Decision and Control Letters,Kobe, Japan, December 1996.
5. S.Samaali, "Commande de Navigation dun dirigeable autonome sous actionn dans les trois plans: Longitudinal, Horizontal et Latral", Ph.D. in 2007 from the University of Evry, France.
6. F.Repoulias and E.Papadopoulos, *Planar Trajectory Planning and Tracking Control Design for Underactuated AUVs*,Ocean Engineering, Vol 34, pp 1650-1667, 2007.
7. H. -H. Zhang, F. Wang, Pavel M. Trivailo. Spin-axis Stabilisation of Underactuated Rigid Spacecraft Under Sinusoidal Disturbance. *International Journal of Control*, **81**, 1901–1909, 2008.
8. Y. Takahashi, S. Nakaura, M. Sampei. Position Control of Surface Vessel with Unknown Disturbances. *Proc of the 46th IEEE Conference on Decision and Control*, New Orleans, LA, USA, 12–14, 2007.
9. S. Samaali, A. Abichou, L. Beji. Longitudinal and horizontal stabilization of an unmanned blimp. *International Journal of Vehicle Autonomous Systems*, **5**, 138–157, 2007.
10. L. Beji, A. Abichou, Y. Bestaoui. Position and attitude control of an underactuated autonomous airship. *Int. J. of Differential Equations*, **8**, 231-255, 2003.

ON-FIELD TEST OF A LOW COST, HIGH PRECISION AUGMENTED GLOBAL NAVIGATION SATELLITE SYSTEM FOR AGRICULTURE APPLICATIONS

F. BONACCORSO, L. CANTELLI, D. LONGO, C.D. MELITA, G. MUSCATO
University of Catania – DIEEI, Viale A. Doria 6
95125 Catania, 95125, Italy – gmuscato@dieei.unict.it

In this paper the architecture developed with the aim of evaluating the performance of a low cost ground based augmented GPS system is presented. Several results will be shown to confirm the effectiveness of the adopted methodology. The proposed architecture could be used in order to enhance the localization of a tracked unmanned ground vehicle for agriculture application.

1. Introduction

The aim of the work presented in this paper is the investigation of the potentialities of a very low cost Global Navigation Satellite System (GNSS) based localization architecture in mobile robotics for agricultural applications.

Despite the appearances, moving a robot using GPS information is a challenging activity; the DIEEI Robotic Group has currently different active projects in the field of autonomous navigation and new GNSS technologies in agriculture, precision farming, for greenhouses or vineyards applications, and volcanic applications [1]. The U-Go Robot [2] is one of the platforms used at DIEEI Robotic Laboratories. The needed accuracy in the position solution is obtained through the adoption of the "Sicili@Net" infrastructure, a free of charge Ground Based Augmentation System (GBAS) offered by the National Institute of Geophysics and Volcanology – INGV. Several test campaigns have been conducted to test both the augmentation and the guidance systems in dynamic conditions and performances have been evaluated in different environments (open spaces, urban environments).

2. GNSS in mobile robotics and agricultural applications

The state of the art in GNSSs shows that Real Time Kinematic (RTK) positioning is able to meet a centimeter level accuracy [3], [4]. This differential technique is now mature and its adoption is growing in more and more dynamic applications. Nevertheless, up to recent models, this technique was only available on dual frequency receivers, at very high cost. Several projects are nowadays dealing with the problem of an accurate, reliable, secure localization and navigation strategy [5], [6], like the famous DARPA Grand Challenge [7], [8].

In the last decade, GPS systems have been widely adopted as tools for automatic guidance of agriculture machines operating in open fields. A Differential GPS (DGPS) system implementation with easy-to-use interface towards operator and machine is provided by John Deere. Drawbacks are that signals coverage is often very limited, that they use non-standard protocols, so dedicated hardware have to be used and that their cost is quite high [9].

The DIEEI Robotic Group has recently been involved in the MOW-BY-SAT project [10]; the priority objective of the project was that of supporting the development of a very low cost satellite based navigation and guidance system to be integrated into an autonomous lawnmower [11]. Moreover it has been involved in a project partially founded by the Italian Agricultural Minister with the aim of developing autonomous machines for precision farming inside greenhouses and different applications with the same purpose but vineyards oriented.

3. Sicili@net Ground Based Augmentation System

The objective of this work is to exploit free of charge augmentation systems that are worldwide available over different kind of media and that allow a standard GPS device to obtain centimeter lever accuracy at a fraction of cost of a complete RTK-DGPS system.

This kind of correction is provided by Sicili@net [12], a ground based augmentation service offered by the "Istituto di Geofisica e Vulcanologia" (National Institute of Geophysics and Volcanology – INGV), section of Catania [13]. Currently, INGV manages more than 80 permanent GPS stations in Sicily and southern Calabria, Italy. This network provides important data on ground deformation in active volcanic areas in Sicily (Etna, Stromboli, Vulcano and Pantelleria) and seismogenic areas of Sicily and southern Calabria ("Area Iblea", "Stretto di Messina").

Thanks to the wide network coverage, the correction data are available virtually everywhere in Sicily and the mean positioning error is minimized thanks to the high density offered by the 80 GPS stations. Data coming from the stations are collected by the control center placed in Catania and processed to provide a correction service over internet via TCP/IP messages by using the RTCM standard, developed by the Radio Technical Commission for Maritime Services [15].

Two operational modalities are available for the on-field users:

- single station mode: the correction is based on the information coming from the single nearest GPS station;
- area correction mode: the corrections are obtained by using the information coming from different GPS stations near the operating area.

The first mode ensures high performance in the proximity of the GPS station and the error increases with the base line between the user receivers and the GPS station. The second mode is suggested when there is not a GPS station near the on-field user or operating area is very wide; in this case the error model is obtained by using the information coming from different GPS stations located around the users, thus decreasing the error due to the long base line with the single station.

3.1. *The experimental setup*

As explained, internet connection by means of a cheap USB internet key, allows to obtain the Sicili@Net data virtually everywhere, using low cost PC and GPS L1 receiver that only support RTCM correction data.

A "GNSS Satellite/Ground Based Augmentation Systems - Performance Evaluation Tool" (GSG PET) has been developed in Matlab to evaluate the performances of different Global Navigation Satellite Systems, Satellite/Ground Based Augmentation Systems, with a specific focus on the topic of autonomous navigation for mobile robotics [11]. The tool allows on-line acquisition and plotting of data coming from the adopted devices and off-line analysis, with the possibility of evaluating the localization algorithms performances.

A Magellan DG14 Sensor Board [16], being capable of accepting RTCM corrections, has been adopted to test the performance of the augmentation system under test. RTCM corrections are provided by INGV over internet via TCP/IP messages; dedicated software has been developed by DIEEI to send RTCM messages over a RS232 wireless link to the DG14 Sensor Board, communicating with the GSG PET through a cabled RS232 bus. The implemented architecture and the data-flow are represented in Figure 1.

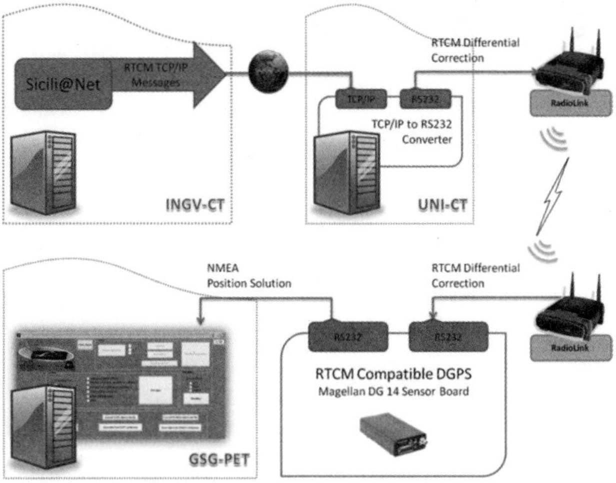

Figure 1. GPS system configuration using Sicili@net correction data.

4. The U-Go Unmanned Ground Vehicle

The U-Go Unmanned Ground Vehicle (UGV) has been used as a platform to test both the localization system and the developed navigation algorithms [14]. The mechanical structure has been designed in order to be compliant to different agriculture requirements. Figure 2 shows the U-Go Robot during some outdoor tests.

The robot control architecture mainly consists of one computer, in which resides software modules used to receive information from the adopted sensors (sonar, laser, GPS etc.). The communication between the different sensors and the on-board PC is through serial port or Ethernet and UDP protocols. Navigation algorithms have been developed using the Microsoft Robotics Developer Studio (MRDS) tool.

Figure 2. The tracked U-Go Robot during a teleoperated outdoor test.

5. Experimental results

Several trials have been performed by executing typical path in the proximity of DIEEI facilities, shown in Figure 3. The track has been selected with the aim of having several effects affecting the receiver performances: tunnels, canyons, multipath, GPS satellites attenuation and total loss of satellites, interruption in the communication between the base station and the mobile devices.

In Figure 4 the localization UTM solutions provided by different satellite based localization systems are shown; the receivers under test during this trial were:

- GPS unit: Garmin GPS MAP 60CSX in autonomous mode [17];
- RTK-DGPS receivers: Ashtech Z-Xtreme RTK-DGPS system[16]; the base station for the RTK-DGPS system was placed in the proximity of the DIEEI facilities and its position is showed in figure 4 with a blue star. A radio link were used to transmit the corrections to the mobile receiver.
- Sicili@Net assisted GPS: Magellan DG14 Sensor Board.

As can be observed GPS receiver is subject to error even in clear-sky conditions, with all satellites in view while RTK-DGPS receiver stops giving the localization solution in presence of tunnels and canyons. Instead, thanks to the received RTCM corrections, Magellan DG14 Sensor Board gives a good solution even in presence of adverse conditions.

These results were confirmed by several test campaigns; the results of one of these tests are shown in Figure 5. During this trial, the RTK-DGPS system and the ground based augmentation service Sicili@Net were compared. The two system have been used at the same time and results reported on Google Earth: the yellow line represents localization data obtained using a RTK-DGPS system and a radio link between fixed and mobile stations while the red line represents data obtained using the Sicili@Net correction. Error sources of the RTK-DGPS system are mainly represented by faults on the radio link between the two stations and by the satellites view obstruction due to buildings and other structures.

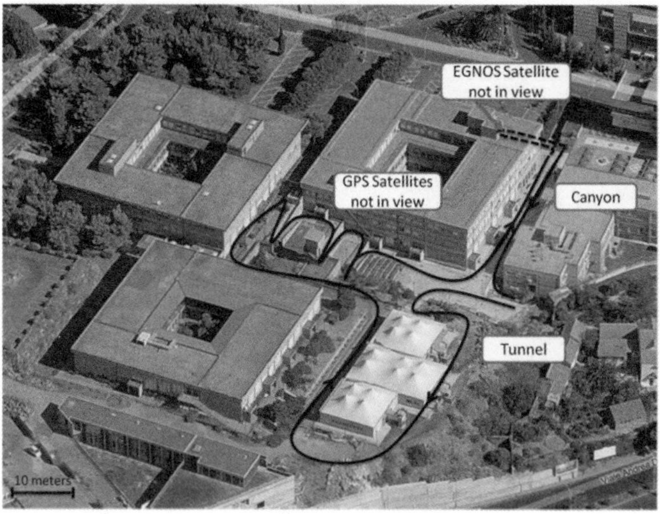

Figure 3. The path selected to test the Sicili@Net augmentation system.

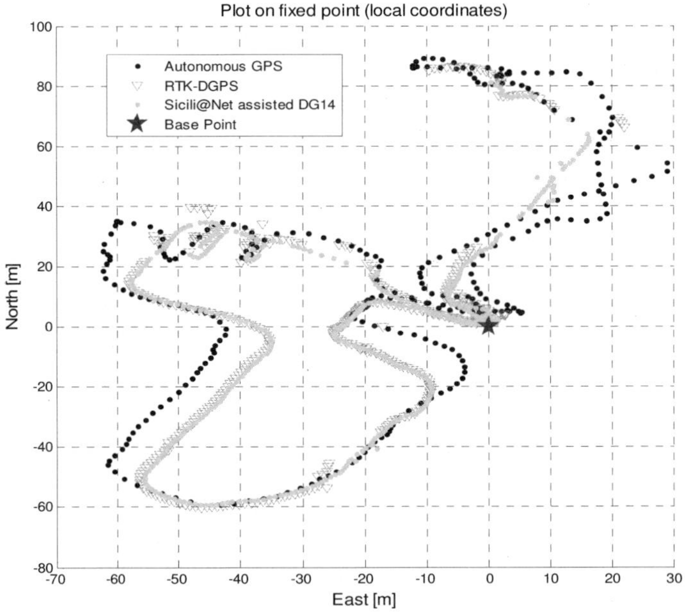

Figure 4. Localization solutions provided by the receivers: GPS positions (black dots), RTK-DGPS (purple triangles), RTCM assisted GPS (green dots).

Figure 5. Comparison between Sicili@Net (in red) and RTK-DGPS (in yellow) systems.

6. Conclusions

In this paper the low-cost, high-performance Sicili@Net Ground Based Augmentation System has been introduced. The architecture developed at the DIEEI laboratories to test the performance of different GNSSs has been introduced. Several trials have been executed and the test campaigns led to the following conclusions:

- single GPS receiver offers poor performance regarding the needs. The position error can often be around 10m or even more. Despite this, a simple GPS receiver can be cheap and availability is good with high sensitive receiver.
- Using RTK needs two GPS receivers and a radio link. But in this case, the phase signal is used instead of pseudo-range. RTK affords great performances and quicker initialization time. Integrity is better since there are twice measures. The drawback is that RTK receivers are expensive regarding the target price of agricultural market. Moreover, a line-of-sight wireless communication link is needed between these receivers.
- The local differential service offered by the Sicili@Net architecture allows to obtain good performance. Moreover, the service offered by INGV is free of charge and the hardware needed is very cheap: just a single RTCM capable receiver is needed and the radio link used for a classical RTK-DGPS system is not needed. The drawback is that an internet connection is needed: it can be achieved by using a simple USB internet key but operational area must be served by UMTS/GPRS signal.

References

1. Service Robots Group - Università di Catania: www.robotic.diees.unict.it
2. S. Balloni, L. Caruso, A. Conti, G. Schillaci, D. Longo, G. Muscato, "Development of an electrical multi-functional autonomous vehicle able to cultivate covered crops and to safe distribute agricultural chemicals inside greenhouses", Proceedings of the XXXIII CIOSTA - CIGR V Conference 2009, pp 355-359, Vol 1, June 17-19 2009, Reggio Calabria, Italy.
3. Grewal, Mohinder S., Weill Lawrence R., Andrews Angus P, "Global Positioning Systems, Inertial Navi-gation, and Integration, 2nd edition", Wiley-Interscience, 2007.
4. Elliott Kaplan, Christopher Hegarty , "Understanding GPS: Principles and Applications, 2nd edition", Artech House Publishers, 2005.
5. D. M. Bevly, Global Positioning System (GPS) A Low-Cost Velocity Sensor for Correcting Inertial Sen-sor Errors on Ground Vehicles, Journal of Dynamic Systems, Measurement, and Control, Vol. 126, June 2004
6. M. B. Gilbreath, M. H. Bruch, G. A. Gilbreath, J. W. Muelhauser, J. Q. Lum, Accurate waypoint navigation using non-differential GPS (2002), AUVSI Unmanned Systems 2002, Lake Buena Vista, FL.
7. DARPA Grand Challenge web-site: www.darpa.mil/grandchallenge/index.asp

8. Thrun S. et al., Stanley: The robot that won the DARPA Grand Challenge: Research Articles, J. Robot. Syst. 23, 9 September 2006, 661-692.
9. www.deere.com/en_US/ProductCatalog/FR/series/ams/guidance/guidance.html
10. The MOW-BY-SAT project homepage: www.mow-by-sat.eu/
11. D. Melita, G. Muscato, M. Poncelet, "Performances evaluation of Global Navigation Satellite Systems and Satellite/Ground Based Augmentation Systems in mobile robotics", ISR 2010 and ROBOTIK 2010, Munich, Germany, 7-9 June 2010. FRANKFURT; pp. 346-353
12. INGV-CT SICILI@NET homepage: http://193.206.223.39:5099/spiderweb/frmIndex.aspx
13. "Istituto di Geofisica e Vulcanologia" section of Catania homepage: www.ct.ingv.it/
14. P. Aranzulla, F. Bonaccorso, C. Bruno, L. Cantelli, G. Lanteri, D. Longo, D. Melita, G. Muscato, A. Pennisi, M. Prestifilippo; "An innovative autonomous outdoor vehicle based on Microsoft Robotic Studio"; CLAWAR 2010: 13th International Conference on Climbing and Walking Robots, Nagoya, Japan, 31 August - 03 September 2010; p. 97, 104; World Scientific Publishing Ltd.
15. Special Committee 104, RTCM Recommended Standards for Differential GNSS (Global Navigation Satellite Systems) Service, Version 3.0, Radio Technical Commission for Maritime Services, Alexandria, VA, February 10, 2004.
16. Ashtech solutions: http://www.ashtech.com/en/
17. Garmin solutions: http://www.garmin.com

MOBILE ROBOT ENIVORNMENT MAP BUILDING USING A RGB-D SENSOR

EVANGELOS GEORGIOU

Centre for Robotic Research, King's College London
London, WC2R 2LS, United Kingdom

JIAN DAI

Centre for Robotic Research, King's College London
London, WC2R 2LS, United Kingdom

This paper presents an alternative approach to implementing a stereo camera configuration for SLAM. The approach suggested implements a simplified method using a single RGB-D camera sensor mounted on a maneuverable non-holonomic mobile robot, the KCLBOT, used for extracting image feature depth information while maneuvering. Using a defined quadratic equation, based on the calibration of the camera, a depth computation model is derived base on the HSV color space map. Using this methodology it is possible to build navigation environment maps and carry out autonomous mobile robot path following and obstacle avoidance, which is validated by some experimental results.

1. Introduction

One of the fundament problems in mobile robot autonomous navigation is map decomposition and building [1]. Building maps as a mobile robot traverses a navigation environment, while using a partially composed map to maintain localization is better known as Simultaneous Localization and Mapping (SLAM). The most common methods of building localization maps are using occupancy grids [2], feature building [3], and raw sensor telemetry [4]. The approach presented by this paper uses the latter approach, utilizing raw sensor telemetry via an RGB-D camera. An RGB-D camera is a projective sensor that measures the bearing of image features and returns an image with depth telemetry, normally using an HSL or HSV color map. Using the depth information from the RGB-D sensor, it is possible to build a navigation map, which in turn can be used for mapping, sensing, localizing, and environment modeling.

An alternative to the RGB-D camera is the implementation of a stereo camera configuration [5], which utilizes two CMOS RGB cameras, to produce 3D reconstructions of the environment in its visual range. The idea behind using a single RGD-D camera is to simplify the stereo camera approach, and reduce the computation time to extract depth information from an image and use this image for SLAM.

2. Navigation Environment Building Using a RGB-D Visual Sensor

2.1. *RGB-D Image Capture using Microsoft Kinect™ Sensor*

In early November 2010, Microsoft released the Kinect™ sensor, which has the capability of RGB and RGB-D image capture. Exploiting the pre-released drivers for this device, it is possible to export and manipulate both the RGB and RGB-D images for mapping, sensing, locating and modeling for SLAM.

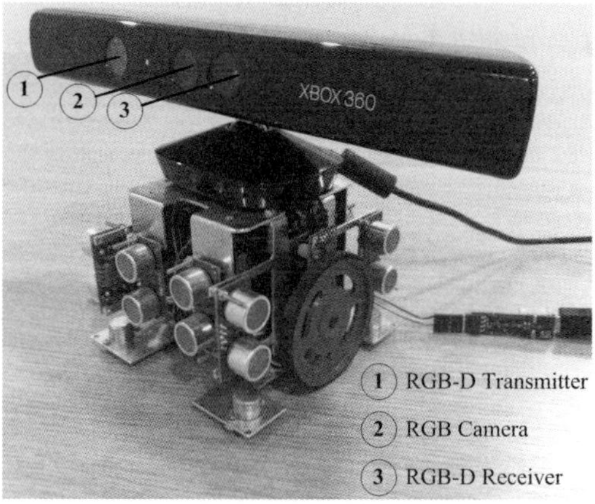

Figure 1. The Kinect™ Mounting Configuration on the KCLBOT.

The Kinect™ depicted in Fig. 1, which is mounted on the KCLBOT, is built around PrimeSense's PS1080 SoC [6]. The PS1080 acts as a controller for the IR light source in order to build the input image with IR light coding depth information. The light coding in this case is mapped on HSV [7]. The device is configured such that a standard CMOS image sensor receives the projected IR light source and transforms the image to produce a depth image. The sensor has a field of view of 58° Horizontal (H), 40° Vertical (V), and 70° Diagonal (D). The sensor allows for a maximum image depth size of VGA, which is a resolution of 640x480 and has maximum image throughput frame rate of 60fps. Finally, the operational range of the sensor is limited to a minimum of 0.8m and to a maximum of 3.5m.

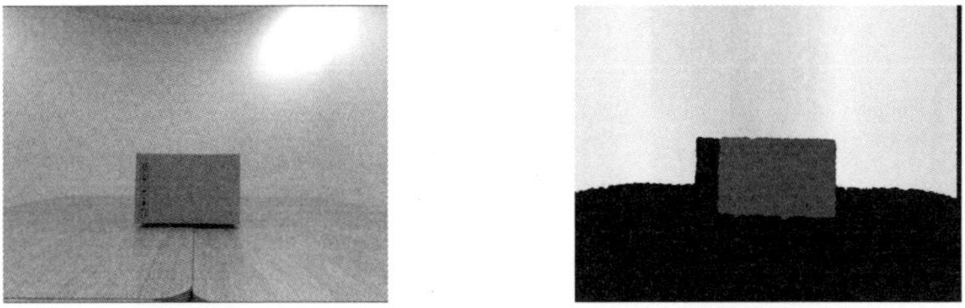

RGB Image RGB-D Image

Figure 2. RGB and RGB-D Sensor Images.

As depicted in Fig. 2, the sensor is able to acquire RGB and RGB-D images, where the RGB-D images are used for environment building.

2.2. *Distance Estimation Using a RGB-D Sensor*

Before the RGB-D images can be used for environment map building, the depth information from the image needs to be computed. To achieve this, a function needs to be derived to convert the RGB color values to Hue. Once the Hue value is acquired, it can be used to compute relative distance estimation. Using the model presented by Joblove and Greenberg [7], the following equation is implemented.

$$H' = \begin{cases} undefined, & if\ C = 0 \\ \frac{G-B}{C} \bmod 6, & if\ M = R \\ \frac{B-R}{C} + 2, & if\ M = G \\ \frac{R-G}{C} + 4, & if\ M = B \end{cases} \tag{1}$$

where (1) describes the transpose of Hue value, and using $C = M - m$, which describes the chroma value and where $M = \max(R, G, B)$ and $m = \min(R, G, B)$ describe the maximum and minimum values of the RGB color, respectively.
The conversion function used to convert the transpose of Hue (21) to the Hue value needed, is as follows.

$$H = 60°.H' \tag{2}$$

Now that a conversion for Hue from RGB has been defined, a function to convert the Hue values to distance is required. To achieve this, sample images are captured from the sensor, using a static object with a known distance from the sensor.

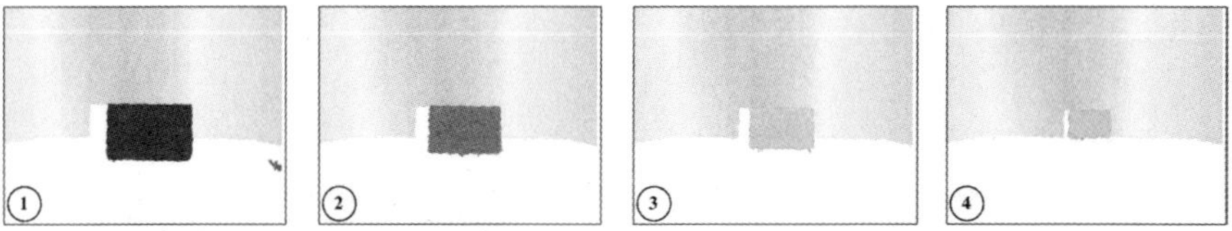

Figure 3. Sample RGB-D Distance Estimation Images.

Using the images presented in Fig. 3, the average RGB values from the static object are extracted and converted to a Hue valuation.

Table 1. Sample RGB-D Distance Estimation Results

Sample	R	G	B	Hue	Distance (cm)
1	0	60	255	225.88°	81.67
2	2	162	255	202.06°	95.60
3	0	249	255	181.41°	110.60
4	0	255	31	127.29°	162.81

The values required to build a Hue to distance function are presented in Table 1.

$$d = 0.0032H^2 - 1.9492H + 359.26 \tag{3}$$

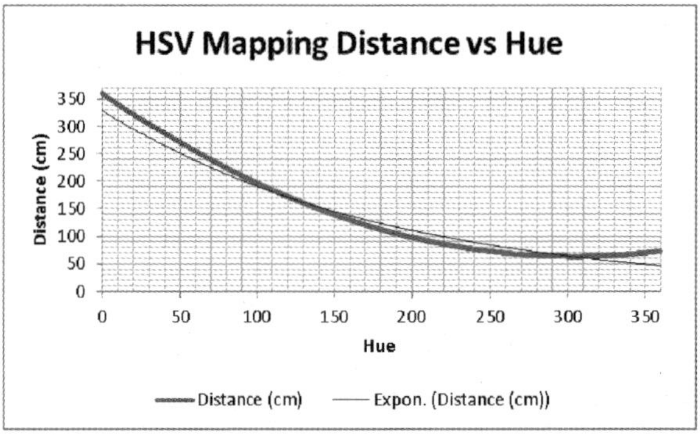

Figure 4. HSV Exponential Mapping of Distance vs. Hue.

Using the quadratic equation (3) that describes distance as a function of Hue, a behavior graph is produced, and presented in Fig. 4. Based on this graph, when distance is less than 80cm or greater than 350cm, the slope of the curve changes more significantly. This conforms to the manufacturer's specifications on the limitations of distance estimation above and below these distance ranges.

2.3. *Environment Building Using a RGB-D Sensor*

Before the autonomous mobile robot can complete any path planning or path following tasks, it requires sufficient information about the environment that it will be traversing. To provide the mobile robot with this information, the detail from the RGB-D camera is used to make a plot of the terrain, plotting the unobstructed space the mobile robot can utilize.

Before the RGB-D image can be used, the noise, resolved as black pixels in the range of #E4E1C0h to #FFFFFFh, needs to be removed from the image. This is achieved by converting the RGB-D image to gray scale [8]. This process is carried out to protect natural colors in the #E4E1C0h to #FFFFFFh range. In the RGB color model, a color image can be represented by the following intensity function:

$$I_{RGB} = (F_R, F_G, F_B) \tag{4}$$

In (4), F_R is the intensity of the pixel (x,y) in the red channel, F_G is the intensity of pixel (x,y) in the green channel, and F_B is the intensity of pixel (x,y) in the blue channel. Using only the brightness information, the color image can be transformed into a gray scale image [8].

The equation that converts a color pixel to a gray scale pixel is presented below (25).

$$I_{GS} = 0.333F_R + 0.5F_G + 0.1666F_B \tag{5}$$

Figure 5. Conversion from RGB-D to Gray Scale.

After the image has been converted to gray scale, as depicted in Fig. 5, the black pixels are filtered out of the image.

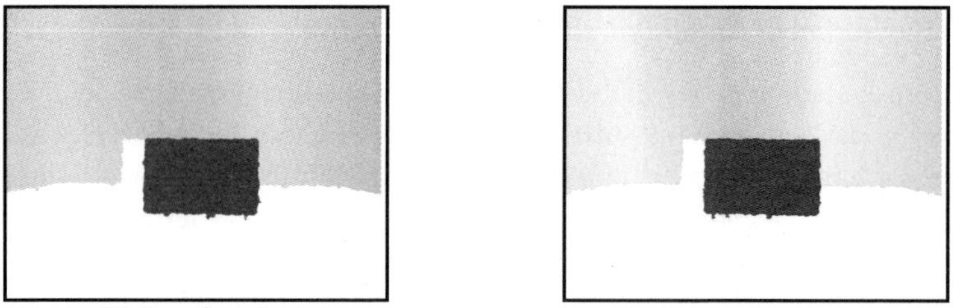

RGB-D to Gray Scale Remapping Color

Figure 6. Conversion from RGB-D to Gray Scale and Color Remapping.

Once the image has been stripped of the black noise pixels, as depicted in Fig. 6, the color detail is required for mapping the traversable environment. The RGB-D depth color information from Fig. 5 is remapped onto the gray scale filtered image; the result is presented in Fig. 6.

$$Input_{pixel} = \begin{bmatrix} x_{pixel} & y_{pixel} & RGB\text{–}D_{pixel} \end{bmatrix} \tag{6}$$

$$R_x(\theta) = \begin{bmatrix} 1 & 0 & 0 \\ 0 & \cos(\theta) & -\sin(\theta) \\ 0 & \sin(\theta) & \cos(\theta) \end{bmatrix} \tag{7}$$

The input matrix (6) is multiplied by the rotation matrix [9] from (7), where $\theta = 90^\circ$, to produce a topological view of the traversable space.

$$Output_{pixel} = \begin{bmatrix} x_{pixel} & RGB\text{–}D_{pixel} & -y_{pixel} \end{bmatrix} \tag{8}$$

Using the first two components of (8), a 2-dimensional traversable space plot is produced, where the $RGB\text{–}D_{pixel}$ is calculated using (3).

$$Input_{pixel} = \begin{bmatrix} x_{pixel} & y_{pixel} & RGB\text{–}D_{pixel} \end{bmatrix} \tag{9}$$

$$R_x(\theta) = \begin{bmatrix} 1 & 0 & 0 \\ 0 & \cos(\theta) & -\sin(\theta) \\ 0 & \sin(\theta) & \cos(\theta) \end{bmatrix} \tag{10}$$

The input matrix (6) is multiplied by the rotation matrix [9] from (7), where $\theta = 90^{\circ}$, to produce a topological view of the traversable space.

$$Output_{pixel} = \begin{bmatrix} x_{pixel} & RGB\text{–}D_{pixel} & -y_{pixel} \end{bmatrix} \tag{11}$$

Using the first two components of (8), a 2-dimensional traversable space plot is produced, where the $RGB\text{–}D_{pixel}$ is calculated using (3).

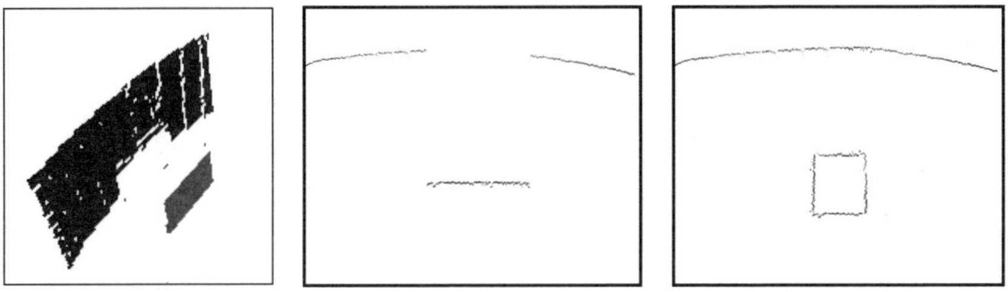

3D Reconstruction   Output Image   Environment Map

Figure 7. Environment Map Generation from RGB-D Image.

The generation of a topological view of the available navigation space, as presented in Fig. 7, provides the mobile robot with sufficient information for path planning and obstacle avoidance. Using a simple Bezier path [10], the KCLBOT is instructed to traverse a defined path and capture sample images using the RGB-D camera. Using the mobile robot's ability to self-localize, implementing while capturing and superimposing the converted images, using (8), an environment map is constructed. The initial results presented in Fig. 7, provide the mobile robot with detailed information of the environment and it can easily estimate the Euclidean distance to all obstacles in range.

3. Conclusion

The introduction of the affordable off-the-shelf RGB-D camera sensor, which captures images with depth information mapped in the HSV range, makes SLAM possible with minimal computation. This is validated by the experimental results presented in Fig. 7. The only concern of note at this stage is the tolerance levels presented by the distance estimation behavior graph in Fig. 4. The limitation is that obstacles 0.8m away from the sensor will not be resolved accurately. This is only a concern for navigation in small spaces but if the navigation area is large and open spaces are available, the sensor will prove to be a valuable tool for SLAM.

References

1. Thrun, S., *Learning Metric-Topological Maps for Indoor Mobile Robot Navigation.* Journal on Artificial Intelligence, 1998. **99**(1): p. 22-71.
2. Elfes, A., *Using Occupancy Grids for Mobile Robot Perception and Navigation.* Computer Journal, 1989. **22**(6): p. 47-57.
3. Leonard, J., Durrant-Whyte, H., and Cox, I, *Dynamic Map Building for an Autonomous Mobile Robot.* The International Journal of Robotic Research, 1992. **11**: p. 286-298.
4. Fang, F., Ma, X., and Dai, X. *A Multi-Sensor Fusion SLAM Approach for Mobile Robots.* in *2005 IEEE International Conference on Mechatronics and Automation*. 2005.
5. Chilian, A., and Hirschmuller, H. *Stereo camera based navigation of mobile robots on rough terrain.* in *International Conference on Intelligent Robots and Systems*. 2009.
6. *The PrimeSensor*. PrimeSense; Available from: http://www.primesense.com.
7. Joblove, G., and Greenberg, D. *Color spaces for computer graphics.* in *5th Annual Conference on Computer Graphics and Interactive Techniques*.
8. Kumar, T., and Verma, K., *A Theory Based on Conversion of RGB image to Gray image.* International Journal of Computer Applications, 2010. **7**(2): p. 7-10.
9. Herter, T., and Lott, K., *Algorithms for decomposing 3-D orthogonal matrices into primitive rotations.* Journal on Computers & Graphics, 1993. **17**(5): p. 517-527.
10. Choi, J., Curry, R., and Elkaim, G. *Path Planning Based on Bézier Curve for Autonomous Ground Vehicles.* in *Proceedings of the Advances in Electrical and Electronics*. 2008.

SECTION–8

HMI, TELE-PRESENCE AND TELE-OPERATED ROBOTS

HUMAN-ROBOT: FROM INTERACTION TO RELATIONSHIP

MARINE GRANDGEORGE
UMR CNRS 6552, Laboratoire Ethologie Animale et Humaine, Université Rennes 1, 263 avenue du Général Leclerc, 35000 Rennes, France

DOMINIQUE DUHAUT
Laboratoire Valoria, Université Bretagne Sud, Campus de Tohannic, 56000 Vannes, France

Our world is widely composed of machines, robots, avatars and so on. We daily interact with them. Thus, it is time to reconsider our interactions between human and robot going from interaction to relationship. To model the relationship, we defined first the notion of *bringing together* which is built on four dimensions: how, where, who and what. Then, we explained the shift to relationship inspired by social science studies, with a focus on human-robot research. Finally, we proposed a general protocol to evaluate the human–robot relationships illustrated by a case study methodology. Now, further research is needed in order to assess how to maximize robot acceptance in the human environment and what type of robot may help develop a positive relationship. For example, many possible application fields exist from education for children to cognitive stimulation and companionship for the elderly.

Human-robot interaction has in recent years been typically studied through the dimensions of cooperation, planning, communication and so on. Since the introduction of the social robot [1, 2] new parameters are investigated: goals, beliefs, feelings … But they are no longer sufficient when considering it with significant human interactions. By significant interactions, we mean long interaction with a robot in an intelligent environment which changes the quality of human life, for instance at home for disabled people. In this case, a relationship is established between the robot and the human; we speak then of human-robot relationships. In this paper, we propose a complete definition of "what a relationship is" based on works from social sciences on human relationships. In a first part, we show what kind of restrictions can be made to consider the human-robot relationship and in a second part how to evaluate it.

1. What is a relationship?

To introduce the notion of relationship, we first need to define the notion of *bringing together* which is built on 4 dimensions: how, where, who and what (Figure 1).

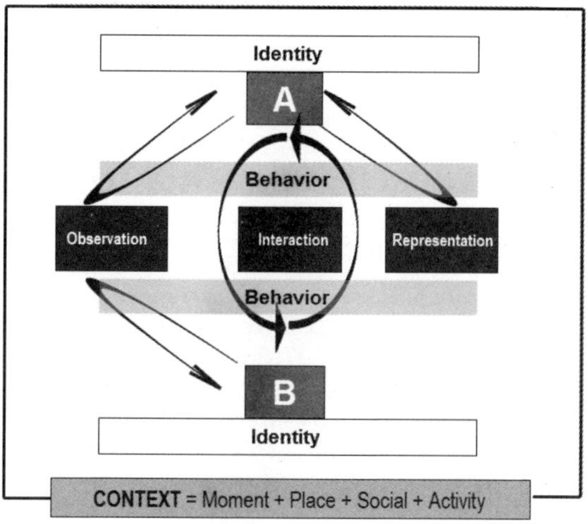

Figure 1. Model of the Bringing together notion, a step to understanding a relationship

A first dimension of *bringing together* is **how** two entities can initiate and develop their relationship. This is given by three different ways. **Representation** is when one can get information on the other one without being "physically" in front. This can appear if you look to something produced by the other: a text, a WEB, an artistic object. **Observation** is when one can see the other partner but does not know that he is being observed. Finally, **interaction** between two individuals involves, at a minimum, individual A showing behaviour X to individual B and B may respond with behaviour Y. There may be a number of repetitions of similar behaviours or not [3]. Both individuals receive information from the other through direct contact either by physical presence or by a mediated object.

The second part focusses on describing **where** the action of *bringing together* is carried out. For this, we define the notion of **context** described with four elements [4]: **Moment** is the time of the presence, for a long or short time, unexpected or expected and so on. **Place** is the location which impacts the behaviour of the individual (*e.g.* a restaurant or a church changes the rules of living between people). **Social** is the presence - or not - of a specific group (which impacts the interpersonal distance, the way to speak and to act with respect to codes of the group). **Activity** or the one-one presence is included in a task to be performed together (in the lower case, the task is only a meeting with no specific goal or outcome).

The third part is to describe **who** is carrying out the *bringing together* process (Figure 2). It is based on three elements [5]: **knowledge** that characterizes the things that an individual shares with the other, the **person** describes the attributes proper to one individual and that does not change in the time (or slowly) and finally the **lifestyle** with is describing the way that an individual acts over the time.

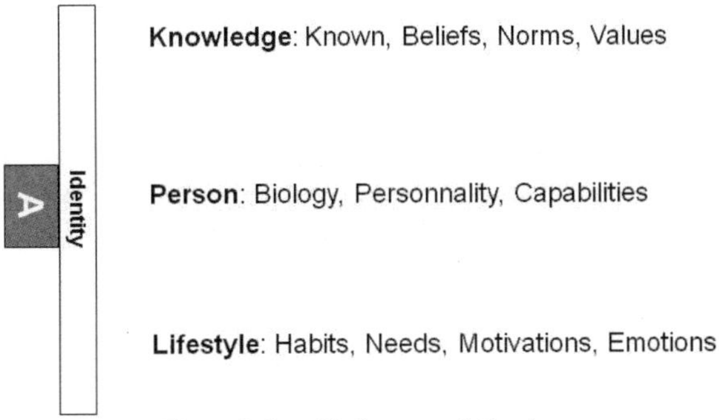

Figure 2. Specific features of identity

The last part is to describe the **what** of the *bringing together*. The **behaviour** is the set of actions performed by an individual with his own **identity** in a **context**. These actions are observed by the other individual with his own identity depending on the state: observation, interaction, representation. It has two main objectives, either a task to solve (act to reach a goal or an outcome in the environment defined by the context) or an exchange with a human (communication to exchange with the other person). The communication is described by its verbal and non verbal components.

2. From Interactions to Relationships

Numerous disciplines study the relationships, such as psychology, ethology, linguistic and so on. Each interaction is influenced by the previous ones in the process of developing a relationship [3]. Hence, individuals have, on the basis of past experience, expectations on the other individual's responses, which can also be modulated by all other previous experiences. Indeed, relationship with one of a category of individuals may be affected by previous interactions with other members of the same category. Thus, the behaviour of a boy with EmI, a companion robot [6], is affected by his past encounters with other companion robots (*e.g.* AIBO [7], Paro [8]) and *vice versa*.

With our definition of *bringing together*, it is possible to give a larger definition of a relationship between two entities. A **relationship** between A & B is the result of modifications in A & B following repeated actions of *bringing together*. The relationship can be analyzed in different ways. The external observer can analyze either the relationship of A and B from the perspectives of A or B, or the relationship between A and B, as well as the behaviours of A or B individually.

3. Relationships between human and robot

From a social point of view, the anthropologist Fiske [9] has suggested four models of relationships in his Relational Models Theory: **communal sharing** (all group members are treated equally and share a common identity), **authority ranking** (people attend to each other's status in a hierarchical order), **equality matching** (the principle of quid pro quo predominates) and **market pricing** (people rationally weigh the utility of their behaviour in achieving outcomes in interaction). In the perspective of a human robot relationship, we have to define the type of relation expected (*i.e.* communal sharing, authority ranking, equality matching, market pricing) and build the robot in consequence. This depends on the context of use and on the identity of the targeted human.

Moreover, we can learn from ten dimensions of relationships previously proposed by Hinde [10]: (1) content, (2) diversity, (3) quality, (4) pattern, (5) reciprocity and complementarity, (6) conflict and power, (7) self-disclosure, (8) interpersonal perception, (9) commitment, and (10) satisfaction.

4. General protocol to evaluate the human–robot relationships: illustration by a case study methodology

Now, we need to investigate how humans consider robots, for example, as a positive, negative or neutral stimulus as the result of previous interactions. Whereas the reverse relationship is also studied in human-human or human-animal relationships, we cannot evaluate how the robot considers humans... The research requires a general protocol design, including tools to evaluate and to measure what impact the experience acquired has, which type of experience may be the best and the more or less long term effects of the interactions. Thus, we use a general protocol design to evaluate Human-Robot Interaction in four parts (Figure 3). We illustrated each part with an example of experience with EmI and preschoolers interacting at school (Figure 4) [11].

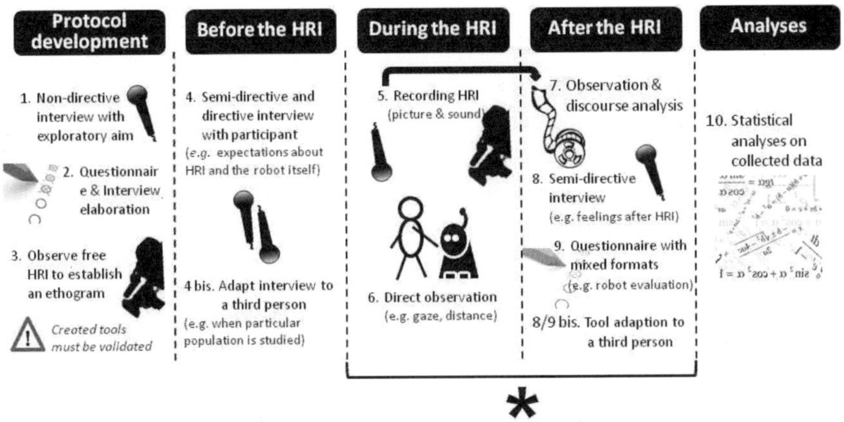

Figure 3. General protocol design to evaluate Human-Robot Interaction (HRI). * corresponds to the experiment repetitions across time to fulfill the need of longitudinal studies.

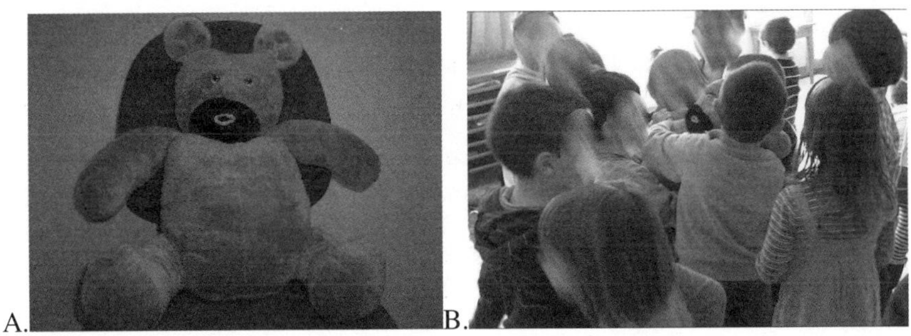

Figure 4. Example of experience with EmI (entire design; A) and preschoolers interacting at school with Emi (B)

4.1. *Protocol development*

Even if this step is not directly dedicated to data collection and human-robot evaluation, it constitutes the reliability of the whole experience. At this step, researchers could use non-directive interviews or observations of free human-robot interactions to create and validate tests that forward relationship evaluation (*e.g.* questionnaires, behavioural tests) based on Hinde's definition [10].

Example: we aimed to study the first encounter between preschoolers at school and a companion robot, EmI [6]. According to literature and parent's reports, we elaborated a short questionnaire for children using simple items.

4.2. *Before the human-robot interactions*

The phase "before the HRI" is a well suited moment to collect information on expectations that the participant may have about the experience and the robot itself. For this, interviews are a good method because they provide access to the participant's mental representations. It may be interesting to duplicate information by interviewing a third person who is familiar with the participant (*e.g.* parent for a child, nurse for an elderly person in an institution) especially for people of low cognitive level whose response reliability is not sure.

Example: a short semi directive interview was only done with the preschooler's teacher (e.g. are the preschoolers eager to encounter EmI?).

4.3. *During the human-robot interaction*

Wherever the experiment takes place, it is important to record verbal and nonverbal interaction between human and robot to keep track for later behavioural evaluation. Direct observation also needs to be conducted and focused on precise items which are difficult to collect later.

Example: each preschooler encountered EmI in an isolated room of the school to study precisely their behaviours in such a situation. All sessions were video-recorded to be later analyzed.

4.4. *After the human-robot interaction*

Except the analysis of a previously collected database, the researchers debriefed the participants about their relationship with the robot using interviews and/or questionnaires (*e.g.* feelings, problems). *Here, each preschooler answers nine short questions about EmI (e.g. would you like to see EmI again?).*

Information could also be duplicated here by interviewing a third person (see above). *Here, the teacher reported some preschooler's reactions.* Finally, all collected data must be analyzed thanks to statistic methods to give robust results.

As the study focus is on relationships and not only on interactions, we need to repeat some experiments over time (*e.g.* observation of human-robot interactions, human interviews about their relationship to the robot). Indeed, to describe and to understand a relationship, researchers need to collect data over a long period of time (*i.e.* longitudinal studies).

5. Some examples of parameters to evaluate the human–robot relationships

As the field of human-robot interactions is quite new, the factors to evaluate them remain relatively scarce. Based on our knowledge about human-animal relationships (*i.e.* "an interspecific couple" such as human and robot), we propose here some theoretical parameters with a possible influence (Table 1). Notice that human culture has a great influence on the human robot interactions and thus on the developed relationships (*e.g.* Japanese people are relatively fond of companion robots - here Paro - whereas European people seem more restrained [12]).

Table 1: Some examples of methods for evaluating parameters of the relationship

Relationship parameters	Methods
General evaluation of the HR relationship	• Likert scales with some qualifying adjectives (*e.g.* useful to not useful, pleasant to unpleasant) • Favourite modalities to interact (*e.g.* vocal, tactile) • Timing of daily interactions
"Social" attraction	• Distance between human and robot at regular intervals • Frequencies of target behaviours (*e.g.* number of interactions with the robot)
Degree of human attention toward robot	• Recording all human gazes towards the robot • Human latency to respond to robot's request
Stress induced by the robot	• Measurement of heart rate and other physiological variables • Human body postures adopted

6. Conclusion

We proposed a model of a human-robot relationship that consisted that consisted in the evolution of the human-robot interaction concept. Real need appears in terms of research in order to improve human-robot relationships. It is highly probable that the way humans "view" their robots has an important influence on the way they use them. Daily human-robot interactions are very varied and include both positive (*e.g.* services) and negative (*e.g.* breakdown) events, leading to more or less balanced relationships. Thus, the robot design must be conceived to reduce the negative events to ensure a relationship as positive as possible that may be beneficial for the human partner. Many possible application fields exist: from education for children to cognitive stimulation and companionship for the elders. Further research is needed in order to assess how to maximize robot acceptance in the human environment - at least in some societies, what type of robot (*e.g.* size, functions) may help develop a positive relationship, what influence human involvement has on the relationship and so on.

References

1. S. Coradeschi, "Human-Inspired Robots," IEEE Intelligent Systems, vol. 21, pp. 74-85, 2006.
2. C. L. Breazeal, Designing sociable robots: MIT Press, 2002.
3. R. Hinde, Towards Understanding Relationships. London: Academic Press, 1979.
4. P. Blanchet, Linguistique de terrain, méthode et théorie (une approche ethnosociolinguistique). Presses Universitaires de Rennes: Rennes, 2000.
5. J. C. Deschamps and P. Moliner, L'identité en psychologie sociale : Des processus identitaires aux représentations sociales: Armand Colin, 2008.
6. S. Saint-Aimé, et al., "EmI – my emotional cuddly companion," presented at the IEEE RO-MAN 2009, 18th International Symposium on Robot and Human Interactive Communication, Toyama, Japan, 2009.
7. M. Fujita, "AIBO: towards the era of digital creatures," International Journal of Robotics Research, vol. 20, pp. 781-794, 2001.
8. T. Shibata, et al., "Mental commit robot and its application to therapy of children," in Advanced Intelligent Mechatronics, 2001. Proceedings. 2001 IEEE/ASME International Conference on, 2001.
9. A. P. Fiske, Structures of Social Life: The Four Elementary Forms of Human Relations. New York: The Free Press, 1991.
10. R. A. Hinde, "Describing relationships," in The diversity of human relationships, A. E. Auhagen and M. von Salisch, Eds., Cambridge: Cambridge University Press, 1996, p. 359 p.
11. S. Saint Aimé, M. Grandgeorge, B. Le Pévédic, D. Duhaut, Comparison of interaction between Emi and young children: alone versus in group, submitted.
12. T. Shibata, et al., Tabulation and Analysis of Questionnaire Results of Subjective Evaluation of Seal Robot in Seven Countries vol. Vols 1 and 2: 17th IEEE International Symposium on Robot and Human Interactive Communication, 2008.

A SOFTWARE ARCHITECTURE FOR EFFICIENT TELEOPERATION OF A SEMI-AUTONOMOUS WALKING ROBOT

Piotr SKRZYPCZYŃSKI

Institute of Control and Information Engineering,
Poznan University of Technology, ul. Piotrowo 3A, PL-60-965 Poznań
E-mail: piotr.skrzypczynski@put.poznan.pl

A software architecture for the semi-autonomous walking robot Messor is described. This software is based on the multi-agent design paradigm with blackboard-type communication between modules. The paper shows how this flexible architecture supports development of an efficient teleoperation system for the walking robot. Use of the teleoperation mode in reconnaissance and search missions is demonstrated with real outdoor scenarios.

Keywords: Walking robot, Software architecture, Blackboard, Teleoperation

1. Introduction

Walking robots are able to access rugged terrains and to climb obstacles without damaging them, which may be very useful in search and rescue missions. However, such missions might be so complex that human intervention is unavoidable. On the other hand, the human operator cannot control a walking robot directly, by defining the movement of each joint. Therefore, an efficient Human-Machine Interface (HMI) is required to operate the robot,[4] but the robot itself should have enough autonomy to free the operator from low-level control tasks, such as generation of feet trajectories, foothold selection, and stability maintenance. Although a higher level of autonomy can make a walking robot more useful, this autonomy comes at the cost of much higher complication of the control software. Therefore, an appropriate software architecture of the perception and control system is required. This architecture should take into account the human operator, which can use his decision and recognition capabilities to support the robot in critical situations.

In Ref. 3 we introduced the Multi-Agent Blackboard (MAB) software architecture for autonomous robots, which addresses the issues of multi-sensor perception, world-modelling, and multi-paradigm control (i.e., deliberative and reactive). The aim of this paper is to present the application of this architecture to a six-legged, semi-autonomous walking robot Messor.[2] We demonstrate how the required integration of the teleoperation mode is accomplished by taking advantage of the flexibility of the blackboard data structure, the modularity of the agent-based framework, and the collaborative human-robot control concept.[7]

2. Software architecture of the Messor robot

The insect-like robot Messor is equipped with a 2D laser scanner Hokuyo URG-04LX, which works as terrain sensor.[9] Another external sensor of the robot is a monocular vision system (a camera), which serves mainly the purposes of SLAM[10] and self-localization based on the visual odometry concept. The robot has also a number of internal sensors: force sensors in the feet, and an Inertial Measurement Unit (IMU), that provides the robot with an instantaneous estimate of its pose. The robot is equipped also with gas sensors, which are able to detect some organic gas substances, and indicate a general direction to the source of such a substance. The control system is layered, with the low-level, real-time functions (e.g., gait execution, kinematics computations) located at the ARM9-based controller of the robot,[2] and the higher-level functions, such like gait generation and foothold selection performed by the on-board PC. This PC running under Linux hosts also the terrain mapping and visual SLAM modules.

The Linux-based part of the control system is characterized by complex, multi-stage data processing, which involves a number of software modules that use sensory data expressed in various formats. Each of the operations can be separately defined as a "black box" with some input and output. Such a kind of data processing can be organized as a blackboard system.[5] Because the software modules are kept separate and independent they can be modelled as software agents.[3] The information needed to arrange control is implemented by means of specialized flags.

The blackboard accommodates data that represent the position and orientation with regard to the global frame, describe the environment and define the motion control parameters in different sub-tasks (Fig. 1). In a robotic system both processing modules and information sources (sensors, communication channels) that are placed around the blackboard are good candidates to be modelled as agents.[3] Therefore, the hardware components of the walking robot are represented by device agents (**ADxxx**):

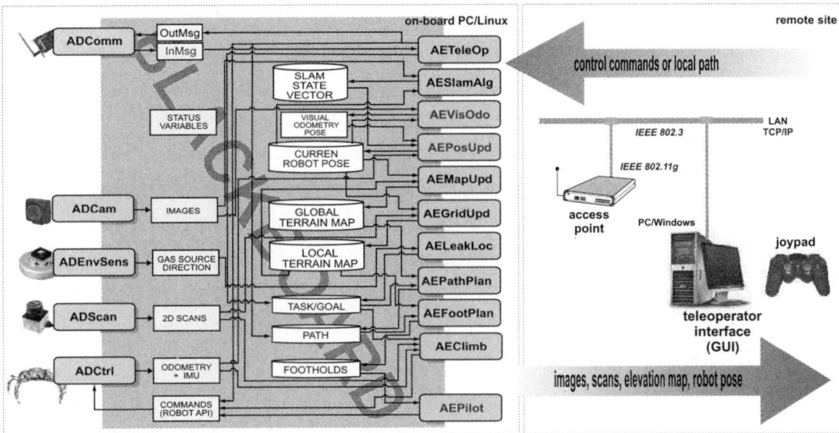

Fig. 1. MAB software architecture of the Messor robot and the teleoperator interface

- **ADScan** – agent of the laser scanner URG-04LX. Processes raw distance measurements, compensates the systematic error, and removes the so-called mixed measurements, as described in Ref. 9.
- **ADCam** – agent of the monocular vision system. Corrects distortions in images that are caused by the wide-angle lenses and writes images to the blackboard.
- **ADEnvSens** – agent of the gas sensors. It detects the given substance and determines the general direction from which the substance flows toward the robot.
- **ADCtrl** – robot controller agent, which writes to the blackboard the internal sensors data (IMU, force sensors in the feet), and computes the pose estimate from the legged odometry. It sends the motion commands provided by the **AEPilot** to the robot motion controller.
- **ADComm** – wireless communication agent. Manages the physical communication channel (IEEE 802.11g) and the input/output queues of messages.

The sensory information processing and motion planning modules are implemented as expert agents (**AExxx**):

- **AESlamAlg** – SLAM agent. It implements the monocular vision-based EKF-SLAM algorithm described in Ref. 10. This agent reads the current robot pose and images from the blackboard, and updates the SLAM state vector (global map of visual features), as well as the robot pose.

- **AEVisOdo** – visual odometry agent. Its role is to estimate the camera ego-motion, and thus the motion of the whole robot analyzing the optical flow from the camera. The input data are images. This agent stores the robot pose estimate on the blackboard.
- **AEPosUpd** – self-localization agent, which updates the current robot pose upon the local estimates currently available on the blackboard: from the visual odometry, from the IMU, and from the legged odometry, performing data fusion.
- **AEGridUpd** – agent that updates the local grid map of the terrain. Implements the elevation map updating scheme described in Ref. 9.
- **AEMapUpd** – agent that updates the global grid map of the terrain upon the current local map and the current robot pose.
- **AEPathPlan** – agent that plans the global path for the robot – from the current robot pose to the goal pose (stored on the blackboard) using a variant of the Rapidly exploring Random Trees (RRT) algorithm.[8]
- **AEFootPlan** – foothold planning agent. Implements the foothold planning method described in Ref. 1. Reads the local grid map, current pose, and the path from the blackboard. The resulting footholds are in turn stored on the blackboard.
- **AELeakLoc** – agent that localizes the gas/leakage source. Requires data stored by the **ADEnvSens** agent, and the current robot pose. Computes new goal for the robot and stores it on the blackboard.
- **AEClimb** – agent that controls the robot motion during climbing obstacles, such like stairs. Implements the algorithm presented in Ref. 11.
- **AEPilot** – agent that generates robot motion commands upon the available pose and footholds. It has a library of gaits and implements co-ordination of legs. The commands are stored on the blackboard and used by **ADCtrl**.
- **AETeleOp** – teleoperation and monitoring agent. Its task is to monitor the whole data and to execute the operator commands, by writing the proper data to the blackboard (i.e. goal pose, path, or direct motion commands – depending on the teleoperation mode).

3. Teleoperation system

The concept of teleoperation can be considered in many different ways – from simple remote control to supervised autonomy.[6] However, our idea is to enable the remote user to interact with the robot at various levels of control – from control of individual joints, through guiding the robot with a simple keyboard/joystick input, to a form of supervised autonomy, when the operator specifies only the general path to follow. Therefore, the teleoperation system should have the following properties: (i) integration with the autonomous functions of the robot; (ii) possibility to issue commands at several different abstraction levels; (iii) multi-modal sensory feedback that gives the operator good situational awareness.

To obtain these properties we implemented the idea of collaborative teleoperation,[7] which assumes that both the operator and the robot control software collaborate on the same decision level, and the operator is viewed as just another module of the control system. This concept is facilitated by our agent-based software architecture of the control system. The multi-agent structure enables to represent the teleoperator in the robot controller as a specialized expert agent **AETeleOp**, which is responsible for executing the commands obtained from the remote operator by writing specific data to the blackboard (i.e. goal pose, path) or by writing direct motion commands. On the other hand, the **AETeleOp** plays also the role of a reporting agent, which reads data from the blackboard and sends them to the operator interface program via the **ADComm** agent.

It is possible to control the robot directly by means of a joypad, which not only enables to change walking direction and speed of the machine, but also makes possible to change the orientation with regard to all the axes (i.e. the roll, pitch or yaw angle), and to change the robot's posture by raising or lowering the trunk. It is also possible to control a single leg or even a single joint (at the level of robot controller's API), which may be useful to execute precise maneuvers, e.g. when one of the legs clamps in a hole and the operator has to withdraw it carefully.

The operator can also give the robot more autonomy by setting only the goal position with regard to the current one. Then, the robot updates its global terrain map, plans a collision-free path to the goal, and starts to follow it, selecting on-line the proper footholds. Alternatively, the operator can send a general 2D path to the robot. This path overrides the one that was planned automatically, but the robot following it still updates the local terrain map and uses the foothold planning algorithm.

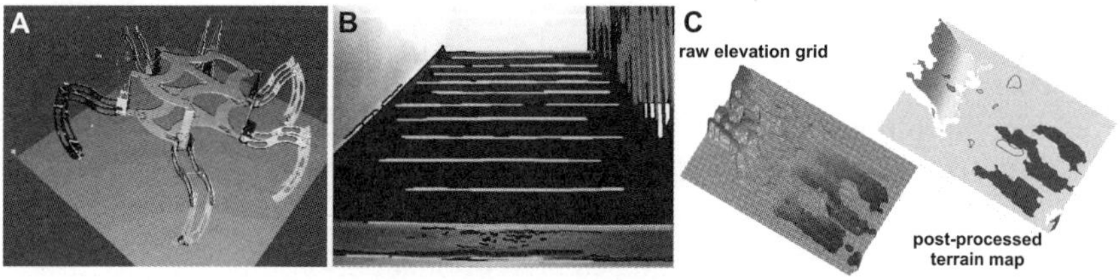

Fig. 2. Examples of the synthetic display concept: robot's frame animation (A), augmented video image (B), and post-processed local elevation map (C)

Because we assume that in most of the missions the remote operator will not have a direct visual contact with the robot, the teleoperation interface should provide enough synthetic data to make the operator fully aware about the environment and the state of the robot. In order to avoid presentation of an overwhelming amount of numerical data a number of synthetic views exploiting the computer animation and the augmented reality concept are implemented. One example is the simplified robot's frame animation, which presents the current posture with regard to the gravity vector, and indicates the support polygon and the state of feet contacts with the ground (by drawing the respective legs in bright green – Fig. 2A). A major part of the screen is occupied by the transmitted images, hence we decided to present other sensory data by overlaying them on the video images. The range data from the laser scanner are presented as a scan line converted to the image coordinates. It is also possible to enhance some image features by extracting them from raw images using standard computer vision algorithms and then displaying them as an overlay. This way object contours, horizontal and vertical lines can be enhanced (Fig. 2B). It is also possible to track objects (e.g. having a distinctive color) over a sequence of images. Each of these functions can be switched off if it is too obtrusive for the operator. The local elevation maps can be displayed to the operator in the raw form or post-processed by a plane segmentation module, which extracts the dominant surfaces in a scene and finds smaller obstacles based on their deviation from these surfaces. The information on main surfaces and the protruding obstacles enriches the information presented to the operator, to whom correct and quick interpretation of the raw grid map can be a problem (Fig. 2C).

4. Experimental results

In this section we present sample results of using the Messor robot in both the supervised autonomy and manual teleoperation modes. In the first experiment the operator specified only a general path to follow, but did not control the robot directly. Results obtained in an area containing randomly scattered rubble (Fig. 3A,B) show that the autonomous functions of the robot work properly. The robot was able to traverse the rubble and to create a reasonably detailed global elevation map of the mission area (Fig. 3C), which is presented to the remote operator.

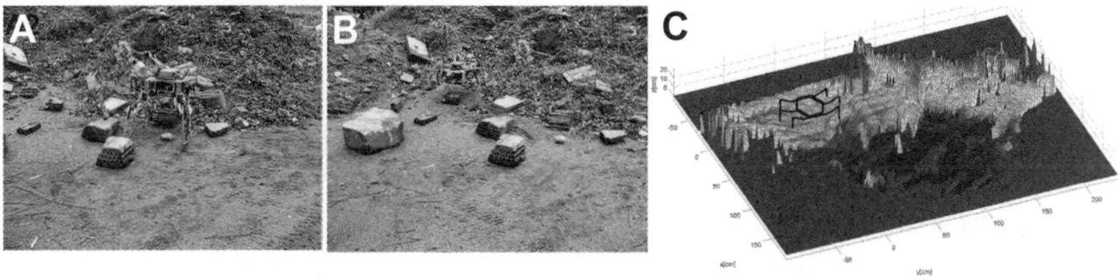

Fig. 3. Messor robot traversing rubble in the supervised autonomy mode

The results of another outdoor experiment performed with the Messor robot (Fig. 4A) illustrate the ability of the map-building and motion-control agents to acquire a map of the terrain with vegetation, and to generate proper footholds in such a demanding environment. The robot created a map of the rough terrain (Fig. 4B), which is presented to the remote operator in an efficient way, enabling him to identify such obstacles as the large root pointed out by the arrow in Fig. 4C.

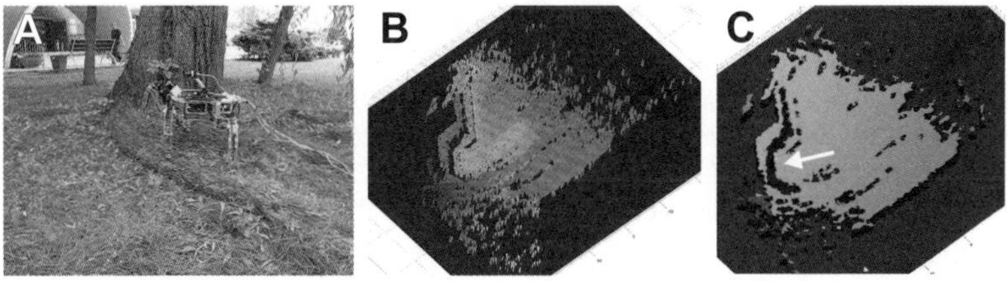

Fig. 4. Outdoor mapping experiment in a vegetation-covered area

An example of inspection-type mission is given in Fig. 5. The task was to find some "suspicious" objects (e.g. explosives) under a car or inside a car. The flexibility of the software, in which the **AETeleOp** agent can override the automatically generated motion commands with commands issued by the remote operator (using a joypad) enabled the robot to change its posture, where it was demanded (Fig. 5A, C). The on-board camera images were sent to the operator (Fig. 5B, D) in order to identify the interesting object, which marked once with a mouse was then tracked automatically.

Fig. 5. Searching for a suspicious object

5. Final remarks

This paper describes the concept and implementation of a multi-agent-based software for a semi-autonomous walking robot. This work capitalizes on our experience with the MAB architecture, which was already used in several autonomous robots built at Poznan University of Technology.

References

1. Belter D., Łabęcki P., Skrzypczyński P. (2010) Map-based Adaptive Foothold Planning for Unstructured Terrain Walking. Proc. IEEE Int. Conf. on Robot. & Autom., Anchorage, 5256–5261.
2. Belter D., Walas K. (2011) Messor – Versatile Walking Robot for Search and Rescue Missions, J. of Autom., Mobile Robotics and Intell. Syst., 5(2), 28–33.
3. Brzykcy G., Martinek J., Meissner A., Skrzypczyński P. (2001) Multi-Agent Blackboard Architecture for a Mobile Robot, Proc. IEEE/RSJ Conf. on Intell. Robots & Systems, Maui, 2369–2374.
4. Estremera J., Garcia E., Gonzalez de Santos P. (2002) A Multi-Modal and Collaborative Human-Machine Interface for a Walking Robot, J. of Intelligent & Robotic Systems, 35, 397–425.
5. Engelmore R., Morgan T., (1988) Blackboard Systems, Boston, Add.-Wesley.
6. Fischer O., Wiersbitzki J., Zöllner M., Dillmann R. (2004) Teleoperating a Six-Legged Walking Machine in Unstructured Environments, Proc. IEEE Int. SSRR Workshop, Bonn, CD-ROM.
7. Fong T., Thorpe C., Baur C. (1999) Collaborative Control: A Robot-Centric Model for Vehicle Teleoperation, AAAI Spring Symp.: Agents with Adjustable Autonomy, Stanford, 210–219.
8. Kuffner J, LaValle S. (2000) RRT-Connect: An Efficient Approach to Single-query Path Planning, Proc. IEEE Int. Conf. on Robot. & Autom., San Francisco, 995–1001.
9. Łabęcki P., Rosiński D., Skrzypczyński P., (2010) Terrain Perception and Mapping in a Walking Robot with a Compact 2D Laser Scanner, Emerging Trends in Mobile Robotics, Singapore, World Scientific, 981–988.
10. Schmidt A., Kasiński A. (2010) The Visual SLAM System for a Hexapod Robot, Computer Vision and Graphics, Berlin, Springer 2010, 260–267.
11. Walas K. (2010) Fully Parametrized Stair Climbing Strategy for a Six-legged Walking Robot, Emerging Trends in Mobile Robotics, Singapore, World Scientific, 777–784.

ON-LINE INCREMENTAL 3D HUMAN BODY RECONSTRUCTION FOR HMI OR AR APPLICATIONS

L. ALMEIDA*†, F. VASCONCELOS†, J. P. BARRETO†, P. MENEZES†, J. DIAS†

†*Institute of Systems and Robotics,*
Department of Electrical and Computer Engineering, University of Coimbra
Polo II, 3030-290 Coimbra, Portugal
E-mail: laa@ipt.pt, {fpv, jpbar, paulo, jorge}@isr.uc.pt

**Department of Informatics Engineering, Institute Polytechnic of Tomar,*
2300 Tomar, Portugal

This research proposes an on-line incremental 3D reconstruction framework that can be used on human machine interaction (HMI) or augmented reality (AR) applications. There is a wide variety of research opportunities including high performance imaging, multi-view video, virtual view synthesis, etc. One fundamental challenge in geometry reconstruction from traditional cameras array is the lack of accuracy in low-texture or repeated pattern region. Our approach explores virtual view synthesis through motion body estimation and hybrid sensors composed by video cameras and a depth camera based on structured-light or time-of-flight. We present a full 3D body reconstruction system that combines visual features and shape-based alignment. The proposed mesh generation algorithm is based on Crust and efficiently adds new vertices to an already existing surface. Modeling is based on meshes computed from dense depth maps in order lower the data to be processed and create a 3D mesh representation that is independent of view-point.

Keywords: Augmented Reality; 3D reconstruction; tele-presence; virtual view synthesis

1. Introduction

This work presents an on-line incremental 3D reconstruction framework that can be used on human machine interaction (HMI) or augmented reality (AR) applications. The project, based on recent low cost depth sensors, intends to create a domestic easy to install 3D acquisition and display system that enables socialization and entertainment. Exploring artificial vision, spatial audio and computers graphics techniques enable us to induce sensations of being physical in the presence of other people useful on several domains like elderly loneness minimization problem, tele-rehabilitation[1,2] education, socialization, 3DTV, entertainment, etc. Phones and internet chat/audio/video conferencing programs (ex: VOIP, NetMeeting, Skype) are not able to create the remote person presence feeling. Means of communications that enable eye contact, gestures reconnaissance, body language and facial expressions are required.

Augmented reality and particularly tele-immersion[34] can provide the technology means that enables users interact remotely and experience the benefits of a face-to-face meeting. In order to aim an incremental on-line 3D human reconstruction solution useful for shared mixed reality workspace[56],[1] we estimate the 3D world information using 2D image sequences and depth information using a depth camera, e.g. a time of flight camera (ToF) or structured light camera. This hybrid approach addresses the geometry reconstruction challenge from traditional cameras array, that is the lack of accuracy in low-texture or repeated pattern regions. The proposed real-time 3D full reconstruction system combines visual features and shape-based alignment. By identifying features in images for which real-world coordinates can be measured, a correspondence between 3D and 2D is set up. Using those annotated 3D points, between consecutive point clouds, it is possible to estimate the motion transformation through a linear, closed form or iterative method, register them on one same referential and create a global model. Correspondence between consecutive image features in images is performed using SURF method.[7] Virtual view synthesis and modeling is based on 3D mesh from dense depth maps in order lower the data to be processed and to create a 3D mesh representation that is independent of view-point.[8]

The proposed mesh generation algorithm is based on an incremental variant of the Crust algorithm,[9] based on the fact that is possible to efficiently add vertices's to a Delaunay Triangulation. The aim is to continuously generate a realistic body model, transfer the model and reconstruct on a remote common display or virtual environment according each users viewpoint by a tracking process. Figure 1 presents an overview of the algorithm.

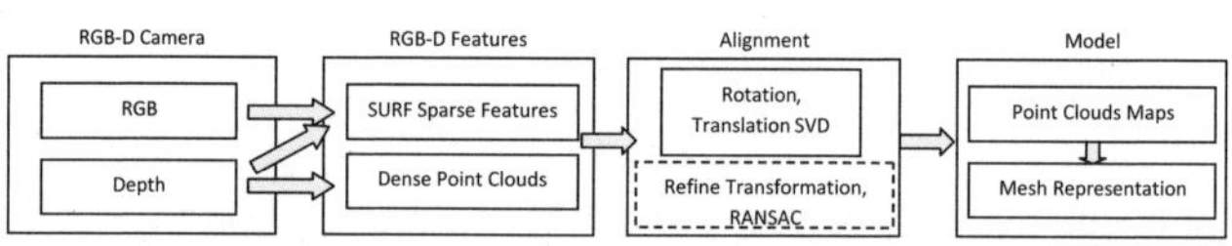

Fig. 1. Algorithm overview. The proposed real-time 3D full reconstruction system combines visual features and shape-based alignment between consecutive point clouds. The model representation is updated incrementally

The reminder of this paper is organized as follows. First some related work, then section 2 describes the suggested methodology and section 3 present some experimental results. Finally, section 4 presents the future work and conclusions.

Background: Virtual view synthesis and modeling are the potential graphic tools to create the eye to eye contact illusion on tele-presence communications.[10] The approach involves surface reconstruction while a basic task for object detection, manipulation and environment modeling. Generally, the object's surface is reconstructed by merging measurements from different views requiring depth data and sensor pose data. When both, pose and depth, are unknown, structure from motion is a solution. Corresponding features in consecutive images are used to estimate the ego-motion of the sensor. Based on this ego-motion information the depth without absolute scale is estimated. Since recent depth cameras also provide RGB data, 2D image processing algorithms are usable. Point feature mapping in RGB images can be improved by the associated depth data obtaining a 3D feature tracking. Most common methods for matching 2D image features are based on the KLT (Kanade-Lucas-Tomasi)[11],[12] SIFT (Scale-Invariant Feature Transform)[13] or SURF (Speeded Up Robust Features)[7] approaches. If only the depth information but no pose is given, i.e. by using a stereo camera or a laser scanner system without inertial sensors, the Iterative Closest Point (ICP) algorithm can be used to register point clouds acquired from different perspectives[14].[15] Finally, if pose and depth are known, the registration procedure is dispensable and the data can simply be merged. Calculating changes in the 3D pose based on these methods have been performed[16][17][18].[19] Our work intends to perform a real-time incremental body modeling.

2. Methodology

We propose a real-time full 3D reconstruction system that combines visual features and shape-based alignment using Xbox Kinect device. Alignment between successive frames is computed by jointly optimizing over both appearance and shape matching.

Registration: Considerer the motion of a rigid body in front of a scanner and the estimation of the rigid transformation (rotation and translation). Assuming that there exist two corresponding 3D point sets $\{\mathbf{x_i^t}\}$ and $\{\mathbf{x_i^{t+1}}\}$, $i = 1..N$, from consecutive t and $t+1$ frames, such that they are related by eq. 1:

$$\mathbf{x_i^{t+1}} = \mathbf{R}\mathbf{x_i^t} + \mathbf{T} + \mathbf{V_i} \quad (1) \qquad \varepsilon^2 = \sum_{i=1}^{N} \left\| \mathbf{x_i^{t+1}} - \mathbf{R}\mathbf{x_i^t} - \mathbf{T} \right\|^2 \quad (2)$$

where $\mathbf{R}$ is a standard 3x3 orthonormal rotation matrix, $\mathbf{T}$ is a 3-D translation vector and $\mathbf{V_i}$ a noise vector. Solving for the optimal transformation $[\mathbf{R}, \mathbf{T}]$ that maps the set $\{\mathbf{x_i^t}\}$ onto $\{\mathbf{x_i^{t+1}}\}$ typically requires minimizing a least squares error criterion given by eq. 2. The minimization of eq. 2 can be based on the singular value decomposition (SVD) of a matrix.[20] Calculating rotation: as a consequence of the least-squares solution to eq. 2, the point sets $\{\mathbf{x_i^t}\}$ and $\{\mathbf{x_i^{t+1}}\}$ should have the same centroid. Using this constraint a new equation can be generated. By defining:

$$\overline{\mathbf{x_i^t}} = \frac{1}{N}\sum_{i=0}^{n}\mathbf{x_i^t}, \quad \overline{\mathbf{x_i^{t+1}}} = \frac{1}{N}\sum_{i=0}^{n}\mathbf{x_i^{t+1}}, \quad \mathbf{x_{ci}^t} = \mathbf{x_i^t} - \overline{\mathbf{x_i^t}}, \quad \mathbf{x_{ci}^{t+1}} = \mathbf{x_i^{t+1}} - \overline{\mathbf{x_i^{t+1}}} \tag{3}$$

Minimizing this equation is equivalent to maximizing $Trace(\mathbf{R}\,\mathbf{H})$, where $\mathbf{H}$ is a 3x3 correlation matrix defined by $\mathbf{H} = \mathbf{x_{ci}^{t+1}}(\mathbf{x_{ci}^t})^{\mathbf{T}}$. If the singular value decomposition of $\mathbf{H}$ is given by $\mathbf{H}$=$\mathbf{UDV}^T$ then the optimal rotation matrix, $\mathbf{R}$, that maximizes the desired trace is $\mathbf{R}$= $\mathbf{U}$ diag(1; 1; det($\mathbf{UV}^T$)) $\mathbf{V}^T$,[20] $\mathbf{R} = \mathbf{UV^T}$.
The optimal translation that aligns the centroid of the sets is $\mathbf{T} = \overline{\mathbf{x_i^{t+1}}} - \mathbf{R}\overline{\mathbf{x_i^t}}$.
Model Mapping: Suppose that the mapping from the world coordinates to one of the scans of the sequence, $\mathbf{^0H_w}$. For any consecutive pair of scans (t, t+1) using tracked points it is possible to measure rotation and translation, $\mathbf{^{t+1}H_t}$, $\mathbf{H} = [\mathbf{R}, \mathbf{T}]$ (single homogeneous matrix 4x4) and compute Eq. 4:

$$\mathbf{^iH_0} = \mathbf{^iH_{i-1}}\mathbf{^{i-1}H_{i-2}}\ldots\mathbf{^1H_0} \qquad \textit{and} \qquad \mathbf{^iH_w} = \mathbf{^iH_0}\mathbf{^0H_w} \tag{4}$$

To update the reconstructed model, each acquired 3D point set is transformed to the world coordinate system using $\mathbf{^iH_w}$. This alignment step adds a new scan to the dense 3D model.*Virtual View Synthesis:* Once having the model, new perspectives views can be presented at each time instant depending on the viewers pose in front of the display (using a head/body tracking module).
Mesh Generation: In order to obtain a surface mesh from the registered 3D point clouds, we use an incremental adaptation of the Crust algorithm.[9] For a given input set of points X, the Crust algorithm determines a set of poles P that lie on the medial axis of the surface. The Delaunay triangulation of $X \cup P$ is then computed, and finally the surface mesh is then obtained by extracting the set of simplices whose vertices belong to X. Given that it is possible to efficiently add vertices to a Delaunay Triangulation,[21] an incremental version of the Crust Algorithm is easy to implement. If the Crust Algorithm was already performed on a set of points X_1, a set of poles P_1 and the Delaunay triangulation of $X_1 \cup P_1$ are available as well. To add a new set of sample points X_2 to the surface mesh, the following steps are performed:

(1) P_2 = poles of X_2
(2) Add $P_2 \cup X_2$ as new vertices of the Delaunay triangulation
(3) Extract triangles whose vertices belong to $X_1 \cup X_2$

This procedure can be repeated for an arbitrary number of point sets X_i. However, in order to avoid an indefinite growth in the number of mesh vertices, points whose

euclidean distance to the closest mesh vertex is lower than a given threshold are deleted from the input cloud before performing the incremental Crust step.
Algorithm:The global model reconstruction algorithm can be described as follow 2.1:

Algorithm 2.1 Model reconstruction algorithm

```
Rg ← Rinit; tg ← tinit
f1 ← undistort(adquire_rgb_image()), f1d ← undistort(adquire_depth_image())
f1xyz ← convert_depth_image_to_xyz_data(f1d), f1r ← map_rgbcolor_to_depth_image(f1xyz, f1)
for (;;) do
    f2 ← undistort(adquire_rgb_image()), f2d ← undistort(adquire_depth_image())
    f2xyz ← convert_depth_image_to_xyz_data(f2d)
    f2r ← map_rgbcolor_to_depth_image(f2xyz, f2)

    (surf1, surf2) ← detect_SURF_features(f1r, f2r)
      matches2D ← SURF_match(surf1, surf2)
      matches3D ← correspond2D3D(matches3D)
      (R,t) ← motion_estimator(matches3D)
      (Rg, tg) ← update_global_transformation(R,t)
      f1r ← f2r; f1xyz ← f2xyz {update_past_data}
      model ← project_points_to_world_coordinates(f2xyz, Rg, tg)
      mesh_modelgeneration
end for
```

3. Implementation and Results

Novel depth sensors like PrimeSense camera or Xbox Kinect can capture video images along with per-pixel depth information and to experimentally test the algorithm we register several 3D point clouds in order to create person model while he is rotating in front of Kinect device. *Calibrations:* The Kinect device combines a regular RGB camera and a 3D scanner, consisting of an infrared (IR) projector and an IR camera as shown in figure 2a) and the calibration is unavoidable. The aim is to undistort the RGB and IR images and map depth pixels with color pixels (see figure 2). Kinect maximal range raw depth is 2^{11}, and it relates to metric depth through a linear approximation $d_m(x_{ir}, y_{ir}) = f(rawdepth(x_{ir}, y_{ir}))$. From the metric depth, the 3D metric position $(X_{ir}, Y_{ir}, Z_{ir})^T$ of the pixel, with the respect to the IR camera, can be computed using the following equation: $(X_{ir}, Y_{ir}, Z_{ir})^T = (\frac{(x_{ir}-c_{xir})*d_m(x_{ir}, y_{ir})}{f_{xir}}, \frac{(y_{ir}-c_{yir})*d_m(x_{ir}, y_{ir})}{f_{yir}}, d_m(x_{ir}, y_{ir}))^T$, where x_{ir}, y_{ir} are the coordinates of the depth pixel in image, f_{xir},f_{yir} the IR camera focal length (pixel size units), c_{xir}, c_{yir} the coordinates of the image center of IR camera, and d_m is depth in meters. The IR and RGB cameras are separated by a small baseline, it is possible to determine the 6 DOF transform between them. Knowing the rotation **R** and translation **T** between the RGB and IR camera, we can then re-project each 3D point on the color image and get its color. The mapping between color image and depth image can be expressed by following equations: $(X_{rgb}, Y_{rgb}, Z_{rgb})^T = \mathbf{R}(X_{ir}, Y_{ir}, Z_{ir})^T + \mathbf{T}$ and $(x_{rgb} = \frac{(X_{rgb} * f_{xrgb})}{Z_{rgb}} + c_{xrgb}$, $y_{rgb} = \frac{(Y_{rgb} * f_{yrgb})}{Z_{rgb}} + c_{yrgb})$, where x_{rgb}, y_{rgb} are the coordinates of the rgb pixel in image, f_{xrgb}, f_{yrgb} the RGB camera focal length, c_{xrgb}, c_{yrgb} the image center, and d_m is depth in meters.

a) *b)* *c)* *d)*

Fig. 2. a) Kinect sensor b) undistorted RGB image c) undistorted depth Image, the body black pixels have unknown depth value, due occlusions or reflective surface material d) Map between undistorted RGB image and depth image.

On figure 3a we present an example of correspondence between consecutive image features in using SURF method (white lines indicate correspondent point). Some matches are undesirable and are related with background static areas. The contribution of erroneous matches is minimized by the number of good matches while using the described minimization method to obtain the transformation. Although SIFT descriptor present better accuracy, we have choose SURF method in order to achieve the real-time characteristic.

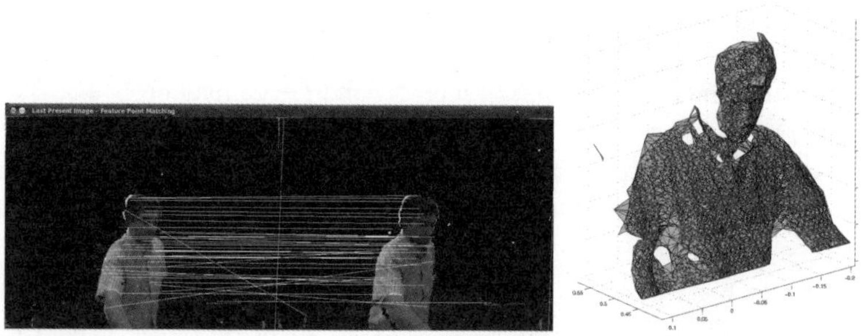

a) *b)*

Fig. 3. a) SURF features matched on consecutive time frames b) Mesh model with 27864 vertices and 31810 faces

An example of mesh generation using the proposed incremental adaptation of Crust algorithm is provided on figure 3b with 27864 vertices and 31810 faces.

Figure 4 depicts a sequence of scans that creates a 3D person model. They result from several 3D point clouds fused after applying successive 3D rigid body transformations. Typically the system has a performance of 2 HZ. The time consuming

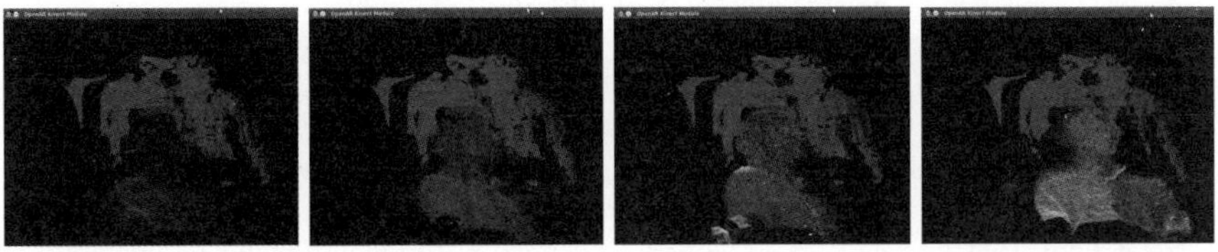

Fig. 4. 3D Model, sequence of point clouds being registered on the same referential, each color represent time sequential scan

stage is related with the surf feature extraction and it takes an average of 300 ms. It depends on the number of detected good feature of the image, although we expect to speed up significantly this step by making use of GPU. The involved number of points also influences the transformation time calculus. On table 1 we present some typically time measure involving some algorithm steps.

Algorithm Steps	**(ms)**
Acquisition	1.55
Undistort Images	10.61
DepthRGB Map and last frame update	36.13
SURF feature extraction	314.853
Matching and transformation calculus	78.0282
Alignment, display and interaction	30.377
Total	471.56 (f=2.12 Hz)

4. Conclusions

There is still a potential for algorithm speedup involving code optimization, GPU CUDA programming and stereo display graphics. Our approach explores virtual view synthesis through motion body estimation and hybrid sensors composed by video cameras and a low cost depth camera based on structured-light, improving the reconstruction accuracy in low-texture or repeated pattern region. We present a full 3D body reconstruction system that combines visual features and shape-based alignment. Modeling is based on meshes computed from dense depth maps in order lower the data to be processed and create a 3D mesh representation that is independent of view-point. This work presents an on-line incremental 3D reconstruction framework that can be used on low cost augmented reality (AR) or HMI applications to enable socialization and entertainment.

References

1. G. Kurillo, T. Koritnik, T. Bajd and R. Bajcsy, *Stud Health Technol Inform* **163**, 290 (2011).
2. A. A. Rizzo and G. J. Kim, *Presence* **14**, 119 (2005).
3. S.-H. Jung and R. Bajcsy, *Journal of Multimedia* **1**, 9 (2006).
4. R. Azuma, Y. Baillot, R. Behringer, S. Feiner, S. Julier and B. MacIntyre, *IEEE Comput. Graph. Appl.* **21**, 34(November 2001).
5. G. Kurillo, R. Vasudevan, E. Lobaton and R. Bajcsy, A framework for collaborative real-time 3d teleimmersion in a geographically distributed environment, in *Multimedia, 2008. ISM 2008. Tenth IEEE International Symposium on*, dec. 2008.
6. B. Petit, J.-D. Lesage, J.-S. Franco, E. Boyer and B. Raffin, Grimage: 3d modeling for remote collaboration and telepresence, in *ACM Symposium on Virtual Reality Software and Technology*, October 2008.
7. H. Bay, T. Tuytelaars and L. V. Gool, Surf: Speeded up robust features, in *In ECCV*, 2006.
8. J. Carranza, C. Theobalt, M. A. Magnor and H.-P. Seidel, *ACM Trans. Graph.* **22**, 569(July 2003).
9. N. Amenta, M. Bern and M. Kamvysselis, A new voronoi-based surface reconstruction algorithm, in *Proceedings of the 25th annual conference on Computer graphics and interactive techniques*, SIGGRAPH '98 (ACM, New York, NY, USA, 1998).
10. F. Isgro, E. Trucco, P. Kauff and O. Schreer, *Circuits and Systems for Video Technology, IEEE Transactions on* **14**, 288 (march 2004).
11. J. Shi and C. Tomasi, Good features to track, in *Computer Vision and Pattern Recognition, 1994. Proceedings CVPR '94., 1994 IEEE Computer Society Conference on*, jun 1994.
12. B. D. Lucas and T. Kanade, An iterative image registration technique with an application to stereo vision, in *Proceedings of the 7th international joint conference on Artificial intelligence - Volume 2*, (Morgan Kaufmann Publishers Inc., San Francisco, CA, USA, 1981).
13. D. G. Lowe, *Int. J. Comput. Vision* **60**, 91(November 2004).
14. P. J. Besl and N. D. McKay, *IEEE Trans. Pattern Anal. Mach. Intell.* **14**, 239(February 1992).
15. J. Lobo, L. Almeida, J. Alves and J. Dias, Registration and segmentation for 3d map building: A solution based on stereo vision and inertial sensors, in *ICRA*, (IEEE, 2003).
16. P. Henry, M. Krainin, E. Herbst, X. Ren and D. Fox, RGB-D Mapping: Using Depth Cameras for Dense 3D Modeling of Indoor Environments (2010).
17. L. G. B. Mirisola, J. L. 0002 and J. Dias, 3d map registration using vision/laser and inertial sensing, in *EMCR*, 2007.
18. A. Akbarzadeh, J.-M. Frahm, P. Mordohai, B. Clipp, C. Engels, D. Gallup, P. Merrell, M. Phelps, S. N. Sinha, B. Talton, L. W. 0002, Q. Yang, H. Stewénius, R. Yang, G. Welch, H. Towles, D. Nistér and M. Pollefeys, Towards urban 3d reconstruction from video, in *3DPVT*, (IEEE Computer Society, 2006).
19. P. Menezes, F. Lerasle and J. Dias, *IVC* **29**, 382(May 2011).
20. K. S. Arun, T. S. Huang and S. D. Blostein, *IEEE Trans. Pattern Anal. Mach. Intell.* **9**, 698(September 1987).
21. L. Guibas, D. Knuth and M. Sharir, *Algorithmica* **7**, 381 (1992), 10.1007/BF01758770.

REMOTE CONTROL SYSTEM FOR RESCUE ROBOT FMT: REMOVAL ALGORITHM OF OBSTACLES IN THE IMAGES

K. TOKUDA * and T. HIRAYAMA

Faculty of Systems Engineering, Wakayama University ,
930 Sakaedani, Wakayama, 640-8510, Japan
**E-mail: tokuda@sys.wakayama-u.ac.jp*

T. KINUGASA ** and T. HAJI

Okayama Univrsity of Science,
1-1, Ridai-cho, Okayama, 700-0005, Japan
***E-mail: kinugasa@mech.ous.ac.jp*

H. AMANO

National Research Institute of Fire and Disaster, Japan,
4-35-3 Jindaiji-higashimachi, Chofu, Tokyo, 182-0012, Japan
E-mail: amano@fri.go.jp

This paper describes a method for removing of obstacles in the images by multiple cameras. Acquiring information of remote location is needed to operate robots. For example, an image by camera is effective in remote control. However, there is a problem that images from camera on front of FMT are hindered by the FMT's crawler. To use them as image for control, removal of crawler in the images is necessary. The proposed method generates the images without crawler by multiple cameras and image processing. Then the method uses edge and difference between some images to detect crawler. We confirmed the availability of the proposed method by experiments.

Keywords: crawler robot, rescue system, image processing

1. Introduction

FMT, Flexible mono-tread mobile track(Fig.1), is a rescue robot system, which is covered by crawler track except the both side surfaces.[1] It is inappropriate to attach a camera on the side of the robot because the camera system structure is easily caught by rubbles. The purpose of this study is to put the camera system onto the front of the robot.

Running track is moving across the face of the front camera system. The images from the camera often catch the crawler track region in the view. The botheration disrupts remote control of the robot system with the nuisance by crawler track region. Removal crawler images is a big problem to realize front camera system.

The specificity of rescue robot require a real-time and simple algorithm. Because the algorithm cannot use knowledge about environment, some methods that apply a background image to clip obstacle areas are effective. Ymashita et al. developed the application algorithm that remove some raindrops in monitoring camera images from by using a common long-distance background image.[2]

To apply short-distance view with big parallax, we tried a technique that extracts big similarity area from 2 camera images. This paper reports the system that remove crawler image from acquired whole images, and shows some experimental results.

Fig. 1. FMT overview and cameras

2. Removal Processing Flow

The experimental system has 2 cameras as minimum requirement(Fig. 2). One image is treated as a base, and another image is treated as a reference. The algorithm processes to the base image by using the reference to remove the crawler region. In the process, (1) it pre-compensate disparity components, (2) detects crawler region, (3) clips the region and generate a patch from the reference. The crawler detection is made by using an image difference calculating algorithm and an edge detection algorithm.

Fig. 3(a) and (b) are example images for description of the processing flow. Fig.3(c) shows a result of calculating definition and gray scaling. The

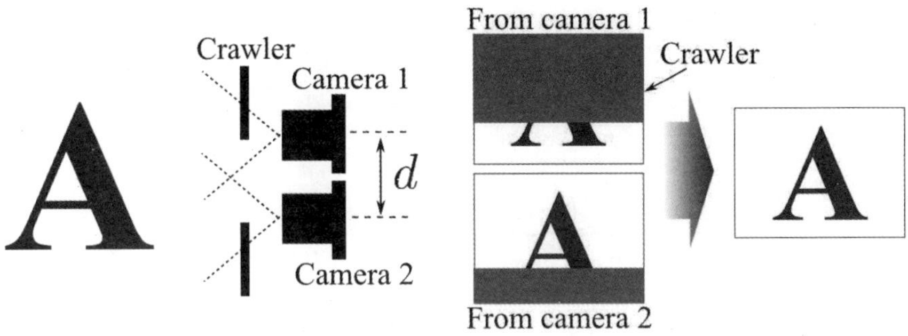

Fig. 2. Removal algorithm concept

gray scale images Fig. 3(d) is an image generated by calculating every region's amount value, shows white area(crawler) and black area(no crawler). Extraction of edges is made by Laplacian filter as Fig. 3(e). Edge image is generated as Fig. 3(f). In the image, edge area is shown as black and no-edge area is shown as white. Removal processing by using a mask(Fig. 3(g)) generate a final image as Fig. 4.

3. Removal Algorithm of Obstacles

Our algorithm uses edges and difference between two images to detect the crawler. This section describes details of the removal algorithm.

3.1. *Parallax Calibration*

We have to process parallax calibrations to acquire 2 images, a base image and a reference image, shown as Fig. 3(a) and (b). In generally, calculating parallax is complex. We use a template matching method for calculating the parallax.

To seek gap during images by template matching, our algorithm clips a center part (1/4 area-ratio) from base image at first. Then, it compares while making a template image move in the reference image(Fig.5). It is using the sum of squares (SSD:Sum of Squared Difference) of the luminance difference which is shown by Eq.(1) for the computation of the similarity. The calculating result moves the reference image and fills the non-common area with black pixels.

$$R_{\mathrm{SSD}}(a,b) = \sum_{i=0}^{H_t-1} \sum_{j=0}^{W_t-1} |I(a+i,b+j) - T(i,j)|^2 \tag{1}$$

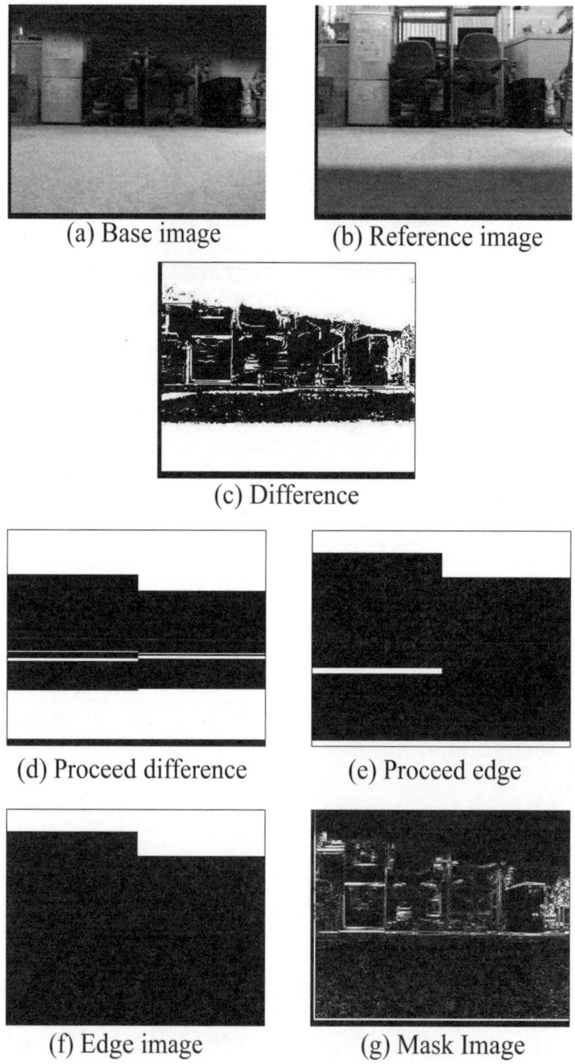

Fig. 3. Image processing frow

3.2. *Region Setting*

The first step is setting region operation in the acquire images as Fig. 6. The region shape is simple enough to trim obstacle areas. Each pixel-value is set during this step.

Fig. 4. Proceeding Result

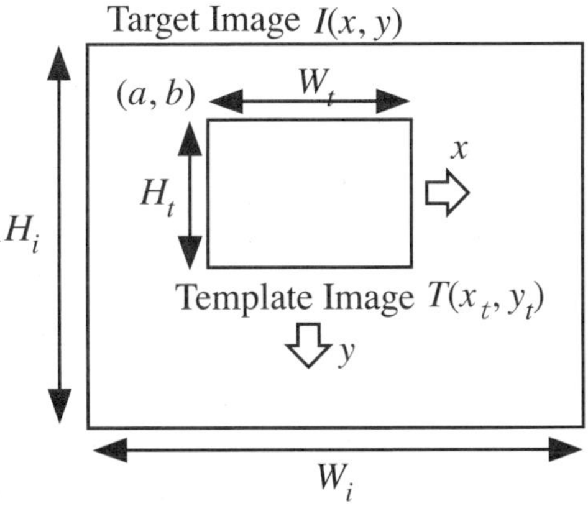

Fig. 5. Parallax Adjustment by Template Matching

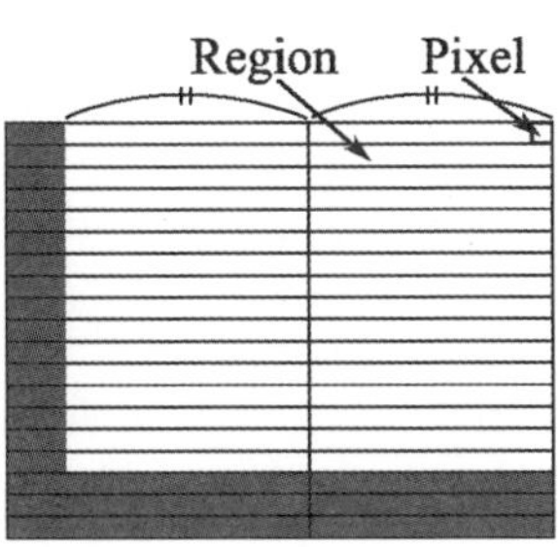

Fig. 6. Region on the image

3.3. *Image Difference Calculating Algorithm*

This algorithm calculates the image difference between base image I_b and reference image I_r. First, the difference every RGB-value between each corresponding pixel in all over the images is computed(Eq. 2). Second, the difference image converts into a gray-scale image, and every pixel-values are set as (255 or 0) by threshold T_d(Eq. 3). (x, y) represents a two-dimensional coordinate on the images, and Gray$(\cdot)$ represents a gray scaling processing.

$$I_{diff}(x,y) = |I_b(x,y) - I_r(x,y)| \tag{2}$$

$$I_{diff-binary}(x,y) = \begin{cases} 255, & \text{Gray}(I_{diff}(x,y)) > T_d \\ 0, & \text{Gray}(I_{diff}(x,y)) \leq T_d \end{cases} \tag{3}$$

Finally, the pixel-value (255 or 0) is given to each region by Eq. 4.

$$I_{diff-region}(region) = \begin{cases} 255, & \sum\limits_{region} I_{diff}(x,y) > T_{dr} \\ 0, & \sum\limits_{region} I_{diff}(x,y) \leq T_{dr} \end{cases} \tag{4}$$

In the generated difference image, white pixel(255) represents difference area, and black pixel(0) points no-difference area.

3.4. *Edge Detection Algorithm*

An edge is defined as point set which pixel-values change sharply. We use edges for detection of crawler on the base image. Extraction of edge on the base image is made by Laplacian filter (which define as Fig.7). The pixel-value (255 or 0) is given to each region by Eq. 5.

0	1	0
1	-4	1
0	1	0

Center pixel

Fig. 7. Laplacian filter

$$I_{edge-region}(region) = \begin{cases} 255, & \sum_{region} I_{edge}(x,y) < T_{er} \\ 0, & \sum_{region} I_{edge}(x,y) \geq T_{er} \end{cases} \tag{5}$$

In the generated image by this process, white pixel (255) represents edge area, and black pixel (0) points no-edge area.

3.5. *Mask Generation Algorithm*

The mask generation algorithm computes a logical AND of the difference image and edge image. In the generated mask image by this process, white pixel(255) represents crawler area, and black pixel(0) points no-crawler area.

4. Experiment

We made a experiment for effectiveness determination by using real FMT robot system. The experiment environment is indoors. We set for the parallax to revise a position difference 2.5m ahead from the lens tip. The camera image size is QVGA(320×240pixel). We set 2 situations for the experiment, in rest and on running.

Fig. 8 shows the result of removal. The image in the figure is the one which 1 frame arranged by the time series by taking out an image as the still image. Some errors of recognition was admitted by the part but it was possible to confirm that the crawler of the standard image was removed about the moving.

5. Conclusion

FMT's crawler interferes a front vision. Our approach in this paper can remove regions of crawler from acquired images. The image processing system is simply constructed by several cameras and a palm size PC. Some experimental results showed that the approach is effective for both scenes at work and at rest. However the experimental system remained some issues that the algorithm is not able to remove the region well when environmental is dark. Crawler's position measurement should resolve the region clipping issues.

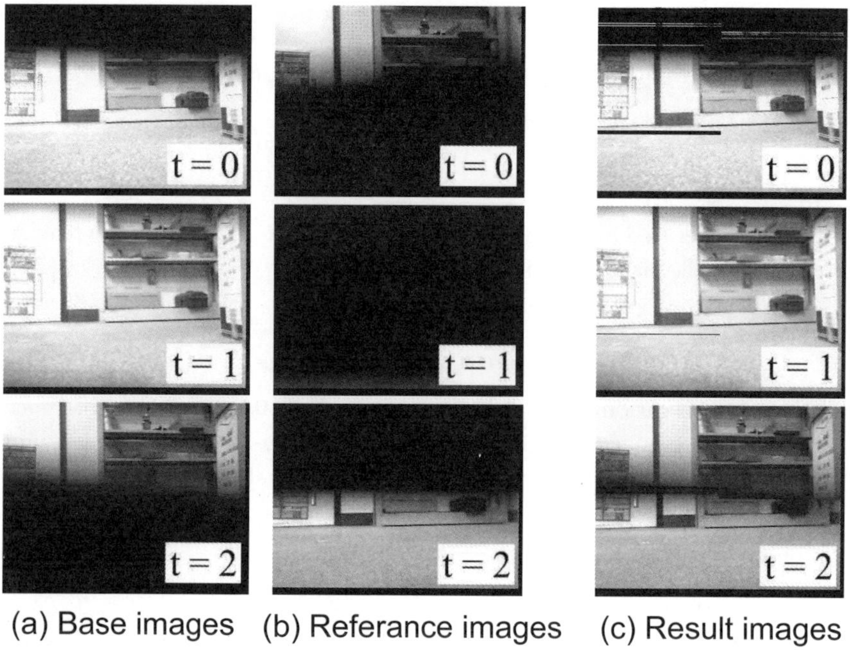

Fig. 8. Experiment result

Acknowledgment

This work was supported in part by the Ministry of Education, Culture, Sports, Science and Technology of Japan through a Financial Assistance Program of the Social Cooperation Study and Chugoku Industrial Innovation Center.

References

1. T. Kinugasa, Y. Otani, T. Haji, K. Yoshida, K. Osuka and H. Amano, *JRSJ* **27**, 107 (2009(in Japanese)).
2. A. Yamashita, M. Kuramoto and T. Kaneko, *IEEJ Trans. on EIS* **127**, 480 (2007(in Japanese)).

MODULAR DISPOSAL TELEOPERATED ROBOT

HERNANDO LEON RODRIGUEZ

Industrial Department, Faculty Engineering, Nueva Granada Military University, Bogotá – Colombia, hernando.leon@unimilitar.edu.co

TARIQ SATTAR

Department of Engineering and Design, London South Bank University, 103 Borough Road, SE1 0AA, London – UK, sattartp@lsbu.ac.uk

The modular disposal teleoperated robot (MDT Robot) can replace man to recognise, remove and deal with explosive bombs or other unknown articles in a dangerous environment. The design uses a modular robot composed of a platform mobile vehicle, an articulated mechanical arm, a teleoperated control system, vision system and a wireless communications system. The MDT Robot is developed to replace a human bomb disposal expert with an apparatus with high manoeuvrability and strong capacity to defeat obstacles, stairs, etc. It can be used in urban areas and in wild environments of sand, grass and soft soil, etc. The robot design is a very low cost robot constructed with several commonly industrial parts, which they can be recycle to further uses in a low cost robot.

1. Introduction

Disposal teleoperated robots have been the focus of research especially in modern society; they require complex algorithms and knowledge in different fields like mechanical design, electronics, image processing, kinetic control, sensor technology and communication. These robots are required to achieve complex missions and some time deal with complicated explosive devices.

There are many tasks which are hazardous to human life which could be conducted remotely using teleoperated robots. A disposal robot is an unmanned machine operated remotely with the intention of removing or making safe of unknown package. Most of these robots are teleoperated machines that can recognise, dispose, climb, search, lift and handle dangerous explosives. [1] [2]

Currently a teleoperated robot is completely operated in manual mode and requires constant human intervention, such as image interpretation, manoeuvrability from one place to another or a simple object manipulation. In the future, most of this robot will be fully autonomous to carry out complex operations in shapeless environments without human intervention. [3] [4] [5]

This paper describes the development of a modular disposal teleoperated robot using recycled technology parts with the purpose of constructing a basic device composed of a mobile crawler platform, 5 degrees of freedom manipulator arm, wireless communications and a vision system. This robot has been developed with a very low cost budget and its aim is to have an expendable machine where if it is destroyed or lost, its cost doesn´t affect the bank balance and the deactivation troops can have even a kamikaze machine.

2. The Robotic Device

The robotic device is composed by four main components, the mobile platform, the manipulator arm, the camera scissor lift mechanics and the control and vision system, see figure 1.The mobile platform is a robot vehicle which has 4 driver motors controlled by one dual motor controller. These motors are mounted in the main structure and provide the lineal, rotational and climbing motion capacity of the system. The manipulator arm is composed by a number of similar modules, where links and joints are attached serially and mounted on top of the robot platform. The manipulator contains 5 revolute joints and one gripper placed at its end, which allows the system to grasp any object of maximum 60 mm width. The main controller and the operator PC/Laptop have a client-server relationship via a graphics user interface (GUI). The operator can access and control wirelessly the mobile platform, the manipulator arm, the camera scissor lift mechanics and the camera itself. The GUI makes the connections straight forward with the main motor controllers permitting brake, move and adjust the camera by individual control of each joint and also obtains error signals from the 9 motors and their corresponding servo drives.

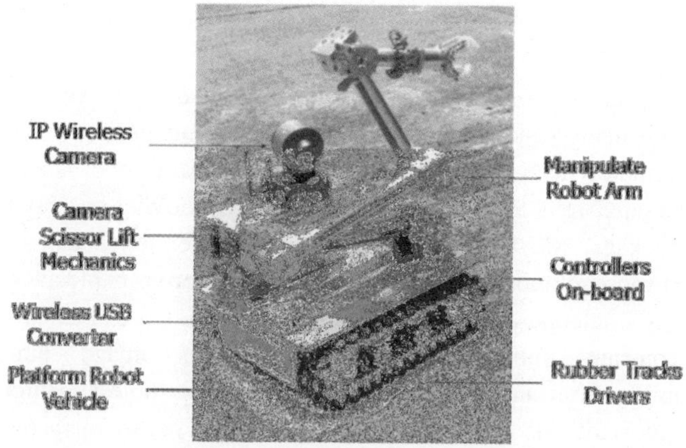

Figure 1. Main parts of the modular disposal teleoperated (MDT) robot.

3. Mobile Platform

The mobile platform of EOD robots are mostly wheeled units, tracked units, or a hybrid of these. This project had develop a tracked robot which has the advantage especially in climbing stairs and crossing mixed terrain environments and high friction contact with the surface but weakness of heavy weight structure.

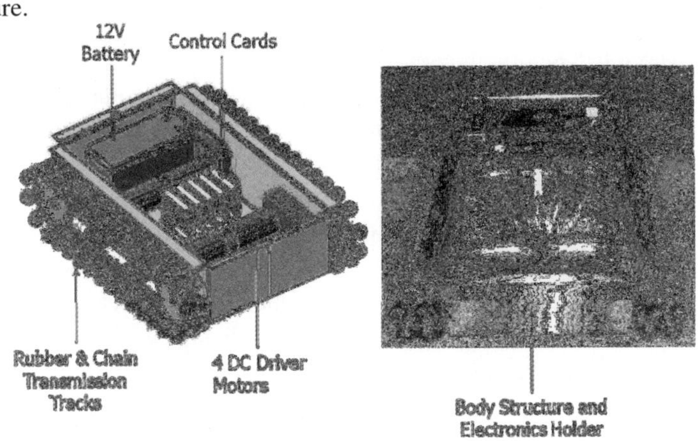

Figure 2: Mobile teleoperated robot platform.

The platform vehicle shows in figure 2 is light weight device of 13 kg; this mobile vehicle is composed by a body structure, four driven motors, two flexible steel chain tracks, four driven pinions and four supported pinions. The body structure was made with stainless steel material and it is a light design to reduce machining problems, decrease the size of the vehicle and also make it stronger to support the driven motors placed on the sides of the vehicle. It uses a dual crawler differential drive; each track is driven by two pinions attached to one geared motor with the intention of increasing power in the vehicle.

On top of the platform are placed a manipulator arm and a camera scissor lift mechanism, which are the main instrument to the teleoperated device. The manipulator arm handled the unknown package and the camera scissor lift mechanism raises/lowers the vision system. These both arms can be controlled by the operator and they can be deploying when is necessary to grasp any object and/or change the vision angle.

4. The Manipulator Arm

The manipulator arm of the disposal system; it has been developed in modular principle in order to provide a great deal of flexibility to reach a given point inside its workspace to collect any unknown object or possible obstacles. The arm is composed by four similar modules on each revolute joint, one rotated joint and one interchangeable gripper. It has redundant joints; its weight is 6 kg including driven motors. Its structure is totally made in profile aluminium; its frames have been made from modified recycled pneumatic cylinders; normally these actuators are thrown away after being used in automated machines or used as support structures, These are extremely light and strong and can be very easily replaced in case of failure or total damage.

The frames are united by the module joints made in aluminium to keep the arm as light as possible, and also these modules are mechanically similar to reduce the cost of the arm and simplify its construction and maintenance. In its fully extended position the manipulator arm can reach about 1.2 metre length and 1.5 metres high, it has the capability to carry 2 kg in its gripper.

The manipulator arm is driven with 5 DC motors which are positioned inside of each frame and they are controlled independently. The description of the developed arm can be displayed using the following parameters of Denavit-Hartenberg algorithm kinematics model [6] [7]. This specific kinematics model for the manipulator represents the transformation from the base coordinate to the gripper coordinate. The result is a matrix, which transforms the first coordinate into the last coordinate by the settings of the joint angles. See figure 3.

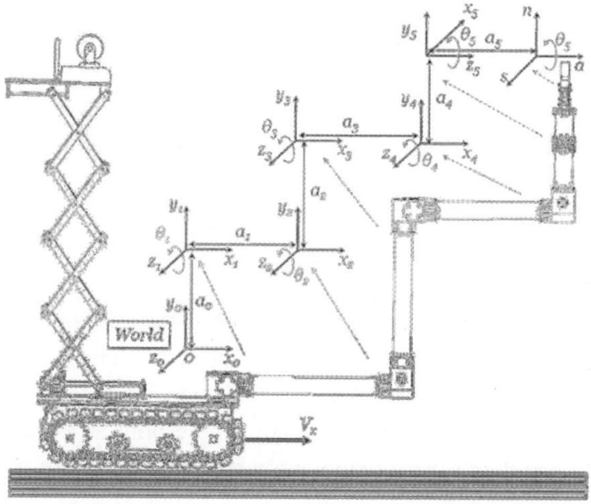

Figure 3: The five degree of freedom kinematics model manipulator arm.

The table 1 shows the parameters and the characteristic representation of the arm and consists of two joint parameters and two link parameters. For an articulated robot whose joints are revolute the joint angle ***(θ$_n$)*** is the only variable parameter. The fixed parameter is the joint angles that varies and lets the robot move its links as preferred. The joint distance ***(a_n)***, the link length ***(d_n)*** and the link twist angle ***(α_n)*** are also fixed parameters.

The link twist angle α is the rotation, which is necessary to transform z_{n-1} to z_n by rotating about x_n. And the joint angle θ is the rotation around the z-axis at will to match the x-axis of the previous coordinate frame x_{n-1} with the one of the following frame x_n.

Table 1. The resulting characteristic table for the revolute manipulate arm.

Axis (Joint)	Joint		Link	
	Angle (θ_n) - variable -	Distance (a_n) [mm] - fixed	Length (d_n) [mm] - fixed	Twist angle (α_n) - fixed
1 (Base)	θ_1	345	0	0
2 (Shoulder)	θ_2	345	0	0
3 (Elbow)	θ_3	310	0	0
4 (Wrist)	θ_4	90	0	0
5 (Roll)	θ_5	0	0	$\pi/2$

The global world coordinate is assigned to the base so that the z-axis represents the axis of rotation. The z-axes are set to align it with the axis of rotation. The x-axes are selected so that they are orthogonal to z_{n-1} and z_n. Because the z-axis and x-axis of the coordinate frame are determined, then the y-axis is defined. The last coordinate frame is set at the gripper tool tip. The origins of the other coordinate frames are located in the point of intersection of the z-axis, so the joint 1, 2, 3, 4 are parallels with the z-axis but joint 5 rotated 90 degrees.

5. The Control and Vision System

The robot device is carrying a several number of control cards which conform to the control system on-board. These cards are dual motor driver controllers composed of five DC servomotor cards and amplifiers embedded on it. These are based on PID control running independently, where all cards are connected to the main driver control card. The system can control velocity and position of each motor by sending PWM signals through the main driver controller. These signals are sent by an USB adapter which communicates with the main control wirelessly, but the video picture between main computer and the robotic system is being sent by Ethernet wireless communications.

The remote control terminal has been programmed in C++, it has graphic user interfaces (GUI) which control the mobile platform, the manipulator arm, camera scissor lift mechanics video system, see figure 4.

The control system is used to send control commands to run the movements and the manipulation of the robotic system. Various commands are created by combining a number of separate buttons and are divided into the control commands as follows; platform backwards/forward, turn left/turn right and stop movements, up/down/stop movements of each joint in the manipulator arm, up/down camera lift mechanics, video system. The operator can modify manually the speed independently of the platform and the manipulator arm.

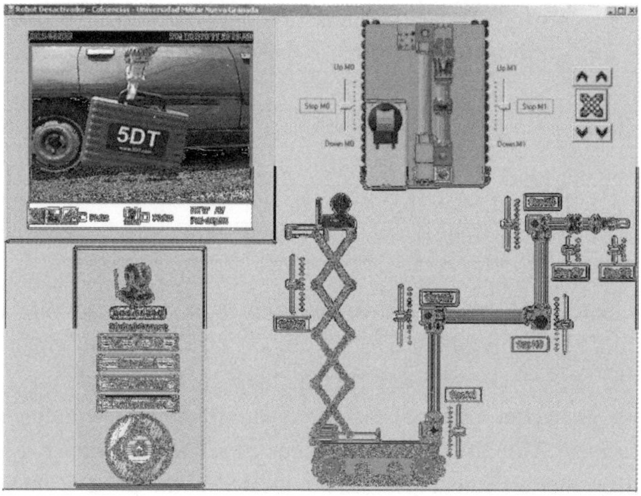

Figure 4: Graphic user interface of the modular disposal teleoperated robot.

The video camera displays a single screen pictures and can be controlled its tilt/pan/zoom; it also has the possibility to send/receive audio from the robot device and some other features of the camera itself like I/O command and darkness. The operator can conduct manual zooming, pan and tilt rotation of the camera wirelessly, to obtain video, images, sound, etc. The operator can also realise actions of the robot's movements, including speed, direction and manoeuvre joint angles of the manipulator arm. The remote control operator gets the information of the robot through the interface, and operates the robot's movements and actions by the main system. The control data and the images are transmitted by the separate wireless transmission module.

Considering the safety features the operator need to be a least 200 meters away of the robot but for absolute security it would be best for them to keep 500 meters away from the unknown package. The wireless communications system of the robot just covers a few metres, but has been increased by around 300 meters with an amplification antenna to achieve the required communication distance. In addition the system has 100 metres cables for a backup communications purposes.

6. Future Work

The prototype robot system has being tested in different environments and surface but most of them can be called lab trials. See figure 5. The robot requires a several tests even in real hazardous environments where mixing pressure is under gone.

Figure 5: Modular disposal teleoperated robot trial.

The maximum distance that the system can operate remotely, accuracy and repeatability of the robot system needs to be constantly evaluated. The system also require some other cameras to increase the view of the operator where multiple cameras gives the operator a clear view of the robot and its surroundings to ensure optimum safety and improve situational awareness, but actual cameras have perform very well in the lab trials.

7. Conclusions

The project has achieved a modular design with recycled materials in the manipulator structure keeping costs low. It has sufficient functionality and the required basic technology to do the required tasks. The robot is cheap enough to be easily expendable and lightweight to be easily transported and deployed.

The robot can be used to recognise inside buildings, sewers, drainpipes, caves and courtyards, perimeter security using on-board motion and sound detectors. Can be used also like bus/train/plane inspector, hostage negotiation and explosive ordnance disposal (EOD)

The robot can replace humans to dispose explosive bombs and remove the excuse of having to use an expensive device robot. The robot design proves that a modular robot can be developed with the purpose of reducing cost but yet having high enough capacity to realise complex tasks, cross obstacles and dispose bombs.

The robot contains some capabilities and features which allows for ready use of a laser pointer and some other accessories that can be attached. In addition, the robot's manipulator is able to support multiple types of instrument for different industrial applications a wide range of detecting sensors.

Acknowledgments

The authors would like to thank the national research council of Colombia Colciencias for sponsored this project.

References

1. M. [1] Boyu Wei, Junyao Gao, Kejie Li, "Design of a Large Explosive Ordnance Disposal Robot", Second International conference on intelligent computation techonology and automation, pp. 403-406, 2009.
2. Gao Junyao, Gao Xueshan, Zhu Jianguo, Zhu Wei, Wei Boyu, Wang Shilin, "Heavy Explosive Removing Robot Control Technique Research" International conferences on intelligent human machines systems and cybernetics, pp. 85-89, 2009
3. Nuttaka Hounsup, terapass Jariyanorawiss, Wiroj Homsup, "A control of A bomb Disposal Robot using a stereoscopic Vision" Ieee, pp. 293-294, 2008
4. Noel Sharkey, "Cassandra or false prophet of domm: al robot and war", IEEE intelligent Systems, pp. 14-17, 2008
5. Zhou Hongfu, Xiao Xinyan, Jiang Liangzhong, "DC servo controlled system design on Bomb-Disposal robot algorithm", the eight international Conferences on Electronics Mesurement and instruments, pp. 510-514, 2007
6. Homeland security newswire, "Snake-like robots dispose of IEDs", Retrieved from: http://homelandsecuritynewswire.com/snake-robots-dispose-ieds, published 28 July 2010, [Visited: March 2011]
7. Alina-A Brenner, T. P. Sattar (2006), Portable 7 DOF Scanning Arm For Non-destructive Inspection Of Objects With Unknown Geometry Proceedings of the 9th International Conference on Climbing and Walking Robots and the Support Technologies for Mobile Machines, September 12 - 14, 2006, Brussels, Belgium, pp 527-532

SECTION–9

INNOVATIVE DESIGN OF CLAWAR

WALL CLIMBING ROBOTS USING ELECTRO-ADHESION TECHNOLOGY

Gareth Griffiths* and Matthew Studley

Bristol Robotics Laboratory, University of the West of England,
Bristol Institute of Technology, Frenchay Campus, Coldharbour Lane, Bristol,
BS16 1QY
**E-mail: gareth.griffiths@brl.ac.uk*
www.brl.ac.uk

This paper reports an investigation in the field of electro-adhesion. The focus has been to develop potential methods which might be applied to wall-climbing robots. Several Electro-adhesive pads were fabricated using easily obtainable materials and we report their effectiveness at holding a high voltage charge and adhering to a glass pane. These initial results show promise and further work will involve optimizing the design of the pads in order to better adhere to a surface, for later use in a wall-climbing robot.

Keywords: Electro-adhesion; Wall Climbing; Electrostatic Adhesion

1. Introduction

In recent years many applications for wall-climbing robots have been suggested, from pipeline inspection and maintenance to large scale window cleaning. *Robug II*[1] used active pneumatic actuators to climb smooth surfaces. The robot was capable of lifting double its body weight and the technology was scalable to larger robots. However pneumatics are regarded as inefficient as a constant vacuum is required to maintain the low-pressure environment on the inside of the suction cup.

Magnetic wall-climbing robots[2] have been developed but share many of the problems of the pneumatic approach, and in addition are limited to climbing ferrous materials. In order to overcome the drawbacks of these methods, researchers investigated a biologically-inspired approach. *Sticky Bot*[3] utilized Van der Waals forces to adhere to walls in a way that simulated the fibrillar setal array found on geckos feet using a structure they termed Directional Polymer Stalks (DPS). *Sticky-bot* could easily climb a large variety of surfaces but had the drawback that the polymer used required high amounts of energy to detach from the surface being climbed.

The common issues that present themselves frequently are either a large detachment force or high power consumption to maintain adhesion. Recent research has demonstrated a new technology for wall-climbing, Electro-adhesion[4,5] was developed from techniques used in the optical manufacturing industry[6] to produce a new form of actuator. A highly compliant substrate is bonded with electrodes that carry a high voltage supply. Positive and negative charges accumulate on alternate electrodes creating an electro-adhesive effect. Prahlad[4] states that the compliance of the pad is key to achieving high adhesion forces.

Electro-adhesion brings many advantages to wall-climbing robots. As the pads can be turned off at will, the force required to detach them from a wall is relatively low, this reduces their power consumption as power is only used to adhere to a wall, which even so is in the order of nanoamps per newton of lateral force.[4]

This paper will investigate the electro-adhesive properties of different combinations of substrate, conductor and adhesive in order to predict their utility in a wall-climbing application.

2. Methods

We investigated combinations of candidate materials which might be used as good substrates, conductors and adhesives. Kapton, Mylar and Paper were used for the substrates as all are known for their good electrical insulating properties and are all easily obtainable. Aluminium Tape and Conductive Paint were chosen for the electrodes as both are frequently used in the electronics industry for static grounding and RFI Shielding, which suggested they are reliable and mechanically flexible electrodes; this latter point being important to achieve the compliance Prahlad suggests is key. Silicone and Latex are versatile adhesives that will adhere to the variety of materials used and both have high dielectric strengths.

A test rig was constructed as shown in fig 2, glass was adopted as the initial test wall as its smooth surface allows for a maximum contact area of the pads and minimises variation in testing. Pads were fabricated with two square electrodes of $40mmx40mm$ with a $15mm$ spacing between them (see fig 1), paying particular attention to keep a uniform pad thickness with all pads to ensure that this factor would not distort the results. This electrode design was used only for initial testing, charges gather in greater quantities at the border between electrodes so future work will implement a comb design to fully exploit this property by creating a larger pad perimeter to surface-area ratio. A series of initial trials were designed to test each

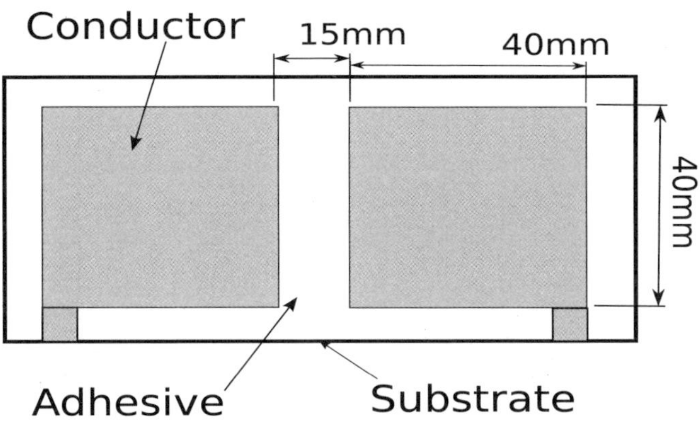

Fig. 1. A test pad

substrate, conductor and glue combination. Each pad was applied to the glass wall and charged using two high-voltage charge pumps[7] at $-10Kv$ and $+10Kv$ for twelve and a half minutes, at which point fifty gram weights were added every thirty seconds until the pad failed. The largest weight successfully held was recorded in table 2.

3. Results

Table 1. Factorial ANOVA of trials.

Source	Degrees of Freedom	F Distribution	Probability>F
Substrate	2	353.4117	$2.2e^{-}16$
Conductor	1	1.0116	0.3171
Adhesive	1	18.1909	$4.719e^{-}05$
Substrate : Conductor	2	0.3750	0.6883
Substrate : Adhesive	2	43.1219	$4.725e^{-}14$
Conductor : Adhesive	1	0.2409	0.6247
Substrate : Conductor : Adhesive	2	10.5355	$7.370e^{-}05$

Table 1 shows the output of a 3x2x3 way factorial Anova of the results shown in Table 2. An outlier was removed as to prevent distortion in the Anova test, this outlier was present in a test using Kapton film, Aluminium Tape and Silicone. The test pad did not show the flexibility of other pads but appeared slightly flatter than other pads, possibly allowing a better substrate contact area with the glass. The Conductor choice had no significant effect on results, this is validated by the conductors Probability>F

Table 2. Results of trials conducted on electrostatic adhesive plates

Substrate **Adhesive** **Conductor**	Kapton Silicone Ink	Kapton Latex Ink	Kapton Silicone Al Tape	Kapton Latex Al Tape	Mylar Silicone Ink	Mylar Latex Ink	Mylar Silicone Al Tape	Mylar Latex Al Tape	Paper Silicone Ink	Paper Latex Ink	Paper Silicone Al Tape	Paper Latex Al Tape
Test No.	Fail Weight (*grams*)											
1	200	150	450[a]	250	150	100	50	150	50	50	50	50
2	350	150	350	200	100	100	50	100	50	50	50	50
3	250	200	250	100	100	100	50	100	50	50	50	50
4	200	200	250	100	100	50	50	100	50	50	50	50
5	200	200	300	150	50	50	50	100	50	50	50	50
6	250	200	250	150	100	100	50	100	50	50	50	50
7	300	200	250	150	100	50	50	100	50	50	50	50
8	250	150	300	200	50	100	50	100	50	50	50	50
9	250	200	250	100	100	100	50	100	50	50	50	50

Note: [a] This outlier was removed from the data.

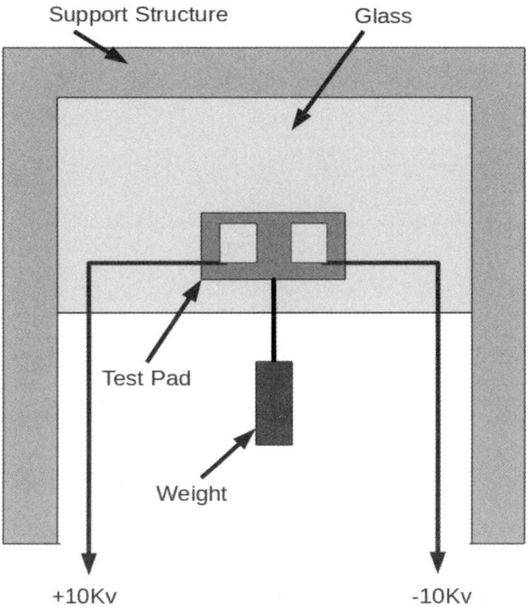

Fig. 2. The test rig

number being greater than 0.05 suggesting that it has no association with the results. There were significant main effects from choice of substrate and adhesive, this is shown by their respective Probability>F values being below 0.05. Of the substrates, Kapton was the most successful, followed by Mylar, while paper was largely unsuccessful and was unable to support the addition of the first weight, Fig 3 shows the distribution of results in respect to the substrate. We assumed that the adhesive only served to adhere a backing plate to the pad to stop deformation of the pad by asymmetric structure,[5] but the Anova tests show that the use of a silicon adhesive produced more successful pads in the trials. It is likely that the adhesive was acting as a strong insulator allowing the build of charge to be biased heavily towards the wall. This is further asserted with the Anova tests when the interaction between sources is examined. Two way interactions with the conductor have no effect. However the interaction with the substrate and adhesive is shown to be highly significant suggesting that the optimum substrate and adhesive have a net improvement on the capability of the pads. With the adhesives, we noted greater variability in the performance with Silicone compared with Latex (see fig 4).

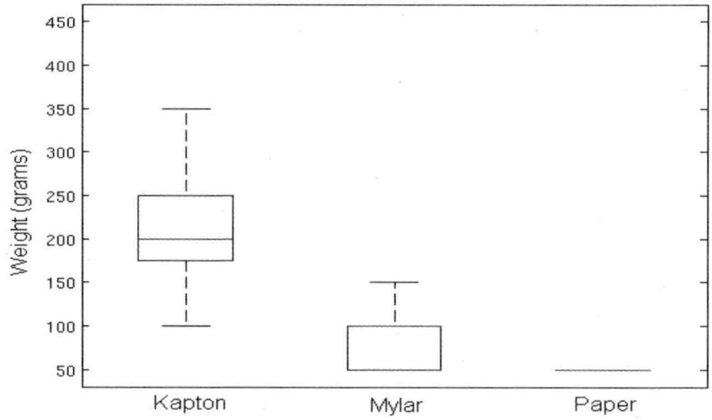

Fig. 3. The distribution of results in respect to the different types of substrate

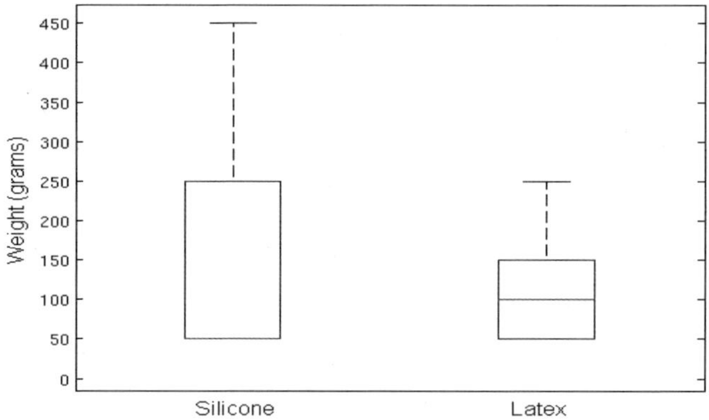

Fig. 4. The distribution of results in respect to the different types of adhesive

4. Discussion

These initial results show that a pad of 105*mm* by 60*mm* is capable of holding four hundred grams of weight without failing. Although this took twelve and a half minutes to charge, Yamamoto *et al*[5] note that when the comb design electrode is implemented, charge time is reduced from minutes to seconds. As the conductor has no significant effect on results, further work will focus on making the electrodes more durable and flexible, as silver conductive paint succumbs to mechanical wear after extended use and the aluminium tape showed little flexibility, reducing the compliance of the pad thus reducing the contact area of the pad. Further work will use highly flexible electrode materials such as gold leaf.

These tests have shown that Kapton film together with a silicone-based adhesive will produce the most effective electro-adhesive pad. Work is currently under way to take the materials from these tests and a new comb-design electrode to produce a series of pads that can act as tank-treads in a novel wall-climbing robot.

References

1. B. Luk, A. Collie and J. Billingsley, Robug ii: An intelligent wall climbing robot, in *Proceedings of the IEEE International Conference on Robotics and Automation*, April 1991.
2. Z. Xu and P. Ma, in *A wall-climbing robot for labelling scale of oil tank's volume*, *Robotica* (Cambridge University Press, New York, NY, USA, March 2002).
3. S. Kim, M. Spenko, S. Trujillo, B. Heyneman, D. Santos and M. Cutkosky, in *Smooth Vertical Surface Climbing With Directional Adhesion*, *IEEE Transactions on Robotics* **24**Feb 2008.
4. H. Prahlad, R. Pelrine, S. Stanford, J. Marlow and R. Kornbluh, Electroadhesive robots - wall climbing robots enabled by a novel, robust, and electrically controllable adhesion technology, in *Proceedings of the IEEE International Conference on Robotics and Automation (ICRA)*, May 2008.
5. A. Yamamoto, T. Nakashima and T. Higuchi, Wall climbing mechanisms using electrostatic attraction generated by flexible electrodes, in *International Symposium on Micro-NanoMechatronics and Human Science*, November 2007.
6. G. Monkman, Electroadhesive microgrippers *Industrial Robot: An International Journal* **30** (MCB UP Ltd, Mechatronics Research Unit, Fachhochschule Regensburg, Fachbereich Elektrotechnik, Regensburg, Germany, 2003).
7. q series dc-dc convertors from emco high voltage corporation online at www.emcohighvoltage.com, (2007), Sutter Creek, CA.

COMPACT CLIMBING ROBOT ROLLING ON FLEXIBLE MAGNETIC ROLLERS, FOR GENERATOR INSPECTION WITH THE ROTOR STILL INSTALLED

WOLFGANG FISCHER, GILLES CAPRARI, ROLAND SIEGWART

Autonomous System Lab, ETH Zürich,
Tannenstrasse 3, 8092 Zürich, Switzerland

ROLAND MOSER

R&D in Inspection Technologies, ALSTOM Power,
Brown Boveri Str. 7, 5401 Baden, Switzerland

This paper describes the design and prototype implementation of a compact climbing robot with magnetic adhesion, which is only 8mm high and allows for moving in any direction on the stators or rotors of power generators. Thanks to a new locomotion system based on flexible magnetic rollers (= magnetic wheels fixed on motorized flexible shafts; patented technology), the robot is able to span over gaps and non-magnetic zones both in axial and circumferential direction while still being flexible enough for adapting to the large curvature of the rotor/stator (~0.5m-2m). With this new design, it is now possible for the first time to access all types of large power generators (>100MW) with only one device – while still keeping the mechanical complexity at a level which allows for realizing the robot in a robust way and at reasonable cost. After a brief description of the industrial environment where it is supposed to be used (inspection of generator stators with the rotor still installed), this paper describes the basic mechanical concept, pointing to the core innovation and the advantages towards previous designs. The paper concludes with a prototype implementation and some test results; and provides an outlook on future extensions in the final industrial version.

1. Introduction

This paper describes the design and prototype implementation of a new robotic crawler for the inspection of stators in generators. These generators are installed in every power plant and have to be checked regularly for failures, to assure that the power plant works safe and reliable. As each day of outage can easily cost up to one million Euros, the most important goal for these inspections is to be as fast as possible. Thus, ideally the most time-consuming work step – the removal of the rotor – should be avoided by using inspection devices of very small size. These devices should to be small enough to fit between stator and rotor but still mobile enough to move along the stator geometry that is normally formed out of several rows of teeth (Fig. 1, c). In some generators, internal obstacles on the stator (zone-rings) can be found – which require an advanced robot mobility to move around them. In the generators with the most narrow entrance gap, the maximum allowed vehicle height is limited to 9mm (Fig. 1, b).

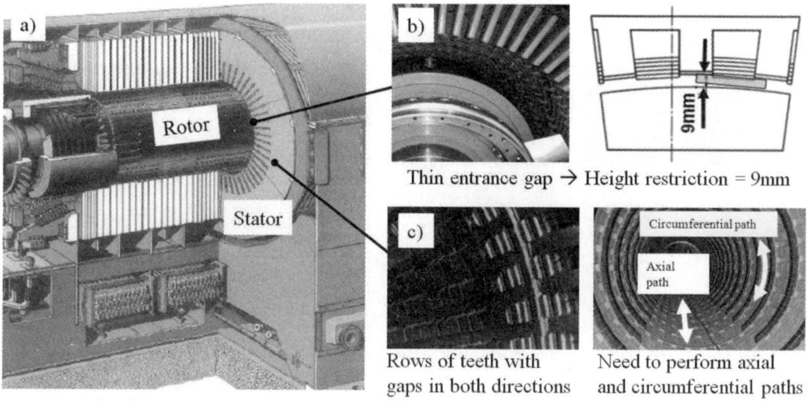

Fig. 1: Sketch of a generator (a), with the entrance geometry that limits the vehicle height (b), and the typical structure of the stator (c) that sometimes contains internal obstacles (zone-rings)

The robot which is presented in this paper is the first one which is able to deal with all these challenges at once, while still keeping a mechanical structure that allows for realizing it in a robust way and at reasonable cost.

The paper is structured is as follows:

Section 2 provides a brief overlook on existing climbing robots for generator inspection and shows their main limitations. The basic locomotion concept of our robot and its advantages towards previous systems is explained in Section 3. This section is followed by the realization of the prototype and the results from the first tests, concluding with an outlook to planned extensions in the final industrial version.

2. State of the art

For inspecting power generators, mainly robots with permanent magnets for adhesion and tracks for locomotion have been realized, such as the DIRIS Flex (Fig. 2-a, [1]) or the Siemens FastGen (Fig. 2-b, [2]). Both are too big for passing the most narrow entrance gaps ([1]: 50mm, [2]: 12.5mm) and cannot move on circumferential paths either. Another competitor product is the GE MAGIC (Fig. 2-c, [3]). This robot achieves axial and circumferential mobility by using three pairs of tracks, with a pneumatic spring in between that allows for spreading between stator and rotor. With this advanced mobility and a height of 12.7mm (0.5"), its application scope is already bigger than the two previously mentioned robots; however it is not yet able to pass the most narrow entrance gaps of only 9mm height. Another recent development is the DIRIS small (Fig. 2-d, [4]), which is a manipulator that pushes a flexible rod into the generator air gap and can move on circumferential paths on the end ring. With this system, it is possible to inspect generators with a 9mm entrance gap and theoretically even smaller, but generators with internal obstacles (zone-rings) cannot be accessed.

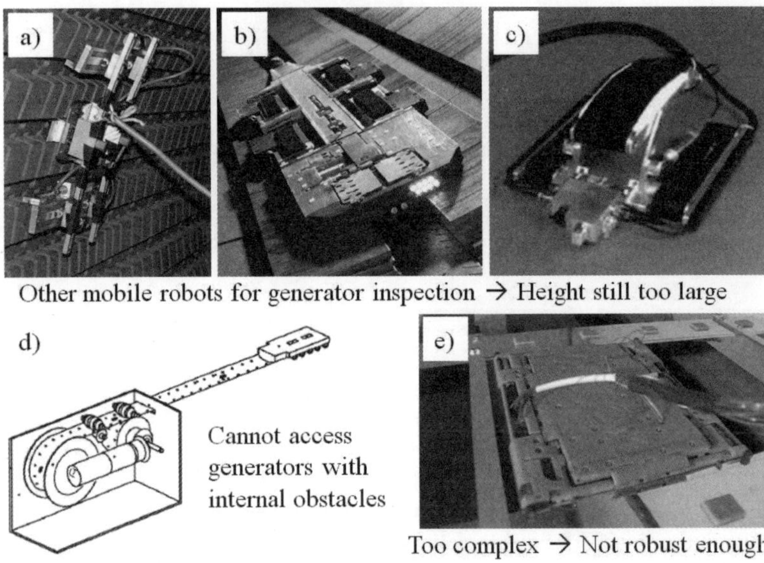

Fig. 2: Previous robotic systems for generator inspection and their main limitations: (a) DIRIS Flex [1], (b) Siemens FastGen [2], (c) GE MAGIC [3], (d) DIRIS small [4], (e) Previous design by our research team [5, 6]

With the goal to achieve the required low height with a mobile robot, a magnetic wheeled climbing robot has been developed by our research team (Fig. 2-e, [5, 6]). The significant reduction of height in comparison to the competitor products could be achieved by using several shafts with magnetic wheels for adhesion and locomotion (instead of magnets in the chassis and tracks), and spur gears for transmitting the torque to the wheels. For passing the gaps in circumferential direction, this robot can transfer its adhesion zone to a frame that runs parallel to the wheel unit and thus move to the next row of teeth in a way that is similar to inchworm-locomotion. The first prototype of this robot performed well in the test-environment; however its robustness was not regarded sufficient enough for real industrial application.

For this reason, the main goal for the here presented robot was set to achieve a mobility in the specified environment which is similar to our previous prototype, but with a mechanical structure that is simple, robust and reliable.

3. Basic mechanical concept

For achieving advanced mobility at very small height and complexity, the new concept is based on a combination of advantageous features taken from previously developed robots: Simple differential-drive-steering like in the GE MAGIC [3] (mobility in all directions at reasonable complexity); but magnetic adhesion for avoiding the spring system that spreads towards the other surface (difficult to realize below 9mm height, patented by GE).

This combination is realized in a new robot structure. It can be bent at a large radius (approximately 0.5-2m) but is yet stiff enough to span over small gaps or non-magnetic zones (up to 30mm). Motorized magnetic wheels are distributed along most of its area for providing adhesion and traction.

The structure consists of a flexible frame which holds two flexible traction units on both sides and the payload in the center – comparable to the structure of a catamaran-boat (Fig. 3, a). The flexibility of the traction units is achieved by flexible magnetic rollers – magnetic wheels glued to flexible shafts and connected to a gear-motor with spur-gears (Fig. 3, b). With this structure, the robot is able to adapt to any curvature between straight and the radius of the stator, while still being stiff enough for spanning over the gaps between the stator teeth and providing enough adhesion force towards the surface for not slipping – both on axial paths (Fig. 3, c, left) and on circumferential paths (Fig. 3, c, right).

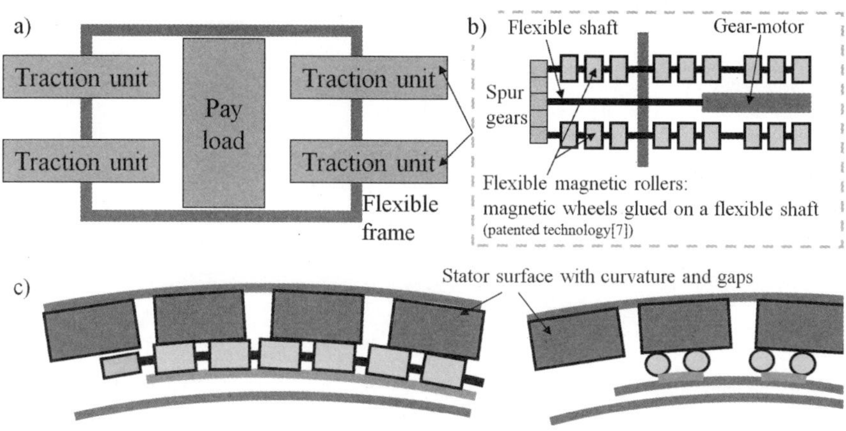

Fig. 3: Basic mechanical concept for the robot: (a) Entire vehicle structure, (b) Detail view of a traction unit with flexible magnetic rollers (patented technology [7], (c) Adaptation of the flexible magnetic roller to a curved ferromagnetic surface with gaps, as it is found in generator stators

With this innovative vehicle structure, a similar mobility as in the GE MAGIC can be achieved (allows for accessing generators with internal obstacles). Furthermore, the mechanical complexity is even lower, which allowa for significantly reduced height (no need for an actively actuated spring system for spreading anymore). Also in comparison to our own previous system [6] the mechanical complexity is reduced significantly.

4. Prototype realization

Based on this concept we realized a prototype for proving its functionality. It achieves a height of only 8mm, at 150mm length, 350mm width, and a mass of approximately 200g (see Fig. 4).

The frame is realized with a combination of thick parts (3-5mm) out of plastics and thin spring-steel bands (0.05…0.3mm) that are clamped between these plastic parts. By changing the free length and the thickness of these steel-bands, the stiffness can be adjusted.

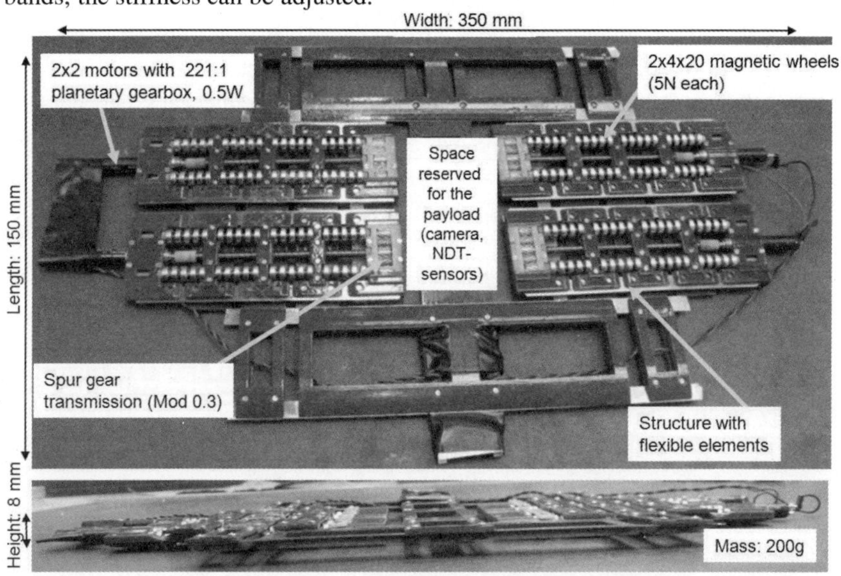

Fig. 4: First test prototype – size, mass and main features

Each traction unit consists of 2 flexible magnetic rollers with 20 ring magnets each (Ø3-6mm, b=2mm, F=5N) that are fixed on an Ø3mm flexible shaft. The torque is provided by Ø6 DC-motors with planetary gearboxes (Maxon RE6 with GP6, 4 stages, i=221:1, P=0.5W, T_nominal=30mNm) and transmitted to the rollers via a flexible shaft and a spur-gear-transmission (2x2 stages without reduction, mod 0.3, 12 teeth, helical shape, brass).

For increasing the stiffness against torsion and bending along an axis normal to the surface (necessary for optimally transmitting the motor torque to the wheels) while still keeping the flexibility to bend around an axis parallel to the driving direction (curvature-adaptation); the structure of the traction units is formed similar to the sandwich-design in skis and snowboards: stiff sections combined with flexible ones, and the intersection lines parallel to the driving direction. Similar to the frame, the stiff sections are manufactured out of relatively thick plastic parts, while the flexible ones are realized with thin spring-steel bands that are clamped between the stiff sections.

For carrying the planned payload (camera + sensors for non-destructive testing, approximately 100g of total mass), a section of approximately 100x50mm is foreseen in the middle of the frame. By changing the geometry of the frame, this space can be increased if necessary. Also a 3rd or 4th pair of traction units could be added, if heavy payloads need to be carried.

In this first prototype, the robot is still manually controlled by the operator – using 2 bi-directional switches that apply either positive or negative voltage on all traction units of each side.

5. Test results

After preliminary tests at ETH Zürich, the robot was then tested on real generators in the ALSTOM-factory in Birr (Switzerland) – on a rotor with 2m diameter (Fig. 5, a), an unpainted stator with approximately 1.5m diameter, and gaps up to 30mm (Fig. 5, b) and a painted stator with approximately 2m diameter, gaps up to 30mm, and paint drops of up to 2mm height (Fig. 5, c).

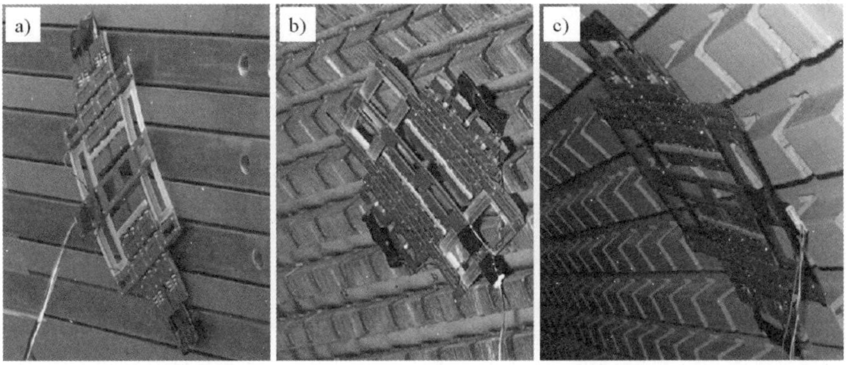

Fig. 5: Tests in real generators: (a) rotor, (b) unpainted stator of small diameter, (c) stator with paint drops up to 2mm height

In all these environments, the prototype was stable in all possible inclinations in respect to gravity and was able to carry twice the specified payload (200g – more was not tested, as there is anyway no additional space in this narrow environment). Regarding its mobility, all zones in the rotor and in the unpainted stator could be accessed without problems. Only on the painted stator the robot sometimes got blocked on some of the bigger paint drops (approximately 2mm height; driving back again was still possible). This limitation has to be mentioned, but should not be severe, as on these painted generators it is possible to place the robot either on the stator- or on the rotor-surface (using a special insertion-stick). When driving on the rotor, the robot cannot move completely to the stator, but is able to access it with its tools by using a small manipulator-arm that covers the distance (20-50mm inside the generator). The detailed design of such a manipulator-arm is planned for the next phase of this project.

The maximum driving speed was measured to approximately 1.5m/min, which is similar to our previous prototypes that use the same gear-motors [5]. With these speed values, the necessary time for scanning a generator of typical size (6m length, 60 rows of teeth) can be calculated to approximately 4-8 hours. This time is significantly lower than the necessary time for removing the rotor (0.5-1 day) and does not bring the risk of damaging it. For further increasing this speed at similar payload-capability, bigger gear-motors than the currently used ones (D=6mm, P=0.5W) would be necessary. Such a bigger version of the robot could be an option for larger generators where the entrance-gap is less narrow.

The inclination of the path (= going straight, upwards or downwards) only plays a minor effect (~10%) on the driving-speed, which is almost proportional to the input-voltage on the motors. For this reason, an almost constant speed can already be achieved without using encoders and motor-controllers.

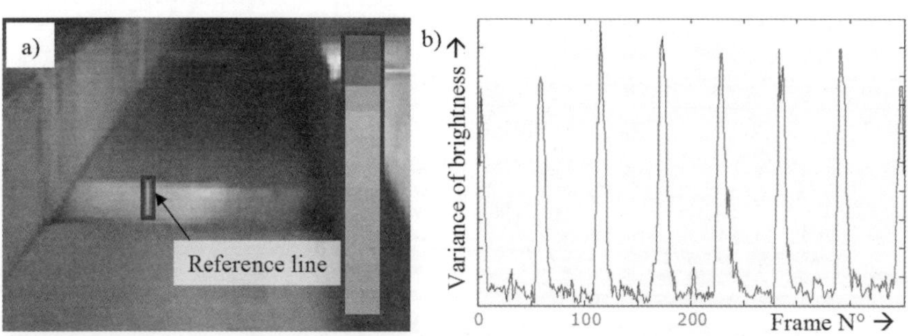

Fig. 6: Detection of the edges directly from the camera-image (a) Typical camera-image with a reference line which is 10 pixels long, (b) Variance of the color-value along the reference line for 400 frames in a movie where the robot is rolling in axial direction at constant speed. The peaks in the graph clearly indicate when the camera is above the edge of a stator tooth.

For the localization of the robot, it is planned to avoid additional sensors that increase its complexity and only use the image of the onboard-camera. These images are usually in very good quality, as there is no other light except the LEDs integrated in the camera. Furthermore, the environment is highly structured and contains many edges that are easily detectable with standard line-extraction-algorithms. Preliminary tests with a very simple program for detecting the edges of a typical camera-image-stream (still taken with the DIRIS small [4]) already showed quite promising results (see Fig. 6).

Ongoing work now consists in developing a program that implements an adapted line-following-algorithm for keeping a straight path, counts all edges for localizing the robot's position and automatically generates a high-resolution-image of the generator where the operator can see potential failures much faster than by looking at the camera-stream alone.

6. Conclusion and outlook to future work

In this paper, we have presented the design and prototype implementation of a compact climbing robot with magnetic adhesion, which is only 8mm high and allows for moving in any direction on the stators or rotors of power generators at a relatively low system complexity, thanks to a new locomotion system based on flexible magnetic rollers. The robot was tested in real environment, where it performed well and could clearly show its advantages towards previous systems.

With this new design, it is now possible for the first time to access all types of large power generators (>100MW) with only one device – while still keeping the mechanical complexity at a reasonable level that allows for realizing the robot in a robust way and at reasonable cost. This ability will make it a valuable tool for most type of generator inspections, and help to significantly decrease the outage-time of power plants during inspection. For this reason, the core innovation was also patented.

Ongoing work towards the industrialization of this robot mainly includes the optimization of the structure for more robustness (Al and stainless steel instead of plastics) and the implementation of the control-program.

Acknowledgments

This work was supported by grant KTI 8435.1 EPRP-IW of the Swiss "Bundesamt für Berufsbildung und Technologie" and ALSTOM Power, Baden, Switzerland. We also want to thank all people at ALSTOM Power and ALSTOM Inspection-Robotics who supported us during the design-phase and the tests, especially Niklaus Hugi, Reinhard Fischer, Markus Wiesendanger and Igor Thommen.

References

1. R. Moser, B. Mark; "Automated Robotic Inspection of Large Generator Stators"; in Proceedings of the IEEE/ASME International Conference on Advanced Intelligent Mechatronics; September 4-7, 2007, ETH Zurich, Switzerland.
2. D. Schoeler, P.O. Miranda; "Improved Unit Reliability & Availability Through Optimised Predictive Maintenance"; Siemens AG, Power Generation; 2003.
3. D.T. Roney, R. J. Zawoyski; "Generator In-Situ Inspections"; GE Power Systems, found on http://www.gepower.com/prod_serv/serv_for/generators/en/testing_insp/ndt.htm
4. R. Moser, N. Hugi, M. Bernhard, P. Isler, S. Honold, J. Erni, "Vorrichtung zur Inspektion eines Spaltes", Patent EP 2 110 678 A1, Alstom Technology Ltd., 14.4.2008.
5. W. Fischer, G. Caprari, R. Siegwart, R. Moser, "Robotic Crawler for Inspecting Generators with Very Narrow Air Gaps", in Proceedings of IEEE-ICM 2009, IEEE-IES International Conference on Mechatronics, Malaga, Spain, April 2009.
6. W. Fischer, G. Caprari, R. Siegwart, R. Moser, "Locomotion System for a Mobile Robot on Magnetic Wheels With Both Axial and Circumferential Mobility but only 8mm Height for Generator Inspection With the Rotor Still Installed", accepted for publication at Transactions on Industrial Electronics, 2011, not published yet.
7. Markus Wiesendanger, Wolfgang Fischer, "DEVICE FOR INSPECTION OF GENERATOR AIR GAPS", Patent US 61/296,199, ALSTOM Technology Ltd., Filing Date: 19.01.2010.
8. W. Fischer, G. Caprari, R. Siegwart, I. Thommen, W. Zesch, R. Moser, " Foldable magnetic wheeled climbing robot for the inspection of gas turbines and similar environments with very narrow access holes", Industrial Robot, Vol. 37, Issue 3, pp. 244-249, 2010.

PARASWIFT – A HYBRID CLIMBING AND BASE JUMPING ROBOT FOR ENTERTAINMENT

L. GEISSMANN, M. DENUDER, D. KEUSCH, L. PFIRTER, D. RÖTHLISBERGER
ETH Zürich, 8092 Zürich, Switzerland

M. RITTER, P. THOMA
ZHAW, 8401 Winterthur, Switzerland

R. SIEGWART, W. FISCHER, G. CAPRARI, J. WEBER
Autonomous Systems Lab, ETH Zürich,
Tannenstrasse 3, 8092 Zürich, Switzerland

P. BEARDSLEY
Disney Research Zürich, Clausiusstrasse 49, 8092 Zürich, Switzerland

This paper introduces Paraswift, a mobile robot that is able to climb an ordinary wall and deploy a paraglider for a remote controlled return to ground. The goal is entertainment and technical education through an unusual, eye-catching robot. Multiple requirements must be met – to provide a mechanism that generates strong adhesion for climbing yet is low weight for flying, to ensure a reliable transition from climbing to flying, and to handle collision forces on landing – in a single compact robot. The climbing technology is vortex adhesion with wheeled locomotion. The paraglider is folded into the robot shell on ascent and deployed at launch time using a novel mechanism based on a 2-DOF manipulator arm. Flight is remote controlled, and the robot has a protective frame of glass fiber reinforced plastic with a hard foam core to absorb collision forces on landing. This paper describes our work on the complete system, starting with the design, simulation, and physical testing of individual components, and culminating in the integration phase with successful climbing and flying on multiple walls of varying characteristics. We believe that Paraswift is the first demonstration of a compact robot that is capable of vertical climbing and passive flying.

Keywords: Climbing Robot, Flying Robot, Vortex Adhesion.

1. Introduction

Paraswift is a climbing and flying robot that provides an eye-catching demonstration for entertainment and technical education. The concept is shown in Figure 1 – the robot climbs a building wall, deploys a paraglider, and launches from the wall to make a remote controlled descent to earth. An onboard wireless camera transmits video to a display at ground level, allowing spectators to see the robot eye view. As an entertainment experience, Paraswift is a base jumping robot that draws on public interest in extreme sports. As a technical education experience, Paraswift demonstrates interesting problems and solutions about how to combine climbing and flying in what we believe is the first compact robot to combine vertical climbing and passive flying in one system.

Figure 1. (a) Climbing an outdoor wall, (b) live video from an onboard camera is transmitted to a screen at ground level, (c) flying under remote control.

In addition, the work has practical potential. The most reliable current technology for climbing robots is magnetic adhesion, which works for ferromagnetic surfaces such as metal tanks and pipes in industrial settings. There is also a variety of research on climbing techniques for ordinary walls, but this is a more difficult problem because the surfaces do not offer a straightforward adhesion mechanism like magnetic force, and ordinary walls have a range of characteristics like indentations and flakiness. Thus falling is a significant risk for an untethered robot that is not using magnetic adhesion. This paper demonstrates a deployable paraglider as a possible safety mechanism for climbing robots. This is not the topic of the paper and there would be many research challenges to address, but it can be proposed as a realistic concept for some applications and settings given the presented results.

2. Related work

There are a number of technologies for creating an adhesion force in order to climb vertical or inclined surfaces. Previous work includes suction [1-5], magnetism [6], artificial gecko hairs and dry adhesion [7-10], and electro-adhesion [11]. Suction can be achieved in two ways, either using a closed vacuum chamber that requires a seal between the robot and the surface [1-3], or using vortex generation that does not require a seal [4, 5]. Magnetic adhesion is a reliable technology that is being used in real world applications such as mobile robot inspection in industrial plants, but it is confined to ferromagnetic surfaces. Gecko hairs and dry adhesion are promising technologies but one of the challenges is that dirt and dust accumulate on the climbing pads and the adhesion force diminishes rapidly. Furthermore, these approaches need a large contact area with the surface e.g. with legs [7], whegs [8, 9] or tracks [10], and simple wheels with only line contact are insufficient. Electroadhesion is a promising approach but is difficult to realize due to limited availability of special materials.

3. Paraswift Robot

3.1 Overview

The first stage of the work was to specify goals and constraints for the robot:

- Climb on unmodified building walls
- Fly under remote-control
- Land on unprepared ground
- Onboard camera
- Stylish appearance and good visual experience for spectators

The principal functions are to climb, to transition between climbing and flying, to fly, and to land on an ordinary ground surface. For the climbing mechanism, we required an adhesion technology that would work on a variety of building surfaces and which was not heavy. Suction was the most suitable choice, along with motorized wheels for locomotion. Suction works on varied surfaces, does not damage or mark the surface, and is an economical choice. This required a further decision between a vacuum chamber and a vortex. Vortex adhesion was the best choice for two reasons – (a) the ratio between adhesion force and mass is superior to vacuum so that the robot can have lower weight, (b) no sealing is required between the robot and the wall so that it is easier to climb on walls with geometric irregularities.

For flying, a paraglider was chosen because of its self-stabilizing character and maneuverability at low velocities, and because it is foldable which allows it to be hidden inside the robot while it is climbing on the wall.

The major unknown in the initial stage was the transition between climbing and flying because this is not a problem with standard solutions and it could not be modeled in simulation. In the rest of the paper, sections 3.2 - 3.5 describe the principal functions in more detail. Section 3.6 describes electrical control and the onboard camera.

3.2 *Climbing*

The climbing mechanism is shown in Figure 2a. Adhesion during the climbing phase is achieved using an impeller that creates a vortex with a low-pressure area at the center as illustrated in the simulation in Figure 2b. A non-sealing carbon fiber case around the impeller keeps the kinetic energy of the vortex inside the system. There is a gap of about 4 mm between the case and the wall, which avoids energy loss due to friction and allows driving over small geometric irregularities. The wheels are standard RC model parts with a friction coefficient of around 1.0 on typical wall surfaces.

The first tests were performed with an impeller from a vacuum cleaner with an outer diameter of 125 mm. Computational Fluid Dynamics (CFD) simulations with ANSYS CFX correlated with these tests and the system was optimized according to our needs. In the implemented climbing mechanism, the case is made of a carbon fiber reinforced polymer structure, weighs 35 g and has size Ø260 mm x 30 mm. The impeller is made of sheet metal and driven by a lightweight brushless motor. The number and form of the rotor blades were determined in the optimization. The impeller weighs 70 g and has size Ø175 mm x 16 mm. The adhesion force is 15 N at a rotational speed of 2900 rpm and energy consumption of 25 W, and 40 N at 7200 rpm and 80 W.

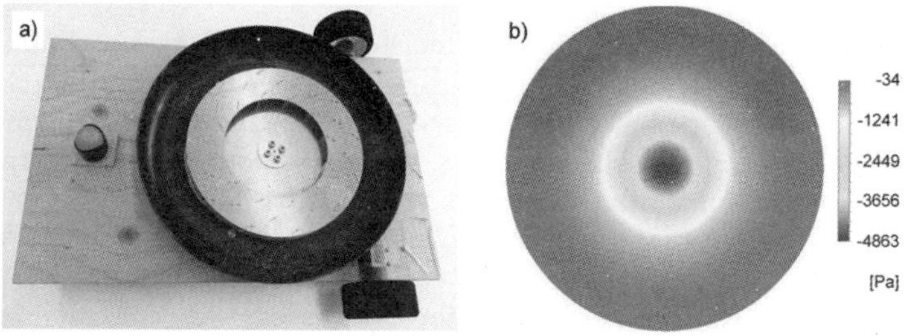

Figure 2. (a) The climbing mechanism of the robot, showing the carbon fiber case, the impeller within the case, and two motorized wheels for locomotion. (b) CFD simulation of the pressure distribution inside the case, with values relative to the environment.

Safety – The case is covered with a grating to prevent accidental contact with the impeller. In addition, the attachment of the impeller to the motor is secured with adhesive, and the robot was handled in operation using safety gloves.

3.3 *Transition from Climbing to Flying*

The paraglider weighs 200 g and has an area of 0.85 m^2 (HobbyKing Paraglider Parafoil 2.15 [12]). The goal of a minimum deployment height of

10 m led to a decision to actively open the paraglider before leaving the wall so that the robot would have more time to reach stable flight.

The paraglider is folded into the robot shell during climbing as shown in Figure 3a. Bendable carbon fiber rods act as flexural springs that are compressed when the paraglider is folded into the shell, and a securing cord holds the compressed rods in place. Deployment is achieved using a lightweight 2-DOF manipulator-arm with two servos as joints. After the manipulator-arm is expanded in a horizontal position, the securing cord is removed and the rods span the paraglider on its entire width, but it is still folded lengthwise around the manipulator-arm as shown in Figure 3b. Now the manipulator-arm can fully expand and by doing so it opens the paraglider completely as shown in Figure 3c. In this position the paraglider lies loosely on the manipulator arm, and nails through eyelets in the paraglider prevent it from slipping off sideways.

As soon as the vortex is turned off the robot quickly falls away from the paraglider because of its lesser air resistance, the lines tauten and the paraglider takes the robot away from the wall. At this point the robot folds the manipulator-arm back in to protect it during the landing.

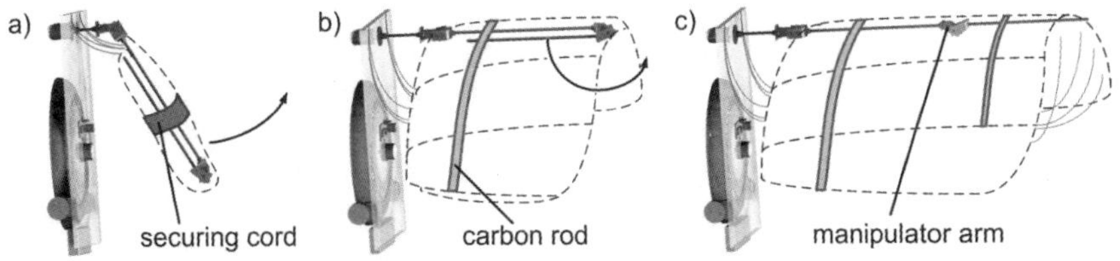

Figure 3. Deployment of the paraglider. (a) The folded paraglider held together by securing cord, (b) the manipulator-arm is half expanded, the securing cord is released and the previously bent carbon rods span the paraglider's entire width, (c) the manipulator-arm is now fully expanded and the paraglider is completely unfolded.

3.4 Flying

The steering mechanism for the paraglider consists of two remotely controlled servos accessing the left and right sets of the paraglider steering lines. This allows control of both banking and the angle of attack of the paraglider before landing.

Figure 4 gives the results for our initial tests on descent velocity versus payload for a paraglider of area 0.85 m^2. This determined that average vertical velocity of the implemented robot would be 4-5 m/s.

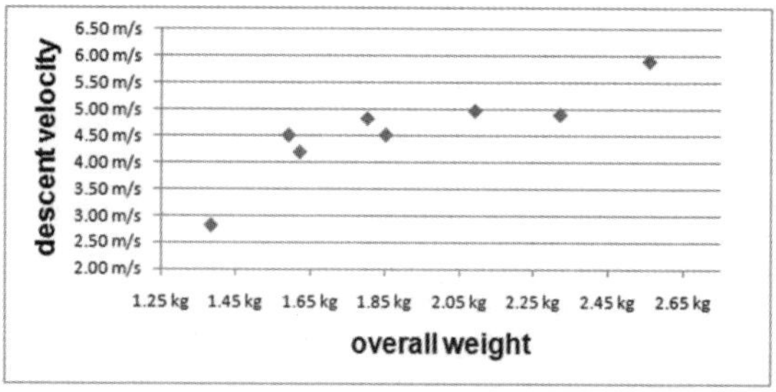

Figure 4. Average vertical velocity of a dummy payload with a paraglider area of 0.85 m^2 for different payload weights.

Safety – Spectator safety during flying is ensured using a safety cord that is attached to the robot and anchored at the bottom of the wall. The rope weighs 15 g so this has no effect on the flight characteristics of the robot (the robot's weight is 1.7 kg).

3.5 Landing

The robot has a protective frame to prevent sensitive components from being damaged on landing. The CAD design is shown in Figure 5a, with the frame being made out of glass fiber reinforced plastic covering a hard foam core. Cushions filled with polystyrene beads provide additional damping for the landing area on the base of the frame (the two struts on the right-hand side in Figure 5a). The frame is also designed to protect the robot in the case of an uncontrolled crash. The frame is covered with the external design shell shown in Figures 5b and 5c, for visual impact. While the robot is climbing, the paraglider is hidden within the design shell. During the launching stage, the front of the shell opens and the paraglider deploys as shown in Figure 5b. The design shell closes again during flight as shown in Figure 5c.

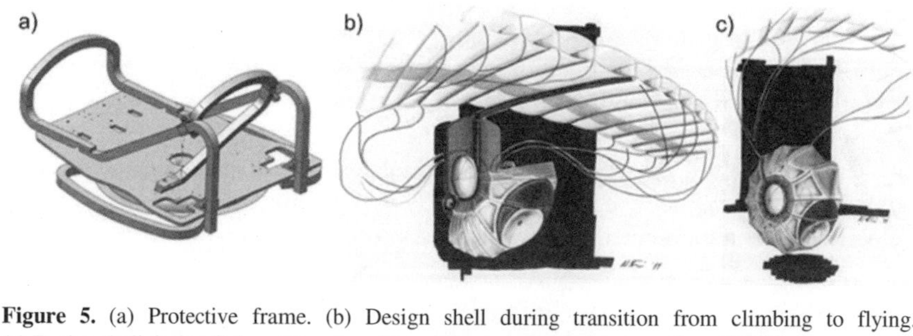

Figure 5. (a) Protective frame. (b) Design shell during transition from climbing to flying. (c) Design shell during flying. (Drawings 5b and 5c made by design team from ZHdK)

3.6 *Electronics and Camera*

The electronics are based on a Skybotix Coax 3 board to provide a microcontroller, IMU, and altitude sensor. The power source is an 11.1 V LiPo battery, which enables three cycles of base jumping with a 15 m climbing height. The robot is steered – both the wheels and the paraglider lines – with a conventional remote control. Signals are received at 2.4 GHz on the Coax board and are transmitted via SPI to the controller. The drive motors are accessed by RS-232.

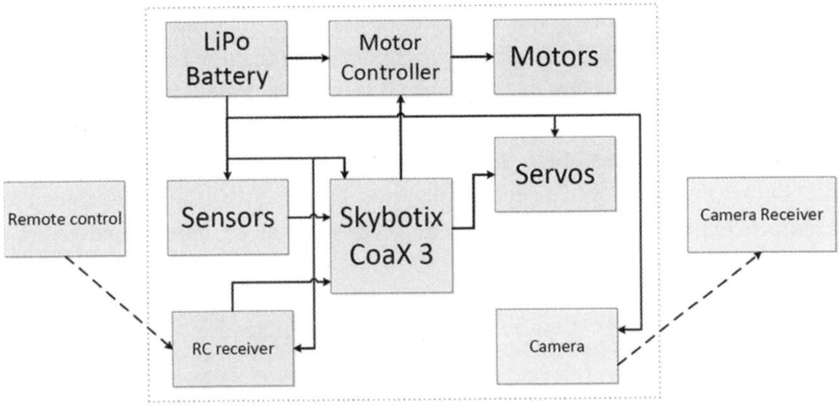

Figure 6. Hardware diagram

The onboard camera is an AXIS M1011-W, mounted on a custom servo-driven pan/tilt mechanism on the robot. The lens has a 20° field of view. The robot is stopped on the wall at a desired height, and the camera is used to capture approximately 200 images for varying pan/tilt angles that are sent by live-stream transmission to a ground laptop. The stitching software PTStitcherNG [13] is used to create a wide-angle panorama of the robot eye view, in typically less than 10 s processing time.

4. Results

The robot is shown in Figure 7 in configurations without the paraglider, with the folded paraglider, and with the design shell. It weighs 1.76 kg without the design shell and is 450 mm long, 300 mm wide and 230 mm high. It weighs 2.18 kg with the design shell. The design shell has been completed as a static shell, and automatic opening and closing of the shell during the transition from climbing to flying was not yet implemented.

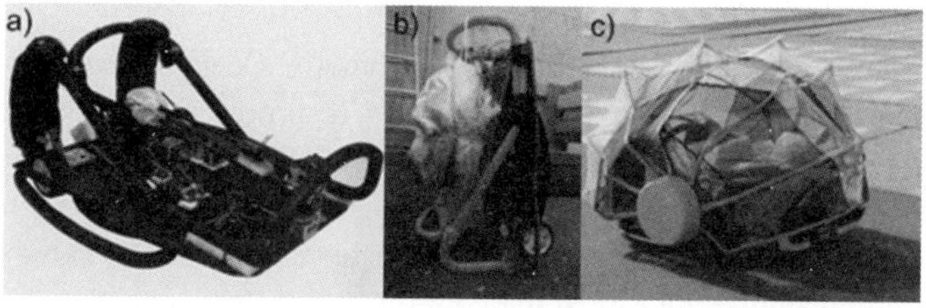

Figure 7. (a) The robot. (b) With folded paraglider. (c) With design shell.

Climbing was successful on multiple surfaces such as brick, concrete, wood and glass. Climbing speed is up to 25 cm/s and the robot can handle rough surfaces and geometric irregularities up to 5 mm. The robot is capable of agile changes of direction during climbing under remote control and can climb down walls backwards or head first. Some examples of climbing on different surfaces are shown in Figure 8. The full cycle from climbing to landing is shown in Figure 9, using a red carpet that was secured from an upper window as a climbing surface.

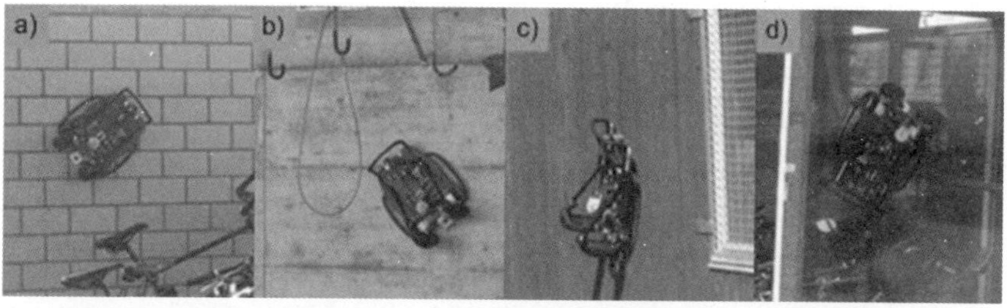

Figure 8. The robot climbing on (a) brick, (b) concrete, (c) wood, (d) acrylic glass.

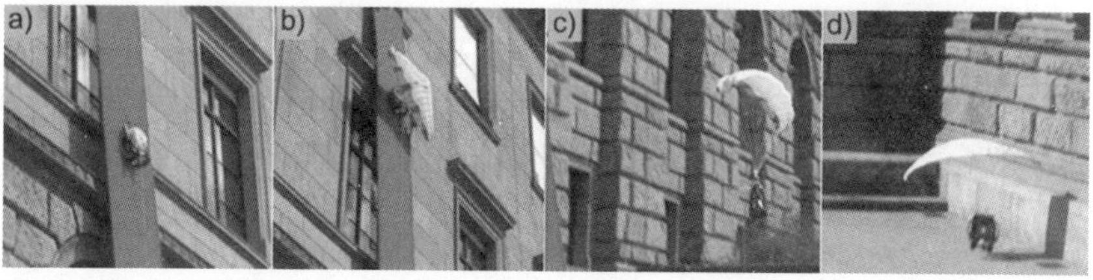

Figure 9. The four key stages of climbing, transition, flying, and landing. The hanging cord visible in (a) and (b) is a lightweight safety cord to constrain the maximum range of the flying robot.

A minimum jump altitude of 10 m was sufficient to allow the robot to reach a stable flight and land safely with a flight time of about two seconds. Steering ability during the flight is limited because the flying speed is low, and attempting to make a large change in direction can cause the paraglider to stall.

But the steering mechanism successfully fulfills the primary goal that it can reliably steer the robot away from the wall.

Most of the testing was done with a safety net but there were also multiple successful landings directly on hard ground surface. One accidental crash of the robot during testing showed that the protective frame was robust enough to prevent damage to the robot for a fall of 12 m onto tarmac, except for small cracks in the frame.

Discussion – As anticipated, the transition from climbing to flying is the key research challenge. Transition has two stages (a) the paraglider is deployed while the robot is still adhering to the wall and (b) the vortex is turned off, the robot falls, and the paraglider starts to fly. Transition is reliable in low wind of up to 5 km/h. Stronger breezes affect stage (a) by pushing the paraglider against the wall, or tilting the whole paraglider. Possible ways to add robustness to transition are to modify the deployment mechanism that supports the paraglider before flight, and to have an anemometer to test that flying conditions are within acceptable parameters. This was not part of the current work.

5. Conclusion

This paper has presented a climbing and base jumping robot with an innovative design that strikingly illustrates the possibilities of recent technology in climbing robots. For the first time, vertical climbing and passive flying are realized in a single lightweight robot that balances the requirements of both functions. With its unique functionality and appealing design, the robot represents an attractive showpiece for entertainment and education about the new generation of mobile robots.

The climbing mechanism of vortex adhesion is lightweight and useful for other applications such as climbing robots that do inspection. The primary research challenge for future work is in the transition from climbing to flying, and ideas for future work were proposed. An interesting possibility is to develop the transition mechanism to handle not just controlled launch of the robot from the wall but also accidental falling.

Acknowledgements

We thank our partners at ZHdK (Zurich University of the Arts) – Nadia Frei, Christa Rüetschi and Thierry Sollberger – for their important input regarding the design. We thank the following people at ETH Zürich for technical support – Stefan Bertschi (Electronics), Marco Steiger & Martin Schütz (CAD-PLM), Markus Bühler & Dario Fenner (Mechanical Workshop), and Roland Haas (Coordinator and Project Coach). We thank Maurizio Nitti of Disney Research Zurich for concept art. This work was sponsored by Disney Research.

References

1. C. Balaguer, A. Gimenez, and CM Abderrahim, *ROMA robots for inspection of steel based infrastructures*, Industrial Robot: An International Journal, 29(3):246–251, 2002.
2. C. Hillenbrand, D. Schmidt, and K. Berns, *Cromsci-A Climbing Robot With Multiple Sucking Chambers For Inspection Tasks*, 11th International Conference on Climbing and Walking Robots and the Support Technologies for Mobile Machines, Coimbra, Portugal, page 311, World Scientific Pub Co Inc, 2008.
3. J. Xiao, A. Sadegh, M. Elliott, A. Calle, A. Persad, H. Chiu, "Design of Mobile Robots with Wall Climbing Capability", Proceedings of the 2005 IEEE/ASME International Conference on Advanced Intelligent Mechatronics Monterey, California, USA, 24-28 July, 2005.
4. L. Illingworth, D. Reinfeld, "Suction cup vortex attractor", Patent US 6881025, signed by Vortex HC, LLC, April 19th 2005. The company website (www.vortexhc.com) does not exist anymore, only old videos can be found (keyword: VRAM).
5. F. Bonaccorso, C. Bruno, D. Longo, G. Muscato, "Structure and model identification of a vortex-based suction cup", Proc. of The 11th International Conference on Climbing and Walking Robots and the Support Technologies for Mobile Machines (CLAWAR), September 2008.
6. F. Tache, W. Fischer, G. Caprari, R. Moser, F. Mondada, R. Siegwart, *Magnebike: A Magnetic Wheeled Robot with High Mobility for Inspecting Complex Shaped Structures*, Journal of Field Robotics, Vol. 26, May 2009, 453-476.
7. S. Kim, M. Spenko, S. Trujillo, B. Heyneman, D. Santos, and M.R. Cutkosky, *Smooth vertical surface climbing with directional adhesion*, Robotics, IEEE Transactions on, 24(1):65–74, 2008.
8. M. Murphy, M. Sitti, "Waalbot: Agile Climbing with Synthetic Fibrillar Dry Adhesives", Proc. of the 2009 IEEE International Conference on Robotics and Automation (ICRA), Kobe, Ja-pan, May 12-17, 2009.
9. H. Tsukagoshi, A.Kitagawa, "Gel-Type Sticky Mobile Inspector to Traverse on the Rugged Wall and Ceiling,", Proc. of the 2009 IEEE International Conference on Robotics and Auto-mation (ICRA), Kobe, Japan, May 12-17, 2009.
10. O. Unver, M. Sitti, "Tankbot: A Miniature, Peeling Based Climber on Rough and Smooth Surfaces", Proc. of the 2009 IEEE International Conference on Robotics and Automation (ICRA), Kobe, Japan, May 12-17, 2009.
11. H. Prahlad, R. Pelrine, S. Stanford, J. Marlow, and R. Kornbluh, *Electroadhesive robots,wall climbing robots enabled by a novel, robust, and electrically controllable adhesion technology*, In Robotics and Automation, 2008. ICRA 2008. IEEE International Conference on, pages 3028–3033. IEEE, 2008.
12. www.hobbyking.com
13. PTStitcherNG, http://webuser.hs-furtwangen.de/~dersch/

CY-MAG^{3DE}: MAGNETIC CLIMBING INSPECTION ROBOT

FREDERIC ROCHAT, PATRICK SCHOENEICH, OLIVIER MARTI, HANNES BLEULER, FRANCESCO MONDADA

EPFL STI IMT LSRO MOBOTS, École Polytechnique Fédérale de Lausanne Lausanne, CH-1015, Switzerland

Cy-Mag^{3De} is a magnetic climbing inspection robot with advanced mobility. The robot is based on an innovative, award-winning and patented magnetic design. Its mobility is enhanced thanks to the addition of a novel active tail. The robot has a dumbbell shape of 80 mm in diameter and 233 mm in length and a weight of 1.6 kg. It is energy autonomous with batteries distributed among the robot subsystems. The robot has an embedded camera and can broadcast wireless video feed-back to the remote user interface. The robot is highly integrated and robust. Cy-Mag^{3De} achieves excellent mobility with remote visual inspection in complex ferromagnetic environment. The resulting mobility and the achieved integration open new possibilities for the application of magnetic climbing inspection robots.

1. Introduction

Complex ferromagnetic environments are common inside power-plants. Some particular areas require regular and compulsory inspection. Currently costly and long-lasting dismantling time is necessary for gaining access for human inspectors. In the future, small sized mobile magnetic climbing robots are expected to carry out those inspections and reduce the shut down time due to dismantling. Such robots need to be sufficiently compact and able to overcome complex obstacles safely.

There already are some rare examples of wheeled magnetic climbing robots with high mobility: the best example is shown in [1], this reference enlists other interesting examples.

In this paper a novel magnetic climbing robot called *Cy-Mag^{3De}* is presented. Its general shape can be seen in Figure 1. The robot has a dumbbell shape with one wheel at both ends. Each wheel embeds its own magnetic adhesion system, actuator and energy. In the next sections, the particular locomotion principle is explained. The properties of the magnetic system, the design of the robot with the addition of the tail to enhance the mobility and the locomotion results are detailed.

Figure 1: (a) the novel Cy-mag^{3De} magnetic climbing robot with its tail and camera for visual inspection surrounded by a LED ring; (b) the Cy-mag^{3D} that previously proved the general mobility concep.

2. Locomotion principle

Magnetic wheels are not the most efficient magnetic systems, as most of their volume does not contribute to the adhesion force, even when they are optimized [2]. Moreover standard magnetic wheels get stuck in inner corners and usually require additional degrees of freedom to overcome such obstacles [1, 3].

A solution to solve both problems simultaneously is to have a magnetic system in the shape of a sector of ring that can rotate inside an outer wheel. At the cost of a larger air-gap due to the outer wheel construction, the part of the magnetic circuit that was not contributing to the adhesion force is suppressed. Passing inner corner can be done by the rotation of the magnetic system. This would require an additional degree of freedom, but its activation requires less force and energy than the mechanical detachment of a wheel in an inner corner, and thus can be done in a more compact manner. This additional degree of freedom can be skipped thanks to the magnetic properties. This is explained in the next section.

The resulting unique working principle for obstacle passing of the robot is shown in Figure 2. The mobility innovation of this configuration is that the holding torque - preventing the magnet to move freely - is magnetic. When the motors activate the wheels forward, the body is hold by the magnetic torque and the robot moves forward until it encounters an obstacle. When the robot is in front of an obstacle - as an inner plan transition for example - the body rotates backward, commuting the magnetic adhesion from one surface to the other. This innovative design allows simple overcoming of inner corner. This principle was already embedded in a miniature robot shown in Figure 1 (b) and proved to have excellent mobility in [4]. In order to design a robot with true inspection capability and larger payload, the next step was to prove the scalability of the adhesion force and validate this unique obstacle passing mechanism for a larger robot.

Simulation and experience shows that a magnetic system with the shape of the sector of a crown exerts a calling torque at its both ends. The results of simulations in force and torque with the system used on the final design of the robot are shown in Figure 3. The torque necessary to reorient the magnetic system in an inner corner is shown in Figure 4.

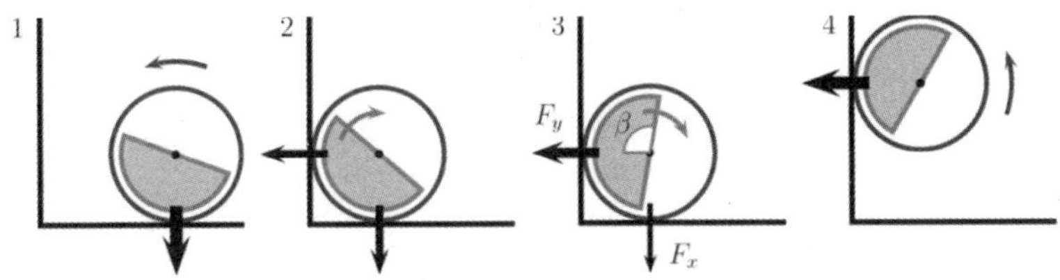

Figure 2 : Description of the unique mechanism for obstacle passing. The motor drives the relative position between the magnet and the wheel. On flat surfaces, the magnetic torque prevents the magnet to rotate and thus the wheel moves forward. Arriving in an inner corner, the wheel is blocked, consequently the magnet rotates in the opposite direction and the force is transferred to the new surface. When the transition is finished, the wheel is free to move again, and the robot can continue its journey.

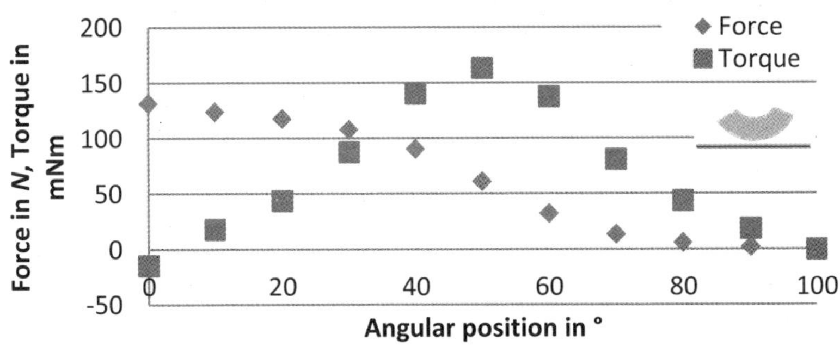

Figure 3: force and torque induced by the magnetic system on a plate. The angular position has its origin with the magnetic system in the symmetric position oriented toward the plate.

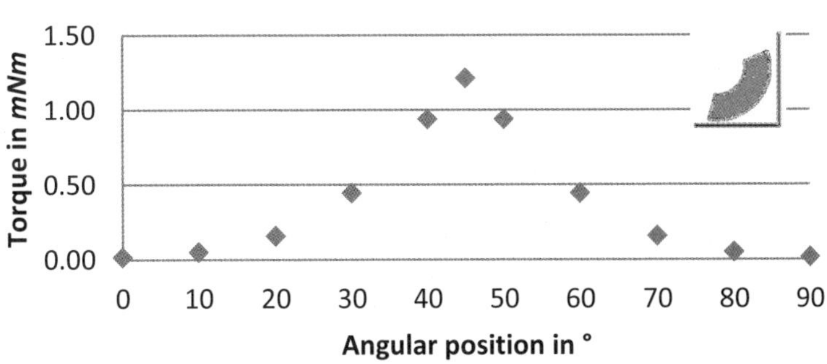

Figure 4: The torque required for the rotation of the magnetic system in a straight inner angle.

3. Design of the wheel module

The design choice was to build a well encapsulated wheel module with the magnetic system and room for control electronics, battery and motorization as seen in *Figure 5* (a). An air-gap is necessary for the implementation of the mobility concept because the magnetic system moves inside an outer wheel. The implemented air-gap is of $3\ mm$. Working with this air-gap results in a lower force, but has the advantage that the force drops less quickly in case of air-gap variation. This allows a robust construction of the wheel, including a rubber layer to improve the friction coefficient and protection against magnetic dust.

The final magnetic adhesion sub-system is shown in *Figure 5* (b). Its angular section in respect to the wheel is $100°$. This value is the most suitable for overcoming an inner straight corner. The magnet has the following dimensions: the external radius is $37\ mm$ and the inner radius is $19\ mm$. The height of the magnet is $14\ mm$. The magnetic circuit is completed with two iron guides placed on each pole of the magnet. Their maximum height is $7\ mm$. Simulations were done using a $3D\ FEM$ analysis software to validate the suitability of the designed magnetic system to pass inner corner and bear the weight of the robot with a sufficient safety margin.

(a)

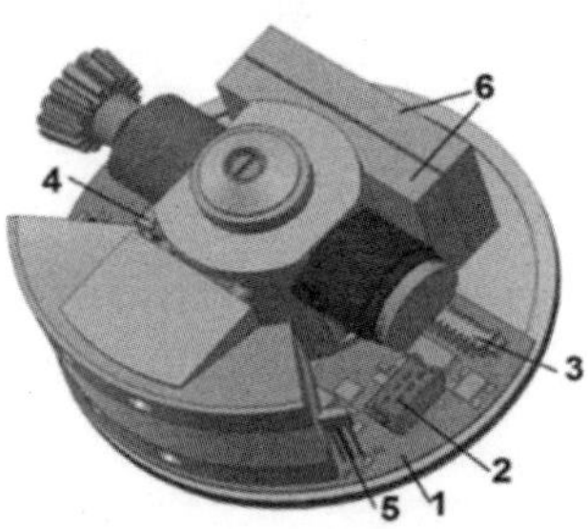

(b)

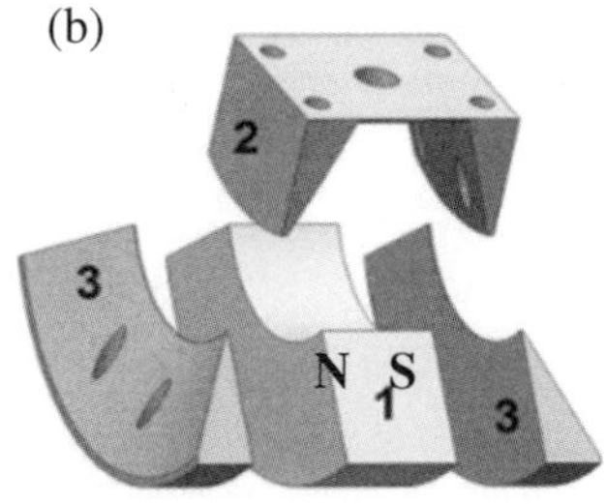

*Figure 5: **(a)** the design of the wheel with the motor, the PCB (1), the connectors for motors (2), battery (3), and system bus (4), Hall sensors for magnetic sensing (5) and the batteries (6); **(b)** the shape of the magnetic system with the axially magnetized magnet (1), the non-magnetic support part (2) and the two flux-guides (3).*

Once the system was built, simulations and measures could be compared for the case of a vertical adhesion force on a 2 mm thick ferromagnetic plate. The results are presented in Figure 6. With the implemented $3\ mm$ airgap, the force is of $125\ N$. Those results demonstrate that measures are very close to the simulations and prove the reliability of the simulations tools to validate the adhesion force of a magnetic system. In *Figure 8*, the force transfer during inner straight corner transition is shown. This shows the reliability and smoothness of inner corner transition, since the force is transferred slowly from one plane to the other and is almost constant.

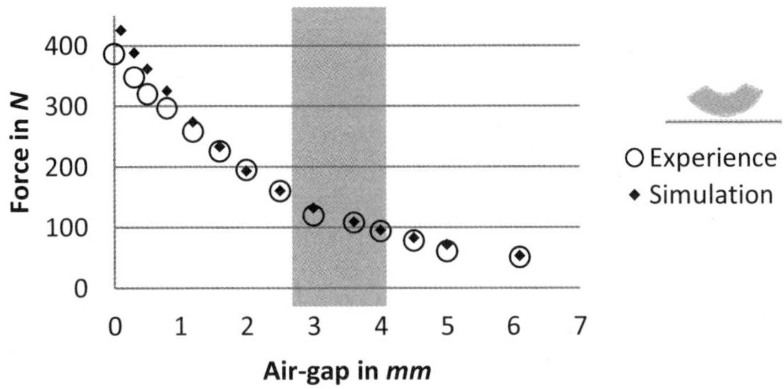

Figure 6: the graphic compares the value of adhesion force for various air-gaps both in the case of experiences and simulations. The grey area shows the area of interest for the robot.

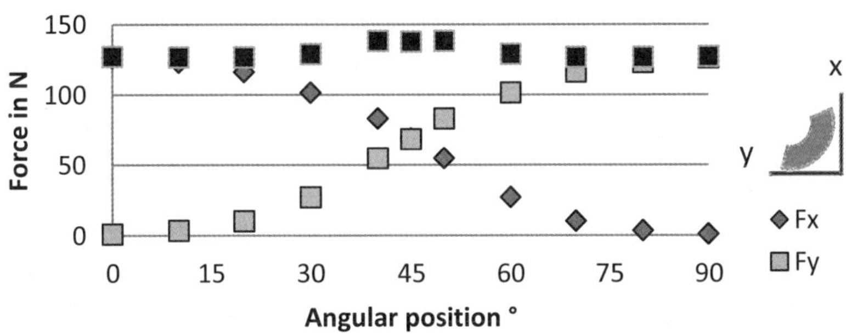

Figure 7: the curves represent simulation results of the transfer of the adhesion force from one to the other surface in case of inner right angle transition. The rotation center of the magnetic system is placed at equidistance of both surfaces, so that the minimum air-gap is 3 mm.

4. Design of the robot

The robot has a dumbbell shape. The goal was to design this robot around a self-propelled and self-powered wheel module that can be plugged at both ends of the central axis. The idea behind this wheel module is to be able to use it in other configurations, for example in a bicycle or tricycle configuration.

In the initial prototype described in [4], the robot weighed 83 g. It was 62 mm in length and 28 mm in diameter. The robot had only one magnetic system in its body. This second prototype is larger and embeds true inspection capability and computational power. The weight of the robot is 1.3 kg without batteries. Due to the large safety margin on adhesion, a payload up to 0.6 kg is possible. As batteries weigh 0.3 kg for an hour of autonomy, this leaves 0.3 kg for inspection equipment.

The robot is composed of two wheel modules and a central body part. The three modules have own batteries but can share energy and information through a common bus. Sharing the energy improves greatly the reliability of the system. The central body, called the turret, can rotate around the central axis that connects the two wheels. This axis is hollow to permit electric connection between the modules. The angular stroke of the turret is limited to 360° by a mobile wiring between the axis and the turrets. The tail is mounted on the driven turret and rotates around the central axis. The tail can rotate freely or be fixed to the turret. Thus, the tail can be used through the turret motion to push the robot when the gravity would otherwise limit obstacle passing. The turret is protected from shocks with a magnetic torque limiter. The rotation of the turret allows the orientation of the camera which allows the inspection and exploration of the environment. The central part embeds a camera, a lighting system and a processing board developed in-house running *Linux*. The design is very integrated as can be seen in *Figure 8*. The control of the robot is simple and intuitive thank to visual feed-back and a joystick interface.

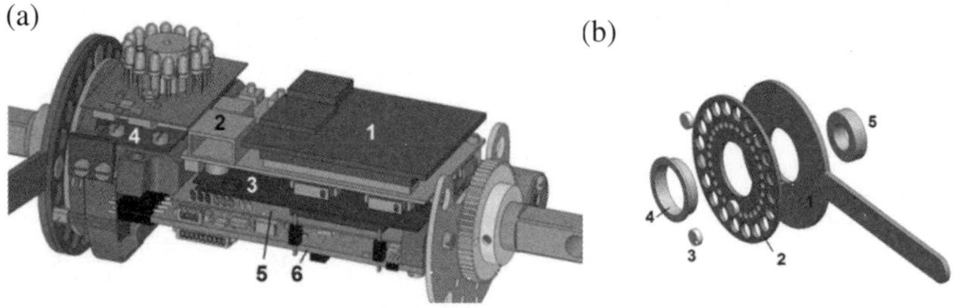

*Figure 8 : **(a)** the central body part with the LINUX board (1) and its motherboard (2), the Wifi module (3), the camera and lighting system (4), and the specific electronics boards (5-6); **(b)** the magnetic torque limiter to damp choc with the tail (1), the disc (2), the magnets (3), a guiding part (4) and bearing (5)*

5. Mobility results

The *Cy-Mag*^{3De} has an impressive mobility thanks to its innovative obstacle passing mechanism and the addition of the active tail. The innovative magnetic configuration allows passing inner straight corners very easily. In the case of outer corner, depending on the orientation of the gravity, two situations can occur. In the first situation the robot is blocked by its mass and cannot pass the obstacle. In that case, the tail is used to push the robot forward and the obstacle can be passed. The second situation occurs when the robot passes an outer corner in the downwards direction. In that case, the tail is placed in front of the robot and is used to damp the shock, as the robot will roll 100° freely due to the angular distance between the two positions where the magnetic torque occurs.

The possible obstacles that the robot can pass are shown in Figure 9. Inner corner transitions are possible in any situation when the value of the angle between the surfaces is between 45° and 315°. Outer corner transitions are possible when the tail is needed either for pushing or for damping, but not when it is needed for both.

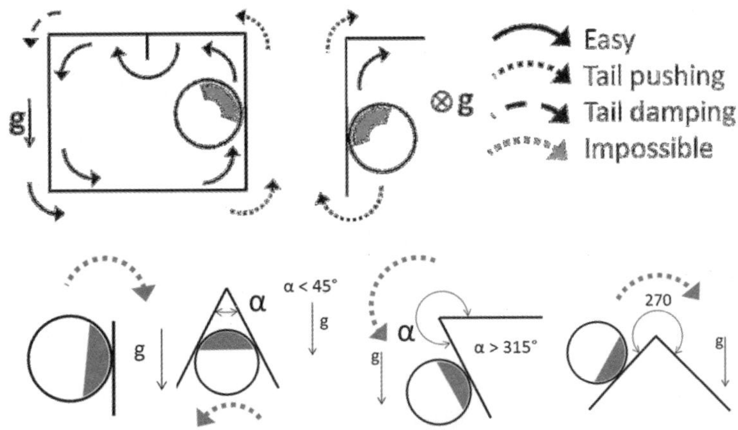

Figure 9: Summary of the mobility of the Cy-mag^{3De} robot.

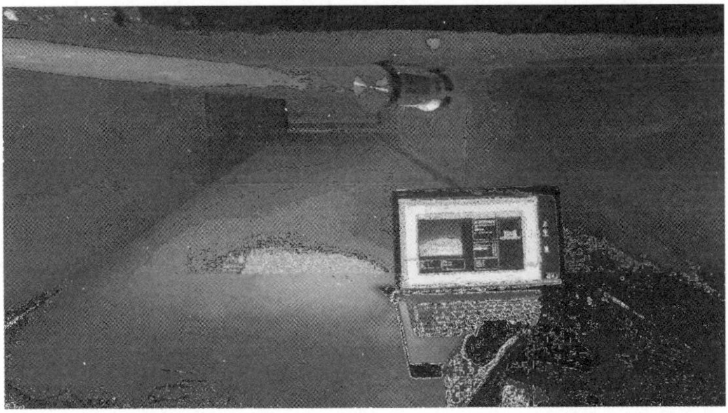

Figure 10: Cy-mag^{3De} inspection tests in an air duct

The described mobility was tested both on steel plates in the laboratory and in an air duct of our buildings. The real situation can be seen in Figure 10. The tests were done with the robot running on batteries. The results fully validate the possibility to scale the mobility concept. The same modular wheel can be used in other configurations enabling other forms of mobility, for example on a bicycle or tricycle. Another potentially interesting possibility for larger and more complex robot is to have two degrees of freedom for each wheel and to activate the outer wheel and the magnetic system separately in respect to the frame. In that case, hall sensors at both ends of the magnetic system would allow detecting magnetic obstacles and allow the implementation of autonomous control.

6. Conclusion

The robot named *Cy-Mag*^{3De} has a cylindrical shape of $80\,mm$ in diameter, $233\,mm$ in length and a weight of $1.6\,kg$. With respect to its size and complexity, it exhibits excellent mobility in ferromagnetic environment beyond the known state of the art. The robot is energy autonomous, has an embedded computer running *Linux,* and can transmit video images wirelessly. *Cy-Mag*^{3De} can be remote controlled with a wireless connection.

The robot has two dedicated degrees of freedom for mobility. Two additional degrees of freedom are for the rotation of the turret and the attachment or release of the tail. The tail allows passing obstacles otherwise impossible to overcome. The turret mobility is convenient for the visual exploration of the environment, allowing a relative movement of the embedded camera.

A better camera equipped with autofocus would be a possible future improvement. An improved video bandwidth is also required for live control. It will be advantageous to implement rotating contacts between the axis and the turret to ease the use. The tail can be an asset to support non-destructive sensors for scanning, alternatively a gripper to retrieve lost part. The final robot design is robust and offers interesting perspective for future industrial applications.

Acknowledgments

This work was possible thanks to the advices, work, collaboration and knowledge of Daniel **Burnier**, Pierre **Noirat**, Philippe **Rétornaz**, Stéphane **Magnenat**, Valentin **Longchamp** and everyone at the faculty workshops. Those results follow on the project funded by the Swiss CTI project: “Highly compact robots for power plants inspections” referenced as CTI 8435.1 EPRP-IW and Alstom.

References

1. Tâche, F., et al. *Adapted magnetic wheel unit for compact robots inspecting complex shaped pipe structures.* in *Proc. of the 2007 IEEE/ASME International Conference*. 2007.
2. Rochat, F., et al. *Magnetic wheels optimization and case study on the Magnebike.* in *Proceedings of the 10th International conference CLAWAR 2009.* 2010. Nagoya.
3. Yukawa, T., H. Okano, and S. Komatsubara, *Mechanisms for the movement of piping inspection robot with magnetic elements.* 2005.
4. Rochat, F., et al. *Cy-mag3D: a simple and miniature climbing robot with advance mobility in ferromagnetic environment.* in *Proceedings of the 10th International conference CLAWAR 2010.* 2010. Nagoya.

CLIMBING ROBOT WITH THERMAL GLUE

FREDERIC ROCHAT, VITALIS HIRSCHMANN, THIERRY BARRAS,

HANNES BLEUER, FRANCESCO MONDADA

EPFL STI IMT LSRO MOBOTS, Ecole Polytechnique Fédérale Lausanne Lausanne, CH-1015, Switzerland

Climbing robots require an adhesion principle in order to overcome gravity. Many physical principles have been applied for this purpose, but few are sufficiently generic. This work explores a new principle based on the use of thermal glue. A miniature thermal glue dispenser has been designed and tested. Experimental results show that the adhesion force depends on the contact surface, but an adhesion force of $8\ N/cm^2$ *was achieved on aluminium and synthetic surfaces. Several ideas were tested to save energy for detaching the glue foot. A climbing inchworm robot using this innovative adhesion principle was designed and tested, illustrating the strength and disadvantage of this technique. This innovative approach offers a very interesting alternative solution to the known adhesion methods.*

1. Introduction

In order to climb, a robot requires a force that compensates the effect of gravity. Many physical principles have already been tested. A very comprehensive state of the art already exists in the literature [1, 2]. The commonly used principles are: pneumatic [3], mechanic [4] or electro-magnetic [5]. Dry adhesive is a popular research topic [6]. The working principles used for grippers in assembly lines are a source of inspiration, for example capillarity [7], Bernoulli levitation [8] or electrostatic chucks [9]. Both physical principles have already been tested on climbing robots [10, 11]. Some alternative solutions are liquid solidification and glue.

The use of glue has not been reported as an adhesive solution for climbing robots. The idea appears to be an interesting research direction. The solution would offer an alternative to exiting solutions with the advantage of a good adhesion force. The chemical bond can be initiated in various manners, for example by humidity, blending of two chemical components, ultra-violet lightning or phase transition. The latter seems to have potential, since it appears simpler to control.

This general concept has been previously explored by using water solidification for connecting a floating modular robot [12]. Thermal glue has been used in micro-manipulation within the design of a tool exchanger [13].

This work presents a prototype of thermal glue dispenser for adhesion and the achieved adhesion force. The feet based on this principle are then integrated into a climbing robot.

2. Prototype

The glue dispenser is similar to a classical glue gun for hobbyists. It is composed by a glue tank, a system to press the glue forward and a heating system. The tank is a square tube. The glue is pushed forward by a spring. Additionally to the functions of a standard glue gun, the designed system requires a contact surface to realize the adhesive bond and support the payload of a climbing robot. This can be done with thick-film technique on a substrate having a matrix of holes. On the substrate are laid heating resistances. The resistances are laid on both sides of the ceramic substrate. On the top side, a PTC resistance is also disposed in order to measure the temperature. The substrate is laser cut with an array of through holes to allow the glue to adhere to the surface. The scheme of the realization can be seen in Figure 1. Nine prototypes were built, exploring different numbers of holes $(14, 22, 52)$ paired with different diameters $(0.7\, mm, 0.4\, mm, 0.2\, mm)$ and thickness of the substrate $(1\, mm, 0.625\, mm, 0.25\, mm)$. The best results were achieved on the configuration with 22 holes and $0.4\, mm$ in diameter. This dimension of the holes let better pass the glue through the holes. The thinner substrate is too fragile and breaks easily. The medium thickness heats faster.

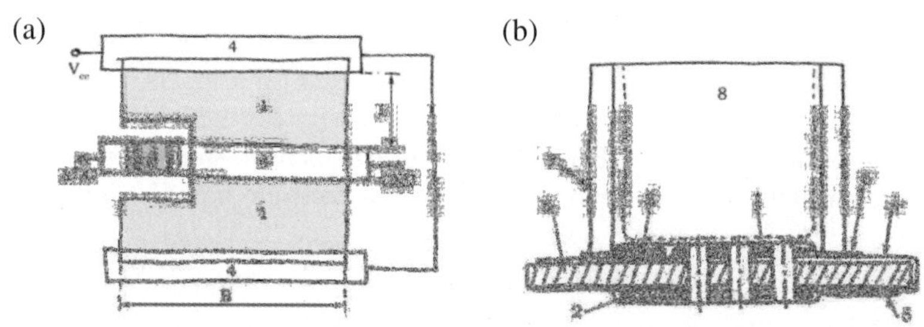

*Figure 1: **(a)** the electronics configuration of the top surface and **(b)** the cross section of the prototype. The top and down resistance (1-2), PTC resistance (3),connection pads (4), vitrified layer for isolation purpose (5), ceramic substrate (6), glue tank (7), thermal glue (8).*

The initial tests were performed with simple paraffin wax as found in candle lights, since it has low liquefaction temperature. Then washable thermal glue, the *Cristal bond 555*, was tested to mitigate the disadvantage of residual foot print. The liquefaction temperature of this glue is low: $54°C$.

The contact superficies of the foot is $1\ cm^2$. The adhesion force was measured to be $8\ Ncm^{-2}$ on aluminium and equally on synthetic POM surfaces, 11 on wood. The resistance to torsion is of $10\ mNm$. The heating time is. But the energy required to melt the glue is high. It takes up to $30\ J$. The cycle time for an adhesion is $14\ s$, including a cooling time of $10\ s$. It is to be noted that the force continues to increase after $10\ s$; the force is even 45% higher after $20\ s$ of cooling time. This can be observed in Table 1.

Table 1: Experimental measures of the vertical adhesion force on aluminum substrate after three various cooling time.

Cooling time	[s]	10	15	20
Vertical force on aluminum	[N]	9.6	12.8	14.3

Several possibilities were tested to detach. The first one is to reheat to thermal glue. This solution uses a lot of energy. The second approach is to break mechanically the bond. The bond is very effective in traction, but weaker in torsion due to the lower resistance to shearing. A prototype was built to detach the bond with a mass attached at the end of an elastic strip. After a deflection, the mass is released and hits a lever attached to the foot. The kinematic energy of the choc breaks the bond. A second mechanical solution is used on the robot in the next section. Both solutions require less energy than melting the glue.

3. Climbing inchworm

A climbing inchworm was built to test the thermal glue as adhesion principle for mobile robotics. An arm links two feet with six degrees of freedom. The degrees of freedom are driven by servo motors. The robot is compact. It can be inserted in a box of $100\ x\ 200\ x\ 60\ mm^3$. The robot was not designed to be energy autonomous and has no battery. The robot can be seen in Figure 2.

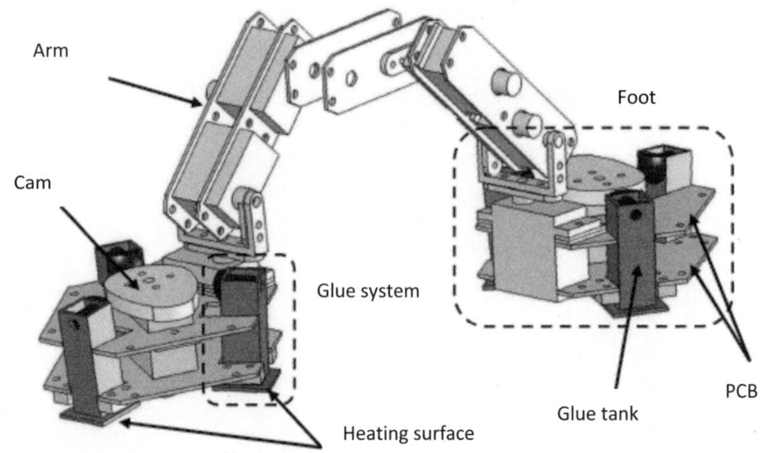

Figure 2: View on the climbing robot that has two feet with tree thermal glue dispenser. In the middle of them, a cam is placed to break mechanically the bond.

Both feet are identical. They have three thermal glue dispensers. They are positioned in a 120° axial symmetry. The thermal dispenser is free to move around an axis tangential to a circle. This allows some compliance for surface variation. In the central axis of the three systems, there is a cam disc. When the cam system is activated, its motion breaks the three bonds one after the other. The mechanism can be seen in Figure 3. The robot was operational, but problem occurred in the adhesion phase due to vibration caused by the servomotors. This weakens or breaks the bond during cooling down of the thermal glue. The prototype can be seen in Figure 4

(a) (b) (c)

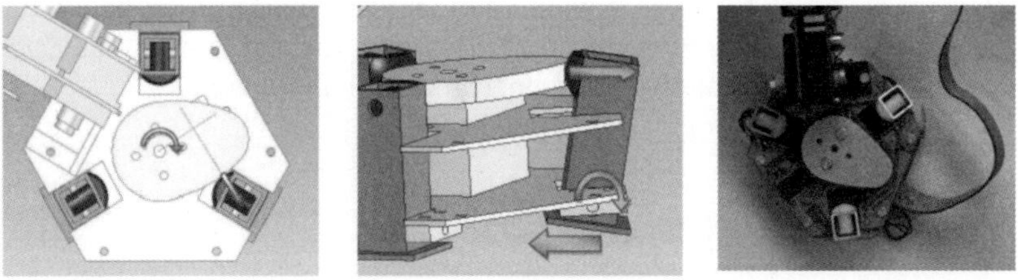

*Figure 3: The cam system **(a)** to detach the adhesive toes of the robot **(b)**, and a top view of the prototype **(c)**.*

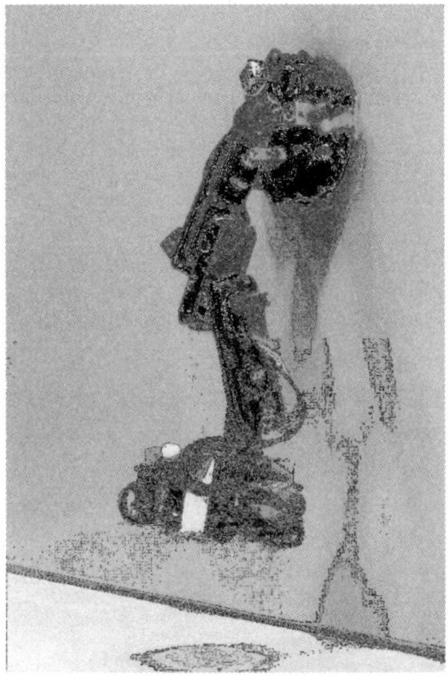

Figure 4: The prototype ***attached to a vertical surface during tests.***

4. Conclusion

In this paper, a prototype of a thermal glue dispenser was built to validate its potential as adhesion principle for a climbing robot. Force, torque, and energy consumption are presented and discussed. An inchworm climbing prototype was implemented using the developed thermal glue dispenser.

Such an adhesion system offers an interesting alternative to other adhesion methods, since it can potentially work on various substrates like ferromagnetic, non-magnetic and even porous ones. The major disadvantages of such a system are the high cycle time and the high energy consumption. Additionally, the quality of the adhesion is very sensitive to motion during the long bonding time. A climbing robot using thermal glue leaves prints behind. This can be a drawback for some applications, even if the used glue is washable with hot water. This drawback is largely mitigated by the larger versatility of the adhesion principle. Such system might also have alternative application in mobile inspection robotics, for example in the implementation of a multipurpose gripper to retrieve lost parts.

Acknowledgments

This work was possible thanks to the work and assistance of Nicolas **Yoakim**, Thomas **Cimprich,** Patrick **Schoeneich**, Thomas **Maeder**, Jean-Marc **Breguet** and Raymond **Clavel**. This project was funded by the Swiss CTI project: "Highly compact robots for power plants inspections" referenced as CTI 8435.1 EPRP-IW and Alstom.

References

1. Silva, M. and J. Machado. *Climbing robots a survey of technologies and applications*. in *Climbing And Walking Robots: Proceedings of the 11th International Conference on Climbing and Walking Robots and the Support Technologies for Mobile Machines (CLAWAR 2008)*. 2008. Coimbra.
2. Longo, D. and G. Muscato. *Adhesion techniques for climbing robots state of the art and experimental consideration*. in *Climbing And Walking Robots: Proceedings of the 11th International Conference on Climbing and Walking Robots and the Support Technologies for Mobile Machines (CLAWAR 2008)*. 2008. Coimbra.
3. Minor, M. and R. Mukherjee, *Under-actuated kinematic structures for miniature climbing robots*. Journal of Mechanical Design(Transactions of the ASME), 2002. **125**(2): p. 281-291.
4. Scheidegger, N., et al., *Bi-pedal Robot for Rescue Operations*, in *International conference on climbing and walking robots, 2006. CLAWAR 2006*. 2006: Brussels.
5. Kotay, K.D. and D.L. Rus. *Navigating 3D steel web structures with an inchworm robot*. in *Intelligent Robots and Systems '96, IROS 96, Proceedings of the 1996 IEEE/RSJ International Conference on*. 1996.
6. Menon, C. and M. Sitti, *A biomimetic climbing robot based on the gecko*. 2006, Elsevier. p. 115-125.
7. Lambert, P. and A. Delchambre. *Design rules for a capillary gripper in microassembly*. in *Assembly and Task Planning: From Nano to Macro Assembly and Manufacturing, 2005. (ISATP 2005). The 6th IEEE International Symposium on*. 2005.
8. Erzincanli, F. and J. Sharp, *Development of a non-contact end effector for robotic handling of non-rigid materials*. Robotica, 1997. **15**(03): p. 331-335.
9. Monkman, G., *Electroadhesive microgrippers*. Industrial Robot: An International Journal, 2003. **30**(4): p. 326-330.
10. Wagner, M., et al. *A novel wall climbing robot based on Bernoulli effect*. in *Mechtronic and Embedded Systems and Applications, 2008. MESA 2008. IEEE/ASME International Conference on*. 2008.
11. Prahlad, H., et al. *Electroadhesive robots - wall climbing robots enabled by a novel, robust, and electrically controllable adhesion technology*. in *Robotics and Automation, 2008. ICRA 2008. IEEE International Conference on*. 2008.
12. Miyashita, S., et al. *Peltier-based freeze-thaw connector for waterborne self-assembly systems*. in *Intelligent Robots and Systems, 2007. IROS 2007. IEEE/RSJ International Conference on*. 2008. Nice.
13. Clevy, C., A. Hubert, and N. Chaillet. *Modeling, identification and control of a thermal glue-based temporary fixing system: application to the micro-robotic field*. in *Robotics and Automation, 2006. ICRA 2006. Proceedings 2006 IEEE International Conference on*. 2006.

TREMO: AN INSPECTION CLIMBING INCHWORM BASED ON MAGNETIC SWITCHABLE DEVICE

FREDERIC ROCHAT, RICARDO BEIRA,

HANNES BLEULER, FRANCESCO MONDADA

EPFL STI IMT LSRO MOBOTS, Ecole Polytechnique Fédérale Lausanne Lausanne, CH-1015, Switzerland

TREMO is a compact magnetic climbing inchworm with the capability to move in complex ferromagnetic environments, like those found in power-plants. The robot fits inside a cylinder of 500 mm *in height and* 120 mm *in diameter for an approximate weight of* 800 *g. It has a modular configuration based on three energy autonomous sub-systems: an arm and two feet. The arm has five degrees of freedom. It has a central articulation with a stroke of 260°. Both ends of the arm have an innovative articulation based on a differential system. This allows unlimited rotation of* 360° *in the heading angles and 180° in the elevation angle. At both ends of the arm a camera is mounted. The camera can be used for visual inspection and remote control of the robot. The feet are donut-shaped and are mounted around the camera. The magnetic adhesion is realized with Magnetic Switchable Devices (MSD). Each foot of the robot embeds three MSD for better compliance and stability. The three MSD of one foot together with their motors weigh only* 32 *g and can hold above* 100 *N.*

1. Introduction

Power industry requires inspection robots with high mobility for complex ferromagnetic environments like the one encountered in coal-fired boiler or generators and its housing. Some existing wheeled robots already have advanced mobility for industrial inspection [1]. However, the mobility of continuous locomotion like wheels, caterpillars or whegs is limited to continuous environment. Point to point locomotion is an alternative to solve the necessary extension of mobility and can pass gaps and complex obstacles with higher versatility. Some examples of climbing inchworm already exist with suction cups [2] or electromagnetic adhesion system [3].

Magnetic adhesion offers the best properties for climbing in ferromagnetic environments in term of force, compactness and energy efficiency. Among other electromagnetic solutions, adhesion with *Magnetic Switchable Device* (*MSD*) has proved to be the most energy-efficient [4]. This is essential for the energy management of an autonomous robot. Based on *MSD* adhesion, a compact magnetic climbing inchworm, called *TREMO*, has been designed and is presented in this work. This paper starts introducing the modular configuration of the robot. Then, the design of the arm is presented with its innovative differential articulation. Two alternative designs of the articulation are presented: using a power transmission with ball crimped endless cable and using a twisted timing belt. Finally, the adhesion feet embedded with *MSD* are presented.

2. General design

The robot is composed of three autonomous modules that are remote controlled independently by a *Bluetooth* connection: the arm and the two feet. In the current configuration, there is no electric connection between the modules, but a rotary contact is foreseen, to let the module share their energy and thus increase system reliability. This configuration is modular. Indeed it permits the independent exchange of the arm or feet in order to adapt to particular situations. Because of the wireless interface, the arm can rotate endlessly around the feet as the illustration of the cinematic configuration shows in Figure 1 (a). This is very advantageous, since it simplifies the operation of the robot. Indeed, if the heading angle is limited, the mobility of the inchworm is limited and the pilot needs to bother about the position of the ends of the angular stroke and thus develop strategies to overcome obstacles; this slows down operation and is not intuitive. The robot has two cameras that are fixed in at both ends of the arm. They are placed in the center of the feet. This position is ideal to explore, move and inspect the environment. The two magnetic feet are donut-shaped and can rotate endlessly around the camera. A CAD view of the prototype can be seen in Figure 1 (b-c).

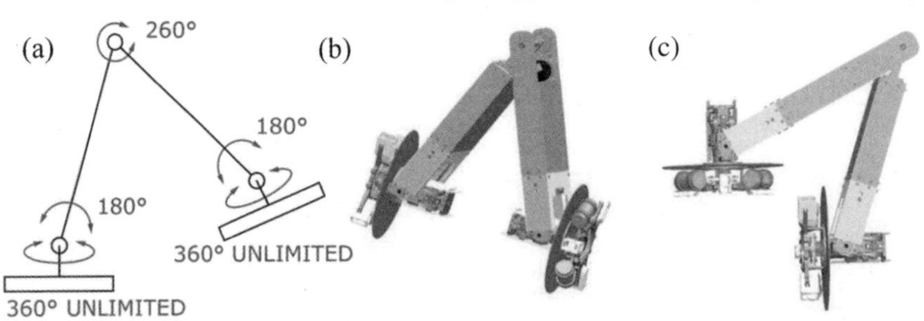

*Figure 1 : **(a)** General cinematic of the magnetic climbing inchworm, called TREMO; **(b-c)** two different CAD views of the developed robot.*

3. Design of the arm

The prototype has an arm with 5 Degrees of Freedom (DoF). This choice de facto limits the final mobility, but reduces the development cost and allows a confirmation of the mobility concepts. The configuration of the DoF is seen in Figure 1 (a). The desired configuration with an unlimited heading angle with the camera positioned in the middle requires the development of specific joints. The suggested solution is based on a differential configuration as seen in

Figure 2 (a). The morphology of this joint can be implemented in various manners. The transmission cannot easily be realized with bevel gears. This solution is heavy and suffers from backlash. Standard cable transmission is not possible due to the desire of endless rotation.

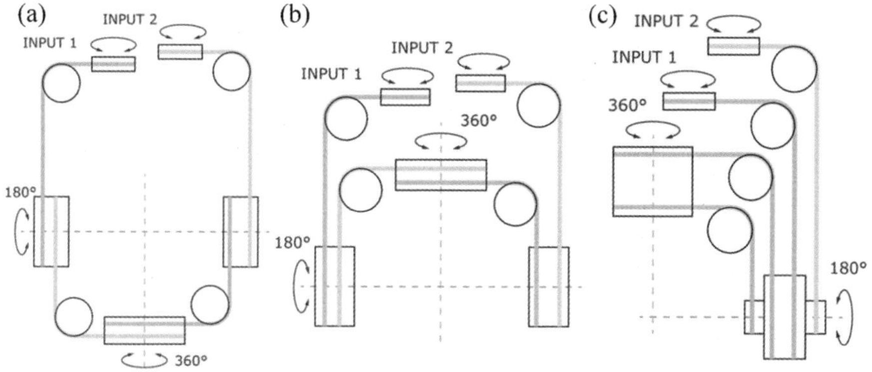

Figure 2 : The joint between arm and feet is based on a differential configuration. The movement is given as input from motors that are placed on the arm (top) and generate a rotation of the foot following two orthogonal axis (bottom). Three different solutions were analyzed, only the first (a) and second (b) have been implemented.

Two different solutions were implemented and tested. The first is based on a closed loop cable with periodic crimped balls. This solution appears ideal for miniaturization, due to the small dimension of the cable and consequently the small thickness of the dedicated tailor-made pulleys. A prototype was built and can be seen in Figure 4. The whole arm-foot joint is composed of four closed loop cables of same length. The cable is made out of tungsten and is 0.4 mm in diameter. The ball are 1.6 mm in diameter and are crimped every 5 mm. The circumference of the loop is 140 mm. This particular transmission demands special pulleys with a circumferential slot and periodic holes for the balls, in order to realize a proper torque transmission. This construction can be seen in Figure 4 (b). The fixed distance between the balls set constraints on the pulleys radius. It also requires very precise alignment of the pulleys in case of right angle transmission. Moreover every loop needs to be tensioned independently. Due to the stringent requirements for alignment around the right angle pulleys, three pulleys need to move at the same time when tensioning the cable. Other configurations are possible where only one pulley moves, but these configurations require additional pulleys and add complexity. The tension mechanism is therefore realized on each loop by a linear movement on three pulleys. It is tensioned with a lateral screw. The torque between the two pairs of pulleys of the elevation axis is transmitter through a linear slider as seen in the middle in Figure 3 (c).

The prototype size is $80\,mm \; x \; 68\,mm \; x \; 12\,mm$. This volume can be reduced further 4 times by shortening the distance between the pulleys and changing the configuration of the tensioning mechanism. On the prototype, the right angle pulleys have a too small radius; this reduces the lifetime of the cable.

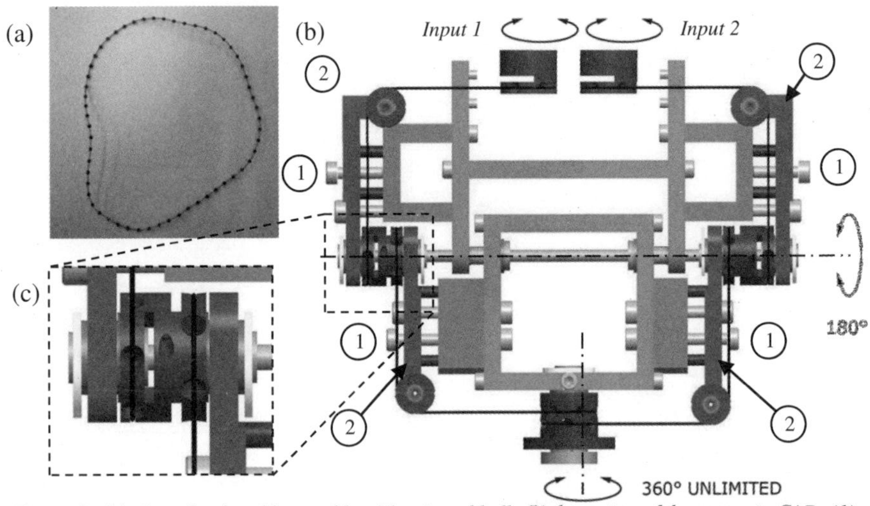

*Figure 3 : **(a)** view of a closed loop cable with crimped ball; **(b)** front view of the system in CAD; (1) the tensioning screws, (2) the tensioning sliders supporting the mobile pulleys; **(c)** detail view on the linear slider transmission mechanism that allows separated tension of each cable loop.*

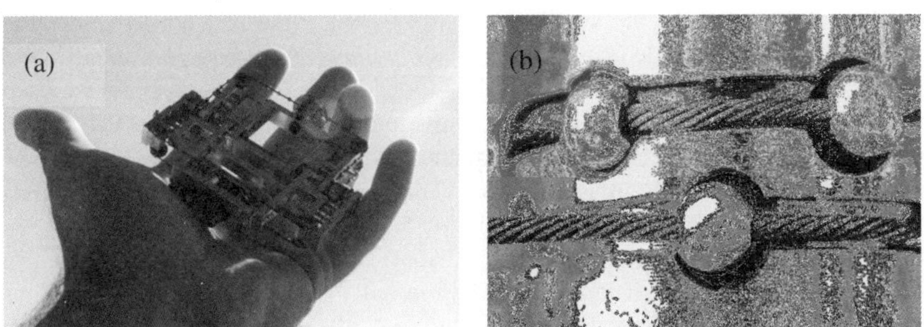

*Figure 4: **(a)** view of the prototype to test the differential articulation with crimped balls cable with hand as size reference; **(b)** detail view of the transmission with crimped balls cables.*

This transmission mechanism is innovative and well adapted to this application. It is generally very interesting for systems having small dimensions; however this design needs special care, since the constraints in the cable are high.

Additionally, the tailor-made cables are expensive. Therefore another solution was designed and implemented. It uses the same differential principle, but with the morphology presented in

Figure 2 (b). The crimped cables are replaced with standard timing belt as seen in Figure 5 (a). In this embodiment various loop lengths are used, since standard components exist. The tensioning mechanism is simpler, since the alignment constraints vanish with timing pulleys. However, this offers less miniaturization possibilities, especially due to the width of the belt and consequently of the pulleys. Moreover the 90° torsion on short distance constrains the timing belt beyond their specifications, which will reduce their lifetime. The deformation reduces the transmission efficiency.

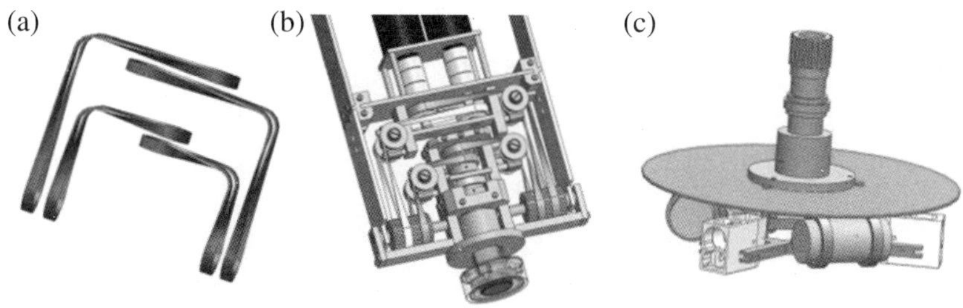

*Figure 5: **(a)** configuration of the twisted timing belt; **(b)** CAD view of the joint without the foot, the camera is seen in the middle, bottom; **(c)** finally the foot with the hollow shaft in the centre to allow the support and connection of the camera fixed to the end of the arm.*

*Figure 6: **(a)** side view of the transmission mechanism with the twisted timing belt; **(b)** front view during assembly of the differential transmission mechanism with the two motors.*

4. Design of the feet

The feet are two autonomous systems with own battery, processing power, proximity sensors and motors drivers. The endless rotation is transmitted through the hollow axis which is seen in *Figure 5* (c). The adhesion is realized with three ***Magnetic Switchable Device*** (***MSD***) systems placed in an 120° axial symmetry. The working principle of an ***MSD*** is based on a ferromagnetic circuit containing permanent magnets where the flux can circulate between different paths. The configuration of the ***MSD*** can be changed by mechanically displacing one component of the circuit. This reroutes the flux in the system and creates or cancels the flux through specific surfaces. Consequently this operation turns *ON* or *OFF* the adhesion force to the nearby ferromagnetic environment. The classical configuration of a ***MSD*** is composed of a fixed ferromagnetic part, the stator, and a moving magnetic part, the rotor, as seen in Figure 7. The rotation of the magnet part by 90° cancels or creates the adhesion.

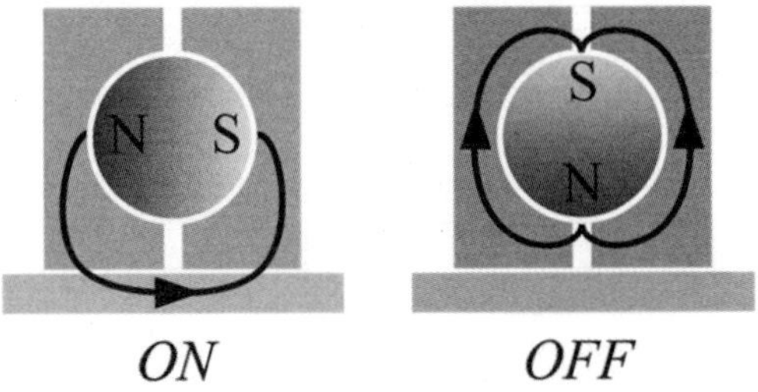

Figure 7: Standard Magnetic Switchable Device (MSD) in ON and OFF position. The arrows represent the magnetic field.

The stators of the six *MSD* that are used in the feet have a size of $5.7 \times 5.7 \times 16\,mm^3$. The rotor and stator weigh together only $4\,g$. One system can hold $38.4\,N$. The switching torque is $1.74\,mNm$. Hence the three *MSD* together composing a foot can hold above $100\,N$ in vertical traction.

MSD are advantageous since their energetic consumption is lower than other solutions. For instance a DC motor can switch the state of this ***MSD*** system with an amount of energy of only4.8 mJ 4.8 m4.8 mJ actuated by a DC motor. The energy required to detach the ***MSD*** mechanically without switching off the adhesion is of $46.2\,mJ$, without counting the mechanical losses. As comparison, the electromagnetic detachment using a coil for a similar force ($36\,N$ of the *Isliker* permanent electro magnet *GTP-20*) requires $10.8\,mJ$. Hence, ***MSD*** are the ideal activation for an autonomous inchworm climbing robot like the *TREMO*.

Each *MSD* is embedded in a casing as seen in Figure 8. It is activated with a DC motor. The transmission is done with gears. The gear driving the magnet is truncated and acts as a mechanical end-stop. Two magnets are glued in it and their position is detected with a Hall sensor for control purpose as seen in Figure 8 (a).

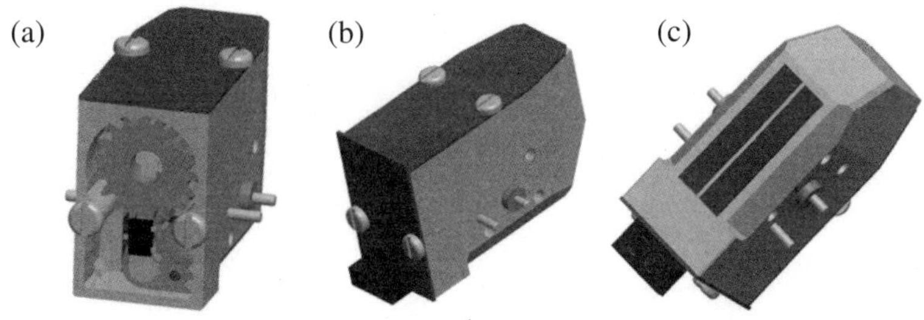

*Figure 8: Different view of the MSD system as used in the foot; **(a)** the gear transmission is seen with the hall sensor, in front of the gears, for detection of ends of stroke; **(b)** side view with the proximity sensors at the bottom; **(c)** bottom view with in the middle the part of the stator that ensures the adhesion with the environment.*

5. Conclusion

A compact magnetic climbing inchworm for visual inspection in complex environments, named *TREMO*, has been designed and can be seen in Figure 9. It fits into a cylinder of $500\ mm$ in height and $120\ mm$ of diameter for about $800\ g$. It uses *Magnetic Switchable Device* (*MSD*) for adhesion. This solution is extremely energy-efficient in respect to existing approaches such as mechanical detachment or use of coils. The inchworm has an innovative modular configuration, based on three autonomous sub-systems. The robot features a differential joint that allows endless rotation of the heading angle. Locomotion tests are in progress on the prototype.

Possible improvements include the addition of rotary contacts to share energy between the modules. The size of the robot can be further reduced with the replacement of timing belts by crimped endless cables. In a future design, mobility can be further improved with additional *Degrees of Freedom* in the arm. The camera is placed ideally to support *Non-Destructive Testing* (*NDT*).

*Figure 9: **(a)** front view of the magnetic climbing inchworm TREMO on a flat surface; **(b)** the robot passing an external straight angle.*

Acknowledgments

This work was possible thanks to the advice, work and knowledge of Daniel **Burnier**, Pierre **Noirat**, Philippe **Rétornaz** and Patrick **Schoeneich**. Those results follow on the project funded by the Swiss CTI project: "Highly compact robots for power plants inspections" referenced as CTI 8435.1 EPRP-IW and Alstom.

References

1. Tache, F.; Fischer, W.; Moser, R.; Mondada, F.; Siegwart, R.; , "Adapted magnetic wheel unit for compact robots inspecting complex shaped pipe structures," 2007 IEEE/ASME international conference on Advanced intelligent mechatronics, pp.1-6, 4-7 Sept. 2007
2. Minor, M. and R. Mukherjee, *Under-actuated kinematic structures for miniature climbing robots.* Journal of Mechanical Design(Transactions of the ASME), 2002. **125**(2): p. 281-291.
3. Kotay, K.D., and Rus, D.L. , "Navigating 3D steel web structures with an inchworm robot," Proceedings of the 1996 IEEE/RSJ International Conference on Intelligent Robots and Systems '96, IROS 96, vol.1, pp.368-375, 4-8 Nov 1996
4. Rochat, F., Schoeneich, P., Bonani, M., Magnenat, S., Mondada, F., Bleuler, H. and Christoph, H. (2010) Design of Magnetic Switchable Device (MSD) and applications in climbing robot. Emerging Trends in Mobile Robotics, pp. 375-382.

DESIGN OF A NOVEL CLIMBING ROBOT - CLIFFRIDER: THE ABSEILING FACE INSPECTION ROBOT

JOEL CAPPELLI, ALI HAYDAR GÖKTOĞAN
Australian Centre for Field Robotics,
School of Aerospace, Mechanical & Mechatronic Engineering
The University of Sydney 2006 NSW, Australia
jcap4010@uni.sydney.edu.au, aligoktogan@acfr.usyd.edu.au

CliffRider is a novel robotic system that has the capability of abseiling steep, crater like terrain. The CliffRider system comprises an autonomous single wheeled platform tethered to two winches mounted on the plateau of the cliff of descent. CliffRider is designed for the navigation of open pit mine faces, providing mining engineers with a portable ground based platform that can be fitted with surface inspection instrumentation. Virtual simulations have shown that, stabilisation of CliffRider is achievable provided specific control strategies are employed. Development of the CliffRider functional prototype and unique control strategies are discussed within this paper.

1. Motivation

Due to the potentially hazardous nature of open pit mining, frequent wall monitoring is required to locate potential failures on site [5]. Alternatively, steep crater walls exposed during planetary exploration offer chronological insight into a planets history [3]. Close inspection of these inclined surfaces reveal rich geological and mineralogical information [6].

This paper will detail the CliffRider concept [4] from high and low level design principles. The CliffRider system is a unique approach for navigation upon steep terrain (Figure 1). CliffRider is a transportation unit for a suite of sensors that can extract useful data from a surface. The CliffRider concept was designed with the following functional and non-functional requirements in mind: (a) Omni-directional motion, (b) inclined rough terrain locomotion capability, (c) 2 kg payload capacity, (d) static stability during inspections, (e) minimal impact upon the cliff face, (f) lightweight and minimal transportation volume.

2. Related Work

One of the first successful cliff descending projects was Dante II. Dante II was an eight legged frame walker robot [2] that was stabilised by a single tether down an Alaskan volcano to perform field tests and extract information. The single tether configuration allowed for only one degree of freedom in motion up/down the cliff face which prevented Dante II from using the support of the tether during an overturn after the rappeller had deviated from the fall line [1].

The design of a co-operative abseiling system (TRESSA) developed to descend the cliffs of Mars is detailed in [7, 8]. TRESSA features a team of three robots – 2 tether anchor robots and a single rappelling robot. This system offers improved stability and control compared to Dante II with the use of 2 independent tether anchors on the cliff plateau. Reference [8] extended the work of [7] demonstrating that the use of an active rappeller requires maintained positive tether tensions otherwise robot appendages will lose mobility. This led to the use of computationally expensive behaviour control techniques that proved to be too demanding for field use.

TRESSA furthered the work from Dante II with an alternative solution demonstrating the effectiveness of an active wheeled rappeller and stability gains of using multiple tether controls. The CliffRider concept was inspired by the previous two related works and further extends the concept to a tethered single wheeled abseiling robotic platform. The simplicity of this new design concept reduces the possible failure modes and increases its robustness for mission critical applications such as its use in planetary exploration and mine face inspection.

3. The CliffRider Concept

The current functional prototype CliffRider system (Figure 1) comprises of two modules: (a) CliffRider platform (Figure 2); (b) two tethering stations (Figure 1). The CliffRider platform is the ground based manoeuvring platform that will carry surface inspection instrumentation. Mounted to each tethering station is a tele-operated winch motor with maximum line speed of 2.9m/min. The tether stations will rappel CliffRider on the cliff face by a single tether respectively.

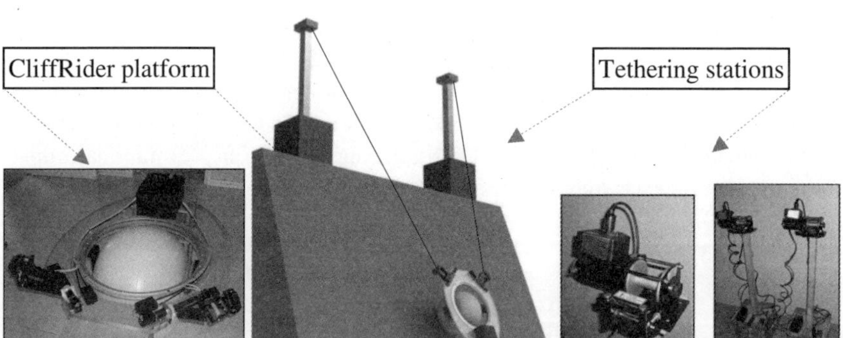

Figure 1: 3D rendering of the CliffRider system and the full scale functional prototype.

The CliffRider platform features a steerable un-driven (free-to-roll) wheel and two tether arms that are connected to the tethers. Each of these subsystems is designed to help stabilise the inherently unstable single wheeled platform. Tether payout is controlled by the operator while the platform is autonomous with an on-board power source, performing self stabilisation techniques.

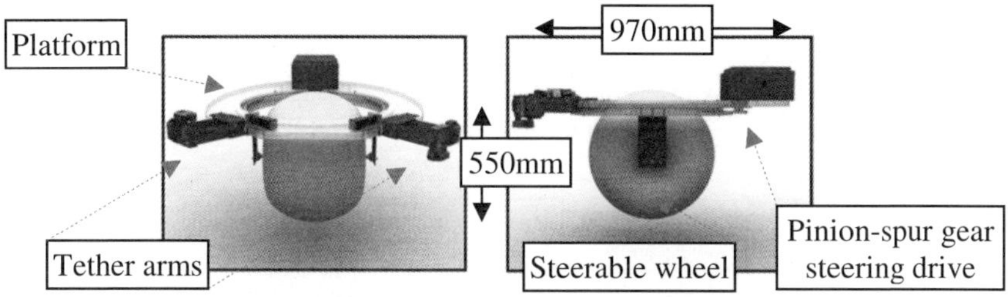

Figure 2: CliffRider platform conceptual design in SolidWorks.

4. Hardware Architecture - Prototype Implementation

4.1. *Mechanical Design*

- **Wheel**

The wheel of CliffRider is a commercial off the shelf (COTS) inflatable fitness ball. It is made of 2.5mm thick polyvinyl chloride thermoplastic with a centre passing 22mm inner diameter poly-propylene sleeve. The wheel can be inflated to an overall diameter of 630 mm with carrying capacity 60 kg. The inflatable wheel is a critical design feature that is beneficial for space applications where weight and volume of payloads before and after deployment have a major impact upon the cost of space missions. The inflatable single wheel concept also provides CliffRider with minimal footprint on the surface and high amount of traction upon rough terrain.

- **Steering mechanism**

The steering mechanism comprises a turntable bearing with a spur gear mounted on its inner race, driven by a pinion gear on the underside of the platform. The turntable is secured into a step on the platform (Figure 3).

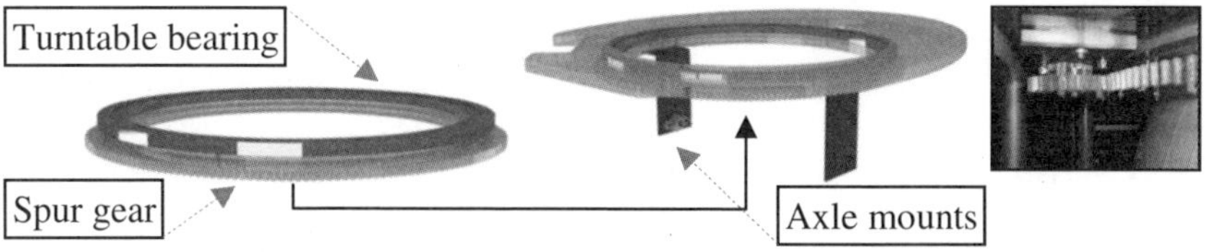

Figure 3: Spur gear mounted onto turntable, assembly with platform and steering pinion.

The spur gear transmits rotary power to the wheel through the axle mounts. The steering mechanism allows for omni-directional motion of the platform and the application of the steering stability control concept.

- **Tether arm mechanism**

The tether arm mechanism employed on CliffRider features a 2 DOF resolver mounted on the end of an actuated arm (mounted on the platform) that can change the position of the tether attachment point (Figure 4). This is to enable the system to measure the tether force direction.

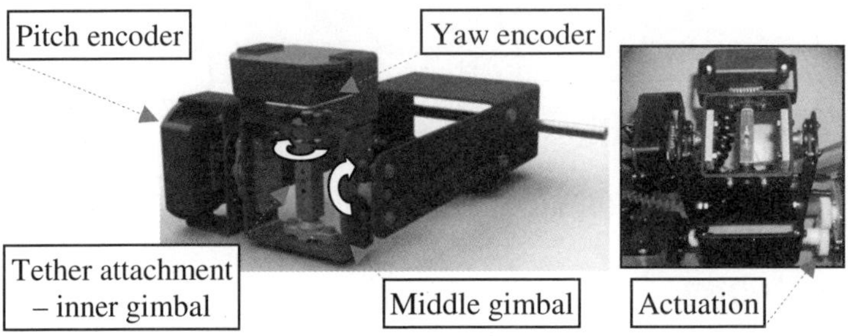

Figure 4: 3D rendering of tether arm assembly and prototype implementation.

4.2. *Electrical Design*

- **Actuation**

A NEMA 11 frame stepper with 100:1 planetary gear box actuates each tether arm, with peak torque of 11.8 Nm. A NEMA 23 frame stepper is used to drive the steering pinion and spur gear (10:1) on the platform underside (Figure 3).

- **State space sensing**

A 5 DOF IMU is required to measure roll rate, pitch rate and X, Y, Z accelerations which can then be combined to compute roll and pitch angles of the platform with respect to the cliff face.

- **Tether sensing**

The tether sensing module requires the use of encoders to measure the angle of rotation of the inner and middle gimbal of the tether arm 2 DOF resolver. Two magnetic encoders are used for each tether arm assembly (Figure 4).

5. Control concepts

The control ideas and principles are unique to the CliffRider system. This section aims to explain the concepts behind the two main sub-system modules of

CliffRider, the steering mechanism and tether arm mechanism. The primary goal of these control concepts is to maintain CliffRider in a state of stability as it is pulled by the tether system.

There are two assumptions made during the development of this work: (a) CliffRider will travel at slow speeds. Therefore, state rates of the system will be minimal; (b) tether tension will be assumed positive and there will be assumed no slack in the tethers over their length of travel.

5.1. *Steering Control*

The steering system is the most important control system of CliffRider. It allows for omni-directional motion on the cliff face and at the same time attempts to maintain system stability, preventing the platform from tilting excessively. The primary objective of the steering system is to 'roll in the direction of the platforms steepest descent'. Much like a bicycle stabilises, the wheel is steered in the direction (within the plane of the surface) that the platform leans most.

5.2. *Tether Arm Control*

The tether arm system will stabilise the platform chassis directly while CliffRider navigates undulating terrain. Tether arm actuation is performed to relocate the tether attachment point with respect to the platform; so that the moments created by the tether forces restore the platform to a state of stability. In order to perform this level of control (Figure 5) the following are required: (a) current location of the tether attachment point ($\vec{r}^{\{A\}}$); (b) line of action of the tether forces ($\hat{F}^{\{A\}}$[a]). This information is obtained through the state space and tether sensing hardware. Combining these in (1) yields the moments created by the tether forces in {A}:

$$\vec{T}^{\{A\}} = \vec{r}^{\{A\}} \times \vec{F}^{\{A\}} = |\vec{F}|(\vec{r}^{\{A\}} \times \hat{F}^{\{A\}}) \tag{1}$$

Where ${T_X}^{\{A\}}$ is the rolling moment of the platform, ${T_Y}^{\{A\}}$ is the pitching moment of the platform and $\hat{F}^{\{A\}}$ is the tether force direction unit vector.

The tether arm control strategy actively alters $\vec{r}^{\{A\}}$ through tether arm actuation (Figure 4) to maximise the absolute pitching moment of the platform (Figure 5). To explain how ${\vec{r}_{desired}}^{\{A\}}$ is computed, we first have to establish a set of transformations that move each tether force unit vector into {A}. The tether arm model is presented in Figure 5. The tether arm frame denoted $\{T_{arm}\}$ $= (O_{arm}, \vec{X}_{arm}, \vec{Y}_{arm}, \vec{Z}_{arm})$ has origin located at the centre of the tether arm

[a] After performing a number of transformations upon $\vec{F}^{\{G\}}$ moving from the gimbal frame to the axle frame of reference.

axle. $\vec{Y}_{arm}$ is fixed to the tether arm shaft, $\vec{X}_{arm}$ is orientated parallel to the tether main arm body. Variable b_{arm} is the control variable representing pitch of the tether arm. The gimbal model is presented in Figure 6. The gimbal frame denoted $\{G\} = (O_G, \vec{X}_G, \vec{Y}_G, \vec{Z}_G)$ has origin located at the centre of the inner gimbal where the tether is attached. Tether direction is measured by the angles of rotation with respect to $\vec{Y}_G$ and $\vec{Z}_G$ by the magnetic encoders (Figure 4). Tether pitch (b_T) denotes angle of rotation about $\vec{Y}_G$. Tether yaw (δ_T) denotes angle of rotation about $\vec{Z}_G$.

The algorithm is to maximise the quantity $\left|T_Y{}^{\{A\}}\right|$ for a suitable $b_{arm,desired}$, the control variable of desired tether arm pitch and therefore the location of the desired tether attachment $O_G{}^{\{T_{arm}\}}$ with respect to each arm.

$\hat{F}^{\{A\}}$ are the unit vectors of the tether forces in $\{A\}$. $\hat{F}^{\{A\}}$ are functions of all state variables necessary to translate $\hat{F}^{\{G\}}$ into $\{A\}$, therefore also a function of $b_{arm,current}$ which is the current tether arm pitch that defines the current tether attachment in $\{T_{arm}\}$.

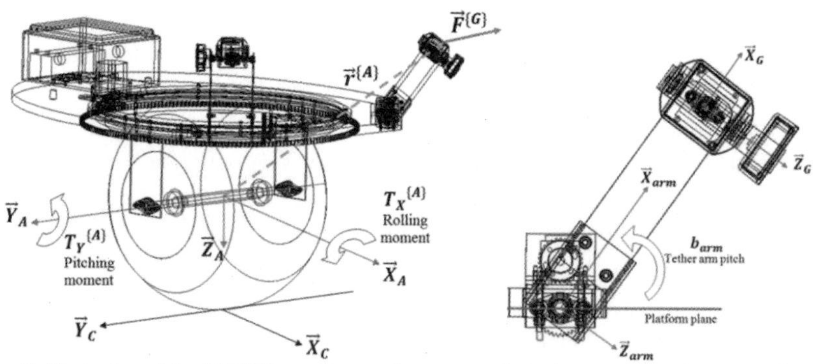

Figure 5: Illustration of the CliffRider system to demonstrate the tether arm control strategy. {A}, {C} and {G} represent the axle, contact surface and gimbal frames of reference respectively.

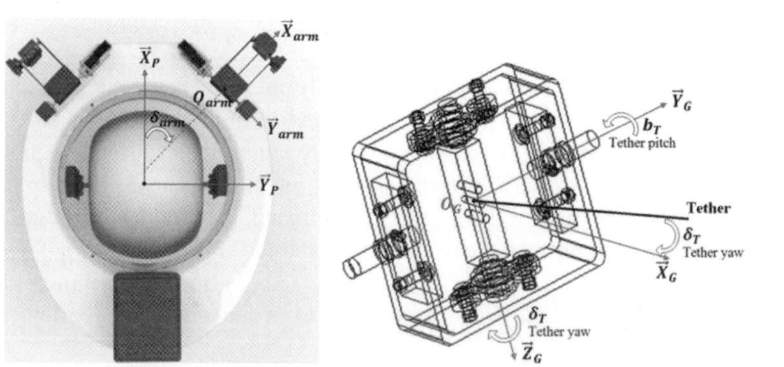

Figure 6: Tether arm control strategy frames of reference; Tether arm with respect to platform representation and the gimbal frame of reference.

Using notation[b] in the tether arm and gimbal model (Figure 5 and Figure 6), the following transformations were performed for each tether arm to translate the current tether forces into {A}.

Converting $\hat{F}^{\{G\}}$ into the tether arm frame $\{T_{arm}\}$:

$$\hat{F}^{\{T_{arm}\}} = [cos(b_T)\, cos(\delta_T) + arm_{length} \quad cos(b_T)\, sin(\delta_T) \quad - sin(b_T)]^T$$

The transformation from the tether arm frame $\{T_{arm}\}$ into the platform frame plane {P}:

$$T^P_{T_{arm}} = \begin{bmatrix} c(\delta_{arm})\, s(b_{arm,current}) & s(\delta_{arm}) & c(\delta_{arm})\, s(b_{arm,current}) \\ -s(\delta_{arm})\, c(b_{arm,current}) & c(\delta_{arm}) & -s(\delta_{arm})\, s(b_{arm,current}) \\ -s(b_{arm,current}) & 0 & c(b_{arm,current}) \end{bmatrix}$$

$$O_{arm}{}^{\{P\}} = [0.2886 \quad \pm 0.246 \quad -0.0077]^T metres$$

The $\pm$ takes into consideration the location of tether arm 1 and 2 respectively.

$$\hat{F}^{\{P\}} = T^P_{T_{arm}} \hat{F}^{\{T_{arm}\}} + O_{arm}{}^{\{P\}}$$

Then further transformations are used to move from {P} into {A} (beyond the scope of this paper). The same approach is taken to develop the equations for $\vec{r}^{\{A\}}{}_{desired}$ by replacing $\hat{F}^{\{T_{arm}\}}$ with $O_G{}^{\{T_{arm}\}}$ and $b_{arm,current}$ with $b_{arm,desired}$.

From quantities derived above,($\vec{r}^{\{A\}}{}_{desired} \times \hat{F}^{\{A\}}$) is computed and the pitching component $T_{Y,desired}{}^{\{A\}}$ is extracted to develop Eq. (2). Eq. (2) is the function to be maximized for single variable; $b_{arm,desired}$ such that the resulting pitching moment $T_Y{}^{\{A\}}$ opposes the platform's current pitch.

$$f\left(b_{arm,desired}\right) = -sign(platform\ pitch) T_{Y,desired}{}^{\{A\}} \tag{2}$$

One method first explored to maximize Eq. (2) was Brent's Method of parabolic interpolation and golden step computations. This proved to be computational expensive for the application. A more elegant solution was derived. Due to the nature in which Eq. (2) was derived; the function can be re-written simply as the summation of periodic trigonometric functions with constants A, B and C (Eq. (3)).

$$f\left(b_{arm,desired}\right) = C + Acos\left(b_{arm,desired}\right) + Bsin\left(b_{arm,desired}\right) \tag{3}$$

Using the auxiliary angle method, Eq. (3) can be transformed into:

$$f\left(b_{arm,desired}\right) = C + \sqrt{A^2 + B^2} cos\left(b_{arm,desired} + tan^{-1}\left(\frac{-B}{A}\right)\right) \tag{4}$$

In order to maximise Eq. (4),

$$b_{arm,desired} = -\, tan^{-1}\left(\frac{-B}{A}\right) \tag{5}$$

This is computed for each respective tether arm. The tether arm control strategy is feasible due to the minimal pitch and roll rates of the system however will need to be explored with respect to actuator dynamics.

---

[b] All rotations follow the right hand rotation rule and axis frames follow the North East Down configuration. T^A_B denotes transformation from frame {B} to {A}.

6. Conclusion and Future Works

CliffRider has the capability of abseiling steep, crater like terrain. Due to this, the CliffRider concept has potential to be utilised in future planetary exploration and open pit mine inspection operations. This work has revealed innovative design of the system and implementation into a functional prototype featuring high level human control. Two critical control concepts to improve inclined high risk terrain manoeuvrability are detailed. These control strategies have been examined through 3D dynamic simulation however field testing will prove fundamental in validating these ideas. Field testing will also allow for the reliability and robustness of design to be investigated in a number of scenarios. Future work exists to extend the CliffRider concept into an autonomous co-operative system removing human-in-the-loop control.

Acknowledgment

This work has been supported and funded by the Rio Tinto Centre for Mine Automation and by the Australian Research Council (ARC) and the New South Wales State Government.

References

1. Apostolopoulos, D., & Bares, J. Locomotion Configuration of a Robust Rappelling Robot. *Robotics Institute*, pp. 280-284. (1995).
2. Bares, J., & Wettergreen, D. Dante II: Technical Description, results and lessons learned. *International Journal of Robotics Research,18(7)*, (pp. 621-649). (1999).
3. Biesiadecki, J.J. *et al.* Mars Exploration rover surface operations: driving opportunity at Meridiani Planum. *Robotics & Automation Magazine, IEEE,* 63-71. (2006).
4. Cappelli, J. *CliffRider: The Abseiling Face Inspection Robot.* Engineering Undergraduate Honours Thesis. USYD. AMME. (2010).
5. Edward, L., & Girard, J. *Evaluating techniques for monitoring rock falls and slope stability.* Spokane, WA: National Institute for Occupational Safety and Health. (2002).
6. Girard, J. & McHugh, E.L. *Emerging technologies and the future of geotechnical instrumentation.* Spokane, WA: NIOSH Spokane Research Laboratory. (2002).
7. Mumm, E. *et al.* Analysis of Behaviour Based Control for Planetary Cliff Descent Using Co-operative Robots. *Autonomous Robots 16 (3)* , 259-272. (2004).
8. Paulsen, G. *et al.* All Terrain Exploration with the Cliff-Bot System. *IEEE Int. Conference on Robotics and Automation.* Spain. (2005).

HARDWARE INDEPENDENT PLATFORM FOR MODULAR ROBOTICS BASED ON PROTOTHREADS AND CONTIKI LOADER*

MICHAL BACHRATÝ

Institute of Control and Industrial Informatics, FEI STU Ilkovičova 3 Bratislava, 812 19, Slovak Republic

PETER HUBINSKÝ,
ANTON VITKO

Institute of Control and Industrial Informatics, FEI STU Ilkovičova 3 Bratislava, 812 19, Slovak Republic

These days there is no modular platform not only for robotics, which would distribute computing power from 8bit low power microcontrollers to 32 or 64 bit high power computers. This paper presents new robotic modular architecture based on knowledge from sensor networks area, where development is rapidly moving forward. The basis of the presented platform is created by functional modules based on protothreads, which communicate in accordance with one common protocol. This gives scalability and independency in the process of creating modular system with code reusability.

1. Introduction

Robotic systems are very complex, starting with hardware systems and ending with software, giving the system intelligence. Universal system for robotic must provide easy way to solve these problems [1]: (i) Rapid development in sensors, actuators and computing technologies leads to increased investigation of robotic capabilities and therefore DSP sensor-actuator algorithms need to be updated easily, (ii) Robotic systems are distributed and all their parts need to be connected into one functional unit, (iii) Components must communicate in an efficient way. (iv) Different algorithms should be incorporated (Obstacle avoidance, mapping, navigation, computer vision ...) and must communicate with each other. The current trend in robotic platform research is focused on developing modular (component) system, which enables easy reuse of

* This work is supported by the grant VEGA 1/0690/09 is duly acknowledged.

components e.g. Kalman filter or navigation algorithms etc. This plug and play concept opens new area in robotics.

These days, there is no modular platform not only for robotics, which would distribute computing power from 8-bit low power microcontrollers to 32 or 64 bit high power computers. The main idea of this paper is to put data processing from sensors or actuator into separate embedded devices, rather than one central processor. Modules located on different devices communicate together with one common protocol. Inside the device, modules communicate only on the level of data referencing and therefore no data copy is needed. Each device can contain more modules each having one special address assigned by configuration module.

The paper presents a new robotic modular architecture based on knowledge from sensor networks area, where development is rapidly moving forwards. The main purpose of sensor networks is to harvest data from wide working area (more than 10000 nodes), with low power energy consumption, over different transport media and then connect data into a human readable form. Sensor network devices have long life on batteries and therefore use processor with constrained memory and computing power. Sensor network have a different purpose in the robotic, but in some aspects the knowledge of communication principles and universal scalable systems independent of hardware can be used. Contiki operating system [2] is portable sensor network operating system based on protothreads [3], stackless threads implementing threads like state machine on different hardware platform. An advantage of Contiky system is possibility of loading and stopping applications, while system is running.

The core is implemented in C language and so it can run on 8bit microcontroller with 1kB RAM and 16kB ROM. The code is compiled with GCC compiler. The code has only a few hardware-depended functions, (communication, timer and loader) and can be easily ported to many platforms. All devices are precisely time synchronized, with 10μs precision. Due to that the sensing of different sensors placed on different devices or different robots can be performed and data from different places can be fused. This advantage can be used for sound processing or sound localization. The devices can also communicate wirelessly. Therefore multi-robot communication and swarm robotic can be easily created.

2. Modular robotic systems

Research in mobile robotic systems – platforms, frameworks is still in progress. Each of the architectures has its own pros and cons and serves to special purpose. Some of them are widely used and are discussed in the next text.

The OROCOS (Open Robot Control Software) platform is designed for industrial robots and machine control. This platform is divided into four libraries: RTT (realtime toolkit), KDL (kynematics and dynamics library), BFL (Bayesian Filtering library) and OCL (Orocos component Library) [4]. Each library is programmed in C++. Orocos uses for communication: in-process, plain C++ method calls, inter- process: socket based messages. Orocos will always get much higher throughput on local machines, but programming must consider memory leaks. If code is buggier, recovery makes worse.

iCub [5], is a system for humanoid robots and it is a part of EU the project roboCub. RoboCub is little humanoid robotic platform composed of 53 motors, that move the head, arms & hands, waist, and legs. It can see and hear, it has the sense of proprioception (body configuration) and movement (using accelerometers and gyroscopes).

ORCA is a framework for developing component-based robotic systems [6]. Communication between components is defined by using strict-defined interfaces powered by ICE (Internet Communication Engine). ICE is contact-based middleware similar in many aspects to CORBA middleware. The Slice language is used to define communication contracts (interfaces) between components at runtime, i.e. an explicit description of services to be provided by other components to others. Ice components can communicate with each other regardless of the language of implementation [7].

Player – a communication platform is an opensource project providing an infrastructure for distributed access to a number of popular robotic hardware devices. Player fits for simple, non-articulated mobile platforms. It was designed to provide easy access to sensors and motors on laser-equipped Pioneers. This architecture runs on many UNIX-like platforms. It is Implemented in C++ and uses POSIX threads or pthread interface for multi-threading. Player supports multiple concurrent client connections to devices, creating new possibilities for distributed and collaborative sensing and control. For simulation of sensors actuators and environment, the Stage platform is used. Stage devices present a Standard Player interface so a few or no changes are required to move between simulation and hardware. Many controllers designed in Stage have been demonstrated to work on real robots [8]. Player/Stage platform is most widely used in the projects.

The ROS (Robot operating system) is a framework for robot software development, providing operating system-like functionality on the top of a heterogenous computer cluster. It provides hardware abstraction, device drivers, libraries, visualizes, message-passing, package management, and more.specification is at the messaging layer. Peer-to-peer negotiation and configuration occurs in XML-RPC, for which reasonable implementations exist in most major languages [9]. To support cross-language development, ROS uses a simple, language-neutral interface definition language (IDL) to describe the messages sent between modules. The IDL uses (very) short text files to describe fields of each message, and allows for composition of messages. Code generators for each supported language then generate native implementations which "feel" like native objects, and are automatically serialized and deserialized by ROS as messages are sent and received. ROS is designed as microkernel, where a large number of small tools are used to build and run the various ROS components. ROS communication layer is not hard real-time compatible, but individual processes within the system can have components operating under a hard realtime operating system, but individual processes within the system can have components operating under a hard realtime operating system.

3. Architecture overview

Basis of the platform is created by functional modules based on the protothreads, which communicate by using one common protocol. That provides scalable architecture. The following block diagram (Figure 1) shows communication structure of modules and devices. Device 1 can be robot A, device 2 small robot B. They communicate through radio. Robot A consists of three hardware boards with embedded sensors and actuators and one software board on the high power computer. Sensors can be for example ultrasonic, infrared or motor driver as actuators. Master board makes bridge between high power computer with operating system and observes on other boards connected to RS485 bus. If the hardware is fully functional, the master board synchronizes time on the slave boards and monitors communication. The Slaves are fully independent from master after data binding and configuration is done. Main software board placed on the high performance computer enables device configuration, address setting, data binding and is always present. For data storage a database module is included. Each software board executes actions - modules in strict order, defined by the thread manager. So module scheduling is defined as data flow programming. Parallel actions can be easily added by new software board.

Software boards and modules can work with a shared memory defined for device. This configuration is useful for visual systems, where algorithms process one picture or region of interest.

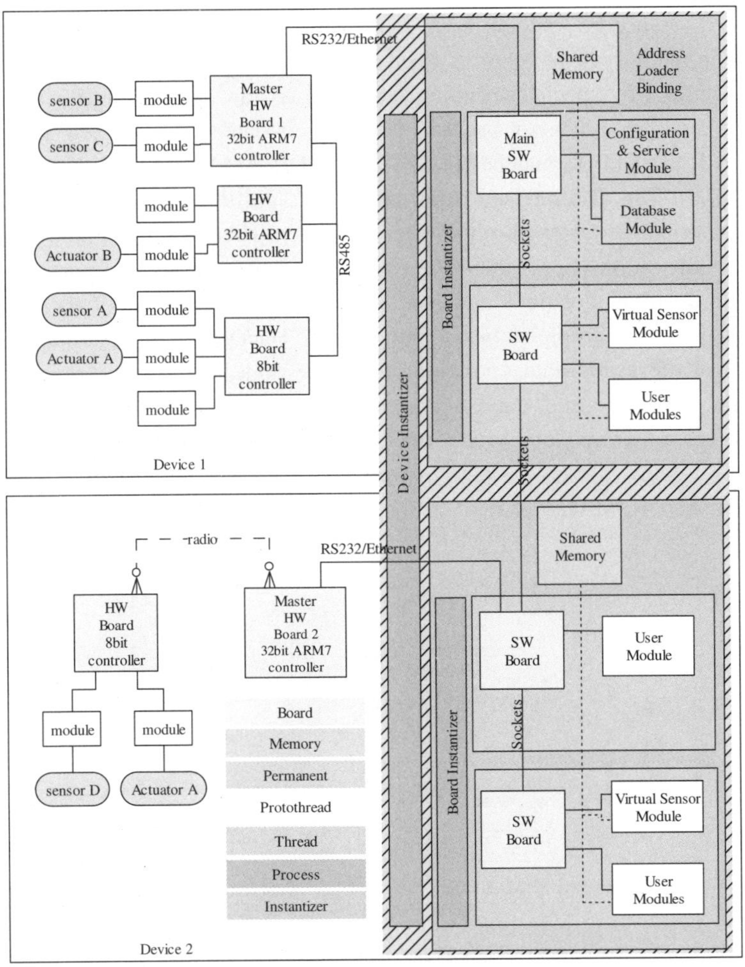

Figure 1. Architecture overview

4. Communication principle and data binding

The described architecture has its own mechanism for communication and data binding. It is similar to Player architecture, where all connected modules - players communicate through broadcast [10]. Modules in presented architecture communicate by sending messages. At first, each module must have set itsr own address. The Address is composed of device address (specific hardware board or software board) and the last 8 bit define specific module. Afterwards binding of module's data and synchronization should to be done. The mask address manager is the core of binding system, which monitors whether there are data assigned to specific module. Each module have DataIn buffer, where data are stored in memory as structure defined by C programming language. These data are described by special definition language and are written in JSON format. Figure 2 describe data set process. Device gets data with specified address of sender. If the address matches in MASK_ADDR_MANAGER, data set process is initiated. The data payload from the specific position of the data stream and the data length is taken from the receiver buffer. This data are copied or set by specific operation (Reset, And, Or, Xor) to DataIn module memory space at specific position. Positions and length are defined in process of data binding with help of configuring software. The variable HasData of module indicate new data and afterwards data operation should be done in module thread. When module wants to send data, a communication packet is created. But when data needs other module inside one microcontroller (or board), there is no need of data packet. Only specific data are transported through MASK_ADDR_MANAGER.

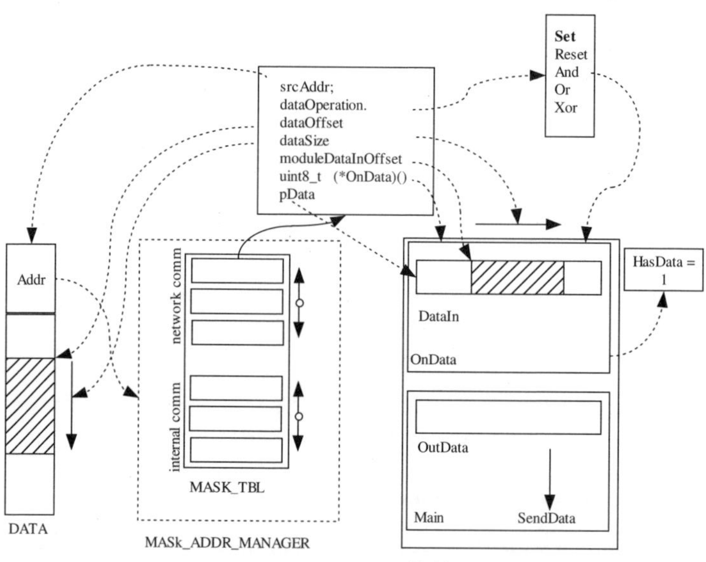

Figure 2. Module communication principle

5. Results

A Simple communication experiment was performed at two different platforms (8-bit AVR microcontroller ATMega32 running at 16MHz and Core2 Duo 1.66GHz running under Windows XP 32-bit. One module performs increments of the received data variable and then sends the result to other modules. The time needed for communication between modules was tested. The first test – communication inside board was performed on both platforms. On AVR platform, 30.26μs was needed to set 4bit data from one module to other, thence 13.0μs for invoking OnData function and 2.6μs for search in mask address table. The second test - on Windows platform, 4.15μs averaged was needed for communication between modules. For inter-process communication between boards, multistream communication is performed. Time needed to create data packet and set data to another module is 7812.0μs averaged.

Comparison with other architectures is problematic, because existing modular robotic systems are mostly configured for work on one processor, where data fusion is done. Presented architecture intends to utilize utilize computing power on small microcontrollers or DSP processors, where data preprocessing or distributed data fusion can be done. Also none of existing platforms has been ported for 8-bit platform and therefore wasn't done.

6. Discussion

Protothreads add a new view to programming modular systems and enable the programming system to be independent of hardware architecture. Currently, the presented system works on AVR (8bit) and x86 computers under Windows operating system. Within in future work we will try to put architecture to ARM platform and write code in C++ for high power computers. There is no support for scripting languages today. Development of a complex robotic system requires contribution of various specialists which contribute with new modules. Comparing to existing robotic platforms, it is scalable multiplatform without need of operating system, time synchronized, with small footprint, without need of central coordination unit. It is safe, redundant and. also can be used for other applications in the industry. Another remarkable aspect which can be remarked is that, this system can be used for swarm robotic, where small microcontrollers are used. Next application can be heterogenous sensor network with moving robots, where the whole devices (sensor nodes, robots) can be composed with this architecture only. Sensor network nodes can provides information on robot localization or specific information needed for enhancing robot precision. This architecture can be also be used for developing complex robotic system, like humanoid robot, though today there exists more complex and powerfull architectures which can do that.

References

1. N. Mohamed, J. Al-Jaroodi, I. Jawhar, Middleware for robotics: a survey, in: Proceedings of RAM, 2008
2. Contiki-a lightweight and flexible operating system for tiny networked sensors, 29th Annual IEEE International Conference on Local Computer Networks [0-7695-2260-2] Dunkels ,2004 pages: 455
3. Adam Dunkels, Oliver Schmidt, Thiemo Voigt, and Muneeb Ali. Protothreads: Simplifying Event-Driven Programming of Memory-Constrained Embedded Systems. In Proceedings of the Fourth ACM Conference on Embedded Networked Sensor Systems (SenSys 2006), Boulder, Colorado, USA, November 2006.
4. H. Bruyninckx. "Open robot control software: the OROCOS project," In IEEE International Conference on Robotics and Automation (ICRA'01) vol. 3, pp 2523–2528, 2001. [Online: http: //www.orocos.org].
5. Beira, R., Lopes, M., Praça, M., Santos-Victor, J., Bernardino, A., Metta, G., Becchi, F., Saltarén, R.: Design of the robot-cub (icub) head. In: International Conference on Robotics and Automation, ICRA, Orlando (May 2006)
6. A. Makarenko, A. Brooks, and T. Kaupp. Orca: Components for robotics. In IEEE/RSJ Int. Conf. on Intelligent Robots and Systems (IROS'06), Beijing, China, 2006.
7. M. Henning and M. Spruiell. "Distributed Programming with Ice". 2006.
8. Matthias Kranz, Radu Bogdan Rusu, Alexis Maldonado, Michael Beetz, and Albrecht Schmidt, "A Player/Stage System for Context-Aware Intelligent Environments". *Proceedings of the System Support for Ubiquitous Computing Workshop (UbiSys 2006)*, at the 8th Annual Conference on Ubiquitous Computing (Ubicomp 2006), Orange County, California, September 2006.
9. M. Quigley, B. Gerkey, K. Conley, J. Faust, T. Foote, J. Leibs, E. Berger, R. Wheeler, and A. Ng. ROS: an open-source Robot Operating System. In International Conference on Robotics and Automation, 2009.
10. YAN Xin-qing, LI Wen-feng, CHEN Ding-fang,: A New Mechanism for Robots Control Based on Player/Stage Proceedings of the 2006 IEEE International Conference on Robotics and Biomimetics, Kunming, China December 17 - 20, 2006.

STATIC ANALYSIS OF A NOVEL BIPED WALKING AND CLIMBING ROBOT

GIORGIO FIGLIOLINI[†] AND PIERLUIGI REA

DiMSAT, University of Cassino, Via G. Di Biasio, 43
03043 Cassino (Fr), Italy

The present paper deals with the static analysis of a novel biped walking and climbing robot, which is provided of a serial-parallel leg mechanism with 6 d.o.f.s, since composed by two 3-RPS parallel modules. Twelve electromechanical linear drives are foreseen to actuate the prismatic joints of both leg mechanisms. Thus, the static analysis of the biped robot is proposed by considering the worst configurations for the walking stability of the robot, during its lateral and forward motions. A specific algorithm for the static analysis of the 3-RPS parallel module is reported and related to the biped robot walking stability, along with first numerical results and validation.

1. Introduction

The mechatronic design of a low-cost biped robot was developed in Cassino and several prototypes of the EP-WAR (Electro-Pneumatic-Walking-Robot) were designed, built and tested since 1995, as reported in (Figliolini and Ceccarelli, 2004). In particular, the last prototype was able to walk along a straight line, turn right and left, climb and descend stairs, because of the effect of suction-cups, which were installed on the underside of each foot. Then, the kinematic model and the gait analysis of a six-legged walking robot were developed, as reported in (Figliolini and Rea, 2007).

Several prototypes of walking and climbing robots were designed, built and tested around the world, but only some of these have inspired the current research project, as discussed in (Figliolini et al., 2010a; 2010b). Moreover, only few walking robot prototypes have been designed and built by using parallel leg mechanisms, as reported in (Hashimoto et al., 2009; Wang et al., 2010) and still more original is the use of serial-parallel leg mechanisms, as adopted in the proposed biped walking and climbing robot. In fact, currently, the mechatronic design of a novel biped walking and climbing robot is in progress at the University of Cassino, as reported in (Figliolini et al., 2010a; 2010b).

This robot is provided of a 2 (3-RPS) leg mechanism with 6 d.o.f.s. and makes use of suction-cups for climbing on sloped and vertical flat surfaces.

[†] Corresponding author, e-mail: figliolini@unicas.it.

A first mechanical design of the whole biped walking and climbing robot was proposed in (Figliolini et al., 2010a) and a leg mechanism prototype that was actuated through six pneumatic cylinders, was built and tested in several significant configurations. Moreover, the stability analysis of this biped walking and climbing robot was proposed in (Figliolini et al., 2010b), in order to perform a dynamic walking on flat or irregular surfaces and a static climbing on vertical or sloped walls, because of suitable suction-cups. Thus, the worst configurations were obtained in order to design each rectangular foot properly.

The present paper deals with the static analysis of this biped walking and climbing robot by considering the worst configurations for its walking stability, with the aim to choose the electromechanical linear drives and as first step for the next dynamic analysis. Thus, a specific algorithm for the static analysis of the 3-RPS parallel module is reported and related to the biped robot walking stability, along with first numerical results and validation.

2. Biped Walking and Climbing Robot

Referring to Fig. 1a, the proposed biped walking and climbing robot shows a leg mechanism with 6 d.o.f.s since composed by two modules of 3-RPS parallel mechanisms, which are connected in series from the robot body to the foot. The transient motion from a horizontal terrain to a vertical wall can be performed through a foot rotation of 90° with respect to the robot body.

In particular, each 3-RPS parallel mechanism shows a prismatic pair P, which is connected to the moving platform through a spherical joint S and to the fixed platform through a revolute joint R. These R and S joints are installed on the fixed and moving platforms at 120° among them.

Figure 3b shows a sketch of the electromechanical linear drive, that is foreseen to actuate the 12 prismatic joints of both leg mechanisms of the robot.

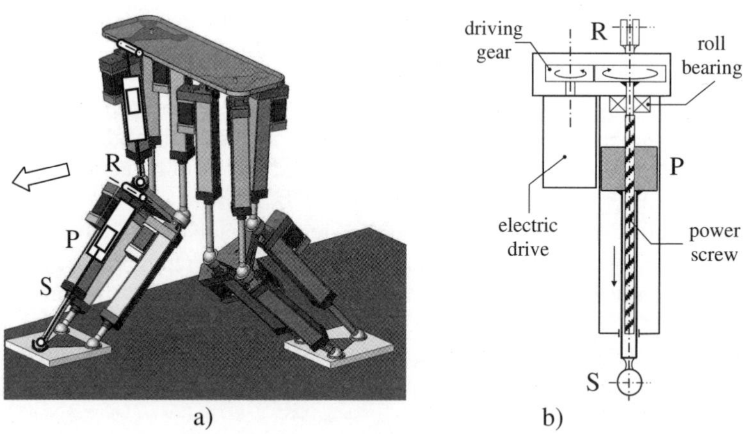

Figure 1. Biped walking and climbing robot: a) SolidWorks model; b) Electromechanical actuator.

3. Walking Stability of the Biped Robot

The static analysis for the walking stability of the biped robot has been carried out, in order to find the worst configurations during its walking motion, as reported in (Figliolini et al., 2010b). The worst configurations were obtained for a lateral walking and a forward walking, where both legs are fully extended, as shown in Figs. 2a and 2b, respectively. In particular, the front side of the biped robot is indicated by the arrow and the position of the projection of the left foot center during its swing phase is referred to the fixed Cartesian frame O_1 X_1 Y_1, which origin O_1 coincides with the center of the right supporting foot.

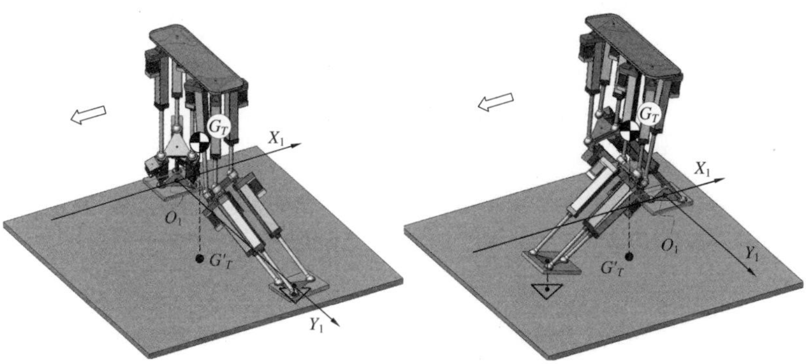

a) b)

Figure 2. Walking stability of the biped robot (the right foot is supposed to be the supporting foot, while the left foot is performing the swing phase, and the arrow indicates the front side of the biped robot): a) Worst configuration for a lateral walking; b) Worst configuration for a forward walking.

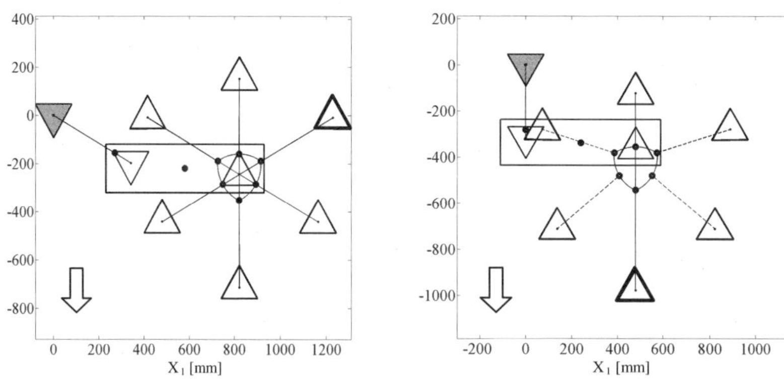

a) b)

Figure 3. Walking stability of the biped robot (the grey triangle is the supporting foot and the dark triangle corresponds to the worst configuration of the other foot): a) Lateral walking (X_1 = 1230 mm and Y_1 = – 8.8 mm); b) Forward walking (X_1 = 479.8 mm and Y_1= – 976.4 mm).

Referring to Figs. 3a and 3b, the grey triangle indicates the supporting foot, while the dark triangle represents the projection of the left foot on the flat floor, which center has Cartesian coordinates: $X_1 = 1230$ mm and $Y_1 = -8.8$ mm for the lateral walking of Figs. 3a; $X_1 = 479.8$ mm and $Y_1 = -976.4$ mm for the forward walking of Figs. 3b. These biped robot configurations can be considered the most dangerous for its stability under a static point of view, when the right supporting foot is supposed to be constrained onto the rigid flat floor through a fixed joint and the left foot is performing the swing phase.

Consequently, the three electromechanical linear drives of the lower parallel module of the supporting leg mechanism can be considered the most loaded, because they have to support the own weight of the other three parallel modules, along with the own weight of the robot body.

Therefore, the constraint force of each prismatic joint of the lower parallel module of the supporting leg mechanism can be evaluated by means of a static analysis, which is formulated in the following. Consequently, according to a safety approach, the twelve electromechanical linear drives for both leg mechanisms can be chosen as equal among them and by referring to the most loaded prismatic joint of the whole biped walking and climbing robot.

4. Static Analysis of the Biped Robot and Results

Referring to Fig. 4, the biped robot has been considered as composed by two different rigid parts: a) the lower parallel module of the supporting leg mechanism; b) the upper part of the robot, along with the other leg mechanism at the extended configuration during the swing phase.

According to this sketch, the supporting foot is supposed to be constrained on the rigid flat floor through a fixed joint, which is able to give a constraint force $\mathbf{R}_\Omega$ that is applied in Ω and a constraint moment $\mathbf{M}_\Omega$ with respect to the pole Ω, in order to ensure the static equilibrium of the whole biped robot on one foot only. Pole Ω coincides with the center of the equilateral triangle among the three spherical joints of the lower platform, which represents the supporting foot.

The upper platform of the same parallel module is loaded through the reaction force $\mathbf{R}_{21}$ that is applied on the center O of the triangle among the three revolute joints and the reaction moment $\mathbf{M}_{21}$ with respect to the pole O. Thus, the static equilibrium of the upper part of the robot is given by the action of its weight $\mathbf{W}$ that is applied in its mass center G and by both reaction force $\mathbf{R}_{12}$ and moment $\mathbf{M}_{12}$, which are exerted by the lower parallel module according to the Principle of Action and Reaction. In particular, $\mathbf{R}_{12}$ and $\mathbf{M}_{12}$ must be equal to $\mathbf{W}$ and to the moment of $\mathbf{W}$ with respect to O, in order to ensure the translational and rotational equilibrium, respectively.

Therefore, the inverse mechanism of the lower parallel module, as shown in the upper-right corner of Fig. 4, can be considered conveniently, in order to formulate a suitable algorithm for the static analysis of the 3-RPS parallel mechanism, which has to give the constraint force acting on each prismatic joint.

In other words, the inverse mechanism shows a fixed platform with three revolute joints R and a moving platform with three spherical joints S. This choice lets to simplify the formulation of the algorithm, according to the approach that was proposed in (Qunfeng and Xiangzhou, 2009), because each link with prismatic joint can only rotate with respect to the fixed platform.

Moving platform of the inverse mechanism of Fig. 4 can be considered as loaded by the force $\mathbf{R}_{\Omega}$ that is applied in the center Ω of the equilateral triangle and by the moment $\mathbf{M}_{\Omega}$ with respect to the pole Ω, which can be given by

$$\mathbf{R}_{\Omega} = -\mathbf{W} \quad \text{and} \quad \mathbf{M}_{\Omega} = -\left(\overline{\mathbf{OG}} + \overline{\mathbf{\Omega O}}\right) \times \mathbf{W} \tag{1}$$

where $\mathbf{W}$ is the weight of the upper part of the robot, while $\overline{\mathbf{OG}}$ and $\overline{\mathbf{\Omega O}}$ are the position vectors of G and O, with respect to O and Ω, respectively.

Referring to the kinematic sketch of Fig. 5, frame $\mathcal{B}$ $(O\ X\ Y\ Z)$ is attached to the fixed platform by showing the origin O coinciding with the center of the equilateral triangle, the X-axis oriented toward the revolute joint B_1 and the Z-axis directed along the vertical axis, while frame $\mathcal{P}$ $(\Omega\ x\ y\ z)$ is attached to the moving platform according to the same position with respect to the platform.

Figure 5a shows that the position vectors $\mathbf{p}_1$, $\mathbf{p}_2$ and $\mathbf{p}_3$ with respect to Ω of points P_1, P_2 and P_3, respectively, of the moving platform, can be expressed as

$$\begin{gathered}[\mathbf{p}_1]_{\mathcal{P}} = [R\ \ 0\ \ 0]^T,\ [\mathbf{p}_2]_{\mathcal{P}} = [-R\,\mathrm{s}(\varphi/2)\ \ R\,\mathrm{c}(\varphi/2)\ \ 0]^T, \\ [\mathbf{p}_3]_{\mathcal{P}} = [-R\,\mathrm{s}(\varphi/2)\ \ -R\,\mathrm{c}(\varphi/2)\ \ 0]^T\end{gathered} \tag{2}$$

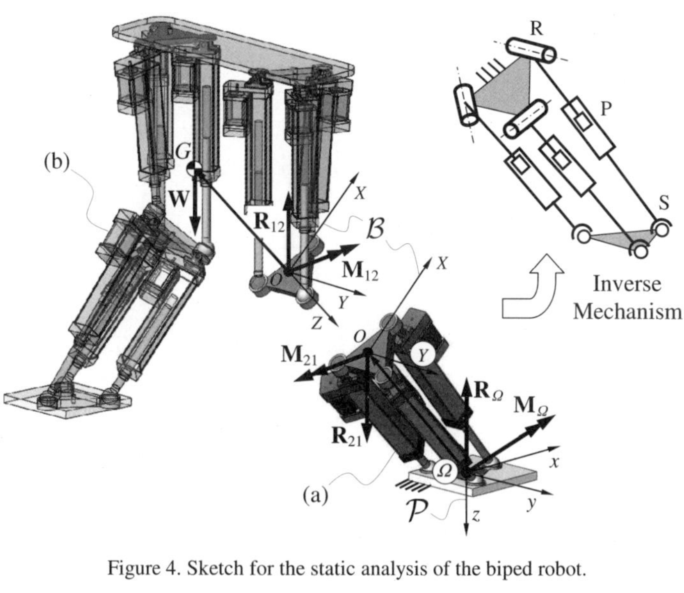

Figure 4. Sketch for the static analysis of the biped robot.

where R is the radius of both circles, through the three spherical joints S and through the three revolute joints R. Functions "sine" and "cosine" are indicated with "s" and "c", respectively, and angle $\varphi = 60°$.

Equations (2) let to obtain the position vectors $\mathbf{b}_1$, $\mathbf{b}_2$ and $\mathbf{b}_3$ of the vertices B_1, B_2 and B_3 with respect to O, respectively, because of the equation

$$[\mathbf{b}_i]_{\mathcal{B}} = [\mathbf{p}_i]_{\mathcal{P}} \cdot \qquad \text{for } i = 1, 2, 3 \qquad (3)$$

The transformation matrix ${}^{\mathcal{B}}[\mathbf{R}]_{\mathcal{P}}$ from frame $\mathcal{P}$ to the frame $\mathcal{B}$ is given by

$$ {}^{\mathcal{B}}[\mathbf{R}]_{\mathcal{P}} = \begin{bmatrix} c\gamma\, c\beta & c\gamma\, s\beta\, s\alpha - s\gamma\, c\alpha & c\gamma\, s\beta\, c\alpha + s\gamma\, s\alpha \\ s\gamma\, c\beta & s\gamma\, s\beta\, s\alpha + c\gamma\, c\alpha & s\gamma\, s\beta\, c\alpha - c\gamma\, s\alpha \\ -s\beta & c\beta\, s\alpha & c\beta\, c\alpha \end{bmatrix} \qquad (4)$$

and, in turn, the transformation matrix ${}^{\mathcal{B}}[\mathbf{R}]_{\mathcal{B}i}$ from $\mathcal{B}_i$ to $\mathcal{B}$ can be expressed as

$$ {}^{\mathcal{B}}[\mathbf{R}]_{\mathcal{B}i} = \left[\, [\mathbf{i}_{\mathcal{B}i}]_{\mathcal{B}} \;\; [\mathbf{j}_{\mathcal{B}i}]_{\mathcal{B}} \;\; [\mathbf{k}_{\mathcal{B}i}]_{\mathcal{B}} \,\right] \qquad \text{for } i = 1, 2, 3 \qquad (5)$$

where the unit vectors $[\,\mathbf{j}_{\mathcal{B}i}\,]_{\mathcal{B}}$ for $i = 1, 2, 3$, are known and given by

$$ [\mathbf{j}_{\mathcal{B}1}]_{\mathcal{B}} = [0 \;\; 1 \;\; 0]^T,\; [\mathbf{j}_{\mathcal{B}2}]_{\mathcal{B}} = [-s(2\pi/3) \;\; c(2\pi/3) \;\; 0]^T, \\ [\mathbf{j}_{\mathcal{B}3}]_{\mathcal{B}} = [-s(4\pi/3) \;\; c(4\pi/3) \;\; 0]^T. \qquad (6)$$

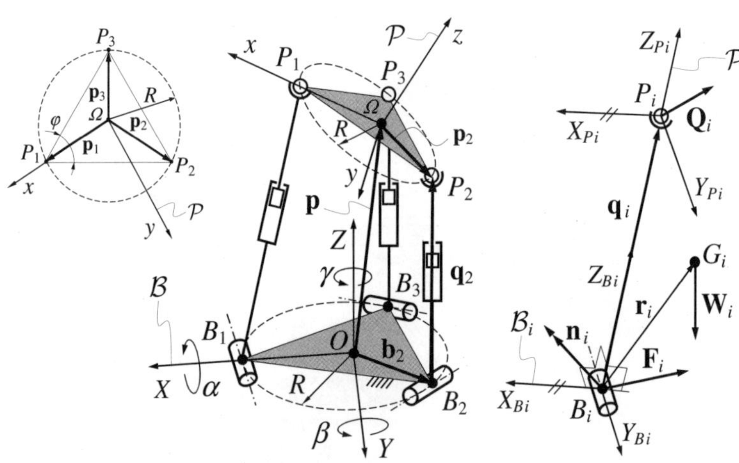

Figure 5. Kinematic sketch of the inverse mechanism of the lower parallel module: a) moving platform; b) 3-RPS parallel mechanism; c) one leg with the sketch of the constraint forces.

Referring to Fig. 5b, the position vector $[\mathbf{q}_i]_{\mathcal{B}}$ of point P_i, with respect to the origin B_i of frame $\mathcal{B}_i$, which Cartesian coordinates are expressed in frame $\mathcal{B}$, can be obtained by the following matrix equation

$$[\mathbf{q}_i]_{\mathcal{B}} = [\mathbf{p}]_{\mathcal{B}} + {}^{\mathcal{B}}[\mathbf{R}]_{\mathcal{P}}[\mathbf{p}_i]_{\mathcal{P}} - [\mathbf{b}_i]_{\mathcal{B}} \qquad \text{for } i = 1, 2, 3 \qquad (7)$$

which gives the unit vectors $[\mathbf{k}_{\mathcal{B}i}]_{\mathcal{B}}$, for $i = 1, 2, 3$, in the form

$$\left[\mathbf{k}_{\mathcal{B}i}\right]_{\mathcal{B}} = \frac{\left[\mathbf{q}_i\right]_{\mathcal{B}}}{\left\|\mathbf{q}_i\right\|} \qquad \text{for } i = 1, 2, 3 \quad (8)$$

where $\left\|\mathbf{q}_i\right\|$ is the magnitude of vector $\mathbf{q}_i$. Moreover, $[\mathbf{i}_{\mathcal{B}i}]_{\mathcal{B}}$ is given by

$$\left[\mathbf{i}_{\mathcal{B}i}\right]_{\mathcal{B}} = \left[\tilde{\mathbf{j}}_{\mathcal{B}i}\right]_{\mathcal{B}}\left[\mathbf{k}_{\mathcal{B}i}\right]_{\mathcal{B}} \qquad \text{for } i = 1, 2, 3 \quad (9)$$

$\left[\tilde{\mathbf{j}}_{\mathcal{B}i}\right]_{\mathcal{B}}$ being the skew-symmetric matrix of the unit vector $[\mathbf{j}_{\mathcal{B}i}]_{\mathcal{B}}$.

The transformation matrix ${}^{\mathcal{P}}[\mathbf{R}]_{\mathcal{B}i}$ from frame $\mathcal{B}_i$ to $\mathcal{P}$ is given by

$${}^{\mathcal{P}}[\mathbf{R}]_{\mathcal{B}i} = {}^{\mathcal{B}}\left[\mathbf{R}^T\right]_{\mathcal{P}} {}^{\mathcal{B}}[\mathbf{R}]_{\mathcal{B}i} \qquad \text{for } i = 1, 2, 3 \quad (10)$$

Finally, the transformation matrix ${}^{\mathcal{P}}[\mathbf{R}]_{\mathcal{P}i}$ is equal to ${}^{\mathcal{P}}[\mathbf{R}]_{\mathcal{B}i}$ because the axes of $\mathcal{P}_i$ are oriented like those of frame $\mathcal{B}_i$, as shown in Fig. 5c.

Thus, the static analysis is formulated by considering each leg of the 3-RPS parallel mechanism as a rigid link with weight $\mathbf{W}_i$ and mass center G_i, which position vector with respect to frame $\mathcal{B}_i$ is given by $[\mathbf{r}_i]_{\mathcal{B}i}$, as shown in Fig. 5c. Constraint force $\mathbf{Q}_i$ of each spherical joint and the constraint force $\mathbf{F}_i$ and moment $\mathbf{n}_i$ of each revolute joint can be expressed in $\mathcal{P}_i$ and $\mathcal{B}_i$, respectively, by

$$\left[\mathbf{Q}_i\right]_{\mathcal{P}i} = \left[Q_{ix} \;\; Q_{iy} \;\; Q_{iz}\right]^T, \; \left[\mathbf{F}_i\right]_{\mathcal{B}i} = \left[F_{ix} \;\; F_{iy} \;\; F_{iz}\right]^T,$$

$$\left[\mathbf{n}_i\right]_{\mathcal{B}i} = \left[n_{ix} \;\; 0 \;\; n_{iz}\right]^T \qquad \text{for i} = 1, 2, 3 \quad (11)$$

The static equilibrium of each rigid link B_iP_i is given by

$$\mathbf{Q}_i + \mathbf{F}_i + \mathbf{W}_i = 0 \qquad \mathbf{n}_i + \mathbf{q}_i \times \mathbf{Q}_i + \mathbf{r}_i \times \mathbf{W}_i = 0 \qquad \text{for } i = 1, 2, 3 \quad (12)$$

and, likewise, for the moving platform with weight $\mathbf{W}_p$, one has

$$\mathbf{R}_{\Omega} + \mathbf{W}_P + \sum\nolimits_{i=1}^{3} \mathbf{Q}_i = 0 \qquad \mathbf{M}_{\Omega} + \sum\nolimits_{i=1}^{3} \mathbf{p}_i \times \mathbf{Q}_i = 0\,. \qquad (13)$$

Equations (12) and (13) have been expressed in Cartesian form by referring to the frames $\mathcal{B}_i$ and $\mathcal{P}$, respectively, by giving a linear system with 24 equations in 24 unknowns, which are Q_{ix}, Q_{iy}, Q_{iz}, F_{ix}, F_{iy}, F_{iz}, n_{ix}, n_{iz} for $i = 1, 2, 3$.

Assuming the starting configuration for $\alpha = \beta = \gamma = 0$ and $[\,\mathbf{p}\,]_{\mathcal{B}} = [0, 0, 0.405]^T$, this algorithm has been tested by considering a vertical load of $[\,\mathbf{R}_{\Omega}\,]_{\mathcal{B}} = [0, 0, -1000]^T$ and a rotation β within the range $[0, -\pi/2]$.

The diagrams of Fig. 6a show the components of $\mathbf{F}_i$ with respect to $\mathcal{B}_i$ or $\mathcal{P}_i$, which are equilibrated by $\mathbf{Q}_i$, for $i = 1,2,3$. Likewise, Fig. 6b shows the diagrams of the components of $\mathbf{n}_i$ with respect to $\mathcal{B}_i$ for $i = 1, 2, 3$. Moreover, referring to the worst configuration of Figs. 2b and 3b, for $[\,\mathbf{p}\,]_{\mathcal{B}} = [-0.014, 0, 0.438\,]^T$ and $\beta = \pi/4$, and considering a load of $[\mathbf{R}_{\Omega}]_{\mathcal{P}} = [0, 0, -460]^T$ and $[\mathbf{M}_{\Omega}]_{\mathcal{P}} = [152, -135, 0]^T$, one has: $\mathbf{F}_1 = [0, -447.86, -1083.06]^T$, $\mathbf{F}_2 = [0, -213.25, -236.92]^T$; $\mathbf{F}_3 = [0, -682.47, 1635.57]^T$, $\bar{\mathbf{n}}_1 = [226.05, 0, 0]^T$, $\bar{\mathbf{n}}_2 = [86.37, 0, 0]^T$, $\bar{\mathbf{n}}_3 = [276.40, 0, 0]^T$, $\mathbf{Q}_1 = [0, 447.86, 1083.06]^T$, $\mathbf{Q}_2 = [0, 213.25, 236.92]^T$, $\mathbf{Q}_3 = [0, 682.47, -1635.57]^T$, which are referred to $\mathcal{B}_i$ or $\mathcal{P}_i$.

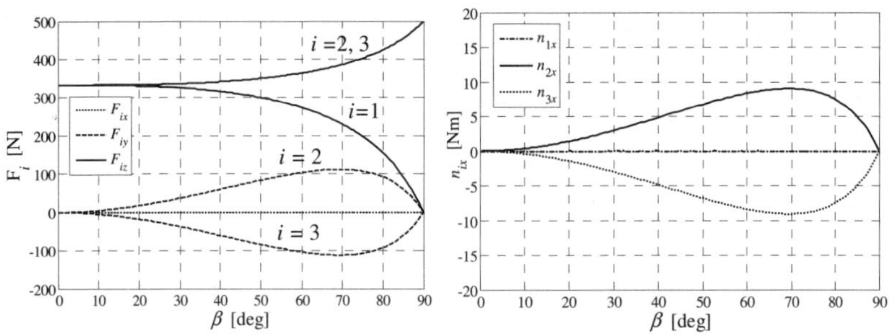

Figure 6. Diagrams of constraint forces and moments for $i = 1, 2, 3$.

Conclusions

The static analysis of a novel biped walking and climbing robot has been formulated by referring to the worst configurations for the walking stability of the robot, during its lateral and forward motions.

References

1. G. Figliolini and M. Ceccarelli, EP-WAR3 biped robot for climbing and descending stairs, Robotica, **22** (4), pp. 405-417, (2004).
2. G. Figliolini and P. Rea, Chapter 1: Mechanics and Simulation of Six-Legged Walking Robots, Climbing & Walking Robots, Towards New Applications, I-Tech Education and Publishing, (2007).
3. G. Figliolini, P. Rea and M. Conte, Mechanical design of a novel biped climbing and walking robot, ROMANSY 18, Robot Design, Dynamics and Control, Eds. V. Parenti-Castelli and W. Schiehlen, Springer, Wien-New York, pp. 199-206, (2010).
4. G. Figliolini, P. Rea and M. Conte, Stability analysis of a novel biped walking and climbing robot, Emerging Trends in Mobile Robotics (13th CLAWAR), Eds. H. Fujimoto, M.O. Tokhi, H. Mochiyama, G. S. Virk, World Scientific, Publishing, New Jersey, pp. 639-647, (2010).
5. K. Hashimoto, A. Hayashi, T. Sawato, Y. Yoshimura, T. Asano, K. Hattori, Y. Sugahara, Hun-ok Lim, A. Takanishi, IEEE/ASME Int. Conf. on Advanced Intelligent Mechatronics, Singapore, pp. 174-179, (2009).
6. H. Wang, Z. Qi, G. Xu, F. Xi, G. Hu, Z. Huang, Kinematics analysis and motion simulation of a quadruped walking robot with parallel leg mechanism, Open Mech. Eng. J. **4**, pp.77-85, (2010).
7. C. Qunfeng, Z. Xiangzhou, Constraint forces applied on limbs of 3-RPS parallel manipulator in static equilibrium, Int. Technology and Innovation Conference, pp. 1-5, (2009).

SECTION–10

INNOVATIVE SENSING AND ACTUATION

HIGH-ACCURACY MEASUREMENT SYSTEM FOR SECURE STEERING OF LARGE VEHICLES

H. MONTES[1, 2], C. SALINAS[1], J. SARRIA[1] and M. ARMADA[1]

[1]*Robot Locomotion and Interaction Group, Centre for Automation and Robotics, CSIC, Spain;* [2]*Faculty of Electrical Engineering, Technological University of Panama, Panama*
hector.montes@ csic.es

A measurement system of high accuracy is of great influence for guidance for automated vehicles. This paper presents the description of the high-accuracy measurement system implemented on a large mobile vehicle, an articulated bus Volvo B10M of 18 m in length. Several experiments have been performed on the test lane, with small radii of curvature, where a guide rail has been embedded. The measurement system consists of a guiding mechanical system, which roll within a guide rail designed for this purpose. The measurement system is capable of sending the bus control system the angular position of the guide rail with respect to the current position of the bus. The measurement of the angular position of the guide rail is obtained by means of two incremental encoders with resolution up to 0.036 degrees. In addition, the measurement system acquires the advance speed data from the bus. In this paper several experimental results with the utilization of the measurement system are presented, with the purpose to develop the lateral and longitudinal control algorithms to be implemented in the bus.

Keywords: High accuracy measurement system, autonomous transport systems, automated vehicles, lateral and longitudinal control.

1. Introduction

The development of automated vehicles has been the subject of important research activities, where measurement systems and control systems play a very relevant role [1]. On the other hand, an increasing interest exists of developing new systems of massive transport safer, more reliable, and with adequate comfort for the users in the field of the Intelligent Transportation Systems [2, 3].

During the last 30 years, several researches in the field of the autonomous transport system have been realized, and several automated transport systems have been constructed, which contain different measurement systems or guiding systems. Thus, BRT systems have several innovative features. Rouen, France, has an optically guided bus that uses the Irisbus Civis dual-mode diesel-electric bus [4, 5]. However, this system has a disadvantage that is difficulty in keeping the pavement striping crisp and clean, for appropriate tracking of the optical system. Essen, Germany, and Leeds, England, have mechanically guided bus. One version of mechanical guidance systems utilizes an arm with a small rubber wheel on one end. The other end of the arm is attached to the bus steering axle such that, when it runs up against a concrete curb, it helps the driver guide the bus closer to the platform edge [6]. In the Netherlands APTS (Advanced Public Transport Systems) developed a new public transport vehicle, the Phileas [7]. This vehicle is equipped with an all-wheel steering docking system, and can drive and steer automatically on dedicated lanes equipped with magnets. Other autonomous transport system is the Translohr, which is a tramway on tires based in a guiding system with rollers. It is equipped with a guiding system consisting of a single central rail [8].

In this paper a measurement system of high-accuracy implemented on guiding mechanical system for research on automatic control of articulated bus is presented. The experiments are performed on an experimental platform developed for this target. Platform consists in an articulated bus, Volvo B10M, where the mechanical system for guidance has been installed, and a ground test area composed of asphalt roads (guide rail embedded) inside CSIC premises [9].

2. Description of the guiding mechanical system

The guiding mechanical system is installed on the axle of the front wheels of the articulated bus and is based on the patent of the RBS [10]. In addition, the mechanical box for guidance is build of steel with the aim of providing high fixing to the guide rail to prevent the derailment of the bus. This provides a value added security to the full system. The mechanical box has a support system (arms) which is fixed on the axle of the front wheel of the bus as shown in Figure 1. The support system allows movement of the guiding mechanical system in the horizontal plane, lateral plane and the plane of advance of the vehicle (yaw, roll and pitch). Besides, inside of the support arms a suspension system is installed, which consist of a spring-damper element used to reduce the vibrations of the guiding mechanical system and, in turn, to force the two vertical wheels of the mechanical box to have contact with the bottom of the guide rail. This provides the correct displacement of the mechanical box into the guide rail.

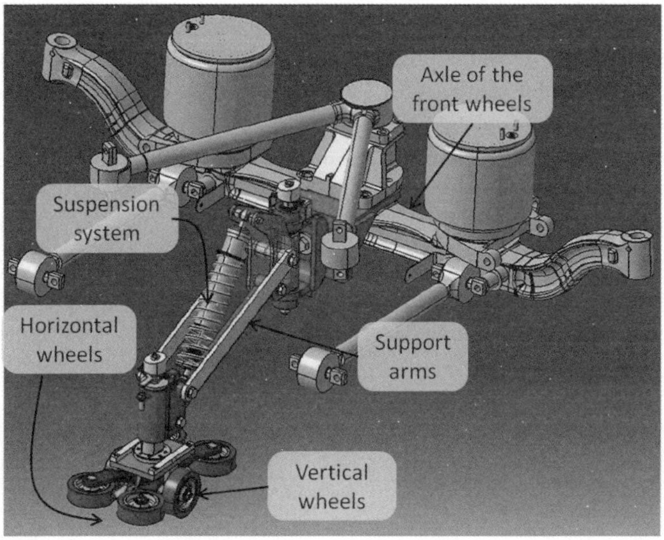

Figure 1. Guiding mechanical system.

3. Description of the high-accuracy measurement system

The guiding mechanical system is fully instrumented to carry out the tracking of the guide rail. With the data provided by the measurement system, several lateral and longitudinal control strategies can be applied to the bus. Figure 2 shows the main elements of the architecture of the measurement system presented in this paper, and besides shows other sensors that not describe in this article.

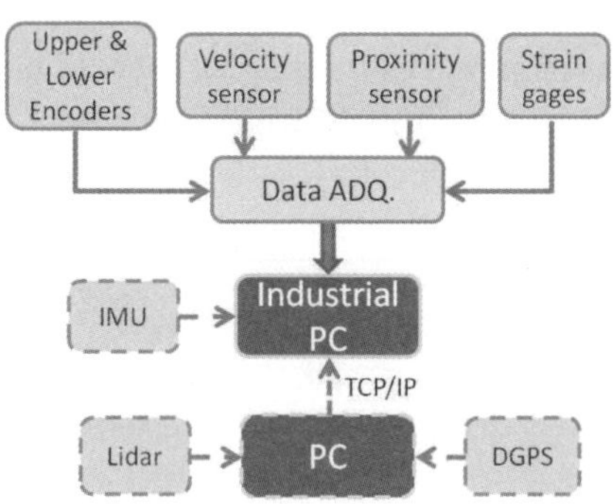

Figure 2. Architecture of the measurement system.

The measurement system consist of two incremental optical encoders, two inductive proximity sensors, strain gauges, and speed sensor, installed on the mechanical system for guidance of the bus. The incremental encoders measure the angular displacements (upper and lower) of the mechanical system on the horizontal plane of the guide rail. These measurements are sent to the control system through the data acquisition board. The inductive proximity sensors are used to initializing the process for guidance of the bus, aligning the mechanical box with the longitudinal axis of the bus. Strain gauges provide information on efforts related to obstruction in the guide rail and possible derailments. Figure 3 shows high-accuracy measurement system installed on the guiding mechanical system of the bus.

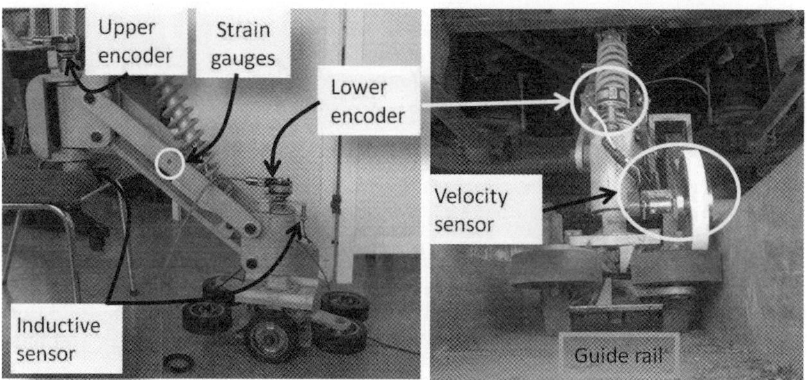

a) b)

Figure 3. Measurement system installed on the guiding mechanical system. (a) Installation process; (b) complete system installed on the axle of the front wheel of the bus.

A classic example of guide rail tracking by mean of the guiding mechanical system is shown in Figure 4. In this schematic diagram are represented the angles measured by the upper and lower encoders (θ_1 and θ_2, respectively), angle of the measurement system (θ_{MS}), the minimum and maximum angles that can have the measurement system (θ_{MSmin}, θ_{MSmax}), and the main lengths of links (l_0, l_1, l_{2a}, l_{2b}, d_2).

The values provided by θ_{MSmax} and θ_{MSmin} are boundary constraints of the measurement system. Therefore, the θ_{MS} supplied to the control system must be within these boundaries. Then, the set of equations of angular displacement of measurement system, which involve the limits (θ_{MSmax} and θ_{MSmin}), are:

$$\theta_{MS} = \text{atan}\left[\frac{2l_1S_1 + 2l_{2b}S_{12}}{2(l_0 + l_1C_1) + l_{2b}C_{12}}\right] \tag{1}$$

$$\theta_{MS\,(\text{min, max})} = \text{atan}\left[\frac{2l_1S_1 + 2l_{2b}S_{12} \mp (l_{2a} + d_2)C_{12}}{2(l_0 + l_1C_1) + l_{2b}C_{12} \pm (l_{2a} + d_2)S_{12}}\right] \tag{2}$$

where, $S_1 = sin(\theta_1)$, $C_1 = \cos(\theta_1)$, $S_{12} = \sin(\theta_1 + \theta_2)$, $C_{12} = \cos(\theta_1 + \theta_2)$; and the links dimensions are $l_0=241.5mm$, $l_1=568.9mm$, $l_{2a}=155mm$, $l_{2b}=338.2mm$, and $d_2=125mm$.

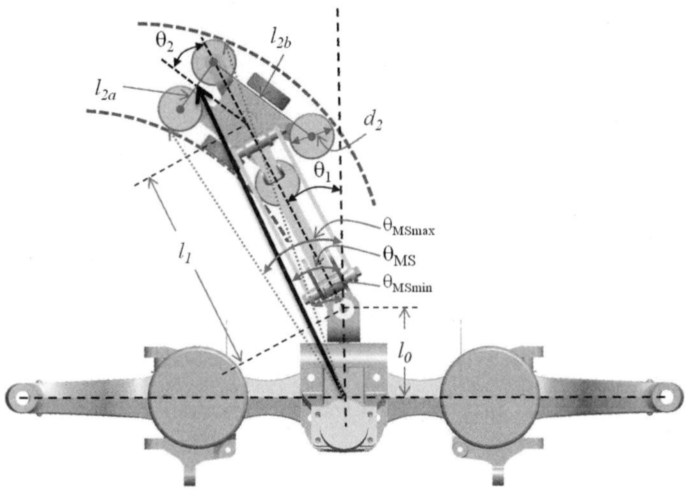

Figure 4. Main variables of the measurement system.

4. Experimental results

Several experiments were carried out, using basic strategies of lateral and longitudinal control on the bus, in order to observe the performance of the measurement system. Figure 5 shows the measurements acquired from the upper and lower encoders (θ_1 and θ_2, respectively) during tracking the trajectory of the test lane. The measurement noise of the encoder 1 (Figure 5) is due to the bus trajectory tracking commanded by the control system. Therefore, this noise is produced by the control system. The excursion of the upper joint is limited to the range $-40^o<\theta_1<40^o$ by the control system, to avoid damage on the "silent block" installed in this joint. Moreover, this constraint prevents undesirable angular displacements of guiding mechanical system.

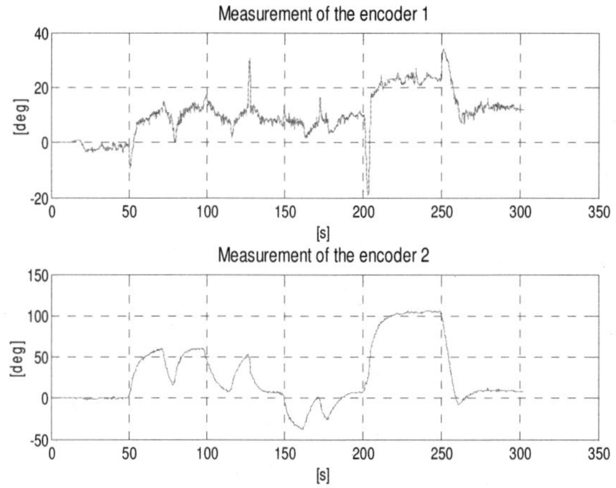

Figure 5. Measurement of the encoders installed on the guiding mechanical system (θ_1 and θ_2).

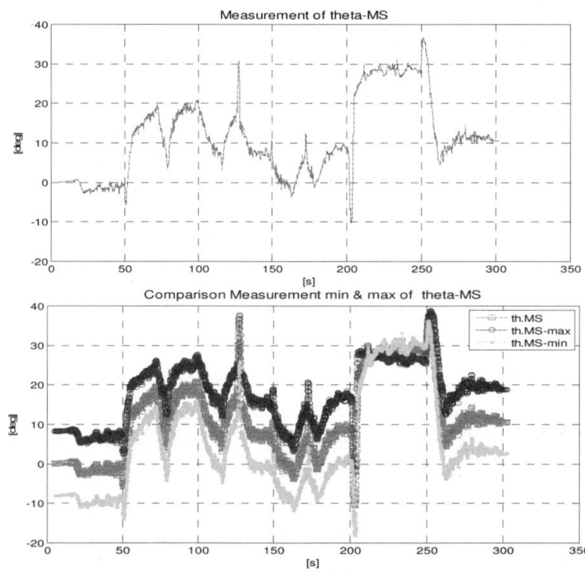

Figure 6. Measurement of the θ_{MS} (top graph) and comparison among boundaries constraints (θ_{MSmin_max}) and θ_{MS} (bottom graph).

Figure 6 shows the functions of θ_{MS}, θ_{MSmin} and θ_{MSmax} obtained with data from the encoders 1 and 2 and the links dimensions of the mechanical system for guidance, using the equations (1) and (2).

Using data supplied by θ_{MS}, taking into account the boundaries constraints given by θ_{MSmin} and θ_{MSmax}, lateral and longitudinal control strategies will be enhanced with respect to those provided in these experiments, which were used to validate the measurement system proposed in this paper.

However, to achieve the benefits described above, the high-accuracy measurement system will be combined with the Lidar System, IMU and DGPS, which will establish complete architecture of the measurement system.

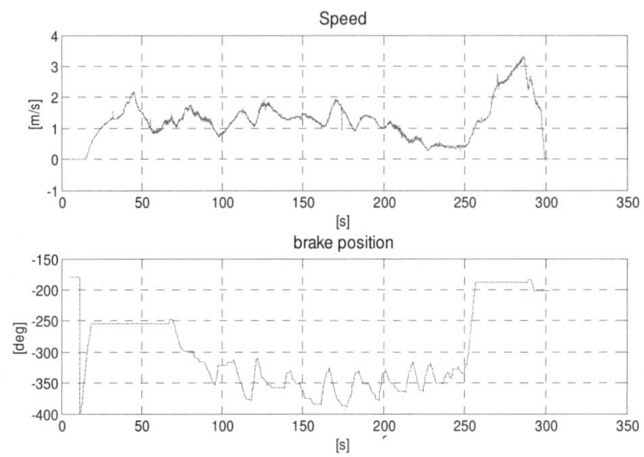

Figure 7. Forward speed of the bus (top graph). Brake pedal position (bottom graph).

To validate the high-accuracy measurement system were carried out trajectory tracking, by mean of the bus, to low speed, within the range 1m/s to 2m/s during most of the test lane, because it has multiple curves with small radii of curvature (10m) in relation to the bus length (18 m). Figure 7 shows the speed that the bus runs during the travel of the test lane. The speed of the bus, most of the trajectory, is between 1m/s to 2m/s, but only at the end of the test lane the speed increases up to 3m/s, because it is a straight line of test lane.

The bottom graph in Figure 7 shows representative values of the position of the brake pedal to maintain the desired speed of the bus. Worth noting that the bus, where the experiments have been carried out, has over 12 years, therefore, it has all the disadvantages of a bus of those years.

5. Conclusions

In this paper a high-accuracy measurement system for guidance of the large vehicles (a Volvo B10M) has been presented. Excellent results have been presented for the detection of the guide rail through of the mechanical box instrumented with several sensors.

The test lane has several curves with small radii of curvature (10m) with relation to the bus length (over 18m). In this case the high-accuracy measurement system is capable to acquire, in real time, the curvature of the test lane. Therefore, the contribution of this measurement system is essential for tracking of any test lane, due to the elements that make up. Mechanical components and electronics devices (sensors) perform an important function in the high-accuracy measurement system that is presented in this article.

6. Acknowledgements

The authors acknowledge partial funding of this research under: Robocity2030 S-0505/DPI-0176, and IMADE PIE/62/2008 (Comunidad de Madrid, ALDESA CONSTRUCCIONES SA, MAXIMASDE); FORTUNA D/030531/10 (Agencia Española de Cooperación Internacional para el Desarrollo, AECID). Dr. Héctor Montes acknowledges support form Universidad Tecnológica de Panamá and from CSIC under grant JAE-Doc.

References

1. R. Fenton, G. Melocik, and K. Olson, IEEE Transactions on Automatic Control, vol. 21, no. 3, 306 (1976).
2. S. Shladover, C. Desoer, J. Hedrick, M. Tomizuka, J. Walrand, W. Zhang, D. McMahon, H. Deng, S. Sheikholeslam, and N. McKeown, IEEE Transaction on Vehicular Technology, vol. 40, no. 1, 114 (1991).
3. W. C. Collier, R. J. Weiland, IEEE SPECTRUM, vol. 31, n. 4, 27 (1994).
4. J-M. Boucheret, Proc. of the Institution of Civil Engineers - Municipal Engineer, vol. 157, Issue 1, 13 (2004).
5. H. S. Levinson, S. Zimmerman, J. Clinger, and J. Gast, Transportation Research Board of the National Academies, Volume 1841, 1 (2003).
6. D. Kantor, G. Moscoe, and C. Henke (2006), Journal of Public Transportation, BRT Special Edition, 89 (2006).
7. Brookhuis, K.A. & De Waard, D, Proceedings IEA2006 congress (CD), Elsevier, (2006).
8. Z. Hamou Amar, P. Chanais and A. Aussem, Proc. of European Symposium on Artificial Neural Networks (ESANN 2006), 659 (2006).
9. H. Montes, C. Salinas, J. Sarria and M. Armada, Proc. Emerging Trends in Mobile Robotics, CLAWAR 2010, 793 (2010).
10. R. Blanco, Integral Train Guide System Comprising A Central Guide Rolling Box, Patent Nº WO/2004/106135, (2004).

THE USE OF VISION SYSTEMS FOR BUILDING TERRAIN MAPS FOR MOBILE ROBOTS*

JANUSZ KOWAL

AGH University of Science and Technology in Krakow, Department of Process Control, Faculty of Mechanical Engineering and Robotics, Al. Mickiewicza 30, 30-059 Krakow, Poland

BOLESŁAW KARWAT

AGH University of Science and Technology in Krakow, Department of Process Control, Faculty of Mechanical Engineering and Robotics, Al. Mickiewicza 30, 30-059 Krakow, Poland

ANDRZEJ SIOMA

AGH University of Science and Technology in Krakow, Department of Process Control, Faculty of Mechanical Engineering and Robotics, Al. Mickiewicza 30, 30-059 Krakow, Poland

This paper discusses the possibility of using three-dimensional vision systems working in the technology of laser triangulation as equipment for mobile robots. Two concepts for the use of a vision system are presented. The first is construction of a terrain map and use of that map to determine the robot's trajectory. The second discusses the possibility of using a vision system by mobile robots with manipulators. On the basis of a three-dimensional image of an object, the manipulator develops a strategy to grab the object or performs measurements on the test object.

1. Introduction

Over time, mobile robots are becoming increasingly more widely used in a growing number of areas. They work in different terrains and carry out a wide range of tasks. Thus, they require equipment capable of meeting such a broad spectrum of activity. One task is navigation in unfamiliar terrain. In mobile robotics, two subclasses of navigation tasks are distinguished: indoor navigation and outdoor navigation. This paper presents a method for building a three-dimensional image of terrain surrounding the mobile robot, using a three-

* Work developed at the Faculty of Mechanical Engineering and Robotics, AGH University of Science and Technology in Krakow

dimensional vision system. The paper also discusses the possibility of using such an image to develop the robot's movement path.

In the case of both indoor and outdoor navigation, the robot's route needs to be set out. If the mobile robot operates autonomously in the field, it must be able to determine the trajectory based on information collected from sensors mounted on the robot. The sensor system should make it possible to receive information from the environment in various weather and lighting conditions. For example, in the case of day and night work.

The paper proposes the use of a vision system equipped with a laser light which enables the construction of a terrain image irrespective of the ambient conditions surrounding the robot.

A model constructed using that method provides information about the actual dimensional parameters of the terrain. Knowledge of the vision system configuration allows designation of the resolution of measurements made on the terrain image in each of the coordinate axes.

2. Configuration and resolution of the vision system

Construction of a three-dimensional model requires proper configuration of the vision system and structural lighting to enable separation and reading of the height profile data of the studied terrain on the image.

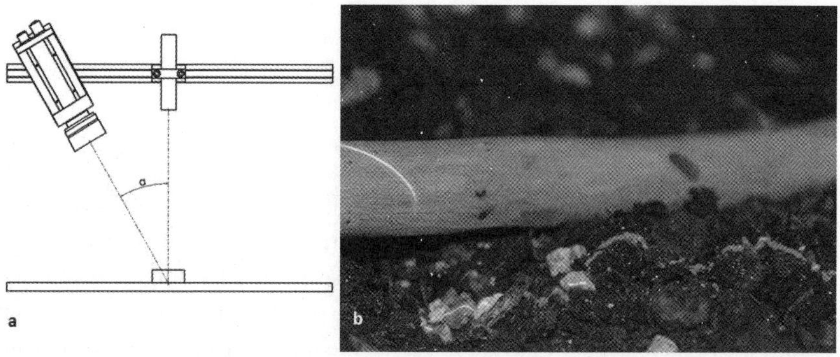

Fig. 1 Geometry of vision system settings (a) and a view of the laser line displayed on the ground (b)

An example of 3D vision system geometry may be a setting in which the laser plane is perpendicular to the surface of the terrain and the camera's optical axis is inclined at angle α to the plane of the laser (Fig. 1). Optical axis is a straight line passing through the centre of a camera's optical components and sensor (e.g. CCD or CMOS).

As a result of using such geometry, the plane of the cross-section of the tested terrain seen in the camera and produced by the laser beam is parallel to axis "Z" in the coordinate system connected with the mobile robot platform. Such a profile requires conversion of the geometry of the cross-section in the image created by the laser beam in order to obtain an actual three-dimensional cross-section and then a three-dimensional image of the terrain (Fig. 1).

In this geometry, on the surface of the terrain, there are surfaces marked with dotted line, which are invisible in the vision system due to the geometry of camera settings (Fig. 2). This phenomenon is known as occlusion, which is associated with the laser beam image being obscured by the shape of the object observed by the camera.

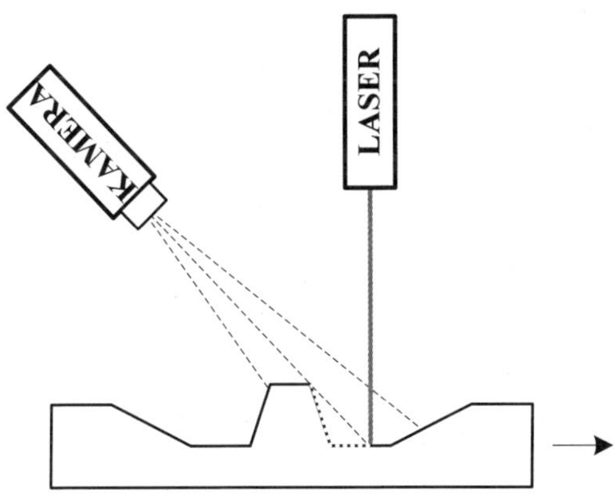

Fig. 2 Lighting the object with a laser beam

The value of points forming a vertical profile on hidden surfaces is equal to 0 (missing data point). In order to improve those parts of the image, one must calculate the height of such points using other points from the surroundings. Using an algorithm that transforms the profile image on surfaces which are not visible in the vision system allows an approximation of the full three-dimensional image of the terrain.

In the next stage of vision system configuration, it is necessary to set the vision system resolution in the adopted configuration. As a result of change in object height, the laser line image moves on the camera sensor. Designation of vision system resolution consists in determining such minimum change in object height at which the laser image moves exactly by one row of pixels on the sensor.

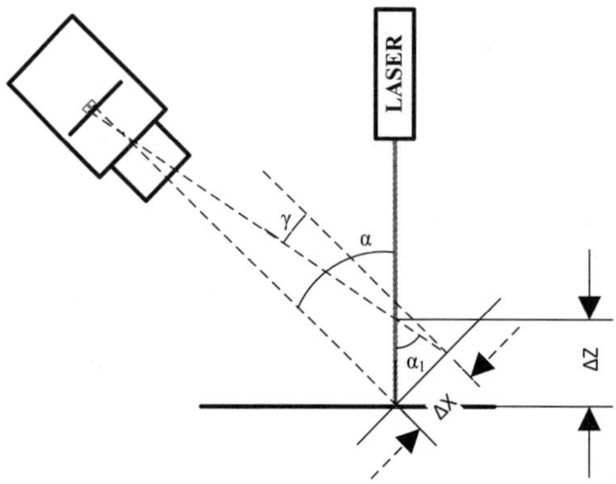

Fig. 3 Designation of vision system resolution in axis Z

On the plane parallel to the sensor plane, resolution ΔX is then determined based on the dimensions of the field of view and sensor resolution given in pixels in axis X. Angle α is between the axis of the laser and the optical axis of the camera.

Resolution ΔY in axis Y is defined as the distance between the generation of an image of successive terrain height profiles. The signal triggering collection of the next terrain height profile can be released from encoders incorporated in the propulsion system of the robot or from an encoder mounted on the mobile platform where the camera and laser are placed.

When calculating the resolution in the configuration presented, an approximation is used where we assume that angle α is equal to angle α_1. In fact, angle $\alpha_1 = \alpha - \gamma$. To simplify the calculation of resolution, it was assumed that angle $\alpha_1 = \alpha$. In addition, the resolution of a vision system is dependent on the field of observation of the terrain by the optical system and the resolution of the sensor used in the vision system.

In the task involving construction of a three-dimensional image of the terrain surrounding the mobile robot, it was assumed that the width of the field of view may be contained within the range of 2000 to 4000 mm.

$$\Delta Z \approx \Delta X / \sin(\alpha) \tag{1}$$

Where: ΔZ – resolution in axis Z, ΔX – resolution in axis X, α – angle between the axis of symmetry of the laser and the optical axis of the camera.

For calculation purposes, a 1536x512 pixel sensor and an optical system allowing observation of a 2000 mm wide object were used (FOV = 2000 mm). In addition, it was assumed that the camera is mounted in the adopted configuration at 45° angle to the plane of the laser. For such data, resolution in the X and Z axes was determined for the object.

$$\Delta X = 2000mm / 1536pixels = 1.3[mm / pixel] \tag{2}$$

$$\Delta Z \approx 1.3 / \sin(45°) = 1.84[mm / pixel] \tag{3}$$

On the plane parallel to the plane of the sensor, with the resolution ΔX in axis X known, resolution in the axis perpendicular to X is taken to be ΔX. Resolution in axis Y direction of the coordinate system is dependent on the shift of the laser line on the ground between the successive image acquisitions by the vision system.

For the geometry in question, there is also change in the resolution in axis Y, which results from the angle of the surface of the terrain in relation to the axis of the laser. As a result, the distance between the profiles created by the laser on the planes of the test object in successive scans will depend on the angle of inclination of the plane.

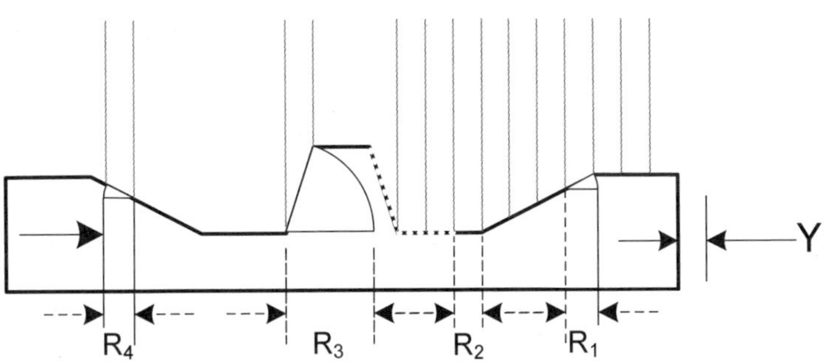

Fig. 4 Changes in resolution in axis Y

In case of considering the use of a vision system by a manipulator mounted on a mobile robot to examine, dimension or reproduce the shape of tested objects, a range of 100 to 400 millimetres can be adopted. Resolution of the system was calculated for a camera with a 1536x512 pixel sensor.

$$\Delta X = 400mm / 1536pixels = 0.26[mm / pixel] \quad (4)$$

$$\Delta Z \approx 0.26 / \sin(45°) = 0.36[mm / pixel] \quad (5)$$

It is also possible to use a different configuration of camera and laser arrangement in a three-dimensional vision system, as shown in the figure below. The choice of vision system arrangement geometry depends on the adopted design of a mobile robot as well as on the mounting system and the tasks to be performed.

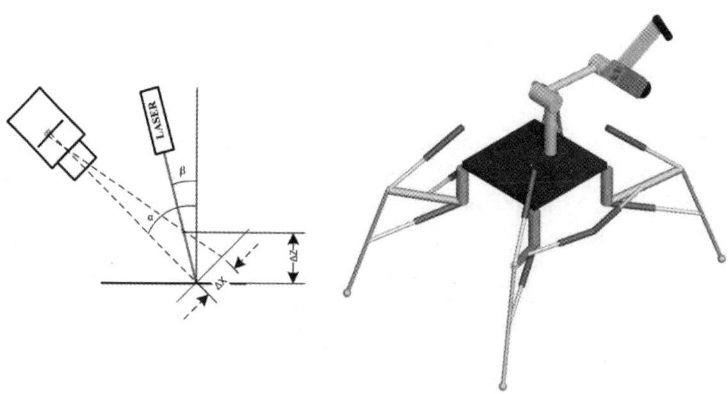

Fig. 5 Example of 3D vision system connection with walking robot

For the calculation of resolution, the following were adopted: angle $\alpha = 60°$ and angle $\beta = 50°$, field of view of the camera 2000 mm (FOV = 2000 mm). With such system parameters, its resolution was determined.

$$\Delta X = 2000mm / 1536pixels = 1.3[mm / pixel] \quad (6)$$

$$\Delta Z \approx \Delta X * \cos(\beta) / \sin(|\beta - \chi|) = 4.82[mm / pixel | \quad (7)$$

Such configuration allows scanning of the terrain at a greater distance from the mobile robot in comparison to the configuration shown in Fig. 3. However, deterioration of the resolutions obtained in the vision system, affecting the accuracy of three-dimensional image construction, is evident.

3. Configuration and resolution of the vision system

The three-dimensional image shown below was produced using the configuration shown in Fig. 1. The image shows the shape of the terrain and objects surrounding the mobile robot. Black spots represent surfaces which have been obscured as a result of occlusion.

Fig. 6 Example of 3D terrain image

In the processing of a three-dimensional image of the terrain, preliminary image filtering should be performed, using e.g. a median filter, and then fill in the missing data in the image (Missing Data Point). The 3D image view after conducting the transformations described above is shown in Fig. 7 It shows filling in of the previously missing surfaces.

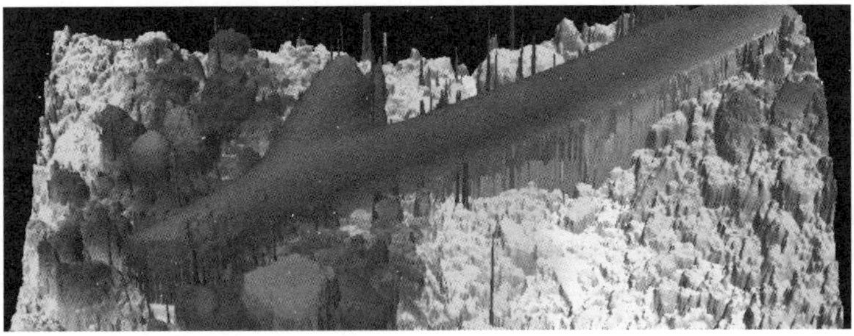

Fig. 7 3D image after transformations to remove the missing data

The 3D image enables measurement of the height of any cross-section. Analysis of the shape and measurements of height in cross-sections allows to develop an algorithm for movement in the terrain. The cross-section shown in the figure below provides an assessment of the depressions and height of obstacles along the route of movement of the mobile robot.

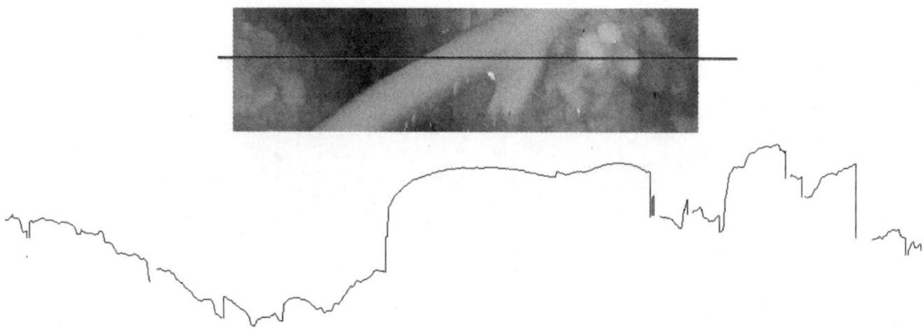

Fig. 8 Example of height profile extract from 3D terrain image

Summary

This paper presents a discussion of the application of vision systems in tasks involving creation of three-dimensional terrain map for a mobile robot. The vision system configurations presented and the image resolutions obtained confirm the possibility of construction of a precise three-dimensional terrain map and using it in the search of path for a mobile robot. In the course of further work, an algorithm for route selection, operating in real time in the robot's control system, will be developed.

References

1. Kowal J., Sioma A. Geometrical parameters product inspection with using 3D vision systems. International Carpathian Control Conference ICCC 2009, Zakopane, 24-27 May 2009
2. Sioma A.: Methods of picture processing in 3D vision systems RESEARCH REPORTS CEPUS PROJECT CII-SK-0030-03-0708 SYSTEMS EQUIPMENT PROCESS SOP2008, Krakow, 7-8 October 2008.
3. Sioma A.: Metody wizyjne – wyższa jakość kontroli procesu technologicznego [Vision Methods – higher quality of process control]. *Inżynieria Jakości*, Gdańsk 2009.

HYBRID FUEL CELL POWER FOR LONG DURATION ROBOT MISSIONS IN FIELD ENVIRONMENTS

Jekanthan Thangavelautham[1], Danielle Gallardo[2], Daniel Strawser[1], Steven Dubowsky[1]

[1]*Mechanical Engineering Department, Massachusetts Institute of Technology 77 Massachusetts Ave., Cambridge, MA, 02139, {jekan, dstrawse, dubowsky}@mit.edu*

[2]*Department of Mechanical, Aerospace and Nuclear Engineering, Rensselaer Polytechnic Institute, 110th 8th Jonsson Engineering Center, Troy New York, 12180*

Mobile robots are often needed for long duration missions. These include search rescue, sentry, repair, surveillance and entertainment. Current power supply technology limit walking and climbing robots from many such missions. Internal combustion engines have high noise and emit toxic exhaust while rechargeable batteries have low energy densities and high rates of self-discharge. In theory, fuel cells do not have such limitations. In particular Proton Exchange Membrane (PEMs) can provide very high energy densities, are clean and quiet. However, PEM fuel cells are found to be unreliable due to performance degradation. This can be mitigated by protecting the fuel cell in a fuel-cell battery hybrid configuration using filtering electronics that ensure the fuel cell is isolated from electrical noise and a battery to isolate it from power surges. Simulation results are presented for a HOAP 2 humanoid robot that suggests a fuel cell powered hybrid power supply superior to conventional batteries.

I. Introduction

Mobile robots, including walking robots are needed to perform long duration missions that are difficult, dangerous and tedious. These include search and rescue, repair, entertainment, sentry, surveillance applications [1, 2]. Continuous operation of these robots, lasting days and weeks (not hours) would be ideal for these applications. Typical power demands for field robots will vary significantly during a mission, often with high peak power demands. These field systems often have constraints on their mass, volume and noise.

Current power supply technology is a key limiting factor for long duration field robotic applications. Internal combustion engines can provide high power for long durations but produce toxic exhaust, noise and strong thermal signatures making them inappropriate for many important applications. Current rechargeable batteries have very low energy densities and high rates of self-discharge, requiring systems to stop and recharge every few hours, making them ineffective for continuous long duration missions. Hence, there is a significant need for a power supply that can provide the high total energy required for long duration missions that is quiet and clean.

Figure 1: (Left) Boston Dynamics Big Dog, a four-legged supply robot. (Right) Robonaut 2 repair robot.

II. Fuel Cell Power for Mobile Robots

Fuel cells are high energy sources of power that have been suggested for robots [7, 8]. They are a promising alternative mobile source of power and have the potential to overcome limitations of current batteries and internal combustion engines. They are simple electrochemical devices that convert chemical energy into electricity (Figure 2). Unlike a battery, fuel cells require a constant supply of fuel and oxidant to produce electricity. Proton Exchange Membrane (PEM) fuel cells are particularly attractive for robotics. These devices consist of simple solid state components sandwiched together as shown in Figure 2. They combine hydrogen fuel, and oxygen (from breathing air) through the most energy releasing reaction known, to produce electricity and water. It has been demonstrated; PEM fuel cells can reach 65-70 % or higher operating efficiencies at room temperature and produce clean water exhaust [9].

III. The Challenges of Fuel Cells for Robots

While PEM fuel cells are simple and sound great in theory, they have three fundamental problems for practical robotics applications. These problems are storage of hydrogen fuel, long-life reliability of fuel cells and low power. Hydrogen fuel due to its high energy content and low density is difficult to store. In our research, we have developed simple, innovative hydrogen storage technologies that promise energy storage densities better than the best batteries of today [5].

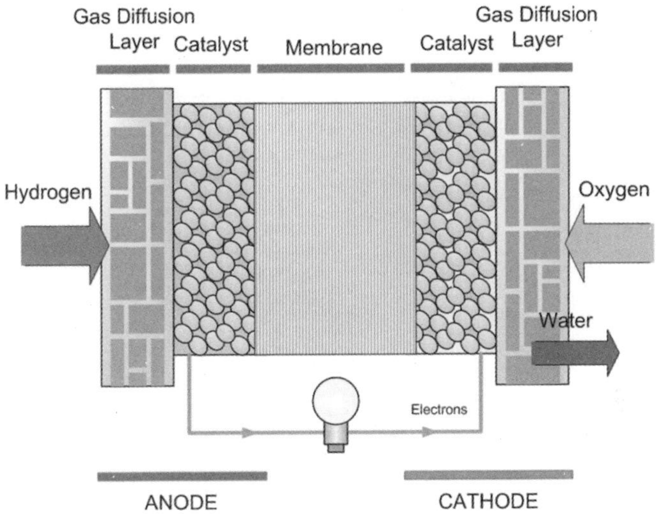

Figure 2: A PEM Fuel Cell Consumes Reactants, Hydrogen and Oxygen to Produce Electricity, Water and Heat.

Second, PEM fuel cells have been found to be unreliable [3]. Our studies of PEM fuel cells show that they are delicate and unreliable due to degradation of their components, resulting in short lives and premature failure [4]. However our physical models and experiments suggests that PEM fuel cells controlled to operate within narrow operating can be made robust, have long lives of 3-5 years or more and high operating efficiencies [4]. Among the factors known to degrade fuel cells, are high operating voltages and electrical noise. As discussed below, mobile field robots operating in unstructured environments are subject to very substantial variation, which, without proper control can result in fuel cell degradation that shortens their lives. A solution to this problem is discussed in the section below.

A third problem with fuel cells is that while they are high energy devices, they have relatively low power. This is a problem for robotics, where typical power requirements can vary substantially over a mission, with low-power rest periods and short bursts at peak power. These varying power demands are known to stress the fuel cells, resulting in short lives. A solution is to use fuel-cells in a hybrid system for mobile robots that maintains a fuel cell at optimal operating conditions to maximize life and efficiency, by protecting it from external electrical load variations, noises and meeting peak power requirement using a battery (see Figure 3).

Fuel cell hybrid systems have been subjected to meet rapid, transient power demands in large and stationary applications, and in robotics, to meet power surges [7, 8]. However these hybrid system designs have not considered the effects of fuel cell degradation.

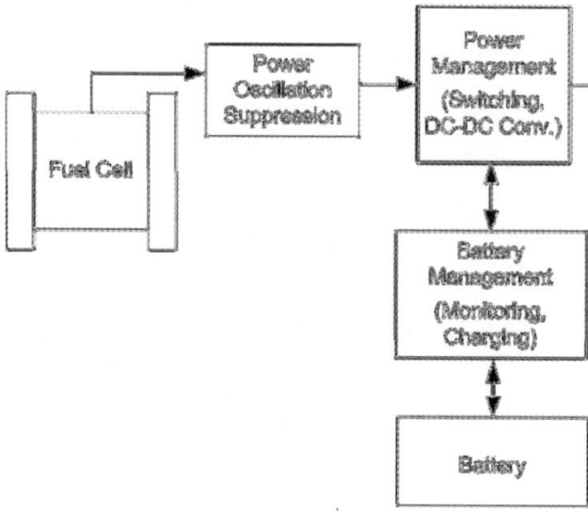

Figure 3: Proposed Fuel Cell-Battery Hybrid Power Supply for Robots.

IV. Research

The research presented here is focused on developing hybrid system design concept for mobile robots that have energy densities that exceed the best battery technology. The hybrid system is designed to meet the required peak power demands and isolate the fuel cell from degrading stresses of high and low frequency noises generated by conditioning circuits required for battery management. Physical models are used to simulate expected conditions and control systems are developed to demonstrate the concept. It is shown that the results are vast improvement over conventional batteries in terms of life, efficiency, energy density and power density.

V. Case Study: Power for Humanoid Walking Robot

Here a hybrid fuel cell power supply for a HOAP-2 humanoid walking robot (Figure 4) developed by Fujitsu is presented. The HOAP-2 is a 7.8 kg robot, with a maximum rated power of 250 W. It contains a 1.2 kg Nickel Metal Hydride rechargeable battery pack by default. The system contains 25 servo actuators, 6 for each leg and 5 for each arm, 2 for the head and 1 one for the waist. The robot has onboard computer equivalent to a PC-104 Pentium III system, a vision system consisting of 2 CCD cameras, onboard accelerometers, gyroscope and pressure sensors on each feet.

Figure 4: (Right) Fujitsu's HOAP 2 Robot (Left) Cyberbotics Webots™ Model of the HOAP-2.

A simulation model of HOAP-2 from Cyberbotics Webots™ [6] is used for our power demand calculations. The robot system consists of three different subsystems for power calculations, namely electro-mechanical system, computer + sensors and power system. For the electro-mechanical system the simulator model provides mechanical power output of the servo motors. The servo motors are assumed to have a 50 % electrical to mechanical efficiency. The computer and sensor system are assumed to be always powered and consume 40 W based on the HOAP-2 specifications. Below, power demand profiles of the robot's walking behavior is shown (Figure 5). For these scenarios, alternative power sources are compared with the default nickel metal hydride battery packs that weigh 1.2 kg.

VI. Fuel Cell Hybrid System

The fuel cell hybrid system consists of a fuel cell stack that provides steady power source and rechargeable lithium ion A123 Nanophosphate™ battery that meets peak power demands. The fuel cells within the stack are operated at constant operating voltage of 0.8, providing a 65 % operating efficiency. Our research into the degradation of fuel cells, based on models and experimental results shows that increased operating voltages exponentially decreases the life of the fuel cell [4]. Operating the fuel cell at constant voltage of 0.8 V or less ensures long-life, while providing sufficiently high operating efficiency. The fuel cell trickle charges the battery during idle times, ensuring the battery is fully charged, to meet power peaks. The Nanophosphate™ battery handles peak demands and can better handle deep battery discharges compared to conventional lithium ion batteries. Ensuring the battery is nearly fully charged, maximizes its life. The battery for the hybrid system is sized based on its specific power density to meet the maximum possible power requirements of the robot.

An oscillation suppression circuit interfaces a fuel cell to the power management system, consisting of power switching circuitry and a DC-DC convertor. The interface circuit effectively extracts the energy from the fuel cell and transfers it into the battery. The oscillation suppression circuits prevents any voltage oscillation from electrical circuits, particularly DC-DC convertors from being noticed by the fuel cell. This ensures the fuel cell operates at steady operating voltage, without any electrical load oscillations.

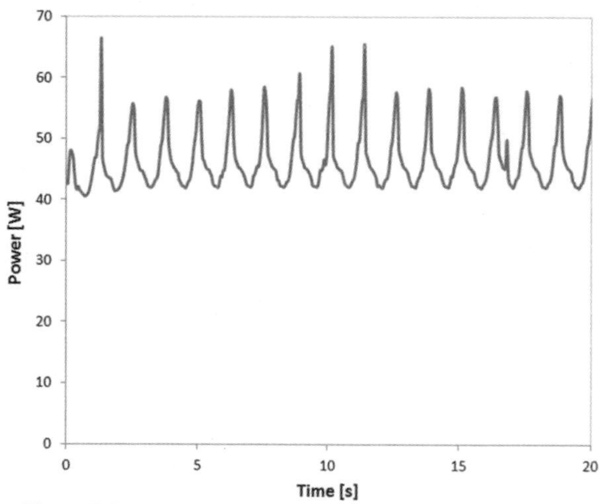

Figure 5: Power Demand of a HOAP-2 Robot, walking at 0.06 m/s.

VII. Hybrid System Sizing

Based on the power demand profiles (Figure 5), the fuel cell will provide a constant steady source of power of 45 W. While the power peaks will be handled by the battery. For 45 W steady supply of power, a fuel cell stack weighing 150 g will be required. For a 250 W peak required by the robot, a 135 g lithium ion Nanophosphate™ battery is required (with specific power of 1,850 W/kg for APR18650 cells) . Another 115 g is allocated for mass of power electronics and other items, leaving 800 g for the Lithium Hydride fuel supply with an energy density of 4,950 Wh/kg.

VIII. Power System Comparison

Four power supply configurations are compared, including a nickel metal hydride and lithium ion battery system, a fuel cell system and a fuel cell hybrid system. See Table 1.

Table 1: Power Supply Comparison for HOAP-2 Humanoid Robot

Power Supply	FC Stack Mass	FC Fuel Mass	Energy Density	System Life	Run-Time
NiMH Battery	-	-	40 Wh/kg	0.3 year	3 hours
Li Ion Battery	-	-	120 Wh/kg	1 year	9 hours
Fuel Cell	300 g	900 g	4950 Wh/kg	5 days	99 hours
Fuel Cell Hybrid	150 g	800 g	4950 Wh/kg	3 years	88 hours

Nickel metal hydride and lithium ion batteries have the lowest energy densities and thus provide short run-times, before requiring recharging. The system life of the batteries is computed based on expected lifetime of 1,000 charge/recharge multiplied by the run-times hours. A direct fuel cell system has the longest run-time. However the life of the system, based on our degradation models is expected to last just 5 days making this option impractical. The fuel cell hybrid system offers a good trade-off between both run-time and system life.

IX. Summary and Conclusions

Based on these results, the fuel cell hybrid system concept offers high energy density, long-life and would meet the required peak power demands of a battery. The key to our hybrid system concept is effective control and design, where a fuel cell and battery are optimally sized to minimize stress on the fuel cell, while enabling a battery to meet power demand peaks. By minimizing stresses on the fuel cell, the system can be operated for long-lives at high operating efficiencies.

X. References

1. Hägele, M., "Contribution to World Robotics 2005," Technical Report, *European Robotics Network*, 2005.
2. Asada, H., et al. "A Roadmap to US Robotics: From Internet to Robotics," Technical Report, Editor: *Computing Community Consortium/Computing Research Association*, 2009.
3. Rubio, M.A., Urquia, A., and Dormida, S. (2009) "Diagnosis of Performance Degradation Phenomena in PEM Fuel Cells," *International Journal of Hydrogen Energy*, pp. 1-5.
4. Thangavelautham, J., Dubowsky, S., (2011) "Controlling Degradation in Fuel Cell Power Supplies for Long-Life Field Applications," Submitted to *Journal of Power Sources*, pp. 1-24.

5. Strawser, D., Cheung, M., Thangavelautham, J., Dubowsky, S., "Application of Lithium Hydride Fed PEM Fuel Cells for Powering Long-Duration Missions in Mobile Robotics," Submitted to *IEEE International Conference on Robots and Systems 2011,* p. 1-8.
6. Michel, O. / Cyberbotics Ltd. (2004) "WebotsTM: Professional Mobile Robot Simulation," *International Journal of Advanced Robotic Systems*, pp. 39-42, Vol. 1 No. 1.
7. Kesner, S. B., Plante, J. S., Boston, P., Fabian, T., Dubowsky, S., "Mobility and Power Feasibility of a Microbot Team System for Extraterrestrial Cave Exploration," *Proceedings of the 2007 IEEE International Conference Robotics and Automation*, Rome, Italy, April 2007.
8. Joh, H., et al., (2010) A direct methanol fuel cell system to power a humanoid robot, *Journal of Power Sources*, Vol. 195, No. 1, pp. 293-298.
9. F. Barbir, *PEM Fuel Cells: Theory and Practice*, Academic Press, 2005.

SELF-CONFIGURING LOCALIZATION ALGORITHMS IN WIRELESS SENSOR NETWORKS: PROBLEMS AND ISSUES IN MOBILE ROBOT APPLICATIONS

MOHAMMAD MEHDI SANAATIYAN

Dep. of Electronic & Computer Engineering, Mazandaran Institute of Technology, Karegar Square., Babol, Mazandran 47166-96519, Iran

HAMID HASSANPOUR

Faculty of Electronic, Computer &IT, Shahroud University of Technology 7 Square, Shahroud, Semnan, Iran

Localization is an essential and fundamental problem in wireless sensor networks (WSNs). The most popular method for localization is using Global Positioning System (GPS) because it increases the consumed cost of sensor nodes particularly for wide environments. To solve the problems caused by nodes using of GPS, different algorithms are proposed for sensor nodes localization which don't need placing GPS on all the nodes. In this case, just a number of nodes exploit GPS for localization and by distribution of their coordinates; they can help other nodes of network in localization. Hence, a useful method is using of a mobile robot. Mobile robot is equipped with GPS can move in the space of network to obtain the place of other nodes. During using of mobile robot, one of the important issues is planning a suitable movement path for network traverse by the robot. In this paper some of the Self-configuring localization algorithms in WSNs are investigated, improved and simulated.

1. Introduction

Generally, "*localization*" or the ability to establish geometric or spatial relationships among objects plays the important goal of localization systems. In fact, determining the location of the sensor nodes with respect to a common reference, or in the context of the object being monitored by them, is an important aspect of WSN applications. Generally, there are two main methods for nodes localization in WSNs; *anchor-based* and *anchor–free method*. A significant number of localization methods use a number of nodes which these nodes know already their positions and play an important role in localization of other nodes. The positions of these nodes are determined by hand or by GPS. These nodes are said *anchor nodes*. As their name refers to it, *anchor-free* methods don't need any *anchor* and in these methods, final place of sensor node will depend on the coordinates of map and the presence or absence of connectivity between different nodes is used for localization. The advantages which *anchor-based* localization methods have compare to the *anchor-free* one include that these methods consume less energy and there is no need already to determine the domain of network deployment and the procedure of nodes distribution but instead of it, they have also some defects that include: these nodes are more expensive and to some extent, they cause to decrease system security.

In this paper, first, path planning of mobile robot is studied using of Graph theory and considering wireless sensor network as non – directional graph; this problem is illustrated by help of covering tree. Therefore, two algorithms of path traverse called as Most Margin Neighborhood First (MMNF) and Least Distance First (LDF) are suggested for covering tree which the proposed algorithm provides more powerful localization for path planning compared to the present algorithms such as BReadth First (BRF) and Back Tracking Greedy (BTG) and also in real conditions and simulation, it exhibits higher efficiency and plays an important role in decreasing the consumed energy by the mobile robot.

2. Related works

In past, several mobile robot based localization method are presented. These methods fall into two general classes. First class are methods in which mobile robot has not its local information whereas second one are methods in which mobile robot has its local position information. This local position information can be available for mobile robot by GPS:

A. First class: In methods in which mobile robot has not its local information, mobile robot considers its beginning point as the coordinates origin and moving along the sensor network domain, it sends its relative coordinates for the other sensors of the network. This method is usually used when we want to use localization information for routing in sensor network. The reason is that sensors' relative information is enough to determine the least distance.

B. Second class: Pathiran and others in [2] used radio waves and mobile robot to perform localization and received signal strength measurement to estimate mobile robot distance, their simulation has been for four sensors and they didn't present a clear path for their mobile robot movement. Corke and others [3,4] used a bird robot which is equipped with GPS for localization and Radio waves strength measurement to estimate the distance up to mobile robot. There are fixed robot in their method which localize a percentage of network nodes. Sichitiu and Remadurai [5,6] have also used GPS and Received signal strength measurement to determine the distance of node up to mobile robot. Ssu and the others [7,8] have presented a distance independent method. In this method, a mobile robot which is equipped with GPS, moves in operation environment and distributes its place everywhere.

To design movement path of mobile robot in which network nodes can localize themselves using of the received signals from it, three standing path of Scan, Double-Scan and Hilbert are proposed in [9]. Then, some methods are proposed to improve robot movement path to increase localization accuracy. In [10], two paths of Circles and S-curves are suggested. Two dynamic algorithms for path planning named as BRF and BTG are proposed which have better performance in respect of the consumed energy of the robot in comparison with the static methods for path planning [11].

1. Distance Measurement: In the study and review of mobile robot path planning, the orientation suggested by Sichitiu and the others [1] is used in which RSSI is used for distance determination. The regulation step for creating PDF table is before calculation step of the given place. This table is saved on all the nodes and writes each RSSI by Probability Distribution Function (PDF). The received RSSI by each node based on Probability Distribution Function is converted into the distance between two nodes.

2. Localization Process: In most localization algorithms, several models for distribution localization are suggested which don't need GPS installation on all the sensors. Localization algorithms proposed in [12,13] use just a limited number of nodes equipped with GPS and these nodes send their coordinates to other nodes to help them in localization. In this paper, a mobile robot is used for sensor networks localization. Based on these algorithms, during passing through the network area periodically and in definite intervals, mobile robot partially distributes beacon packages in width of network. When the node received beacon package, it refers to PDF table to obtain its distance from the anchors which distribute beacon packages. This process will be repeated for each beacon package which is received. When the node receives more than three beacon packages from the robot, it can calculate its coordinates.

3. Path planning for mobile robot

Main goal to design a path for mobile robot is to create suitable efficiency in terms of cost and precision in order to localization for network nodes. In fact, these methods are seeking to design a suitable path for mobile robot movement along the sensor network area such that most nodes of the network shall be localized and the robot shall consume less energy for this path traversing. Among the existing methods to design mobile robot path for localization of unknown nodes, BTG algorithm has a better performance to the other existing algorithms. This algorithm along with BRF algorithm have been proposed and evaluated in paper [11]. In this algorithm, for planning movement path of mobile robot, it was conducted in this way that during path traversing, for selecting the next node among the neighbors, the robot would pay attention to the marginal node which had a less weight. In this paper, for robot consumed cost decrement in path planning for mobile robot and also for localization of a more number of nodes by it, we suggest some methods which are highly more efficient and more inexpensive according to the performed evaluations. The proposed methods are within two algorithms named as LDF and MMNF which have been formed like previous algorithms that are BTG and BRF based on graph theory. The following hypotheses have been considered to design mobile robot path by implementing Graph Theory:

1. WSN nodes use multi-directional antennas.
2. Communication area is circular-shaped and by radius of R_c. communication radius of robot is also R_m.
3. All nodes of wireless sensor network have similar technical characteristics.
4. All wireless communications between the nodes are bi-directional.

Definition 1- Neighbor-Nodes: If during visiting time of robot with node v_i, the distance between the node v_n and robot is less than r_m, the node V_n is called as neighbor-Node of V_i.

Definition 2- Interior-Node: If during visiting time of robot with the node V_i, the distance between V_n an the robot is less than 2 R_m / 3, the node V_n is called as Interior-Node of V_i.

Definition 3- Margin-Node: If during visiting time of robot with the node V_i, the distance between the node V_n and robot is less than r_m and more than $2R_m$ / 3, the node V_n is called as Margin-Node of V_i.

Definition 4- Non–directional node: In graph G=(V(G),E(G)),V(G) and E(G) are the nodes of wireless sensor networks and communication directions between the nodes, respectively, which are called the vertices and edges of non-directional graph.

Definition 5- vertices weight in non-directional graph: If the number of unknown neighbor nodes v_i is n, the weight of node v_i is regarded as 1/n.

Definition 6- spanning tree: spanning tree is a sub-tree which includes all vertices of graph G and enough number of edges for the construction of the connective sub-graph but has not any cycle.

A. ***LDF algorithm:*** The first proposed dynamic algorithm is Least Distance First algorithm. As it is evident by algorithms name, the basis of mobile robot movement in this algorithm is the least distance of robot from the Neighbor-Nodes. In fact, this algorithm for selecting the next node of mobile robot in path planning algorithm, addresses the current distance of robot from the Neighbor-Nodes and we select a node which has the least distance. A justification which exist to select the least distance is that firstly, by selecting shorter distances, we can increase localization precision for network nodes because at less distribution distance of beacon, mobile robot can often distribute its place, therefore at less distances, we expect better localization, secondly, by increasing nodes visited by the robot, more beacons will be distributed in the network, this causes to

Step 1

A margin vertex v_0 regard as the root of the Spanning Tree.

The ROBOT starts to visit vertex v_0

Step 2

For all the Neighbour-Nodes of vertex v_i visited currently

If Neighbour-Node v_n is Interior-Node of vertex v_i

Delete the vertex v_n

Else

compute ***Distance*** of margin-node v_n

End

End

If (***Distance*** of vertex v_n is ***minimal***)

Take vertex v_n for the current visited vertex;

Go to Step 2;

End

Step 3

If (there are WSN nodes that are not localized)

Take the nearest vertex v for the current visited vertex;

Go to Step 2;

End

End of algorithm

Algorithm1: Least Distance First (LDF)

increase the probably of receiving more beacons by each node in the network and consequently, the number of nodes which can be localized, will be increased. However, regarding to increase in precision due to short distribution distances by mobile robot, its consumed energy will be increased which a tradeoff between localization precision of nodes and the consumed energy of mobile robot is observed.

B.*MMNF algorithm:* The fundamental problem of LDF algorithm is that traversing steps of the robot is relatively high and increases its consumed energy. This means that time and energy is not desirably used in some way. To remove LDF weak point, we suggest an algorithm named as Most Margin Neighborhood First. In this algorithm, for selecting the next node during path traversing in network graph, we select a node that has more margin-neighbors.

Step 1

A margin vertex v_0 regard as the root of the Spanning Tree.

The ROBOT starts to visit vertex v_0

Step 2

For all the Neighbor-Nodes of vertex v_i visited currently

If Neighbor-Node v_n is Interior-Node of vertex v_i

Delete the vertex v_n

Else

compute ***number of Margin-Neighborhoods*** of margin-node v_n

End

End

If (***number of Margin-Neighborhoods*** of vertex v_n is ***maximal***)

Take vertex v_n for the current visited vertex;

Go to Step 2;

End

Step 3

If (there are WSN nodes that are not localized)

Take the nearest vertex v for the current visited vertex;

Go to Step 2;

End

End of algorithm

Algorithm2: Most Margin Nodes First (MMNF)

4. Simulations

The performance of the proposed algorithms (i.e the order of visiting nodes) in constant conditions includes number of nodes, network dimensions and radius of communications is shown in figures 1 and 2. Simulation environment is 35m*35m in size, nodes number is 50 and communication radius is 10m.

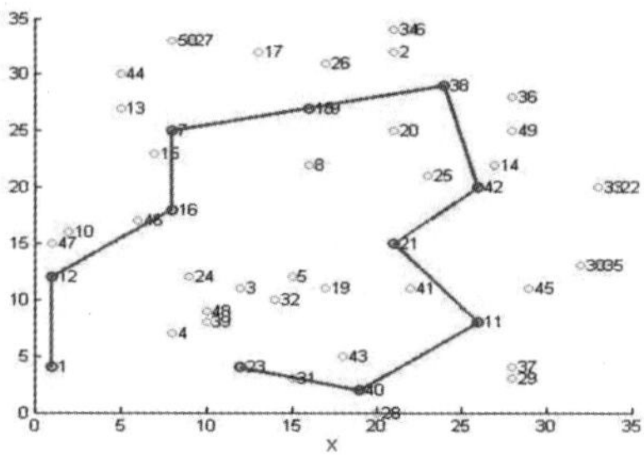

Figure1. The order of nodes visiting in LDF algorithm

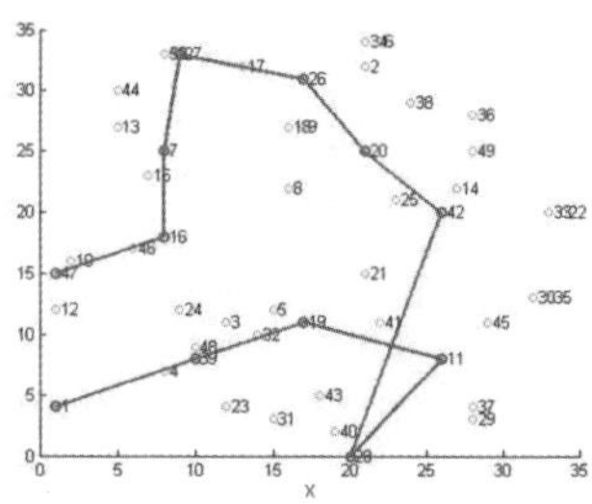

Figure2. The order of nodes visiting in MMNF algorithm

In the above figures, green nodes represent network nodes and red nodes represent nodes visited by mobile robot and robot movement path in any of the above-mentioned algorithms is shown by red. For comparison between the proposed algorithms and the previous dynamic algorithms, several simulation scenarios are conducted which we point out to two cases of these scenarios, and then we will analyze the obtained results. In simulation scenarios, we have studied three criteria of the number of visited nodes, the traverse distance and the number of localized nodes by robot against three factors of change in nodes number, change in simulation dimensions and change in communication radius, respectively.

5. Conclusion

In this paper, path planning for mobile robot in wireless sensor networks has been studied. WSNs are introduced as non-directional connective graph and the issue of path planning has been converted in to covering tree and traversing graph. Two algorithms of path traversing called as MMNF and LDF are proposed in distance statistics which are obtained from simulation show that these two algorithms have been adapt very well for wireless sensor networks applications in bigger scales. These two algorithms obtain relatively high efficiency for localization and perform localization by more effective use of energy. Also, due to shorter distribution distance, LDF algorithm increases the probability of the required information reception to determine the situation of unknown nodes of the network and as a result, it improves the factor of localized nodes number in the network. This is while the proposed MMNF algorithm because of addressing margin neighbors which have more distance to interior neighbors of robot, exhibit better performance in terms of energy consumption and path traversing time. This is because which in MMNF algorithm, robot visits less nodes number and consequently traverses less distance to cover all nodes of the network. Therefore, it spends less energy and cost for doing this. According to the statistics which was obtained by simulation, in 63% of cases, MMNF algorithm has a better efficiency in respect of the traverse distance to the other algorithms, In 21% of cases, LDF algorithm and just in 16% of cases, BTG algorithm have been efficient in these respects. In respect of the localized nodes number according to diagram1, in 91% of cases, LDF algorithm has better efficiency in respect of the localized nodes number to the other algorithm. In 9% of cases, BTG algorithm has been efficient and in none of cases, MMNF algorithm has a better performance to the other two algorithms in this case. In the case of the number of visited nodes, MMNF algorithm acts always better than BTG algorithm and BTG algorithm acts always better than LDF algorithm.

References

[1] M. L. Sichitiu , V. Ramadurai, "Localization of Wireless Sensor Networks with a Mobile Beacon" In Proc. of MASS, September 2004.

[2] P. Pathirana, A. Savkin, S. Jha, and N. Bulusu, "Node localization using mobile robots in delay-tolerant sensor networks", IEEE Trans. Mobile Computing, Vol. 4 ,No. 3, pp. 285-296, 2005.

[3] P. Corke, R. Peterson, and D. Rus, Networked robots, "Flying robot navigation using a sensor net", in Proc. 11th ISRR, Siena, Italy, Nov. 2008

[4] M. Bahi, A.Makhoul and A.Mostefaoui," A Mobile Beacon Based Approach for Sensor Network Localization", Third IEEE International Conference on Wireless and Mobile Computing, Networking and Communications (WiMob), 2007.

[5] M. Sichitiu and V. Ramadurai," Localization of wireless sensor networks with a mobile Beacon", in Proc. 1st IEEE MASS, Fort Lauderdale, FL, pp. 174-183, Oct.2004

[6] K.-F. Ssu, C.-H. Ou, and H. C. Jiau," Localization with Mobile Anchor Points in Wireless Sensor Networks", IEEE Transactions on Vehicular Technology, vol. 54, no.3, 1187-1179, 2005 .

[7] K. Ssu, C. Ou, and H. Christine," Localization with Mobile Anchor Points in Wireless Sensor Networks", IEEE Trans. Vehicular Technology, Vol. 54, No. 3 ,May 2005.

[8] Peter Corke, Ron Peterson, Daniela Rus, "Localization and Navigation Assisted by Cooperating Networked Sensors and Robots", International Journal of Robotics Research 24(9):786-771, October,2005.

[9] D. Koutsonikolas, S. M. Das, and Y. C. Hu, "Path Planning of Mobile Landmarks for Localization in Wireless Sensor Networks", in Proc. of ICDCSW, 2007 .

[10] R. Huang, G. V. Zaruba," Static Path Planning for Mobile Beacons to Localize Sensor Networks", In IEEE PerComW, 2007.

[11] Hongjun Li, Jianwen Wang, Xun Li and Hongxu Ma, "Real-time Path Planning of Mobile Anchor Node in Localization for Wireless Sensor Networks", The IEEE International Conference on Information and Automation, pp.384-389, 2008.

[12] Shang Y,Ruml W,Zhang, "Localization from Mere Connectivity", MobiHoc'03[C],Annapolis, Maryland, USA, 2003.

[13] Niculescu , B. Nath, " Ad Hoc Positioning System(APS)", In Proc. Of Globecom, November 2001.

SECTION–11

LOCOMOTION

A SPARE METHOD FOR OBSTACLE SENSING IN LEGGED LOCOMOTION

Q. BOMBLED* and O. VERLINDEN

Department of Theoretical Mechanics, Dynamics and Vibrations,
University of Mons, Place du Parc, 20 – Mons, 7000, Belgium
**E-mail: quentin.bombled@umons.ac.be*
http://mecara.fpms.ac.be/

This article assesses the motor current sensing as a "backup" of force sensors for the ground detection in legged locomotion. The general method is first described: it is simply based on the difference between the real current measurement and a real-time estimation in the case of a free motion. The estimator is based on the dynamic model of a leg, where joint velocities and accelerations are derived from position measurements on the real robot. The ground detection algorithm is implemented in a free-gait framework with posture correction on the real robot. Tests in real environment show that the algorithm is sufficiently reliable for ground detection. Finally, a discussion establishes the potential interests of the developed method.

Keywords: dynamic modeling, friction, real-time control, current sensing, posture control

1. Problem Formulation and Motivation

Walking robots are designed to move in an unstructured environment, where ground is quite uneven and inaccessible to classical wheeled robots. But their control is more complex, especially with regard to the management of the feet swing/support sequences, which must ensure the robot stability and drive it according to a desired trajectory. The adequate sequence of swing(s)/support(s) can only be determined by knowing that leg is effectively on the ground. Therefore, a reliable ground detection method is of first importance.

The most basic solution consists in equipping the feet with mechanical switches, which deliver a simple on/off information. The most widespread solution is to insert force sensors at leg tip[1–3] so that normal forces (and eventually the tangential forces) coming from ground contact can be measured. But, in addition to these sensors located at feet, it could be interesting to add redundancy at joint level, for two main reasons:

- the importance of ground detection in a gait algorithm is such that any failure at feet sensors could have disastrous consequences;
- with information at joint level, an impact located anywhere on the leg can be detected.

The redundant information we propose is the current measured in the DC motors of the joints. The key idea is to compare the real joint force/torque derived from the current sensing, with an estimation computed from a leg moving freely; therefore a difference important enough reflects that an "unexpected" load is encountered by the leg. The current sensors have the advantage of being readily embeddable, inexpensive and easy to implement in the hardware driver of DC motors.

In this paper we develop a general methodology for obstacle detection based on the estimation of joint forces/torques from currents measurement and a leg dynamic model. The developments are made to be applied to any leg structure, provided that there is no over- or under-actuation. We implement the approach for the ground sensing in the case of a robot evolving on hard and natural terrain. The walking machine AMRU5 is used for the experimental validation. Its leg mechanism is represented in Fig. 1. A picture of the actuation is shown on the left, while a CAD view of the pantograph mechanism is depicted on the right. The position of each joint is given by q, while the force/torque acting on it is denoted by τ.

The methodology is detailed in Sec. 2. The assessment of the joint estimation mechanism is made in Sec. 3. Results for the ground detection algorithm are detailed in Sec. 4. The rough implementation of ground detection leads to a drift in the robot posture during walking, so a posture correction is proposed in Sec. 5. The behavior of the robot in natural environment is examined in Sec. 6, before concluding in Sec. 7.

2. Methodology

In the joint space of one leg ($\underline{\boldsymbol{\tau}} = (\tau_r, \tau_v, \tau_h)$ of dimension $3{\times}1$), the joint force balance is written:

$$\underline{\boldsymbol{\tau}} = \underline{\boldsymbol{\tau}}_{obs} + \underline{\boldsymbol{\tau}}_{iner} + \underline{\boldsymbol{\tau}}_{grav} + \underline{\boldsymbol{\tau}}_{fric} \tag{1}$$

where

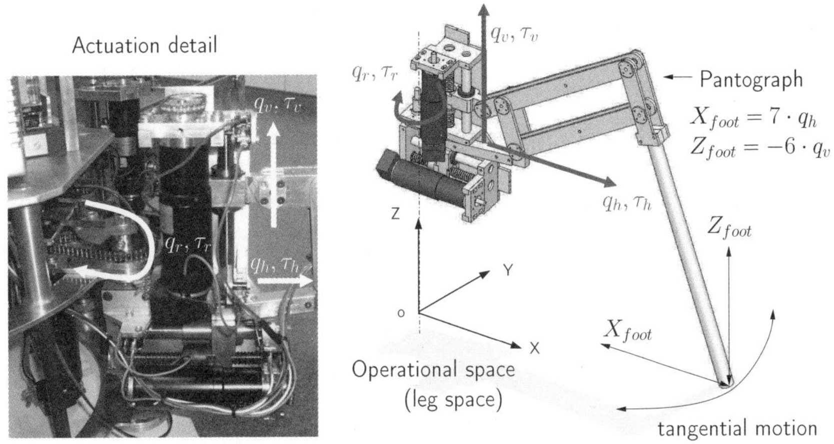

Fig. 1. AMRU5 leg and joint estimation mechanism for q_v

- $\underline{\tau}$ are the motor torques;
- $\underline{\tau}_{obs}$ are the contributions due to obstacle contact;
- $\underline{\tau}_{iner}$ and $\underline{\tau}_{grav}$ are the contributions of inertia (including Coriolis and centrifugal effects) and gravity;
- $\underline{\tau}_{fric}$ are the contributions of friction (in a general sense, that means any effect against the motion).

A dynamic model of the leg, developed with the EasyDyn[4] C++ library, estimates the joint force/torque vector $\hat{\underline{\tau}}$:

$$\hat{\underline{\tau}} = \hat{\underline{\tau}}_{iner} + \hat{\underline{\tau}}_{grav} + \hat{\underline{\tau}}_{fric}. \tag{2}$$

in real-time (while the leg is moving), from real joint positions, and an estimation of their velocity and acceleration. The inertial and gravity terms are deduced from the virtual power principle, where the kinematics of all the bodies composing a leg has been expressed according to a minimal coordinate approach. The friction force at joint i is directly computed with a classical dry/viscous model of the form:

$$\hat{\tau}^i_{fric} = F^i_c + f^i_v \dot{q}^i. \tag{3}$$

Dry and viscous parameters, F_c and f_v respectively, have been identified by imposing constant velocity profiles to each joint separately, such that inertia effects vanish. If the leg model is accurate enough, difference between Eqs. 1 and 2 leads to:

$$\underline{\tau} - \hat{\underline{\tau}} \approx \underline{\tau}_{obs} \tag{4}$$

where the motor torques $\underline{\tau}$ are directly measured through the currents. It is possible to find the contact forces in the cartesian space $\underline{\boldsymbol{F}}_{obs}$ according to the duality principle between joint and operational space, we have:

$$\underline{\boldsymbol{F}}_{obs} = \left(\underline{\boldsymbol{J}}(\underline{\boldsymbol{q}})^T\right)^{-1} \cdot \underline{\tau}_{obs} \tag{5}$$

where $\underline{\boldsymbol{J}}(\underline{\boldsymbol{q}})$ represents the geometric Jacobian of the leg. The proposed method is hence a redundancy regarding force sensors, but it could be useful if they fail.

3. Performance of the joint estimation mechanism

A sawtooth and a sine profile have been assigned separately to the three joints of a leg. During the motion, the different contributions of Eq. (2) are computed in real-time. Joint velocities and accelerations are not directly measured, but calculated with numerical derivatives from the real joint positions. This derivation increases the noise in the estimations, that's why the position of the joint is low-pass filtered at 3 Hz before the derivative operations.

Fig. 2(a) shows the real vertical joint force τ_v and its estimation for a sawtooth profile. In this case, friction has the most important contribution, the inertia effects vanish and the gravity effect is quite insignificant. The ground detection occurs in a similar case, because the foot descent is made with constant velocity. Ground detection is detailed in Sec. 4.

The sine profile illustrated in Fig. 2(b) makes more clear the usefulness of a dynamic leg model. Here the inertial effects can not been neglected anymore, as they can represent up to 30 % of the total joint force. Results are similar for the other three joints but are not detailed in the scope of this article.

4. Ground detection

The previous force estimation algorithm has been successfully tested on the three joints of a leg, by imposing sawtooth and sine motions. From this, algorithm of Eq. (5) can be particularized to the the vertical joint in the case of a ground detection.

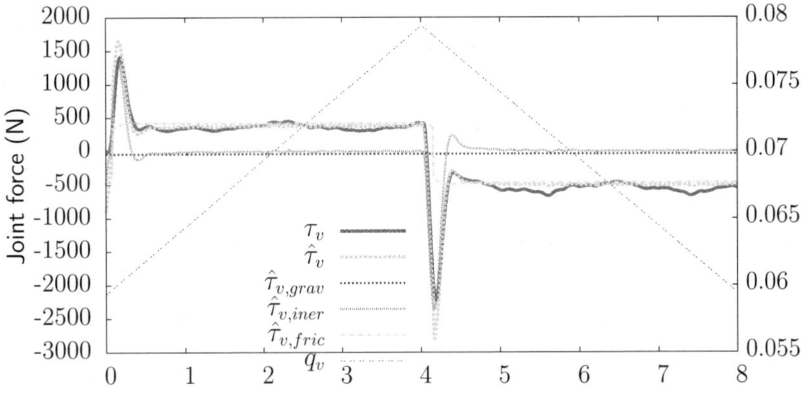

(a) Sawtooth motion

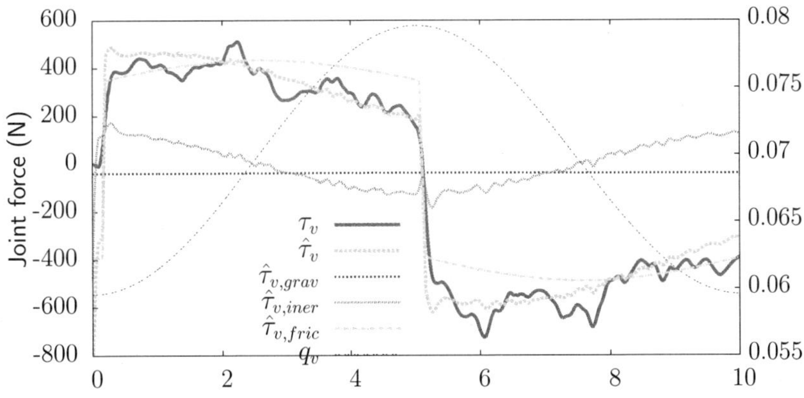

(b) Sine motion

Fig. 2. Comparison between real joint force τ_v and its estimation $\hat{\tau}_v$

Practically, contact occurs if:

$$F_{gnd} \geq F^* \tag{6}$$

where F_{gnd} is computed from Eq. (5) and F^* is a threshold proper to the robot. This value is set to ensure a "sufficient" ground contact force. Tests have been made on the six legs, by giving sawtooth motions to each foot, in such a way that they encounter the ground on which an HBM-U9B force transducer has been placed. When Eq. (6) is verified, motor voltage related to the vertical motion is cut off. Joint position q_v stops while the vertical

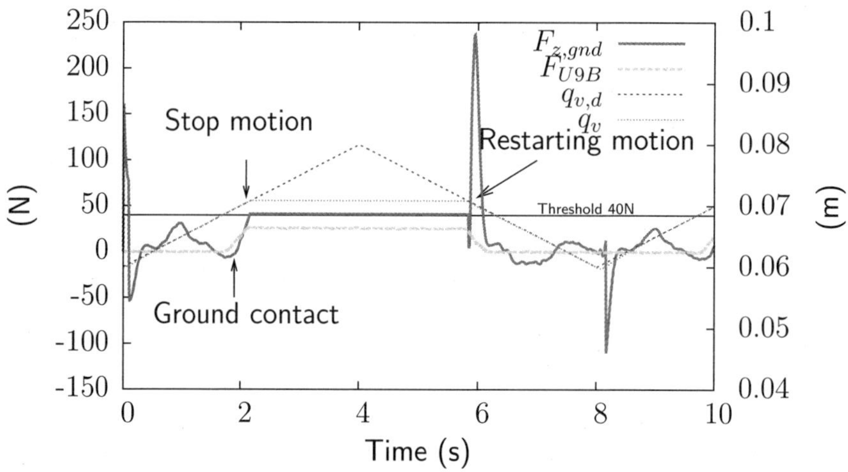

Fig. 3. AMRU5 leg and joint estimation mechanism for q_v

joint reference $q_{v,d}$ still evolves. Motion begins again when the foot reference goes back up. This situation is shown in Fig. 3, where ground contact occurs at t=2 s. It can be seen that the threshold of 40 N is not really reached, the ground contact force F_{U9B} is rather at 25 N. Effectively, some power is lost in the deformation of the mechanism (flexibilities). Moreover, the threshold can not be too low, because transmission mounting defects create important variation in the motor current, thus in computed joint force/torque.

The method has been validated by testing thresholds of 40, 50 and 60 N. The ground detection is correctly obtained but the resulting ground contact force estimation has errors going up to 40 %, depending on the studied leg. As expected, the results strongly depend on the mechanical quality of the transmissions.

5. Posture correction

Even if it walks on a flat ground, the use of the ground detection algorithm alone produces a drift in the robot posture, because of unavoidable modeling errors in Eq. (2). With force sensors, the posture of the robot can be adapted with force control. But with current sensing, the measure is not possible when joint is not moving, so the posture adaptation we implemented is a simplification of the Porta and Celaya's one.[5] The input of the algorithm are the local roll $\Delta\psi$, pitch $\Delta\theta$ and altitude Δz of the robot. The criterion to minimize is given by:

$$D = \sum_{i=1}^{n_{leg}} (p_z^i - r_z^i)^2 \tag{7}$$

where r_z^i is the distance of foot i with respect to a plane parallel to the robot body in an adequate reference posture, and p_z^i computed by:

$$\begin{aligned} p_z^i = & -\sin(\Delta\theta) \cdot p_x^{i,0} + \cos(\Delta\theta)\sin(\Delta\psi) \cdot p_y^{i,0} \\ & + \cos(\Delta\theta)\cos(\Delta\psi) \cdot p_z^{i,0}. + \Delta z \end{aligned} \tag{8}$$

with $\underline{\boldsymbol{p}}^{i,0}$ the initial position of foot i in the body frame. Practically, Eq. (7) is minimized each time a new foot touches the ground. Also note that n_{leg}=6 : the minimization concerns legs in support phase, but also legs in transfer, for which the last contact point is used. Keeping the robot close to a reference position automatically corrects the posture drift caused by the ground detection algorithm.

6. Walking in natural environment and discussion

The robot has been tested under two environments: the lab where ground is very rigid, and outside, on a grass ground moderately hard. In both case, the ground detection algorithm coupled with the posture correction gives satisfying results. The robot never tips over because of the combination of its hexagonal configuration and the use of strictly stable gait.[5] These tests under real conditions prove that current sensing in the motor can be employed for ground detection.

We recall here that the proposed method is not intended to replace force sensing at foot. The aim of this test is to consider a global failure of foot sensing, and to show that motor current sensing can temporarily take over. Besides, current sensing complements the force sensing of the legs at joint level, which is of practical importance in the case of obstacle located elsewhere than at foot, or for transmission failure.

7. Conclusion

The ground detection is an important keypoint in the free gait generation. A ground detection algorithm based on current sensing has been investigated in this paper. The current is used to calculate the real force at joint level. In parallel to the measurement, a force estimator predicts the joint force in the case of a free motion of the foot. The ground is detected when the difference between both forces exceeds a certain threshold fixed by the user.

The force/torque estimator has been designed for the three joints of a leg, in a very general way, such that the obstacle detection could be implemented in addition to the ground sensing, and for any leg structure. The method requires:

(1) a current sensor in the motor circuitry(ies);
(2) an accurate dynamic model of the leg (or at least of one joint, in the case of a ground detection alone);
(3) a posture correction, not based on force informations, due to the unavoidable drift propagating during the walking.

However, some advantages deserve to be mentioned:

(1) the motor current is an interesting redundant information for ground detection, even if it is less accurate than force sensors;
(2) whatever the impact location of the obstacle, an increase of the measured joint force/torque is detected and the leg can react adequately, which is impossible with force sensors at foot;
(3) the comparison with the force estimation is an indicator of the transmission deterioration: it could prevent severe damages at joint level which would not be detected with only force sensors.

8. Ackowledgements

The authors want to thanks the Royal Military Academy, and especially the Prof. Y. Baudoin, for the loan of AMRU5.

References

1. B. Gaßmann, K.-U. Scholl and K. Berns, Locomotion of LAURON III in rough terrain, in *IEEE/ASME Int. Conf. on Advanced Intelligent Mechatronics*, (Como, Italy, 2001).
2. P. Gonzalez de Santos, E. Garcia, J. Estremera and M. Armada, *International Journal of Systems Science* **36**, 545 (2005).
3. A. Scheider and U. Schmucker, Force sensing for multi-legged walking robots, in *Mobile Robot, Moving Intelligence*, ed. J. Buchli (ARS/plV, 2006) pp. 447–470.
4. O. Verlinden, K. Kouroussis and C. Conti, Easydyn: A framework based on free symbolic and numerical tools for teaching multibody systems, in *Eccomas Thematic Conference*, (Madrid, Spain, 2005).
5. J. Porta and E. Celaya, *Robotics and Autonomous Systems* **47**, 187 (2004).

MODULAR-JOINT DESIGN FOR EFFICIENT WALKING AND MODELING OF A BIPEDAL ROBOT TPINOKIO

TECK-CHEW WEE[†], A.ASTOLFI

EEE/C&PG, Imperial College London, South Kensington Campus London, SW7 2AZ, UK

MING XIE

MAE Department, Nanyang Technological University, 50 Nanyang Avenue S(639798), Singapore

This paper presents the work done in designing and building of a modular-joint teen-size bipedal humanoid robot. Due to its modular-joint form, the robot is able to be modeled as a multi-body point-mass system without considering the links' inertia, this is an alternative approach in improving the accuracy of the robot's kinematics and dynamic model. Most of the joint components and parts are inter-changeable, as such, the robot is cost effective and is suitable for both education and engineering research purposes. This paper will focus mainly on the robot's mechanical structural design and the kinematics modeling.

1. Introduction

The development of bipedal robot locomotion began more than 30 years ago, some well-known robots are WABOT, ASIMO, HRP, HUBO, JOHNNIE, LOCH and QRIO. The success of ASIMO triggered a worldwide research fever on humanoid robot. Stable walking gait pattern generation and real-time on-line control remains a challenging task in humanoid robot research, it requires precision modeling of the robot dynamics and kinematics including mass distribution, location of center-of-mass (CoM) and inertia of each links. Therefore, obtaining the correct and accurate robot model is critical. Other challenges most of the researchers faced including the difficulty in designing a low cost robot with robust mechanical structure. With this in mind, the mechanical structure of TPinokio is in modular form, all the masses are concentrated at the link joints, and the links are made of light-weight aluminum, as such, the moment and inertia of the links are negligible compare to the joints' mass, this results in a simplified multi-body point-mass system model. TPinokio platform is designed for both educational learning and engineering research applications, the prototype is as shown in Fig. 1.

[†]author contact email : t.wee10@imperial.ac.uk

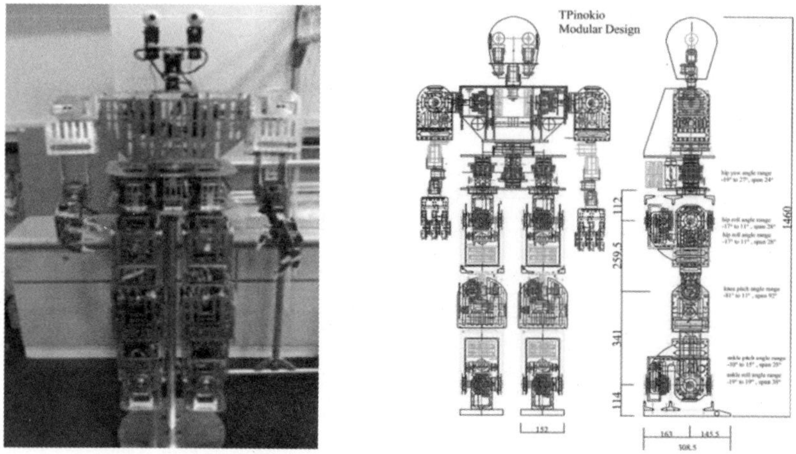

Figure 1. TPinokio Modular design and point-mass distribution stick diagram.

2. Design Concept

TPinokio is modular, light weight and low cost ;

- Compact modular-joint design, can be easily modeled as 'point-mass'.
- Simple kinematics.
- Teen-size human-like more than 28 D.O.F.
- Zero backlash, robust mechanical structure for easy maintenance.

3. Overview of design

The CAD drawing of the TPinokio design is as shown in Fig. 1, the quality, accuracy and performance of the components used for the upper body range from low-end to mid-range, this is a trade-off between cost effectiveness and performance requirements.

3.1. *Modular-joint design*

The joint-shaft is driven by a DC motor combined with harmonic drive and pulley-belt system. The sensors, electronics and wiring harness are all assembled in a compact housing, as depicted in Fig. 2. Most of the parts and components are standardized with 85% similarity in their specifications. As such, the modular-joint can be easily swapped, reconfigured, or removed for repair and maintenance. The communication between each joint is through CANbus interface with a single five-cores cable (3-wires for CANbus protocol and 2-wires for power supply), thus, messy cabling is eliminated, as depicted in Fig. 2.

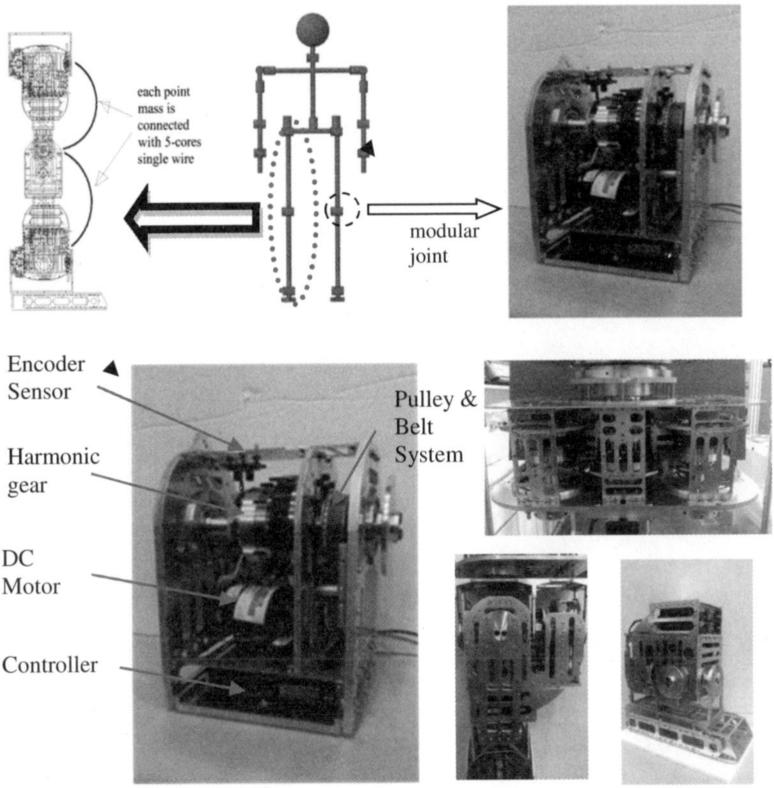

Figure 2. Modular-joint design

3.2. *Lower Body Design*

The lower-body has 13 D.O.F. All the joints are modular and compact. The encoders, pulley-belt system are standardized for all the joints, the main parts are listed in Table 2.

Table 1. TPinokio lower-body actuators part list.

Position	Joint	HD Gear Size/ratio	Pulley-Belt	DC motor
Pelvis	yaw	17/ 120:1	2:1	150w
Hip	roll/pitch/yaw	20/120:1	2:1	150w
Knee	pitch	17/120:1	2:1	200w
Ankle	roll/pitch	14/100:1	2:1	150w
		20/120:1	2:1	150w

4. Kinematics

4.1. *Simple Kinematics*

TPinokio's joints mechanism and structure are designed to produce simple kinematics model. The robot is modeled as having point-mass located at the distal end and the links to each joints are fabricated with light-weight aluminum. The mass and inertia of the link are negligible compared to the weight of the modular-joint housing, as shown in Fig. 3.

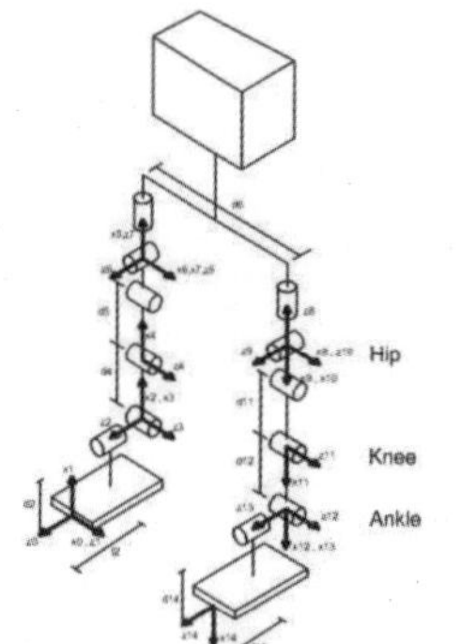

TPinokio Kinematics stick diagram

Parameters	Value
Head	0.5kg
Shoulder (yaw)	2.4kg
Torso (shoulder pitch)	5.8kg
Waist (yaw) MT	11.6 kg
Hip (M5,M9)	10.7 kg
Knee (M3 , M10)	4.1 kg
Ankle + feet (M1, M12)	7.1kg

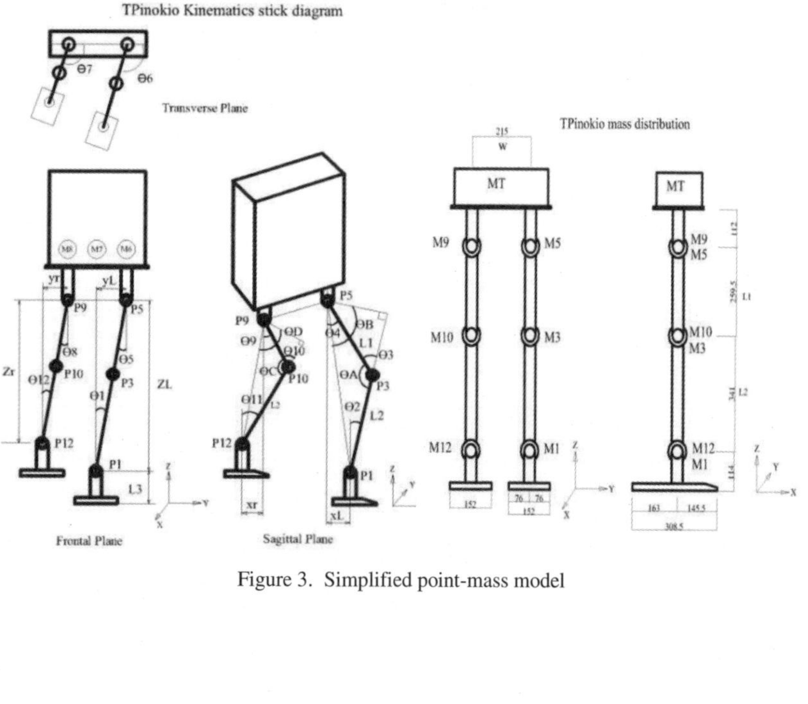

Figure 3. Simplified point-mass model

The CoM ($x_{com}, y_{com}, z_{com}$) is calculated with the following Equations 1,

$$x_{com} = \frac{\sum_i m_i P_{ix}}{\sum_i m_i} = \frac{m_3 P_{3x} + m_5 P_{5Lx} + m_T \left(\frac{P_{5Lx} + P_{5rx}}{2}\right) + m_9 P_{5rx} + m_{10} P_{10x} + m_{12} P_{12x}}{m_1 + m_3 + m_5 + m_T + m_9 + m_{10} + m_{12}}$$

$$y_{com} = \frac{\sum_i m_i P_{iy}}{\sum_i m_i} = \frac{m_3 P_{3y} + m_5 P_{5Ly} + m_T \left(\frac{P_{5Ly} + P_{5ry}}{2}\right) + m_9 P_{5ry} + m_{10} P_{10y} + m_{12} P_{12y}}{m_1 + m_3 + m_5 + m_T + m_9 + m_{10} + m_{12}}$$

$$z_{com} = \frac{\sum_i m_i P_{iz}}{\sum_i m_i} = \frac{m_3 P_{3z} + m_5 P_{5Lz} + m_T \left(\frac{P_{5Lz} + P_{5rz}}{2}\right) + m_9 P_{5rz} + m_{10} P_{10z} + m_{12} P_{12z}}{m_1 + m_3 + m_5 + m_T + m_9 + m_{10} + m_{12}} \tag{1}$$

4.2. *Inverse Kinematics*

We assume that the upper body mass of the robot are concentrated at a point-mass located at the centre of the pelvis (MT). The lower body of the robot has 6 D.O.F for both legs, with reference from Goswami Dip et al. [3], the inverse kinematic parameters are as shown in Equations 2 :

$$x_L = P_{1x} - P_{5x} \;;\; y_L = -P_{5y} \;;\; z_L = P_{1z} - P_{5z}$$
$$x_r = P_{12x} - P_{5x} \;;\; y_r = -P_{5y} \;;\; z_r = P_{12z} - P_{5z}$$

$$\theta_A = cos^{-1}\left[\frac{L_1^2 + L_2^2 - (x_L^2 + y_L^2 + z_L^2)}{2L_1L_2}\right] \;;\; \theta_B = cos^{-1}\left[\frac{L_1 \sin(\theta_A)}{\sqrt{x_L^2 + y_L^2 + z_L^2}}\right]$$

$$\theta_C = cos^{-1}\left[\frac{L_1^2 + L_2^2 - (x_r^2 + y_r^2 + z_r^2)}{2L_1L_2}\right] \;;\; \theta_D = cos^{-1}\left[\frac{L_1 \sin(\theta_C)}{\sqrt{x_r^2 + y_r^2 + z_r^2}}\right] \tag{2}$$

The values of the joints' angle [$\theta_1, \theta_2, \theta_3, \ldots \theta_{12}]^T$, can be found from these four angular parameters (θ_A, θ_B, θ_C, θ_D), by solving trigonometric Equations .

4.3. *Direct Measurement of ZMP*

Direct measurement of ZMP without knowledge of kinematics is also possible by reading the force sensors feedback values. There are four force sensors on each TPinokio's foot sole. The voltage output of the four force sensors are read via A/D converter. The position of ZMP is calculated by using Equation 3.

$$x_{zmp} = \frac{(f_2 + f_3)x_1 - (f_1 + f_4)x_2}{f_1 + f_2 + f_3 + f_4} \;;\; y_{zmp} = \frac{(f_1 + f_2)y_1 - (f_3 + f_4)y_2}{f_1 + f_2 + f_3 + f_4} \tag{3}$$

5 CONTROL ARCHITECTURE AND GAIT ALGORITHM

5.1. *Simplified Dynamic*

During the swing phase, the Centre of Mass (CoM) of the robot may be modeled as point-mass and it is connected to the stance foot like an inverted pendulum as shown in Fig. 10a. In order to define the relationship of the ZMP and the CoM trajectories, a simplified ZMP $(x_{zmp}, y_{zmp}, 0\ )$ with Inverted Pendulum Model (IPM) is computed [1][3], as shown in Fig. 10b.

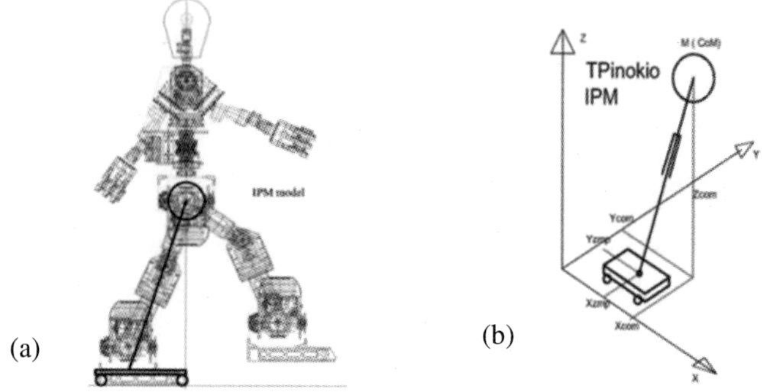

Figure 4. TPinokio simplified dynamic model

The simplified ZMP equations are given as ,

$$x_{zmp} = x_{com} - \frac{z_{com}}{g}\ddot{x}_{com} \quad ; \quad y_{zmp} = y_{com} - \frac{z_{com}}{g}\ddot{y}_{com} \tag{4}$$

The above Equations 4 are decoupled, and they can be express independently, and with the introduction of an extra variable u (system jerk),

$$u_x = \frac{d}{dt}\ddot{x}_{com} = \dddot{x}_{com} \quad ; \quad u_y = \frac{d}{dt}\ddot{y}_{com} = \dddot{y}_{com} \tag{5}$$

Equation 4 can be expressed as a proper dynamical system [3].With sampling time T, the discretized form are,

$$\hat{x}_{k=}\begin{bmatrix} x_{com}(kT) \\ \dot{x}_{com}(kT) \\ \ddot{x}_{com}(kT) \end{bmatrix} ; \dddot{x}_k = u_x(kT) ; z_k^x = x_{zmp}(kT) \quad ; \quad \hat{y}_{k=}\begin{bmatrix} y_{com}(kT) \\ \dot{y}_{com}(kT) \\ \ddot{y}_{com}(kT) \end{bmatrix} ; \dddot{y}_k = u_y(kT) ; z_k^y = y_{zmp}(kT)$$

$$\hat{x}_{k+1} = \begin{bmatrix} 1 & T & T^2/2 \\ 0 & 1 & T \\ 0 & 0 & 1 \end{bmatrix}\hat{x}_k + \begin{bmatrix} T^3/6 \\ T^2/2 \\ T \end{bmatrix}\dddot{x}_k \quad ; \quad \hat{y}_{k+1} = \begin{bmatrix} 1 & T & T^2/2 \\ 0 & 1 & T \\ 0 & 0 & 1 \end{bmatrix}\hat{y}_k + \begin{bmatrix} T^3/6 \\ T^2/2 \\ T \end{bmatrix}\dddot{y}_k$$

$$z_k^x = [1 \quad 0 \quad -z_{com}/g]\hat{x}_k \quad ; \quad z_k^y = [1 \quad 0 \quad -z_{com}/g]\hat{y}_k \tag{6}$$

5.2. *Interpolation and gait trajectory pattern generation*

In real-time operation, biped locomotion CoM/ZMP path planning is achieved by sending the desired foot position to the controller so that the joints' angle can be calculated by inverse kinematics formulations. The Cartesian path of the feet movement is expressed as a function of time, an observer may be designed to determine its velocity and acceleration [4]. In order to achieve a path start and stop with zero jerk, seven degree polynomial is used for TPinokio path planning. A zero jerk start-stop path of a joint moving from 10° to 45° (t=0 to 1s) is as shown in Fig. 5a. For a IPM model [1][3], the robot foot landing position can be obtained by off-line setting of the trajectories parameters of the CoM, the walking cycle S, the step height H, the step time T and the frontal-shift F as shown in Fig. 6a. The 3D simulation for a straight walking path is as shown in Fig. 6b.

Step trajectory (Fig. 5b) :

$$x(t) = \frac{S}{2}(1 - cos\left(\frac{\pi t}{T}\right) \quad ; \quad z(t) = \frac{H}{2}\left[1 - \cos(\frac{2\pi t}{T})\right] \quad ; \qquad for\ 0 \le t \le T \tag{7}$$

CoM trajectory (with start /stop jerk = 0):

$$q(t) = a_0 + a_1(\frac{t}{T}) + a_2(\frac{t}{T})^2 + a_3(\frac{t}{T})^3 + a_4(\frac{t}{T})^4 + a_5(\frac{t}{T})^5 + a_6(\frac{t}{T})^6 + a_7(\frac{t}{T})^7 \tag{8}$$

$$\begin{bmatrix} 1 & 0 & 0 & 0 & 0 & 0 & 0 & 0 \\ 0 & 1 & 0 & 0 & 0 & 0 & 0 & 0 \\ 0 & 0 & 2 & 0 & 0 & 0 & 0 & 0 \\ 0 & 0 & 0 & 6 & 0 & 0 & 0 & 0 \\ 1 & 1 & 1 & 1 & 1 & 1 & 1 & 1 \\ 0 & 1 & 2 & 3 & 4 & 5 & 6 & 7 \\ 0 & 0 & 2 & 6 & 12 & 20 & 30 & 42 \\ 0 & 0 & 0 & 6 & 24 & 60 & 120 & 210 \end{bmatrix} \begin{bmatrix} a_0 \\ a_1 \\ a_2 \\ a_3 \\ a_4 \\ a_5 \\ a_6 \\ a_7 \end{bmatrix} = \begin{bmatrix} q_0 \\ 0 \\ 0 \\ 0 \\ q_f \\ 0 \\ 0 \\ 0 \end{bmatrix} \; ; \quad \begin{pmatrix} for\ CoMx: \ [q_0, q_f] = [0,\ 0.5S] \\ for\ CoMy: [q_0, q_f] = [-0.5F, 0.5F] \end{pmatrix}$$

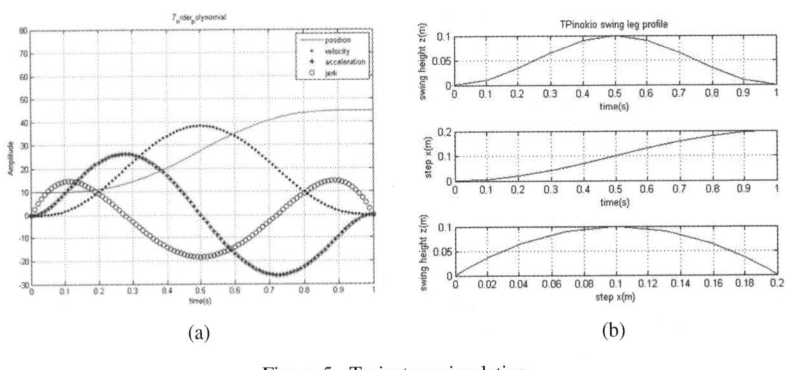

(a) (b)

Figure 5. Trajectory simulation

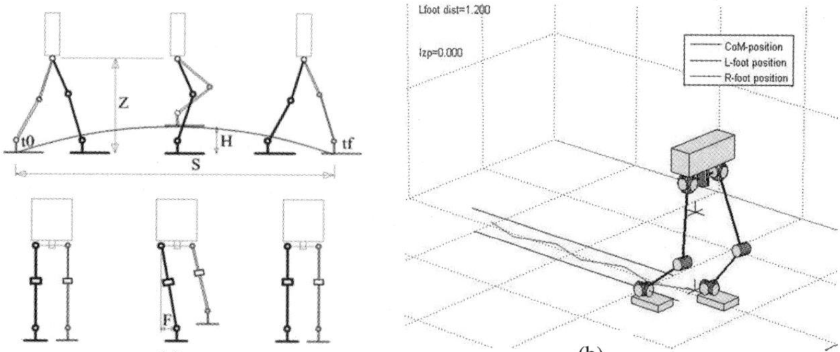

(a) (b)

Figure 6. Walking path simulation

6 Conclusion

This paper presented and focused on the modular-joint and a point-mass design concept of a teen-size humanoid robot TPinokio. The modular-joint design provides an alternative method for point-mass modeling. It also results in the creation of a cost effective and easy maintenance robot for both research test-bedding and educational learning. The continuous development of this project includes improvement of the mechanical structure, development of the dynamic walking path, optimization of the control algorithm and analysis of the stability margin. These results shall be presented in future papers.

References

1. M. Xie, Z. W. Zhong, L. Zhang, L. B. Xian, L. Wang, H. J. Yang, C. S. Song and J. Li, *A Deterministic Way of Planning and Controlling Biped Walking of LOCH Humanoid Robot*, **International Journal of Industrial Robot** 36/4 (2009), pp 314–325.
2. Goswami Dip, Vadakkepat Prahlad, Phung Duc Kien, *Genetic algorithm-based optimal bipedal walking gait synthesis considering tradeoff between stability margin and speed*, **Robotica (2009) volume 27**, pp. 355-365.
3. Kajita, S., Kahehiro, F., Kaneko, K., Fujiwara, K., Harada, K., Yokoi, K., Hirukawa, H., **Biped Walking Pattern Generation using Preview Control of the Zero-Moment-Point**, Proceedings of IEEE International Conference on Robotics and Automation, vol.2, 1620 -1626 (2003)
4. Astolfi, A.; Ortega, R.; Venkatraman, A., **Global observer design for mechanical systems with non-holonomic constraints** , American Control Conference (ACC), 2010, pp. 202 – 207

HIGH COMPLIANT SERIES ELASTIC ACTUATION FOR THE ROBOTIC LEG SCARL*ETH*

MARCO HUTTER, C. DAVID REMY, MARK A. HOEPFLINGER, ROLAND SIEGWART*

Autonomous Systems Lab, Institute of Robotics and Intelligent Systems, Swiss Federal Institute of Technology (ETHZ), Zurich, Switzerland, mahutter@ethz.ch

This paper presents the actuation system of the robotic leg Scarl*ETH*. It was developed specifically for a quadrupedal robot and is designed to achieve fast position control as well as accurate joint torque control. It introduces strong passive dynamics to create an efficient running behavior. High spring compliance with low damping in combination with a cascaded, motor velocity based, control structure was successfully tested in simulation and experiments. Final tests with the entire leg demonstrate that the system can perform a hopping motion providing only positive actuator power.

1. Introduction

Series elastic actuation (SEA) is a favored actuation approach when it comes to dynamic maneuvers in legged locomotion [1]. The elastic decoupling of the actuator from the joint is not only beneficial due to its mechanical low-pass filter characteristic that inherently increases safety for the environment [2] and that additionally protects the actuator from peak loads at landing impact [3]. Using the spring as a force element also allows performing torque control by controlling the deflection of the spring. The principle "the stiffer, the better" dominated classical robotic applications for a long time, and a similar tendency can be observed in the field of series elastic actuation: Such actuators are usually evaluated regarding force control bandwidth, which makes stiff springs clearly preferable due to the smaller motor travel distances that are necessary. In contradiction thereto, we would like to make the point that a great benefit of **high** compliance in combination with low damping is that it allows storing and releasing a substantial amount of energy during ground contact phases. These principles are adapted from nature, where tendons and elastic elements in

* This research was supported by the Swiss National Science Foundation through the National Centre of Competence in Research Robotics and the Hans-Eggenberger-Foundation.

muscles largely contribute to the bouncing behavior of dynamic maneuvers such as running. Provided that the stiffness is well chosen, this approach significantly increases the energy efficiency and reduces the power consumption of the actuators - a crucial element when building legged systems for highly dynamic maneuvers.

In the present paper, we will introduce the series elastic actuation developed for Scarl*ETH* (**S**eries **c**ompliant **ar**ticulated **r**obotic **l**eg), a single leg prototype for a quadruped running robot. The hardware as well as the controller is designed to achieve accurate torque control with a good reference action as well as good disturbance rejection and fast position control including active damping. In addition, with properly tuned joint stiffness, this setup allows to optimally exploiting the natural dynamics of the system such that efficient running is possible.

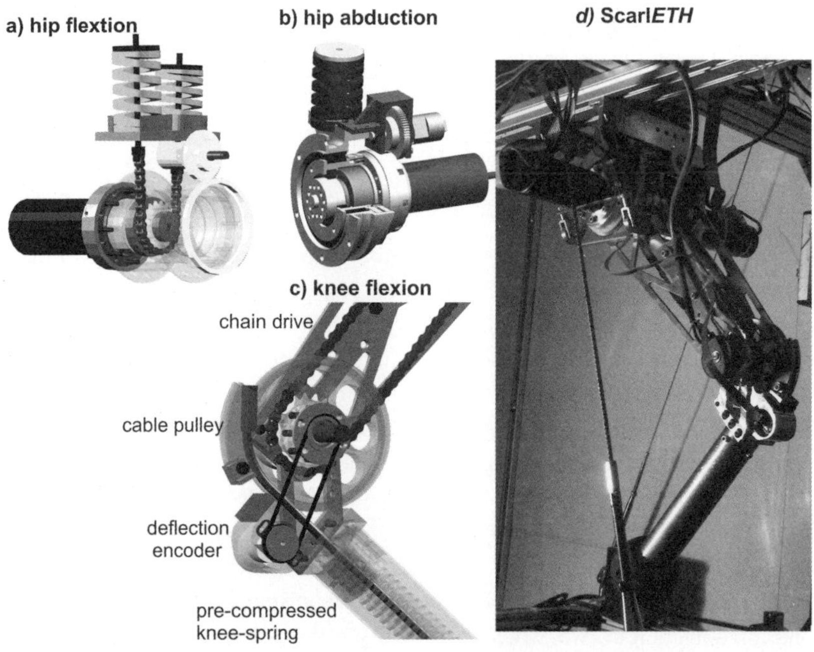

Figure 1 Scarl*ETH* (d) is driven by 3 high compliant series elastic actuators for hip flexion (a), hip abduction (b), and knee flexion (c). A chain/cable pulley design in combination with compression springs in all joints allows placing all motors at the hip joint such that the segments are lightweight (beneficial for fast motion, impact loss, and CoG close to the hip joint) and that a huge range of motion is achieved (complete knee retraction $\varphi_{knee}^{max} = 170°$ and hip flexion $\varphi_{hip}^{max} = \pm 85°$). All joints are actuated by 200 watt EC-motors in combination with harmonic drive units. Accurate force control is possible due to high linearity in the spring characteristics and very accurate deflection (80'000 qc/rev) and motor position (160'000 qc/rev) encoding.

2. System description

The fully actuated leg (Fig. 1) consists of three series elastic actuators for knee and hip flexion/extension as well as hip adduction/abduction. Powerful maxon motors (RE35 90W, EC4pole30 200W[†]) in combination with harmonic drive units (CSG14, 1:100 reduction[‡]) are connected over a chain and a steel cable pulley design with linear compression springs. This allows placing all actuators directly at the hip joint (Fig. 1) and facilitates not only a huge range of motion but also minimizes the inertia of the thigh and shank segment and hence the impact collision losses. The harmonic drive gearbox, the output shaft, as well as the springs are included in a highly integrated design that was optimized for weight (total leg weight of 4.5kg).

All springs are pre-loaded, resulting in an entirely backlash free design. A series of equal sized springs with different stiffness allow an optimal system tuning not only in simulation but also in experiments. In the present work, the knee stiffness is approximately 36Nm/rad with a corresponding maximal joint deflection of 60°, the hip joint about 70Nm/rad with a maximal deflection of 30°.

Each motor is controlled by a maxon EPOS controller that features current and velocity control. They are connected by a CAN bus to the low level FPGA controller on a compactRIO[§] system of National Instruments where joint position as well as torque controllers are implemented. The torque is measured using an incremental deflection encoder[**] with high resolution (80'000 qc/turn) connected to a sensor board.

3. Control

Controlling a legged robotic system generally poses two major, but inherently different control problems. On one side, during non-contact phases, accurate and fast position control is required for leg positioning. Due to the high joint compliance, this requires active damping: In Scarl*ETH*, we solved this problem by an LQR controller design based on motor as well as joint position and speed measurements. A more detailed description can be found in [4], which is directly compatible with the presented control structure in this paper.

[†] maxon motor AG, www.maxonmotor.ch

[‡] Harmonic Drive AG, www.harmonicdrive.de

[§] compactRIO, National Instruments, www.ni.com/compactrio/

[**] AEDA-3300, AVAGO Technologies, www.avagotech.com

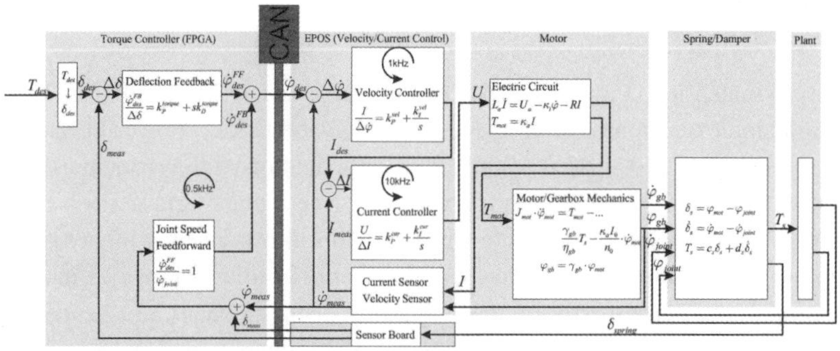

Figure 2 SEA controller structure: a cascaded structure with current control and velocity control on a maxon EPOS controller and a torque as well as a position controller on a NI FPGA. For high compliant springs with large deflections this velocity based torque control is beneficial.

The second control strategy, namely accurate joint torque control, is required during support phase. This facilitates virtual model control techniques [5], operational space control [6], or low gain walking control [7], which is clearly beneficial when dealing with uncertainties in the environment. In the standard SEA setup, a PID torque controller with a subordinate current controller is used [8]. In contrast thereto, some studies have demonstrated that a cascaded structure with additionally closing the motor velocity loop is beneficial [9],[10], especially to compensate undesired gearbox effects. In our application of the SEA, high compliance with low damping enforces large deflections and hence long motor shaft travel distances. Therefore, and because we have a setup with a high performance velocity controller at hand, we propose the cascaded control structure depicted in Fig. 2.

3.1. *Setup*

The linearized continuous time model of the entire actuation system is summarized in Fig. 2. System identification of the spring characteristic showed nearly perfect linearity with low damping and virtually no hysteresis in all joints [4], hence torque control remains equivalent to spring deflection control (damping is negligible). In this cascaded control structure the actual torque controller is implemented on the FPGA where it runs in a 500 Hz loop and creates the set points for the subordinate velocity and current controllers which are running with 1 kHz and 10 kHz respectively. The entire structure was rebuilt in simulation including saturation, discretization, and filtering effects. Detailed analysis of the individual loops in different model identification tests showed great equivalence to the simulation (Fig. 3). Both, current and velocity gains were tuned on a simplified test bench, resulting in a closed loop velocity bandwidth of about 100 Hz (b) and a step response time of about 5ms (a).

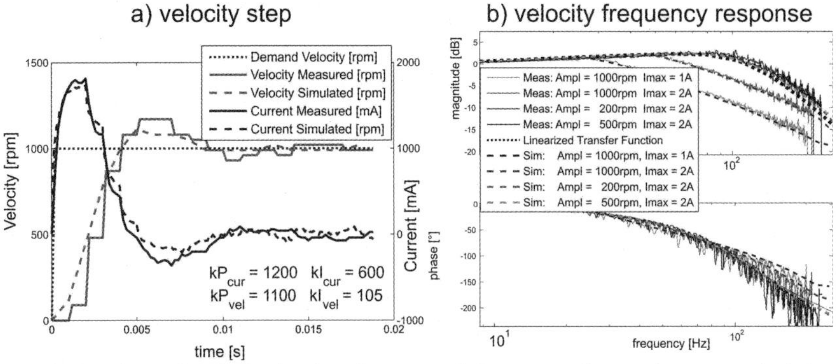

Figure 3 Experimental and simulated performance of the presented control structure depicting a nearly perfect matching of simulation and experiment (a) and a velocity control bandwidth of about 100Hz (b).

3.2. *Torque control design*

Based on detailed simulations as well as experiments, the torque controller was tuned with regard to two design criteria:

Firstly, fast reference signal-following must be achieved; i.e, a high closed loop torque bandwidth is required in combination with high robustness. Fig. 4a shows a step response time of about 30ms for an input of 3 Nm. The closed loop transfer function (Fig. 4b) emphasizes the drawback of high compliance: with increasing amplitudes, the bandwidth substantially drops due to saturation effects. Nevertheless, the TF estimation with the entire model and the validated control parameters varies between 28Hz for 1Nm, 15Hz for 5Nm and 11Hz for 10Nm. The closed loop system is robust with a phase margin of more than 85°.

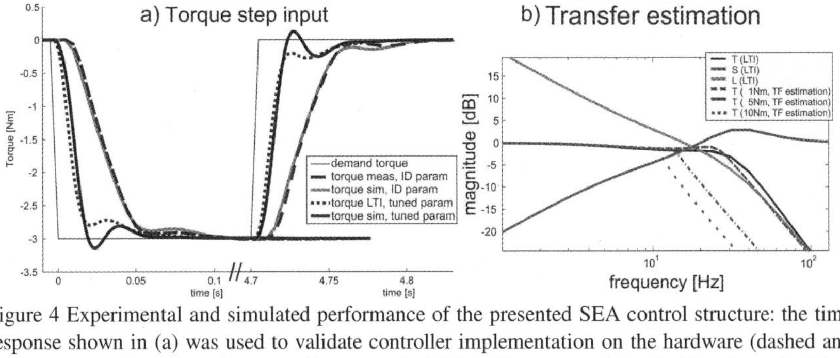

Figure 4 Experimental and simulated performance of the presented SEA control structure: the time response shown in (a) was used to validate controller implementation on the hardware (dashed and gray curve match well) and to tune the system for high performance. A bandwidth shown in (b) of about 28Hz is achieved for small amplitudes; about 11Hz remain when saturation effects occur. Keeping the integrator gains small results in a negligible overshoot (T_{max}<1dB) and good robustness with a phase margin of more than 85°.

Due to the high compliance, the resulting bandwidth is poor compared to systems with stiffer springs [11], still they are within the range of Pratt's electrical SEAs (5-25Hz [1]) and above the frequency range that can be found in human motion (4-8Hz [12]).
Secondly, next to a fast reference signal response, a good disturbance rejection is crucial. This can be tested in a zero torque behavior with modulated external joint deflection. Using only a PI controller on the deflection leads to substantial torque ripples[9], mainly caused by a phase delay (dash-dotted line in Fig. 5a) between the joint and motor angles. As depicted in Fig. 5b, the sensitivity of the PI controller has a flat ascent requiring for higher integral gains to keep the torque ripple small. This would clearly influence the reference following behavior, having a negative influence on the robustness.

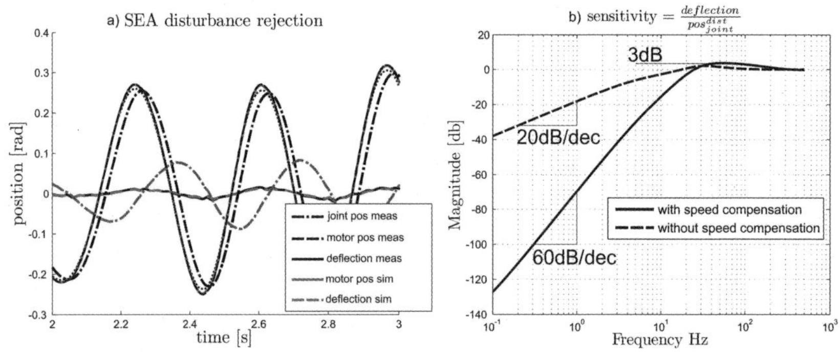

Figure 5 Significant improvement regarding disturbance rejection is achieved by disturbance compensation based on joint velocity measurement. Without that, high integral gains that also influence the closed loop reference signal behavior would be needed to achieve a good zero torque behavior.

Thanks to the backlash free mechanical design (pre-compressed springs and harmonic drive gearboxes) and the accurate measurements of motor position and spring deflection, we have very direct means of measuring the actual disturbance. Because of the high resolution (160'000 counts/rev at the motor, 80'000 counts/rev for the deflection) encoders show minimal discretization noise and hence provide accurate joint speed estimation. Using these inputs, a disturbance compensator is included. This compensator has no influence on the properties of the closed loop controller, such that the previously achieved stability and robustness properties are conserved. Due to the low-pass behavior of the velocity controlled motor, the joint velocity input can be directly used as a feet forward path, inherently ensuring that the output value is independent of the input value for the stationary case since $\dot{\varphi}_{mot}(s=0)=\dot{\varphi}_{joint}(s=0)$. The sensitivity is still bounded by 3dB but shows a much better behavior for lower frequencies (at which the system is usually operated) (Fig. 5b). The zero torque behavior was successfully tested in experiments. The measured joint disturbance was then used as simulation input and compared against the measurements. A significant improvement to the controller without disturbance compensation can be observed. At this point, the high compliance clearly helps, since the time constant for the required reaction is lower than in a system using stiff springs.

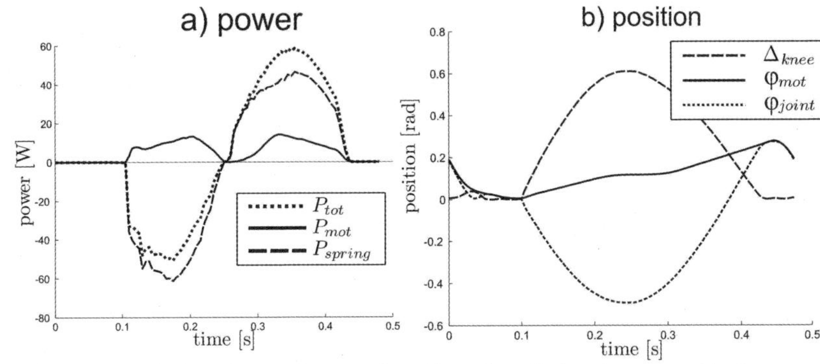

Figure 6 A modified Raibert style controller allows for efficient hopping with the presented hardware. During the whole stance phase, the knee motor is expanded (b) and provides exclusively positive power to the knee joint. When comparing the actuator power with the energy that is being stored in the spring (a) it becomes evident that the motion is dominantly driven by the natural dynamics. The motor only has to compensate for energetic losses and constitutes to less than 20% of the total joint work.

4. Efficient running

The main advantage of using high compliances with low damping becomes evident when it comes to perform highly dynamic maneuvers. In the present setup, knee joint deflections of 60° allowed to store up to 30 Joules of energy. This corresponds to a vertical excursion of the total system's mass by 0.5m. A detailed description of the controller for 2D running can be found at [4]. During stance phase, we additionally ensured that the knee motor which is the main energy source delivers only positive power by an expansion speed $\dot{\varphi}_{mot} > 0$. Fig. 6 shows the power distribution as well as the joint angles during a stance period in steady state ($\dot{x}_{Hip} = 0$) hopping. The motor itself provides exclusively positive power during the deceleration and acceleration phase. Approximately 20% of the total energy is introduced by the motor. The remaining fluctuations are created passively.

5. Conclusion and Outlook

In this paper we presented the actuation system of the prototype leg Scarl*ETH* that was built for a quadruped robot. This system combines all aspects needed for a versatile, fast, and efficient legged platform: high damped position control for precise leg positioning, accurate torque control during stance phase, and high compliant energy storage to generate efficient and fast running behaviors.

References

1. Pratt, J.E. and B.T. Krupp, *Series elastic actuators for legged robots.* Unmanned Ground Vehicle Technology Vi, 2004. **5422**: p. 135-144, 618.
2. Wyeth, G. and G. Walker, *Assessing the safety of a velocity sourced series elastic actuator using the head injury criterion*, in *Proceedings of Australasian Conference on Robotics and Automation 2007*, M. Dunbabin and M. Srinivasan, Editors. 2007.
3. Pratt, G.A., et al., *Stiffness Isn't Everything*, in *The 4th International Symposium on Experimental Robotics IV*. 1997, Springer-Verlag.
4. Hutter, M., et al., *ScarlETH: Design and Control of a Planar Running Robot*, in *in prep. for IROS 2011*. 2011: San Francisco, USA.
5. Pratt, J., et al., *Virtual model control: An intuitive approach for bipedal locomotion.* International Journal of Robotics Research, 2001. **20**(2): p. 129-143.
6. Sentis, L., *Compliant Control of Whole-Body Multi-Contact Behaviors in Humanoid Robots*, in *Motion Planning for Humanoid Robots*. 2009, Springer Global Editorial.
7. Kalakrishnan, M., et al. *Fast, Robust Quadruped Locomotion over Challenging Terrain*. in *ICRA 2010*. 2010. Anchorage, AK.
8. Robinson, D.W., et al. *Series elastic actuator development for a biomimetic walking robot*. in *International Conference on Advanced Intelligent Mechatronics (AIM)*. 1999. Atlanta, USA.
9. Vallery, H., et al. *Passive and accurate torque control of series elastic actuators*. in *Intelligent Robots and Systems, 2007. IROS 2007. IEEE/RSJ International Conference on*. 2007.
10. Robinson, D.W., *Design and Analysis of Series Elasticity in Closed-loop Actuator Force Control*, in *Department of Mechanical Engineering*. 2000, Massachusetts Institute of Technology: Cambringe.
11. Melnykov, A., et al., *Linear Elastic Actuator of a Biped Robot "Rotto"*, in *CLAWAR 2011*. 2010: Nagoya, Japan. p. 588-595.
12. Kyoungchul, K., B. Joonbum, and M. Tomizuka, *Control of Rotary Series Elastic Actuator for Ideal Force-Mode Actuation in Human–Robot Interaction Applications*. Mechatronics, IEEE/ASME Transactions on, 2009. **14**(1): p. 105-118.

STEERING CONTROL FOR A 3D BIPED ROBOT CONSIDERING GROUND CONTACT AND STABILITY

Ting Wang, Christine Chevallereau

L'Institut de Recherche en Communications et Cyberntique de Nantes (IRCCyN)
CNRS, Ecole Centrale de Nantes, France
E-mail: ting.wang@irccyn.ec-nantes.fr, Christine.Chevallereau@irccyn.ec-nantes.fr

This paper proposes a method to steer a 3D robot while it is walking on a flat surface. A control law consists of ZMP controller, swing ankle rotation controller and partial joint angles controller is presented at first. The stability of the walking gait under closed loop control is evaluated with the linearization of the restricted Poincaré map of the hybrid zero dynamics. As a result, the robot can achieve the stable and periodic walking along a straight line. Next, a stride-to-stride feedback control is supplemented to the original walking control law to adjust the net yaw rotation of the robot over a step in order to steer the robot along paths with mild curvature. Finally, some examples of the robot walking along different paths are given.

1. Introduction

The robot is required to move all over the working place when it works in human environment. Therefore, we need the precise steering control not only the walking straight simply. There are few studies about direction control. Ref. 1 and Ref. 2 studied the turning motions for biped with ZMP-based footstep planning. Ref. 3 proposed the robust direction control system that used gyro sensor feedback under environment with disturbance. In Ref. 4, motion stability during turning is ensured by adjusting the swing leg center of mass (COM) and hip position trajectories in a trial and error fashion. This work is extended from the previous studies for a 3D underactuated biped robot in Ref. 5 . A stride-to-stride feedback controller proposed in Ref. 6 is integrated with the original control law to regulate the walking direction through the net yaw motion about the stance foot over a step. Different from other methods, the stability during steering is maintained. In addition, an interesting feature of this work is that one is able to control the robot's motion along a path with limited curvature using only a predefined reference motion.

2. Original control for stable walking along a straight line

The studied robot is comprised of a torso, two identical legs and feet. Each hip and ankle consists of a revolute joint with 3 DOF and the knee is a 1 DOF revolute joint, and each DOF is independently actuated. As a result, it has 14 DOF in the single support (SS) phase and 14 actuators. The vector $q = [q_1, \ldots, q_{14}]^T$ describes the shape and orientation of the biped. The walking gait consists of SS phases on flat stance foot separated by impacts. The original control for the walking along a straight line consists of a ZMP controller, a swing ankle rotation controller and a partial joint angles controller. The former two controllers are used to insure stable contact with the ground. The third controller implements a joint path control and it is partial since the torques are already partially defined by the former two controllers. The objective of three controllers is to obtain:

$$\begin{cases} ZMP = ZMP^d \\ \xi_{xy} = 0 \\ u = Mq = Mq^d \end{cases} \tag{1}$$

Where the first line of (1) describes 2 positions of ZMP in the horizontal plane have to be regulated to their desired values, that can offer 2 dynamic equations about the joint angle, its velocity and acceleration. Thus, only other $n-2$ outputs can be defined (where $n = 14$ is the number of actuated joints), that means not all the joints can be controlled. Because all of our work is done with a hypothesis that the ground is flat, therefore, if the motion of all the joints track the desired trajectory very well, the swing foot touch the ground with flat-foot. As a result, the support foot in the next step will be flat too, then the control law will be valid. The swing ankle rotation controller described as the second line of (1) is used to satisfy this condition, in which the roll and pitch angle of the swing foot are controlled to zero, that can also offer 2 dynamic equations. Consequently, for a 3D robot with n actuated joints, since 2 positions of ZMP and 2 rotation angles of the swing ankle are controlled only other $n - 4$ controlled outputs can be chosen. For simplicity, the vector of the $n - 4$ controlled outputs, which is written as u, can be defined as a linear combination of the n actuated joints, i.e., $u = Mq$. The partial joint angles controller is applied to make u track its desired value u^d, as shown in the third line of (1). As a result, the three controllers can offer n dynamic equations to resolve the acceleration of n actuated joints, then the input torque can be obtained by the inverse dynamics.

Because the first equation of (1) which offers 2 outputs has relative degree 0, so the overall control law induces a zeros dynamics of degrees 4 and its stability must be studied. It is evaluated with the linearization of the restricted Poincaré map of the hybrid zero dynamics such as Ref. 7. The Poincaré map linearized about a fixed-point x^* gives rise to a linearized system,

$$\delta x_{k+1} = A^z \delta x_k, \tag{2}$$

Ref. 7 has proposed that a fixed-point of the restricted Poincaré map is locally exponentially stable if and only if the eigenvalues of A^z have magnitude strictly less than one. Using this criteria, the controlled partial joints or M can be chosen judiciously to obtain stable walking along a straight line.

3. Direction control of the yaw motion

In order to describe the yaw motion of the robot, the walking direction angle q_0 is supplemented to the configuration variables of the robot, there is $q^e = [q^T, q_0]^T$. q_0 defines the orientation of the robot in an absolute frame, it does not belong to the shape variable q. As shown in Fig. 1, defining q_l and q_r are the direction angle relative to x_s-axis of left foot and right foot respectively, the walking direction angle of the whole body q_0 is given by:

$$q_0 = \frac{q_l + q_r}{2}. \tag{3}$$

Where q_l and q_r can be obtained using the position of two feet in the $x_s y_s$ plane of the absolute frame.

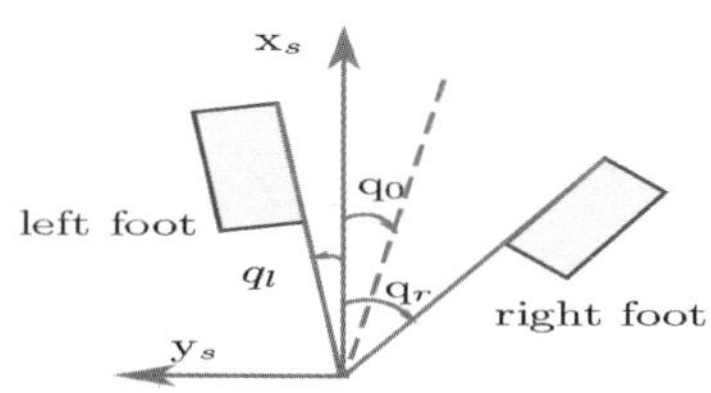

Fig. 1: Description of walking direction

Next, the stability of the extended system can be studied as the previous section, the corresponding linearized extended restricted Poincaré map at the fix point is written as:

$$\delta x^e_{k+1} = A^e \delta x^e_k, \tag{4}$$

where $x^e = [x, q_0]^T$ and $\delta x^e_k = x^e_k - x^{e*}$. Since the dynamic model is invariant with respect to q_0, the walking direction doesn't influence the behavior of the original closed-loop system. As a consequence, the last eigenvalue of A^e is 1 and the others are the same as the eigenvalues of A^z in (2). Because the ZMP control and swing ankle rotation control are always desired, so the feedback control is only added to the partial joint angles controller, then the desired property of the overall controller shown in (1) is rewritten as:

$$\begin{cases} ZMP = ZMP^d \\ \xi_{xy} = 0 \\ Mq = Mq^d + u_s(\beta) \end{cases} \tag{5}$$

where the supplemented term $u_s(\beta)$ is introduced from the moment of Poincaré section and it is defined as a linear polynomial such that:

$$u_{sf} = \beta \tag{6}$$

where u_{sf} denotes the value of u_s at the end of one step. Linearizing the Poincaré map about the fix point x^{e*} and the nominal parameter value $\beta^* = \mathbf{0}_{10\times 1}$ leads to

$$\delta x^e_{k+1} = A^e \delta x^e_k + F\delta\beta_k, \tag{7}$$

where $\delta\beta_k = \beta_k - \beta^*$ and F is the Jacobian of P^z with respect to β. Designing a feedback matrix

$$\delta\beta_k = -K\delta x^e_k \tag{8}$$

such that the eigenvalues of $(A^z - FK)$ have magnitude strictly less than one will exponentially stabilize the fixed point x^{e*}.

In (7), F is calculated numerically, analogously to the calculation of A^e. The (10×4) gain matrix K is calculated via DLQR technique so that the state-feedback control law (8) minimizes the cost function $\sum (\delta x^{e'}_k \delta x^e_k + r\delta\beta'_k \delta\beta_k)$.

Since the feedback control (8) can stabilize the system at fixed point with the walking direction angle $q_0{}^*$, it can also stabilize the system at the other fixed point $q_0{}^* + C$. Therefore, the robot can converge to a motion with a desired direction of travel $q_0{}^* + C$, and the direction of the robot is steered by changing C.[5]

Next, an example of the robot passing through a door is given. The initial distance between the midpoint of two feet and the goal along y_s

direction is denoted by y^d. Our principle is regulating the walking direction to eliminate this distance step by step, and then controlling the walking direction angle to 0.

According to the previous analysis, the steering control will be implemented through changing the desired fixed point ${q_0}^*$ to ${q_0}^* + C$. Supposing the walking direction is steered by changing C_k at the kth step, it is chosen as:

$$C_k = \begin{cases} Q_{sat} \ , \ k_1(y^d - y_k) > Q_{sat} \\ -Q_{sat} \ , \ k_1(y^d - y_k) < -Q_{sat} \\ k_1(y^d - y_k) \ , \ otherwise. \end{cases} \tag{9}$$

where y_k is the current position of the midpoint of two feet at the kth step. k_1 is the control gain and Q_{sat} is a saturation of turning that must be chosen appropriately.

As soon as C_k is determined, ${q_0}^* + C$ will replace ${q_0}^*$ to be used in (8) as the last element of x^{e*}. Supposing $y^d = -0.5$ and choosing $k_1 = 0.6$, $Q_{sat} = 5°$, the simulation results are shown in Fig. 2 and Fig. 3. Fig. 2 indicates that the walking path converges to the desired one $y^d = -0.5$. Fig. 3 describes the direction angle of two feet q_l, q_r and the walking direction angle q_0. It shows that q_0 converges to $-5°$ (because $Q_{sat} = 5°$) at first to obtain $y^d = -0.5$, then it converges to 0 so the robot walking along the x_s axis in Fig. 2.

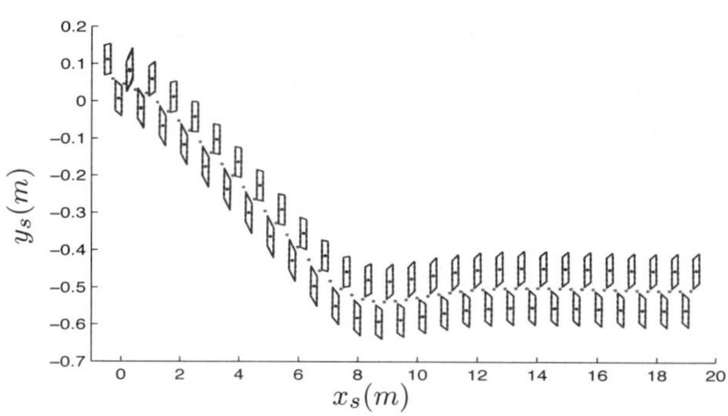

Fig. 2: The position of the feet at impact for *Example* 1, where the red points denote the midpoint of two feet. The walking path converges to the desired one $y^d = -0.5$.

It is well known that the robot cannot turn with large direction during one step because of the constrain of torques. Next, an example of a solution

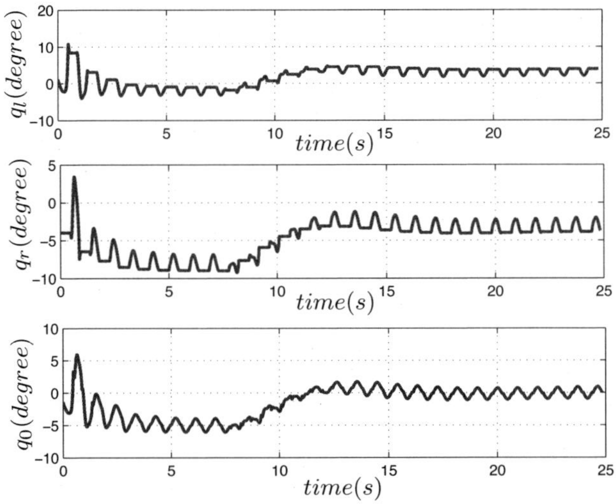

Fig. 3: The direction angle of left foot q_l, right foot q_r, and the walking direction angle q_0, where the straight lines in figures of q_l and q_r denote the single support phases. q_0 converges to $-5°$ (because $Q_{sat} = 5°$) at first to obtain $y^d = -0.5$, then it converges to 0 so the robot walking along the x_s axis in Fig. 2.

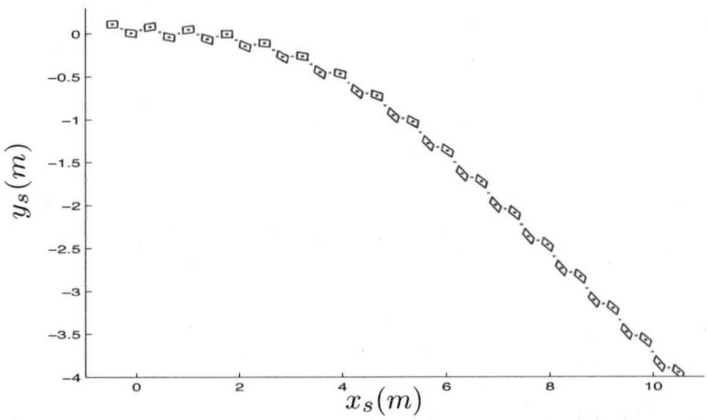

Fig. 4: The position of the feet at impact for *Exemple* 2, where the red points denote the midpoint of two feet. The walking path is steered $-30°$ then the robot walks along this direction.

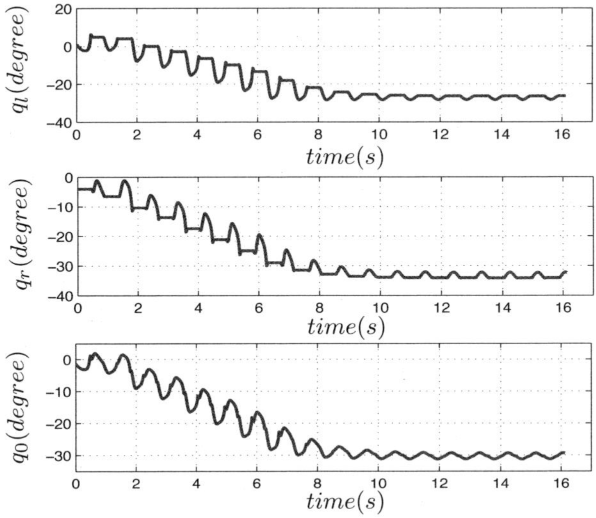

Fig. 5: The direction angle of left foot q_l, right foot q_r, and the walking direction angle q_0, where the straight line in figures of q_l and q_r denote the single support phase. It shows q_0 converges to the desired value ${q_0}^d = -30°$

to this problem is given. In order to avoid the saturation of torques, the control quantity is changed stepwisely. Supposing the walking direction is steered by changing C_k at the kth step, it is chosen as:

$$C_k = C_{k-1} + \alpha_k \tag{10}$$

with

$$\alpha_k = \begin{cases} Q_{sat} \ , \ k_2({q_0}^d - C_{k-1}) > Q_{sat} \\ -Q_{sat} \ , \ k_2({q_0}^d - C_{k-1}) < -Q_{sat} \\ k_2({q_0}^d - C_{k-1}) \ , \ otherwise. \end{cases} \tag{11}$$

where k_2 is the control gain and C_{k-1} denotes the value of C at $(k-1)^{th}$ step, and if $k = 1$, $C_0 = 0$.

Supposing ${q_0}^d = -30°$ and choosing $k_2 = 0.6$ and $Q_{sat} = 2°$, the simulation results are shown in Fig. 4 and Fig. 5. Fig. 5 indicates that the steering control of direction control of two feet is stepwise and the walking direction angle q_0 converges to the desired value ${q_0}^d = -30°$.

4. Conclusions

This paper proposed a walking control strategy for a 3D biped robot with flat-feet. A computed-torque control method is implemented. Its objective is controlling the partial joints angles, the ZMP and the swing ankle rotation simultaneously. It is important that the choice of controlled partial joints is based on stability analysis using Poincaré method. Finally, a supplemental stride-to-stride feedback controller was designed to stabilize the yaw motion. By adjusting the set point of the feedback controller, it is possible to steer the direction of the robot, and even to direct the walking along a given path.

References

1. K. Nishiwaki, S. Kagami, J. Kuffner, M. Inaba and H. Inoue, Online humanoid walking control system and a moving goal tracking experiment, in *Proceedings of the 2003 IEEE International Conference on Robotics and Automation*, (Taipei, Taiwan, 2003).
2. R. Kurazume, T. Hasegawa and K. Yoneda, The sway compensation trajectory for a biped robot, in *Proceedings of the 2003 IEEE International Conference on Robotics and Automation*, (Taipei, Taiwan, 2003).
3. K. Matsumoto and A. Kawamura, The direction control of a biped robot using gyro sensor feedback, in *Proceedings of the 11th IEEE International Workshop on Advanced Motion Control*, (Nagaoka, Niigata, 2010).
4. M. Yagi and V. Lumelsky, Synthesis of turning pattern trajectories for a biped robot in a scene with obstacles, in *Proceedings of the 2000 IEEE/RSJ International Conference on Intelligent Robots and Systems*, 2000.
5. C. Chevallereau, J. Grizzle and C. Shih, Steering of a 3d bipedal robot with an underactuated ankle, in *Proceedings of the 2010 IEEE/RSJ International Conference on Intelligent Robots and Systems*, (Taipei, Taiwan, 2010). To be appeared.
6. J. Grizzle, Remarks on event-based stabilization of periodic orbits in systems with impulse effects, in *Proceedings of the Second International Symposium on Communications, Control and Signal Processing*, 2003.
7. E. Westervelt, J. Grizzle, C. Chevallereau, J. Choi and B. Morris, *Feedback Control of Dynamic Bipedal Robot Locomotion*Control and Automation, Control and Automation (CRC Press, Boca Raton, June 2007).

PASSIVE DYNAMIC QUADRUPED WALKER WITH FLAT FEET AND ANKLE SPRINGS

T. KINUGASA$^\diamond$, H. KOMATSUBARA$^\diamond$, K. YOSHIDA$^\diamond$, S. FUJIMOTO$^\diamond$, M. IRIBE *, and K. OSUKA †

$^\diamond$ *Okayama University of Science,*
1-1, Ridai-cho, Kita-ku, Okayama 700-0005, Japan
E-mail: kinugasa@mech.ous.ac.jp

It is well known that passive dynamic walking is attractive in the field of robotics research. Some researchers are focusing on passive dynamic multiped walking. The existence of passive dynamic multiped walking seems one locomotion essence of multiped creatures. Previous researches are indicating that passive dynamic quadruped walking has various gait patterns such as 'pace' and 'trot'. However, stride of the quadruped walker developed by Osuka was small relative to length of legs and feet. This means that its gait is not enough dynamic. It is, moreover, important to prove the fact that various passive dynamic multiped walking exist. The purpose of the paper is, therefore, to develop a passive dynamic quadruped walker of which foot is flat feet and ankle is connected by springs and a ball joint. As a result, the quadruped realize more dynamic gait with relatively wide stride, 54% of its leg length and distance, 66% of a 3.6 m long slope.

1. Introduction

It is well known that passive dynamic walking[1] is attractive in the field of robotics research. Almost all of the researches are using a biped model not only for stability analysis but also experimental tests. On the other hand, some researchers are focusing on passive dynamic multiped walking. For example, Berkemeire[2] analysed stability of quadrupedal walker with a rigid body and Remy[4] showed that a quadruped walker with elastic connection between legs and a body has some gait patterns through some simulations in two dimensional space. It was not until the result by Osuka et al.[3] that a passive dynamic quadrupedal walker in three dimensional plane is realized and existence of its various locomotion such as 'pace' and 'trot' was proved.

*Osaka Electro-Communication University, Japan.
†Osaka University, Japan.

In addition, Osuka et al.[5] realized passive dynamic multiped walkers. We expect that passive dynamic multiped walking is the essence of multiped locomotion in common with passive dynamic bipedal walking. However, stride of the quadruped walker developed by Osuka[3] was small relative to length of legs and feet. This means that its gait is not enough dynamic, thus the result seems weak to prove that the locomotion of passive quadruped is analogous to quadruped creatures, especially to large quadruped mammals. Feet of the previous quadruped walker was composed of spherical surface only. It is important to show various passive dynamic multiped walking exist, as passive walker is strongly affected by physical property such as shape of sole which is a dominant factor for stability and leg length and spring stiffness which affect natural frequency.

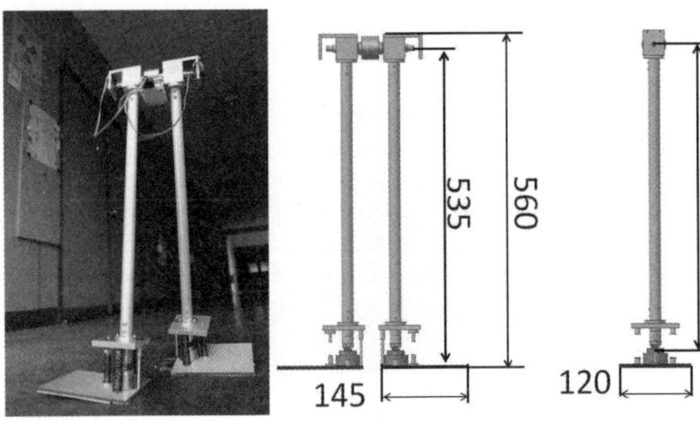

Fig. 1. Ridai passive Walker N°2 (RW02)

We have developed some three dimensional passive dynamic biped walker of which foot is composed of flat sole and connected to ankle by a ball joint and coil springs. The flat feet instead of usual arc sole for passive dynamic walker give larger friction force against the ground. The largest stride of our walker was about 60% of leg length and 200% of foot, which means that total centre of mass of the biped leaves from supporting polygon. Unfortunately, the larger the stride was, the more difficult the walker walked. We cannot use the unstable walker for developing a new passive quadruped walker, straightforwardly.

The purpose of the paper is, therefore, to develop a passive dynamic quadruped walker with flat feet and ankle springs using more stable three dimensional passive dynamic biped and to realize more dynamic gait.

2. Passive Dynamic Quadruped Walker

2.1. *Ridai Passive Walker N° 2 (RW-02)*

The most important point for developing a passive dynamic quadruped walker is to realize a stable three dimensional passive dynamic biped walker. We have developed various 3-D passive dynamic biped walkers called 'Ridai

Fig. 2. Ballast for RW-02 (left) and ankle joint with springs

Passive Walker (RW)'. RW-01[7] and RW-02[6] have flat feet and ankles connected by springs and a ball joint. The flat feet gives relatively large friction between the sole and the ground, and stabilize standing still position. The

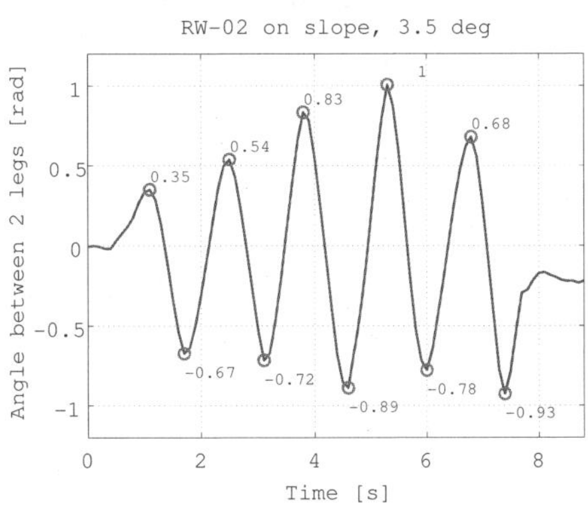

Fig. 3. Biped locomotion of RW-02

results of our researches led to the conclusion that ballast attached to legs was effective for stability of RW-01[7] and stiffness of springs affected stride of gait of RW-02.[6] In order to realize wider stride and more stable biped walking, RW-02 are improved using ballast, 500g (in Fig.2), for each leg and spring, 10.3 N/mm, for pitch axis of the ankle joint. Figure 3 shows relative angle between the both leg of RW-02 on 3.5 [deg] slope. In this case, natural frequency of legs (average) is 4.7 rad/s, and walking frequency was 4.4 rad/s. When RW-02 walks on a 1.8 m long slope, completeness rate[7] was 17 % without the ballast and 74 % with the ballast. The completeness rate is defined by the number of event in case that the walker walks on the slope completely per the number of trial. In case of the improved RW-02, there is a 32 % probability that RW-02 could completely walk on 3.6 m long slope, and its average stride and the number of steps were 360 mm (72 % of the leg length, 500 [mm], and 300% of the foot length, 120 [mm]) and 10, respectively. An average relative angle between both legs at double support phase was 0.74 [rad]. Especially, the stride is three times longer than the foot length, and thereby projection of total centre of mass (CoM) to the ground surface widely leaves from supporting polygon of the foot at stance phase.

2.2. *Ridai Passive Walker 02 Duo (RW-02D)*

Fig. 4. RW-02D (left) and 'pace' gait

Next, Ridai Passive Walker 02 Duo (RW-02D) is developed (the left figure of Fig.4). RW-02D is composed of two RW-02 with the ballast and a 360 mm long rod connecting to hip joint shafts of the both RW-02. The ballast is attached to each leg at 30 mm high from ankle joint. Relative angles between each leg and the rod and acceleration of three axis can be measured by potentiometers and accelerometers at the both end of the rod. The data is logged every 100 [msec] by WAA (Wireless Tech.) through Bluetooth.

3. Experimental result

3.1. *'Pace' Gait*

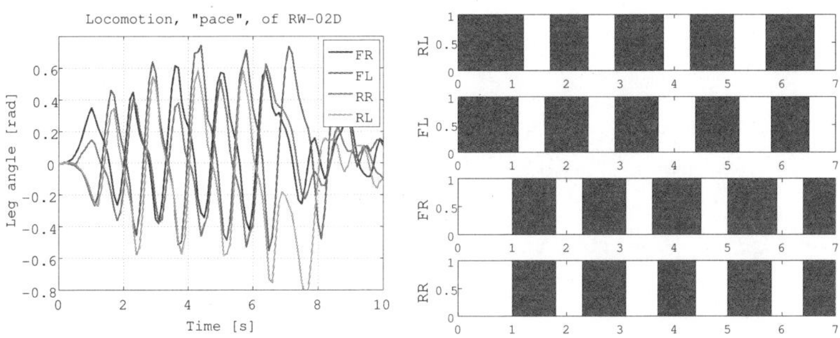

Fig. 5. Quadruped locomotion, 'pace', and gait chart of RW-02D

The left figure of Fig.5 shows an example of quadruped locomotion of RW-02D on slope 4.5 deg as the relative angle of each leg. It can be found that RW-02D walked 10 steps with an average stride of 300 mm long. The rear left leg (RL in Fig.5) hit the front left leg (FL) after 10 steps, then the walker stopped accordingly. Walking frequency was 4.52 rad/s in this case. Though natural frequency about roll axis is depending on roll angle, we can predict that it is corresponding to the walking frequency through our experience. The obtained gait is categorised into 'pace'. Two feet of one side (left or right) strike the slope at roughly the same time. We can also find the 'pace' gait from the right figure of Fig.4.

In order to clarify the fact, a gait chart is shown in the right figure of Fig.5. In the figure, green area indicates the stance phase, and white the swing phase. The abscissa axis means time [s]. The figure shows that the gait has interval of double support phase, the front and the rear legs of each side are same phase, and the left and the right pair has 180 [deg] phase difference. As a result, we can conclude that the gait was 'pace'.

3.2. *Gait w.r.t. Slope Angle*

RW-02D can walk on 3.6 mm long slope, as has been verified by ten times experiment in cases of three slope angles, 3.5, 4.0 and 4.5 degrees. The gait of all experiments were 'pace'. Figure 6 shows normalised stride for the leg length, 500 mm, and distance for the slope length, 3.6 [m]. The normalised stride is stride length divided by leg length, 500 [mm] and the normalised distance is walking distance divided by 3240 [mm], the slope length excepting the rod length. From the figure, it can be found that the stride was about 50% of the leg length and more than 200% of the foot length. Moreover, the gait is stable in the sense of walking distance, 66% of a 3.6 m long slope. Since the rate of distance gradually decreases when

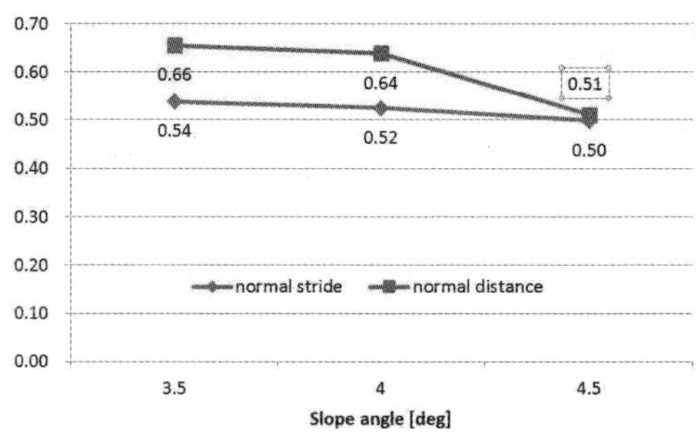

Fig. 6. Normalised stride and distance

the slope angle increases, the slope of 3.5 degrees is optimal. The rate of stride slightly decreases because there were some unstable narrow steps on the relatively steep slope.

3.3. *Dynamic Effect for Walking*

Finally, dynamic effect of walking gait is taken into consideration. Dynamic gait index is introduced so that we can evaluate the effect. Let d be distance between toe and line through hip joint and total CoG, l leg length, d_f length between toe and leg, and and d_s stride (see Fig.7). The dynamic walking index is defined by the following formula:

$$i_{dw} = \frac{d}{l} \approx \frac{d_s - d_f}{l} \tag{1}$$

If the index is negative, then the gait is static. If the index is positive and large, then we can regard that the dynamic effect is strong. The obtained gaits of our biped and quadruped are evaluated by the index i_{dw}. Table 1 shows the index and other parameters of the walkers. 'Human' in Table 1 is basically calculated by Winter[8] and including the case of abnormally long stride gait measured by an author. Stride of human in case of steady walking is from 50% to 60% of leg length, and from 200% to 300% of foot length, and then the index of steady human gait is distributing from 0.3 to 0.4, accordingly. From the table, i_{dw} of RW-01 was negative, then the gait was static. RW-02 are analogous to human, and thereby the gait of RW-02 is enough dynamic. Leg joints of RW-02D are free and the rod is a rigid

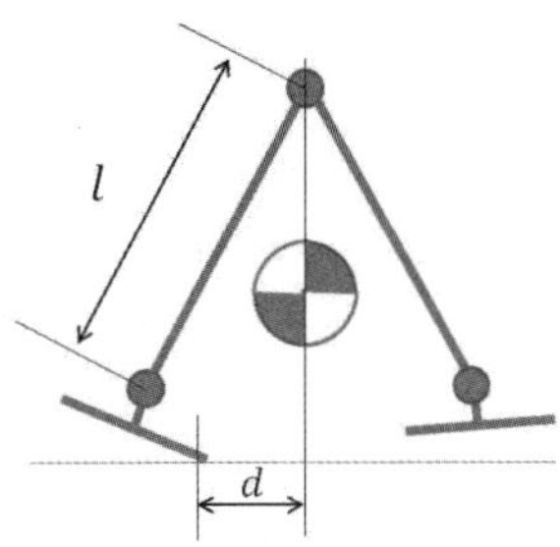

Fig. 7. Dynamic walking index (dw-index)

Walker	l[mm]	stride[%]	d_f[%]	i_{dw}
RW-01	420	7∼15	21	-0.14∼-0.07
RW-02	500	60∼72	12	0.48∼0.6
RW-02D	500	60	12	0.48
Human(an author)	910	50∼80	20	0.3∼0.6

link, thus dynamic model of RW-02D is similar to that of the biped RW-02 except mass of the rod as long as gait of the quadruped is pace. Hence i_{dw} of RW-02D is also given by Eq.1. The gait of the quadruped had relatively wide stride, 54% of its leg length. As a result, the gait of the quadruped, RW-02D was dynamic in the sence that the dynamic walking index was as large as that of human.

4. Conclusion

The purpose of this research has been to develop a passive dynamic quadruped walker. To that end, stable 'pace' gait with relatively long stride of the quadruped has been achieved. We have presented some experimental tests examined the influence of slope angle. Moreover, the gait was evaluated by the dynamic walking index. As a result, the existence of passive quadruped gait that is enough dynamic was proved.

In the future, we will confirm existance of other gait patten such as 'trot' with respect to physical structure and realize passive dynamic gait close to quadruped animals.

References

1. T. McGeer, *CSS-IS TR*, **88-02**, (1988).
2. M. D. Berkemeier and A. C. Smith, *ICRA*, 34 (1997).
3. K. Nakatani, Y. Sugimoto and K. Osuka, *Adv. Robotics*, **23**(5) 483 (2009).
4. C. C. Remy, K. Buffinton and R. Siegwart, *IJRR*, **on line**, Aug. 14 (2009).
5. H. Yoshioka, Y. Sugimoto and K. Osuka, *Conf. of SICE SI div.*, 1335 (2010).
6. T. Kinugasa and K. Yoshida, *Proc. of Intl. Symp. on Mobiligence*, 425 (2009)
7. T. Kinugasa et al., *Proc. of CLAWAR*, 721 (2010).
8. D. A. Winter, *Biomechanics and Motor Control of Human Gait*, (1991).

GAIT GENERATION FOR THE HUMANOID NAO BASED ON HUMAN WALKING PARAMETERIZATION

J.-Ch. PALYART LAMARCHE[1,2,*],O. BRUNEAU[3] and J.-G. FONTAINE[1]

[1] *Italian Institute of Technology, Via Morego, 30 - 16163 Genova, Italy,*
[2] *Università degli studi di Genova, Via Balbi, 5 - 16126 Genova, Italy,*
[3]*Laboratoire d´Ingenierie des Systèmes de Versailles, Université de Versailles, 10/12 Avenue de l´Europe, 78140 Velizy, France*
**E-mail: jean.palyart@iit.it*

The work presented here focus on humanoid walking. More precisely we deliberately divide the control architecture in two distinct parts, coordination and stability. The goal is to design robot coordination mimicking the human one, by reducing the latter's complexity to a minimal set of relations. In our previous work, polynomial approximations were done and implemented into a kinematic simulation of an anthropomorphic humanoid model. This simulation showed smooth biomimetic walking at different speeds. This paper presents the real-world implementation of our methodology using the humanoid robot NAO. The results are promising for applying this entire work on more complex humanoid robots, other gaits, and gait transitions.

Keywords: Biomimetism; Parameterization; Humanoid locomotion; Motion Capture; NAO

1. Introduction

Reproduction of human locomotion by humanoid robots is at cross-road between the fields of robotics, animation, and biomechanics.[1] The principal results were produced with off line optimization of the motion according to stability criteria.[2,3]

Here, we propose to divide gait generation, in two parallel control loops: coordination and stability. Our research highlights coordination to reduce humanoid kinematic complexity, making a favorable basis for stability control, that thereby we hope would be easier. Our previous papers[4,5] describe how from human motion captured data we generate humanoid coordinated movements. From a single spatial input and a set of spatial functions we compute the feet and pelvis coordinated motion. We validated this time-independent coordination method using a kinematic simulation of a virtual humanoid. We thus were able to reproduce smooth and natural three dimensional human gaits.

In this paper, we extended this validation to real world by implementing our method in the humanoid robot NAO, from Aldebaran Robotics company

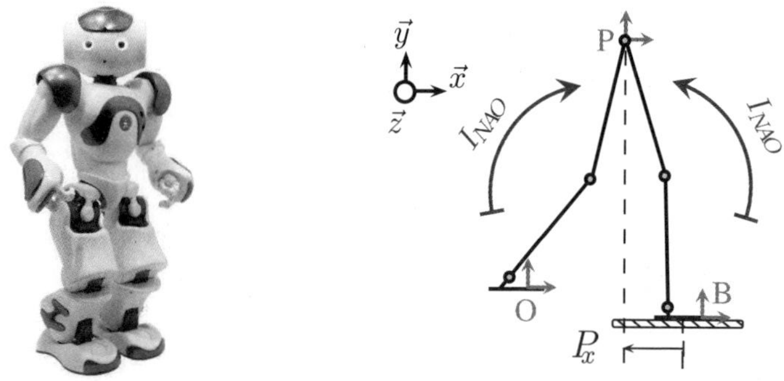

(a) (b)

Fig. 1. Humanoid NAO (a), and inverse kinematics applied on NAO's legs (b).

(Fig. 1(a)). This robot was chosen for several reasons. Leg joint configuration is close to the great majority of current humanoid robots.[6] It also has force foot feedback which is needed for leg coordination and equilibrium. The robot is easy to use, because of its size and weight, and has a user friendly programming environment.

2. Method

Here in the first section, we give a brief outline of the parameterization of human motion we proposed in the previous papers.[4] Then, we describe how we adapt the obtained coordination relations to he robot NAO, and finally how the stability loop controlling the robot's balance is simplified.

2.1. *Our coordination approach*

While a subject (26 years old man, $1.75\,m$, and $70Kg$) was walking on flat ground, we used six infrared cameras to record the position of reflective targets stuck on its skin. From the Cartesian positions of the reflective targets, we computed the positions and orientations of subject's pelvis, right and left feet. The length along the movement axis between the foot and pelvis ($\overrightarrow{BP}\,.\vec{x}$), called P_x (Fig. 1(b)) was chosen as the gait parameter of our parameterization.[4] Others coordinates, called gait descriptors, were expressed as function of the parameter P_x value. These functions, $f(P_x)$, were obtained though polynomial interpolations of ten samples of two walking cycles (see an example in Fig. 2(a)).

By overlapping right and left functions, we were able to position and orient feet and pelvis frames, thus leg joint angles by inverse kinematic model, generating walking motions indefinitely.[4]

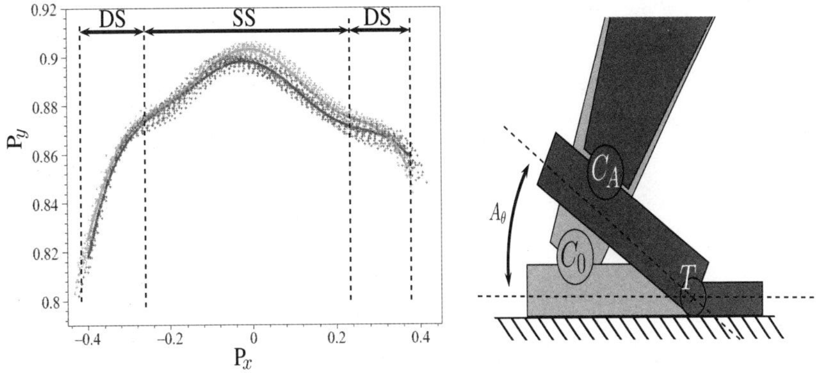

(a) (b)

Fig. 2. Example of left (gray) and right (black) human P_y gait descriptor functions (altitude of the pelvis center in the stance foot reference frame) during one Single Support (SS) and two Double Support (DS) phases (a). Figure (b) shows two feet, one with a MTP joint angle of A_θ (dark gray), the other is flat on the ground (light gray). $||\vec{C_0 C_A}||$ is the distance between the ankle center C_0 and C_A.

2.2. *Adaptation to NAO*

In the previous works[4,5] we validated feet and pelvis coordinated motion on a virtual humanoid that has the same sizes as the subject and seven Degrees of Freedom (DoF) per leg. We now present how we dealt with the DoF and size reductions implied by the use of NAO.

2.2.1. *Degrees of freedom*

The locomotor apparatus of NAO robot has 11 degrees of freedom for both legs:

- five for both hips (two pitch, two roll, and one coupled yaw/pitch)
- one per knee
- two per ankle (pitch and roll)

Another important difference between NAO robot and human is the lack of toe joints, which is very important in human locomotion[7] Then robot's feet must be flat on the ground and the reproduction of heel strike and specially toe off are very complex. The maximal pelvis altitude is lowered because of the distance $\| \overrightarrow{C_0 C_A} \|$ lost between the ankle centers of a flat foot and a bended foot (Fig. 2(b)). For example, when the subject walks this projected distance on the vertical axis is equals $0.13\,m$, which is about 15% of the maximum vertical distance between the toe joint and pelvis centers. In order to stay in the accessible leg domains, gait descriptor positions related to the vertical axis are reduced.

2.2.2. *Size scaling*

To obtain NAO's swing foot (O) and stance foot (B) coordinates from human motions we applied size coefficients on the spatial relations $f(P_x)$. These coefficients are determined by using the ratio (noted K_i), between segmental lengths of the human (${}^{HUMAN}l_i$) and those of NAO (${}^{NAO}l_i$),

$$K_i = \left(\frac{{}^{NAO}l_i}{{}^{HUMAN}l_i} \right) \tag{1}$$

These ratio K_i are applied on the gait parameter,

$${}^{HUMAN}P_x \equiv \frac{{}^{NAO}P_x}{K_i} \tag{2}$$

and then on their respective gait descriptor functions,

$${}^{NAO}f = {}^{HUMAN}f \left(\frac{{}^{NAO}P_x}{K_i} \right) \times K_i \tag{3}$$

One difficulty is the fact that segment lengths of leg and foot do not have the same length ratio. Some coefficients are listed in the Eq. 4.

$$\begin{cases} K_{tibia} = 0.25 \\ K_{femur} = 0.25 \\ K_{foot} = 0.50 \end{cases} \tag{4}$$

2.2.3. *Inverse kinematic model*

We use the embedded inverse kinematic model I_{NAO}, which is able to position and orient an end-effector in the torso reference frame. I_{NAO} is applied on both feet positions and orientations (Fig. 1(b)) to compute leg joint angles.

2.2.4. *Foot contact transition*

The transition between the left and the right systems are coordinated by a state machine. As contact with the ground is absent in the kinematic simulation we performed before experimenting on the real robot (see Sec. 3), the transitions are implemented by defining a threshold on the control parameter P_x. This value corresponds to the instant when step length is near the maximum, and both feet altitude are equivalent. Thus, end effector trajectories, and joint angles q_i via inverse kinematics, are continuously and indefinitely generated.

In the case of actual robot, the state machine triggers transitions when a new double foot contact with the ground is detected, that is to say when the swing foot

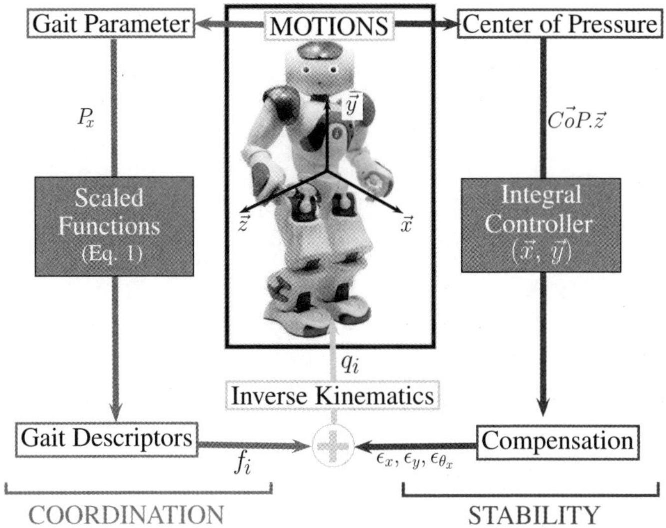

Fig. 3. Gait generation by the on-line parallelization of the coordination and stability loops.

strikes the ground and when the force sum under this foot overtakes a limit. At that instant, the swing foot O and the stance foot B are switched.

2.3. *Stability*

In the case of the real robot, we need to add a balance control in parallel to the coordination generation. To this end, we implemented an equilibrium control based on an integral controller principle. This control is only done in the frontal plane (defined by $\vec{y}$ and $\vec{z}$ axis) using the center of pressure (CoP) under the stance foot. The CoP is computed, if contact exists, in each foot reference frame with its Force Sensitive Resistor (FSR) sensors. In the case where the CoP moves over a limit to the interior or exterior of the foot along $\vec{z}$ axis, the controller adds or removes a cumulative constant ϵ_i (every fixed period) that pushes the pelvis desired position (P_y, P_z) and orientation (P_{θ_x}), in the opposite side, respectively exterior or interior, until the CoP returns into the middle zone (Fig. 3).

$$\epsilon_i^{next} = \epsilon_i^{current} \pm C_i \quad \text{with } C_i \text{ predetermined constants} \tag{5}$$

3. Results

3.1. *Test with Choregraphe*

Before experimenting our walking generation on NAO robot, we tested the coordination part using Choregraphe, the NAO's kinematic simulator. The gait parameter

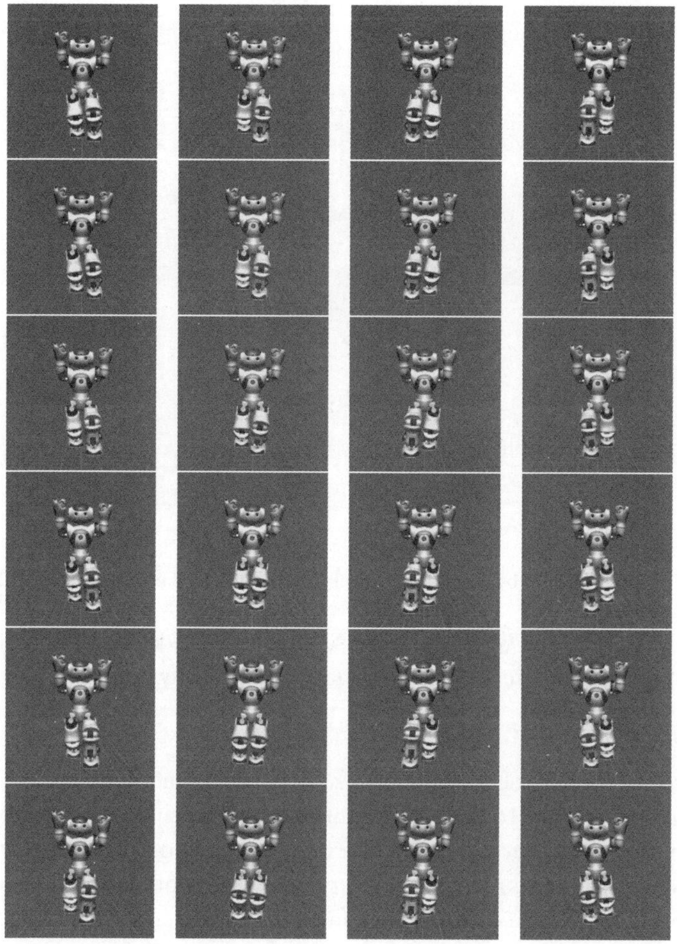

Fig. 4. Kinematic simulation of NAO walking (one walking cycle) at a constant speed (the model is hold by the pelvis). Read from up to down, then left to right, snapshot rate is 8 Hz.

is computed from its measurements by the following equation,

$$P_x{}_{next}^{desired} = P_x{}_{current}^{measured} + \dot{P_x}^{desired} \times \Delta \tag{6}$$

with Δ the sampling period.

We scaled and implemented the coordination relations (Eq. 3), from human motion captures, to the virtual model of NAO. In this kinematic simulation the model was able to perform walking motion from 5 to 15 $cm.s^{-1}$ (Fig. 4).

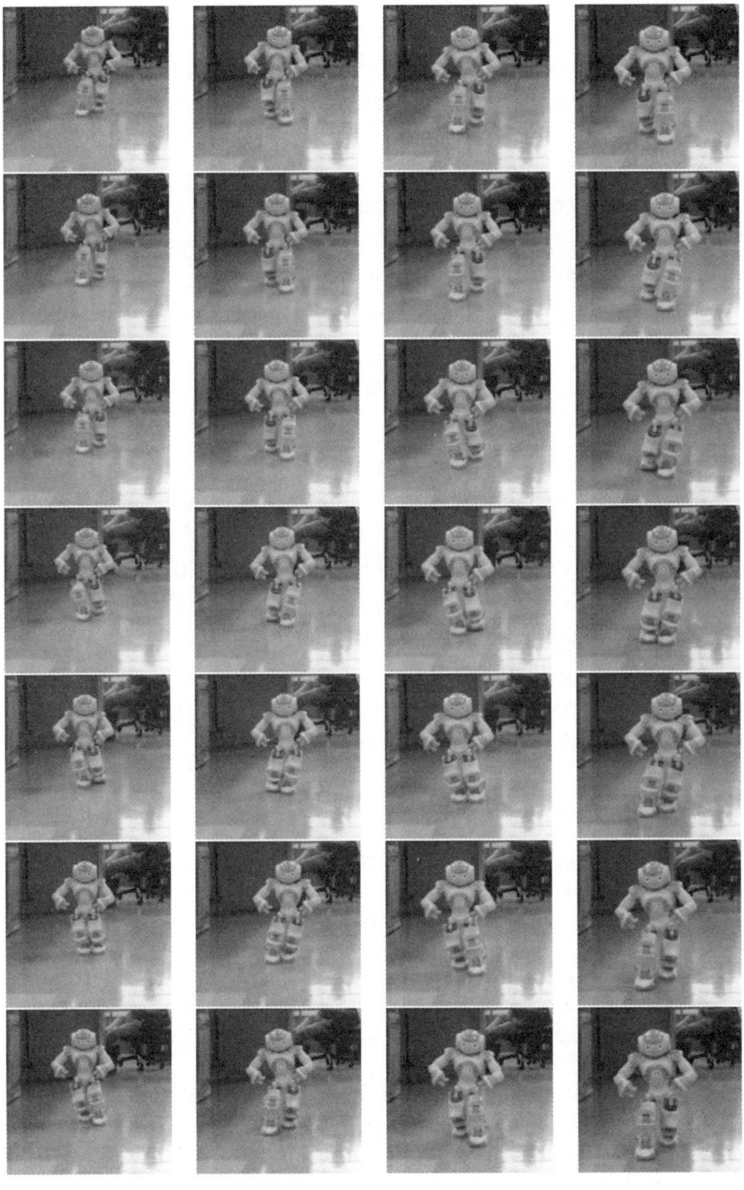

Fig. 5. (Actual humanoid NAO walking ($6\,cm.s^{-1}$) with three dimensional time-independent relationships depending on the spatial parameter P_x. Read from up to down, then left to right, snapshot rate is $4\,Hz$.

3.2. *Experimentation on NAO*

The controlled robot was able to perform walking motions at $6\,cm.s^{-1}$ (Fig. 5). Despite this slow speed (default Aldebaran's walking speed[8] is $12\,cm.s^{-1}$), we obtained a very good stride length (around $16\,cm$ against $4\,cm$ for Aldebaran's walking[8]), in which the robot is able to put its leg in front of the other. We also observed a bending knee walking, because of the absence of toe joints. Moreover the vertical scaling coefficient (relative to the leg) are twice the horizontal scaling coefficient (relative to the feet), thus this creates a deformation of the gait descriptors, and the walking shape. They are more compressed vertically than horizontally.

4. Conclusion

In this paper we adapted our gait generation methodology to the robot NAO. The robot walked with a biomimetic coordination (three dimensional) and stability requires only a basic control strategy (plane defined by $\vec{y}$ and $\vec{z}$). The coordination is based on the study of human motion capture relation with regard to the spatial parameter P_x. The lateral stability is based on integral compensation, a cumulative constant which is a function of the pressure center position.

Our future work will focus on the stability criteria in order to increase the walking speed and versatility. Furthermore, we want to extend to more complex humanoids, other gaits, and gait transitions.

References

1. C. L. Vaughan, *Journal of Biomechanics* **36**, 513 (2003).
2. S. Nakaoka, A. Nakazawa, F. Kanehiro, K. Kaneko, M. Morisawa, H. Hirukawa and K. Ikeuchi, *The International Journal of Robotics Research (IJRR)* **26**, 829(August 2007).
3. L. Boutin, A. Eon, S. Zeghloul and P. Lacouture, *Robotica* **29**, 325 (2011).
4. J.-Ch. Palyart Lamarche and O. Bruneau and J.-G. Fontaine, Humanoid walking coordination through a single spatial parameter, in *The International Conference on Humanoid Robots (Humanoids)*, December 2010.
5. J.-Ch. Palyart Lamarche and O. Bruneau and J.-G. Fontaine, Biomimetic human coordination based on spatial relationships, in *International Conference on Robotics and Biomimetics (ROBIO)*, December 2010.
6. S. Kajita and B. Espiau, *Legged Robots* (Springer Handbook of Robotics, 2008), ch. 16, pp. 361–389.
7. L. Ren, D. Howard, L. Ren, C. Nester and L. Tian, *Journal of Biomechanics* **43**, 194 (2010).
8. D. Gouaillier, C. Collette and C. Kilner, Omni-directional closed-loop walk for nao, in *International Conference on Humanoid Robots (Humanoids)*, 2010.

COMPLIANT JOINTS INCREASE THE ENERGY EFFICIENCY OF BIPEDAL ROBOT

A. Haq, Y. Aoustin and C. Chevallereau

LUNAM, Institut de Recherche en Communications et Cybernétique de Nantes
UMR CNRS 6597, CNRS, Université de Nantes,
1 rue de la Noe, 44321 Nantes Cedex 3, France
E-mail: Abdul.Haq@irccyn.ec-nantes.fr, Yannick.Aoustin@irccyn.ec-nantes.fr
Christine.Chevallereau@irccyn.ec-nantes.fr

The energetic effects of knee locking and addition of linear elastic members to different joints of a seven-link fully actuated planner bipedal robot were studied. The focus was on the reduction of energy consumption during walking. An impactless walking gait was studied and the energetic cost of walking was determined without joint stiffness and knee locking as a baseline for comparison. The gait trajectory was then optimized by adding spring to different joints, energetic cost of walk was then calculated at different walking speeds. Support knee was then mechanically locked and gait was optimized to find the walking cost. The energetic cost of walking determined for the above two cases was then compared to the baseline cost. It was observed that addition of torsional springs at both hips can reduce the walking cost up to 38%, support hip up to 40% with spring stiffness as an optimization parameter for both cases while mechanically locking the support knee can reduce the cost of walking up to 92% at slow walking speeds with gait and knee locking angle optimized.

Keywords: biped, energetic efficiency, passive springs, compliance, impactless gait

1. Motivation

Energetic efficiency and stability are the fundamental criteria that can improve autonomy and task performance capabilities of humanoid robots. Researchers are trying to generate dynamically stable and energetically efficient bipedal gaits as close as possible to human walk.[2,6,11] Studies shows that the legs of the humanoid robots consume more energy in the stance phase than in the swing phase.[3] This difference in energy consumption is because of the demand of high torques to support the robot weight on the ground. Therefore, there is a space of energy optimization in the support leg.

The scope of this work is to investigate the energetic effects of knee locking and addition springs to different joints of a planner bipedal robot. The focus is on the reduction of energy consumption during walking.

2. Introduction

Recently, springs have been used to recover the lost energy, decrease the energy consumption and stabilize the walking gait. In most of the cases, springs are added to the ankle of the robot to store energy and use it when needed mainly during ankle push-off just before heel strike. Studies indicate that there is a direct trade-off between the toe-off impulse from the trailing leg and the rotational torque between the legs.[5] Using the toe-off impulse alone to power gait is four times less energetically expensive then using the hip torque alone.[4,5] Another method of reducing energy consumption is to mechanically lock the support knee and release the lock during swing phase.[9] The knee-lock with active release mechanism is found to be technologically simple and energetically efficient.[8] However, neither the combined effects of knee locking and addition of springs have been explored nor the effects of compliance on energy consumption have been presented in function of walking speed.

In this paper, we carried out four different studies, One on the robot with no springs and locking mechanism, second with spring in different articulations, third with knee locking during support phase and fourth by adding the spring to the joint which economize maximum energy and then locking the support knee in parallel with springs. We assumed that the support knee can be mechanically locked and the torque required is provided by the mechanical lock.

3. Presentation of the Biped

Our seven-link bipedal robot is a planer bipedal robot, composed of two identical legs and a torso. Each leg consists of a thigh, shin and a rigid flat foot. All joints are revolute, frictionless and can only move in the sagittal plane. Our biped is 2D equivalent of humanoid robot HYDROiD[1] having body mass and length similar to those of a human.

The desired gait consists of single support phases separated by instantaneous impactless phase (the swing foot comes in contact with ground with zero velocity). Right and left feet are supposed to be the stance and swing foot respectively. The walking step starts with a single support phase and ends with impactless flat foot contact on the ground where the feet change their role *i.e.* the stance foot becomes swing foot and vice versa.

Following different types of studies were carried out on the biped:

case A. The robot trajectories were optimized and energetic cost of walk was calculated without adding springs or blocking the knee.

case B. Springs were added to the hip, knee or ankle joints of the robot.

- **case B1.** Spring was added only to the support leg joints *i.e.* support ankle, knee or hip.
- **case B2.** Identical springs were added to both joints of the ankle, knee or hip.

In all the cases where a spring was added to any of the joint, the spring constant K was optimized along with the gait and the spring offset or bias angle θ_0 was kept zero.

case C. Support knee was mechanically locked at transition between single support without adding springs at any of the joint. The knee remained locked during the entire support phase.

- **case C1.** The knee locking angle (β) was an optimization parameter for trajectory optimization.
- **case C2.** Based on the numerical values obtained in case **C1**, a constant value of β was selected and then the gait was optimized.

case D. Support knee was mechanically locked and identical springs were added to both hip joints. The knee locking angle β and spring stiffness K were optimized with gait trajectory.

4. Model of the Biped and Optimization

4.1. *Dynamic Model in Single Support*

The dynamic model can be developed by using the Lagrange formulation. In single support phase, the dynamic model, considering an implicit liaison of the stance foot with the ground, *i.e.* there is no take off and no sliding during the single support phase, can be written as in (1) for simple biped and in (2) for a biped with torsional springs.

$$\mathbf{A}(\mathbf{q})\ddot{\mathbf{q}} + \mathbf{C}(\mathbf{q}, \dot{\mathbf{q}})\dot{\mathbf{q}} + \mathbf{G}(\mathbf{q}) = \mathbf{B}\mathbf{\Gamma} \tag{1}$$

$$\mathbf{A}(\mathbf{q})\ddot{\mathbf{q}} + \mathbf{C}(\mathbf{q}, \dot{\mathbf{q}})\dot{\mathbf{q}} + \mathbf{G}(\mathbf{q}) + \mathbf{\Gamma}_s = \mathbf{B}\mathbf{\Gamma} \tag{2}$$

where $\mathbf{A}(q) \in \mathbb{R}^{6\times6}$ is the positive definitive inertia matrix; $\mathbf{C}(q,\dot{q}) \in \mathbb{R}^{6\times6}$ contains the Coriolis and centrifugal forces, $\mathbf{G}(q) \in \mathbb{R}^{6\times1}$ is the vector of gravity forces, $\mathbf{B} \in \mathbb{R}^{6\times6}$ is the actuation matrix consisting of zeros and ones and $\mathbf{\Gamma}_s \in \mathbb{R}^{6\times1}$ is the vector of spring torque.

4.2. *Optimization Criterion*

In optimization problem, reference trajectory for a walking step is generated and then optimized to minimize a criterion shown in (3) under constraints. We are using a polynomial of order four for the evolution of actuated joints motion in function of time to generate reference trajectory.

$$C_\Gamma = \frac{1}{d}\int_0^T \mathbf{\Gamma}^t\mathbf{\Gamma}dt \tag{3}$$

The above criterion was optimized using MATLAB function *fmincon* under inequality constraints of max. joint torques, max. joint velocities, max. Joint angles, Zero Moment Point (ZMP) and no slipping and take-off of support foot during swing phase.

5. Results

Our results show that the most effective way to reduce energy consumption during walking is to add springs to the support joints only and the most economical is the support knee joint (see fig. 2) which economizes up to 92% of energy at 0.30 m/sec.

Adding springs to the support leg only is possible by having a variable spring stiffness mechanism[11] capable of producing spring stiffness from zero to required value, however this mechanism will add an additional mass and the robot's parameters will change. In our simulation tests, we assumed that this mechanism is massless and the energy consumed by the mechanism is negligible. The time of activation and deactivation of the variable spring is very important. If the spring in not activated at the right time, energy is lost. To avoid this variable stiffness mechanism, we restrained ourselves to the simple option and studied the effects of adding identical torsional springs to both legs.

Figure 1 presents the evolution of criteria in function of walking speed for our biped in case **A**, case **B1** where spring was introduced to the support leg and case **B2** where identical springs were added to both of the legs. It is to be noted that identical springs were added to the same joint of both

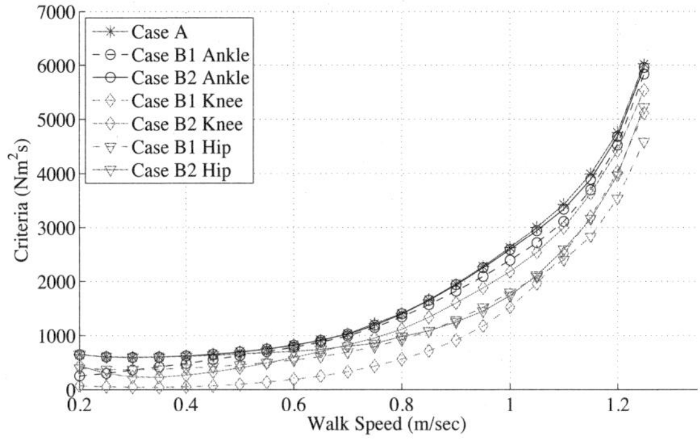

Fig. 1. Evolution of criterion in function of walk speed (solid lines for springs at both legs and dashed lines for springs at support leg)

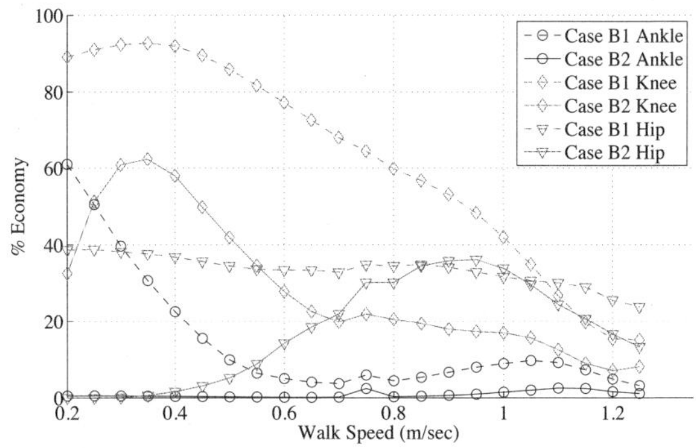

Fig. 2. Evolution of % economy in function of walk speed (solid lines for springs at both legs and dashed lines for springs at support leg)

legs. Figure 1 shows that the optimization criteria is significantly reduced after introducing identical springs to both knee or hip joints in parallel with existing actuator.

Figure 2 case **B2** indicates that knee springs are effective at low walking speeds while hip springs are effective for hight walking speeds. The two curves cross at walking speed of 0.70m/sec which means that if the application area is below this poit then use knee springs otherwise use hip springs. It was however observed that adding springs to both ankles were not effective for our robot. This is contrary to Ref. 7 and Ref. 10 where they found ankle springs useful for stability and energetic efficiency. This is because we considered a flat foot impactless walking gait and there is no rotation of support foot during entire swing phase and in addition we kept the spring offset angle zero which can improve the role of ankle springs.

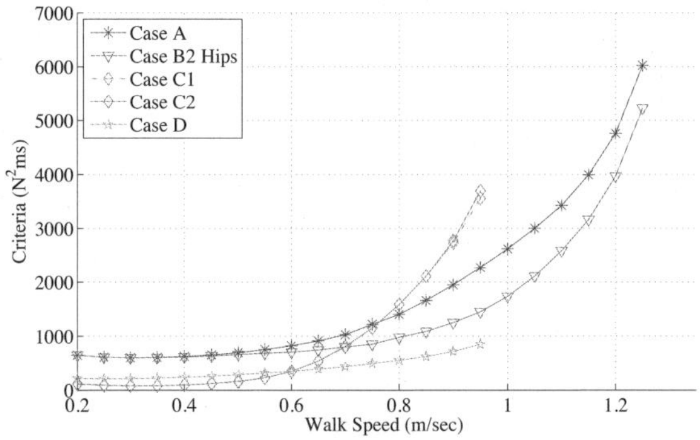

Fig. 3. Evolution of criterion in function of walk speed

Figure 3 gives comparison of criteria curves in function of walking speed in case **A**, case **B2** (identical springs on both hips), case **C1** where knee locking angle β was optimized, case **C2** where β of 3 degree was obtained from optimization results in case **C1** and for a biped in case **D** with an identical spring on both hips and support knee locked. Hip springs are selected here for comparison and eventual addition along with knee locking because it is not possible to study spring effects on the knee joint while the joint is locked.

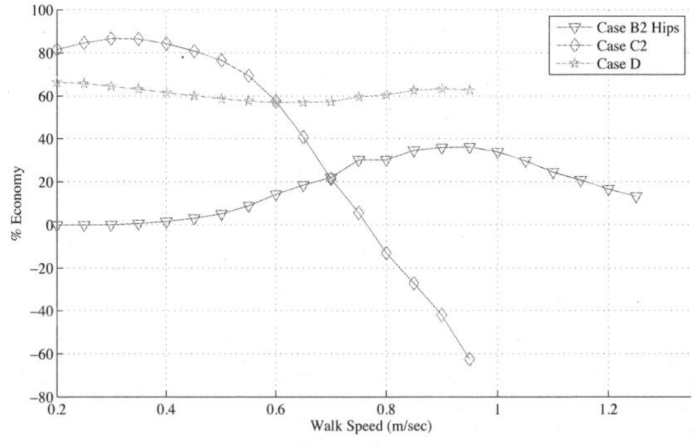

Fig. 4. Evolution of % economy in function of walk speed

Our simulation results show that only knee locking for an impactless bipedal walking gait is economical at lower walking speeds but more costly at higher speeds. The two curves in case **C1** and **C2** are superposed which clearly indicate that knee can be locked at a constant angle for all possible feasible walking speeds. It was also observed that no feasible walk trajectory was found above 0.95m/sec while without locking and with springs added, the biped was able to walk at 1.25m/sec.

The energetic efficiency was dramatically improved when hip springs were added to the biped in case **C1**. Figure 4 shows that the energetic efficiency in case **D** is almost constant for a large range of walking speeds and is always greater than that of a biped with only hip springs. However, only knee locking is more effective upto 0.6m/sec and for speeds above 0.95m/sec, the only choice is the hip springs.

6. Conclusion and Perspectives

In this paper impactless trajectories were generated for a seven-link bipedal robot with three different strategies, one by adding torsional springs to different joints, second by mechanically locking the knee and third by combining the two strategies *i.e.* locking the support knee and adding identical springs to hip joints. The key focus of the study was to economize the energy consumption during walking.

We conclude from this study that the energy consumption of a biped was significantly reduced by adding identical torsional springs at the hip joint and mechanically locking the knee however the maximum attainable walking speed was reduced by a small amount. The energetic efficiency of blocking the knee alone at low speeds and adding springs only to the knee or hip joints is also noticeable. Practical implimentation of this strategy will significantly improve the energetic efficiency as well as the autonomy of our biped.

In line with the previous research, our study reinforces the idea of using passive joint stiffness to improve energetic efficiency of the biped especially on the hip joints in our case. However, contrary to previous work, ankle springs were not effective in our study.

In perspective of this study, the next step is to explore the effects of spring offset or bias angle and the ankle springs with foot rotation on the energetic efficiency of the biped. This can also be extended to study effects on a 3D bipedal robot.

Acknowledgment

This work is supported by ANR grants for the R2A2 project and by Higher Education Commission (HEC) of Pakistan under overseas scholarship program.

References

1. S. Alfayad, F.B. Ouezdou, and F. Namoun. New three dof ankle mechanism for humanoid robotic application: Modeling, design and realization. In *Proceedings of the IEEE/RSJ International Conference on Intelligent Robots and Systems (IROS 2009)*, pages 4969–4976, 2009.
2. S.H. Collins and A. Ruina. A bipedal walking robot with efficient and human-like gait. In *Proceedings of the 2005 IEEE International Conference on Robotics and Automation (ICRA 2005)*, pages 1983–1988, 2005.
3. Heinz Ulbrich Daniela Frg, Martin Frg. A bipedal robot model with elastic actuation. In *Proceedings of the 1st Joint International Conference on Multibody System Dynamics*, May 25-27 2010.
4. Arthur D. Kuo. Energetics of actively powered locomotion using the simplest walking model. *Journal of Biomechanical Engineering*, 124(1):113–120, 2002.
5. Cara L. Lewis and Daniel P. Ferris. Walking with increased ankle pushoff decreases hip muscle moments. *Journal of Biomechanics*, 41(10):2082–2089, 2008.
6. Xiuhua Ni, Weishan Chen, and Junkao Liu. A comparison between human walking and passive dynamic walking. In *Proceedings of the 4th IEEE Conference on Industrial Electronics and Applications (ICIEA 2009)*, pages 2552–2555, 2009.
7. T. Schauss, M. Scheint, M. Sobotka, W. Seiberl, and M. Buss. Effects of compliant ankles on bipedal locomotion. In *Proceedings of the IEEE International Conference on Robotics and Automation (ICRA '09)*, pages 2761–2766, 2009.
8. K. Trifonov and S. Hashimoto. Active knee-lock release for passive-dynamic walking machines. In *Proceedings of the IEEE International Conference on Robotics and Biomimetics (ROBIO 2007)*, pages 958–963, 2007.
9. Gijs Oort van, Raffaella Carloni, Dian J. Borgerink, and Stefano Stramigioli. An energy efficient knee locking mechanism for a dynamically walking robot. In *IEEE International Conference on Robotics and Automation, ICRA 2011*, USA, May 2011. IEEE Robotics and Automation Society.
10. M. Wisse, D.G.E. Hobbelen, R.J.J. Rotteveel, S.O. Anderson, and G.J. Zeglin. Ankle springs instead of arc-shaped feet for passive dynamic walkers. In *Proceedings of the 6th IEEE-RAS International Conference on Humanoid Robots*, pages 110–116, 2006.
11. J. Yamaguchi, D. Nishino, and A. Takanishi. Realization of dynamic biped walking varying joint stiffness using antagonistic driven joints. In *Proceedings of the IEEE 1998 IEEE International Conference on Robotics and Automation*, volume 3, pages 2022–2029 vol.3, 1998.

ON-BOARD ENERGY CONSUMPTION ESTIMATION FOR A SIX-LEGGED WALKING ROBOT

A. Roennau, F. Sutter, T. Kerscher, R. Dillmann

FZI Research Center for Information Technology,
Interactive Diagnosis- and Servicesystems (IDS),
Haid-und-Neu-Str. 10–14, D-76131 Karlsruhe, Germany
E-mail: {roennau, sutter, kerscher, dillmann}@fzi.de, www.fzi.de

Walking robots have a great number of active joints, which makes them very agile and able to operate in rough and unstructured terrain. But all these joints have to be powered individually. Therefore it is not surprising that walking robots require much energy, which reduces their operational range. The energetic optimisation of walking robots can increase this range and thus their autonomy. In this paper we present an on-board energy consumption estimation for a six-legged walking robot, which provides detailed data for each joint and is the first step to a sophisticated locomotion analysis. The developed estimation is based on an online and on-board joint current measurement method. Experiments with LAURON show the high precision of the received current and energy measurement results.

Keywords: six-legged walking robot, hexapod, on-board energy consumption

1. INTRODUCTION

Many walking robots have great climbing capabilities and can overcome impressive obstacles. This makes them suitable for rough and unstructured terrain, which can usually not be traversed by wheeled robots. But their manoeuvrability comes with a drawback: the great number of active joints. Each joint has to be actuated separately, which requires energy in each of these joints and makes walking less energy efficient compared to wheeled locomotion. Therefore, precise energetic measuring methods are required for a detailed energy analysis. With this data it is possible to better understand all influences on the walking process and to optimise the amount of energy needed.

Energy efficiency is an important criterion for the optimisation of walking robots concerning the evaluation, choice of gait patterns and mechanical design. The consumed energy can be calculated from the torques and angular velocities[1] or from the currents of the motors[2]. It is possible to use the received energy consumption for extended measurements like the transport cost[3]. Recent research on energy efficiency in walking robots tends to take biological, mathematical or simulation results into account rather than experimental analysis[45]. Only few approaches measure the consumed energy of real robots and make use of on-board circuits to determine the currents needed while walking and calculate the energy actually consumed[6]. The results of a sophisticated locomotion analysis can be used during the design process of the next generation of walking robots and help to develop energy optimised control strategies.

Fig. 1. Bio-inspired six-legged walking robot LAURON IVc with 20 active joints.

In this work we will present an on-board and online measurement method to determine the motor currents of each joint of the six-legged walking robot LAURON. These currents are combined and used to estimate a precise overall system current. With the help of the integral calculus we are able to identify the energy consumption of each joint and consequently of the entire robot. Several experiments will demonstrate the high precision of the system current measurement and the developed energy consumption estimation.

The bio-inspired, six-legged walking robot LAURON has 20 active joints (see Fig. 1) and weighs about 27 kg with its accumulators. Like its ancestors, the mechanical design of LAURON IVc was derived from the stick insect Carausius morosus.[7] Each leg has three rotational joints and the head is equipped with a pan tilt unit (2 joints). All joints are controlled by custom made UCoM control units (UCoM: Universal Control Module[8]). These UCoMs use a PID controller with a velocity feedback loop to move the joints to the desired angles. All joint angles are measured by optical encoders and high resolution motor encoders. The ground contact forces are estimated by a strain gauge 3D force sensor and a spring-damper system. Several sensor systems, like camera systems, an IMU etc. complete LAURON's perception capabilities. With its robust hardware setup, all sensor systems and its behaviour-based control system LAURON is able to walk autonomously in rough and unstructured terrain.

2. APPROACH

2.1. *On-board Current Measurement*

For walking robots it is not only interesting to determine the overall system energy consumption, but also the energy needed by each joint. With DC motors as actuator for a robot's joint, it is necessary to measure the energy of each motor separately. The consumed energy can be retrieved by summing up the needed power over time (integral calculus). In this approach we measure the electric power ($P = U \cdot I(t)$) of each motor (with $P(t) :=$ electric power, $U :=$ input voltage, $I(t) :=$ motor current). All motors are controlled by PWM-based voltages, which cause a reasonable amount of noise in the currents. The PWM-based voltages cannot be interpreted as DC constant input voltage for the electric motor. Depending on the PWM value the input voltage can vary in wide range from -12V to +12V. The voltage drop of a reference resistor in the motor driver circuits of the UCoM is used to measure the current $I(t)$. The noise in the current is reduced by a combination of a median and mean filter. Then a non-linear curve with an offset is fitted to reference data. The reference data was collected in experiments with a PWM-based motor controller implemented on the UCoM, which allowed us to set fixed PWM values and then measure the resulting current. With the help of this fitting we are able to consider and compensate the effects of the PWM-based non-constant motor input voltage in our measurement. This non-linear curve is parameterised and then optimised for each leg respectively, which enables us to determine the motor currents very precisely. Because LAURON is moved by conventional DC motors, it is easy to convert the motor currents to corresponding motor torques. The overall leg current is the sum of the three motor currents with an offset for the UCoM control unit. Finally, all leg currents are combined to one overall system current. Of course, the supply currents for all remaining electronic components were estimated and added as a fixed offset. In several exemplary measurements we were able to identify a very low mean error of the overall system current, which confirms the high precision of the identified motor currents.

2.2. *Detailed Energy Consumption*

Given the varying value for the current of a motor or the overall system current over time, the energy can be determined by the integral calculus. The motor input voltage is created with a PWM modulation of a 24V supply voltage. Due to the non-linear fitting of the current in the on-board current measurement method, we are able to approximate the input voltage U by a constant voltage of 24V.

$$E = \int_0^T P(t)dt = \int_0^T U \cdot I(t)dt = \int_0^T 24V \cdot I(t)dt$$

The discrete integral calculus is realised by the *multimeter behaviour* in the behaviour-based control system of LAURON. This behaviour sums up the current values in discrete time steps of about 20 ms (50 Hz). Due to its online character and measurement speed, this estimation can also be used to gather actual information about the system status and detect errors in the electronics. With the high precision of the current measurement, we receive a high precision in our energy estimation. This precision was confirmed in multiple experiments.

3. RESULTS AND EXPERIMENTS

3.1. *Joint and Overall System Current*

The first presented experiment illustrates the results of a single motor current measurement while walking with the robot. In this experiment (see top of Fig. 2) the beta currents are dominant, because these joints have to compensate the biggest portion of the gravitational forces.

The second experiment shows the effect of adding 15kg payload to the robot. Despite this high payload of approximately half of the total robot weight, the increase of LAURON's system current was rather small. Only additional 900mA were needed to carry the extra weight. The result of this experiment is shown in the bottom of Fig. 2.

In order to evaluate the precision of the on-board current measurement, we decided to use an external reference measurement, which determines the

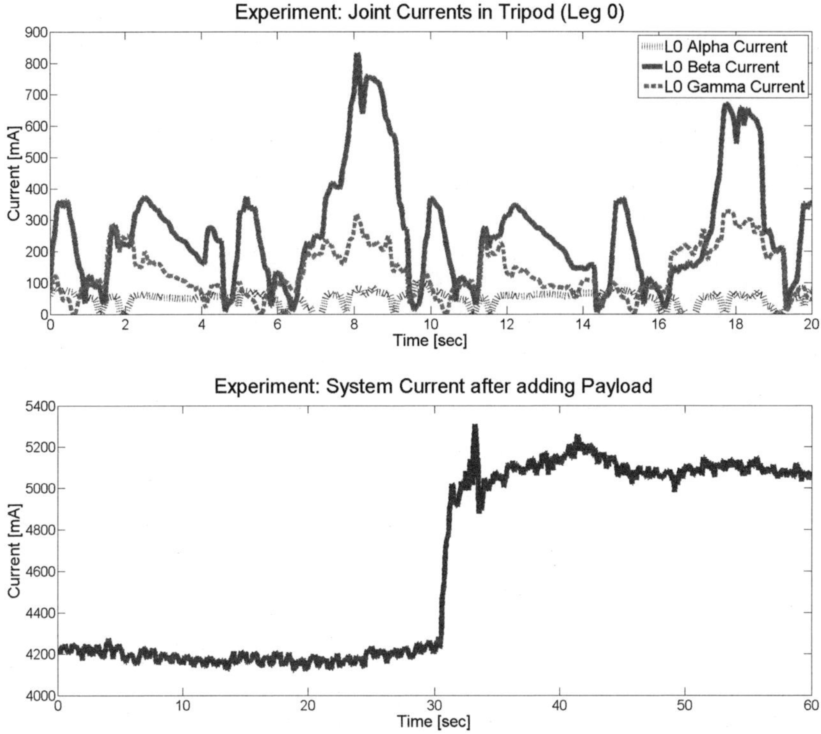

Fig. 2. Experiment: (top) motor currents of Leg 0, (bottom) system current after adding 15kg payload.

current (at 24V output voltage) supplying the robot. The reference current was compared to our estimated overall system current. Figure 3 shows both currents and the resulting error of the developed on-board measurement. Although, the plot shows some visible differences at larger current peaks, the mean error of the on-board measurement was at the low value of only 68.8 mA in this experiment. This low mean error was confirmed in several other experiments (mean system current error below 100mA).

3.2. *Energy Consumption Estimation*

The precision of the developed electrical energy consumption estimation was evaluated in a long run experiment. LAURON's energy consumption was measured for about 60 min. The result of the on-board energy estimation can be found in Fig. 4. According to our developed method, LAURON

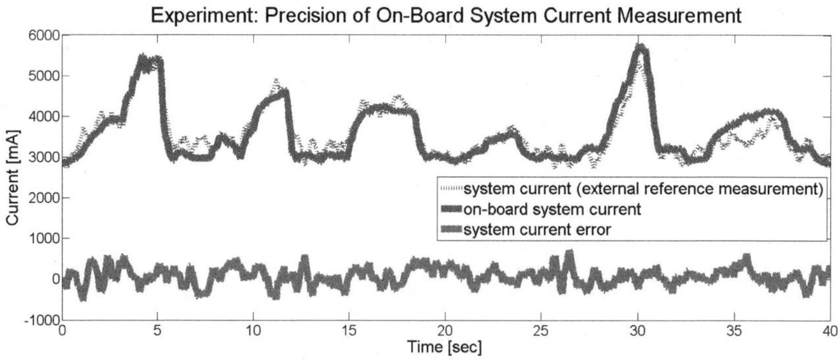

Fig. 3. Experiment: Comparing on-board system current estimation with reference current, mean current error: 68.8mA, mean absolute current error: 176.1mA

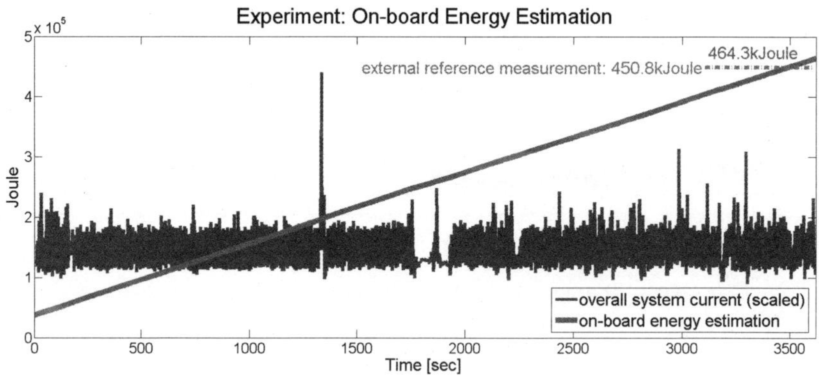

Fig. 4. Experiment: Comparing on-board energy consumption estimation with external reference measurement: Peak in system current was the result of a collision and a safety behaviour reaction.

consumed 464.3 kJ during this long run experiment. Besides the estimated energy, we also illustrated the overall system current (scaled to fit into the plot). This value shows, that the robot was not only standing still, but was very active during the experiment. A high system current peak after approx. 1300 sec was the reaction to a collision of the robot. As a reference energy measurement we used a *Voltcraft Energy Check 3000*, which was plugged in front of the 24V DC power supply of the robot. The input/output efficiency of the DC power supply (*Voltcraft HPS-13030*) was determined by multiple measurements at the average current output level (measured efficiency: 84 percent). At the end of this long run experiment the *Voltcraft Energy*

Check 3000 showed the value 0.149 kWh. After converting this to Joule and taking the DC efficiency of the power supply into account we receive 450.8 kJ. This means we have an error smaller than 2.99 percent, which is very good for such a long run experiment.

The last presented experiment was conducted to analyse the coherence of the walking width and the overall system energy consumption. In three runs LAURON's walking width was increased from 780mm to 880mm and then finally set to 980mm. In each run LAURON walked for 5 minutes using its tripod gait. The system energy consumptions can be found in Fig. 5.

Energy Consumption

step width	after 100 sec	after 200 sec	after 300 sec
780mm	12.61 kWs	25.16 kWs	37.91 kWs
880mm	12.93 kWs	25.41 kWs	38.26 kWs
980mm	13.88 kWs	27.58 kWs	41.24 kWs

Fig. 5. Experiment: Energy Consumption depends on LAURON's step width.

4. CONCLUSIONS AND FUTURE WORKS

4.1. *Conclusions*

In this paper we have presented a method to analyse the detailed energy consumption of a six-legged walking robot. The precise estimation is based on an exact on-board current measurement method performed for each joint individually. A non-linear function was fitted to a great number of reference measurements and multiple filters were implemented, in order to achieve this high precision. By summing up all motor currents and adding well identified offsets, it is possible to estimate an overall system current. Experiments have shown the precision of this estimation (mean error smaller than 100mA). With the help of the integral calculus and the exact motor currents, we are able to estimate an overall energy consumption of the six-legged walking robot. It is even possible to determine the energy consumed in each joint separately. Several experiments have shown the good results of the on-board energy estimation approach. The influence of the walking width on the energy consumption was analysed and showed that a narrow walking width is less energy consuming.

4.2. *Future Works*

The ability to measure the motor currents of each joint separately enables us to determine the generated torques. Now, we can also investigate mechanical stress and find mechanical coupling problems. The results of these analyses can be used to design new energy efficient walking robots and develop efficient control strategies. In the future, the presented energy consumption estimation approach will be used to optimise LAURON's locomotion, compare walking patterns to each other and estimate the remaining battery capacity. This will improve LAURON's autonomy and walking capabilities in rough and unstructured terrain.

References

1. F. Yamasaki, K. Hosoda and M. Asada, An energy consumption based control for humanoid walking, in *IEEE/RSJ International Conference on Intelligent Robots and Systems, IROS*, 2002.
2. Y. Ogura, T. Kataoka, Aikawa and et al., Evaluation of various walking patterns of biped humanoid robot, in *Proceedings of the 2005 IEEE International Conference on Robotics and Automation, ICRA*, 2005.
3. S. Collins, A. Ruina, R. Tedrake and M. Wisse, Efficient bipedal robots based on passive-dynamic walkers, in *Science 18 February 2005: Vol. 307 no. 5712 pp. 1082-1085*, 2005.
4. M. F. Silva, J. A. T. Machado and A. M. Lopes, Energy analysis of multi-legged locomotion systems, in *Proceedings of 4th International Conference on Climbing and Walking Robots, CLAWAR*, 2001.
5. J. Nishi, Gait pattern and energetic cost in hexapods, in *Proceedings of the 20th Annual International Conference of the IEEE Engineering in Medicine and Biology Society*, 1998.
6. R. Kurazume, A. Byong-won, K. Ohta and T. Hasegawa, Experimental study on energy efficiency for quadruped walking vehicles, in *IEEE/RSJ International Conference on Intelligent Robots and Systems, IROS*, 2003.
7. A. Roennau, T. Kerscher and R. Dillmann, Design and kinematics of a biologically-inspired leg for a six-legged walking machine, in *2010 3rd IEEE RAS and EMBS International Conference on Biomedical Robotics and Biomechatronics (BioRob)*, 2010.
8. K. Regenstein, Modulare, verteilte hardware-software-architektur fuer humanoide roboter, in *PhD Thesis on Modular, Distributed Hardware-Software Architecture for Humanoid Robots, FZI Karlsruhe*, 2010.

KINEMATIC AND DYNAMIC ANALYSIS OF A HEXAPOD PRONKING GAIT ROBOT AND CONTROL ACTIONS

SERVET SOYGUDER

Mechanical Engineering, Firat University, Elazig, 23279,Turkey

HASAN ALLI

Mechanical Engineering, Firat University, Elazig, 23279,Turkey

This in this study, A hexapod robot was carried out walking, running and pronking with Spring Loaded Inverted Pendulum (SLIP) model in stance phase. This is solving a natural problem in which the main goal is to enable the robot to walk, run and pronking in a stable condition regardless of the environmental conditions. In order to solve this problem, kinematic and dynamic analysis, and control actions of a hexapod robot were realized for a pronking gait in this study. In addition, the stance and flight phase dynamic structures were solved in a sequential closed loop to obtain the equation of motion for pronking gait. The hexapod robot with pronking gait was controlled by PD control algorithm. As a result, all computer simulations have shown that the proposed control actions and method are more effective and make the system control quite easy and successful.

1. Introduction

In recent years, numerous studies on legged robots have been reported [1-10]. Scientists and engineers have taken an interest in constructing legged robots with biologically-based designs [4, 8]. The aim of this study is to realize a dynamic stable pronking gait of a hexapod robot. First of all, the dynamic model of the hexapod robot with pronking gait has been formed and then, the simulation of the system has been realized. To obtain the numerical simulation of the equation of motion for pronking gait, the various dynamic structures have been solved in a sequential closed loop (stance, flight phase). SLIP has been used as a dynamic model to simplify the simulation of the system and ease to understand the pronking gait. Different from the studies in the literature [9, 10], the number of the performed control actions during the phases of pronking gait in this study is smaller than those of the studies in the literature in the limit of authors knowledge. Therefore, the number of the performed control actions in this study makes the system control quite easy. In addition, the manufacturing of robot was done for experimental works. Fig. 1 shows the manufactured robot. In addition, the manufacturing of robot was done for future experimental works.

Fig. 1 shows the manufactured hexapod robot ROBOTURK SA-3. It is used as both the quadrupedal and hexapod robot. The fixed middle legs hexapod robot is used as quadrupedal robot. In previous studies, we carried out the designed and manufactured six legged robot named ROBOTURK SA-1 [11] and eight legged robot named ROBOTURK SA-2 [12] by reducing the number of actuators to simplify the control problem. ROBOTURK SA-1 has been driven by using only two actuators while ROBOTURK SA-2 has been driven by using only one actuator.

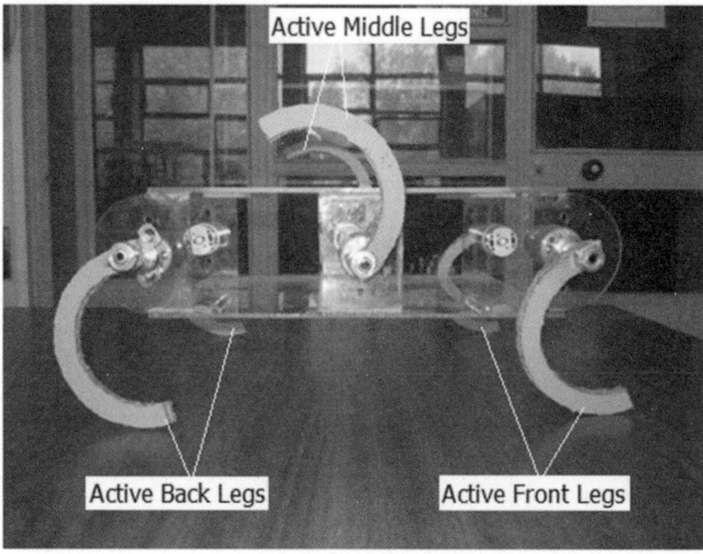

Figure 1. Photograph of the hexapod robot ROBOTURK SA-3.

2. The Coordinate Transformations of The Kinematic Equations

The only way acting external force on the body is the six legs by means of their interactions with the ground. That's why, the position and orientation of the legs with respect to th body frame are the most important parameters in this manner [13].

According to B body frame, we can define:

$$\vec{\mathrm{I}}_i = R^T(\alpha).(\vec{f}_i - \vec{b}) - \vec{a}_i \tag{1}$$

$$\vec{\mathrm{I}}_i = \begin{vmatrix} \cos(\alpha)\sin(\alpha) \\ -\sin(\alpha)\cos(\alpha) \end{vmatrix} . \begin{vmatrix} f_{iy} - b_y \\ f_{iz} - b_z \end{vmatrix} - \begin{vmatrix} \partial_{iy} \\ \partial_{iz} \end{vmatrix} \tag{2}$$

If the radial leg forces are rearranged, we get

$$\tau_i = F_i.\rho_i \tag{3}$$

$$F_i = \tau_i / \rho_i \tag{4}$$

$$F_{iy} = \cos(\alpha + \phi_i).(\tau_i / \rho_i) + \sin(\alpha + \phi_i).F_{ri} \tag{5}$$

$$F_{iz} = \sin(\alpha + \phi_i).(\tau_i / \rho_i) - \cos(\alpha + \phi_i).F_{ri} \tag{6}$$

Furthermore, the power consumption for all virtual leg can be written as,

$$P_{hexapod\_total} = \dot{\phi}_{front\_leg}.\tau_{front\_leg} + \dot{\phi}_{middle\_leg}.\tau_{middle\_leg} + \dot{\phi}_{back\_leg}.\tau_{back\_leg} \tag{7}$$

3. Equations of Motion of The System

The dynamics of the planar hexapod robot are obtained based on a planar rigid body affected by the leg forces. Therefore, the equations of motion can be obtained as follows:
If we use Newton's second law for rigid body, we have

$$m(\ddot{b} + g) = \sum_{i=1}^{6} s_i F_i \tag{8}$$

Again from Newton's second law, the following equations are obtained:

$$\mathrm{I}\ddot{\alpha} = \sum_{i=1}^{6} s_i (f_i - b) \times F_i \tag{9}$$

If leg masses are considered, the equations of motion can be expressed as follow:

$$m_t.\ddot{f}_i = (1 - s_i)F_i \tag{10}$$

$$m_t.\ddot{f}_i = (1 - s_i)\begin{bmatrix} \cos(\alpha + \phi_i) & -\sin(\alpha + \phi_i) \\ \sin(\alpha + \phi_i) & \cos(\alpha + \phi_i) \end{bmatrix} . \begin{bmatrix} \tau_i / \rho_i \\ -F_{ri} \end{bmatrix} \tag{11}$$

4. Control Actions

Most of the control action is applied during the flight phase since the actuator torques are limited during the stance phase [9, 14]. Fig. 7 indicates the control action methods in two-phases of the system. At least, one of the virtual legs must not contact with the ground during the flight phase. In this case, ϕ^{td} angle is always controlled. Moreover, the angle of the leg should become the desired angle when the virtual legs touch the ground after the flight phase. This action is generally performed during the flight phase. PD control algorithm is assigned to obtain the desired touchdown angle by moving the legs to the desired positions. The required torque is obtained by the actuators mounted the hips of each virtual leg. As a result, ϕ^{td} angle can be taken as a control variable of the flight phase. [15].

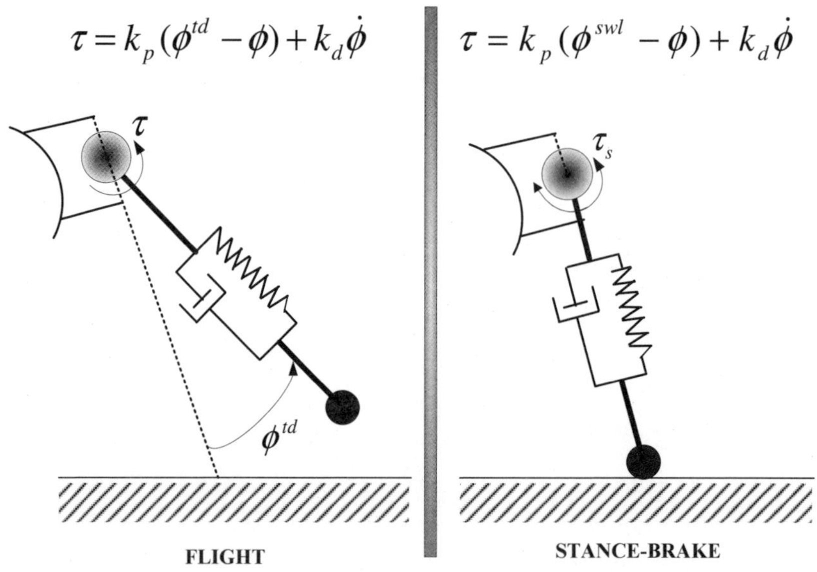

Figure 2. Control Actions During The Phases of The Pronking Gait.

5. Simulation Results

In this study, numerous simulation walking, running and bounding gait of the hexapod robot have been performed. Fig. 3 shows the simulation of the bounding gait of the hexapod robot in the sagittal plane.

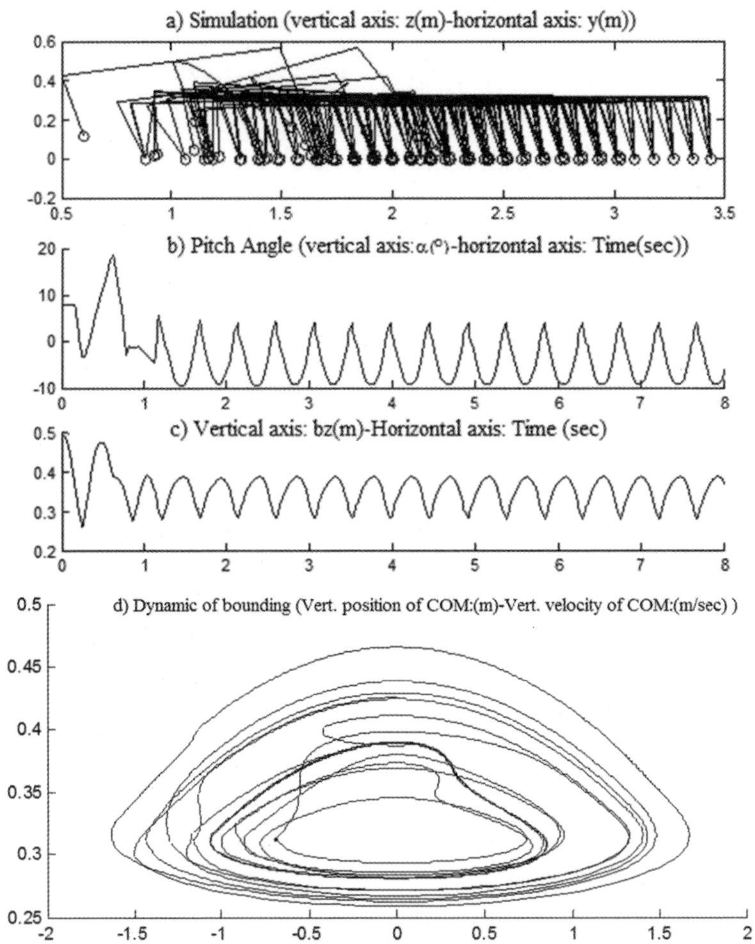

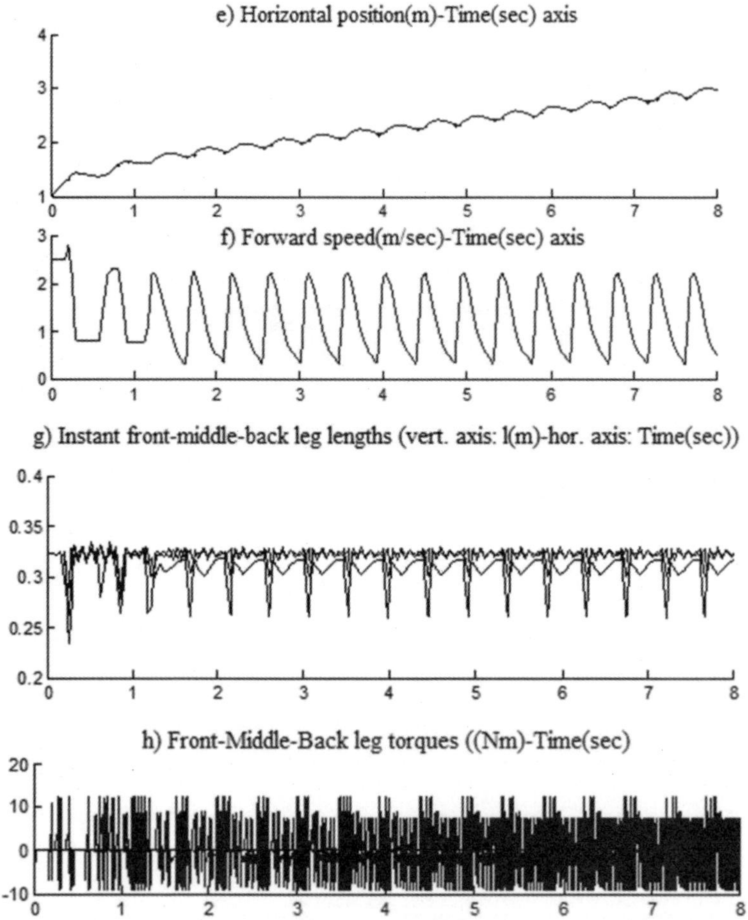

Figure 3. Simulation of the pronking gait of the hexapod robot in the sagittal plane.

6. Conclusion

In this study, the proposed controllers are simpler than those of Raibert proposals because the body-state feedback and leg-length control are not necessary. The idea of the proposed controller has been successfully used in the studies of different robots with bounding gait. [13, 16-17]. In addition to that, the number of the performed control action during the phases of bounding gait is smaller than those of the studies in the literature in the limit of authors knowledge [9-10]. This showed that, the proposed control action method in this study makes both the system control easy and the system performance increase by decreasing the run time for each loop. Consequently, the SLIP model has been successfully used as a dynamic model of bounding gait of the hexapod robot in this study as also those of walking [6, 14], running [18], flipping [13] and climbing gaits [15] in the literatures.

References

1. A. Sato, A Planar Hopping Robot with One Actuator: Design, Simulation, and Experimental Results, McGill University, Montreal, Canada, (2004).
2. P. Gonzalez De Santos, M.A. Jimenez, Path tracking with quadruped walking machines using discontinuous gaits, Computers & Electrical Engineering, Vol. 21(6), pp. 383-396 (1995).
3. V. A. Nishi, Development of wall-climbing robots, Computers & Electrical Engineering, Vol. 22(2), pp. 123-149 (1996).
4. D. Son, D. Jeon, W.C. Nam, D. Chang, T.W. Seo, J. Kim. Gait planning based on kinematics for a quadruped gecko model with redundancy, Robotics and Autonomous Systems, Vol.58(5), pp. 648-656 (2010).
5. Y. Sooyeong, Reliable gait planning and control for miniaturized quadruped robot pet, Mechatronics, Vol.20(4), pp. 485-495 (2010).
6. J.M. Yang., Two-phase discontinuous gaits for quadruped walking machines with a failed leg, Robotics and Autonomous Systems, Vol. 56(9), pp. 728-737 (2008).
7. T. Ho, and S. Lee, Piezoelectrically Actuated Biomimetic Self-Contained Quadruped Bounding Robot, Journal of Bionic Engineering, Vol.6(1), 2009, pp. 29-36.
8. R.D. Quinn, G.M. Nelson, R.J. Bachmann, D.A. Kingsleg, J. Offi, R.E. Ritzmann, Insect design for improved robot mobility, Proceedings of 4th International Conference on Climbing and Walking Robots, Berns and Dillmann eds, Prof. Eng. Pub, pp. 69-76 (2001).
9. I. Poulakakis, J.A. Smith, M. Buehler, Modeling and Experiments of Untethered Quadrupedal Running with a Bounding Gait: The Scout II Robot The International Journal of Robotics Research, vol. 24, pp. 239 – 256 (2005).
10. I. Poulakakis, E. Papadopoulos M. Buehler, On the Stable Passive Dynamics of Quadrupedal Running, Proc of the IEEE Int. Conf. on Robotics and Automation, Taiwan, pp. 1368-1373 (2003).
11. S. Soyguder, and H. Alli, Design and prototype of a six-legged walking insect robot, Industrial Robot –An International Journal, Vol: 34(5), pp. 412-422 (2007).
12. S. Soyguder, and H. Alli, Motion Mechanism Concept and Morphology of A Single Actuator Tetrapod Walking Spider Robot: The ROBOTURK SA-2 Robot, Industrial Robot –An International Journal, (2011), In Press.
13. U. Saranli, Dynamic locomotion with a Hexapod Robot, Michigan University, Michigan, USA, (2002).

14. R. Altendorfer, N. Moore, H. Komsuo¹glu, M. Buehler, H.B. Brown Jr., D. McMordie, U. Saranli, R. Full, and D.E. Koditschek. RHex: A biologically inspired hexapod runner. Au- tonomous Robots, 11:207{213, (2001).
15. Talebi, S., Buehler, M., and Papadopoulos, E., Towards Dynamic Step Climbing for a Quadruped Robot with Compliant Legs, 3rd Int. Conf. on Climbing and Walking Robots, Madrid, Spain, (2000).
16. D. Campbell and M. Buehler, "Preliminary Bounding Experiments in a Dynamic Hexapod", In Bruno Siciliano and Paolo Dario, editors, Experimental Robotics VIII, Springer-Verlag, (2003).
17. Yamamoto Y., Fujita M., De Lasa M., Talebi S., Jewell D., Playter R. and Raibert M., "Development of Dynamic Locomotion for the Entertainment Robot – Teaching a New Dog Old Tricks, 4th Int. Conf. on Climbing and Walking Robots (CLAWAR), pp. 695-702, (2001).
18. Buehler M., "Dynamic Locomotion with One, Four and Six-Legged Robots", in J. of the Robotics Society of Japan, 20 (3), pp. 15-20, (2002).

ADAPTIVE WALKING GAIT FOR LOCOMOTION ON TERRAIN WITH NON-UNIFORM SLOPE

Carlos Neves* and Rodrigo Ventura** and João Sequeira***

Institute for Systems and Robotics
Instituto Superior Técnico
1049-001 Lisboa, Portugal
**E-mail: cneves@isr.ist.utl.pt*
*** Visiting Scholar, Computer Science Department, Carnegie Mellon University*
E-mail: rodrigo.ventura@isr.ist.utl.pt
**** E-mail: jseq@isr.ist.utl.pt*

The autonomous locomotion of a legged robot presents several challenges, such as stability and sensitivity to the ground slope. Complex sensory systems, accurate physical models, and demanding computational resources are often required to maintain balance, that together with precise servo control leads to high energy consumption levels. This paper addresses the problem of controlling the locomotion of a humanoid robot, using measurements from inertial sensors to create adaptive walking gaits. These gaits are based on the interpolation of walking gait templates, weighted by the estimated ground slope. A supervised learning method was also implemented to perform a faster recognition of the ground slope when in a static pose. This system provides a simple, yet effective approach to humanoid locomotion on sloped terrain, thus freeing computational resources for other tasks. The approach was validated in simulation, using a realistic model of a Bioloid Humanoid platform.

Keywords: Humanoid robots, Legged Locomotion, Extended Kalman Filter.

1. Introduction

One of the challenges in legged locomotion of humanoid robots is to maintain a stable gait, even on a non-uniform slope terrain. Unless the walking gate is appropriately adjusted, once the ground is no longer horizontal, balance is often lost. In humans, walking makes use of our complex muscular system dynamics, such as the heel strike or the leg swing, exploiting the physics of the body, and thus reducing the energy consumption. On robots, this process is hard to mimic, although extensive work on passive walkers can be found in the literature.[1–4] Other common approaches to humanoid locomotion include robots like Honda's P series[5] and ASIMO[6] using ground reference points,[7] like the Zero Moment Point (ZMP),[8] in order to plan trajectories for the center of mass (CoM), from which gaits are adjusted using Inverse Kinematics. However, these methods require complex and precise modelling of the robot in order to be feasible in practice. The use of pre-programmed gaits for known slopes is, therefore, a simpler, alternative solution.

This paper addresses the design of an adaptive walking gait controller by combining previously developed walking gaits, according to the ground slope estimation. Stochastic filtering of inertial sensor measurements is used for this estimation, with a small computational workload. This allows robots to dedicate less computational resources on walking, enabling them to focus on other tasks.

2. Adaptive Walking

In order to create an adaptive walking gait, the controller requires an estimation of the ground slope. An Extended Kalman Filter (EKF) was implemented following the work on Refs. 9, 10 and 11, allowing the estimation of the rotation transform between the World frame and the Bioloid frame. Since this transform is equivalent to the composition of the rotation from the World frame to the ground slope, with the rotation from the robot feet to its body (Robot frame), knowing the robot joint configuration it is possible to recover the ground slope.

This method was validated using a simple balancing controller: placing the robot on an orientable platform, the controller compensates the slope of the platform by actuating on the joints in order to maintain the robot body vertical with respect to the World Frame. The robot was able to maintain its balance on slopes up to 20°, both in simulation and in the real robot.

The adaptive walking gait is based on three handcrafted template gaits:

- gait **H**, to cope with a horizontal ground;
- gait **F**, to cope with a ground plane rotated 15° around an horizontal axis perpendicular to the robot saggital plane (herein called *frontal* slope); and
- gait **L**, to cope with a ground plane rotated 10° around an horizontal axis paralel to the robot saggital plane (*lateral* slope).

Each gait contains a sequence of poses, as servo set points and associated timestamps.

The resulting gait is obtained by interpolating walking gait **H** with either **F** or **L**, depending on the orientation of the current slope. The interpolation is linear, weighted by the amount of slope detected, and is performed in a per-servo and per-timestamp fashion.[a] The position θ_i of servo i for an estimated slope of γ is given by Eq. (1), where m and b are obtained with Eqs. (2) and (3). θ_{ig} and γ_g refer to the servo position and slope of a gait where g is the gait being used.

$$\theta_i = m \cdot \gamma + b \tag{1}$$

$$m = \frac{\theta_{ig} - \theta_{i\mathbf{H}}}{\gamma_g - \gamma_{\mathbf{H}}} \tag{2}$$

$$b = \theta_{ig} - m \cdot \gamma_S \tag{3}$$

In the case of gait **L**, a non-linear correction $\boldsymbol{\rho}$, explicited in Eq. (4), is to be added to the servos of the leg that needs to be bend (either servos 11, 13 and 15 or 12, 14 and 16) to ensure the robot maintains its balance. θ_{foot} is the position of servo 17, if the right leg is bent, or 18, if the left leg is bent. The servos localization can be seen in Figure 1.

$$\boldsymbol{\rho} = -11.159 \cdot |\theta_{foot}|^2 + 3.855 \cdot |\theta_{foot}| \tag{4}$$

The walking algorithm can be summarized as:

```
estimate initial slope
generate initial walking gait files
execute motion until left single support pose
FOR step from 0 to total_steps:
  execute motion until right single support pose
  estimate slope
  IF abs(estimated_slope - old_slope) > threshold THEN
    generate walking gait file for current slope
  END IF
  execute motion until left single support pose
  estimate slope
  IF abs(estimated_slope - old_slope) > threshold THEN
    generate walking gait file for current slope
  END IF
END FOR
execute motion until double support pose
finish
```

[a]Note that the motion files have to be properly synchronized in time.

The walking gait controller estimates the slope of the ground and generates the corresponding walking gait at each step. Two methods are used to estimate the slope of the ground: (1) the *balancing method*, using the balancing controller described above, and (2) using a *learnt table*, described next. The first method runs the balancing controller until the robot body is aligned with the Frontal plane, everytime the robot is on a static single support pose. The second method uses a lookup table mapping a set of pair (*slope, gait*), to the corresponding positions of the Center of Mass (CoM), as estimated by the EKF. This table is initially learnt using a set of known slopes. Everytime the static single support pose is achieved, the robot uses this table to estimate, by interpolation, the current slope, given the current gait and the estimated CoM position. This results in a correction of the gait. Even if this estimation is not accurate once, the re-estimation of the slope at the next time the single suuport pose is achieved leads the algorithm to converge to the appropriate gait (assuming that there is enough data so that the vertical stance is a fixed point of the table).
For instance, in Table 1 is a line taken from the estimation table, corresponding to using a a gait for 10.0° frontal slope. When the gait causes the robot to fall, the corresponding cell is filled with '*'. In this case, the robot falls down when using this gait for slopes of 0.0°, 2.0°, 4.0° or above 12°. If after taking a step using this gait the estimation results in $-1.36cm < \hat{x}_{CoM} < 0.49cm$, then a linear interpolation is made to estimate the current slope. If $\hat{x}_{CoM} \approx -0.58cm$ the slope remains unchanged at 10.0°.

	0.0°	2.0°	4.0°	8.0°	10.0°	12.0°
10.0°	*	*	*	0.0049m	-0.0058m	-0.0136

The advantage of the learnt table method with respect to the balancing one is that a table lookup is faster than running the balancing controller.
The estimation of the slope is performed by:

```
read servo measurements and update EKF
IF current_gait exists in table
  find lowest recorded CoM value higher than estimated CoM
  find highest recorded CoM value lower than estimated CoM
  use the (current_gait, slope) pairs to obtain slope
    estimation by linear interpolation
ELSE
```

```
  WHILE robot not aligned with Frontal plane:
    correct servo positions
  END WHILE
  write new data in the learnt table
END IF
```

3. Results

The simulations presented here were obtained using the Webots™ software, where a realistic model of the Bioloid Humanoid robot developed by the authors was used. The similarities between the Bioloid robot and the model can be seen in Figure 1.

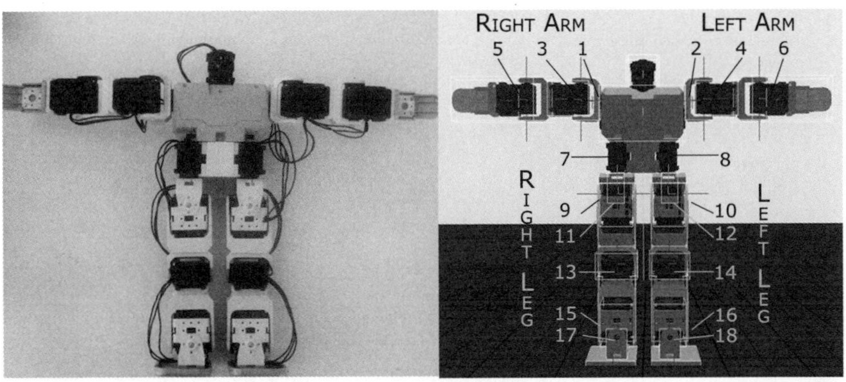

Figure 1. The real Bioloid, in the left image, and the simulated Bioloid, with servo numbering, in the right image

The method of interpolating walking gaits was tested on platforms with either a frontal or a lateral constant slope, with good results on slopes up to 20°. The gaits generated for platforms with rotations around horizontal axes different than these two are very unstable and a better mapping of the corrections that need to be done to the servos is necessary.

The plots in Figure 2 show the results for the adaptive walking gait, in terms of the CoM position (X_{CoM} and Z_{CoM})[b], as the robot travels the test environment depicted in Figure 3, where the slope of the ground increases progressively, from 0° to 20°, in discrete steps of 2°. Even when the CoM estimate is not accurate, but still close to the real value, the generated

[b]The X and Z axis correspond to the horizontal axels, perpendicular and paralel with the robot saggital plane, respectively.

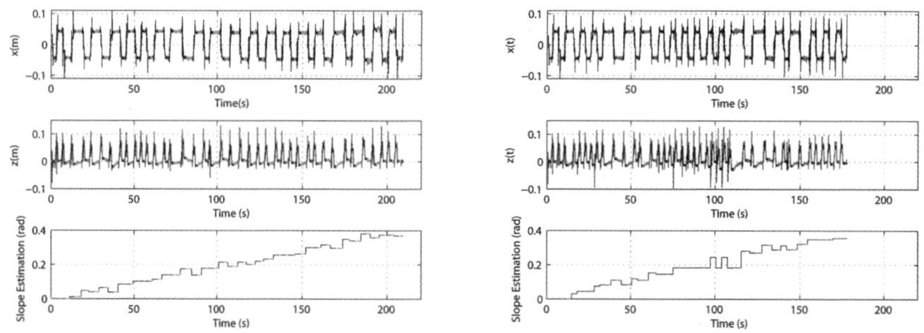

(a) balancing method (b) learnt table method

Figure 2. Simulation results using the two presented methods: (1) balancing, and (2) learnt table. Top - X_{CoM}; Middle - Z_{CoM}; Bottom - Estimated slope,

Figure 3. Snapshots taken of the simulation running.

gait allows the robot to maintain balance. The results in Figure 2(a) was obtained using the balancing method, while Figure 2(b) was obtained using the learnt table method. As expected, using the learnt table method allowed the robot to walk faster, taking 11% less time to walk the same distance than using the balancing method.

Robustness of these methods to noisy measurements was evaluated by artificially adding noise to the inertial sensor measurements. This trial is of extreme importance, since the goal is to implement this controller on a Bioloid Humanoid and there will be noise sources not accounted for in the simulations (the plastic frame is not rigid, the ground will not be completely smooth, etc.). In this case, the robot is placed on the floor, with no slope, and walks up to a platform with 3° of slope. The plots in Figure

4(a) show its progress when using normal noise values, and it is clear the transition from the plain floor to the platform at the 31st second, when it takes him longer to balance. As in the previous chapter, it is important to know how the controller handles higher noise rates. In this case, the value of $\sigma^2_{max} = 20\sigma^2_{real}$ was the highest possible to ensure the robot did not stumble or lose its walking direction. The estimation results are shown in Figure 4(b), where it can be seen that are much more unstable than on the previous plots. It is also of note that the estimations tend to stray from the real positions of $\boldsymbol{x}_{CoM}$ after the first 40 seconds of walking.

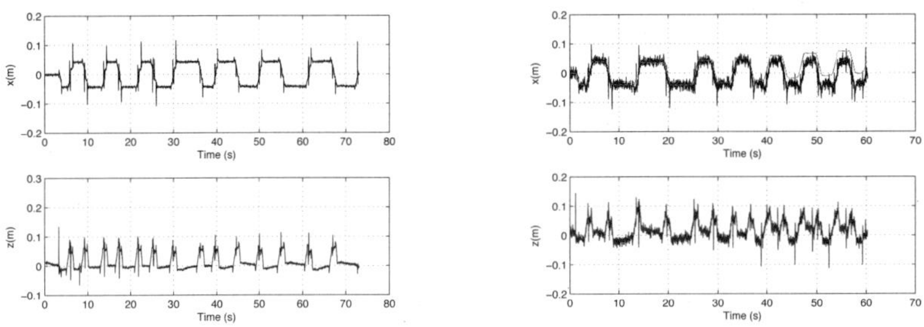

(a) normal noise variances

(b) $\sigma^2_{max} = 20\sigma^2_{real}$

Figure 4. $\boldsymbol{x}_{EKF}$(t) when the robot adjusts to a new slope with normal and increased noise variances

4. Conclusions

The walking gait controller proposed in this paper enables the Bioloid Humanoid robot to have an adaptive walking gait in the presence of non-horizontal terrains, using only inertial sensory data. The simulation results are promising, showing the effectiveness of the proposed methods. The generation of walking gaits by linearly interpolating previously designed gaits is easily applicable to other humanoid robots, since there is no need of a kinematic model of the robot. Future work includes the creation of gaits to handle ground slopes around an arbitrary axis.

Acknowledgments

This work was supported by the FCT (ISR/IST plurianual funding) through the PIDDAC Program funds. We also want to thank Ricardo Oliveira, Limor Schweitzer and the Robosavvy Forum users, who gave invaluable help with the Bioloid platform.

Bibliography

1. S. H. Collins, A bipedal walking robot with efficient and human-like gait, in *Proceedings of IEEE International Conference on Robotics and Automation*, 2005.
2. M. Wisse and A. L. Schwab, First steps in passive dynamic walking, in *Proceedings of the 7th International Conference CLAWAR 2004*, 2004.
3. M. Wisse, C. G. Atkenson and D. K. Kloimwieder, Swing leg retraction helps biped walking stability, in *Proceedings of 2005 5th IEEE-RAS International Conference on Humanoid Robots*, 2005.
4. M. Wisse, *International Journal of Humanoid Robotics* **2**, 459 (2004).
5. K. Hirai, M. Hirose, Y. Haikawa and T. Takenaka, The development of Honda humanoid robot, in *Procedings of the 1998 II-1-1 International Conference on Robotics and Automation*, 1998.
6. H. Masato and T. Tooru, *Honda R&D Tech Rev.* **13**, 1 (2001).
7. M. B. Popovic, A. Goswani and H. Herr, Ground reference points in legged locomotion: Definitions, biological trajectories and control implications, in *International Journal of Robotics Research*, (12)2005.
8. M. Vukobratocić, B. Borovac and D. Šurdilocić, *USC Neuroscience* (2004).
9. E. Lefferts, F. Markley and M. Shuster, *Journal of Guidance, Control and Dynamics* , 417 (1982).
10. N. Trawny and S. I. Roumeliotis, *Indirect Kalman Filter for 3D Attitude Estimation - A Tutorial for Quaternion Algebra*, tech. rep., Department of Computer Science&Engineering - University of Minnesota (March 2005).
11. A. I. Mourikis, N. Trawny, S. I. Roumeliotis, D. M. Helmick and L. Matthies, *The International Journal of Robotics Research* , 737 (2007).

ENERGY CONTROL FOR DESIGN OF OSCILLATIONS OF BYPED IN THE FRONTAL PLANE

A. TELESH, F.PALIS, A. RUDSKYY, A. MELNYKOV, T. DYNNIK,
University of Magdeburg, Department Electrical Energy Systems, P.B. 4120, 39106 Magdeburg, Germany

M. KONYEV, U. SCHMUCKER
Fraunhofer IFF, Department Virtual Engineering, Sandtorstr. 22, 39106 Magdeburg, Germany

In this paper we propose a method of organizing the frontal plane oscillations of a feetless biped mechanism based on the total mechanism energy management, which also allows indirect assess and control of the basic parameters of oscillations – amplitude and period. Also a mathematical model of a flat mechanism with taking into account a completely inelastic impact is considered. Proposed methods of control are based on both intuitive considerations, as well as on the classical methods of oscillation theory.

1. Introduction

Many papers discuss different ways of solving walking problem designing the control system only in the "main" - sagital-plane. This approach is rather widespread, despite the fact that the obtained motion properties could significantly restrict future possibilities of generation of 3D walking. In this paper we consider a biped mechanism with inelastaic legs without feet (see figure 1a,c). The torque is applied to the hinge between hips and trunk and realized by a power drive. [2]. The lower body ("legs") is attached to the hinge at the point *O*. It consists of three rods welded together: the left *ON*, the right *LF* legs and the hip joint *NBL*. The legs are parallel to each other and perpendicular to the hip joint. We assume that lengths, masses and moments of inertia of the legs and the hip joint are known and the center of mass of these three bodies (point *C*) can be determined. The "trunk" is pivotally attached to the lower body at the point B. *OB* and *OC* -_ distances from the origin *O* to the point *B*, where the upper body is attached, and to the center of mass C of the lower body, respectively. γ_0 – angle formed by the lines *OC* and *BC*, γ_1 – angle formed by the lines *OB* and *BC*. . φ – inclination of the legs from the vertical line, α – inclination of the trunk *BD* from the line *BC*. *D* – center of mass of the "trunk", *C* – center of mass of the lower body.

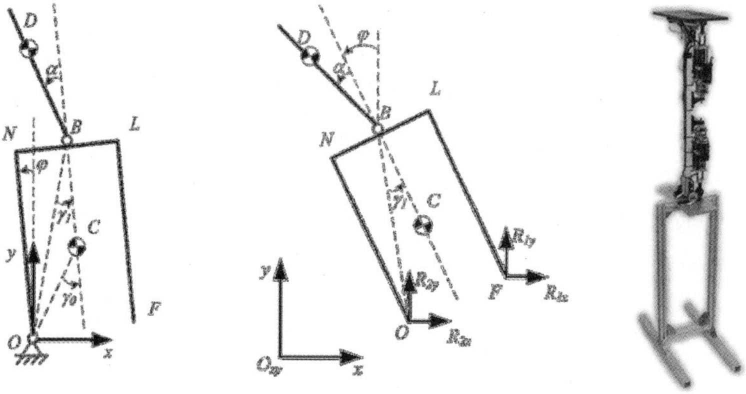

Fig. 1 –Two-link mechanism: a) b) c)

According to the classical theory of oscillations such mechanism belongs to a class of uncontrolled mechanical systems characterized by the appearance of forced oscillations. As known, these systems can be design using a Lyapunov function based on the total energy of the system.

2. Mathematical model of mechanism

The equations of motion of the two-link mechanism are solved using the Lagrange equations of second kind [3]

$$\begin{aligned} & a_{11}\ddot{\varphi}+a_{12}\left(2\ddot{\varphi}+\ddot{\alpha}\right)\cos\alpha-a_{12}\left(2\dot{\varphi}+\dot{\alpha}\right)\dot{\alpha}\sin\alpha+a_{22}\left(\ddot{\varphi}+\ddot{\alpha}\right)- \\ & \qquad -b_{11}\sin\left(\varphi-\gamma_0\right)-b_{12}\sin\left(\varphi-\gamma_1\right)-b_{13}\sin\left(\varphi+a-\gamma_1\right)=M_1 \\ & a_{12}\ddot{\varphi}\cos\alpha-a_{12}\dot{\varphi}\dot{\alpha}\sin\alpha+a_{22}\left(\ddot{\varphi}+\ddot{\alpha}\right)+a_{12}\dot{\varphi}\left(\dot{\varphi}+\dot{\alpha}\right)\sin\alpha- \\ & \qquad -b_{13}\sin\left(\varphi+\alpha-\gamma_1\right)=M_2 \end{aligned} \tag{1}$$

Here $a_{11}=(I_1+m_2OB^2)$, $a_{12}=m_2OB\cdot BD$, $a_{22}=I_B$, $b_{11}=m_1gOC$, $b_{12}=m_2gOB$, $b_{13}=m_2gBD$, M_1 and M_2 – torques applied at joints O and B.

3. Mathematical realization of an absolutely inelastic impact

The mathematical description and studying of double support phase are important for the problem of organizing walking for two-legged mechanisms. The double support occurs at the instant when setting one foot on the surface. At this moment an impact occurs and the second leg either stays on the surface or leaves it. In this paper is considered, that the second leg leaves the surface and the time of double support phase is infinitely small. At the moment of impact the velocities in the mechanism are redistributed [4].

Computation of an absolutely inelastic impact is based on the method, described in [4]. According to this the final equations yield

$$\begin{aligned}\dot{\varphi}(+0) = \dot{\varphi}(-0)\big[& I_1 I_2 + BC^2 I_2 m_1 + BD^2 I_1 m_2 + \\ & BC^2 BD^2 m_1 m_2 - 2BD^2 OB^2 m_2{}^2 cos(2\psi) + \\ & \left(I_2 OB^2 m_1 + I_2 OB^2 m_2 + BD^2 OB^2 m_1 m_2 - BD^2 OB^2 m_2{}^2\right) cos(2\gamma_1) - \\ & 4OBm_1 BC\left(I_2 + BD^2 m_2\right) cos(\gamma_1)\big] / \\ & \big[I_1 I_2 - BD^2 OB^2 m_2{}^2 + BC^2 I_2 m_1 + BD^2 I_1 m_1 + I_2 OB^2 m_1 + \\ & I_2 OB^2 m_2 + BC^2 BD^2 m_1 m_2 + BD^2 OB^2 m_1 m_2 - \\ & 2BD^2 OB^2 m_2{}^2 cos(2\gamma_1 - 2\psi) - \\ & 4BCI_2 OBm_1 cos(\gamma_1) - 4BD^2 BCm_1 OBm_2 cos(\gamma_1)\big];\end{aligned} \tag{2}$$

$$\begin{aligned}\dot{\alpha}(+0) = (BDm_2 OB\dot{\varphi}(-0)\big[& (2BC^2 m_1 + OB^2 m_1 + OB^2 m_2 + 2I_1) cos(\gamma_1 + \psi) - \\ & 2(I_1 + BC^2 m_1) cos(\gamma_1 - \psi) - (m_1 + m_2) OB^2 cos(3\gamma_1 - \psi) + \\ & 4BCm_1 OB(cos(2\gamma_1 - \psi) - cos(2\gamma_1 + \psi))\big] / \\ & \big[I_1 I_2 + BD^2 I_1 (m_1 + m_2) + I_2 OB^2 (m_1 + m_2) + \\ & (BC^2 m_1 + OB^2 m_1 - OB^2 m_2) BD^2 m_2 - \\ & 2BD^2 OB^2 m_2{}^2 cos(2\gamma_1 - 2\psi) - 4BCm_1 OB(I_2 + BD^2 m_2) cos(\gamma_1)\big] \\ & + \dot{\alpha}(-0) + \dot{\varphi}(-0) - \dot{\varphi}(+0);\end{aligned} \tag{3}$$

Here $\psi = \alpha + \varphi$ - inclination of upper body against the vertical.

4. Synthesis of control system

The model of the two-link mechanism includes the following parts – dynamic equations of motion, computation of the absolutely inelastic impact and the controller of total energy of mechanism. The mathematical model for the synthesis of the controller can be represented in the following form

$$\dot{x} = f(x) + G(x)u \tag{4}$$

Here *f(x)* – function describing the conservative part of the system, *G(x)* – continuous vector functions, u – vector of inputs, $x^T = [\varphi \quad \alpha \quad \dot{\varphi} \quad \dot{\alpha}]$ – vector of states.

For the mechanism presented in the figure 1 the total energy is calculated using the following expressions:
Potential energy

$$V = V_0 + b_{11}\cos(\varphi - \gamma_0) + b_{12}\cos(\varphi - \gamma_1) + b_{13}\cos(\varphi + \alpha - \gamma_1) \tag{5}$$

Here, V_0 – offset of the level of potential energy
Kinetic energy

$$T = \frac{1}{2}\cdot\left[a_{11}\dot{\phi}^2 + 2a_{12}\dot{\phi}\left(\dot{\phi} + \dot{\alpha}\right)\cos\alpha + a_{22}\left(\dot{\phi} + \dot{\alpha}\right)^2\right], \tag{6}$$

The control strategy is based on a control method of a one-link parametrical pendulum [5, 6, 7]. This method uses length control of the radius vector of the center of mass relative to the supporting point for the increment (decrement) of total energy per oscillation period.

In this case the inclination of the upper body in fall direction (opposite direction) is enough to make incrementation (decrementation) of energy. The fall direction of the system can be obtained using the sign of the velocity $\dot{\varphi}$. The gradient on an input is described by following equation

$$\frac{\partial H}{\partial u} \equiv -\alpha_{\max} sign(\dot{\varphi}) \tag{7}$$

The sign of φ can be used for small velocity

$$\frac{\partial H}{\partial u} \equiv -\alpha_{\max} sign(\varphi) \tag{8}$$

The inclination of the upper body is proportional to the energy error - difference between desired and actual energy level. The resulting form of the control law is equivalent to the method of the speed-gradient law.

$$\alpha_s = u = \gamma(H_S - H)\frac{\partial H}{\partial x} = -\gamma(T + V - H_S)\cdot(-\alpha_{\max} sign(\varphi)) \tag{9}$$

The control loop of the upper body inclination is constructed using the pole placement method in state space form. The integrator with the reinitialization state port is included in the model to take into consideration the impact equations.

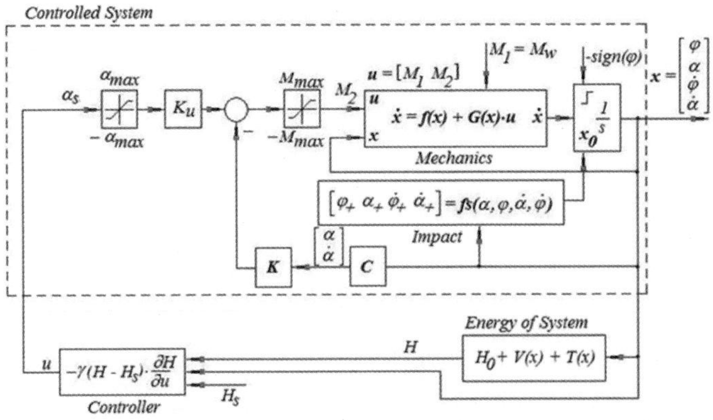

Fig. 2 – Control system of two-link mechanism

5. Simulation of Control System

The mathematical model and the experimental set-up have been considered with the following numerical parameters: Mass $m_1 = 2.2\ kg$, $m_2 = 1.8\ kg$, moment of inertia about the center of mass $I_2 = 0.0559\ kg{\bullet}m^2$; moment of inertia link *ONLP* relative to the point 0 $I_1 = 0.315\ kg{\bullet}m^2$, location of centers of mass $BD = 0.63\ m$, $OC = 0.8\ m$, $OB = 0.714\ m$; $\gamma 0 = 8.3\ grad$, $\gamma 1 = 9.3\ grad.$

In Figure 3, the graphs of actual energy H, desire energy H_S, angles φ, α and angular velocity $\dot{\varphi}$ as function of time are shown.

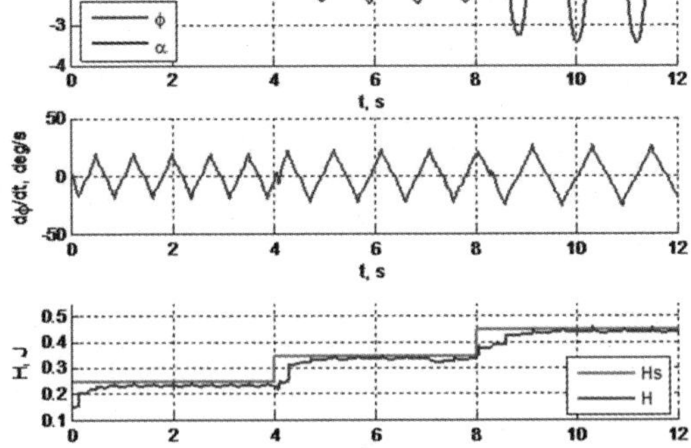

Fig 3 - The transitional processes of energy control with absolutely inelastic impact.

The figure shows that the system loses energy at each impact and the various levels of desired energy correspond to steady-state oscillations with different frequency and amplitude. The system has a static error because the control algorithm is based on static control over the energy.

Figure 4 shows phase portraits of the two-link mechanism, from which the orbital stability of the system can be seen.

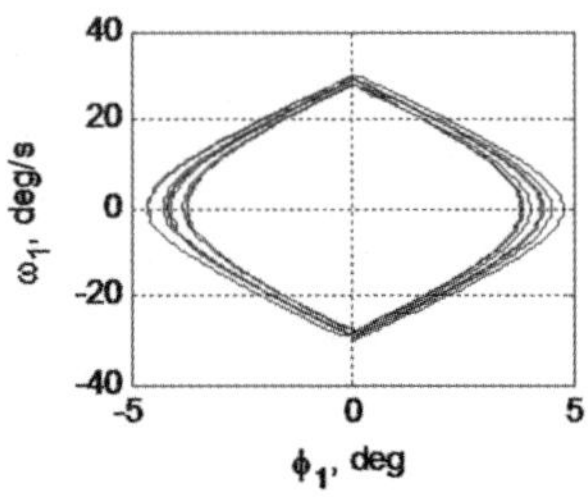

Fig 4 – Phase portraits of the oscilation of two-link mechanism.

6. Experimental Results

The experimental set-up have been investigated for various surfaces: wood, carpet, rubber and rubber foam. All these surfaces have different properties of elasticity and damping.

The figure 5 shows, that the energy of the experimental set-up has more significant fluctuations in comparison to the ideal one.

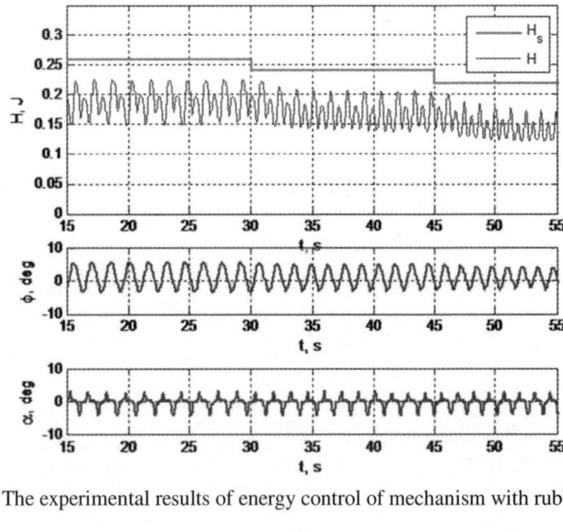

Fig 5 – The experimental results of energy control of mechanism with rubber foam

These differences are caused by large losses of energy at impact, the presence of oscillations in the perpendicular plane (not the working plane), the complexity of measuring of the angle and angular velocity at the support point.

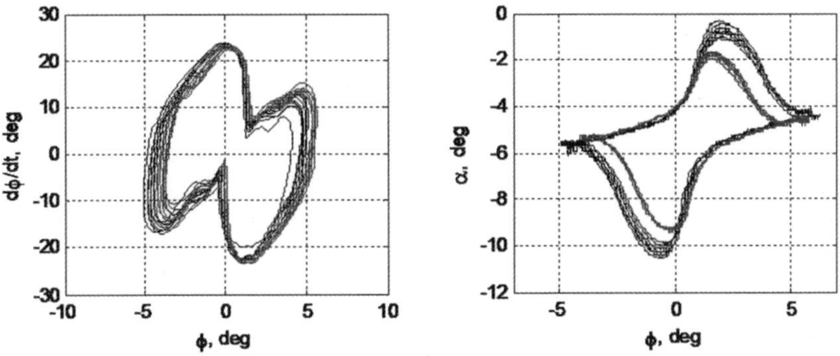

Fig 6 – Experimental phase portraits of the oscilation of two-link mechanism..

Figure 6 shows the steady state oscillations of the mechanism on wood (black) and woven (blue) surfaces. The periodic motion of the upper body executes the incrementation of energy. Stochastic processes of impact cause small differences between the trajectories. The phase portraits of two types of surfaces have different amplitudes and periods of oscillations.

7. Conclusion

On the basis of total energy control of an underactuated mechanical system good dynamic qualities of study stable oscillation are obtained. Experiments have shown significantly bigger losses of energy in comparison to theoretical calculations for absolutely inelastic impact. Different properties of contact surface may change the character of oscillation but they remain stable. The obtained results can be used to control biped walking mechanisms in the sagital plane.

References

1. http://www.uni-magdeburg.de/ieat/robotslab
2. Linear elastic actuator of a biped robot "ROTTO"/ Melnykov, Andriy; Konyev, Mikhailo; Palis, Frank; Schmucker, Ulrich/ In: Emerging trends in mobile robotics . - Singapore [u.a.] : World Scientific, ISBN 981-432797-2, S. 588-595, 2010 Kongress: CLAWAR; 13 (Nagoya) : 2010.08.31-09.03
3. Landau, L. D. and Lifschitz, E. M. Mechanics, 3rd ed. Oxford, England: Pergamon Press, 1976.

4. Peremeshhenie antropomorfnykh mekhanizmov. Formal'skij A.M. - M.: Nauka. Glavnaja redakcija fiziko-matematicheskojj literatury, 1982. - 368 s. (in Russian).
5. Kurt Magnus .Schwingungen. Eine Einfuhrung in die theoretische Behandlung von Schwingungsproblemen. Tubner, 1986, s. 251 . (in German)
6. Global stabilization of a double inverted pendulum with control at the hinge between the links /A. M. Formal'skii //Mechanics of Solids, Volume 43, Number 5, 687-697.
7. On the Design of Optimal Feedback Control for Systems of Second Order / Alexander M. Formalskii // Applied Mathematics Vol.1 No.4, October 2010 ISSN Print: 2152-7385

LIMIT CYCLES WALKING IN SAGITAL PLANE BASED ON THE TOTAL ENERGY CONTROL AND VIRTUAL CONSTRAINTS

A. TELESH, F.PALIS, A. RUDSKYY, A. MELNYKOV, T. DYNNIK,

University of Magdeburg, Department Electrical Energy Systems, P.B. 4120, 39106 Magdeburg, Germany

M. KONYEV, U. SCHMUCKER

Fraunhofer IFF, Department Virtual Engineering, Sandtorstr. 22, 39106 Magdeburg, Germany

The article describes research works for dynamic walking of a biped robot without feet. The control system is synthesized using oscillation theory and virtual constrains to generate the feet movement. The proposed control method is based on the following decomposition. First: the servo-drive realizes a special equation $\alpha=f(\varphi)$ of angle between feet and the angle of the support leg against the surface. This equation was obtained to provide properties of free, passive, ballistic motion (step). Second: the increment (decrement) of energy which is important to fullfill a step is compensated using the torso.

1. Introduction

Recent research works [1] are related to biped walking, allowing to analyze and synthesize dynamic and stable walking pattern. In the presented paper a dynamic walking control system of the biped social robot Rotto, which was designed by the research group RobotsLab [2], is investigated. Torso and legs of the robot are described in earlier papers [3]. Now arms, shoulders and head are added as shown in fig. 1.

The research tasks is divided into four levels, which can be linked together. The first level includes robot feet, equipped with a 6-component force sensor. Its control tasks are: contact with the surface and the distribution of support of the masses (forces) [4]. Tasks of static and dynamic walking are solved at the second level. Solutions of the first level can be used for static walking. Several solutions are offered for the task of dynamic walking, such as control in accordance with the theory of virtual forces [5], ballistic control [6] and others. This article offers energy-control for a mechanical system. The third level includes manipulation of objects (arms) [7], the increase of dynamics using additional actuators[8], and social problems. The fourth level includes Social-emotion robot-human Interaction.

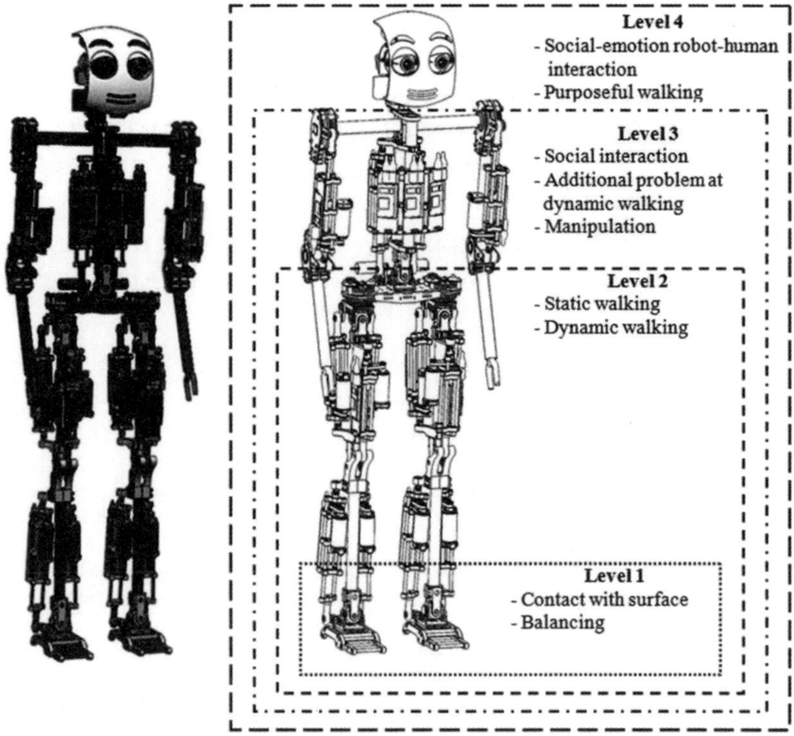

Fig. 1 – Four levels of problems for anthropomorphous robot ROTTO

2. Virtual constraint for step control

Obtaining of virtual constraints [9,10] for linked leg movements (interlink angle and the angle of the support leg against surface) is based on the ballistic method to construct free motion of a mechanism without friction in joints. The mathematical model of a 2D two-mass mechanism (Fig. 2a) is considered. It can be described using the Lagrange equations of 2nd order, its linearized model is:

$$\begin{gathered}[b_1+2b_2(a_2+cg_{2y})]\ddot{\varphi}+[b_3+b_2(a_2+cg_{2y})]\ddot{\alpha}- \\ -m_2g((a_2+cg_{2y})(\varphi+\alpha)+a_1\varphi)-m_1g(a_1+cg_{1y})\varphi=0 \\ [b_3+b_2(a_2+cg_{2y})]\ddot{\varphi}+b_3\ddot{\alpha}-m_2g(a_2+cg_{2y})(\varphi+\alpha)=0\end{gathered} \tag{1}$$

where a_1 and a_2 - lengths of links, m_1 and m_2 - masses of links, I_1 and I_2 - moments of inertia of links with respect to the centers of mass, centers of mass of links OB and BD are located at a distance cg_{1y}, cg_{2y} from the points B and D in the links coordinate systems, $b_2 = m_2 a_1$, $b_3 = I_2 + m_2(a_2 + cg_{2y})^2$, $b_1 = I_1 + I_2 + m_1(a_1 + cg_{1y})^2 + m_2((a_2 + cg_{2y})^2 + a_1^{\ 2})$.

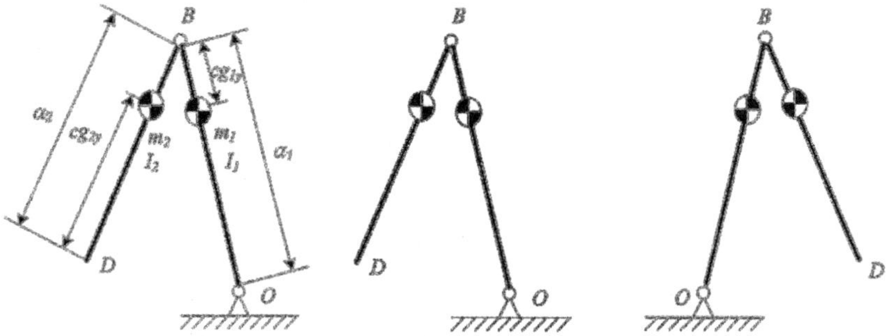

a) b) c)

Fig. 2 – Two-mass mechanism, its initial and final position

Equation (1) can be transformed into matrix form:

$$B\ddot{z} + Az = 0. \tag{2}$$

An analytical solution of the system of differential equations can be obtained using the transition matrix R to convert equations to a normal form, by the method proposed in [11]

$$\ddot{x}_1 + w_1^2 x_1 = 0,\ \ddot{x}_2 - w_2^2 x_2 = 0. \tag{3}$$

where w_i^2 - the roots of (2). Analytical solution of the equations (3) is:

$$\begin{aligned} x_1 &= [x_1(T)\sin w_1 t + x_1(0)\sin w_1(T-t)]/\sin w_1 T \\ x_2 &= [x_2(T)\,\mathrm{sh}\, w_2 t + x_2(0)\,\mathrm{sh}\, w_2(T-t)]/\,\mathrm{sh}\, w_2 T \end{aligned} \tag{4}$$

Here $x(0) = R^{-1}\left[\varphi(0) \quad a(0)\right]^T$ and $x(T) = R^{-1}\left[\varphi(T) \quad a(T)\right]^T$ - vectors of initial and final conditions for system (3). The boundary problem for the unknown velocities $\dot{\varphi}(0)$, $\dot{\varphi}(T)$ and step time T is to solve using equations (4). The boundary conditions, which are shown on Fig. 2b (initial position) and Fig. 2c (final one), can be presented as:

$$\begin{aligned} x(0) &= R^{-1}[\varphi(0) \quad a(0)]^T, \quad \dot{x}(0) = R^{-1}[\dot{\varphi}(0) \quad \dot{a}(0)]^T, \\ x(T) &= R^{-1}[\varphi(T) \quad a(T)]^T, \quad \dot{x}(T) = R^{-1}[\dot{\varphi}(T) \quad \dot{a}(T)]^T. \end{aligned} \tag{5}$$

The obtained trajectories should be realizable. The velocity in the interlink joint must be equal to zero up to the impact moment in order to "freeze" the mechanism. Blocking of joint B (Fig. 2) during the impact is important for entering and leaving the contact at a given configuration. According to [11] for the symmetric boundary value problem boundary conditions are:

$$x(0) = R^{-1}[\varphi(0) \quad a(0)]^T, \qquad \dot{x}(0) = R^{-1}[\dot{\varphi}(0) \quad 0]^T,$$
$$x(T) = R^{-1}[-\varphi(0) \quad -a(0)]^T, \quad \dot{x}(T) = R^{-1}[\dot{\varphi}(0) \quad 0]^T. \tag{6}$$

The resulting equations α=f(φ), and the total energy as function of the step length are shown in Fig. 3.

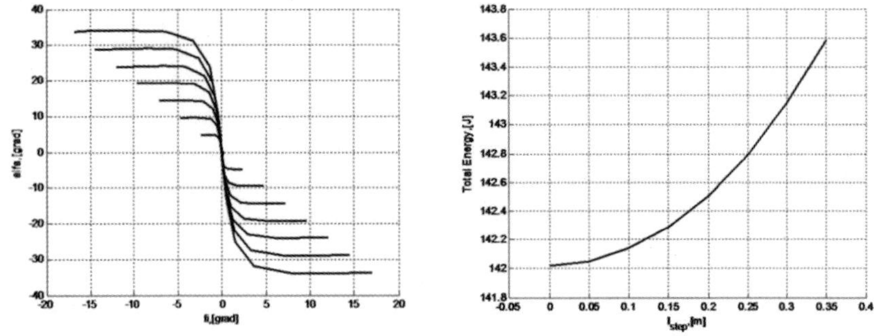

Fig. 3 Resulting equations

It should be noted that the obtained solutions for this task have the following properties: for every step length, the time of the step is constant, the dependence of $\alpha=f(\varphi)$ is scalable (Fig. 3a), depending on step length. At the same time to maintain a constant step time, system must have more initial energy for longer steps, as it can be seen in Fig. 3b.

3. Walking control

To solve the walking control problem a 5-mass mechanism with attached ground reaction is considered (Fig. 4a).

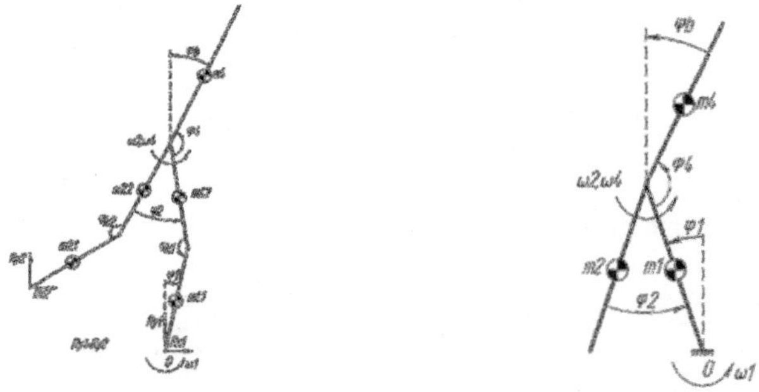

a) b)

Fig. 4 - Anthropomorphic mechanism, the sagittal plane.

Angles are denoted as φ_1 – angle of support leg respect to the vertical, φ_2 – interlink angle, φ_4 – body position respect to support leg, φ_b – angle of the body respect to vertical, φ_{k1} – angle of the knee joint of support leg, φ_{k2} – angle of the second knee joint.

The original mechanical system, which is described as a system of first order non-linear differential equations in matrix form, has the state vector $\boldsymbol{x}$ and input vector $\boldsymbol{u}$:

$$\begin{aligned}\mathbf{x}^T &= [\varphi_1\ \varphi_2\ \varphi_4\ \varphi_{k1}\ \varphi_{k2}\ \omega_1\ \omega_2\ \omega_4\ \omega_{k1}\ \omega_{k2}\ x_O\ y_O\ \dot{x}_O\ \dot{y}_O],\\ \mathbf{u} &= [M_1\ M_2\ M_4\ M_{k1}\ M_{k2}\ R_{x1}\ R_{x2}\ R_{y1}\ R_{y2}]\end{aligned} \tag{7}$$

where M_i – torque at the i-th joint; R_{x1}, R_{y1} – forces acting on a body in the fulcrum; R_{x2}, R_{y2} – forces acting on a body at the end of the second leg; x_O, y_O – global coordinates of the fulcrum O.

Contact processing is implemented according to [12]. Due to the position control loops, realized in all joints except the supporting one, as well as the presence of internal links in the system, the vectors of inputs and considered states can be rewritten as: $\mathbf{x}^T = [\varphi_1\ \varphi_2\ \varphi_4\ \varphi_{k1}\ \varphi_{k2}\ \omega_1\ \omega_2\ \omega_4\ \omega_{k1}\ \omega_{k2}]$ $\mathbf{u} = [\varphi_{2s}\ \varphi_{4s}\ \varphi_{k1s}\ \varphi_{k2s}]$.

The system under control is artificially introduced as feedback system (expression for interlink angle as a function of the support leg supplemented with adjustments related to the function of the knee). Final expressions are:

$$\begin{bmatrix}\varphi_{k1s}\\ \varphi_{k2s}\end{bmatrix} = \mathbf{fk}(\phi_2, \omega_1) = \begin{bmatrix}\pi \\ \begin{cases}\pi - \varphi_{k\max}, & \omega_1 \cdot \varphi_2 \le 0\\ \pi - \varphi_{k\max}\dfrac{\varphi_{2\max} - |\varphi_2|}{\varphi_{2\max}}, & \omega_1 \cdot \varphi_2 > 0\end{cases}\end{bmatrix}. \tag{8}$$

$$\varphi_{2s} = \begin{cases} fsri(\varphi_1) + \dfrac{1.1}{2}(\pi - \varphi_{k2s}),\ forward \\ fsri(\varphi_1) + \dfrac{0.9}{2}(\pi - \varphi_{k2s}),\ backward\end{cases} = fsr(\varphi_1, Direction, \varphi_{k2s}) \tag{9}$$

Where *frsi* – virtual constraint for step control. Introducing such feedback, the system is obtained with the following state and input vectors: $\mathbf{x}^T = [\varphi_1\ \varphi_2\ \varphi_4\ \omega_1\ \omega_2\ \omega_4]$ $\mathbf{u} = [\varphi_{4s}]$.

The system can be reduced to the mechanism presented in Fig. 4b. For this mechanism the total energy is calculated as H = V + T and intuitive control is built, that implements the increment (falling) of energy of the system over the period of oscillations (a step) through inclination of the body with respect to the vertical. The setpoint of the angle of the body with respect to the support leg is:

$$\varphi_{4s} = \pi + \frac{\varphi_2}{2} - \Delta\varphi_{4s} \tag{10}$$

Such a setpoint means, that for $\Delta\varphi_{4s} = 0$, both legs deviate from the body at the same angle. Symmetric lifting, moving of leg and placing of leg on a surface will cause a symmetric oscillation of the body with respect to the vertical. Setting any other $\Delta\varphi_{4s}$ will lead to asymmetric oscillations of the body, with a forward or backward inclination of the body. Analogously to the method of the energy increment for an one-mass parametric pendulum (the change of the radius vector of the center of mass [13]), the influence of an increment of the total energy of the oscillation per period on the input Δφ4s is approximately obtained as:

$$\frac{\partial H}{\partial u} \equiv \Delta\varphi_{4\max}\, sign(\omega_1) \tag{10}$$

The angle $\Delta\varphi_{4s}$ is calculated proportionally to the error of the energy, thus avoiding switching mode. This leads to a controller form [15, 16], similar to the method of speed-gradient:

$$\Delta\varphi_{4s} = u = -\gamma\left(H - H_S\right)\frac{\partial H}{\partial u} = -\gamma\left(T + V - H_S\right)\cdot\Delta\varphi_{4\max}\, sign(\omega_1) \tag{10}$$

The resulting block diagram of control system is shown in Fig. 5.

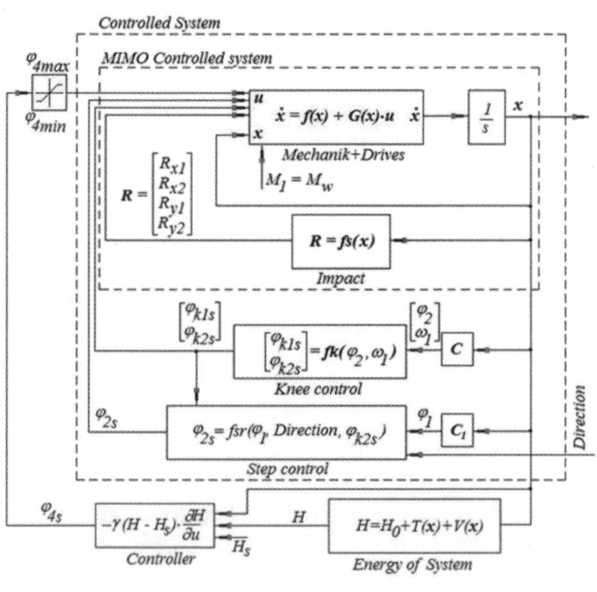

Fig.5 - Resulting block diagram of control system

4. Simulation Results

The parameters of mechanism (Fig. 4a) are: masses of links $m_{1.1} = m_{1.2} = m_{2.1} = m_{2.2} = 2.2$ kg, $m_4 = 12$ kg; moments of inertia of links respect to centers of mass $J_{1.1} = J_{1.2} = J_{2.1} = J_{2.2} = 0.02$ kg·m^2, $J_4 = 0.001$ kg·m^2; lengths of links $a_{1.1} = a_{1.2} = a_{2.1} = a_{2.2} = 0.35$ m, $a_4 = 0.5$ m; centers of mass of links CG1, CG2 and CG4 are located at a center of link. Simulation results are shown in Fig.6, phase diagrams in the steady-state behavior are shown in Fig. 7.

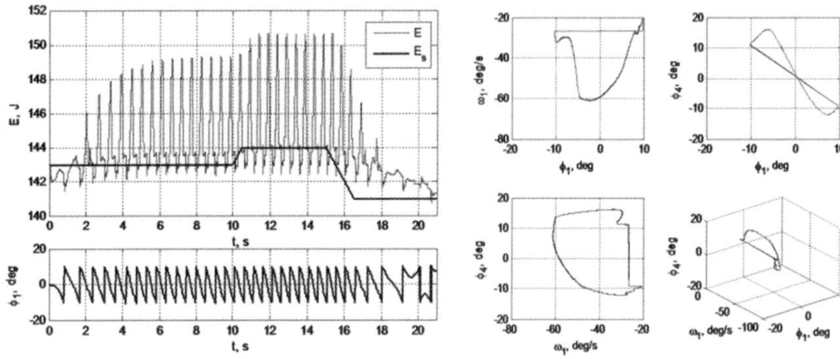

a) b)

Fig. 6 – a) Transients while walking forward, b) phase diagrams in the steady-state behavior 12-15s

Plots were obtained by walking forward with constant step length of φ2max =20°. Mechanism starts to move from a vertical position slowly gaining energy, coming up to the periodic mode. Step time is 0.54 seconds (1.85 Hz step frequency, by step length of 0.24m velocity is 0.45 m / s = 1.6km / h). System is noticeably faster gaining energy after increase of set point. This can be explained by increase of energy during the period - the mechanism makes more steps over the same time with increasing frequency. After increase of the energy level step time becomes 0.49s (frequency of step is 2.04 Hz, for step length 0.24m velocity is 0.49 m / s = 1.75km / h). When the energy level set point is less than required to perform step, mechanism reduces the speed and at the 19th second starts to make steps on the spot. From the phase trajectory (Fig 6a) can be concluded that the system works in a stable oscillation regime in the period between 12 and 15 sec.

5. Conclusion

The proposed method of introducing a feedback in the system implementing virtual constraints, can simplify the system to control the total energy of the system. The change of the energy set point causes a change of the attractor set in the systems state space and leads it to a new level of energy. Considered system is a system with deficit control inputs, and its modelling as an oscillating system allows to implement periodic walking, based on free motion of the system and using the features of the system.

References

1. Matthias Hackel Humanoid Robots, Human-like Machines/Daan G.E. Hobbelen and Martijn Wisse, Limit Cycle Walking. 2007 I-Tech Education and Publishing, 642 pages.
2. www.uni-magdeburg.de/ieat/robotslab/
3. Presentation of a new biped robot "ROTTO". Konyev, F. Palis, Y. Zavgorodniy, A. Melnykov, A. Rudskyy, A. Telesh, U. Schmucker// MOBILE ROBOTICS Solutions and Challenges (pp 551-558).
4. Shneider A.Yu.: Force control in locomotion of legged vehicle over rigid and soft surfaces – Gorinevsky - 1990
5. 5Jerry Pratt, Peter Dilworth, Gill Pratt, "Virtual Model Control of a Bipedal Walking Robot", In Proceedings of the 1997 International Conference on Robotics and Automation, ICRA '97
6. Y. Zavgorodniy, „Konstruktion und Steuerung von Schreitrobotern mit ballistischem Laufverhalten", Ph.D. Thesis. Magdeburg, 2009
7. 7.Aaron Ladd Edsinger, „Robot Manipulation in Human Environments",Ph.D. Thesis, Massachusetts Institute of Technology, January 2007
8. Masahiro Mori. Bukimi No Tani (The Uncanny Valley). Energy 7(4), 1970, S.33–35
9. Virtual-Holonomic-Constraints-Based Design of Stable Oscillations of Furuta Pendulum: Theory and Experiments, Anton S. Shiriaev, Leonid B. Freidovich, Anders Robertsson, Rolf Johansson, and Anders Sandberg IEEE TRANSACTIONS ON ROBOTICS, VOL. 23, NO. 4, AUGUST 2007
10. A Passive 2DOF Walker: Finding Gait Cycles Using Virtual Holonomic Constraints Leonid B. Freidovich, Uwe Mettin, Anton S. Shiriaev, Mark W. Spong. Proceedings of the 47th IEEE Conference on Decision and Control Cancun, Mexico, Dec. 9-11, 2008
11. Peremeshhenie antropomorfnykh mekhanizmov. Formal'skij A.M. - M.: Nauka. Glavnaja redakcija fiziko-matematicheskojj literatury, 1982. - 368 s. (in Russian).

PROTOTYPE PASSIVE WALKER DEVELOPMENT WHICH HAS CAPABILITIES TO MEASURE THE STATE PARAMETERS OF ITS OWN

MASATSUGU IRIBE, HIROKI FURUKAWA†

Div. of Mechanical and Control Engineering, Osaka Electro-Communication University, 18-8 Hatsu-cho, Neyagawa, Osaka, 572-8530, Japan

YASUHIRO SUGIMOTO, KOICHI OSUKA

Dept. of Mechanical Engineering, Osaka University, 2-1 Yamadaoka, Suita, Osaka, 565-0871, Japan

TETSUYA KINUGASA

Dept. of Mechanical Engineering, Okayama University of Science, 1-1 Ridai-cho, Kitaku, Okayama, 700-0005, Japan

In the previous researches on Passive Dynamic Walking, there is an obvious remaining problem, whether the adopted conservative system model really shows the behaviors of Passive Dynamic Walking in the real world or not. We therefore try to verify the adopted conservative system model by using our developed new compass-type prototype robot which has capabilities to measure the state parameters of its own.

1. Introduction

It is well-known that walking behaviors of Passive Dynamic Walking is natural, efficient and very attractive as a new method of legged walking. Therefore, there have been several researches on Passive Dynamic Walking. And those researches seem to be classified by two types after surveying.

One is the experimental research. The researchers who make a choice of this method develop the prototype robot actually, and do the walking experiment by using their prototypes. They make an observation about the prototype robot's behaviors carefully and try to develop the legged walking technologies by appropriating the essence of behaviors [1][2][3][4].

The other is the theoretical and numerical simulative research. The researchers who make choice of this method try to clarify the principle and

† Moved to NIPPON MECHATECHNO CO, LTD on Apr. 2011.

behaviors of Passive Dynamic Walking by computer simulation [5][6][7][8]. On these researches, the researchers adopt the conventional dynamical system model which consists of several mechanical links and the leg-transition law which shows the swing leg and support leg's transition based on the theorem of angular momentum conservation.

However, it has not verified yet that the adopted dynamical system model really shows the essential behaviors of Passive Dynamic Walking in the real world. We therefore cannot help having the doubt that the simulative results really agree in experimental result. For example, the simple vibration system shown in upper side of Figure 1 is well analyzed by two ways, theoretical way and experimental way, and the results from these ways agree with each other. Therefore the theoretical model is reliable enough.

On Passive Dynamic Walking research, it is also advisable that the theoretical research results agree with the experimental research results. On the other hand, there is no research to verify the adequacy between theoretical results and experimental results. We therefore developed the compass type passive dynamic walking prototype robot with the expansion-contraction mechanism and the posture sensing devices to enable to walk stably and reliably, and to compare the theoretical and experimental results. This prototype robot has capabilities to measure the state parameters of the dynamical system model and to keep walking stably and passively.

In this paper we report the comparison results between the walking experiment and theoretical examination by using the prototype robot, and then try to show the adequacy of the theoretical model.

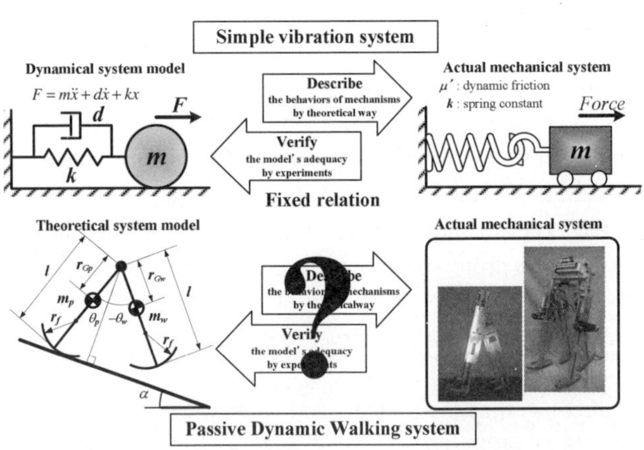

Figure 1. Basic concept of our research, "Does PDW's theoretical model really show actual behaviours ?"

2. Development of the compass-type prototype walking robot

2.1. *Overview*

Figure 2 shows a conservative dynamical model which is generally adopted by past researches [7]. As shown in the figure, we consider the support leg angle is θ_p and the swing (non-supported) leg angle is θ_w, and the slope angle parameter α. The dynamic equation derived using the well-known Euler-Lagrange approach is

$$\mathbf{M}(\boldsymbol{\theta})\ddot{\boldsymbol{\theta}}+\mathbf{N}(\boldsymbol{\theta},\dot{\boldsymbol{\theta}})\dot{\boldsymbol{\theta}}+\mathbf{G}(\boldsymbol{\theta},\alpha)=\mathbf{0} \tag{1}$$

where $\mathbf{M}(\boldsymbol{\theta})$ is the inertia matrix, $\mathbf{N}(\boldsymbol{\theta},\dot{\boldsymbol{\theta}})$ is the centrifugal and Coriolis term, and $\mathbf{G}(\boldsymbol{\theta},\alpha)$ is the gravity term. Because torque input is ignored, the right side of (1) becomes zero.

Assuming that the support and swing legs change instantly and the swing leg impacts with the ground inelastically and without sliding, the equation of the transition at impact becomes

$$\mathbf{P}_b(\beta)\dot{\boldsymbol{\theta}}^-(t)=\mathbf{P}_a(\beta)\dot{\boldsymbol{\theta}}^+(t) \tag{2}$$

by using the conservation of angular momentum, where $\dot{\boldsymbol{\theta}}^-$, $\dot{\boldsymbol{\theta}}^+$ are pre- and post-impact angular velocities, respectively.

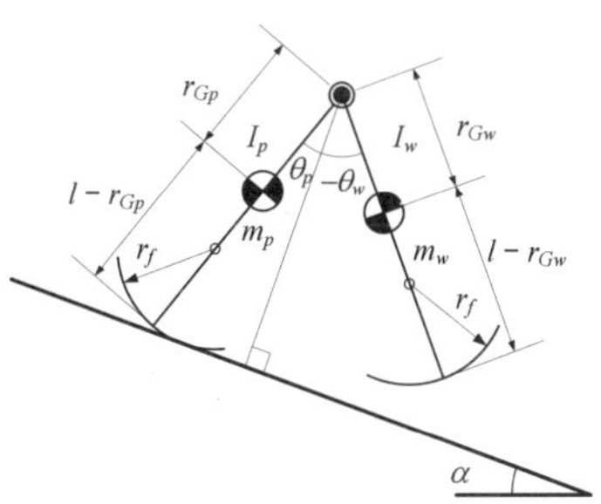

Figure 2. Theoretical model of the conservative Passive Dynamic Walk; support leg's mass: m_p [kg], swing leg's mass: m_w [kg], support leg's moment of inertia: Ip [kg-m^2], swing leg's moment of inertia: Iw [kg-m^2], leg length: l [m], radius of the circular arc foot: r_f [m], distance between hip center and support leg's centroid: r_{Gp} [m], distance between hip center and swing leg's centroid: r_{Gw} [m], G-forces: g = 9.8[m/s^2], slope angle: α [rad], support leg angle: θ_p [rad], swing leg angle: θ_w [rad].

Figure 3 shows overview of the prototype robot which realizes the conservative dynamical model described the above, and the specifications are described in Table 1. The developed prototype robot is two legged compass type robot to reduce the degree of freedom to make analysis easy. The robot consists of two legs; the outer leg and inner leg. Foot part of the robot has arch shape whose curvature radius is same length between the center of rotation and end of the leg. And there are elastic sponges attached to the sole part of arch foot to absorb the heel strike shock.

In the case of the compass type walking robot development, the foot part strikes to the slope by each step because the outer and inner leg's lengths are equal. We therefore developed the 5 link mechanism attached to each leg to expand and contract to prevent the stumbling to the slope.

The leg expansion and contraction is controlled by the 16bit microcomputer and its peripheral circuits, power supply, motor driving, signal processing of the sensors, are contained in Control box which is attached to the upper side of outer leg.

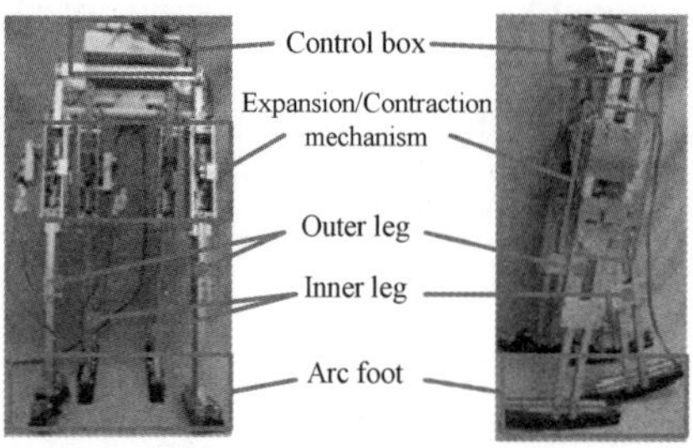

Figure 3 Developed Compass type 2-legged Passive Dynamic Walking robot with new expansion and contraction mechanism.

Table 1 Specifications of the robot

Height	780 mm
Width	330 mm
Leg length	600 mm
Weight	9.5 kg
Curvature radius of foot	600 mm
Expansion stroke	13 mm
Thickness of sole rubber	5 mm

2.2. *Leg Expansion-Contraction mechanism*

Figure 4 shows the expansion and contraction mechanism. The actuation unit consists of a DC-motor and reduction gears, and the reduction gears also consist of spur gears and rack and pinion mechanism. There is 5-link mechanism which moves the lower leg part to extract and contract, and mechanism Slider G is driven by rack and pinion mechanism as shown in Figure 4and 5. And shaft parts E and E' shown in figure 5 intercept the impact shocks from heel strike.

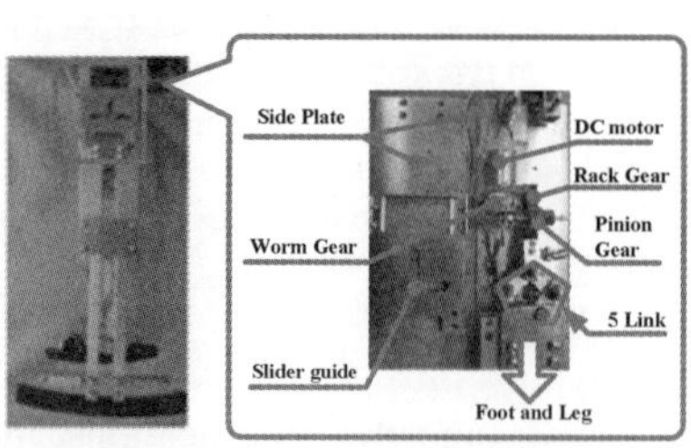

Figure 4. Developed Expansion-Contraction mechanism, the stroke is 10mm.

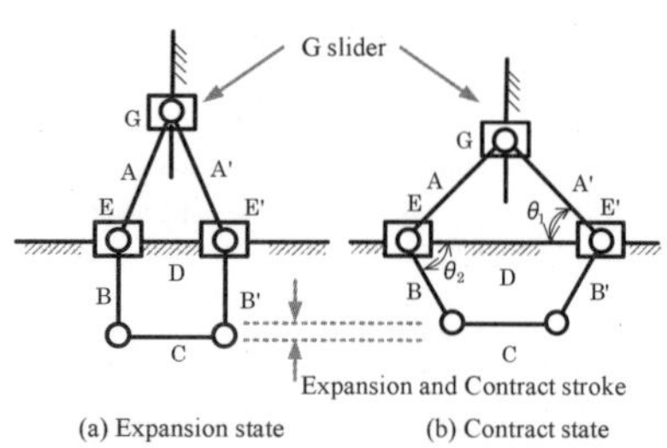

Figure 5. 5-link mechanism operation ; the impact shock from heel strike is intercepted by the shaft parts E and E'.

Several past researches adopt ball-screw devices to realize the extract and contract mechanism. However the ball-screw device tends to break down easily because the device is shocked by the heel strike [9][10]. On the other hand, our developed 5-link mechanism is able to protect the actuation mechanism such as reduction gears form impact shocks of heel strikes. This is a feature point of this mechanism

2.3. *State parameters measurement system*

The acceleration sensors (MMA7361, Freescale Co. LTD.) which measure the acceleration including the gravity- force are attached to the outer and inner leg. The acceleration signals from each leg are imported to the Micro-SD memory by 10-bit resolution and the sampling time is 10ms. The block diagram of the signal processing is shown in Figure 6. And as the walking behaviors of Passive Dynamic Walking is basically generated by the gravity force and only the gravitational acceleration elements works when the prototype robot walks passively, so the leg angles are measured certainly.

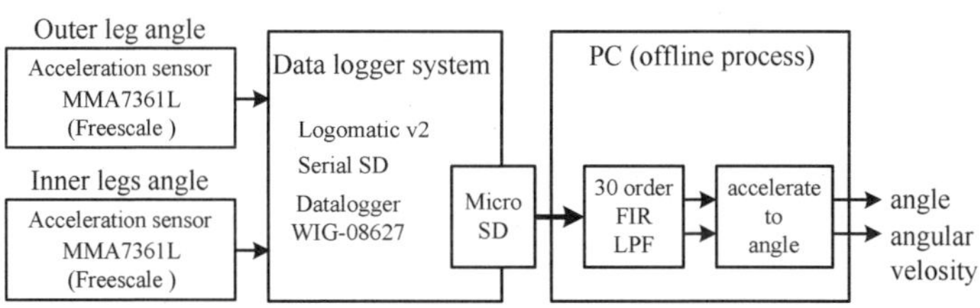

Figure 6. Block diagram of the measurement system. Measured signals are converted to the leg angles θ_p , $\dot{\theta}_p$ and θ_w, $\dot{\theta}_w$ via 30 order FIR type Low Pass Filters and some transformation equations.

However, the high-frequency noise superimposed to the circuit and the impact shock element caused by heel strike are detected inevitably. We therefore adopt the elastic sponges attached to the sole part of arch foot to absorb the heel strike shock, and the 30-order FIR digital filter whose cut-off frequency is 1.5Hz. As the countermeasures works well, the leg angles become to be measured.

2.4. *Leg Expansion-Contraction control*

To control the leg expansion and contraction motion, the prototype robot applies the outer and inner leg's angle and angular velocity. Figure 6 shows the hardware block diagram, and figure 7 shows the state transition diagram.

This simple control, which only iterates the leg's extract and contract according to the sign of leg's angular velocity and state switch's signal of extract or contract, is very robust and reliable.

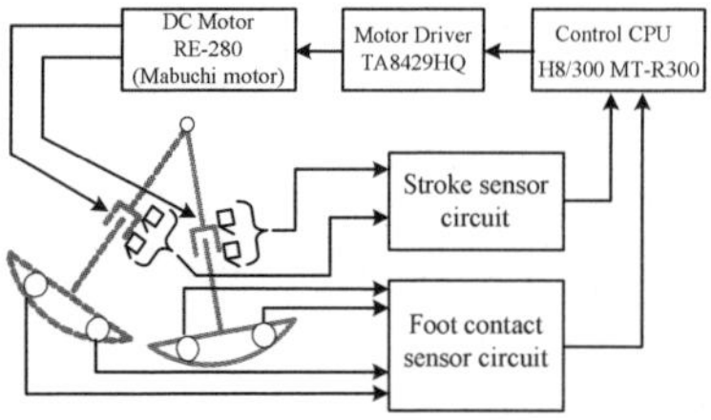

Figure 7. Block diagram of the mechanism control

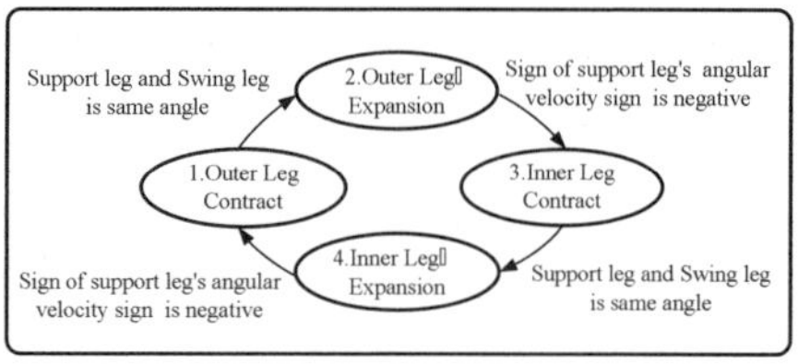

Figure 8. State transition diagram of the mechanism controller

3. Walking experiments and analysis

3.1. *Experiments via the prototype robot*

We conducted the walking experiment by applying the above-mentioned prototype robot. On the 1.1 deg slope angle, we achieved 11steps walk passively as shown in Figure 9. Figure 10 shows the leg angles of impact point of leg-transition and Figure 11 shows the configuration space trajectories for analyzing the time-series trajectories in detail.

As shown in Figure 9 and 10, the prototype robot walks by 2-periodic gait. It seems that there are differences between the outer and inner leg, such as mass, moment of inertia, gravity point, so the 2-periodic gait appears. Although the prototype robot walks by 2-periodic gait, as shown in Figure 11, the trajectories in configuration space settle to a constant orbit.

3.2. *Stability analysis via Linearized Poincare map*

Then we analyze the experimental result via the linearized Poincare map described in [8], and Figure 12 shows the analytical results.

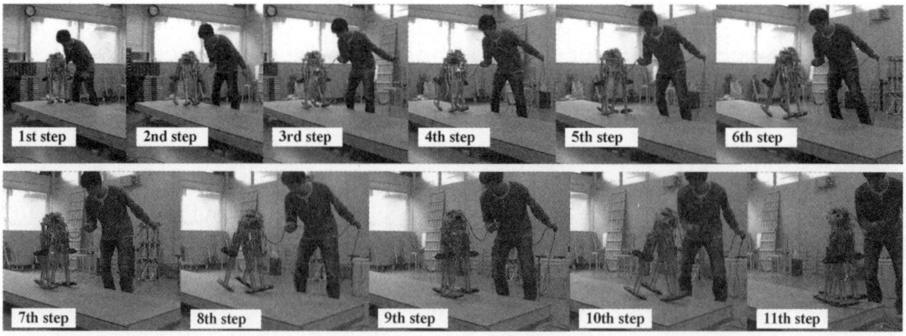

Figure 9. Walking Experiment; the developed compass type prototype robot walks 11 steps stably.

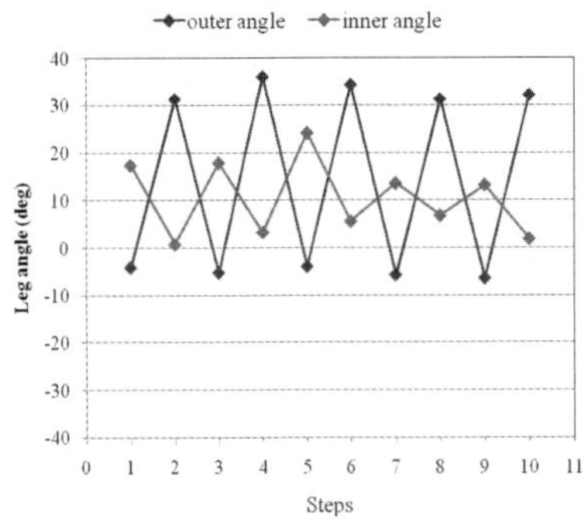

Figure 10. Leg angles of impact points of walking experiment. Walking gaits show 2-periodic gait

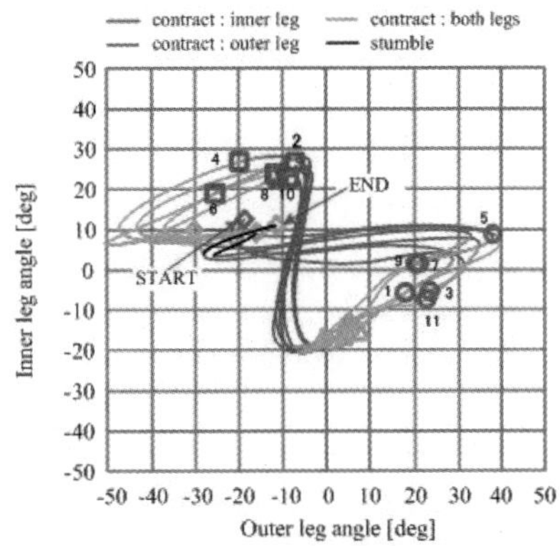

Figure 11. Configuration space trajectories of leg angle θ_{out} and θ_{in}, the trajectories settle to a constant orbit.

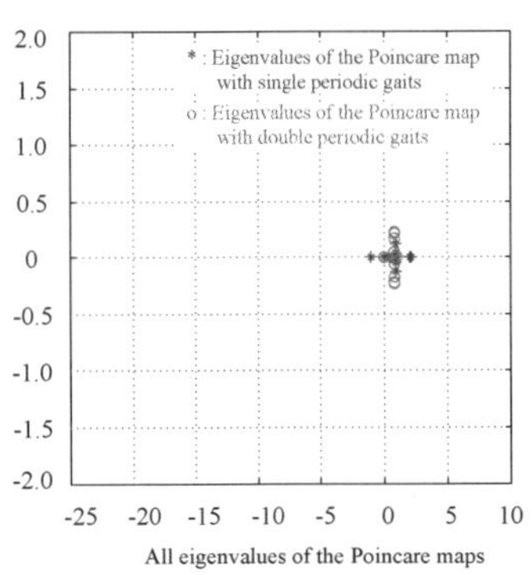

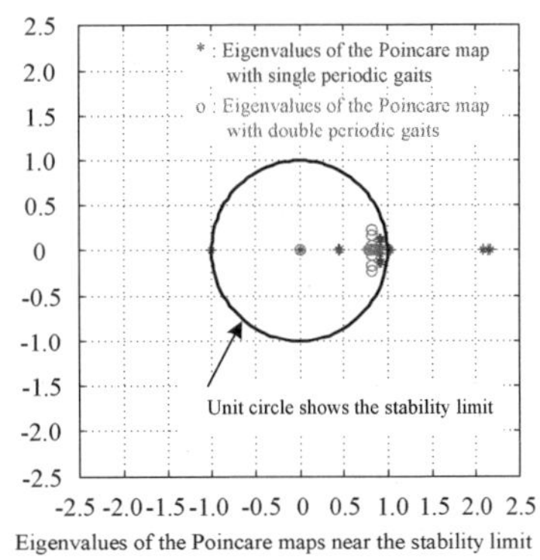

Fig.12 Eigenvalues of Linearized Poincare maps

The configuration space trajectories settle to a constant orbit as shown in Figure 11, and the absolute maximal eigenvalues of 1-periodic gait become more than 1, so the walking behavior seems to be unstable. However, absolute maximal eigenvalues of 2-periodic gait become less than 1, so the result shows this walking behavior becomes stable when the unit of walking gait is 2-periodic. It is very interesting that these experimental results agree with the simulative result in [8][11].

According to the above and our past research [12], behaviors of Passive Dynamic Walking can be explained by the conservative system model which has adopted and described in (1) and (2). And addition to it, as our theoretical research results described in [7][8] agrees with actual experimental results, our adopted system model of Passive Dynamic Walking can be a strong design tool of legged walking robots by applying the principle of Passive Dynamic Walking.

4. Conclusion

In this paper we tried to develop the prototype Passive Dynamic Walking robot, which has capabilities to keep walking passively and to measure the state parameters for analyzing its walking behaviors, to verify the adequacy of the conservative system model. Then we conducted walking experiments by using the prototype robot and verified the adequacy of conservative system model.

From the above results, it is thought that the results of theoretical researches can be strong design tool of legged walking robots by applying the principle of Passive Dynamic Walking. For the future work, we are going to try to develop Semi-Passive Dynamic Walking robot by applying the design tool which now we have.

References

1. T. McGeer, The Int. J. of Robotic Research, Vol.9, No.2, pp.62-82 (1990).
2. M. Wisse, A. L. Schwab, R. Q. van der Linde, F. C. T. van der Helm, IEEE TRANSACTIONS ON ROBOTICS, VOL.21, NO.3, pp.393-401 (2005).
3. Y. Ikemata, A. Sano, H. Fujimoto, Proc. of ICRA 2006, pp.836-841, May, 2006.
4. K. Osuka, K. Kirihara, JRSJ, Vol.18, No.5, pp.121-126(2000).(In Japanese)
5. A. Goswami, B. Thuilot, B. Espiau, Technical Report 2996, INRIA (1996).
6. M. Garcia, A. Chatterjee, A. Ruina and M. Coleman, ASME J. of Bio-mechanical Engineering, Vol.120, No.2, pp.281-288 (1998).
7. M. Iribe, K. Osuka, J. of Robotics and Mechatronics, Vol.19, No.4, pp.402-408 (2007).
8. Y. Sugimoto, K. Osuka, J. of Robotics and Mechatronics, Vol.20, No.4, pp. 559-566 (2008).
9. T. Kinugasa, S. Miwa, K. Yoshida, J. of Robotics and Mechatronics, Vol.20, No. 1, pp.98-105(2008).
10. T. Hayashi, K. Kaneko, F. Asano, Z.W. Luo, Advanced Robotics, Vol.25, Nos.1-2, pp.273-287(2011).
11. S. Morita, H. Fujii, T. Kobiki, S. Minami, T. Ohtsuka, Proc. of IROS 2004, pp.2825-2830, 2004.
12. M. Iribe, T. Kinugasa, Y. Sugimoto, K. Osuka, Proc. of SICE 2010, pp.2800-2803, Taipei, 2010.

MODELING OF QUADRUPEDAL RUNNING GAITS

KENNETH J. WALDRON

Faculty of Engineering and Information Technology, University of Technology, Sydney, 15 Broadway, Ultimo, NSW 2007, Australia

In this work we correct a previous ballistic model of the gallop cycle that could not be reconciled with experimental evidence. One of the assumptions on which that model was based proved to be incorrect. Dropping that assumption and taking a different approach to the development of the governing equations, we are able to develop expressions for the durations between footfalls during the stride cycle that are consistent with available experimental data. We also suggest a reason why some fast running animals use a rotary gallop, rather than a transverse gallop.

1. Introduction

Despite years of research there are still many mysteries to be solved before we can claim to fully understand the mechanics of dynamic quadrupedal running. Mathematical modeling has proven to be difficult to interpret, even when drastically simplified models are used [1]. While modeling of the dynamic interactions of leg and body during stance has a long history, and reasonably good agreement among researchers [2, 3], a complete understanding of the process depends on combining the stance model with a ballistic model of the behavior of the system between stances. This combination has received very little attention.

Recent work by the author pursuing this objective has shown that the interpretation of the previous model of the ballistic behavior of the system was erroneous. The unavoidable conclusion is that one or more of the assumptions upon which that model was based were invalid [4]. In particular, the attempt to use the motion about the roll axis to explain footfall timing must be abandoned because it leads to results that conflict both with empirical observation, and with the essential dynamics of motion in the sagittal plane that are obviously dominant in terms of the overall locomotion of the system. Although the fundamental dynamic laws govern the motion about the roll axis, the ability of a running animal to rotate the pelvic axis relative to the shoulder axis allows the timing of the rear footfalls relative to those of the front feet to be varied to fit the sagittal plane dynamics, rather than constraining the timing. Thus, the flexibility of a running animal about the roll axis allows the roll dynamics to be partially decoupled from sagittal plane and pitch dynamics. Of course, a robot with a rigid body frame cannot decouple the dynamics in the same way. In that case it is not valid to assume, as was done in [1], that the impulses delivered by the two front, or the two rear feet are identical. There is evidence from observations of biological systems that supports this assumption [5], although other observations conflict with it [6].

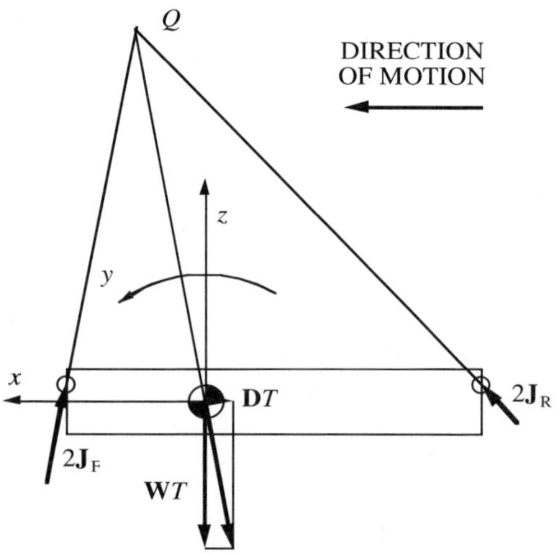

Figure 1: Impulsive equilibrium over the duration of a stride for a bound. **W** is the weight of the system, and **D** is the drag. T is the stride duration. $2\mathbf{J}_F$ is the net impulse imparted by the front feet, and $2J_R$ is that imparted by the rear feet.

The asymmetry between the gathered and extended flight phases during the gallop gait is a product of the differences in the vertical components of the impulses produced at the front and rear feet respectively, and of the pitch motion in the sagittal plane. That difference, in turn, is a direct result of the location of the center of mass of the system closer to the shoulder joints than to the hips. Why this should be so has never been explained, but it is an inherent architectural feature of every mammal that employs a gallop as its high-speed gait.

In this paper a different approach to the ballistic model of reference [1] provides a mechanically consistent model of the gallop gait that is consistent with available experimental data.

Following references [1] and [4]. We observe that the work done by horizontal components of the forces of contact between the feet and the ground are small relative to the horizontal kinetic energy of a fast running system. We therefore choose to focus on motions in the vertical direction.

It has been shown that the loss of energy from the system due to impacts between the feet and the ground is reduced by distributing those impacts as evenly as possible over the stride cycle [9, 1]. The dynamics of a quadruped in a single foot gait does not permit the intervals between foot impacts to be equal. Splitting the paired feet of a bound is beneficial from this point of view. A gallop can be viewed as a bound in which the front and rear foot pairs are split into individual footfalls. That is the approach taken in this paper. From this point of view, extending the durations of the double stances to be as long as possible is important.

An overall stride cycle model that can be compared to empirical observations is presented below. The well-known asymmetry of the gallop cycle is partially explained by means of the differences between the impulses delivered by the front and the rear feet. The greater vertical impulse delivered by the front feet, as compared to the rear explains the ubiquity of the gathered flight phase when galloping, and its greater duration as compared to the spread flight phase. The extended duration of the stance combines with the ballistic model to produce a theoretical gallop cycle with strong resemblances to those observed in running mammals [7, 13].

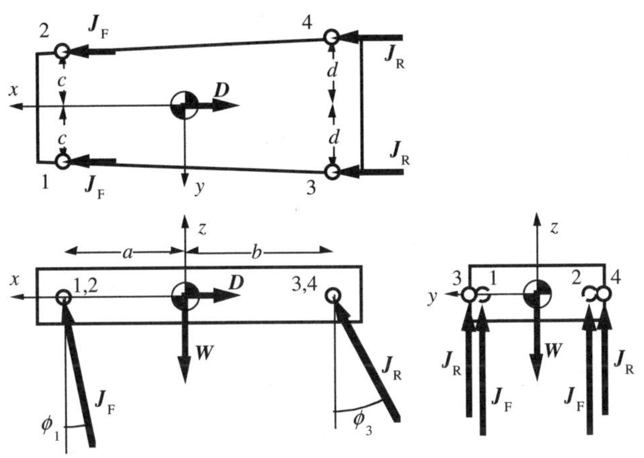

Figure 2: Free body diagram. The center of mass is assumed to be coplanar with the shoulder and hip joints. It can be argued that location of the center of mass a small distance above, or below the plane of the joints makes little difference to the dynamics. The shoulder and hip joints are assumed to be pairs of intersecting revolutes that are, respectively, parallel to the x axis of the body and orthogonal to the plane of the leg. The body reference frame is centered on the center of mass and aligned as shown. The principal axes of inertia are assumed to coincide with the x, y, z axes.

The function of pelvic rotation during a gallop to control the durations of the intervals between the front and rear foot stances promises to resolve another mystery: namely the occurrence of two types of gallop. As was observed in [1] there is an important dynamic distinction between a transverse gallop and a rotary gallop since in the former case the system performs two roll oscillations per stride, while in the latter it performs only one. At the same time, it is evident that some very fast runners like cheetahs and greyhounds employ pronounced pelvic rotation relative to the shoulder [7, 8] In this paper we will explore the implications of those observations.

Another component of this picture, which motivates some modification of the SLIP stance model [2], is the behavior of a foot when it impacts the ground, maintains contact over the duration of the stance, and then pushes off [9]. An important observation is that the feet of biological quadrupeds do not bounce or

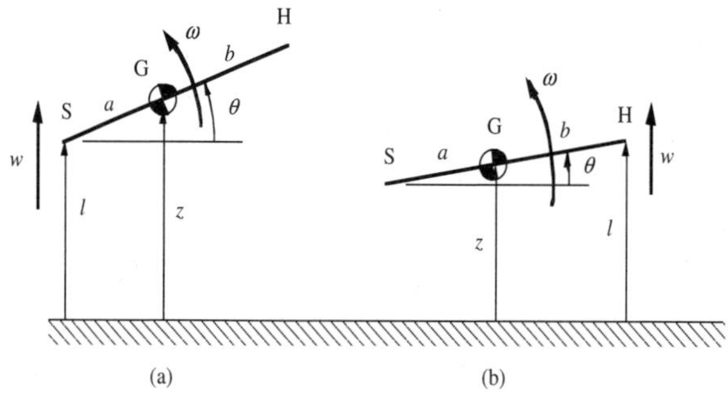

Figure 3: Coordinate senses during (a) front foot stance, and (b) rear foot stance. Point S represents the shoulder joint, and point H represents the hip joint. G is the center of mass.

chatter when brought into contact with the ground, although this is commonly observed behavior in artificial running systems [10]. We have studied the mechanics of foot/ground interaction in considerable depth, in order to be able to predict dynamic foot behavior [11], and design leg and foot structures that behave like biological feet.

2. Analysis

We choose to work in the impulse-momentum domain allowing us to consider an impulse balance over the entire stride period, *T*. We also work in a reference frame fixed to the body. This does not imply that we regard the stance period as being instantaneous. Rather, we integrate the contact force during stance with respect to time. This is analogous to replacing a spatially distributed force system by its resultant. In that case we integrate over the area of action of the force system. In the present case, the force is temporally distributed and we use the product of it's time average with the duration of action. We will discuss the effects of replacing each contact force by its impulse in some detail below when interpreting the results of the theoretical models based on these assumptions.

If the net impulse imparted to the system by the front feet is $2\mathbf{J}_F$, and that imparted by the rear feet is $2\mathbf{J}_R$, **W** and **D** are the weight of the system and the drag respectively, the system becomes a three-force system. Therefore the lines of action of the front and rear impulses must be concurrent with that of the impulse of the resultant of the weight and drag as shown in Figure 1.

Note that we assume that the impulses delivered by the two members of a leg pair are identical. That assumption is well supported in the case of galloping dogs [5], but is more problematic in the case of horses [6]. Please note that, if there is a difference between the vertical impulses delivered by the front feet, there must be a proportionate but opposed difference between those of the rear feet, or else the values of a and b must be varied, in order for the system to return to the same angular velocity about the x axis at the end of a stride.

The equivalent dynamic equilibrium equations are:

$$2J_{xF} + 2J_{xR} = DT \tag{1}$$

$$2J_{zF} + 2J_{zR} = WT \tag{2}$$

$$aJ_{zF} = bJ_{zR} \tag{3}$$

3. Bounding

As in reference [1], we start by considering a bound. For convenience, we will recapitulate a little of the mathematics from that reference, with some modifications.

Let the vertical velocity of the center of mass of the system immediately before front foot impact be u_F, regarded as being positive in the upward direction; and that immediately after the impact be u'_F. The vertical velocity immediately before rear foot impact is u_R, and that immediately after is u'_R. If the vertical impulse delivered by a front foot is J_{zF}, and that delivered by a rear foot is J_{zR} then applying impulse-momentum across the impacts gives

$$u'_F = u_F + \frac{2J_{zF}}{M}, \quad u'_R = u_R + \frac{2J_{zR}}{M} \tag{4}$$

where ω_{yFR} is the pitch angular velocity after the front foot stance and before the rear foot stance, and ω_{yFR} is the corresponding angular velocity after the rear foot stance. M is the total mass of the system and I_y is its pitch moment of inertia about G.

As was pointed out in reference [4], the pitch moment of inertia is significantly different when pitching from rear to front to that when pitching front to rear [12]. On the other hand, the model of reference [1] indicates that there is very little, or no pitching during the front and rear double stances of a gallop, so the changes in pitch inertia are easily handled by using different values for I_y during the rear-to-front flight (I_{yFR}) to that during front-to-rear flight (I_{yFR}). However, this does complicate the mathematical development below. Since we seek to understand the behavior of the system, rather than perform accurate calculations we choose to retain $I_{yFR} = I_{yFR} = I_y$. Hence

$$\omega_{yFR} = \omega_{yRF} - \frac{2aJ_{zF}}{I_y} \tag{5}$$

Looking at the velocity changes during the two ballistic trajectories (hops)

$$u_R = u'_F - g\tau_{FR}, \quad u_F = u'_R - g\tau_{RF} \tag{6}$$

Here τ_{FR} is the time period from the front foot stance to the rear foot stance, and τ_{RF} is that between the rear stance and the front. Obviously

$$\tau_{FR} + \tau_{RF} = T \tag{7}$$

where T is the stride period.

Applying the condition that the vertical height of the center of mass at the end of the stride must be identical with that at the beginning

$$\left(u_{\mathrm{F}}'+u_{\mathrm{R}}\right)\tau_{\mathrm{FR}}+\left(u_{\mathrm{R}}'+u_{\mathrm{F}}\right)\tau_{\mathrm{RF}}=0 \tag{8}$$

The condition that the system must return to the same pitch angle after a complete stride gives

$$\omega_{y\mathrm{FR}}\tau_{\mathrm{FR}}+\omega_{y\mathrm{RF}}\tau_{\mathrm{RF}}=0 \tag{9}$$

Combining Equations 4, 6 and 9 gives

$$\begin{aligned} u_{\mathrm{F}}' &= \frac{gT}{2(a+b)}(2ap-a+b), \quad u_{\mathrm{R}}' = \frac{gT}{2(a+b)}(b+a-2bp), \\ u_{\mathrm{F}} &= \frac{gT}{2(a+b)}(2ap-a-b), \quad u_{\mathrm{R}} = \frac{gT}{2(a+b)}(b-a-2bp) \end{aligned} \tag{10}$$

Here $p = \tau_{\mathrm{FR}}/T$, $1 - p = \tau_{\mathrm{RF}}/T$.

Note that we can apply Equations 2 and 3 above to get

$$J_{\mathrm{F}z}=\frac{bMgT}{2(a+b)}, \quad J_{\mathrm{R}z}=\frac{aMgT}{2(a+b)} \tag{11}$$

These expressions are also used in the derivation of Equations 10 above.

Similarly, Equations 5 and 9 can be combined to give

$$\omega_{y\mathrm{FR}}=\frac{\kappa gT}{a+b}(p-1), \quad \omega_{y\mathrm{RF}}=\frac{\kappa gT}{a+b}p \tag{12}$$

where

$$\kappa=\frac{Mab}{I_y} \tag{13}$$

The parameter, κ, actually measures the relative magnitudes of the impulsive moment about the pitch axis, and the impulse of the inertial moment about that axis. Thus, it quantifies the importance of pitch rotation in the model. Note that for real mammals $\kappa > 1$. It is typically of the order of 1.5 or 2.0.

At this point we diverge from reference [1] to directly seek expressions for the durations of the two flight phases. We make the assumption that the front and rear leg pairs each have preferred working heights, l_{F} and l_{R} respectively, and that, in the absence of disturbances, the respective legs will be placed at the right times in the stride cycle to work at these heights. This assumption is well supported by biological studies. Most robotic legs, also, have a designed working length, and the gait is adjusted so they can operate at that length.

Referring to Figure 3, l_{F} and l_{R} are related to the respective heights of the center of mass, z_{F} and z_{R} at the effective instants of application of the front and rear impulses by

$$l_{\mathrm{F}}=z_{\mathrm{F}}-a\theta_{y\mathrm{F}}, \quad l_{\mathrm{R}}=z_{\mathrm{R}}+b\theta_{y\mathrm{R}}. \tag{14}$$

Here θ_y is the pitch angle, as defined on Figure 3. Note that the Equations 14 imply a small angle assumption. This assumption is valid for the relatively small pitch angles that occur in a gallop. It is not valid for bounding with extreme pitch angles, as observed in the case of some antelopes, or for the pitch angles employed when leaping across obstacles.

Also

$$z_{\mathrm{R}} = z_{\mathrm{F}} + u'_{\mathrm{F}} Tp - \frac{gT^2 p^2}{2}, \quad \theta_{y\mathrm{R}} = \theta_{y\mathrm{F}} + \omega_{\mathrm{FR}} Tp \tag{15}$$

Combining Equations 13 and 14, and letting $l_{\mathrm{F}} = l_{\mathrm{R}}$ for simplicity we get

$$p^2 - p + Q = 0 \tag{16}$$

where

$$Q = \frac{2(a+b)^2 \theta_{y\mathrm{F}}}{gT^2(2\kappa b - a + b)} \tag{17}$$

Solution of Equation 16 gives

$$p = \frac{1 \pm \sqrt{1-4Q}}{2} \tag{18}$$

The two solutions embodied in Equation 18 correspond to p and $p - 1$, and hence to both τ_{FR} and τ_{RF}. As can be seen, the form of the equation places an upper limit on $\theta_{y\mathrm{F}}$ beyond which there is no real solution:

$$\theta_{y\mathrm{F}} \le \frac{gT^2(2\kappa b - a + b)}{8(a+b)^2} \tag{19}$$

However, values of θ_y approaching this would invalidate the small angle assumption.

The larger of the two solutions given by Equation 18 corresponds to a bound for which $\tau_{\mathrm{FR}} > \tau_{\mathrm{RF}}$. As may be seen, however,

$$\frac{\tau_{\mathrm{FR}}}{\tau_{\mathrm{RF}}} \ne \frac{J_{z\mathrm{F}}}{J_{z\mathrm{R}}}$$

which would have been true if pitch rotation was ignored.

4. Galloping

As was demonstrated in reference [1] using a similar model, the pitch rotations during the front and rear double stances are ideally zero. In a double stance, the relationship between the downward vertical velocity of the center of mass at impact of the first foot placed and the upward vertical velocity when the second foot lifts off is the same as given by Equations 4. In fact, all of the algebra above can be reinterpreted as applying to the flight phases of the gallop. Thus, τ_{FR} becomes τ_{23}, and τ_{RF} becomes τ_{41} for a transverse gallop. The subscripts denote the feet numbered as in Figure 2. However, the total stride duration must be replaced by the combined durations of the two flight phases: $T - \tau_{12} - \tau_{34}$, that is, the stride duration less the combined durations of the double stances. Thus, Equation 17 becomes

$$Q = \frac{2(a+b)^2 \theta_{y2}}{g(T - \tau_{12} - \tau_{34})^2 (2\kappa b - a + b)} \tag{20}$$

and Equation 18 then gives

$$\tau_{23} = \frac{1+\sqrt{1-4Q}}{2}(T - \tau_{12} - \tau_{34}), \quad \tau_{41} = \frac{1-\sqrt{1-4Q}}{2}(T - \tau_{12} - \tau_{34}) \tag{21}$$

For a rotary gallop the equations are exactly the same except τ_{34} becomes τ_{43} and τ_{23} and τ_{41} become τ_{24} and τ_{31} respectively.

The durations of the double stance phases, τ_{12} and τ_{34} (or τ_{43}) are completely unconstrained in this model, since we have abandoned the use of roll dynamics to constrain the periods between the foot impulses as was done in reference [4]. As discussed above, rotation of the hips relative to the shoulders about the longitudinal axis of the body can be used to adjust the times at which the respective legs reach their optimal working heights and are placed on the ground. However, the whole system still rolls in response to the foot impulses, and that roll motion can be expected to play a role in determining the duration between foot placements in the double stances. It is noticeable that the front double stance duration is typically longer than that at the back. We might speculate that if foot 1 (the left front foot) is placed first, the foot impulse will tend to roll the whole system in the clockwise direction about the x axis resulting in a very short interval τ_{12} before it is necessary to place foot 2. That might be countered by torqueing the hindquarters relative to the forequarters in the clockwise direction about the x axis. The reaction exerts a counter-clockwise torque on the forequarters slowing the rotation of the shoulders to permit a longer τ_{12}. After the roll rotation is arrested by the impact of foot 2, and starts to reverse, that movement has placed foot 3 higher than foot 4 so foot 4 is in position to be placed before foot 3 resulting in a rotary gallop. After foot 4 impacts increasing the counter-clockwise roll the forequarters might be torqued in the counter-clockwise direction to slow the roll of the hindquarters allowing increased τ_{43}. In photographic sequences of fast runners using rotary gallops like greyhounds or cheetahs [7, 11] strong rotation of the hindquarters toward the first front foot placed is visible. The reverse rotation of the forequarters after the first hind foot is placed is less visible. Since the roll inertia of the forequarters is typically substantially larger than that of the hindquarters this is not unexpected, but it would seem to run counter to the disparity in the durations of the front and rear double stances cited above. This issue needs to be further examined by means of more detailed models and experiments.

5 Conclusions

In this paper we have presented a new model of the gross dynamics of a galloping quadruped that does not exhibit the discrepancies when compared to experiment of the model of reference [1]. The model presented here predicts that the duration of the gathered flight phase must always be somewhat greater than that of the spread flight phase, and predicts what that difference should be for a given system. The durations of the flight phases and the double stances in the model are the intervals between the instants of application of the resultant vertical impulses of the foot impact forces. Since the stance durations during fast running are of the order of 20% of the stride duration, the durations of the double stances as seen by a stationary observer need to be increased by ~ $0.2T$, and the apparent durations of the flight phases reduced by a similar amount. Thus, it can be seen why the spread flight phase is not observed in a slow gallop, although the gathered flight phase is always present.

We present here, for the first time, a possible reason why fast runners with ability to substantially rotate the hindquarters relative to the forequarters should be expected to favor a rotary gallop over a transverse gallop. That suggestion needs to be pursued by further modeling and experimentation.

Acknowledgement

The work described here was partially supported by grant number LP100200750 from the Australian Research Council and the Roads and Traffic Authority of New South Wales.

References

1 Waldron, K.J., Estremera, J., Csonka, P.J., Singh, S.P.N., "Analyzing Bounding and Galloping Using Simple Models", *ASME Transactions J. Mechanisms and Robotics*, Vol. 1, No. 1, February 2009, pp 011002-1 to 011002-11.

2 Raibert, M. H., 1986, Legged Robots That Balance, MIT Press, Cambridge, Massachusetts.

3 Abdallah, M.E., Waldron, K.J., "The Mechanics of Biped Running and a Stable Control Strategy", *Robotica*, ,2009, volume 27, pp. 789–799.

4 Waldron, K.J., Singh, S.P.N., "Resolving the Paradox of Asymmetry in the Gallop Gait," 13th World Congress in Mechanism and Machine Science, Guanajuato, México, 19-25 June, 2011 (to appear).

5 Walter, R.M. and Carrier, D.R., "Ground Forces Applied by Galloping Dogs", *J. Experimental Biology*, 210, 2007, pp. 208-216.

6 Merkens, H.W., Schmhardt, H.C., van Osch, G.J., Hartman, W., "Ground reaction force patterns of Dutch Warmbloods at the canter," *Am. J. Vet. Res.*, Vol. 54, pp. 670-674.

7 Gambaryan, P.P., 1974, How Mammals Run: Anatomical Adaptions, John Wiley & Sons, New York.

8 Estremera, J., Waldron, K.J., "Thrust Control, Stabilization and Energetics of a Quadruped Running Robot", *International Journal of Robotics Research*, Vol. 27, 2008, pp. 1135-1151.

9 Ruina, A., Bertram, J.E.A., Srininvasan, M., "A collisional model of the energetic cost of support work qualitatively explains leg sequencing in walking and galloping, pseudo-elastic leg behavior in running and the walk-to-run transition," *Journal of Theoretical Biology*, Vol. 237, 2005, pp. 170-192.

10 Jacobs, D.A., Waldron, K.J., "Application of Hunt-Crossley Viscoelastic Nonlinear Contact Model to Elastic and Plastic Collisions," submitted to *Journal of Applied Mechanics.*

11 Muybridge, E., Animals in Motion, Dover Publications, New York, 1957.

12 Schmiedeler, J.P., Siston, R., and Waldron, K.J., 2002, "The Significance of Leg Mass in Modeling Quadrupedal Running Gaits," ROMANSY 14: Theory and Practice of Robots and Manipulators, ed. G. Bianchi, J.C. Guinot, C. Rzymkowski, Springer, pp481-488.

13 Singh, S.P.N., Waldron, K.J., "A Stance Period Approach for Simplified Observation of Galloping as Applied to Canines", *Robotica*, 2011, to appear.

ANALYZING ENERGY-EFFICIENT KINEMATIC CONFIGURATIONS IN HEXAPOD ROBOTS

D. Sanz-Merodio, E.Garcia, and P. Gonzalez-de-Santos

Centre for Automation and Robotics, CSIC
La Poveda, 28500 Madrid, Spain
**E-mail: daniel.sanz@car.upm-csic.es*
www.iai.csic.es/dca

Reducing energy consumption in walking robots is a topic of huge importance to increase mission time for a given power supply. This paper addresses the problem of improving energy efficiency in statically stable walking machines by comparing two leg configurations, insect and mammal, on the hexapod robotic platform SILO6. Dynamic simulation of this hexapod is used to develop a set of rules that optimize energy expenditure in both configurations. Later, through a theoretical analysis of energy consumption and the experimental measurements on the real platform SILO6, a configuration is chosen.

Keywords: Energy efficiency, hexapod robots, insect and mammal configurations

1. Introduction

Detection and location of antipersonnel landmines is a slow process that requires endurance and energy efficiency using robots. This research is intended to determine the most energy-efficient kinematic configuration for a hexapod robot, developed for humanitarian demining missions.

This work is motivated be the real need to increase the energy autonomy of the SILO6 walking robot. The SILO6 is a six-legged robot developed as a mobile platform for humanitarian demining activities (see Fig. 1).

This paper shows the comparative analysis of the power consumption of two classic leg configurations on the robotic platform SILO6. It's widely accepted that the mammal configuration consumes less energy than the insect one.[5] Through a theoretical analysis based on dynamic simulation and real experiments this paper analyzes the role of leg dynamics in the energy consumption of hexapod robots in the two configurations. Section 2 introduces both leg configurations. Section 3 presents an optimization of the

Fig. 1. SILO6 walking robot

chosen critical parameters. Sections 4 and 5 show the results of theoretical and experimental analysis. Finally, section 6 summarizes the main results and presents some conclusions.

2. Leg configurations

In order to determine the most efficient configuration for a hexapod, two configurations have been implemented in the SILO6 robot , the insect and the mammal one, shown in figure 2. These leg configurations are used for simulation and theoretical analysis and are implemented in the experimental platform SILO6 to compute the energy consumption.The insect and the mammal configuration are

First, a theoretical analysis is performed making use of a model of power consumption and a dynamic simulation package in order to determine the sources of energy expenditure. Later, the two configurations are implemented in the real robot, power consumption is acquired during real experiments and results are analyzed.

3. Theoretical approach

In this theoretical approach, the two configurations are analyzed in order to compare their energy efficiency. A non-dimensional parameter is considered for comparison, evaluation and optimization purposes, which is the specific resistance. The specific resistance is a feature used in any kind of vehicle and it is defined as:[3]

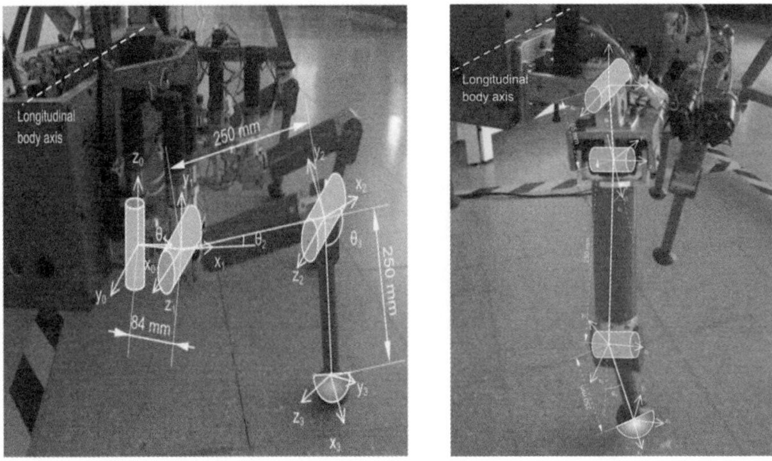

(a) (b)

Fig. 2. Leg configurations, joint-axis distribution and dimensions. (a) Insect ; (b) Mammal

$$\varepsilon = \frac{E}{WL} = \frac{P}{Wv} \tag{1}$$

where E is the energy required to travel a distance L, by a vehicle with weight W. P is the power consumed and v is the speed of the vehicle. The specific resistance considers the power consumed per unit of mass and unit of speed. Therefore, it allows to compare vehicle performance with independence of weight and speed. For example, in a hexapod robot, the mammal configuration is slower than the insect one.

A walking robot can perform a variety of gaits featuring very different properties. It is well known that a hexapod performing an alternating-tripod gait can achieve its highest speed[4] and, due to the fact that statically stable walking robots are intrinsically slow, alternating tripods becomes the most usual gait by both natural and artificial hexapods. Thus, we focus our study on an alternating-tripod gait.

In order to compare the energy consumption of the two configurations a set of rules that optimize the gait will be established. This set of rules to optimize the energy consumption are based on three critical parameters of the gait (see Fig. 3):

(1) The distance of the footholds to the origin of the leg reference frame L_i.

(2) The leg stroke R_x.
(3) The body height of the robot, H_i.

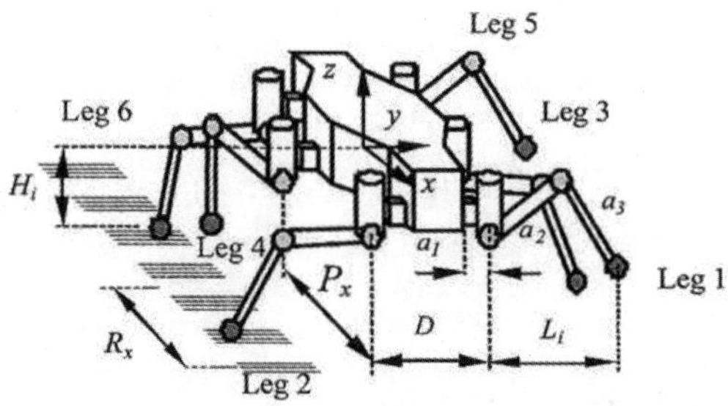

Fig. 3. Gait parameters in a six legged robot.

These three parameters can change joint speed and torque, and therefore, the power consumption. As a first approach, to establish the optimum values of the mentioned parameters for a given configuration, it is considered that the mechanical power consumed is:

$$P = \frac{\int_0^T \sum_{i=1}^{6} \sum_{i=1}^{3} (\omega_{ij}(t)\tau_{ij}(t))dt}{T} \tag{2}$$

where $\omega_{ij}(t)$ is the angular speed and $\tau_{ij}(t)$ is the torque in the joint j of the leg i. Using the dynamics simulation package Yobotics! SCS,[6] joint speed and torque for each leg are obtained and inserted in equation (2) to compute the power consumed.

3.1. *Parameter optimization in the insect configuration*

Marhefka and Orin[2] set up some rules to optimize the gait in an insect configuration. Here these rules have been adapted to the critical parameters, as follows:

Insect rule 1 (Footholds and stroke length) The footholds should be set at points on an arc of a circle about the vertical hip axis, whose radius is determined so that the leg distal link is nearly vertical (see Figure 4).

Insect rule 2 (Body height) To generate the required support forces with a minimum of joint torques (at joint 2), the body should be lowered so as to raise the knee above the hip.

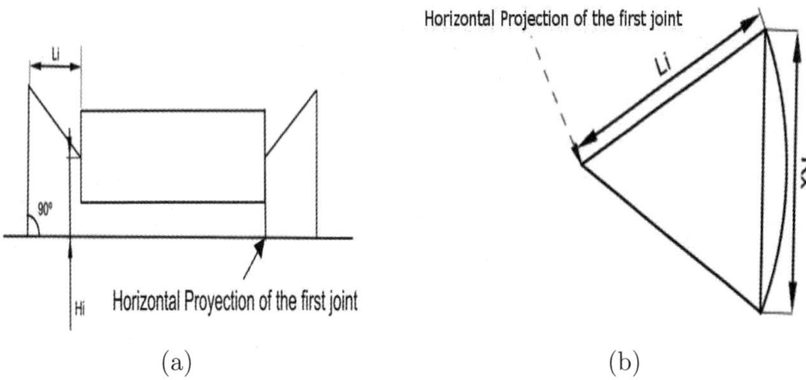

Fig. 4. (a) Leg distal link is nearly vertical (b) Points on an arc of a circle about the vertical hip axis

3.2. *Parameter optimization in the mammal configuration*

The mammal configuration is not frequently used in hexapod robots, and there is not literature about rules to optimize the gait in this configuration. By analyzing the power consumption on the dynamics simulation the following set of rules are here established.

Mammal rule 1 (Footholds) The footholds that minimize power consumption are those on the horizontal projection of the first joint on the ground. Thus, the lateral length should be L_i=0.

Mammal rule 2 (Stroke length) The foot-ground reaction force should pass through the hip to minimize joint torque. Also, stroke length must increase as the speed does.

Mammal rule 3 (Body height) Power consumption is reduced if the foot-ground reaction force passes through the knee, which is achieved if the robot walks as high as possible.

Following the above rules, optimum gait parameters have been applied to simulation and to real experiments. Table 1 shows the optimum critical values gathered after applying these mentioned rules and the specific resistance computed in a first approach using equation (1) and (2).

Table 1. Specific resistance of insect and mammal configuration at the first theoretical approach.

Configuration	Lateral length L_i (m)	Stroke length R_x (m)	Body height H_i (m)	Specific Resistance ε
Insect	0.2	0.25	0.2	1.04
Mammal	0	0.3	0.52	1.37

As it is seen, the insect configuration seems to be a 25% more efficient that the mammal one walking along a straight line. However, notice that power consumed at the motors has not been considered in this initial approach. The electro-mechanical model of power consumption will be addressed in the next section.

4. Electro-mechanical analysis of power consumption

The energy consumption in a robot is given by the sum of the energy consumed at every joint plus the energy consumed by electronic equipment (computers, drivers, analog I/O, etc.). The contribution of this last term is out of the scope of this work.

If a joint is driven by an electric motor and taking into account the well-known electrical model of a motor,[1] the consumed energy is given by:

$$E(t) = \int u_{app}(t)i(t)dt = \int (\frac{k_E}{k_M}\omega(t)\tau(t) + \frac{R}{k_M^2}\tau^2(t))dt \tag{3}$$

where u_{app} is the applied voltage, $i(t)$ is the current through the motor, k_E is the back electromotive constant, k_M is the torque constant, R is the electric resistance of the motor, $\omega(t)$ is the motor speed and $\tau(t)$ is the motor torque. The relation between k_E and k_M for the motors in SILO6 is as follows:

$$k_E = k_M(\frac{2\pi}{60}) \tag{4}$$

In equation 3 the term $\frac{k_E}{k_M}\omega(t)\tau(t)$ is related with the mechanical energy and the term $\frac{R}{k_M^2}\tau^2(t)$ corresponds to electrical losses in the motor.

Taking the contribution of all joints in the hexapod into consideration and under the assumption that the negative mechanical energy is lost, because an electrical motor cannot store energy, the total energy becomes:

$$E = \int_0^T \sum_{i=1}^{6} \sum_{i=1}^{3} (\Delta(\frac{k_E}{k_M} \omega_{m_{ij}}(t) \tau_{m_{ij}}(t) + \frac{R_j}{k_{M_j}^2} \tau_{m_{ij}}^2(t)))dt \tag{5}$$

where the function $\Delta(x)$ is defined as:

$$\Delta(x) = \begin{cases} x \; if \; x > 0 \\ 0 \; if \; x \leq 0 \end{cases} \tag{6}$$

Joint speeds and torques are directly obtained from the dynamic simulation, but equation (5) refers to the speeds and torques of the motors. Joint speed and torque and motor speed and torque are related through the mechanical and electrical parameters of the motor-gear system (motor-gear parameters for SILO6 are shown in table 2). The equations that relate both joint speed and torque with motor speed and torque are:

$$\omega_m = N\omega_s \qquad \tau_m = JN\alpha_s + BN\omega_s + \frac{\tau_s}{\eta N} \tag{7}$$

where J is the rotor inertia, N is the gear ratio, B is the friction constant, α_s in the joint's angular acceleration, ω_s and τ_s are joint angular speed and torque respectively and η is the mechanical efficiency. Using the values obtained of α_s, ω_s and τ_s in dynamic simulation, the energy consumption is computed. Also, from the terms in right hand side of (3), it's possible to distinguish between mechanical consumption and electrical losses.

Table 2. SILO6 joint parameters.

Leg joint	Motor		Gear	
	Torque constant $k_M(Nm/A)$	Resistance $R(\Omega)$	Ratio N	Efficiency $\eta(\%)$
1	0.0333	0.62	246	63
2	0.0333	0.62	881.5	42
1	0.0333	0.62	881.5	42

Table 3 shows results of applying the values obtained in dynamic simulation of the joint's angular acceleration, angular speed and torque to equations (5) and (7) for 0° and for 20° crab gait.

Through the observation of these tables some interesting conclusions are stated:

Table 3. Mechanical, electrical and total consumed power in the configuration of insect and mammal walking along a straight line (0° crab gait) and in a 20° crab gait.

Configuration	Crab Angle	Mechanical Power (W)	Electrical Power (W)	Total Power (W)	Specific Resistance ε
Insect	0°	2.28	37.96	40.25	5.31
Mammal	0°	5.50	23.81	29.32	3.65
Insect	20°	3.06	38.51	41.58	5.48
Mammal	20°	5.50	24.5	30.01	3.73

(1) Electrical consumption dominates over mechanical power; it takes an 81% of the total power on the mammal configuration and a 92% on the insect one.
(2) Regardless mechanical power expenditure is still higher in the mammal configuration, electrical consumption is larger in the insect one. And, as the electric consumption is more relevant than the mechanical one, the largest total power consumption occurs in the insect configuration.
(3) When the crab gait changes from a 0° to a 20° crab angle, the mechanical power increases a 34.2% in the insect configuration while in the mammal one there is not an important increase.

5. Experimental analysis

With the above theoretical analysis of energy consumption in hand, both configurations have been implemented in the robotic platform SILO6. The mechanical structure of the leg has been changed (see Fig. 2a y 2b) to fit with the two configurations and the optimum gait parameters have been applied to compare power expenditure. The SILO6 robot is driven by 18 DC motors (three per leg). There are also onboard electronics (control cards, power amplifiers, sensors, etc) that require some power. In these experiments only the actuator's power consumption is measured , while the consumption of the onboard computer and sensors are assumed to be constant. Output current and voltage from the actuators power supply is captured making use of a custom-made power measurement unit (PMU). These two analog measurements have been acquired through a PCMCIA National Instruments DAQCard-A1-16XE-50 in a portable PC. The acquisition sampling time is 1ms.

In order to estimate each term of the power consumption, three measurements have been accomplished:

(1) Power consumption of SILO6 stopped and suspended on the air (without touching the ground).

(2) Power consumption of SILO6 walking on the air along a straight line in both configurations the insect and the mammal one.
(3) Power consumption of SILO6 walking on flat, rigid ground along a straight line and in a 20° crab gait in both configurations, the insect and the mammal one.

Experiment 1 provides the power losses in the power electronic cards. Experiment 2 provides power consumption for the legs movement. Finally experiment 3 would give the power consumed to support the body. Table 4 shows the different sources of power consumption obtained in the experiment.

Table 4. Mechanical, electrical and total consumed power in the configuration of insect and mammal walking in a 0° and 20° crab gait

Configuration	Gait	Power consumed at the power electronic cards(W)	Power consumed for the leg motion(W)	Power consumed to support of the body(W)	Consumption at the actuators (W)	Total power consumed (W)
Insect	Crab 0°	40.89	17.36	6.9	24.26	65.15
	Crab 20°	40.89	18.89	12.29	31.18	72.07
Mammal	Crab 0°	40.89	20.02	3.67	23.89	64.58
	Crab 20°	40.89	22.23	5.55	27.79	68.67

Analyzing this result considering the dynamic model of a leg structure, it is observed that the power consumed at leg motion is related with the leg dynamics during its transfer phase, following the expression $D(q)\ddot{q} + H(q,\dot{q}) + C(q)$. Where $D(q)$ is the 3x3 mass matrix of the leg, q is the vector of generalized coordinates, H is a 3x1 vector of centrifugal and Coriolis terms, and $C(q)$ is a 3x1 vector of gravity terms. While the measured power consumed to support the robot weight is related with the term $J^T F$. Where F is the vector of the ground-contact forces and J is the Jacobian matrix.

Analyzing and comparing the experimental results it can be deduced that the insect configuration has less capacity to support static loads, but conversely the power consumption due to leg dynamics is lower in the insect configuration than in the mammal one.

6. Results and conclusions

It is widely accepted that the mammal configuration in statically stable walking machines is better to support high loads while the insect one is considered to improve mobility. This paper presents a theoretical and experimental analysis of power consumed by an hexapod robot in both configurations. To allow for a thorough analysis, a non-dimensional parameter has been used for comparison of power consumption. Also, rules for optimizing the two configurations have been proposed in order to compare the insect and the mammal configuration in their most efficient disposition. After the results showed in this paper, it is stated that this affirmation is valid if only static loads are taken into account, that is, the power consumption for supporting the body and its pay-load is larger in an insect configuration. However, as it is shown in this paper the leg dynamics is not negligible even in a statically-stable locomotion and its contribution to power consumption is even larger than the static-load one. Also it is important to remark that the larger contribution to the power consumed in a walking robot is due to the electric term (larger than 80%).

Finally, taking the effect of the gait crab angle into account, the mammal configuration features lower consumption in a crab gait. In a natural environment, like the one found in demining applications, a robot has to change its trajectory constantly so based on the point of view of the energy consumption, the better configuration for this applications is the mammal one. The robot energy consumption could be also improved if the motor's efficiency does so.

7. Acknowledgments

This work has been funded by the Spanish Ministry for Innovation and Science through grant DPI2010-18702.

References

1. P. Gonzalez de Santos, E. Garcia, R. Ponticelli, and M. Armada. Minimizing energy consumption in hexapod robots. *Advanced Robotics*, 23:681–704, 2009.
2. Duane W. Marhefka and David E. Orin. Gait planning for energy efficiency in walking machines. *IEEE Robotics and Automation Magazine*, 97(April):474–478, 1997.
3. B. Siciliano and O. Khatib. *Springer Handbook of Robotics*. Springer, London, 2008.
4. S.M. Song and K.J. Waldron. *Machines that walk: The adaptive suspension vehicle*. The MIT Press, Cambridge, Massachusetts, 1989.
5. D. J. Todd. *Walking machines: An introduction to legged robots*. Kogan Page, Ltd., 1985.
6. Yobotics Inc., Boston, MA. *Yobotics! Simulation Construction Set: Users Guide*, 2002. Available: `http://www.yobotics.com/`.

BIO-INSPIRED LOCOMOTION CONTROLLER DESIGN FOR LEGGED ROBOTS

RIADH ZAIER

Department of Mechanical and Industrial Engineering, College of Engineering, Sultan Qaboos University, P.O. Box 36, Al Khod 123, Sultanate of Oman

To control a humanoid locomotion, much work can be found in the literature that has been focused on the Zero Moment Point approach. More recently, biologically inspired control strategies such as Central Pattern Generators have been proposed to generate autonomously adaptable rhythmic movement. Despite the extensive research works in this area, suitable autonomous control system that can adapt and interact safely with the surrounding environment while delivering high robustness is yet to be discovered. In this paper, therefore, we deal with the design of oscillatory neural network for bipedal motion pattern generator and locomotion controller. Locomotion pattern will be generated by a neural network that represents the lower layer of the overall control system of the humanoid. The neural network will be augmented by neural controllers with sensory connections to maintain the stability of the system. Moreover, we investigate the stability of the system simplified as an inverted pendulum and we propose analytical method on how to decide the parameters of the motion To validate the theoretical results, we use the humanoid robots "HOAP-3".

1. Introduction

Motion control of humanoid robots has attracted considerable attention during the last decade. Much work has been focused on the dynamics of the robot using the zero moment point (ZMP) approach [1-3]. Huang et al. [4] proposed a method for planning a walking pattern, where the reference trajectory is designed offline for given constraints on the foot and ground, and satisfying particular ZMP constraints using third order spline functions. More recently, biologically inspired control strategies have been proposed to generate autonomously adaptable rhythmic movement. These are based on a neuronal network, termed a central pattern generator (CPG) [5-7], that is capable of generating a rhythmic pattern of motor activity in the absence of sensory input signals. Taga [6], [7] has demonstrated that bipedal locomotion can be realized as a global limit cycle generated through entrainment between a neural network consisting of a neural oscillator and the physical system.

In this paper, instead of using nonlinear dynamics models and solving complex equations, our control strategy is based on piecewise linear oscillators with few parameters that can be easily obtained. Moreover, the motivation is to provide much flexibility to the motion pattern generator so that combination of reflexes during locomotion can be realized without complexity or re-design of the motion pattern generator [8]. The stability of the system is assured by a gyro feedback loop, while terrain irregularities are compensated by adjusting the pitching motion using a virtual spring-damper systems. In our previous work [8] it has been shown that the proposed method is straightforward with respect to three design parameters namely, the slope of the piecewise function, the time delay of the neuron, and the rolling amplitude of the hip and ankle joints, which has been determined experimentally. In this paper, we investigate the stability of the system simplified as an inverted pendulum and we propose analytical method on how to decide the parameters of the motion.

The rest of this paper is organized as follows. Section 2 presents our motivation; Section 3 explains the system behavior and stability; Section 4 presents the proposed generation method of walking motion; Section 5 presents the experiment result, and Section 6 is conclusion.

2. Motivation

To control the motion of a humanoid robot we consider the general model of the neural network as follows;

$$\begin{aligned} \frac{dx}{dt} &= -Dx + T\,g(x) + Sv \\ y &= Cx \end{aligned} \tag{1}$$

where $x \in \boldsymbol{R}^n$ is the neurons' outputs vector, $y \in \boldsymbol{R}^m$ depicts the motors' commands vector, D is an n x n diagonal matrix with strictly positive real numbers, T is an n x n neurons' interconnection matrix, $g(x)$: $\boldsymbol{R}^n \rightarrow \boldsymbol{R}^n$ is the nonlinear vector activation function, C is an m x n output matrix, $v \in \boldsymbol{R}^m$ is a sensory input vector, and S is an n x m matrix. Note that the right side of (1) can be considered as decomposition into linear part "Sv" and a nonlinear one "Tg(x)". To investigate the stability of (1), we consider the behavior of a linearized model of system at the equilibrium points with Tg(x)=0. Indeed, a motion can be described topologically as a homotopy mapping at each equilibrium point as well as switching, or jump mapping between these points.

3. System Behavior and Stability

The proposed control strategy is illustrated in Fig.1, where the neural network (1) has to control the humanoid robot locomotion as continuous commutations

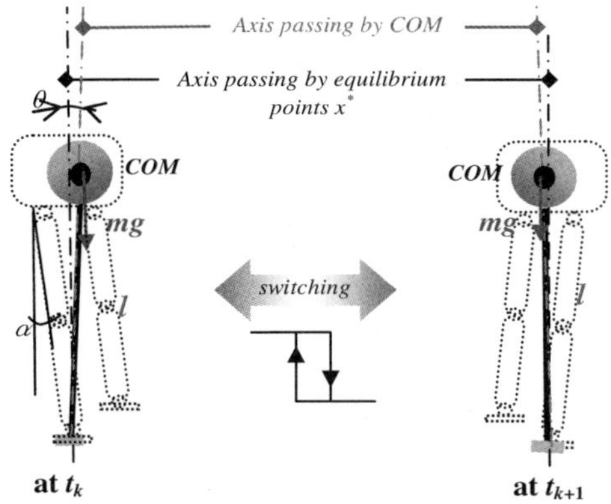

Figure 1 Switching the system state between equilibrium points defined at single support phases.

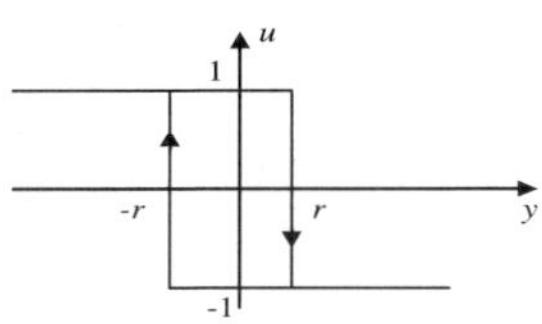

Figure 2 Relay with hysteresis property.

between two equilibrium points defined at single support phases. Then based on the output of the switched system, we can define the motion pattern generator output to the rolling joints, while maintaining the stability of the equilibrium points during locomotion by gyro feedback and adjusting the pitching motion with virtual dumper-spring system. To begin with, we investigate the stability of the robot simplified as an inverted pendulum at each equilibrium point, which can be expressed by;

$$\frac{d^2\theta}{dt^2} + 2\varsigma\frac{d\theta}{dt} - \mu\theta = u(\theta) \tag{2}$$

where $\theta \in \boldsymbol{R}$ is the counterclockwise angle of the inverted pendulum from the vertical, $\zeta \in \boldsymbol{R}$ is damping ratio, $\mu = g/l \in \boldsymbol{R}$, and $u \in \boldsymbol{R}$ is the control input to the system. To control the inverted pendulum as illustrated in Fig. 1 and Fig. 2, we propose the control input as follows;

$$u(\theta) = u_g(\theta) + u_{sw}(\theta) \tag{3}$$

where $u_g(\alpha)$ is the gyro feedback signal stabilizing the inverted pendulum around $\theta = \theta_s$, and $u_{sw}(\theta)$ is a relay type feedback signal with a hysteresis property defined by;

$$u_{sw}(y(t)) = \begin{cases} +1 & if \ y(t^-) > -r \\ -1 & if \ y(t^-) < r \\ \{-1,1\} & if \quad t = k\tau \ and \ if \ y(t) = \pm r \end{cases} \tag{4}$$

where t^- is the time just before commutation. Notice that the switching law (4) depends not only on the output $y(t)$ but also on the switching time constant τ. In our design method we choose a fixed switching time, which reduces the impact map to a linear one. For the feedback signal (4), the trajectories of the system will follow the invariant manifolds of only two equilibrium points of (2). To investigate the stability of (2) for θ close to θ_s, reconsider the system in the absence of control input. Then at each equilibrium point the system has the same eigenvalues as follows;

$$\lambda_{1,2} = -\varsigma \pm \sqrt{\varsigma^2 + \mu} \tag{5}$$

Since $\zeta \ll \mu$, the equilibrium points (5) are saddle ones. To generate periodic movement as explained in Fig. 1 using (4), we first stabilize the equilibrium points by a feedback control $u_g(\theta) = k_1\theta + 2k_2 d(\theta)/dt$ using the angular velocity of the inverted pendulum measured by the gyro sensor. Therefore, (2) with (4) becomes as

$$\frac{d^2\theta}{dt^2} + 2(\varsigma + k_2)\frac{d\theta}{dt} + (k_1 - \mu)\theta = u_{sw}(\theta) \tag{6}$$

Let $\varsigma' = \varsigma + k_2$, and $\mu' = k_1 - \mu$. The equilibrium points becomes stable if $k_2 > -\varsigma$ and $k_1 > \mu$. The state space trajectory will follow the invariant manifold of a stable equilibrium point but a switching will occurs before reaching this point, then the trajectory will be governed by the invariant manifold of the other stable virtual equilibrium point and so on so forth. This consecutive switching will result in a limit cycle in the phase space. Without loss of generality we can assume that the time origin is when the state space is at M_0 (see Fig. 3). It can be shown that the state space trajectory is as follows;

$$x(t) = \begin{cases} \dfrac{V_0}{\gamma} e^{-\varsigma' t} \sin \gamma t + u_{sw}(x_1) \\ e^{-\varsigma' t} (V_0 \cos \gamma t - \varsigma' \dfrac{V_0}{\gamma} \sin \gamma t) \end{cases}, \tag{7}$$

where $\gamma = \sqrt{\mu' - \varsigma'^2}$ and the constants c_1 and c_2 depend on the initial conditions $x(0)=(-1,V_0)$. The position $x_1(t)$ for $\gamma t \in [0, \pi/2]$ is shown in Fig. 4.

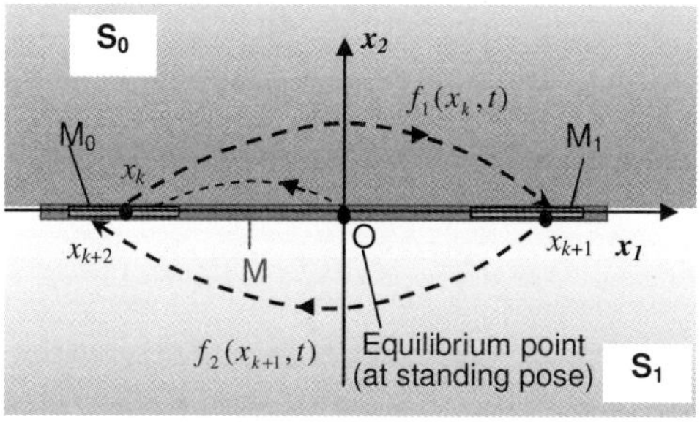

Figure 3 Phase space and Poincare map.

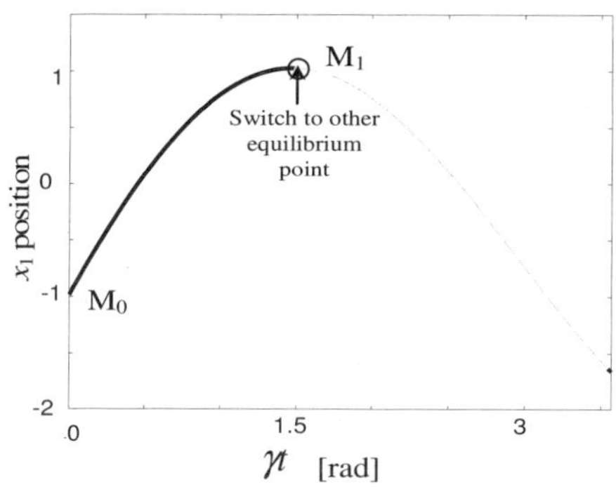

Figure 4 Plot of x_1 position for $V_0 = 2.326, \varsigma' = 0.1$ and $\gamma = 0.995$

The smallest time τ the state trajectory may take when the robot is commutating from one leg to another is such that $\gamma\tau = \pi/2$ in (7) and we will have;

$$\tau = 0.5\pi / \gamma, \tag{8}$$

where $\gamma = \sqrt{\mu' - \varsigma'^2}$, ζ' and μ' are parameters of the inverted pendulum.

4. Walking Pattern Generator

4.1. *Approximate Solution of Rolling Motion*

For the sake of reducing the computing cost and ease the implementation of the controller, instead of the exact solution of the state trajectory (7), we propose an approximate solution $\theta(t)$, using a unit piecewise linear function defined as;

$$u(t_i, \omega_i) = \omega_i (t - t_i)\left[u_s(t - t_i) - u_s(t - t_i - 1/\omega_i)\right] + u_s(t - t_i - 1/\omega_i), \tag{9}$$

where $u_s(.)$ is the unit step function, $\omega_i > 0$ is the slope of the function $u(.)$ between t_i and (t_i+1/ω_i). The approximate solution $\theta(t)$ can be formulated, therefore, as a function of time delay ε, joint angular velocity ω, walking period T, and the rolling amplitude θ_s.

$$\theta = f(\varepsilon, \omega, T, \theta_s), \tag{10}$$

Using (9), the trajectory in Fig.4 for $\gamma t \in [0, \pi/2]$ can be approximated as $\varepsilon\, d\theta(t)/dt + \theta(t) = 2(u(0,\omega_r)-1$, with $\theta(0)$=-1 and $\varepsilon = \zeta$. The commands $\theta^r_{am}(t)$ and $\theta^r_{hm}(t)$ to be sent respectively to the hip and ankle joints are given by,

$$\theta^r_{am}(t) = -\theta^r_{hm}(t) = \theta(t) + \theta^r_{fb}(t), \tag{11}$$

where $\theta^r_{fb}(t)$ is the feedback signal stabilizing (2).

4.2. *Swing Motion*

The dynamic of the swing leg can be considered the same as that of a pendulum, and hence it is inherently stable without any compensator. Due to the switching control approach, this motion is generated in a similar fashion as that of the rolling motion, and is expressed as follows;

$$\varepsilon \frac{d\theta_l(t)}{dt} + \theta_l(t) = A_l[u(t_{l0}, \omega_l) - u(t_{l1} + nT, \omega_l) + u(t_{l2} + nT, \omega_l) - u(t_{lf}, \omega_l)], \tag{12}$$

where θ_l is the lifting motion, A_l is the amplitude of lifting, and t_{l0} and t_{l2} are the switching time for the first and second lifting, t_{l1} and t_{lf} are the switching times for the first and the last landing, and ω_l is the joints' angular velocity generating the lifting motion. On the other hand, we assume that the landing of each leg is accomplished with flat foot on a flat ground, and the thigh and shank of the robot have the same length. Therefore, with respect to the angles definition in [9], this condition can be satisfied as follows.

$$\begin{cases} \theta^p_{am}(t) = \theta_l(t) + \theta^l_{fb}(t) \\ \theta_{km}(t) = -2\theta_l(t) \\ \theta^p_{hm}(t) = \theta_l(t) - \theta^l_{fb}(t) \end{cases}, \tag{13}$$

5. Experiment

At first, we conducted experiment where the virtual damper-spring system is added to the system as in [9]. Hence, Fig. 5 demonstrates how the robot uses gravity for landing and the spring energy for lifting.

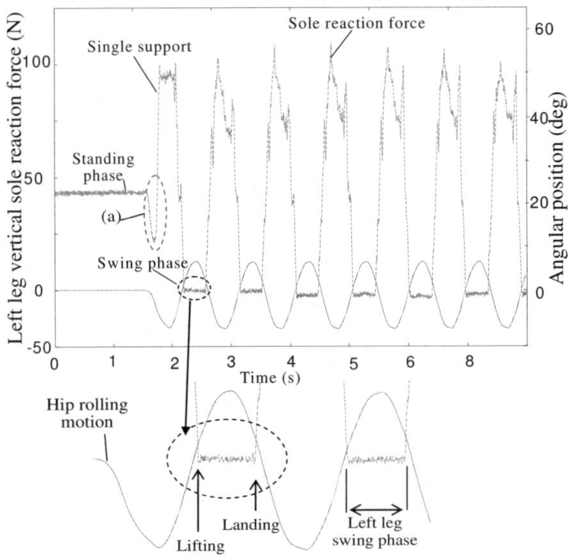

Figure 5 Hip rolling joint output and sole reaction force acting on the left leg; the robot uses spring energy for lifting and gravity for landing.

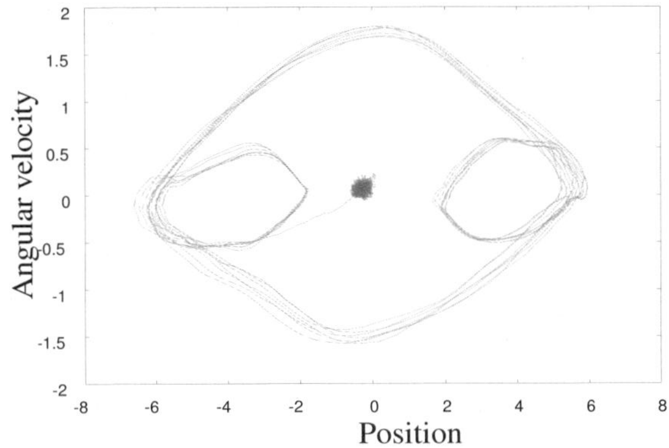

Figure 6 Phase portrait of the ZMP in the lateral plane (Experiment)

Fig. 5 shows that the lifting phase starts when the angular position of the rolling joints is almost zero. In other words, the actuators of the hip and ankle rolling joints do not contribute in moving the ZMP to the supporting leg. These joints, therefore, can be considered as locked joints. It demonstrates also how smooth the approximate solution is, using the proposed pattern generator with sensory feedback. The phase portrait is shown in Fig.6, which demonstrates that the system exhibits stable limit cycle with three periods.

6. Conclusion

In this paper a qualitative analysis of a proposed piecewise linear patter generator of walking motion with sensory feedback for humanoid robots was presented. To improve the robustness against lateral and frontal disturbances a gyro feedback loop was implemented, while terrain irregularities were compensated by adjusting the pitching motion using virtual spring-damper systems. Moreover, we obtained the rolling profile parameters analytically and the approximate solution was implemented giving much modulation flexibility to the motion pattern generator to include reflexes. The locomotion controller becomes so adaptive that the robot is enabled to walk on floor, carpet, and slope. In order to demonstrate the effectiveness of the proposed system, we conducted experiment using the humanoid robot HOAP-3.

Acknowledgments

The experimental work was conducted at Fujitsu Laboratories Limited, Japan.

References

1. M. Vukobratovic and D. Juricic, "Contributuion to the Synthesis of Biped Gait", Bio-Medical Engineering, IEEE Trans. On Biomedical engineering, vol BME-16, No. 1, pp. 1-6, (1969)
2. H. Miura, I. Shimoyama, "Dynamic walk of a biped", *The International Journal of Robotics Research*, vol. 3 no. 2, pp.60-74(1984).
3. S. Kajita, O. Matsumoto, "Real-time 3D walking pattern generation for a biped robot with telescopic legs", *Proceeding of the 2001 IEEE International Conference on Robotics & Automation*, pp.2299-2306 (2001).
4. Q. Huang, K. Yokoi, S. Kajita, K. Kaneko, N. Koyachi, H. Arai, and K. Tanie, "Planning walking patterns for a biped robot," in *IEEE Trans. Robotics and Automation*, vol. 17, no. 3, pp. 280–289, (2001).
5. S. Grillner, "Neurobiological bases of rhythmic motor acts in vertebrates", *Science*, no. 228, pp143-149 (1985).
6. G. Taga, "A model of the neuro-musculo-skeletal system for human locomotion, I. Emergence of basic gait", *Boil. Cybern*, no.73, pp.97-111 (1995)
7. G. Taga, Y. Yamaguchi, and H. Shimizu, "Self organized control of bipedal locomotion by neural oscillators in unpredictable environment", *Biological Cybernetics*, vol. 65, pp.147-159 (1991).
8. R. Zaier, and S. Kanda, "Piecewise-Linear Pattern Genarator and Reflex System for Humanoid Robots", *IEEE Int. Conf. on Robotics and Automation,*(Roma, Italy), p.2188-2195, (2007).
9. HOAP-3, Fujitsu Automation Ltd. Available: http://www.automation.fujitsu.com/group/fja/services/hoap/

A DISTRIBUTED HYBRID CONTROLLER FOR WALKING OVER UNEVEN TERRAIN

Luther Palmer III* and Mayur Palankar

Department of Computer Science & Engineering, University of South Florida, Tampa, Florida 33620, USA
**E-mail: palmer@cse.usf.edu*
www.cse.usf.edu/~palmer

A Hybrid Impulse-Position (HIP) controller is presented for legged locomotion over uneven terrain. The HIP controller decouples the control of each leg such that the multi-leg Jacobian does not need to be computed, is robust to terrain elevations without sensing information about the ground, and does not require vestibular/inertial feedback to remain stable. The algorithm is applied to a simulated hexapod walking on uneven terrain, and an experimental system is being built to further investigate the approach.

Keywords: Legged locomotion; Biologically-inspired robotics; Hybrid Impulse-Position Control

1. INTRODUCTION

Legged robots will be particularly useful for planetary exploration, time-critical search and rescue, military reconnaissance and a wealth of other tasks in which the terrain is complex or rife with uncertainty. To effectively maneuver in these arenas, legged systems need to have mobility comparable to biological systems such as squirrels, cockroaches and even humans. Animal legs appear in many arrangements and morphologies, and enable multiple locomotion modes such as walking, running, jumping, climbing and even swimming. The ability of an animal to utilize multiple mobility modes to circumvent obstacles or gaps in the terrain far outpaces the capabilities inherent to wheels and tracks that can nominally just move over near-horizontal terrain.

Even within a single mobility mode, such as running, legs can be coordinated to transition between gaits, such as trotting and galloping, in order to increase stability or conserve energy. This level of versatility can be achieved because of the high number of degrees of freedom (DoFs) apparent in many legs. Some researchers have achieved remarkable mobility in a single mode by implementing extracted principles of legged locomotion on systems with significantly reduced DoFs.[1–4] The resulting leg-like mechanisms often do not resemble any biological systems and do not have the potential for multi-mode locomotion. Achieving full animal-like mobility requires the pursuit of high DoF biomorphic systems and the versatility of mode and gait that these systems will afford.

A large number of DoFs can be very difficult to coordinate and control. To make this control problem more manageable, gaits have been isolated and studied individually.[5–9] Although legged systems with articulating joints have been developed to perform singular mobility tasks,[10–12] many of the underlying mechanisms and control algorithms for these systems are gait or task specific, and no link to multi-modal or multi-gait mobility is laid out.

This paper presents a Hybrid Impulse-Position (HIP) controller for legged systems that regulates forward velocity and stabilizes body tilt axes without measuring these motions. The HIP controller decouples the control of each leg such that no multi-leg Jacobian is computed. Leg control is completely localized in that only a feedforward CPG signal and intraleg feedback is utilized within a single leg. The CPG signal is static in this implementation, but the algorithm can also be implemented with dynamic CPG algorithms.[13] Also, the HIP strategy stabilizes hexapod locomotion over uneven terrain without sensing any information about the ground. Figure 1 shows the hexapod system in simulation.

Fig. 1. Hexapod in simulation.

By decoupling the leg controllers, each leg can be thought of as a distributed system, each autonomously achieving a certain performance criteria. This performance criteria can be preprogrammed for a particular morphology and gait, and achieved based entirely upon localized feedback. This algorithm has broad-reaching potential for walking, running, climbing and jumping legged systems, as the task becomes learning the desired performance criteria for each maneuver.

2. HYBRID IMPULSE-POSITION CONTROLLER

Biologists comparing legged systems of differing morphology have recorded foot force profiles during the walk, run, jump and climb. Joint positions during these motions have also been archived for a significant number of biological systems. These leg force and position trajectories will serve as targets for the HIP controller.

2.1. *Position Control*

For many years, position control has been used to control walking systems.[14] Once joint angles are chosen for the leg touchdown position and the desired body motion during stance is known, the desired joint angles throughout the stance phase can be computed kinematically. Proportional-derivative (PD) control is used to compute a desired foot force, $f_{position}$ based upon position error of the foot. Position control inherently stabilizes body motions on flat terrain, but success is limited on highly irregular terrain because of inconsistent ground contact. The results will show that the position control element of the HIP controllers helps to maintains stable body movements on uneven terrain as well.

2.2. *Impulse Control*

The top subplot of Fig. 2 shows a representative vertical ground reaction force profile for a single leg during two steps. The force trajectory can also be represented as the impulse, drawn in the bottom subplot. The impulse, or area under the force curve, is used as the reference trajectory for the HIP controller. Step 2 in Fig. 2 shows how the algorithm is designed to respond when ground contact is delayed due to an unexpected drop in the terrain. After contact, the force required to eliminate the impulse error is computed. This force, $f_{impulse}$, is large and cannot be achieved instantaneously so there is some delay in the time required to match the desired impulse. This commanded force must ultimately grow to be larger than the initial desired force trajectory.

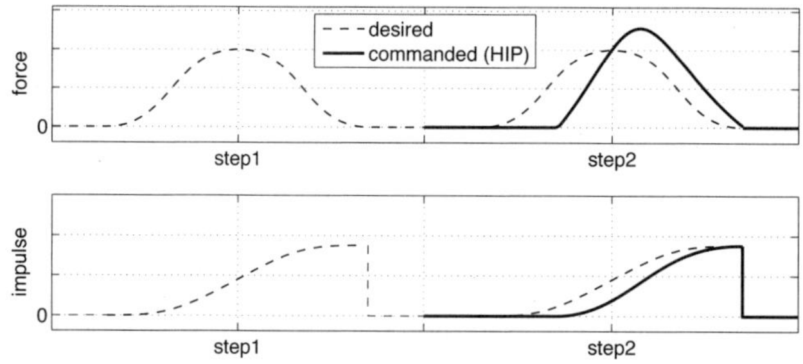

Fig. 2. Nominal vertical force and impulse for a single leg during the stance phase of a step. Step 2 shows the desired response of a controller which attempts to eliminate impulse error after a late foot touchdown. The top subplot shows that larger forces are required to achieve the desired impulse when the foot contacts the ground late.

By seeking to eliminate impulse error, the HIP controller requires legs that contact the ground late to still generate the desired impulse. Vertical impulse is responsible for lifting the body, and producing less than the desired impulse will result in inadequate body height for walking over rough terrain.

2.3. *Hybrid Integration*

Once the position-based and impulse-based forces have been computed, $f_{position}$ and $f_{impulse}$ respectively, joint outputs, τ_{out}, are computed by:

$$f_{out} = \alpha_1 f_{impulse} + \alpha_2 f_{position}, \text{ and then}$$

$$\tau_{out} = J^T f_{out}.$$

where J^T is the transpose of the single leg Jacobian and α_1, α_2 are the hybrid weights.

The HIP controller is similar in architecture to the hybrid force-position (HFP) controller presented many years ago.[15] This strategy was initially described on manipulator arms, but has also been adapted for walking. The implementation of the HFP controller, however, is very different compared to HIP controller as it either uses the position control error or the force control error to calculate the joint torques. By contrast, the HIP algorithm produces a cumulative result which capitalizes on the advantages of both position and force control.

3. SIMULATION AND RESULTS

Simulation of the HIP controller was done using the RobotBuilder[16] simulation environment, which is built upon the DynaMechs software package. System losses were modeled as damping in the compliant ground. Ground spring and damping coefficients was taken as $75kN/m$ and $2kN/m/s$ respectively. Ground static and kinetic friction coefficients were 0.75 and 0.6 respectively, matching the properties of rubber on concrete. The hexapod executes an alternating tripod gait during the below tests.

Each leg of the hexapod has three degrees of freedom, as shown in Fig. 3. The hexapod weighs a total of 3.05 kg, and the body dimensions of the simulated system are given in Table 1. Each leg link is modeled as a 0.1 kg slim rod with geometrically-centered mass. Leg mass combined to make up 59% of the complete system mass, each leg making up approximately 9.8%.

Product	Centimeters
Body Width	7.8740
Body Length	35.5346
Body Width	2.5400
Leg Length 1 (l_1)	3.7211
Leg Length 2 (l_2)	4.5009
Leg Length 3 (l_3)	2.2504
Leg Length 4 (l_4)	6.9850

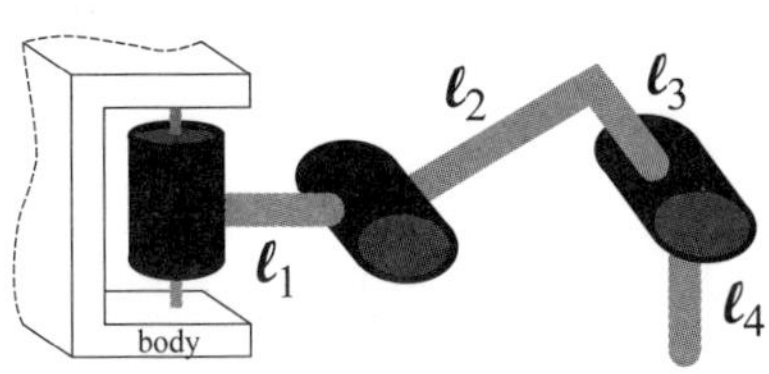

Fig. 3. Each leg has three actuators and is described by four leg lengths.

Figure 4 contains two subplots which show the forward velocity and height of the hexapod walking over uneven terrain. As a reminder, there is no feedback to the control system of either forward velocity or body height. As in animal locomotion,[17] the forward velocity undergoes oscillations due to the intermittent nature of the ground contact and the prescribed leg positions. Selecting the optimal impulse and position trajectories is an area that will require a significant amount of future work. For now though, the average velocity over this uneven terrain maintains at approximately 0.5 body lengths per second. The body also undergoes expected vertical oscillations during each step, and the changing height of the terrain is evident as the body's average height trends upward and then downward during the trial.

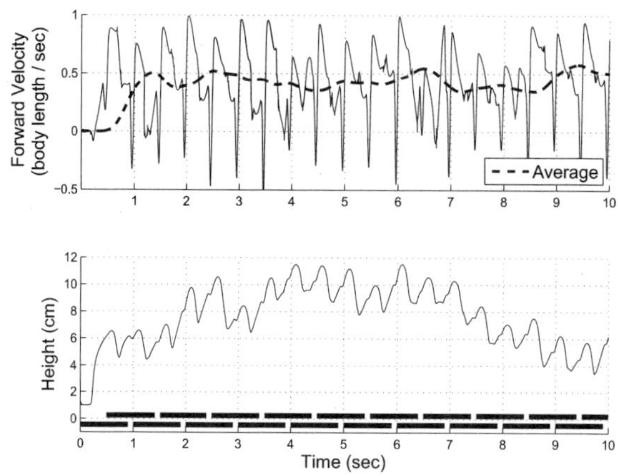

Fig. 4. Body forward velocity and height during hexapod walking on uneven terrain. Although there is no feedback of velocity and height, the HIP controller is able to stabilize these motions. The oscillations are expected[17] and result from the biologically-inspired position and impulse profiles.

Figure 5 shows the comparison between the HIP and the feedforward position controller as the hexapod walks over uneven terrain for 20 tripod steps. The displacement is shown in body lengths (Table 1). A 3D randomly generated uneven terrain was used for this simulation. The terrain consists of square tiles of random elevation with a normal distribution centered at $0cm$ with a deviation of $\pm 5cm$. At multiple times, as the feedforward position controller tries to achieve the desired leg position with respect to the body, the foot misses or lightly engages the terrain. As a result, the hexapod moves comparatively slowly and is prone to becoming stuck. In contrast, the HIP controller generates forces at the foot that seek the ground, create a sufficient vertical impulse, and result in forward motion in each step.

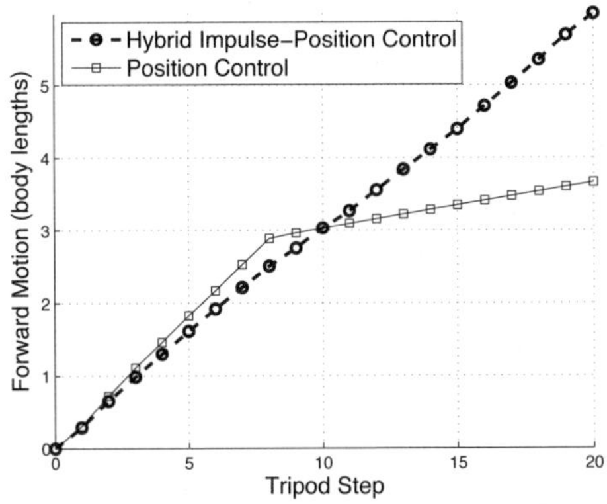

Fig. 5. Comparison between the HIP controller and the position controller on uneven terrain. Position control of the legs results in terrain misses that result in sluggish motion on uneven terrain. HIP control of the hexapod legs forces the foot to engage the ground and propel the body.

4. DISCUSSION AND FUTURE WORK

A Hybrid Impulse-Position (HIP) controller for legged locomotion over uneven terrain has been presented. The HIP controller for legged systems stabilizes the body tilt axes without measuring these motions or relying on the multi-leg Jacobian. Results of the simulation of a hexapod walking on uneven terrain using the HIP controller have been presented. Comparisons to the position controller have revealed advantages which could be very useful in uneven terrain.

This HIP controller has broad-reaching potential for walking, running, climbing and jumping legged systems and could be adapted to biped or quadruped locomotion. In the future, testing will be done on different gaits and smooth transition between them. Future work will also include turning or steering of the body based on biological data regarding leg forces and kinematics. Adaptive CPGs would also be tested to further increase robustness. An experimental hexapod is being built to test the real-world limitations of the this algorithm.

References

1. M. Kovac, M. Fuchs, A. Guignard, J.-C. Zufferey and D. Floreano, A miniature 7g jumping robot, in *Proc. of the IEEE International Conference on Robotics and Automation*, may 2008.
2. T. Allen, R. Quinn, R. Bachmann and R. Ritzmann, Abstracted biological principles applied with reduced actuation improve mobility of legged vehicles,

in *Proc. of the IEEE/RSJ International Conference on Intelligent Robots and Systems*, october 2003.

3. N. Neville, M. Buehler and I. Sharf, A bipedal running robot with one actuator per leg, in *Proc. of the IEEE International Conference on Robotics and Automation*, may 2006.
4. U. Saranli, M. Buehler and D. Koditschek, Design, modeling and preliminary control of a compliant hexapod robot, in *Proc. of the IEEE International Conference on Robotics and Automation*, 2000.
5. M. Berkemeier, Approximate return maps for quadrupedal running, in *Proc. of the IEEE International Conference on Robotics and Automation*, april 1997.
6. R. Kurazume, S. Hirose and K. Yoneda, Feedforward and feedback dynamic trot gait control for a quadruped walking vehicle, in *Proc. of the IEEE International Conference on Robotics and Automation*, 2001.
7. K. N. Murphy and M. H. Raibert, Trotting and bounding in a planar two-legged model, in *In Theory and Practice of Robots and Manipulators, Proc. of RoManSy'84*, eds. A. Morecki, G. Bianchi and K. Kedzior (MIT Press, Cambridge, 1985).
8. H. M. Herr and T. A. McMahon, *International Journal of Robotics Research* **19**, 566(June 2000).
9. H. M. Herr and T. A. McMahon, *International Journal of Robotics Research* **20**, 26(January 2001).
10. K. Autumn, M. Buehler, M. Cutkosky, R. Fearing, R. J. Full, D. Goldman, R. Groff, W. Provancher, A. A. Rizzi, U. Saranli, A. Saunders and D. E. Koditschek, Robots in scansorial environments, in *Proc. of SPIE*, eds. G. Gerhart, C. Shoemaker and D. Gage (Bellingham, WA, 2005).
11. R. Niiyama, A. Nagakubo and Y. Kuniyoshi, Mowgli: A bipedal jumping and landing robot with an artificial musculoskeletal system, in *Proc. of the IEEE International Conference on Robotics and Automation*, april 2007.
12. M. Buehler, R. Playter and M. Raibert, Robots step outside, in *Proc. of the International Symposium on Adaptive Motion in Animals and Machines (AMAM)*, (Ilmenau, Germany, 2005).
13. Y. Fukuoka, H. Kimura and A. Cohen, *International Journal of Robotics Research* **22**, 187 (2003).
14. S. M. Song and K. J. Waldron, *Machines that Walk* (MIT Press, Cambridge, MA, 1989).
15. J. J. Craig and M. H. Raibert, A systematic method of hybrid position/force control of a manipulator, in *Proceedings of the IEEE Computer Society in Computer Software and Applications Conference*, November 1979.
16. S. J. Rodenbaugh, RobotBuilder: a graphical software tool for the rapid development of robotic dynamic simulations, Master's thesis, The Ohio State University (2003).
17. R. T. Schroer, M. J. Boggess, R. J. Bachmann, R. D. Quinn and R. E. Ritzmann, Comparing cockroach and whegs robot body motions, in *Proc. of the IEEE International Conference on Robotics and Automation*, April 2004.

ON THE STEP HANDLING OF VIRTUAL SLOPE WALKING

MINGGUO ZHAO[†]

State Key Laboratory of Robotics and System (HIT)
Department of Automation, Tsinghua University, Beijing 100084, China
mgzhao@mail.tsinghua.edu.cn

Virtual Slope Walking is a new realization of powered walking based on Passive Dynamic Walking. It is effective in generating fast walking and has a good performance in the disturbance rejection. The step-handling walking with maximum step height of 10% leg length is realized on a planar bipedal robot Stepper-2D using this method. This paper theoretically studies the disturbance rejection of Virtual Slope Walking by introducing the ground step perturbation. It is theoretically proved that the step perturbation can be transformed to the disturbance of initial state and successful step handling walking comes from the system's cyclic stability.

1. Introduction

McGeer[1] demonstrated that a Passive Dynamic Walker can walk down a shadow slope with no control and actuation in the early of 1990. Then the concept of Passive Dynamic Walking has been used as a starting point for designing powered walkers on the level ground. Collins, et al[2][3][4] demonstrated several realizations of powered walking based on kinematic energy complement. Asano et al [5][6] introduced the parametric excitation for potential energy restoration. In our previous work [7][8], we proposed Virtual Slope Walking by introducing the leg length modulation and achieved a relative speed of 4.48leg/s on a planar bipedal robot Stepper-2D.

Disturbance rejection, defined as the ability to deal with unexpected disturbances [9], is considered as one of the fundamental performances for bipedal walker. There exists several ways to measure the disturbance rejection for a PDW based walker theoretically, such as Floquet multipliers, Basin of Attraction and the Gait Sensitivity Norm. But the most commonly used experimental measure is the ground step perturbation that a walker can handle without falling. Wisse[10] realized powered walking under a step height of 2% leg length, Pratt [11] realized 9% and Geng[12] achieved 4% leg length. We

[†] Work partially supported by State Key Laboratory of Robotics and System (HIT) (No.SKLRS2010-ZD-01) and the National Nature Science Foundation of China (No.60875065).

[13][14] have realized a powered walking under a step height of 10% leg length on Stepper-2D based on Virtual Slope Walking.

In this paper, we theoretically study the disturbance rejection of Virtual Slope Walking and present its stabilizing mechanism. Based on the asymptotic expression of stride function and the fixed point, we theoretically proved that ground step perturbation can be transformed to the disturbance of initial system state and the successful step handling walking comes from the fixed point's stability. The necessary and sufficient condition of recovering is presented based on the analysis of the minimum initial state in the transition phase.

The remainder of this paper is organized as follows. In Section II, the model of Virtual Slope Walking is presented. In Section III, the ground step disturbance handling is illustrated, and the performance of disturbance rejection is analyzed in Section IV. Section V presents the experimental results and Section VI is the conclusion and future work.

2. Model of Virtual Slope Walking

2.1. *Model Description*

The model of Virtual Slope Walking is shown in Fig.1. It has two telescopic massless legs and a point mass body at the hip. The stance leg is actuated for extending from r_s to r_e following a smooth leg length trajectory r(t). The swing leg is actuated for shortening from r_e to r_s in one step. The length shorten ratio is then defined as $\beta=r_s/r_e$. Since the swing leg is assumed massless, it can be swung arbitrarily quickly to the position with constant inter-leg angle φ_0 before heel strike. The impact of the swing leg with the ground is assumed to be fully inelastic (no slip, no bounce) and instantaneous.

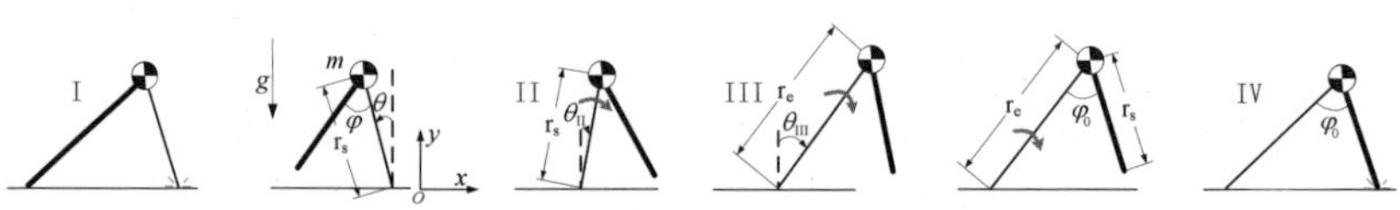

Fig. 1 Model of Virtual Slope Walking

We define a walking step starts at the new stance leg (lighter line) contact with the ground (instant I). The stance leg swings to the position at $\theta=\theta_{II}$ (instant II) for the beginning of extension. And the stance leg extension ends at $\theta=\theta_{III}$ (instant III). The swing leg (heavier line) is shortened and swings to the position with constant inter-leg angle φ_0 just before heel strike, and hits the ground at instant IV. Then, the stance leg and swing leg exchange subsequently, and the walking cycle repeats continuously.

2.2. *Governing Equations*

The governing equations of the system consist of nonlinear differential equations for the swing phase and algebraic equations for the transitions of heel strike.

(1) Swing phase from **I** to **II**: Using *Lagrangian Equations*, the equation of motion is given below with the constant length r_s and the dimensionless time variable $\tau = \sqrt{g/r_e}t$

$$\ddot{\theta}(\tau) = \frac{1}{\beta}\sin\theta(\tau) \tag{1}$$

(2) Stance leg extension From **II** to **III**: The stance leg acts as an inverted pendulum with variable length r(τ).

$$\begin{cases} \ddot{\theta}(\tau) = \dfrac{r_e}{r(\tau)}\sin\theta(\tau) - \dfrac{2}{r(\tau)}\dot{r}(\tau)\dot{\theta}(\tau) \\ \ddot{r}(\tau) = r(\tau)\dot{\theta}^2(\tau) - r_e\cos\theta(\tau) + r_e\dfrac{F}{mg} \end{cases} \tag{2}$$

Where *F* is the stance leg extension force on the center of mass.

(3) Swing phase from **III** to **IV**: Similar to the Swing phase from **I** to **II**, the equation of motion for the stance leg with the constant length r_e can be written as

$$\ddot{\theta}(\tau) = \sin\theta(\tau) \tag{3}$$

(4) Heel strike transition from **IV** to **I** of the subsequent step: The heel strike from step *n* to the subsequent step *n*+1 occurs when the geometric condition is met

$$\begin{cases} \theta_{\mathrm{I}}(n+1) = -(\varphi_0 - \theta_{\mathrm{IV}}(n)) \\ \beta\cos\theta_{\mathrm{I}}(n+1) = \cos\theta_{\mathrm{IV}}(n) \end{cases} \tag{4}$$

Where φ_0 is a constant. Eq. (4) also reflects a change of names for the two legs.

From the conservation of angular momentum about the swing foot contact point at heel strike, we obtain the following transition equation

$$\omega_{\mathrm{I}}(n+1) = \frac{\cos\varphi_0}{\beta}\omega_{\mathrm{IV}}(n) \tag{5}$$

Eq. (1)-(5) construct the dynamic equations of this hybrid system.

2.3. *Stride Function and Fixed Point*

Our Poincaré section is at the instant **I** in Fig. 1. Given the state of the system at instant **I,** the Poincaré map **f** determines the state just after the next heel strike. Note that in the geometric collision condition Eq. (4), the stance leg angle θ_I is constant with inter-leg angle φ_0

$$\theta_{\mathrm{I}} = -\arctan\frac{\beta-\cos\varphi_0}{\sin\varphi_0} \tag{6}$$

So the heel strike transition reduces this problem from 2D state space $\{\theta_I, \omega_I\}$ to a 1D map **f**, only consisting of angular velocity ω_I. While the system has only one independent initial condition, we need to specify ω_I at the start of walking step n to fully determine the subsequent motion at steps n+1, n+2, ….

We have proved that under the **Equivalent Definition**, the Trajectory Leg Extension (TLE) can be equivalently transformed to the Instantaneous Leg Extension (ILE) [15]. By defining a new variable $\mathbf{q}=\boldsymbol{\omega}^2_{\mathbf{I}}$ as the system state, the stride function **f** can be analytically written as

$$\mathbf{f}(q) = \beta^2\cos^2\varphi_0 q + 2\cos^2\varphi_0[\cos\theta^*_{\mathrm{II}}(\frac{1}{\beta^2}-\beta) - \cos\theta_{\mathrm{I}}(\frac{1}{\beta}-\beta)] \tag{7}$$

Where θ^*_{II} is the equivalent extension angle. The fixed point of the stride function is defined as $\mathbf{f}(q^f) = q^f$. From Eq. (7), the fixed point can then be obtained as follows

$$q^f = \frac{2\cos^2\varphi_0[\cos\theta^*_{\mathrm{II}}(1-\beta^3) - \cos\theta_{\mathrm{I}}(\beta-\beta^3)]}{\beta^2(1-\cos^2\varphi_0\beta^2)} \tag{8}$$

3. Ground Step Disturbance Handling of Virtual Slope Walking

3.1. *Transition Walking*

After entering into the periodic state in Virtual Slope Walking, once the system is perturbed by a single step, there will be a transition phase in the subsequent steps. The system state will approach the fixed point asymptotically the same as the condition of initial state's disturbance. We will illustrate such transition walking by introducing the single step-up and step-up-down perturbations in the following section.

3.2. *Single Step-Up Perturbation*

We assume that the system is in the periodic state in step *n*-2, and a single step-up perturbation occurs at the end of step *n*-1. The subsequent transition walking step n is shown in Fig. 2. There exist perturbations not only on the system state q, but also on the stance leg angle θ_I and θ_{IV}. After the transition phase, only perturbations on the system state q exists.

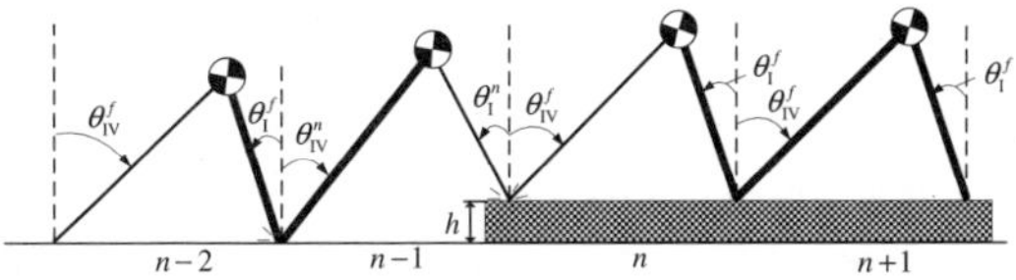

Fig. 2 Transition walking of the Single Step-Up disturbance

Let h be the step disturbance height, and $h_r=h/r_e$ be the relative value. Then, the perturbation on the stance leg angle θ_I and θ_{IV} resulted from the step height disturbance in the transition walking step n can be obtained as follows

$$\begin{cases} \theta_{IV}^n + \theta_I^n = \varphi_0 \\ \cos\theta_{IV}^n = \beta\cos\theta_I^n + h_r \end{cases} \tag{9}$$

The system state at the start of step n can be considered as the output of the stride function with perturbed θ_{IV} of step n-1. Let θ^f_I and θ^f_{IV} be the constant stance leg angle which is corresponded to the fixed point, then $q(n)$ can be represented as

$$q(n) = \beta^2\cos^2\varphi_0 q^f + 2\cos^2\varphi_0[\cos\theta_{II}^*(\frac{1}{\beta^2}-\beta) - (\frac{1}{\beta^2}\cos\theta_{IV}^n - \beta\cos\theta_I^f)] \tag{10}$$

There exists the perturbation on θ_I in transition walking step n (Fig. 2), so q(n+1) can be considered as the output of the stride function with perturbed θ_I of step n

$$q(n+1) = \beta^2\cos^2\varphi_0 q(n) + 2\cos^2\varphi_0[\cos\theta_{II}^*(\frac{1}{\beta^2}-\beta) - (\frac{1}{\beta^2}\cos\theta_{IV}^f - \beta\cos\theta_I^n)] \tag{11}$$

θ_I and θ_{IV} returns to the constant value from step n+1, then q(n+2) can be represented as

$$q(n+2) = \beta^2\cos^2\varphi_0 q(n+1) + 2\cos^2\varphi_0[\cos\theta_{II}^*(\frac{1}{\beta^2}-\beta) - \cos\theta_I^f(\frac{1}{\beta}-\beta)] \tag{12}$$

The transition phase ends after step n, and the system state starts approaching the fixed point asymptotically with the initial state of $q(n+1)$. It can be concluded from Eq. (12) that after the transition phase ends, the step perturbation can be totally transformed to the disturbance of initial system state.

3.3. *Single Step-Up-Down Perturbation*

We assume that the system is in the periodic state in step *n*-2, and a single step-up-down perturbation occurs at the end of step *n*-1. The subsequent transition walking steps *n* and *n*+1 are shown in Fig. 3. The same as the step-up perturbation, in the transition phase, there exists the perturbations on the system state *q*, the stance leg angle θ_I and θ_{IV}, which is constant in normal walking, resulting in the variation of the stride function. After the transition phase ends, only disturbance on the system state *q* exists.

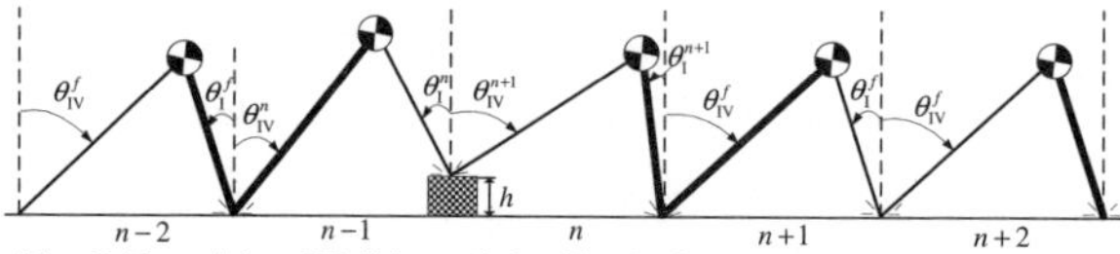

Fig. 3 Transition Walking of the Single Step-Up-Down Perturbation

The perturbation on the stance leg angle θ_I and θ_{IV} resulted from the step height disturbance in the transition walking step *n* and *n*+1 can be obtained as follows

$$\begin{cases} \theta_{IV}^n + \theta_I^n = \varphi_0 \\ \cos\theta_{IV}^n = \beta\cos\theta_I^n + h_r \end{cases}, \begin{cases} \theta_{IV}^{n+1} + \theta_I^{n+1} = \varphi_0 \\ \cos\theta_{IV}^{n+1} = \beta\cos\theta_I^{n+1} - h_r \end{cases} \tag{13}$$

The system state at the start of step *n* can be considered as the output of the stride function with perturbed θ_{IV} of step *n*-1, so as the subsequent steps.

$$q(n) = \beta^2\cos^2\varphi_0 q^f + 2\cos^2\varphi_0[\cos\theta_{II}^*(\frac{1}{\beta^2} - \beta) - (\frac{1}{\beta^2}\cos\theta_{IV}^n - \beta\cos\theta_I^f)] \tag{14}$$

$$q(n+1) = \beta^2\cos^2\varphi_0 q(n) + 2\cos^2\varphi_0[\cos\theta_{II}^*(\frac{1}{\beta^2} - \beta) - (\frac{1}{\beta^2}\cos\theta_{IV}^{n+1} - \beta\cos\theta_I^n)] \tag{15}$$

$$q(n+2) = \beta^2\cos^2\varphi_0 q(n+1) + 2\cos^2\varphi_0[\cos\theta_{II}^*(\frac{1}{\beta^2} - \beta) - (\frac{1}{\beta^2}\cos\theta_{IV}^f - \beta\cos\theta_I^{n+1})] \tag{16}$$

θ_I and θ_{IV} returns to the constant value from step *n*+2, then q(*n*+3) can be represented as

$$q(n+3) = \beta^2\cos^2\varphi_0 q(n+2) + 2\cos^2\varphi_0[\cos\theta_{II}^*(\frac{1}{\beta^2} - \beta) - \cos\theta_I^f(\frac{1}{\beta} - \beta)] \tag{17}$$

The transition phase ends after step *n*+1, and the system state starts approaching the fixed point asymptotically with the initial state of q(*n*+2). It can be concluded from Eq.(17) that after the transition phase ends, the step perturbation can be totally transformed to the disturbance of initial system state.

4. Analysis of Disturbance Rejection of Virtual Slope Walking

4.1. *Maximum Relative Step Height*

There exists the maximum relative step height h_r^{max} when $q(n)=q^z(n)$ holds as h_r increases, which describes the performance of disturbance rejection of Virtual Slope Walking. Therefore, h_r^{max} can be obtained as follows

$$h_r^{max}=\beta\tan^2\varphi_0\cos\theta_I^n+\begin{cases}\dfrac{1}{2}\beta^4 q^f+\cos\theta_{II}^*(1-\beta^3)+\beta^3\cos\theta_I^f-\dfrac{\beta}{\cos^2\varphi_0},\theta_{II}^*>0\\[2ex]\dfrac{1}{2}\beta^4 q^f+\dfrac{1-(1-\beta^3)(1-\beta^2\cos^2\varphi_0)\cos\theta_{II}^*}{\beta^2\cos^2\varphi_0}+\beta^3\cos\theta_I^f,\theta_{II}^*\le 0\end{cases}\tag{18}$$

where h_r^{max} and θ_I^n satisfy

$$\begin{cases}\theta_{IV}^n+\theta_I^n=\varphi_0\\ \cos\theta_{IV}^n=\beta\cos\theta_I^n+h_r^{max}\end{cases}\tag{19}$$

It can be concluded from Eq. (18) & (19) that the maximum relative step height h_r^{max} is determined by the model parameters. We will illustrate the influence of model parameters on the disturbance rejection in the following section.

4.2. *Influence of Model Parameters*

4.2.1. *Effect of the Length Shorten Ratio β*

The maximum relative step height h_r^{max} is shown as a function of β in Fig.4 with four values of φ_0. It is indicated from Fig.4 that h_r^{max} decreases with an increase in β. The main conclusion from this graph is that **a decrease in β leads to a greater h_r^{max} and a larger disturbance rejection in Virtual Slope Walking**.

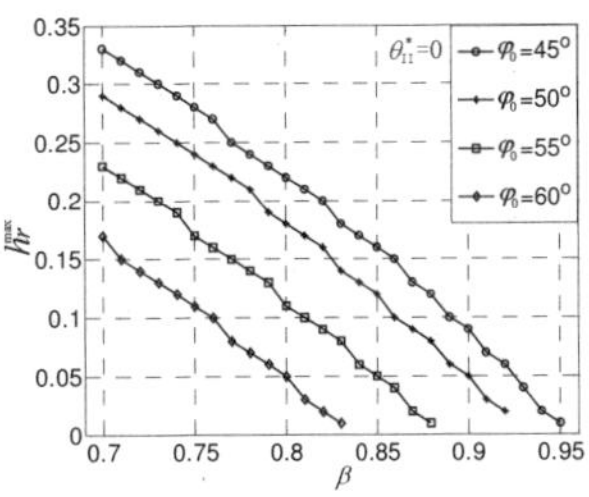

Fig. 4 Trajectory of maximum relative step height h_r^{max} versus length shorten ratio β

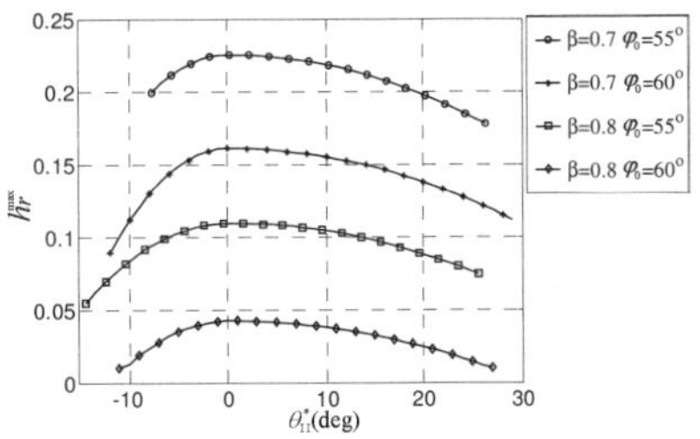

Fig. 5 Trajectory of maximum relative step height h_r^{max} versus equivalent extension angle θ^*_{II}

4.2.2. *Effect of the Equivalent Extension Angle θ^*_{II}*

The maximum relative step height h_r^{max} is shown as a function of θ^*_{II} in Fig. 5 with two values of β and φ_0 respectively. Fig. 5 shows a second order relationship between h_r^{max} and θ^*_{II}. As θ^*_{II} approaching zero from both side, h_r^{max} increases and reaches a maximum value at $\theta^*_{II} = 0^o$. It can be concluded from this graph that **extending the stance leg more close to mid-stance will result in a greater h_r^{max} and a larger disturbance rejection in Virtual Slope Walking**.

4.2.3. *Effect of the Inter-leg Angle φ0*

The maximum relative step height h_r^{max} is shown as a function of φ0 in Fig. 6 with four values of β. As shown in Fig. 6, h_r^{max} decreases with an increase in φ_0. The dissipation energy E_r increases as φ_0 increases. So we can conclude from this graph that **a smaller φ_0 results in greater h_r^{max} and a larger disturbance rejection in Virtual Slope Walking**.

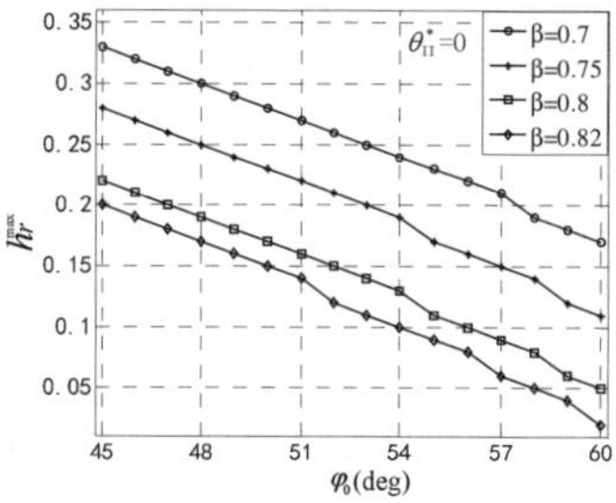

Fig. 6 Trajectory of maximal relative step height h_r^{max} versus inter-leg angle φ_0

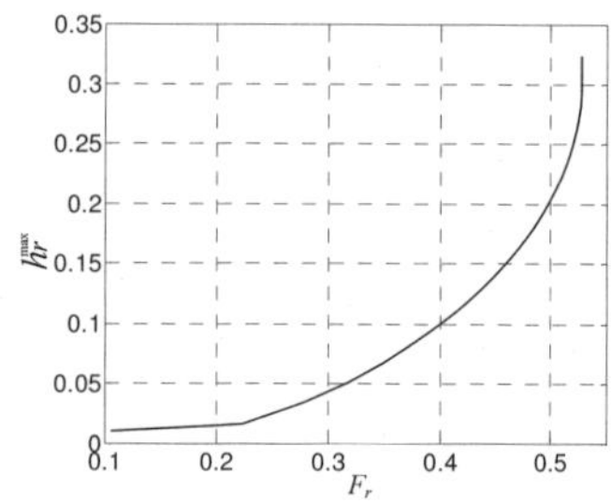

Fig. 7 Trajectory of maximal relative step height h_r^{max} versus walking speed F_r

4.2.4. *Adjoint Relationship of Walking Speed and Disturbance Rejection*

The walking speed described by the Froude Number F_r is also determined by the model parameters β, θ^*_{II}, and φ_0 [16]. Therefore, there exists an adjoint relationship between the walking speed and disturbance rejection. The maximum relative step height h_r^{max} is shown as a function of F_r in Fig. 7. It is indicated that **h_r^{max} increases with an increase in Fr**. Such conclusion suggests that achieving fast walking speed always accompanying with large disturbance rejection.

5. Experiment

5.1. *Planar Bipedal Robot Stepper-2D*

Planar bipedal robot Stepper-2D is used as a test bed of disturbance rejection as shown in Fig.8. It is mounted on a boom to constrain the motion in the sagittal plane. Stepper-2D's leg length is 250mm and weights 390g.

Fig. 8 Planar Bipedal Robot Stepper-2D with Point Foot.

The leg with the point foot is actuated in the hip and knee joint by digital servo motors. The telescopic leg motion is realized by bending and unbending the knee joint. And the swing leg motion is achieved by hip motor actuation [15].

5.2. *Experimental Results*

Stepper-2D successfully recovers from a maximum single step perturbation of 25mm (h_r^{max}=10% leg length). Fig.9&10 presents the image sequences of the walking experiment under a single step-up and step-up-down perturbation of Stepper-2D respectively. Videos about the walking experiments could be found on the website http://v.youku.com/v_show/id_XMjA3ODM5OTcy.html. The comparison of relative step height of with other walkers is shown in table 1.

Fig. 9 Image sequence of a single step-up perturbation experiment (h_r^{max}=25mm)

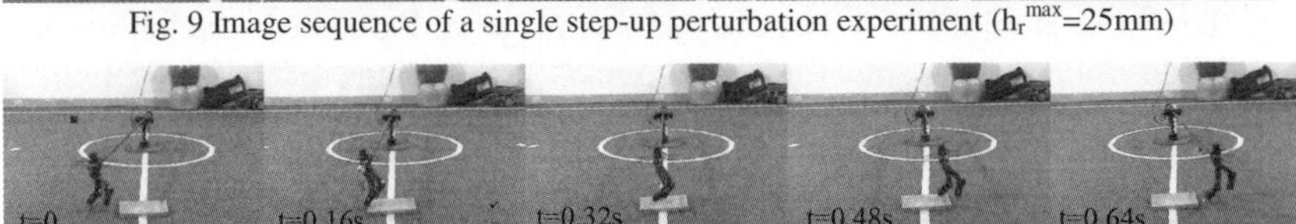

Fig. 10 Image sequence of a single step-up-down perturbation experiment (h_r^{max}=25mm)

Table 1 Comparison of maximum relative step height

Type of dynamic walker	Maximum Relative Step Height
Stepper-2D	10%
Flaminge [17]	9%
Runbot [18]	4%
Mike [15]	2%

6. Conclusion

In this paper, we analytically study the disturbance rejection of Virtual Slope Walking by introducing the ground step perturbation. We theoretically prove that ground step perturbation can be transformed to the disturbance of initial system state and the successful step handling walking comes from the system's cyclic stability. We illustrate the effect of leg length shorten ratio β, equivalent extension angle θ^*_{II} and inter-leg angle φ_0 on the maximum relative step height h_r in Virtual Slope Walking. The step handling walking experiment of Stepper-2D verifies the theoretical analysis results.

Reference

1. McGeer, T., 1990a, Passive dynamic walking, International Journal of Robotics Research, 9(2), pp.62–82.
2. S. Collins, A. Ruina, R. Tedrake, et al. Efficient Bipedal Passive-Dynamic Walkers. Science, Vol.307, No. 5712, pp.1082-1085, 2005.
3. M. Wisse. Three additions to passive dynamic walking; actuation, an upper body, and 3d stability, Proc. of Int. Conf. on Humanoid Robots, pp. 1113-132, 2004.
4. D.G.E. Hobbelen, M. Wisse. Ankle Actuation for Limit Cycle Walkers. International Journal of Robotics Research, 27(6):709-735, 2008.
5. F. Asano, Zhi-Wei Luo. Energy-Efficient and High-Speed Dynamic Biped Locomotion Based on Principle of Parametric Excitation, IEEE Trans. Robot., vol. 24, no. 6, pp. 1289–1301, 2008.
6. Toyoyuki Honjo, Zhi-Wei Luo, Akinori Nagano. Parametric excitation of a biped robot as an inverted pendulum," Proceedings of IEEE/RSJ 2008 International Conference on Intelligent Robots and Systems, pp. 3408-3413, 2008.
7. Mingguo Zhao, Ji Zhang, Hao Dong, Yu Liu, Liguo Li, and Xuemin Su. Humanoid Robot Gait Generation Based on Limit Cycle Stability, Proc. of the RoboCup Symposium, pp. 403-413, 2008.
8. Hao Dong, Mingguo Zhao, Ji Zhang, Naiyao Zhang. Hardware design and gait generation of humanoid soccer robot Stepper-3D. Robotics and Autonomous Systems, 57:828-838, 2009.

9. D. G. E. Hobbelen, M. Wisse. A Disturbance Rejection Measure for Limit Cycle Walkers: The Gait Sensitivity Norm. IEEE Transactions on Robotics, Vol. 23, No. 6, pp.1213-1224, 2007.
10. M. Wisse. J. van Frankenhuyzen. Design and construction of mike; a 2d autonomous biped based on passive dynamic walking. Adaptive motion of Animals and machines, pp.143-154, 2006.
11. Pratt J. Exploiting Inherent Robustness and Natural Dynamics in the Control of Bipedal Walking Robots. PhD Thesis, Massachusetts Institute of Technology, 2000.
12. Geng T, Porr B, Wörgötter F. Fast biped walking with a sensor driven neuronal controller and real time online learning. International Journal of Robotics Research, Vol. 25, No. 3, pp. 243 -259, 2006.
13. Hao Dong, Mingguo Zhao, Naiyao Zhang. High Speed and Energy Efficient Biped Locomotion Based on Virtual Slope Walking. Autonomous Robots, Vol.30, No. 2, pp.199-216, 2011.
14. Mingguo Zhao, Hao Dong, and Naiyao Zhang. The Instantaneous Leg Extension Model of Virtual Slope Walking, Proc. of the IEEE/RSJ International Conference on Intelligent Robots and Systems, pp.3220–3225, 2009.

DYNAMIC OBSTACLE-CROSSING OF A WHEELED ROVER WITH DOUBLE-WISHBONE SUSPENSION

J.C. FAUROUX, B.C. BOUZGARROU

Clermont University, French Institute for Advanced Mechanics (IFMA), EA3867, FR TIMS / CNRS 2856, Mechanical Engineering Research Group (LaMI), BP 10448, F-63000

In order to evaluate the feasibility of fast terrestrial inspection, this work presents an experimental study of the longitudinal stability of a rapid wheeled rover crossing an obstacle at 10m/s. The considered rover uses wheels suspended on double wishbones. Results show that a stability front can be traced, after which tip-over occurs systematically. Impact force measurements and vehicle analytic modeling complete this work and trace future improvement on suspension architecture including additional mobilities.

1. Introduction

Many mobile all-terrain robots are used for safety and inspection applications but most of them are slow (less than 3m/s, Fig. 1). The low speed allow them to use special modes of locomotion such as obstacle climbing modes for military applications (I-Robot Packbot [1]) or industrial inspection (M-Tecks Arthron [2]). Low speed is also suitable for gaming / home use (Meccano Spykee [3]).

Fig. 1: a) I-Robot Packbot [1]. b) M-Tecks Arthron [2]. c) Meccano Spykee [3].

It appears that many outside applications could benefit from high speed. Inspecting a vast area such as an airport would require less terrestrial robots it they are fast enough. Fast inspection is also more dissuasive. Even if unmanned aerial vehicles and flying drones become popular, they may be dangerous for the human environment (risk of fall) and have limited autonomy because of mass constraints. In this case, terrestrial drones keep their full interest. Fast terrestrial drones could have many other applications in agriculture (weeding, seeding), transport of small parts and casualty detection during natural catastrophes.

One major difficulty for fast terrestrial robot in natural environment is to manage obstacle-crossing at high speed. Relatively few works describe what happens when a mobile robot crashes against an obstacle. Some robots can be thrown above obstacles by an operator: they may be equipped by a suspension, such as the 3-spring star-configured suspension of [4] or may be very light with no suspension, such as the DragonRunner [5]. The dynamic crossing of an obstacle is studied in [6] but restricted to low obstacles (e.g. pavement edge).

Fig. 2: a) Throwable robot with 3-spring star-configured suspension [4].
b) Ultra-rugged DragonRunner robot without suspension [5].

This work is part of the FAST research project, where the considered fast terrestrial robots use wheel locomotion in all-terrain up to 10m/s. This paper will evaluate the longitudinal stability of a classical vehicle (with double wishbone suspensions) crossing a high obstacle at high speed. The violent shock and transient fast phenomena to study led us to use small scale vehicles and to directly experiment before modeling. Section 2 introduces the vehicle and experimental measurements and results. Section 3 develops a simplified analytical model fitting the experimental results.

2. Experimenting obstacle crossing at high speed

2.1. *Experimental protocol*

Experiments are performed with an unmodified electric vehicle at scale 1:10 (Fig. 3, E-Maxx #3903, Traxxas Inc.), weighing 5.16kg for a total length of 518mm (track-width = 330mm, wheelbase = 335mm). With two brushed motors in parallel, the vehicle can reach a maximal speed of 14m/s, which exceeds our objective speed of 10m/s.

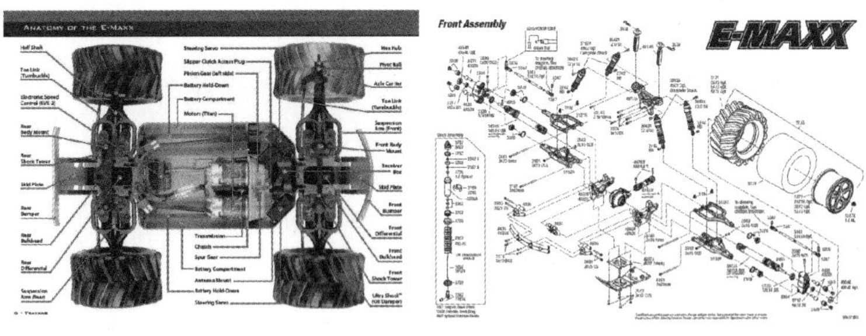

Fig. 3: a) Overview of vehicle architecture. b) View of the front double-wishbone suspension [7].

The obstacle is made of a square section steel bar with adjustable height along vertical rails (Fig. 4). The rails are screwed on brackets, that are also screwed on 3-component piezoelectric force sensors (9257B, Kistler Inc.).

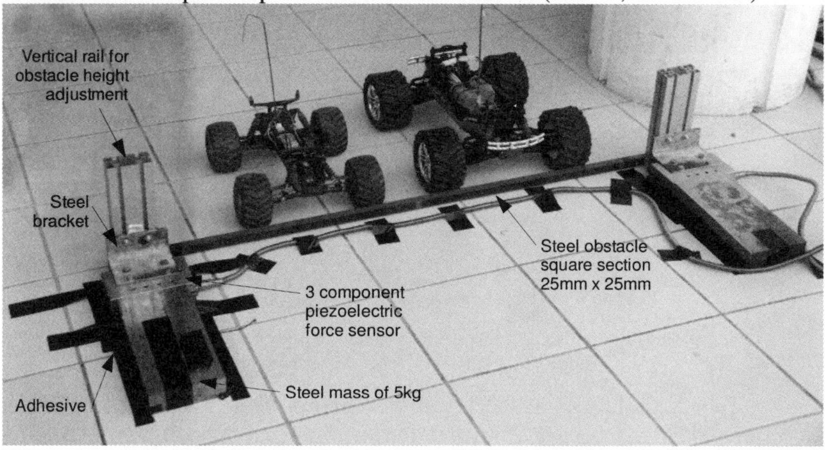

Fig. 4: Adjustable obstacle mounted on dual 3-component force-sensors.

Speed measurements are made by vision with a 30Hz camera located above of the impact zone. The vehicle runs on a tiled floor with a periodic pattern of 300mm. Only the last two images before impact are considered to calculate the instantaneous speed (Fig. 5a).

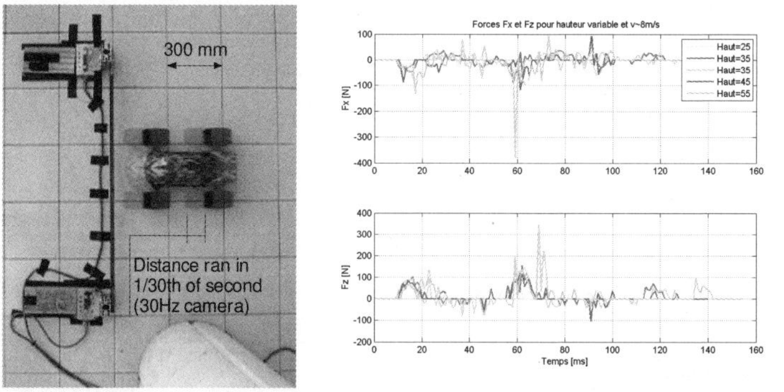

Fig. 5: a) Vision measuring of the instantaneous speed before impact (30 Hz video).
b) Evolution of horizontal *Fx* and vertical *Fz* impact forces according to obstacle height [8].

2.2. *Experimental results*

The experimental force measurements at 1kHz confirm that the impact force increases with the obstacle-height. For high obstacles, peaks can reach 400N and the horizontal force *Fx* has the same order of magnitude as the vertical one *Fz* (Fig. 5b). This suggests that vehicles dedicated to fast obstacle crossing should be also designed with a horizontal suspension, as modeled in [9].

A total of 77 experiments were performed and summarized in Fig. 6. For high speeds or high obstacles, longitudinal stability is compromised and the vehicle crashes by tip-over (red dots). A stability front that separates experiments with or without tip-over can be traced. It has a decreasing non-linear shape. Improvement of the stability zone can be expected from the development of a future innovative suspension with 2DDL (vertical+horizontal).

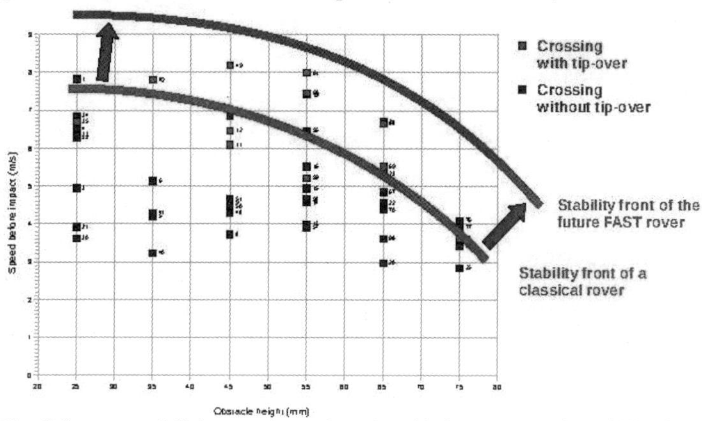

Fig. 6: Summary of all the experimental results with the presence of a stability front.

3. Simple analytical model

A simplified model of the vehicle can be set up, based on experimental results and considering only the vertical wheel translations and the pitch angle of the chassis. Therefore, it is proposed to represent the vehicle by a bicycle model with three rigid bodies: two wheels and the chassis. The wheels and the chassis are connected by spring-damper systems. Vehicle suspension mechanisms authorize planar motion of the wheels relatively to the chassis.

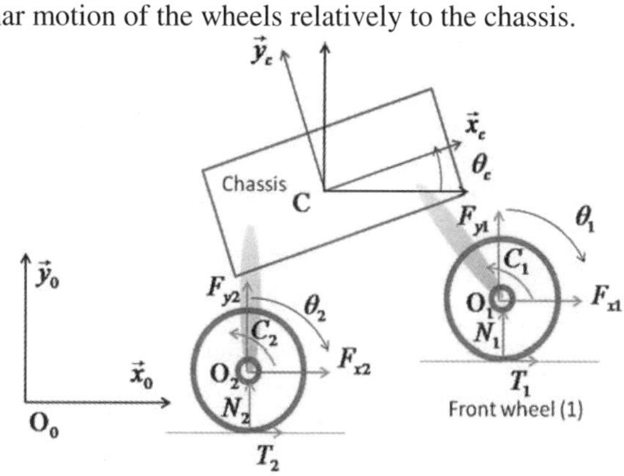

Fig. 7: Parametrization of the vehicle bodies and applied forces.

3.1. *Parameterization*

We consider a planar motion of the in the (O_0, x_0, y_0) plane where $R_0(O_0, x_0, y_0, z_0)$ is the fixed reference frame (Fig. 7). A frame is attached to each body at its center of mass: $R_C(O_C, x_C, y_C)$ for the chassis, $R_1(O_1, x_1, y_1)$ for the front wheel and $R_2(O_2, x_2, y_2)$ for the rear wheel.

The positions of R_C relatively to R_0, projected in R_0, is given by $\overrightarrow{O_0 O_C} = X_C \vec{x}_0 + Y_C \vec{y}_0$ and its orientation is given by the rotation angle $\theta_C = \sphericalangle(\vec{x}_0, \vec{x}_C) = \sphericalangle(\vec{y}_0, \vec{y}_C)$. These parameters are grouped in the position vector (X_C, Y_C, θ_C). In the same way, the positions and orientations relatively to R_0 of frames R_1 and R_2 attached to the wheels are given by (X_1, Y_1, θ_1) and (X_2, Y_2, θ_2).

3.2. *Equations of motion*

By isolating separately vehicle solids, three dynamic equations can be written for each one (Newton-Euler equations). The chassis is submitted to wheel forces and torques as well as gravity (Fig. 7). The following equations can be written:

$$m_c \ddot{X}_c = F_{x1} + F_{x2} \tag{1a}$$

$$m_c \ddot{Y}_c = F_{y1} + F_{y2} - m_c g \tag{1b}$$

$$I_{cz} \ddot{\theta}_c = M_1 + M_2 - C_1 - C_2 \tag{1c}$$

where F_{xi} and F_{xi} $(i=1,2)$ are the joint forces applied by the wheel (i) on the chassis. These forces depend on suspension loading. C_1 and C_2 are the torques applied by the wheel motors. M_1 and M_2 denote the moments of the wheel force at the chassis center of mass point C. m_C and I_{CZ} are respectively the mass and the moment of inertia about z_C of the chassis.

The following equations can be written for each wheel:

$$m_i \ddot{X}_i = T_i + R_{xi} - F_{xi} \tag{2a}$$

$$m_i \ddot{Y}_i = N_i + R_{yi} - F_{yi} - m_i g \tag{2b}$$

$$I_{zi} \ddot{\theta}_i = C_i + rT_i \tag{2c}$$

where N_i and T_i are respectively the normal and the tangential wheel – ground contact forces (Fig. 7). R_{xi} and R_{yi} are the additional obstacle impact forces. m_i and I_{zi} are respectively the mass and the moment of inertia of the wheel (i).

3.3. *Wheel ground contact*

In vehicle dynamics, the wheel – ground contact modeling is a critical issue since contact forces are dependent on the vehicle motion that they generate. The normal contact forces are given by:

$$N_i = \begin{cases} k(r - Y_i) - c\dot{Y}_i & (Y_i \le r) \\ 0 & (Y_i > r) \end{cases} \tag{3}$$

where k and c are respectively the stiffness and damping of the wheel – ground unilateral contact. The tangential contact force is modeled by equation (4).

$$T_i = N_i \min\left(g_r \mu - (1-g_r)\frac{C_i}{r N_i},\ g_r \mu \right) \tag{4}$$

This formula takes into account the relative slipping g_r, given by equation 5, and uses one friction parameter μ, which can be identified experimentally. Contrarily to classic Pacejka like formulae, this formula can be used even in static equilibrium and allows rolling without slipping condition.

$$g_r = \frac{-r\dot{\theta}_i - \dot{X}_i}{\max(|r\dot{\theta}_i|,|\dot{X}_i|)} \tag{5}$$

Finally, the impact force, applied by the obstacle to the wheel is modeled by the action of a linear spring-damper (Fig. 8). This force is proportional to the wheel – obstacle penetration e and oriented from the contact point to the wheel center.

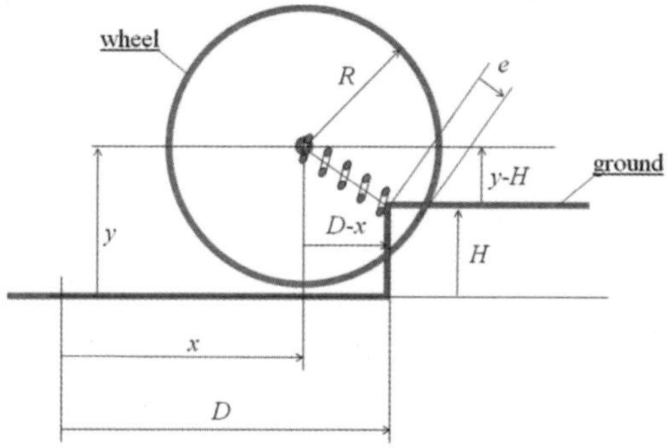

Fig. 8: Impact force modeling.

The spring-damper model was chosen after analyzing a wheel impact with a high speed camera at 10000Hz (Fig. 9). Our reduced car is equipped with rubber tires wrapping a foam layer. Tires have no pressure (decompression holes in the rim). The video clearly showed two steps: in a first step, tire compression; in a second step, hard shock on the rim of the wheel (images 3-4 from the left). In this model, only step 1 was represented. Future work could also include step 2.

Fig. 9: High speed camera analysis at 10000 Hz of wheel impact.

3.4. *Simulation results*

By using the elaborated model, crossing operations can be simulated for several values of impact velocity and obstacle height. As expected in reality, a tip-over failure is obtained for relatively low speed and high obstacle or high speed and low obstacle. These simulations are quite realistic and allow analyzing crossing conditions. They also enable design optimization by tuning suspension stiffness and damping. A simulated motion is represented in Fig. 10. In this case, the impact speed is 5m/s and the obstacle height is 5 cm (2/3 of the wheel radius). The vehicle succeeded to cross the obstacle but the trajectory contains ballistic phases and two bounces.

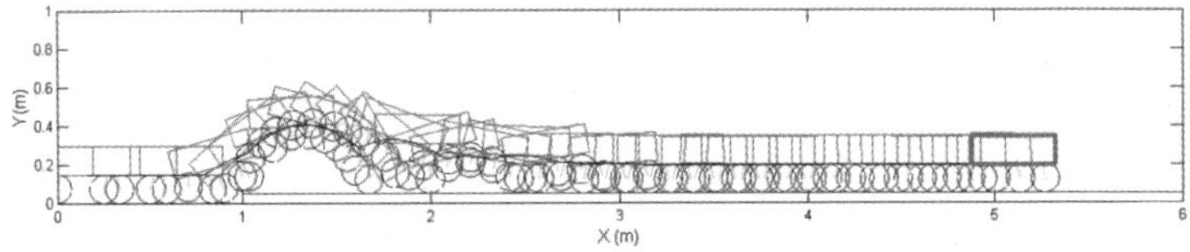

Fig. 10: Trajectory simulation.

In order to analyze the obstacle crossing capacity of the vehicle, a numerical design of experiment (DoE) was performed with two variables: the impact speed and the obstacle height. With only two variables, a full factorial DoE was achievable with ten levels for each variable. Fig. 11 shows the result of the DoE on one hundred of virtual crossings. Each simulation can lead to four possible situations, combination of two factors : obstacle crossing (yes or no) and stability (tip-over or not).

We can identify a successful and stable crossing zone under the blue bold circle points. A precise boundary of this zone can be obtained by a finer meshing of the crossing parameters around these points. It can be noticed the existence of some isolated zones or points which can't be exploited safely in a real application.

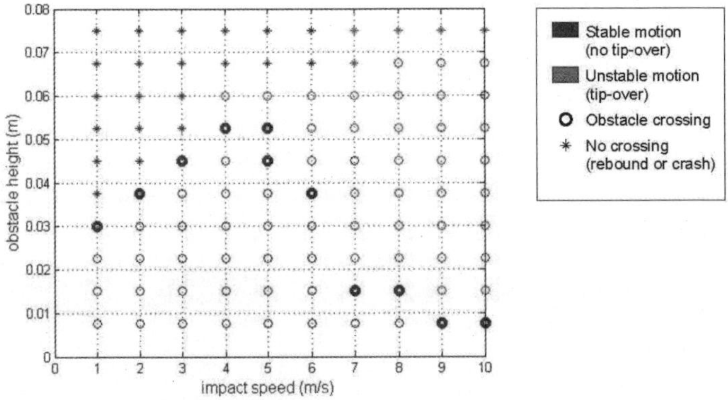

Fig. 11: Numerical characterization of crossing capacity and stability.

4. Conclusion

In this work, 77 experiments of dynamic obstacle-crossing were performed with an electric vehicle at scale 1/10. It allowed showing that there exists a stability limit after which the vehicle crashes systematically by tipping over. This limit can be expressed as a non-linear relation between the vehicle speed and the obstacle height. The vehicle can cross low obstacles at high speed or high obstacles at low speed. This means that the speed limit for keeping the vehicle controllable can be directly inferred from sensing obstacle height.

This behaviour was corroborated by an analytical model based on a bicycle model and simple equations of contact forces. The model was tested in a design of experiment containing one hundred virtual crossings.

Experimental measurement of impact forces showed that, for steep obstacles that are typical of all-terrain use, the horizontal component of the contact force can even exceed the vertical component. This traces a major perspective of improvement for dynamic obstacle-crossing: creating innovative suspensions that are capable to also absorb horizontal shocks.

Future work include developing a new kinematics of suspension to absorb horizontal shocks; refining the impact model by adding a non constant stiffness; finding the analytical expression of the frontier between stable and unstable behaviour; and last but not least: choosing a control strategy for optimal obstacle crossing.

Acknowledgments

This work is supported by the French National Research Agency and is included into the ANR FAST project. The authors wish to thank Richard COUSTURIER for help in experiments, Miguel CANO for piloting and Jamil DAKHLALLAH for post-processing.

References

1. iRobot Inc. 510 Packbot robot specifications. http://www.irobot.com
2. M-Tecks EAC Inc. Arthron-R robot. http://www.m-teckseac.com
3. Meccano Inc. Spykee robot. http://www.meccano.com/models/spykee
4. D. O'Halloran, A. Wolf and H. Choset. Design of a high-impact survivable robot. *Mechanism and Machine Theory.* **40**, 1345–1366 (2005).
5. H. Schempf, W. Crowley, C. Gasior and D. Moreau. Ultra-rugged Soldier-Robot for Urban Conflict Missions. *Unmanned Systems 2003 Conference - AUVSI 30th Annual Symposium and Exhibition*, Baltimore, MD, July 15-17 (2003). http://www.rec.ri.cmu.edu/projects/dragonrunner/tech
6. H.B. Pacejka, *Tyre And Vehicle Dynamics*, Butterworth-Heinemann, 2005.
7. Traxxas Inc. E-Maxx#3903, 1/10 scale, 4WD electric Monster truck, Reference manual and Part list. http://traxxas.com/products/models/electric/3903emaxx-downloads
8. J. Dakhlallah, Design of mechatronics systems for innovative suspensions dedicated to high speed off-road mobile robots, Post-doctoral report (2010).
9. J.C. Fauroux, J. Dakhlallah, B.C. Bouzgarrou. A New Concept of FAST Mobile Rover with Improved Stability on Rough Terrain. *Proc. of* HUDEM'2010*, 8th IARP Workshop on Robotics and Mechanical assistance in Humanitarian De-mining*, Sousse, Tunisia. 16 p, 10-12 May (2010).

DEVELOPMENT OF A PERSONAL MOBILITY VEHICLE FOR ROUGH TERRAIN

Shuro Nakajim

The Dept. of Advanced Robotics, Chiba Institute of Technology, Narashino, Chiba 275-0016, Japan. Email: shuro.nakajima@it-chiba.ac.jp, www.nakajima-lab.it-chiba.ac.jp

We have developed a personal mobility vehicle for rough terrain, called RT-Mover P-type, which shows adequate mobility performance on targeted types of rough terrain. It has four drivable wheels and two leg-like axles but only five active shafts. A strength of it is that it realize both a leg mode and a wheel mode in a simple mechanism.

In this paper, the mechanical design concept of this robot is described, and fundamental leg motions are evaluated by experiences.

The proposed vehicle can move on discontinuous, rough terrain while maintaining its carrier in a horizontal position.

Keywords: Personal Mobility Vehicle, Leg-wheel Robot, Rough Terrain

1. Introduction

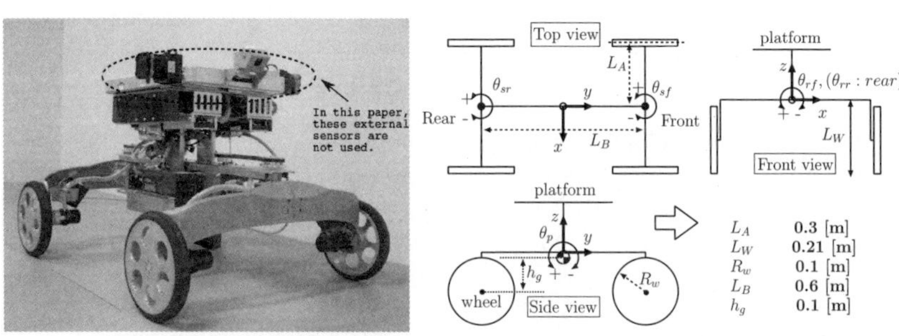

Fig. 1. A small type of RT-Mover

The already proposed robot, called "RT-Mover[1]" (Fig. 1), has a simple mechanism and adequate mobility with a stable posture position for the following target environments: 1. an indoor environment including a step and an uneven ground surface; 2. an artificial outdoor environment with an uneven ground surface and a bump; and 3. natural terrain such as a path in a forest. RT-Mover's mechanism is different from that of conventional mobile robots: four wheels are mounted at the tip of every leg but it has four active wheels and only five other active shafts. With an emphasis on minimizing the number of drive shafts, the mechanism is designed for a four-wheeled mobile body that is widely used in practical locomotive machinery. RT-Mover can move like a wheeled robot and also walk over a step like a legged robot, despite the simplicity of the mechanism.

The robot can move on discontinuous, rough terrain while maintaining its platform in a horizontal position. Therefore, it can be used as a stable carrier for an object.

In this paper, we develop a personal mobility vehicle as a series of RT-Mover, shown in Fig.3.

2. Mechanical Design of RT-Mover P-type

The target of this paper is a practical mobile robot that carries people. It is necessary to keep the boarding seat (hereinafter, the platform) horizontal in order to carry people. We also aim to develop a practical robot with high cost performance by reducing the number of drive shafts to the extent possible and employing a simple mechanism.

A mobile robot used for general purposes should have high speed on a paved road and good ability of locomotion over rough terrain. Although there is no mechanism superior to a wheel for high speed and energy efficiency, a wheeled robot can move only on continuous rough terrain. To move on discontinuous terrain, a leg mechanism is better than a wheel mechanism. Therefore, to perform all the essential functions of mobile robots in rough terrain, both wheel and leg mechanisms are needed.

The target application environment such as an urban environment is basically a paved surface upon which transportation by wheel is possible, with several steps that necessitate a leg function in order to reach the destination. Therefore, it is advantageous to reduce the complexity of the leg mechanism to a minimum and to limit each leg's motion space.

In this paper, we take a four-wheeled mobile body often used in practice as the starting point in considering the mechanism of the proposed robot, and from there we develop the proposed mechanism. In addition, this discussion is based on a four-wheeled mobile body that drives each wheel independently.

After listing possible mechanisms for each function, the mechanism with the minimum number of drive shafts is chosen for the proposed robot with consideration of the possible combinations.

2.0.1. *Steering Function*

Practical mechanisms to achieve a steering function for a four-wheeled mobile body are shown in Fig. 2(a). Although there are other mechanisms to achieve steering, the following discussion targets highly practical representative mechanisms. The mechanism of 1-1 is the Ackermann steering system. That of 1-3 is a mechanism that rotates the center of a shaft to steer. It is possible to attach a steering mechanism to the rear wheels of both 1-1 and 1-3. However, this case is the same as 1-1 and 1-3 and is omitted from following discussion. Those of 1-2 and 1-4 are 4-Wheel Steering (4WS) systems.

2.0.2. *Suspension Function in Wheel Mode*

The wheels are required to have active vertical travel according to terrain in order to keep the platform horizontal when moving on gently varying irregular terrain. Systems to fulfill this requirement are shown in Fig. 2(b). The mechanism of 2-1 provides an up-down function to each wheel. The necessary number of drive shafts is four, and the platform can be kept horizontal on rough terrain by the vertical movement of each wheel according to the ground surface.

The mechanism of 2-2 is a bogie mechanism for the left and right wheel that uses a rotating shaft as a drive shaft. The horizontal degree of roll direction of the platform can be maintained by the active control of the rotating shaft of front and rear bogie mechanism in response to the ground surface. The horizontal degree of pitch direction of the platform is maintained by attaching an independent pitch control shaft. The number of necessary drive shafts is three.

2.0.3. *Suspension Mechanism in Leg Mode*

The two factors are necessary for traveling in leg mode over steps and other obstacles: 1.lifting, moving forward, and landing for each leg, 2.supporting the body with three other legs.

The up-and-down function of each leg is achieved by a wheel mode suspension function. The mechanisms to move each wheel back and forth are shown in Fig. 2(c). The mechanism of 3-1 is a method to affix a back-

and-forth drive function to each wheel, and the number of required drive shafts is four. That of 3-2 is the same as 1-4, with the wheels moving back-and-forth through rotation of the center shaft. Both of them can locate the center of gravity of the body in the support polygon and thus support the body by adjusting positions of supporting legs.

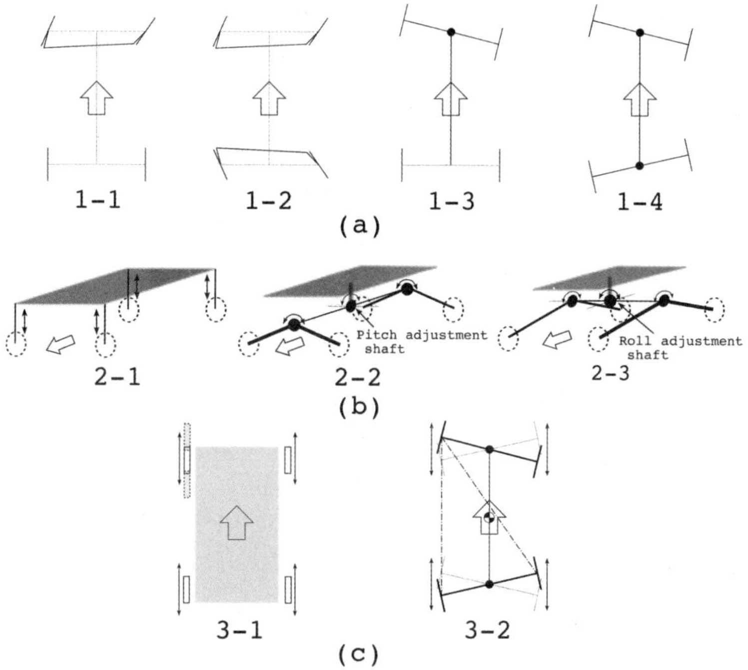

Fig. 2. Mechanism for each function. (a)Steering mechanism. (b)Suspension mechanism for wheel mode. (c)Suspension mechanism for leg mode (Top view).

2.0.4. *Mechanism of Proposed Robot*

The combinations of each mechanism are listed in Table 1. Impossible or unavailable combinations are indicated by "-".

We take an example to explain how to see Table 1. As the mechanism of 3-2 is the same as that of 1-4, it is also a steering mechanism. Therefore, 3-2 cannot be combined with 1-1, 1-2, and 1-3 because this would result in unnecessarily duplicating steering functions in equipping the drive shaft.

The combination of 1-4, 2-2, and 3-2 gives five drive shafts, which is a minimum. Therefore, the mechanism of the proposed robot consists of four drive wheels and five drive shafts, as shown in Fig. 3.

Table 1. Combinations of each mechanism

Steering	Suspension (wheel mode)	Suspension (leg mode)	Number of drive shafts
1-1	2-1	3-1 / 3-2	9 / -
	2-2	3-1 / 3-2	8 / -
	2-3	3-1 / 3-2	8 / -
1-2	2-1	3-1 / 3-2	10/ -
	2-2	3-1 / 3-2	9 / -
	2-3	3-1 / 3-2	9 / -
1-3	2-1	3-1 / 3-2	7 / -
	2-2	3-1 / 3-2	6 / -
	2-3	3-1 / 3-2	- / -
1-4	2-1	3-1 / 3-2	- / 6
	2-2	3-1 / 3-2	- / 5
	2-3	3-1 / 3-2	- / -

The specifications of the robot are listed in Table 2. When rotating the roll-adjustment axis through 30[deg] such that the wheel on one side is in contact with the ground, the other wheel attached to a 650[mm] steering arm can rise 325[mm]. Therefore, the movement range is sufficient for the targeted terrain. Likewise, moving 325[mm] in the front and rear directions is possible by moving the steering from 0[deg] to 30[deg], and holes of 325[mm] can be crossed. With regards to locomotion on a slope, back-and-forth movement and traversal of a slope of up to 30[deg] is possible.

In fact, additional motors are attached to the robot, for example, for adjusting footrest mechanism. Those are, however, not essential functions for moving on rough terrain, so they are not discussed in this paper.

3. Assessment of Ability of Locomotion in Leg Mode

Evaluations are performed through experiments taking a step-up gait and a step-down gait as examples. The same methodology in the paper[1] is used for those gaits. At the current stage, road shapes are known in advance.

Fig.4 shows data of the step-up walking experiment over a 0.15[m]-high step. The robot can get over a 0.15[m] step with a person riding on it while maintaing the horizontal position of its platform within ±5[deg]. The main conditions are the followings. The angular velocity of a roll-adjustment shaft when lifting and landing the wheel are 0.2[rad/s] and 0.1[rad/s] respectively, that of a steering shaft to put forward the lifted leg is 0.2[rad/s], and the

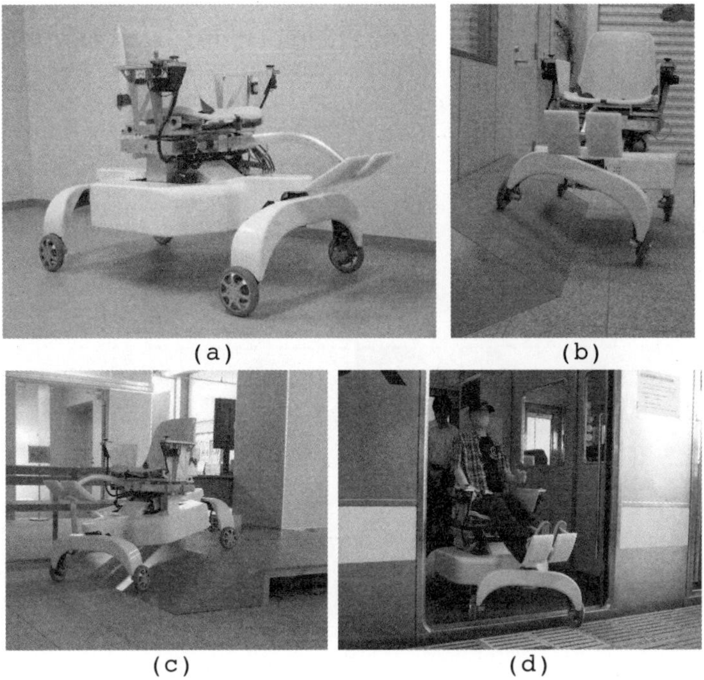

Fig. 3. (a)RT-Mover P-type. (b)On a bank. (c)On a slope. (d)Getting off a train.

Table 2. Main specifications

Dimensions	Length 1150[mm](excluding footrest); Width 696[mm] (Tread 650[mm]); Height to seat 577.5[mm]; Height to bottom 170[mm]
Wheel	Radius:150[mm]; Width:30[mm]
Weight	80[kg] (including batteries at 20[kg])
Motor	maxon brushless motor 100[W] $\times$ 9
Gear ratio	100 (each wheel, front and rear steering); 820 (pitch-adjustment shaft); 2400 (roll-adjustment shaft)
Sensor	Encoder (each motor); Current sensor (each motor); Posture angle sensor (roll and pitch of platform)
Angle limit	$\pm$30[deg] (steering, roll-adjustment shaft, and pitch-adjustment shaft)
Max speed	4.5[km/s]
Power supply	48[V] lead accumulator

forward velocity of the body in wheel mode is 0.12[m/s]. The configured value of each process velocity in leg mode is obtained experimentally from a velocity that gives static leg motion. There are plans to address high-speed leg processes in near future.

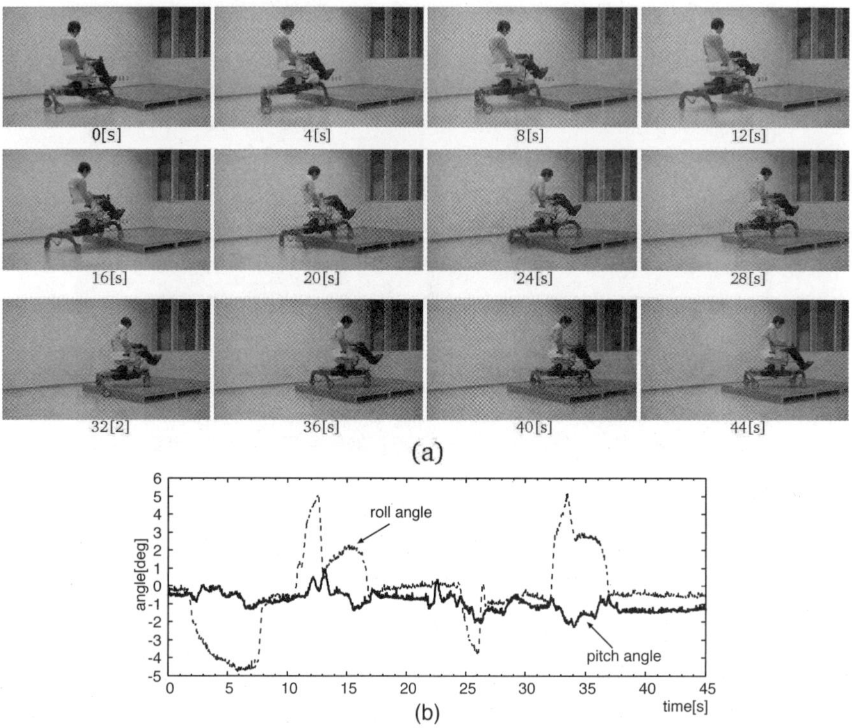

Fig. 4. Experimental result of the step-up gait. (a)Snapshots. (b)Posture angles of the platform.

Fig.5 shows data of the step-down walking experiment down a 0.15[m]-high step. The main conditions are basically the same as the step-up gait experiment. The robot can decline a 0.15[m] step with a person riding on it while maintaing the horizontal position of its platform within ±4.5[deg].

4. Conclusion

In this paper, a personal mobility vehicle for rough terrain was develoed and its fundamental mobility performance was evaluated. As future works,

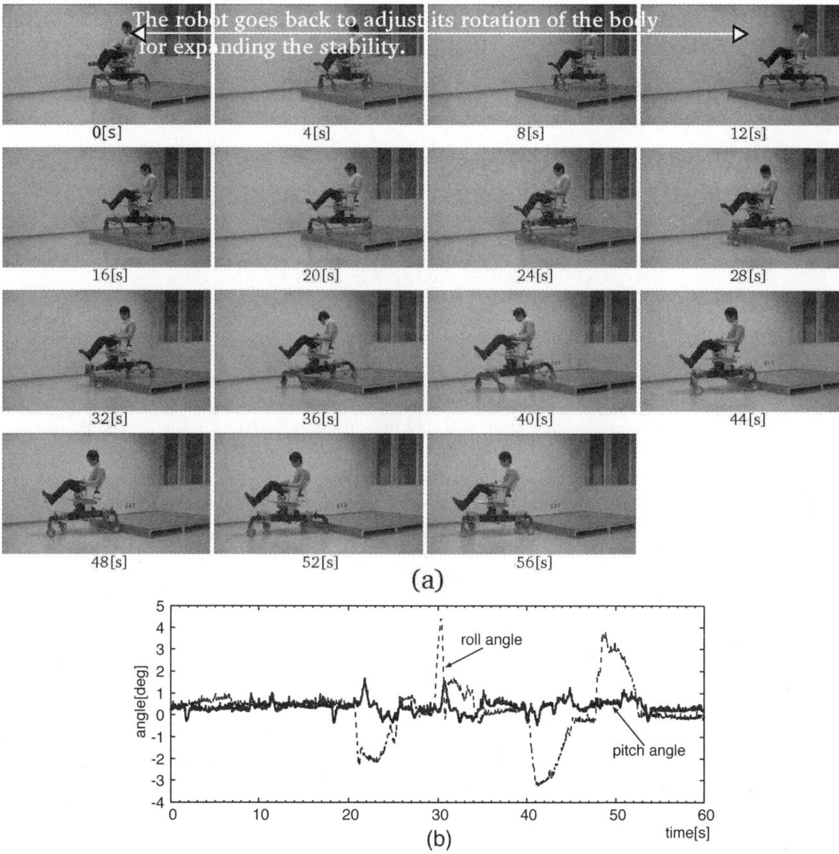

Fig. 5. Experimental result of the step-down gait. (a)Snapshots. (b)Posture angles of the platform.

we have to improve its mobility performance regarding high speed motion and high stability.

Acknowledgement: This work was funded by the joint research with East Japan Railway Company.

References

1. S. Nakajima"RT-Mover: a rough terrain mobile robot with a simple leg-wheel hybrid mechanism"International Journal of Robotics Research, accepted.

DESIGN, SIMULATION AND EXPERIMENTAL TESTS OF A HYBRID ROVER FOR OVERPASSING OBSTACLES

GIANNI CASTELLI AND ERIKA OTTAVIANO

DiMSAT, University of Cassino, via Di Biasio 43
Cassino, 03043, Italy

In this paper the design and simulation are presented for THROO: Tracked Hybrid Rover for Overpassing Obstacles. The proposed mobile robot has 3-DOFs and it is capable of straight motion, turning ability and two operations, namely rover-like motion with tracks on flat terrain and walking-like motion with legs to overpass obstacles on uneven terrain. A prototype has been built and the reported experimental results show its ability in overpass obstacles, which size is comparable or greater than the track high.

1. Introduction

Mobile robots can be classified according to the type of locomotion they perform as wheeled, tracked, legged or hybrid robots [1]. Wheeled and tracked ones can perform fast and robust motion on flat terrain, but they are ineffective in overpass obstacles that are greater, or even with similar size, than the radius (dimension) of the wheel (track). Theoretically, legged robots can go where wheeled robots cannot, so they have a potentiality that justifies the extra effort required to control their locomotion [2]. It is obvious by reported works that legged robots have favorable characteristics for obtaining flexible and robust motion, but suffers of some drawbacks, such as energy consumption, stability and locomotion speed, both if we are dealing with two, four, six or more legs. Hybrid systems were developed to exploit the terrain adaptability of legs in rough terrain and simpler control as well as high speed related with wheels.

Research efforts have been made toward hybrid mobile robots that can drive quickly and smoothly on flat terrain and can stably negotiate natural or artificial uneven terrain. Hybrid mobile robots can be classified into two main groups: in the first one legs and wheels work independently where articulated legs can provide traction for the obstacle-climbing when wheels cannot. This basic idea is used for example in [3, 4]. Innovative prototypes of wheelchairs capable of climbing stairs have been proposed in [5, 6].

The second type of hybrid locomotion involves wheels installed on each leg end-tips in particular, the articulated leg is used to place the wheel in such a way to avoid slopes or obstacles that cannot be surpassed. This design solution was adopted for example by [7-10].

The goal of this research is to develop a robot that can be used in industrial sites, especially in hazardous areas such as nuclear power plants. Other application areas are related to surveillance and rescue operations for military and civil tasks. Most of these robots, however, cannot be easily adapted to environments designed for human beings. If such devices are to be used at industrial sites, they will require special arrangements such as ramps to allow them to move around. Since these places are originally designed for human beings, it is desirable to have robots capable to surpass obstacles and adapt to the environment. In this paper the design of a new prototype of hybrid rover is proposed as shown in Fig. 1 with low-cost easy-operation features. In order to obtain a low-cost easy operation walking machine the following requirements have been considered: compactness, light weight, reduced number of DOFs.

The leg structure shown in Figs. 1 and 2 has one DOF and it is possible to combine more legs by sharing the same input crank. This solution allows the use of just one actuator for moving at same time four or more legs in order to design a robot capable of overcoming a large number of obstacle types [11]. Dynamic simulations and experimental tests have been performed to validate the operation of the proposed robot [a].

Figure 1. A built prototype of the proposed hybrid rover named as THROO.

2. Kinematics and synthesis of the leg mechanism

According to the kinematic scheme of the leg in Fig. 2, the position of point E, with coordinates (x, y) can be described with respect to AXY fixed frame, as a function of kinematic parameters of the mechanism ABCDE in the form letters for table footnotes (Table I).

Since the curve traced by any point of the coupler link of a planar four-bar linkage is algebraic of sixth degree, the coupler curve of a four-bar linkage

[a] You can obtain these files from: http://www.youtube.com/user/RoboticaLab

intersects a line at a maximum of six points. As a consequence, the coupler curve cannot have straight segments of finite length. Therefore, the required trajectory can have only approximate straight line segments.

In particular links AD and BC can be evaluated with respect to the AXY frame in the form

$$\underline{AD} = \begin{bmatrix} x + u\cos\theta - v\sin\theta \\ y + u\sin\theta + v\cos\theta \end{bmatrix}; \ \underline{BC} = \begin{bmatrix} x - p + (u + c)\cos\theta - v\sin\theta \\ y - q + (u + c)\sin\theta + v\cos\theta \end{bmatrix} \tag{1}$$

Furthermore, by considering

$$\|\underline{AD}\|^2 = d^2 ; \ \|\underline{BC}\|^2 = m^2 \tag{2}$$

and substituting Eq.(1) in Eq.(2), one can obtain two equations, which are function of design parameters and x, y and θ. The desired implicit function can be obtained by eliminating θ, and further algebraic manipulations.

A parametric study has been carried out involving design parameters and coupler curve shape [11]. In particular, the trajectory of point E is required to have an approximate straight line trajectory (during the support phase) with a sufficient high to overpass obstacles. Specifically, point E trajectory should have a suitable size in terms of high (comparable with the body robot size) for the flight phase and length (half length of the track size) during the support phase. Results of the above mentioned study are reported in Fig. 3 when the kinematic parameters values are reported in Table I. In particular, Fig. 3 shows a planar scheme of the robot for which the front and rear legs are actuated by a single motor installed at point B.

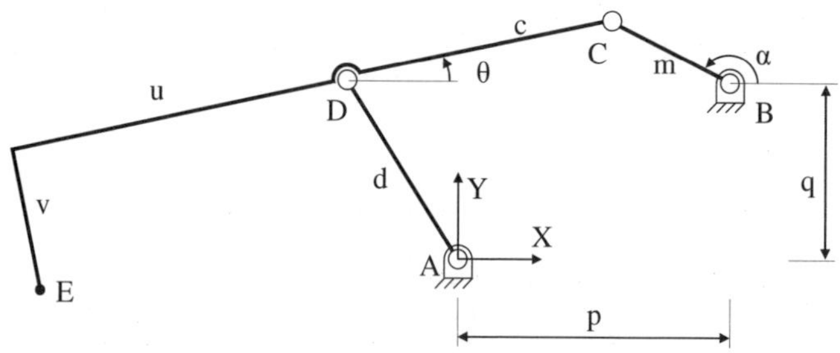

Figure 2. A kinematic scheme for the 1-DOF leg.

Table. I. Kinematic parameters of the leg in Fig. 2.

Parameter	d	m	c	u	v	p	q
Dimension [mm]	75.0	45.0	104.0	160.5	63.5	100.0	60.0

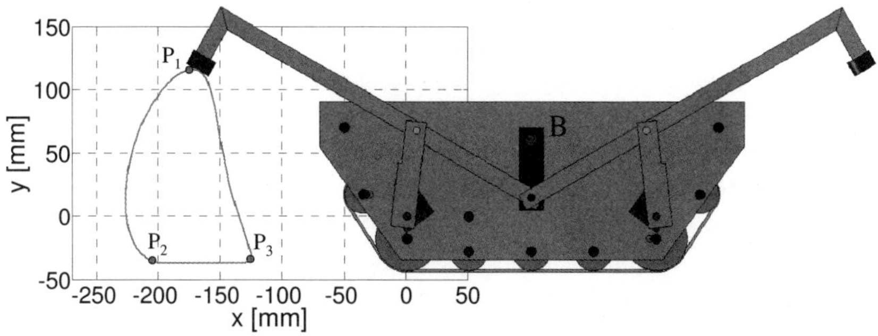

Figure 3. A scheme for the hybrid robot with coupler curve traced by point E.

The common crank for the two legs is depicted in dark grey in Fig. 3 and it is fixed to the chassis at point B.

3. Dynamic simulation

A suitable model in SolidWorks environment has been developed in order to test the operation of the hybrid rover. In particular, several paths with obstacles have been tested and simulated, by considering either regularly shaped obstacles but also irregular terrain with obstacles.

Figure 4 shows the snapshots of the developed dynamic simulation for the robot movement during its motion in overpassing a step obstacle of 90 mm. It is worth noting that Fig.4 shows a lateral view (in which just one rear leg and one front leg are shown) but the hybrid rover has been modeled and simulated in 3D to test the lateral stability also.

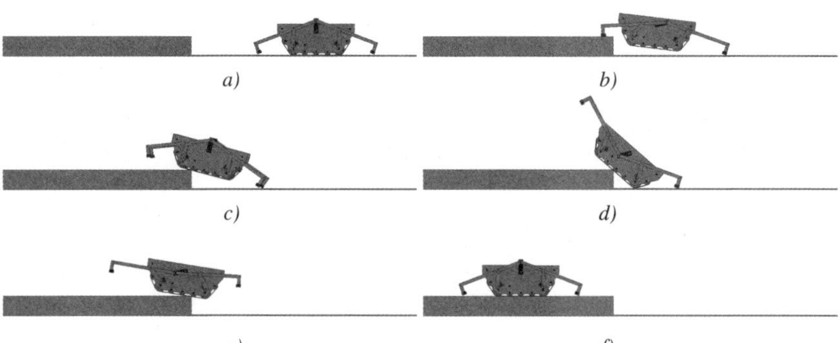

Figure 4. Dynamic simulation snapshots for the robot movement in overpassing an obstacle: a) 1 sec; b) 3.2 sec; c) 5 sec; d) 5.6 sec; e) 6.7 sec; f) 9 sec.

In particular, to analyze the robot robustness with respect to possible overthrowing, several dynamic simulations have been carried out. It has been verified that the legs can be used to stabilize the hybrid robot during the motion and prevent longitudinal and lateral overthrow in climbing an obstacle.

The operation of the robot can be summarized as the following: 1) rover-like motion on flat or smooth terrain (Fig.4a and Fig.4f), which is the most efficient type of locomotion under this condition; 2) walking-like motion when approaching and overcoming an obstacle (Figs. 4b-e). In particular, the robot has been provided by rear and front legs to stabilize it during the obstacle crossing. Moreover, the rear leg is used as propulsive element in overpass the obstacle. The main goal to pursue is therefore to use the front legs to approach the obstacle and rear legs to help the robot in overpassing the obstacle.

In Figure 5 numerical results of the dynamic simulation have been reported, namely the rover CM position (Fig. 5a) and CM velocity (Fig. 5b).

Furthermore, the dynamic simulation allows to evaluate the torques required to give suitable power to the legs and track. By observing the simulation results it is worth to note that on a flat surface the torque required for the legs is less than 0.3 Nm and the one needed for the track is less than 0.1 Nm. When the hybrid rover is overpassing the obstacle in Fig. 4, it is observed that the maximum torque required for the legs is 4 Nm and it is less than 1 Nm for the track.

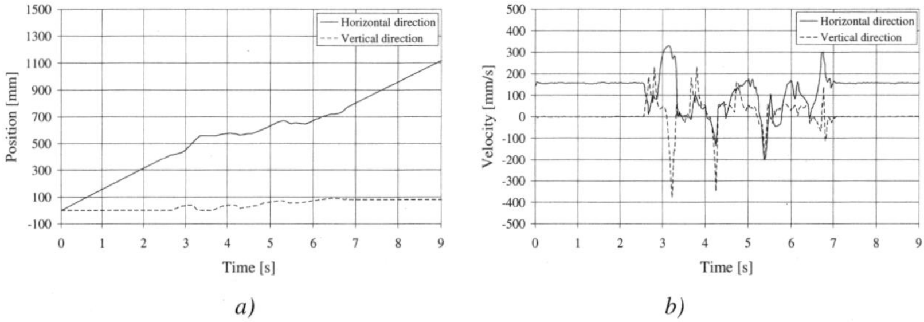

a) *b)*

Figure 5. Results of the dynamic simulation in Fig. 4 for the robot movement overpassing an obstacle: a) rover CM position; b) rover CM velocity.

4. A prototype of the hybrid rover

According to the results obtained from the dynamic simulation, a prototype of the hybrid rover has been designed and built considering the following constrains: minimum number of DOFs; low-cost mechanical design; easy operation; modularity of the legs.

4.1. *Mechanical design*

Figure 6 shows a mechanical design of the robot with 3 DOFs, two of them are used for the track giving the ability of straight motion and turning ability.

An additional motor is used to actuate the four legs, each couple of rear and front legs share the same crack and work providing a specular trajectory of the leg end-point. A common shaft is used for left and right cranks. Belt transmission is used to drive the motion from motors to the tracks and to the input cracks of the legs. Motors have been installed to the robot chassis in order to distribute uniformly their masses. According to the proposed mechanical design and suitable mass distribution, the center of mass of the robot lies in the longitudinal plane of robot body, while its y-coordinate varies according to the legs' configuration. Its average relative y-coordinate is 55 mm, with respect to the leg reference frame shown in Figure 2.

It is important to point out that as a design requirement the robot should work in both locomotion directions having the same ability of overpassing obstacles. This is to increase the robot's operation flexibility.

According to initial design idea and simulation results during the rover-like motion the legs should be inactive, and in the walking-like motion both legs and track should be active. It worth to note that even if the legs and tracks operate simultaneously on flat surface, the leg end-points do not interact with the terrain, as it is shown in the scheme of Fig.3 and in the simulation results, which are reported in Figs. 4 and 5.

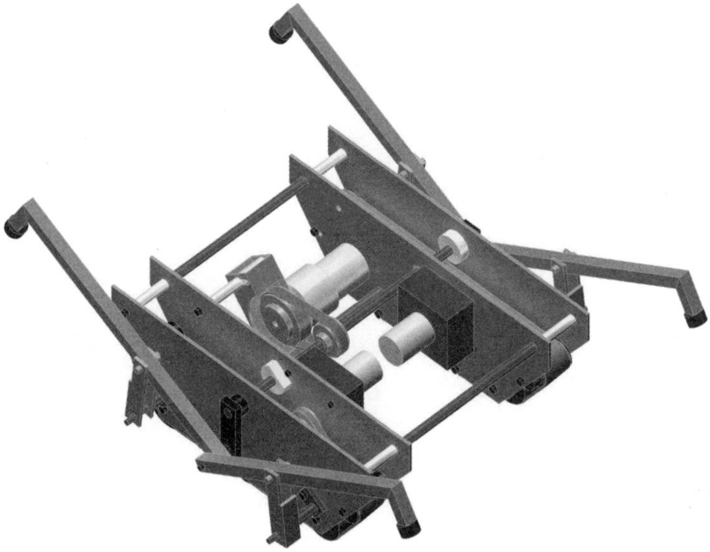

Figure 6. Mechanical design of the prototype of the hybrid rover.

4.2. *Experimental tests*

A prototype of the THROO hybrid rover has been built according to the results of the dynamic simulation proposed in the previous Section. Its 3-DOFs are currently remotely controlled and future development of the robot is its sensorization in order to automatically activate the legs just in proximity of obstacles. The prototype size is 690 x 450 x 210 mm, all mechanical components are made of aluminum and actuation and transmission systems are of commercial type. The overall mass of the prototype is 4.5 kg (batteries not included). It is worth to note that its dimensions vary according to the legs' configuration, here maximum values are given. The actuation system is composed by 2 DC motors for the track with following features: power supply of 24 V, max torque of 5 Nm, nominal power of 3.9 W. The motor used to actuate the legs has a power supply of 24 V, max torque of 12 Nm, nominal power of 24 W.

The THROO hybrid robot has been tested in several operating conditions. A comparison between numerical and experimental tests is performed and it gives a good match of expected results. Several experimental tests have been carried out in indoor and outdoor environment. In particular, it has been experimentally tested the ability of the hybrid robot in overpass obstacles.

Figure 7 shows snapshots of the hybrid robot overpassing an obstacle which is greater than the track size. The high of the obstacle is 100 mm, which represents the 125 % of the high of the robot track.

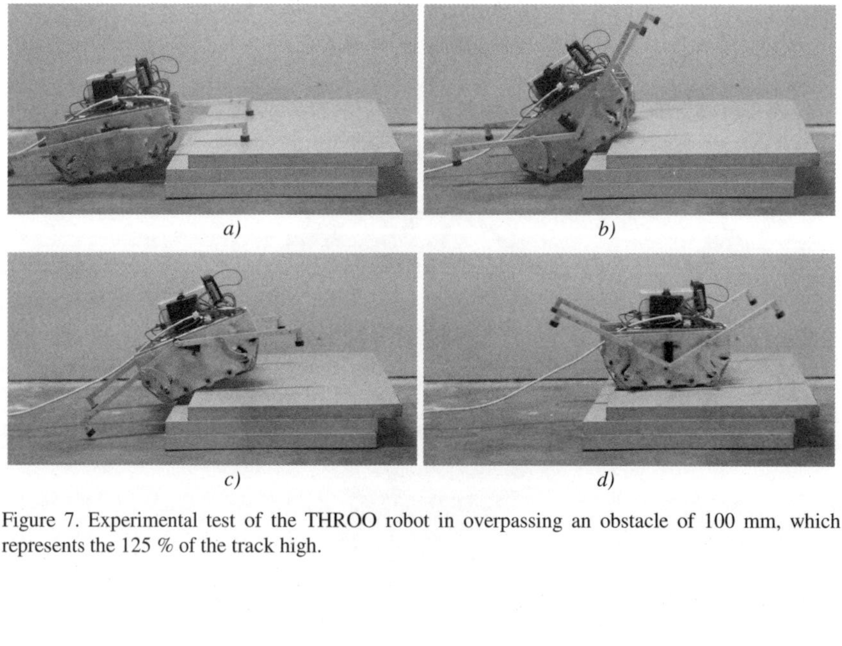

a) *b)* *c)* *d)*

Figure 7. Experimental test of the THROO robot in overpassing an obstacle of 100 mm, which represents the 125 % of the track high.

5. Conclusion

The design and construction of the THROO hybrid rover have been presented and discussed in this paper. The proposed hybrid rover is capable of overpassing obstacles, which are greater than the dimension of its track. Future development will concern a suitable sensorization of the robot to provide the ability of automatically switch from rover-like to walking-like modes and vice-versa in the proximity of an obstacle. Further developments are also concerning with robot control strategies to be implemented and used by a PLC, which can be even installed on board.

References

1. P. E. Sandin, *Robot Mechanisms and Mechanical Devices Illustrated*, McGraw-Hill, (2003).
2. R. Siegwart, I. R. Nourbakhsh, *Introduction to Autonomous Mobile Robots*, MIT Press, (2004).
3. S. Guccione and G. Muscato, "Control Strategies Computing architectures and Experimental Results of the Hybrid Robot Wheeleg", *IEEE Robotics & Automation Magazine*, **10** (4), pp. 33-43, (2003).
4. C. Tavolieri, E. Ottaviano, M. Ceccarelli, "Design and Problems of a New Leg-Wheel Walking Robot", *10th Int. Conf. on Climbing and Walking Robots CLAWAR 2007*, pp. 319-328, (2007).
5. P. Wellman, V. Krovi, V. Kumar, W. Harwin, "Design of a wheelchair with legs for people with motor disabilities", *IEEE Trans. on Rehabilitation Engineering*, **3** (4), pp 343-353, (1995).
6. R. Morales, V. Feliu, A. González, P. Pintado, "Kinematic Model of a New Staircase Climbing Wheelchair and its Experimental Validation," *The Int. Jnl of Robotics Research*, **25** (9), pp. 825-841, (2006).
7. C. Grand, F. Benamar, F. Plumet, P. Bidaud, "Stability and Traction Optimized of a Reconfigurable Wheel-Legged Robot", *The International Journal of Robotics Research*, **23** (10-11), pp. 1041-1058, 2004.
8. P. Aarnio, K. Koskinen, S. Salmi, "Simulation of the hybtor robot", *3rd Int. Conf. on Climbing and Walking Robots*, Madrid, pp. 267-274, 2000.
9. S. Hirose, H. Takeuchi, "Study on Roller-Walk. Basic Characteristics and its Control", *Proc. IEEE International Conference on Robotics and Automation*, Minneapolis, pp. 3265-3270, 1996.
10. R. Siegwart, P. Lamon, T. Estier, M. Lauria, R. Piguet, "Innovative Design for Wheeled Locomotion in Rough Terrain", *Journal of Robotics and Autonomous Systems*, **40**, pp. 151-162, 2002.
11. G. Castelli E. Ottaviano, "Design And Simulation Of A New Hybrid Mobile Robot For Overpassing Obstacles", *Proceedings of EuCoMeS 2008*, pp. 101-108, 2008.

SENSOR-BASED CONTROL OF A SKID-STEERING MOBILE ROBOT COMBINED WITH FEET FOR MOVING ON ALL TERRAIN

T. OKADA*, S. TEZUKA and Y. SASAKI

Faculty of Engineering, Niigata University,
Niigata, 950-2181, Japan
**E-mail: okada@eng.niigata-u.ac.jp*
http://okada.eng.niigata-u.ac.jp

This article proposes a mechanism and control of a skid-steering mobile robot combined with six feet for each wheel. Wheels and feet in front and back on one side are mechanically synchronized, but odd numbered and even numbered feet groups are independently controlled. Likewise, other side has two feet groups too. Therefore, the robot has 6 degrees of freedom (DOF) to drive all the wheels and feet. We consider sensor-based robot controls for progressing and steering by rolling and walking since cooperative controls of wheels with feet generate various movements to cope with obstacles. Particularly, angular values of hip and ankle joints are discussed for stepping up or over obstacles. The robot PEOPLER-IV, demonstrates various movements on a practical field for verifying usefulness of the proposed controls using the gravity direction.

Keywords: Skid-Steering, Wheel with Feet, Wheelchair, Gravity Sensor

1. Introduction

Skid-steering vehicles (SSVs) and mobile robots (SSMRs) composed of multiple wheels have no steering mechanism and no ability of recognizing where it is, and which direction it moves, due to sliding, skidding and slipping. But its robustness in various road conditions are high. Therefore, SSVs and SSMRs are beneficial on rough terrain when appropriate commands for moving are given from a driver or a remote operator based on the knowledge interpreted using visual information.

The wheel-feet hybrid mechanism locates feet frontward as parts of links connecting two same wheels in size for enabling a climbing.[1] Duplicated eccentric shaft driving mechanism makes segmented wheel rims change their angles and directions.[2] The robot Wheeleg, exchanges the front wheels of a vehicle with legs.[3] The robot Wheg, has discontinuous treads to use a rotating unit as not only a wheels but also split spokes, i.e. legs.[4] Our robot PEOPLER-II, serves not only as a wheel-type, but also as a leg-type.[5,6] Two legs of each wheel are synchronized mechanically, but the next version PEOPLER-III, drives the two legs independently by augmenting the degrees of feedom from 4 to 6. Also, the newest version PEOPLER-IV, has 6 short feet for improving performance of mobility. Reasonable number of the feet, foot length, and feet control rules for standing/sitting with minimal energy consumption are discussed so far.

This paper describes sensor-based controls of a SSMR combined with multiple feet for moving by wheels, feet, and mixture of these aiming for advancing mobility of a wheelchair. We introduce basic idea for generating functions of motion that are important for developing a robust electric wheelchair. A gravity sensor detects robot's pitch and roll angles directly. This angular information makes it possible to recognize a progressing direction of the robot from the top of a road, in addition to the road inclination angle. We demonstrate basic motions of the SSMR using the PEOPLER-IV in actual field, and show successful movements of stepping up/over and climbing stairs by sensor-based controls of wheels and feet. Experimental results verify that the proposed controls are applicable on all terrain.

2. Specification of the SSMR combined with feet

Fig.1 shows a sketch of our four-wheeled SSMR. Each wheel of radius R rotates around its axis, i.e., a hip joint J_h. In addition, an ankle joint J_a is attached on the periphery of each wheel with equal distance between neighbors. J_a rotates around J_h with radius r ($r < R$). Detailed design is similar to those of PEOPLER-II,[6] III by supposing that J_a is supposed to be a knee joint J_k.

From now we focus on motions of wheels and feet located at a front right position for simplicity of explanation. Also, we define parameters of θ and β as control variables of wheels and feet, respectively. That is, θ is an angular shift from front horizontal for progressing front (see Fig.2). β is also an angular shift from gravity direction to make a foot touch the ground ahead. It takes a negative value when the foot touches the ground behind.

All the feet have shorter length ($=\ell$) than r. And the robot can move using only feet without any contact of wheels on the ground when $\ell > R - r\cos(\pi/m)$, where m is the number of feet in each wheel (see Fig.2). Two inclinometers are attached on the robot to sense pitch; θ_x and roll; θ_y directly against the gravity direction. Specifications of motors, reduction

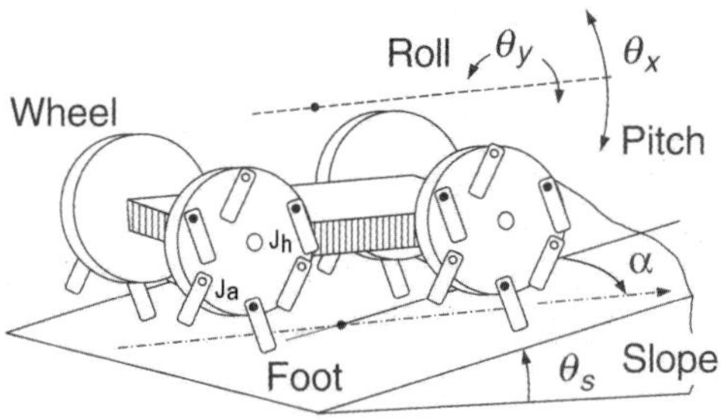

Fig. 1. Sketch of the PEOPLER-IV climbing a slope.

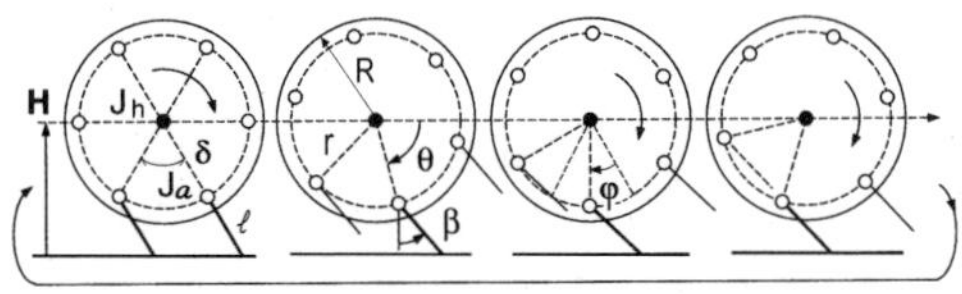

Fig. 2. Motion sequence of a smooth walking for suppressing vertical swing.

Table 1. Specifications of the hybrid robot, PEOPLER-IV.

Items	Detailed expressions
Body	110cm×74cm with ℓ=15cm, r=14cm, R=18cm. Weight;40kg
Wheel	Two aluminum plates with thickness 1.5mm are combined to form a wheel of outer diameter 36cm & tread 5cm.
Feet	Solid aluminum of 15cm long, 1cm thickness & 5cm width.
Gravity sensor	1-axis inclinometers (AccuStar, Pacico) ±60° for pitch & roll.
Controller	PC (VersaPro-NEC VY16M/XF-4) with 2 PC card slots.
Driver	MCUSB4sd (HPtec)
Motor & gear	DC motors Maxon 273753, 24V, 90W giving 106.26NM(max) with gear ratio 1/1150. 6 units.
Rotary encoder	MG-20-500P-5V.
Interface	PC Card(ADA16-32/2(CB)F; CONTEC) for gravity sensor & IOUSB32sd (HPtec) for Rotary encoder.
Battery(remote)	30A19RL 24V, 21Ah, 15Kg (Panasonic).
Performance	Max. Payload; 10Kg. Velocity; 2.0km/h(max). Step-up height; 15cm(Max).

gear ratios, encoders and etc. are listed in Table 1.

In general, ℓ is limited to be smaller than the distance between two neighboring J_as to keep free rotation without bumping. Therefore, it is evident that $\ell < 2r\sin(\delta/2)$. Irrelevantly to the dimension and number of feet of a robot under such that all the J_a keep the feet angles same to β, a center angle δ is written as

$$\delta = 2\cos^{-1}\{(R - \ell\sin\beta)/r\} \tag{1}$$

3. Environmental condition detected using a gravity sensor

In order to step up obstacles, feet need to keep their directions downward and continue to move forward so that some foot get a chance to stamp obstacles while wheels are rolling. Therefore, the sense of a gravity direction and a front horizontal direction of a robot is important. Notice that θ_s and α express angles of the ground inclination and the deviation of the robot front from a hilly direction, respectively. Then these are extracted uniquely using θ_x and θ_y in Fig.1 as follows:

Let us suppose that the robot aligns initially on the $x - y$ plane with its body center at the origin of a world coordinate system Σ_0, and with its body front toward the direction of y-axis. Also, suppose that $\mathbf{V_f}$ and $\mathbf{V_r}$ stand for unit direction vectors pointing frontward and rightward directions of the robot in Σ_0 having axial unit vectors **i**, **j** and **k**. Then, we have

$$\mathbf{V_f} = \sin\theta_x \sin_y \ \mathbf{i} + \cos\theta_x \ \mathbf{j} + \sin\theta_x \cos\theta_y \ \mathbf{k} \tag{2}$$

$$\mathbf{V_r} = \cos\theta_y \ \mathbf{i} - \sin\theta_y \ \mathbf{k} \tag{3}$$

Accordingly, we get the plane λ_b on which the robot moves. Actually, it is expressed as the plane including $\mathbf{V_f}$ and $\mathbf{V_r}$.

$$(\cos\theta_x \sin\theta_y)x - \sin\theta_x y + (\cos\theta_x \cos\theta_y)z = 0 \tag{4}$$

Noting that θ_s is the cross angle between λ_b and $x - y$ plane yields

$$\theta_s = \cos^{-1}(\cos\theta_x \cos\theta_y) \tag{5}$$

Also, we have a unit direction vector $\mathbf{V_p}$ pointing the hilly top of the ground in Σ_0

$$\mathbf{V_p} = -(cos^2\theta_x \sin\theta_y \cos\theta_y)/c \ \mathbf{i} + (\sin\theta_x \cos\theta_x \cos\theta_y)/c \ \mathbf{j} + c \ \mathbf{k} \tag{6}$$

where

$$c = \sqrt{1 - \cos^2\theta_x \cos^2\theta_y} \tag{7}$$

Therefore

$$\begin{aligned} \alpha &= \cos^{-1}(\mathbf{V_f} \cdot \mathbf{V_p}) \\ &= \cos^{-1}\{(\sin\theta_x \cos\theta_y)/c\} \end{aligned} \tag{8}$$

4. Cooperative Control of wheels and feet

Each wheel with odd and even numbered feet is arranged at a corner of a rectangular body. All the wheels and feet are rotated to move forward or backward, independently by the same driving mechanisms. However, those in one side behave always similar due to mechanical connections by chains. That is, all the front and back wheels are always same in angular phase. Likewise, all the odd numbered feet in front and back are always same in angular phase. Also, the even numbered feet are same in angular phase in the same manner. In the following, we deal with θ_s=0, otherwise designated.

4.1. *Feet control to change grip force on the ground*

Nails of a spike tyre spread to radial directions. In stead, each wheel enables all the odd numbered or even numbered feet to change their directions. Wider toe area prevents wheels from getting into the ground like *Kanjiki* in Japanese that is a boot attachment for widening contact area on the snow. The robot, PEOPLER-IV serves as a wheeled SSMR by setting all the feet toward upward with $\beta = \pm\pi$ radian, for instance. On the other hand, for the moving on frozen surface or soft ground like ice and sand, the robot can change the value of β intentionally for augmenting grip force on the ground to prevent pivot feet from sliding and skipping including slipping. The ends of feet operate as if they are spikes or pickels. Therefore, the direction control of feet are promising to move on snowy road, sandy beach, farm road, muddy road, pasture, and etc..

4.2. *Suppression of a body height swing*

For the purpose of keeping a robot's body at a designated height, feet landing is beneficial. Fig.2 shows a fixed height H when β is changed so that the following relations are satisfied under m=6, according as θ increases or decreases for frontward or backward traveling, respectively.

$$H = r\cos(\pi/6 - \psi) + \ell\cos\beta \tag{9}$$

$$\psi = \theta - n\pi/3 \tag{10}$$

where ψ is the angular displacement within the range δ, and n is the maximal natural number satisfying $0 < \theta - n\pi/3$.

Obviously, β decreases according to the increase of H. Postures of (a) in the left of Fig.3 show the motion that J_h moves right by repeating the control sequences from the top 1) to the bottom 5).

4.3. *Change of right and left strides for steering*

The SSMR can turn its traveling direction by changing the rotation speed on both sides as not only a wheel type but also a foot type. In addition, it can turn by changing strides between both sides under a same rotation

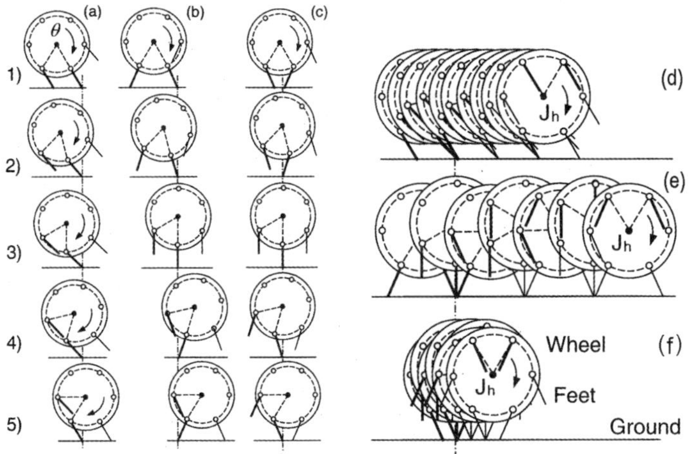

Fig. 3. Variation of feet controls to keep the body height same and to change strides (Left). Comparison of three walking related to the vertical swing and stride (Right).

speed of both side wheels. For instance, the stride becomes long and short by opening and closing adjoining feet, respectively. Simple control of only wheels with β=0 makes all the feet stamp the ground vertically. Therefore, the control is effective to step up or climb an object so that the object does not roll when a foot stamps its corner. We call this stride a *standard stride.* This stride is equal to the distance between two neighboring of J_as.

Postures (b) and (c) in Fig.3 show the moving with long and short strides, respectively. In both cases, all the feet happen to become vertical.[5] Posture 3) in (b) and (c) shows this occasion. In these walking, long and short strides appear alternately. Also, the robot body swings vertically, contrary to (a). In order to compare both stride and vertical swing range, transitional postures 1)-5) in (a), (b) and (c) are overwritten to have the respective figures of (e), (f) and (g) by aligning initial pivot foot at a same position. Figures from (e) to (g) clarify that the robot can proceed by selecting landing positions actively.

4.4. *Stepping on and over obstacles and climbing a stairs*

Feet are desirable to stamp the ground vertically with much power for driving both wheels and feet. Therefore, we use a word *traverse-ability* to evaluate a performance of stepping over/on obstacles concerned with β including its sign. Fig.4 shows different feet control types from (a) to (e) for moving right according as θ increases in the order from 1 to 5. In (a), the robot moves with standard stride. Odd numbered feet stretch frontward horizontally in (b), for instance. But all the feet stretch in (c) to move by

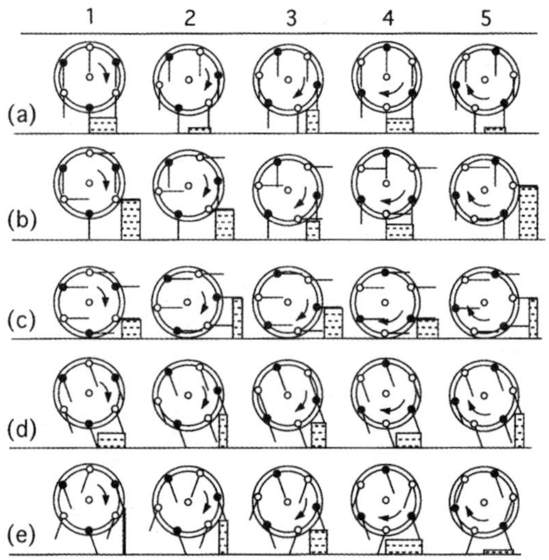

Fig. 4. Various feet directions to cope with stepping on and over obstacles.

wheels. In (d), all the feet lean frontward, but odd and even numbered feet lean opposite in (e) without changing β. Other controls might be considered as a mixture of (a) and (d), for instance. However, the controlling methodologies are summarized into the five in this study. Dotted rectangular boxes show the area where the foot can stamp. In this viewpoints, the posture in (b)-5 has an ability to climb a higher bump. The posture in (c) has also the same ability. However, it moves always by wheels, and the lower toe has a tendency to push the bump, or block the movement before starting a stepping up by an upper foot. This interruption happens in (d). Postures (d)-2 and (e)-2 are practically in the same situation. Likewise, postures (d)-3 and (e)-3 are same. Postures (d)-5 and (e)-2 are also same.

On the other hand, big power is necessary to step up the bump in (b) and (c). This is caused by the long distance between J_h and the toe contact position. Notice that each foot comes down by the rotation power of the wheel. Then, a foot pushing displacement is desirable to be small. This clarifies that (b) and (c) are inappropriate. Then, we select a better control that makes the pushing area minimal while some foot comes down.

Afore-mentioned considerations give us a reasonable selection that the control (e) is superior to others when some foot is positioned appropriately before the bump to start a stepping up motion. Otherwise, (d) and (e) are acceptable usually in this order.

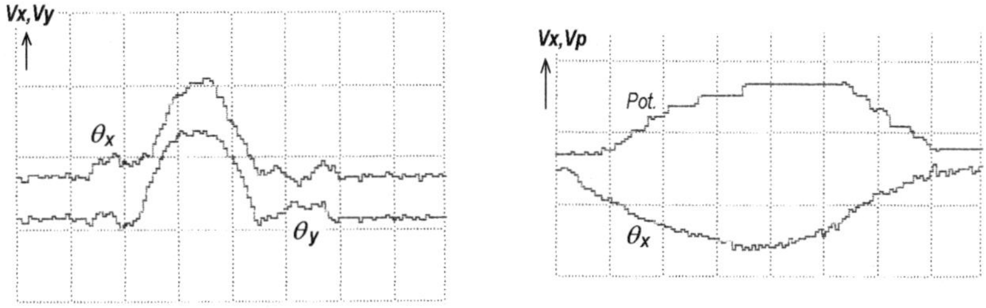

Fig. 5. Voltages from two inclinometers and their use for making feet directions vertical. Digital storagescope is set to 0.5V/div and 2s/div in a roll mode sweep.

5. Experiments of rolling, walking and climbing

5.1. *Sensor-based feet direction control*

The inclinometers produced voltages of Vx and Vy that are almost linear to θ_x and θ_y, respectively. Left part of Fig.5 shows the data collected in the movement from a flat lower level to a flat higher level through a slant looking a hillside to the left. Right part shows adaptation response to control feet directions vertical to any of θ_x. Clearly, the signal Pot. from the potentiometer increases when θ_x decreases to cancel the affect of θ_x. It is observed that there is a time-delay (approximately 1.8sec.) in the direction control but no stationary error.

5.2. *Indoor experiments*

We demonstrated fundamental movements by controlling wheels and feet on an indoor p-tile. Actually, only wheels or feet rotated for moving straight forward/backward. Also, cooperative controls of wheels and feet are examined. For the robot turn, we changed angular speeds of wheels and feet between right and left. These movements including the motion combining wheels and feet for suppressing the robot's vertical swing are carried out successfully, however the traveling speed is out of control at this moment.

Moreover, the robot traveled on a chaotic floor that was made by piled wooden plates of $91\times9\times2\text{cm}^3$(=L×W×H) on duckboards of $85\times45\times7\text{cm}^3$ with a slant angle 20° and blocks of $90\times9\times9\text{cm}^3$ located on ahead. The traveling performed by the control type (e) in Fig.4 with β=13° is overviewed in left of Fig.6. In a transient state, pivot foot was suspended due to absence of suspension mechanisms. Even though the robot traveled without stopping and falling by dragging power cables from a remote battery.

Fig. 6. PEOPLER-IV that stepps over the mock-up(left) and rolls on a concrete slope(right).

Fig. 7. Demonstration of a climbing on a concrete stairs having 2 steps.

5.3. *Outdoor experiments*

For the feasibility study of the SSMR, we found a practical site where concrete slope and stairs are established with joining each other. Usually, pedestrians use the site for coming and going between their office building and a parking area. The stairs have two steps of 1.46m width. That is, the height and depth, say $H_s \times D_s$, of the first and second steps are $9.7{\times}59\text{cm}^2$ and $11{\times}59\text{cm}^2$, respectively. Length and width of the slope surface is 2.76m and 1.5m, respectively. Inclination angles of the stairs and slope are $10.5°\{=\tan^{-1}(H_s/D_s)\}$ and $4.6°$, respectively. Right of Fig.6 shows the scene of the SSMR rolling up the slope by wheels with feet directed upward. The driver is a dummy.

For climbing up the stairs, the control type (e) is applied with $|\beta|=13°$ that is controlled automatically by the help of the two inclination sensors. Left and right parts of Fig.7 show the scene of climbing the first and second steps, respectively. Going downstairs was carried out successfully by changing only the wheel rotation direction opposite. The movie is attached in this paper for the observation of this climbing up/down.

6. Conclusion

For the purpose of improving mobility and performance of SSMRs on rough terrain, we proposed a wheel mechanism installing six ankle joints at circumferential positions of a wheel for landing by the toe of each foot one by one. Particularly, we considered the traverse-ability to evaluate a performance of stepping up/over obstacles concerned with β including its sign. Then, we disclosed a beneficial control by integrating traverse-ability and smallness of driving torques. That is, the control to make odd and even numbered feet face opposite directions about the gravity direction so that the SSMR can travel steadily by stepping up/over obstacles on the ground or climbing stairs. In the experiments, we could verify that the PEOPLER-IV can manage to travel on rough terrain by using θ_s and α that are calculated from the data of θ_x and θ_y. The SSMR might be useful in hazardous environment for accessing as a robust carrier, too. Finally, this study was supported by KAKENHI(C21560259).

References

1. K. Taguchi : "Enhanced Wheel System for Step Climbing", Advanced Robotics, vol. 9, no.2, pp137-147, 1995.
2. T. Iwamoto, K. Shibuya: "Concept and Mechanism of Robot Vehicle Using Variable Configuration Segmented Wheels", JSME J., vol.C71, no.701, pp.171-177, 2005 (in Japanese).
3. M. Lacagnina and G. Muscato, R. Sinatra: "Kinematics, Dynamics and Control of a Hybrid Robot Wheeleg", J. Robotics and Autonomous Systems, Vol.45, no.3, pp.161-180, 2003,
4. A.S. Boxerbaum, M.A. Klein, R.B. Roger Quinn, R. Harkins, R. Vaidyanathan: "Design of a Semi-Autonomous Hybrid Mobility Surf-Zone Robot", IEEE/ASME Int. Conf. on Advanced Mechatronics, pp.974-979, 2009.
5. T. Okada, W. Botelho, T. Shimizu: "Motion Analysis with Experimental Verification of the Hybrid Robot, PEOPLER-II, for Reversible Switch between Walk and Roll on Demand", Int. J. of Robotics Research, vol.29, no.9, pp.1199-1221, 2010.
6. T. Okada, W. Botelho, T. Shimizu: "Driving Mechanism of a Legged and Wheeled Mobile Robot with Minimal DOFs and Control for Reversible Switching between Walking and Rolling", J. of RSJ, vol.29, no.3, pp.306-316, April, 2011 (in Japanese).

OBSTACLE CROSSING WITH THE WHEEL-LEGGED ROBOT HYLOS BASED ON A ROBUSTNESS CRITERION ON THE FRICTIONAL CONTACTS

Pierre Jarrault, Christophe Grand and Philippe Bidaud

Institut des Systemes Intelligent et de Robotique
Universite Pierre et Marie Curie(CNRS - FRE 2507)
4, place Jussieu 75252 Paris cedex 05, France

This paper focuses on the control of "wheel-legged" mobile robot. We present a stability criterion based on the measure of the sustainable forces at the wheel-ground contacts. Then we propose a control law for obstacle crossing using the system capacities both in terms of reconfigurability and forces distribution. The goal of the control algorithm is to improve the robustness relatively to the errors induced, by noise and friction, in the control of the contact forces. The criterion is also used for analysis of obstacle crossing and we show that a reconfiguration planning is needed due to discontinuities of the optimal solution. A simple planning algorithm is presented to guarantee sufficient smoothness during the obstacle crossing.

Keywords: Kinematic reconfigurability, Obstacle crossing, Mobile robot

Introduction

Hybrid wheel-legged robots are systems designed to increase crossing capacities on uneven terrains. Systems like the HyLoS robot[1] are hybrid locomotion systems that use polyarticulated chains as active suspensions, ended by driven and steerable wheels. Such robots have the ability to change the position of their center of mass (CoM) and to modify the distribution of their contact forces. This paper presents a control algorithm that use, at the same time, the reconfigurability inherent to polyarticulated robots and the forces distribution in order to maximize the robustness, relatively to modeling error affecting the contact forces control, of both the contact stability and the tip-over avoidance.

Previous work on rover stability mostly use geometrical stability criteria based on the CoM[2] or ZMP[3] projection on the support polygon formed by the contact points. These methods do not consider the friction in contacts and thus are only suited for tip-over avoidance. Extensive work has been done on the problem of forces distribution with frictional contacts in the grasping research field. They express the combination of the constraints in the wrenches space associated with the CoM of the grasped object. As a result, this gives all the wrenches that are resistible if they are applied on the CoM. Different criteria derive from this representation.

Some authors[4] uses the smallest wrench applicable on the CoM as a criterion to choose a grasp configuration. Others[5] solves the forces distribution problem by maximizing the residual ball radius in this space, making sure that the robot will be able to add forces in order to support the largest possible disturbance. However, as these methods do not consider individually each contact force, the stability of some contacts may be reduced for the sake of global stability, making modeling errors affecting the contact force control more likely to cause slippage.

Thus we propose in this paper a stability criterion based on the smallest perturbation allowed at the contact level, we then present formulation of the optimization problem using this criterion. Simulation results are presented and shows that a planning of the CoM position is needed. A simple path planning algorithm is presented using the stability map computed with the criterion.

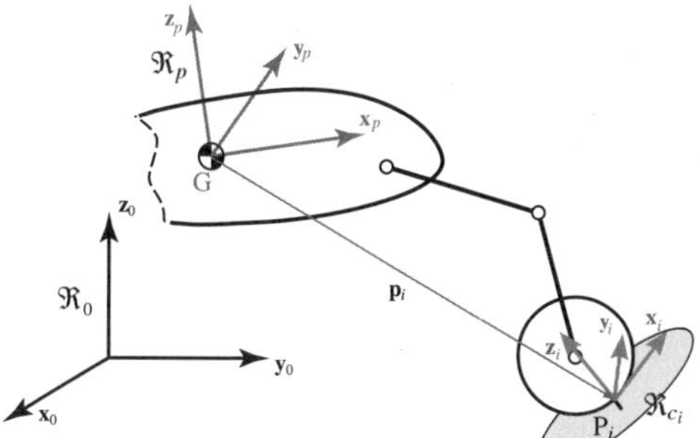

Fig. 1. Model of the system

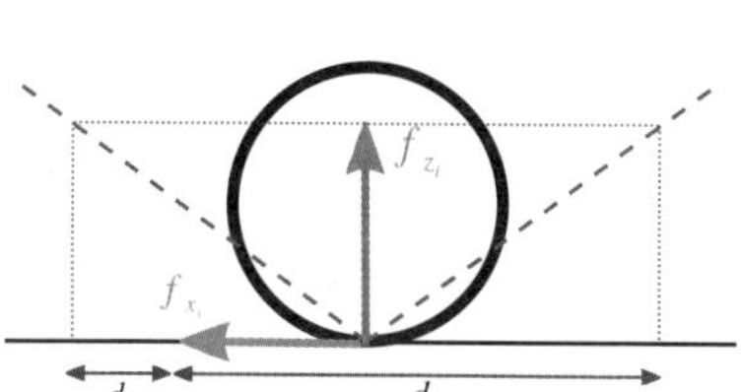

Fig. 2. Residual force on a contact

1. Force distribution equations

We consider a system supported by n legs (Fig. 1). The ith leg is in a frictional contact with the ground at point P_i which coordinates in the frame $\Re_p$ attached to the CoM are $\mathbf{p_i}$. The contact force of the ground on the leg is denoted $\mathbf{f_i} = \begin{bmatrix} f_{x_i} \; f_{y_i} \; f_{z_i} \end{bmatrix}^T$ where f_{x_i}, f_{y_i} and f_{z_i} are the components of the force along the contact frame's axis $\Re_{c_i}$, $\mathbf{z}_i$ is the contact normal and $\mathbf{x}_i$, $\mathbf{y}_i$ are the tangential directions.

The equations describing the equilibrium of the system are given by:

$$G\,\mathbf{f} = \mathbf{F} \tag{1}$$

where $\mathbf{f} = \left[\mathbf{f_1}^T \ldots \mathbf{f_n}^T\right]^T$ is a $[3n \times 1]$ vector containing all contact forces and $\mathbf{F}$ is the set of external and inertial wrench applied on the platform. G is a $[6 \times 3n]$ matrix giving the equivalent wrench to the contact forces at the center of mass in the frame $\Re_p$.

In addition to satisfy (1), the contact forces must respect some constraints:

- actuators limits:

$$\begin{aligned} J^T\,\mathbf{f} &< \boldsymbol{\tau}_{max} \\ -J^T\,\mathbf{f} &< \boldsymbol{\tau}_{max} \end{aligned} \tag{2}$$

  where $J = blockdiag(J_i)$, J_i being the Jacobian matrix of the ith leg and $\boldsymbol{\tau}_{max}$ is the torque limit vector.
- contact constraints defined by Coulomb's friction law which, expressed in a pyramidal form[6] are given by:

$$A_i f_i < 0 \tag{3}$$

  with

$$A_i = \begin{bmatrix} 0 & 1 & \frac{\mu_i}{\sqrt{2}} \\ 0 & -1 & \frac{\mu_i}{\sqrt{2}} \\ 1 & 0 & \frac{\mu_i}{\sqrt{2}} \\ -1 & 0 & \frac{\mu_i}{\sqrt{2}} \end{bmatrix}$$

2. Stability criterion

In order to maximize robustness with respect to modeling errors affecting the contact force control, we need a metric giving information about the sustainable perturbation at the contact level. Let us define vector $\mathbf{d}$ as:

$$\mathbf{d} = A\,\mathbf{f} \tag{4}$$

where A is a matrix defined by $A = blockdiag(A_i)$.

The elements of $\mathbf{d}$ represent the tangential force that can be added to the contact force of a leg in each tangential direction without breaking the contact stability (Fig. 2). By maximizing the smallest element of $\mathbf{d}$, we maximize the perturbations that can be sustained by the robot. The optimization criterion is then:

$$\phi(\mathbf{f}) = \min(A\,\mathbf{f}) \tag{5}$$

where $\min(\mathbf{x})$ is the function returning the smallest element of the vector $\mathbf{x}$. If $\phi(\mathbf{f})$ is negative, at least one contact is sliding or broken.

3. Control problem formulation

3.1. *Compact form of the force distribution problem*

Considering the elements given in the previous section, the force distribution problem is written as maximizing (5) subject to the constraints (1) and (2). We transform it into his compact primal form:[8]

$$\begin{array}{ll} \text{Find } \mathbf{x_f} \text{ maximizing} & \min\left(A[G^{+}\mathbf{F} + N_f\ \mathbf{x_f}]\right) \\ \text{Subject to:} & \\ & \begin{bmatrix} N_f^T\ J & -N_f^T\ J \end{bmatrix}^T \mathbf{x_f} < \mathbf{a_f} \end{array} \tag{6}$$

where:

- $G^{+}\mathbf{F}$ represents the particular solution of (1), G^{+} being the weighted pseudo-inverse of G.
- $N_f\ \mathbf{x_f}$ represents the homogeneous solution of (1), N_f being the matrix containing the basis vectors of the null space of G
- $\mathbf{a_f} = \left[(\boldsymbol{\tau}_{max} - J^T\ G^{+}\ \mathbf{F})^T\ (\boldsymbol{\tau}_{max} + J^T\ G^{+}\ \mathbf{F})^T \right]^T$

3.2. *Optimization of the CoM position*

We define $\mathbf{x_p} = [X \quad Z]^T$ as the vector containing the longitudinal and vertical coordinates of the CoM position in the global frame (considering the global frame is chosen to have the Yaw of the robot equal to zero). The null space of G is entirely defined by the contact geometry, so for fixed contact points, the null space of G is not modified by a change of X or/and Z. Furthermore for a frontal crossing, *ie* the rotation between the ground and the contact's frame of the obstacle is around the lateral axis of the robot, the particular solution can be stated as a combination of X and Z:

$$G^{+}\mathbf{F} = \mathbf{c_x}\ X + \mathbf{c_z}\ Z + \mathbf{f_0} \tag{7}$$

Thus, the CoM position can be added to the optimization variables:

$$\begin{array}{l} \text{Find } \mathbf{x} \text{ maximizing: } \min\left(A\left[\mathbf{f_0} + N\ \mathbf{x}\right]\right) \\ \text{Subject to} \\ \qquad J_p^T\ \mathbf{x} < \mathbf{a} \\ \qquad \|\mathbf{p_i}\| < D_{max} \end{array} \tag{8}$$

with $N = \begin{bmatrix} N_f\ \mathbf{c_x}\ \mathbf{c_z} \end{bmatrix}$ and $\mathbf{x} = \begin{bmatrix} \mathbf{x_f}^T\ \mathbf{x_p}^T \end{bmatrix}^T$, $J_p = \begin{bmatrix} N^T\ J & -N^T\ J \end{bmatrix}^T$ and $\mathbf{a} = \left[(\boldsymbol{\tau}_{max} - J^T\ \mathbf{f_0})^T\ (\boldsymbol{\tau}_{max} + J^T\ \mathbf{f_0})^T \right]^T$

4. Simulation

The control algorithm is evaluated on the HyLoS2 robot, an actively actuated polyarticulated system consisting of 4 legs supporting the chassis. Each leg is composed of 2 segments and a wheel with four actuated degrees of freedom (DoF): two for the leg movement, one for the steering and one for the traction.

The robot speed is sufficiently low to neglect the inertial effects on the CoM. The task for problem (8) is only to compensate for gravity ($F = \begin{bmatrix} 0 & -Mg & 0 & 0 & 0 \end{bmatrix}^T$) and the displacement of the robot is created by an additional torque on the wheels. Considering this, the only position's parameter which is optimized is X. During the simulation, the force optimization (6) is performed every 10 time steps of the simulation (1ms) while the position optimization (8) is performed every 100 steps. The null space's basis N_f is given by a singular value decomposition of the matrix G and the minimax optimization is performed by Dutta's algorithm.[9] Given a new desired configuration, the position of the CoM is controlled by adding a force to the task vector F in the force distribution problem (6). This force is computed by a PD controller. The distance between two wheels on the same side is fixed and the static friction coefficient is set to 0.8.

The robot is asked to climb a step with a height representing 25% of the distance between the wheels. The stability margin defined in (5) is plotted on Fig. 3 and the actual and desired horizontal position are given Fig. 4.

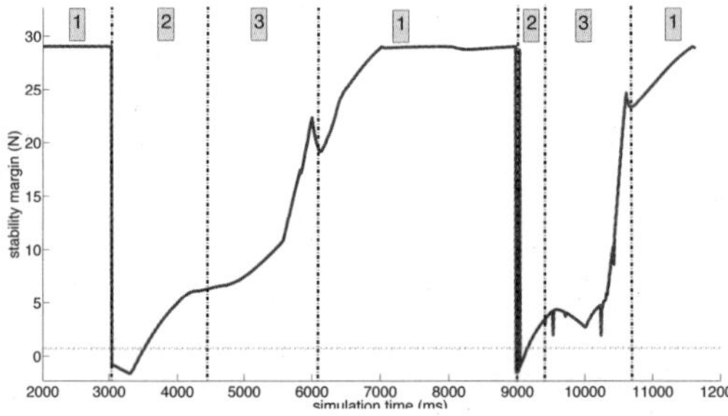

Fig. 3. Stability margin during a step crossing

During phase 1, the robot is on a horizontal flat ground. The front wheels touch the step and it enters in phase 2, the stability margin drops below zero, indicating that it can not produce the necessary forces to start the climbing motion. The desired position is calculated and the robot reaches

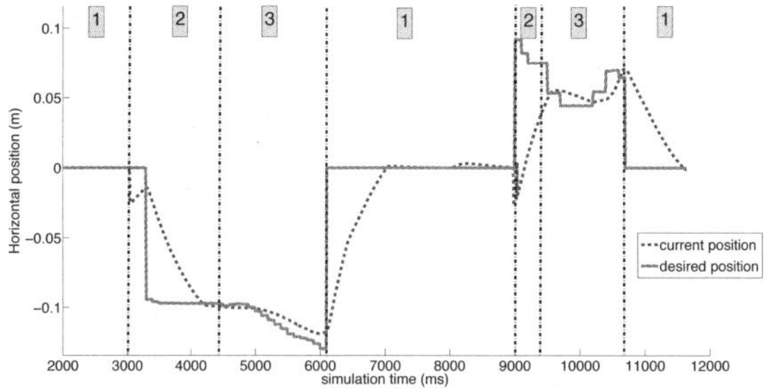

Fig. 4. Desired and actual horizontal position of the CoM

the optimal configuration. During this phase, the robot still uses the ground to sustain its weight. Once the stability margin has increased enough, the robot enter in phase 3 and starts to climb over the step. While climbing, the optimal position of the CoM is updated, avoiding possible tip-over as the robot is rising. Once the front wheels have reached the top of the step, the robot is once again in a phase similar to phase 1 where all the wheels are on horizontal planes. And the same phases follows: the rear wheels touch the obstacle, the robot reaches the optimal configuration and the wheels climb over the step.

5. Posture planning

The reactive algorithm presented above has some limits. To illustrate them, we draw, by solving problem (6) for each available posture and for each phase of the crossing, a stability map (Fig. 5) for a step to climb up followed by a step to climb down. For comprehension of the figure, unstable and impossible configuration are lowered and the optimal position is drawn in white.

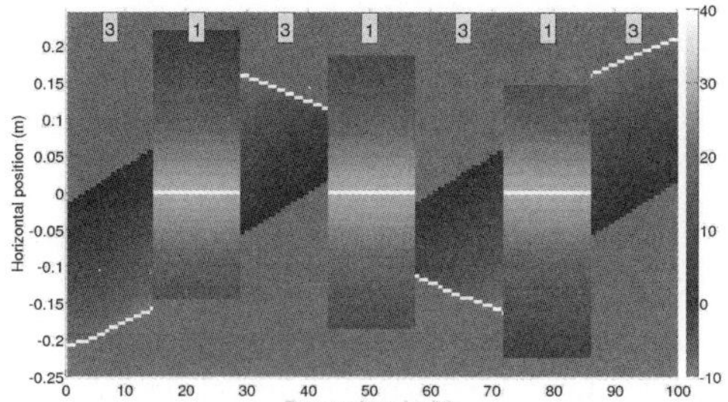

Fig. 5. Map of the stability margin depending on the robot's situation

The optimal posture trajectory is discontinuous, the forces applied to follow the optimal configuration can critically reduce stability. Furthermore, when the contact angles change suddenly, the optimal configuration at a given time can become an unstable one just after (57% of the crossing). It is thus necessary to plan a smooth configuration trajectory insuring stability at all time and considering the effects of the posture control.

To find this trajectory, we run a path planning algorithm using the stability map as a graph where each pixel of the map is a node. Each node from a column of the map are connected to all nodes of the next column.The weight of the link between two nodes is chosen as the lowest value among:

- the stability of the start node
- the stability of the goal node
- the stability of the start node modified by the posture control

We consider that response time of the posture control is lower than the time step of the graph. The maximum force applied by the posture control is the force computed at the beginning of the reconfiguration:

$$\mathbf{F_p} = K(\mathbf{X_e} - \mathbf{X_s}) \tag{9}$$

with K the proportional gain of the posture control, $\mathbf{X_s}$ and $\mathbf{X_e}$ are the starting and ending configuration.

To compute the optimal stability with this new task force, we need to solve problem (6) again, however the time required to do this through the graph is huge. Instead, we consider only the particular solution $\mathbf{f_c}$ of (1) and add it to the contact forces computed during the map drawing. The stability measure obtained from this process is conservative.

The value of each node is the lowest stability encountered to get to the node. When several nodes have the same value because the path leading to them is through the same low stability node, we differentiate them by the path starting from this node. Results of the algorithm on the map is given Fig. (6)

6. Conclusion

A criterion describing the robustness of the contacts stability for polyarticulated mobile robots has been proposed. It is based on the measure of the smallest perturbation sustainable at the contacts level. A control algorithm that compute both the best forces distribution and the best configuration of the robot relative to this criterion is described. Simulations show good results of the control scheme for a step crossing, but an analysis of the

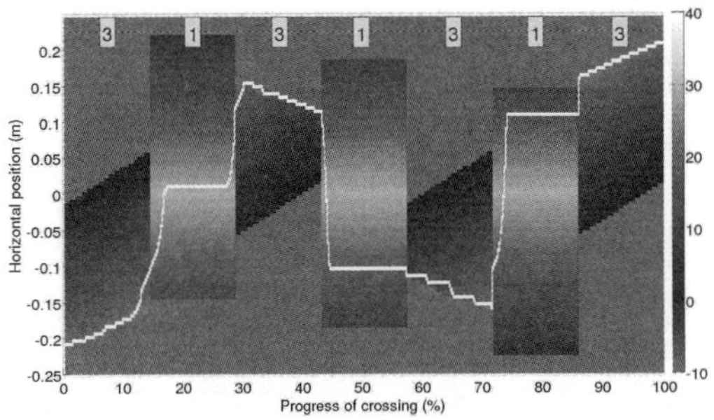

Fig. 6. Posture trajectory given by the breadth-first algorithm

posture trajectory shows that it needs to be planned in order to insure smoothness. A path planning algorithm is then presented. It uses the stability map obtained from the stability criterion to find a smooth trajectory of the CoM. Future works will focus on the implementation of the control algorithm on the actual platform HyLoS2.

References

1. C. Grand, F. B. Amar and F. Plumet, Motion kinematics analysis of wheeled-legged robot over 3D surface with posture adaptation *Mechanism and Machine Theory* **45**, 477(March 2010).
2. R.B. McGhee and A. A. Franck,On the Stability Properties of Quadruped Creeping Gaits *Mathematical Bioscience* **3**, 331 (1968).
3. M. Vukobratovic, A. A. Frank and D. Juricic, On the Stability of Byped Locomotion. *IEEE Transaction on Biomedical Engineering* **17**, 25 (1970).
4. D. Kirkpatrick, B. Mishra and C. Yap, Quantitative Steinitz's Theorems with Applications to Multifingered Grasping *Discrete and Computational Geometry* **7**, 295 (1992).
5. T. Yoshikawa and K. Nagai, Evaluation and determination of Grasping Forces for Multi-fingered Hands *IEEE Conference on Robotic & Automation* (1988).
6. J. Kerr and B. Roth, Analysis of multifingered hands *The International Journal of Robotic Research* **4**, 3 (1986).
7. A. Bicchi and V. Kumar, Robotic Grasping and Contact : a Review *IEEE International Conference on Robotics & Automation* (2003).
8. F.T. Cheng and D.E. Orin, Efficient Algorithm for Optimal Force Distribution - The Compact-Dual LP Method *IEEE Transaction on Robotics and Automation* **6** (1990).
9. S. R. K. Dutta and M. Vidyasagar, New algorithms for constrained minimax optimization *IEEE Mathematical Programming* **13** 140 (1977).

THE SIMPLEST PARKOUR MODEL: EXPERIMENTAL VALIDATION AND STABILITY ANALYSIS

A. W. Long*, R. D. Gregg, and K. M. Lynch

Department of Mechanical Engineering, Northwestern University, Evanston, IL 60208, USA

**E-mail: awlong@u.northwestern.edu*

We describe and experimentally validate the Simplest Parkour Model (SPM) for the ParkourBot, a planar dynamic climbing robot equipped with two springy BowLegs. By controlling the leg angles and injected energy at impact, the ParkourBot is capable of climbing up and down in a rigid chute on an inclined air table. The SPM consists of a point mass and two massless legs. The legs are assumed to be infinitely stiff, resulting in an instantaneous stance phase and a closed-form solution of the hybrid dynamics. In this paper, we show that the SPM is a good predictor of the actual experimental behavior. Using the SPM we compute the fixed points, stability and basins of attraction of period-1 limit cycles.

1. Introduction

Parkour is a form of dynamic locomotion that exploits body dynamics and reaction forces from the environment to efficiently access places that are otherwise difficult to reach.[1] Inspired by the flexibility of parkour, Carnegie

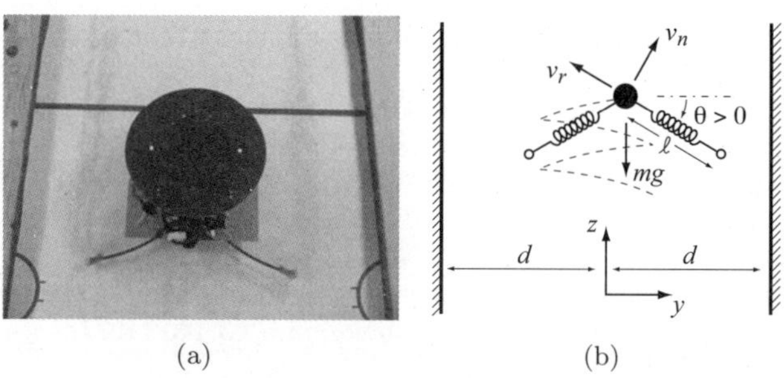

(a) (b)

Fig. 1. (a) Image of ParkourBot, a planar biped climbing robot, on an inclined air table to reduce gravity. (b) A cartoon of a simplified model of the ParkourBot ascending a chute.

Mellon University, in collaboration with Northwestern University, developed the ParkourBot[2] (see Fig. 1(a)), a dynamic climbing robot equipped with two springy BowLegs. Potential energy is stored in the legs during flight and is quickly converted to kinetic energy at impact. By controlling the leg angles and injected energy, our eventual goal is to control the ParkourBot to reliably move through a complex environment of footholds. Our first step is to control the ParkourBot to dynamically climb in a chute on an inclined air table with reduced gravity as shown in the cartoon in Fig. 1(b).

This paper describes and experimentally validates the Simplest Parkour Model (SPM), a simple model with closed-form hybrid dynamics. With the SPM, we compute fixed points, stability and basins of attraction (BOA) of period-1 limit cycles.

In Section 2 we place this work in context of previous work. Section 3 briefly summarizes the mechanical design of the ParkourBot. The SPM is described in Section 4 with experimental validation in Section 5. We conduct stability analysis of this model in Section 6.

2. Related Work

A common model for dynamic running and hopping legged robots is the spring-loaded inverted pendulum (SLIP).[3] The SLIP model has been used mainly to model robots traversing horizontal terrain such as Raibert's monoped,[4] Brown and Zeglin's BowLeg hopper[5] and RHex.[6] The nonlinear differential equations of the finite-time stance phase for the SLIP model cannot be solved analytically, so limit cycles and stability properties are computed numerically. Recently, Degani *et al.* modeled the ParkourBot as a two-legged SLIP and numerically analyzed its stability properties.[2] They observed theoretically and experimentally that climbing up at constant speed (with a larger energy injection at impact) is more stable with a larger BOA than climbing down.

By analogy to the simplest walking model[7] of passive dynamic walking[8] and the simplified one-dimensional hopping robot,[9] we investigate a simpler version of the SLIP model, which arose from personal communication with Degani. We refer to this model as the Simplest Parkour Model (SPM). The SPM is a special case of the SLIP model in that it has an instantaneous stance phase, enabling us to derive a closed-form solution of the hybrid dynamics and explicit expressions for fixed points of period-1 limit cycles.

The analytical nature of the SPM is particularly useful for motion planning applications based on sequential composition of asymptotically stable movements, known as Lyapunov funneling. Estimates of BOA can be used to stably compose distinct controllers, such as for sequential paddle-and-ball batting maneuvers.[10] A set of asymptotically stable walking gaits has been used as motion primitives for a bipedal robot,[11] where walking paths are planned as sequences of controllers from the set of primitives. The same procedure could be used with the analytical SPM to plan provably stable transitions between climbing velocities.

3. Mechanical Design of ParkourBot

The ParkourBot is equipped with two springy BowLegs which are low weight energy-efficient springs that operate similarly to an archer's bow.[5] During flight, a servo motor retracts a string attached to the foot, storing energy in the bow. At impact the string goes slack and the leg extends quickly, converting the stored potential energy to kinetic energy of the ParkourBot's body. Thus a low-power servo motor can store energy in the leg during the relatively long flight phase, and much of this energy is recovered in a high power event at impact. Another servo motor is used to independently change the leg angle. By controlling the leg angle and the stored potential energy, the robot is capable of climbing up and down as well as bouncing in place. Note that the robot does not adhere to or grab the wall, but instead uses the springy leg to kick off the wall. The robot is also equipped with a gyrostabilizer to minimize rotation in the plane of the air table. With the gyrostabilizer and low weight legs, a reasonable simple model of the ParkourBot is a point mass with massless legs. For a more complete description of the mechanical design, see Degani *et al.*[2]

4. Simplest Parkour Model

The SPM consists of a point mass m and two massless legs of length ℓ. The position of the point mass is $q = (y, z)$ as shown in Fig. 1(b). The springy legs are modeled as infinitely stiff, leading to an instantaneous stance phase. The inertial frame is centered between the two vertical walls separated by a distance $2d$. An impact occurs at $y = \pm a$, where $a = d - \ell \cos\theta$ and θ is the leg angle. At impact, kinetic energy KE is governed by

$$(KE)^+ = \epsilon(KE)^- + \sigma, \tag{1}$$

where $\epsilon \in (0, 1)$ is a restitution coefficient representing energy loss, σ is energy injected at impact and the superscripts $-$ and $+$ refer to just before and after impact. Positive σ corresponds to converting the energy stored during the flight phase into kinetic energy where as negative σ corresponds to storing more energy in the spring at impact. The current ParkourBot is not capable of negative σ, but future designs may be based on this idea.

The continuous dynamics are ballistic motion given by $[\ddot{y}, \ddot{z}]^T = [0, -g]^T$. At impact, an impulse is applied along the leg through the mass. We represent the robot's velocity at impact as $[v_n, v_r]^T = R(\theta - \pi/2)[\dot{y}, \dot{z}]^T$, where $R(\cdot)$ is the standard counterclockwise rotation matrix and (v_n, v_r) represent the normal (perpendicular to the leg) and radial (along the leg) velocities, respectively, as shown in Fig. 1(b). Since the impulse can only influence the radial velocity, the normal velocity remains constant through the impact. The impact map of the SPM is

$$q^+ = q^-, \quad v_n^+ = v_n^-, \quad v_r^+ = \sqrt{\epsilon(v_r^-)^2 + \frac{2\sigma}{m}}, \tag{2}$$

where the controls for the ParkourBot are $\xi = (\theta, \sigma)$.

We compose the integrable flight dynamics and the impact map (2) to find a closed-form solution from one post-impact state to the next. We do not write the solution here due to its length. Because the dynamics are invariant to the robot's height z, and because y at impact is determined by the leg angle θ, we restrict our analysis to the post-impact velocities.

By symmetry, we can assume the robot is always impacting the right wall; the position and velocity are reflected to the left wall immediately after impact. The velocity update map is $x_{k+1}^+ = P(x_k^+, \xi)$, where $x^+ = [\dot{y}^+, \dot{z}^+]^T$ is the velocity vector immediately after an impact. If the post-impact velocity $\dot{y}_{k+1}^+$ is negative, then the impulse did not reverse the sign of the horizontal velocity, and the point mass continues toward impact with the wall. We call this failure mode"tripping," resulting in the robot falling. We can solve analytically for the region of the $(\dot{y}^+, \dot{z}^+)$ space that leads to tripping on the next impact.

5. Experimental Validation of the SPM

Using a high-speed PhotonFocus TrackCam mounted above the air table, we recorded the positions of two LEDs fixed on the robot at a rate of 500 Hz to compute the position and orientation of the robot's center of mass. We recorded the trajectory through impact for many impacts with various leg retract controls and fixed leg angles of $\theta = \pi/6$. The leg retract controls correspond to different injected energies σ.

A recording for a single impact is shown in Fig. 2(a). As shown in Fig. 2(b), the body angle stays approximately constant throughout the trajectory due to the gyrostabilizer, allowing us to ignore orientation in

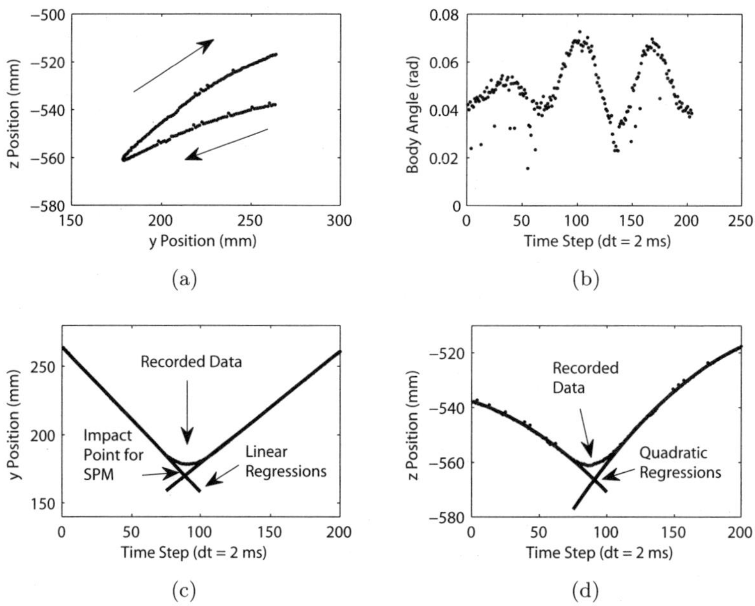

Fig. 2. (a) Recorded trajectory of ParkourBot for one impact. (b),(c), and (d) show the body angle, horizontal position y and vertical position z as a function of time, respectively, of the ParkourBot for one impact. (c) and (d) have linear and quadratic regressions overlaid for pre-impact and post-impact flights.

the SPM. During flight, the horizontal and vertical positions respectively follow linear and quadratic trajectories as predicted by ballistic motion, shown in Fig. 2(c) and Fig. 2(d). We fit linear and quadratic regressions to the pre- and post-flight positions. The impact is estimated to occur at the intersection of the linear regressions. With this method, we do not estimate the robot's state during the stance phase and focus on finding an instantaneous impact state that is compatible with the two flight phases. With the regressions, we estimate the pre- and post-impact states for the SPM. To compute the normal and radial velocities, we calculate the effective angle θ with the wall by combining the angles of the leg and the body.

The kinetic energy and normal velocity for post- versus pre-impact for three sample energy controls for many impacts can be seen in Fig. 3(a) and 3(b), respectively. The slopes of linear regressions of the energy data were averaged over many controls to produce $\epsilon = 0.7$, which is comparable to energy loss measured with the single leg BowLeg hopper.[5] The energy intercept gives an approximately linear relationship between the leg retract

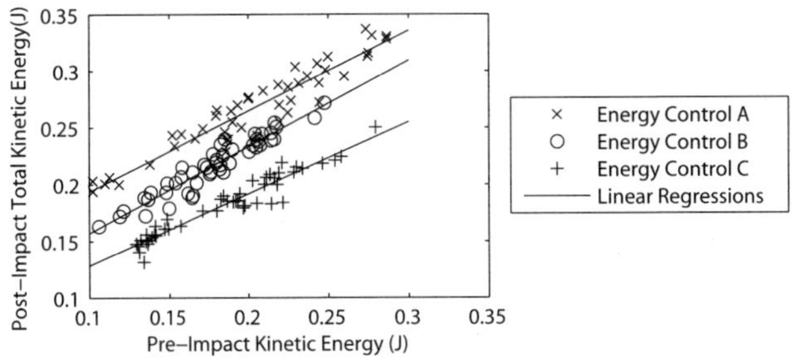

(a)

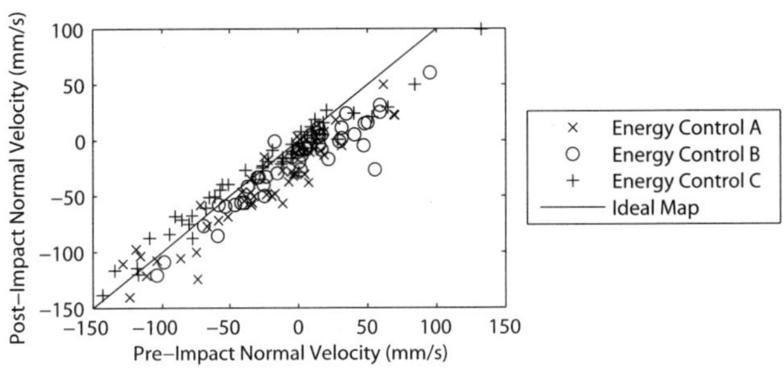

(b)

Fig. 3. (a) Post- vs. pre-impact kinetic energy with linear regressions. (b) Post- vs. pre-impact normal velocity with ideal map ($v_n^+ = v_n^-$).

control and the injected energy. Other parameters were measured to be $a = 0.1$ m, $m = 1.5$ kg, and $g = 0.98$ m/s^2. These experiments show that the SPM can be used to approximately model the ParkourBot.

6. Fixed Points and Stability Analysis

A fixed point x_* (corresponding to a period-1 limit cycle) satisfies $x_* = P(x_*, \xi)$. The velocities of the period-1 fixed points can be computed as

$$\begin{bmatrix} \dot{y}_*^+ \\ \dot{z}_*^+ \end{bmatrix} = \begin{bmatrix} 1 \\ \tan\theta - \frac{2}{(1-\epsilon)\sin 2\theta}\left(1+\epsilon-\sqrt{4\epsilon+(1-\epsilon)\frac{\sigma}{mga}\sin 2\theta}\right) \end{bmatrix} \sqrt{ag\cot\theta}. \tag{3}$$

The change in height per bounce h for the fixed point is given by

$$h = -\frac{4a}{(1-\epsilon)\sin 2\theta}\left((1+\epsilon) - \sqrt{4\epsilon + (1-\epsilon)\frac{\sigma}{mga}\sin 2\theta}\right). \tag{4}$$

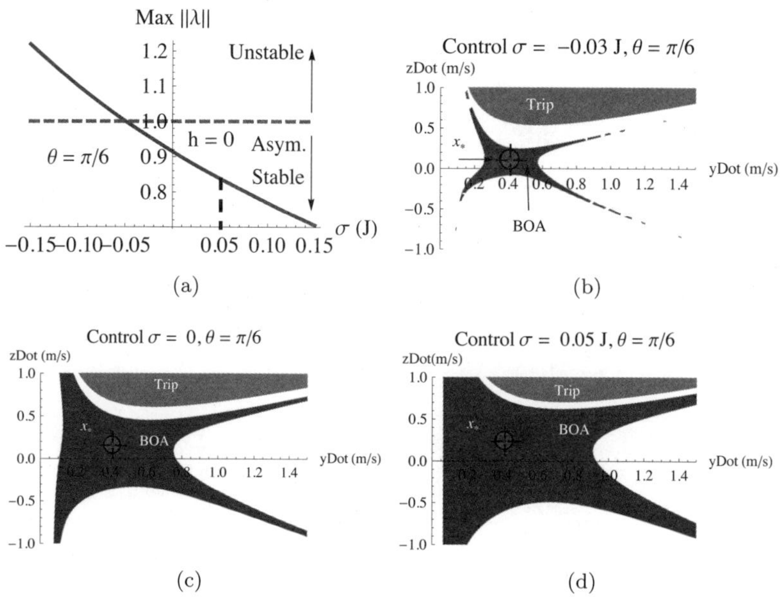

Fig. 4. (a) Maximum eigenvalue $||\lambda_{\max}||$ of the linerized discrete dynamics as a function of energy control σ for fixed leg angle $\theta = \pi/6$. (b) - (d) BOA in the velocity $(\dot{y}, \dot{z})$ space for three controls with their respective fixed points. The immediate trip region corresponds to points that do not reverse the horizontal velocity after impact.

We can solve (4) for the control σ as

$$\sigma = mga\left(\frac{(1-\epsilon)}{\sin(2\theta)} + \frac{(1+\epsilon)h}{2a} + \frac{(1-\epsilon)h^2 \sin 2\theta}{16a^2}\right), \tag{5}$$

which allows us to re-parameterize the fixed point with h as

$$\dot{y}_*^+ = \sqrt{ag\cot\theta}, \quad \dot{z}_*^+ = \left(\tan\theta + \frac{h}{2a}\right)\sqrt{ag\cot\theta}. \tag{6}$$

With this parameterization, we can more easily create motion plans than with the numerical results of the SLIP model.

If the maximum eigenvalue of $\partial P/\partial x$, evaluated at the fixed point x_*, has magnitude less than one, the system is asymptotically stable. Using the calibrated parameters and a leg angle of $\theta = \pi/6$, we computed the maximum eigenvalue as shown in Fig. 4(a), which shows climbing up is more stable than descending.

For asymptotically stable fixed points, the BOA is the set of points that will converge to the fixed point as the number of impacts approaches ∞. We estimate the BOA by simulating a sample grid in the velocity space and determining which points converge to the fixed point. Fig. 4(b) - 4(d) show the fixed point, the BOA and an immediate trip region for energy controls $\sigma = -0.03, 0$, and 0.05 J and fixed leg angle $\theta = \pi/6$. The immediate trip region is shown in red. The intermediate white space contains points that will eventually trip. The area of the BOA for the SPM decreases with decreasing energy levels, which was also observed with the SLIP model.

7. Conclusion

We describe the Simplest Parkour Model, experimentally validate the model and analyze the stability and basin of attraction of period-1 fixed points. The stability and BOA of the SPM qualitatively match those of the SLIP model. An advantage of the SPM is the closed-form hybrid dynamics, allowing for analytical expressions of the fixed points. The analytical nature of the fixed points can more easily be utilized in motion plans.

Acknowledgments

This work was supported by NSF IIS-0803734 and the NDSEG fellowship. The authors thank Adam Barber, Tsahai Phillip, Nelson Rosa, Ben Brown, Amir Degani and Siyuan Feng for discussions about the ParkourBot.

References

1. http://en.wikipedia.org/wiki/parkour
2. A. Degani, S. Feng, H.B. Brown Jr., K. Lynch, H. Choset, and M. Mason, The ParkourBot - A Dynamic BowLeg Climbing Robot, in *IEEE Int. Conf. Robotics and Automation*, 2011.
3. R. Blickhan, *J. of Biomechanics* **22**, 1217 (1989).
4. M. Raibert, *Legged Robots that Balance* (The MIT Press, Cambridge, MA, 1985).
5. G. Zeglin, The bow leg hopping robot, PhD thesis, The Robotics Institute, Carnegie Mellon University1999.
6. R. Altendorfer, N. Moore, H. Komsuoglu, M. Buehler, H. Brown, D. McMordie, U. Saranli, R. Full and D. Koditschek, *Autonomous Robots* **11**, 207 (2001).
7. M. Garcia, A. Chatterjee, A. Ruina and M. Coleman, *J. Biomech. Eng. Trans. ASME* **120**, 281 (1998).
8. T. McGeer, *Int. J of Robotics Research* **9**, p. 62 (1990).
9. D. Koditschek and M. Bühler, *Int. J. of Robotics Research* **10**, p. 587 (1991).
10. R. Burridge, A. Rizzi and D. Koditschek, *Int. J. of Robotics Research* **18**, 534 (1999).
11. R. D. Gregg, T. W. Bretl and M. W. Spong, Asymptotically stable gait primitives for planning dynamic bipedal locomotion in three dimensions, in *IEEE Int. Conf. Robotics and Automation*, (Anchorage, AK, 2010).

SECTION–12

MEDICAL AND BIOMEDICAL ROBOTS

ROBOTIC ASSISTED INTRAMEDULLARY NAILING OF FEMORAL SHAFTS

E. Cerny*, K. Bouazza-Marouf and D. Kerr

Mechanical and Manufacturing Engineering, Loughborough University, LE11 3TU, UK

**E-mail: E.Cerny@lboro.ac.uk*

In surgical procedures for femoral shaft fracture treatment, current techniques for locking the distal end of intramedullary nails, using two screws, rely heavily on the use of two-dimensional X-Ray images to guide three-dimensional bone drilling processes. Therefore, a large number of X-Ray images are required, as the surgeon uses his/her skills and experience to locate the distal hole axes on the intramedullary nail. The long-term effects of X-Ray radiation and their relation to different types of cancer still remain uncertain. Additionally, mistakes in locating the holes may lead to multiple attempts being made, weakening the fixture. Therefore, there is a need to develop a surgical technique that can limit the use of X-Rays during the distal locking procedure and allow for the distal locking holes to be drilled correctly, first time, every time. The various robotic and non-robotic methods that have been attempted to achieve this are discussed, along with barriers to adoption and potential improvements.

1. Introduction

Closed locked intramedullary nailing is now the preferred treatment for long bone fractures. Intramedullary nails are stainless steel tubes or rods which are inserted down the centre of the femur (medullary cavity) through a point of introduction at the proximal end of the bone. Locking screws fix the nail to the proximal and distal ends of the bone and provide rotational and axial stability.[1]

As it is a minimally invasive operation, there are fewer post-operative complications and shorter recovery times compared to conventional surgical methods. In almost all cases of intramedullary nailing, it is required to lock both ends of the nail to the bone, to prevent fragment rotation or bone shortening (which would traditionally be prevented by a long period of traction).

The process of intramedullary nailing consists of first making an incision in the soft tissue at the hip to gain access to the proximal end of the femur.

A hole is then drilled into the femur's medullary cavity to be able to insert the intramedullary nail. Following this, the nail is inserted and the fracture is reduced (the fractured sections of the femur are manipulated to return them to their original positions). Finally, after the nail is fully inserted the bone is locked to the nail with one or two screws at each end. A diagram is shown in Figure 1.

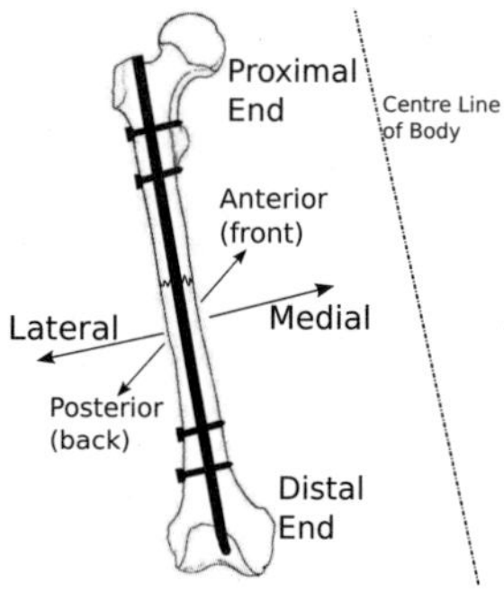

Fig. 1. Diagram of intramedullary nailed femur, including anatomical terms

When referring to different ends of the femur, or drilling from different directions, the medical anatomical terms lateral, medial, anterior, posterior, proximal and distal are used. These terms are shown in Figure 1 for the right femur. The lateral and medial directions are defined with respect to the centre of the body, so these would be mirrored for the left femur; the other terms are the same.

2. Non-Robotic Methods

The locking of an intramedullary nail at the proximal end is relatively straightforward, as a drill guide can be attached to the exposed thread in the proximal end of the nail. Drilling the distal locking holes is however far more complicated. The intramedullary nail may undergo bending during insertion,[2] and a small angular deviation of the drill guide, attached to the proximal end of the nail, would result in misalignment of the drill guide hole with respect to intramedullary nail distal hole.

There have been many different non-robotic methods for drilling distal locking holes, but their advantages have never outweighed their disadvantages in terms of cost, complexity and size. Hence the recommended method is still the freehand method.[3,4]

2.0.1. *Freehand Method*

The traditional method of drilling the distal locking holes involves the radiographer positioning the C-Arm so that the distal locking holes appear as perfect circles (from the lateral direction). The accuracy of the rest of the procedure depends on the accurate positioning of the C-arm, however this often takes many images to achieve and the C-Arm is very difficult to position accurately. The surgeon will then align a tool such as a scalpel, with X-Ray guidance, over the perfect circles image to locate the centre of the distal locking hole, as shown in Figure 2.

The surgeon can then create the stab incision down to the lateral cortex. This process is repeated to locate the starting position for the drill. The radiographer then must continuously take X-Ray images while the surgeon adjusts the drill position based on lateral and anterior-posterior X-Ray images. This results in a high exposure to X-Rays for the patient, and also for the surgeon. In addition to this, the surgeon will often have to re-drill the hole as it was not correctly aligned, further weakening the femur. Furthermore, each consecutive attempt becomes more difficult as the drill bit is more likely to slip into one of the existing incorrectly positioned holes. In the cases where the drill contacts the nail, the shape of the locking hole will not be correct and the locking may not be as robust. Throughout all of this, the surgeon will be exposed to indirect scattered radiation.

For experienced surgeons, the time taken to drill the distal locking holes will be less, fewer attempts will be required, and accuracy will be better, but this requires a long learning curve while operating (practising) on live patients. Additionally, as a large part of the process relies on the surgeon being able to visualise 3D positions from two orthogonal (and therefore dissimilar) 2D images, some surgeons may never be very good at the procedure, as visuospatial ability is largely inherent and difficult to improve.[2]

2.1. *Stryker S2™ IM Nail Distal Targeting Device for reduction of radiation*

In this method, the intramedullary canal must be over-reamed to 2mm greater diameter than normal, to allow easier insertion and retraction of the nail. The intramedullary nail is then inserted by an extra 10mm distally and a lateral image is used to select which of the drill guide holes in the calibration frame to use to drill a pilot hole.[5] After drilling, the targeting device is removed and the hole is cleared of debris. The alignment of the pilot hole is then confirmed by the interference of a bevelled probe with the

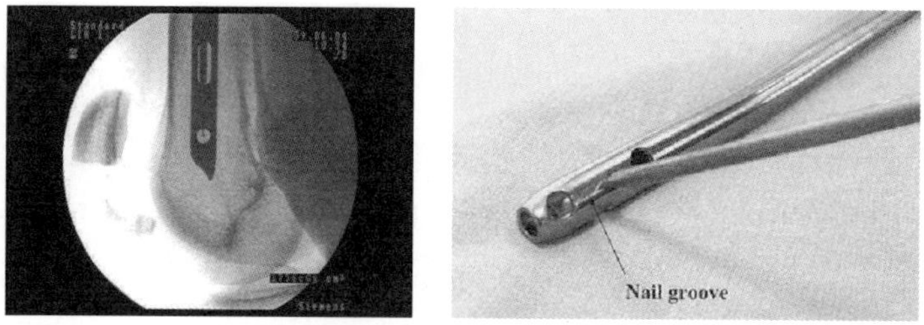

Fig. 2. Left: In the free-hand method, a scalpel is used to locate the distal locking hole[3] Right: Stryker S2 Targeting Device - Bevelled probe is used to locate the nail groove[5]

groove of the nail, see Figure 2. If the alignment is not correct, the nail is rotated to bring the groove under the pilot hole. The nail is then retracted until the nail distal locking hole is under the pilot hole. This is confirmed by the probe passing into the distal locking hole and subsequently the surgeon being unable to rotate the nail about it's long axis. A fixation sleeve is fitted over the probe to secure the location of the distal locking hole. The probe is locked in this position while the proximal holes are drilled and then a drill guide is attached to the probe to allow the drilling of the distal holes.

During testing on 127 patients, locking was unsuccessful in 5 cases, and the surgeon had to revert to the free-hand method. The average operation times and radiation exposure were considerably lower than the free-hand method.[5,6] The disadvantages of the method are the requirement to over-insert and retract the nail, meaning a shorter nail must be used, and the over-reaming of the medullary canal. Both of these will make the final osteosynthesis weaker than necessary, and also, excessive reaming can contribute to pulmonary complications from embolised medullary fat.[3,7]

3. Robotic Methods

The aim of the robotic assisted orthopaedic surgery system is to accurately locate the distal locking holes and either position a drill guide in the correct position (for a non-invasive system), or to drill the distal locking holes (in the case of an invasive system). For the non-invasive case there are also methods that require the surgeon to position the drill, but provide some form of visual feedback to guide the surgeon to the correct position, this will be referred to from here on as the assisted navigation method.

There are many possible ways to accomplish this, and all the different methods have different advantages and disadvantages. In all cases, the process consists of using fluoroscopic imaging to locate the distal locking holes, a method to identify the 3D position of the holes relative to some other known reference position(s) and finally a method to position a robotic manipulator of some kind in the position required to drill the holes (or for the assisted navigation method, to track the position of the drill in real-time and provide feedback for the surgeon).

3.1. *MARS Robot with distal locking attachment*

The system consists of a sterilizable robotically controlled targeting device, an image calibration ring for the fluoroscopic X-Ray unit, a video frame grabber, and a PC with a monitor. The targeting device consists of a miniature robot (MARS), a base to attach it to the bone, and a targeting guide mounted on the robot. The MARS robot is a $50 \times 50 \times 70$mm, 150-g, 6-degree-of-freedom parallel manipulator whose work volume is several centimetres. Its positional accuracy is better than 0.1 mm and when locked, it is very rigid and can withstand several kilograms of loading.

This implementation uses a small combination drill guide and calibration frame ($40 \times 55 \times 15$mm) with 28 3mm stainless steel ball bearings. The ball bearings are mounted in two planes, one plane on each side of the drill guide. The MARS robot can be attached directly to the bone with self tapping screws, or can be mounted to the proximal end of the nail.[8,9]

The C-Arm is aligned (under software guidance) to obtain full circles for the distal locking holes. This typically takes 2-6 images. Alternatively, the authors state that using a more generic method, the distal locking can be achieved with one lateral image only (not aligned to obtain full circles).

The whole image is distortion calibrated using a frame fixed to the X-Ray image intensifier as described in earlier work.[10] This may be necessary due to the very small size of the drill guide / calibration frame: errors in the calculation of the focal point by as much as 50mm still allows for tracking with an accuracy of 0.5mm in translation and 0.65° in rotation.[11]

Using the "Fronto-Parallel" method (where the C-Arm is aligned so that the distal locking holes appear as full circles), the image diameter of the holes is used to calculate the distance of the nail from the focal point. This is not very accurate due to the size of the nail holes in the image, but as the nail frontal plane is almost parallel to the XRII in this specific setup and the drill guide / calibration frame is very close to the nail, the effect of angular errors is small.

Using the "generic" method, the size and position of the distal locking holes is used to calculate the three-dimensional position of the nail (without having to align the X-arm to view the full circles). While this is definitely possible mathematically, inaccuracies are introduced due to the low resolution of the images (the distal locking holes are approximately 20 pixels in diameter).

The results from testing show average errors of 3.0mm translational and 1.3° rotational (maximum 4.5mm and 2°) for the Fronto-Parallel method, meaning that this method may fail occasionally and require re-drilling (assuming a nail hole diameter of 7mm). The generic method is currently not accurate enough to successfully locate the distal locking holes every time.

3.2. *Harbin Institute of Technology Robotic Assisted Orthopaedic Surgery system (HIT-RAOS)*

This is a general orthopaedic surgery system, designed to assist an othopaedic surgeon throughout an operation, from fractuce reduction to nailing and locking. The leg is attached to a parallel fracture reduction robot, fixed at the knee and at the foot. To drill the locking holes, this system uses a serial 6 degree of freedom robot, with a radio-transluscent calibration box attached to it. This contains 8 fiducials, 4 on each plane. This is a low number considering the amount of non-linear distortion typical of X-Ray images, however to counteract this, a custom "precision" C-Arm was manufactured to reduce the amount of image distortion. Linear interpolation is then used to calculate the position of the holes. Zhijiang et al[12] state that the accuracy from this is sufficient to complete the drilling of locking holes.

3.3. *Braunschweig & Hannover robot-assisted long bone fracture reduction system*

This is a complete fracture reduction and intramedullary nailing system.[13] The fracture reduction is calculated from 3D X-Ray imaging, with respect to two Dynamic Reference Bases (DRBs) attached at either end of the femur, and one attached to the manipulator of a 6 DOF robot. The robot used is a large Stäubli RX 90 with a reach of 1 metre, which is floor mounted.The same DRBs are used to calculate the distal locking hole trajectory, but with two 2D X-Ray images - lateral and anterior-posterior. A line of sight surgical navigation system is used to calculate the relative positions or all the DRBs.

The results from testing give a mean translational error of 0.96mm (min 0.42 / max 1.57) and a mean rotational error of 2.70 degrees (min 1.21 / max 4.5). The angular error is the greatest error in this method, as it relies on the accuracy of the radiographer in positioning the C-Arm so that the distal locking holes appear as full circles.

3.4. *Bi-planar robot navigation system - Beihang, China*

In this system,[14] a very small number of fiducials are used - only 4 per plane. The surgeon designates the locking path by identifying the entry point and axis of the nail hole on the X-Ray image. Their tests have shown an average distal locking error of 0.28mm with a maximum error of 0.44mm. The paper does not explain how the robot is fixed relative to the nail so it may just be a proof of concept and accuracy at this stage. As the calibration frame / drill guide must be placed around the leg, it will be less practical to use compared to other methods. Its advantages are: potential for higher accuracy as fiducials are places both sides of the nail, The robot is safe by design, as it does not have the degrees of freedom to interfere with the patient, and it has high stiffness and low weight due to its geometry.

3.5. *Hull Computer Assisted Orthopaedic Surgery System*

In this system, a line of sight registration method is used to get a transformation between the C-Arm and manipulator coordinate systems. The manipulator is a drill guide attached to a passive lockable arm. The transformation between the distal locking holes and the manipulator is achieved through the use of a "registration phantom", an H arrangement of metal balls that is attached to the manipulator and must be positioned in both lateral and AP X-Ray images. For this method, X-Ray calibration needs to be performed monthly, to calculate the undistortion characteristics to be used during surgery. It should be noted at this point that a significant amount of distortion is experienced simply from the orientation of the C-Arm to the earth's magnetic field,[15] so the C-Arm can never be completely calibrated pre-operatively.

After taking lateral and AP images, the surgeon identifies key points on the image and marks them using the mouse. This allows the computer to process the data and automatically calculate the pose of the nail relative to the manipulator. The surgeon must then align the drill guide using computer feedback based on the optical tracking of the manipulator.[16,17]

3.6. *Brussels Orthopaedic Surgery System*

This method[18] uses two Lateral images separated by an acute angle, but where the nail holes can be seen in both images. A calibration frame of two sheets of radio-transluscent polycarbonate containing radio-opaque fiducials is attached to the X-Ray image intensifier. This is used to calculate the intrinsic and extrinsic parameters for the projection - i.e. the focal length and distortion. An optical tracking reference marker is used to compute the actual 3D position of this calibration frame and therefore the C-Arm position. The intersections of corresponding points in each image is calculated, modelling the X-Ray as a point source. Online modelling of the nail is then used to compute the angle of the nail with respect to the C-Arm (via line of sight tracking). Mean errors were 1.5mm translational and 1° rotational. The computational time for this method is around 4 minutes.

The surgeon drills the distal locking holes using either a drill with an optical tracking reference frame attached, or a passive drill guide with a reference frame attached and an ordinary drill. A lateral profile of the nail is displayed on a screen, with two circles representing sections of the drill. In order to drill through the calculated trajectory, the surgeon must align the two circles concentrically over the centre of the distal locking hole. In their experiments, Leloup *et al* noted that the surgeon found it easier to align the referenced drill guide then drill while holding the drill guide, rather than to align the referenced drill on its own.

3.7. *Loughborough Orthopaedic Assistant System*

The Loughborough Orthopaedic Assistant System (LOAS) has been developed in an attempt to remove some of the barriers to adoption that are present in other systems. The procedure involves the external attachment of a Γ-shaped calibration frame to the nail using existing attachments with an adapter. The Γ-calibration frame which overlaps the nail in both lateral and AP X-Ray images has two planes of radio-opaque fiducials embedded in radio-translucent plates on each arm (i.e 2 AP planes and 2 lateral planes). This allows the calculation of the focal point of each (AP and lateral) uncalibrated image, and hence the 3D position of any point in the images. The AP and lateral images are then transferred to the system computer. The shape of the distal locking holes in the lateral image and the nail orientation in AP image are then automatically used to derive the drilling trajectory and thus the pose of the drill guide attached to a semi-active robotic device. The surgeon then removes the Γ-calibration frame and replaces it by the small robotic manipulator.

This method has the advantage that the robotic manipulator is autoclaveable, relatively very small, lightweight and easy to use/implement. The accuracy of the vision system planned trajectory has been verified using theodolites[19] and the accuracy of the whole system (vision system & robotic alignment) is being assessed using artificial bones and animal bones. However medical trials have not been undertaken yet.

Table 1. Comparison of distal locking methods

Method	X-Ray Images Required	Mean Entry/ Exit Error/ mm	Max Entry/ Exit Error/ mm	Mean Angular Error/ °	Max Angular Error/ °	Line of sight navigation	Large robot required
Braunschweig	3-5	0.94	1.57	2.54	3.75	•	•
Brussels	2	1.5	-	1	-	•	•
Hull	2-4	0.5	2.6	0.2	-	•	-
Harbin	-	-	-	-	-	-	•
Beihang	-	0.28[a]	0.44[a]	-	-	-	-
Jerusalem[b]	2-6	3.0	4.5	1.3	2	-	-
Jerusalem[c]	1	3.4	6	2.7	8	-	-
Athens	2	-	-	-	-	-	-
Loughborough[d]	2	0.81	1.18	1.01	1.25	-	-
Loughborough[e]	2	-	0.95	-	0.6	-	-

Note: [a]Error is defined as distance of hole centre from drilling centre, measured in the lateral plane only [b]Fronto-Parallel method [c]Generic method [d]Vision accuracy [e]Drilling accuracy

4. Conclusion

Despite a very large amount of research having been conducted over the last 20 years, including a lot of clinical trials in some cases, the free-hand method is currently still the recommended method for the distal locking of intramedullary nails. As can be verified in the qualitative summary of results in Table 1, most of the reviewed robotic assisted surgery methods have the necessary accuracy to correctly drill the distal locking holes first time, with no need to re-drill and with far less X-Ray exposure than the traditional method. The lack of uptake so far of these methods is clearly due to other factors.

Cost is of course a large factor in the uptake of any system, and the price of the majority of the reviewed systems cannot be justified with respect to the targeted surgical procedure. Another major factor is that of the use of large industrial designed robots — safety issues, sterility compliance and electromagnetic emission compatibility compliance make it almost impossible to adopt a system using a robot of this kind. Finally, a system that requires the surgical operation to be carried out very differently to traditional methods will always be difficult to implement, as the change-over to such a system would be relatively complex. In conclusion, there is still a need for a relatively cheap (cost-effective) and easy to implement osteosynthesis assistance robot, that does not drastically change the rest of the surgical procedure.

References

1. I. Kempf, A. Grosse and G. Beck, *J Bone Joint Surg Am* **67**, 709 (1985).
2. I. Browbank, Intraoperative registration for roboticassisted orthopaedic surgery, PhD thesis, Loughborough University 2001.
3. P. V. Giannoudis, H. C. Pape and M. Schuetz, *AO Surgery Reference* (Davos, 2011), ch. Femur Shaft.
4. V. Buehren, J. D. DiCicco III and T. G. DiPasquale, *T2 Femoral Nailing System - Operative Technique* (Stryker Trauma GmbH, 2006).
5. G. Anastopoulos, P. Ntagiopoulos, D. Chissas, A. Papaeliou and A. Asimakopoulos, *Clinical Orthopaedics and Related Research* **466**, 216 (2008).
6. G. Anastopoulos, P. G. Ntagiopoulos, D. Chissas, G. Loupasis, A. Asimakopoulos, E. Athanaselis and P. Megas, *Injury* **39**, 1210 (2008).
7. M. P. M. Klein, B. A. Rahn, R. Frigg, S. Kessler and S. M. Perren, *Archives of Orthopaedic and Trauma Surgery* **109**, 314 (1990).
8. Z. Yaniv and L. Joskowicz, *IEEE Trans. on Medical Imaging* **24**, 624 (2005).
9. L. Joskowicz, C. Milgrom, M. Shoham, Z. Yaniv and A. Simkin, *International Congress Series* **1256**, 485 (2003).
10. H. Livyatan, Z. Yaniv and L. Joskowicz, *MICCAI* **2489**, 60 (2002).
11. A. Jain, R. Kon, Y. Zhou and G. Fichtinger, *MICCAI* **3749**, 639 (2005).
12. D. Zhi-jiang, K. Min-xiu, F. Li-xin and S. Li-ning, A novel fluoroscopy-guided robot-assisted orthopaedic surgery system, in *Proc. IEEE Int. Conf. Robotics and Biomimetics ROBIO '06*, 2006.
13. R. Westphal, S. Winkelbach, F. Wahl, T. Goesling, M. Oszwald, T. Huefner and C. Krettek, *Int. J. Rob. Res.* **28**, 1259 (2009).
14. H. Lei, L. Sheng, W. Manyi, W. Junqiang and L. Wenyong, *Int. J. Med. Robotics Comput. Assist. Surg.* **6**, 61 (2010).
15. I. Browbank, K. Bouazza-Marouf and J. Schnabler, *Proc. of the IMechE, Part H: J. of Eng. in Medicine* **214**, 165(January 2000).
16. R. Phillips, A. Mohsen, W. Viant, S. Malek, Q. Li, N. Shah, M. Bielby and K. Sherman, *MICCAI* **3217**, 621 (2004).
17. S. Malek, R. Phillips, A. Mohsen, W. Viant, M. Bielby and K. Sherman, *Int. J. Med. Robotics Comput. Assist. Surg.* **1**, 34 (2005).
18. T. Leloup, W. El Kazzi, F. Schuind and N. Warzee, *IEEE Transactions on Medical Imaging* **27**, 1202 (2008).
19. F. Junejo, K. Bouazza-Marouf, D. Kerr, A. Taylor and G. Taylor, *Proc. of the IMechE, Part H: J. of Eng. in Medicine* **221**, 365(January 2007).

FORCE CHARACTERISATION OF ONE AXIS FOR AN ADHESIVE FOOT IN A CLIMBING ROBOT FOR MINIMALLY INVASIVE ABDOMINAL SURGERY

ALFONSO MONTELLANO LÓPEZ*, MOJTABA KHAZRAVI, ROBERT RICHARDSON, ABBAS DEHGHANI, RUPESH ROSHAN, TOMASZ LISKIEWICZ, ARDIAN MORINA, ANNE NEVILLE

School of Mechanical Engineering, University of Leeds, Woodhouse Lane Leeds, West Yorkshire, LS9 2JT, United Kingdom

DAVID G. JAYNE

Academic Surgical Unit, St. James's University Hospital Leeds, West Yorkshire, LS9 2JT, United Kingdom

This paper presents the characterisation of the force generated by a miniature piezo-electric motor (SQUIGGLE®) integrated into a mechanism to move a micro-structured adhesive pad. The characterisation focuses on the amount of force the configuration is able to exert along its axis. These pad and mechanism are part of an adhesion-reliant robot for intra-abdominal operation against gravity. Control of the adhesive behaviour of the pad will be enabled by precise knowledge of the force that the motor is able to apply. Firstly, an introduction presents a summary of state-of-the art technologies in robot-assisted surgery. Secondly, the purpose, environment and mechanical design of the robot are introduced. Next, results from the force characterisation show the mechanism is able to deliver sufficient force for the control of the adhesive foot. Further tests to determine the way force and moment are transmitted through the robot are to be performed to optimise the design of the hardware and software.

1. Current developments, challenges and trends in robot-assisted surgery

The number of robots used for surgical interventions has steadily increased since the 1980's [1] including procedures where robotic assistants move a camera, cut, suture, drive needles and drill bones for a variety of operations. With these devices, a properly trained surgeon is able to perform a more precise and shorter intervention, reducing the risk of post-operative complications as compared to open surgery and conventional (non-robotic) laparoscopy [1, 2]. However, current surgical robots are costly, cumbersome and have limited accessibility

* Corresponding author: mnaml@leeds.ac.uk

and visual feedback. To overcome these difficulties, the new trend is to design miniature robots able to intelligently interact with biological tissue.

Amongst the projects concerned with the miniature size in robotic surgery, the following are particularly remarkable: HeartLander from Carnegie Mellon University [3], where a remotely actuated robot crawls using vacuum suction cups over the beating heart; Nebraska Wheeled Robot [4]: a robot that moves over the abdominal organs driven by two DC motors; the ARAKNES project [5] in University of Pisa with other European partners, aimed to build an array of micro robots to access the stomach and perform image-guided operations controlled by a console.

One of the challenges of designing a miniature robot is the actuation technology to be used. For selection of a commercially available motor in the miniature size, four main actuation technologies are currently available [6, 7]. These are: piezo-electricity, brushless DC micro-motors (BLDC), Shape Memory Alloys (SMA) and Electro-active Polymers (EAP). EAP applications currently exert very low forces (normally less than 10gf) and their control is still a very complex task. The control of SMA is currently more advanced and the force higher, but their integration in a compact design is difficult. Besides, the high heat dissipation of these materials is a significant downside for an intra-corporeal robot. DC micro-motors and piezo-electric motors are the most mature technologies. Compared to DC motors, the main disadvantage of piezo-electric motors for medical applications is the high voltage they require.

Another major challenge is the manufacturing of the mechanism where the motor is integrated. Due to the difficulties of working on the miniature scale, it is specially important to characterise the resulting mechanism.

In line with the former research trend, this paper presents a new miniature surgical robot for abdominal surgery which operates inside the body without external fixtures. In the following section the purpose of the robot within its specific environment is presented. Section 3 introduces the design of the robot and the last section shows the results of the force characterisation.

2. Purpose, environment and requirements of the robot

This robot is designed to go inside the abdomen to assist a surgeon performing a laparoscopic operation. During this procedure, the abdomen is insufflated with carbon dioxide to create an operating space. The robot will be placed on the peritoneum and after this initial attachment will have to walk upside down over the surface. In this way the robot will enable the transportation of a camera to explore the abdominal cavity. In the future, it could also carry surgical tools. Figure 1 shows the environment of the robot and its position within it.

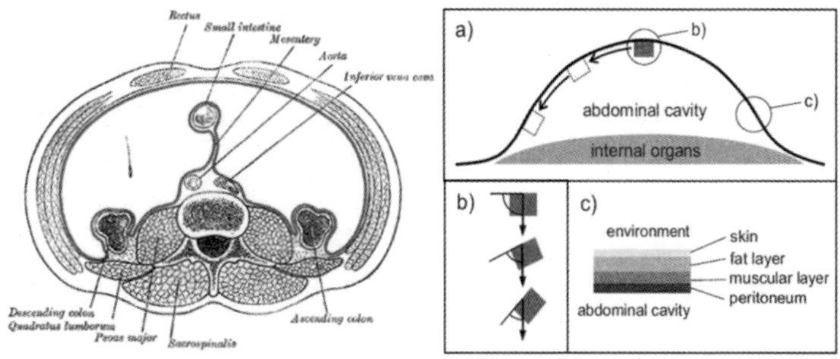

Figure 1. Transverse section of the abdomen from [8] on the left. The peritoneum is represented by a dark red line, enclosing the operating space created by insufflation with carbon dioxide. On the right, diagram showing (a) the device in the operating environment, (b) working at different angles, (c) the composition of the abdominal wall [9].

The robot is therefore required to be positioned anywhere on the peritoneum and operate against gravity, keeping contact with the tissue at all times. It also has to detach smoothly to minimise tissue damage and be as safe and compact as possible to enable intra-corporeal mobility. Although the robot will operate at all possible angles inside the abdomen, namely from 90° to around 60° between gravity and the surface of the peritoneum, this initial study assumes an angle of 90°. It is also assumed that the contact surface between the robot and the peritoneum is small enough not to be affected by the curvature of the abdominal cavity.

3. Conceptual design of the robot

The starting point for the design of the robot is the adhesive material for the feet. Such material is a bio-inspired surface developed in the University of Leeds. Drawing inspiration from beetles, geckoes and tree frogs [10-12] the micro-structure on this surface uses Van der Waals and capillary forces to provide wet adhesion to biological tissue [9]. The handling of this adhesion force is the main purpose of the mechanism of the robot, together with its own locomotion. This is achieved by applying forces to the pad, so that it can be attached, detached and moved. The selected motor to apply these forces is a ultrasonic motor from the travelling wave family. Although piezoelectric actuators normally use very high voltages, recent developments (Squiggle RV® [13]) use as little voltage as DC micro-motors and are more energy-efficient for the small scale.

For their simplicity and ease of integration into a compact design, two main locomotion paradigms were considered:

- Adhesive wheels or tracks: allow the use of just two rotary motors for steering ability. However, there is the risk of skidding or straining the tissue while moving over it.
- Adhesive feet: need a more complex actuation systems but offer the advantage of a smoother motion.

The final choice for the conceptual design is a Cartesian walker with four individually actuated adhesive pads (feet). Although four-legged robots have typically a narrower static stability margin as compared to six-legged, upside-down operation and the use of broad, adhesive feet enables stable robot locomotion. Locomotion is obtained by transportation of the feet as shown in Figure 2. This figure also shows how the robot reaches a target position subdividing the trajectory.

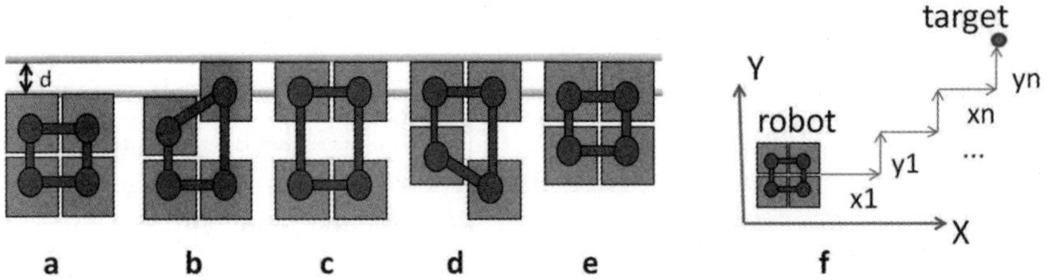

Figure 2. On the left (a-e), four-pads robot taking one step, travelling a distance *d*: (a) initial position, (b) top-right pad has travelled a distance *d*, (c-d) the same for the other three pads until (e) the whole robot reaches the final position. On the right (f), the robot reaches a target position by taking several steps alternating direction X and Y.

No radial force is allowable on the shaft of the selected motor [14]. A structure is required to transmit motion from the motor to the adhesive pad and connect it to the rest of the feet. Once motor and pad are connected, the rest of the mechanism to obtain independent motion of each pad in the three spatial dimensions can be assembled. Figure 3 shows different views of the CAD design of the structure to actuate one adhesive pad using three linear motors in (a), (b) and (c). In (d), the mechanism of the whole robot inter-connecting the four pads and providing them with motion in the three spatial dimensions is presented.

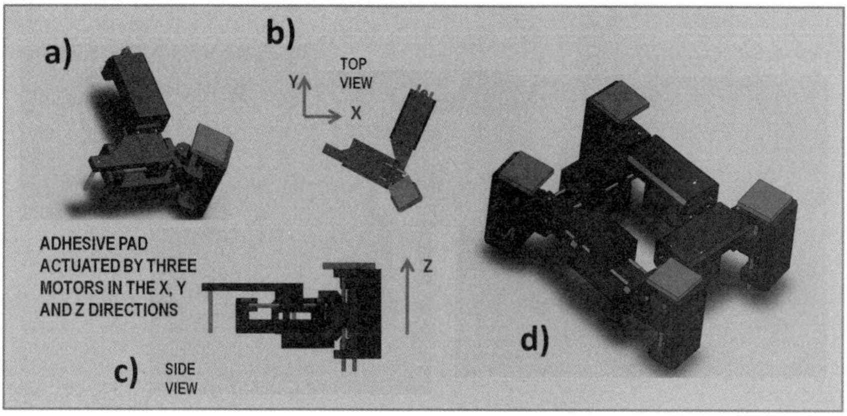

Figure 3. Adhesive pad actuated by three motors (a), in top (b) and side (c) view. And the whole robot (d).

4. Force characterization of one axis for the adhesive foot

For attachment and detachment of the foot a vertical force is applied from the vertically-positioned motor (F_{motor} in Figure 4). This is the force that the robot directly controls to obtain motion and increase or decrease adhesion.

The force exerted by the motor depends on the driving voltage, power of operation (which determines speed) and load [15]. When integrated into a mechanism like the one previously shown, the force of the assembly depends on the mechanical performance of such mechanism. Specially, the friction that may appear along the distance to travel. To assess this performance is particularly relevant in this case considering the miniature scale of the design and the precision required for its manufacturing.

In Figure 4, the system proposed to move the pad is shown. The system is composed of: the motor moving the adhesive surface away from the peritoneum and applying F_{motor}, the connection of the motor to the supporting structure (the rest of the robot) and a magnetic encoder measuring the movement of a magnet fixed to the supporter of the adhesive pad. In this figure, the points considered for the characterisation, as explained in the next section, and the force to be characterised F_{motor} are highlighted.

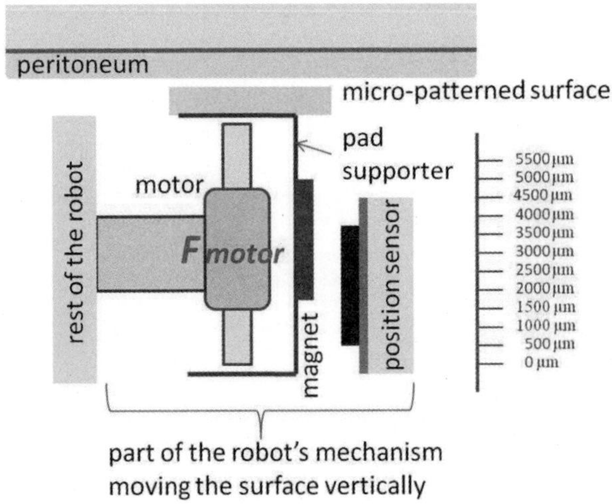

Figure 4. Scheme of the system proposed for the attachment and detachment of the adhesive pad.

4.1. *Experimental procedure and set-up*

In order to measure how uniform the performance of the motor is along the distance travelled (6 mm), the axis is divided into 12 points, one point each 500 ± 200 μm (see Figure 4). The driving voltage is set to 3.3 V. For this test, a load of $15 \pm 1.5\, gf$ is initially set. The force is measured by setting the driving voltage, moving the axis to the corresponding point, blocking the movement with an intial load from the load cell (Transducer Techniques, GS0-100) and applying the driving signal. The driving signal is set to 70% of the maximum power that the motor can deliver for the given initial load.

Figure 5 shows the CAD design of the mechanism and the manufactured parts integrated into the experimental set-up. The supporting pieces corresponding to the moving and the static part were manufactured using micro-machining in Delrin Acetal® 500 AF. Then assembled including the two iron rods and fixed to an aluminium structure. Between the end of the assembly applying the force and the load cell stem a layer of foam is placed to favour full, soft contact. In Figure 5, (a) shows the supporter and guiding rods assembled around the motor, (b) shows the former assembly plus the platform that holds the motor housing with and slotted extension to fix it to the test rig and (c) shows the actual assembly in the rig, ready to be tested. Below picture (c) is shown the part in the full robot design to which the assembly corresponds.

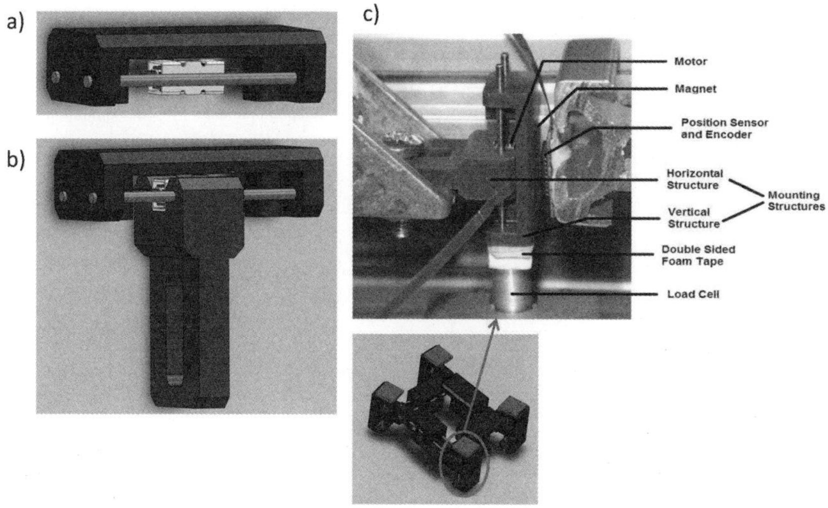

Figure 5. Prototype of the one-axis mechanism and set-up for force characterisation.

4.2. *Results*

This section presents the results from the experiments to determine the force the mechanism is able to exert. The load during operation of the robot will be given by the weight the mechanism has to move and the adhesive force it has to handle. A quarter of the robot is expected to weigh between 5 and 10 gf. The force obtained from the micro-structured pattern during some initial experiments with 113 mm^2 of adhesive area was between 8 and 9 gf [9]. To consider this mechanism design successful, a force increment of at least the maximum expected adhesive force is desirable.

Figure 6 shows the measurements of force obtained during the tests along the axis of the motor divided into 12 points as explained in section 4.1. To compare force at different points along the axis the initial load has been normalized to zero. The actual values of initial load together with the real initial position for each point along the axis and the force increment can be found in Table 1.

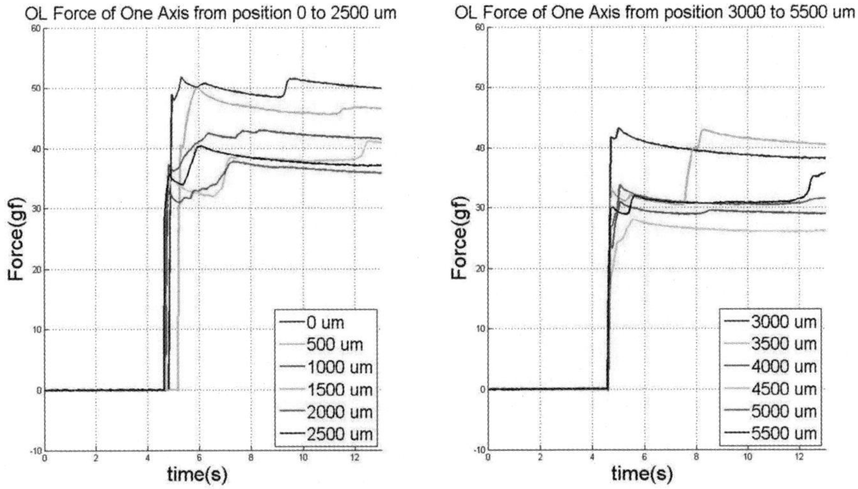

Figure 6. Force recorded during the experiment. The plot on the left-hand side shows the force from points 0 to 2500 μm and on the right from 3000 to 5500 μm.

Point (μm)	Initial position (μm)	Initial load (gf)	Force increment (gf)
0	60	14.1159	51.8195
500	-468	15.0540	50.0387
1000	-900	14.8473	42.9630
1500	-1500	14.6406	41.1980
2000	-1902	13.9251	37.8430
2500	-2456	15.5946	40.3871
3000	-3036	15.5787	43.2333
3500	-3408	16.4533	42.9948
4000	-3848	15.3879	31.0058
4500	-4378	15.4674	28.1120
5000	-5076	14.8950	34.0269
5500	-5388	14.4657	37.3024

Table 1. This table shows the theoretical initial position (Point), the initial position, initial load and force increment for each of the 12 points selected along the axis of the mechanism.

5. Conclusions

Following the trend for miniaturisation in medical robotics an adhesion-reliant robot has been designed to operate inside the abdominal cavity hanging upside-down from the peritoneum. Results on the force characterisation for a configuration which moves the adhesive foot in one direction, have been presented. These results show that the force increment obtained from the mechanism, between 28 and 52 gf depending on the point of operation along the axis, is sufficient to control the force the robot has to handle (8-9 gf).

6. Future work

Experiments with the adhesive pad interacting with rat (human-like) peritoneum are to be performed in order to determine the force of adhesion obtained from the micro-structured pad integrated into the motor's mechanism.

The testing and characterisation will be extended to a two-axes configuration, investigating the transmission of force and moment to the whole structure of the robot. The outcome of all these experiments will be fed into the final design of the hardware and controller.

Acknowledgments

This work was supported by the New and Emerging Applications of Technologies Programme.
The authors would like to thank Ling Yee for her work on the software to control the motor.

References

[1] P. Dasgupta, "Advanced laparoscopic surgery using tha da Vinci robotic system," in *Robotics in Surgery. State of the Art.*, London, 2010.
[2] J. Gilbert, "Robots in laparoscopic surgery," in *Robotics in Surgery. State of the Art.*, London, 2010.
[3] N. Patronik, T. Ota, M. Zenati, and C. Riviere, "A miniature mobile robot for navigation and positioning on the beating heart," *Robotics, IEEE Transactions on,* vol. 25, pp. 1109-1124, 2009.
[4] M. Rentschler, J. Dumpert, S. Platt, K. Iagnemma, D. Oleynikov, and S. Farritor, "An in vivo mobile robot for surgical vision and task assistance," *Journal of Medical Devices,* vol. 1, p. 23, 2007.
[5] http://www.araknes.org/home.html, Accessed Nov. 2010.
[6] J. Pons, *Emerging actuator technologies: a micromechatronic approach*: John Wiley & Sons Inc, 2005.
[7] P. Dario, R. Valleggi, M. Carrozza, M. Montesi, and M. Cocco, "Microactuators for microrobots: A critical survey," *Journal of Micromechanics and microengineering,* vol. 2, p. 141, 1992.
[8] www.theodora.com/anatomy, Accessed November 2010.

[9] G. Taylor, A. Neville, D. Jayne, R. Roshan, T. Liskiewicz, A. Morina, and P. Gaskell, "Wet adhesion for a miniature mobile intra-abdominal device based on biomimetic principles," *Proceedings of the Institution of Mechanical Engineers, Part C: Journal of Mechanical Engineering Science,* vol. 224, pp. 1473-1485, 2010.

[10] S. Gorb, M. Sinha, A. Peressadko, K. Daltorio, and R. Quinn, "Insects did it first: a micropatterned adhesive tape for robotic applications," *Bioinspiration & Biomimetics,* vol. 2, p. S117, 2007.

[11] K. Autumn and A. Peattie, "Mechanisms of adhesion in geckos," *Integrative and Comparative Biology,* vol. 42, p. 1081, 2002.

[12] W. Federle, W. Barnes, W. Baumgartner, P. Drechsler, and J. Smith, "Wet but not slippery: boundary friction in tree frog adhesive toe pads," *Journal of The Royal Society Interface,* vol. 3, p. 689, 2006.

[13] http://www.newscaletech.com/, Accessed Nov. 2010.

[14] http://www.newscaletech.com/app_notes/DesignNote_QuickTips-for-Integrating-Squiggle-Motors.pdf, "Design Note: Quick Tips for Integrating SQUIGGLE Motors," New Scale Technologies Inc., Accessed Nov. 2010.

[15] http://www.newscaletech.com/doc_downloads/SQL-RV-1-8_datasheet.pdf, "SQL-RV-1.8 Reduced Voltage SQUIGGLE RV linear drive system data sheet," New Scale Technologies Inc., Accessed Nov. 2010.

PHASE-DEPENDENT KNEE MODEL IDENTIFICATION IN GAIT MOTION USING AN ACTIVE KNEE REHABILITATION ORTHOTIC DEVICE

Ludovic Saint-Bauzel , Ozer Unluhisarcikli
Student Member, IEEE, and Constantinos Mavroidis, Member, IEEE

Abstract— This paper presents a new approach in modeling human motions using robotic rehabilitation devices. A phase dependent identification method is proposed that is based on applying a Gaussian white noise around the knee joint through the use of a rehabilitation robotic exoskeleton. The input-output relationship is analyzed and the concept to represent the walking motion as a phase-dependent action is described. The method takes advantage of the time independent property of Gaussian noise and shows great accuracy in modeling human limb motions when compared to classical (i.e. global) models. The new modeling method for human limb motions is important in the control and planning of rehabilitation robotic devices and in evaluating human biomechanics using a portable gait measuring device.

1. INTRODUCTION

Rehabilitation robotics that has seen great growth during the last years can be classified into end-effector devices and wearable exoskeleton devices. The end effector design approach is both applied on upper limb (e.g. MIT Manus [1],Handle Device[2])) and on lower limb rehabilitation (Haptic Walker [3]). Similarly, exoskeleton devices could be used on upper limb (Able[4], 2RobotArm[5]) and lower limb (Lokomat[6],PAM-POGO[7], KneeDevice[8]). In both types of rehabilitation devices the controllers are similar. However due to the additional safety concerns when wearing a an exoskeleton, its controller should guarantee safe operation of the device..

In general, the main control aims of a rehabilitation robotic devices are that the controller should be able to a) implement accurately the prescribed rehabilitation treatment protocol and b) eliminate unwanted (e.g. hazardous) effects on the subject. In order for the controller to be able to achieve these aims, it should be able to take into account the changing dynamics of the human body and its environment using a dynamic model.

One way to develop a dynamic model of the human body interacting with its

---

This work was supported in part by the U.S. National Science Foundation under Grant CBET-0828772. Any opinions, findings, conclusions, or recommendations expressed in this publication are those of the authors and do not necessarily reflect the views of the National Science Foundation.

L. Saint-Bauzel is with the Université Pierre et Marie Curie-UPMC "Institut des Systèmes Intelligents etde Robotique" ISIR, Paris, F-75014 France (corresponding author to provide phone: (+33)144-276-210; fax: (+33)144-275-145; e-mail: saintbauzel@isir.upmc.fr).

O. Unluhisarcikli and C. Mavroidis are with the Department of Mechanical and Industrial Engineering, Northeastern University, Boston MA 02115 USA (phone: 617-373-4121; fax: 617-373-2921; e-mail: mavro@coe.neu.edu).

Corresponding authors: L. Saint-Bauzel and C. Mavroidis

environment for use in the control of a rehabilitation device, is to use a global model that takes into account the whole body. However, This global modeling leads to a very large number of parameters to be identified using experimental data and a complex identification process [4,5 or 6] It is also computationally very expensive that makes its real time implementation on the device almost impossible. [14], [15] and [16]. But this bio-mechanical models that aim to understand and present a realistic model of the joint. In our case we want to represent properties during defined motion. As describe below the walking lead to non-linearities when the foot impact the ground, many unwanted motion are appearing because of the robot attachment. So the issue is that one don't know if the knee properties while walking motion can be presented as a linear parameter variable system. So this paper will evaluate if it is accurate to represent the knee mechanical properties as a LPV system and present a ensemble data method to identify the parameters. It will be applied on knee motion equipped of ANDROS Device that present the property to be a one DOF exoskeleton and is therefor a good experimental tool to apply the method.

2. System Overview - ANDROS V1

ANdROS V1 is designed as a wearable and portable assistive tool for gait rehabilitation and monitoring. The exoskeleton frame consists of two braces: an actuated brace attached to the impaired leg via rigid straps, and a sensorized brace attached to the healthy leg via elastic straps (Figures 1 & 2). The power and torque specifications of ANdROS were determined based on the reference trajectories obtained from tests conducted on 70 adult healthy subjects at Spaulding Rehabilitation Hospital in Boston, MA. The specifications obtained from these tests are summarized in Table 1.

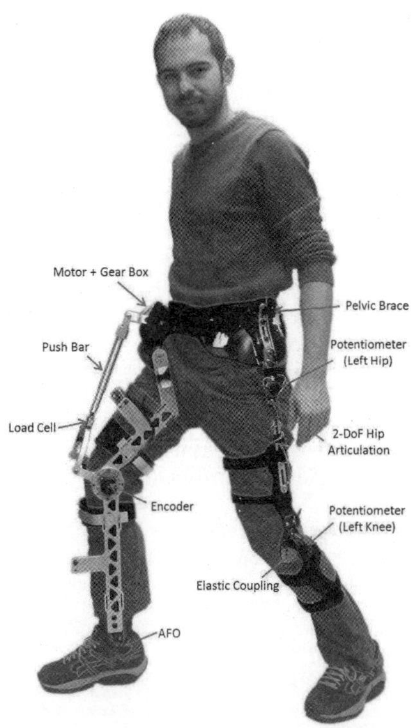

Figure 1 : ANdROS V1: The new Active Knee Rehabilitation Orthotic System.

The actuated brace is driven by a brushless DC motor (MCG IB23000-E1) coupled to a gear-box (Anaheim Automation GBPH-0602-NP-040, 40:1 gear ratio) which is located next to the pelvis, to reduce the moving mass and keep the center of mass (CM) of the exoskeleton close to the wearer's CM. The motor has been selected to provide at least 50% assistance. The peak torque at the gearbox output is 22.88 Nm and rated power is 179 W. The torque generated by the motor is transferred to the knee joint by means of a push bar that features a quick disengagement mechanism. By disengaging the lower part of the exoskeleton from the power train, easy adjustment during donning/doffing is possible. The push bar mechanism of ANdROS is basically a four bar linkage. Kinematic analysis has been carried out to convert the torque at the motor shaft to the torque at the knee, but has been excluded herein due to space limitations.

TABLE 1: TECHNICAL SPECIFICATIONS

	10% of Gait Cycle	90% of Gait Cycle
Torque[a] (Nm)	54 (peak)	27
Velocity (rad/sec)	2	7 (peak)
Power (W)	108	189

[a] The torque values are calculated for a 135 kg individual

A load cell at the end of the push bar measures the interaction forces. The load cell is located proximal to the knee joint to maximize the amount of mass between the motor and itself, whose apparent inertia is reduced by force-feedback. An incremental encoder and a rotary potentiometer are placed on either side of the knee joint, where the potentiometer is used to initialize the incremental encoder's position. The rotary encoder provides high accuracy (0.088°) and noise free measurement of the knee angle, which are critical properties in calculating digital derivative for velocity estimation.

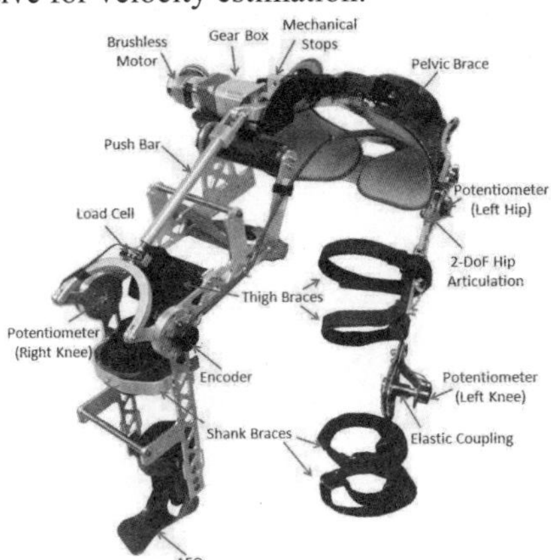

Figure 2: Close-Up View of ANdROS V1 and its Components.

An exoskeleton should be attached to the body so that there is minimal play (migration) with respect to the human tissue, yet still be comfortable to wear. The actuated brace of ANdROS is attached to the user's body at four locations (i.e. hip, thigh, shank, and ankle) to prevent the migration effect. The weight of the exoskeleton and the user is transferred to the ground through an AFO. The symmetrical structure of the actuated brace applies torque to the knee from both medial and lateral sides of the leg, thus eliminating the twisting effect which would occur in a single-sided design.

The brace worn on the unimpaired leg measures the knee and hip flexion/extension angles via high precision potentiometers.

3. Models

3.1. *Phases of Gait Cycle*

Figure 3: One Stance Cycle Schematic Representation.

Figure 3 shows the four phases in a gait cycle. The stance is supposed to begin when the foot is close to leave the ground. During Phase-1 (from 0 to 20% of the gait cycle) one can see that the right foot is in the air. Then Phase-2 is a high torque phase corresponding to the ground hitting pick (between 20 to 40% of the gait cycle). Then a second opposite pick is observed corresponding to the main knee action to move the other leg in Phase-3 (40-60% of the gait cycle). The last phase, Phase-4, is more or less a constant part of the walking motion for the knee. It is when the leg is straight and turns around the ankle while the other foot is in the air and goes to the front of the subject.

3.2. *Mechanical Models of the Knee*

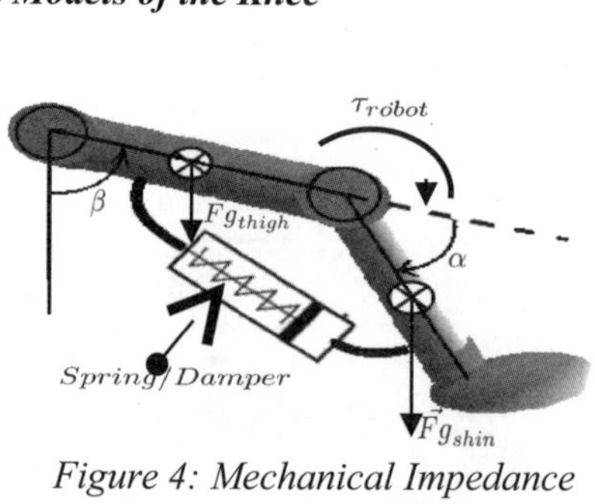

Figure 4: Mechanical Impedance Representation of the Leg

The biomechanical model of the knee shown in Figure 4 leads to Equation (1).

$$\tau_{muscle} + \tau_{robot} + \tau_{gravity} = I\ddot{\omega} \tag{1}$$

where τ_{muscle}, τ_{robot} and $\tau_{gravity}$ are respectively the torque coming from the leg muscle, the robot actuator and the gravity on the leg, I is the inertia of the

system. But one hypothesis is that the actuators are not perfect so they have some impedance (where k represents spring parameter and b stands for damper actions) in the motion. U is the initial torque.
Consequently τ_{robot} and τ_{muscle} will be written a impedance torques :

$$\tau_{muscle} = U_{muscle} + k_{muscle}\Delta\alpha + b_{muscle}\Delta\dot{\alpha} \quad (2)$$

$$\tau_{robot} = U_{rob} + k_{rob}\Delta U + b_{rob}\Delta\dot{U} \quad (3)$$

So the previous equation becomes :

$$\begin{aligned} &U_{muscle(t)} + k_{muscle(t)}\Delta\alpha + b_{muscle(t)}\Delta\dot{\alpha} \\ &+U_{rob} + k_{rob}\Delta U_{rob} + b_{rob}\Delta\dot{U_{rob}} \\ &+\tau_{gravity} + \Phi_w = (I_{rob} + I_{shin})\ddot{\omega} \end{aligned} \quad (4)$$

where ω is the global orientation of the shin, α is the angle between the shin and the thigh. And I is decompose of inertia property of the shin I_{shin} and inertia of the robot I_{rob}. One can notice that these equation are representing the dynamics of the leg without external forces coming from the ground during the walking motion. As a reminder, time varying parameters are identified in this equation. In this paper one would like to represent this ground action as a part of time varying impedance. So as a time varying impedance, it has been added to the previous equation and is called Φ_w.

The idea of this work is to represent walking motion as a time varying impedance. So it seems that we may have to represent Φ_w as a function of time t and α. The problem is that a big part of walking motion is done with α close to a blocked position (during the swing of the other Leg). And the action of the robot could be useful during this motion. So the proposed solution is to include Hip angle (β) in the time varying model of the walking motion.
So the time varying of dynamics of walking can be presented as follow :

$$\Phi_{w(t,\alpha,\beta)} = U_{(t)} + k_{w1(t)}\alpha + b_{w1(t)}\dot{\alpha} + k_{w2(t)}\beta + b_{w2(t)}\dot{\beta} \quad (5)$$

where $U(t)$ will be considered as a time varying wanted torque.
Having an identification of the human joint mechanical parameters is not the aim of this paper. The aim is to have a model of the leg connected to the robot. Consequently knowing the $k_{muscle(t)}$ and $b_{muscle(t)}$ parameters for themself is not relevant. In this paper, it will be added in (6) as follow :

$$\Phi_{wm(t,\alpha,\beta)} = \Phi_{w(t,\alpha,\beta)} + k_{muscle(t)}\alpha + b_{muscle(t)}\dot{\alpha} \quad (6)$$

Consequently, the parameters that are dependent to α and its differentiates can be aggregated and will be called :

$$k_{wm1(t)} = k_{w1(t)} + k_{muscle(t)} \ (7), \quad b_{wm1(t)} = b_{w1(t)} + b_{muscle(t)} \ (8)$$

3.3. *Phase-dependent representation*

In this paper, the idea is to take advantage of periodicity of walking motion. So an hypothesis of this work is that there is some periodicity in the properties of the leg between each stance. Thanks to this property, the time varying

parameters will be considered as phase-dependent parameter. So one can introduce a new parameter Ω that represents the phase of the motion. This parameter is only a position of the phase. It is a real scalar value between 0 and 1 that represents a percent of the phase.
The difficulty is that this numerical representations are not representing the proximity between a parameter close to 0 and a parameter close to 1. In order to obtain a realistic relation, it has been decided to use a lookup table that represent a piece of data centered on the percent presented by Ω and a time window of W. As for example, $k_{wm1(\Omega,W)}$ is computed with every data at time $t_\Omega \pm W * \frac{T_{stance}}{2}$ where t_Ω are every moment that are in Ω position of the stance motion, T_{stance} is the duration of the stance and W is a percent of this duration. Consequently, $k_{wm1(0,W)} = k_{wm1(1,W)}$ and continuity is kept. Another property of this notation is one can represent a fixed in time parameter as robot parameters for example:
$k_{rob} = k_{rob(x,1)}$, $b_{rob} = b_{rob(x,1)}$, $\forall x$
From equation (1) with impedance parameters (*k* for spring action, *b* for damper action parameter), Inertia matrices decomposed by their origin $(I_{robot} + I_{shin})$, the system can therefore be rewritten as follow :

$$\begin{aligned} &U_{muscle(\Omega)} + U_{rob(t)} + k_{rob}\Delta U_{rob} + b_{rob}\Delta \dot{U}_{rob} \\ &+k_{wm1(\Omega,W)}\Delta\alpha + b_{wm1(\Omega,W)}\Delta\dot{\alpha} + k_{w2(\Omega,W)}\Delta\beta \\ &\quad +b_{w2(\Omega,W)} + \tau_{gravity} = (I_{robot} + I_{shin})\ddot{\omega} \end{aligned} \tag{9}$$

where all the Ω dependent parameters, *k* and *b* robot parameters and $(I_{robot} + I_{shin})$ are unknown and are the parameters that are aimed to be identified in this paper.

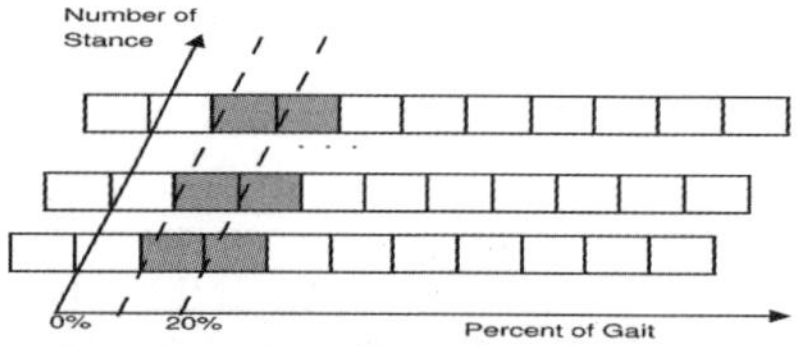

Figure 5 : data(Ω, W) *with* $\Omega = 0.2$ *and* $W = 0.2$

The idea behind this variable model is to take a small set of data around the working point and to find the best model parameters by converging to the local minimum. The data selection is phase dependent and is collected as explained before for the parameters and as presented in Figure 5.

This approach is also a way to reduce inter-trial variations, and as far as time evolutions is not modified it is respecting dynamics properties of the motion.

4. Results & Discussion

4.1. *Experimental setup*

The comparison is done between a model $Model_{wi(-,1)}$ with constant parameters and the model with phase varying parameters is called

$Model_{wi(\Omega,0.2)}$ with $i \in [m1, 2]$ and because the evaluated window size is 20% of stance.

4.2. *Results*

If one compares the proposed models as they estimate the knee torque applied by the human when walking with AKROD, we can observe that the phase varying parameter model is closer to the constant parameter model.
As presented in Fig. 6, the black line that represent the phase model torque evaluation seems closer to grey line than green dashed line, the most relevant parts are around sample 200. and around the sample 100.
A better way to evaluate the accuracy of the phase varying model is to look at the measured quadratic error along the stance as it is done in Fig. 7. According to this graphic, the phase varying model is always under the constant model, It

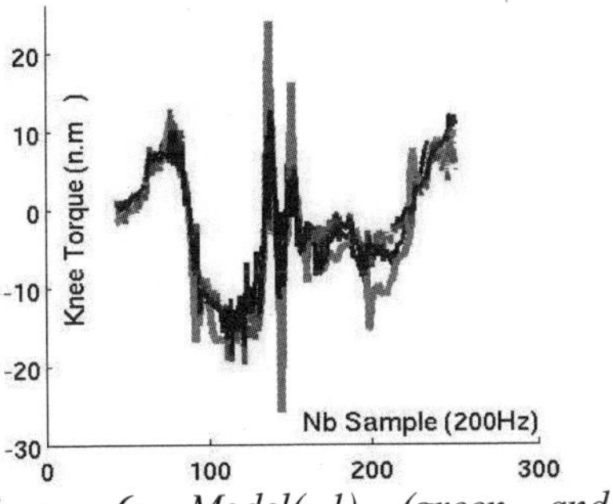

Figure 6: Model(-,1) (green and dashed line), Model(Ω, 0.2) (Black line), and Measured data(grey line) of Knee Torque during one stance.

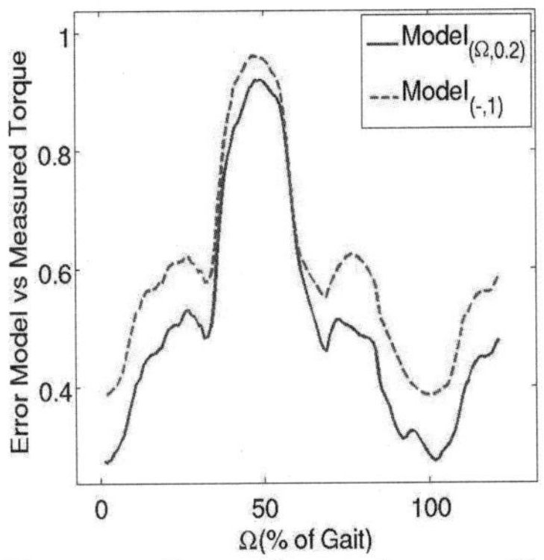

Figure 7: Comparison Phase dependent model ($Model_{(\Omega,W)}$) vs global model ($Model_{(-,1)}$) error along motion

means that the errors between model predicted torque and measured torque.

4.3. *Discussion*

This study shows a phase-dependent model that fits to a human leg with good accuracy. The methodology based on random noise is an interesting approach in obtaining a dynamic model. However, as explained in the introduction, one of the goals of this work is to create a model of the patient. Obviously, introducing random noise to the patient while he/she is in motion can be a problem. Therefore, one limitation of this method is the patient being subjected to perturbations as he/she walks during the identification process.
This method is also interesting in showing the different fields of accuracy of the proposed model. Since this is a phase dependent algorithm, it is possible to propose more accurate models in some parts of the motion.
It should be noted that this study is conducted on a single subject. This is a proof of concept that must be confirmed with statistical analysis on multiple

subjects. However, this larger study will have to pay attention to the subject-dependency of this method.

References

[1] Hogan, N.; Krebs, H.; Charnnarong, J.; Srikrishna, P. & Sharon, A.,Robot and Human Communication, 1992. Proceedings., IEEE International Workshop on : 161 -165(1992).

[2] Khanicheh, A.; Mintzopoulos, D.; Weinberg, B.; Tzika, A. & Mavroidis, C.,Rehabilitation Robotics, 2007. ICORR 2007. IEEE 10th International Conference on : 883 -889(2007).

[3] Schmidt, H.; Hesse, S.; Bernhardt, R. & Krüger, J.,ACM Transactions on Applied Perception 2 : 166-180(2005).

[4] Garrec, P.; Friconneau, J.; Measson, Y. & Perrot, Y.,Intelligent Robots and Systems, 2008. IROS 2008. IEEE/RSJ International Conference on : 1483 -1488(2008).

[5] Culmer, P.; Jackson, A.; Richardson, R.; Bhakta, B.; Levesley, M. & Cozens, A.,Rehabilitation Robotics, 2005. ICORR 2005. 9th International Conference on : 61 - 65(2005).

[6] Bernhardt, M.; Frey, M.; Colombo, G. & Riener, R.,Rehabilitation Robotics, 2005. ICORR 2005. 9th International Conference on : 536-539(2005).

[7] Aoyagi, D.; Ichinose, W.; Harkema, S.; Reinkensmeyer, D. & Bobrow, J., 15 : 387 -400(2007).

[8] Weinberg, B.; Nikitczuk, J.; Patel, S.; Patritti, B.; Mavroidis, C.; Bonato, P. & Canavan, P.,Robotics and Automation, 2007 IEEE International Conference on : 4126 -4133(2007).

[9] Jarrasse, N.; Paik, J.; Pasqui, V. & Morel, G.,Robotics and Automation, 2008. ICRA 2008. IEEE International Conference on : 2134 -2139(2008).

[10] Zhang, L.-Q.; Nuber, G.; Butler, J.; Bowen, M. & Rymer, W. Z.,Journal of Biomechanics 31 : 71 - 76(1997).

[11] Kiss, R.; Kocsis, L. & Knoll, Z.,Med.Eng Phys. 26 : 611-620(2004).

[12] Sok, K. W.; Kim, M. & Lee, J.,SIGGRAPH '07: ACM SIGGRAPH 2007 papers : 107(2007).

[13] Sasaki, O.; Gagey, P.; Ouaknine, M.; Martinerie, J. andVan Quyen, M.; Toupet, M. & L'Heritier, A.,Neuroscience Research 41 : 185-192(2001).

[14] Xu, Y. & Hollerbach, J.,Biomedical Engineering, IEEE Transactions on 45 : 1051 -1060(1998).

[15] Xu, Y. & Hollerbach, J.,Biomedical Engineering, IEEE Transactions on 46 : 409 -419(1999).

[16] Konczak, J.; Brommann, K. & Kalveram, K.,Motor Control 3(4) : 394-413(1999).

SECTION–13

MODELLING AND SIMULATION OF CLAWAR

MODELLING AND ANALYSIS OF AN AUTO-RECONFIGURING MODULAR MICRO MOBILE ROBOT*

JONATHAN HODGINS, DAN ZHANG†

University of Ontario Institute of Technology, 2000 Simcoe Street North
Oshawa, Ontario L1H 7K4/Canada

This paper presents a green solution for a field-reconfigurable modular micro mobile robot. The robot is small enough that it can be used for searching in tight spaces, where humans cannot easily reach and the robots can be adapted to many situations by attaching extra modules to enhance the robot's capabilities. When the module is plugged in and the robot is restarted, it automatically recognises the new module and incorporates the module into its code. By using only the modules needed for a given situation and sharing some modules between robots using IR communication, the robots are greener because they do not have unnecessary parts to add weight and increase power consumption, resulting in reduced cost and maximised battery life.

1. Introduction

Many people are affected by robots on a daily basis, whether it is the latest toy or an industrial robot at work, there is a need for robots to do more and more.

One area of development is to shrink robots to fit in smaller spaces as well as increase the precision of their manipulation. This also allows the robots to work together in a swarm and precisely manipulate small objects around them. This is advantageous because small robot swarms can enter and quickly search where humans can't, aiding in rescue operations.

The paper Modularity in small distributed robotics describes a 6cmx6cmx6cm robots called Millibots, each with limited capabilities but using add-ons and working together they can be adapted to a situation and create a map of an area. [4].

Some reconfiguring modular robots have already been developed, such as Cubelets by Modular Robotics LLC. They have recently developed an education robotics construction set with self reconfiguring modules. The modules change their programming based on the type of modules present as well as their configuration. [3].

* This work is supported by the Natural Sciences and Engineering Research Council of Canada (NSERC) and Canada Research Chairs prograqm..

† Corresponding author: Dr. D. Zhang, tel: 905 7218668 ext. 5721, fax: 905 7213370, email: dan.zhang@uoit.ca.

This paper describes a novel design for an auto reconfiguring robot that automatically incorporates any added modules into its code when the additional module is attached and the robot is restarted. This allows for the swarm to only have the parts needed reducing cost and maximising battery life.

2. Hardware

The hardware of the robot consists of different modules, each with a dedicated purpose. The primary module that coordinates all of the functions is the main module. The main module is programmed using the programming module, which simply allows the programming headers to connect to the microcontroller. The locomotion modules are what move the robot around while the sensor modules record and report data from various sensors to the main module.
The communication modules can report the data back to a monitoring station as well as receive instructions. The manipulator modules are the robots to interact with their environment.

2.1. *Main Module*

The main module consists of an ATMEGA328 microcontroller with basic support hardware required to function, including buffering capacitors, crystal, battery charger and voltage regulator. This microcontroller was chosen because it is also used in the Arduino platform, making it compatible with their programming software (arduino) as well as the manufacturer's software (AVR studio 4).

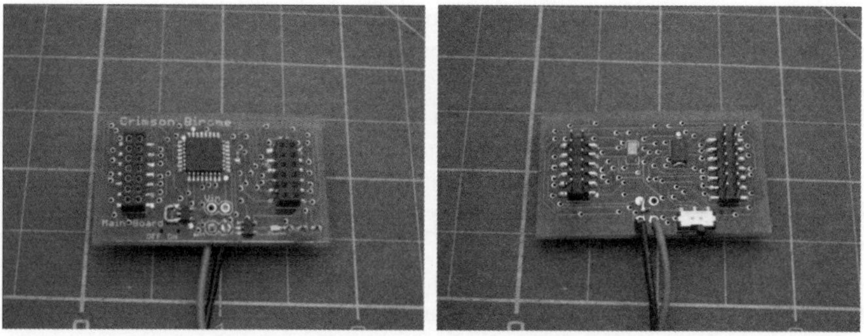

Figure 1 – Main robot module top and bottom

The main module can be programmed using an AVR in system programmer (AVR ISP) or TTL serial communications by attaching the programming board and connecting the appropriate cable.

2.2. *Locomotion Modules*

The reasoning behind the robot having a separate locomotion module from the main board is so, unlike other robots that sacrifice versatility for simplicity [2], it can be easily adapted to multiple situations. For example, if the robot needed to climb through ductwork it could be given magnetic base [6] or if it needed tank-style treads another base could be designed [1].

The only base designed and prototyped to date is a simple 2 wheel design as seen in Figure 2.

Figure 2 – 2 wheel locomotion base

2.3. *Sensor Modules*

The sensor modules allow the robot to perceive its current surroundings. The basic sensor module is a board with 4 IR emitter/detector pairs, one pointed in each direction (forward, backward, left and right). Each of the sensors are connected to and analog input on the main board and each of the IR emitters are connected to a digital output, allowing for communication between robots.

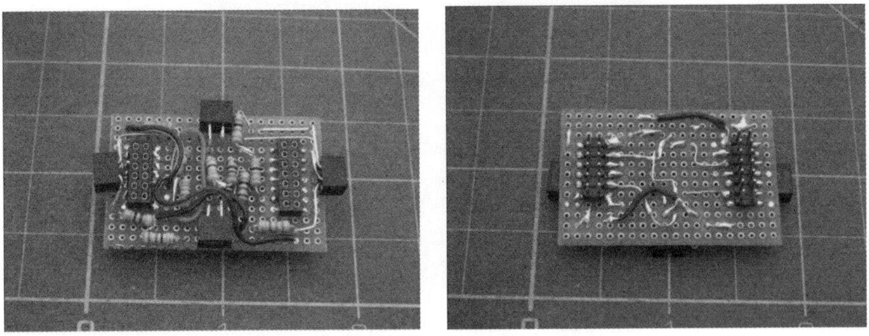

Figure 3 – IR Sensor Board top and bottom

2.4. *Communication Modules*

The communication modules allow for the robots to report back to a base station as well as receive commands for actions to perform. Because the Main module uses standard serial communications, there are a couple options for wireless communication.

Option 1 is point to point Bluetooth communication as seen in figure 4. The Bluetooth module can be paired with either a computer or cell phone for a mobile base station.

Figure 4 – Bluetooth communication module

Option 2 is a Zigbee module (seen in figure 5) which can be arranged in point to point networks or mesh networks to include communication between robots.

Figure 5 – Wireless Zigbee module

2.5. *IMU Module*

The IMU module (Inertia measurement unit) consists of a 3 axis accelerometer, gyroscope and magnetometer all connected to a microcontroller. The microcontroller can read the sensor's data and interpret it to calculate the robot's position in 3D space.
The accelerometer measures the acceleration along each axis, which can then be derived twice to achieve an estimation of the robot's position (there is some compounding error associated with this. The gyroscope measures the angular velocity of all 3 axis and it is then derived once to achieve an estimation of the angular position. The magnetometer measures the magnetic field around the robot, which should be the earth's magnetic field, and it corrects any error in the angular position calculations. The IMU module with only the microcontroller and accelerometer attached is shown in figure 6.

Figure 6 – IMU module with only microcontroller (shown on the underside in the left picture) and the accelerometer (in the right picture)

3. Software

The software is structured so that on start-up the robot has a way to check if any possible module is attached.
The exceptions to this are the locomotion modules and the communication modules. If the locomotion module is removed, the robot will just not move and the code will continue unaffected, which is useful for debugging purposes. If the communication module is removed the robot will just not send or receive any data and again the behaviour of the code is unaffected.

The software begins by checking for the presence of the IR module. This is done by reading the analog values from the sensors, if the module is removed they will all be around 0 (out of a possible 1024) so the variable that remembers if the IR module is present is set to false. The same procedure is performed for any additional modules (such as an IMU board) by using a digital input pin on the motherboard that the attached module pulls high to signal that it is connected.
Because the robot is highly modular, the process intensive modules, such as the IMU board, have a dedicated processor that sends any relevant data back to the

main board upon request using a 9600 baud serial connection to maintain a fast overall processing speed.

Once the main board has established what modules are connected it begins to run the main program. The main program loop consists of a pseudo-parallel program structure, where each sub function runs at set intervals. First, any sensors attached to the main module are read and the values are stored in variables and any data is requested and read from the attached modules. The second function computes how to control the motor based on the inputs and the robot's goal. The third function updates the motors and the 4th is responsible for communication (both IR and Bluetooth/zigbee).

4. Conclusion

This report was motivated to create a simple and versatile platform for further study of mobile swarm robotics. The first task outlines and prototyped is the hardware modules of the robot. This includes the main module which makes decisions based on all of the data collected from the other modules and controls the locomotion module. The locomotion module is a simple 2 wheel design with the ability to be easily swapped for another style of base such as a walking base. The second task outlined and created is the robot programming. It first detects which of the possible modules are connected and incorporated their code into the program. Once the program is running, it uses a pseudo-parallel task structure to streamline the code flow for including/excluding the extra code pieces.

References

1, Caprari, G., & Siegwart, R. (2003). Design and Control of the Mobile Micro Robot Alice. In Proceedings of the 2nd International Symposium on Autonomous Minirobots for Research and Edutainment, (pp. 23-32).

2, Caprari, G., & Siegwart, R. (2005). Mobile micro-robots ready to use: Alice. IEEE/RSJ International Conference on Intelligent Robots and Systems, (pp. 3295 - 3300).

3, Modular Robotics LLC. (2010). Cubelets. Retrieved 02 17, 2011, from Modular Robotics: http://www.modrobotics.com/cubelets

4, Navarro-Serment, L. E., Grabowski, R., Paredis, C. J., & Khosla, P. K. (Sept. 19-22, 1999). Modularity in small distributed robots. Proceedings of the SPIE conference on Sensor Fusion and Decentralized Control in Robotic Systems II. Boston, MA.

5, P. Valdastria, P. C. (2006). Micromanipulation, communication and swarm intelligence issues in a swarm microrobotic platform. Robotics and Autonomous Systems , 789-804.

6, Shen, W., Gu, J., & Shen, Y. (2005). Proposed wall climbing robot with permanent magnetic tracks for inspecting oil tanks. IEEE International Conference Mechatronics and Automation, (pp. 2072 - 207 vol. 4).

MODELLING OF TWO-WHEELED ROBOTIC WHEELCHAIR WITH MOVING PAYLOAD

ABDULLAH M ALMESHAL and M O TOKHI

Department of Automatic Control and Systems Engineering, University of Sheffield, Sheffield, UK cop10ama@shef.ac.uk

KHALED GOHER

Department of Mechanical and Industrial Engineering, College of Engineering, Sultan Qaboos University, Oman, kgoher@squ.edu.om

This paper considers modeling of a novel two-wheeled robotic wheelchair with moving payload. The design of the vehicle is based on double inverted pendulum system model with a linear actuator at the second link. The linear actuator is able to lift up the payload to a demanded height. The vehicle model presented is highly non-linear with five degrees of freedom. Lagrangian modeling approach is used to derive the system equations of motion. The developed model is a multi-input multi-output (MIMO) type with highly nonlinear and coupled dynamics. A control strategy was developed and implemented in Matlab Simulink. Simulations and results showing successful control of the systems are presented in this paper.

1. Introduction

The inverted pendulum system has always been of interest to control system engineers and researchers due to its nonlinearity and under actuated nature [1]. There are many variations of the inverted pendulum system. These include the linear inverted pendulum system, mobile inverted pendulum on cart, and rotational inverted pendulum. These have enabled researchers to study various applications including gait of humanoid robots [2], personal transporters [3], and self-balancing wheelchairs [4]. In recent years, Inverted pendulum systems have been extended to have multiple links, such as double inverted pendulum and triple inverted pendulum systems [5]. This extension has increased the degrees of freedom and complexity of the system providing models for more complex applications.

In this paper, a model of double inverted pendulum system with five degrees of freedom is developed. Equations of motion describing system dynamics are derived and presented. The model developed will be used as a foundation block for ongoing research of self-balancing wheelchairs with ability to maneuver on irregular and uneven surfaces with moving payload.

2. System Description and analysis

A schematic diagram of the vehicle considered in this paper is shown in Figure 1. The associated variables and physical parameters are described in Table 1. The vehicle is designed based on the double inverted pendulum system model with novel modifications [1]. The vehicle consists of two links and a cart driven by two DC motors that in turn drive the entire system. In addition, the vehicle has a third DC motor to drive the second link. These motors will help stabilizing the system in the upright position by applying an appropriate control signal. The second link consists of two co-axial rods connected by a linear actuator to enable lifting up the payload to a demanded height. Therefore, the system has five degrees of freedom; translational motion with the right and left wheels, first and second links and linear actuator on the second link.

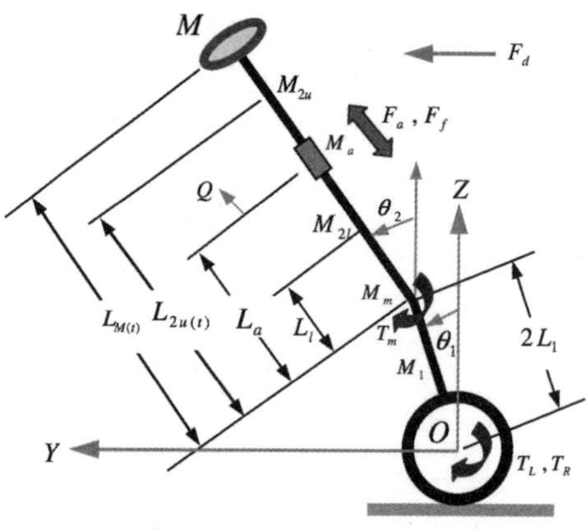

Figure 1. Schematic description of the vehicle

Table 1. Nomenclature

Variable	*Description*	*Unit*
$L_{M(t)}$	Distance to the COM of the payload	m
$L_{2u(t)}$	Distance to the COM of the upper part link 2	m
L_a	Position of the linear actuator	m
L_l	Half length of link 1	m

H	Distance between wheels along X axis	m
Q	Displacement of the linear actuator	m
M_1	Mass of link 1	kg
M_m	Mass of motor driving link 2	kg
M_{2l}	Mass of the lower part of link 2	kg
M_a	Mass of the linear actuator	kg
M_{2u}	Mass of the upper part of link 2	kg
M	Payload mass	kg
T_R, T_L	Right and left wheels driving torques	N.m
T_m	Motor torque	N.m
F_a	Linear actuator force	N
F_f	Frictional force in the linear actuator	N
F_d	External disturbance force	N
θ_1	Angular position of link 1 to the positive Z axis	rad
θ_2	Angular position of link 2 to the positive Z axis	rad
ϕ	Yaw angle of the vehicle around the Z axis	rad
δ_R, δ_L	Angular position of the right and left wheels	rad
J_1	Mass moment of inertia of link 1	$kg.m^2$
J_{2u}	Mass moment of inertia of the upper rod of link 2	$kg.m^2$
J_a	Mass moment of inertia of the actuator	$kg.m^2$
J_M	Mass moment of inertia of the payload	$kg.m^2$
J_w	Mass moment of inertia of the wheels	$kg.m^2$
J_{IB}	Mass moment of inertia of the intermediate body	$kg.m^2$
J_{2L}	Mass moment of inertia of the lower rod of link 2	$kg.m^2$

3. System Dynamic model and equations of motion

In this paper, Euler-Lagrange approach is used to derive the equations of motion of the double inverted pendulum system. The Lagrangian approach employs the system kinetic and potential energies to derive the dynamic model of the system. The Lagrange equation is defined as:

$$\frac{d}{dt}\left(\frac{\partial L}{\partial \dot{q}_i}\right) - \frac{\partial L}{\partial q_i} = Q_i \tag{1}$$

where

$L = T_T$ - V_T *Lagrangian function*
Q_i = *Generalized force vector*
q_i = *Generalized coordinate vector*
T_T = *Total kinetic energy of the system*

V_T = *Total potential energy of the system*

The system has five degrees of freedom and therefore the coordinates of the system are selected as

$$q_i = [\delta_L \quad \delta_R \quad \theta_1 \quad \theta_2 \quad Q]^T \tag{2}$$

and the force vector is defined as

$$Q_i = [T_{LT} \quad T_{RT} \quad 0.5(T_{LT} + T_{RT}) \quad T_{MT} \quad F_{aT}]^T \tag{3}$$

where T_{LT}, T_{RT}, T_{MT}, F_{aT}, are the total moments based including Coulomb's friction model of the left wheel joint, right wheel joint, intermediate joint between the pendulum links, and the force in the actuator respectively. Thus, it is required to find the system total kinetic energy and potential energy. The total kinetic energy of the system is expressed as

$$T_T = T_1 + T_{2l} + T_{2u} + T_m + T_a + T_M + T_c + T_\phi \tag{4}$$

and the total potential energy of the system is expresses as

$$V_T = V_1 + V_{2l} + V_{2u} + V_m + V_M + V_a \tag{5}$$

The kinetic energies of system components are defined as follows:

$$T_1 = \frac{1}{2} M_1 \left(\left(\frac{R_w}{2} (\dot{\delta}_L + \dot{\delta}_R) + L_1 \dot{\theta}_1 \cos\theta_1 \right)^2 + \left(L_1 \dot{\theta}_1 \sin\theta_1 \right)^2 \right) + \frac{1}{2} J_1 \dot{\theta}_1^2 \tag{6}$$

$$T_m = \frac{1}{2} M_M \left(\left(\frac{R_w}{2} (\dot{\delta}_L + \dot{\delta}_R) + 2L_1 \dot{\theta}_1 \cos\theta_1 \right)^2 + \left(2L_1 \dot{\theta}_1 \sin\theta_1 \right)^2 \right) + \frac{1}{2} J_m \dot{\theta}_1^2 \tag{7}$$

$$T_{2l} = \frac{1}{2} M_{2l} \left(\begin{array}{l} \left(\frac{R_w}{2} (\dot{\delta}_L + \dot{\delta}_R) + 2L_1 \dot{\theta}_1 \cos\theta_1 + L_{2l} \dot{\theta}_2 \cos\theta_2 \right)^2 \\ + \left(2L_1 \dot{\theta}_1 \sin\theta_1 + L_{2l} \dot{\theta}_2 \sin\theta_2 \right) \end{array} \right) + \frac{1}{2} J_{2l} \dot{\theta}_2^2 \tag{8}$$

$$T_a = \frac{1}{2} M_a \left(\left(\frac{R_w}{2} (\dot{\delta}_L + \dot{\delta}_R) + 2L_1 \dot{\theta}_1 \cos\theta_1 + 2L_{2l} \dot{\theta}_2 \cos\theta_2 \right)^2 + \left(2L_1 \dot{\theta}_1 \sin\theta_1 + 2L_{2l} \dot{\theta}_2 \sin\theta_2 \right)^2 \right) + \frac{1}{2} J_a \dot{\theta}_2^2 \tag{9}$$

$$T_{2u} = \frac{1}{2} M_{2u} \left(\dot{Q}^2 + \left(\frac{R_w}{2} (\dot{\delta}_L + \dot{\delta}_R) + 2L_1 \dot{\theta}_1 \cos\theta_1 + 2L_{u(t)} \dot{\theta}_2 \cos\theta_2 \right)^2 + \left(2L_1 \dot{\theta}_1 \sin\theta_1 + L_{2u(t)} \dot{\theta}_2 \sin\theta_2 \right)^2 \right) + \frac{1}{2} J_{2u} \dot{\theta}_2^2 \tag{10}$$

$$T_M = \frac{1}{2} M \left(\dot{Q}^2 + \left(\frac{R_w}{2} (\dot{\delta}_L + \dot{\delta}_R) + 2L_1 \dot{\theta}_1 \cos\theta_1 + 2L_{M(t)} \dot{\theta}_2 \cos\theta_2 \right)^2 + \left(2L_1 \dot{\theta}_1 \sin\theta_1 + L_{2M(t)} \dot{\theta}_2 \sin\theta_2 \right)^2 \right) + \frac{1}{2} J_M \dot{\theta}_2^2 \tag{11}$$

$$T_c = (M_w R_w^2 + J_w)(\dot{\delta}_L^2 + \dot{\delta}_R^2) \tag{12}$$

$$T_\phi = \frac{1}{2}(2J_w + J_{IB})\dot{\phi}^2 \tag{13}$$

Summing equations (6)-(13) and simplifying yields the total kinetic energy of the system as:

$$\begin{aligned} T_T &= C_{21}(\dot{\delta}_L^2 + \dot{\delta}_R^2) + C_{22}\dot{\delta}_L \dot{\delta}_R + \frac{1}{2} C_8 \dot{Q}^2 + \frac{1}{2} C_{16} \dot{\phi}_2 + C_{18} \dot{\theta}_1^2 \\ &+ \frac{1}{2}(C_{20} + C_{12}Q + C_8 Q^2)\dot{\theta}_2^2 + C_9 \frac{R_w}{2} L_1 \dot{\theta}_1 (\dot{\delta}_L + \dot{\delta}_R)\cos\theta_1 \\ &+ \frac{R_w}{2}(C_{10} + C_8 Q)(\dot{\delta}_L + \dot{\delta}_R)\dot{\theta}_2 \cos\theta_2 + 2L_1(C_{10} + C_8 Q)\dot{\theta}_1 \dot{\theta}_2 \cos(\theta_1 - \theta_2) \end{aligned} \tag{14}$$

The potential energy of each system component is defined as

$$V_1 = M_1 g L_1 \cos\theta_1 \tag{15}$$

$$V_m = 2M_{m1} g L_1 \cos\theta_1 \tag{16}$$

$$V_{2l} = M_{2l} g (2L_1 \cos\theta_1 + L_{2l} \cos\theta_2) \tag{17}$$

$$V_{2u} = M_{2u}g(2L_1\cos\theta_1 + 2L_{2u(t)}\cos\theta_2) \tag{18}$$

$$V_a = M_a g(2L_1\cos\theta_1 + 2L_{2l}\cos\theta_2) \tag{19}$$

$$V_M = Mg(2L_1\cos\theta_1 + 2L_{2M(t)}\cos\theta_2) \tag{20}$$

where

$$L_{2u(t)} = 2L_{2l} + L_{2u} + Q \tag{21}$$

$$L_{M(t)} = 2L_{2l} + 2L_{2u} + Q \tag{22}$$

Summing equations (15)-(20) and simplifying yields the total potential energy of the system as:

$$V_T = C_3 g\cos\theta_1 + (C_{15} + C_8 Q)g\cos\theta_2 \tag{23}$$

Using equations (14) and (23) the system Lagrangian function can be obtained as follows:

$$\begin{aligned}L &= C_{21}(\dot{\delta}_L^2 + \dot{\delta}_R^2) + C_{22}\dot{\delta}_L\dot{\delta}_R + \frac{1}{2}C_8\dot{Q}^2 + \frac{1}{2}C_{16}\dot{\phi}_2 + C_{18}\dot{\theta}_1^2\\&+\frac{1}{2}(C_{20} + C_{12}Q + C_8Q^2)\dot{\theta}_2^2 + C_9\frac{R_w}{2}L_1\dot{\theta}_1(\dot{\delta}_L + \dot{\delta}_R)\cos\theta_1\\&+\frac{R_w}{2}(C_{10} + C_8Q)(\dot{\delta}_L + \dot{\delta}_R)\dot{\theta}_2\cos\theta_2 + 2L_1(C_{10} + C_8Q)\dot{\theta}_1\dot{\theta}_2\cos(\theta_1 - \theta_2)\\&-C_3g\cos\theta_1 + (C_{15} + C_8Q)g\cos\theta_2\end{aligned} \tag{24}$$

System equations of motion can be derived by solving equation (1) for each system coordinate defined in coordinate vector of equation (2). Solving equation (1) for each coordinate yields the following five non-linear differential equations describing the system dynamics:

$$\begin{aligned}&2C_{21}\ddot{\delta}_L + C_{22}\ddot{\delta}_R + C_9\frac{R_w}{2}L_1\ddot{\theta}_1\cos\theta_1 - C_9\frac{R_w}{2}L_1\dot{\theta}_1^2\sin\theta_1\\&+\frac{R_w}{2}(C_{10} + C_8Q)\ddot{\theta}_2\cos\theta_2 - \frac{R_w}{2}(C_{10} + C_8Q)\dot{\theta}_2^2\sin\theta_2\\&+\frac{R_w}{2}C_8\dot{Q}\dot{\theta}_2\cos\theta_2 = T_L - T_{fL}\end{aligned} \tag{25}$$

$$\begin{aligned}&2C_{21}\ddot{\delta}_R + C_{22}\ddot{\delta}_L + C_9\frac{R_w}{2}L_1\ddot{\theta}_1\cos\theta_1 - C_9\frac{R_w}{2}L_1\dot{\theta}_1^2\sin\theta_1\\&+\frac{R_w}{2}(C_{10} + C_8Q)\ddot{\theta}_2\cos\theta_2 - \frac{R_w}{2}(C_{10} + C_8Q)\dot{\theta}_2^2\sin\theta_2\\&+\frac{R_w}{2}C_8\dot{Q}\dot{\theta}_2\cos\theta_2 = T_R - T_{fR}\end{aligned} \tag{26}$$

$$2C_{18}\ddot{\theta}_1+C_9\frac{R_w}{2}L_1(\ddot{\delta}_L+\ddot{\delta}_R)\cos\theta_1-C_9\frac{R_w}{2}L_1(\dot{\delta}_L+\dot{\delta}_R)\dot{\theta}_1\sin\theta_1$$
$$+2L_1(C_{10}+C_8Q)\ddot{\theta}_2\cos(\theta_1-\theta_2)-2L_1(C_{10}+C_8Q)\dot{\theta}_1\dot{\theta}_2\sin(\theta_1-\theta_2)$$
$$+2L_1(C_{10}+C_8Q)\dot{\theta}_2^2\sin(\theta_1-\theta_2)+2L_1C_8\dot{Q}\dot{\theta}_2\cos(\theta_1-\theta_2)$$
$$+C_9\frac{R_w}{2}L_1\dot{\theta}_1^2(\dot{\delta}_L+\dot{\delta}_R)\sin\theta_1+2L_1(C_{10}+C_8Q)\theta_1^2\dot{\theta}_2\sin(\theta_1-\theta_2)$$
$$-C_3g\dot{\theta}_1\sin\theta_1=\frac{1}{2}(T_{LT}+T_{RT}) \tag{27}$$

$$C_{20}\ddot{\theta}_2+(C_{12}\dot{Q}+2C_8Q)\dot{\theta}_2+(C_{12}Q+C_8Q^2)\ddot{\theta}_2$$
$$+\frac{R_w}{2}(C_{10}+C_8Q)(\ddot{\delta}_L+\ddot{\delta}_R)\cos\theta_2-\frac{R_w}{2}(C_{10}+C_8Q)(\dot{\delta}_L+\dot{\delta}_R)\dot{\theta}_2\sin(\theta_2)$$
$$+C_8\frac{R_w}{2}\dot{Q}(\dot{\delta}_L+\dot{\delta}_R)\cos\theta_2+2L_1(C_{10}+C_8Q)\ddot{\theta}_1\cos(\theta_1-\theta_2)$$
$$-2L_1(C_{10}+C_8Q)\dot{\theta}_1^2\sin(\theta_1-\theta_2)+2L_1(C_{10}+C_8Q)\dot{\theta}_1\dot{\theta}_2\sin(\theta_1-\theta_2)$$
$$+2C_8L_1\dot{\theta}_1\dot{\theta}_2\cos(\theta_1-\theta_2)+\frac{R_w}{2}(C_{10}+C_8Q)(\dot{\delta}_L+\dot{\delta}_R)\dot{\theta}_2^2\sin\theta_2$$
$$-(C_{15}+C_8Q)g\dot{\theta}_2\sin\theta_2-2L_1(C_{10}+C_8Q)\dot{\theta}_1\dot{\theta}_2^2\sin(\theta_1-\theta_2)=T_M-T_{FM}-L_dF_d \tag{28}$$

$$C_8\ddot{Q}-\frac{1}{2}(C_{12}+2C_8Q)\dot{\theta}_2^2-C_8\frac{R_w}{2}\theta_2(\dot{\delta}_L+\dot{\delta}_R)\cos\theta_2$$
$$-2L_1C_8\dot{\theta}_1\dot{\theta}_2\cos(\theta_1-\theta_2)+C_8g\cos\theta_2=F_a-F_{fa} \tag{29}$$

4. Control Strategy

The system described in this paper is a Multi-Input Multi-Output (MIMO) system. To control the system, five control loops are used in which two of them for angular displacement of the wheels, two for the pendulum links and one for the payload linear actuators displacement. The inputs to the system are the driving torques of each wheel, torque driving the motor activating the second link, and the linear actuator force. As a first step to test the system controllability, a combination of Proportional-Derivative (PD) and Proportional-Integral-Derivative (PID) controllers were used to assess the system response. The PD controllers are used to control the angular displacement of the wheels and the tilt angle of the first link. While PID controllers are used to control the tilt angle of the second link and the displacement of the payload.

5. Simulation and Results

The derived model was implemented and simulated in Matlab Simulink environment. The desired values for the system are $\delta_{Rd} = \delta_{Ld} = 0.8$ m and $\theta_1 = \theta_2 = 0\,^{o}$ while the payload actuator should follow a predefined input signal shown in Figure 2. Figure 3 represents the open-loop system response. The system is clearly unstable with outputs reaching infinity. With the proposed control strategy, PD and PID controller were added to the closed-loop system and tuned heuristically with the gains described in Table 3. The closed loop controlled system response is shown in Figure 4 where it is clear that the system was stable and converge to the desired set values.

The simulation parameters are $L_{2l}=0.25$ m, $L_{2u}=0.25$ m, $L_l=0.11m$, $M_l=3$ kg, $M_m=0.3$ kg, $M_{2l}=1.5$ kg, $M_a=0.8$ kg, $M_{2u}=1.5$ kg, $M=70$ kg, $J_l=0.003025$ $kg.m^2$, $J_{2u}=0.005\ kg.m^2$, $J_w=0.225\ kg.m^2$, $J_{2L}=0.005\ kg.m^2$, δ_{Rd}, $\delta_{Ld}=0.8m$.

Table 3. PID loops gains

Loop	*PID controller Gains*		
	K_P	K_I	K_D
Left wheel	1000	-	600
Right wheel	1000	-	600
First pendulum link	50	-	50
Second pendulum link	80	30	60
Payload linear actuator	6000	100	1200

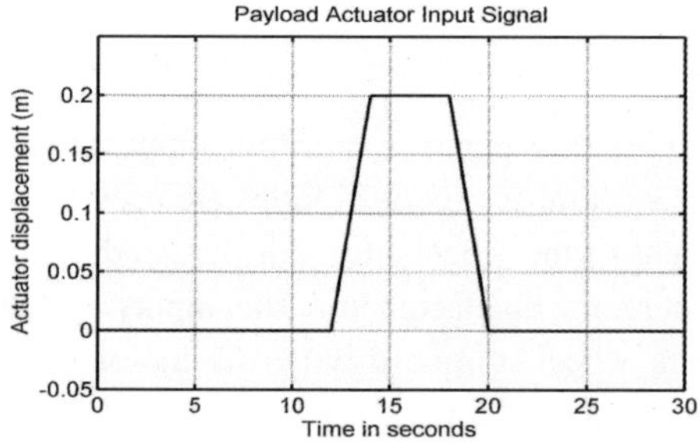

Figure 2. Predefined input signal for the Payload linear actuator

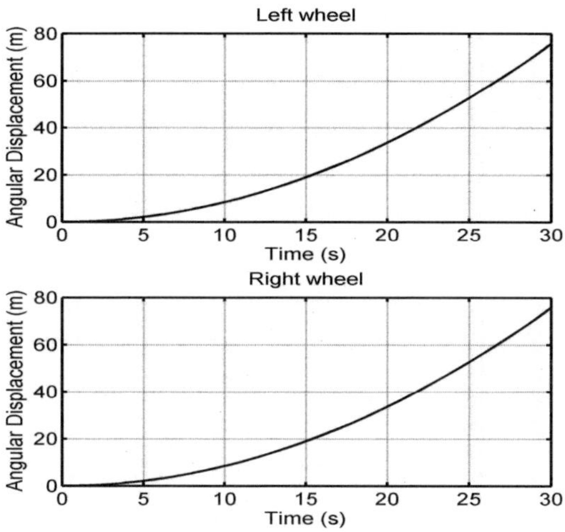

(a) Left and right wheels angular displacement

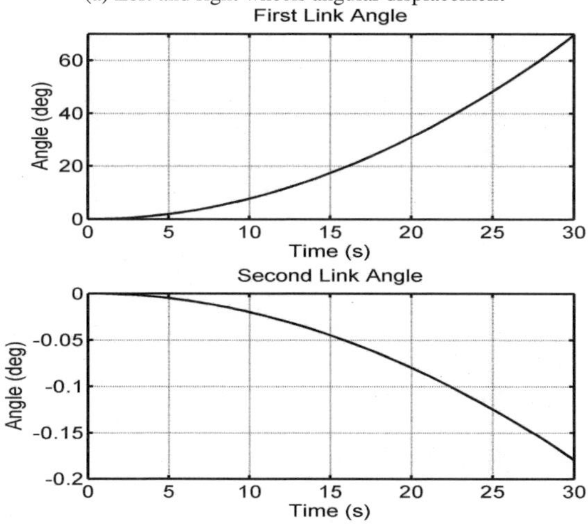

(b) First and second pendulums tilt angles.

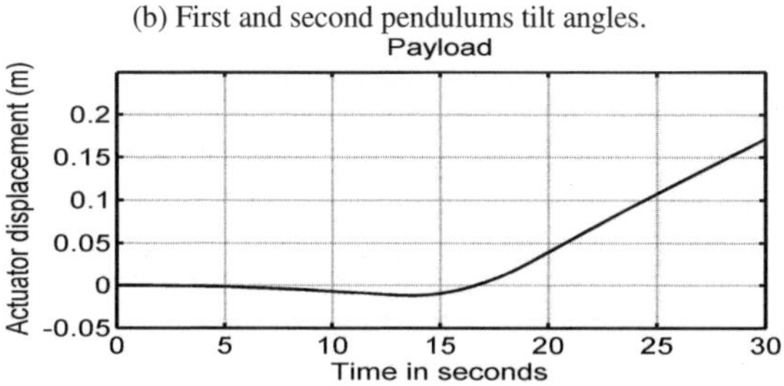

(c) Displacement of Payload linear actuator.

Figure 3. Uncontrolled open loop system response

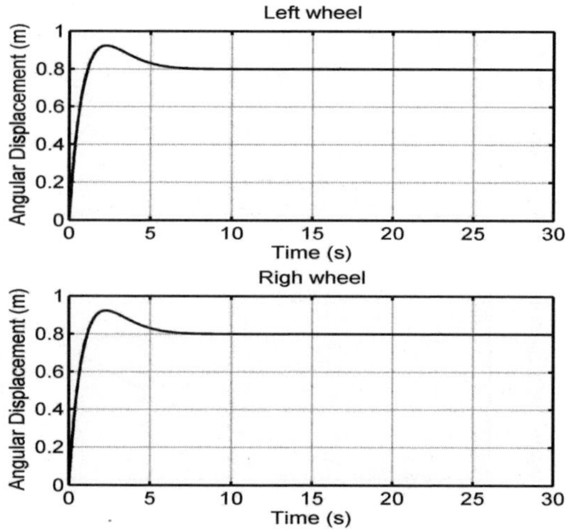

(a) Left and right wheels angular displacement

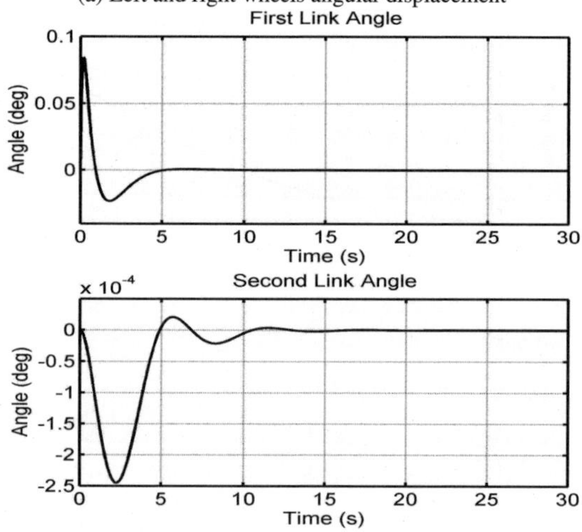

(b) First and second pendulums tilt angles.

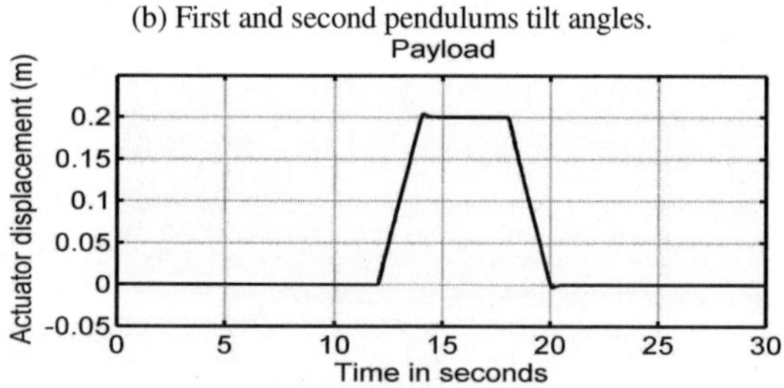

(c) Displacement of Payload linear actuator.

Figure 4. PID Controlled closed loop system response

6. Conclusion and future work

The development of a model of a novel two-wheeled robotic machine with moving payload has been presented. Lagrangian dynamic approach has been used to derive the dynamic equations of motion of the vehicle while maintaining the nonlinear terms. The model developed comprises five highly nonlinear and coupled differential equations. A control algorithm was developed and implemented on the system model in Matlab Simulink environment using a combination of PD and PID controllers. Simulations showing a successful and stable balancing behaviour were presented.

References

1. K. Goher ,S. Ahmad, M.O. Tokhi, "A new configuration of two wheeled vehicles: Towards a more workspace and motion flexibility", Proceedings of 4th Annual IEEE conference on Intelligent Systems, San Diego, CA, USA, 5-8 April 2010.
2. M. Takahashi, T. Narukawa, K. Yoshida,"Intelligent transfer and stabilization control to unstable equilibrium point of double inverted pendulum," SICE 2003 Annual Conference, vol.2, pp.1451-1456, 4-6 August 2003.
3. D.S. Nasrallah, J. Angeles, and H. Michalska, "Velocity and orientation control of an anti-tilting mobile robot moving on an inclined plane", IEEE International Conference on Robotics and Automation, 2006. ICRA 2006, pp.3717-3723, 15-19 May 2006.
4. S. Ahmad S., M.O. Tokhi, "Modeling and Control of a Wheelchair on Two Wheels" Second Asia International Conference on Modeling & Simulation, 2008. AICMS 08. ,pp.579-584, 13-15 May 2008.
5. K.G. Eltohamy, and C.Y. Kuo, "Real time stabilisation of a triple link inverted pendulum using single control input" , IEE Proceedings on Control Theory and Applications , vol.144, no.5, pp.498-504, Sep 1997.

DYNAMIC STABILITY IMPROVEMENT OF WALKING ROBOTS[1]

LUIGE VLADAREANU

Institute of Solid Mechanics of the Romanian Academy, Bucharest, Romania

ION ION, ADRIAN CURAJ

Politechnica University of Bucharest, Romania

STEFAN A. DUMITRU

Institute of Solid Mechanics of the Romanian Academy, Bucharest, Romania

The paper presents ZPM dynamic control of walking robots, developing an open architecture real time control multiprocessor system, for dynamics stability improvement of MERO walking robots. In order to increase mobility and stability in real conditions and to obtain superior results relating to the possibility of moving walking robots on terrains with close configuration to real situations such as slope walking, overtaking or avoiding obstacles, the paper presents research regarding the integration of compliant control and fuzzy control into the hybrid position force control system architecture for hexapod walking robots. The results obtained shows that this allows the robot to adapt to rough terrain through a real time control with increased stability during walking.

1. Introduction

The walking robots motion control is included in the category of systems with a high degree of automation. The mechanical system must be equipped with a large number of degrees of free (DOF), in order to form complex synergies and to achieve coordinated movement of the legs. The action of such disturbances like an additional load, changes in weight, position of center of gravity and the robot platform inertial moments and may be a cause of significant deviations from the robot prescribed motion [1,2,3]. A number of compliant control techniques are known for obtaining high performance in robot trajectory control, some of which include a dynamic model in the control loop: solved acceleration loop (Luh, Walker and Paul), operational space method (Khatib),

[1] * This work is supported by the research project no. 263/2007-2010, ID 005/2007, in the framework of Grants of CNCSIS Program IDEAS, PN II (National Program for Scientific Research and Innovation Technologies), Romania.

impedance control (Hogan, Kazerooni, Sheridan) and other which do not include dynamic models: hybrid control (Railbert si Craig) and rigidity control (Salisbury). Stability analyses and experimental implementations have shown not only that using dynamic models leads to a more precise control, but also that using an inadequate dynamic model can sometimes lead to unstable force control [4,5,6]. A dynamic position-force control system is presented by integrating the multi-stage fuzzy method with solved acceleration in position-force control and dynamic control loops through the ZMP method. It is also noted that in addition to hybrid position-force control, three other main tasks are added, resulting from the analysis of the robot's walking cycle, conforming to movement characteristics: real time balance control, walking scheme control and predictable control of the walking robot movement [7,8,9].

Starting from similar research applied to humanoid robots, in order to increase mobility and stability of walking robots movement, this paper approaches the new concepts of applied mathematics in modeling and hybrid position-force control for walking robots, developing an open architecture real-time control multiprocessor system. The main strategy to the dynamic control of the movement of walking robots lies in the Zero Movement Point (ZMP) method by processing inertial information of force, torque and tilting and by implementing intelligent high level algorithms.

2. Strategy for fuzzy dynamic walking robot control using ZMP and inertial information

The walking robot is considered as a set of articulated rigid solids, which are standing as a platform and leg elements. With the increasing number of walking robot legs, the driving and control system is becoming more complicated. Moreover, due to the increased number of points of support, the static and quasi-static displacement becomes more stable. As noted above, the walking robot movement is stable only under certain and quite restrictive conditions [5,9]. The static stability problem is solved by calculating the extremity of each leg position according to the system of axes attached to the platform, with origin at the center of gravity of it.

In order to carry out new capabilities for walking robots, such as walking down the slope, going by overcoming or avoiding obstacles, it is necessary to develop high-level intelligent algorithms, because the mechanism of walking robots stepping on a road with bumps is a complicated process to understand, being a repetitive process of tilting or unstable movements that can lead to the overthrow of the robot. The chosen method that adapts well to walking robots is the zero moment point ZMP (Zero Moment Point) method. During research there was developed a new strategy for the dynamic control for walking robot stepping using ZMP and inertial information through . The control diagram includes pattern generation of compliant walking, real-time ZMP compensation in one phase - support phase, the leg joint damping control, stable stepping control and stepping position control based on angular velocity of the platform. In this way, the walking robot is able to adapt on uneven ground, through a real time control, without losing its stability during walking.

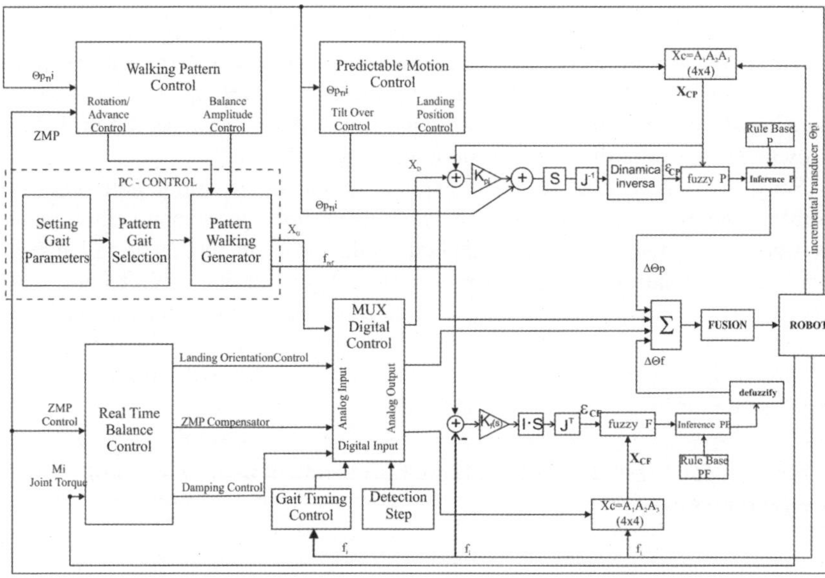

Figure 1. Robots' real time control method through virtual projection

A dynamic position-force control system is presented in figure 1 by integrating the multi-stage fuzzy method with solved acceleration in position-force control and dynamic control loops through the ZMP method. It is also noted that in addition to hybrid position-force control, three other main tasks are added, resulting from the analysis of the robot's walking cycle, conforming to movement characteristics: real time balance control, walking scheme control and predictable control of the walking robot movement. The first main task, balance control, leads to a control model that periodically modifies the walking scheme, depending on the sensory information received from the robot transducers. Real time balance control contains 4 types of reactive loops: damping control, ZMP compensation control, walk timing control and walk orientation control. The second main task, walking scheme control, represents a real time control of the robot equilibrium using the reactions of inertial sensors. The walking control scheme can be changed periodically in accordance with the information received from the inertial sensors during the walking cycle, by processing them into two real time loops: platform swing amplitude control and platform rotation/advance control. The third main task, predictable movement control, represents the control of predictable movement based on a fast decision from previous experimental data. **ZMP compensation control strategy.** The damping loop is not sufficient to maintain a stable walking motion because of the ZMP movement. So, there was developed a ZMP compensator in a phase, respectively in single support phase (FSU), where the platform will move back and forth according to ZMP dynamics. Control strategy consists in mathematical modeling of ZMP compensator through the springloaded inverted pendulums. **Gait timing control strategy**. Timing control at landing avoids instabilities on the robot motion at landing. This module generates a signal to block the walking diagram if the foot doesn't touch the ground at the end of phase 2 and 4. This way the command to move the leg is stopped, until the leg touches the ground, as a result of the inertia movement. **The walking pattern control,** may be changed periodically according to the information received from an inertia transducer during each walking cycle. The control pattern of the walking model contains two control loops in real time: the amplitude control for platform balance and advance/rotate control of the platform. For **modeling of platform balance** amplitude was used inverse pendulum method according to lateral balance. In terms of control strategy, periodically adjust the lateral balance amplitude of the platform in order to move ZMP, corrected by measuring ZMP during each walking cycle. Changing the platform balance amplitude is achieved by offsetting the position of the legs which are in contact with the ground so that actual ZMP amplitude (measured) can converge to the average reference ZMP.

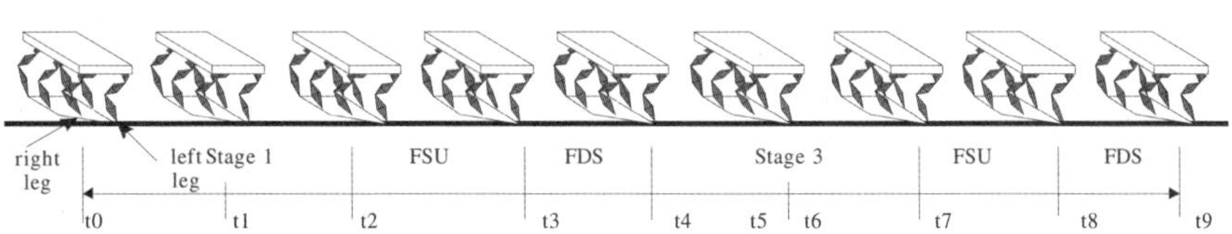

Fig. 2 Walking phases and time cycle

Rotation/advance platform control strategy, driven at the walking on a slope, allows the central position of the platform to move in the opposite direction to the inclined transverse plane so that the swinging movement can be well balanced. To measure the platform motion, the command system processes on-line the analog signal from an inertia transducer mounted on the robot platform. Position vector of the platform is obtained through an iterative relation of integration for angular error vector of the platform between the unique support phases (FSU) 1 and 3, amplified by a gain coefficient determined experimentally. In the walking pattern control, the gait phases related to the cycle time have been taken into consideration (fig.2),, where: t0-t1 is starting phase 1, t0-t2 is stage 2 having t2-t3 FSU and t3-t4 FDS, t4-t7 is stage 3 respectively t7- t9 stage 4 with FSU and FDS. **The predictable motion control** is based on generating probable robot movements achieved by processing earlier movements, in order to avoid overthrowing the walking robot. Anticipating robot movements is obtained through experimental statistic information, taking into account the margins of stability, prevention of abnormal conditions and processing of additional movements. It has two online control loops.

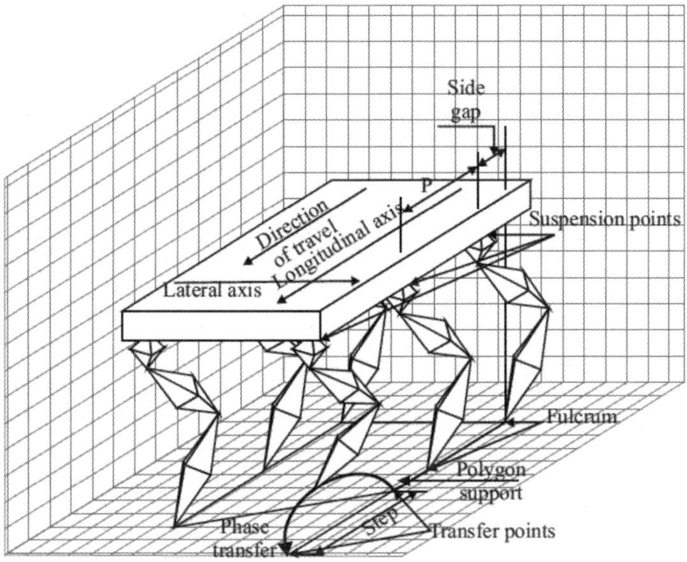

Fig. 3 The safety limits of walking robot

The limits of safety have been defined, according to the value imposed to the limit of stability for a system of walking and/or sizes (fig. 3) depending on the reaction forces of the contact points of the feet with the supporting surface. Bringing the projection center of gravity of walking robots in the area of stability, meaning inside the polygon of support is achieved by horizontal movement of the platform, controlled by the block maintaining stability.

Another phase in research development was to **establish programming strategies** for control schemes reported in control loops and phases of walking robot, as follows: in balance control is achieved damping control, ZMP compensation, position control on landing phases 1, 2 FSU, 3 and 4 FSU; within walking pattern control it's achieved the platform balance amplitude control and the controller for advance/rotation in phases 2 FDS and 4 FDS.

Within predictable motion control we have the tilt over robot controller, in phases 1 and 3, respectively the control of landing position in phases 2 FSU and FDS, 4 FSU and FDS. Choosing the movement period of the robot to 3.7 s , we've obtained its natural frequency of 0.27 Hz. Double support phase portion in a walking cycle is approximately 10% to 20% for humanoid robot, while for the HFPC robot was experimentally determined as only 5% because HFPC has no joint in the palm of the robot. Lateral balance movement of the platform is essential to moving the ZMP on each foot while walking.

3. Results and conclusions

The research evidences that stable gaits can be achieved by employing simple control approaches which take advantage of the dynamics of compliant legs. This allows a decentralization of the control system, through which a central command establishes the general movement trajectory and local control laws presented in the paper solve the movement stability problems, such as: damping control, ZMP compensation control, gait timing control, walking pattern control, predictable motion control. A distributed control system architecture was integrated into the HFPC architecture so that it can be controlled with high efficiency and high performance.

The command for walking robots is achieved by a control system with three levels. The first level is to produce control signals for motor drive mounted on leg joints, ensuring the robot moving in the direction required with a given speed. The language for this level is that of differential equations. The second level controls the walking, respectively it coordinates the movements and provides the data necessary data to achieve progress. At this level, work is described in the language of algorithms types of walking. The third level of command defines the type of walking, speed and orientation. At this level, the command may be provided by an operator who can use the control panel, in pursuit of its link with the robot, to specify the type of running and passing special orders, for the definition of the vector speed of movement. To maintain the platform in a horizontal position, information provided by the horizontality or verticality transducers is used, that senses the walking robots deviation platform to the horizontal position.

In order to achieve greater mobility and an increased stability in real life conditions and for obtaining superior movement performances on terrains with a configuration closer to reality, systematic and thorough studies have been performed regarding trajectory control, knowledge of the coordinates of the robot leg fulcrum in relation to its body, as well as the coordinates of the body's position during walking in relation to the fulcrum. For simplification purposes the existence of a point contact between the leg and the support surface is admitted. The results of modelling have shown that the semi-dynamic walk, as well as the step, may be stabilized by using the absolute system of coordinates solely for the angle of rotation and its corresponding angular speed.

For experimental testing and real time robot control we used the virtual projection method (L. Vladareanu, et. all, "Real time control method and device for robot in virtual projection", EU patent no. EPO-09464001, 2009). The results obtained have allowed the graphical determination of the walking robot model's reply to round the coordinate axes rotation and have led to an improvement in the response time to disturbances with up to 20%, and to tracking the motion trajectory with higher than 30% precision in conditions of high stability.

The research showed the efficiency of the proposed method for controlling the walk of a walking robot with applicability to the modular MERO robot. More than this the results obtained shows that this allows the robot to adapt to rough terrain through a real time control with increased stability during walking. In the end, a multi-microprocessor architecture was designed with multi-tasking control that allows a fast feedback loop for real time robot control, improving stability and flexibility performance.

Acknowledgments

The authors wish to express their gratitude to the Research and Educational Ministry (MEC) and to the Romanian Academy for their support of the program of work reported herein. The work took place as part of research project no. 263/2007-2010, ID 005/2007, in the framework of Grants of CNCSIS Program IDEAS, PN II (National Program for Scientific Research and Innovation Technologies).

References

1. Jung-Yup Kim, Ill-Woo Park, Jun-Ho Oh, Walking Control Algorithm of Biped Humanoid Robot on Uneven and Inclined Floor, Springer Science, J. Intell Robot Syst (2007) 48:457–484, DOI 10.1007/s10846-006-9107-8
2. L. Vladareanu1, G. Tont, I. Ion, V. Vladareanu, D. Mitroi, Modeling and Hybrid Position-Force Control of Walking Modular Robots, ISI Proceedings, Recent Advances in Applied Mathematics, Harvard University, Cambridge, USA, 2010, pg. 510-518, ISBN 978-960-474-150-2, ISSN 1790-2769
3. Sandru, OI; Vladareanu, L; Sandru, A, Grasping objects with mobile robots, Proceedings of the 9th wseas international conference on applications of electrical engineering, 2010, Penang, Malaysia, Pages: 210-215, ISSN: 1790-5117, ISBN: 978-960-474-171-7
4. J. Rummel, A. Seyfarth, Stable Running with Segmented Legs, The International Journal of Robotics Research 2008; 27; 919, DOI: 10.1177/027836490895136

5. Vladareanu L., Ion I., Velea L.M., Mitroi D., Gal A., The Real Time Control of Modular Walking Robot Stability, Proceedings of the 8th International Conference on Applications of Electrical Engineering (AEE '09), Houston, USA,pg.179-186, ISSN: 1790-5117 , ISBN: 978-960-474-072-7
6. An C.H., Hollerbach J.M., The Role of Dynamic Models in Cartesian Force Control of Manipulators, The International Journal of Robotics Research,Vol.8,No.4, August 1989,pg. 51-71
7. L. Vladareanu, I.Ion, I. Simionescu; The Reaction Forces in Real-Time Control for Modular Walking Robot Legs; Proceedings of the 13th International Conference on Climbing and Walking Robots and the Support Technologies for Mobile Machines, Nagoya Inst. Technology, Japan 2010; ISBN-13 978-981-4327-97-8; ISBN-10 981-4327-97-2.
8. Tont G; Vladareanu L; Munteanu R.A.; Tont D.G.; Fuzzy Petri Net-Based Approach in Modelling Simultaneous Task Assignment for Robotic System; Proceedings of the American Conference on Applied Mathematics; Pag:409-414; 2010 Harvard Univ, Cambridge, MA; ISBN: 978-960-474-150-2.
9. I. Ion, I. Simionescu, L. Vladareanu, A. Curaj, V. Aurelian; Reconfiguring MERO Modular Walking Robots, Proceedings of the 13th International Conference on Climbing and Walking Robots, Nagoya Ins. Technology, Japan 2010; ISBN-13 978-981-4327-97-8; ISBN-10 981-4327-97-2.

COMPLIANCE FOR A CROSS FOUR-BAR KNEE JOINT

A. Hamon and Y. Aoustin
IRCCyN, UMR CNRS 6597,
Nantes, 1 rue de la noe, France
E-mail: [arnaud.hamon],[yannick.aoustin]@irccyn.ec-nantes.fr

We propose a mechanical design for the knee joints of a planar bipedal robot. Each knee joint is composed of cross four-bar with springs. The dynamic model of a planar bipedal robot with cross four-bar knees and a parametric optimization problem are presented to produce a set of optimal reference trajectories. We use these trajectories to compare the performance of the bipedal robot with respect to different physical characteristics of the knee joints.

Keywords: Bipedal robot. Compliance. Four-bar knee joint. Optimization process.

1. Introduction

The biomechanics researchers have done important progress in the comprehension of the human lower limb and especially on the knee joint[1] and the ankle joint.[2] Indeed, these two joints have a complex structure formed by non symmetric surfaces. They can produce more complex movements than a simple revolute joint. In addition to the flexion in the sagittal plane, there are an internal rotation with displacement of the Instantaneous Center of rotation(ICR) of the knee joint and a posterior translation of the femur on the tibia. These motions cannot be represented by one or two revolute joints. Different studies have confirmed these results by an observation of the movements of the human knee in the 3D space.[3]

From these studies, a new kind of prosthetic knee is appeared, called polycentric knee. A classical polycentric knee is the four-bar linkage,[4] which used in most prosthetic knees. This structure forms a closed mechanism, which allows a combined rotation and translation of the knee joint in the sagittal plain without using of artificial ligaments to keep the rigidity of the mechanism. The dimensions of the four-bar structure can be chosen by measurement, on a real subject, of the length of the anterior and posterior ligaments, and the position of the cross ligament attachments on the tibia and the femur, projected in the sagittal plain in the maximum extension position.[5] This choice of the dimensions produces similar motions in the sagittal plain of the knee than those obtained with the human knee.[6]

Our objective is to improve the bipedal robot performance by a new conception of the knee joint. Several papers are devoted to bipedal robots equipped with complex knees, like G. Gini et al.[7] which used knee joints based on the human knee surfaces. F. Wang et al.[8] developed a bipedal robot with two different joints, a rotoide joint and a four-bar joint. However, a four-bar structure presents a configuration of singularity which limits the flexion of the knee contrary to a cross four-bar solution for which the singular position is out of the range of use of the knee joint. In Ref. 9, we have proved a cross four-bar joint is a better solution for the knee than a rotoide joint in term of energy consumption. In this paper, we improved the study by adding a compliance in the knee with different solutions of springs.

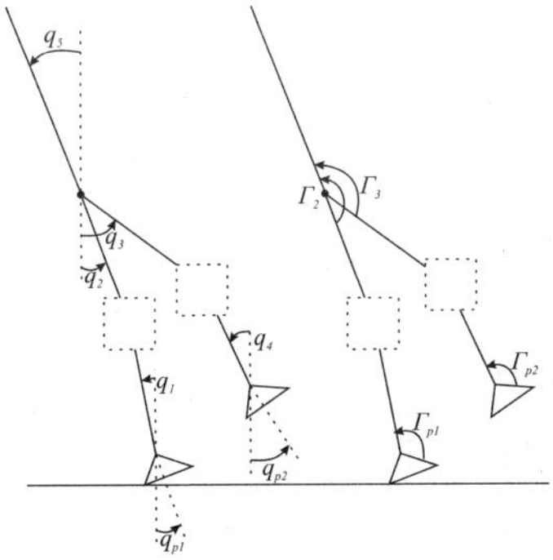

Fig. 1. Diagram of a planar bipedal robot. Absolute angular variables and torques.

2. Dynamic model

2.1. *Reduced model in single support*

The aim is to propose a dynamic model with an implicite liaison of the stance foot with the ground to calculate the torques during the optimization process, with the knowledge of the reference trajectories of the generalized coordinates. This reduced dynamic model is only valid if the stance foot does not take off and there is no sliding during the swing phase.

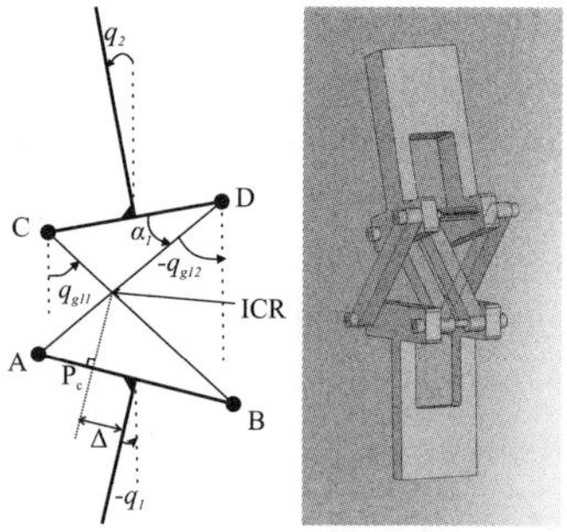

(a) (b)

Fig. 2. Details of the cross four-bar joint and position of the Instantaneous Center of Rotation (ICR)

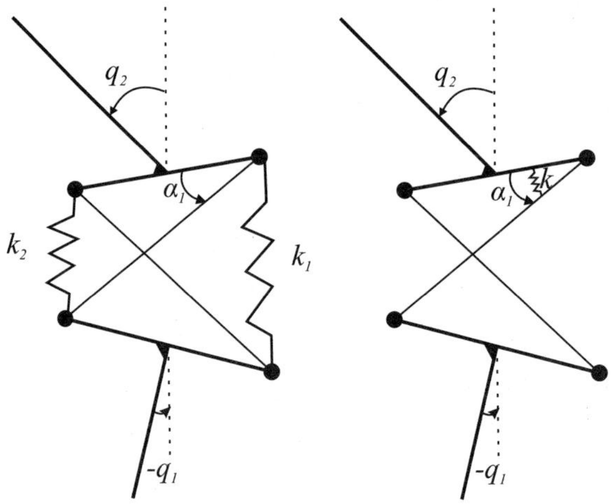

(a) (b)

Fig. 3. Cross four-bar knee with extension springs and with torsion spring

Then, during the single support phase, the stance foot is assumed to remain in flat contact on the ground, *i.e.*, no sliding motion, no take-off, no rotation ($q_{p_1} = 0$). So, we can use a reduced dynamic model in single support, based on vector $q = [q_{p_2}, q_1, \cdots, q_5, q_{g_{11}}, q_{g_{12}}, q_{g_{21}}, q_{g_{22}}]^t$, which is independent from the reaction force of the gait on the stance foot.

The bipedal robot is equipped of two closed structures. Let us introduce the constraint equations solving the dynamic model.[10] For the sake of clarity

Table 1. Physical parameters of the bipedal robot.

	Masse (kg)	Length (m)	Inertia ($kg.m^2$)	Center of mass (m)
foot	0.678	$L_p = 0.207$ $H_p = 0.059$	0.002	0.0235
shin	2.188	0.392	0.027	0.223
thigh	5.025	0.392	0.066	0.2234
trunk	24.97	0.403	1.036	0.281
four-bar knee	$m_b = 0.3$ $m_d = 0.3$		$l_a = AB = 0.029$ $l_b = BC = 0.030$ $l_c = CD = 0.015$ $l_d = AD = 0.025$	

let us consider only knee 1 to detail the principle of the calculation:

$$\begin{aligned} l_a\cos(q_1) - l_b\sin(q_{g_{11}}) + l_c\cos(q_2) + l_d\sin(q_{g_{12}}) = 0 \\ l_a\sin(q_1) + l_b\cos(q_{g_{11}}) - l_c\sin(q_2) - l_d\cos(q_{g_{12}}) = 0 \end{aligned} \quad (1)$$

These geometrical relations, their first and their second derivatives in times yields to the constraints in velocities and accelerations of the closed structure of the knee 1. By using the virtual work principle, these constraints equations can be expressed in the dynamic model by adding the Lagrange multipliers $J_{1_e}\lambda$.

Here J_{1_e} is the 2x11 Jacobian matrix calculated from the geometrical relations (1) and $\lambda = F_{c_1} = [F_{x_1}, F_{z_1}]^t$ defines the two Lagrange multipliers vector associated to the constraint of the closed structure. These multipliers represent the exerted forces on point A of the knee joint (Fig. 2). We apply the same principle for the knee joint 2 to obtain the dynamic model of the bipedal robot with the cross four-bar knees:

$$A(q)\ddot{q} + H(q,\dot{q}) = \begin{bmatrix} D_\Gamma \; J_1^t \; J_2^t \end{bmatrix} \begin{bmatrix} \Gamma_m \\ F_c \end{bmatrix} \quad (2)$$

2.2. *Adding spring effect in the dynamic model*

In a general case, the dynamic model of a robot using spring is depicted by :

$$A(X)\ddot{X} + H(X,\dot{X}) + Q_{spring}(X) = D(X) \quad (3)$$

Where $A(X)$ is the matrix of the inertial effect, $H(X,\dot{X})$ is the vector of the centrifugal, Coriolis and gravity effects. D is the vector of the torque effect and the external force. $Q_{spring}(X)$ is the effect of the spring which obtained by the derivation of the potential energy of the spring :

$$Q_{spring} = \frac{\partial U_{spring}}{\partial X} \tag{4}$$

with :

$$U_{spring} = \frac{k}{2}(\beta - \beta_0)^2 \tag{5}$$

Where k is the stiffness coefficient of the spring. β can be the current length of the springs or its angular position and β_0 is the equilibrium state of the spring.

3. Gait optimization for the cyclic walking

To design a cyclic walking gait, the behaviour of the actuated joint variables are prescribed using cubic spline functions. The set of the optimization parameters are used to calculate these cubic spline functions, taking account the properties of continuity between each step. The objective is to calculate a cyclic gait by minimizing a criterion based on the energy consumption. Physical conditions of contact between the feet and the ground and limits on the actuators define non-linear constraints of this optimization process. The cyclic walking motions are composed of single support phases with impulsive impacts.

By parametrizing the joint motion in terms of cubic spline functions, the optimization problem is reduced to a constrained parameter optimization problem of the form:

$$\begin{array}{c}\text{Minimize } \mathbf{C}\ (P) \\ \text{subject to } \mathbf{g_j}\ (P) \leq 0 \text{ for } j = 1, 2, ...l\end{array} \tag{6}$$

where P is the set of optimization variables. $\mathbf{C}(P)$ is the criterion to minimize with l inequality constraints $\mathbf{g_j}(P) \leq 0$ to satisfy. The criterion and constraints are given in the following sections.

We used the SQP method (Sequential Quadratic Programming),[11 12] with the *fmincon* function of Matlab ® to solve this problem.

$$C_W = \frac{1}{d}\int_0^{\mathrm{T}} \Gamma_m^{\mathrm{t}} \Gamma_m d\tau \tag{7}$$

where T is the step duration and Γ_m is the vector of the joint torque. During an optimization process the step length d is an optimization variable and the walking rate v is fixed, such as the step duration is directly given by the relation $T = d/v$.

Two types of constraints are used to obtain a realistic gait.

- The necessary constraints, which ensure a valid walking gait. The first constraint ensures the support leg tip does not take off or slide on the ground. So, the ground reaction force is inside a friction cone, defined by the coefficient of friction f.
  To ensure the non rotation of the support foot we introduce a constraint on the ZMP during the single support phase and at the instant of the impact.
  Just after the impact, the velocity of the taking-off foot should be directed upward. In consequence, the positivity of the vertical component of the velocities for the heel and the toes is added to the set of constraints.
  The last constraint allows ensuring the non penetration of the swinging foot in the ground.
- The unnecessary constraints, which ensure a technological realistic gait. We introduced mechanical stops on the joint variables. Moreover, we limited the torques with a constraint, which sets a template of the maximum torque of the motor relatively to the velocity.

4. Effect of the springs on the knee joints on the energy consumption.

In this paper, we compare the effect of a cross four-bar knees with spring units on the energy consumption of the biped during a cyclic walking gait. We use different springs : a torsion spring in parallel of the actuator of the knee or two extension springs on the two side of the cross four-bar knee. Indeed, several papers pay interest to the effect of springs equipping the bipedal robot joints.[13] We produce a set of optimal trajectories for the bipedal robot where the stiffness coefficient is an optimization parameter. The stiffness coefficient of the springs depends on the walking velocity and can be different for the support leg and for the swing leg. This choice implies the spring units used, have a variable stiffness. We can note, the work of S. Wolf in Ref. 14, which designs a rotoide joint with a control of the stiffness coefficient. Different types of springs can be used for the knee joint. We can see on Fig.4, the energy consumption according to the walking velocity of the bipedal robot equipped of cross four-bar knees without springs and with the two solutions of springs. We observed the biped equipped of cross four-bar knees with extension springs is more efficient than the biped equipped of simple cross four-bar knees for the walking velocity lower than 2.5 Km/H. Moreover, the biped equipped of cross four-bar knee with torsion spring have a similar energy consumption than the biped with cross four-bar knees with extension springs for the walking rates lower than 2 Km/H. For the higher walking velocities the cross four-bar knees with torsion springs is the better solution of knee.

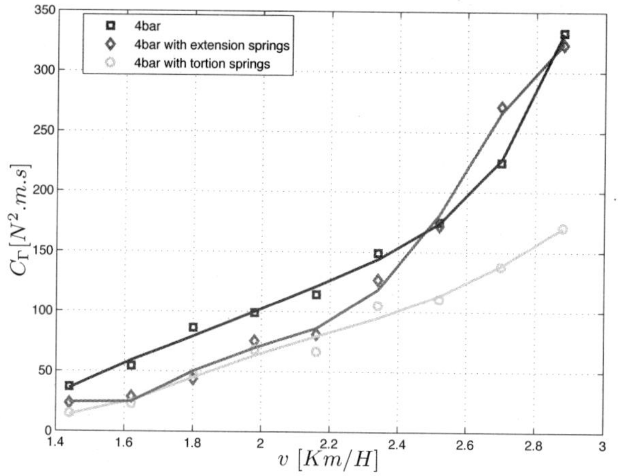

Fig. 4. Evolution of the energy consumption according to the walking rate for the cross four bar knee without springs and with extension springs and torsion springs.

5. Conclusions

We have proposed a mechanical structure for the knee of a planar bipedal robot called cross four-bar knee and different solutions of actuation with spring units. We have produced with a parametric optimization method a set of optimal reference walking trajectories for a bipedal robot for different solutions of knee structures. We have compared the energy consumption of the bipedal robot for the different structure of the knees. This study proved that the cross four-bar knee with spring units can reduce the energy consumption of the biped. The perspective of this study is to extend this work for a 3D bipedal robots. Moreover a measure of the evolution of center of rotation of the human knee during a walking gait will be used to choose more precisely the dimension of the cross four-bar mechanism.

References

1. D. R. Wilson, J. D. Feikes and J. O'Connor, *Journal of Biomechanics* **31**, 1127 (1998).
2. A. Leardini, J. O'Connor, F. Catani and S. Giannini, *Journal of Biomechanics* **32**, 585 (1999).
3. B. Landjerit and M. Bisserie, *Acta Orthopaedica Belgica* **58**, 147 (1992).
4. S. A. Gard, D. S. Childress and J. E. Uellendahl, *Journal of Prosthetics and Orthotics* **8**, 34 (1996).
5. J. D. Feikes, J. J. O'Connor and A. B. Zavatsky, *Journal of Biomechanics* **36**, 125 (2003).
6. F. K. Fuss, *American Journal of Anatomy* **184**, 165 (1989).
7. G. Gini, U. Scarfogliero and M. Folgheraiter, Human-oriented biped robot design : insights into the development of a truly antropomophic leg, in *IEEE International Conference on Robotics and Automation*, 2007.
8. F. Wang, C. Wu, Y. Zhang and X. Xu, Design and implementation of coordinated control strategy for biped robot with heterogeneous legs, in *IEEE International Conference on Mechatronics and Automation*, 2007.
9. A. Hamon and Y. Aoustin, Cross four-bar linkage for the knees of a planar bipedal robot, in *2010 IEEE-RAS International Conference on Humanoid Robots*, 2010.
10. W. Khalil and E. Dombre, *Modeling, identification and control of robots* (Hermes Sciences Europe, 2002).
11. P. Gill, W. Murray and M. Wright, *Practical optimization* (Academic Press, London, 1981).
12. M. Powell, *Variable metric methods for constrained optimization*, in *Computing Methods in Applied Sciences and Engineering*, Lecture Notes in Mathematics Lecture Notes in Mathematics, (Springer Berlin / Heidelberg, 1977), pp. 62–72.
13. T. Yang, E. Westervelt, J. Schmideler and R. Bockbrader, *Autonomous robots* **25**, 317 (2008).
14. S. Wolf and G. Hirzinger, A new variable stiffness design : matching requirments of the next robot generation, in *2008 IEEE International Conference on Robotics and Automation*, 2008.

WALKING DIRECTION CONTROL OF A SIX-LEGGED ROBOT BY OPTIMAL SERVO SYSTEM*

HIROAKI UCHIDA†, TAKUYA KAMEYAMA

Department of Mechanical Engineering, Kisarazu National College of Technology, 2-11-1, Kiyomidai-higashi, Kisarazu, Chiba, 292-0041, Japan

This paper dealt with a control method for an achieving walking not to follow the desired angle of the rotating links between supporting phase. In this study, we proposed a new control method to control the body in the walking direction. Concretely, mathematical model in the walking direction was derived, optimal servo system was composed, and the 3D simulations were done.

1. Introduction

The general control method of a six-legged robot is to control the angle of the leg joint to follow the desired orbit. However, on irregular terrain, it is necessary to recognize the walking environment beforehand, and, to design the desired orbit that is appropriate for the environment. For such a problem, it is well known that force control and posture control are valid control methods for the adjustment to the walking environment. Hirose et al. designed and marketed a spread type quadruped robot with a leg consisting of links to generate the moving speed in the rotating part and to generate the power for supporting the weight of the body in the thigh part[1]. We've focused on this mechanism to control the posture (pitching angle, rolling angle, and height of the body) of the six-legged robot by controlling the thigh link by the three supporting legs. And the validity of the proposed method were verified by 3D simulations and the walking experiments[2]. However, the report that examined the control method for controlling the position on the walking direction and the yawing angle of the body at the same time hasn't been performed for a six-legged robot. In this study, we examine the control method of controlling the walking both direction (x, y) and the yawing angle of the body at the same time for the six-legged robot having the above mentioned mechanism. In this method, it isn't necessary to design the desired orbit of the rotating link at the supporting leg beforehand, and it is able to correspond also to the walking speed control of the multi-legged robot as well as the speed control of a car. Six-legged robot is able to realize the

---

*

†

flexible walking to correspond to the walking environment. Moreover, it becomes possible to control the six-axes of the body by applying the proposed control method with the posture control method we've proposed. The six-legged robot can realize a walking to be suitable for the walking environment. In this study, first, the mathematical model is derived. Next, optimal servo system is designed. Finally, the proposed control method is examined by 3D simulations.

2. Six-legged robot

Figure 1 shows the 3D CAD model of a six-legged robot. One leg of the six-legged robot consists of three joints that are a rotating part, a thigh part and a shank part. The advantages of the leg mechanism of the six-legged robot are that the pitching angle, the rolling angle and the height of the body are controlled by controlling the thigh links, the walking direction (*x*, *y*) and yawing angle of the body are controlled by controlling the rotating links. Table 1 shows the dimension of the six-legged robot using in this study.

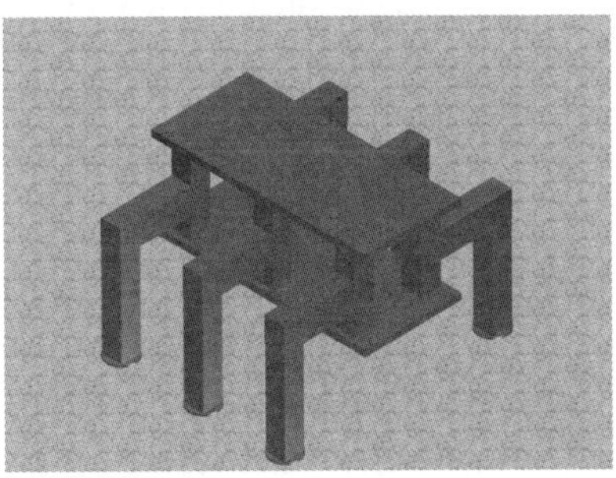

Fig.1 3D model of 6-legged robot

Table 1 Dimension of robot

Height	500mm	Length of Thigh	168mm
Length	666mm	Length of Shank	312mm
Width	710mm	Moment of Inertia I_z	1.966 Kgm^2
Weight	24.9kg		

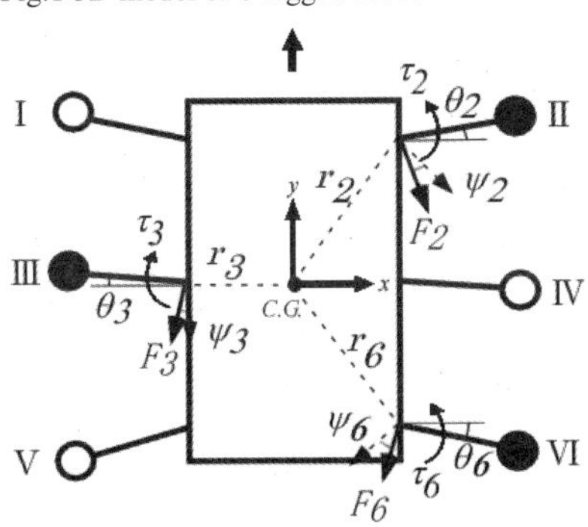

Fig.2 Mathematical model of walking direction in the 3 support legs

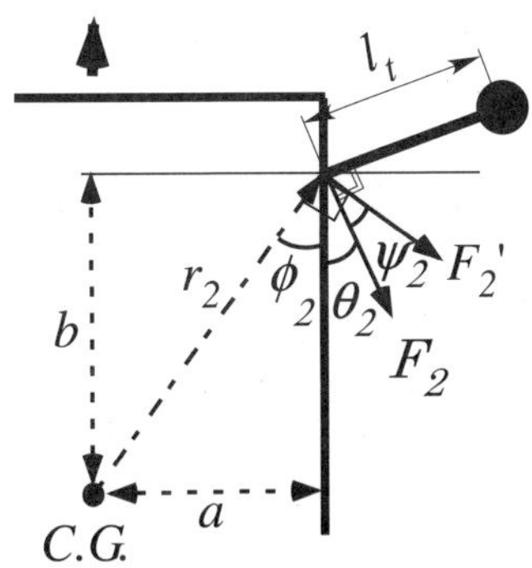

Fig.3 The relationship of ϕ , θ and ψ .

3. Walking Planning

The effectiveness of the proposed method is verified by the walking pattern of three supporting legs and three swinging legs. Walking period is 12sec.

4. Walking Direction Control Method

4.1. *Mathematical Model of the Body in the Walking Direction*

Figure 2 shows the relation of the force working at the rotational part in the 3 support legs. Here, support legs are Leg II, III and VI, and swing legs are Leg I, IV, and V. Moreover, the support legs are assumed to support the body firmly, and not to slip. y is the walking direction, x is the vertical direction to y, θ_z is the angle around z axis. Moreover, the force $F_i(i=1,\cdots,6)$ from the support leg acts on the rotational axis in the support leg. M is the weight of the body, I_z is the inertia around z axis, $\theta_i(i=1,\cdots,6)$ is the angle of the rotating link, $r_i(i=1,\cdots,6)$ is the distance between the center of gravity of the body and to the center of the rotation in the rotating link. The motion equation of x-direction, y-direction and around θ_z are shown as follows:

$$\begin{cases} M\ddot{x} = F_2\sin(-\theta_2) + F_3\sin\theta_3 + F_6\sin(-\theta_6) \\ M\ddot{y} = F_2\cos(-\theta_2) + F_3\cos\theta_3 + F_6\cos(-\theta_6) \\ I_z\ddot{\theta}_z = F_2 r_2\cos\psi_2 - F_3 r_3\cos\psi_3 + F_6 r_6\cos\psi_6 \end{cases} \tag{1}$$

By defining state vector $x = [x, y, \theta_z, \dot{x}, \dot{y}, \dot{\theta}_z]^T$, input vector $u = [\tau_2, \tau_3, \tau_6]^T$, following state equation and output equation are obtained. Here, input matrix $B(t)$ becomes time variant matrix.

$$\dot{x}(t) = Ax(t) + B(t)u(t) \tag{2-a}$$

$$y(t) = Cx(t) \tag{2-b}$$

where,

$$C = \begin{bmatrix} 1 & 0 & 0 & 0 & 0 & 0 \\ 0 & 1 & 0 & 0 & 0 & 0 \\ 0 & 0 & 1 & 0 & 0 & 0 \end{bmatrix}, \quad \tau_i = l_t F_i(i=1,\cdots,6),$$

$$A = \begin{bmatrix} 0 & 0 & 0 & 1 & 0 & 0 \\ 0 & 0 & 0 & 0 & 1 & 0 \\ 0 & 0 & 0 & 0 & 0 & 1 \\ 0 & 0 & 0 & 0 & 0 & 0 \\ 0 & 0 & 0 & 0 & 0 & 0 \\ 0 & 0 & 0 & 0 & 0 & 0 \end{bmatrix}, \quad B(t) = \begin{bmatrix} 0 & 0 & 0 \\ 0 & 0 & 0 \\ 0 & 0 & 0 \\ \frac{\sin(-\theta_2)}{l_t M} & \frac{\sin\theta_3}{l_t M} & \frac{\sin(-\theta_6)}{l_t M} \\ \frac{\cos(-\theta_2)}{l_t M} & \frac{\cos\theta_3}{l_t M} & \frac{\cos(-\theta_6)}{l_t M} \\ \frac{r_2 \cos\psi_2}{l_t I_z} & -\frac{r_3 \cos\psi_3}{l_t I_z} & \frac{r_6 \cos\psi_6}{l_t I_z} \end{bmatrix}$$

4.2. *Optimal Servo System*

The servo system that the system shown by Eq. (4) follows to the desired value is designed.

$$\begin{cases} \dot{z}(t) = r(t) - Cx(t) \\ \dot{x}(t) = Ax(t) + B(t)u(t) \end{cases} \tag{3}$$

Equation (3) is described in equation form as follows:

$$\dot{x}_g(t) = A_g x_g(t) + B_g(t)u + f_g r \tag{4}$$

The control input to the actuator driving the rotating link is obtained to minimize the following equitation.

$$J = \int_0^{\infty} \left[x_g(t)^T Q x_g(t) + u(t)^T R u(t) \right] dt \tag{5}$$

Here, $Q(n \times n)$ and $R(m \times m)$ are the weighting matrix given by the design specification, and $Q \geq 0, R > 0$. The control input to minimize Eq.(5) is shown as following:

$$u_b^o(t) = -R^{-1} B_g^T(t) P(t) x_g \tag{6}$$

Here, $P(t)(n \times n)$ is the solution of the following Ricatti equation.

$$P(t)A_g + A_g^T P(t) - P(t)B_g(t)R^{-1}B_g^T(t)P(t) + Q = 0 \tag{7}$$

5. 3D Simulations

The walking simulations were done in Case I and Case II. Case I was that the body of the robot was controlled to follow the desired function ($y_r = (step \times t)/T,\ T = 12s,\ \ step = 0.12m$) in reference to *y*-direction and the constant value (0) in reference to the x-direction and yawing angle by using 3DCAD model shown in Fig.1. Figure 3 shows the simulation results. The walking period is 12s, Fig.3(b) shows the angles of rotating links, and Fig.3 (c) shows the control inputs to the rotating links between support phases. In Fig.3, the solid line shows the 3D simulation results, the dashed line shows the reference signals. The weighting matrix of Eq.(5) was applied as $Q = diag($ $1000,1000,1000,0,\cdots,0)$, $R = diag(1,1,1)$. In Fig.3(b), the rotating link was controlled by the proposed control method in the supporting phase and the position control to follow the desired orbit in the swing phase, respectively. In this case, the control inputs grew at the support leg, and the case to exceed the range of the movement in the rotating link was appeared. Then, the range of the movements in the rotating link was $-30 \leq \theta_i \leq 30$ [deg] $(i = 1,\cdots,6)$. And, when the rotating link exceeded it, the control input to the rotating link was zero. The reference angles of the thigh links and the shank links were obtained from the desired orbit by using Inverse Kinematics when the toe orbit was designed by a step 0.12m, and each angle of the thigh and the shank was controlled by PD control to follow the desired orbit. In this case, there was a possibility that slipping was caused at the foot. However, in this study, even if slipping was caused, the simulations were continued without change because it has aimed to confirm the effectiveness of the proposed control method. From Fig.3(a), it is seen that the 3D simulation results of y-direction shows the good performance following the reference signal without error. And, in the x-direction and the yawing angle, though some vibrations occurred from the 3rd cycle to the 4th cycle, the body of the robot was controlled around zero of the reference signal. In Fig.3(b), the dashed line shows the reference signals obtained by Inverse Kinematics. Because of the position control to follow the reference signal, it is understood that the PD control in the swing phase corrects the error of the angles in the rotating links that was occurred by the proposed control method at the supporting phase. Moreover, the phase that the control inputs vibrated in the front and rear legs (I, II, V, and VI) of the body at the three supporting legs appeared from Fig.3(c), and the response of the rotating link vibrated. The angles of the rotating link in Leg I and II reached the limitation of the movement in a certain phase. On the other hand, it is seen that the control inputs becomes stable in Leg III and IV. Figure 4 shows the 3D simulation results of Case II where the periodic function ($ref = 0.05\sin(2\pi t / T),\ T = 12s$) for the yawing angle of the body is added to the reference signals of Case I. In this case, the phase where the control input vibrated appeared as well as Case I, and the response of the rotating links vibrated. However, it is seen that the 3D simulation results of y-direction shows the good performance as well as Case I, and, the body of the robot was controlled to follow the periodic reference signal of the yawing angle with a little error. Six-legged robot could realize a stable walking by the proposed control method.

6. Conclusions

As a walking direction control method of a six-legged robot, the motion equation of walking direction was derived by the relation of the force in the three supporting phase. And, optimal servo system was designed to this model. By using 3D model of the six-legged robot, the effectiveness of the proposed control method was examined in the case that the walking velocity in y-direction was constant and the yawing angle of the body was the periodic function. As a result, the validity of the proposed control method was verified by 3D simulations.

References

1. Hirose S. and Arikawa K., Development of Quadruped Walking Robot TITAN–VIII for Commercially Available Research Platform, Journal of Robotic Society of Japan, Vol. 17, No.8 (1999), pp1191-1197. (Japanese)
2. H. Uchida and K. Nonami, *Climbing & Walking Robots Towards New Application, I-Tech,* ***pp*** 299-312 (2007).

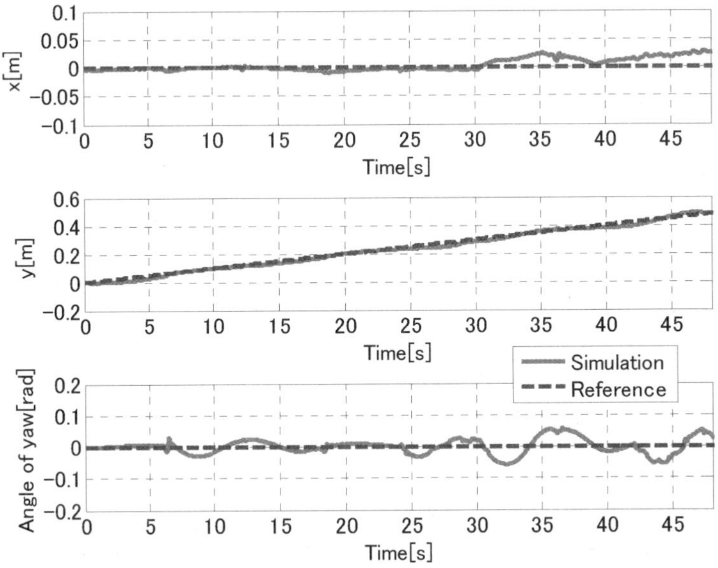

(a) x, y and yawing angle of the body

Fig.3 Simulation results of Case I

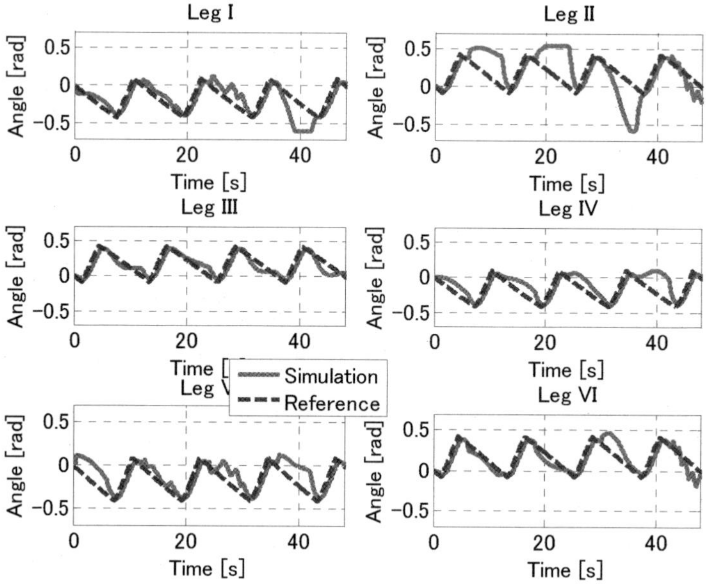

(b) Rotating angles

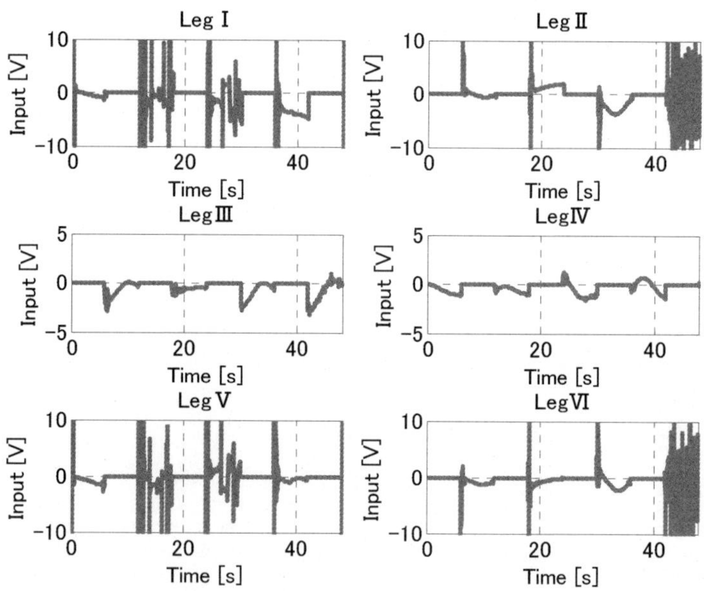

(c) Control inputs between support phase

Fig.3 Simulation results of Case I

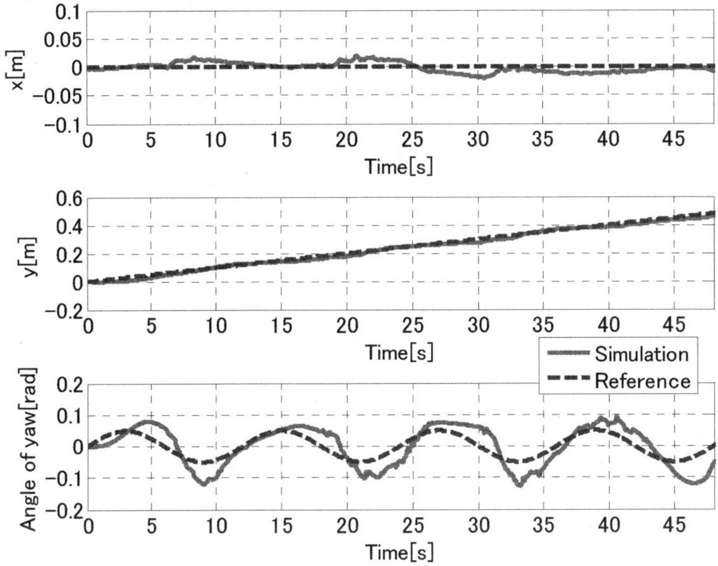

(a) x, y and yawing angle of the body

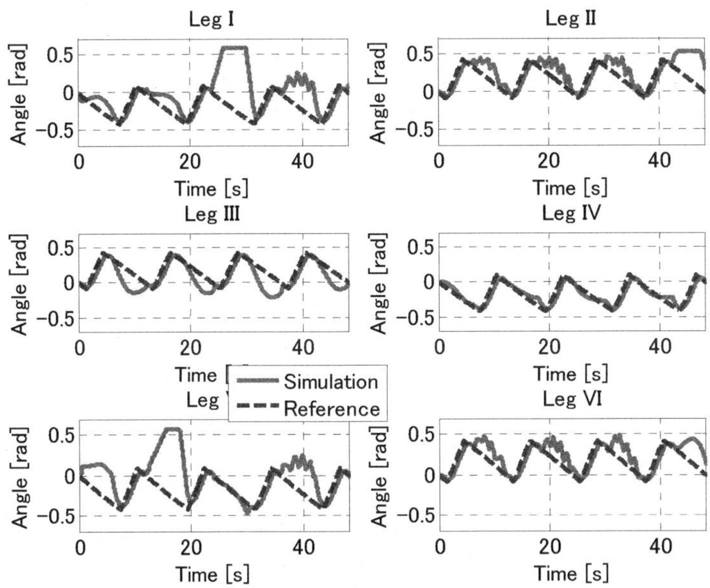

(b) Rotating angles

Fig.4 Simulation results Case II

VIRTUAL MODELLING, PROTOTYPING AND SIMULATION OF TRANSFEMORAL ACTIVE PROSTHETIC LIMB

ZHEN WEI LUI[1], MOHAMMED AWAD[1,†], ABBAS A. DEHGHANI[1], NEIL MESSENGER[2]

[1] *Institute of Engineering Systems and Design, School of Mechanical Engineering, University of Leeds, Leeds, West Yorkshire, LS2 9JT, United Kingdom*

[+]*Mechanical Engineering, Faculty of Engineering, Ain Shams University, Cairo, Egypt.*

[2] *Centre for Sport and Exercise Sciences, Faculty of Biological Sciences, University of Leeds, Leeds, West Yorkshire, LS2 9JT, United Kingdom*

An articulated ankle-foot prosthetic model, with two types of operating modes; passive and active controlled, and a single axis prosthetic knee model that has three operating modes; fully active, semi-active and passive, are being studied. The measures to determine the effectiveness of the model were taken from a reference model; human subject. Data for level ground walking was collected by using photogrammetric methods to improve the accuracy in data collection. These data were used in simulation software, MD ADAMS 2010, where a transfemoral amputee was simulated. The simulated results have shown that the prosthetic knee is able to reduce the electrical energy consumption by decreasing the activation time of the actuator by approximately 50%. Furthermore, the active ankle has successfully demonstrated the effects of push-off during terminal stance which helped reduce the BCOM additional vertical excursion with respect to the reference model.

1. Introduction

Adaptability has always been the main concern on contemporary lower limb prosthesis design. It has been reported that these prostheses have high flexibility, improve gait smoothness and reduce metabolic cost during walking [1]. Human lower limb biomechanics during level ground walking have previously been studied, especially the effects of critical gait events. The six determinants of gait, initially proposed by Saunders et al. [2], and later described by Whittle [3] are pelvic rotation, pelvic obliquity, knee flexion in stance phase, ankle plantar-dorsiflexion mechanism, foot mechanism, and lateral displacement of body, were believed to reduce energy expenditure by contributing to a reduced excursion of the body centre of mass (BCOM). However, it has been subsequently claimed that not all six determinants contribute to lowering the energy consumption but the occurrence of these determinants during normal walking cannot be neglected [4, 5]. The pelvic rotation has also been reported to play a minor role which only contributed approximately 10% BCOM vertical displacement reduction [5]. Instead, it is suggested that the ankle mechanism, specifically heel-rise from around mid-stance, is the main factor in reducing the energy consumption [5, 6]. This would suggest that an articulated prosthetic ankle-foot would effectively contribute in reduction of energy consumption over conventional non-articulated prosthetic foot.

Conventional single-axis prosthetic knee joint is comparable simpler to polycentric prosthetic knee mechanism thus requires minimal maintenance and increases engineering efficiency. However, such simple knee mechanism has been reported to have caused limitation on amputee's gait as these knee prostheses can only swing properly for a certain cadence despite its joint friction has been finely tuned [7]. This limitation has later been treated with heavy and costly hydraulic or pneumatic knee prosthetic. These improvements have, undoubtedly, increased the bandwidth of amputee's walking speed as well as the overall maintenance cost. Thus, it is postulated that an electrically-actuated single-axis prosthetic knee could fulfil all of the mentioned criteria: simplicity, durability, and efficiency.

In this paper, a total of six configurations of operating modes for both prosthetic knee and prosthetic ankle have been proposed. Joint kinematic data from healthy subjects performing level ground walking, collected by using photogrammetric methods, are used as a design reference in the modelling process of the prosthesis. The comparisons of simulation results with the reference are analyzed to observe the adaptability of the prosthesis kinematic responses. Thus, we hypothesized that should the prostheses kinematic behaviour is similar to that of a healthy subject's walking gait, the vertical displacement of the BCOM will have a similar trajectory which will in turn reduce the energy consumption of the amputee wearing conventional prosthesis.

1.1. *Motivation*

The main motivation of this work is to determine the optimum lower limb prosthesis by studying the human lower limb biomechanics during level ground walking. Thus by analyzing the behaviour of the modelled prosthesis, its dynamic responses could be evaluated and improvements could be made towards the model during the development stage of the prosthesis.

In order to evaluate the behaviour of a prosthetic on an amputee, 3D virtual modelling and simulation of the proposed prosthetics has been chosen to model its performance towards the dynamically changing behaviour of a subject. It has been reported that an amputee's adaptive behaviour towards prosthetics is close to impossible to model mathematically [8]. Not only that, the interfacing factors between the amputee and prosthetic have also been reported to be extremely difficult to model. However, there is a need to use virtual modelling and simulation tools so as to improve engineering efficiency and to reduce product development cost. Moreover, using simulation provides system validation for the design before implementation.

1.2. *Methodologies*

Instead of modelling the amputee's input responses on the prosthetic mathematically, computer simulation software has been proposed to evaluate the performance of a novel lower limb prosthetic design. This is due to the aforementioned difficulties in the predictions of an amputee's adaptation ability as well as the input responses on the prosthetic limb. Rodriguez [9] and Kim [10] have also used similar approach in determining the effectiveness of their research model.

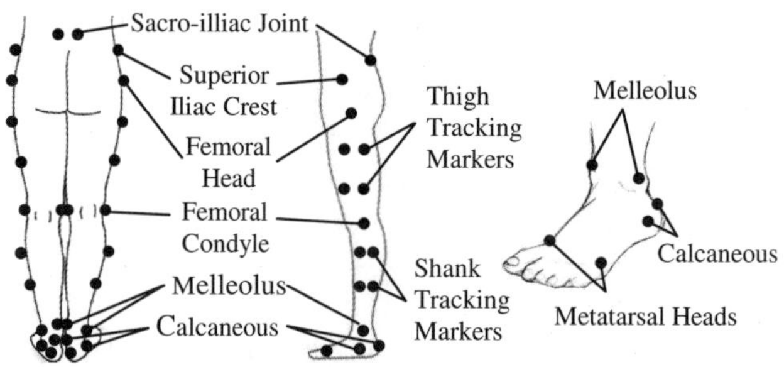

Figure 1. Reflective markers placement used in gait collection.

Gait analysis experiments were carried out to obtain kinematic joint data during human level ground walking by using a real time 3D motion capture system: Qualisys ProReflex MCU240 and Track Manager (QTM). These motion data were later exported into C-Motion Visual3D V4, which was used to define pelvis, thigh, shank, and foot body segments for both lower limbs using the marker system illustrated in Figure 1. Four healthy subjects volunteered to participate in this experiment. Each subject was requested to walk with self-selected normal walking speed along a straight walking path with a total of 6 steps. Two force-plates were used to determine critical gait events, i.e. heel-strike and toe-off.

Next, the model of each part of the prosthetics; knee and ankle-foot, were modelled by using 3D CAD modelling software. The geometry of the model, together with an amputated human model, were imported into the simulation software: MD ADAMS 2010, where simulations, based on the lower limb joint kinematic data collected during gait analysis experiments on normal walking gait, were carried out. The difference between the kinematic responses of the prosthesis and the reference were taken into consideration and improvements were made to the prosthesis model.

2. Virtual Model Description

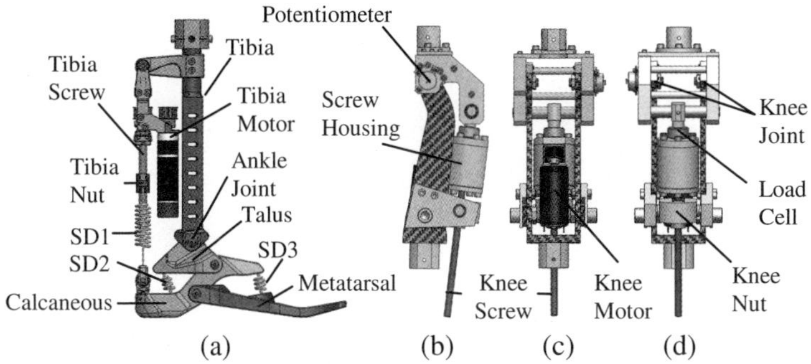

Figure 2. CAD of prosthetic model. (a) Prosthetic ankle-foot side view, (b) prosthetic knee side view, (c) prosthetic knee read view, and (d) prosthetic knee front view.

The proposed prosthetic ankle-foot model in this study had an articulated single-axis prosthetic ankle, which it is postulated, has better versatility over non-articulated prosthetic ankle in terms of manoeuvrability and adaptability. Figure 2(a) shows the CAD of the biological inspired prosthetic ankle-foot model constituting four main parts, namely Tibia, Talus, Calcaneous and Metatarsal, which represents the bone structure, and three passive components, spring-damper pair (SD), which imitate the tendons and ligaments of the human foot. This prosthetic ankle-foot model allows contraction of the foot during stance phase allowing energy to be stored during the first half of stance and released during termination of stance. A motor is used to drive the screw in order the change the position of the nut connecting SD1 that indirectly drives the ankle joint angle.

A single-axis prosthetic knee, which has a maximum flexion angle of 105.1° is proposed. This large range of operating angle is sufficient for normal walking as well as for sit to stand from a low chair with a maximum angle of 104.9° [11]. This prosthetic knee model consists of a motor with a power transmission unit connected to a screw. This was modelled so as to impose a certain degree of damping of the knee joint during non-actuated swing phase. The screw and nut of the prosthetic knee was proposed to be a non-self-locking ball screw unit that allows the knee joint to move freely while not actuated. This back-drivable mechanism allowed the prosthetic knee to have semi-active and passive operating modes that help reduce electrical energy consumption for continuous usage of the prosthesis. Figure 3 shows the human model wearing the proposed transfemoral prosthetic limb. Notice that the left foot of the modelled amputee had been replaced with a similar modelled prosthetic foot due to the inability of a rigid and solid foot to demonstrate critical gait events during initial stance and terminal stance.

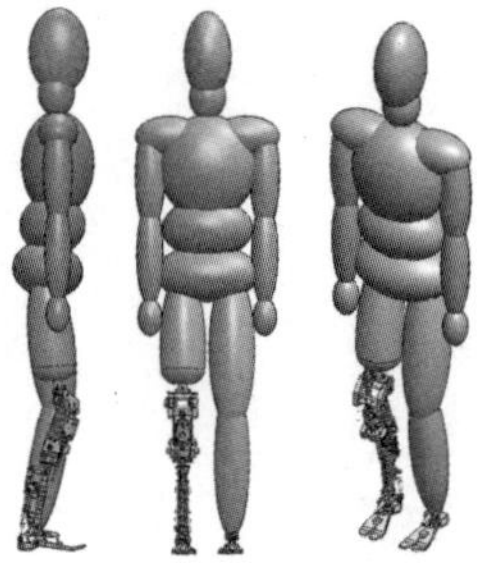

Figure 3. Amputated human model for simulations.

3. Simulation Configurations

The simulations carried out on the human model are based on the gait data collected from the gait analysis experiments. The motion of the human model in the simulation has been constrained to the sagittal plane. The sound limb was driven by motion at the hip, knee, and ankle joint while the residual limb is driven by a motion only at the hip joint. These motions represent a healthy person's lower limb joint motion during walking.

Table 1. Configuration number for different knee and ankle operations.

Configurations	Passive Ankle	Active Ankle
Active Knee	1	2
Semi-active Knee	3	4
Passive Knee	5	6

Apart from these motions applied on the human model, there are 3 different prosthetic knee controls: active, semi-active, and passive, and 2 different ankle operating modes: passive and active. Each combination of configuration has been simulated to determine the dynamic response of the lower limb prosthetic model. Referring to Table 1, active knee control is simply driving the knee motor to move the knee angle in a similar manner to that of a collected knee angle data from a healthy person. Semi-active and passive knee operating modes both deactivate the knee motor during terminal stance phase of a gait to allow the knee to flex freely before it swings. However, the difference between semi-active and passive is that semi-active knee control will resume the motor's motion while passive knee would stop the knee joint from rotating by braking the motor, where the actuator can be easily replaced with a braking mechanism. Figure 4(a) shows the proposed state diagram of the prosthetic knee model.

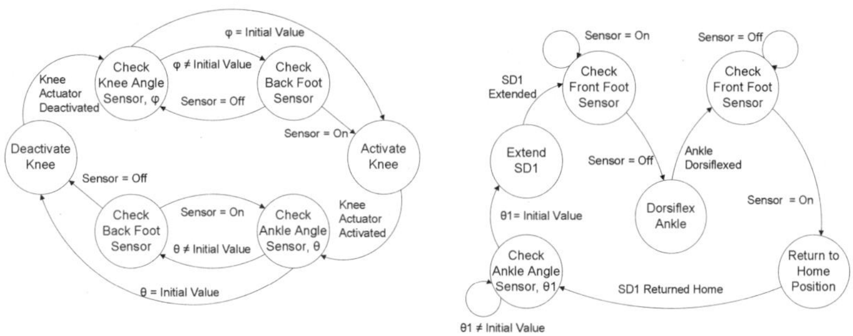

(a) (b)

Figure 4. (a) State diagram for prosthetic knee model semi-active and passive operating modes, (b) state diagram of tibia nut position control for active ankle operating mode.

In the passive ankle, the position of the nut at the Tibia would be fixed at its initial position which allows storing of energy during heel-strike and releasing of energy at toe-off. On the other hand, the active ankle changes the maximum storable energy by altering the position of the nut during stance phase. Figure 4(b) shows the control strategy of the active prosthetic ankle.

Results

The most significant difference in the context of dynamic response between Configuration 1 and Configuration 2, as depicted in Figure 5, is the ankle response during swing phase. The prosthetic ankle-foot, acting as a passive foot in Configuration 1, did not plantar flex. Instead, it returns to its neutral position after the foot is lifted off the ground. On the contrary, active ankle mode controls the position of the tibia nut to allow SD1 to have the ability to store additional mechanical energy between the intervals of foot-flat to heel-off. This effect can be clearly seen during terminal stance as the ankle plantar flexes similar to what a healthy limb would do before toe-off. This gait event, also known as push-off, which occurs between heel-off and toe-off that is believed to assist with forward progression.

The prosthetic knee angle response for Configuration 5 and Configuration 6 are also shown in Figure 5. Notice that in passive operating modes, the prosthetic knee is able to maintain its kinematic response with the reference despite the fact that the time that the knee motor is actuated had been reduced by approximately 50% of a gait cycle. This reduction can also be observed in Configuration 3 and 4. The drawback of passive operating modes for the prosthetic knee is that it behaves like a stump during stance which does not flatten the BCOM trajectory.

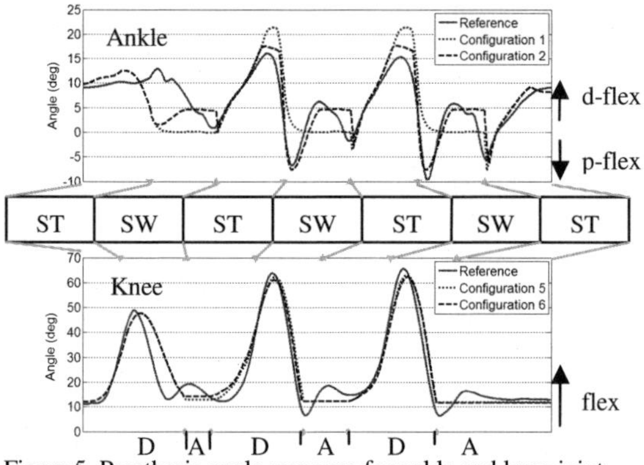

Figure 5. Prosthesis angle response for ankle and knee joint.

Another observation is on the vertical displacement of BCOM throughout the course of walking. In determining the error of the BCOM displacement, the complete joint data has been imported to move each major segments of the lower limb to obtain the reference. The error computed marks the difference in the BCOM trajectory which could be interpreted as the additional energy used to maintain the BCOM trajectory with respect to the reference while walking. As tabulated in Table 2, RMS value of error for active prosthetic ankle-foot configuration has been considerably lowered, which in turn produce high similarity with a healthy subject's walking pattern. This reduction in error also agrees that the ankle-foot mechanism, the sixth determinant of gait, does contribute in the behaviour of the BCOM trajectory.

Table 2. RMS value of pelvis vertical displacement error.

Prostheses Configuration		Ankle	
		Passive	Active
		RMS (mm)	RMS (mm)
Knee	Active	2.2	1.5
	Semi-active	2.3	1.3
	Passive	3	2.5

4. Conclusions

This paper has presented some investigations of the modelled transfemoral amputee during level ground walking. These proposed prosthetic models and configurations are able to adapt to a particular gait model, while maintaining the movements of the BCOM.

Acknowledgements

The collaborating author (Mohammed Awad) would like to acknowledge the Egyptian Government and Egyptian Cultural Bureau in the UK for the financial support.

References

1. Johansson, J.L., et al., *A Clinical Comparison of Variable-Damping and Mechanically Passive Prosthetic Knee Devices.* American Journal of Physical Medicine & Rehabilitation, 2005. **84**(8): p. 1 - 13.
2. Saunders, J., V. Inman, and H. Eberhart, *The major determinants in normal and pathological gait.* Journal of Bone & Joint Surgery,1953.**35**:p.543–558.
3. Whittle, M.W., *Gait Analysis: an introduction*. 2006.
4. Childress, D.S. and S.A. Gard, *Commentary on the six determinants of gait*, in *Human Walking*, J. Rose and J.G. Gamble, Editors. 2006, Lippincott Williams & Wilkins: Philadelphia, USA. p. 19-21.
5. Croce, U.D., et al., *A refined view of the determinants of gait.* Gait & Posture, 2001. **14**(2): p. 79-84.
6. Kerrigan, D.C., et al., *A refined view of the determinants of gait: Significance of heel rise.* Archives of Physical Medicine and Rehabilitation, 2000. **81**(8): p. 1077-1080.
7. Staros, A. and E. Murphy, *Properties of fluid flow applied to above-knee prostheses.* Bulletin of Prosthetics Research, 1964. **10**(1): p. 40-65.
8. Grimes, D.L., *An Active Multi-Mode Above-Knee Prosthesis Controller* in *Department of Mechanical Engineering*. 1979, MIT. p. 160.
9. Rodriguez, N.E.N., *Design and Simulation of a New Low-Cost Easy-Operation Humanoid Robot*, in *Department of Mechanical, Structure, Environment and Territory*. 2006, University of Cassino. p. 177.
10. Kim, C., J.-J. Lee, and X. Xu, *Design of Biped Robot with Heterogeneous Legs for advanced prosthetic knee application*, in *SICE-ICASE International Joint Conference*. 2006: Busan, Korea. p. 1852-1855.
11. Rowe, P.J., et al., *Knee joint kinematics in gait and other functional activities measured using flexible electrogoniometry: how much knee motion is sufficient for normal daily life?* Gait and Posture, 2000. **12**: p. 143-155.

MOBILE VIBRATING ROBOT FOR MOVEMENT ON METAL SURFACE*

JATSUN S.F.

Mechatronical Department, South-West State University, 50 let Oktyabrya, 94, Kursk, 305040, Russia

CHEREPANOV A.A.

Mechatronical Department, South-West State University, 50 let Oktyabrya, 94, Kursk, 305040, Russia

RUBLEV S.B.

Mechatronical Department, South-West State University, 50 let Oktyabrya, 94, Kursk, 305040, Russia

YATSUN A.S.

Mechatronical Department, South-West State University, 50 let Oktyabrya, 94, Kursk, 305040, Russia

The following paper presents the mathematical model of movement of the vibration driven robot that moves on a vertical metal surface by means of variable pressing of the case to a surface. The robot is equipped with an electromagnet drive and the rotation drive with debalance masses. The results of numerical modeling of motion are presented.

1. Introduction

Nowadays the robotics area directed at creation and research of mobile robots develops intensively, including mobile systems research in which new principles of movement are applied. Many authors show the classification of vibrating mobile systems and some prototypes [1-6] have been created. However, all developed models have one common drawback – movement can occur only on a horizontal surface or on a surface with small incline angle. Research of movement of vibrating robots on vertical surfaces therefore is of great interest.

* This work is supported by state contract №П2228 and №П2285 of the Russian Foundation of Basic Research 10-08-00769.

To enable movement on vertical ferromagnetic surfaces it is necessary to equip the robot with adhesive device which is capable of keeping mobile robot on a metal surface. For metal surfaces there can be used an electromagnet. By operating voltage submitted on an electromagnet, it is possible to control pressing force of a case to a metal surface and to generate variable friction force.

2. Description of vibrating mobile system.

Robot moves under the influence of a friction force arising on contact of case and a rough metal surface. Control over value and direction of friction force is carried out by means of rotation of internal inertial masses and variable effort of case pressing onto a rough metal surface.

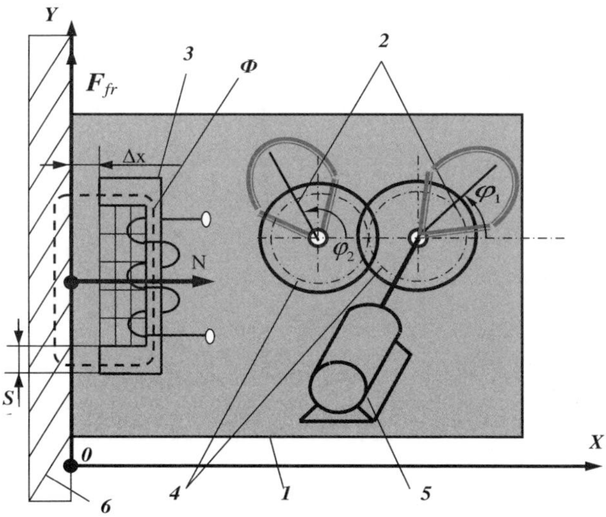

Fig.1. Robot general scheme.

Robot consists of following parts: case – 1, rotational masses – 2, electromagnet – 3, gear wheel with transfer ration (synchronizer) – 4, DC-motor – 5.

3. Mathematical model of robot motion.

Let us consider inertial masses and robot case as material points, and the inertial masses are identical. Robot moves in XOY system of coordinates. Given mechanical system can be described by the following equation system:

$$\begin{cases} \varphi_1 = \pi - \varphi_2 \\ x = 0 \end{cases} \tag{1}$$

$$\begin{cases} \ddot{\varphi}_1 = \dfrac{k_\omega I_Я - 2 \cdot \ddot{y}_3 \cdot m_1 \cdot l \cdot \cos(\varphi_1) - \mu_{\varphi 1} \cdot \dot{\varphi}_1 - 2m_1 \cdot g \cdot l \cdot \cos(\varphi_1)}{2 \cdot m_1 \cdot l^2 + J_Я} \\ \ddot{y}_3 = \dfrac{2 \cdot m_1 \cdot \left(\dot{\varphi}_1\right)^2 \cdot l \cdot \sin(\varphi_1) - \mu_{y3} \cdot \dot{y}_3 - 2 \cdot m_1 \cdot \left(\ddot{\varphi}_1\right) \cdot l \cdot \cos(\varphi_1) - m_3 g - 2m_1 g + F_{TP}}{(m_3 + 2 \cdot m_1)} \\ \dot{\Phi} = U_M(t) - R_M \dfrac{\Phi \cdot \Delta x}{\mu_0 \cdot S \cdot z^2} \\ \dot{I}_Я = \dfrac{U_Я(t) - R_Я I_Я - k_\varepsilon \dot{\varphi}_1}{L_Я} \end{cases} \tag{2}$$

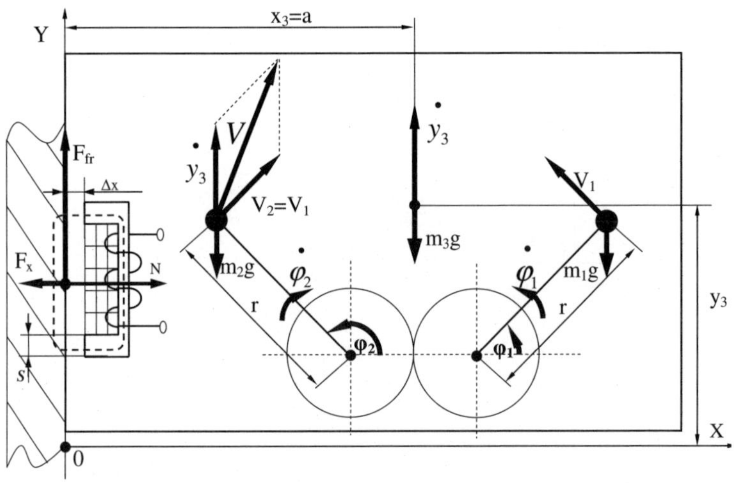

Fig. 2. Calculating scheme of robot.

Let us write the Eq. (2) in a dimensionless form. For this purpose we will accept the following scales: τ – is dimensionless time, y_0 – is scale of length, φ_0 – is scale of an angle rotation, U_0 – is scale of electric voltage, I_0 – is electric current scale, Φ_0 – is scale of a magnetic flux.

Let's substitute dimensionless variables in Eq. (1). Considering accepted designations, system equations can be written on the next form:

$$\begin{cases} \ddot{\overline{\varphi}}_1 = -\gamma_2 \cdot \ddot{\overline{y}}_3 \cdot \cos\left(\varphi_0 \cdot \overline{\varphi_1}\right) - \gamma_3 \cdot \dot{\overline{\varphi}}_1 - \gamma_4 \cdot \cos\left(\varphi_0 \cdot \overline{\varphi_1}\right) + \gamma \overline{I}_Я \\ \ddot{\overline{y}}_3 = -\chi_2 \cdot \dot{\overline{y}}_3 - \chi_3 \cdot \ddot{\overline{\varphi}}_1 \cdot \cos\left(\varphi_0 \cdot \overline{\varphi_1}\right) + \chi \cdot \left(\dot{\overline{\varphi}}_1\right)^2 \cdot \sin\left(\varphi_0 \cdot \overline{\varphi_1}\right) + \chi_4 \cdot \overline{F_{TP}} - 1 \\ \dot{\overline{\Phi}} = \theta \cdot \overline{U_M} - \theta_2 \cdot \overline{\Phi} \\ \dot{\overline{I}}_Я = \xi \cdot \overline{U_Я} - \xi_2 \cdot \overline{I_Я} - \xi_3 \cdot \dot{\overline{\varphi}}_1 \end{cases} \quad (3)$$

Eq. (3) is nonlinear, therefore for decision it is expedient to apply numerical methods, for example, Verlet integration method [7].

Due to the step-type time calculation, the moment at which case speed is equal to zero is challenging to define. The probability of defining the condition, in which $(\dot{\overline{y}}_3)_i = 0$, at fixed $\Delta\tau$, is not significant.

Therefore it is necessary to define the moment of time at which condition is satisfied $\left(\dot{\overline{y}}_3\right)_i = 0$. For this purpose the algorithm correction of an integration step, shown on fig. 3, is applied. The idea is to detect the transition of speed through a zero. If inequality $sign\left[\left(\dot{\overline{y}}_3\right)_i \cdot \left(\left(\dot{\overline{y}}_3\right)_i + \Delta\tau \cdot \left(\ddot{\overline{y}}_3\right)_{i+1}\right)\right] < 0$, is true, it means that on a current step of integration there was a transition of speed through a zero.

I.e. equality is carried out:

$$(\dot{\overline{y}}_3)_i + \Delta\tau\_adapt \cdot (\ddot{\overline{y}}_3)_{i+1} = 0 \, . \quad (4)$$

From Eq. (3) it is possible to define an adaptive step of integration at which speed becomes equal to zero on a following step of integration:

$$\Delta\tau\_adapt = -\frac{(\dot{\overline{y}}_3)_i}{(\ddot{\overline{y}}_3)_{i+1}} \, . \quad (5)$$

Developed algorithm allows to change a step of integration time that essentially raises accuracy of received decision and minimizes time of computing experiment.

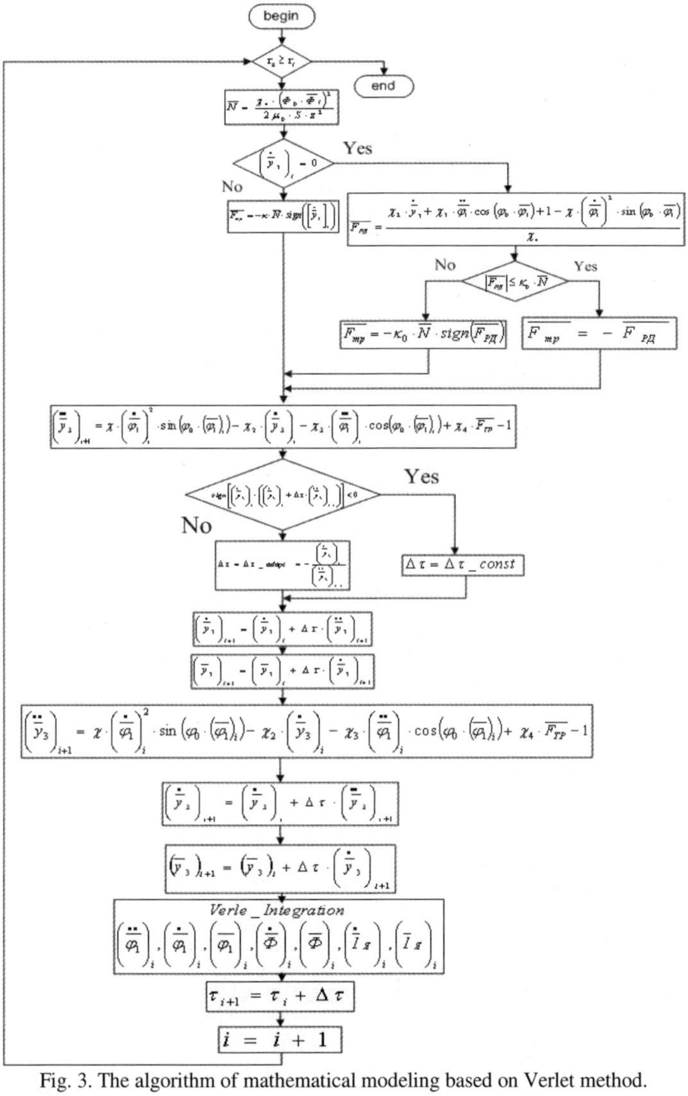

Fig. 3. The algorithm of mathematical modeling based on Verlet method.

Let us accept the following control voltage law $U_M\,(t)$:

$$U_M(t) = U_1 + \begin{cases} U_0 & if \quad \sin(\omega \cdot t + \alpha) > 0 \\ 0 & if \quad \sin(\omega \cdot t + \alpha) \le 0 \end{cases}, \tag{6}$$

where U_1 – is constant component of voltage submitted on an electromagnet, and where

$$\omega = \left(\dot{\varphi}_1\right)_{CP},\ \left(\dot{\varphi}_1\right)_{CP} = \frac{1}{T}\int_0^T \dot{\varphi}_1\,dt\,.$$

The following values of parameters of electromechanical system were chosen for modeling: $k = 0.7$, $\omega = 25(1/s)$, $m_1 = 0.1$ (kg), $m_3 = 0.1$ (kg), l=0.05 (m), z=1000, $\Delta x = 0.001$ (m), $U_я(t) = 32.5$ (V), U_1=0 (V). Electromechanical system modeling was carried out with zero initial conditions.

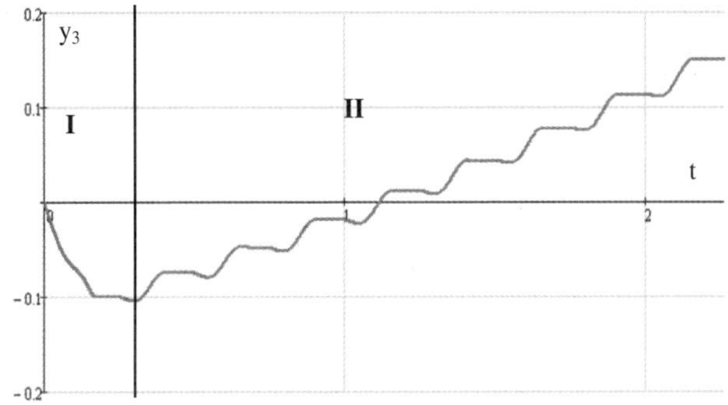

Fig. 4. Displacement of robot case on a vertical surface depending on time.

As shown on fig.4 (I), first period of robot motion begins from falling, that can be explained by transient process occurring in system. Further movement of robot case is shown on field (II) and is caused by vertically upwards established movement.

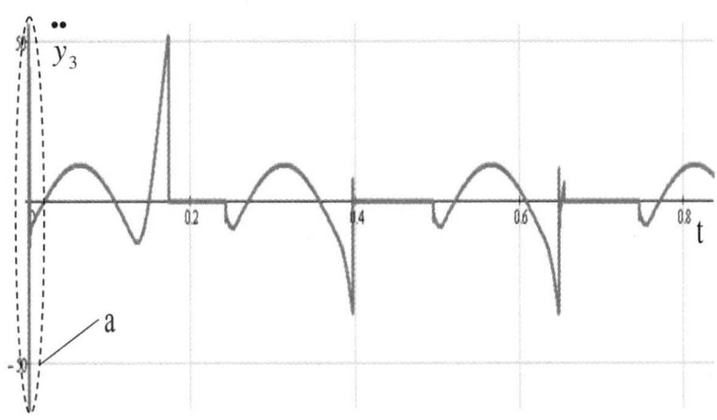

Fig. 5. Acceleration of robot case depending on time.

On the fig. 5 that shows acceleration it is visible that during initial moment of time system has great values of accelerations influence. Let us combine diagrams of case acceleration and angular acceleration of DC-motor shaft. Zoomed dependences are shown on fig. 6.

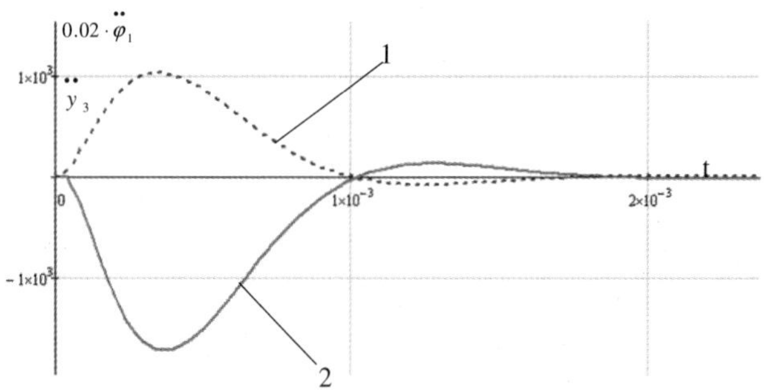

Fig. 6. Diagram of case acceleration and angular acceleration of DC-motor shaft (zoomed).

Diagrams of case displacement on vertical surface show that falling is caused by engine underrun in performing demanded rotation frequency. Hence, great values of a tangential inertia force entering $\chi_3 \cdot \overline{\varphi_1} \cdot \cos\left(\varphi_0 \cdot \overline{\varphi_1}\right)$ into system of equations (3).

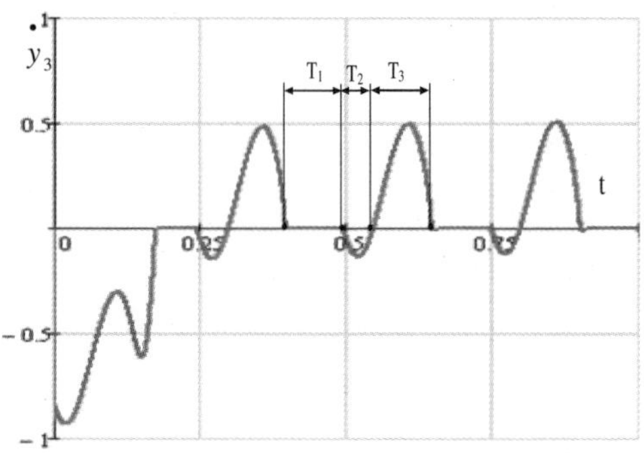

Fig.7. Robot case speed depending on time.

Established mode of robot speed, shown on figure 7, consists of 3 phases. During a movement phase T1 robot case is immobile. Further at phase T2 it slides downwards, because friction force is not enough to compensate gravity force applied to case. At T3 phase mobile robot rises upwards under the influence of forces produced by inertial masses.

Robot prototype has been developed for experimental research. The robot photo is shown on a fig. 8. During prototype construction the greatest attention was paid to weight minimization. Robot case consists of cardboard boxes, therefore it possesses big rigidity and minimum weight. The role of inertial masses is carried out by coins that have the desired weight and a small size. Electromagnets used in the prototype, have been taken from the automobile electromechanical relay.

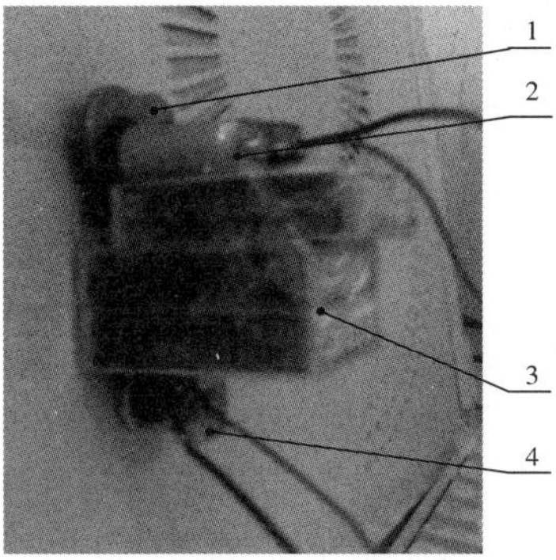

Fig. 8 General view of the prototype (1 – inertial masses, 2 – DC- motor, 3 – case, 4 - electromagnet)

Conclusions

The principle of movement described in this paper allows to expand functionality and area of application of mobile vibrating robots. Further research of movement of such system will consist in more detailed modeling of starting modes, and revealing of influence of robot parameters on movement. It can also be interesting to perform research of such robots in real life applications.

References

1. Vartholomeos P., Papadopoulos E. Dynamics, Design and Simulation of Novel Microrobotic Platform Employing vibration Microactuators // *Journal of Dynamics System, Measurement and Control*. **Vol. 128**. (2006)
2. Jatsun S., Dyshenko V., Yatsun A., Malchikov A. Modelling of Robots Motion by Use of Vibration of Internal Masses // *Proceedings of EUCOMES*. Pp.267-274. (2008)
3. Zimmermann K., Zeidis I. and Pivovarov M. Dynamics of a nonlinear oscillator in consideration of non_symmetric Coulomb dry friction // Fifth Euromech. *Nonlinear Dynamics Conf. Book of Abstracts*. P. 308. (2005)
4. Gradetsky, V.G.; Knyazkov, M.M. et al. Microsensor control of motion of compact robots inside tubes (in Russian), *Mikrosistemnaya Tekhnika [Microsystem Engineering*], **No. 8**, P. 11-19. (2002)
5. Bolotnik N. N., Jatsun S. F, Jatsun A. S., Cherepanov A. A. Automatically controlled vibration-driven robots, Proceedings. *International Conference on Mechatronics ICM* P.438-441. (2006)
6. Aoshima S., Tsujimura T., Yabuta T. A miniature mobile robot using piezo vibration for mobility in a thin tube, *Journal of Dynamic Systems, Measurement, and Control* **Vol. 115**, pp. 270-278 (1993)
7. Verlet L. Computer experiments on classical fluids. I. *Thermodynamic properties of Lennard-Jones molecules. Phys. Rev. 1967.* **V.159**. (1967)

QUADRUPED ROBOT OPTIMIZATION USING A GENETIC ALGORITHM

Manuel F. Silva

Department of Electrical Engineering, Institute of Engineering of Porto
Porto, Portugal
**mss@isep.ipp.pt*
http://ave.dee.isep.ipp.pt/ mss/

During the last two decades the research and development of legged robots has grown steadily. Legged systems present major advantages when compared with "traditional" vehicles, because they allow locomotion in terrain inaccessible to vehicles with wheels and tracks. However, the robustness of legged robots, and specially its energy consumption, among other aspects, still lag being mechanisms that use wheels and tracks. Therefore, in the present state of development, there are several aspects that need to be improved and optimized. One of these aspects relates to the optimum parameters that these machines should adopt in order to minimize the energy consumption. Due to the large number of parameters involved in this optimization process, one way to achieve meaningful results is using evolutionary strategies. Genetic Algorithms are a way to "imitate nature" replicating the process that nature designed for the generation and evolution of species. This paper presents a genetic algorithm, running over a simulation application of legged robots, which allows the optimization of several parameters of a quadruped robot model.

Keywords: Legged Robots, Gait, Optimization, Genetic Algorithms.

1. Introduction

Several walking robots have been developed up to date.[1] Compared with traditional vehicles having wheels and tracks, their major advantage is the fact of allowing locomotion in terrain inaccessible to other type of vehicles, because they do not need a continuous support surface. A frequent approach to design and build legged locomotion robots is to make a mechatronic mimic of the animal that is intended to replicate.[1]

However, in the present state of development, there are several aspects that need to be improved and optimized in these machines. With this idea in mind, different optimization strategies have been proposed and applied to these systems, either during its design and construction phases, or during its operation. One possibility to implement this idea makes use of genetic algorithms (GAs) as the engine to generate robot structures. In some cases it is performed a GA modular approach to the robot design. There are also works on which evolutionary strategies are used to optimize the structure of a specific robot. Other authors proposed the use of GAs for the simultaneous generation of the mechanical structure and the robot controller, for distinct types of robots.[2]

Bearing these ideas in mind, the objective of this paper is to present a GA, running over a simulation application of legged robots, which allows the optimization of several parameters of a quadruped robot model.

The remainder of this paper is organized as follows. Section two presents the robot model and its control architecture. Sections three and four present the implemented GA, and some preliminary simulation results, respectively. Finally, section five outlines the main conclusions of this study.

2. Robot Model and Control Architecture

2.1. *Kinematics and Trajectory Planning*

We consider the model of a walking system (Figure 1, left) with $n = 4$ legs, equally distributed along both sides of the robot body, having each one two rotational joints (*i.e.*, $j = \{1, 2\} \equiv \{\text{hip, knee}\}$).[3]

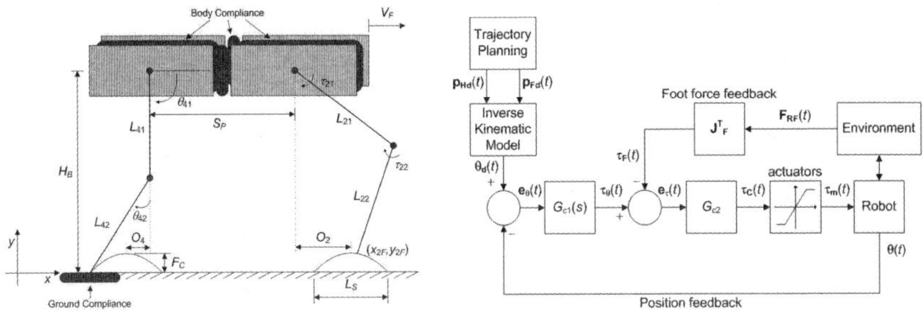

Fig. 1. Quadruped robot model (left) and control architecture (right)

Motion is described by means of a world coordinate system. The kinematic model comprises: the cycle time T, the duty factor β, the transference time $t_T = (1 - \beta)T$, the support time $t_S = \beta T$, the step length L_S, the stroke pitch S_P, the body height H_B, the maximum foot clearance F_C, the ith leg lengths L_{i1} and L_{i2} (being the total length of each robot leg equal to 1 m) and the foot trajectory offset O_i ($i = 1, \ldots, n$). The robot body, and by consequence the legs hips, is assumed to have a desired horizontal movement with a constant forward speed $V_F = L_S \,/\, T$. Regarding the feet trajectories, the trajectory of each foot is periodic and, for each cycle, the desired trajectory of the foot of the swing leg is computed through a cycloid function.[3]

In this work we consider three walking gaits (Walk, Chelonian Walk and Amble), two symmetrical running gaits (Trot and Pace) and five asymmetrical running gaits (Canter, Transverse Gallop, Rotary Gallop, Half-Bound and Bound). These are the gaits usually adopted by animals moving at low, moderate and high speed, respectively, being their main characteristics presented in Silva (2010).[4]

2.2. *Robot Dynamic Model*

The model for the robot inverse dynamics is formulated as:[3]

$$\mathbf{\Gamma} = \mathbf{H}\left(\mathbf{\Theta}\right)\ddot{\mathbf{\Theta}} + \mathbf{c}\left(\mathbf{\Theta}, \dot{\mathbf{\Theta}}\right) + \mathbf{g}\left(\mathbf{\Theta}\right) - \mathbf{F_{RH}} - \mathbf{J}^{\mathbf{T}}(\mathbf{\Theta})\mathbf{F_{RF}} \tag{1}$$

Moreover, the joint actuators are not considered ideal, exhibiting a saturation, being the maximum torque that each actuator can supply τ_{ijMax}.

The dynamic model for the hexapod body and foot-ground interaction (Figure 1) considers a compliant robot body, divided in n identical segments (each with mass $M_b n^{-1}$, while making the total mass of the robot equal to 100 kg) and a linear spring-damper system is adopted to implement the intra-body compliance.[3]

The contact of the ith robot foot with the ground is modelled through a non-linear system (Figure 1) with linear stiffness and non-linear damping.[5]

2.3. *Control Architecture*

The general control architecture of the multi-legged locomotion system is depicted in Figure 1 (right), being $G_{c1}(s)$ a PD controller and G_{c2} a simple P controller.[5] The trajectory planning is held in the Cartesian space, but the control is performed in the joint space, which requires the integration of the inverse kinematic model in the forward path. The control algorithm considers an external position and velocity feedback and an internal feedback loop with information of foot-ground interaction force.

3. Developed Genetic Algorithm

GAs are adaptive methods which may be used to solve search and optimization problems.[6] By mimicking the principles of natural selection, GAs are able to evolve solutions towards an optimal one. Although the optimal is not guaranteed, the GA is a stochastic search procedure that, usually, generates good results. The GA maintains a population of candidate solutions (the individuals). Individuals are evaluated and fitness values are assigned based on their relative performance. They are then given a chance to reproduce, i.e., replicating several of their characteristics. The offspring produced are modified by means of mutation and/or recombination operators before they are evaluated and reinserted in the population. This is repeated until some condition is satisfied.

3.1. *Structure of the Used Chromosome*

The chromosome used in the developed GA presents 34 genes (i.e., 34 robot parameters). The genes are organized as presented in Table 1: the first gene (L_s) contains information regarding the step length and the last gene (Kd_{22}) contains the derivative gain of joint 2 of the robot rear legs. These values are coded directly into real numbers (value encoding).[4]

3.2. *Measure for the Fitness Evaluation*

For the fitness function is used the mean absolute density of energy per travelled distance E_{av}.[7] This index is computed assuming that energy regeneration is not available by actuators doing negative work (by taking the absolute value of the power). At a given joint j (each leg has $m = 2$ joints) and leg i (since we are adopting a quadruped it yields $n = 4$ legs), the mechanical power is the product of the motor torque and angular velocity. The global index E_{av} is obtained by averaging the mechanical absolute energy delivered over the travelled distance d:

$$E_{av} = \frac{1}{d} \sum_{i=1}^{n} \sum_{j=1}^{m} \int_0^T \left| \tau_{ij}(t)\, \dot{\theta}_{ij}(t) \right| dt \quad [\mathrm{Jm}^{-1}] \tag{2}$$

The performance optimization is achieved through the minimization of the index E_{av}.

Minimum Value	Variable	Maximum Value
$0 <$	L_s	$\leq$ 10 m
$0 <$	H_B	$\leq$ 1 m
$0 <$	F_C	$\leq$ 1 m
$0 <$	L_{11}	$\leq$ 1 m
$0 <$	L_{12}	$\leq$ 1 m
$0 <$	L_{21}	$\leq$ 1 m
$0 <$	L_{22}	$\leq$ 1 m
$0 <$	O_1	$\leq$ 10 m
$0 <$	O_2	$\leq$ 10 m
$0 <$	M_b	$\leq$ 100 kg
$0 <$	M_{11}	$\leq$ 10 kg
$0 <$	M_{12}	$\leq$ 10 kg
$0 <$	M_{21}	$\leq$ 10 kg
$0 <$	M_{22}	$\leq$ 10 kg
$0 <$	K_{xh}	$\leq$ 10000 Nm
$0 <$	K_{yh}	$\leq$ 10000 Nm
$0 <$	B_{xh}	$\leq$ 10000 Nms^{-1}
$0 <$	B_{yh}	$\leq$ 10000 Nms^{-1}
$-400 <$	τ_{11min}	$\leq$ 0 Nm
$0 <$	τ_{11Max}	$\leq$ 400 Nm
$-400 <$	τ_{12min}	$\leq$ 0 Nm
$0 <$	τ_{12Max}	$\leq$ 400 Nm
$-400 <$	τ_{21min}	$\leq$ 0 Nm
$0 <$	τ_{21Max}	$\leq$ 400 Nm
$-400 <$	τ_{22min}	$\leq$ 0 Nm
$0 <$	τ_{22Max}	$\leq$ 400 Nm
$0 <$	Kp_{11}	$\leq$ 10000
$0 <$	Kd_{11}	$\leq$ 1000
$0 <$	Kp_{12}	$\leq$ 10000
$0 <$	Kd_{12}	$\leq$ 1000
$0 <$	Kp_{21}	$\leq$ 10000
$0 <$	Kd_{21}	$\leq$ 1000
$0 <$	Kp_{22}	$\leq$ 10000
$0 <$	Kd_{22}	$\leq$ 1000

4. Simulation Results

The main objective of this study was to find the optimal values for the robot model and controller parameters, considering that the robot was moving with $V_F = 1\ ms^{-1}$, while adopting the gaits Walk (1), Chelonian Walk (2), Amble (3), Trot (4), Pace (5), Canter (6), Transverse Gallop (7), Rotary Gallop (8), Half-Bound (9) and Bound (10).

Running the GA, with the parameters described in,[4] the algorithm converged to the results that are described in the sequel. There was one exception: for the Chelonian Walk gait the GA did not converge, although several attempts (distinct runs of the GA) were made.

Figure 2 depicts two charts with the optimum values of the Step Length (L_S) (left) and Body Height (H_B) (right), for the distinct gaits under consideration, determined by the GA. It is seen that the robot should move with a value of $L_S \approxeq 1.1$ m, except for the Canter and Bound gaits, for which case it should be $L_S \approxeq 1.3$ m. Concerning the value for the Body Height, analyzing the chart presented in the left part of this figure, one can conclude that the robot should adopt a value of $H_B \approxeq 0.8$ m.

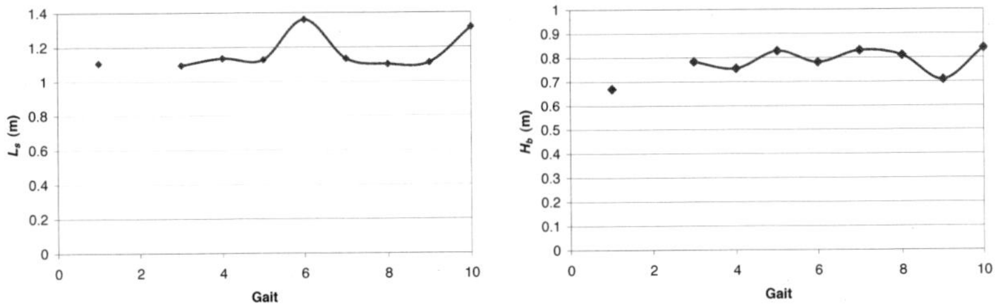

Fig. 2. Optimum values of the Step Length L_S (left) and Body Height H_B (right), for the gaits under study

Figure 3 depicts two charts with the optimum values of the front legs links lengths L_{11} and L_{12} (left) and of the rear legs links lengths L_{21} and L_{22} (right), for the distinct gaits under study, determined by the GA. Analyzing the results presented in these figures it should be referred that the length of the upper segment of the leg should be smaller than the corresponding length of the lower segment. The relation between the lengths of both segments is $L_{i1}/L_{i2} \approxeq 0.3/0.7$.

There are only two exceptions to these general results; for the case of the front legs, when the robots adopts the Amble gait both segments should be of similar lengths and when the quadruped adopts the Bound gait the results obtained by the GA indicate that the lower segment of the leg should be smaller than the corresponding upper segment.

In Figure 4 are presented two charts with the optimum values of the foot trajectory offset O_1 and O_2 (left) and of the Body Mass (M_B) (right),

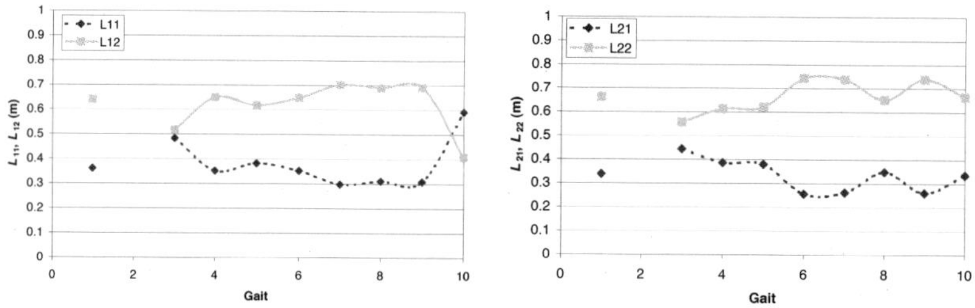

Fig. 3. Optimum values of the front legs links lengths L_{11} and L_{12} (left) and of the rear legs links lengths L_{21} and L_{22} (right), for the gaits under study

determined by the GA, for the gaits under study. Concerning the foot trajectory offset, the results presented in this chart indicate that the robot should move with the feet trajectory displaced to the rear of the hip trajectory (in the moving direction), as indicated by the values of the parameters O_i. Regarding the robot mass distribution, the body should concentrate most of its value (it is assumed that the total mass of the robot is equal to 100.0 kg) being $M_B > 70$ kg for all gaits under study, except for the Half-Bound.

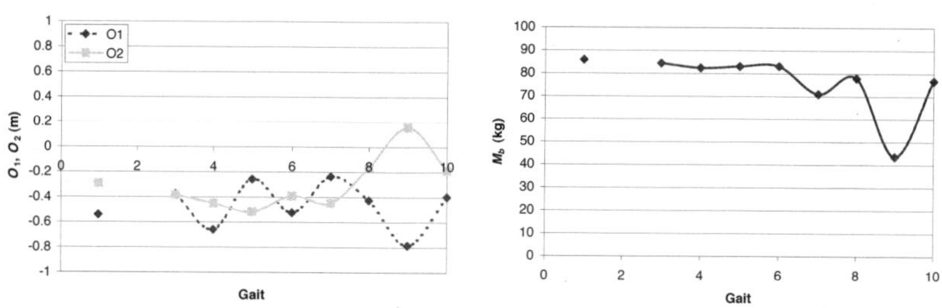

Fig. 4. Optimum values of the foot trajectory offset O_1 and O_2 (left) and of the Body Mass M_B (right), for the gaits under study

Finally, Figure 5 shows two charts with the optimum values of the front legs link masses M_{11} and M_{12} (left) and of the rear legs link masses M_{21} and M_{22} (right), determined by the GA, for the gaits under study. The left chart indicates that the lower segment of the front legs should be heavier than the upper segment ($M_{12} > M_{11}$), except for the Half-Bound gait.

In a similar manner, the right chart indicates that the lower segment of the rear legs should be heavier than the upper segment ($M_{22} > M_{21}$), except for the Transverse Gallop gait. These results seem to agree with the ones presented in Figure 3, since longer legs links segments are heavier.

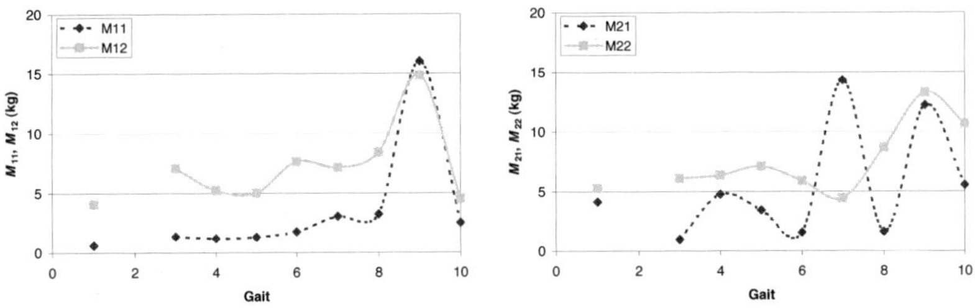

Fig. 5. Optimum values of the front legs link masses M_{11} and M_{12} (left) and of the rear legs link masses M_{21} and M_{22} (right), for the gaits under study

5. Conclusions

This paper presented a GA developed for the optimization of quadruped robot parameters, while walking with distinct gaits at $V_F = 1\ ms^{-1}$. This GA runs over a simulation application of legged robots (developed in the C programming language), which allows the optimization of several parameters of the robot model and of its gaits for different locomotion speeds. Based on this GA, the author plans to develop several simulation experiments to find the parameters that optimize the robot locomotion, from the viewpoint of the index E_{av}, for different values of V_F in the range $0.1 \leqslant V_F \leqslant 10.0\ ms^{-1}$.

Acknowledgments

The author would like to thank Sérgio Carvalho, for implementing the GA, and GECAD, of ISEP–IPP, for their financial support to this work.

References

1. M. F. Silva and J. A. T. Machado, *J. of Vibration and Control* **13**, 1447 (2007).
2. M. F. Silva and J. A. T. Machado, *J. of Vibration and Control* (Accepted for Publication).
3. M. F. Silva, J. A. T. Machado and A. M. Lopes, *ROBOTICA* **23**, 595 (2005).
4. M. F. Silva, A genetic algorithm for the optimization of quadruped robot parameters, in *Proc. of the MME 2010 - Int. Conf. on Mathematical Methods in Engineering*, (Coimbra, Portugal, 2010).
5. M. F. Silva, J. A. T. Machado and A. M. Lopes, *Machine Intelligence and Robotic Control* **5**, 33 (2003).
6. P. J. Fleming and R. C. Purshouse, **IFAC Professional Brief** (2002).
7. M. F. Silva and J. A. T. Machado, *ROBOTICA* **26**, 19(January 2008).

STABILITY AUGMENTATION OF SLIP-LIKE LEGGED LOCOMOTION EXPLOITING HIP ACTUATION

SOHA POUYA, RICO MOECKEL, FRANK PEUKER[‡],
ANDRE SEYFARTH[‡], AUKE JAN IJSPEERT

Biorobotics Laboratory, Institute of Bioengineering, Ecole Polytechnique Fédérale de Lausanne (EPFL), Switzerland
[‡]*Locomotion Laboratory, Friedrich-Schiller University of Jena, Jena, Germany*

Stable locomotion that tolerates parameter variations is an important feature for legged robots. In this paper we introduce a locomotion control framework for legged robots that combines the well-known spring-loaded inverted pendulum (SLIP) with active hip actuation through Central Pattern Generators. Using this framework we present studies suggesting that compliant in contrast to rigid hip actuation and the addition of a simple feedback scheme can highly enhance the robustness of the robot locomotion against parameter changes.

1. Introduction

For legged robots a high robustness to parameter variability is of great advantage. Parameters can vary between different legs, among different robots e.g. caused by imperfections in the fabrication of mechanical parts and can change over time due to aging. Despite this variability the robot is desired to perform stable locomotion. In this paper we introduce a control and optimization framework that is combining the Spring-Loaded Inverted Pendulum (SLIP) locomotion model with hip actuation that is controlled through a Central Pattern Generator (CPG). The periodic actuation scheme of the CPG is found through an optimization algorithm called Particle Swarm Optimization (PSO). We analyze two mechanisms in hip actuation for enhancing the robustness of locomotion stability to parameter variation: (1) the addition of compliance in series to the actuator and (2) the implementation of a simple feedback that triggers a pre-optimized locomotion pattern when a touch down of the leg is sensed.

Locomotion previously has been analyzed at different levels of abstraction from very simplified models to sophisticated neuro-mechanical models (Ref. 1). For animal and human gaits, the Spring-Loaded Inverted Pendulum (SLIP) is an archetypal model, predicting correctly ground reaction forces and center of mass trajectories (Ref. 2 and Ref. 3). While the standard SLIP model represents the stance phase with passive dynamics, the swing phase is abstracted by a ballistic curve. Locomotion patterns are the result of the intercoupling (through different types of proprioceptive and exteroceptive feedback) between the CPG, the musculo-skeletal system, and modulation signals from higher-level brain centers (Ref. 2. Similarly the SLIP model being a valid *mechanical* template, coupled phase oscillators have been proven to be valid *neural* template for the legged locomotion mechanism (Ref. 2). In the following Section we introduce the details of the locomotion control and optimization framework while Section 3 presents experimental results. Section 4 concludes and describes future work.

2. Locomotion Control and Optimization Framework

Our control and optimization framework contains three major elements. (1) A modified SLIP model with hip actuation models the passive dynamics of the legged robot while (2) a Central Pattern Generator controller generates the periodic locomotion pattern of the hip controller. (3) Finally we use the Particle Swarm Optimization algorithm to optimize the actuation pattern.

2.1. *SLIP Model with Hip Actuation*

The SLIP model is a two-state hybrid dynamical system consisting of a point mass attached to a massless leg spring. One gait cycle of running with the SLIP model follows the typical sequence: touch-down, stance phase, lift-off and flight phase (Fig.1). Touch-down occurs when the foot point of the extended leg reaches ground. The angle between the leg and the ground at touch-down ($\alpha = \phi(t_1)$) is an important control parameter for stable locomotion (Ref. 6). The foot point is constant during stance phase; no slipping or foot point translation is included in the model. During leg compression, axial forces are generated pointing from foot point to point mass. The magnitude of this force is calculated according to Hook's law: $F_{leg} = k \cdot (L_0 - L(t))$ where k is the spring constant while $L(t)$ and L_0 are the spring length at time t and at rest, respectively. When the leg spring reaches its rest length again, the leg lifts off and detaches from ground. During flight phase, the model is driven only by gravitational force. The SLIP model represents the passive leg behavior and depends on both model parameters (body mass, leg length, leg stiffness) and initial conditions (CoM position and velocity). The standard SLIP model does not contain any hip spring nor it includes any continuous active actuation scheme. Only the touch-down angle α is controlled. Otherwise all locomotion is achieved purely due to passive dynamics. We extend this standard SLIP model (Ref. 3,8) by

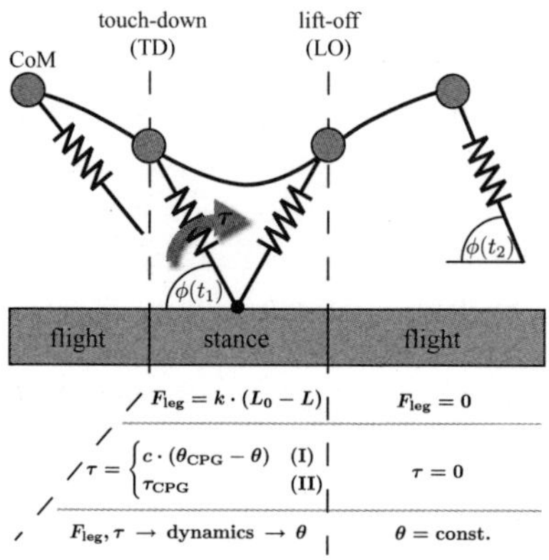

Fig. 1: Hip-actuated SLIP model. During stance phase, a torque is applied to the model either by a serial combination of motor and spring (I) or by direct actuation (II). The leg angle θ emerges from dynamics according to leg force F_{leg} and hip torque τ. In contrast, no torque is applied during flight phase and the leg angle θ is set to a constant value.

adding active hip actuation during stance phase. This is achieved either by applying a torque τ generated by changing the hip spring's rest angle θ_{CPG} ($\tau(t) = c \cdot (\theta(t) - \theta_{\mathrm{CPG}}(t))$ or by actuating the hip joint directly ($\tau(t) = \tau_{\mathrm{CPG}}(t)$). For both cases a periodic locomotion pattern is generated with a CPG (either θ_{CPG} or τ_{CPG}). The value of the hip spring stiffness c is an open parameter subject to optimization. In the SLIP model, no trunk or leg is present that could provide hip torque internally. Therefore, in a technical realization, the source of hip torque (e.g., motor) has to be placed in the robot environment. One example is shown by Andrews et al. (Ref. 10) who mounted the motor onto a boom that at the same time was restricting robot movement to 2D (hopping in a circle). To apply hip actuation to the SLIP point mass model, torque needs to be transferred to force. In 2D, the force magnitude can be derived as $F_{hip} = \tau/L$ where the force direction is perpendicular to leg axis. We formulate the equations within the cartesian world referential composed from x-axis (horizontal) and y-axis (vertical). Given the world referential projections F_{leg_x}, F_{leg_y}, F_{hip_x} and F_{hip_y} of leg force magnitude F_{leg} and hip actuation force F_{hip}, the equations are:

$$m \cdot a_x = \begin{cases} F_{leg_x} + F_{hip_x} & \text{stance phase} \\ 0 & \text{flight phase} \end{cases}$$

$$m \cdot a_y = \begin{cases} F_{leg_y} + F_{hip_y} - mg & \text{stance phase} \\ -mg & \text{flight phase} \end{cases}$$

Here, the gravitational constant ($g = 9.81 m/s^2$) and the accelerations of the center of mass along x-and y-axis (a_x, a_y) have been introduced.

2.2. *Central Pattern Generator Controller*

We use an extend phase oscillator CPG Controller with a non-linear output filter to generate the actuation patterns for the rest angle of hip spring ($\theta_0(t)$) in complaint joints or the patterns for the direct torque ($\tau(t)$) in rigid joints. The CPG controller can be described by a set of differential equations where the output of oscillator number i is derived from the function $\Gamma(t)$ that takes as argument the oscillator's phase $\phi_i(t)$ and that is multiplied by the amplitude $r_i(t)$:

$$\theta_{0_i}(t) = r_i(t) \cdot \Gamma(\phi_i(t)) + g_{\theta_i}(s(\vec{t})) \tag{1}$$

$$\text{where:} \begin{cases} \dot{r}_i(t) & = a_i(R_i - r_i(t)) \\ \dot{\phi}_i(t) & = 2\pi \cdot \nu_i + K_i(t) \\ K_i(t) & = \sum_j w_{ij} \cdot r_j(t) \cdot \sin(\phi_j(t) - \phi_i(t) - \psi_{ij}) \end{cases} \tag{2}$$

where ν_i is the frequency of the ith oscillator, $g_{\theta_i}(s(\vec{t}))$ allows incorporating sensory feedback into the state equations and a_i is a positive constant affecting the rise time of the state variable $r_i(t)$ to reach the desired amplitude R_i. Open parameters subjected to optimization are the shape of the function $\Gamma(t)$ and the value of parameter ν_i.

For the single-leg SLIP model presented in this paper a single oscillator is used. However, for actuating multi-legged models the CPG controller can generate coordinated rhythmic patterns using a network of oscillators (used for biped and quadruped gaits). In such a network, the coupling term $K_i(t)$ is extracted by using the parameters w_{ij} and ψ_{ij}; respectively the weight and the phase bias of the coupling between the ith and j^{th} oscillators. For the future works we exploit this property to control multi-legged models.

2.3. *Optimal Trajectory Generation*

As described in Section 2.2 the output function $\Gamma(t)$ and the frequency ν_i of the CPG (Equation 1) together with the hip spring stiffness c are used to generate different candidate solutions for the hip actuation pattern. The function $\Gamma(t)$ is used to encode the profile (with normalized values in $[0 \quad 1]$ interval) while the frequency ν_i are used to squeeze or stretch the output trajectories for θ_0 or τ in time. Different waveforms for $\Gamma(t)$ are generated via a piecewise cubic Hermite interpolating polynomial (PCHIP) method by using four control points. The x and y coordinates of these points in Cartesian space are optimized by a well-known stochastic optimization method called Particle Swarm Optimization (PSO, see Ref. 9).

For finding stable locomotion solutions that are robust to parameter variations we chose the following fitness criterion: The stability is measured by counting the number of steps that the SLIP model can perform before falling, a numerical method suggested in previous work (Ref. 7). By sweeping the SLIP model parameters touch-down angle α and spring constant k and assigning the corresponding step count value a so-called *steps-to-fall-map* (see Fig. 2) can be generated. For each candidate solution, the ratio of the stable region which includes those k-α-pairs where the SLIP model performed at least 50 steps without falling over the whole map size is computed:

$$\textit{fitness value} = \frac{\text{size of stable region}}{\text{total size of } k - \alpha \text{ map}} \tag{3}$$

Every actuation pattern is verified for the whole $k-\alpha$ space and this results in gaining optimized control trajectories that lead to stable locomotion for a variety of k-α parameter pairs and that e.g. do not need to be re-optimized when the k or α parameter is changing within a maximum range.

3. Experiments and Results

With the presented SLIP-CPG-PSO control and optimization framework we are exploring the following questions: (1) How does the k-α steps-to-fall map and thus the robustness of the locomotion to variations of these parameters change when hip actuation is added in comparison to the non-actuated standard SLIP model? (2) How does the addition of compliance in the hip change the robustness in comparison to a rigid actuation? (3) How can robustness further benefit from the addition of a simple feedback mechanism that detects when touch-down occurs? To answer these questions we are running four groups of experiments. Taking into account the stochastic nature of the optimization method each experiment is repeated five times. Each run is initialized with a random initial population of 50 PSO-particles that are exploring the parameter space extracting optimal trajectory, frequency and hip stiffness (in compliant joints) over 100 iterations. For each

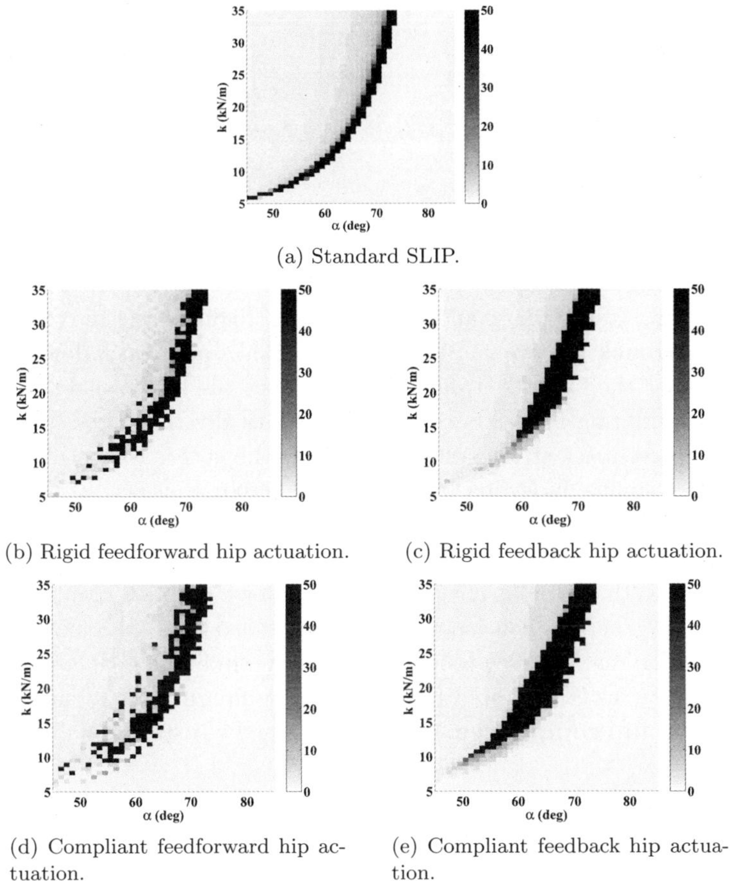

Fig. 2: k-α maps showing the number of steps to fall for different values of leg spring constants k and angles of attack at touchdown α for different SLIP model variations. (a) Standard SLIP model without hip actuation. (b)-(e) Four different combinations of the hip actuated SLIP model with/without touch-down detection and hip compliance.

of the groups the results with the best k-α map out of the five trials are shown in Fig. 2b to 2e. Fig. 2a is showing k-α steps-to-fall map for the **standard SLIP model** without hip actuation or hip compliance. The result reproduces the well-documented J-shape for the stability region in this map (Ref. 6) and are used as the control or base-line results. The fitness of this solution according to Equation 3 is 0.051. To explore simple feedback we distinguish between two groups of experiments. For the **feed-forward group** no feedback mechanism is included in the control scheme but we

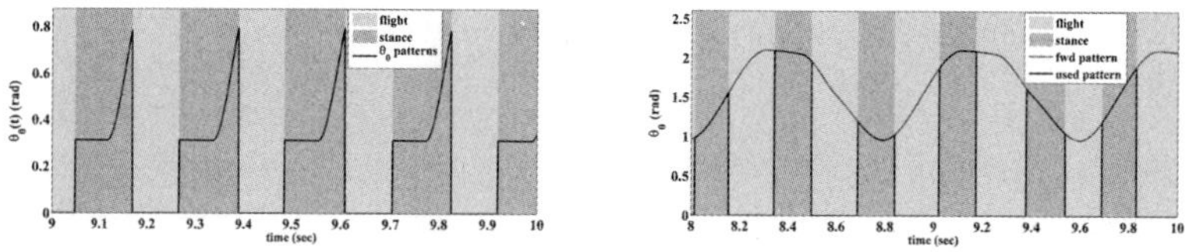

Fig. 3: CPG actuation patterns using touchdown feedback (left) and using a feed-forward scheme (right). In both schemes the output signals during flight phase is zero.

are optimizing a continuous control trajectory that is used to control hip torque during stance phase and ignored during flight phase. In comparison for the **feedback group** an event is generated at every touch-down that is used to reset the CPG controller and to trigger the hip torque trajectory that is zero during fight phase. Fig. 3 schematically shows the used feed-forward vs. feedback strategies. In both cases the goal is to control the hip only during stance. In feedback scheme CPG actuation is activated once receiving touchdown event feedback, thus resulting in the synchronization of the CPG outputs with the stance dynamics. In the feed-forward scheme the assumption is that CPG is not aware of touchdown events to adjust its output. Thus it generates continuous patterns (dashed lines) in a feed-forward manner which make the leg touch the ground with different θ_0 adjustments in different cycles (solid lines). Finally we are comparing hip actuation with and without **hip compliance**. The most optimal k-α steps-to-fall maps for

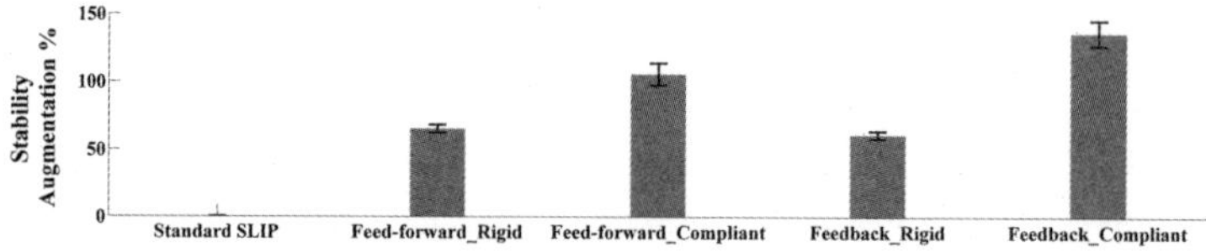

Fig. 4: Augmentation of fitness measure and thus robustness for the hip-actuated SLIP model with optimized CPG control in comparison to the standard SLIP model without hip actuation. Error bars show the standard deviation among the five simulation trials.

the different groups of experiments are shown in Fig. 2b to 2e. A comparison of the augmentation of the robustness to parameter changes is shown in Fig. 4. The bar plots show the average between the five optimization trials of each group. Error bars are indicating the standard deviation. The results are given in comparison to the fitness of the standard SLIP model. They are suggesting that by adding hip actuation the robustness to parameter changes can be increased by at least 50%. However according to the k-α maps in Fig. 2, feed-forward strategies result in sparse stable regions.

This means that the feed-forward solutions are sensitive to the parameters variation. In contrast, the solutions with using feedback both increase the size of stable region and keep the map continuity. Furthermore, adding the comparsion between the compliant and rigid actuations suggests that the best results are achieved by including both hip compliance and feedback through touch-down detection which can result in stability augmentation of 143% while keeping the map continuity.

4. Conclusion and Future Work

We presented a framework for studying the robustness of gait stability to parameter changes for legged robots. This framework is combining a standard 2-D SLIP model for exploring the passive dynamics of a hopping or running robot, a CPG scheme for hip actuation and an optimizer for finding optimal hip actuation parameters and trajectories. We are exploring different actuation strategies including feedback through touch-down detection and compliant hip actuation. Simulation results are suggesting that compliant hip actuation scheme together with using touchdown feedback can increase the stability region of hopping or running robots in comparison to a purely passive robot by up to 143 %. Future work includes more detailed studies of hip actuation schemes that are e.g. using more complex and continuous feedback to the CPG controller as well as the application of our control and optimization framework to monoped, biped and quadruped robot hardware.

Acknowledgments

This project was supported by the European project Locomorph under the grant agreements No. 231688. We gratefully acknowledge the contributions by Jesse van den Kieboom for developing the optimization framework and the support and inputs by Sten Grimmer and Martin Gross.

References

1. R.J. Full, D. Koditschek *J Exp Biol*, **202**, 3325-3332, (1999).
2. P. Holmes, R.J. Full, D. Koditschek, J. Guckenheimer, *SIAM Rev.* **48**, (2006).
3. R. Blickhan *J. of Biomechanics*, **22**, 1217–1227 (1989)
4. A.J. Ijspeert, *Neural Networks* **4**, 642–653 (2008).
5. S. Grimmer, M. Ernst, M. Gunther, and R. Blickhan, *J Exp Biol*, **211**, (2008).
6. A. Seyfarth, H. Geyer, M. Guenther, R. Blickhan. *J. of Biomechanics* (2002) **35**, No.5, pp. 649-655.
7. J. Seipel and P. Holmes, *Regular and Chaotic Dynamics*, **12**, No.5, (2007).
8. A. Seyfarth, Hartmut Geyer and Hugh Herr, *J. Exp. Biol.*, **206**. (2003).
9. J. Kennedy and R. Eberhart, *Proceedings of ICNN* , 1942–1948 (1995).
10. B. Andrews, B. Miller, J. Schmitt, J.E. Clark, *Bioinspiration and Biomimetics*, **6**, No.2 (2010).

SECTION–14

PERCEPTION AND SENSOR FUSION

SENSORY SYSTEM EVALUATION METHODOLOGY FOR AUTOMATIC FRUITS HARVESTING*

C. SALINAS[1], R. FERNANDEZ[1], H. MONTES[1-2], AND M. ARMADA[1]

[1]Robot Locomotion and Interaction Group, Centre for Automation and Robotics, CSIC-UPM, Spain; [2]Faculty of Electrical Engineering, Technological University of Panama, Panama carlota.salinas@car.upm-csic.es

This paper presents the study of a wide collection of activities focused on automatic fruit harvesting, in order to understand the main challenges of this research area. A methodology to evaluate the fusion and the performance of several sensor systems has also been addressed. The purpose of the proposed methodology is to successfully deal with the occlusions caused by leafs, trunks, stamps and the similarities between the fruits and the background during the detection, localisation and characterisation of fruits.

1. Introduction

Precision agriculture oriented to automatic fruits harvesting tasks requires non-destructive, accurate and efficient techniques to be adequately addressed, and to improve and preserve the products quality. Therefore embrace efficient resource consumption and profit-making procedures. That mainly leads the researchers to investigate sensory systems based on the electromagnetic spectrum properties of reflection and absorption.

A common solution for sensing systems, regarding the environments perception, involves answers that mimic the human vision by using cameras. Some important drawbacks of working only with these systems lie in their sensitivity to variations in the illumination conditions and to harvesting modifications. Nevertheless, in line for great technology advances it is possible to perceive more information that the one acquired by the human eye (visible wavelength). During the last decade, remarkable works have shown the capabilities of new solutions, such as the multispectral or hyperspectral systems created by the combination of devices that are sensitive to different regions of the electromagnetic spectrum.

In this work we present a general overview of the sensory system for fruit harvesting and we introduce a methodology to evaluate the sensory systems

* This work is supported by Grant Agreement number 246252

under field conditions for detection, localisation and characterisation of fruits in greenhouses and plantations.

2. Overview of research areas of fruits harvesting

There are a large number of studies in precision agriculture related to fruits harvesting; in this paper we have reviewed some works where the sensory system exposes new interesting solutions and gathered them in three groups: fruit detection, diseases detection and post-harvesting tasks.

2.1 Fruit detection

The scenarios for fruit detection can be defined as complex environments, since they usually have variable illumination conditions which are not easily controlled, and the fruits surroundings are invaded with leaves, stalks and trunks. In [1] the authors have dealt with these problems by combining colour images with reflectance and range data, and shape-base data analysis has been implemented to detect oranges (orange and green colour). Those strategies are more independent of the colour features but also more time consuming. The authors claim that there were not false alarms and the correct detection was over 80%. The main drawbacks are enclosed in the occlusions. A cherry-harvesting robot was presented in [2]; this work explores the combination of red light and infrared beams acquired by 2 PSD, a multispectral device. The IR data is used to measure the 3D shape of cherry trees while the spectral-reflection difference is used to detect cherries, because the red light is not well reflected in unripe red fruits, stalks or leaves. Again the occlusions and closed groups of cherries provoke complications, but the ambiguities between fruits and trunks are solved by the specula phenomena of the cherries.

The harvesting of green-like fruits increases the complexity of the scenarios. For instance, in [3] (recognition and cutting system of sweet pepper), the system is strongly dependent on illumination conditions (glares on the fruit and leaves surfaces) since only colour images are used. Alternatively, in [4] the authors explored characteristics of hyperspectral cameras and their results denote potential of these devices. By applying a simple pixel discrimination procedure and morphological operations, a good detection success rate of over 70% was achieved.

2.2 *Disease detection*

A great deal for disease detection is to identify the primary symptoms of the diseases in plantations. In [5] the author's goal was to introduce a ground-based real time remote sensing for detecting diseases under field conditions and in an early stage, before they can be visibly detected. They combine hyperspectral reflection images (450 to 900 nm) and fluorescence imaging. The disease variable was set, by the characterization of pixels fraction in fluorescence images recognized as diseased. Then images of healthy and not healthy plants were taken with a hyperspectral camera and the Normalized Difference Vegetation Index (NDVI) was calculated in order to select only the plant specific spectra. An initial approach fuses the two measures by Quadric Discrimination with a result of 95% of healthy discrimination. A further approach achieves a 99% by using Self Organization Map. Another illustration of multispectral analysis by NIR and visible red was presented in [6], where the authors propose the combination of linear relation (N, NIR, NIR/R, NIR-R, NDVI) to create 5-index images. The approach of clustering classification of the 5-dimensional feature space can identify the late blight (LB) when the disease reaches stage 3 or above, which helps to reduce the economic losses. The accuracy in anthracnose (or canker, are general terms for a large number of plant diseases) detection increased in 86%, and the green mould detection from 65% to 94%.

The researchers of [7] and [8] show the complexity of using visible information to detect or classify the injured region of the plants. In [7] a grape leafs disease detection is presented, by using colour images and artificial intelligent techniques. Since the first step of leafs extraction is pre-processed, the method is very sensitive to changes in the illumination conditions and the harvesting. In [8] the authors propose an indoor set-up based on texture analysis, colour camera and structured light with fluorescent tube and polarized filters. Although the result is over 93% of success ratio and the model is able to identify new unpredictable defects, it is conditioned to only one specific scenario.

2.3 *Post-harvesting: quality*

In quality control activities, the environments are usually highly controlled, including structured lights, fluorescence lights, polarized filters, etc. Initially only video cameras were used. Nowadays the multispectral sensing has increased its popularity.

In 2001 an early study for demonstration of hyperspectral imaging capabilities was proposed by [9], combining hyperspectral reflectance and fluorescence imaging, the system was calibrated and normalized. The authors performed several experiments to show not only the versatility of these systems, but also the clearly visible differences amount normal apples and apples with fungal contamination and bruised spot.

A very popular method for fruit-shape detection is the combination of colour cameras and structured lights. The researchers of [10] used reflection patterns (strips) to analyse apples region in order to localize its stem and calyx. The technique was an early stage, though they could approximate the region in which the stem or calyx were placed.

2.4 Discussion

A wide collection of activities was taken into account, in order to deeply understand the problem of fruit harvesting task from the sensing point of view. The sensory systems has to be capable of dealing with dynamic environments, the poor differences between background and targets, the variable harvesting conditions and the changeable shapes and colour of fruits.

From the sensors point of view:

- Single type of sensors provides good results but they are sensitive to the specific conditions of the environment.
- Image fusion techniques increase the good performance of the system and decrease the sensitivity to changes in the environment. Remarkable works have shown the capability of working with these techniques.
- Visible range sensors (colour and monochrome) provide plenty of information but dependent on the illumination conditions.
- Multispectral reflectance sensors have a great potential because of its wide range of wavelength detection. Therefore the identification of the specific spectrum of different targets in the same environment can be computed.

From physical characteristics:

- Occlusions caused by leaves are a huge problem for fruits and vegetables detection.
- Green coloured fruits are difficult to be detected because of their similarities with leaves and foliage.
- The thermal temporal variations can contribute to misinterpretation of the information provided by thermal cameras.
- The problem of localisation of fruits can be addressed as a problem of detecting small objects in large environments.
- The classification of different targets with similar spectrums in the same environment can be troublesome.

3. Methodology for the sensory system data acquisition in the validation stage

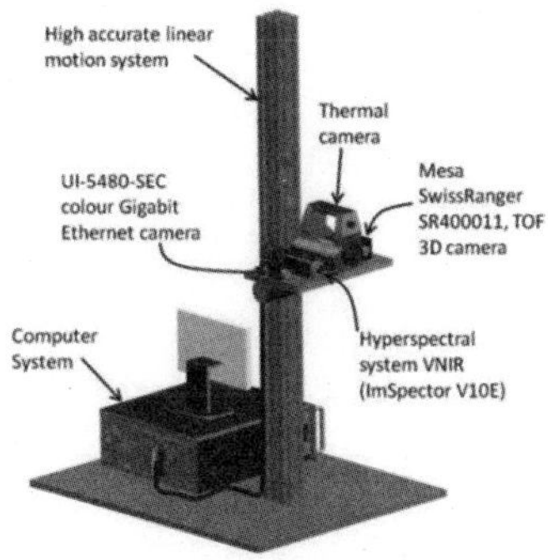

Figure 1. Testbed for validation of the sensory system for detection of fruits.

In order to validate the feasibility of a sensory system for the detection, localisation and characterisation of fruits in different scenarios and with variable conditions, a testbed has been designed. It consists of a computer and an electrically driven high accurate linear motion system where a progressive colour camera, a TOF camera, a hyperspectral system and a thermal camera are installed. Although a priori the thermal camera will not be constituent part of the sensory system for detection and localisation of fruits, it will be used in the preliminary experiments with the aim of analysing the amount of energy emitted by the fruits, leaves and peduncles, and determining if this information could be useful for this or other subsequent task (such as the health evaluation of the crops). With the proposed configuration (see Figure 1), a target could be scanned with linear velocity of 0.5mm/s to 500mm/s. The linear motion system provides a position accuracy of ±0.1mm and can support a load of 1500N.

In the validation stage, the data acquisition procedure is divided into two principal phases (Figure 2). The first one is focused on the acquisition scanned and analysis of an entire stationary working area, defined by the field of view of the spectral system and the height of the plant, or the tree to be harvested. The second phase address the acquisition and analysis of local areas denoted as zones with a high probability of fruit presence. This paper addresses the experimentation and results of the first phase procedures. In the first phase, initialisation settings are defined by the fruit harvesting specifications, as the dimension of the plants (height and width), the characteristics of the fruit (colour and average size) and the devices parameters (light source, working distance, focal length, resolution, spatial position, field of view and type of output data).

The scanning parameters of the working area are dependent of the initialisation settings and defined as the number of snaps to cover the scene and the vertical pushbroom displacement of the equipment. This scanning task will be carried out from the bottom to the top of the crop. The data analysis of the working area will provide a probability map of the possible zones with fruits onto the image plane, delivered from the hyperspectral image analysis.

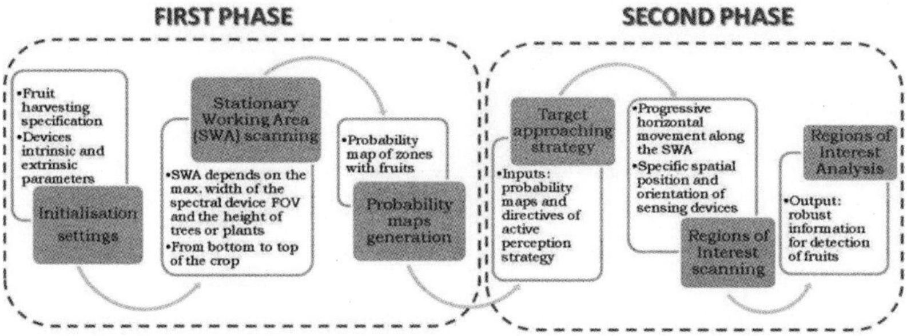

Figure 2. Outline of the methodology for the sensory system data acquisition in the validation stage.

During the second phase, the active perception strategy should decide the effective trajectory for target approaching. For instance, as primary and simplest solution, a progressive horizontal movement along the width of the working area, from top to bottom direction, could be adopted. Combining the probability maps of the zones with fruits and the directives of the active perception strategy, the spatial position and the orientation of the sensing devices (colour camera, TOF, Hyperspectral system) for scanning and analysing the local regions of interest (zones with high probability of fruits/targets location) will be provided. The analysis procedure in this second phase is focused on producing robust information for extracting the image regions corresponding to the fruit between leaves and stalks. This robust information, denoted as conditioned data, entails the data flow provided by the device itself together with the devices parameters and the sensorial fusion conversion parameters.

4. Early stage experimentation

In order to analyse the system capability and feasibility, two groups of experiments have been performed. The first group addressed the performance of the TOF camera under laboratory and field conditions for sweet pepper (green, yellow, read) and leaf of grapes acquisition. Three important parameters have been tested: the illumination conditions, the camera depth resolution and the optimal working distance. Figure 3 shows the results of the clustering of similar regions into the scenes. It is possible to detect depth variation of 5-10 mm between surfaces, if the working distance does not exceed the 1500 mm. The uncontrolled illumination condition is not critical for this device; the noise caused by glares from the light source can be filtered.

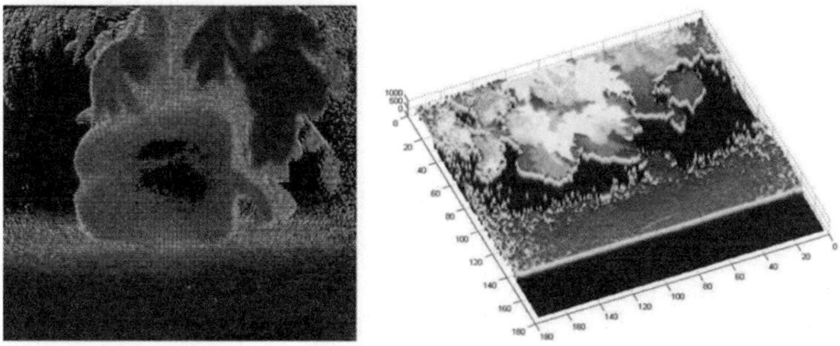

(a) (b)

Figure 3. Results of TOF camera: (a) Sweet peppers with leafs; (b) Leaf of a vineyard

The second group of the experiments was focused on the acquisition and analysis with the hyperspectral system. These tests were carried out with the support of the TOF camera to compute the spatial resolution of the hyperspectral images. A line of five LEDs with different wavelengths {534, 565, 583, 626, 920}nm were scanned, analysed and represented with the goal of demonstrating the ability of the system to detect objects with very close wavelength and to preserve the two-dimensional spatial relation of these objects, and finally the acquisition and characterization of green and red sweet pepper is presented in Figure 4.

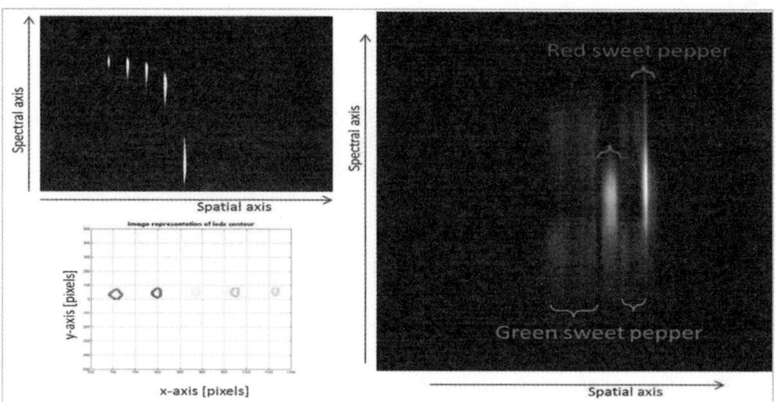

(a) (b)

Figure 4. Results of the Hyperspectral System: (a) Leds line; (b) Sweet peppers.

5. Discussion

A study of several works of precision agriculture, focused on fruit harvesting tasks was accomplished. A wide collection of activities was taken into account, in order to deeply understand the problem of fruit harvesting from the sensory system point of view. The sensory system has to be capable of dealing with complex environments (occlusions, gleams, etc.), the poor differences between background and target, the variable harvesting conditions and the changeable shapes and colour of fruits.

A sensory system evaluation methodology has been proposed for automatic fruit harvesting. This proposed methodology involves an electrically driven high accurate linear motion system where a progressive colour camera, a TOF camera, a hyperspectral system and a thermal camera are installed. An analysis procedure for the evaluation stage has been proposed. This procedure deals with the occlusions caused by leafs, trunks, stamps and the similarities between the fruits and the background.

Acknowledgments

The authors acknowledge funding of this work under Crops Projects - Grant Agreement N° 246252 of the 7FP. Carlota Salinas acknowledges support from Robocity2030 S-0505/DPI-0176. Dr. Roemi Fernández acknowledges support from CSIC under grant JAE-Doc. Dr. Héctor Montes acknowledges support from Universidad Tecnológica de Panamá and from CSIC under grant JAE-Doc.

References

1. A.R. Jimenez, R. Ceres, J.L. Pons, *Trans. of ASAE*, vol. **43(6)**, 1911 (2000).
2. K. Tanigaki, T. Fujiura, A. Akase, J. Imagawa, *Comp. Elec. in Agric.*, vol. **63(1),** 65 (2008).
3. S. Katamura, K. Oka, *IEEE Conf. on Mech. and Automation*, 1807 (2005).
4. H. Okamoto, W. Lee-Suk, *Comp. Elec. in Agric.*, vol. **63,** 201 (2009).
5. D. Moshou, C. Bravo, R. Oberti, J. West, L. Bodria, A. McCartney, *RT Imaging*, **11**, 75 (2005).
6. M. Zhang, Z. Qin, X. Lui, *Prec. Agr.,* **6**, 489 (2005).
7. A. Meunkaewjinda, P. Kumsawat, K. Attakitmongcol, A. Srikaew, *Proc. Of ECTI-CON*, 513 (2008).
8. F. López-García, G. Andreu-García, J. Blasco, N. Aleixos, J.M. Valiente, *Comp. Elec. in Agric.*, vol. **71,** 189 (2010).
9. M.S. Kim, Y.R. Chen, P.M. Mehl, Trans. of the ASAE, vol. 44(3), 721 (2001).
10. D.W. Penman, Comp. Elec. in Agric., vol. 33(1), 7 (2001).

A DISTANCE-DEPENDENT PEOPLE DETECTION FOR A WHEELED MOBILE DOMESTIC ROBOT

X. WANG and X. CLADY

Institut des Systèmes Intelligents et de Robotique,
Université Pierre et Marie Curie, CNRS UMR 7222,
cc 173, 4 place Jussieu, 75252 Paris Cedex
E-mail: wang/clady@isir.upmc.fr
http://www.isir.upmc.fr

In this paper, we present a people detection system embedded on a wheeled service robot. The probabilistic outputs of laser and vision based detectors are merged considering explicitly the human-robot distance. A grid based approach allows to combine these probabilities in Gaussian Mixture Models. Our generic approach is particularly efficient because it benefits from their strong distance-dependent complementarity. The correct detection rate is increased, for example, from 0.5 to 0.8 with the false alarm rate fixed at 0.1.

Keywords: wheeled mobile domestic robot; people detection; human-robot distance

1. Introduction

Recent days, mobile robots are more and more involved in our daily lives. They can be employed in various missions like cleaning the room, taking care of the patient, accompanying the elders, playing with kids, etc.. People detection is an important skill providing these robots the ability to fulfill more complex tasks within the human society. Many wheeled domestic robots are equipped with 2D laser scanners and cameras. The laser scanner provides accurate distance estimation while the camera gives discriminative information. The advantage of merging the information from both sensors has been clearly shown. In a previous study combining a vision based whole body detector with a laser based leg one, we noticed that the complementarity of these two sensors depends strongly on the human-robot distance.

ACKNOWLEDGMENTS: this work has been supported by European Union and French National Research Agency (ANR) through AAL program (project DOMEO n° ANR-08-AALI-001-02).

In this paper, we propose to merge the probabilistic outputs of laser and vision based detectors considering explicitly this distance.

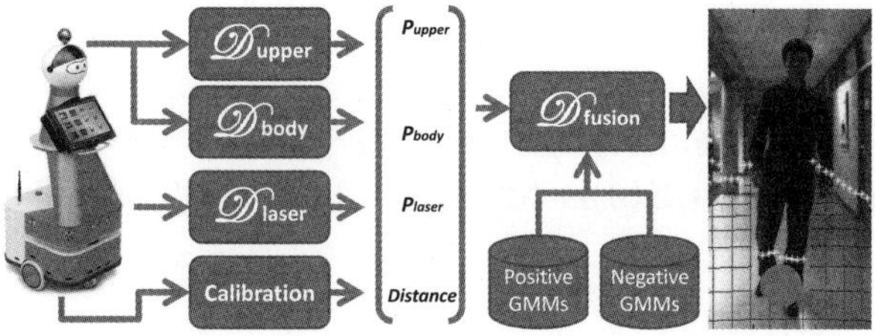

Fig. 1. Synopsis of the detection system

2. Related Work

Embedded on wheeled mobile robots, laser scanner(s) and camera(s) are used respectively for legs detection and for face or body detection.

In *Domestic Robots*, some studies[1–3] use the information provided by the laser rangefinder to focus a vision based algorithm on an interest region of the image, to identify a moving object detected in the image sequence[4] or to reject the false alarms[5] provided by the visual sensor. Most of the recent studies[6–10] merge the information provided by the two sensors via the tracking process. In these scheme, vision based observations are often used to confirm more or less regularly the laser based tracks. Concerning the leg detection using laser scans, several algorithms have been proposed in the literature. Some[11,12] are based on the search of local minima. Other approaches[13,14] can only detect moving persons, normally searching for differences between two or more scans. Bellotto[8–10] *et al.* propose a structural approach to identify typical patterns. Recently machine learning approaches[15] have been proposed. Concerning the vision based algorithms, some papers[2,16] propose skin-colored based segmentation. Many use the OpenCV's[17] face detector. Similar appearance based approaches[5,9,10,18] have been used to detect the upper-body or the whole body. Schiele[5] *et al.* have presented recently an interesting overview of other promising techniques for visual people detection.

In *Intelligent Vehicles*, there is a wider range[19] of used sensors: visible light imaging, infrared imaging, time-of-flight sensors (RADAR, Laser,...),... The most common software architecture can be decomposed into two steps:

a hypothesis generation step and a hypothesis validation step. The 3D sensors (Radar, Laser,...) are often used in the first step to provide potential positions of pedestrians in a simple and rapid way resulting in a reduced search area. Hypotheses generated in the earlier step are verified by using some vision based complex algorithms and false alarms are eliminated. Gerónimo[20] *et al.* present a recent survey in this domain. As reported, the sensor fusion for pedestrian detection is an open area of research. The fusion between laser scanners and cameras is particularly studied because the first ones provide accurate distance estimation and the second ones discriminative information. Bayesian fusion or filtering[21–23] are generally used to combine vision and laser based information. Some authors[24,25] concatenate all the features provided from the two sensors and apply machine learning algorithms. Another strategy has been proposed by Premebida[25,26] *et al.*: some learning machine process are tested to combine the outputs of two boosted classifiers; each classifier was learned on the feature outputs from each sensor.

The proposed people detection system deals essentially with the complementarity of this two sensors. It is particularly exacerbated by the sensor configuration for domestic robots. They are commonly equipped with a 2D laser scanner (mounted approximately $30cm.$ above the floor) and a camera (located approximately $1m.$ above the laser or the floor and often embedded on a Pan-Tilt head). In a previous study[27] combining a vision based whole body detector and a laser based legs one, we can notice that the human-robot distance is an important issue. At close distances ($< 1.4m.$), only the laser based detector works indeed. It is due to the camera view-range: the floor (and the whole body) can be viewed in the image only between $2m.$ and $9m.$. At farther distances ($> 3m.$), the body detector is more reliable and the legs one works badly. It is because the probability that a laser ray intersects a leg decreases drastically farther than $3m.$ At middle distances, a merging process is needed to exploit the redundancy of the detectors.

3. Detection Algorithm

Three detectors are employed: a laser based leg detector, a body detector and an upper-body detectors both based on vision. Using a grid based approach and Gaussian Mixture Models (GMM), their output probabilities are combined as illustrated in the Figure 1.

Probability Grids. In order to do a consistent fusion of laser and camera data, we build for the two sensors the same type of probabilistic grid.

Our grids are based on a polar split of the space. The cells have different sizes with respect to the distance from the robot. Each cell is identified by its polar coordinates (ρ, θ). For the laser grid, the space is divided in 270 radius (one for each degree). The distance increment is given by: $d_{i+1} = d_i \cdot (1 + E_r) + E_a$, where E_r could be considered as a relative error and E_a as an additional absolute error in distance. We take $E_r = 5\%$, $E_a = 5cm.$ and $d_0 = 0.2cm.$. A similar grid is defined for the camera on the floor according to the same equation in distance, but with a smaller angular range (according to the field of view of the camera). As we can see in the figure 2.a, the grid is more dense close to the robot (where we need a more accurate position estimation). In the Personal Space (between $0.45m.$ and $1.2m.$), the relative mean error in distance due to our space sampling (Figure 2.b) is quite similar as the results obtained by Oskoei *et al.* In the same way, they have recently presented a work[28] on the evaluation of the accuracy performance of a state-of-art Autonomous Proxemics System[8] to keep an acceptable Human-Robot proxemics (HRP) distance during interaction with a human. This experiment was conducted with two subjects in the University of Hertfordshire "Robot House". Their analysis of the relative error means (approximately around 10%) between the desired and the obtained HRP distances, lead them to conclude that the proposed system works reliably.

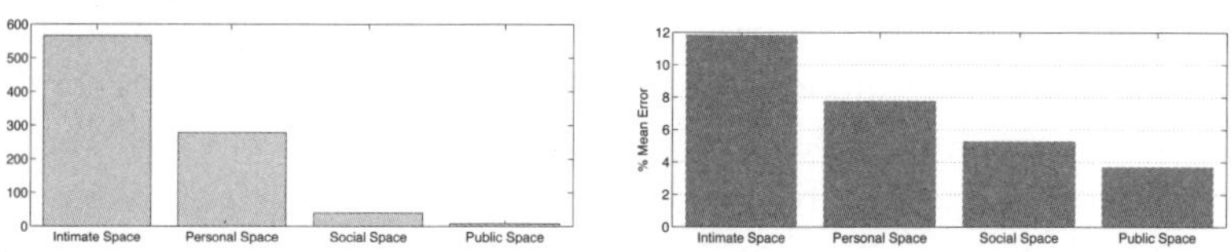

(a) Density of cells (in number of cells/m^2) (b) Mean relative error (%) in distance

Fig. 2. Grid statistics related to proxemics[29]

Detectors. Each detector is learned using Real Adaboost. The laser features[15] are inspired from previous works. In the two vision based detectors, two types of features are employed: Haar-Like (Haar-like features or Haar features) and HoG (Histogram of Oriented Gradients). These two features are frequently used in the domain of object detection and recognition. It[30,31] has been proved the combination of the two types of features produces a more efficient detection system. Haar-like features are derived from the wavelet decomposition (using Haar wavelets). They provide information about the gray-level distribution between two adjacent regions. Their outputs characterize the contrasted regions in the subimage. The histogram of oriented gradients is a histogram of neighborhood pixels according to their gradient orientation and weighted by their gradient magnitude. Subdivided and head-tail connected, HoG features[32,33] are particularly adapted to people detection: they provide compact and discriminative information on the contours (essentially on the body shape).

Grid Fusion. The merging process is composed of two steps. Firstly, the geometric relation between the camera and the laser scanner is estimated in order to match the two grids. The position and the orientation of the camera related to the floor is estimated with a test pattern (a known planar checkerboard) and using the Bouguet's[34] calibration toolbox. The geometric relation between the two sensors has been obtained following a calibration method proposed by Couverture[35] *et al.* For the second step, as proposed by Premebida[25,26] *et al*, Gaussian Mixture Models are learned in a vectorial space grouping the outputs of the three detectors. To take explicitly account of the human-robot distance, we add it as another dimension of the vectorial space. The likelihoods with both positive and negative GMMs are computed; a log operator is then applied. A score is given by the difference between the positive and negative log-likelihoods. Applying a threshold to this score, we can estimate if there is a person in the cell.

4. Experimental Results

The Kompaï robot used for our experiments embeds a laser sensor Sick S300 and a camera on its head. During the experiments, the robot has been remotely controlled. We have asked fifteen persons (one or two persons at a time) to walk naturally in front of the robot. We have collected 271 data (image+laser frame): 131 with one person and 140 with two persons. They represent 411 potential observations of a human in the environment by the robot.

In Figure 3.a, the fusion's curve rises above other individual detectors because of the great complementarity between two sensors. The correct detection rate is increased from 0.5 to 0.8 with the false alarm rate fixed at 0.1. The results of the vision based detectors are similar as they are clearly redundancy with this sensor configuration. Figure 3.b shows that the performances can be greatly improved by considering explicitly the human-robot distance in the detection system. It is particularly true for a standard domestic robot which the complementarity of the embedded

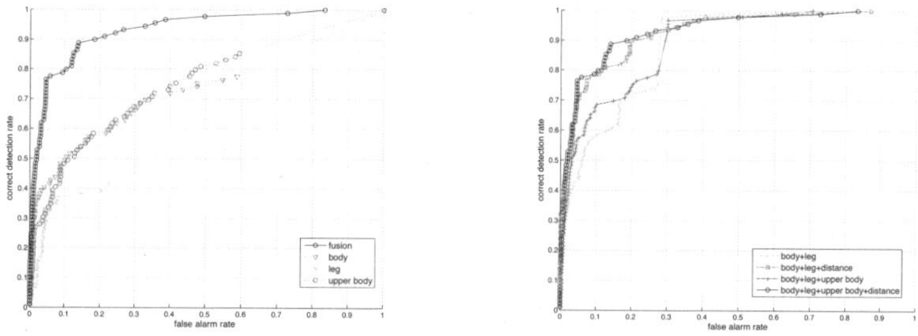

(a) Comparison with the 3 detectors (b) Comparison with different fusion process

Fig. 3. ROC curves

sensor is depending on the distance. Figure 4 illustrates the obtained results.

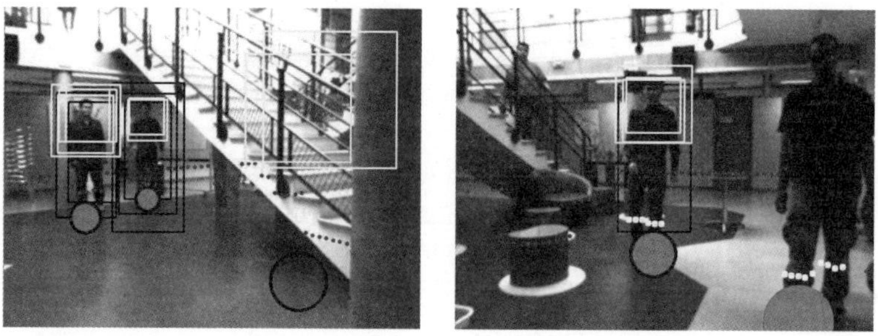

Fig. 4. Examples of detection results (Body: black rectangular; Upper-body: yellow square; Leg: white points; Fusion with distance: red disk; Fusion without distance: black circle)

5. Conclusions

In this paper, we have presented an embedded human detection system for a wheeled mobile domestic robots. The used methodology is based on the fusion of three detectors using two embedded sensors: a laser range finder and a camera. They are dedicated to leg, body and upper-body detections. A grid based approach allows to combine the resulting human presence probabilities in Gaussian Mixture Models. The experimental results show that the explicit consideration of the human-robot distance in the detection system improves clearly the performances. This generic method make easier for integrating other sensors (i.e. Time-Of-Flight Cameras[36] or the Kinect). In future works, we would extend our database and our method for including sitting or occluded people.

References

1. J. Blanco, W. Burgard, R. Sanz and J. Fernandez, Fast face detection for mobile robots by integrating laser range data with vision, in *Conf. Adv. Robotics*, 2003.
2. Z. Byers, M. Dixon, K. Goodier, C. Grimm and W. Smart, An autonomous robot photographer, in *IEEE Conf. Robots Syst.*, 2003.
3. M. Scheutz, J. McRaven and G. Cserey, Fast, reliable, adaptive bimodal people tracking for indoor environments, in *IEEE Conf. Robots Syst.*, 2004.
4. Z. Zhang and K. Kodagoda, Multi sensor approach for people detection, in *ISSNIP*, 2005.
5. B. Schiele, M. Andriluka, N. Majer, S. Roth and C. Wojek, Visual people detection: Different models, comparison and discussion, in *IEEE ICRA workshop on People Detection and Tracking*, 2009.
6. Z. Zivkovic and B. Krose, Part based people detection 2d range data and images, in *IEEE/RSJ IROS*, 2007.
7. C. Lin, C. Yang, C. Wang, K. Song and J. Hu, A new design on multi modal robotic focus attention, in *IEEE Int. Symposium on Robot and Human Interactive Communication*, 2008.
8. N. Bellotto and H. Hu, Multisensor based human detection and tracking for mobile service robots *IEEE Trans. on Systems, Man, and Cybernetics: Part. B* 2009.
9. N. Bellotto and H. Hu, Computationally efficient solutions for tracking people with a mobilerobot: an experimental evaluation of bayesian filters *Autonomous Robots* 2010.
10. N. Bellotto and H. Hu, A bank of unscented kalman filters for multimodal human perception with mobile service robots *Int J Soc Robot* 2010.
11. J. Fritsch, M. Kleinehagenbrock, S. Lang, T. Plotz, G. A. Fink and G. Sagerer, Multimodal anchoring for human robot interaction *Robot. Auton. Syst.* 2003.
12. N. Bellotto and H. Hu, Multisensor integration for humanrobot interaction *IEEE Journal of Intelligent Cybernetic Systems* 2005.
13. M. Lindstrom and J. Eklundh, Detecting and tracking moving objects from a mobile platform using a laser range scanner, in *IEEE/RSJ IROS*, 2001.
14. P. Chakravarty and R. Jarvis, Panoramic vision and laser range finder fusion for multiple tracking, in *IEEE/RSJ IROS*, 2006.
15. S. Juriĉ-Kavelj and I. Petroviĉ, Experimental comparison of adaboost algorithms applied on leg detection with different range sensor setups, in *RAAD*, 2010.
16. M. Kleinehagenbrock, J. F. S. Lang, F. Lomker, G. A. Fink and G. Sagerer, Person tracking with a mobile robot based on multimodal anchoring, in *IEEE ROMAN*, 2002.
17. OpenCV, Open computer vision library, [Online] Available: http://sourceforge.net/projects/opencvlibrary.
18. R. C. Luo, Y. J. Chen, C. T. Liao and A. C. Tsai, Mobile robot based human detection and tracking using range and intensity data fusion, in *IEEE Workshop Adv. Robot. Social Impacts*, 2007.

19. T. Gandhi and M. Trived, Pedestrian protection systems: Issues, survey, and challenges *IEEE Trans. ITS* 2007.
20. D. Geronimo, A. M. Lopez, A. D. Sappa and T. Graf, Survey of pedestrian detection for advanced assistance systems *PAMI* 2010.
21. H. Cheng, N. Zheng, X. Zhang, J. Qin and H. van de Wetering, Interactive road situation analysis for driver assistance and safety warning systems: Framework and algorithms *IEEE Trans. ITS* 2007.
22. L. Spinello and R. Siegwart, Human detection using multimodal and multidimensional features, in *IEEE ICRA*, 2008.
23. L. N. Pangop, R. Chapuis, S. Bonnet, S. Cornou and F. Chausse, A bayesian multisensor fusion approach integrating correlated data applied to a real-time pedestrian detection system, in *IEEE/RSJ IROS*, 2008.
24. B. Douillard, D. Fox and F. Ramos, A spatiotemporal probabilistic model for multi-sensor object recognition, in *IEEE/RSJ IROS*, 2007.
25. C. Premebida, O. Ludwig and U. Nunes, Lidar and vision-based pedestrian detection system *Journal of Field Robotics* 2009.
26. C. Premebida, O. Ludwig, M. Silva and U. Nunes, A cascade classifier applied in pedestrian detection using laser and image-based features, in *IEEE ITSC*, 2010.
27. X. Wang, X. Clady and C. Granata, A human detection system for proxemics interaction, in *IEEE/ACM HRI (Late Breaking Results)*, 2011.
28. M. A. Oskoei, M. L. Walters and K. Dautenhahn, An autonomous proxemic system for a mobile companion robot, in *AISB: New Frontiers in Human Robot Interaction*, 2010.
29. E. Hall, Proxemics *Current Anthropology* 1968.
30. P. Geismann and G. Schneider, A two-staged approach to vision-based pedestrian recognition using haar and hog features, in *IEEE IV*, 2008.
31. P. Negri, X. Clady, S. M. Hanif and L. Prevost, A cascade of boosted generative and discriminative classifiers for vehicle detection *EURASIP Journal on Advances in Signal Processing (Special Issue : Machine Learning in Image Processing)* 2008.
32. I. Laptev, Improving object detection with boosted histograms *Image Vision Comput.* 2009.
33. P. Dollar, C. Wojek, B. Schiele and P. Perona, Pedestrian detection: A benchmark, in *CVPR*, 2009.
34. J.-Y. Bouguet, Camera calibration toolbox for matlab, [Online] Available: http://www.vision.caltech.edu/bouguetj/.
35. C. Couverture, C. Rosenberger, G. Mourioux, C. Novales and G. Poisson, Matching between telemetric measures and images for mobile robotics, in *Int. Conf. on Image and Graphics*, 2007.
36. C. Zong, X. Clady and M. Chetouani, An embedded human motion capture system for an assistive walking robot, in *IEEE ICORR*, 2011.

OBTAINING GEOMETRY OF STAIRS USING MONOCULAR VISION SYSTEM

K. WALAS

Institute of Control and Information Engineering, Poznan University of Technology, Poznan, Poland
E-mail: krzysztof.walas@put.poznan.pl

Modern walking robots are able to negotiate rough terrain. However there are still open topics, especially when there is a need to climb or descend an obstacle. This article presents the perception system for descending stairs with the six-legged walking robot. The perception system for stair descent differs much from that used for stair climbing, while during the descent most of the surfaces of the step suffer from occlusions. Author describes the solution to this problem. The perception system consists of two sensors. Namely a video camera and a force sensor at the tip of the leg for active haptic sensing. The monocular vision system gives the scale of the stairs and the active haptic sensing system provides the additional information to obtain the real geometry of the obstacle. In the presented paper experiments were performed on the walking robot Messor. The article describes the monocular vision system and the method for obtaining the geometry of stairs. Next the ways of improving the accuracy of the system are presented.

Keywords: Walking Robots, Perception, Stair Descending

1. Introduction

Modern walking robots are able to negotiate rough terrain. However there are still open topics, especially when there is a need to climb an obstacle for example stairs. In most cases the attention of the researchers is focused on climbing stairs. The list of references concerning this topic could be found in Ref.8. The problem of descending stairs has less attention. The list of papers concerning this particular topic is rather short Ref. 1–3. The author is aware that it could be non-exhaustive. The reason for this disproportion could be an assumption that the stair descent is the same problem as the stair ascent. However these problems differ especially in the type of the perception required to obtain the information about the shape of the obstacle. While climbing stairs the robot is able to easily built the map of the stairs because all the surfaces of the step are visible for its vision system. The

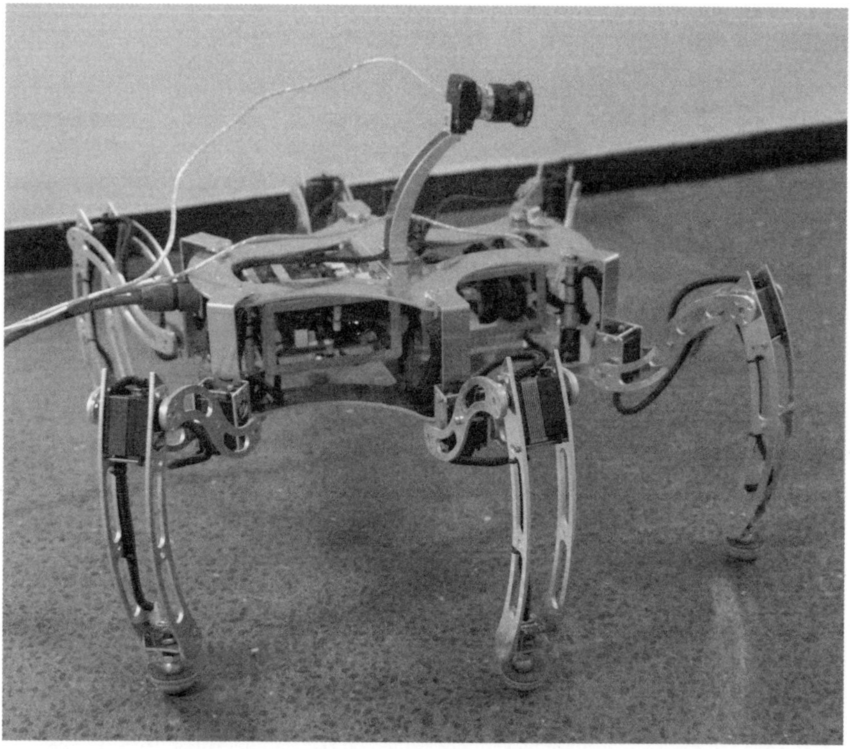

Fig. 1. Messor robot.

exemplary perception system for stair climbing was presented in Ref. 9. Whereas during the descent most of the surfaces of the step suffer from occlusions. This article presents the solution to this problem. The perception system consists of two sensors. Namely a video camera and a force sensor at the tip of the leg for active haptic sensing. The monocular vision system gives the scale of the stairs, and the active haptic sensing system provides the additional information to obtain the real geometry of the obstacle. The use of the active haptic sensing was first presented in Ref. 6. This type of perception is based on the observations made in nature as it was described in Ref. 4. The recent research made on this topic is presented in Ref. 5 where the active haptic sensing is used for terrain classification.

In the presented paper experiments were performed on the walking robot Messor. It is a six-legged walking robot with three degrees of freedom in each leg. Its trunk is 26 cm wide and 30.6 cm long. The segments of the

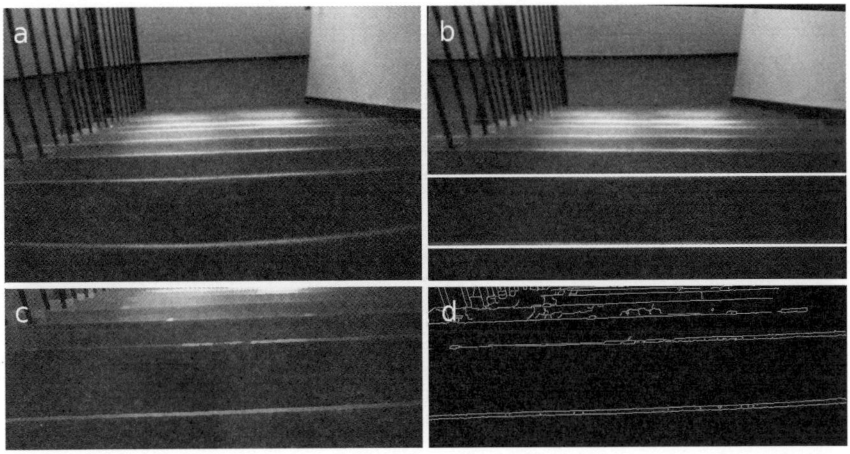

Fig. 2. Image processing phases: input image(a); output image with edges of the steps marked with white line (b); corrected and cropped image after Kuwahara filtering(c), binary image of edges found with Canny detector(d).

leg are: coxa 5.5 cm, femur 16 cm, tibia 23 cm. It weights 4.3 kg. Further information could be found in Ref.10. The robot is shown in Fig. 1.

The content of the article is as follows. At the beginning monocular vision system is described. Then the method for obtaining the geometry of stairs is presented. Next the ways of improving the accuracy of the system are described. At the end concluding remarks are given, followed by the future work plans.

2. Monocular vision system

First phase in obtaining the dimensions of stairs in presented approach is based on the visual perception. The image of the stairs obtained from the camera mounted on the robot is shown in Fig. 2a. Due to the distortion caused by the lens of the camera, correction with appropriate calibration tool was needed. Next the image was cropped and its upper part was removed. It made image processing easier, because superfluous edges were discarded. The lower part of the image is sufficient to establish the geometry of the stairs while the obstacle is regular and the dimensions of the first two steps provides the information on dimensions of the whole stairs.

In the subsequent step texture of the floor had to be removed. The Kuwahara filter[7] was used. It smooths the image but preserves the edges. The Kuwahara filter is rather slow and it takes about two seconds to process presented image. However this delay in the stair descent procedure, which is a deliberative process, is allowed. The filtering is performed while it add robustness to the line detection algorithms. The result of distortion correction, image cropping and Kuwahara filtering is shown in Fig. 2c.

Next the Canny edge detection was performed. The result of this step is shown in Fig. 2d. Obtained binary image containing all the edges was processed using Hough line transform to get straight lines. Then the line with the highest value in the accumulator plane is chosen. Later the image is rotated to align the edges of the stairs horizontally. In the following step the Hough transform is performed one more time. Now two horizontal lines were chosen: closest to the robot and second closest to the robot. The distance in pixels between these two lines ought to be above certain value. The threshold was calculated from the constraint on the least physical distance between subsequent stairs. This prevents choosing two line representing the edge of the same stair.

The result of the whole algorithm is presented in in Fig. 2b. The edges of two steps were found and were marked with white line. Image processing procedure has provided the distance in pixels. Next section describes the process of obtaining the real distance between edges of the step.

3. Obtaining the geometry from monocular vision system

The image processing algorithm described in the previous section provided information on the position of the projection of the stair edges on the camera imager. It was marked as the points q_1 and q_2 as shown in the Fig. 3a. The physical dimensions of the stairs could be obtained when at least one geometrical parameter of the stairs is known. In this case the height of the step $(y_{c_2} - y_{c_1})$. This dimension is obtained using the leg of the robot as a sensor. For this purpose the leg is equipped with the contact sensor in its foot. The robot while standing at the edge of the stairs, lowers its leg until the foot reaches the surface of the step below. The height of the step is calculated using robot kinematics.

The information from the vision system and active haptic sensing system has to be merged to obtain the depth of the step $(z_{c_2} - z_{c_1})$. To do this the coordinates of the points y_{c_1} and y_{c_2} have to be transformed to the camera coordinate system. It was done by using Eq. 1.

$$A_2^{-1} A_1^{-1} * \mathbf{a_i} = \mathbf{c_i} \tag{1}$$

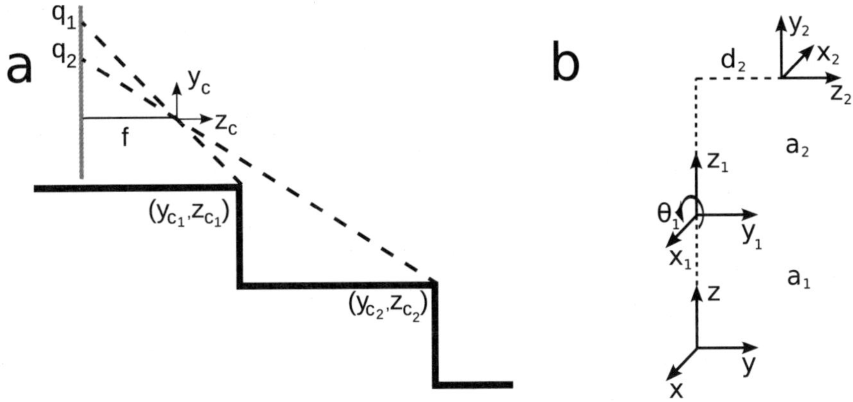

Fig. 3. Perception system: schematic view of the vision perception(a); robot coordinates frames(b).

where:

$$A_1 = \begin{bmatrix} 1 & 0 & 0 & 0 \\ 0 & cos(\theta) & -sin(\theta) & 0 \\ 0 & sin(\theta) & cos(\theta) & a_1 \\ 0 & 0 & 0 & 1 \end{bmatrix}, \ A_2 = \begin{bmatrix} -1 & 0 & 0 & 0 \\ 0 & 0 & 1 & d_2 \\ 0 & 1 & 0 & a_2 \\ 0 & 0 & 0 & 1 \end{bmatrix}, \ \mathbf{a_i} = \begin{bmatrix} 0 \\ y_{a_i} \\ z_{a_i} \\ 1 \end{bmatrix} \ \mathbf{c_i} = \begin{bmatrix} 0 \\ y_{c_i} \\ z_{c_i} \\ 1 \end{bmatrix}$$

The coordinate frames of the system are shown in Fig. 3b. The global reference frame has no indices. The index 1 is related to the robot local reference frame, and index 2 is related to the camera local reference frame. In this set-up a_1 and θ_1 are the variables. The robot is able to lift its body and to lean it forward – the rotation around x_1 axis. This two movements are required due to the restricted field of view of the camera. In the ideal case shown in Fig. 3a the field of view of the camera is large enough that, there is no need of lifting the robot and leaning it forward. Equation 2 provide all the points of interest in the camera reference frame. Now using the pinhole camera model Eq. 3, the depth of the step $(z_{a_2} - z_{a_1})$ could be obtained according to the Eq. 4. In this equation all the coordinates with subscript a are referring to coordinates in global frame and these with subscript c are referring to coordinates in the local camera frame.

$$\mathbf{c_i} = \begin{bmatrix} 0 \\ y_{c_i} \\ z_{c_i} \\ 1 \end{bmatrix} = \begin{bmatrix} 0 \\ -y_{a_i} sin(\theta) + z_{a_i} cos(\theta) + a_1 cos(\theta) - a_2 \\ y_{a_i} cos(\theta) + z_{a_i} sin(\theta) + a_1 sin(\theta) - d_2 \\ 1 \end{bmatrix} \tag{2}$$

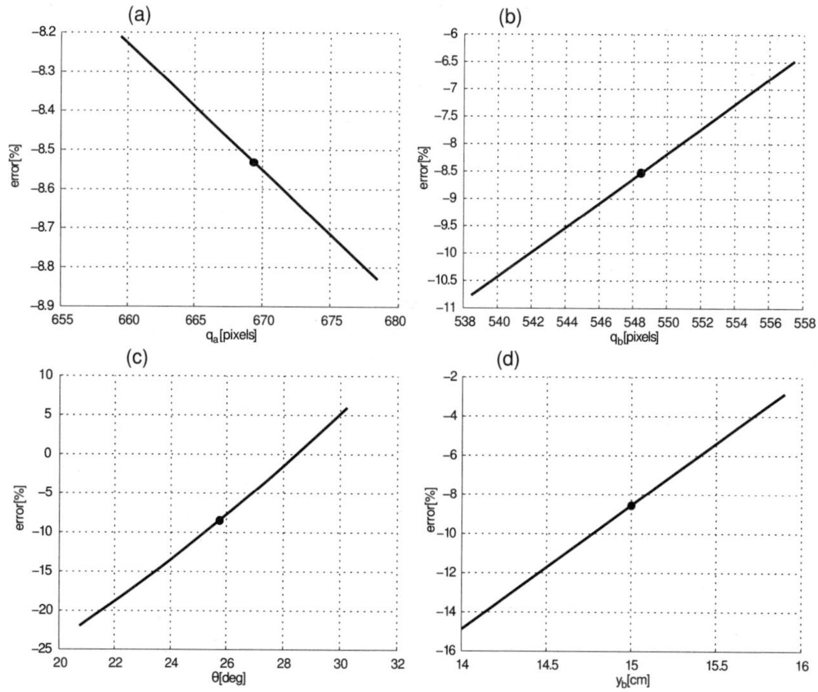

Fig. 4. The change of the relative error of the depth of the stair measurements caused by: position of the first edge of stairs on the image (in pixels) – q_a (a); the position of the second edge of the stairs on the image (in pixels) – q_b (b); the tilt angle of the robot (in deg) – θ (c); the height of the stair – y_b (d).

$$\frac{y_{c_i}}{z_{c_i}} = \frac{q_i}{f_y} \tag{3}$$

$$z_{a_i} = \frac{q_i * (y_{a_i} cos(\theta) + a_1 sin(\theta) - d_2) - f(-y_{a_i} sin(\theta) + a_1 cos(\theta) - a_2)}{f cos(\theta) - q_i sin(\theta)} \tag{4}$$

In the experiment presented in Fig. 3 the dimensions of stair were: height 15 cm, depth 30 cm. The depth of the stair obtained with the presented system was equal to 27.4 cm. Which gives a relative error of -8.5%.

4. System precision improvement

The result presented in subsequent section is not satisfactory. In order to improve it four possible sources of measurements errors were identified and checked. Namely, the position of first edge of stairs on the image (in pixels) – q_a, the position of the second edge of the stairs on the image (in pixels) – q_b, the tilt angle of the robot (in deg) – θ and the height of the stair – y_b.

As it is shown in Fig. 4a and in Fig. 4b the change of the position of q_a and q_b causes linear growth of the depth determination relative error. The change of the position of the line on the image was in the range of ± 5 pixels around the measured value (marked by dot). The increment of change was equal to one pixel. The difference of one pixel caused the change of relative error value by 0.03%.

The relation between the tilt angle change and the relative error of the stair depth measurement is non-linear it was shown in Fig. 4c. The increment of the change of the angle was equal to 0.5 deg. This increment caused the average change of error equal to 1.6 %.

In the next chart Fig. 4d the relation between the change of the height measurement and the relative error of depth measurement is shown. The function is linear. The increment of the height was equal to 0.1 cm. This increment caused the change of the relative error equal to 0.6 %.

As it was presented above the highest influence on the precision of the system has the change of the tilt angle and the height measurement. The least influence on the precision of the whole system has the accuracy of the vision system. There are two more parameters which could be changed in order to obtain better results. These are the calibration parameters which are: distance of the camera axis from the robot local coordinates system along z-axis and along the x-axis. These parameters has to be calibrated experimentally while the determination of the exact position of the local camera coordinates frame is hard to obtain.

The improvement of the height of the step measurement is hard to obtain while this sensing is done using the leg and the robot kinematics. The possible improvement could be done in tilt angle measurements. Moreover the better experimental calibration of the camera position is possible.

5. Conclusions

The paper presented the monocular vision system which supported by active haptic sensing with the leg of the robot provides the information on the real dimensions of the stairs. The obtained geometry of the stairs is further used to perform stair descent algorithm which is based on the strategy presented in Ref.8. The geometry of the stairs is measured with high value of error so the system gives only the rough position of the edge of the stairs. To assure safe descent the strategy requires active haptic sensing of the edge of the stair when robot approaches this very edge. The robot could move fast by the most of the descent. However when it is approaching the edge of the stairs it has to switch to the low-speed cautious walk. As the future work better calibration of the system has to be done. Moreover its integration with the descent strategy is required. Additionally the investigations on speeding-up the Kuwahara filter by using parallel computing are in progress.

References

1. P. Ben-tzvi, S. Ito, A. Goldenberg, A mobile robot with autonomous climbing and descending of stairs, Robotica 27(2), 2009, pp. 171 – 188.
2. D. Campbell, M. Buehler, Stair Descent in the Simple Hexapod RHex, in Proc. IEEE Int. Conf. on Robotics and Autom. (ICRA'10), Taipei, Taiwan, 2003, pp. 1380 – 1385.
3. G. Figliolini, M. Ceccarelli, EP-WAR3 biped robot for climbing and descending stairs, Robotica 22(4), 2004, pp. 405 – 417.
4. W.C. Flannigan, G.M. Nelson, R.D. Quinn, Locomotion controller for a crab-like robot, in *Proc. IEEE Int. Conf. on Robot. and Aut.(ICRA'98)*, (Leuven, Belgium, 1998), pp. 152 – 156.
5. M. Hoepflinger, C. Remy, M. Hutter, L. Spinello, R. Siegwart, Haptic Terrain Classification for Legged Robots*Proc. IEEE Int. Conf. on Robot. and Aut.(ICRA'10)*, (Anchorage, Alaska, 2010), pp. 2828 – 2833.
6. R. Hoffman, E. Krotkov, Perception of rugged terrain for a walking robot: true confessions and new directions, in *Proc. IEEE/RSJ Int. Workshop on Intell. Robots and Syst.(IROS'91)*, (Osaka, Japan, 1991), pp. 1505 – 1510.
7. M. Kuwahara, K. Hachimura, S. Eiho, M. Kinoshita, Processing of RI-angiocardiographic images, in Digit. Process of Biomedical Images, New York: Plenum, 1976. pp. 187-202.
8. K. Walas, Fully parametrized stair climbing strategy for a six-legged walking robot, in *Proc. Int. Conf. on Climb. and Walk. Robots and the Support Technol. for Mob. Mach. (CLAWAR'10)*, (Nagoya, Japan, 2010), pp. 777 – 784.
9. P. Labecki, K. Walas, Multisensor perception for autonomous stair climbing with a six legged robot, in *Proc. Int. Conf. on Climb. and Walk. Robots and the Support Technol. for Mob. Mach. (CLAWAR'10)*, (Nagoya, Japan, 2010), pp. 1013 – 1020.
10. K. Walas, D. Belter, Messor – Versatile Walking Robot for Search and Rescue Missions, J. of Autom., Mob. Robotics & Intell. Sys., vol. 5(2), pp. 2834, 2011, www.jamris.org.

LASER SCAN MATCHING FOR SELF-LOCALIZATION OF A WALKING ROBOT IN MAN-MADE ENVIRONMENTS

Piotr SKRZYPCZYŃSKI

Institute of Control and Information Engineering,
Poznan University of Technology, ul. Piotrowo 3A, PL-60-965 Poznań
E-mail: piotr.skrzypczynski@put.poznan.pl

This paper describes a novel application of the well-established 2D laser scan matching technique for self-localization of a walking robot. The presented system combines two scan matching algorithms to deal with poor range data from a compact laser scanner and to provide robust self-localization in various types of environments. Proprioceptive sensors and some simplifying assumptions holding in man-made environments are exploited to compensate for the changing attitude of the walking robot.

Keywords: Walking robot, Self-localization, Scan matching, Attitude correction

1. Introduction

Legged robots are very suitable for performing missions in complex environments, because they are able to negotiate many types of obstacles and climb stairs. An example of such a mission are urban search and rescue scenarios. However, due to the inherent properties of legged locomotion some tasks that can be easily carried out by wheeled robots are hard to solve if a legged robot is being used. One of such tasks is self-localization, which is one of the central problems for autonomous robots. Many of the assumptions of classic self-localization algorithms are invalid in the case of a legged robot. The most common assumption is that the robot moves on a plane and the sensory readouts (from range sensors in particular) are parallel to that plane. This is not true for most of the walking robots, which cannot maintain their attitude (pitch φ and roll ψ angles) stable enough to keep a sensor perpendicular to the gravity vector all the time. Another problem is the reliable dead reckoning assumption – in most walking robots reliable odometry is not available. The situation is worsened by the limited payload of most legged robots, which cannot carry large or heavy exteroceptive sensors. Thus, in the case of range sensors, miniature devices have to be used, such like the Hokuyo URG-04LX, which yields short-range and noisy data comparing it to the Sick LMS family of laser sensors typically used on wheeled robots.

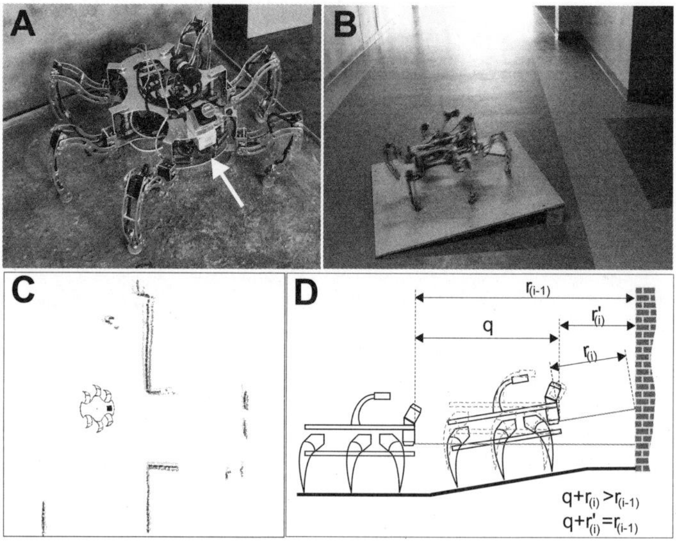

Fig. 1. Messor the robot (A), scanning in non-flat environment example (B,C), and the attitude correction concept (D)

This paper addresses the issues of reliable incremental self-localization for the six-legged walking robot Messor (Fig. 1A), which is intended for urban search and rescue missions. The terrain mapping system of this robot uses a tilted down URG-04LX laser scanner, which sweeps the ground ahead of the machine, enabling it to sense the terrain profile.[6] The mapping procedure relies on an external pose estimate with regard to 6 degrees of freedom (d.o.f.). The Inertial Measurement Unit (IMU) is insufficient for positioning due to the drift. If only the proprioceptive sensors are used for positioning, the pose uncertainty grows quickly. Therefore the pose estimate is provided by a visual SLAM system.[8] However, this system is unreliable in environments with few visually salient features. Self-localization in 6 d.o.f. can be accomplished by matching point clouds from a 3D laser scanner,[1] but such a sensor cannot be carried by our robot due to the tight mass and energy limits. In Ref. 7 a lightweight Time-of-Flight (ToF) camera is applied to create a local terrain model, which is then used by a fast histogram based registration method, providing a form of odometry. However, ToF cameras are expensive, and because this technology is still in its infancy, they have some drawbacks, most notably high dependence on environment conditions, which make them less suitable in our application. On the other hand, known methods for estimating the attitude of a walking robot from sparse range measurements of inexpensive sensors observing the terrain cannot be used here because they work under a flat ground plane assumption.[9]

However, the urban search and rescue missions are usually carried out in man-made environments characterized by many approximately flat vertical surfaces (walls – Fig. 1B,C). Therefore, a 2D laser range scanner mounted horizontally on the robot (not tilted) can be used to obtain environment profiles (2D scans) that are suitable for scan matching. This concept was already applied to learn grid maps with a humanoid robot equipped with a 2D laser scanner.[11]

Scan matching in 2D is a well-established self-localization technique, which can be used to obtain either globally consistent maps or as an alternative form of odometry, yielding the relative displacement $\mathbf{q}_i = [\Delta x_i \ \Delta y_i \ \Delta\theta_i]^T$ (translation and rotation) for a pair of scans $s(i-1)$ and $s(i)$. There are many methods for matching 2D scans known from the literature. Recently, a robust and precise scan matching algorithm was proposed in Ref. 3. This algorithm is provided with a covariance estimation procedure and is very suitable for our purposes. But using a 2D scan matching procedure on a walking robot requires solving several problems that are specific to this platform: (i) compensation of the varying pitch and roll angles in order to correct the range readings that hit the observed objects at different heights in consecutive scans (Fig. 1D), (ii) estimation of the instantaneous motion of the robot between the consecutive scans to have an initial estimate for scan matching, (iii) dealing with the poor quality of range data from the URG-04LX scanner.

2. Scan matching algorithms

A scan matching algorithm aims at finding the best alignment between two laser scans. The most used approach to scan matching is the Iterative Closes Point (ICP), which attempts to find the displacement $\mathbf{q}_i$ that minimizes the Euclidean distances between the points of a given scan, rotated and translated by $\mathbf{q}_i$ and the points of another scan, treated as a reference. The point-to-point matching approach does not require the environment to contain predefined features. We apply a particular modification of this approach, the Point-to-Line ICP (PLICP),[3] which uses a distance provided by the normal to the locally defined line between scan points, and applies an exact closed-form solution for minimizing this distance. The PLICP is interesting for its robustness to outliers and speed of convergence. However, PLICP is less robust to large initial displacement errors, particularly with regard to the rotation, as shown in Fig. 2A and C. This is a problem in a walking robot application, as the robot often quickly changes the orientation of its trunk (the yaw angle) negotiating obstacles, while the internal sensors provide only very rough estimates with regard to the yaw angle and the horizontal displacement.

Because of the abovementioned issues of the PLICP algorithm we implemented for our system also another scan matching method, called Polar Scan Matching (PSM),[4] which avoids searching for point associations by simply matching points with the same bearing. The matching bearing rule makes PSM very fast, while it has also a large average area of convergence.[4] A drawback of PSM is a limited orientation estimation accuracy, as illustrated by experimental data in Fig. 2B and D.

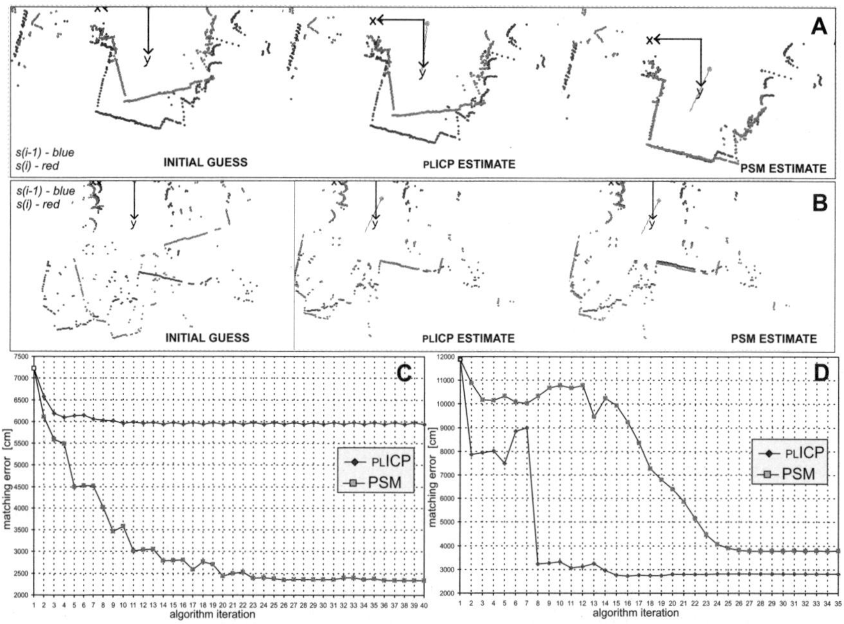

Fig. 2. Different properties of the PLICP and PSM algorithms (see text for explanation)

One can note that the strengths and weaknesses of the PLICP and PSM algorithms compensate each other, giving rise to the main idea of our scan matching system: use the PSM to compute an initial guess for $\mathbf{q}_i$, and then switch to PLICP for higher precision and robustness. Therefore, the scan matching based self localization works according to the following scheme:

(1) correction of the laser scanner data to account for the sensor-specific errors in range measurements, and elimination of the most unreliable range readings;
(2) correction of the range readings r to compensate the non-zero φ and ψ angles of the robot:

$$r' = r\sqrt{\cos^2\alpha\,(\cos\psi + \sin\varphi\sin\psi)^2 + \sin^2\alpha\cos^2\varphi}, \tag{1}$$

where r' is the corrected range, and α its corresponding bearing in the scan;
(3) scan pre-processing, which prior to matching removes clutter and remaining outliers by using the clustering approach described in Ref. 6, and eliminates occlusions by scan projection, as in Ref. 4;
(4) computation of the first guess of the robot displacement between the two poses at which the scans $s(i-1)$ and $s(i)$ were taken by using dead reckoning and data from internal sensors;
(5) computation of the improved rotation and translation estimate $\hat{\mathbf{q}}_i^{\mathrm{PSM}}$ by using the PSM algorithm with a fixed maximal number of iterations;
(6) computation of the final rotation and translation estimate at the i-th time instant $\hat{\mathbf{q}}_i$ by using the PLICP algorithm, starting from the estimate given by the PSM method;
(7) estimation of the displacement covariance matrix $\mathbf{C}_{q_i}$ by using the method from Ref. 2;
(8) computation of the robot's 2D pose estimate $\hat{\mathbf{x}}_{R(i)}$ with uncertainty $\mathbf{C}_R$ by compounding the local displacement estimates head-to-tail:

$$\hat{\mathbf{x}}_{R(i)} = \hat{\mathbf{x}}_{R(i-1)} \oplus \hat{\mathbf{q}}_i, \quad \mathbf{C}_{R(i)} = \mathbf{J}_{1\oplus}\mathbf{C}_{R(i-1)}\mathbf{J}_{1\oplus}^T + \mathbf{J}_{2\oplus}\mathbf{C}_{q_i}\mathbf{J}_{2\oplus}^T, \tag{2}$$

where $\oplus$ is the compounding operator, and $\mathbf{J}_{1\oplus}$, $\mathbf{J}_{2\oplus}$ are the Jacobians of this operation.

3. 2D scan matching adopted to a walking robot

The scan matching procedure is implemented on the insect-like hexapod robot Messor. It is equipped with various sensors, including a 6-d.o.f. ADIS 16350 IMU, which provides acceleration, rate-of-turn, and 3-axis angular orientation. The Hokuyo URG-04LX laser scanner for self-localization is mounted in an inverted position, in the front part of the trunk (pointed by an arrow in Fig. 1A), which, however, limits somewhat the angular field of view of the sensor. The URG-04LX suffers from dependency between the measured range accuracy and the received light intensity. Therefore, the range measurements are corrected by using a calibration model.[6] But the calibration does not eliminate corrupted range measurements, which are not suitable for scan matching.[10] Some of these errors are caused by the so-called mixed pixels, and they are removed by clustering the range measurements and analyzing discontinuities in the scanned sequence.[6] Erroneous range readings are also caused by pitch black, highly reflective or translucent surfaces. They can be eliminated by taking into account the reflection intensity output available in the URG-04LX sensor. Like in Ref. 5 we experimentally identified the range-intensity characteristics of the most common erroneous readouts, and we use thresholding in the range-intensity space to remove them, e.g., by throwing away measurements of very low intensity values or high intensity values with short range. We also discard maximum range readings.

Unlike a wheeled robot the legged machine does not maintain a constant attitude while walking, even on a flat surface. Therefore, the surrounding vertical surfaces are observed at slightly different heights in consecutive scans, which changes the range readings, as seen in Fig. 1C, where the current scan is shown in green, the previous one in blue, and the previous one without correction in red (these scans were taken in the setup shown in Fig. 1B). However, in a man-made environment we can assume that the surrounding walls have the same shape for all heights. Using this assumption we can correct the range measurements accounting for the changing attitude (cf. Fig. 1D). The attitude is obtained from the IMU, which is used merely as a two-axis inclinometer, because the yaw angle estimate is unreliable (there are no means to internally compensate its drift). Thus, the initial estimate of the robot displacement has to be obtained from the prediction of the motion based on the last executed command, but in a walking robot this method provides only a very rough estimate due to feet slippages. One approach to overcome this problem is to use the last pose estimate computed by the PLICP algorithm as the initial guess in the next matching cycle. However, this approach works well only if the displacement between the scans is small. Therefore, we use the last pose estimate from scan matching to reset the yaw angle in the IMU (to compensate the drift), and then use the IMU-based orientation estimate in the next cycle. This approach is facilitated by using the PSM algorithm, which tolerates much bigger initial orientation errors than the PLICP .

4. Experimental results

Figure 3 presents results of matching scans in a lab room, where the teleoperated robot followed an arbitrary trajectory. In this experiment the robot moved on a flat floor. If the PLICP algorithm was used without attitude correction (only the basic geometric scan pre-processing was applied to remove outliers) the results are not satisfactory, as seen from the poorly overlapping scans in Fig. 3A. Our method, combining two matching algorithms with more sophisticated scan pre-processing yielded much better results, as shown by Fig. 3B, also converging much faster (Fig. 3C).

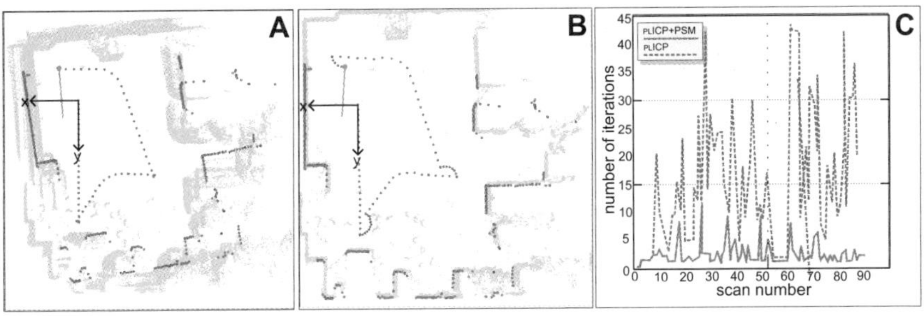

Fig. 3. Comparison of results for the PLICP algorithm and our method in a lab room

Figure 4 presents results of a self-localization experiment in a corridor-like environment, in which the teleoperated robot collected 310 laser scans. Small ramps were used to imitate to some degree the uneven terrain characteristics (cf. Fig. 1B). The trajectory computed only from the legged odometry and IMU measurements is practically useless, due to large orientation errors that arose when the robot was turning (Fig. 4A). Our scan matching algorithm enables to correct these errors, but if the changing attitude is not corrected large residual errors are present in the pose estimates, as seen in Fig. 4B, where the consecutive scans do not overlap in some areas. The proposed attitude correction decreases these errors significantly (Fig. 4C). Also the robot's trajectory (shown in blue) is estimated more precisely. In Fig. 4B some parts of the trajectory (example indicated by an arrow) are shortened, because the range data were corrupted by the changed robot attitude. The average time required to match two scans in this experiment was 0.02 s on a 1.8 GHz PC.

5. Conclusions

This paper presents a practical implementation of the 2D scan matching paradigm to the self-localization problem of a walking robot. Although

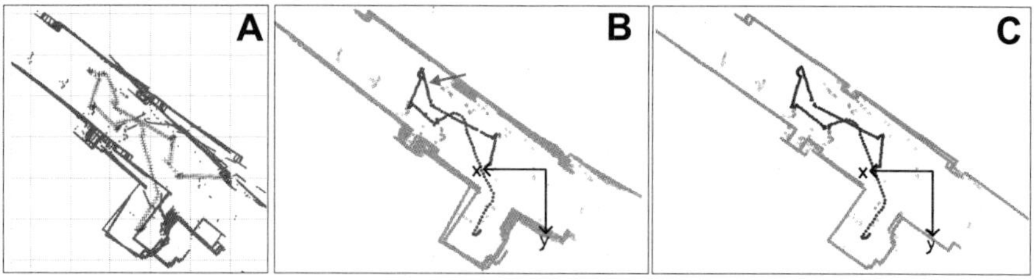

Fig. 4. Influence of the attitude correction on scan matching results

the localization algorithms used here are not new, the contribution lies in combining them in order to overcome problems specific to a legged platform, using only common affordable hardware. Our further work will focus on experiments in outdoor environments, with more realistic scenarios.

References

1. Borrmann D., Elseberg J., Lingemann K., Nüchter A., Hertzberg J. (2008) Globally Consistent 3D Mapping with Scan Matching, Robotics and Autonomous Systems, 56, 130–142.
2. Censi A., An Accurate Closed-Form Estimate of ICP's Covariance (2007) Proc. IEEE Int. Conf. on Robotics and Automation, Rome, 3167–3172.
3. Censi A., An ICP Variant Using a Point-to-Line Metric (2008) Proc. IEEE Int. Conf. on Robotics and Automation, Pasadena, 19–25.
4. Diosi A., Kleeman L. (2007) Fast Laser Scan Matching Using Polar Coordinates, Int. Journal of Robotics Research, 26(10), 1125–1153.
5. Hara Y., Kawata H., Ohya A., Yuta S. (2006) Map Building for Mobile Robots using a SOKUIKI Sensor – Robust Scan Matching using Laser Reflection Intensity, Proc. Int. Conf. SICE-ICASE, Busan, 5951–5956.
6. Łabęcki P., Rosiński D., Skrzypczyński P., (2010) Terrain Perception and Mapping in a Walking Robot with a Compact 2D Laser Scanner, Emerging Trends in Mobile Robotics, Singapore, World Scientific, 981–988.
7. Roennau A., Kerscher T., Ziegenmeyer M., Schamm T., Zoellner J., Dillmann R. (2009) Localisation of a Six-Legged Walking Robot in Rough Rerrain by ToF-Odometry, Proc. IARP Robotics for Risky Interventions and Environmental Surveillance (RISE), Brussels.
8. Schmidt A., Kasiński A. (2010) The Visual SLAM System for a Hexapod Robot, Computer Vision and Graphics, Berlin, Springer 2010, 260–267.
9. Singh S., Waldron K. (2005) Attitude Estimation for Dynamic Legged Locomotion Using Range and Inertial Sensors, Proc. IEEE Int. Conf. on Robotics & Automation, Barcelona, 1663–1668.
10. Skrzypczyński, P. (2008) On Qualitative Uncertainty in Range Measurements from 2D Laser Scanners, JAMRIS, 2(2), 35–42.
11. Stachniss C., Bennewitz M., Grisetti G., Behnke S., Burgard W. (2008) How to Learn Accurate Grid Maps with a Humanoid, Proc. IEEE Int. Conf. on Robotics & Automation, Pasadena, 3194–3199.

DYNAMIC CALIBRATION OF A GYROSCOPE USING A COMPOUND PENDULUM

KIAN SEK TEE[*], MOHAMMED AWAD[*+], ABBAS DEHGHANI[*]

[*]*Institute of Engineering Systems and Design, School of Mechanical Engineering, University of Leeds, Leeds, LS2 9JT, United Kingdom.*

[+]*Mechanical Engineering, Faculty of Engineering, Ain Shams University, Cairo, Egypt.*

DAVID MOSER, SAEED ZAHEDI

Chas A Blatchford & Sons Ltd, Lister Road, Basingstoke, Hampshire, RG22 4AH, United Kingdom.

A novel method of dynamic calibration of a MEMS type gyroscope is presented in this paper. A compound pendulum installed with a potentiometer as angular and angular speed reference was used to justify the scale factor and zero bias of a gyroscope. Two gyroscopes, i.e. IDG300 and IDG500, were studied. It was found that scale factor of IDG300 matched with the manufacturer's datasheet while IDG500 did not match.

1. Introduction

Microelectromechanical systems (MEMS) type accelerometers and gyroscopes are getting popular in all sort of motion relevant applications [1-4] such as virtual sport games, gait analysis, aviation etc. There are a number of papers reporting different techniques of accelerometer static calibration [5-10] such in-field calibration and conventional rotary table. No researchers have reported specifically the dynamic calibration of a gyroscope except expensive laboratory gimbal [1, 11]. Two important parameters are the scale factor and zero bias. In normal practice, since gyroscope is non-ratiometric, manufacturer scale factor is used without further scrutinizing the accuracy of the specification while zero bias is the average voltage during static. In this paper, two gyroscopes are compared to a reference angular speed in a test rig built of a pendulum, a potentiometer as an electronic protractor as well as tachometer and a customized data logger sampling at 200Hz.

2. Methodology

2.1. *Test Rig*

The test rig as shown in Figure 1 consists of a compound pendulum installed with a potentiometer as angular and angular speed transducer, a customized data logger, an IMU-5DOF and a stopper to initiate consistent release angle. A electronic spirit level is used to align the test rig platform horizontally.

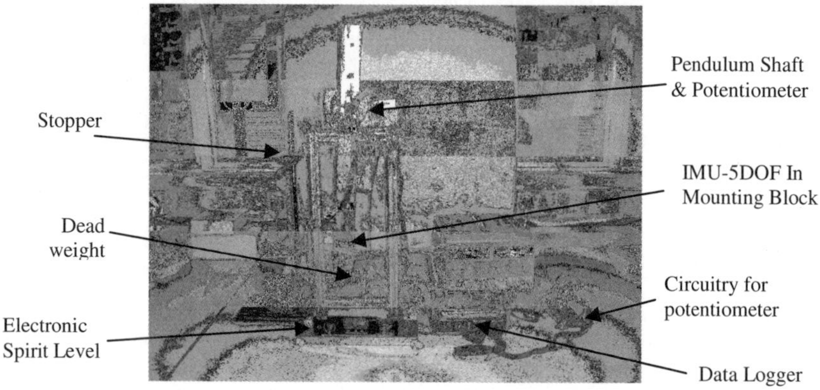

Figure 1: System Test Rig

2.2. *Gyroscopes*

Two types of IMU-5DOF (SparkFun, Inc.) in the form of PCB-breakouts were used as shown in Figure 2. In this paper, gyroscopes (XR-axis, YR-axis), i.e. IDG300[12] and IDG500[13] (InvenSense, Inc.) were studied and compared.

(a) (b) (c)

Figure 2: (a). ADXL330 & IDG300, (a). ADXL335 & IDG500, (c). Axes Assignment

2.3. *Experiment Protocol*

A differential op-amp was used to build the main circuitry for the potentiometer. Its output was fed into the data logger. Firstly, the potentiometer is calibrated as an electronic protractor over the angle range -40^0 to +40^0. A regression equation of voltage versus angles was modeled. For each trials, the gyroscope was aligned in a specific plane such as Y-Z plane for XR axis or Z-X plane for YR axis. For convenient purpose, Z-X plane (YR axis) is explained throughout the paper. The pendulum was released at -40^0 specified by the stopper. Data logging starts from the release angle until the pendulum stops gradually. Repetitive trials were taken. The same procedures were repeated for both gyroscopes in two planes.

3. Results

The potentiometer was statically calibrated and the average result of five trials were plotted as shown in Figure 3. Maximum standard deviation, 0.011178 was observed throughout the trials, indicating highly repeatability in the device.

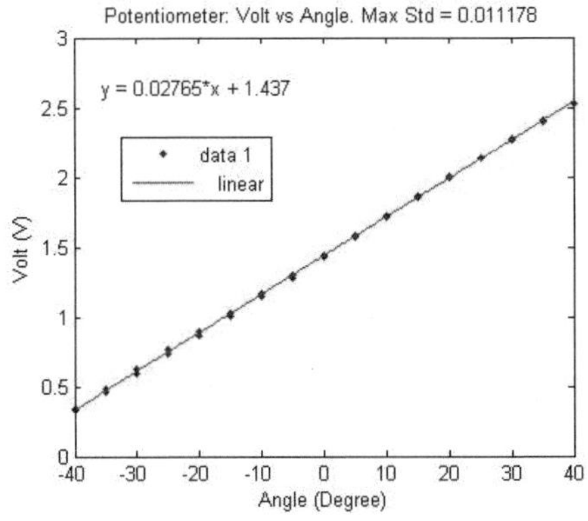

Figure 3: Static Calibration of Potentiometer

The gyroscope output of IDG300 and IDG500 in volt and its corresponding pendulum angular speed in deg/s are shown in Figure 4. For comparison purpose, zero phase delay FIR low pass filter (ZPLP) was applied to pendulum angular speed. For the same release angle, IDG500 exhibits inverse output and nearly half the amplitude as compared to IDG300.

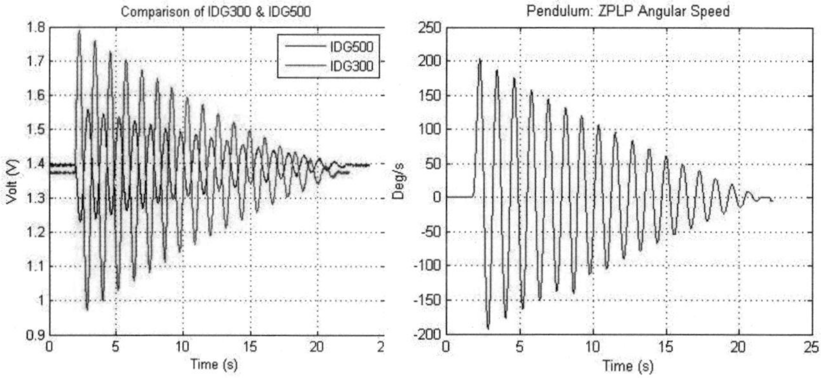

(a) (b)

Figure 4: (a) Angular Speed – YR, (b) Angular Speed - Pendulum

The gyroscope in any axis is formulated as Eq. (1).

$$V = K_c\omega_p + ZB_{av} \tag{1}$$

where V is gyroscope output in volt, K_c is scale factor (mV/^{0}/s) to be determined, ω_p is pendulum angular speed in unit of ^{0}/s and ZB_{av} is average voltage of the gyroscope output. As shown Figure 5(a), scale factor, K_c was calculated and plotted. However, K_c is not symmetrically distributed due to vibration at the start and the end. A frequency distribution method is proposed. The scale factor, K_c is located at the maximum frequency.

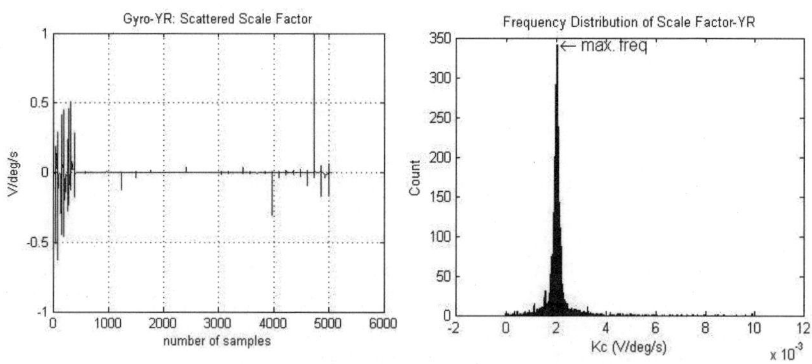

(a) (b)

Figure 5: (a). Gyro-YR scale factor in a trial, (b) Frequency Distribution

Table 1: Comparison of Scale Factor (mV/°/s)

Specification	Calibrated IDG300	Calibrated IDG500
0.002	0.0019	0.0.00083

Calibrated scale factor, K_c of IDG300 and IDG500 were compared to the manufacturer specification. IDG300 performs near to the given specification while IDG500 is less than half of the specification.

4. Discussion

The range of the pendulum angular speed can be slightly increased by releasing at a bigger angle. Adding dead weights at the end of the pendulum would reduce the damping ratio, meaning that the pendulum will swing longer at slower damping rate. The pendulum is working on the gravity and it is not easy to acquire higher speed range without major mechanical design changes. It is most convenient and cheap for calibration of low range of angular speed such as for the application of gait analysis.

5. Conclusion

MEMS type gyroscope is popular in many applications. A simple dynamic calibration was proposed using a compound pendulum to identify its scale factor and zero bias. Two gyroscopes are calibrated and their differences with respect to the manufacturer's specification were compared. It was found that IDG500 did not match the specification and a new scale factor was proposed.

Acknowledgments

The collaborating author (M. Awad) would like to acknowledge the Egyptian Government and Egyptian Cultural Bureau in the UK for their financial support.

References

[1] D. H. Titterton and J. L. Weston, "Strapdown Inertial Navigation Technology," 2 ed: Institution of Engineering and Technology 2004.

[2] M. Brodie, A. Walmsley, and W. Page, "Fusion motion capture: a prototype system using inertial measurement units and GPS for the biomechanical analysis of ski racing," *Sports Technology,* vol. 1, pp. 17-28, 2008.

[3] J. Favre, B. M. Jolles, R. Aissaoui, and K. Aminian, "Ambulatory measurement of 3D knee joint angle," *Journal of Biomechanics,* vol. 41, pp. 1029-1035 2008.

[4] J. Rueterbories, E. G. Spaich, B. Larsen, and O. K. Andersen, "Methods for gait event detection and analysis in ambulatory systems," *Medical Engineering & Physics,* vol. 32, pp. 545-552, 2010.

[5] W. T. Fong, S. K. Ong, and A. Y. C. Nee, "Methods for in-field user calibration of an inertial measurement unit without external equipment," *Measurement Science and Technology,* vol. 19, pp. 1-11, 2008.

[6] J. C. Lötters, J. Schippe, P. H. Veltink, W. Olthuis, and P. Bergveld, "Procedure for in-use calibration of triaxial accelerometers in medical applications," *Sensors and Actuators A: Physical,* vol. 68, pp. 221-228, 1998.

[7] I. Skog and P. Handel, "Calibration of a MEMS Inertial Measurement Unit," in *XVII IMEKO WORLD CONGRESS, Metrology for a Sustainable Development, September, 17-22, 2006 Rio de Janeiro, Brazil,* 2006.

[8] F. Ferraris, U. Grimaldi, and M. Parvis, "Procedure for Effortless In-Field Calibration of Three-Axis Rate Gyros and Accelerometers," *Sensors and Materials,* vol. 7, pp. 311-330, 1995.

[9] J. C. Hung, J. R. Thacher, and H. V. White, "Calibration of accelerometer triad of an IMU with drifting Z -accelerometer bias," *Aerospace and Electronics Conference, 1989. NAECON 1989., Proceedings of the IEEE 1989 National,* vol. 1, pp. 153 - 158, 1989.

[10] A. Kim and M. F. Golnaraghi, "Initial calibration of an inertial measurement unit using an optical position tracking system," in *Position Location and Navigation Symposium, 2004. PLANS 2004*: IEEE, 2004, pp. 96 - 101

[11] K. R. Britting, *Inertial Navigation Systems Analysis*: Wiley-Interscience, 1971.

[12] IDG300, "Integrated Dual-Axis Gyro," InvenSense, Inc., 2006.

[13] IDG-500, "Integrated Dual-Axis Gyro," InvenSense, Inc., 2008.

A PROBABILISTIC MODEL OF EQUILIBRIUM PERCEPTION

S. N'Guyen*, J. Droulez and A. Berthoz
UMR 7152 Collège de France - CNRS
75005, Paris, France
**E-mail: steve.nguyen@college-de-france.fr*

The objective of this work is to show how sensors can be fused to detect perturbations on a humanoid robot. We propose a probabilistic model taking into account uncertainties in sensors readings and in the model. Under basic assumptions, redundancy becomes a linear statistical correlation, and unexpected situations can be revealed by large (statistically significant) deviations. We then describe a series of experiments conducted to learn the correlation coefficients and to validate the ability to detect perturbations. Finally, we will discuss several solutions to further improve this model.

1. Introduction

To detect and avoid equilibrium perturbations is of primary importance in robotics and especially in humanoid robotics. Bipeds are fundamentally unstable and one obviously would like to prevent any fall. To do so, some indicators have been used such as the famous Zero Moment Point (ZMP), but majority of ZMP-based controllers for balance maintenance only relies upon joint and ground reaction force feedback assuming an horizontal, stiff and non-slippy ground[1,2] while humans extensively relies upon multiple sensors and can deal with very wide range of perturbations.[3,4]

Some bio-inspired multi-sensory model for postural control have been tested in both simulation[5,6] and robotic platforms[7,8] being able to quite well replicate humans behavior.

Following Mergner et al. (2009) we propose an external disturbance estimation model intended to be used to orchestrate further disturbance rejection. Moreover, uncertainties on sensor readings and on model imperfections are rarely taken into account. So the aim of this work is to introduce a probabilistic framework taking advantages of sensory redundancies in order to detect perturbations.

2. Deterministic model

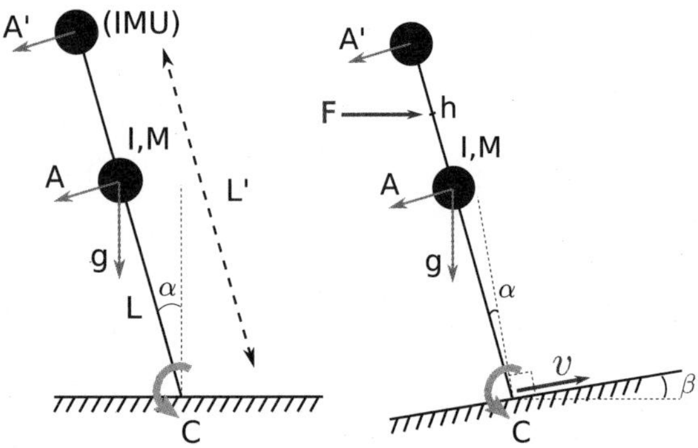

Fig. 1. Inverted pendulum model. C is the torque, α the angle of the pendulum, A and A' are respectively center of mass (of mass M and inertia momentum I) and Inertial Measurement Unit (IMU) accelerations. Under external perturbations (right), F is a force directly applied at height h, β is the angle of the slanted ground and v is the slip velocity.

Considering the robot as a simple perfectly rigid inverted pendulum (cf. Fig. 1) dynamics can be described by the following equation:

$$(I + ML^2)\ddot{\alpha} = C + M.g.L.sin(\alpha) \tag{1}$$

As the linear acceleration of the IMU A' is given by $A' = L'\ddot{\alpha} - g.sin(\alpha)$, we can express the coupling between A', C and α as following:

$$(I + ML^2)A' - L'C + [I - (L' - L)ML]g.sin(\alpha) = 0 \tag{2}$$

Notice that all these equations can be extended to the two dimensional case by replacing the scalar values by corresponding vectors.

3. Unexpected situation

In a "perfect" world (i.e. without any perturbation) equation 2 is always true. However in a "real" world perturbations can occur, then violating this coupling.

In case of horizontal external force, slanted ground or slippy surface, an extra torque will appear.

3.1. *External force*

Supposing that an external horizontal force F is exerted on the robot at height h, thus producing an extra torque Fh and an additional acceleration (cf. Fig. 1, right). We now have:

$$(I + ML^2)A' - L'C + [I - (L' - L)ML]g.sin(\alpha) = FhL' \qquad (3)$$

Unfortunately it is impossible to determine the height h without any additional sensor (e.g. tactile).

3.2. *Slanted ground*

It was previously stated that α was the angle of the robot with respect to the gravity (horizontal ground). But angle sensor here can only measure angle of the robot with respect to a the direction normal to the ground. If ground is now slanted by an angle β (cf. Fig. 1, right), body angle with respect to the gravity becomes $\alpha + \beta$ involving:

$$(I + ML^2)A' - L'C + [I - (L' - L)ML]g.sin(\alpha + \beta) = 0 \qquad (4)$$

Thus, β can be extracted in order to make any corrective movement:

$$\beta = arcsin\Big(\frac{L'C - (I + ML^2)A'}{(I - (L' - L)ML)g}\Big) - \alpha \qquad (5)$$

3.3. *Slippy surface*

Finally, if a slipping motion occurs, body acceleration A would not be anymore proportional to $\ddot{\alpha}$. Considering v as the slip velocity (with $v > 0$ for feet moving to the right, cf. Fig. 1, right) we have $A' = L'\ddot{\alpha} - g.sin(\alpha) - \dot{v}.cos'(\alpha)$, thus giving from equation 2:

$$\dot{v}.cos(\alpha) = \frac{L'}{L}.\frac{C + M.g.L.sin(\alpha)}{I + ML^2} - A' - g.sin(\alpha) \qquad (6)$$

4. Probabilistic model

Unfortunately, equation 2 is never really true due to sensor noise and model imperfections, producing a "residual torque" R. Probabilistic framework is a natural way to take into account these uncertainties.

Then the general model can be described by the following joint probability:

$$P(C)P(A')P(\alpha)P(R|C, A', \alpha)P(Feet|C)P(Imu|A')P(Enc|\alpha) \qquad (7)$$

giving us a general probabilistic model with $Feet$, Imu and Enc being the observations of C (torque given by feet pressure sensors), A' (total head acceleration given by IMU) and α (ankle angle given by joints encoders) respectively.

We can now decide to parametrize this model.

4.1. *Sensor models*

Knowledge one can extract about a relevant physical variable (state variable) from sensor reading (observation) can be modeled as probability distribution.

Assuming Gaussian distributions, we obtain the following likelihood functions:

- $P(Enc|\alpha) = N(\alpha, \sigma_\alpha)$
- $P(IMU|A') = N(A', \sigma_{A'})$
- $P(Feet|C) = N(C, \sigma_C)$

4.2. *Detection of perturbations*

Of course as stated previously, in the real world, even without any perturbation eq. 2 is never perfectly exact. The "residual torque" R can be formalized in the model:

$$R = C - \frac{I + ML^2}{L'}A' - \frac{[I - (L' - L)ML].g}{L'}sin(\alpha) \tag{8}$$

with $P(R|C, A', \alpha) = N(\epsilon, \sigma_R)$.

After marginalization and small angle approximation we obtain $P(R|Enc, IMU, Feet) = N(\mu, \sigma)$ with:

$$\mu = \epsilon + Feet - \frac{(I + ML^2)}{L'}Imu - \frac{[I - (L' - L)ML].g}{L'}Enc$$

$$\sigma^2 = \sigma_R^2 + \sigma_C^2 + \frac{(I + ML^2)^2}{L'^2}\sigma_{A'}^2 + \frac{[I - (L' - L)ML]^2.g^2}{L'^2}\sigma_\alpha^2$$

thus giving a linear correlation between the observations plus an additional bias ϵ.

Assuming the correlation coefficients computed or estimated (according to previous equations), we can now easily compute the torque residue in real time:

$$r(Feet, Imu, Enc) = \epsilon + Feet - k_{IMU}Imu - k_{Enc}Enc \tag{9}$$

This residue can then be used to detect "abnormal" situations while in a perfect inverted pendulum case (without any perturbation) this residue should be equal to 0.

5. Results

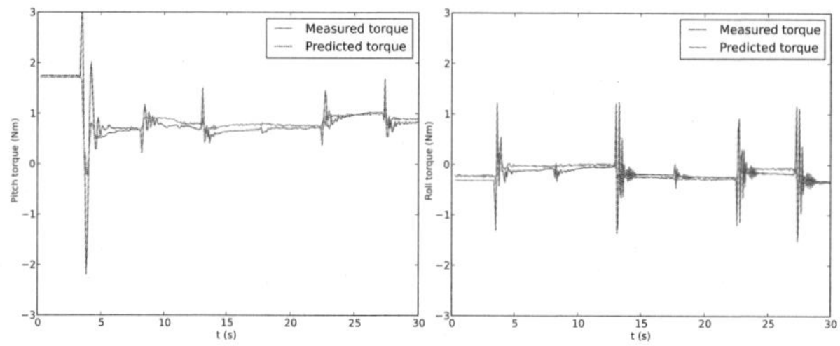

Fig. 2. Comparison between measured torque (using FSR) and predicted torque during random movements, using IMU and ankle angle for Pitch (left) and Roll (right).

We tested these principles on the humanoid robot Nao from Aldebaran Robotics. We first learned correlation coefficients by simply producing random movements (in a limited range to avoid falls) of the ankle for Pitch and ankle+hip for Roll (keeping a parallelogram constraint for foot and pelvis). The following table shows the result of the correlation.

For pitch movement R^2 and the RMSE of regression indicate a good correlation between sensors with a good prediction of linear regression. Roll movements produce a worse R^2 indicating that the linear regression doesn't really "explain" all the variance, while keeping a relatively low RMSE. One explanation of this behavior can be seen on Figure 2, where a time plot of predicted and measured torques are compared. We can see for the Roll that high amplitude high frequency oscillations during movements are not "captured" by the prediction. Experimentally, these variations seems to be provoked by small foot-ground detachment. For the following postural experiments these transient variations can be ignored.

	R^2	RMSE (Nm)
Pitch	0.721424	0.1780744
Roll	0.243210	0.2098518

5.1. *External force*

According to equation 9 we can compute the residual torque r and then detect any perturbations such as external "pushes". Knowing from section 5

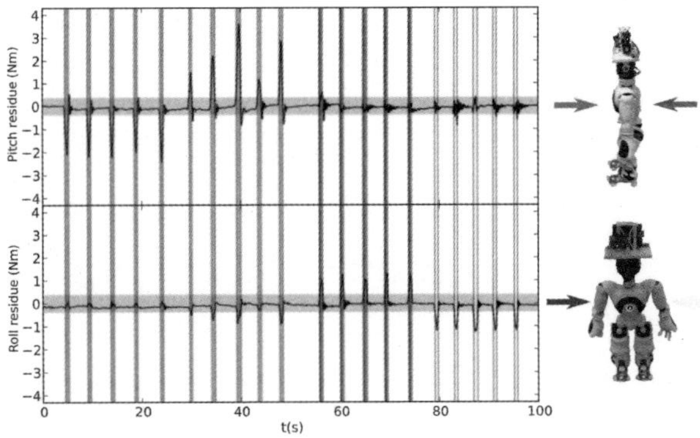

Fig. 3. Evolution of the measured residue while applying external "pushes". Colored vertical bars indicate the detected "push" events. Horizontal gray area represents 2σ around the mean residual value.

the standard deviation of r from 0 (σ), one can consider a "push" event with a simple *ad-hoc* threshold when r crosses $\pm 2\sigma$ (95% chance of perturbation).

Figure 3 shows example of detected perturbations with different orientations in Pitch and Roll.

5.2. *Slanted ground*

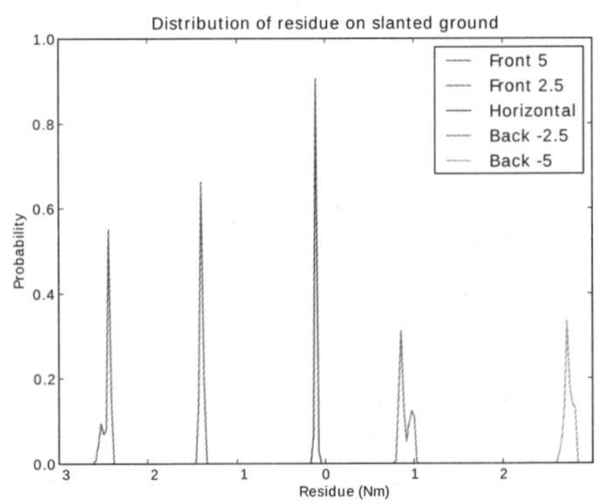

Fig. 4. Measured distribution of residue while the robot is placed on different slopes.

Ground slope detection and estimation is also a possibly very useful feature for postural stability. Accumulation of the residual torque during a period of time (or looking at the DC value) should inform of any "constant" slope of the ground.

Figure 4 shows obtained distributions of the residue on different slopes on the Pitch direction (both forward and backward). Knowing all the parameters of the model one should be able to compute an estimation of the slope angle β thanks to equation 5.

Notice that indicated slope values are only approximations due to imprecision of the experimental setup.

5.3. *Slippy surface*

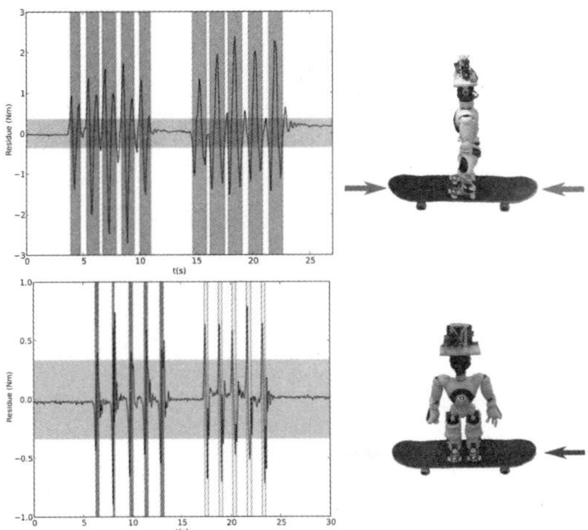

Fig. 5. Evolution of the measured residue while simulating slippy surface along the antero-posterior (top) and lateral (bottom) directions. Colored vertical bars indicate the detected "slide" events. Horizontal gray area represents 2σ around the mean residual value.

Finally, we simulated effects of a slippy surface by placing the robot on a skateboard. Rapid displacement of the skateboard induces a velocity of the robot's foots somewhat similar to a slip velocity.

Figures 5 shows measured residual torque for antero-posterior (top) and lateral (bottom) simulated slip.

6. Discussion and future work

Qualitative results shown in this work seem to validate the usability of a very simple mechanism for disturbance detection. As we only considered the inverted pendulum simplification, this system cannot account for knees or trunk movements hence it has now to be extended.

We introduced a probabilistic framework able to detect perturbations, but using this formalism we should also be able to deal with priors in order to estimate their origins. This disambiguation could be done using some assumptions about occurrence of perturbation. E.g. ground surface (slope and sliding) can be considered as relatively stationary (between each step), eventual external forces can be detected by vision. Future work in this direction could enhance robots capabilities and help to explain the so-called sensory reweighting[5,10] observed in humans.

Finally, the proposed perception model can be integrated with a probabilistic control model in order to take into account uncertainties in perception and consequences of actions.

Acknowledgments

This work has been funded by project ROMEO (French government) and by EU project RoboSom (ICT-2009.2.1).

References

1. K. Kaneko, F. Kanehiro, S. Kajita, M. Morisawa, K. Fujiwara, K. Harada and H. Hirukawa, Slip observer for walking on a low friction floor, in *Intelligent Robots and Systems, 2005.(IROS 2005). 2005 IEEE/RSJ International Conference on*, (IEEE, 2005).
2. M. Morisawa, F. Kanehiro, K. Kaneko, N. Mansard, J. Sola, E. Yoshida, K. Yokoi and J. Laumond, Combining suppression of the disturbance and reactive stepping for recovering balance, in *Intelligent Robots and Systems (IROS), 2010 IEEE/RSJ International Conference on*, (IEEE, 2010).
3. R. J. Peterka, *Journal of physiology, Paris* **103**, 149 (2009).
4. T. Mergner, *Annual Reviews in Control* **34**, 177(December 2010).
5. T. Mergner, C. Maurer and R. J. Peterka, *Progress in brain research* **142**, 189(January 2003).
6. C. Maurer, T. Mergner and R. J. Peterka, *Experimental brain research* **171**, 231(May 2006).
7. A. Mahboobin, P. J. Loughlin, M. S. Redfern, S. O. Anderson, C. G. Atkeson and J. K. Hodgins, *Neural networks : the official journal of the International Neural Network Society* **21**, 621(May 2008).
8. K. a. Tahboub, *Journal of physiology, Paris* **103**, 195 (2009).
9. T. Mergner, G. Schweigart and L. Fennell, *Journal of physiology, Paris* **103**, 178 (2009).
10. R. Peterka, *Journal of Neurophysiology* , 1097 (2002).

AN AUTONOMOUS SYSTEM FOR GROUND-BASED CLOUD TRACKING USING THE MEAN-SHIFT ALGORITHM

M.D. O'Toole*, S.A. Wormald, D. Kerr, J. Coupland

Wolfson School, Loughborough University,
Leicestershire, LE11 3TU, UK
** M.D.OToole@lboro.ac.uk*

A.P. Sandford

Campbell Scientific Ltd, Loughborough, Leicestershire, LE12 9GX, UK

Increasing desire for automation in weather reports generated by air-traffic control centers and meteorological observatories has created the opportunity for autonomous system to identify and track clouds as they move across the sky. Essential to the operation of the apparatus is a robust algorithm capable of identifying and isolating individual clouds and tracking their motion. In this paper, we describe a procedure for the tracking of clouds using the mean-shift algorithm and the Hampel filter. The procedure is demonstrated on a sequence of infrared images, taken at our facility, of a cumulus cloud traversing the sky.

Keywords: mean-shift, autonomous tracking, meteorology, remote sensing

1. Introduction

Clouds play an influential role on the dynamics of the global climate by modulating the amount of incoming an outgoing radiation. Consequently, knowledge of cloud conditions is a key factor in accurate climate modeling and prediction. The characteristics of the clouds in any given locality are also indicative of local weather phenomena, or significantly, weather phenomena that is likely to occur.[1] The identification of cloud formations and sky-scenes can provide important supporting information for weather forecasting over short-ranges, or nowcasting. Airports are required to report cloud vertical extent, cloud cover, and the presence of cumulonimbus and towering cumulus clouds to all local air traffic via METARS. The reporting of cumulonimbus and towering cumulus clouds is particularly important. These clouds develop from powerful atmospheric updrafts and downdrafts, which can pose a hazard to incoming or outgoing aircraft.

There are many macroscopic properties that are of interest when observing clouds, such as cloud cover, cloud-type, cloud-base height, cloud height, cloud formation, shape, and wind-speed at differing heights. Some of these properties can be obtained directly from satellite coverage. However, satellites can only return images with a low-spatial resolution, and high-altitude objects, such as high cloud layers, can occlude formations at lower altitudes. Ground-based sensors such as ceilometers can supplement satellite images with direct atmospheric measurements, however, the sensors cannot return a complete picture of the sky-conditions. For the most part, local ground-based cloud observation must be performed by a network of human observers, which are expensive and can result in inconsistent and subjective information insufficient for numerical weather models.

Autonomous observation systems using visual or infrared cameras to capture the key characteristics of a sky-scene are a potential alternative to human observation. For example, cloud classification systems have been proposed using texture based pattern recognition algorithms.[2,3] Autonomous nowcasting systems have been proposed that use satellite data to generate short-term storm forecasts.[1,4] However, their sucess-rate has been limited thus far. Cloud tracking has been demonstrated to obtain wind-field vectors by cross-correlating sucessive infrared images[5] and by using photogrammetric techiques with visible cameras.[6]

In this paper, we demonstrate an autonomous system for tracking cloud motion. This is the first component of research towards an autonomous observation system that aims to identify, characterize and quantify a sky-scene by producing a continually updating, three-dimensional virtual-map of the local cloud formations. Cloud motion is useful for local short-term weather prediction, for quantizing and formalizing data on the evolution of the sky-scenes, and for determine wind field velocity vectors at different heights.

2. Instrumentation

Experiments are performed using our infrared camera system based on site at Campbell Scientific Ltd, Loughborough, UK. The system consists of a two-degree-of-freedom (azimuth and elevation) pan-tilt unit (IDesignC-CTV, UK) mounted on a 10m high mast. The pan-tilt unit holds an infrared camera (Avio, NEC) operating in the 8-14μm range, with 22.1^o (horizontal) and 16.2^o (vertical) field of view. The unit and camera are controlled via an Ethernet connection from a work-station running Matlab (MathsWorks, Inc).

3. Method

In this section, we describe our method for tracking of cloud formations using the infra-red camera system. An infrared image is taken of a target portion of sky. The image is segmented into separate cloud and sky segments using a temperature threshold. The cloud segments are projected (mapped) onto a three-dimensional real-world coordinate system, referred to henceforth as the cloud-space. Velocity vectors are computed by tracking the clouds using the mean-shift algorithm and by using a Hampel filter[7] to filter the results.

Some notation is introduced: Denote $I(u, v)$ as an $M \times N$ image and (u, v) as the pixel position, taken from a reference frame with origin in the top left corner of the image, with u going positively downwards and v going positively rightwards. Define (x, y, z) as a set of spatial coordinates on a 'real-world' coordinate frame with z as height.

3.1. *Cloud-Space Transform*

The image I is projected onto a cloud-space - that is, an approximate 3-dimensional real-world coordinate system. Each pixel is assumed to be a representative packet of air. The height of the air packet z is calculated using using the adiabatic lapse rate and pixel temperature,

$$z = (T_A - T) / (\partial T / \partial z) \tag{1}$$

where $(\partial T/\partial z)$ is the adiabatic lapse rate, T is the pixel temperature, and T_A is the ambient ground temperature. The 3-dimensional spatial position (x, y, z) of the air packet is calculated using its height and the known angular position (azimuth and elevation), inferred from the pan-tilt system sensors and the pixel location in the image. The projection is performed on all pixels in the image to populate the cloud-space.

3.2. *Mean-Shift Algorithm*

The mean-shift algorithm is simple, robust, and versatile technique for finding local modes of a data-set and for clustering data-points into membership sets. In this section, we briefly describe the mean-shift algorithm, following the method proposed by Comaniciu and Meer[8,9] with some variations. We then discuss the adaption of this algorithm to the tracking of clouds across an approximate construct of a real-world space.

Let us denote a data set of n data points $\mathbf{x}_i \in \mathbb{R}^d, i = 1, 2, ..., n$, where d is the dimension of the space. For some point $\mathbf{y}_j$ in the d-dimensional space, the successive mean-shift coordinate $\mathbf{y}_{j+1}$ is computed by,

$$\mathbf{y}_{j+1} = \frac{\sum_{i=1}^{n} \mathbf{x}_i w_i g\left(\left\|\frac{\mathbf{y}_j - \mathbf{x}_i}{h}\right\|^2\right)}{\sum_{i=1}^{n} w_i g\left(\left\|\frac{\mathbf{y}_j - \mathbf{x}_i}{h}\right\|^2\right)} \tag{2}$$

where $g(x)$ is the Epanechnikov kernel profile defined as:

$$g(x) = 1 - x^2 \tag{3}$$

w_i is some weighting parameter, and h is some chosen scalar bandwidth value. The center of a local density maxima $\mathbf{y}_k$ is computed by successive iteration of equation 2 for $j = 1, 2, ..., k$ until $y_k \approx y_{k+1}$.

For the present work, it is sufficient to set the dimension of the space to three ($d = 3$). We determine the set $\{\mathbf{x}\}_N = \{\mathbf{x}_i, i = 1, 2, ..., n\}$ by finding the nth nearest neighbors to $\mathbf{y}_j$. The nearest neighbors are found efficiently using a kd-tree structure.[10] The bandwidth parameter is chosen as $h = \max\left(\|\{\mathbf{x}\}_N\|\right)$. Using nearest neighbors as opposed to a fixed spatial range guarantees good performance of the mean-shift algorithm, even in sparsely populated regions of the space.

To track the cloud, the mean-shift algorithm weighting parameter is set according to the cloud texture, or more specifically, the spectral profile of the image returned by a band-pass filter. Denote $f(I(u, v))$ as the band-pass filtered image. The pixels in $f(I(u, v))$ are each associated with a data-point in the cloud map. Consider the nearest neighbor set $\{\mathbf{x}\}_N$ obtained at each iteration of the meanshift algorithm. Every data-point in $\{\mathbf{x}\}_N$ is associated with a value in $f(I(u, v))$. We create a discrete histogram of the values of $f(I(u, v))$ for the points in $\{\mathbf{x}\}_N$,

$$q_c = \frac{\sum_{i=1}^{n} \delta\left[b(\mathbf{x}_i) - c\right]}{n}, \quad c = 1, 2, ..., m \tag{4}$$

We refer to this as the target model. Now consider the same point in space after some small time step. Using the same histogram bins, we compute a second histogram p_c for the new nearest neighbor set of the new filtered image. The weights for the meanshift algorithm are function of the similarity between the two histograms:

$$w_i = \sum_{c=1}^{m} \delta\left[b(\mathbf{x}_i) - c\right] \sqrt{\frac{q_c}{p_c}} \tag{5}$$

The derivation of equation 5 is given Comanciu et al.[9] Thus with the weights included, the mean-shift algorithm is extended so that it will move towards the point of maximum density most similar to the target model.

We propose the following process for cloud tracking using the weighted mean-shift algorithm: A set of k data points are chosen randomly from the data-points contained in the largest segment on the cloud-map and the mean-shift weights are initialized to unity. Each of the k data points is mean-shifted to find a local point of maximum density. We refer to this point as a cluster. Any of the data-points with identical or similar clusters are culled, so that the remaining clusters represent an even spread of local maximum densities. We denote the set of clusters $C_- = \{\mathbf{x}_{l,-} \in \mathbb{R}^3, q_{c,l} \in \mathbb{R}^m, l = 1, 2, ..., k\}$ where $\mathbf{x}_{l,-} = \left[x \ y \ z\right]^T$ is the spatial coordinate and $q_{c,l}$ is the respective spectral histogram of each cluster, as described by equation 4.

A subsequent cloud-space is produced from the image taken with the camera pointing in the same position but some time δt later. The positions and histogram of the clusters in the previous image $\mathbf{x}_{l,-}$ are used as the starting positions for the mean-shift operation to find a new set of clusters (of maximum local density) in the new image. We denote the new set of clusters $C_+ = \{x_{l,+} \in \mathbb{R}^3, q_{c,l} \in \mathbb{R}^m, i = 1, 2, ..., k\}$ resulting from repeated mean-shift operation. A velocity vector is obtained for each cluster $v_l = (\mathbf{x}_{l,+} - \mathbf{x}_{l,-})/\delta t$. We then set $\mathbf{x}_{l,-} = \mathbf{x}_{l,+}$ and repeat the process for the next image.

3.3. *Filtering*

In ideal circumstances, the velocity vectors for the clusters would be sufficient to give a velocity vector for the cloud. However in practice, the computation of velocity vectors is complicated by image noise, mast wobble induced by the wind, and the occasional jostling and leaping of the clusters in regions with even density or minimal texture. We propose a regime of data-treatment to obtain a robust single velocity vector at each time-step:

Before mean-shifting, each pixel in the image sequence is filtered in the time domain. A second-order Butterworth digital filter is used with the z-domain form:

$$I_f(u, v) = \frac{0.0675z + 0.1349z^{-1} + 0.0675z^{-2}}{z - 1.143z^{-1} + 0.4128z^{-2}} I(u, v) \tag{6}$$

where $I(u, v)$ is the original image obtained from the camera and $I_f(u, v)$ is the filtered image used in the mean-shift tracking.

The velocity vectors $v_l(\tau)$ of each cluster at each time-step τ are filtered using a Hampel filter; a nonlinear filter which can smooth signals and reject outliers caused by sudden jumps or leaps of the clusters between regions of high density. The filter is based on the median distance from the median. Define $\tilde{V}_l = \{v_l(\tau - N_w), v_l(\tau - N_w + 1), ..., v_l(\tau + N_w)\}$ as a window of size $2N_w$ taken from the sequence of velocity vectors of a single cluster. The filter takes the form of the following inequality for the τth sample,

$$\left|v_l(\tau) - \text{median}(\tilde{V}_l)\right| \geq \alpha\, \text{median}\left(\left|v_l(\tau) - \text{median}(\tilde{V}_l)\right|\right) \qquad (7)$$

where α is some design constant or function. If the condition is satisfied then the filtered velocity vector becomes $\hat{v}_l(\tau) = \text{median}(\tilde{V}_l)$ otherwise $\hat{v}_l(\tau) = v_l(\tau)$. An overall filtered velocity vector $V(\tau)$ is obtained by taking the median of the filtered velocity vectors of the multiple clusters, i.e. $V(\tau) = \text{median}(\{\hat{v}_l(\tau), l = 1, 2, ...N\})$.

4. Results and Discussion

Cloud tracking was tested on a sequence of images taken with a sampling rate of 2Hz at an elevation angle of twenty degrees from the horizontal. The image sequence is shown in figure 1. The sequence contains one large cumulus cloud traveling across the camera. Wind conditions were moderate to high. The following parameters were chosen for the tracking procedure: The adiabatic lapse rate for the cloud-space transform is set to the aviation standard of 6.8. The assumption of a constant lapse rate is generally legitimate for the low-altitude clouds in the image sequences. More accurate heights and spatial coordinates can be obtained by modulating the adiabatic lapse rate according to atmospheric models or auxiliary sensor input - see for example.[5] The mean-shift algorithm uses three cluster points and a kernel size of 300 data points. The Hampel filter parameters are $\alpha = 0.7$ and window size $N_w = 5$.

The initial cluster positions are shown projected onto the image in figure 1 a). The mean-shift tracking procedure shifts the cluster positions with each new update of the cloud-space computed from the next image in the sequence. In other words, the clusters are moved to the same relative point on the cloud in the clouds new spatial position. The cluster positions are projected onto the final image in the sequence in figure 1 d). The precise position of the initial clusters (their location determined in the first image) on the cloud in the last image in the sequence is somewhat subjective, due among other reasons to the amorphous and evolving structure of the cloud. It is clear however, that most of the cluster points have been able to move with the cloud over the sequence time period. There is a signficant sudden leap towards the left-side of the cloud made by each of the clusters within the first 25 seconds (figure 1 a) to b)). After this (figure 1 b), c) and d)),

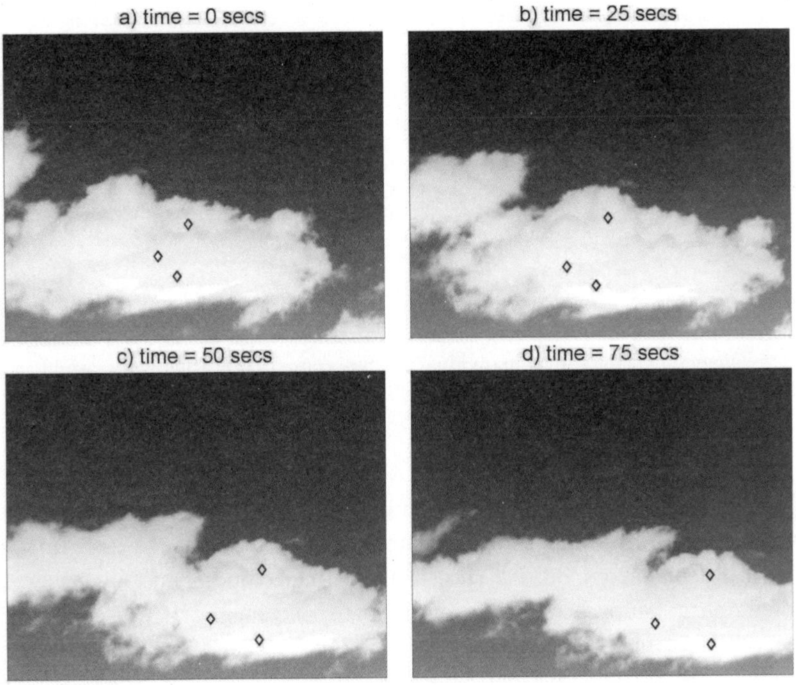

Fig. 1. Image sequence of a cumulus clouds taken at an angle of 20° and a sampling rate of 2Hz. The cluster positions are shown as diamonds

the clusters appear to retain an approximate lock on their specific region of the cloud.

Figure 2 shows the computed velocity vectors in the x and y direction for each cluster, as well as the overall Hampel filtered result. Cluster leaps and jostling are evident from the noise in the data. There are two significant leaps made by different clusters at 13 seconds and 13.5 seconds. Numerically differentiating the leap results in two significant outliers on the y-direction plot at 556.5m/s and 230.6m/s respectively. These are not shown in figure 2 b) as to do so would require an uninformative scaling of the plot axis. The resulting filtered velocity vector is mostly unaffected by the leaps caused by cluster tracking failure. The Hampel filter is able to identify leaps as outliers and reject them when smoothing the data. Velocity vectors rejected as outliers are replaced by the median value of the window. This approach smooths the noisy data. The remaining low-frequency oscillations are mostly due to mast wobble and intermittent high-winds during the observation period.

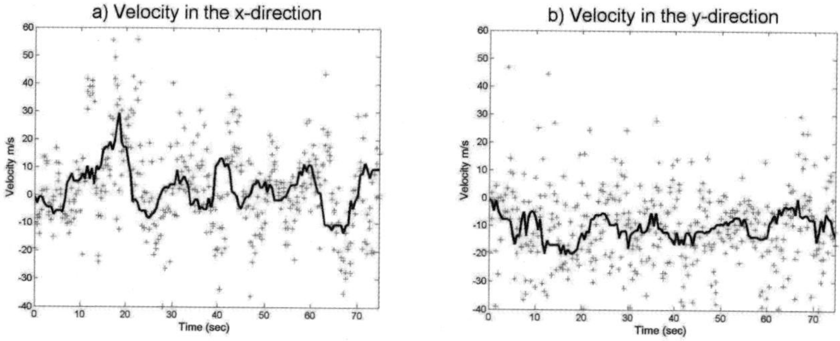

Fig. 2. Velocity vectors of the individual clusters computed at each time-step in the a) x-direction and b) y-direction. The overall filtered velocity vector is shown in black

5. Conclusion

In this paper, we have demonstrated a procedure for tracking clouds across a sequence of infrared images in order to compute a velocity vector. Infrared images are projected onto a three-dimensional cloud-space, by using the known relationship between air temperature and height. It was shown that the regions of maximum density in the cloud-space, which correspond to the structure of the cloud, can be successfully tracked using the mean-shift algorithm. The effects of cluster jostling in regions of even density and leaps between local density regions are easily suppressed by using a Hampel filter.

References

1. C. Mueller, T. Saxen, R. Roberts, J. Wilson, T. Betancourt, S. Dettling, N. Oien and J. Yee, *Weather and Forecasting* **18**, 545 (2003).
2. J. Calb and J. Sabburg, *Journal of Atmospheric and Oceanic Technology* **25**, 3 (2008).
3. M. Singh and M. Glennen, *Pattern Analysis amp; Applications* **8**, 258 (2005).
4. J. W. Wilson and C. K. Mueller, *Weather and Forecasting* **8**, 113 (mar 1993).
5. E. Brocard, M. Schneebeli and C. Matzler, *Geoscience and Remote Sensing, IEEE Transactions on* **47**, 3319 (oct. 2009).
6. G. Seiz, J. Shields, U. Feister, E. P. Baltsavias and A. Gruen, *Int. J. Remote Sens.* **28**, 2001 (May 2007).
7. R. Pearson, Control Systems Technology, *IEEE Transactionson* **10**, 55 (jan 2002).
8. D. Comaniciu and P. Meer, *IEEE Transactions on Pattern Analysis and Machine Intelligence* **24**, 603 (2002).
9. D. Comaniciu, V. Ramesh and P. Meer, *Computer Vision and Pattern Recognition, IEEE Computer Society Conference on* **2**, p. 2142 (2000).
10. A. Moore, A tutorial on kd-trees Extract from PhD Thesis, (1991), Available from http://www.cs.cmu.edu/simawm/papers.html.

OBSTACLE DETECTION USING INTEGRATION OF OMNI-DIRECTIONAL CAMERA AND INERTIAL SENSOR

Yan Guo, Sio-Hoi Ieng, Miguel Romero, Ryad Benosman, Bruno Gas

Institut des Systemes Intelligent et de Robotique, Universite Pierre et Marie Curie (CNRS - UMR 7222)

4, place Jussieu 75252 Paris cedex 05, France

**E-mail: firstname.lastname@upmc.fr*

Multi-sensors system is important in the research of unmanned surface vehicle for the environmental perception. The ASAROME project * (Autonomous SAiling Robot for Oceanographic MEasurements) is focused on an autonomous sailboat to make measurements and observations with a multi-sensor system in the marine environment for long term.

Keywords: Multi-Sensor; Sensors integration; Omni-directional camera; Obstacle detection; USV.

1. INTRODUCTION

In the research of the autonomous surface vehicles (ASV), the goal is the obstacle detection and avoidance.[1–3] The obstacle detection is one of the most important tasks to fulfill to ensure the safety of the boat and the possibility to define the most adequate manoeuver commands. These tasks are possible only if the handling of the sensors informations are highly accurate. ASAROME embedded several sensors to provide this information: an omnidirectional camera, an inertial measurement unit , two hydrophones and a sonar.

In this paper we are focusing on the integration of vision sensor and Inertial Measurement Unit (IMU) to address the problem of obstacles detection. The paper is organized as follows. First, the calibration of sensors is presented. Second, estimations of the boat initial pose and the pose in real time are described. Third, the obstacle detection using a colorimetric criterion is presented. Finally, we have tested the system's ability to detect and locate the obstacle position.

*This work is funded under the project ANR ASAROME (Num. ANR-07-ROBO-0009)

2. THE ALGORITHM

2.1. *Calibration of camera and IMU*

Sensors data need firstly to be calibrated correctly, i.e., the mappings between measurements and the real world metric must be estimated and all sensors must be related to each other. In our system, the IMU is used just for measurements of orientation change. We do not need to calibrate its related position to camera. A catadioptric system is used and the relative pose of each component is estimated with a variation of the method described in,[4] based on an homography between two judiciously chosen planes according to.[5] In our system, these planes are the image plane and the one defined by the upper edge of the mirror. The calibration is complete and it is able to map any pixel to a 3D line going through the point on the mirror.

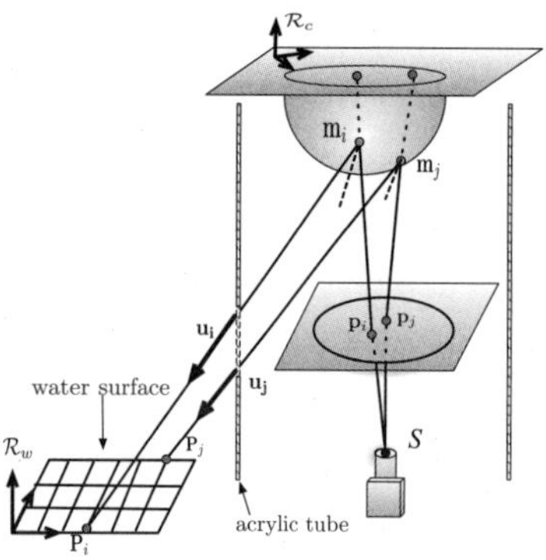

Fig. 1. A complete calibration of the catadioptric sensor provides the mapping of any image pixels $\mathbf{p}_i$ to a 3D line $(\mathbf{m}_i, \mathbf{u}_i)$ and its intersection with the surface defined by the water. $\mathbf{P}_i$ can be computed if the relative pose is known.

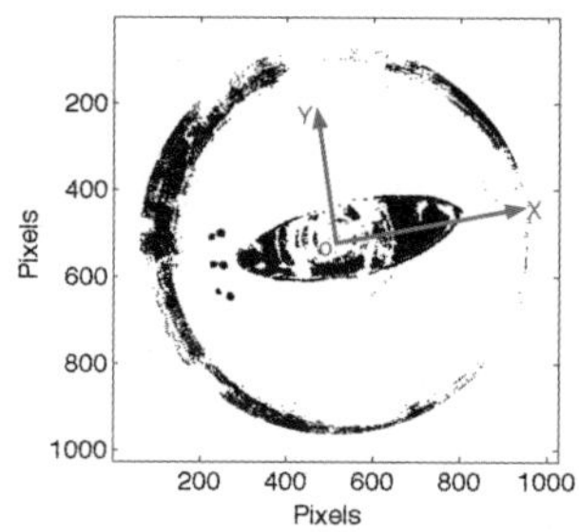

Fig. 2. The ship's axes are used to define a global coordinate frame assuming the camera is placed perpendicular to it.

2.2. *Boat initial pose estimation*

A global coordinate frame has to be set at first time, which is defined by the axes of the boat. Any structure or scene object will be referenced within this frame in this paper (Fig. 2). As we did for the camera pose with respect to the mirror estimation, we can estimate the omnidirectional sensor initial pose by detecting projections of the calm sea plane (the sea is considered as a perfect plane) on the catadioptric image according to.[6] To achieve this task we place into the water several easy-to-detect buoys that are acting as seamarks to underline the sea plane (Fig. 3). If the structure defined by the buoys is known, i.e., the relative distances of the buoys are known, then enough geometric constraints can be provided in addition to the coplanarity one for the plane estimation. The water plane decision is made with a Hough-based method which is tested by Guo et al. in.[7] A standard parametrization of a plane we use is:

$$ax + by + cz + d = 0 \tag{1}$$

The initial orientation of the water surface $\mathbf{n_I^w}$ in the 3D coordinate system of camera is presented as:

$$\mathbf{n_I^w} = (a, b, c)^T \tag{2}$$

With the change of reference, the boat orientation $\mathbf{n_I^b}$ can be obtained by $\mathbf{n_I^w}$(Fig. 5).

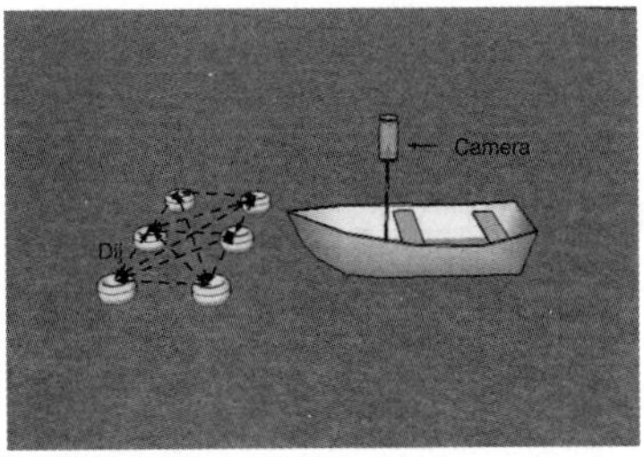

Fig. 3. Seamark used for the pose estimation.

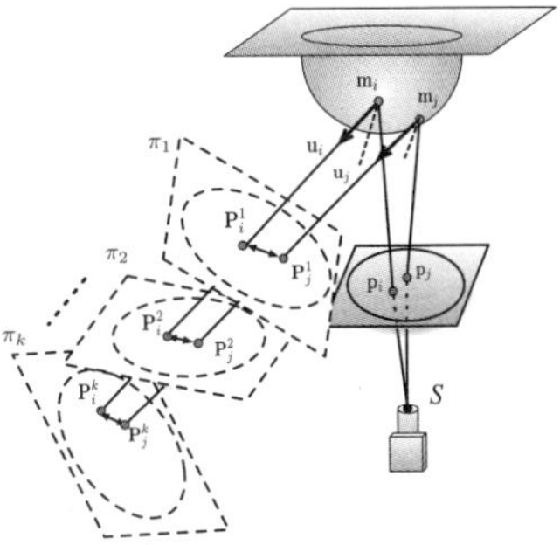

Fig. 4. Fitting plane on the calibrated sensor.

With the possibility to determine the camera pose with respect to the sea surface, we can map any pixel $\mathbf{p}_i$ to a 3D point $\mathbf{P}_i$ on the water plane (Fig. 4). The distance of each pixel to the boat in the image can be computed in the global coordinate frame introduced earlier.This latter result allows to built a resolution map of the sensor on the sea surface since we can project all the pixels on it with the mast normal to the water plane (calm sea state), however an oscillating sea is more likely expected, implying to recompute the resolution map for different orientations of the ship with respect to the sea surface. To avoid a too large amount of computation, we introduce the IMU to detect the orientation of the ship in real time.

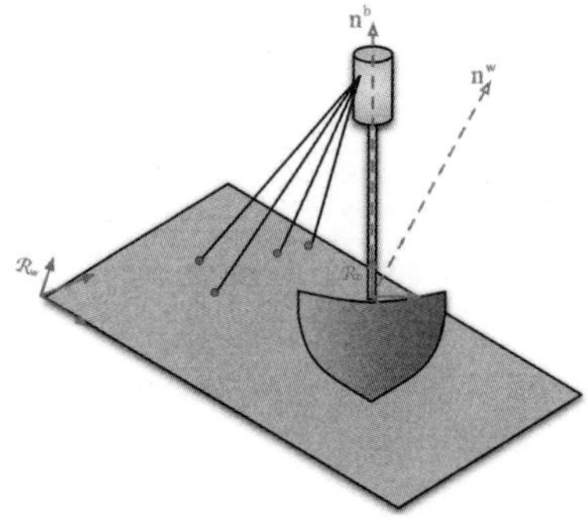

Fig. 5. $\mathcal{R}_w$ is the water surface referenced coordinate system, n^b is the boat orientation in this system. $\mathcal{R}_b$ is the boat referenced coordinate system, n^w is the water surface orientation in this system

2.3. *Boat pose estimation in real time*

In this part, we integrate the IMU for the pose estimation in real time. The detected data of IMU cannot be used directly for the orientation calculation of the boat, but the IMU and sailboat have the same orientation change during the measurement, i.e. the same rotation matrix. As we have already estimated the boat initial pose, with the measurement of orientation change by IMU, the sailboat pose in real time can be then estimated.

The rotation matrix R^{IMU} can be obtained from the IMU data. The 3D orientation output of our IMU is defined as the orientation between the body-fixed coordinate system and the earth-fixed co-ordinate system using the earth-fixed coordinate system as the reference. We define $\mathbf{E_r}$ as the earth-fixed reference vector of IMU, with the detection we have:

$$\begin{cases} \mathbf{n_I^{IMU}} = R_I^{IMU}\mathbf{E_r} \\ \mathbf{n_T^{IMU}} = R_T^{IMU}\mathbf{E_r} \\ \mathbf{n_T^{IMU}} = R_G^{IMU}\mathbf{n_I^{IMU}} \end{cases} \Rightarrow R_G^{IMU} = R_T^{IMU}(R_I^{IMU})^{-1} \tag{3}$$

Where $\mathbf{n_I^{IMU}}$ and $\mathbf{n_T^{IMU}}$ represent the initial and instant orientation vector of IMU, R_G^{IMU} is their global matrix rotation.

In our system, the movements of IMU and boat are synchronized, their rotations matrices are identical, the pose of the boat $\mathbf{n_T^b}$ in real time can be computed:

$$\mathbf{n_T^b} = R_G^{IMU}\mathbf{n_I^b} \tag{4}$$

And the orientation of the water surface $\mathbf{n_T^w}$ in the 3D coordinate of camera can be calculated from the pose of the boat $\mathbf{n_T^b}$ by changing the reference of coordinate system (Fig. 5). Similar to the estimation of boat initial pose, when the water surface is defined by using the data of IMU, we can map any pixel to a 3D point on the water plane in real time.

2.4. *Obstacle detection*

The obstacle detection is done with a colorimetric criterion which is used to segment objects in the image. A color signature of the sea is first computed by selecting samples of regions representing the water. Each pixel of these regions are reprojected to the three planes of the RGB coordinate frame according to their color components. This operation produces a cluster in each plane and is representative of the color distribution of the background (i.e sea). To segment an object pixel from the background, we project it into the RGB components planes and it is considered being an obstacle if at least one of its RGB component does not belong to the clusters defined earlier. The obstacles can be properly detected and segmented.

The Fig. 6 show an example of segmentation of a seamark formed by six yellow buoys using the combination of image difference andcolorimetric signature of the water.

3. EXPERIMENTATION

We conducted this experimentation in a large place with its surface defining a plane. A special seamark with 6 buoys (1 m long x 2 m wide) is used for this test. The camera is fixed on a 1.5 m high tripod. The seamark position is changed between each record and the distance between the camera and the center of the seamark is manually measured. The distance tested ranges from 2 to 10 m. For each distance, we have tested several poses of the camera to simulate the different poses of the boat. There are two principal types of pose: the static poses and dynamic poses. The static pose is stable and without movement. The dynamic pose means that the camera is in the process of moving in arbitrary directions (Fig. 7).

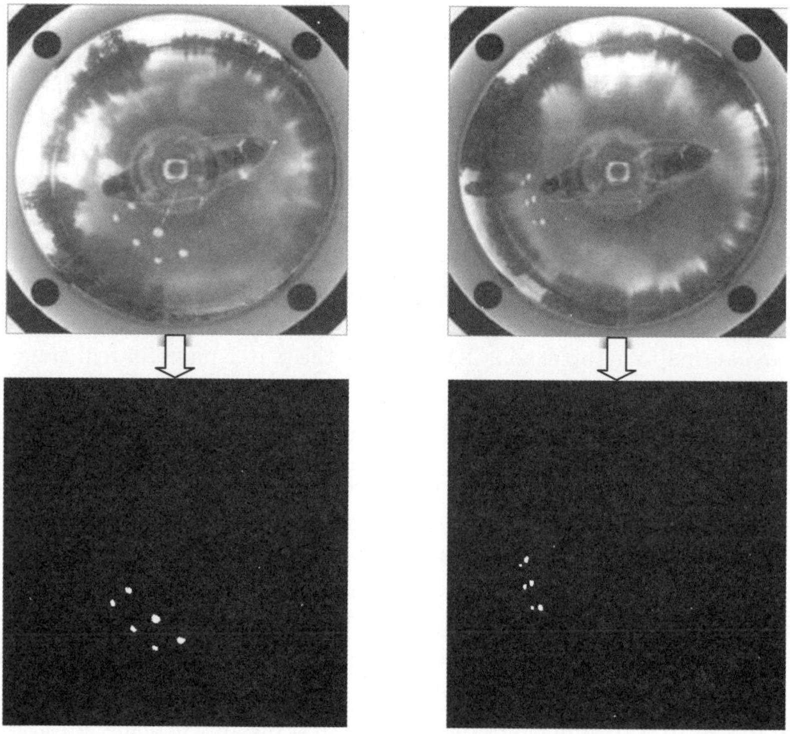

Fig. 6. Seamark detections using colorimetric signature.

Table 1. Comparison of acoustic for frequencies for piston-cylinder problem.

Mesured Distance (m)	Estimated Distance Static Pose (m)	Estimated Distance Dynamic Pose (m)	Distance Estimated Error Rates (%)
2	2.10	2.06	1.33
4	4.17	4.39	4.67
6	6.21	6.10	5.00
8	7.93	8.89	8.00
10	10.57	10.49	8.83

From table 1, our algorithm seemed to have excellent performance with error rates within 9%. The distances estimated are smaller than the measured distances. The maximum value of estimated error is 0.89 m for a distance of 8 m, and the minimum is 0.06 m for a distance of 2 m.

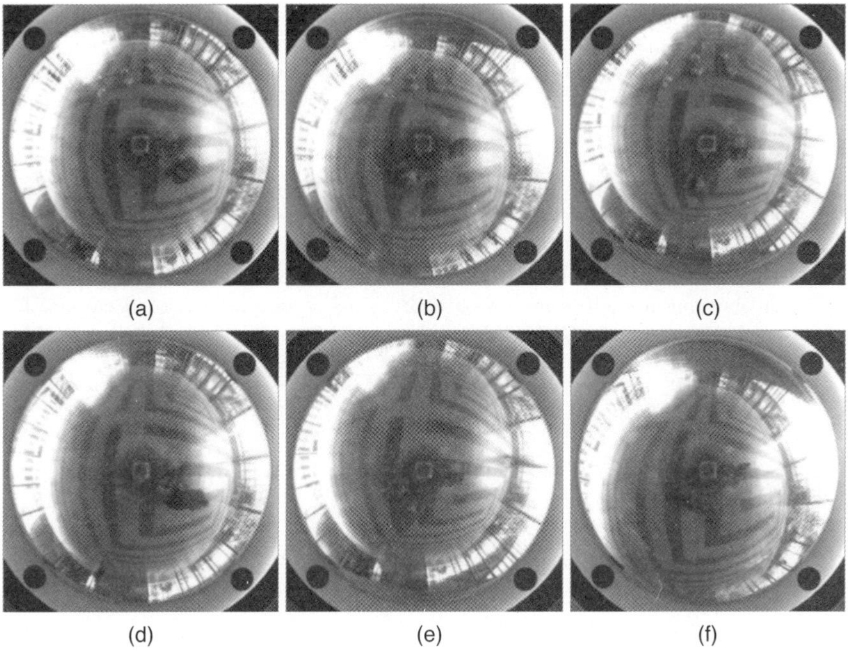

Fig. 7. The photos are taken by the panoramic camera with various poses in different distances. (a)-(c) are the photos taken with the buoys at 2 m. (d)-(f) are the detection at 6 m. (a) and (d) are the static poses and others are the dynamic poses.

4. CONCLUSION

We have described in this paper a method of obstacle detection by integrating the information from visual and Inertial sensor. Experiments results have shown the expected performances for obstacle detection. In the final paper, we will present more detailed results and descriptions about our simulation.

References

1. T. Pastore and V. Djapic, *Journal of Field Robotics* (2010).
2. M. Dunbabin, A. Grinham and J. Udy, An autonomous surface vehicle for water quality monitoring, in *Australasian Conference on Robotics and Automation (ACRA)*, 2009.
3. M. Tall, P. R. ans J.M. Lorio and K. von Ellenrieder, Visual-based navigation of an autonomous tugboat, in *OCEANS 2009, MTS/IEEE Biloxi - Marine Technology for Our Future: Global and Local Challenges OCEANS 2009, MTS/IEEE Biloxi - Marine Technology for Our Future: Global and Lo-*

cal Challenges OCEANS 2009, MTS/IEEE Biloxi - Marine Technology for Our Future: Global and Local Challenges, 2009.
4. J. Fabrizio, P. Tarel and R. Benosman, Calibration of panoramic catadioptric sensors made easier, in *Proceedings of the IEEE Workshop on Omnidirectional Vision*, june 2002.
5. P. Sturm, Algorithms for plane based pose estimation, in *Proceedings of the IEEE International conference on Computer Vision and Pattern Recognition*, 2000.
6. R. Hicks and R. Bajcsy, *Image and Vision Computing* (2001).
7. Y. Guo, M. Romero, S.-H. Ieng, F. Plumet, R. Benosman and B. Gas, Reactive path planning for autonomous sailboat using an omni-directional camera for obstacle detection, in *IEEE International Conference on Mechatronics*, 2011.

SECTION–15

PLANNING AND CONTROL

BEHAVIOR-BASED ADHESION CONTROL SYSTEM FOR SAFE ADHERENCE OF WALL-CLIMBING ROBOTS

D. SCHMIDT* and K. BERNS

Robotics Research Lab, University of Kaiserslautern,
Kaiserslautern, Germany
** E-mail: dschmidt@cs.uni-kl.de*
http://agrosy.cs.uni-kl.de/cromsci

Safe adhesion is a large research area in the field of climbing robots. Many robotic systems use well-known magnetic adhesion or mechanical connections like claws, some others use classic closed-loop controllers in combination with negative pressure adhesion to ensure operation safety. This paper will introduce important safety aspects of the climbing robot CROMSCI which has a negative pressure system, an adaptive inflatable sealing and an omni-directional drive. We will illustrate the novel behavior-based closed-loop controller architecture which has different control levels — starting from individual chamber pressure controllers up to an overall force control. Experimental results will show the performance of the complete behavioral network.

Keywords: Climbing Robot, Behavior-Based Control, Adhesion Control, Negative Pressure Adhesion

1. Motivation and State of the Art

Safety is a main requirement for all kind of robots. In most cases the developers pay attention to the operation safety to ensure that the system will not harm people or damage something.[1] Regarding climbing robots the most endangered object is the robot itself. In general the threat caused by climbing robots[2–7] is nearly zero because they are slow and do not interact with humans. Only a drop-off of the robot is dangerous and should be avoided under all circumstances.

Therefore, a behavior-based adhesion controller has been developed which consists of individual, but interacting components to reduce or avoid robot slip and tilt – which are main threats for climbing robots. By this structure, we are able to use the advantages of a behavior-based network in combination with several closed-loop controllers and operation modules:

The control system can be built and enhanced bottom-up, the behavior meta data can be used as additional virtual sensor information.

As test platform we use the climbing robot CROMSCI (fig. 1) which has been developed for inspections of large concrete structures. The system combines several innovative techniques like a negative pressure adhesion system with multiple chambers and a reservoir, adaptive sealings in the shape of spokes and a drive system with integrated load cells.[8] For inspection tasks a manipulator arm can be mounted to carry the inspection sensors. An umbilical cable connects the robot to a supply station.

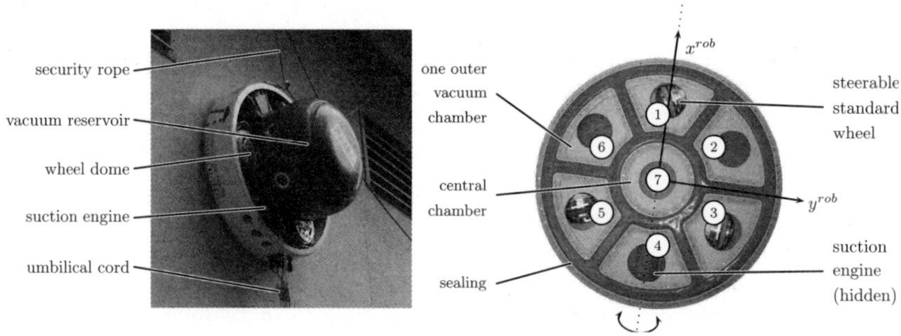

Fig. 1. Climbing robot CROMSCI in action (left) and the setup of the seven-chamber adhesion system (bottom view, right)

2. Behavior-Based Control Architecture iB2C

The negative pressure control system as presented by Hillenbrand *et al.*[9] has been transfered into a behavior-based network in which several individual controllers, limiters and calculation units interact. Each of these behaviors follows the design of the iB2C[a] architecture by Proetzsch *et al.*[10]

In general, one differs between behavior and fusion modules (see fig. 2). Behaviors are algorithmic elements generating outputs from given inputs. Fusion modules combine these outputs, if two or more behaviors try to control the same resource. Each behavior generates control data $\vec{u}$ based on an arbitrary transfer function $F(\vec{e}, \iota)$ which uses the given input vector $\vec{e}$ and its *activation* $\iota = s \cdot (1 - max_{1..n}(i))$ (maximum possible activity). A standardized interface allows to influence other modules and to signal the current behavior state: *stimulation* $s \in [0, 1]$ (activates the behavior),

[a]iB2C stands for *Integrated Behaviour-Based Control* and can be downloaded on `http://rrlib.cs.uni-kl.de/mca2-kl/libraries/ib2c/`

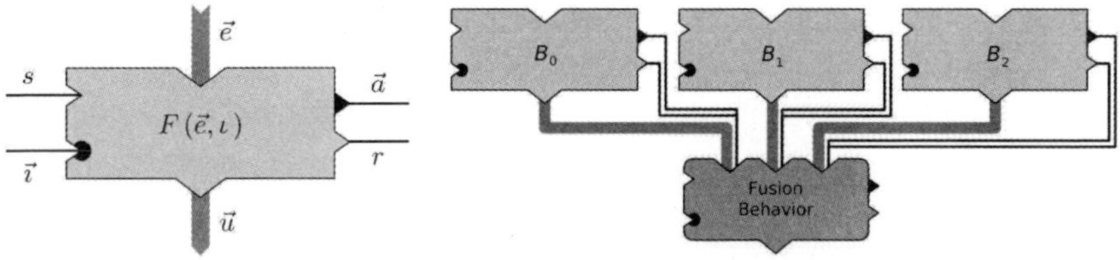

Fig. 2. Behavior module in the behavior-based architecture iB2C (left) and an exemplary fusion of the outputs of three behaviors by taking meta data into account (right)

inhibition $\vec{\imath} \in [0,1]^n$ (reduces stimuli), *activity* $\vec{a} \in [0,1]^m$ (real amount of action it performs) and *target rating* $r \in [0,1]$ (dissatisfaction in the current situation). The advantage of this architecture is that additional meta data will be received which can be used for behavior interaction and data fusion to implement a very complex system behavior based on many individual units. Behaviors and fusion modules can be encapsulated to behavior groups for a better survey. It has to be mentioned that the iB2C architecture scales from the reactive level up to high-level functions.

3. Control Behaviors

The novel behavior-based adhesion control system consists of different control levels as illustrated in fig. 3. The following list shows, which behaviors are used to control the adhesion and how the meta data *activity* and *target rating* for behavior interaction are calculated.

Chamber Controller (CCi) A basic pid closed-loop pressure controller for one working chamber $i \in \{1, 2, ...7\}$ which exists seven times within the network. The *activity* $a_i^{CC} = \iota_i^{CC} \cdot A_i^{act} / A^{max}$ depends on the actual valve area A_i^{act} and its maximum. In colloquial speech, this points out that the behavior is more active if the valve is opened. The *target rating* $r_i^{CC} = \left|p_i^{act} - p_i^{des}\right| / \Delta p^{max} + \left(1 - \iota_i^{CC}\right)$ uses the difference of desired p_i^{des} and actual chamber pressure p_i^{act} and therefore is unhappy if the desired pressure value can not be reached or if the behavior is not activated.

Chamber Deactivator (CDi) A deactivator surveys one working chamber and inhibits the corresponding CCi in cases of high leakages. By this means the chamber is cut off from the adhesion system to prevent a global fail. Additionally, the deactivator tests the leak chamber before it is reintegrated. The *activity* $a_i^{CD} = \iota_i^{CD} \cdot s_i^{act}$ is determined by the chamber state $s_i^{act} \in \{0, 1\}$ which depends on an internal state and the estimated leakage. It is active if the leakage limit has been exceeded and a waiting or testing cycle is started. The *target rating* $r_i^{CD} = s_i^{act} + \hat{A}_i^{leak}/A^{max}$ also depends on the estimated leakage $\hat{A}_i^{leak}$. Therefore, it is dissatisfied if the leakage area tends towards the valve area or if it has to be active.

Chamber Deactivator Control (CDC) This behavior coordinates the individual chamber deactivators to prohibit simultaneous testing of chambers. Additionally, it inhibits the inactive CDi behaviors if the maximum number of inactive chambers nic^{max} has been reached to prevent the cut-off of more chambers. The *activity* $a^{CDC} = \iota^{CDC} \cdot nic/nic^{max}$ increases with the amount of inactive chambers nic. The *target rating* increases with the amount of unhandled leakage: $r^{CDC} = max_{i\in\{1,2,\ldots 7\}}(r_i^{CD} \cdot \iota_i^{CD})$.

Manual Chamber Control (MCC) Via this behavior the seven chambers can be activated and controlled individually by hand. These manual values are fused with the calculated ones from the CPC. *Activity* and *target rating* are $a^{MCC} = \iota^{MCC}$ and $a^{MCC} = 0.5$.

Chamber Pressure Calculator (CPC) This behavior calculates individual chamber pressures based on a desired total downforce. Here, the *activity* is equal to the *activation* $a^{CPC} = \iota^{CPC}$, the *target rating* depends on the possible and desired downforce values and on the position of the downforce point: $r^{CPC} = max\left(\left|F_z^{pos} - F_z^{des}\right|/\Delta F^{max}, \left|\vec{P}_F^{pos} - \vec{P}_F^{des}\right|/\Delta P_F^{max}\right)$. Thus, the behavior is dissatisfied if the desired downforce value can not be reached by the adhesion system. This can happen in cases of leakages or 'unusual' settings.

Manual Force Control (MFC) This behavior is used to set a desired total downforce for the negative pressure system. Again, the *activity* is set to the internal *activation* $a^{MFC} = \iota^{MFC}$, whereas the *target rating* is calculated similar to the CPC: $r^{MFC} = max\left(\left|F_z^{act} - F_z^{des}\right|/\Delta F^{max}, \left|\vec{P}_F^{act} - \vec{P}_F^{des}\right|/\Delta P_F^{max}\right)$ by comparing the actual and desired adhesion force values.

Force Controller (FC) This is the high-level closed-loop controller for regulating the downforces at the three wheels which takes the load cell values into account. *Activity* is set to $a^{FC} = \iota^{FC}$, *target rating* again depends on the actual and desired values — but now determined by the load cells (LC): $r^{FC} = max\left(\left|F_{LC,z}^{act} - F_{LC,z}^{des}\right|/\Delta F_{LC}^{max}, \left|\vec{P}_{LC,F}^{act} - \vec{P}_{LC,F}^{des}\right|/\Delta P_{LC,F}^{max}\right)$.

4. Behavior-based Adhesion Control

As important as the individual components is the interaction of them. Figure 3 shows the connections between the different control and fusion behaviors within our iB2C framework. The bottom-layer consists of the *chamber control group* with the seven *chamber control* (CCi) modules and the *chamber deactivator group* with the *chamber deactivator control* (CDC) and *chamber deactivator* (CDi) behaviors. The controller outputs of the CCi behaviors are the desired valve positions to control the air flow between negative pressure reservoir and working chamber. The CDC output is connected to the inhibition inputs of the CDi modules. The CDi *activity* outputs again are used to inhibit the corresponding CCi. All these different controlling behaviors work in parallel.

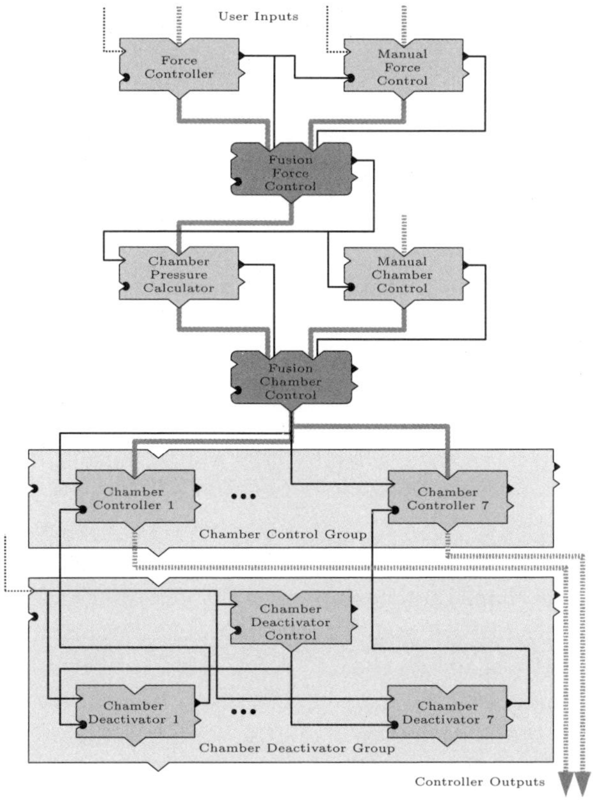

Fig. 3. Overall behavior structure of the negative pressure and force control system. The user can activate additional functionalities by stimulating the behaviors *force controller*, *manual force control* and the *chamber deactivator group*.

In the middle of fig. 3 the *manual chamber control* (MCC) and *chamber pressure calculator* (CPC) are located, a fusion module combines their controller outputs. The MCC can be used for additional manual control settings like a position control of the chamber valves. The CPC receives the desired force values from the *fusion force control* on top. This fusion module stimulates the CPC and inhibits the MCC via its *activity*. The incoming force values are taken either from the *force controller* (FC) which uses the load cell values or from the *manual force control* (MFC) which is inhibited by FC's *activity*. It is obvious that this structure allows only one of the higher behaviors to control the adhesion because of the different user commands which are possible: The FC works on desired loadcell force values, the MFC takes the desired adhesion force and the MCC either uses individual valve positions or chamber pressures. The user can now activate FC, MFC and CDC depending on the desired system functionalities and set the corresponding control values.

5. Experiments and Test Results

Figure 4 gives an overview on how the individual components interact. On the left, *activity* (darkgray) and *activation* (lightgray) are given. On the right, one can find the *target rating* of each behavior (because of page limits the individual behaviors of chambers 4-7 are not shown). In this setup the simulated robot drives downward a wall facing a deep crack.

At the beginning, the MCC is active with the same desired chamber pressure for all chambers. Therefore, robot tilt is not balanced out which results in a x-position of the downforce point $P^{act}_{LC,x}$ of $0.042[m]$ (bottom right graph). Then the user stimulates MFC and FC. Finally, the FC is active and uses the desired downforce $F^{des}_{LC,z} = 2200[N]$ while MFC and MCC are inhibited ($a^{MFC} = a^{MCC} = 0$). Now the robot tries to overcome the crack but the leakages are too high: CD1, CD2 and CD6 become active and shut down the frontal chambers (compare fig. 1) by inhibiting CC1, CC2 and CC6 – the light gray *activation* graphs go down to zero. Because of the maximal amount of inactive chambers the CDC is full active ($a^{CDC} = 1$) and inhibits CD3, CD4, CD5 and CD7, no more chambers can be shut down. After some time the CDC starts a testing cycle of CC1 which fails because the chamber is still influenced by the crack. The minimum downforce in this period is at about $1760[N]$ provided by the four remaining chambers. The x-coordinate of the downforce moved to $-0.082[m]$ which is still within the stability triangle – the robot does not tilt. Now the robot is driven backward, the CDi behaviors become inactive again and the system can recover.

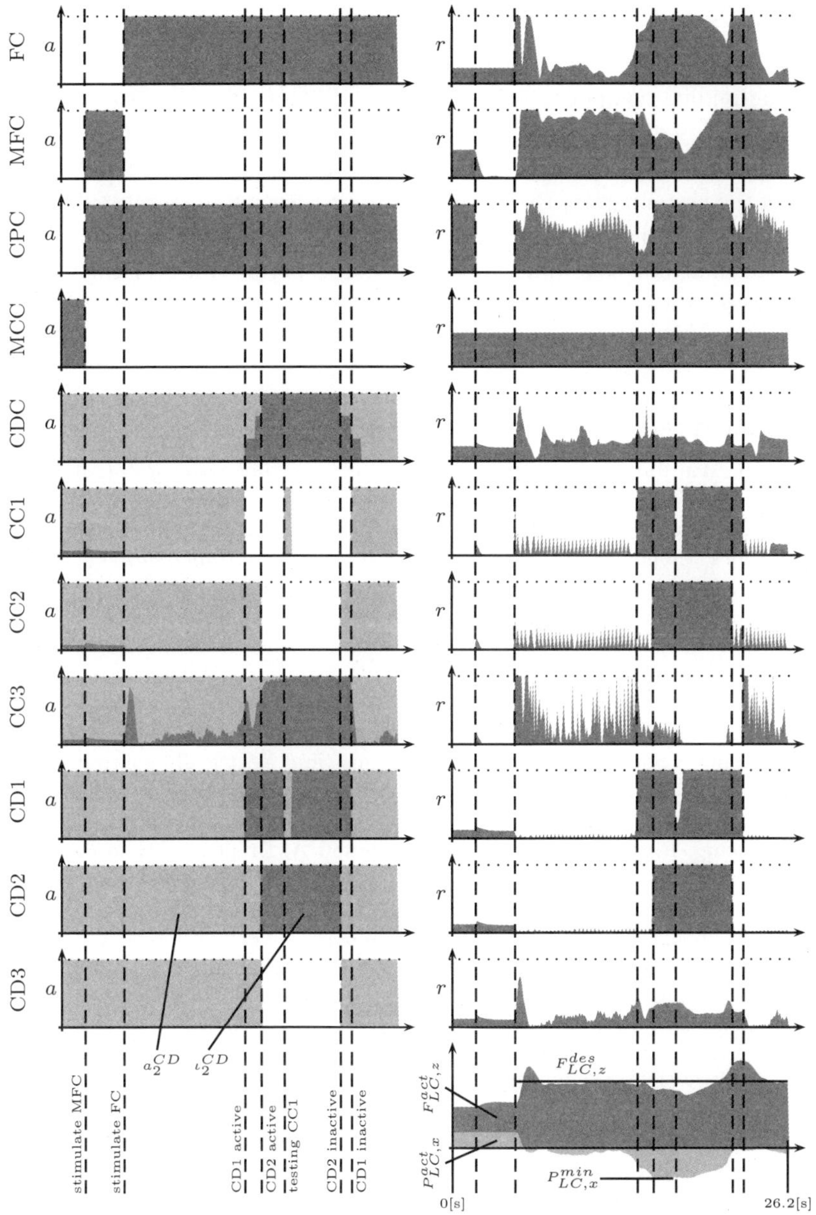

Fig. 4. Plot of behavior *activities a* (left, gray), *activation* (left, light gray) and *target ratings r* (right). On bottom right one can find the resulting downforce $F_{LC,z}^{act}$ (dark gray graph) with a desired force of $F_{LC,z}^{des} = 2200[N]$. The light gray graph indicates the x-position of the downforce point which reaches a minimum of $P_{LC,x}^{min} = -0.082[m]$.

6. Conclusion and Outlook

We presented the novel behavior-based network for climbing robots which uses individual components based on the iB2C architecture to perform a cascaded closed-loop adhesion control. Meta data are used for behavior interaction and will be used in future for additional controlling and evaluation of the system state. By this generality it is easy to integrate additional components to the control system, to reuse them in other contexts and to perform analyses. The functionality and the performance of this system have been proved by experiments. The future work is now focused on an evaluation of the behaviors for risk estimation and prediction.

References

1. G. R. Ward and S. R. G.Went, Robot safety, in *Industrial Robot: An International Journal*, (1) 1995 pp. 10–13.
2. F. Bonaccorso, D. Longo and G. Muscato, Modelling of an innovative actuator for climbing robot adhesion, in *12th International Conference on Climbing and Walking Robots (CLAWAR)*, September 2009.
3. T. Hayakawa, T. Nakamura and H. Suzuki, Development of a wave propagation type wall-climbing robot using a fan and slider cranks, in *12th International Conference on Climbing and Walking Robots (CLAWAR)*, September 2009.
4. H. Prahlad, R. Pelrine and S. Stanford et al., Electroadhesive robots - wall climbing robots enabled by a novel, robust and electrically controllable adhesion technology, in *International Conference on Robotics and Automation (ICRA)*, May 19-23 2008.
5. J. Shang, B. Bridge, T. P. Sattar, M. S and A. Brenner, Development of a climbing robot for inspection of long weld lines, in *Climbing and Walking Robots*, July 2007.
6. M. J. Spenko, G. C. Haynes, J. A. Saunders, M. R. Cutkosky, A. A. Rizzi, R. J. Full and D. E. Koditschek, *Journal of Field Robotics* **25** (2008).
7. H. X. Zhang, W. Wang and J. W. Zhang, High stiffness pneumatic actuating scheme and improved position control strategy realization of a pneumatic climbing robot, in *IEEE International Conference on Robotics and Biomimetics (ROBIO)*, February 2009.
8. D. Schmidt, C. Hillenbrand and K. Berns, Omnidirectional locomotion and traction control of the wheel-driven, wall-climbing robot, cromsci, in *Robotica Journal (First View)*, (Cambridge University Press, 11th April 2011).
9. C. Hillenbrand, D. Schmidt and K. Berns, Cromsci - development of a climbing robot with negative pressure adhesion for inspections, in *Industrial Robot Volume 35 Issue 3*, (Emerald Group Publishing Ltd., May 13 2008).
10. M. Proetzsch, T. Luksch and K. Berns, Development of complex robotic systems using the behavior-based control architecture iB2C, in *Journal of Robotics and Autonomous Systems*, January 2010 pp. 46–67.

STOCHASTIC CONTROLLERS FOR ROBUST UNDERWATER MOBILE MANIPULATION

FABIO BONSIGNORIO
Heron Robots, V.R.Ceccardi, 1 18
Genova,16121 ,Italy

RoboticsLab, University Carlos III de Madrid, Av. Universidad, 30
Leganes, 28911 Madrid, Spain

In this paper the technological underpinnings of a novel robust underwater manipulation controller platform currently under development by Heron Robots are briefly described. Underwater visual servoed manipulation performed by a dexterous arm connected to a ROV (Remotely Operated Vehicle) or an AUV (Autonomous Underwater Vehicle) can be seen as a (hard) mobile manipulation problem, as the arm base is rigidly connected to a platform whose motion is stochastically independent from the object motion, from a mechanical standpoint. In our design the 'reaching behaviors' are not be pre-designed, their instantiation builds and expands on recently proposed techniques using belief space planning maximum likelihood observations. They will be made more efficient by a high level functional representation of the configuration space of the mobile manipulation task, based on invariant and constraint properties of displacement (Lie) groups. The adoption of such a canonical representation of both the kinematics and the dynamics of the manipulation action will allow the accurate parameterization of the hand-object relations to characterize the phase space portrait of the overall system and reduce the search space of the planning algorithm. We are implementing a probabilistic kinematic planning of the manipulation task involving the modeling of nonholonomic movement of the manipulator platform: the platform moves naturally for the water streams and moves in a partially actuated way to ease the manipulation. The key scientific objective is the development of robust planning and control algorithms for the underwater mobile manipulation and robust mobile manipulation in general.

1. Introduction

The broad aim is to achieve (or to do significant step towards) mobile manipulation planning and control algorithms robust and computationally efficient enough to enable the mobile manipulation systematic utilization in open ended underwater environment.

Belief Space Planning, assuming maximum observation likelihood, has been analyzed with promising results in simplified examples, in simulation, of the proposed experiment: the laser-grasp domain and the light-dark domain, [1]. They are interesting as they are based on a stochastic theoretical framework, that, in principle, may more likely lead to robust real world solutions. On the other end methods based on the subgroup classification of SE(3) have been proposed for grasping [4] and mobile robotics [2].

In [2,10] it is argued that the recognition of the Lie group structure of the mobility space may help planning methods based on searching in configurations space. Perception methods for mobile manipulation and grasping based on (active) stereo vision have been proposed for more complex task than that targeted by this experiment, [3].

2. System Concepts

The controller integrates these three approaches, and the two scenarios of paper [1], implementing them in parallel GPU (Graphical Processing Unit) or FPGA (Field Programmable Gate Array) computing architectures. Mobility of the manipulation platform is used for two purposes: 1) the platform moves to ease the manipulation 2) to allow (maybe trivially) the movement from a working place to a closer one. In both cases high speed it is not required for the mobile platform, ROV (Remotely Operated Vehicle) or AUV (Autonomous Underwater Vehicle) in this case.

There will be a stereo vision camera mounted on the vehicle. A number of occlusion and random obstacle of growing severity will be tested. All scientifically relevant software developed in the project will be developed on the 'yarp' (yet another robotics platform), [9], platform and open sourced (although the company is developing non free and non open source product engineered software based on this approach, and fully integrated soft-arm systems, in the next future). The development of such technologies is scientific relevant as it may pave the way to flexible and robust manipulation methods suitable for disruptive applications in many contexts including the shop floor where it may enable new logistic organizations. For example disruptive and industrially successful solutions for warehouse logistics have been proposed by Kiva, [12]. This technology goes in the direction of providing movable manipulation units substituting the human intervention.

The mobile manipulation system we develop should help moving in that directions: as it aims at a robust plug and play manipulation platform: self localizing and based on stereo vision, instead of laser range.

We implement a Belief Space Planning maximum observation likelihood planning framework.

We:

1) merge the laser-grasp domain and the light-dark domain for the given setting, in simulation.

2) modify the resulting algorithms exploiting the relevant subgroup classification of SE(3) have been proposed for grasping, [4], and mobile robotics, [2].

3) simplify and apply perception methods for mobile manipulation and grasping based on (active) stereo vision to the simpler setting of this experiment, [3]

4) Integrate in a common framework (based on yarp)

5) Define proper benchmarking metrics and experiment replication procedures

The degree of success of the system will be measured by the capability of managing the given task with growing level of occlusions and obstacles with a given level of the In particular we are interested to test on field by which degree the Belief Space Planning with and without the proposed Lie-group algebra aware modification are effective and robust. Mapping the limit will allow to get a better understanding of scientific bottlenecks to perform this kind of tasks and to select the most promising of the basket of methods of our 'recipe' and to change the recipe itself.

3. Belief Space Planning

In [1] new approaches to the problem of belief space planning are proposed and show interesting results in a toy grasping problem where it is necessary to acquire information before acting.
The problem of manipulation and grasping in an underwater environment can be seen as a control problem in a partially observable environment as the scene reconstruction allowed by the robot sensors and cameras are limited by dirty waters and streams and by the stochastic movement of the platform (AUV or ROV) to which the arm is connected. A widespread approach to cope with partially observable control problems is to model them as POMDP (Partially Observable Decision Process. This approach leads to PSPACE complete problems and as such is not viable for practical applications.

The solution proposed in [1] is based on the idea of generating plans in the 'belief space' of the manipulator, i.e. the space of all possible distributions over the state space.
The controller decides actions on the basis of current estimate of the state and of the available information or more precisely the 'beliefs' on the distributions. Some of actions are actually information gathering actions.
There are a number of challenges. Even in the case of coarse-grained approximation of the belief space the planning has to performed in a much higher dimensional state than the state space, moreover the belief state dynamics is non linear, stochastic and inherently and significantly under actuated (as the number of control input is significantly lower than the number of the dimensions of the belief space).
The main contribution of [1] is that it shows that by assuming maximum likelihood observation, i.e. it is assumed for planning purposes that the state of the system is the most likely state according to the past observations, it is possible to exploit standard and effective methods developed for the planning and control of high-dimensional non linear under actuated systems with Gaussian noise.
It is shown that it is possible to apply linear quadratic regulation (LQR) to calculate policies based on a linearization of local (belief space) dynamics.
It is also shown that this kind of linearization is optimal in linear Gaussian process assumptions and generates reasonable behaviors in a neighboring of the linearization points.
The method can be generalized to more noisy settings when the systems is forced away from the locally stabilized belief space region by means of re-planning from the new belief state (updated space of possible probability distributions over the state space).
We suggest to check [1] for the details, here below the method applied is summarized.
The observations $z_t \in Z$ are modeled as a non linear stochastic function of the vector of dimension d $x_t \in X$ the, not directly observed, state of the system and the environment,

$$z_t = g(x_t) + \omega \tag{1}$$

where g is a deterministic function of the measurement and ω is a zero mean Gaussian noise with covariance W_t dependent on the state.
The deterministic function f links the new state to the older one under the control action u_t

$$x_{t+1} = f(x_t, u_t) \tag{2}$$

Both f,g are assumed to be differentiable functions of x_t and u_t.

We assume that the controller knows the state through a probabilistic density function *P(x)*.
The parameters of this distribution is what we refer to as 'belief state'.
This belief state can be update through the following equation:

$$P(x_{t+1}) = \eta P(z_{t+1}|x_{t+1}) \int_x P(x_{t+1}|x,u_t)\, P(x) \tag{3}$$

If we assume linear Gaussian belief state dynamics we have:

$$x_{t+1} \approx A_t(x_t - m_t) + f(m_t, u_t) \tag{4}$$

$$z_t \approx C_t(x_t - m_t) + g(f(m_t, u_t)) + \omega \tag{5}$$

where m_t is the mean of the belief state and A_t and C_t are the Jacobian matrices $A_t = \frac{\partial f}{\partial x}(m_t, u_t)$, $C_t = \frac{\partial g}{\partial x}(m_t)$.

The (Gaussian) distribution is then given by:

$$\Sigma_t : P(x) = N(x|m_t, \Sigma_t) \tag{6}$$

In this hypotheses it can be proved that assuming maximum likelihood the control action has the form, see [1]:

$$u^* = -\left(\tilde{B}_t^T \tilde{S}_{t+1} \tilde{B}_t + R_t\right)^{-1} \tilde{B}_t^T \tilde{S}_{t+1} \tilde{A}_t b_t \tag{7}$$

4. Perception using Active Stereo

In [3] a perception system for manipulation and grasping task based on stereo vision. This system exploits dense stereo point clouds by means of projected light textures. Stereo vision is faster than laser panning as it allows 30-60 Hz scene acquisition while a laser pan requires. The main disadvantage is that the stereo depth map point cloud is much more noisy. The method proposed there is claimed to have obtained 98.27% accuracy on a dataset of IKEA objects (made of a random assortment of about 20 different kinds of glasses and cups). The method exploits a FPFH (Fats Point Feature Histograms). The method first estimate the FPFH and then passes them to a Conditional Random Field which labels local surfaces chunks with geometric primitives. Then the scene is modeled by reconstructing and decoupling surfaces and simplified by mesh relaxation.

The FPFH method is a simplified version of a previously developed for laser panning. It builds a multi-dimensional feature histogram space representing estimated surfaces and generalized local curvatures, on the basis of point pairs, p_i, p_j and their estimated surface normals $\boldsymbol{n}_i$, $\boldsymbol{n}_j$, see [3] for details.

Model learning is based on CRF (Conditional Random Fields). CRFs are discriminative probabilistic graphical models. They model a conditional probability distribution $P(y|x)$ avoiding, w.r.t. generative models, to provide a model of the observations x.

The CRF is trained to assign ('tag') points to a geometric class.

In the published experiments the scene is first segmented by means of planar fitting with robust estimators and Euclidean clustering methods. This leads to a clustering in groups if about 600 points within a radius of 3 cm.

Then the FPFH is produced for all the clusters and the points are labeled by means of a trained CRF model. The original surfaces are reconstructed by means of decoupled mesh triangulation and a simple mesh relaxation.

The resolution of the triangular mesh is related to the grasping and manipulation purposes, in our case (underwater manipulation and simple gripping) the needed resolution is lower than in [3], we have, on the other end, more noisy observations.

5. Our system approach

In our solution design we develop the belief space planning methods in [1] in a 3D world, reconstructed by a simplified and parallel version of the algorithms described in [3]. With respect to [3] we essentially apply a coarser grain mesh to much simpler although more noisy environmental setting. We respect to [1] the environment is more complex and the belief space includes an even coarser but still big triangular than that used for static perception.

The manipulation is performed in 3 phases: (1) Scene reconstruction at high mesh resolution (although with a significantly coarser grained mesh than in [3]); (2) Calibration at high resolution of the triangular mesh; (3) Manipulation at low mesh resolution.

Lie subgroups symmetry is exploited to reduce the search space of the planning algorithms, see for example the analysis in [10]. Lie Algebras of the tangent space in the neighboring of the stabilization point obtained by belief space planning allow to derive a linear expression similar to (6), this will be analyzed in detail in a future publication.

6. Experimental validation

The experimental validation of the approach will be performed in three campaign:

1) Simulation
2) Underwater pool manipulation
3) Undersea manipulation

The initial reference simulation scenario is pictured in Fig. 1. The origin of the reference axis is sampled by a bivariated Gaussian distribution. There is light gradient on the field. The force constraints (force closure) are translated into kinematic constraints (form closure) Only when >98% accuracy and repeatability in a multivariated and statistically significant randomized set of tests will be achieved it will be possible to move to the next campaign level.

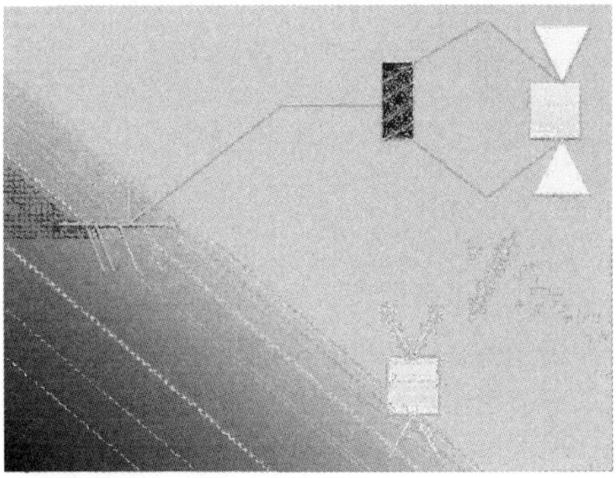

Fig. 1. A concept picture of the simulation scenario.

7. Conclusions and Future Work

The main contribution of this paper is to provide summary information on an industrial research project aiming to provide unprecedentedly dexterous and robust underwater and open-ended environment mobile manipulation capabilities. This project aims to exploit cutting edge basic research in manipulation, mechanics and stochastic kinematics to obtain breakthrough features in much needed novel design soft-arm mobile manipulation systems. Some preliminary simulation results are shown and discussed. The mathematical models developed in this project may in a less short time frame allow to mimic the climbing behaviors of animals such as the gibbons which actually exploit their dynamics while climbing ('brachiating'), see [11].

Acknowledgments

This work is partially funded by Regione Liguria and the European ERDF (European Regional Development Fund) within the POR CRO Regione Liguria 2007/2013 program. This project involves scientific cooperation with CNR-ISSIA, CNR-IMATI and the Dept. of Applied Mechanics and Machine design of the University of Genova.

References

1. R. Platt, R. Tedrake, L.P. Kaelbling, and T. Lozano-Perez. Belief space planning assuming maximum likelihood observations. In Proceedings of Robotics: Science and Systems, 2010.
2. G.S. Chirikjian. Information Theory on Lie-groups and Mobile Robotics Applications, IEEE ICRA 2010, Anchorage (AK), 2010
3. Radu Bogdan Rusu, Andreas Holzbach, Rosen Diankovy, Gary Bradski, Michael Beetz. Perception for Mobile Manipulation and Grasping using Active Stereo. IEEE-RAS International Conference on Humanoid Robots (Humanoids09), Paris, 2009
4. W.Nederbragt, B. Ravani. Design of Tactile Fixtures for Robotics and Manufacturing, ASME J. Mechanical Design, 119, pp. 204-211, (1997)
6. Y.Wang and G. Chirikjian. Error Propagation on the Euclidean Group With Applications to Manipulator Kinematics. *IEEE Transactions on Robotics*, Vol. 22(4): 591-602, (2006)
8. G.S. Chirikjian. Stochastic Models, Information Theory, and Lie Groups, Vol. 2, Birkhauser,(2010), in press
9. Yarp website. http://eris.liralab.it/yarp/
10. F.P. Bonsignorio. On the stochastic stability and observability of controlled serial kinematic chains, ASME ESDA 2010, Istanbul, 2010
11. M. Tavakoli, L. Marques, F. Bonsignorio. Biomimetic Inspiration in Climbing Robots, Clawar 2010, Nagoya, 2010
12. www.kivasystems.com

PLANNING OF MOTION TRAJECTORIES FOR THE CONTROL OF TORQUE-UNIT MANIPULATOR

K. YOSHIDA*, Y. KATAYAMA and T. KINUGASA

Department of Mechanical Systems Engineering, Okayama University of Science, 1-1 Ridai-cho, Okayama 700-0005, Japan
**E-mail: k_yoshida@mech.ous.ac.jp*

K.FUKUDA

Freelance
E-mail: k5-fukuda@mvd.biglobe.ne.jp

K. OSUKA

Department of Mechanical Engineering, Osaka University 2-1, Yamadaoka, Suita, Osaka 565-0871, Japan
E-mail: osuka@mech.eng.osaka-u.ac.jp

A new design concept of space manipulator has been proposed, which is called Torque-unit Manipulator (TUM). Each joint of TUM is free and a device which is called "torque unit" is equipped on each link. The "torque unit" can be made easily of a DC motor and a disc, for example. Position-controllability of the links of TUM has been shown with conventional control laws for manipulators, however, when such control laws are used, positions of the links are controlled to desired positions but the angular velocities of the discs usually do not come to zero. The reason is that TUM is a kind of nonholonomic system hence the angular velocity of each disc at each time depends on the trajectories of the links. In this paper, a scheme is presented for control of positions of the links and angular velocities of the discs by plannning motion trajectories of the links for 2-dimensional N d.o.f. TUM .

Keywords: Planning Motion Trajectory; Space manipulator; Torque-unit Manipulator; Nonholonomic system.

1. Introduction

We have proposed a new design concept of space manipulator which is called "Torque-unit Manipulator (TUM)"[1] . Figure 1 shows a schematic picture of TUM. The TUM also has an open-loop kinematic chain as the typical traditional types of manipulators, however the differences are as follows.

(i) Each joint of the TUM is free joint which is 1 degree-of-freedom.

(ii) Each link has a "torque-unit" on an arbitrary position. The "torque-unit" applies torque to the link on the position. The "torque-unit" can be made, for example, of a rotary actuator such as AC-servo motor and a disc which is connected to the rotating shaft of the actuator.

Packing an actuator and a disc into one "unit", we can make a torque-unit and attach it on an arbitrary position of each link of a kinematic chain whose joints are all free. Thereby, we would be able to construct a Torque-unit Manipulator very easily. The TUM would have merits, for example, **an easy maintenance**, **alterability of dynamical feature**. See Ref. 1 in detail. The easy maintenance is very important if we use a manipulator in such a cruel environment as the space. TUM is an interesting sample of Multibody Systems.

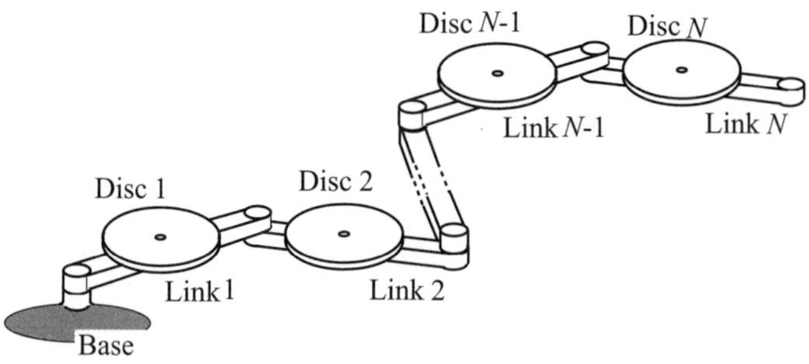

Fig. 1. TUM: A New Design Concept of Space Manipulator

We have shown the position-controllability of the links for general type of N d.o.f. TUM by using a conventional control law for manipulators.[2] However there arises a problem in the position control of the TUM with the control law. When we use it, we only focus on the movement of links and ignore the behavior of the discs in torque-units. Hence, we only ensure that positions of the links are controlled to desired positions and that the angular velocities of the discs do not go infinity even asymptotically. The angular velocities of the discs usually do not come to zero and result to be constant when the links come to desired positions. Because of this problem, TUM might get out of control. The residual angular velocity might give bad influence to the actuator. The problem arose because we did not take TUM as a system with non-holonomic constraints. Taking the all variables to represent the state of TUM into account, i.e., taking the movement of the links and also the discs into account, we have found that TUM is a system with non-holonomic constraints: a Non-holonomic system.

Taking TUM as a non-holonomic system, Osuka, *et al.*[3] have made it clear that the angular velocity of each disc at each time depends on the trajectories of the links: how the links moved until the time from the beginning. Based on the condition, they have proposed a scheme to make the discs stop when the links come to desired positions for N d.o.f. TUM which can move in a 2-dimensional plane.

Generalizing the problem in Ref. 3, we have considered the possibility to control the position of each link to desired position and the angular velocity of each disc to desired constant value which is not zero. (When the all discs have constant angular velocities, the links stay in the same positions.) As results of the consideration, we have given the followings:[4]

- We can control the angular velocities of $N-1$ discs; angular velocity of one disc is dependent on the angular velocities of the other discs.
- We show a set of trajectories of the links which lead the positions of the links to desired positions and the angular velocities of $N-1$ discs to desired constant values.

We have shown a set of trajectories of the links for the control of positions of linkes and angular velocities of the discs, however, in planning the trojectories we need the exact values of some physical parameters of the TUM, e.g., mass, mass center, inertial moment of the links. It is difficult to obtain exact values of such physical parameters. By using the uncertain parameter values we cannot plan the trajectories of the links good enough. We present in this paper a scheme to control the position of each link and to make the angular velocity of each disc to come to a desired value asymptotically in enough accuracy by planning trajectories of the links of the TUM, even in case that the values of the physical parameters are uncertain.

2. *N* D.O.F. TUM

In this section, we describe a N degree-of-freedom TUM which we treat in this paper. Each "torque-unit" is made of a rotational motor and a disc.

Figure 2 shows a schematic picture of the TUM. All the joint axes are parallel and rotational axis of each disc is parallel to the joint axes. The TUM can move only in a plane. First, we attach a coordinate system for each link and each disc to describe the motion of the TUM.

Let $(o_i; \boldsymbol{x}_i, \boldsymbol{y}_i, \boldsymbol{z}_i)$ denote the coordinate system i, where o_i denotes the origin of the coordinate system. All the $\boldsymbol{z}_i$ are aligned with each other. $\boldsymbol{x}_i$ is aligned with longitudinal direction of link i. Thereby we can define the joint variable. Let θ_i for $i = 1, 2, \cdots, N$ denote the joint angle. Let $\boldsymbol{\theta} = \begin{bmatrix} \theta_1 \ \theta_2 \ \cdots \ \theta_N \end{bmatrix}^T$, where superscript T denotes transposition. We attach a coordinate system $(o_{di}; \boldsymbol{x}_{di}, \boldsymbol{y}_{di}, \boldsymbol{z}_{di})$ to the disc i such that $\boldsymbol{z}_{di}$ is aligned with the rotational axis of the disc i and o_{di} is at the center of mass of the disc i. The rotational displacement of the disc i is denoted by φ_i which is the angle between $\boldsymbol{x}_i$ and $\boldsymbol{x}_{di}$ and let $\boldsymbol{\varphi} = \begin{bmatrix} \varphi_1 \ \varphi_2 \ \cdots \ \varphi_N \end{bmatrix}^T$. For link i, let l_i denote distance from o_i to o_{i+1}, m_i mass, and I_i inertial moment around the center of mass, respectively. For disc i, let m_{di} denote mass and I_{di} inertial moment around the center of mass. Here, we assume that the center of mass of link i and o_{di} are on the line from o_i to o_{i+1}. Then, let r_i denote distance from o_i to center of mass of link i and l_{di} distance from o_i to o_{di}, respectively.

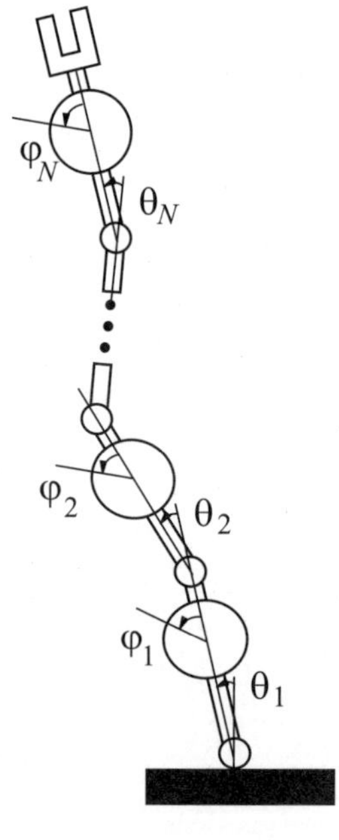

Fig. 2. 2-dimensional TUM

Then the dynamic model for the links and discs of the TUM is obtained as follows, see Ref. 2 in detail,

$$\boldsymbol{H}_{11}(\boldsymbol{\theta})\ddot{\boldsymbol{\theta}} + \boldsymbol{H}_{12}(\boldsymbol{\theta})\ddot{\boldsymbol{\varphi}} + \dot{\boldsymbol{H}}_{11}(\boldsymbol{\theta})\dot{\boldsymbol{\theta}} - \frac{\partial}{\partial \boldsymbol{\theta}}(\frac{1}{2}\dot{\boldsymbol{\theta}} \cdot \boldsymbol{H}_{11}(\boldsymbol{\theta})\dot{\boldsymbol{\theta}}) = \boldsymbol{0}, \qquad (1)$$

$$\boldsymbol{H}_{12}^{T}(\boldsymbol{\theta})\ddot{\boldsymbol{\theta}} + \boldsymbol{H}_{22}\ddot{\boldsymbol{\varphi}} = \boldsymbol{\tau}_{disc}, \qquad (2)$$

where, $\boldsymbol{H}_{11}(\boldsymbol{\theta})$ is $n \times n$ symmetry matrix that is the inertial matrix of the open-loop kinematic chain with frozen discs.

$\boldsymbol{H}_{12}$ is an upper triangular matrix whose elements are I_{di}, $1 \leq i \leq N$. $\boldsymbol{H}_{12}$ is a diagonal matrix whose elements are I_{di}, $1 \leq i \leq N$. and $\boldsymbol{\tau}_{disc} = \begin{bmatrix} \tau_{disc1} & \tau_{disc2} & \cdots & \tau_{discN} \end{bmatrix}^T$ denotes the vector of applied torque to the discs.

If we focus on only the movement of links and ignore the behavior of the discs in torque-units, the dynamic model of the TUM is the same as that of conventional manipulators, hence, we can use conventional control laws for the TUM. However, if we use such conventional control laws, the angular velocities of the discs do not usually come to zero and result to be constant when the links come to desired positions. The problem arose because we did not take TUM as a system with non-holonomic constraints. Taking the all variables to represent the state of TUM into account, i.e., taking the movement of the links and also the discs into account, TUM is a Non-holonomic system.

Here, we state a problem.

Problem 1 Suppose the TUM stays at an arbitrary attitude (i.e., joint angles of the links are arbitrary constants and angular velocities of the links are zero). and angular velocities of the discs are arbitrary constants (The links do not move as far as the discs have constant angular velocities.) Then, are there sets of trajectories of the links which lead the links to desired positions and let the angular velocities of the discs to be desired constant when the links get to the desired positions in time interval T?

Suppose the links of TUM stays in arbitrary positions and angular velocities of the discs have arbitrary constants as initial state. It has been shown that there exist a set of trajectories of the links which lead the links to desired positions in time interval T and the angular velocities of the discs come to be the same as the initial constant values when the links come to the desired positions. Hence, we can give a solution to **Problem 1** if we can give a solution to **Problem 2**:

Problem 2 Suppose the links of TUM stay in arbitrary positions and angular velocities of the discs have arbitrary constants as initial state. Then, are there sets of trajectories of the links which lead the links back to the initial positions and let the angular velocities of the discs to be zero when the links get back to the initial position.

We have given the following proposition[4] as a solution to **Problem 2**

Proposition1 There is at least a set of trajectories of the links which lead the links back to the initial positions and let the angular velocities of N-1 discs to be desired constants. It is inevitably impossible to let the angular velocities of all the discs to be desired constants.

We have given a set of trajectories of the links for the control of positions of linkes and angular velocities of the discs in **Proposition1**, however, in planning the trojectories we need the exact values of some physical parameters of the TUM, e.g., mass, mass center, inertial moment of the links. It is difficult to obtain exact values of such physical parameters.

Then, we present in the next section a scheme to control the position of each link and to make the angular velocity of each of N-1 discs to come to a desired value asymptotically in enough accuracy by planning trajectories of the links of the TUM, even in case that the values of the physical parameters are uncertain.

3. Trajectory Planning for Control of TUM in Case of Uncertain Parameter Values

We give the following proposition.

Proposition2 Suppose that each link of N d.o.f. TUM is in a stationary state and each disc is rotating with a constant angular velocity as an initial state. Then we plan motion trajectories for the joint angles of the links as follows:

$$\begin{cases} \dot{\theta}_{i-1} = B\sin\omega t \\ \dot{\theta}_{i+1} = -\dot{\theta}_{i-1} \\ \dot{\theta}_j = 0 \qquad \text{for } j \neq i-1, i+1, \end{cases} \tag{3}$$

where, $\omega = \dfrac{2\pi}{T_1 - T}$, $T_1 > T$ and θ_j is the joint angle of link j,

$$B = \sqrt{a}\sqrt{|\dot{\phi}_{di} - \dot{\phi}_i(T)|}, \tag{4}$$

a is a constant, $\dot{\phi}_{di}$ is a desired angular velocity for disc i, and $\dot{\phi}_i(T)$ is the angular velocity for disc i at time T. If the TUM moves according to the trajectories from T to T_1, we have

$$|\dot{\phi}_{di} - \dot{\phi}_i(T_1)| < |\dot{\phi}_{di} - \dot{\phi}_i(T)| \tag{5}$$

as far as a is within a range. And the angular velocities of the disc j $(i < j \leq N)$ have the same constant values as those at time T and the joint angles of the links have the same values as those at time T.

Proof Integrating Eq.(1) from T to T_1, we obtain

$$\int_T^{T_1} \boldsymbol{H}_{12}(\boldsymbol{\theta})\ddot{\varphi}dt = \int_T^{T_1} \frac{\partial}{\partial \boldsymbol{\theta}}(\frac{1}{2}\dot{\boldsymbol{\theta}} \cdot \boldsymbol{H}_{11}(\boldsymbol{\theta})\dot{\boldsymbol{\theta}}) \tag{6}$$

If the TUM moves according to the trajectories above, we obtain

$$\dot{\varphi}_i(T_1) - \dot{\varphi}_i(T) = \frac{-l_{i-1}R_i}{2I_{di}} \sin\theta_i B^2(T_1 - T) \tag{7}$$

for the i-th equation from the top of Eq.(6). In the equation, R_i is defined as

$$R_i = m_i r_i + M_{i+1} l_i + m_{di} l_{di}, M_i = \sum_{j=i}^{N}(m_j + m_{dj}). \tag{8}$$

Then, substituting B into Eq.(7), we obtain

$$\dot{\varphi}_i(T_1) = \frac{-l_{i-1}R_i}{2I_{di}} \sin\theta_i a(T_1 - T)|\dot{\varphi}_{di} - \dot{\varphi}_i(T)| + \dot{\varphi}_i(T). \tag{9}$$

Taking a norm of the difference between $\dot{\varphi}_i(T_1)$ and $\dot{\varphi}_{di}$, we obtain

$$|\dot{\phi}_{di} - \dot{\phi}_i(T_1)| < |\dot{\phi}_{di} - \dot{\phi}_i(T)| \tag{10}$$

as far as a is within a range according to $\sin\theta_i$ which is a constant. □

Hence if the TUM repeats to move according to the trajectories with B which would be updated at each planned trajectories, the angular velocity of disc i approaches to the desired angular velocity asymptotically. The angular velocities of the disc$j$ $(i < j \leq N)$ have the same constant values as those at time T and the joint angles of the links have the same values as those at the initial stationary state. Beginning with the N th disc, we try to let angular velocities of discs to be desired constant value one by one, i.e., first, we try to let angular velocity of the N th disc to be desired value, next that of N-1 th disc, $\cdots$, finally that of 2nd disc. Then, we can control the position of each link and make the angular velocity of each of N-1 discs to come to a desired value asymptotically in enough accuracy.

4. Conclusions

If we focus on only the movement of links and ignore the behavior of the discs in torque-units, the dynamic model of the TUM is the same as that of conventional manipulators, hence, we can use conventional control laws for the TUM. However, if we use such conventional control laws, the angular velocities of the discs result to be constant when the links come to desired positions. The problem arose because if we take the all variables to represent the state of TUM into account, TUM is a Non-holonomic system.

Extending the result by Osuka *et al.*,[3] we have given a consideration of the posibility to control the posion of each link and the angular velocity of each disc to desired constant values which are not zero by planning trajectories of the links for 2-dimensional N d.o.f. TUM. As results of the consideration, we have shown that there exists a set of trajectories of the links which lead the positions of the links to desired positions and the angular velocities of $N-1$ discs to desired constant values. We have also shown a set of trajectories of the links for the control of positions of linkes and angular velocities of the discs,[4] however, in planning the trojectories we need the exact values of some physical parameters of the TUM. Hence, we have presented in this paper a scheme to control the position of each link and to make the angular velocity of each of N-1 discs to come to a desired value asymptotically in enough accuracy by planning trajectories of the links, even in case that the values of the physical parameters are not exact. We will show simulation results of the presented scheme.

References

1. K.Osuka, K.Yoshida, and T.Ono, New Design Concept of Space Manipulator, in *Proc. of The 33rd IEEE CDC*, pp.1823–1825 (1994).
2. K.Osuka, K.Yoshida, and T.Ono, Proposal of Torque-unit Manipulator, *Journal of the Robotics Society of Japan*, **14**, 7, pp.1018–1025 (1996) (in Japanese).
3. K.Osuka and T.Nohara, Attitude Control of Torque Unit Manipulator via Trajectory Planing, *Journal of the Robotics Society of Japan*, **18**, 4, pp.612–615 (2000) (in Japanese).
4. K.Yoshida and K.Osuka, On Motion Trajectories for the Control of All State Variables for Torque-unit Manipulator, *Journal of the Robotics Society of Japan*, **25**, 7, pp.1108–1117 (2007) (in Japanese).

A GOAL DRIVEN PERSPECTIVE TO GENERATE COMPLEX HUMANOID MOTION SYNTHESIS

Joseph Salini, Vincent Padois, Aurélien Ibanez, Philippe Bidaud
Université Pierre et Marie Curie,
Institut des Systèmes Intelligents et de Robotique - CNRS UMR 7222
Pyramide Tour 55, BC 173 — 4 Place Jussieu, F-75005 Paris
E-mail: {salini,padois,ibanez,bidaud}@isir.upmc.fr

Axel Buendia
SpirOps AI,
8 passage de la bonne graine, F-75011 Paris
CNAM CEDRIC,
292 rue St Martin, F-75003 Paris
E-mail: axel.buendia@spirops.com

Keywords: Humanoid Control; Dynamic; Linear Quadratic Problem; Goal-Driven Perspective.

1. Introduction

In the paper of Salini *et al.*,[1] the authors proposed a generic controller which performs the synthesis of complex humanoid whole-body behaviors. The core of this low-level controller is based on a Linear Quadratic Program (LQP) where the minimum of a quadratic cost function is computed, subject to linear constraints. The cost function represents a weighted sum of tasks to be achieved by the robot and the constraints the physical limitations of the system. Similar controllers are used in the works of Abe *et al.*[2] and Colette *et al.*[3]

Although multi-criteria (weights based) controllers cannot give a meaningful physical interpretation of the criterion to optimize, the approach is competitive in terms of tasks accomplishment compared to hierarchies based methods (see Kanoun *et al.*,[4] and Escande *et al.*[5]). It is computationally more efficient and offers greater flexibility by allowing complex task blending. However, the choice of the weights/hierarchies is left to the programmer, which can limit the missions/goals achievement.

This issue can be bypassed with a higher level controller which reactively elects tasks of interest based on contextual information. These tasks and their associated relative weights can then be accounted for at a lower, physics related, control level.

A recent work in this domain is proposed in Philippsen *et al.*[7] which introduces an architecture combining a Whole Body Control (low) level and a Reactive Symbolic Planning (high) level through what is called a Mobile Manipulation Database and whose goal is to deal with information which is relevant to all components of the system.

The contribution of the proposed work has to be interpreted from an "action selection" perspective, where no explicit plan is constructed, and where a mission is described in terms of high level goals, associated to contextual information (percepts) and rules, which allow the triggering of high level actions — blends of lower level tasks. Hence it illustrates the use of such a high level controller in combination with a powerful task-level controller where complex whole-body humanoid behaviors cannot be planned in a deterministic manner.

2. Method

2.1. *Task level control*

2.1.1. *The Constraints*

The task-level controller has been extensively described in the paper of Salini *et al.*[1] The robot is subject to many constraints, such as the equation of motion

$$M(\boldsymbol{q})\ddot{\boldsymbol{q}} + \boldsymbol{n}(\boldsymbol{q}, \dot{\boldsymbol{q}}) = J_{\chi}(\boldsymbol{q})^T \boldsymbol{\chi} + \boldsymbol{g}(\boldsymbol{q}). \tag{1}$$

where $\boldsymbol{q}, \dot{\boldsymbol{q}}, \ddot{\boldsymbol{q}}$ are respectively the generalized coordinates, velocity and acceleration vectors, $M, \boldsymbol{n}, \boldsymbol{g}$ the generalized inertia matrix, the non-linear effects vector (Coriolis and centrifugal) and the gravity vector. $J_{\chi}(\boldsymbol{q})^T = \left[J_{c,1}(\boldsymbol{q})^T \ \dots \ J_{c,n_c}(\boldsymbol{q})^T \ S^T \right]$ is the generalized external wrench jacobian composed of the jacobians of contacts $J_{c,i}(\boldsymbol{q})$ and of the actuation matrix S. $\boldsymbol{\chi} = \left[\boldsymbol{w}_{c,1}^T \ \dots \ \boldsymbol{w}_{c,n_c}^T \ \boldsymbol{\tau}^T \right]^T$ is called the **action variable** of the system, with $\boldsymbol{w}_{c,i}$ the external wrenches of contact and $\boldsymbol{\tau}$ the input torque vector.

The robot has also several limitations, which can be interpreted as internal or external constraints. The internal ones are typically the joint actuation limits, as the torque (Eq. 2) and its derivative (Eq. 3), the position (Eq. 4), the velocity (Eq. 5) and the acceleration (Eq. 6), expressed as follows:

$$\boldsymbol{\tau}_{min} \leq \tau \leq \boldsymbol{\tau}_{max} \tag{2}$$
$$\dot{\boldsymbol{\tau}}_{min} \leq (\tau - \tau_{prec})/dt \leq \dot{\boldsymbol{\tau}}_{max} \tag{3}$$
$$\boldsymbol{q}_{min} \leq \boldsymbol{q} + h_2\dot{\boldsymbol{q}} + \frac{(h_2)^2}{2}\ddot{\boldsymbol{q}} \leq \boldsymbol{q}_{max}, \tag{4}$$
$$\dot{\boldsymbol{q}}_{min} \leq \dot{\boldsymbol{q}} + h_1\ddot{\boldsymbol{q}} \leq \dot{\boldsymbol{q}}_{max} \tag{5}$$
$$\ddot{\boldsymbol{q}}_{min} \leq \ddot{\boldsymbol{q}} \leq \ddot{\boldsymbol{q}}_{max} \tag{6}$$

where τ_{prec} is the previous torque vector and h_1, h_2 are anticipation coefficients set to predict the future value of the state $(\boldsymbol{q}, \dot{\boldsymbol{q}})$ given the generalized acceleration $\ddot{\boldsymbol{q}}$. The external constraints typically describe the interactions of the robot with its environment. For example frictional contacts are characterized by two constraints, one which equals the acceleration at contact point to 0 (Eq. 7) and the other which bounds the force into the linearized friction cone $C_{f,i}$ (Eq. 8).

$$J_{c,i}(\boldsymbol{q})\ddot{\boldsymbol{q}} + \dot{J}_{c,i}(\boldsymbol{q}, \dot{\boldsymbol{q}})\dot{\boldsymbol{q}} = \boldsymbol{0} \tag{7}$$
$$C_{f,i}\boldsymbol{w}_{c,i} \leq \boldsymbol{0} \tag{8}$$

Kinematic loops, which are also considered as external constraints, link two frames of the tree-structure system, and result in the following equation $\boldsymbol{t}_i - \boldsymbol{t}_j = \boldsymbol{0}$ where $\boldsymbol{t}_i$ and $\boldsymbol{t}_j$ are respectively the twists of frames i and j. Expressed in terms of task space acceleration, this equation becomes

$$(J_i(\boldsymbol{q}) - J_j(\boldsymbol{q}))\,\ddot{\boldsymbol{q}} + (\dot{J}_i\,(\boldsymbol{q}, \dot{\boldsymbol{q}}) - \dot{J}_j\,(\boldsymbol{q}, \dot{\boldsymbol{q}}))\dot{\boldsymbol{q}} = \boldsymbol{0}. \tag{9}$$

2.1.2. *The Tasks*

Given the **variable** of the system $\boldsymbol{\chi}' = \begin{bmatrix} \ddot{\boldsymbol{q}}^T & \boldsymbol{\chi}^T \end{bmatrix}^T$, a task can be defined in a general manner as the minimization

$$\min_{(\boldsymbol{\chi}')} \; T_j(\boldsymbol{q}, \dot{\boldsymbol{q}}, \boldsymbol{\chi}') = \left\| E_j(\boldsymbol{q})\boldsymbol{\chi}' + \boldsymbol{f}_j(\boldsymbol{q}, \dot{\boldsymbol{q}}) \right\|. \tag{10}$$

In the case of a desired task space acceleration $\dot{\boldsymbol{t}}_j^{des}$, the task is written $\left\| J_j(\boldsymbol{q})\ddot{\boldsymbol{q}} + \dot{J}_j(\boldsymbol{q}, \dot{\boldsymbol{q}})\dot{\boldsymbol{q}} - \dot{\boldsymbol{t}}_j^{des} \right\|$. With D_a defined such that $\ddot{\boldsymbol{q}} = D_a\boldsymbol{\chi}'$, the formulation of a task space acceleration leads to $E_j(\boldsymbol{q}) = J_j(\boldsymbol{q})D_a$ and $\boldsymbol{f}_j(\boldsymbol{q}, \dot{\boldsymbol{q}}) = \dot{J}_j(\boldsymbol{q}, \dot{\boldsymbol{q}})\dot{\boldsymbol{q}} - \dot{\boldsymbol{t}}_j^{des}$.

In the case of a desired task space wrench $\boldsymbol{w}_j^{des}$, the task is written $\left\| \boldsymbol{w}_j - \boldsymbol{w}_j^{des} \right\|$. With $D_{w,j}$ defined such that $\boldsymbol{w}_j = D_{w,j}\boldsymbol{\chi}'$, the formulation of a task space wrench leads to $E_j(\boldsymbol{q}) = D_{w,j}$ and $\boldsymbol{f}_j(\boldsymbol{q}, \dot{\boldsymbol{q}}) = \boldsymbol{w}_j^{des}$.

These generic tasks allow to control many parts of the robot, as frames and contact forces, but also its joints in acceleration and torque, its center of mass (CoM), etc.

The desired values of the tasks $\dot{\boldsymbol{t}}_j^{des}$ and $\boldsymbol{w}_j^{des}$ are computed separately with any controller of the literature, for example a simple proportional derivative controller $\dot{\boldsymbol{t}}^{des} = \dot{\boldsymbol{t}}^{goal} + K_p\boldsymbol{\epsilon}_p + K_d$ where K_p, K_d are respectively the coefficient of stiffness and damping, and more complex one like a preview controller to control the CoM of the robot for walking tasks.[10]

$$\dot{\boldsymbol{\epsilon}}_p \tag{11}$$

2.1.3. *The Resolution: Weighting and Hierarchy*

Two strategies exist to solve a multi-tasks problem, the weighting or the hierarchy. Given a set of n_t tasks, in the first case each task is related to a weight, $(T_i(\boldsymbol{q}, \dot{\boldsymbol{q}}, \boldsymbol{\chi}'), \omega_i)$ $i \in [1..n_t]$, and the minimization problem is defined as:

$$\begin{aligned} &\min_{(\boldsymbol{\chi})} \tfrac{1}{2}\left(\textstyle\sum_{i=1}^{n_t}((\omega_i.T_i(\boldsymbol{q}, \dot{\boldsymbol{q}}, \boldsymbol{\chi}'))^2) + (\omega_0.T_0(\boldsymbol{q}, \dot{\boldsymbol{q}}, \boldsymbol{\chi}'))^2\right) \\ &s.t. : \quad G\boldsymbol{\chi} \preceq \boldsymbol{h} \\ &\qquad\quad A\boldsymbol{\chi} = \boldsymbol{b} \end{aligned}$$

where $\omega_0 \ll \omega_i$ $\forall i$ and G, A, $\boldsymbol{h}$ and $\boldsymbol{b}$ are the concatenated representations of the constraints.

In the second case, each task is related to a level, $(T_i(\boldsymbol{q}, \dot{\boldsymbol{q}}, \boldsymbol{\chi}'), i)$ $i \in [1..n_t]$, and the minimization problem is solved recursively as following:

for i=1..n_t:

$$\begin{aligned} &\min_{(\boldsymbol{\chi})} \frac{1}{2}\left((T_i(\boldsymbol{q}, \dot{\boldsymbol{q}}, \boldsymbol{\chi}'))^2 + (w_0.T_0(\boldsymbol{q}, \dot{\boldsymbol{q}}, \boldsymbol{\chi}'))^2\right) \\ &s.t. : \; G\boldsymbol{\chi} \preceq \boldsymbol{h} \\ &\qquad\;\; A_i\boldsymbol{\chi} = \boldsymbol{b}_i \end{aligned}$$

$$\begin{aligned} A_{i+1} &\leftarrow A_i \cup E_i \\ \boldsymbol{b}_{i+1} &\leftarrow \boldsymbol{b}_i \cup (E_i\boldsymbol{\chi}_i^*) \end{aligned}$$

As the general purpose is to synthesize the humanoid motion, transitions may occurs in the set of importances and constraints. These transitions may involve sharp evolutions of the actuation vector, so in order to avoid this problem the authors prefer to use the weighing strategy, because weights (floating values) can be modified in a continuous way, whereas in a hierarchy strategy levels (discrete values) change in a discontinuous way.

2.2. *Goal driven control*

To blend the different tasks and perform good behavior synthesis, one needs to change the tasks weights efficiently. This can be done manually, as shown in the paper of Salini *et al.*[1] for simple missions, but it has its limitation when unknown dynamic external events occur. The robot must be controlled with goals in a high-level description language as "grab the red box" or "go to this location", but also with the contextual information which describes the robot perceptions, i.e. "box is there", "location is reachable", "goal has been performed", etc.

The lowest level includes a set of n_t **tasks** T (defined above) that the robot can perform. It is composed of posture tasks, i.e. sitting, crouching, and standing, and frame tasks, i.e. hands, feet, head, pelvis, bottom and CoM. The blending of all these tasks is done by the LQP controller and the weighting strategy.

The intermediate level includes a set of n_a **actions** $\mathbb{A}$ which defines predefined tasks combination using **blending factors** β_{ij} between the action $\mathbb{A}_j$ and the task T_i.

Finally, at the highest level, a set of n_g **goals** G is defined in accordance to the objective of the mission. Each goal G_k refers to an action $\mathbb{A}_j$ and compute its **Desire** of completion $\delta_{jk} = \alpha_{jk}.\theta_{jk}$ according to its **adequacy** $\alpha_{jk} \in [0,1]$ ("I want"), and its **opportunity** $\theta_{jk} \in [0,1]$ ("I can"). Notice that one action can be related to many goals, for example "running" related to "attack" and "flee".

The decision making engine selects for each action the goal with the highest desire δ_j^*, which leads to the weight ω_i of task T_i

$$\delta_j^* = \max_k(\delta_{jk})$$
$$\omega_i = \sum_{j=1}^{n_a} \beta_{ij}\delta_j^*. \tag{12}$$

This high level goal driven controller is performed with Spirops,[11] which provides an open environment to easily and incrementally design autonomous decision processes, thanks to a graphical editor which edits independent goals. Each of them can be seen as extended 0+ fuzzy rules,[12] and the decision making engine computes the interest and the parameters

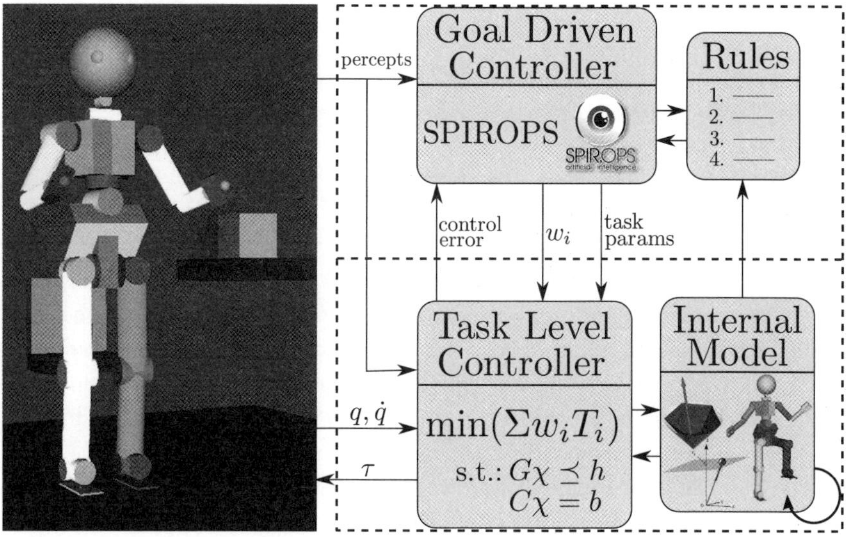

Fig. 1. Overall control architecture

for each action to produce. The general architecture of the whole controller is shown in Fig. 1.

3. Results

This architecture has been used in several simulated experimental results where a humanoid robot iCub moves in an unknown environment *a priori* and has to perform some high level goals, for example "to grab a box" and "to drop it".

3.1. *A changing World*

The first experiment shows the iCub robot in some situations with different initial conditions. It may sit or stand, and its objective is to grab a box which can be in its range or not. The situation is evaluated by the high level decision making engine which decides what set of actions it has to perform, and leads to the control of low level tasks. In Fig. 2, the pictures on the left side show that the scenario starts with 2 different initial conditions, And the robot successfully grabs the box. On the right of the figure, the graph shows the evolution of the tasks weights according to its perceptions.

In the second experiment, the robot opens a door many times in the same sequence. It pushes the door at the maximum, and when the task

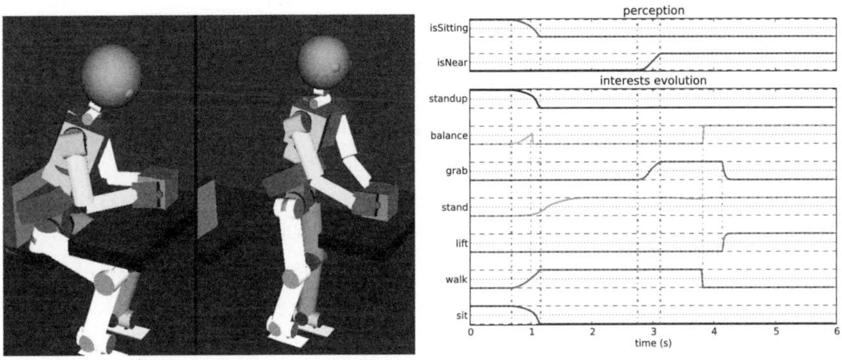

Fig. 2. Experiment 1 – Left: Two scenarios with different initial configurations and one objective, to grab the box – Right: Evolution of the actions interests according to perceptions. Bold dashed red lines show the correlation with perception, the thin ones highlight the relations between actions.

cannot be performed correctly, the robot adapts its position and posture, uses the closest hand to grab the knob and pushes the door again. The Fig. 3 illustrates this experiment.

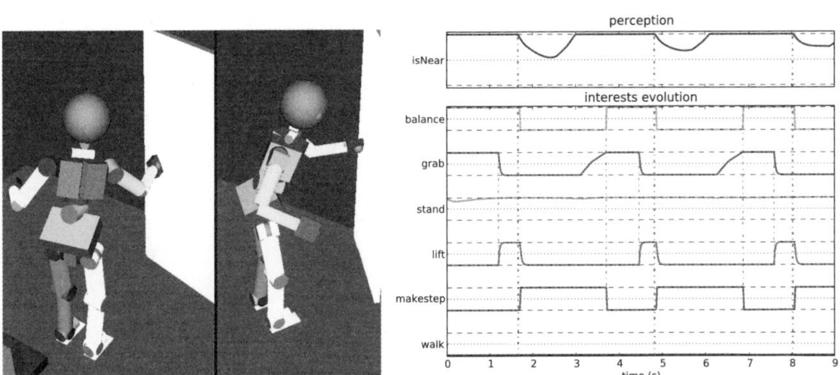

Fig. 3. Experiment 2 – Left: The robot opens a door many times. It uses its left or right hand, depending on the distance between the hands and the knob. – Right: Evolution of the actions interests according to perceptions. Bold dashed red lines show the correlation with perception, the thin ones highlight the relations between actions.

3.2. *Dynamic Adaptation*

The last experiment describes the adaptation of the robot subject to dynamic disturbances. The objective is to displace a box from a location to another. This box can be light or heavy. During the displacement, the center of mass of the system "robot + box" get closer to the base of the support defined by the convex hull of the feet, and if it reaches a threshold, the robot drops the box and moves to a closer location to achieve its objective. Fig. 4 gives the configuration of the robot displacing a heavy box. On the right of the figure, the evolution of the center of mass, the box and the action interests are plotted. Notice the peaks in the evolution of the center of mass, due to the actions of grabbing and dropping.

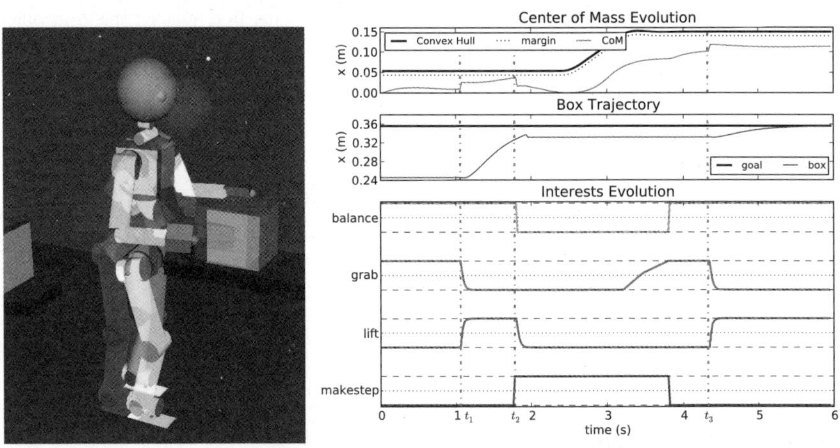

Fig. 4. Experiment 3 – Left: The robot has to displace a $2Kg$ box from one side of the table to another – Right: Evolution of the position of the center of mass and the box, and the interest of the actions. The robot needs to drop the box once and make a step to achieve its goal.

4. Conclusion

This paper shows the combination of a low level controller based on weighted tasks mixed with a LQP, and a high level controller based on goals and actions. This association is useful for non deterministic scenarios, and it is illustrated trough three simulations.

In the future, the authors will focus on the consistence of the tasks to get a physically meaningful interpretation of the weighted sum of tasks in the LQP. Furthermore, a new formalism of the problem will be explored to reduce the computational time of the control loop.

References

1. J. Salini, V. Padois and P. Bidaud, Synthesis of complex humanoid whole-body behavior: a focus on sequencing and tasks transitions, in *Proc. of the IEEE ICRA*, (Shanghai, China, 2011).
2. Y. Abe, M. da Silva and J. Popovic, Multiobjective control with frictional contacts, in *Proc. of the Symposium on Computer Animation*, 2007.
3. C. Collette, A. Micaelli, P. Lemerle and C. Andriot, Robust balance optimization control of humanoid robots with multiple non coplanar grasps and frictional contacts, in *Proc. of the IEEE ICRA*, (Pasadena, USA, 2008).
4. O. Kanoun, F. Lamiraux, P.-B. Wieber, F. Kanehiro, E. Yoshida and J.-P. Laumond, Prioritizing linear equality and inequality systems: Application to local motion planning for redundant robots, in *Proc. of the IEEE ICRA*, (Kobe, Japan, 2009).
5. A. Escande, N. Mansard and P.-B. Wieber, Fast resolution of hierarchized inverse kinematics with inequality constraints, in *Proc. of the IEEE ICRA*, (Anchorage, USA, 2010).
6. W. Li, Optimal control for biological movement systems, PhD thesis, University of California, (SanDiego, USA, 2006).
7. R. Philippsen, N. Nejati and L. Sentis, Bridging the gap between semantic planning and continuous control for mobile manipulation using a graph-based world representation, in *1st Int. Work. on Hybrid Control of Autonomous Systems held in conj. with the Int. Joint Conf. on AI*, (Pasadena, USA, 2009).
8. E. Yoshida, I. Belousov, C. Esteves and J.-P. Laumond, Humanoid motion planning for dynamic tasks, in *Proc. of the IEEE/RAS Int. Conf. on Humanoid Robots*, (Tsukuba, Japan, 2005)
9. S. Cambon, R. Alami and F. Gravot, *The Int. Jour. of Robotics Research* **28**, 104(January 2009).
10. S. Kajita, F. Kanehiro, K. Kaneko, K. Fujiwara, K. Harada, K. Yokoi and H. Hirukawa, Biped walking pattern generation by using preview control of zero-moment point, in *Proc. of the IEEE ICRA*, (Taipei, Taiwan, 2003).
11. A. Buendia, Spirops AI, a behavior production tool.
12. P. D. Magnus, *forallx An Introduction to Formal Logic* (University at Albany, State University of New York, 2005).

SECTION–16

SERVICE ROBOTS

A LEG DESIGN FOR A BIPED HUMANOID SERVICE ROBOT

TAO LI

LARM: Laboratory of Robotics and Mechatronics-DiMSAT, University of Cassino Cassino(Fr), Via Di Biasio 43-03043, Italy

MARCO CECCARELLI

LARM: Laboratory of Robotics and Mechatronics-DiMSAT, University of Cassino Cassino(Fr), Via Di Biasio 43 03043, Italy

A leg mechanism has been proposed specially for biped humanoid service robots as based on the analysis from aspects of working environment, movement performances, energy efficiency, and coexistence with human. Specifications have been elaborated for a suitable leg design. The key characteristic of this leg mechanism is that it has feet and wheels at the leg extremity simultaneously, with the aim to enable it to have advantages of both legged and wheeled robots. Basic motion modes have been illustrated along with its motion capability and operation feasibility.

1. Introduction

Robots working in households, offices, and amusement parks are examples of so called service robots. They should be characterized by performance to permit an efficient adjustable operation within living environments of human beings as pointed out in [1]. These performances are mobility, portability and adaptability to widely varying operations and environmental conditions, so that mobility is the most indispensable for a service robot.

Service robots are most commonly wheeled, but they also include legged ones with two or more legs, like humanoids, or resembling animals. In general, wheeled robots have better stability, reliability and higher efficiency than legged ones. Furthermore, wheeled robots are more flexible in changing directions; and they make less noise while moving because there is no big impact between robot extremities and ground. Because of these advantages, in the past few years, many wheeled robots have been developed in institutes, universities and corporations as research platforms or even commercial products in market [2-4]. They represent good performances in households and offices environment because of flat ground. However, there are not only flat ground but also obstacles and stairs in these places. Thus wheeled robots are not perfect solutions for service robots.

On the other hand, legged robots are more suitable in an unstructured environment than wheeled ones. When referring to service robots, biped humanoid robots are most acceptable from an ethical and psychological point of view [5], and they are endowed with better potential abilities in obstacle overcoming, stair ascending and descending. In [6-9], some representative humanoid robots are presented with different kinds of biped leg mechanisms.

At LARM: Laboratory of Robotics and Mechatronics in University of Cassino, a research line has been dedicated to leg mechanisms since the early of 1990s'. Research works have been carried out both on design solutions and design procedures for leg mechanisms and their applications in walking machines. Several kinds of leg mechanisms have been developed based on linkage mechanisms as reported in [10-12]. In [13], a short account has been reported to illustrate the activity and its results at LARM in the aspects of designing and experiencing leg mechanisms for walking machines.

However, also legged robots have drawbacks, such as low energy efficiency and unstable with respect to wheeled robots. In order to combine advantages of both wheeled and legged robots, hybrid systems have been considered in the design of mobile robots. A hexapod robot and a quadruped robot have been presented in [14] and [15], respectively. They can move on wheels or climb stairs through cooperation of legs. But these robots are not specially designed as service robots for working in human's day life environments.

In this paper, a novel leg mechanism has been proposed, which is supposed to be a suitable solution for biped humanoid service robots.

2. Design Considerations for Service Robots

Characteristics for service robot design depend of the requirements for the applications not only in terms of task goals but also as function of the environment constraints. When considering household and office (Fig. 1) as examples, it can be seen that the space is filled with furniture. Thus a service robot should not have very large dimensions. Furthermore, the dimensions should be similar to an adult in order to make it well adaptable to an environment filled with objects and tools which have been designed and produced according to human-size. In addition, human-sized service robots can be well accepted from a psychological point of view [5]. Besides furniture, there are also thresholds and stairs. Thus service robots should have the abilities of overcoming obstacles and climbing stairs.

During a task, the path between a robot's start location and its destination probably will not be straight. Thus the robot should be able to change directions.

Sometimes a robot may move into a "U" shaped space (in Fig. 1, the table and sofa form "U" shaped areas) where there is no space for changing directions thus it should be able to move backward. In addition, it is better for service robots to have rapid reaction ability and fast speed. Meanwhile, robust stability is required due to the rapid reaction ability and fast speed. Furthermore, it is not suitable for service robots dragging a power supply cable while working or moving around. A built-in battery will be a good choice, which has already been widely used in the design of robots. However, this solution brings a problem with it, which is the less battery duration. Thus a good design proposal should be frugal with electric power otherwise the robots need to charge frequently during working which will also lead to time-consuming.

Based on the abovementioned analysis, it is desirable that the following aspects will be considered in the design of service robots:

- Dimensions should be similar to an adult;
- Ability of overcoming obstacles, ascending and descending stairs;
- Ability of changing directions;
- Ability of moving backward or turning with zero radius;
- Rapid reaction ability and fast speed;
- Robust stability and safety issue in human interactions;
- Energy-saving.

These abovementioned characteristics will be considered as specifications, which play an important role during the design process. A good design will not only provide the desired functionalities but also will generally reduce the control effort, and will present fewer problems during operation.

3. Design of a Biped Leg Mechanism

3.1. *DOF Configuration*

For service robots, they should be able to move in 3 dimensions. Namely the robots must be physically capable of changing directions, climbing up and down stairs, and some other tasks instead of just making straight walking. To be able to do these things, the robots need to have sufficient DOFs. In order to fulfill these requirements, each leg is preliminarily endowed with 6 DOFs, in which three are for hip joint, one is for knee joint, and two are for ankle joints.

3.2. *Actuator Selection*

Normally, hydraulic actuators, pneumatic actuators and electric motors are the three most common types of actuators. But for service robots, electric motors are preferred because both hydraulic and pneumatic actuators are not feasible since both require bulky pumps or compressors, which only can been seen in factories or other situations except in houses.

3.3. *Foot/Wheel Mechanism Design*

A leg mechanism, which has both foot and wheel at the extremity, is proposed. It is designed as a combination of foot and wheel so that the leg has two basic working modes. One is foot mode and the other is wheeled mode as shown in Fig. 2(a) and Fig. 2(b), respectively. A mechanism for buffering and energy-saving has been designed in Fig. 1(c) which will be described in 3.4.

This design allows the robot to choose a suitable motion mode when facing a situation. Namely it can work under wheeled mode on flat ground, with the aim to enable the robot move fast from place to place and change directions with zero turning radius. When needing to overcome thresholds or climb stairs, the robot adopts foot mode. Working in these two modes at suitable situations can make the robot more time and energy efficient, less noisy and more reliable.

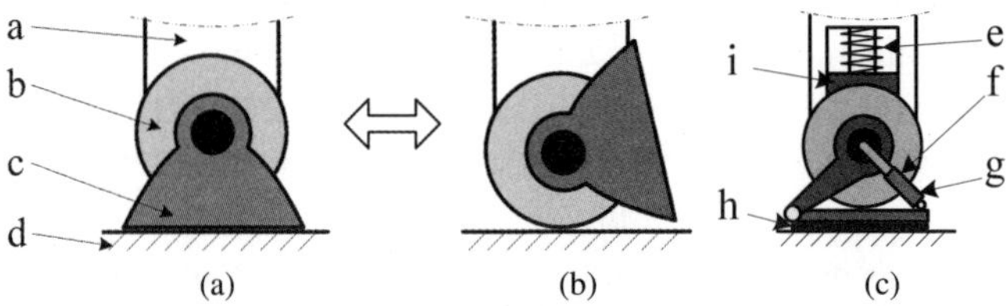

Figure 1. The foot/wheeled leg: (a) foot mode; (b) wheeled mode; (c) buffering and energy-saving solution (a- shank; b- wheel; c- foot; d- ground; e- spring; f- pneumatic spring; g-rubber).

3.4. *Buffer and Energy-saving Device Design*

Impacts between foot and ground have great influence on robot stability and reliability. In Fig. 2(c), one spring (e) and two pneumatic springs (each at one side of the wheel, the one that can be seen in is named as g, the other cannot be seen) are used. Slider i moves along the shank to compress or release the spring for buffering and energy saving. In addition, rubber h also is installed to the bottom side of the foot for absorbing impact generated during a walking.

3.5. *Linkage and Joint Design*

One proposal for the leg design in Fig. 3(a) has been chosen as suitable to fulfill the requirements. At point A, hip Yaw and hip Roll, which are the joints connecting to waist, are omitted. The thigh is composed of two triangles BCD and BEF. Sides CD and BE, which are composed of two linear motors, are length-alterable. Motor 1 is the actuator of hip Pitch, while motor 2 is the actuator of the knee pitch. The shank is a triangle composed of sides EF, EG and FG. When the axis of motor 1 is elongates or curtails, the thigh will make clockwise or anticlockwise turn around joints B, respectively. When the axis of motor 2 is elongates or curtails, the shank will make clockwise or anticlockwise turn around joint F, respectively.

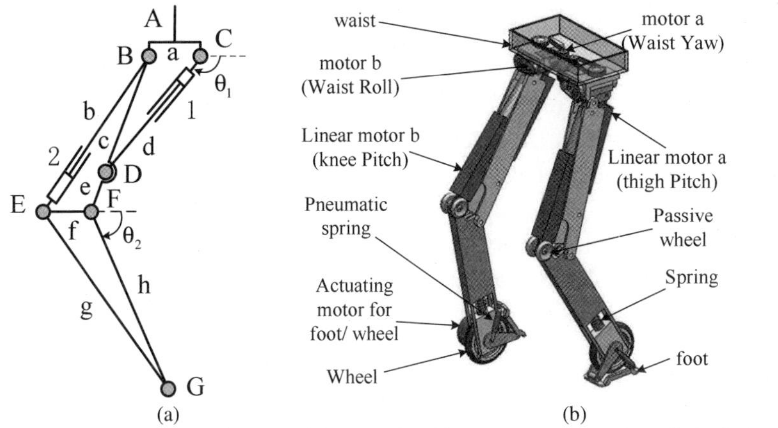

Figure 2. The proposed leg mechanism with translational joints: (a) a scheme; (b) A CAD design

A CAD design of a combination of all the parts proposed in section 3 is shown in Fig. 3(b). Two legs are connected to the waist with two revolute joints respectively. These two joints are Yaw and Roll of the waist. Pitch of the thigh and shank are actuated by two linear motors respectively. In addition, two passive wheels are installed at the knee joints. Motion ranges for the joints are modeled with reference to human joints. Table I shows the typical motion ranges for a human and the robot for the joints of the hip, knee and ankle, [16].

Table 1. Range of motion of human and robot joints, [16]

	human		robot	
Joint angle(deg)	Min.	Max.	Min.	Max.
Hip Yaw	-20	65	-180	60
Hip Roll	-25	40	-50	50
Hip Pitch	-10	135	-45	100
Knee Pitch	-135	5	-125	0
Ankle Roll	-60	30	-45	45
Ankle Pitch	-50	30	-50	30

4. Performance Characteristics

4.1. *Transmission Efficiency*

Transmission efficiency of the linear motors has been studied through numerical simulations carried out in ADAMS environment. Suitable modes are first constructed in SolidWorks environment and then they are imported into ADAMS for a dynamic simulation. The efficiencies of the two motors are evaluated individually, which means when one is under evaluating the other is fixed. The transmission efficiency of a leg mechanism can be defined as

$$K_{ei} = \Delta\theta_i / \Delta l, (i = 1, 2) \tag{1}$$

in which θ_1 is the angle between thigh and horizontal line; θ_2 is the angle between shank and horizontal line in Fig. 3(a). Values of length listed in table 2 have been given to the linkages. These values have been designed with reference to human size (but in a reduced scale). Motions have been given to the thigh and shank with the same velocity of 20mm/s. Two plots are obtained via simulations as shown in Fig. 4. The computed minimum values of the slope shown in Fig. 4(a) and Fig. 4(b) are approximately to 20deg/s and 20.2deg/s, respectively. Thus values of K_{e1} and K_{e2} are approximately to 1deg/mm. These values indicate that the leg mechanism has good transmission efficiency.

Table 2. Sizes for a design solution as in Fig. 3 (in millimeter)

a	b	c	d	e	f	g	h
30	125	150	110	32	140	120	40

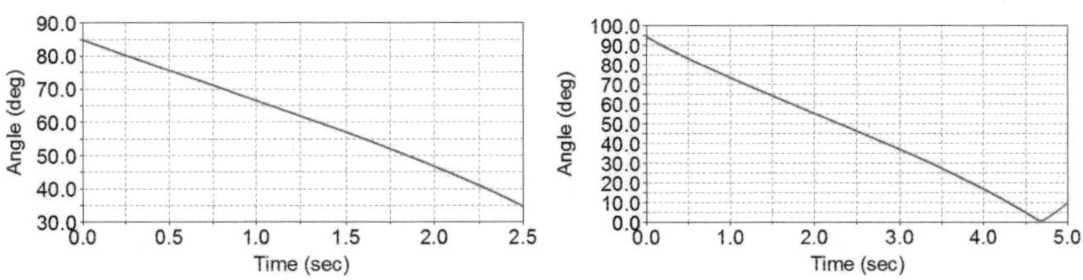

(a) (b)

Figure 3. Computed values of angles vs. time: (a) result of thigh; (b) result of shank.

4.2. *Working Modes*

With two passive wheels installed at the knee joints mentioned in section 3, the leg mechanism can have three basic motion modes as shown in Fig. 5. The motion mode in Fig. 5(a) can be used when the ground is flat. When working under this mode, all joints are locked except the two at the ankle joints. In this situation, control of this mode can be seen as control of an inverted pendulum, which has been implemented in many literatures. The motion mode in Fig. 5(b) can be used when the robot meets stairs or obstacles. Control of this mode can be realized based on ZMP theory. Fig. 5(c) shows a four-wheeled working mode. Under this mode, the robot kneels on ground to provide higher stability and can carry more payload than the others. It also can be used when the robot is idle.

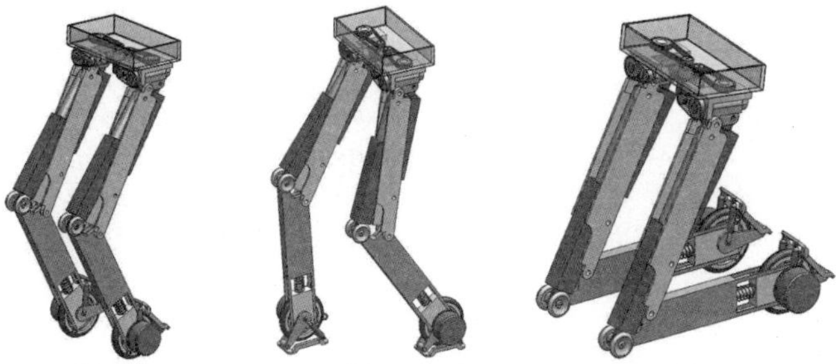

(a) (b) (c)

Figure 4. Three main motion modes: (a) A two-wheeled motion mode; (b) A two-foot motion mode; (c) A four-wheeled motion mode with bended knees.

5. Conclusions

A new leg mechanism is proposed specially for biped humanoid service robots as based on the consideration of the design specifications. The key characteristic of this leg mechanism is that it has feet and wheels at the leg extremity simultaneously, which make it has both advantages of legged and wheeled robots. Feasibility of its three basic motion modes has been verified through theoretical analysis. While working with different modes in different suitable situations, the robot could be more stable, more reliable, less time consuming and have higher energy efficiency.

Acknowledgments

The first author likes to acknowledge Chinese Scholarship Council (CSC) for supporting his PhD study and research at LARM in the University of Cassino in Italy for the years 2010-2012.

References

1 M. Ceccarelli, *New challenging applications for service robots: problems and experience*, proceedings of IARP Workshop, Beijing, 2009, invited paper, No.2.

2 Toyota. Co. JP., *Toyota Partner Robot-announced an overview of the project to develop partner robots designed to function as personal assistants for humans*, http://www.toyota.co.jp/en/special/ robot (2011).

3 RobotShop, *Robot Vacuums*, http://www.robotshop.com/eu/robot-vacuums.html (2011).

4 Information and Robot Technology Research Initiative, *Robot Housekeeper*, December, http://www.youtube.com/watch?v=1zD 45oO0ZO4 (2008).

5 H.F.M. Van der Loos, *Ethics by Design: A Conceptual Approach to Personal and Service Robot Systems*, ICRA Roboethics Workshop, Rome, Italy: IEEE (2007).

6 K. Kaneko, F. Kanehiro, M. Morisawa, K. Miura, and S. N. S. Kajita, *Cybernetic human hrp-4c*, in Humanoids, pp.7–14 (2009).

7 Ill-W. Park, J.Y. Kim, J. Lee, and J.H. Oh, *Mechanical Design of the Humanoid Robot Platform, HUBO*, Journal of Advanced Robotics, Vol. 21, No. 11 (2007).

8 K. Hirai, M. Hirose, Y. Haikawa, and T. Takenaka, *The Development of Honda Humanoid Robot*, Proc. IEEE Int. Conference on Robotics and Automation, pp. 1321-1326 (1998).

9 H. Ulbrich, T. Buschmann, S. Lohmeier, *Development of the Humanoid Robot LOLA*, Applied Mechanics and Materials, Volumes 5-6, pp. 529-540 (2006).

10 M. Ceccarelli, G. Carbone, *A Study of Feasibility for a Leg Design with Parallel Mechanism Architecture*, IEEE/ASME Conference on Advanced Intelligent Mechatronics AIM'09, Singapore, paper no. 131 (2009).

11 M. Ceccarelli, *An Historical Perspective of Robotics Toward the Future*, Fuji International Journal of Robotics and Mechatronics, Vol.13, No.3, pp.299-313 (2001).

12 G. Carbone, M. Ceccarelli, *Legged Robotic Systems*, Cutting Edge Robotics ARS Scientific Book, Wien, pp. 553-576 (2005).

13 M. Ceccarelli, G. Carbone, E. Ottaviano, C. Lanni, *Leg Designs for Walking Machines at LARM in Cassino*, Workshop Robotica per esplorazione lunare unmanned, Roma, July (2009).

14 G. Carbone, M.. Suciu, M. Ceccarelli and D. Pisla, *Design and Simulation of Cassino Hexapode Walking Machine, International Journal of Mechanics and Control*, Vol. 10, n. 2, pp. 27-34, December (2009).

15 A.G. Gonzalez-Rodriguez, A. Gonzalez-Rodriguez, A.J. Nieto, R. Morales, *Design and simulation of an easy operating leg for walking robots*, Malaga, April 14-17 (2009).

16 E. Choong, C. Chew, A. Poo, G. Hong, *Mechanical design of an anthropomorphic bipedal robot,* First Humanoid, Nanotechnology, Information Technology, Communication and Control Environment and Management (HNICEM) International Conference March 27-30, Manila, Philippines (2003).

TRALOC – TELEOPERATED SNAKE-ROBOT WITH TRACKS ON EACH SEGMENT, FOR SEARCHING VICTIMS IN COLLAPSED BUILDINGS

T. BLEIKER, J. DARBRE, S. DROLL, F. HEER, D. HUNZIKER, T. WINANDY
ETH Zürich, 8092 Zürich, Switzerland

W. FISCHER, R. SIEGWART
Autonomous System Lab, ETH Zürich,
Tannenstrasse 3, 8092 Zürich, Switzerland

This paper presents the "Traloc", a mobile remote-controlled snakelike robot, which is designed to search victims in the debris after earthquakes. It consists of 5 similar elements with motorized tracks on each side; 4 joints with 2 DOF each in between, a "head-unit" with camera and the cable which connects it to the control unit. Thanks to the actively driven tracks in each element, the mobility of the snake is significantly higher than comparable systems, where only the joints are actuated. With this advanced mobility, our robot is already able to overcome steps and gaps which are approximately the size of one segment – with only 5 elements in total. By implementing an innovative actuation principle for the joints which is based on a direct chain-drive, the entire robot could be realized very lightweight and without the need of pneumatic air-supply – which is crucial for reliability and future upgrades towards a wireless version running on batteries.

Keywords: Field Robot, Mechanism Design, Search and Rescue, Hybrid Track Snake robot,

1. Introduction

Our goal is to provide a system which is capable to move through the debris resulting after an earthquake and to support the first responders in localizing buried victims. The situation on site is very difficult due to the expectations of the local population, the problematic logistics and the environmental influences. This is in an increase of pressure and delay in the rescue. Because of the dramatically decreasing chance of survival after 72 hours, it is indispensable that the rescue is as much facilitated as possible.

Our response to this situation is a robot called Traloc, which supplements a more precise and faster possibility to locate victims to the already existing methods (dogs, borescopes, etc.). The target is to develop and construct such a robot which is able move through the debris after an earthquake and handle the specifications that are a result of the challenging terrain.

Figure 1. Typical environment in a collapsed building
(a) Photo with a classical rescue-team at work, (b) current search instrument, (c) Detail-view of typical obstacles, with the most difficult challenges for robot locomotion highlighted.

After analyzing the requirements in more detail in collaboration with our partners from the Swiss Military Disaster Relief, the following specifications for such a robot were defined (see Figure 1-c):

- Need to localize a victim precisely
- The system should be easy to control from the outside of the debris
- Should be transportable as standard-luggage (<30kg)
- A gap of up to half a meter must be overcome
- A step of up to half a meter must be overcome
- Holes with a minimal size of 20x20cm must be passed
- The robot should be able to pass a slope of up to 45°
- A minimum action time of about an hour is required
- The path which the robot needs to travel can be up to 30m

2. State of the art

Currently the standard method for localizing buried victims is searching with dogs. However, the animals tend to be too large to directly reach the buried victims. For this reason technical support aids, such as telescopic cameras and acoustic sensors, are normally used to verify the approximate position of the victims indicated by the dog. Unfortunately all currently available technical equipment in this field is insufficient. Different robotic approaches are being investigated to address the existing problems.

One possible solution is the development of airborne robots [1] with the main advantage of overcoming any type of obstacle regarding the terrain. However the payload is limited and the spatial requirements and energy consumption are too high.

Robots with wheels [2] or tracks [3] and active joints also have a very good ability to overcome large obstacles, but still tend to be too large for accessing the most narrow gaps in the debris (20x20cm are specified).

Snakelike robots inspired by nature [4], which are more flexible and potentially smaller than common vehicles, are very difficult to control due to the high number of required elements and the complex gait patterns of a snake.

A combination of a snakelike system with wheels [5] or tracks [6, 7] allows achieving similar mobility with a smaller number of elements and significantly simplified control. While a system with wheels is easier to realize regarding the mechanical design, tracks guarantee a higher mobility in difficult terrain. A combination of a snakelike robot with tracks therefore seems to be the most promising solution, given some changes compared to the previous designs.

The main problem in the most advanced previous system [7] is the conversion from electrical energy into pressure for pneumatics. For this reason the main focus in our work consisted in developing a joint with an innovative actuation principle which only needs electrical power, but is almost as lightweight as the pneumatic solution.

3. Basic concept

To sum up, the goal of this project was set to realize an innovative snake-robot with tracks on all sides of its segments (similar to [7]), with an actuation mechanism for the joints which does not need pneumatic supply; and to realize this robot in a robust and lightweight design and prove its functionality in realistic test-environments.

The basic mechanical concept of the robot can be seen in Figure 2. It consists of 5 similar, 30 cm long elements with motorized tracks on each side; 4 joints with 2 DOF each in between, a "head-unit" with camera and the cable which connects it to the control unit. For controlling it, we use angles given by a remote control interface. The electronics implemented in the system control the motors in the defined positions.

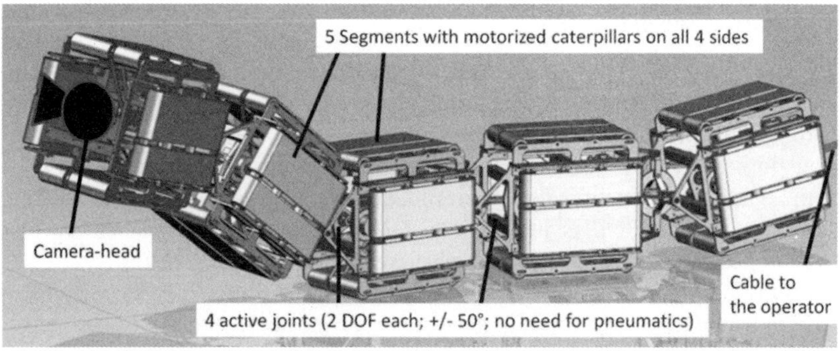

Figure 2. Overall mechanical design with the main assemblies.

4. Detailed design

Within the detailed design, the following challenges had to be solved: Realizing a direct-driven joint-mechanism with 2 DOF and at least +/-45°, driving the tracks in each segment with only one motor and a robust and lightweight torque-transmission, controlling all motors including the camera-head, and providing an intuitive user-interface to the operator.

4.1. Joints

An important part of the system is the joint mechanism that connects the elements of the snakelike system. According to the goals that Traloc should climb over steps and gaps in the range of 0.5 m, the joint has to carry high loads. After studying several load cases, the expected turning moment on the joint was specified as 21 Nm and the sheer force as 120 N; while the maximum total mass for each joint with +/-45° minimum-angle in 2 directions was defined to be less than 1.5 kg (40% of the total weight per element).

We have considered various concepts to fulfill these specifications. Nevertheless, no satisfying solutions based on pneumatic or hydraulic actuation could be found. The reasons are that commercially available compressors with enough pressure would be too big for fitting on the robot; and externally generating the pressure and feeding it to the robot via a cable (as implemented in the Omni-Tread [7]) was regarded as unsuitable for this application. To solve this problem, we developed an innovative mechanism based on the principle of a cardan-joint that is able to generate high forces and reaches a bending angle of up to 50° between two elements. The two degrees of freedom are mechanically decoupled and each of them is steered independently. All elements include two similar units with a gear-motor (Maxon EC-max25 with GP 32C 1:190, 6Nm) and a secondary transmission with two chain-drives (1:15), resulting in 42Nm of maximum torque on the joint and a turning-time of approximately 2.5 seconds for 45°.

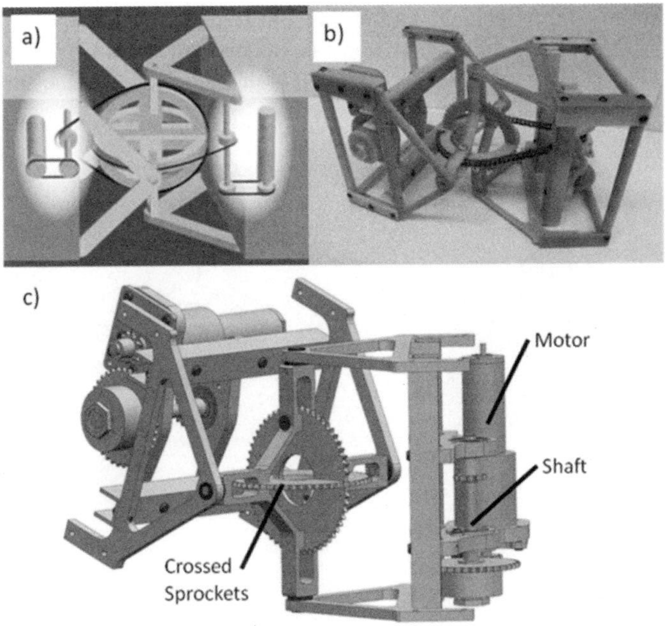

Figure 3. Joint actuation principle with direct chain-drive:
(a) Conceptual drawing of the actuation-principle, (b) Picture of a preliminary plastic model
(c) Detailed Cut-view showing the main components (chains not displayed).

4.2. Drive-train

The drive train is essential to move the system with a reasonable number of elements. The decision to place tracks on all 4 sides of the element has been taken to increase the propulsion rate (= powered surface / total surface) and to assert the capability of driving on each side. Moreover the probability of getting stuck decreases by increasing the powered surface.

All tracks of one element are driven by one motor placed at the center of it, similar as in [7]. To transmit the power from the motor to the tracks we developed a distribution gearbox which consists of two stages combining a worm and a belt gear. As there are four tracks, the worm transfers the torque to four worm wheels which are connected to toothed belt wheels. With these four belt wheels, the connection between the drive at the center and the outer tracks is established. The distribution gear has an overall gear transmission ratio of 1:44. As Motor we chose the EC 32 flat with integrated electronics from Maxon Motor. It has a maximal rotational speed of 6000 rpm and nominal torque of 34mNm. Put in relation with the overall gear transmission ratio this results in a maximal moment of about 1.5 Nm and 140 rpm for the tracks. The apparent asymmetry of the tracks results from the internal construction of the gear. The tracks consist of four axes whereof two can be flexibly positioned to enable the pre-stressing of the tracks. The other two are fixed in the middle of the drive train to ensure that the timing belt won't be bended too much when driving over an edge. The tracks are manufactured of double-side polyurethane timing belts in trapezoidal shape.

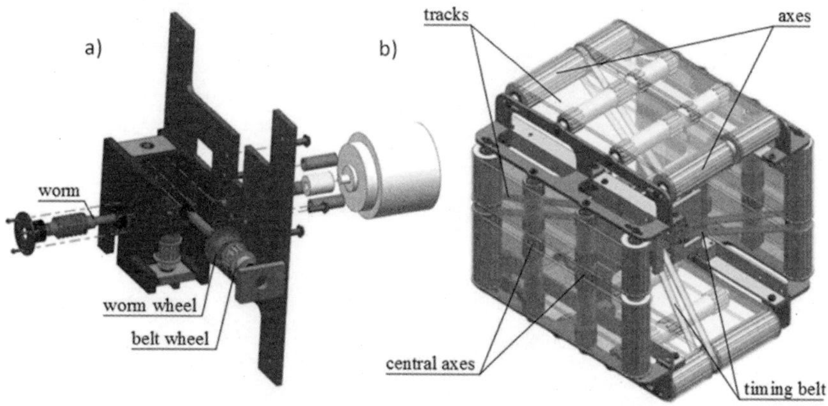

Figure 4. Drive-train for axial motion, (a) detail view of distribution gear, (b) overview with all four tracks.

4.3. Control and user-interface

With respect to the electronics, each element is built the same way: There are two brushless DC-motors to drive the joints and a third one to power the train-drive. Furthermore, a magnetic rotary encoder in each joint measures the current angle. With this information, a processor controls the angle in the two joints as well as the drive speed.

This concept gives the freedom to treat each element separately and to operate the snake with a different number of elements. If elements are put together, the control units in each element can communicate with one another and exchange information. Then, one of the elements operates as master and can be connected to a remote controller from where desired reference values for the joint angles can be received and passed to all other elements.

For additional control, an IMU equipped with a gyroscope, a compass and acceleration sensors is placed in one element in order to get information about the current orientation.

Two approaches are realized for the steering of the snake. The first one is based on a small model of the snake with encoders in each joint. The robot follows the movements the person does with the model. The second variant is to drive trajectories in the plane between several given points.

5. Tests

The first test with two elements proved that the concept of the joints and the drive train is working. The robot was able to drive curves as well as to lift one element. Further tests showed that Traloc is able to overcome a step of 40 cm and a gap of 60 cm. Also the functionality of the steering with the model and the movable stereovision camera in the head has been demonstrated.

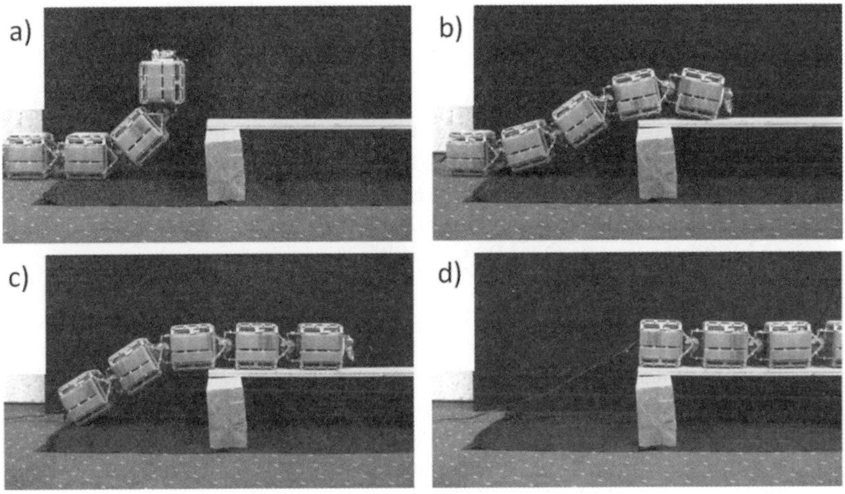

Figure 5. Sequence of the robot passing a step.

6. Conclusion and Outlook

The snake-robot with tracks on its elements, which was presented in this paper, illustrates an innovative prototype-design for a field-robot with high mobility in search-and-rescue-applications. Thanks to its directly-driven joints, its lightweight drive for the tracks and its intuitive control, it achieves outstanding performance and should become an important, additional tool in future missions for searching victims in collapsed buildings.

Future improvements in a final industrial version will mainly stress on the resistance against dust and water, the energy supply with batteries or a cable reel on the robot itself and the controllability of the system.

Acknowledgement

We want to thank the technical support-team at ETH Zurich (Marco Steiger, Martin Schütz, Jean-Luc Emery: CAD-FEM; Markus Bühler, Dario Fenner: Mechanical workshop; Stefan Bertschi, Thomas Baumgartner: Electronics; Roland Haas: coordination and coaching). Further on we thank our sponsors in particular Maxon Motor, Qualicut, LW-Basel, ABB, Ruag and Armasuisse.

References

1. D. Schafroth, S. Bouabdallah, C. Bermes, R. Siegwart, "From the Test Benches to the First Prototype of the muFly Micro Helicopter", Journal of Intelligent and Robotic Systems, 2008.
2. M. Lauria, Y. Piguet, R. Siegwart, "Octopus - An Autonomous Wheeled Climbing Robot", Proc. of The Fifth International Conference on Climbing and Walking Robots (CLAWAR), 2002.
3. A. Masayuki, T. Takayama, S. Hirose, "Development of 'Souryu-III': Connected Crawler Vehicle for Inspection inside Narrow and Winding Spaces", Proceedings of 2004 IEEORSJ International Conference on Intelligent Robots and Systems, September 28 - October 2, Sendai, Japan, p. 52-57, 2004
4. C. Wright, A Johnson, A. Peck, Z. McCord, A. Naaktgeboren, P. Gianfortoni, M. Gonzalez-Rivero, R. Hatton, H. Choset, "Design of a Modular Snake Robot", IEEE RSJ International Conference on Intelligent Robots and Systems, October 29 – November 2, San Diego, USA, p. 2609-2614
5. K.L.Paap T.Christaller F.Kirchner, "A Robot Snake to Inspect Broken Buildings", Proceedings of the 2000 IEEE/RSJ International Conference on lntelligent Robots and Systems, p. 2079-2082
6. T. Kamegawa, T. Yamas, H. Igarashit, F. Matsuno. "Development of The Snake-like Rescue Robot 'KOHGA'." Proceedings of the 2004 IEEE International Conference on Robotics *8* Automation, New Orleans, LA April 2004, p. 5081-5086
7. G. Granosik, M. Hansen, J., "The OmniTread Serpentine Robot for Industrial Inspection and Surveillance", Industrial Robot: An International Journal, Band 32, p. 139-148, 2005.

PANGOLIN : A VARIABLE GEOMETRY TRACKED VEHICLE WITH INDEPENDENT TRACK CONTROL

MI YUSOF[1,2] and TONY DODD[1]

[1] *Department of Automatic Control and System Engineering, The University of Sheffield, Sheffield S1 3JD, United Kingdom.*

[2] *Universiti Kuala Lumpur, 1016 Jalan Sultan Ismail, 50250 Kuala Lumpur, Malaysia.*

E-mail: {cop08mb,t.j.dodd}@shef.ac.uk

Variable Geometry Tracked Vehicle (VGTV) mobile robot configuration has successfully demonstrated its capability to traverse on unstructured ground surface in many military and disaster rescue activities. This paper presents a geometric analysis and control strategy for a new VGTV with independent flippers which is developed to traverse on uneven terrain especially in urban search and rescue (USAR) operations. A simple leadscrew-flipper mechanism control scheme that considers the effect of track deformation is presented. The behaviour of track length is studied in simulation. It is shown that with a knowledge of leadscrew operation and flipper angle relationship, a nearly constant track length is achievable.

1. Introduction

The implementation of mobile robot in USAR operation requires a platform that can autonomously adapt itself to disaster area and work robustly in complex and highly unstructured environments.

Tracked platforms, in particular VGTV offers several advantages such as better stability and high surmountability on uneven terrain. For example, Marques et al. proposed RAPOSA[3] that is able to climb higher obstacles by tilting the front-body part to provide climbing angle. Prior to this, Iwamoto et al.[1] patented a VGTV using single-tracked mechanism that is able to modify its geometrical shape to pass over an obstacle. Meanwhile, Kim et al.[2] designed similar type of VGTV with different method of controlling the track's tension. These type of VGTV can overcome various types of obstacles with least complex control method.

In this paper, we propose PANGOLIN, a new VGTV configuration with two sets of independent flippers. PANGOLIN is able to autonomously modify its geometrical shape based on sensor input while independent flipper enhances stability of its body when traversing along uneven terrain. The track's tension variation is controlled by a simple leadscrew mechanism. Additionally, it has self-righting capability due to a symmetrical design.

2. Methodology

Commands that indicate basic PANGOLIN movements are under minimal command by an operator. In order to minimise the operator load, PANGOLIN autonomously adapts the track configuration to the terrain. The behaviour of the robot when it encounters obstacles is shown in Figure 1b-1d, whereby Figure 1a shows the location of obstacle detection sensors.

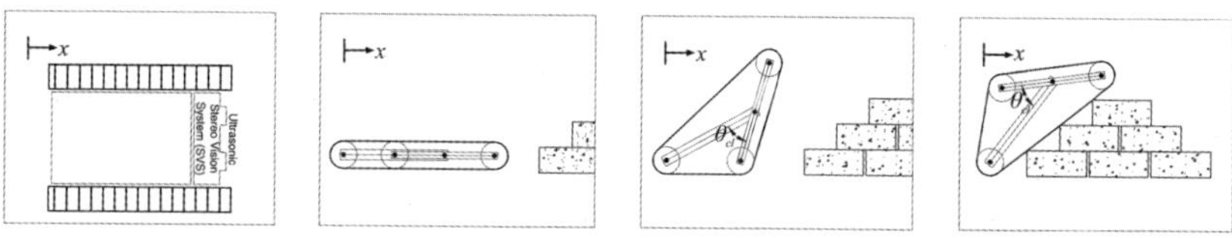

(a) Top view. (b) Flat surface. (c) Start to climb. (d) Climbing.

Fig. 1: Obstacle climbing principle.

2.1. *Control System Concept*

The overall control system structure (Figure 2) of PANGOLIN can operate in three layers of autonomy and operator may monitor or change current mission through *Human-Robot Interface* (HRI). ***Mission Control*** is the highest layer in overall control system hierarchy which determines robot mission. For example, the operator commands PANGOLIN to move from point A to point B with constant velocity, ν. At the same time, the system monitors the mission and keeps updating the status. Meanwhile, ***Robot Control*** layer provides control layer to monitor robot motion parameters including; (a) main body platform orientation (ψ_{roll}, ψ_{pitch} and ψ_{yaw}); (b) linear displacement (x, y, and z); and (c) linear and angular velocity (ν, ω). In addition, this is the layer where stereo vision system (SVS) and ultrasonic sensors interact with robot controller. In this paper, we highlight ***Track Control*** layer which is controls the tracks as well as flipper behaviour based on 'look-ahead' algorithm by obstacle detection sensor to provide obstacle parameter, θ_{cl}. It is necessary to adjust the flippers' angle to satify θ_{cl} so that it will be able to pass over obstacles. For some type of terrain, robot may loosen the tracks tension to enhance its grip to surface.

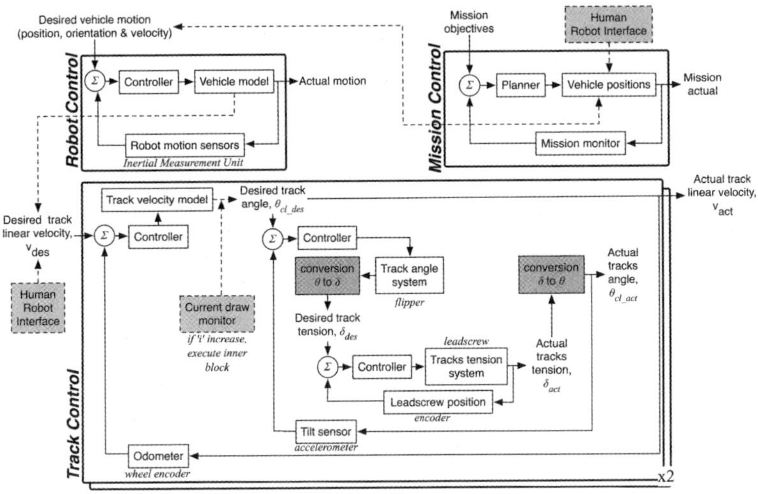

Fig. 2: Control Block Diagram.

2.1.1. *Mechatronic System Structure*

The prototype in Figure 3a, is composed of three main parts: (1) main body, (2) independent left and right flippers, and (3) stereo vision system (SVS) casing. Unlike existing rescue robots, PANGOLIN is able to perform symmetrical climbing angle on both sides of flippers (Figure 3b). Two sets of leadscrew mechanisms (Figure 3c) are used to control tracks' tension whenever the flippers change their angles to adapt with terrain.

Further, PANGOLIN is equipped with multiple sensors, command system and wireless communications as illustrated in Figure 4.

2.1.2. *Flipper Control Strategy*

The behaviour of the robot when it encounters an obstacle is shown in Figure 5. Control rules are based on obstacle parameters defined by SVS and ultrasonic sensor: (a) θ_{cl}, angle of servo to initiate flipper rotation and measured by tilt sensors (Analog Devices' ADXL335); (b) δ, leadscrew linear travel distance which controls track tension with respect to θ_{cl}; (c) T_R

(a) Full body (b) Independent flipper (c) Leadscrew mechanism

Fig. 3: The first prototype of PANGOLIN.

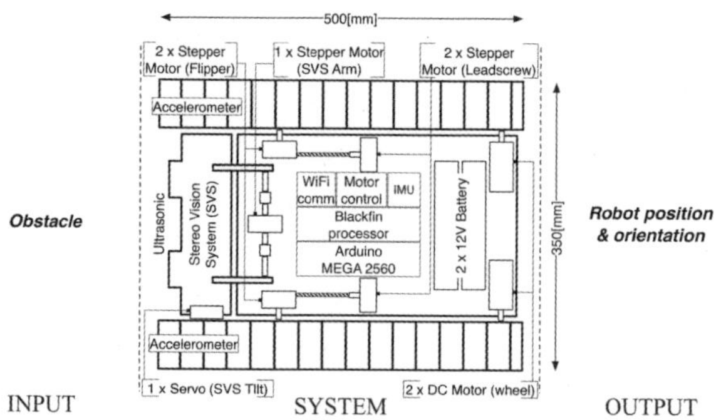

Fig. 4: Overview of PANGOLIN Embedded system.

and T_L, minimum torque required by driving wheels to overcome obstacles; and (d) l_{obs}, distance to obstacle, determined by SVS triangulation method with an ultrasonic sensor as feedback. To overcome obstacles, the sequence starts with the infrared sensor detecting an obstacle within range of $20cm \leqslant l_{obs} \leqslant 150cm$. Then the obstacle's image is captured by SVS and processed to produce a geometric representation to determine the appropriate θ_{climb} for flipper rotation. Meanwhile, the robot controller computes and sends the value of δ to the leadscrew actuator. Finally, main body platform orientation (ψ_{roll}, ψ_{pitch} and ψ_{yaw}) is measured using an inertial measurement unit (Sparkfun 9DOF Razor IMU).

2.2. *Track Geometry Formulation*

The desired track length, tl_{des}, is given when platform body is in initial position ($\theta_{cl} = 0$), Figure 6a, and calculated by;

$$tl_{des} = tl_{flat} = 2x_1 + s_2 + s_3$$

However, if the flippers' pivot point, O, is fixed (Figure 6b), value of tl increases with the flipper angle ($\theta_{cl} > 0$). Figure 7 shows the elongation of tl when $0 < \theta_{cl} < \pi/2$ which is computed from;

$$tl_{0<\theta<\frac{\pi}{2}} = x_2 + a + b + s_1 + s_2 + s_3 \tag{1}$$

which clearly illustrates that $tl_{flat} \neq tl_{0<\theta<\frac{\pi}{2}}$ (assume ratio of $m:n$ is $1:1$ for ease of computation). Therefore, in order to eliminate or minimise the variation in tl, it is necessary to adjust the track's tension via parameter δ (Figure 6c).

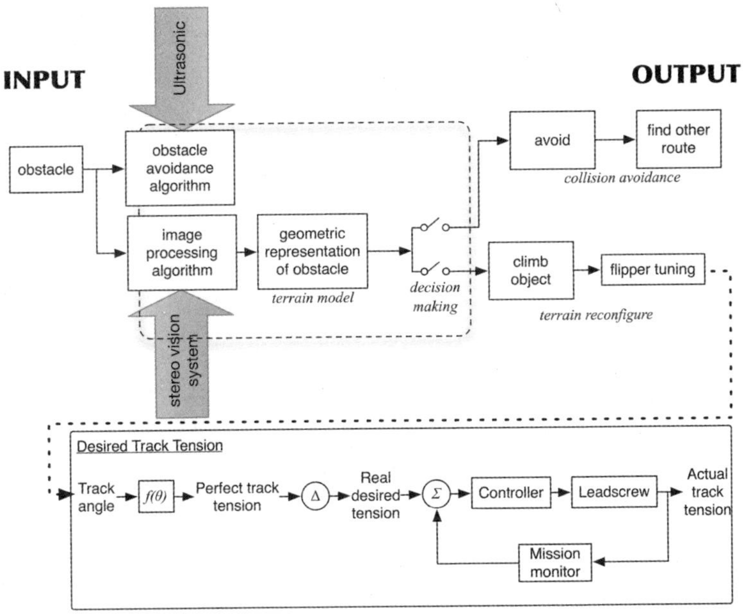

Fig. 5: Control flow for terrain navigation and flipper control strategy.

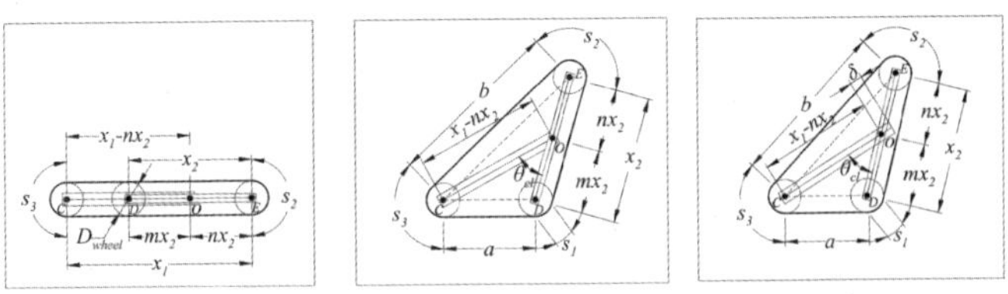

(a) flat ($\theta = 0$). (b) climb ($0 < \theta < \frac{\pi}{2}$). (c) climb with δ.

Fig. 6: 2D view of basic transformation stages.

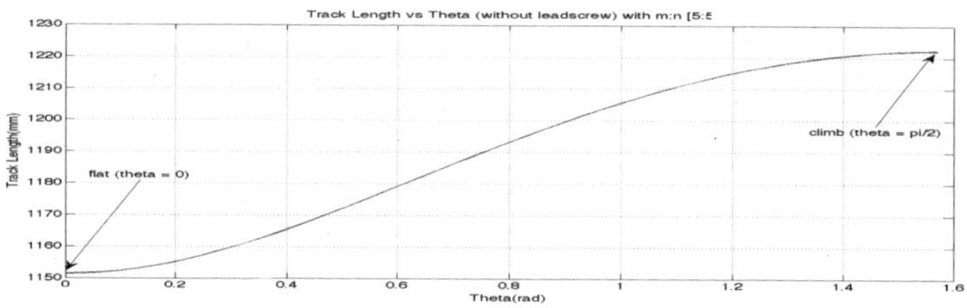

Fig. 7: Track length for $0 \leqslant \theta \leqslant \frac{\pi}{2}$.

3. Track Length Analysis

We considered two methods to solve the issue of tl as θ varies from 0 to $\pi/2$. One of them is by introducing δ which can control the track tension despite increase of tl. Another method is by determining an approriate distance between O to flipper end,[2] OD and OE that is indicated by the ratio of $m : n$ (Figure 6).

3.1. *Parameter Validation*

The derivation of additional parameters (Figure 6b and Figure 6c) to fullfil Eq.1 are given by;

$$a = \sqrt{(mx_2)^2 + (x_1 - nx_2)^2 - 2(mx_2)(x_1 - mx_2)\cos(\theta)} \tag{2}$$

$$b = \sqrt{(nx^2) + (x_1 - nx_2)^2 - 2(nx_2)(x_1 - nx_2)\cos(\pi - \theta)} \tag{3}$$

$$s_1 = \frac{D_{wheel}}{2}\left[\pi - \arccos\left(\frac{a^2 + (mx_2)^2 - (x_1 - nx_2)^2}{2amx_2}\right)\right] \tag{4}$$

$$s_2 = \frac{D_{wheel}}{2}\left[\pi - \arccos\left(\frac{b^2 + nx_2^2 - (x_1 - nx_2)^2}{2bnx_2}\right)\right] \tag{5}$$

$$s_3 = \frac{D_{wheel}}{2}\left[\pi - \arccos\left(\frac{a^2 + b^2 - x_2^2}{2ab}\right)\right] \tag{6}$$

From Eq.1, we have calculated various value of tl as θ varies from 0 to $\pi/2$, Figure 8. In theory, an approximate value of m and n giving the least difference between tl_{max} and tl_{min}. In the calculation, we get $20mm$ as the least difference where $60mm$ is the distance of OD and $90mm$ is the distance of OE which is given by the ratio $2 : 3$. However from the plot, we can see that tl does not monotonically increase with θ. Therefore, we chose $[1 : 1]$ as combination of parameters m and n gives a plots nearly proportional

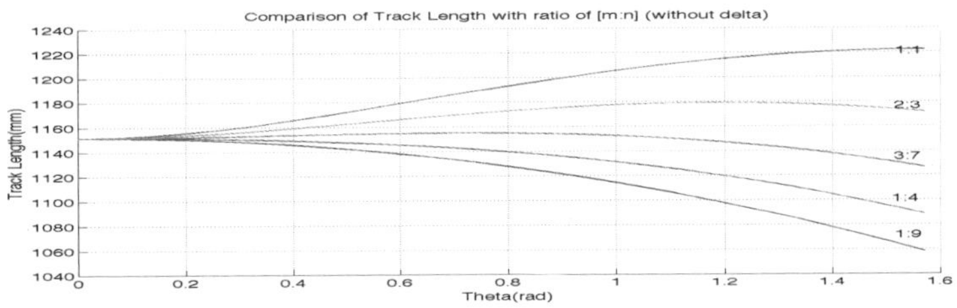

Fig. 8: Comparison of tl value with combination of $m : n$ without variation in parameter δ.

result of tl for $0 \leq \theta \leq \pi/2$ with difference of $71mm$ $(tl_{max} - tl_{min})$ where OD and OE are $75mm$.

3.2. *Mathematical Modeling*

The physical track length is fixed to desired length tl_{des}, but as the physical track angle, θ_{cl}, varies, the required total track length varies. The leadscrew distance, δ, is used to control the required track length to within acceptable bounds to avoid undue tension on the physical track. Simulations were used to determine the required relationship between θ_{cl} and δ. We show that a constant value of tl is achievable to satisfiy the value of $tl_{des} = 1151mm$ throughout track transformation, within the range of $0 \leqslant \theta_{cl} \leqslant \pi/2$. In order to eliminate track's material deformation and achieved tl_{des}, leadscrew linear travel distance, δ needs to be tuned automatically. Figure 9a shows the track length required for combinations of θ_{cl} and δ. Figure 9b gives the relation between θ_{cl} and δ by slicing through Figure 9a at tl_{des}. Then, an approximate curve fitting method is used to obtain a polynomial relation that is used to determine the tuning parameter of δ when flipper angle is θ_{cl}. The final result of the simulation is shown in Figure 9c which clearly shows that tl_{act} is approximated as tl_{des} with error $\pm 0.1mm$. The function $\delta(\theta)$ is given by;

$$\delta(\theta) = 4.12\theta^5 - 8.93\theta^4 - 22.68\theta^3 + 58.69\theta^2 - 1.08\theta + 0.04 \qquad (7)$$

4. Conclusions and further work

In this study, we proposed a control strategy to overcome unknown obstacle with simple leadscrew mechanism to control θ_{cl}, δ, tl and track tension. One of the main features is to control both flippers independently and traverse

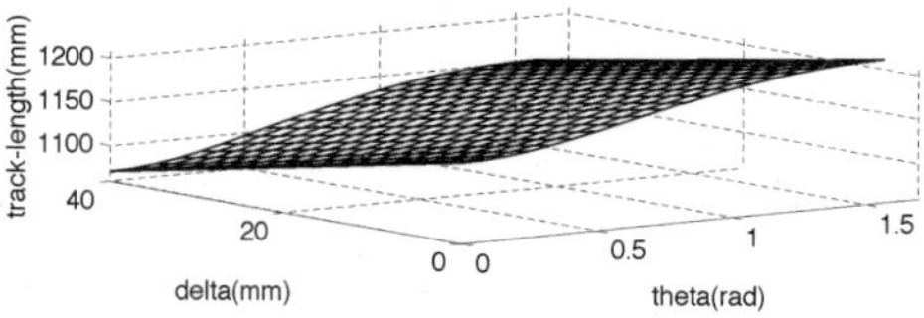

(a) Relation of θ, δ & TL.

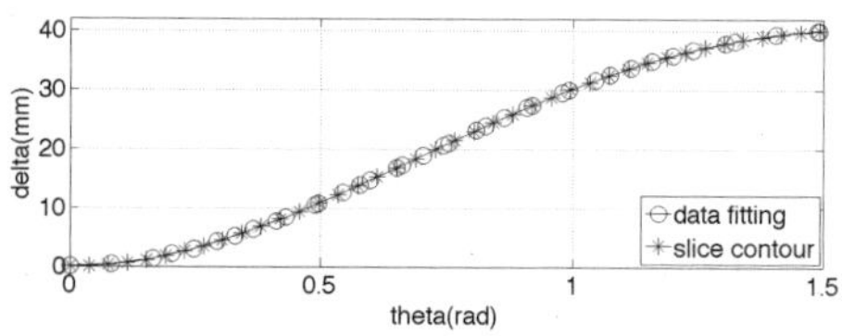

(b) Relation θ and δ.

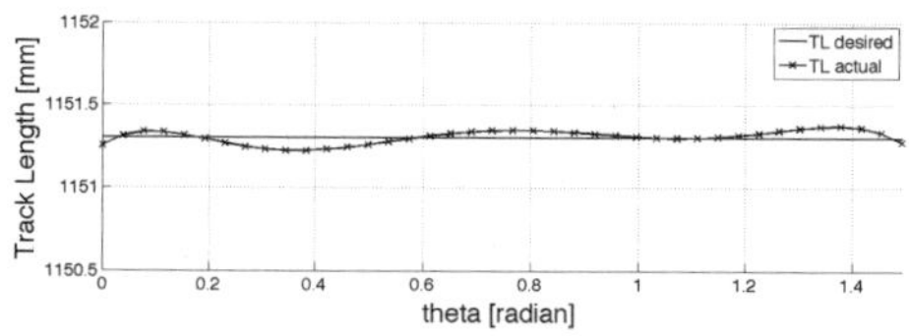

(c) Track replot.

Fig. 9: Track length geometry modeling.

seemlessly on any uneven terrain using information from obstacle avoidance sensors. In addition, both flippers are very useful to gain better stability to prevent 'tip-over'. Future work will focus on the practical implementation of the autonomous track configuration and collision avoidance system.

Acknowledgement

The authors gratefully acknowledge the financial support from Universiti Kuala Lumpur, Malaysia.

References

1. T. Iwamoto and H. Yamamoto. Mechanical design of variable configuration tracked vehicle. *Journal of Mechanical Design*, 112(3):289–294, 1990.
2. J. Kim and C. Lee. Variable transformation shapes of single-tracked mechanism for a rescue robot. In *International Conference of Control, Automation and System*, pages 1057–1061, Seoul, Korea, 2007.
3. C. Marques, Crist, Jo o, P. Alvito, P. Lima, Fraz, I. Ribeiro, and R. Ventura. A search and rescue robot with tele-operated tether docking system. *International Journal of Industrial Robot*, 34:332–338, 2007.

DESIGN OF MOBILE BASE FOR A SELF-RECONFIGURABLE INTELLIGENT SWARM FIXTURE SYSTEM*

LI XIONG†

DIMEC, University of Genoa, Via All'opera Pia 15a
Genoa, GE 16145, Italy

MATTEO ZOPPI

DIMEC, University of Genoa, Via All'opera Pia 15a
Genoa, GE 16145, Italy

REZIA MOLFINO

DIMEC, University of Genoa, Via All'opera Pia 15a
Genoa, GE 16145, Italy

LUIS MIGUEL DE LEONARDO GIRARD

DIMEC, University of Genoa, Via All'opera Pia 15a
Genoa, GE 16145, Italy

A self-reconfigurable fixture system, primarily targeted at the aerospace industry, is being developed within the inter-European SwarmItFIX project. The system is composed of one mobile base, one parallel mechanism and one adjustable head. This paper describes the mobile base, the subsystem which ensures rapid and precise locomotion, as well as secure docking, of the fixturing agent during the machining process. Several candidate designs with different locomotion methods are considered. A final solution is selected and the structure design is proposed.

1. Introduction

Thin-sheet and composite components are being increasingly adopted in the aerospace, automotive, and other industries. This is in keeping with the current trends towards life-cycle design, sustainable production, and geometrical complexity. The effect on manufacture equipment is a gradual shift towards more flexibility, reconfigurability, and automation. The same tendency affects

* This work is supported by the Seventh Framework Program (Collaborative Project FP7-214678) of Europe Commission.

† lixiong@dimec.unige.it

the development of fixtures, which are the devices used for locating and clamping the workpieces.

The current flexible fixture systems (FFSs) can be classified into modular flexible fixture systems (MFFSs), single structure flexible fixture systems (SSFFSs), and robotic fixtureless assemblies (RFAs) [1, 2]. The basic elements of a MFFS are: a base plate, locators, clamps, and connections, which can be combined into various configurations [3-5]. SSFFSs include "bed-of nails" systems [5], adaptive FFSs using internal mechanisms [7], and phase-change FFSs [8]. However, all of these systems need human intervention and lack of automation flexibility. The high initial cost and the time-consuming reconfiguration are other weak points. RFAs [9, 10] have limitations due to their high complexity, the relatively small number of robots that can realistically be used and the high dependence on software.

A novel concept of a fixture system combining flexibility, self-reconfigurability, automation, and swarm collaboration is put forward. The project SwarmItFIX, funded by European commission, is developing a self-reconfigurable intelligent swarm fixture system. The focus is on large sheet metal machining, especially in the manufacturing of parts for the aerospace industry, such as fuselage sections, aerofoils, and other panels. The proposed solution involves, mobile autonomous fixturing agents, periodically repositioning in order to be able to support the workpiece area near the moving machine tool.

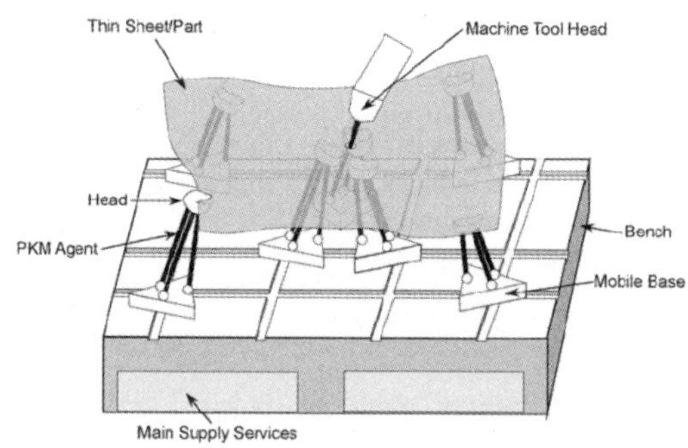

Figure 1 General scheme of the SwarmItFIX system

The system consists of a bench and several actuated fixtures (agents), which collaborate, without human interference, to support a thin-sheet workpiece, as illustrated in Figure 1. Beside its primary target application, the aerospace industry, this system could be beneficial in the future to other secondary sectors as shown in Figure 2. Each agent consists of one mobile base, one parallel kinematic mechanism (PKM), and one adjustable head.

In this paper, we mainly focus on the locomotion design and analysis of the mobile base. The mobile base has two main groups of functions: it carries the parallel robot that supports the workpiece and provides the repositioning of the whole agent on the bench. The significant specification is the time of repositioning independently from the type of locomotion adopted. The required positioning accuracy of the head is ± 0.5 mm including the positioning errors of the base and of the parallel robot.

In this paper, after a brief outline of the SwarmItFIX fixturing system, we focus on the development of the mobile base of the fixturing agent. Different locomotion and docking methods have been considered. Herein, the corresponding design solutions are outlined, analyzed, and compared. Finally, the best overall design is presented.

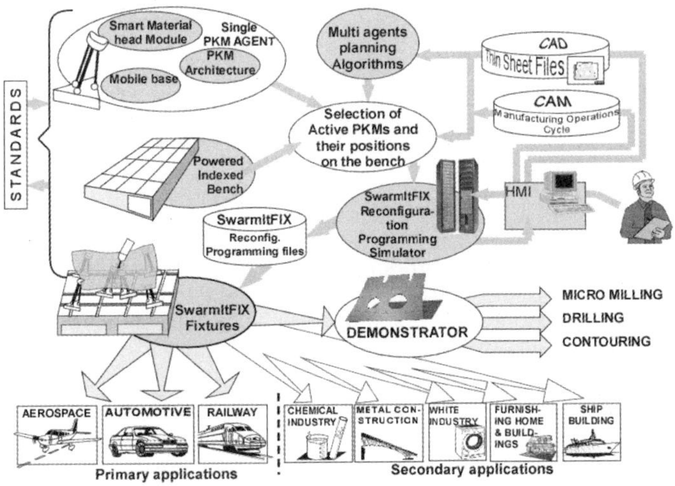

Figure 2 SwarmItFIX system design and development: primary and secondary beneficial sectors

2. Mobile Base Architecture

Based upon the design requirements velocity of forward motion, dwell time during locomotion, positioning accuracy and stiffness; alternative methods of locomotion are possible. Five types are proposed: sliding, rolling in grooves, rolling with tracks, rolling directly on the bench, and swinging around a vertical axis. Mechanisms with various architectures are designed and analyzed to implement these methods.

2.1. *Locomotion method – sliding*

The mobile base crawls on the bench by pulling itself with sliding arms.

A possible design (mechanism (M1)) with four pairs of rack and pinion transmissions with pins is shown in Figure 3(a). A grid of holes is available on the surface of the bench. The initial configuration of a step cycle is indicated by the solid line in the figure. The gears operating the sliders parallel to the direction of locomotion rotate to drive their corresponding racks. The racks move forward (dashed line) and stop. The pins (red dots in figure 3(a)) at the extremities of the racks are inserted into the bench holes. The gears rotate backward and the base is pulled forward. After that, the pins are pulled out and a new cycle can start. Figure 3(b) shows the use of a timing belt replacing the gears used in Figure 3(a).

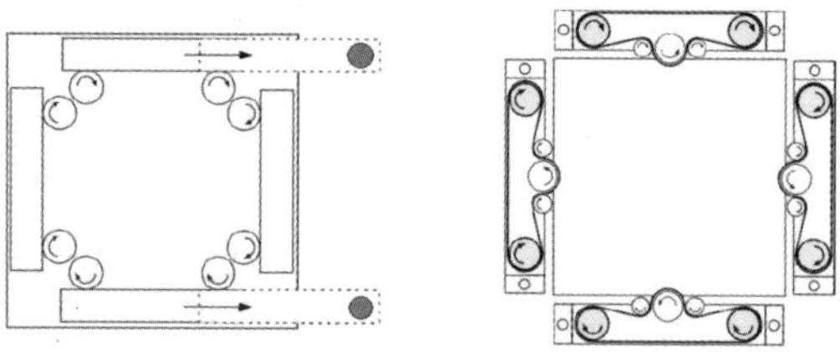

(a) Using rack and pinion transmission (b) Using timing belt transmission

Figure 3 Schematics of the sliding method (M1)

2.2. *Locomotion method – rolling in grooves*

In this method, the mobile base is equipped with wheels and the bench has a grid of grooves. The wheels roll in the grooves which ensure rectilinear motion of the base and help in its accurate positioning.

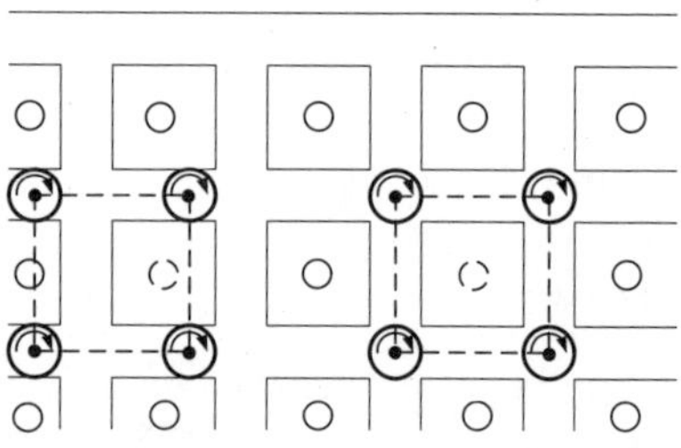

Figure 4 Scheme of rolling in grooves mechanism (M2)

The driving force is provided by the friction between the wheels and walls of the grooves. By coordinating the motion of the wheels, the mobile base can move at any orthogonal direction on the bench as shown in Figure 4.

2.3. *Locomotion method – tracks*

In this mechanism (M3) tracks with conical pins are used to generate thrust and move the agent on the bench, Figurc5. The pins can be rigid or compliant. If compliant pins are used, the accurate positioning can be obtained with additional pins sliding vertically in the bench during the docking cycle.

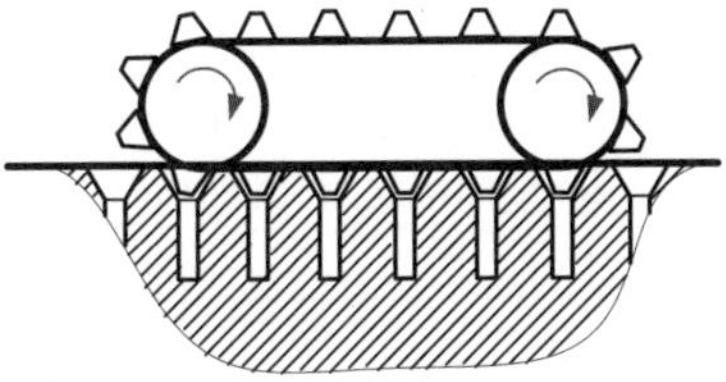

Figure 5 Track with pins locomotion (M3)

2.4. *Locomotion method – direct rolling on the bench*

The agent rolls to the target location on the bench (M4). A design solution comprises a frame with two rollers. The frame rotates about a vertical axis. Four pins sliding vertically are used to dock to the bench. When the agent changes direction, the frame with rollers is slightly lifted up until rollers detach from the bench. The frame is rotated to the desired angle and then lifted down with the rollers contacting the bench, Figure 6. The rollers rotate and the agent moves. The initial condition is with the four pins inserted into the holes and the rollers system lifted up.

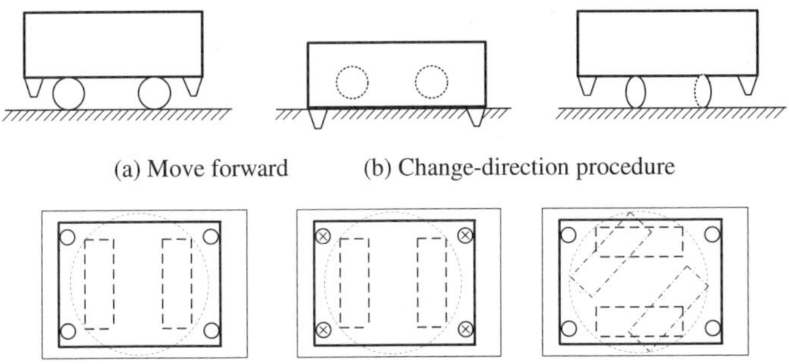

(a) Move forward (b) Change-direction procedure

(c) Top view of the change-direction procedure

Figure 6 Schematics of Locomotion with rollers (M4)

2.5. *Locomotion method – swing*

In the swing locomotion design (M5), the mobile base is equipped with two pins moving up and down and rotating. The two pins are inserted alternatively in the bench. The base rotates about the inserted pin while the other pin is up. The base rotates at an angle and stops with the pin which is up right above a free bench hole. The pin was up slides down and the one which was down slides up. The presence of one pin always inserted in the bench guarantees the equilibrium of the agent.

During the manufacturing of the workpiece both pins are inserted in the bench to increase the docking stiffness.

A sequence of steps and a schematic of the base architecture are shown in Figs. 7 and 8 respectively. The geometric accuracy of pins and holes must be high.

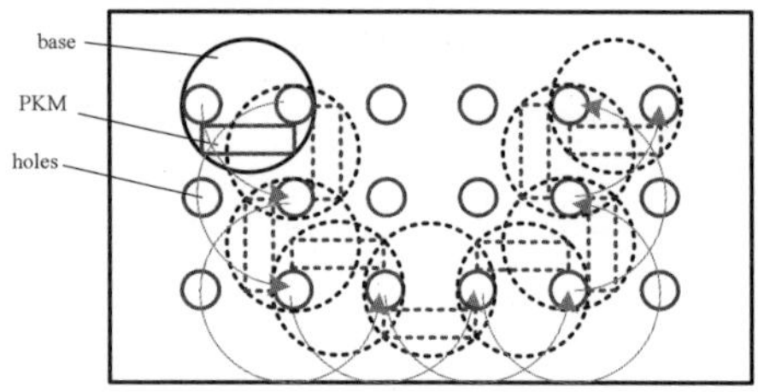

Figure 7 Swing method (M5)

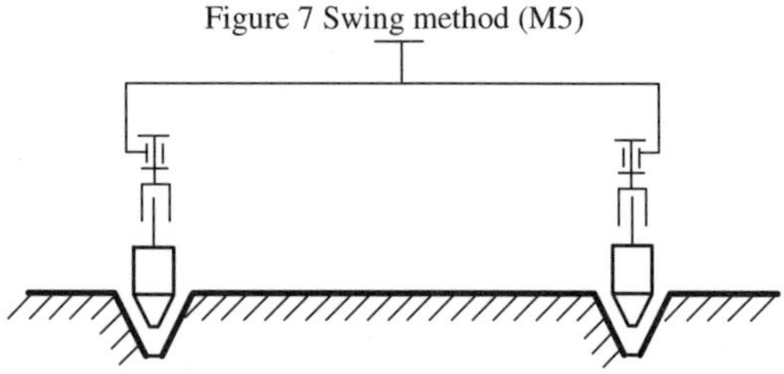

Figure 8 Architecture of swing method (M5)

3. The Final Solution and Structure Design

Considering the velocity of forward motion, dwell time during locomotion, positioning accuracy and stiffness, the "swing" method is adopted. Based upon the architecture of M5, the corresponding structure of the mobile base is designed as illustrated in Figure 9 (a) and (b). Two gear motors are embedded into the pin modules for pin rotation. Harmonic Drive units are adopted with the following functional characteristics: rated speed 4500 r/min and maximum output torque and brake holding torque 82~Nm. Two pairs of air cylinders are used to move the pins up and down. The pins for positioning and docking are commercial part Schunk NSA 120, with repeated accuracy < 0.005 mm, draw in force 18 kN per module, holding force 75 kN per module. Power transmitted from the bench to the agent uses contact brushes placed at the bottom of the agent. An air compressor with a tank is placed onboard.

All locomotion components are placed in the bottom space and the electronics and control equipment is hosted in the upper space of the base. The whole considered fixture system is illustrated in Figure10.

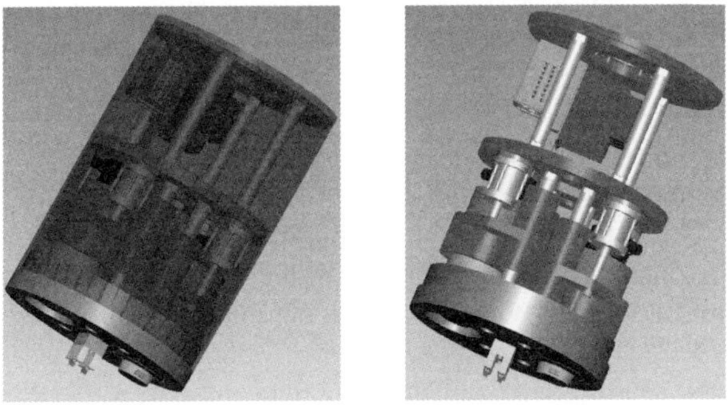

(a) Overview (b) View with cover removed

Figure 9 Design of the mobile base

Figure 10 Structure of the whole system

4. Conclusion

The paper presents the concept of a swarm fixture and outlines the design solutions adopted for the agent body and the bench.

The focus is on the selection of the locomotion principle. Different solutions are presented and critically compared. One is selected. The embodiment of this solution is briefly outlined. The technical feasibility of the concept of swarm fixturing is proven. The agent and the system are under development and a prototype of the first swarm fixture will be available in May of 2011.

Acknowledgments

The research is developed within the SwarmItFIX project funded under the Seventh Framework Program (Collaborative Project 214678). We acknowledge the assistance of European Commission and the other partners of this project: Warsaw University of Technology (Poland), Piaggio Aero Industries (Italy), Exechon (Sweden), ZTS VVU Kosice a.s. (Slovakia), and Centro Ricerche FIAT (Italy)

References

1. Rezia Molfino, Matteo Zoppi, Dimiter Zlatannov. "Reconfigurable Swarm Fixture". ASME/IFToM ReMar, 730 (2009)
2. Bi, Z.M., Zhang, W.J. "Flexible Fixture Design and Automation Review, issues and future directions". International Journal of Production Research, 39(13), 2867 (2001)
3. Asada, H., Andre, B.B., "Kinematic analysis of workpiece fixture for flexible assembly with automatically reconfigurable fixtures". IEEE Journal of Robotics and Automation, 1(2), 86 (1985)
4. Shirinzadeh, B., Tie, Y. "Experimental investigation of the performance of a reconfigurable fixture system". International Journal of Advanced Manufacturing Technology, 10(5), 330 (1995)
5. Sela, M.N., Gaudry, O., Dombre, E., Benhabib, B. "A reconfigurable modular fixturing system for thin-walled flexible objects". International Journal of Advanced Manufacturing Technology, 13(9), 611(1997)
6. T. Aoyamaa., Y. Kakinumaa. "Development of Fixture Devices for Thin and Compliant Workpieces". CIRP Annals-Manufacturing Technology, Volume 54, Issue 1, 325 (2005)
7. Wallak, A.S., Canny, J.F. "Planning for modular and hybrid fixtures". Proceedings of the IEEE Int. Conference on Robotics and Automation, 520(1994)
8. Plut, W. J., Bone, G. M. "Grasping of 3-D sheet metal parts for robotic fixtureless assembly". Proceedings of the CSME Forum, 13th Symposium on Engineering Applications of Mechanics, 221 (1996)
9. Backesl F., Frankel V., Geigerl M. "Concurrent manufacturing of parts and tools for the sheet-metal industry". Journal of Intelligent Manufacturing, Springer Netherlands, 9(4), 1998
10. Kang, Y., Rong, Y., Yang, J., Ma, W. "Computer-aided fixture design verification". Assembly Automation, 22(4), 350 (2002)

MODULARITY AND PARALLEL KINEMATICS: AN ORIGINAL DESIGN METHODOLOGY APPLIED TO HIGH PRECISION

MURIELLE RICHARD, PROF. REYMOND CLAVEL

Laboratoire de Systèmes Robotiques (LSRO), Ecole Polytechnique Fédérale de Lausanne (EPFL), Station 9, CH-1015 Lausanne, Switzerland

This paper introduces a new methodology to design modular industrial ultra-high precision robots; it aims at significantly reducing both the complexity of their design and their development time. This modular concept can be considered as a robotic Lego®, where a finite number of building bricks is used to quickly design the robot and to easily change its mobility. The core of the concept is the thorough conceptual solution catalogue, which is independent from any mechanical design. This paper will first present the methodology and the techniques to establish this solution catalogue. Then, its application to high precision will include the formulation of hypotheses and a practical example of a 5-degree of freedom ultra-high precision robot design.

1. Motivation

The current trend in numerous industrial domains is to miniaturize products, mainly microelectronic, optic and biomedical devices; the design of industrial robots capable of performing micromanipulation and micro-assembly tasks with a sub-micrometric precision is thus becoming a crucial need [1]. At the present time, the R&D process to design and build such ultra-high precision machines is still costly. Therefore, design methodologies have been developed in order to reduce the time-to-market, for example in [2]. However, if the industrial specifications of the robot are modified, requiring for instance a change in its mobility, the whole design process has to be restarted from the beginning.

This paper thus presents an original concept of modular kinematics which allows to rapidly design a parallel robot and to modify only a small part of the kinematics to change its mobility. This approach can be compared to a robotic Lego®, where a finite number of conceptual bricks can be chosen and assembled within a small amount of time to create the desired structure, thanks to a conceptual solution catalogue [3]. In parallel with the conceptual aspects, the mechanical design of the building bricks is investigated, making use of flexure hinges: this type of joints is indeed compulsory to meet the sub-micrometric precision requirements [4]. Consequently, the robots resulting from this methodology achieve similar high performances as more traditionally designed machines, with a significantly shortened development time.

This paper will first present the modular design methodology, especially the steps of the orderly process generating the conceptual solution catalogue. Then, its application to high precision robots will be detailed: hypotheses linked to the mechanical flexure-based design of the bricks will be introduced, and an ultra-high precision 5-DOF (Degrees of Freedom) robot design will illustrate the practical use of the methodology.

2. Modular design methodology

2.1. *Theoretical bases and terminology*

This methodology consists in designing a parallel robot composed of one to three kinematic chains, which are orthogonally arranged. This robot is symbolized by a cube: the chains are disposed along its faces, while the end-effector is located on one of its corners. The design of this robot makes use of a finite number of conceptual building bricks, which present two states: active and passive. Each kinematical chain of the modular robot consists in the serial arrangement of an active brick with a passive one.

As their name indicates, the role of the active bricks is to actuate from one to three DOF; the motions which are not actuated are blocked. As for the passive bricks, their function is to link the output of the active bricks to the end-effector of the robot, thus achieving a transmission of the motion. Their DOF are either passive, *i.e.* free to move but not actuated, or blocked. A generic notation uniformly represents the building bricks by symbolizing their two main features (see figure 1). First, it indicates their mobility: T represents a translation and R stands for a rotation. Uppercase letters symbolize actuated motions, whereas lowercase letters stand for passive ones. Then, the notation also specifies the directions of the DOF in relation to the face of the cube on which the brick is located: $\perp$ indicates that the motion is along an axis which is orthogonal to the plane of the face, whereas a motion along one of the directions of that plane is represented by $\parallel$. Likewise, the subscripts 1 and 2 are used to explicitly express the directions of several DOF along both possible axes of that plane.

The methodology transforms the desired mobility of the robot into an exhaustive list of all possible combinations of active and passive bricks fulfilling it, with the help of a conceptual solution catalogue. Then, the most suitable kinematics for the considered problem is selected depending on the robot specifications, as well as on mechanical design considerations.

The generation of the exhaustive conceptual solution catalogue comprises two main steps: first, the thorough list of the active bricks arrangements actuating the desired mobility is established; then, for each of them, all combinations of passive bricks performing the transmission of the motions without overconstraints are itemized. The next paragraphs will detail both of these steps.

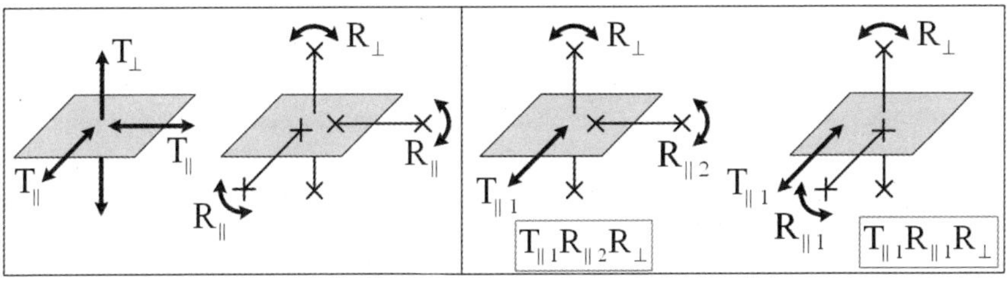

Figure 1. Illustration of the building bricks notation system (left) and examples of correspondence between brick mobility and notation (right)

2.2. *First step: active bricks arrangements*

The generation of the catalogue first consists in listing all possibilities of arranging one to three active bricks on the cube, so that they actuate the motions which the robot has to perform. It is to note that the permutations of the same bricks along the cube faces are listed as one solution. An intuitive approach first consists in establishing the different families of solutions which perform the required motions. For instance, a kinematics performing a planar joint motion (two translations and a rotation along three different axes) present the following ones:

1. *1 active brick*, actuating 2 translations and 1 rotation
2. *2 active bricks*, the first actuating 2 translations, the second 1 rotation
3. *2 active bricks*: the first actuating 1 translation and 1 rotation, the second 1 translation
4. *3 active bricks*, two of them performing 1 translation, the third 1 rotation

Then, for each of these families, all bricks actuating the required motions are combined: nonetheless, the consistency of these arrangements with the concept definition must be verified. Indeed, the second family of the previous example admits four bricks: $T_\parallel T_\perp$ and $T_{\parallel 1}T_{\parallel 2}$ to actuate two translations, $R_\parallel$ and $R_\perp$ for the rotation. A maximum of four combinations is thus possible, among which only three are compatible with the concept: the $T_{\parallel 1}T_{\parallel 2}$ and $R_\perp$ solution is impossible, as it would require both bricks to be located on the same face of the cube (see figure 2).

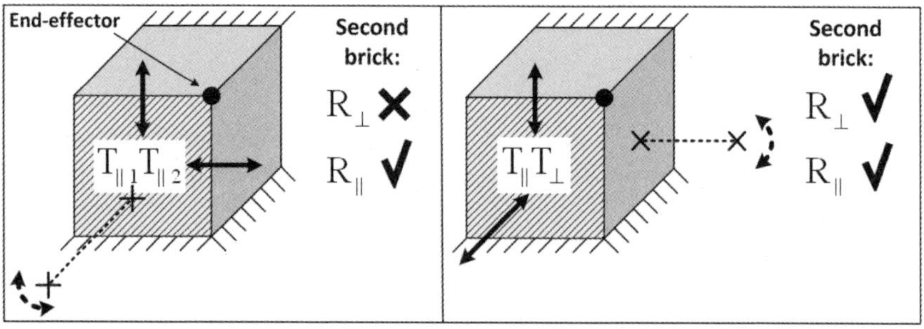

Figure 2. Second solution family of the plane joint mobility: the first brick, performing both translations, is arbitrarily placed on one face of the cube; the second brick has to actuate the remaining rotation, symbolized by a dashed line.

This first approach has allowed to establish the conceptual solution catalogue; moreover, a second method has been developed to automatically generate the solutions, making use of combinatory algorithms to produce the arrangements: first, a list of all active brick combinations is established. Then, three dismissing algorithms are applied: the first suppresses the solutions actuating more than once the same DOF, whereas the second deletes the combinations which are not compatible with the concept, for instance the ones where two bricks should be located on the same face of the cube. Last, all permutations of the same arrangement are dismissed. These methods have been implemented on Matlab, which has allowed to prove the thoroughness of the solution catalogue established with the first approach.

2.3. *Second step: passive bricks arrangements*

Once the exhaustive active bricks arrangements are listed, the second step consists in determining which passive bricks should be added to each of them to obtain the required robot DOF. The mobility of each passive brick is established as following: first, to perform the transmission of the actuated motions, the passive brick has to block the DOF which are actuated by the serially arranged active brick. Then, the motions which are actuated by the other active bricks of the robot (if any) must be free to move. Last, the remaining motions must avoid overconstraints, *i.e.* must ensure that the position of each blocked DOF is imposed by only one passive brick. In other words, the movements must be blocked on one passive brick and remain passive on the others. Figure 3 illustrates the determination of the passive bricks for a specific solution of the planar joint example (see 2.2).

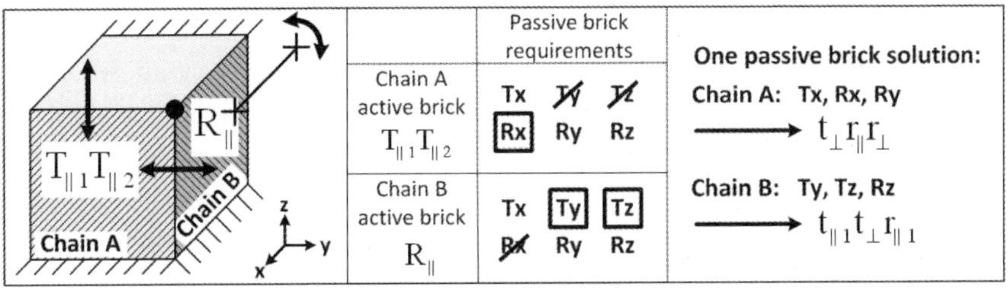

Figure 3. Example of the determination of the passive bricks solutions: planar joint mobility, active bricks: $T_{\parallel 1}T_{\parallel 2}$ and $R_{\parallel}$. The crossed motions represent the DOF which are actuated by the corresponding active brick, whereas the framed DOF stand for the ones which are actuated by the other active brick.

The total number of solutions of the exhaustive conceptual solution catalogue exceeds 3000, which makes it hardly applicable to a real design context. Consequently, simplifying hypotheses must be formulated: these are based on the robot requirements, especially on the mechanical design of the building bricks. The next paragraphs will illustrate possible assumptions for the specific case of ultra-high precision robots.

3. Application to ultra-high precision

Ultra-high precision robots refer to machines which are capable of performing micromanipulation and micro-assembly tasks with a resolution of less than 10 nanometers. To achieve this requirement, their mechanical design must be without friction, backlash and wear: the use of flexure hinges is thus mandatory. These joints are based on the elasticity of the material itself and can be monolithically machined by Wire Electro-Discharge Machining. The mechanical design of the bricks and some machining considerations thus allow the formulation of simplifying hypotheses.

3.1. *Hypotheses for high precision*

For high precision applications, the active bricks are assumed to be planar flexure-based mechanisms, which are motorized by linear actuators only, the rotations being obtained by the differential motions of the actuators of the same active brick. Consequently, only the conceptual active bricks which can shrewdly be designed with flexure joints are considered.

Likewise, the passive bricks are assumed to be designed as planar or axisymmetric structures. As the shrewdness of their integration in a high precision robot highly depends on the combination of the active and passive bricks within the kinematic chain, all conceptual passive bricks are allowed for.

These hypotheses first transform the complex 3D robot design problem, which results in scarcely modular structures, into several 2D ones, which are simpler and well-mastered. Then, the aforementioned dismissal of some active bricks allows to establish a simplified conceptual solution catalogue for high precision, thus dividing by two the number of kinematical possibilities. More details on the mechanical design of the building bricks can be found in [5].

4. Application of the methodology to a 5-DOF robot

The choice of the 5-DOF mobility is motivated by the typical industrial applications of high-precision robots, for instance micromanipulation tasks. These often require three translations (Tx, Ty, Tz) and two rotations (Rx, Ry), the rotation around the axis of the robot tool being blocked. This kinematics is complex to design and manufacture, which explains the scarcity of industrial prototypes. The hypotheses which are formulated for this problem are the following:

1. *Active bricks*: all bricks presenting the same mobility with different orientations relatively to the face of the cube share the same design; the physical plane of the brick does thus not necessarily match the face of the cube.
2. The robot presents a cubic arrangement of its kinematic chains.
3. *Passive bricks*: the mechanical design of each brick must belong to a plane which is parallel to the one of the corresponding active brick.
4. The number of different building bricks must be minimized.
5. The robot must be easily transformable into a 4-DOF or a 6-DOF one.

4.1. *Choice of the kinematical solution*

The use of the simplified conceptual solution catalogue results in only one cubic active brick arrangement for each family, as the first hypothesis causes solutions which are conceptually different to result in the same mechanical design. Then, one passive brick solution is selected for each active brick arrangement: its choice is based on the machinability of the selected mechanical designs of the passive bricks and on the third hypothesis. Figure 4 illustrates the four remaining kinematics: solution 4 is preferred as it minimizes the number of building bricks (hypothesis 4). Figure 5 illustrates the principle of the robot design, as well as a double-scaled mock-up which has been built with assembled blades.

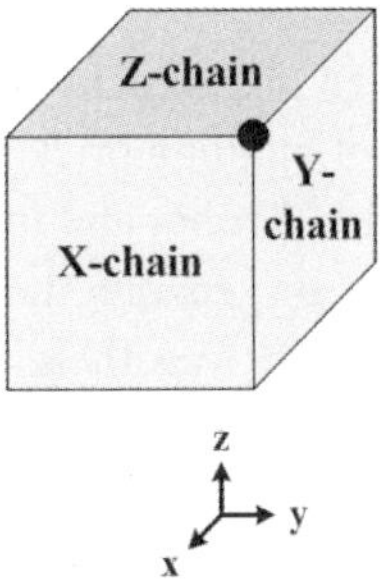

	X-chain	Y-chain	Z-chain	
Solution 1	$T_{\parallel 1}T_{\parallel 2}R_{\perp}$	$T_{\parallel}R_{\perp}$	Ø	4 different bricks
	$t_{\perp}r_{\parallel}$	$t_{\parallel 1}t_{\perp}r_{\parallel 1}r_{\parallel 2}$	Ø	
Solution 2	$T_{\parallel 1}T_{\parallel 2}R_{\perp}$	$R_{\perp}$	$T_{\parallel}$	6 different bricks
	$t_{\perp}r_{\parallel 1}r_{\parallel 2}$	$t_{\parallel 1}t_{\parallel 2}t_{\perp}r_{\parallel 1}r_{\parallel 2}$	$t_{\parallel 1}t_{\perp}r_{\parallel 1}r_{\parallel 2}$	
Solution 3	$T_{\parallel}R_{\perp}$	$R_{\perp}$	$T_{\parallel 1}T_{\parallel 2}$	6 different bricks
	$t_{\parallel 1}t_{\perp}r_{\parallel 1}r_{\parallel 2}$	$t_{\parallel 1}t_{\parallel 2}t_{\perp}r_{\parallel 1}r_{\parallel 2}$	$t_{\perp}r_{\parallel 1}r_{\parallel 2}$	
Solution 4	$T_{\parallel}R_{\perp}$	$T_{\parallel}R_{\perp}$	$T_{\parallel}$	3 different bricks
	$t_{\parallel 1}t_{\perp}r_{\parallel 1}r_{\parallel 2}$	$t_{\parallel 1}t_{\perp}r_{\parallel 1}r_{\parallel 2}$	$t_{\parallel 1}t_{\perp}r_{\parallel 1}r_{\parallel 2}$	

Figure 4. The four remaining kinematics of the 5-DOF robot

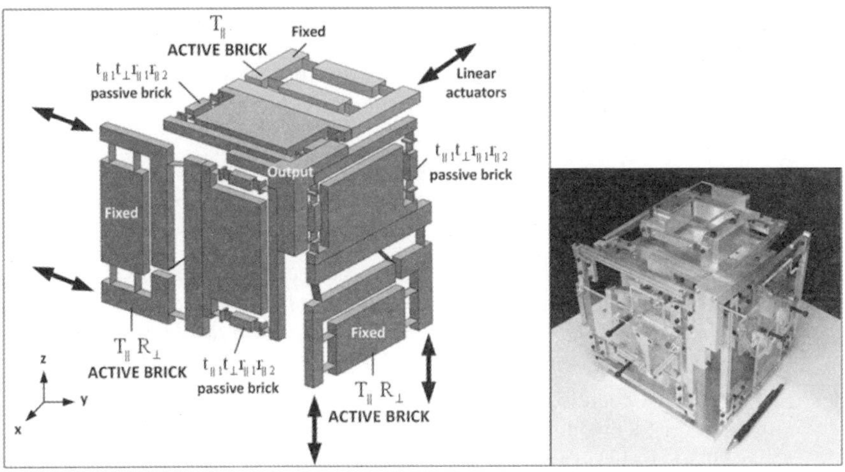

Figure 5. Design principle (left) and double-scaled mock-up (right) of the 5-DOF Legolas 5 robot

The fifth hypothesis is also fulfilled as the mobility of this kinematics can straightforwardly be modified: the substitution of the $T_{\parallel}$ brick by a $T_{\parallel}R_{\perp}$ indeed results in a 6-DOF robot, whereas the opposite operation creates a 4-DOF one. This application of the methodology has thus generated a new family of ultra-high precision robots, named Legolas: with only the four building bricks of the 5-DOF robot (thus named Legolas 5) and a simple passive one, all 19 possible DOF arrangements can be performed. This family notably includes the orthogonal version of the Delta kinematics [6, 7].

5. Conclusion

This paper has introduced an original methodology to design high precision modular robots. This robotic Lego®, based on a finite number of building bricks and on a conceptual solution catalogue, allows to rapidly develop a robot and to easily modify its mobility. This paper has detailed the methods to generate the exhaustive conceptual catalogue, as well as possible simplifying hypotheses based on high precision specific requirements. These have been applied to a practical example consisting in the development of a 5-DOF robot, which has highlighted the creation of a new family of ultra-high precision parallel machines. The design of an industrial prototype of the Legolas 5 is currently in progress; its characterization will allow to prove that the performances of the robots resulting from this new methodology are at least similar to the ones of more classically designed kinematics.

References

1. D. Tolfree, ‘Commercialising Nanotechnology, concepts – products – markets’, *International Journal of Nanomanufacturing*, vol. 1, n°1, pp. 117-133, 2006.
2. P. Helmer, ‘Conception systématique de structures cinématiques orthogonales pour la microrobotique’, Ph.D. thesis n° 3365, Swiss Federal Institute of Technology Lausanne (EPFL), 2006.
3. M. Richard, R. Clavel, ‘A new concept of modular kinematics to design ultra-high precision flexure-based robots’, *Proceedings for the joint conference of the 41st International Symposium of Robotics and the 6th German Conference on Robotics*, Munich, pp. 940-947, 2010.
4. S. Henein, ‘Conception des guidages flexibles’, Presses Polytechniques et Universitaires Romandes, Collection Meta, Lausanne, 2001.
5. M. Richard, R. Clavel, ‘Concept of modular flexure-based mechanisms for ultra-high precision robot design’, *Mechanical Sci*ences, vol. 2, 99-107, 2011.
6. J.-P. Bacher, S. Bottinelli, J.-M. Breguet, R. Clavel, ‘Delta3: a New Ultra-high Precision Micro-Robot: Design and Control of a Flexure Mechanism’, Journal Européen des Systèmes Automatisés, vol. 36, n°9/2002, ed. Lavoisier: France, pp. 1263-1275, 2002.
7. S. Bottinelli et al., ‘Movement transmission unit and movement transmission apparatus employing the same’, *US Patent US006453566B1*, September 24th, 2002.

CLAWAR SYSTEMS AS ENABLERS OF INTELLIGENT BUILDINGS OPERATION, CONSTRUCTION AND DESIGN: A HOLISTIC VIEW

FABIO BONSIGNORIO

RoboticsLab, University Carlos III de Madrid, Av. Universidad, 30
Leganes, 28911 Madrid, Spain

Heron Robots, V.R.Ceccardi, 1 18
Genova,16121 ,Italy

In this paper the state of the art of robotics applications in constructions is shortly reviewed. A holistic approach for robotics construction technology is discussed. The massive urbanism of the past century has led to city landscapes, in Europe, in North America, Japan and the industrialized part of the other continents, which are often far from being optimal in term of quality of life and architectural and urban design. Similar processes are ongoing in Asia, Latin America and Africa today. The need for a cheap and fast mass deployment of buildings to lodge a quickly growing mass of new inhabitants forces to build chaotic, inefficient, unpleasant and at various degrees unhealthy huge suburbs which usually may be defined, with some understatement, largely suboptimal with respect to quality of life and efficient resource exploitation. It is thought that agile manufacturing concepts, robotics science and technology and cognitive science, allowing craftwork like flexibility and creativity at mass production cheap costs, may enable a novel construction technology paradigm. We will analyze the issues and opportunities entailed by such a scenario.

1. Introduction

This work has been inspired by the work of British architect Richard Rogers, [16]. In the late 60s Rogers proposed the 'Zip-up house', see Figure 1, meant as an aesthetically cool house at low cost coming from an industrialized manufacturing process and assembled on site with a very short lead time. This idea of 'industrial' prefabricated houses was popular in America already in the

XIX century and become even more popular with the advent of 'series production', 'production line' and the related 'economy of scale'. Nevertheless this construction approach has had limited impact so far, although there is a growing tendency to incorporate in new buildings prefabricated components, in particular in industrial buildings. Despite remarkable design results, see Figure 1,2 and the catalogue of the 2008 MOMA exhibition on prefabricated houses, [11].

Figure 1. Richard and Su Rogers. *Zip-Up Enclosures No. 1 and 2*, 1968-71 Model. On behalf of Rogers Stirk Harbour + Partners, from [11]

Figure 2. KieranTimberlake Associates, Stephen Kieran and James Timberlake. Cellophane House (Exterior). 2008, from [11]

Prefabrication seems to constrain too much the architect (and structural engineer). With some simplification the application of robotics and automation in constructions has been in the form of factory automation integrated by 'standard' industrial robotics in the prefabrication phase, and to some robotics application (mainly in Japan) in the construction site, for example teleoperated cranes, [4,5,6,8,9]. More extensive robotization has been considered for constructions in space, [10]. At research level many climbing robots have been proposed for construction (e.g. wall painting) and maintenance (e.g. window cleaning). The design of new construction machines has been an integral part of architecture from the very beginning: Vitruvius and L.B. Alberti already in their classical treatises, [31,32], dedicated entire chapters to the construction machines. Today the potential for construction practice of recent service robotics and cognitive sciences advances might be overlooked. In [3,15] the potential area of exploitation of robotics and automation technologies is summarized as in Figure 3. A 'holistic' approach and the exploitation of service robotics and cognitive sciences research results in the different phases of the building process, may help a sensible improvement in the construction activities: a) Ambient Innovation, ambient intelligence, a 'robot enabled house'; b) Industrialization, agile manufacturing and cognitive factory; c) Site Automation, cognitive robotics on the construction site; d) Robotic Deconstruction/ Decommissioning.

In all these phases, and namely in a), c), d), remarkable progress might be enabled by an integrated concurrent design approach, the systematic utilization of CLAWAR (Climbing and Walking Robots), with adequate manipulation and perception and cognitive capabilities, and, above, all a 'design for X' approach, where 'X' is in general robotization and CLAWAR utilization.

Figure 3(a-d). a) Ambient Innovation; b) Industrialization; c) Site Automation; d) Robotic Deconstruction, from [15]

2. The Potential Benefits of a Holistic Perspective

2.1 Ambient Innovation: the house as a robot network

The first area where the exploitation of already obtained research results may improve significantly the quality of life is the ambient innovation.
Domotics applications such as managing lights, window shutters, heating and air conditioning in a coordinated way and from a central control panel by some degree already common in new buildings. For example this is commonly

exploited in order to minimize energy consumption sometimes in conjunction with special materials like transparent photovoltaic panels integrated into the windows, as in a recent redevelopment project in the Chicago Sears' Towers. More ambitious projects even consider to move different sections of the building. In Fig. 4 several examples of network robots applications to the home environment are shown. These examples might be complemented (with proper distribution in the environment of hanging stations by climbing systems such as ASIBOT, [21]. Although these systems are usually designed for assistive purposes, as their prices will lower nothing prevents their usage in common living environments. Other systems evolved from system such as Care-o-bot or Armar,[27,28], might become mature in 10-15 years from now. Other applications may include moving the walls and piece of furniture according to the changes in the external environment and the user needs.
The external walls of buildings and their internal corridors and pipes might be designed in the perspective of using CLAWAR robot systems for cleaning, painting and in general maintenance and supervision, for example with steel ways if you want to use a magnetic climbing systems, or even a proper set of hooks or special ways with fit surface materials. These simple precautions in the design phase of building would allow a massive utilization of already developed climbing systems reducing the maintenance cost of building and paying back the equipment costs.

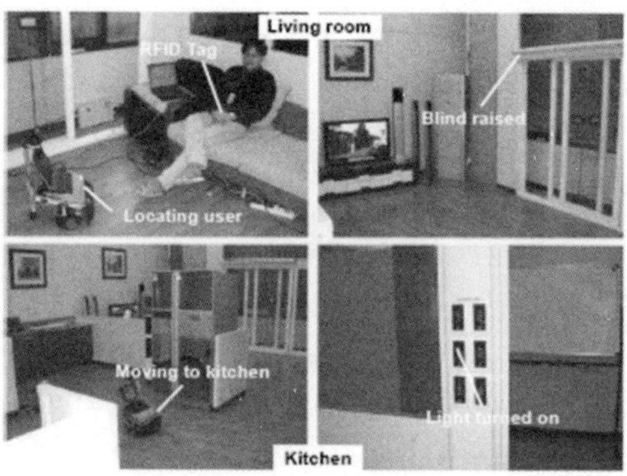

Fig. 4. Examples of Network Robot usage (courtesy KAIST)

2.2 Industrialization

There are hints that in a 5-15 year timeframe significant changes in the full cycle of consumer service-products from raw materials to disposal may occur, [28].
These changes are enabled by the continuous progress in the price/performance ratio of computing equipment, the transition from a 'browsing' internet to internet computing, 'cloud computing', 'grid computing' etc. , happening together with the parallel continuous reduction in the cost of manufacturing –

towards personal fabrication - and the expected progress in cognition sciences and robotics.

We should look at a highly connected and highly distributed WW (Manufacturing)/Supply system seen as a massive loosely coupled network of networks merging seamlessly 'information' and material management/ procurement/transformation and energy production/distribution. The manufacturing facilities should range from personal autonomous fabricators based on tiny autonomous personal factories, merging 3D printers, algorithmic assemblers, personal robots etc.), to massive adaptive 'server' factory for the mass scale production of 'components', namely chips or generic nanoengineered films/tissue, including information management like server farms ('a la google', but more close to strong AI), from distributed factories for big items like airplanes/car 'skeletons', to local (to the 'market') assembling units customizing at personal level anything from cars to fridges.

A totally new sustainable ecology of service/products might emerge.

This, together with the systematic utilization of service robotics solutions and namely mobile manipulation [22, 23] will allow 'mass customization' or in other terms tailor made service-products at transfer line costs. For example the approaches applied in [17] to clothing manufacturing can be transferred to construction components and subsystems, while the 'final assembly' will occur on the field.

This will allow a huge design freedom without compromising cost effectiveness.

2.3 Site Automation

Assuming that in a comparatively close future it will be possible to have industrial cost 'make to order' components and subsystems for constructions, the most technically tough challenges remain the 'final assembly' on the construction site. Current service robotics solutions lack flexibility and robustness when they have to cope with open ended environments.

Yet there are promising progresses in AI and Cognition.

It is thought that the strong autonomous service robots paradigm enabled by the progress in the modeling design and development of networks or swarms of 'embodied cognitive agents' should allow the design and implementation of a new generation of autonomous constructions robots with advanced cognitive capabilities for a new generation distributed highly connected building paradigm.

In the short term mobile manipulation techniques for assembly and manipulation and autonomous navigation vehicle technologies for construction site logistics might be used, together with advance Human Robot Interaction techniques for teleoperation. In the long term cognitive and humanoid technologies might also be exploited. With currently available service robotics technologies the components and subassembly designs must be constrained in order to allow their subsequent robotized assembly on the construction site and the massive utilization of CLAWAR systems for those tasks.

2.4 Robotic Deconstruction (Decommissioning, Recycling)

Current view of manufacturing includes decommissioning, disassembly and recycling. The design of a new building should consider from the beginning the methods to decommission it and to recycle its component and materials. This implies that the design of the machines (including CLAWAR systems) used for the decommissioning and the structure of the buildings should be concurrently engineered.

3. Preliminary Examples

The purposes of this section is essentially that of triggering intuition as many new devices and procedures need to be designed in order to be able to put into practice what it is envisioned here. Specific examples will be detailed elsewhere in more specific analysis. Think about a building like that in Fig. 2. It actually represents a wide class of buildings, as it is a composition of parallelepiped units. It may look like you only have to prefabricate a set of rectangular panels and then assemble them on site. Yet if these panels are an output of a conventional manufacturing automation process the possible variations of the prefabricated panels will be constrained for example by the water/power supply networks in the all building and the single panels subnetworks. A robotized flexible production plant would allow to have a huge variability in the panl dimensions and layout. Moreover a direct remote connection from the designer office to the factory would allow lot-of-one production and just in time, virtually reducing work in progress and 'sitting warehouse' to zero. The onsite assembly (with proper caution in the structure and panel co-design) might be reduced to a cooperative manipulation problem of (now) recent but standard industrial grade (with scaled up dimensions). The finishing (i.e. wall painting/cleaning) maintenance cost of such a building might be reduced and the technological issues scaled down, if, in the design phase, in between concrete walls, properly preset ways (i.e. trivially ferromagnetic, less trivially with ad hoc surface and shape/texture design to facilitate adhesion or clinging) are foreseen.

4. Conclusions and Future Work

From the above discussion it should be apparent that the concurrent design approach from industrial production of components and subassembly of the buildings to their final decommissioning might help the cost effectiveness while keeping flexible enough the building process and improving the new building quality and comfort.

The advances in Robotics and Cognitive Sciences enable a new breed of architectural design which may reconcile esthetics, flexibility and cheap and fast construction needs. Also the 'machine for living', 'machine a habiter' in French, which was envisioned by Le Corbusier at the dawn of the past century, [29], might be closer to a literal embodiment than what we may think. This might have a remarkable impact on urban quality of life and maybe eventually overcome some older criticisms to the supposed 'ugliness' of 'contemporary' cities, [30].

The robotics in construction as a research topic and a technology for construction practice has been more popular in a recent past and is now comparatively stagnating. The discussion above should suggest why and give some hints about how it could be revitalized.

The successful exploitation of service robotics research results requires a paradigm change in construction design and the implementation of a newly designed supply chain for constructions. This is a costly challenge which probably requires a long term public commitment. The return of such an investment might be dramatic.

References

1. The Urban Task Force. Towards and Urban Renaissance, Queen's Printer and Controller of HMSO, London, (1999)
2. Rem Koolhaas, B. Mau and H. Werlemann. S M L XL, Monacelli Press, (2007)
3. T. Bock. Construction robotics, Auton. Robots, 22, 3, 201-209, (2007)
4. C. Balaguer and M. Abderrahim. FutureHome: An integrated construction automation approach. IEEE Robotics & Automation magazine. Vol. 9. No. 1. pp.55-66, (2002)
5. C. Balaguer and E. Gambao, Robotics and automation in construction, IEEE Robotics & Automation magazine, Vol. 9, No. 1, pp.4-6, (2002)
6. C. Balaguer and M. Abderrahim (eds.). Robotics and Automation in Construction, (2008)
7. J. Wawerla, G. S. Sukhatme and M. J. Mataric. Collective Construction with Multiple Robots, IROS 2002, Lausanne (2002)
8. http://www.iaarc.org/
10. http://www-robotics.jpl.nasa.gov/applications/applicationArea.cfm?App=9
11. K. Tadashi Oshima, R. Waern (authors), B. Bergdoll and P. Christensen (eds). Home Delivery, The Museum of Modern Art, New York, (2008)
12. S. Kieran and J. Timberlake. Refabricating Architecture: How Manufacturing Methodologies are Poised to Transform Building Construction, McGraw-Hill Professional, (2003)

13. L. Iwamoto. Digital Fabrications: Architectural and Material Techniques (Architecture Briefs), Princeton Architectural Press, (2009)
14. M. Meredith (author, editor), Aranda-lasch, M. Sasaki (eds). From Control to Design: Parametric/Algorithmic Architecture, Actar, (2008)
15. T. Block. TARSA, Teaching Automation, Robotics and Services to Architects, (2010)
16. K. Powell. Richard Rogers Complete Works, Phaidon Press, (2008)
17. F.P. Bonsignorio and R. Molfino. An object based virtual reality simulation tool for design validation of a new paradigm cloth manufacturing facility. In: Information Technology for Balanced Manufactured Systems (Shen W. ed.), Springer, New York, (2006)
18. N. Gershenfeld. FAB: The Coming Revolution on Your Desktop--From Personal Computers to Personal Fabrication. Basic Books, New York, (2005)
19. R. Brooks. A Robust Layered Control System for A Mobile Robot, IEEE Journal of Robotics and Automation, (1986)
20. M. Lungarella, F. Iida , J. Borgard and R. Pfeifer (Eds.). 50 Years of AI , LNCS/AI, Springer, Berlin-Heidelberg, (2007)
21. R.Correal, A.Jardón, S.Martínez, R.Cabas, A.Giménez, C.Balaguer. Human-Robot Coexistence in Robot-Aided Apartment. The 23rd International Symposium on Automation and Robotics in Construction (ISARC 2006), Tokyo, Japan, (2006)
22. Kuka web site. http://www.kuka-robotics.com/NR/exeres/AF1FBA31-2D7E-4771-B450-E8B605863928
23. Kiva Systems web site. www.kivasystems.com
24. URUS project website, http://www-iri.upc.es/groups/urus/
25. ViewFinder project web site http://www.shu.ac.uk/mmvl/research/viewfinder/
26. Care-o-bot web site. http://www.care-o-bot-research.org/
27. T. Asfour, K. Berns, and R. Dillmann. The Humanoid Robot ARMAR: Design and Control, In. Proceedings IEEE-RAS International Conference on Humanoid Robots, Boston, USA, (2000)
28. F. Bonsignorio. Towards a Self Organizing Swarming Distributed Personal Manufacturing Environment. RAAD 2008, Ancona, Italy, (2008)
29. Le Corbusier (Charles-Edouard Jeanneret). Vers une architecture, (Toward an Architecture), (1923)
30. J. Ruskin. The Stones of Venice, 1851-1853
31. Vitruvius (Marcus Vitruvius Pollio). De Architectura (The Ten Books of Architecture), 15 BC
32. L. B. Alberti. De re aedificatoria (The Ten Books of Architecture), Florence, (1452)

TUBULO II: MAGNETIC CLIMBING INCHWORM FOR INSPECTION OF BOILER TUBES

FREDERIC ROCHAT, PATRICK SCHOENEICH,
OLIVIER TRUONG-DAT NGUYEN
HANNES BLEULER, FRANCESCO MONDADA
EPFL STI IMT LSRO MOBOTS, École Polytechnique Fédérale de Lausanne
1015 Lausanne, Switzerland

TUBULO II is a miniature climbing inspection robot for the inner visual inspection of boiler tubes. Its locomotion principle is based on magnetic attachment to the walls of the tube coupled with an inchworm-based displacement. The magnetic ON-OFF adhesion is realized with Magnetic Switchable Devices (MSD). The robot is able to operate inside a tube with a diameter of 25 mm with bends as tight as 150 mm in radius and soldering joints up to 2 mm in height. Those environmental features are commonly found in coal-fired boiler tubes layout. The innovative design allows a robust construction. Hence, the robot can withstand the harsh environment of corroded boiler tubes, including water and ferromagnetic particles, which are very abrasive. The robot embeds a camera and lighting for visual inspection. It is energy autonomous. The robot is modular, which allows the future addition of other inspection modules. The robot achieves a mean speed of $18\ cm/min$.

1. Introduction

In power and chemical plants, there are kilometres of tubes used to transport water, oil, steam, heat, etc. from a place to another. To ensure a good efficiency of the plants, and to avoid failure due to defects or damages, most of those pipes require to be regularly inspected. One example is the preventive maintenance and inspection of water wall tubes inside coal-fired boiler. They usually have an inside diameter of $25\ mm$ with bends up to $150\ mm$ and with some solder joints up to $2\ mm$ in height. Their length is above $25\ m$.

The tubes are ferromagnetic, but can be covered with a layer of rust and other defaults caused by corrosion and aging. This surface is very abrasive and ferromagnetic particles detach easily. Most of the tubes are vertical, but portions can be in any inclination against gravity. This means that a climbing robot needs to attach to the tube in order to climb. Actual inspection methods are done with inspection sensors that are pushed and pulled on the top of rods. Standard *Non-Destructive Testing* (*NDT*) sensors such as ultrasonic, Eddy current or magnetic flux leakage method are mostly used [1]. As ultrasonic inspection requires a coupling liquid, and as the environment is often wet after outage, the design was orientated toward submersible solution.

There are two major reasons to research and study other solutions than current endoscopic solutions. The first is that a huge force is required to push the rod due to friction. This limits the inspection distance especially if several bends have to be passed. Hence, the distance that can be covered by endoscopes is limited, sometimes insufficient. Finally it requires a lot of space at the entrance of the tube to push and pull the rod. It is not always simple or even possible to bring cumbersome equipment in those often stringent size restricted areas. This constraint also limits the possibility for parallel tube inspection. Such strategy would be an excellent opportunity to save time during overhaul. There are mobile robotics solutions with wheels [2] or inchworm [3], but they lack robustness for the harsh conditions. Those findings were the starting point for the development of an autonomous robot dedicated for mobility inside rusty ferromagnetic tubes typically found in the layout of coal-fired boilers.

2. Design of *Tubulo II* with magnetic inchworm locomotion

A first prototype, called *Tubulo*, was built in a modular and train-like configuration [4]. This first study prototype can be seen in Figure 1. The robot is rolling on magnetic wheels oriented in the direction of motion. A compact electronic embeds the control of the robot. Due to the bends in the tube layout, a flexible shape was necessary. This feature was achieved with the development of several modules linked by an electronic bus for the exchange of energy and the communication. This modular configuration is open to future developments, like the addition of other inspection methods. The resulting prototype proved the feasibility of an autonomous robot for boiler tube inspection, but mobility was limited to straight and low curvature tubes. The main problem of this prototype is the wheels. Magnetic wheels are not compatible with the harsh environment due to ferromagnetic particles that stick on the wheel and disturb the driving mechanism, blocking the gear drive. Therefore another approach was necessary to achieve sufficient robustness within the target industrial applications.

Figure 1: Top view of the prototype of Tubulo. This robot has a train-like configuration with four wagons. Its locomotion is based on magnetic wheels.

The robot *Tubulo II* is shown in Figure 2. The robot design keeps the modular construction. This allows the independent development of inspection modules that can be later added or removed in order to fit the required specifications. For example, a second energetic module can be added to increase the autonomy of the robot, or a second vision module can be attached at the end of the robot looking backwards, in order to visualize both sides of solder joints. This makes the robotic platform more versatile and adaptive to various inspection scenarios.

The modules communicates through a *CAN* bus. The robot has an energy module and is therefore energy autonomous. It also has an embedded camera surrounded by LED. The head of the robot with camera and LED can be seen in Figure 3. Since ferromagnetic tubes damper standard wireless communication, currently the video signal is transmitted through cables, using a serializer / deserializer *(SerDes)* interface. The advantage of the *SerDes* interface is to reduce the number of cables to a single twisted pair. Alternatively, the inspection data could be stored on-board. The wireless transmission of the inspection data will require the development of a specially designed communication board working at a higher frequency, which depends on the tube diameter.

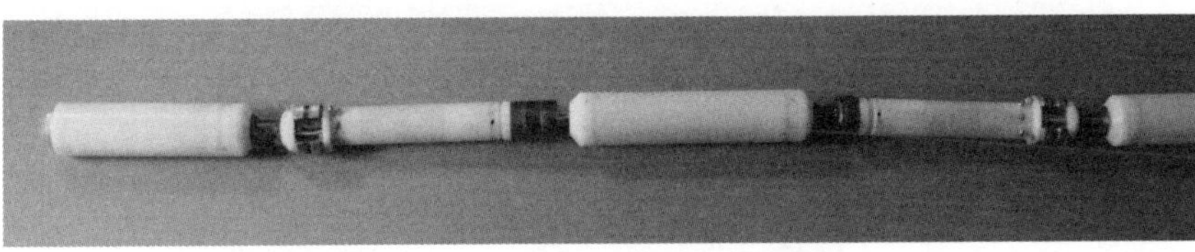

Figure 2: Overview of the Tubulo II inspection robot. Five modules are connected by a CAN and energy bus through mechanically flexible joints.

Figure 3: The head of Tubulo II at the entrance of a tube of 25 mm in diameter.

In order to improve the mobility and achieve better reliability and robustness in this harsh environment, the design is based on an innovative inchworm locomotion principle based on *Magnetic Switchable Device* (*MSD*) [5]. The inchworm locomotion applied to this robot works as following (Figure 4): an extendable module is placed between two modules responsible for adhesion. One adhesion module is activated while the extension module elongates. When the full extension is reached, the second adhesion module is activated. The first one can then be deactivated and the extension module can contract. If the operation is repeated, it is possible to move forward. By inverting the sequence, the robot is also able to move backwards.

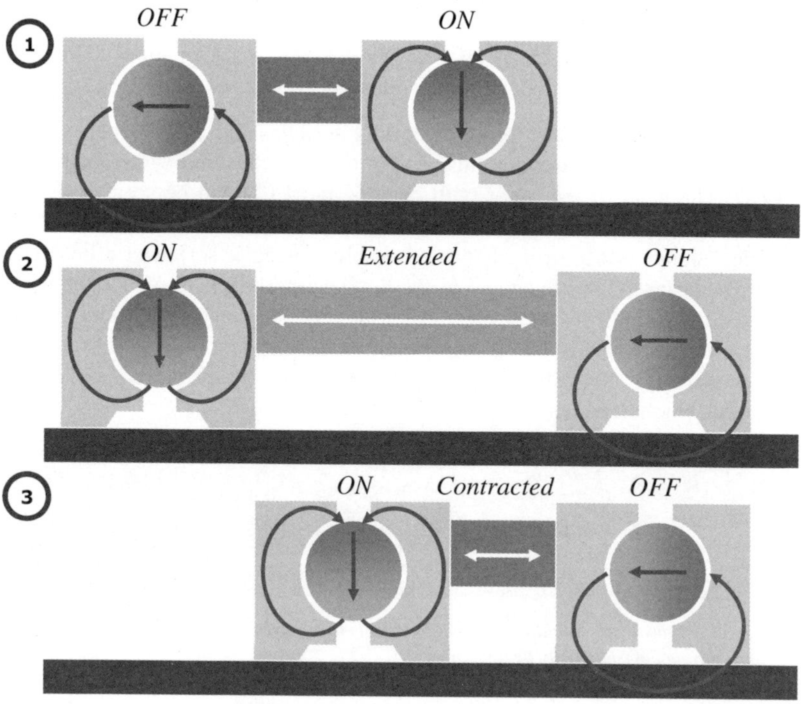

Figure 4: Scheme of the inchworm locomotion implemented with Magnetic Switchable Device (MSD). In a MSD, the adhesion is activated or cancelled by the change of the orientation of the cylindrical magnet.

A major advantage of this locomotion principle for the propulsion of an inspection robot is the possibility to use one adhesion module as a reference point and then use the extension module to scan the environment with *NDT* sensors. Additionally, this locomotion principle seems very promising in order to be robustly implemented and made water-resistant. The drawback of the current embodiment is that its motion is slow: $3mm/s$. The slow speed is due to the stroke and speed of the extension module. The stroke limitation is due to the need for passing sharp bends.

Possible improvements of this critical aspect include the use of a motor with a higher speed. Another solution to mitigate this disadvantage is with the serialization of several extension modules. Hence their strokes would sum up and increase the speed. Such solution is easy to implement with the existing modular approach. A final alternative would be the implementation of flexible rack rail transmission.

The *MSD* solution is ideal to act as an ON OFF adhesion device as required for the construction of an inchworm robot inside a ferromagnetic tube. The *MSD* devices have proved to be energy-efficient, and thus perfect to ensure a good autonomy. As the ferromagnetic system is in direct contact with the environment, it provides large adhesion force since the air-gap is null. Additionally the design using *MSD* allows the construction of an adhesion system with no external moving part. This is an asset for the design of a robust and waterproof robot.

The working principle of the *MSD* is known [5], but a dedicated design is required in order to fit the particular constraints of the current environment. The used configuration is described in Figure 5. Due to the tubular environment, it is ideal to have the motor for the activation of the *MSD* placed in the axial direction of the tube. The motor, gearbox and MSD can be seen in a half view in in Figure 6 (a). An adhesion force is required to act on the side of the tube wall. The *MSD* must have a shape able to fit to tubes, curvature and solder joints as found in industrial environments.

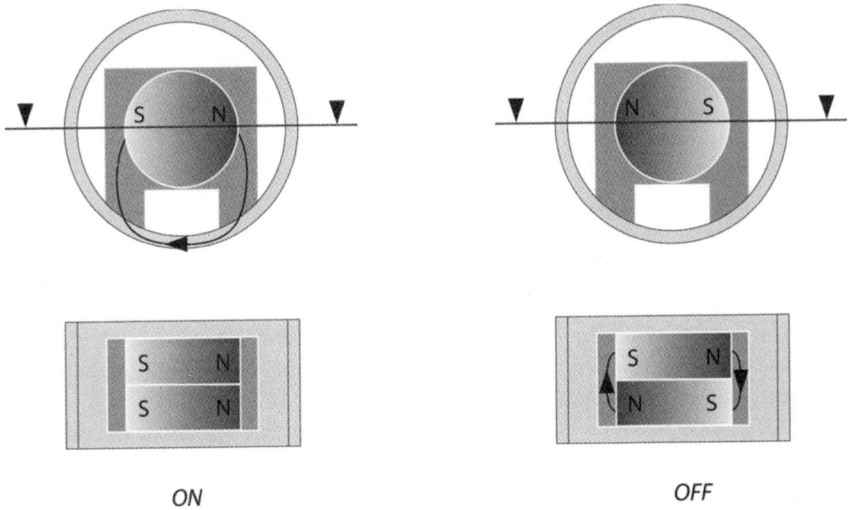

Figure 5: Scheme of the configuration of the Magnetic Switchable Device (MSD) in ON and OFF mode. From top to down: front view and cut view, following the cutting line described above.

Additionally the MSD needs to have a shape that can be adapted to a water resistant construction. Two side growths on both sides of the magnets serve as iron guides and come in contact with the tubular surface. The top of their shapes follow a 25 *mm* diameter, in order to perfectly fit the tube bore. Those growths are chamfered at both ends to adapt to concave curvatures of the tube consequent of bending. A chamfered slot is machined in the middle of the growth to facilitate the adaptation to solder joint or convex curvatures. All these aspects results in a complex surface that can be seen in Figure 6 (b). The validation of the design of the *MSD* was done using *3D FEM* simulation tools. The result can be seen in Figure 7.

(a) (b)

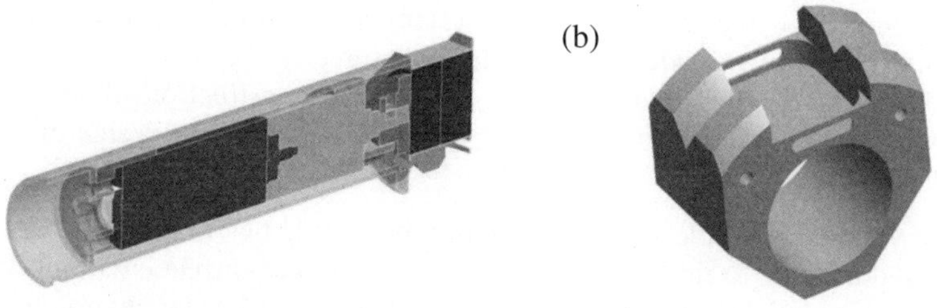

Figure 6: On the left, the MSD-wagon with the motor, gearbox, two magnets and stator. On the right, the stator with its adapted shape.

(a) (b)

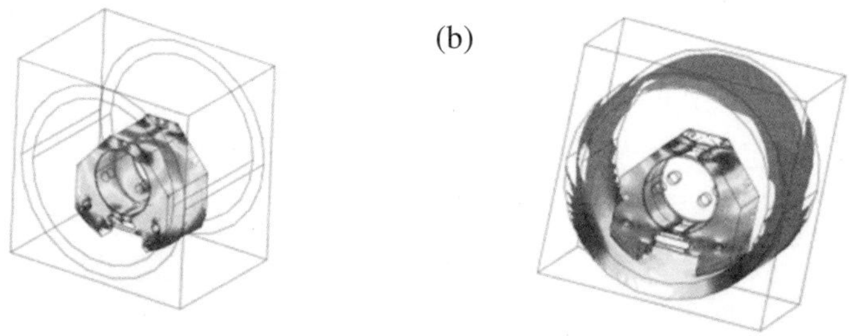

Figure 7: 3D simulations of the MSD inside a tube, (a) in OFF-mode, where it can be seen that the magnetic field is closed within the robot, generating no adhesion to the tube; (b) in ON-mode the magnetization can be seen in the tube, generating adhesion.

The built *MSD* has two magnets. Their diameter is $10\ mm$ and their height is 5 mm. The general outer dimension of the *MSD* is $16\ mm\ x\ 16\ mm\ x\ 10\ mm$. Its adhesion force is $50\ N$ measured inside a clean tube of $25\ mm$ in diameter, whose thickness is $2\ mm$. The adhesion force of the *MSD* is sufficient to attract the *MSD* from a distance of up to $8\ mm$. This means that whatever the position of the *MSD* inside the tube, once it is activated, the *MSD* is going to adhere properly to the tube surface and no elastic module is required to position the *MSD* against the wall surface during motion. The maximum activation torque of the MSD is $40\ mNm$. The activation is done with a Ø10mm Faulhaber DC motor. Its output torque is $1.28\ mNm$, with a gearbox of reduction $64:1$. In this configuration, the rotation required to activate or deactivate the *MSD* takes $0.22\ s$. The mechanical stops are done with a pin. The stator of the *MSD* is integrated into the tube construction of the module with silicon joints. The construction shape allows room to make the electric connection to pass through it. The working principle of the *MSD* allows turning *ON* or *OFF* the magnetic adhesion force against the tube and has been verified experimentally. This solution proved its suitability for tube locomotion. The robustness has been proven in real application-like environments.

3. Conclusion

Tubulo II is an innovative inchworm robot using magnetic *ON OFF* adhesion achieved with specially designed *Magnetic Switchable Device* (*MSD*). This embodiment is robust even in the harsh environment of boiler tubes. Ferromagnetic particles fall easily off the *MSD* when it is deactivated, making the cleaning easier and the system robust. The robot allows visual inspection and is energy autonomous.

The robot is not fully submersible, but this is achievable by adding a gasket or silicon joint to the extension module. The property of having no external moving parts helps in making the whole system waterproof. Thanks to the modularity of the robot, advanced inspection modules for *NDT* inspection can be added in the future. The innovative design with *MSD* offers a very interesting solution for inchworm locomotion inside tube layouts.

Acknowledgments

This work was possible thanks to the advice, work and knowledge of Daniel **Burnier**, Pierre **Noirat** and Philippe **Rétornaz**. This project was funded by the Swiss CTI project: "Highly compact robots for power plants inspections", referenced as CTI 8435.1 EPRP-IW, and Alstom.

References

1. Bickerstaff, R., Vaughn, M., Stocker, G., Hassard, M., Garrett, M. (2002) *Review of Sensor Technologies for In-line Inspection of Natural Gas Pipelines.* Sandia National Laboratories, Albuquerque, NM.
2. Suzumori, K.; Miyagawa, T.; Kimura, M.; Hasegawa, Y. (1999) Micro inspection robot for 1-in pipes. IEEE/ASME Transactions on Mechatronics, vol.4, no.3, pp.286-292.
3. Bertetto, A.M.; Ruggiu, M. (2001) In-pipe inch-worm pneumatic flexible robot. Advanced Intelligent Mechatronics, 2001. Proceedings. 2001 IEEE/ASME International Conference on , vol.2, no., pp.1226-1231 vol.2.
4. Schoeneich, P., Rochat, F., Nguyen, O.T.-D., Caprari, G., Moser, R., Bleuler, H. and Mondada, F. (2010) Tubulo – a train-like miniature inspection climbing robot for ferromagnetic tubes. in *1st International Conference on Applied Robotics for the Power Industry*. Montreal, Canada.
5. Rochat, F., Schoeneich, P., Bonani, M., Magnenat, S., Mondada, F., Bleuler, H. and Christoph, H. (2010) Design of Magnetic Switchable Device (MSD) and applications in climbing robot. Emerging Trends in Mobile Robotics, pp. 375-382.

SAFETY FOR EMERGING SERVICE ROBOTS

GURVINDER SINGH VIRK

Faculty of Engineering and Sustainable Development, University of Gävle, Gävle, Sweden, and CLAWAR Association Limited, Sheffield, United Kingdom.

SEUNGBIN MOON

Department of Computer Science, Sejong University, Seoul, Republic of Korea.

The paper presents an update on the international standardization activities in emerging areas of service robotics; two areas of activity are described, namely, safety of personal care robots, and medical robot safety and essential performance. The personal care robot safety work has reached the DIS balloting stage, and the standard considers issues that must be addressed by robot designers and manufacturers to reduce risk; these include hazards due to robot shape, robot motion, incorrect autonomous actions, etc. The medical robot standardization work focuses on existing safety regulations for medical electrical equipment, and how these may be extended to cover robots intended to perform surgery, rehabilitation, medical diagnosis and treatment. The key issues that medical robots bring to the medical electrical equipment area are their autonomous capabilities which make the task of formulating basic safety and essential performance requirements quite unique and challenging.

1. Introduction

Although robotics is an established field since Unimation sold the first robot in the 60s, the interest was largely in industrial manufacturing applications involving tasks such as welding, painting, assembly, pick and place, packaging and palletizing, product testing, etc. Over the years, an active industrial robot market was developed and some of the current leading companies are ABB, Kuka, Motoman, Fanuc and Panasonic. This industrial robot sector has been stable for over thirty years, during which time industrial robots have become sophisticated, reliable systems having high endurance, speed and precision. They are also extremely powerful machines able to carry heavy loads, and hence are deemed to be dangerous for humans when the robots are operational. Because of their potential for causing significant harm, strict safety regulations have been formulated and imposed by authorities, together with the adoption of other standards on vocabulary, characteristics and specifications for manipulators and end effectors/grippers to facilitate international trade. ISO (International Organization for Standardization, www.iso.org), under its Technical Committee TC 184 (Automation systems and integration), and Sub-committee SC 2, has been largely responsible for developing the international standards needed. SC 2 was created in 1983 for this purpose, although its scope was initially limited to "Robots for manufacturing environment". This was expanded to "Robots for industrial environment" in 2003, and then further widened in 2006 to its current scope "Robots and robotic devices" as new applications for robots broadened to new sectors which do not seem to have any clear restrictions.

The main concern raised by the new emerging robot systems for these new sectors is the trend towards providing "services to humans" rather than performing the "traditional manufacturing tasks in an industrial environment" which they have been doing for many years. This has created the need to focus on safety issues arising from the new service robots' intended purposes because of this requires (generally) close proximity to humans to perform their tasks, whereas the previous industrial robots have been designed on the basis of keeping humans and robots separated while the robots are operating. These close robot-human interaction issues that need addressing have been discussed in [1]-[4] due to the interest of the CLAWAR community in this area to widen the application base of robotics; in fact the recent surge in the international robot safety standardization activities was initiated for this purpose in 2005 by the EC funded Network of Excellence on Climbing and walking robots (CLAWAR), when detailed proposals were made to ISO which led to the creation of the new standardization work groups.

The areas of personal care robot safety and medical robot safety have been receiving the most attention recently, and the work has been carried out within the following ISO workgroups:

- ISO TC 184/SC 2/WG7 (Personal care safety); Chair: UK (GS Virk), Members: France, Germany, Hungary, Japan, Korea, Sweden, Switzerland and USA
- WG7's Study Group on Medical care robots; Chair: UK (GS Virk), Members: France, Germany, Japan, Korea, Romania, Switzerland and USA

The two groups have been extremely busy over the past few years and considerable work has been carried out in the face-to-face meetings and the large number of e-meetings which have been organised. This paper presents a summary of the work and the latest results that have been achieved to date. The details of the personal care robot safety work are presented in section 2, and section 3 describes the medical robot safety standardization activities. Conclusions and future activities are highlighted in section 4.

2. Personal care robot safety

As discussed in Virk [4] the key difference between traditional industrial robots and the new emerging personal care robots is the close robot-human interactions which must to be included to allow for the provision of the needed assistance to persons. It is because of this that the safety standard for personal care robots is being formulated so that close robot-human interactions, as well as robot-human contacts, are allowed while the robot is operating. It is clear that, as with industrial robots, the new personal care robots will still be regarded as "machines", and as such will be regulated in Europe under the Machinery Directive [5] which requires all machines to be "safe"; how such close robot-human interactions will be allowed while maintaining human safety is a major challenge.

The approach adopted in formulating the new international standard has used information in line with the ISO 12100 [6] standard, together with the approach proposed in the ISO 13849 [7] and IEC 62061 [8] standards, to formulate ISO 13482 [9] which specifies the conditions for physical human-robot contact in personal care robot applications. The method uses the three step process presented in ISO 12100, which is the following:

- Step 1: Adapt inherently safe design measures
- Step 2: Implement safeguarding and complementary protective measures
- Step 3: Provide appropriate information for use

to ensure that the desired level of safety is achieved for specific types of service robots.

Personal care robots have been classified into three classes (mobile servants, physical assistant robots, and person carriers) whose typical tasks have been defined in the ISO/DIS 13482 standard to be the following:

- mobile servant robots: travelling in buildings while avoiding collisions with objects in their environment to perform pose-to-pose motions or full area coverage, interaction with humans including object exchange, handling of small and medium sized objects for grasping, manipulating, transporting, placing tasks and handling large objects possibly having constraints, e.g., opening a door/window, which may include travelling to extend the workspace.
- physical assistant robots: exoskeleton or other strap-on robotic devices for applying cooperative control forces on human limbs designed in order to do the following: (i) to help comfortable walking, (ii) to help in carrying heavy loads, (iii) to reduce loads on parts of the body, (iv) allow fine manipulation, etc.

- person carrier robots: physically transporting a person from one location to another autonomously in various environments (flat 2D surfaces, 3D unstructured areas, etc).

Risk assessment exemplars for different personal care robots have been investigated in detail, but it has been decided that the ISO 13482 standard should only present generic requirements at this stage, leaving the detailed risk reduction work to be carried out by robot designers for their specific applications, and supporting their work by appropriate experimental testing. Hence the three step risk reduction process has been applied to the defined tasks for personal care robots in a generic manner, with the view that future versions of the standard may be extended to present more detailed particular safety requirements for specific types of robots, e.g., for exoskeleton assistant robots or for person mobility carrier robots. It is clear that the different personal care robot applications will have different safety requirements, e.g., speeds of operation, maximum forces which can be applied to various parts of the body, etc., although it is too early to state these at the current time. It is also clear that such safety requirements will also need to include more complete safety-related numeric information for different categories of people (children, elderly persons, pregnant women, etc), which will need major international cooperation to determine if it is to be acceptable normative design data for personal care robots.

The DIS (Draft International Standard) balloting standard for personal care robot safety has been recently finalized and distributed to ISO members for balloting and commenting. This ISO/DIS 13482 standard has considered the different types of hazards as presented in ISO 12100 and specific hazards have been indentified by giving particular consideration to:

- unexpected travel surface conditions in the case of mobile robots;
- uncertainty of objects to be handled in the case of mobile servant robots;
- conformity to the human anatomy and its variability in the case of physical assistant robots.
- normal but unexpected movement of the personal care robot
- unexpected movement of humans, animals and other objects
- unintended movement of the personal care robot

In this way, clauses have been formulated to reduce risk due to the following reasons: hazards related to charging battery, energy storage and supply, hazards due to robot shape, emissions, electromagnetic interference, stress, posture and usage, hazards due to robot motion, insufficient durability, incorrect autonomous actions, and contact with moving components. Each hazardous area has been looked at via the three step risk reduction process to ensure an acceptable level of safety is achieved.

Safety-related control systems for personal care robots have also been considered, especially the stopping functions since these need to comply with more complex safety requirements than industrial robots so that hazards are not caused by the stopping functions themselves. Another interesting clause that has been developed considers different operational modes of the personal care robot (manual, autonomous and some mixture), and their implications on safety.

As already mentioned, the DIS balloting standard has been recently distributed and comments are awaited from the international community on how to move forward in producing a Final Draft International Standard (FDIS), and finally publishing the International Standard (IS).

3. Medical robot safety and essential performance

Virk [4] has described standardization work for medical robot safety, which commenced in October 2009 within WG7's Study Group on Medical care robots in collaboration with IEC TC 62 (Electrical Equipment in Medical Practice), and especially with its Sub-committee SC 62A (Common aspects of electrical equipment used in medical practice). This work has brought together medical and robot experts to investigate the open and wide ranging issues which need to be explored before the safety requirements for medical robots can be formulated. It is clear that medical robots are medical devices and as such, they must be regulated under the Medical Device Directive [10] rather than the Machinery Directive; this poses interesting questions which need to be addressed if an acceptable safety standard is to be formulated. Some of these questions, which have arisen in the discussions, include the following:

- What classifies a robot as either "a medical device" or "a machine"?
- What are the differences between existing medical electrical equipment and medical robots?
- When can a (non-medical) personal care robot become a medical robot?
- etc

The study group's early work on medical robots has focused on defining the general aims for medical robot standardization; what should be the key areas to study, what are the differences in the Machinery Directive and the Medical Device Directive, formulation of key definitions, classifications of medical robots, and what should be the structure of the new safety standard. In the detailed discussions to date, it has been strongly felt that a general standard for medical robots is urgently needed because harmonisation of medical device and machine safety issues will lead to considerable clarification of the complex issues which have been exposed. The medical robot standardization work would in fact, assist in defining the risk-benefit balance for patients, versus the risk to operators/third parties in using medical robots to perform intended tasks. It would also help in the formulation of acceptable risk control measures covering the unique aspects of medical robots having appropriate levels of autonomy.

The concept of autonomy for medical robots, when they are defined as medical electrical equipment, is felt to be one of the key issues that need to be investigated thoroughly. This has raised many further questions and the need to possibly revisit some of the key definitions (robot, robotic device, autonomy, etc) that have been proposed by ISO TC 184/SC 2/WG 1 (Vocabulary on robots and robotic devices) in the revising and updating of the growing vocabulary for this area of technology [12].

From the beginning of the study group's work, it has been clear that the medical robot standard should fit the IEC 60601 family of standards for medical electrical equipment. The view from IEC TC62 (who is responsible for producing the 60601 standards) has been that a collateral standard entitled, "Additional general requirements for basic safety and essential performance of medical robots" would be the most appropriate to formulate. This collateral standard should specify general requirements for medical robots that are in addition to those of the general IEC 60601-1 standard [11], and to serve as the basis for the particular standards for different types of medical robots (e.g., for surgery or for rehabilitation, etc).

This general agreement to produce a medical robot standard fitting into the IEC 60601 family has raised the question, "what does basic safety and essential performance mean when applied to medical robots?" Up to now, robots have been designed on the basis of fulfilling safety requirements only, and the concept of "essential performance" is quite new and needs investigating properly in order to fit into the medical electrical equipment arena.

The potential boundaries between non-medical and medical robot applications have also been discussed in detail, however this still remains an open question, and other issues such as "what is the intended use?" are critically important; namely is the purpose, "non-medical" or "medical", or "does the application involve invasive procedures?". In addition, "is the state of health of a person considered?" in the application, or "where (home, medical centre, etc) and how (by lay person or a medical professional) is the robot to be used", etc. These are all important issues when considering the boundaries between medical and non-medical robot applications.

It is clear from the above that the standardization work for the medical robots is in its initial phases, and there are more questions than answers at present. However, the work to ensure safety of medical robots is felt to be essential in allowing these types of robots to be further developed. The questions being raised are intriguing, and it is imperative that clinical experts work with robot experts to discuss the details before the requirements for basic safety and essential performance are finally formulated.

The latest update to this work is that ISO TC 184/SC 2 has proposed to IEC TC 62 to set up a Joint Working Group (JWG) on Safety for medical devices using robotic technology, using the study group's work as the starting point for the JWG. This proposal has been balloted and accepted by IEC, and hence JWG9 has been set up in April 2011 for this purpose and IEC has nominated additional experts to join the ISO experts in forming the JWG.

JWG9 has the following structure; Chair: UK (GS Virk), Members: Canada, France, Germany, Japan, Republic of Korea, Romania, Switzerland, The Netherlands, Sweden and USA, and Liaisons: ISO/TC 121, ISO/TC 173, IEC/SC 62D.

The work task given to JWG 9 is the following: *Develop general requirements and guidance related to the safety of medical electrical equipment and systems that utilize robotic technology. (i.e., medical robots). The work would encompass medical applications (including aids for the disabled) covering invasive and non-invasive procedures such as surgery, rehabilitation therapy, imaging and other robots for medical diagnosis and treatment.*

The detailed work for medical robot standardization is due to commence soon, and will be reported in future publications. Experts in robotics and medical sciences are warmly invited to participate in this valuable work to both facilitate the development of the medical robot standard and to ensure it will be robust and acceptable to the whole stakeholder community.

4. Conclusions

The paper has presented an update on the ISO/IEC robot standardization work of ISO TC 184/SC 2/WG 7 (Personal care safety), which is developing the safety standard for personal care robots as well as investigating the safety requirements for medical robots under its study group. In this respect, the personal care safety standard (ISO/DIS 13482) has progressed to the DIS balloting phase, and comments are awaited to progress it to FDIS and final publication of the IS in late 2012 or early 2013 is expected. The medical robot standardization work has formally started by the creation of JWG 9 under IEC admin lead is due to have its first meeting in June 2011. The starting point will be the wide ranging work that has been carried out by WG7's study group.

Acknowledgments

The authors would like to express their gratitude to the international experts on the ISO TC 184/SC 2/WG 7 and WG7's Study group on Medical care robots which are doing excellent work to produce the new safety standards for personal care robots and for medical robots. The first author is partially supported by the UK DTI under the AITS travel scheme for attending the international ISO/IEC meetings; this financial support is gratefully acknowledged.

References

1. Virk GS, New standards for new robots, Proceedings of the 10th International Conference on Climbing and Walking Robotics (CLAWAR 2007), pp 698-707, Singapore, 16-18 July 2007.
2. Virk GS, Moon S and Gelin R, ISO standards for service robots, Proceedings of the 11th International Conference CLAWAR 2008, Coimbra, Portugal, pp 133-138, 2008.
3. Virk GS, Safety standard for personal care robots, Proceedings of the 12th International Conference CLAWAR 2009, Istanbul, Turkey, pp 147-154, 2009.
4. Virk GS, Personal care robot safety, Proceedings of the 13th International Conference CLAWAR 2010, Nagoya, Japan, pp 1333-1339, 2010.
5. EC Machinery Directive, Directive 2006/42/EC of the European Parliament and of the Council of 17 May 2006 on Machinery, and amending Directive 95/16/EC (recast).
6. ISO 12100:2010, Safety of machinery – General principles for design – Risk assessment and risk reduction.
7. ISO 13849-1:2006/Cor 1:2009, Safety of machinery – Safety-related parts of control systems – Part 1: General principles for design; and ISO 13849-2:2003 Safety of machinery – Safety-related parts of control systems – Part 2: Validation.
8. IEC 62061 ed1.0; Safety of machinery – Functional safety of safety-related electrical, electronic and programmable electronic control systems.
9. ISO/DIS 13482; Robots and robotic devices — Safety requirements — Non-medical personal care robot.
10. Medical Device Directive; Directive 2007/47/EC of the European parliament and of the Council of 5 September 2007 amending Council Directive 90/385/EEC on the approximation of the laws of the Member States relating to active implantable medical devices, Council Directive 93/42/EEC concerning medical devices and Directive 98/8/EC concerning the placing of biocidal products on the market.
11. IEC 60601-1:2005; Medical electrical equipment Part 1: General requirements for basic safety and essential performance.
12. ISO/DIS 8373, Robots and robotic devices – Vocabulary, 2010.

AUTHOR INDEX

Abichou A. 311
Agouri S. A. 53
Ahmed N. 233
Akagi T. 275
Aldebrez F. 292
Algazaly R. 267
Alli H. 555
Almeida L. 355
Almeshal A. M. 731
Amano H. 275, 363
Amar F. B. 78
Aoustin Y. 539, 750
Armada M. 455, 801
Astolfi A. 499
Awad M. 35, 766, 833

Bachratý M. 437
Barras T. 415
Barreto J. P. 355
Beardsley P. 397
Beira R. 421
Beji L. 311
Benckendorff C. 45, 142
Bender J. 196
Benosman R. 855
Berns K. 149, 865
Berthoz A. 839
Bidaud P. 11, 676, 889
Birkenhofer C. 182
Bleiker T. 910
Bleuler H. 407, 415, 421, 950
Blumel M. 196
Bohn S. 182
Bombled Q. 491
Bonaccorso F. 321
Bonsignorio F. 873, 942
Bouazza-Marouf K. 695
Boumghar R. 215
Bouzgarrou B. C. 642
Briere Y. 102
Bruneau O. 531
Buendia A. 11, 889

Cameron S. A. 257
Cantelli L. 321
Cappelli J. 429
Caprari G. 388, 397, 658
Ceccarelli M. 901
Cerny E. 695
Cherepanov A. A. 774
Chevallereau C. 515, 539
Chugo D. 27
Clady X. 809
Clavel R. 934
Coupland J. M. 847
Culverhouse P. 135
Curaj A. 62, 742

Dai J. 329
Darbre J. 910
Dehghani A. 35, 705, 833
Dehghani A. A. 766
Denuder M. 397
de-Santos P. G. 605
Dias J. 355
Díaz-Hernández C.-A. 303
Dillmann R. 182, 233, 547
Dodd T. 918

Dogramadzi S. 190
Droll S. 910
Droulez J. 839
Dubowsky S. 471
Duhaut D. 339
Dumitru S. A. 742
Dynnik T. 571, 579

Fauroux J. C. 642
Fernández R. 801
Figliolini G. 445
Fischer W. 388, 397, 910
Fontaine J.-G. 531
Fujimoto S. 523
Fukuda K. 881
Furukawa H. 587

Gallardo D. 471
Garcia E. 605
Gas B. 855
Geissmann L. 397
Georgiou E. 329
Girard L. M. D. L. 926
Goher K. 731
Goher K. M. 53
Göktogan A. H. 429
Granata C. 11
Grand C. 676
Grandgeorge M. 339
Gregg R. D. 684
Griffiths G. 381
Guo Y. 855

Haji T. 275, 363
Hamon A. 750
Haq A. 539
Harbaum T. 233
Hassanpour H. 479
Hayakawa T. 169
Hayashi R. 275
Heer F. 910
Hermann A. 233
Higuchi T. 27
Hild M. 45, 142
Hirayama T. 363
Hirshmann V. 415
Hodgins J. 725
Hoepflinger M. A. 94, 507
Horsfield I. 190
Hubinský P. 437
Hunziker D. 910
Hutter M. 94, 507

Ibanez A. 889
Ieng S.-H. 855
Ijspeert A. J. 790
Ion I. 62, 742
Iribe M. 275, 523, 587

Jarrault P. 676
Jatsun S. F. 774
Jaulin L. 110
Jayne D. G. 705
Júnior C. L. N. 249

Kagawa T. 3
Kameyama T. 758
Karwat B. 463
Katayama Y. 881
Kerr D. 695, 847
Kerscher T. 547
Keusch D. 397
Khazravi M. 705
Kinugasa T. 275, 363, 523, 587, 881
Komatsubara H. 523
Konyev M. 571, 579
Koos S. 70
Kos-Grabar T. 233
Kowal J. 463
Krid M. 78
Kubisch M. 45, 142
Kurz G. 233

Lacroix S. 215
Lamarche J.-C. P. 531
Lewinger W. 196
Li T. 901
Liskiewicz T. 705
Liu D. 174
Long A. W. 684
Longo D. 321
Lopez A. M. 705
López-Coronado J. 303
Lozano A. 174
Lui Z. W. 766
Luige V. 62
Luksch T. 149
Lynch K. M. 684

Ma J. 257
Madani K. 207, 241
Marti O. 407
Mavroidis C. 715
Md Zain B. A. 157
Melita C. D. 321
Melnykov A. 571, 579
Menezes P. 355
Messenger N. 766
Moeckel R. 790
Molfino R. 926
Mondada F. 407, 415, 421, 950
Montes H. 455, 801
Moon S. 957
Morina A. 705
Moser D. 35, 833
Moser R. 388
Mouret J.-B. 70
Muñoz-Lozano J.-L. 303
Muscato G. 321
Myoraku S. 275

N'Guyen S. 839
Nakajima S. 650
Nakamura T. 127, 169
Neves C. 563
Neville A. 705
Nguyen O. T.-D. 950
Niknam S. A. 283

O'Toole M. D. 847
Oberländer J. 233
Oda Y. 3
Ogawa T. 127
Okada T. 666
Omar M. 292
Osuka K. 523, 275, 587, 881
Ottaviano E. 658

Padois V. 889
Palankar M. 623
Palis F. 571, 579
Palmer III L. 623
Patel T. K. 223
Pearson M. J. 190
Peters G. 174
Petres C. 86
Peuker F. 790
Pfirter L. 397
Pfotzer L. 182
Plumet F. 86
Pouya S. 790

Quinn R. 196

Rahideh A. 283
Ramik D. M. 207, 241
Rea P. 445
Remy C. D. 94, 507
Richard M. 934
Richardson R. 705
Ritter M. 397
Ritzmann R. 196
Rochat F. 407,. 415, 421
Rochat F. 950
Rodriguez H. L. 223, 371
Roennau A. 547
Romero M. 855

Romero-Ramirez M. A. 86
Rönuauand A. 233
Rooney T. 190
Roper D. T. 135
Rosendo A. C. 53
Roshan R. 705
Röthlisberger D. 397
Rublev S. B. 774
Rudskyy A. 571, 579
Rutter B. 196

Sabourin C. 207, 241
Saint-Bauzel L. 715
Sakaida Y. 27
Salinas C. 455, 801
Salini J. 889
Samaali S. 311
Sanaatiyan M. M. 479
Sandford A. P. 847
Santos J. L. D. 249
Sanz-Merodio D. 605
Sarria J. 455
Sasaki Y. 666
Sattar T. 371
Sattar T. P. 223
Schmidt D. 865
Schmucker U. 571, 579
Schoeneich P. 407, 950
Sequeira J. 563
Sewell R. 190
Seyfarth A. 790
Shaheed M. H. 283
Sharma S. 135
Siedel T. 45
Siegwart R. 94, 388, 397, 507, 910
Silva M. F. 782
Simionescu I. 62
Sioma A. 463
Skrzypczynski P. 347, 825
Sliwka J. 110
Soyguder S. 555
Strawser D. 471
Studley M. 381
Sugimoto Y. 587
Supriyono H. 157
Sutter F. 547
Sutton R. 135

Takase K. 27
Taylor B. 196
Tee K. S. 35, 833
Telesh A. 571, 579
Tezuka S. 666
Thangavelautham J. 471
Thiele C. 45
Thoma P. 397
Tokhi M. O. 53, 157, 267, 292, 731
Tokuda K. 275, 363

Uchida H. 758
Unluhisarcikli O. 715
Uno Y. 3

Vasconcelos F. 355
Vasile A. 62
Ventura R. 563
Verlinden O. 491
Virk G. S. 957
Vitko A. 437
Vladareanu L. 742

Walas K. 817
Waldron K. 174
Waldron K. J. 595
Wang T. 207, 515
Wang X. 809
Weber J. 397
Wee T.-C. 499
Welsby J. 190
Winandy T. 910

Wormald S. A. 847

Xiao K. 110
Xie M. 499
Xiong L. 926

Yamamoto M. 127
Yatsun A. S. 774
Yokota S. 27
Yoshida K. 275, 523, 881
Yusof M. 918

Zahedi S. 35, 833
Zaier R. 615
Zhang D. 725
Zhao J. 149
Zhao M. 631
Zoppi M. 926